WITHDRAWN

Ecological Studies, Vol. 128

Analysis and Synthesis

Edited by

M.M. Caldwell, Logan, USA
G. Heldmaier, Marburg, Germany
O.L. Lange, Würzburg, Germany
H.A. Mooney, Stanford, USA
E.-D. Schulze, Bayreuth, Germany
U. Sommer, Kiel, Germany

Ecological Studies

Volumes published since 1992 are listed at the end of this book.

Springer
*New York
Berlin
Heidelberg
Barcelona
Budapest
Hong Kong
London
Milan
Paris
Santa Clara
Singapore
Tokyo*

Robert A. Mickler   Susan Fox
Editors

The Productivity and Sustainability of Southern Forest Ecosystems in a Changing Environment

With 221 Illustrations, 30 in Full Color

Springer

Robert A. Mickler
ManTech Environmental
  Technology, Inc.
Raleigh, NC 27606
USA

Susan Fox
USDA Forest Service
Southern Research Station
Raleigh, NC 27606
USA

Library of Congress Cataloging-in-Publication Data
The productivity and sustainability of southern forest
  ecosystems in a changing environment / [edited by]
  Robert A. Mickler and Susan Fox.
     p.   cm. — (Ecological studies ; 128)
  Includes bibliographical references and index.
  ISBN 0-387-94851-1 (hc : alk. paper)
  1. Forest productivity—Climactic factors—Southern
States. 2. Climactic changes—Southern States.
3. Forest ecology—Southern States. I. Mickler, Robert A.
II. Fox, Susan (Susan A.) III. Series: Ecological
studies ; v. 128.
SD390.7.C55 P76   1997
577.3—dc21   97-10648

Printed on acid-free paper.

© 1998 Springer-Verlag New York, Inc.
All rights reserved. This work may not be translated or copied in whole or in part without the written permission of the publisher (Springer-Verlag New York, Inc., 175 Fifth Avenue, New York, NY 10010, USA), except for brief excerpts in connection with reviews or scholarly analysis. Use in connection with any form of information storage and retrieval, electronic adaptation, computer software, or by similar or dissimilar methodology now known or hereafter developed is forbidden.
The use of general descriptive names, trade names, trademarks, etc., in this publication, even if the former are not especially identified, is not to be taken as a sign that such names, as understood by the Trade Marks and Merchandise Marks Act, may accordingly be used freely by anyone.

Production coordinated by Chernow Editorial Services, Inc., and managed by Natalie Johnson; manufacturing supervised by Jacqui Ashri.
Typeset by Agnew's, Inc., Grand Rapids, MI.
Printed and bound by Walsworth Publishing Co., Marceline, MO.
Printed in the United States of America.

9 8 7 6 5 4 3 2 1

ISSN 0070-8356
ISBN 0-387-94851-1 Springer-Verlag New York Berlin Heidelberg SPIN 10523482

Acknowledgments

The studies described in this book represent the first five years of research conducted by the Southern Global Change Program (SGCP). The SGCP is one of five regional research cooperatives, which comprise the Forest Service Global Change Program (FSGCRP) and are designed to provide a sound scientific basis for making regional, national, and international management and policy decisions regarding forest ecosystems in the context of global change challenges. Through participation in the U.S. Department of Agriculture's Global Change Research Program, the FSGCRP is part of the U.S. Government's Global Change Research Program (USGCRP). The USGCRP has been developed under the direction of the Office of Science and Technology Policy in the Executive Office of the President, through the Federal Coordination Council on Science, Engineering, and Technology and its Committee on Earth and Environmental Sciences.

We acknowledge the assistance given by the SGCP's Research Task Group, which provided scientific guidance to the program. We especially thank James Barnett, Prosanto K. Biswas, Patricia Brewer, Nicholas Comerford, Thomas Dell, Gary Foley, Walter Heck, John Hodges, Mike Rogers, David Shriner, Eric Vance, and Wayne Swank.

The research described in this volume represents five years of study by scientists, research associates and technicians, and support staff of the USDA Forest Service, the U.S. Environmental Protection Agency, the U.S. Department of Energy, other federal and state agencies, and major universities throughout the southern United States. Their individual efforts as part of a research cooperative, the Southern Global Change Program, have contributed to our understanding of the growth and physiology processes present in forest ecosystems in the southern United States.

The Southern Global Change Program is primarily supported by the USDA Forest Service. The assistance of William Sommers, Director of the Forest Fire and Atmospheric Sciences Research Staff; Lamar Beasley, former Director of the Southeastern Forest Experiment Station; and Tom Ellis, former Director of the Southern Forest Experiment Station was invaluable in the establishment of the research program. The support of the research staff of the Southern Research Station, Station Director Peter Roussopoulos, and Assistant Directors Nancy Herbert, Samuel Foster, and Jimmy Reaves was essential to the continuing development and implementation of the research program and in the preparation and production of this book. We acknowledge the support and research direction provided by the USDA Forest Service Research staffs. We thank Deputy Chief Jerry Sesco, Richard Smythe, H. Fred Kaiser, Jr., Calvin Bey, Nelson Loftus, Jr., James Stewart, and Ann Bartuska. A special thanks is owed to Elvia Niebla, National Coordinator of the FSGCRP, for her encouragement and guidance of the regional research programs.

There were more than 100 scientists who reviewed the manuscripts in this volume. The rigor of their commentary contributed to the content and clarity of the scientific information presented in this volume.

We thank Chuck Gaul, graphic arts coordinator, for the cover design.

This book has not been subject to policy review by the USDA Forest Service and should not be construed to represent the policies of the agency. The National Council for Air and Stream Improvement provided financial support for individual studies which increased our understanding of forest growth and physiological processes that contribute to the productivity of the region's forest resource. All the research findings from studies described in this book have not been subject to scientific review by the National Council for Air and Stream Improvement.

Robert A. Mickler
Susan Fox

Contents

Acknowledgments v
Contributors xiii

Section 1. An Introduction to Southern Forests in a Changing Environment

1. **Southern Forest Ecosystems in a Changing Chemical and Physical Environment** 3
   Robert A. Mickler

2. **General Circulation Model Scenarios for the Southern United States** 15
   Ellen J. Cooter

3. **Developing Policy-Relevant Global Climate Change Research** 55
   J. Christopher Bernabo

Section 2. Global Change Impacts on Tree Physiology and Growth

4. **Influence of Drought Stress on the Response of Shortleaf Pine to Ozone** 73
   Richard B. Flagler, John C. Brissette, and James P. Barnett

Contents

5. **Effects of Elevated Carbon Dioxide on the Growth and Physiology of Loblolly Pine** ... 93
   Makonnen Alemayehu, Douglas R. Hileman, Gobena Huluka, and Prosanto K. Biswas

6. **Environmental Stresses and Reproductive Biology of Loblolly Pine (*Pinus taeda* L.) and Flowering Dogwood (*Cornus florida* L.)** ... 103
   Kristina F. Connor, Timothy C. Prewitt, Franklin T. Bonner, William W. Elam, and Robert C. Parker

7. **Interactions of Elevated Carbon Dioxide, Nutrient Status, and Water Stress on Physiological Processes and Competitive Interactions Among Three Forest Tree Species** ... 117
   John W. Groninger, John R. Seiler, Alexander L. Friend, Paul C. Berrang, and Shepard M. Zedaker

8. **Effects of Elevated Carbon Dioxide Levels and Air Temperature on Carbon Assimilation of Loblolly Pine** ... 131
   Robert O. Teskey

9. **An Investigation of the Impacts of Elevated Carbon Dioxide, Irrigation, and Fertilization on the Physiology and Growth of Loblolly Pine** ... 149
   Phillip M. Dougherty, H. Lee Allen, Lance W. Kress, Ramesh Murthy, Chris A. Maier, Timothy J. Albaugh, and D. Arthur Sampson

10. **Effects of Elevated Carbon Dioxode, Water, and Nutrients on Photosynthesis, Stomatal Conductance, and Total Chlorophyll Content of Young Loblolly Pine (*Pinus taeda* L.) Trees** ... 169
    Thomas C. Hennessey and Venkatesh K. Harinath

11. **Ecophysiological Response of Managed Loblolly Pine to Changes in Stand Environment** ... 185
    Mary A. Sword, Jim L. Chambers, Dennis A. Gravatt, James D. Haywood, and James P. Barnett

12. **Dynamic Responses of Mature Forest Trees to Changes in Physical and Chemical Climate** ... 207
    Samuel B. McLaughlin and Daryl J. Downing

13.	Productivity of Natural Stands of Longleaf Pine in Relation to Competition and Climatic Factors Ralph S. Meldahl, John S. Kush, Jyoti N. Rayamajhi, and Robert M. Farrar, Jr.	231
14.	The Impacts of Acidic Deposition and Global Change on High Elevation Southern Appalachian Spruce-Fir Forests Samuel B. McLaughlin, J. Devereux Joslin, Wayne Robarge, April Stone, Rupert Wimmer, and Stan D. Wullschleger	255
15.	The Influences of Global Change on Tree Physiology and Growth Robert O. Teskey, Phillip M. Dougherty, and Robert A. Mickler	279

Section 3. Modeling the Biophysical Effects of Global Change

16.	Modeling Nutrient Uptake as a Component of Loblolly Pine Response to Environmental Stress J. Michael Kelly and Ruth D. Yanai	293
17.	A Linked Model for Simulating Stand Development and Growth Processes of Loblolly Pine V. Clark Baldwin, Jr., Phillip M. Dougherty, and Harold E. Burkhart	305
18.	MAESTRO Simulations of the Response of Loblolly Pine to Elevated Temperatures and Carbon Dioxide Wendell P. Cropper, Jr., Kelly Peterson, and Robert O. Teskey	327
19.	Projections of Growth of Loblolly Pine Stands Under Elevated Temperatures and Carbon Dioxide Harry T. Valentine, Timothy G. Gregoire, Harold E. Burkhart, and David Y. Hollinger	341
20.	Modeling the Potential Sensitivity of Slash Pine Stem Growth to Increasing Temperature and Carbon Dixode Wendell P. Cropper, Jr.	353
21.	An Index for Assessing Climate Change and Elevated Carbon Dioxide Effects on Loblolly Pine Productivity David Arthur Sampson, H. Lee Allen, and Phillip M. Dougherty	367

Contents

22. **Predictions and Projections of Pine Productivity and Hydrology in Response to Climate Change Across the Southern United States** 391
    Steven G. McNulty, James M. Vose, and Wayne T. Swank

23. **Scaling Up Physiological Responses of Loblolly Pine to Ambient Ozone Exposure Under Natural Weather Variations** 407
    Robert J. Luxmoore, Scott M. Pearson, M. Lynn Tharp, and Samuel B. McLaughlin

24. **Intregrating Research on Climate Change Effects on Loblolly Pine: A Probability Regional Modeling Approach** 429
    James E. Smith, Peter B. Woodbury, David A. Weinstein, and John A. Laurence

25. **Projected Impacts of Global Climate Change on Forests and Water Resources of the Southeastern United States** 453
    Jeffrey G. Borchers and Ronald P. Neilson

26. **Summary of Simulated Forest Responses to Climate Change in the Southeastern United States** 479
    David A. Weinstein, Wendell P. Cropper, Jr., and Steven G. McNulty

Section 4. The Effects of Climate Change on Forest Soils

27. **Simulated Effects of Atmospheric Deposition and Species Change on Nutrient Cycling in Loblolly Pine and Mixed Deciduous Forests** 503
    Dale W. Johnson, Richard B. Susfalk, and Wayne T. Swank

28. **Influence of Microclimate on Short-Term Litter Decomposition in Loblolly Pine Ecosystems** 525
    B. Graeme Lockaby, Arthur H. Chappelka, Mary A. Sword, and Allan E. Tiarks

29. **Soil Organic Matter and Soil Productivity: Searching for the Missing Link** 543
    Felipe G. Sanchez

30. **Effects of Soil Warming on Organic Matter Decomposition and Soil-Nitrogen Cycling in a High Elevation Red Spruce Stand** 557
    J. Devereux Joslin, Mark H. Wolfe, and Charles T. Garten

31.	**Effects of Soil Warming, Atmospheric Deposition, and Elevated Carbon Dioxide on Forest Soils in the Southeastern United States** J. Devereux Joslin and Dale W. Johnson	571

Section 5.   Disturbance Interactions With Global Change

32.	**Environmental Effects on Pine Tree Carbon Budgets and Resistance to Bark Beetles** Richard T. Wilkens, Matthew P. Ayres, Peter L. Lorio, Jr., and John D. Hodges	591
33.	**Predictions of Southern Pine Beetle Populations Using a Forest Ecosystem Model** Steven G. McNulty, Peter L. Lorio, Jr., Matthew P. Ayres, and John D. Reeve	617
34.	**Soil Effects Mediate Interaction of Dogwood Anthracnose and Acidic Precipitation** Kerry O. Britton, Paul C. Berrang, and Erika Mavity	635
35.	**Effects of Temperature and Drought Stress on Physiological Processes Associated With Oak Decline** Theodor D. Leininger	647
36.	**Effects of Global Climate Change on Biodiversity in Forests of the Southern United States** Margaret S. Devall and Bernard R. Parresol	663
37.	**Regional Climate Change in the Southern United States: The Implications for Wildfire Occurrence** Warren E. Heilman, Brian E. Potter, and John I. Zerbe	683
38.	**Detecting and Predicting Climatic Variation from Old-Growth Baldcypress** Gregory A. Reams and Paul C. Van Deusen	701
39.	**Modeling the Differential Sensitivity of Loblolly Pine to Climatic Change Using Tree Rings** Edward R. Cook, Warren L. Nance, Paul J. Krusic, and James Grissom	717

40. Global Change and Disturbance in Southern Forest
    Ecosystems 741
    Matthew P. Ayres and Gregory A. Reams

Section 6. Socioeconomic Impacts of Global Change

41. Evaluation of Effects of Forestry and Agricultural Policies
    on Forest Carbon and Markets 755
    Ralph J. Alig, Darius M. Adams, and Bruce A. McCarl

42. Economic Dimensions of Climate Change Impacts on
    Southern Forests 777
    Diana M. Burton, Bruce A. McCarl, Claudio N.M. de Sousa,
    Darius M. Adams, Ralph J. Alig, and Steven M. Winnett

43. Assessing Present Biological Information for Valuating
    the Economic Impacts of Climate Change on Softwood
    Stumpage Supply in the South 795
    James T. Gunter, Donald G. Hodges, and James L. Regens

44. An Integrated Assessment of Climate Change on Timber
    Markets of the Southern United States 809
    Joseph E. de Steiguer and Steven G. McNulty

45. Integrating Local and Global Objectives in Forest
    Management: A Value-Based, Multiscalar Approach 823
    R. Gordon Dailey, Jr. and Bryan G. Norton

46. Sensitivity of Protection Value to Forest Condition in the
    Southern Appalachian Spruce–Fir Forest 837
    Dylan H. Jenkins, Jay Sullivan, Niki S. Nicholas,
    Gregory Amacher, and Dixie Watts Reaves

47. Economics and Global Climate Change 855
    David N. Wear

Index 865

Contributors

Darius M. Adams
Department of Forest Resources, Oregon State University, Corvallis, OR 97333, USA

Timothy J. Albaugh
Department of Forestry, North Carolina State University, Raleigh, NC 27695, USA

Makonnen Alemayehu
School of Agriculture and Home Economics, Tuskegee University, Tuskegee, AL 36088, USA

Ralph J. Alig
USDA Forest Service, Corvallis, OR 97331, USA

H. Lee Allen
Department of Forestry, North Carolina State University, Raleigh, NC 27695, USA

Gregory Amacher
Department of Forestry, Virginia Polytechnical Institute and State University, Blacksburg, VA 24061, USA

Matthew P. Ayres	Department of Biological Science, Dartmouth College, Hanover, NH 03755, USA
V. Clark Baldwin, Jr.	USDA Forest Service, Pineville, LA 71360, USA
James P. Barnett	USDA Forest Service, Pineville, LA 71360, USA
J. Christopher Bernabo	Science and Policy Associates, Washington, DC 20005, USA
Paul C. Berrang	USDA Forest Service, Klamath Falls, OR 97601, USA
Prosanto K. Biswas	School of Agriculture and Home Economics, Tuskegee University, Tuskegee, AL 36088, USA
Franklin T. Bonner	USDA Forest Service, Starkville, MS 39760, USA
Jeffrey G. Borchers	USDA Forest Service, Corvallis, OR 97331, USA
John C. Brissette	USDA Forest Service, Orno, ME 04473, USA
Kerry O. Britton	USDA Forest Service, Athens, GA 30602, USA
Harold E. Burkhart	Department of Forestry, Virginia Polytechnical Institute and State University, Blacksburg, VA 24061, USA
Diana M. Burton	Department of Forestry, Texas A&M University, College Station, TX 77843, USA
Jim L. Chambers	Department of Forestry, Louisiana State University, Baton Rouge, LA 70803, USA
Arthur H. Chappelka	School of Forestry, Auburn University, Auburn, AL 36849, USA

Contributors

Kristina F. Connor	USDA Forest Service, Starkville, MS 39760, USA
Edward R. Cook	Lamont-Doherty Earth Observatory, Palisades, NY 10964, USA
Ellen Cooter	National Oceanic and Atmospheric Administration, Research Triangle Park, NC 27711, USA
Wendell P. Cropper, Jr.	Rosential School of Marine and Atmospheric Science, University of Miami, Miami, FL 33149, USA
R. Gordon Daily, Jr.	Battelle Pacific Northwest Lab, Washington, DC 20024, USA
Claudio N.M. de Sousa	Department of Agricultural Economics, Texas A&M University, College Station, TX 77843, USA
Joseph E. de Steiguer	USDA Forest Service, Raleigh, NC 27695, USA
Margaret S. Devall	USDA Forest Service, Stoneville, MS 39759, USA
Phillip M. Dougherty	Westvaco, Summerville, SC 29483, USA
Daryl J. Downing	Oak Ridge National Laboratory, Oak Ridge, TN 37831, USA
William W. Elam	Department of Forestry, Mississippi State University, Mississippi State, MS 39762, USA
Robert M. Farrar, Jr.	USDA Forest Service, Mississippi State, MS 39762, USA
Richard B. Flagler	543 Walden Way, Ft. Collins, CO 80526, USA
Susan Fox	USDA Forest Service, 1509 Varsity Drive, Raleigh, NC 27606, USA

Alexander L. Friend	Department of Forestry, Mississippi State University, Mississippi State, MS 39762, USA
Charles T. Garten	Oak Ridge National Laboratory, Oak Ridge, TN 37831, USA
Dennis A. Gravatt	Department of Forestry, Louisiana State University, Baton Rouge, LA 70803, USA
Timothy G. Gregoire	Department of Forestry, Virginia Polytechnical Institute and State University, Blacksburg, VA 24061, USA
James Grissom	Department of Forestry, North Carolina State University, Raleigh, NC 27695, USA
John W. Groninger	Department of Forestry, Virginia Polytechnical Institute and State University, Blacksburg, VA 24061, USA
James T. Gunter	Tulane University, Spatial Analysis Research Laboratory, New Orleans, LA 70112, USA
Venkatesh K. Harinath	Department of Forestry, Oklahoma State University, Stillwater, OK 74078, USA
James D. Haywood	USDA Forest Service, Pineville, LA 71360, USA
Warren E. Heilman	USDA Forest Service, East Lansing, MI 48823, USA
Thomas C. Hennessey	Department of Forestry, Oklahoma State University, Stillwater, OK 74078, USA
Douglas R. Hileman	School of Agriculture and Home Economics, Tuskegee University, Tuskegee, Al 36088, USA

Donald G. Hodges	Tulane University, Spatial Analysis Research Laboratory, New Orleans, LA 70112, USA
John D. Hodges	Mississippi State University, Mississippi State, MS 39762, USA
David Y. Hollinger	USDA Forest Service, Durham, NH 03824, USA
Gobena Huluka	Tuskegee University, Tuskegee, AL 36088, USA
Dylan H. Jenkins	Department of Forestry, Virginia Polytechnical Institute and State University, Blacksburg, VA 24061, USA
Dale W. Johnson	Desert Research Institute, University of Nevada-Reno, Reno, NV 89506, USA
J. Devereux Joslin	Tennessee Valley Authority, Norris, TN 37828, USA
J. Michael Kelly	Department of Forestry, Iowa State University, Ames, Iowa 50011, USA
Lance W. Kress	USDA Forest Service, RTP, NC 27709, USA
Paul J. Krusic	Lamont-Doherty Earth Observatory, Palisades, NY 10964, USA
John S. Kush	School of Forestry, Auburn University, Auburn, AL 36849, USA
John A. Laurence	Boyce Thompson Institute, Ithaca, NY 14853, USA
Theodor D. Leininger	USDA Forest Service, Stoneville, MS 38776, USA
B. Graeme Lockaby	School of Forestry, Auburn University, Auburn, AL 36849, USA
Peter L. Lorio, Jr.	USDA Forest Service, Alexandria, LA 71360, USA

Robert J. Luxmoore	Oak Ridge National Laboratory, Oak Ridge, TN 37831, USA
Chris A. Maier	USDA Forest Service, Research Triangle Park, NC 27709, USA
Erika Mavity	USDA Forest Service, Gainesville, GA 30501, USA
Samuel B. McLaughlin	Oak Ridge National Laboratory, Oak Ridge, TN 37831, USA
Bruce A. McCarl	Department of Agricultural Economics, Texas A&M University, College Station, TX 77843, USA
Steven G. McNulty	USDA Forest Service, Raleigh, NC 27606, USA
Ralph S. Meldahl	School of Forestry, Auburn University, Auburn, AL 36849, USA
Robert A. Mickler	ManTech Environmental Technology, Inc., Raleigh, NC 27606, USA
Ramesh Murthy	Department of Forestry, North Carolina State University, Raleigh, NC 27695, USA
Warren L. Nance	USDA Forest Service, Saucier, MS 39574, USA
Ronald P. Neilson	USDA Forest Service, Corvallis, OR 97331, USA
Niki S. Nicholas	Tennessee Valley Authority, Norris, TN 37828, USA
Bryan G. Norton	School of Public Policy, Georgia Institute of Technology, Atlanta, GA 30332, USA
Robert C. Parker	Department of Forestry, Mississippi State University, Mississippi State, MS 39762, USA

Bernard R. Parresol	USDA Forest Service, Asheville, NC 28802, USA
Scott M. Pearson	Biology Department, Mars Hill College, Mars Hill, NC 28754, USA
Kelly Peterson	USDA Forest Service, Pineville, LA 71360, USA
Brian E. Potter	USDA Forest Service, East Lansing, MI 48823, USA
Timothy C. Prewitt	Department of Forestry, Mississippi State University, Mississippi State, MS 39762, USA
Jyoti N. Rayamajhi	School of Forestry, Auburn University, Auburn, AL 36849, USA
Gregory A. Reams	USDA Forest Service, Asheville, NC 28802, USA
Dixie Watts Reaves	Department of Forestry, Virginia Polytechnical Institute and State University, Blacksburg, VA 24061, USA
John D. Reeve	USDA Forest Service, Pineville, LA 71361, USA
James L. Regens	Tulans University, Spatial Analysis Research Laboratory, New Orleans, LA 70112, USA
Wayne Robarge	Department of Soil Science, North Carolina State University, Raleigh, NC 27695, USA
David Arthur Sampson	Department of Forestry, North Carolina State University, Raleigh, NC 27695, USA
Felipe G. Sanchez	USDA Forest Service, Research Triangle Park, NC 27709, USA

John R. Seiler	Department of Forestry, Virginia Polytechnical Institue and State University, Blacksburg, VA 24061, USA
James E. Smith	USDA Forest Service, Portland, OR 97208, USA
April Stone	University of Tennessee, Knoxville, TN 37996, USA
Jay B. Sullivan	Department of Forestry, Virginia Polytechnical Institute and State University, Blacksburg, VA 24061, USA
Richard B. Susfalk	Desert Research Institute, University of Nevada-Reno, Reno, NV 89506, USA
Wayne T. Swank	USDA Forest Service, Otto, NC 28763, USA
Mary A. Sword	USDA Forest Service, Pineville, LA 71360, USA
Robert O. Teskey	School of Forest Resources, University of Georgia, Athens, GA 30602, USA
M. Lynn Tharp	Oak Ridge National Laboratory, Oak Ridge, TN 37831, USA
Allan E. Tiarks	USDA Forest Service, Pineville, LA 71360, USA
Harry T. Valentine	USDA Forest Service, Durham, NH 03824, USA
Paul C. Van Deusen	National Council of the Paper Industry for Air and Stream Improvement, Inc., Medford, MA 02155, USA
James M. Vose	USDA Forest Service, Otto, NC 28763, USA
David N. Wear	USDA Forest Service, Research Triangle Park, NC 27709, USA

David A. Weinstein	Boyce Thompson Institute, Ithaca, NY, 14853, USA
Richard T. Wilkens	Department of Ecology and Evolutionary Biology, University of Connecticut, Storrs, CT 06269, USA
Steven M. Winnett	US Environmental Protection Agency, Washington, DC 20024, USA
Rupert Wimmer	University of Agriculture, Vienna, Austria
Mark H. Wolfe	Tennessee Valley Authority, Oak Ridge, TN 37831, USA
Peter B. Woodbury	Boyce Thompson Institute, Ithaca, NY 14853, USA
Stan D. Wullschleger	Oak Ridge National Laboratory, Oak Ridge, TN 37831, USA
Ruth D. Yanai	College of Environmental Science and Forestry, State University New York, Syracuse, NY 13210, USA
Shepard M. Zedaker	Department of Forestry, Virginia Polytechnical Institute and State University, Blacksburg, VA 24061, USA
John I. Zerbe	USDA Forest Service, Madison, WI 53713, USA

1. An Introduction to Southern Forests in a Changing Environment

1. Southern Forest Ecosystems in a Changing Chemical and Physical Environment

Robert A. Mickler

For much of Earth's history, terrestrial vegetation evolved in a carbon dioxide (CO_2) atmosphere that saturated photosynthesis and enhanced the growth of C_3 plants. Estimates of atmospheric CO_2 for 420 millions years ago suggest that the first terrestrial plants grew in CO_2 concentrations 16-times higher than those present today (Yapp and Poths, 1992). During the period from 50 to 100 million years ago, atmospheric CO_2 concentrations have been estimated at 1,000 to 3,000 μLL^{-1} (Budyko et al., 1987, Ehleringer et al., 1991). In contrast, during the last 160,000 years atmospheric CO_2 concentrations have been atypically low, ranging from 190 to 280 μLL^{-1} as measured from air trapped in the Vostok ice cores (Barnola et al., 1994), until stabilizing at about 280 μLL^{-1} CO_2 after the last glacial period. The atmospheric CO_2 record obtained from the Siple Station ice core indicates that the pre-industrial atmospheric CO_2 concentration ca.1750 was 280 μLL^{-1} and increased to 345 μLL^{-1} in 1984, from anthropogenic sources (Neftel et. al., 1994). Beginning in the nineteenth century, CO_2 concentration began to rise in a logarithmic manner to the 1992 annual mean value of 356 μLL^{-1}. The National Oceanic and Atmospheric Administration's (NOAA) Climate Monitoring and Diagnostic Laboratory (CDML) flask data from Mauna Loa documents an increase in annual CO_2 concentration from 325.3 μLL^{-1} in 1970 to 356.4 μLL^{-1} in 1992. NOAA/CMDL flask data report an annual increase of 1.41 μLL^{-1} for the Mauna Loa site and an annual global increase of 1.43 μLL^{-1} for all sampling sites over the 22-year period (Conway et al., 1994). At the current rate of increase, a typically managed loblolly pine (*Pinus taeda* L.) plantation with a rotation of thirty years will experience an approximately 10% increase in atmospheric CO_2 concentrations from planting to harvest. Trees from long-lived species in natural forest ecosystems will encounter over a 100% increase in atmospheric CO_2 during their life spans into the middle of the next century when atmospheric CO_2 is projected at 530 to 600 μLL^{-1} (Trabalka et al., 1986, Watson et al., 1990).

Although today's warnings from the global scientific community of the biological importance of a doubling in atmospheric CO_2 seems extreme or is nonexistent in the minds of many in the ongoing global change political debate, from the ecological perspective the rate of change is more significant than the degree of change. Terrestrial ecosystems are experiencing changes in their chemical and physical environments at heretofore unprecedented rates. Present day CO_2 and O_2 concentrations restrict most vegetation to 60 to 70% of its photosynthetic potential as a result of kinetic constraints imposed by rubisco in the photosynthetic carbon reduction cycle (Bowes, 1993). Setting aside the uncertainty associated with whatever climatic changes are occurring in association with rising CO_2 concentrations, most of the scientific community agree that rising CO_2 levels will have substantial direct and indirect impact on terrestrial ecology. One compilation of literature studying the effects of a doubling of atmospheric CO_2 found that the average growth stimulation among 156 plant species was 41% for C_3 plants, 22% for C_4 plants, and 15% for CAM species (Poorter, 1993). An analysis of the response of 39 tree species reported by Gunderson and Wullschleger (1994) found that trees grown at elevated atmospheric CO_2 had an average photosynthetic enhancement of 44%. Findings from these and additional forest ecosystem studies indicate the importance of rising atmospheric CO_2 concentrations and other environmental resources and ambient levels of pollutants in modifying the response of forests and associated plant communities.

Nutrients, water, irradiance, temperature, and ambient levels of atmospheric pollutants are a few of the known factors interacting with CO_2 enrichment at the leaf, canopy, stand, and forest scales. In addition to direct impact, there are indirect effects from disturbance, land-use change, and altered vegetation distribution. These interactions confound our ability to understand the mechanisms necessary to predict the magnitude of forest ecosystem responses to a rapidly changing physical and chemical environment.

Interaction of CO_2 With Other Environmental Factors

The interaction of elevated CO_2 with other factors that affect plant growth has already been demonstrated to occur in agricultural crops and managed and natural forest ecosystems. Several literature reviews by Eamus and Jarvis (1989), Bazzaz (1990), Musselman and Fox (1991), Strain and Thomas (1992), Rogers et al. (1994), Gunderson and Wullschleger (1994), Ceulemans and Mousseau (1994), and Idso and Idso (1994) have all shown that rising CO_2 will alter the competitive interaction that influences forest ecosystems by direct effects on plant growth and development. Although a substantial body of evidence has been accumulating which points to increases in growth in an enhanced CO_2 environment, debate continues in the science and policy communities.

Environmental stresses and limiting resources are frequently identified as potential factors that currently restrict growth in forest ecosystems and which may limit or eliminate any promotion of growth by elevated CO_2. Several studies have reported that the full potential of forest ecosystems to increase net primary productivity (NPP) in a rising CO_2 environment will not be achieved because of nutrient and water limitations (Kramer, 1981; Allen et al., 1990; Thomas et al., 1994). Nutrient limitations, such as the effects of low phosphorus supply on limiting ribulose-1,5-bisphosphate carboxylase/oxygenase (rubisco) activity and the regeneration of 1,5-bisphosphate, may reduce photosynthetic capacity (Lewis et al., 1994), but it has also been observed that enhanced growth can be maintained when environmental parameters are colimiting in a CO_2 enriched environment (Gifford, 1992; Idso and Idso, 1994). The benefits of increased photosynthetic water use efficiency and nutrient use efficiency observed in trees growing in a rising CO_2 environment, when other environmental parameters may be limiting, have important implications for long-term forest ecosystem productivity and sustainability. On a regional, national, and world-wide scale, estimates of the direction and magnitude of changes to NPP and the possible enhancement of temperate forest ecosystem carbon accumulation are being made with increasing confidence.

Nutrients

Many forest soils throughout the southern United States have been characterized as having inherent nutritional problems characterized by low cation exchange capacity, relatively few weatherable primary minerals, and low organic matter as a result of past agricultural practices (Richter and Markewitz, 1995). The early observations of Norby, O'Neill, and Luxmore (1986) showed that enhanced growth is not precluded under the coexisting conditions of elevated CO_2 and nutrient deficiencies. A literature survey of experiments conducted to study the interactions of mineral nutrition and elevated CO_2 (Ceulemans and Mousseau, 1994) found that the largest increases in biomass were found in high nutrient treatments, but biomass did increase in low nutrient treatments. Several studies have shown that plant growth at elevated CO_2 and low nutrient concentrations was not enhanced (Thomas et al., 1994), greater than expected (Silvola and Alholm, 1992), greater at low versus high nutrient treatments (Tissue et al., 1993), or adjusted from low- to higher-growth rates with time (Conroy et al., 1990).

This range of plant response reported in the literature shows that the link between elevated CO_2 and plant nutrition is not clear. The physiological and biochemical mechanisms for observed plant responses range from "acclimation" (Gunderson and Wullschleger, 1994) to short-term regulation (Sage, 1994). Eamus and Jarvis (1989) and Gunderson and Wullschleger (1994) have reviewed the tree acclimation literature. They use the term "acclimation" to address any biochemical and physiological changes in the plant that result from growth in elevated CO_2 over time. They conclude that photosynthetic acclimation will occur in trees in response to CO_2, and nutrient limitations and sink-source strength appear to effect photosynthetic acclimation. However, additional data will be needed to fully understand the underlying mechanisms. In contrast, Idso and Idso (1994) reviewed ten years of research on the interactions of CO_2 and environmental stresses and concluded that atmospheric CO_2 enrichment enables plants to cope with soil nutrient deficiencies. The concentrations of CO_2 required to observe growth enhancement were higher under nutrient limiting conditions than when adequate nutrients were available to the plant. The mechanisms for observed growth promotion include: increased efficiency in extracting soil nutrients from enlarged root systems of CO_2 enriched plants; enhancement of soil rhizosphere organisms from increased photosynthetic rates; increased availability of carbohydrates for metabolic processes; and the stimulation of biological nitrogen fixation. Sage (1994) reviewed the theories of acclimation and short-term regulation in response to stress and changes in resource levels. He concludes that the gas exchange literature does not support the acclimation theory and that natural environments may exhibit enhanced photosynthesis per unit leaf area following CO_2 enrichment. These conclusions are supported by analytical modeling (Comins and McMurtrie, 1993; Kirschbaum et al., 1994; Loehle, 1995) and in other field and modeling research reported in this volume.

Water

The response of plants to elevated CO_2 under well-watered and water-stressed growing conditions has repeatedly shown that water stress is ameliorated by increasing concentrations of CO_2 in C_3 and C_4 species. This response has been observed in herbaceous plants (Smith et al., 1987; Nijs et al., 1989a), agricultural crops (Rogers et al., 1984; Cure and Acock, 1986), and trees (Eamus and Jarvis, 1989; Ceulemans and Mousseau, 1994). An increase in CO_2 concentrations decreases stomatal density, reduces the stomatal opening, and decreases stomal conductance, which result in a reduction in the rate of transpiration and an increase in the water use efficiency (WUE) (moles CO_2 assimilated per mole water transpired) (Woodward and Bazzaz, 1988; Eamus and Jarvis, 1989; Paoletti and Gellini, 1993). Several studies have shown that plants grown at preindustrial and current ambient atmospheric CO_2 concentrations had lower WUE than plants grown at elevated CO_2 (Woodward, 1987; Norby and O'Neil, 1989, 1991). Increases in WUE have been reported for several plant species in the absence of changes in stomatal conductance as a result of enhanced CO_2 stimulation of photosynthesis (Gunderson et al., 1993).

In routinely drought-stressed environments, like the southern United States, CO_2 enrichment may enable plants to ameliorate drought stress and delay its onset. Plants will be able to reduce their requirements for ground water through increased WUE and better exploit available water with a typically enlarged root system in elevated CO_2 environments (Idso et al., 1988; Tschaplinski et al., 1993). Although growth is reduced under drought-stress conditions, on a relative plant growth response basis the overall response in CO_2 enhanced treatments is higher when water is limiting than when adequate water is present: 62 ± 11% vs. 31 ± 6% at 300 μLL^{-1} CO_2 and 219 ± 66% vs. 51 ± 17% at 600 μLL^{-1} CO_2 (Idso and Idso, 1994).

Thus, drought stress is unlikely to limit the promotion of plant growth effects from increases in atmospheric CO_2. This effect will increase at higher concentrations of CO_2 enrichment as plants increase WUE at the individual leaf and canopy levels and increase their ability to exploit larger soil volumes for soil water at the root level.

One additional trend important to coastal and near-coastal forest ecosystems is that enhanced CO_2 may ameliorate the adverse effects of water salinity on plant growth (Schwartz and Gale, 1984; Bowman and Strain, 1987). The mechanisms for this response are thought to be linked to reductions in uptake of salt through increases in WUE or by increases in rubisco efficiency. Increases in salt tolerance may be an important plant response in coastal forest ecosystems that are experiencing increased salinity levels and more frequent salt water intrusions from storm events and in near-coastal areas where salt water intrusion into ground water is becoming more frequent and longer in duration.

Light

The interaction of light and enhanced CO_2 on photosynthesis was first observed on agricultural crops. Brun and Cooper (1967) developed photosynthetic response curves to CO_2 concentrations over a range of irradiance for soybean. They observed that photosynthesis becomes light saturated at increasingly higher intensities with increasing CO_2 concentrations. The response curves differed with cultivars, which suggested to the authors that the selection of genotypes that are better able to utilize enhanced CO_2 concentrations could be utilized to increase yield performance.

Almost all recent C_3 crop plant studies have shown that enhanced CO_2 increases photosynthetic rates and plant growth, and that this increase occurs even under limiting irradiance (Sionit et al., 1982; Cure and Acock, 1986; Allen, 1990; Gifford, 1992). On a relative enhancement basis, growth may be greater under limiting irradiance versus high irradiance. Valle et al. (1985) estimated quantum yield in soybean plants grown at 330 μLL^{-1} and 660 μLL^{-1} CO_2. They observed a 60 to 75% enhancement in quantum yield at the higher CO_2 concentration and that this increase was relatively higher across a range of low irradiance. They concluded that photosynthesis may be stimulated to a greater degree at the bottom of plant canopies. This response has also been observed for sedge (Long and Drake, 1991) and ryegrass (Nijs et al., 1989b).

The number of tree studies reporting on the interaction of CO_2 and over a range of irradiance is limited. Most studies have been conducted with plants grown at a single irradiance level in controlled-exposure environments. In forest ecosystems, photosynthesis generally occurs at irradiance levels which limit the rate of photosynthesis. Observations from several studies have shown that tree photosynthesis and growth at low irradiance levels is enhanced with elevated CO_2 concentrations (Tolley and Strain, 1985; Bazzaz el al., 1990; Wullschleger et al., 1992). Bazzaz et al. (1990) examined the response of seven co-occurring tree species to elevated CO_2 and observed that the increase in total biomass of shade tolerant trees was greater than shade intolerant tree species. Similar results were reported by Williams et al. (1986) for shade tolerant hickory (*Carya* sp.) relative to shade intolerant tulip poplar (*Liriodendron tulipifera* L.). In general, tree studies utilizing one irradiance level have observed higher quantum yield at low irradiance in elevated atmospheric CO_2 (Eamus and Jarvis, 1989; Wullschleger et al., 1992). In the few reported studies in which trees were grown at multiple irradiance levels, Reekie and Bazzaz (1989) showed no difference in tree response to low and high irradiance at elevated CO_2, and Tolley and Strain (1985) showed high irradiance and elevated CO_2 treatments delayed tree stress responses to drought. Thus, crop and tree studies have shown that at the plant and canopy levels, low levels of irradiance do not constrain the plant response to enhanced atmospheric CO_2. The general trend for most plants is that they exhibit the largest relative response under limiting irradiance and this response is greatest at higher CO_2 concentrations (Idso and Idso, 1994). The limited number of reported studies, however, increases the uncertainty associated with predicting an ecosystem photosynthetic response to limiting irradiance and elevated CO_2 based on the range of responses observed for individual plant species and cultivars.

Temperature

Global circulation model predictions for the southeastern United States show a 3.6 to 8.0 °C rise in the mean annual temperature and an increase in high- and low-temperature extremes. One of the long-term effects of potential increases in global temperatures is a plant species migration that on average moves plant distributions 100 to 160 km northward and 100 m in elevation for each 1 °C warming (Davis, 1989; Dobson et al., 1989). A more immediate and major consequence of increases in air temperature in an enhanced CO_2 environment is a shift upward in the optimum temperature ranges for carbon sequestration (Berry and Bjorkman, 1980; McMurtrie and Wang, 1993). Observations of the interaction of elevated CO_2 and temperature have shown that plant growth and physiology in an enriched CO_2 environment is strongly temperature dependent. In general, C_3 plants have a lower optimum temperature for photosynthesis than C_4 plants because the CO_2/O_2 specificity of rubisco favors oxygenase at higher temperatures. Enhanced atmospheric CO_2 increases the optimum photosynthetic temperatures for C_3 plants and shifts the range nearer to that of C_4 plants (Pearcy and Bjorkman, 1983; Long, 1991). C_3 plants may benefit preferentially at higher temperatures as photorespiration increases with increasing temperature and is increasingly inhibited by rising CO_2 concentrations (Idso et al., 1987). Although individual species differ in their responses to CO_2 and temperature, positive interactions have been reported for the limited number of studies addressing direct and indirect effects for both C_3 and C_4 plants.

Pearcy and Bjorkman (1983) reported a 2.5 times increase in photosynthetic rates in the C_3 desert shrub *Larrea divaricata* resulting from an increase in CO_2 from 330 μLL^{-1} to 1000 μLL^{-1} and a shift of optimum temperature from 35 to 40 °C to near 44°C. The temperature response curves for *L. divaricata* and a C_4 shrub *Tridestronia oblongifolia* were similar in the 1000 μLL^{-1} treatment. A linear increase from 30 to 56% in biomass growth was observed for five C_3 crops and submerged plant species over a temperature range of 19 to 34 °C and elevated CO_2 (Idso et al., 1987). They also observed a reduction in plant growth at elevated CO_2 and cool daily mean temperatures (<18.5 °C). A replication of this study confirmed that the effects of elevated CO_2 are temperature dependent (Idso and Kimball, 1989). Cure (1985) reported a 40% increase in biomass for rice, maize, and soybean grown under a doubling of CO_2 and an increase of day/night temperatures from 23/20 °C to 28/23 °C. In addition to increases in growth, enhanced CO_2 has been observed to increase chilling survival of okra [*Abelmoschus esculentus* (L.) Moench, cv. Clemson Spineless] at decreased day/night temperatures (Sionit et al., 1981). Potvin (1985) studied the C_4 grasses *Echinocloa crusgalli* and *Elusine indica* grown over a temperature range and elevated CO_2. Plants grown at elevated CO_2 has less of a reduction in photosynthetic rates and stomatal conductance when exposed to chilling than plants grown at ambient CO_2. In contrast, Baker and Allen (1994) reviewed canopy level observations for cotton, rice, and soybean. They state that there were only small differences in the photosynthetic rate responses across a wide range of air temperatures with increasing CO_2 which they attribute to increased evaporative cooling of leaves at increasing air temperatures.

The interaction of increasing CO_2 and temperature on tree growth and physiology has been reported for only a small number of studies. Callaway et al. (1994) measured growth parameters on ponderosa pine (*Pinus ponderosa* Dougl. ex Laws) seedlings at 350 μLL^{-1} and 650 μLL^{-1} CO_2 and night/day temperatures of 10/25°C and 15/30°C. Although biomass was higher after the first month in the elevated CO_2 and temperature treatment, after two months the relative growth rates did not differ among treatments. Observations of mature trees in the field also suggest a limited growth response to elevated CO_2 (Callaway et al., 1994). The authors conclude that shifts in biomass allocation and gas exchange in response to temperature and atmospheric CO_2 may not result in increased carbon sequestration for western forests. In a study of ponderosa pine and loblolly pine (*Pinus taeda* L.) responses to temperature, CO_2, and nitrogen, ponderosa pine showed a 105% increase in total root biomass which mirrored the whole plant response (King et al., 1996). The species was not responsive to temperature and nitrogen, which was thought to reflect the conservative growth strategy of this late-successional species. Loblolly pine accumulated more root biomass at higher temperatures, within nitrogen treatments at higher temperatures and elevated CO_2, and at low nitrogen concentrations. The species appears to have a greater capacity to reduce carbon losses from increased respiration and photorespiration at high temperatures and to exploit low soil nitrogen sites. The results indicate that both species have the potential to increase belowground biomass in response to enhanced CO_2 and that the response of loblolly pine is sensitive to temperature and nitrogen (King et al., 1996). In contrast, Wang et al. (1995) reported a reduction in net carbon assimilation for Scots pine (*Pinus sylvestris* L.) at elevated CO_2 and higher air temperatures. These findings are consistent with the data presented by Teskey in this volume.

Ido and Idso (1994) plotted the results of CO_2 and temperature interaction studies for a ten-year period in an attempt to generalize plant responses. Their results show that relative growth increases with enhanced CO_2 over the range of higher temperatures. This effect is attributed to the carbon exchange rate efficiency reducing the process of photorespiration and the inhibition of photorespiration at higher atmospheric CO_2 concentrations. The effect of both processes is to raise the optimum temperature for plant growth (McMurtrie et al., 1992; McMurtrie and Wang, 1993).

Air Pollutants

Regional air pollutants have been shown to suppress growth and yield of some crops (Miller, 1987; Heagle, 1989) and forest tree species (see review by Berrang et al., 1996). A recent review by Allen and Gholz (1996) characterized air quality and atmospheric deposition for the southeastern United States. Although the focus of national monitoring networks is ozone, nitrogen oxides, and sulfur dioxide, monitoring in the southeastern United States had shown that concentrations of sulfur dioxide and nitrogen oxide are rarely sufficient in duration and concentration to cause visible injury to plants and trees. In contrast, ozone has been linked to visible injury and reduction in growth and yield of crops and forest trees in the region. These finding have been summarized for the National Acidic Precipitation Assessment Program's (NAPAP) National Crop Loss Assessment Program (NCLAN) (Heck et al., 1988) and the Forest Response Program (FRP) (Fox and Mickler, 1996). One current topic of interest is the interaction of ozone and elevated concentrations of atmospheric CO_2 and its impact on plant growth and yield.

An early study on tobacco demonstrated that the addition of 500 μLL^{-1} CO_2 to ambient air reduced visible injury when tobacco plants were exposed to acute concentration of ozone (Heck and Dunning, 1967). Similar greenhouse studies with a doubling of ambient concentrations of CO_2 and ozone exposures demonstrated reduced suppression of growth or biomass in wheat (*Triticum aestivum* L.) (Mortensen, 1990), tomato (*Lycopersicon esculentum* L.) (Mortensen, 1992), radish (*Raphanus sativus* L.) (Barnes and Pfirrmann, 1992), white clover (*Trifolium repens* L.) (Heagle et al., 1993), and soybean (*Glycine max* L. Merr) (Reinert and Ho, 1995). These studies were conducted in controlled environment chambers. In the only reported field study, soybean was exposed in open-top chambers to 350, 400, and 500 μLL^{-1} CO_2 and 20, 40, and 80 μLL^{-1} ozone. The CO_2 enrichment has a similar ameliorating effect on plant photosynthesis and growth. The limited number of crop studies and the lack of studies on trees illustrates that the potential interaction of ozone and carbon dioxide have not been adequately described in the literature for field studies and over a range of concentrations of CO_2 and ozone.

References

Allen LH Jr. (1990) Plant responses to rising carbon dioxide and potential interactions with air pollutants. J Environ Qual 19:15–34.

Allen HL, Dougherty PM, Cambell RG (1990) Manipulation of water and nutrients—practice and opportunity in Southern U.S. pine forests. Forest Ecol Manage 30: 437–453.

Allen ER, Gholz HL (1996) Air quality and atmospheric deposition in southern U.S. forests. In Fox S, Mickler RA (Eds) *Impacts of air pollutants on southern pine forests.* Springer-Verlag, New York.

Baker JT and Allen LH, Jr. (1994) Assessment of the impact of rising carbon dioxide and other potential climate changes on vegetation. Environ Pollut 83:223–235.

Barnes JD, Pfirrmann T (1992) The influence of CO_2 and O_3, singly and in combination, on gas exchange, growth and nutrient status of radish (*Raphanus sativus* L.). New Phytol 121:403–412.

Barnola JM, Raynaud D, Lorius C, Korotkevich YS (1994) Historical CO_2 record from the Vostok ice core. In Boden TA, Kaiser DP, Sepanski RJ, Stoss FW (Eds) *Trends '93: A Compendium of Data on Global Change.* ORNL/CDIAC-65. Carbon Dioxide Information Analysis Center, Oak Ridge National Laboratory, Oak Ridge, Tenn., 7–10.

Bazzaz FA (1990) The response of natural ecosystems to the rising global CO_2 levels. Ann Rev Ecol Syst 21:167–196.

Bazzaz FA, Colemen JS, Morse SR (1990) Growth responses of seven major co-occurring tree species of the northeastern United States to elevated CO_2. Can J For Res 20:1479–1484.

Berrang P, Meadows JS Hodges JD (1996) An overview of responses of southern pines to airborne chemical stresses. In Fox S, Mickler RA (Eds) *Impacts of air pollutants on southern pine forests.* Springer-Verlag, New York.

Berry J, Bjorkman O (1980) Photosynthetic response and adaptation to temperature in higher plants. Ann Rev Plant Physiol 31:491–543.

Bowes G (1993) Facing the inevitable: Plants and increasing atmospheric CO_2. Annu Rev Plant Physiol Plant Mol Biol 44:309–332.

Bowman WD, Strain BR (1987) Interaction between CO_2 enrichment and salinity stress in the C_4 non-halophyte *Andropogon glomeratus.* Plant Cell Environ 10:267–270.

Brun WA, Cooper RL (1967) Effects of light intensity and carbon dioxide concentration on photosynthetic rate of soybean. Crop Sci 7:451–454.

Budyko MI, Ronov AB, Yanshin, AL (1987) History of the Earth's atmosphere. Springer-Verlag, New York.

Callaway RM, DeLucia EH, Schlesinger WH (1994) Biomass allocation of montane and sesert ponderosa pine: An analog for response to climate change. Ecology 75(5):1474–1481.

Callaway RM, DeLucia EH, Thomas EM, Schlesinger WH (1994) Compensatory responses of CO_2 exchange and biomass allocation and their effects on the relative growth rate of ponderosa pine in different CO_2 and temperature regimes. Oecologia 98:159–166.

Ceulemans R, Mousseau M (1994) Effects of elevated atmospheric CO_2 on woody plants. New Phytolog 127:425–446.

Comins HV, McMurtrie RE (1993) Long-term biotic response of nutrient-limited forest ecosystems to CO_2-enrichment; equilibrium behavior of integrated plant-soil models. Ecol Appl 3: 66–681.

Conroy JP, Milham PJ, Mazur M, Barlow EWR (1990) Growth, dry weight partioning, and wood properties of *Pinus radiata* D. after 2 years of CO_2 enrichment. Plant Cell Environ 13: 329–337.

Conway TJ, Tans PP, Waterman LS (1994) Atmospheric CO_2 record from sites in the NOAA/CMDL air sampling network. In Boden TA, Kaiser DP, Sepanski RJ, Stoss FW (Eds) *Trends '93: A Compendium of Data on Global Change.* ORNL/CDIAC-65. Carbon Dioxide Information Analysis Center, Oak Ridge National Laboratory, Oak Ridge, Tenn., 41–119.

Cure JD, Acock B (1986) Crop response to carbon dioxide doubling: a literature survey. Agric For Meterol 38:127–145.

Cure JD (1985) Carbon dioxide doubling response: A crop survey. In Strain BR, Cure JD (Eds) *Direct effects of increasing carbon dioxide doubling on vegetation.* DOE/ER-0238. U.S. Department of Energy, Carbon Dioxide Res. Div., Washington, DC, 99-116.

Davis MB (1989) Lags in vegetation response to greenhouse warming. Clim Change 15:75.

Dobson A, Jolly A, Rubenstein D (1989) The greenhouse effect and biological diversity. Trends Ecol Evol 4:64.

Eamus D, Jarvis PG (1989) The direct effects of increase in the global atmospheric CO_2 concentration on natural and commercial temperate trees and forests. Adv Ecol Res 19:1–55.

Ehleringer JR, Sage RF, Flanagan LB, Pearcy RW (1991) Climate change and the evolution of C_4 photosynthesis. Trends Ecol Evol 6:95–99.

Fox S, Mickler RA (Eds) (1996) *Impacts of air pollutants on southern pine forests.* Springer-Verlag, New York.

Gifford (1992) Interaction of carbon dioxide with growth-limiting environmental factors in vegetation productivity: implications for the global carbon cycle. In Stanhill G (Ed) *Advances in Bioclimatology.* Springer-Verlag, New York, 24–58.

Gunderson CA, Norby RJ, Wullschleger SD (1993) Foliar gas exchange responses to two deciduous hardwoods during 3 years of growth in elevated CO_2: no loss of photosynthetic enhancement. Plant Cell Environ 16:797–807.

Gunderson CA, Wullschleger SD (1994) Photosynthetic acclimation in trees to rising atmospheric CO_2: a broader perspective. Photosynthesis Res 39:369–388.

Heagle AS (1989) Ozone and crop yield. Ann Rev Phytopath 27:397–423.

Heagle AS, Miller JE, Sherill DE, Rawlings JO (1993) Effects of ozone and carbon dioxide mixtures on two clones of white clover. New Phytol 123:751–762.

Heck WW, Dunning JA (1967) The effect of ozone on tobacco and pinto bean as conditioned by several ecological factors. J Air Pollut Control Assoc 17:112–114.

Heck WW, Taylor OC, Tingey DT (Eds) (1988) *Assessment of crop loss from air pollutants.* Elesevier Applied Science, London.

Idso KE, Idso SB (1994) Plant responses to atmospheric CO_2 enrichment in the face of environmental constraints: a review of the past 10 years' research. Agric For Meterol 69:153–203.

Idso SB, Kimball BA (1989) Growth response of carrot and radish to atmospheric CO_2 enrichment. Environ Exp Bot 29:135–139.

Idso SB, Kimball BA, Anderson MG, Mauney JR (1987) Effects of atmospheric CO_2 enrichment on plant growth: the interactive role of air temperature. Agric Ecosyst Environ 20:1–10.

Idso SB, Kimball BA, Mauney JR (1988) Effects of atmospheric CO_2 enrichment on root:shoot ratios of carrot, radish, cotton, and soybean. Agric Ecosyst Environ 22:293–299.

King JS, Thomas RB, Strain BR (1996) Growth and carbon accumulation in root systems of *Pinus taeda* and *Pinus ponderosa* seedlings as affected by varying CO_2, temperature, and nitrogen. Tree Physiol 16:635–642.

Kirschbaum MUF, King DA, Comins HN, McMurtrie RE, Medlyn BE, Pongracic S, Murty D, Keith H, Raison RJ, Khanna PK, Sheriff DW (1994) Modelling forest responses to increasing CO_2 concentration under nutrient-limiting conditions. Plant Cell Environ 17:1081–1099.

Kramer PJ (1981) Carbon dioxide concentration, photosynthesis, and dry matter production. BioScience 31:29–33.

Lewis JD, Griffin KL, Thomas RB, Strain BR (1994) Phosphorus supply affects the

photosynthetic capacity of loblolly pine grown in elevated carbon dioxide. Tree Physiol 14:1229–1244.
Loehle C (1995) Anomalous responses of plants to CO_2 enrichment. OIKOS 73:181–187.
Long SP (1991) Modification of the response of photosynthetic productivity to rising atmospheric CO_2 concentrations: Has its importance been underestimated? Plant Cell Environ 14:729–739.
Long SP, Drake BG (1991) Effect of the long-term elevation of CO_2 concentration in the field on the quantum yield of photosynthesis of the C_3 sedge, *Scirpus olneyi*. Plant Physiol 96:221–226.
McMurtrie RE, Comins HN, Kirschbaum MUF, Wang YP (1992) Modifying existing forest growth models to take account of effects of elevated CO_2. Aust J Bot 4:657–677.
McMurtrie RE, Wang YP (1993) Mathematical models of the photosynthetic response of tree stands to rising CO_2 concentrations and temperatures. Plant Cell Environ 16:1–13.
Miller JE (1987) Effects on photosynthesis, carbon allocation, and plant growth. In Heck WW, Taylor OC, Tingey DT (Eds) *Assessment of crop loss from air pollutants* Elesevier Applied Science, London.
Mortensen LM (1990) Effects of ozone on growth of *Triticum aestivum* L. at different light, air humidity and CO_2 levels. Norwegian J Agric Sci 4:343–348.
Mortensen LM (1992) Effects of ozone concentration on growth of tomato at various light, air humidity and carbon dioxide levels. Scientia Horticulturae 49:17–24.
Musselman RC, Fox DG (1991) A review of the role of temperate forests in the global CO_2 balance. J Air Waste Manage Assoc 41(8):798–807.
Neftel A, Friedli H, Moor E, Lotscher H, Oeschger H, Siegenthaler U, Stauffer B (1994) Historical CO_2 record from the Siple Station ice core. In Boden TA, Kaiser DP, Sepanski RJ, Stoss FW (Eds) *Trends '93: A Compendium of Data on GlobalChange*. ORNL/CDIAC-65. Carbon Dioxide Information Analysis Center, Oak Ridge National Laboratory, Oak Ridge, Tenn., 11–14.
Nijs I, Impens I, Behaeghe T (1989a) Effects of long-term elevated atmospheric CO_2 concentrtions on *Lolium perenne* and *Trifolium repens* canopies in the course of a terminal drought stress period. Can J Bot 67:2720–2725.
Nijs I, Impens I, Behaeghe T (1989b) Leaf and canopy responses of *Lolium perenne* to long-term elevated atmospheric carbon-dioxide concentrations. Planta 177:312–320.
Norby RJ, O'Neill EG (1989) Growth dynamics and water use of seedlings of *Quercus alba* L. in CO_2-enriched atmospheres. New Phytol 111:491–500.
Norby RJ, O'Neill EG (1991) Leaf area compensation and nutrient interactions in CO_2-enriched seedlings of yellow-poplar (*Liriodendron tulipifera* L.) New Phytol 117:515–528.
Norby RJ, O'Neil EG, Luxmoore RJ (1986) Effects of atmospheric CO_2 enrichment on the growth and mineral nutrition of *Quercus alba* seedlings in nutrient-poor soil. Plant Physiol 82:83–89.
Paoletti E, Gellini R (1993) Stomatal density variation in beech and holm oak leaves collected over the last 200 years. Acta Oecologia 14:173–178.
Pearcy RW, Bjorkman O (1983) Physiological effects. In Lemon ER (Ed) *Carbon Dioxide and Plants: The Response of Plants to Rising Levels of Atmospheric Carbon Dioxide*. Westview Press, Boulder, CO, 65–105.
Poorter H (1993) Interspecific variation in the growth response of plants to an elevated ambient CO_2 concentration. Vegetatio 104/105:77–97.
Potvin C (1985) Amelioration of chilling effects by CO_2 enrichment. Physiol Veg 23:345–352.
Reekie EG, Bazzaz FA (1989) Competition and patterns of resource use among seedlings of five tropical trees grown at ambient and elevated CO_2. Oecologia 79:212–222.
Reinert RA, Ho MC (1995) Vegetative growth of soybean as affected by elevated carbon dioxide and ozone. Environ Pollut 89:89–96.
Richter DD, Markewitz D (1995) Atmospheric deposition and soil resources of the south-

ern pine forest. In Fox S, Mickler, RA (Eds) *Impact of Air Pollution of Southern Pine Forests*. Springer-Verlag, New York, 315–336.

Rogers HH, Runion GB (1994) Plant responses to atmospheric CO_2 enrichment with emphasis on roots and their rhizosphere. Environ Pollut 83:155–189.

Rogers HH, Sionit N, Cure JD, Smith JM, Bingham GE (1984) Influence of elevated carbon dioxide on water relations of soybean. Plant Physiol 74:233–238.

Sage RF (1994) Acclimation of photosynthesis to increasing CO_2: the gas exchange perspective. Photosynthesis Res 39:351–368.

Schwartz JR, Gale J (1984) Growth response to salinity at high levels of carbon dioxide. J Exp Bot 35:193–196.

Silvola J, Alholm U (1992) Photosynthesis in willow (*Salix* x *dasyclados*) grown at different CO_2 concentrations and fertilization levels. Oecologia 91:208–213.

Sionit N, Hellmers H, Strain BR (1982) Interaction of atmospheric CO_2 enrichment and irradiance on plant growth. J Agron 74:721–725.

Sionit N, Strain BR, Beckford HA (1981) Environmental controls on growth and yield of okra. I. Effects of temperature and CO_2 at cool temperature. Crop Sci 25:533–537.

Smith SD, Strain BR, Sharkey TD (1987) Effects of CO_2 enrichment on four Great Basin grasses. Func Ecol 1:139–143.

Strain BR, Thomas RB (1992) Field measurements of CO_2 enhancement and climate change in natural vegetation. Water Air Soil Pollut 64:45–60.

Tissue DT, Thomas RB, Strain BR (1993) Long-term effects of elevated CO_2 and nutrients on photosynthesis and rubisco in loblolly pine seedlings. Plant Cell Environ 16:859–865.

Thomas RB, Lewis JD, Strain BR (1994) Effects of leaf nutrient status on photosynthesis capicity in loblolly pine (*Pinus taeda* L.) Seedlings grown in elevated atmospheric CO_2. Tree Physiol 14:947–960.

Tolley LC, Strain BR (1985) Effects of CO_2 enrichment and water stress on gas exchange of *Liquidamber styraciflua* and *Pinus taeda* seedlings grown under different irradiation levels. Oecologia 65:166–172.

Trabalka JR, Edwards JA, Reilly JM, Gardner RH, Reichle DE (1986) Atmospheric CO_2 projection with globally averaged carbon cycle models. In Trabalka JR, Reichle DE (Eds) *The Changing Carbon Cycle: A Global Analysis*. Springer, New York, 534–560.

Tschaplinski TJ, Norby RJ, Wullschleger SD (1993) Responses of loblolly pine seedlings to elevated CO_2 and fluctuating water supply. Tree Physiol 13:283–296.

Valle R, Mishoe JW, Cambell JW, Allen LH Jr. (1985) Photosynthetic responses of 'Bragg' soybean leaves adapted to different CO_2 environments. Crop Sci 25:333–339.

Wang K, Kellomaki LK (1995) Effects of needle age, long-term temperature, and CO_2 treatments on the photosynthesis of Scots pine. Tree Physiol 15:211–218.

Watson RT, Rohde H, Oeschger H, Siegenthaler U (1990) Greenhouse gases and aerosols. In: Houghton JT, Jenkins GJ, Ephraum JJ (Eds) *Climate Change: the IPCC Scientific Assessment*. Cambridge University Press, Cambridge, 1–40.

Williams WE, Garbutt K, Bazzaz FA, Vitousek, PM (1986) The response of plants to elevated CO_2 IV. Two deciduous forest communities. Oecologia 69:454–459.

Woodward FI (1987) Stomatal numbers are sensitive to increases in CO_2 from preindustrial levels. Nature 327:617–618.

Woodward FI, Bazzaz FA (1988) The responses of stomatal density to CO_2 partial pressure. J Exp Bot 39:1771–1781.

Wullschleger SD, Norby RJ, Hendrix DL (1992) Carbon exchange rates, chlorophyll content, and carbohydrate status of two forest tree species exposed to carbon dioxide enrichment. Tree Physiol 10:21–31.

Yapp CJ, Poths H (1992) Ancient atmospheric CO_2 pressures inferred from natural goethites. Nature 355:342–44.

2. General Circulation Model Scenarios for the Southern United States

Ellen J. Cooter

The "Southern United States" extends roughly from 75° to 100° west longitude and from 30° to 37° north latitude (Figure 2.1). Elevations within the region range from near sea level along the Gulf and Atlantic coasts to more than 1800m in the Appalachian Mountains. The climate of locations within the southeastern United States is determined primarily by latitude, by proximity to the Gulf of Mexico and the Atlantic Ocean, and by altitude. Overall, the climate is temperate, becoming largely subtropical near the coast. Summers are long, hot, and humid, with little day-to-day temperature change. Late June to mid-August receive local afternoon thundershowers. The coldest months are December, January, and February, during which the region is subject to frequent shifts between exposure to moist, mild, Gulf air and cool, dry, continental air. Severely cold weather seldom occurs. Except at higher elevations, temperatures of -17.8 °C or lower are rare and occur only when there is snow on the ground. The last spring freeze (T < -2.2 °C) generally occurs between March 15 and April 15. The first fall freeze (T < -2.2 °C) generally occurs between October 15 and November 15 (Koss et al., 1988).

Precipitation is nearly all in the form of rain and varies greatly from year to year. Snow can occur frequently at higher elevations, but is usually reported only once or twice a year at most locations. Interannual rainfall patterns range from bimodal in the southern coastal regions of Mississippi, Alabama, and Georgia, to nearly uniform in North Carolina and Virginia. Nearly all precipitation is from local thundershowers, which most frequently occur in the afternoon. During late August and in September, summer atmospheric temperature and moisture condi-

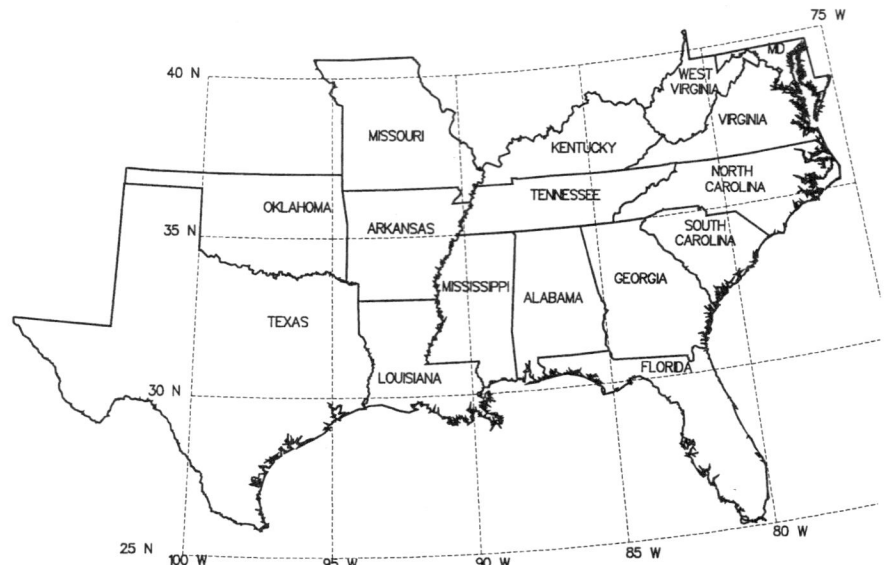

Figure 2.1. Map of the Southern Global Change Program (SGCP) study region.

tions persist, but thundershowers become less frequent. Late night and early morning thundershowers, characteristic of late summer on the coast, continue until mid-September. Rains during October are nearly always from showers or thundershowers occurring ahead of temperature drops. Such changes become more frequent and more pronounced as winter approaches. Dry, sunny weather prevails during September and October, but from August through early October, heavy rainfall may occur when tropical disturbances or hurricanes move inland from the Gulf of Mexico or Atlantic Ocean.

Droughts may occur any time during the growing season, from late April through October. Relatively long periods with little or no rain are more likely to occur in late summer and autumn than at any other time, while a secondary maxima of such periods occurs in May and June. Severe local droughts occur nearly every year, but severe region-wide droughts are relatively rare.

A brief description of the historical database and methodologies employed to project changes from these historical conditions are presented. This is followed by a summary of region-wide and within-region changes projected by four General Circulating Models (GCM's) and applied to the southern U.S. historical database.

A Historical Climate Database for Southern U.S. Applications

Temperature and Precipitation

A discussion of the current state of biological modeling of forests is presented in Dale and Rauscher (1994). With few exceptions, the studies described in this volume, which include several of these models, focused on relatively small spatial (less than 10^4 ha) and temporal (10 to 10^2 years) scales. To support such modeling studies, a historical climate database for the southern United States, which represents areas on the order of .1° latitude/longitude rectangular resolution and contains long, continuous time series of daily maximum temperature, minimum temperature, precipitation, humidity, and solar radiation, is desirable. The database should also adequately capture naturally occurring spatial and temporal intercorrelations between time series, and seasonal and interannual variability with time series.

The Richman–Lamb historical database meets many of these requirements. The data (daily maximum temperature, minimum temperature, and precipitation) are spatially and temporally consistent, continuous through time (1949 through 1993), are of the greatest spatial resolution practicable using observational data (a nearly regular 1° latitude by longitude grid), and are easily accessed. Research using these data indicates that they are of sufficient quality and are appropriate for regional impact assessment applications (Richman and Lamb, 1985; Richman and Montroy, 1996), but should not be used for the detection of subtle, small-scale changes or local long-term climate trends.

Solar Radiation

One of the most important and least available climate variables required by biological models is solar radiation. The most frequently reported radiation component is global radiation. Global radiation is the sum of incoming direct and diffuse shortwave radiation received by (incident upon) a unit horizontal surface. If this surface is a leaf, then global radiation can be partitioned into that which is absorbed (photosynthetically active, PAR), that which is reflected, and that which is transmitted (Bannister, 1976). Most biological calculations of leaf energy budgets begin on the basis of a plane, isolated, and horizontal leaf in the air. These estimates can then be adjusted for plant-specific physiology and canopy architecture.

Global radiation observations on the same spatial scale as the temperature and precipitation database were not available, and so several stochastic models of solar radiation were considered. The model reported in Hodges et al. (1985) was selected. The original model parameterizations were developed for use in regional crop assessments with the CERES-Maize physiological model (Hodges et al., 1987). This combination was later used for regional climate change impact assessment studies (Smith and Tirpak, 1990; Cooter, 1990). It is well suited for regional time series studies because of its minimal input requirements and its response to precipitation events and interannual variability.

The Hodges model was reparameterized for the southeastern United States using observed and modeled global radiation data from the Solar and Meteorological Surface Observation Network (SAMSON; NREL, 1992) which improved its representation of spatial and temporal patterns of historical solar radiation time series. Details of the model revision and performance in two maize process models are presented in Cooter and Dhakhwa (1996). Each historical time series of temperature and precipitation for the SGCP research area was then used as input to the revised radiation model, and daily global radiation time series for the 1949 through 1993 time period were generated.

Vapor Pressure Deficit

Modelers often use stomatal response to vapor pressure deficit (VPD) to simulate the direct effects of atmospheric humidity on forest productivity. Indirect effects are often modeled through environmental stresses such as the restricted availability and transport of moisture and nutrients.

Vapor pressure deficit (VPD) was chosen to represent atmospheric humidity in this database. Following Murray (1967), VPDs were obtained from

$$VPD = e_s - e \tag{2.1}$$

Where VPD is day-average vapor pressure deficit (mb). The vapor pressure (e, mb), is determined from the dewpoint temperature (SDEW, °C) according to

$$e = 6.1078 * e^{(17.269 * SDEW)/(237.3 + SDEW)} \tag{2.2}$$

and saturation vapor pressure, (e_s, mb) was obtained from SDEW and the average temperature for the day (STEMP, °C) as

$$e_s = 6.1078 * e^{(17.269 * STEMP)/(237.3 + STEMP)} \tag{2.3}$$

Dewpoint temperature estimates are not widely available beyond First Order Meteorological Station sites, and so a necessary assumption is that the dewpoint is approximately equivalent to the 24-hour minimum temperature. Friend (1996) found that this assumption provided good estimates of regional patterns of humidity in mid-latitudes (Europe and North America), but not in arid portions of Africa. Running et al. (1987) also found that the correlation between minimum temperature and dewpoint temperature deteriorated in arid environments. Following up on this finding, Kimball et al. (unpublished manuscript) examined the impact of this assumption on estimated humidity at First Order Meteorological Station sites throughout the United States. They found that, for data from humid subtropical regions (which includes all of the SGCP area, including sites as far west as Tulsa, Oklahoma and Austin, Texas), the use of minimum daily temperature versus reported dewpoint temperature resulted in vapor pressure differences of less then .1 kPa (1 mb). Given these results, the dewpoint/minimum temperature assumption was accepted for the present assessment and daily time series of VPD for the SGCP study region, which are temporally and spatially consistent with temperature observations could be generated..

GCMs—Tools to Explore Change

In 1990 the Intergovernmental Panel on Climate Change (IPCC) announced that, while recognizing the considerable uncertainty in the entire issue of climate

2. Circulation Models for the Southern United States

change, several statements concerning anticipated climate change in response to increases in CO_2 and other trace gases could be made with some certainty (Table 2.1) (IPCC, 1990). These include statements concerning changes in temperature, precipitation, soil moisture, snow, and sea ice.

In 1992, the IPCC issued a supplement to the original report in which current findings concerning observed climate variability and change were reported. A partial list of findings, which includes the observation of surface and mid-tropospheric warming, is provided in Table 2.2. The executive summary concludes:

> It is still not possible to attribute with high confidence all, or even a large part of, the observed global warming to the enhanced greenhouse effect. On the other hand, it is not possible to refute the claim that greenhouse-gas-induced climate change has contributed substantially to the observed warming. (Folland et al., 1992).

GCMs are widely considered one of the best tools available for the exploration of future climate conditions under great uncertainty. Although GCMs are far from being acurate tools of prediction, it has been demonstrated that (1) some are able to simulate the important large-scale features of the climate system well, including seasonal, geographical, and vertical variations of forcing and dynamics in space and time; (2) many climate changes are consistently projected by different models in response to greenhouse gases and aerosols and can be explained in terms of physical processes which are known to be operating in the real world; and (3) GCM results exhibit "natural" variability on a wide range of time- and spacescales which is broadly comparable to that observed (IPCC, 1996).

GCMs are three-dimensional (latitude, longitude, height) models of the climate system. They simulate the physical processes that determine large-scale climate. All GCMs attempt to solve a fundamental set of physical equations representing the conservation of mass, momentum, and energy as well as equations of motion, state, and radiative transfer; predictive equations for water vapor and heat energy balances at the earth's surface are also included. The source terms in these equations include numerical representations of turbulent transfer at the groundatmosphere boundary, cloud formation, condensation or rain, and transport of heat by ocean-boundary currents. All these process simulations are ultimately driven by the spatial and temporal distribution of solar radiation. The selection of which processes to model explicitly, to parameterize, or to eliminate—as well as input and output variables and time aggregation—vary with the model and model version.

The first widely released generation of GCMs (circa early- to mid-1980s) estimated conditions over relatively few, very large grid cells and for equilibrium double CO_2 conditions. The results of these models have been made widely available to scientific and policy communities. As a result, their advantages, disadvantages, caveats, and assumptions are fairly well known. Some of the most notable areas of uncertainty included coupling of the ocean and atmosphere, transient (time dependent) greenhouse gas changes, explicit regional changes, and climate feedback and sensitivity.

Table 2.1. Main Equilibrium Changes in Climate Due to Doubling CO_2 Deduced from a Suite of GCMs

Rating	Description
Temperature	
*****	The lower atmosphere and Earth's surface warm
*****	The stratosphere cools
***	Near the Earth's surface, the global average warming lies between +1.5 °C and +4.5 °C, with a "best guess" of 2.5 °C
***	The surface warming at high latitudes is greater than the global average in winter but smaller than in summer. (In time dependent simulations with a deep ocean, there is little warming over the high latitude southern ocean.)
***	The surface warming and its seasonal variation are least in the tropics
Precipitation	
****	The global average increases (as does that of evaporation); the larger the warming, the larger the increase
***	Increases at high latitudes throughout the year
***	Increases globally by 3 to 15% (as does evaporation)
Soil moisture	
***	Increases in high latitudes in winter
Snow and sea-ice	
****	The area of sea ice and seasonal snow-cover diminish

Items listed were assigned three or more *s on a five * scale. Five *s indicate virtual certainties, one * indicates low confidence (Mitchell et al., 1990).

Table 2.2. Selected Findings of the IPPC Concerning Observed Climate Variability and Change

Temperature

- Continuing research into the nineteenth century ocean temperature record has not significantly altered our calculation of surface temperature warming of 0.45 ± 0.15 °C since the late nineteenth century.
- A new analysis of radiosonde data confirms that mid-tropospheric warming has occurred over the past several decades.
- Microwave Sounding Unit (MSU) data provide a more complete satellite-based global data set for tropospheric and stratospheric mean temperatures, but the record is still too short for a meaningful assessment of trends.

Precipitation

- Precipitation variations of practical significance have been documented in a number of regions on many time and space scales. Because of data coverage and inhomogeneity problems, however, we cannot yet say anything new about global-scale changes.
- Evidence continues to support an increase in water vapor in the tropical lower troposphere since the mid-1970s, though the magnitude is uncertain. However, we cannot say whether the changes are larger than natural variability.

From Folland et al., 1992.

A second generation of GCMs emerged during the late 1980s and early 1990s which address many of these uncertainties. A summary of their findings are reported in Gates et al. (1992). These findings are updated yet again in IPCC (1996) to reflect the now widespread use of transient, fully coupled atmosphere-ocean GCMs. Many of the general findings reported in IPCC (1990) and Gates (1992) remain unchanged, but the latest generation of GCMs now estimate a slightly lower mean global temperature change: +1.5 °C to +4.5 °C (circa. 1990), +1 °C to +3.5 °C by the year 2100 (c. 1996). The 1996 estimate reflects model advances that address the role of aerosols in the pre-1990 radiative forcing history, a revised understanding of the carbon cycle, and other important, but previously missing or overly simplified atmospheric processes and attributes (Kattenberg et al., 1996). In spite of notable model advances, confidence in regional change projections remains low. The key factors that affect the regional performance of global-coupled models are their horizontal resolution and their physical parameterizations. The characterization of cloud and radiative effects on the hydrologic balance over land and of heat flux at the ocean surface remain significant sources of overall model uncertainty (Gates et al., 1996).

With the large number of GCM projections available to applications research scientists (more than twenty in 1993), how does one identify "which model is the best?" A common evaluation approach is to compare model results to historical climate records, but this must be done carefully. Comparative performance in simulating current and past climate varies from region to region and season to season. One initiative, designed specifically to facilitate such evaluations, has been organized by the World Meteorological Organization's (WMO) World Climate Research Programme (WCRP) and is called the Atmospheric Model Intercomparison Project (AMIP). AMIP evaluates the ability of atmospheric GCMs to simulate the global climate of the decade 1979 to 1988 (Gates, 1992). Twenty-five diagnostic subprojects have been examining various aspects of the simulations. Reports have been prepared that compare, for example, radiation, cloudiness, soil moisture, humidity, precipitation, and vegetation parameterization performance. Readers should contact the WMO-WCRP for a complete list of AMIP reports and journal articles.

The GCM Output

The SGCP assessment required the development of a set of climate change scenarios that parallel in space and time resolution, those for the historical period. A database representing a range of future conditions, rather than a single outcome, was developed using monthly equilibrium summary output arrays obtained from the National Center for Atmospheric Research (NCAR) archives for the NASA Goddard Institute for Space Studies (GISS; Hansen et al., 1983), the Geophysical Fluid Dynamics Laboratory (GFDL; Manabe and Wetherald, 1987), the Oregon State University (OSU; Schlesinger and Zhao, 1989), and the United Kingdom Meteorological Office (UKMO; Wilson and Mitchell, 1987) GCMs. The general

Table 2.3. Attributes of Four General Circulation Models Used for SGCP Scenario Development

Model	GFDL	GISS	OSU	UKMO
Date of output generation	1988 (Q-flux)	1982 (Q-flux)	1984–1985	1986
Numerical solution technique	Spectral (R-15)	Finite difference	Finite difference	Finite difference
Horizontal resolution (Lat × Long)	4.5° × 7.5°	7.8° × 10.0°	4.04° × 5.0°	5.0° × 7.5°
Vertical resolution layers	9	9	2	11
Initial CO_2 concentration (ppm)	300	315	326	320
Solar constant (W/m^2)	1467	1367	1354	1395
Diurnal cycle	No	Yes	No	Yes
Surface characterization	Uniform	Fractional	Uniform	Uniform
Land cover	6 vegetation types	8 vegetation types	Snow, sea ice, 6 veg. types	Sea, sea-ice, snow, snow-free land
Convective parameterization	Moist adiabatic	Penetrating convection	Penetrating convection	Penetrating convection
Cloud cover	On/off, RH>99%	On/off, RH>100%	On/off, RH>100%	Fractional
Ocean parameterization	60-meter slab, prescribed	65-meter slab, prescribed	60-meter slab, prescribed	50-meter slab, prescribed

characteristics of these second generation models are presented in Table 2.3.

Baseline (1 × CO_2) and double CO_2 (2 × CO_2) temperature, precipitation, humidity, and solar radiation fields produced by these models were retrieved. Unfortunately, the temporal and spatial scales of these fields were inadequate for SGCP applications (Cooter et al., 1993; Dale and Rauscher, 1994) and so, scenario generation techniques were applied to provide more useful climate change information.

Change Scenarios for the Southern United States

The relative certainty expressed in Tables 2.1 and 2.2 all but disappears when the implications of such large-scale findings for smaller regions such as the southern United States are explored. Scenario development techniques must be employed. Robinson and Finkelstein (1991) define a scenario as one possible set of future climate conditions. A scenario should be internally consistent, developed using sound scientific principles, and have no specific probability of occurrence attached.

Giorgi and Mearns (1991) describe a range of possible scenario generation techniques that have been used in the past. A semiempirical approach was adopted for the SGCP scenario database. Semiempirical approaches attempt to translate large-scale, GCM information into local statistics by using empirically derived relationships between large-scale and local-surface variables. The basic assumption underlying this approach is that the inaccuracy from the coarse GCM resolution is reduced when GCM-produced baseline and 2 × CO_2 field differences are applied to higher-resolution observed data. The classic semiempirical methodology applied here lacks sophistication but has provided useful results (e.g., Smith and Tirpak, 1990; Cooter, 1990). The application of this approach to each historical time series along with appropriate assumptions and caveats are described later. A more detailed discussion is found in Cooter et al. (submitted). One example of more sophisticated statistical "down-scaling" scenario techniques is found in von Storch et al. (1993) and Gyalistras et al. (1994).

Temperature

Two temperature scenario options considered were equal day and nighttime warming and differential warming. For equal warming, SGCP 2 × CO_2 temperature time series were constructed by adding the difference between the 1 × CO_2 and 2 × CO_2 GCM mean monthly temperatures to daily maximum and minimum historical temperatures. Equal warming scenarios were applied by Smith and Tirpak (1990) and Cooter (1990). Differential warming scenarios do not assume that day and nighttime temperatures warm by the same amount. To construct such scenarios requires that the GCM include diurnal temperature projections. Table 2.3 indicates that only the GISS and UKMO models include diurnal cycles and so a full suite of GCM-directed differential warming scenarios could not be developed.

A second diurnal temperature scenario, based on the analysis of historical time series, was proposed in Karl et al. (1993). They determined that minimum temperature (nighttime) warming has been approximately three times that observed during the day (maximum temperature) and that the diurnal temperature range is approximately equal to the mean monthly temperature change. Implementation of this scenario was not possible for the SGCP assessment, but preliminary applications have been explored.

The equal warming scenario was used for the SGCP. Temporally, the mean intraannual climate change reflected in each daily time series is dictated entirely by changes between $1 \times CO_2$ and $2 \times CO_2$ GCM output fields. Physically reasonable spatial detail is added by maintaining the original empirical relationships of the historical time series. This approach assumes that spatial relationships characterized by the set of historical time series do not change under double CO_2 conditions.

Several limitations are imposed by this scenario approach. Adding GCM differences equally to temperatures on all days transposes the overall distribution of temperature (maximum and minimum) upward (warmer). Within a month, for a given grid cell, diurnal temperature ranges and standard deviations remain unchanged. Climate change differences estimated for some months are quite large and, occasionally, result in daily temperatures above historical climate bounds. Some biological models, developed exclusively from historical environmental conditions, may produce unreasonable results using these scenarios and modifications, or upper response limits may need to be added before being used in climate change assessment applications. Historical and GCM-directed region-wide scenarios are summarized in Tables 2.4 through 2.6. Within-region SGCP scenarios are illustrated in Figures 2.2 through 2.5.

Precipitation

For the precipitation scenarios, each daily rainfall total in the historical record was multiplied by the ratio of mean monthly $1 \times CO_2$ to $2 \times CO_2$ GCM results. The mean intraannual climate change reflected in each resulting daily time series is dictated entirely by changes between $1 \times CO_2$ and $2 \times CO_2$ GCM output fields. Spatial relationships contained within the historical data base are maintained, thus increasing spatial detail in a physically plausible fashion. This approach assumes that spatial relationships characterized by the set of historical time series does not change under double CO_2 conditions. The principle limitation of this scenario approach is that no change in rainfall frequency or autocorrelation is assumed. Further, by multiplying each event by a constant GCM ratio, the variance of the intensity process is altered by the constant ratio, squared (Mearns et al., 1996). Historical and GCM-directed region-wide precipitation scenarios are summarized in Tables 2.4 through 2.6. Within-region SGCP scenarios are illustrated in Figures 2.6 through 2.7.

Table 2.4. Early Growing Season (April, May, June) Historical and Climate Change Conditions Across the Southern Global Change Program's Study Region

	Maximum temperature (°C)		Minimum temperature (°C)		Precipitation (mm)		Vapor pressure deficit (mb)		Solar radiation (W/m^2)	
	Mean	S.D.	Mean	S.D.	Mean	S.D.	Mean	S.D.	Mean	S.D.
History	27.4	1.8	14.4	2.5	329.7	41.9	8.33	.9	239	10.1
GISS	31.5	1.7	18.4	2.4	372.3	54.8	10.4	1.1	244	9.1
GFDL	31.2	1.9	18.2	2.6	386.1	84.2	10.3	1.2	244	11.9
OSU	31.1	1.7	18.0	2.5	302.8	38.7	10.2	1.1	240	9.3
UKMO	33.5	1.7	20.4	2.2	350.7	68.5	11.6	1.4	245	9.7

Table 2.5. Late Growing Season (July, August, September) Historical and Climate Change Conditions Across the Southern Global Change Program's Study Region

	Maximum temperature (°C)		Minimum temperature (°C)		Precipitation (mm)		Vapor pressure deficit (mb)		Solar radiation (W/m^2)	
	Mean	S.D.	Mean	S.D.	Mean	S.D.	Mean	S.D.	Mean	S.D.
History	31.6	1.7	19.2	2.1	327.7	91.3	10.2	1.3	230	15.0
GISS	35.8	2.0	23.3	2.1	379.3	84.1	12.7	1.9	234	12.7
GFDL	35.9	1.7	23.4	1.9	308.5	114.8	12.8	1.8	234	15.4
OSU	35.2	1.9	22.7	2.3	398.6	111.8	12.4	1.7	231	13.4
UKMO	38.3	2.2	25.8	2.2	318.5	109.4	14.5	2.2	229	12.6

Table 2.6. Non Growing Season (January, February, March, October, November, December) Historical and Climate Change Conditions Across the Southern Global Change Program's Study Region

	Maximum temperature (°C)		Minimum temperature (°C)		Precipitation (mm)		Vapor pressure deficit (mb)		Solar radiation (W/m^2)	
	Mean	S.D.	Mean	S.D.	Mean	S.D.	Mean	S.D.	Mean	S.D.
History	17.0	2.9	4.1	3.2	581.8	143.6	4.8	.8	138	10.5
GISS	21.5	2.7	8.7	2.8	518.8	130.8	6.3	.9	137	10.1
GFDL	21.1	2.7	8.0	3.0	642.9	164.7	6.1	.9	136	10.7
OSU	20.5	2.7	7.5	3.0	540.3	128.7	5.8	.9	138	9.7
UKMO	23.7	2.6	10.7	2.8	550.8	141.8	7.0	1.1	138	10.1

Solar Radiation

GCMs are, ultimately, driven by spatial and temporal distributions of solar radiation. Several studies have been conducted which compare radiation algorithms across GCMs. For instance, Cess et al. (1990) and Randall et al. (1992) compared surface energy flux and feedback, and Fouquart et al. (1991) compared shortwave radiation flux under an international initiative called the Intercomparison of Radiation Codes in Climate Models (ICRCCM) project which predates AMIP. Each analysis indicates a wide range of responses (sensitivity) to carefully specified environmental changes. Fouquart et al. (1991) conclude that ". . . many radiation algorithms could have inherently unknown errors that may significantly affect the conclusions of the studies in which they are used. This is true for climate modeling and weather forecasting studies as well as for other applications, such as inferences from satellite observations." Algorithmic improvements have been made and our understanding of radiation processes improved since these analyses, but Cess et al. (1992) and Fouquart (1991) reflect radiation algorithm performance for the generation of GCMs used in the SGCP.

A few comparisons of GCM results with field observations have been made. One of the earliest is Mearns et al. (1989). They reported that for the Great Lakes, Great Plains, and the southeastern and northwestern United States, GCMs were able to simulate the annual solar cycle, but did not simulate effectively the magnitude of solar (incident) radiation. A second study, Brazel et al. (1993) reported similar results for the southwestern United States. Although GCM projections of incident solar radiation should not be relied upon for applications research, these studies indicate that changes in average radiation patterns might be meaningful.

Two radiation scenario approaches were considered: GCM-directed and SGCP-modeled. GCM-directed scenarios were generated by modifying the baseline SGCP radiation time series by the ratio of GCM baseline to $2 \times CO_2$ projections. The SGCP scenario approach assumes that historical statistical and physical relationships represented by the model described earlier remain intact under changed climate. Temperature-dependent parameterizations for the model were recalculated using the GCM-directed temperature scenarios and new radiation time series were produced. These two radiation scenarios were then compared. We found that mean annual global radiation estimates generated by the two approaches are not statistically different and so the time series of global radiation obtained via the SGCP radiation model was accepted into the applications data base. By doing so, internal consistency across climate variables assigned to an SGCP grid cell is maintained, consistency between baseline and changed climate scenarios is maintained, physical consistency between grid cells and among variables is maintained, and GCM internal and physical consistency is recognized and maintained through the use of GCM-directed temperature scenarios to drive the radiation time series. Other implications of this scenario choice is explored in greater detail in Cooter et al. (submitted). Historical and SGCP-generated region-wide change scenarios are summarized in Tables 2.4 through 2.6. Within-region SGCP scenarios are illustrated in Figures 2.8 through 2.9.

Vapor Pressure Deficit

The suitability of direct GCM estimates of humidity for SGCP applications versus simulation via Eqs. (2.2) and (2.3) was considered. A recent AMIP evaluation of water vapor (humidity) simulations across 28 GCMs suggests that estimates are within 5 to 15% of observed historical values for the SGCP region at large spatial and temporal analysis scales (4° grid spacing and decadal median annual, seasonal and monthly estimates) (Gaffen et al., 1997). Several sources of estimate bias have been identified, but firm conclusions could not be made.

In light of this analysis and the need to provide climate change scenarios that are spatially and temporally consistent with the historical time series, SGCP 2 × CO_2 humidity scenarios were generated using the relationships described by Eqs. (2.2) and (2.3). However, humidity changes simulated as being indirectly induced by increased atmospheric CO_2 must be interpreted carefully. For instance, the SGCP scenario approach assumes that daily maximum and minimum temperature are increased equally. The vapor pressure and saturation vapor pressure, in turn, increase proportionately [see Eqs. (2.2) and (2.3)]. VPD increases (i.e., the air is further from saturation), but the relative humidity (approximated by the ratio of e to e_s) remains constant and the mixing ratio (the ratio of the mass of water vapor in the air to the mass of dry air with which the water vapor is associated) increases. Therefore, depending on the measure chosen, humidity under $2 \times CO_2$ conditions can increase, decrease, or remain unchanged from historical conditions. Direct GCM output indicates increased mixing ratios in most cases which is in general agreement with the empirically derived results presented here. Implications of these humidity scenarios for loblolly pine stand evapotranspiration applications are presented in Cooter et al. (submitted). Historical and SGCP-generated region-wide scenarios are summarized in Tables 2.4 through 2.6. Within-region SGCP scenarios are illustrated in Figures 2.10 through 2.11.

Critical Findings

Region-wide

Temperature

- Mean regional temperature increases range from

Early growing season[1]	+3.7 °C (OSU) to +6.1 °C (UKMO)
Late growing season[2]	+3.6 °C (OSU) to +6.7 °C (UKMO)
Non-growing season[3]	+3.5 °C (OSU) to +6.7 °C (UKMO)

- Nongrowing (October through March) and early (April through June) growing seasons show decreased spatial variability. Late growing season temperatures show increased spatial variability.

---

[1] April, May, and June.
[2] July, August, and September.
[3] October through December, and January through March.

Precipitation

- Most[4] models considered suggest increased growing season precipitation and decreased nongrowing season rainfall
- Patterns of change within the growing season vary with model

	Early growing season	Late growing season
GISS	+	+
GFDL	+	−
UKMO	+	−
OSU	−	+

Global solar radiation

- All models considered indicate slightly increased early growing season global radiation, with most showing decreased spatial variability.
- Most models suggest increased mean late season global incident radiation. Spatial variability decreases in most cases.

Vapor pressure deficit

- Vapor pressure deficit increases across all time periods and change scenarios
- Greatest average vapor pressure deficit change occurs during late growing season months

Within-region

Temperature

- Historical early growing season temperature spatial patterns are maintained in all scenarios. In most cases, magnitudes are increased 2 °C to 4 °C.
- Most models maintain late season temperature spatial patterns and values in North Carolina, South Carolina, and Tennessee, with 2 °C to 4 °C increases elsewhere.
- Area of greatest temperature change from historical conditions are projected for Louisiana, Mississippi, and Alabama.

Precipitation

- There is no model agreement concerning drying or intensification of the early growing season moisture corridor that, historically, extends from the Louisiana coast, northward.
- Historical late growing season spatial patterns remain essentially unchanged. GFDL and UKMO scenario rainfall totals are also unchanged. GISS and OSU models indicate more moist conditions throughout the region.

---

[4]In this discussion, "most" is used only when three of the four GCMs considered in this analysis are in agreement.

Global solar radiation
- The models agree that although the spatial pattern of mean growing season solar radiation will remain unchanged, values may increase slightly.

Vapor pressure deficit
- Most models project early growing season VPD spatial patterns and values to remain unchanged in the Northern one third of the region and VPD increases in the interior of the region. All models project no VPD changes along the Gulf Coast.
- Late growing season spatial patterns are maintained. Magnitudes along the Gulf Coast are maintained by all models and most others project uniform deficit increases on the order of 2 mb.

Summary

Historical and future climate conditions for the southern United States have been assembled and presented. Although some statements concerning possible global changes under double CO_2 conditions can be made with varying levels of certainty, the regional expression of these changes remains highly uncertain.

The regional results derived for the SGCP suggest agreement regarding warming in the South, with sensitivity tending towards the upper end of the IPCC range (Table 2.1). The area containing Louisiana, Mississippi, and Georgia could experience the greatest degree of change within the region. Although projected annual precipitation increases could result in a more favorable biological environment for the South, changes in the within-growing season distribution of precipitation, as well as uncertainty concerning changes in spatial patterns of precipitation, make the ultimate impact of these changes on biological productivity less certain. Uncertainty in the GCM parameterization of cloud formation and cloud radiation feedback make statements regarding changes in humidity and solar radiation highly speculative.

References

Bannister P (1976) Introduction to physiological plant ecology. John Wiley, New York, 36–61.

Brazel AJ, McCabe GJ Jr, Verville HJ (1993) Incident solar radiation simulated by general circulation models for the southwestern United States. Clim Res 2(3):177–181.

Cess RD, Potter GL, Blanchet JP, Boer GJ, Del Genio AD, Deque M, Dymnikov V, Galin V, Gates WL, Ghan SJ, Kiehl JT, Lacis AA, Le Treut H, Li Z-X, Liang X-Z, McAvaney BJ, Meleshko VP, Mitchell JFB, Morcrette J-J, Randall DA, Rikus L, Roeckner E, Royer JF, Schlese U, Sheinin DA, Slingo A, Sokolov AP, Taylor KE, Washington WM, Wetherald RT, Yagai I, Zhang M-H (1990) Intercomparison and interpretation of climate feedback processes in 19 atmospheric general circulation models. J. Geophysical Rsch. 95 (D10): 16601–16615.

Cooter EJ (1990) The impact of climate change on continuous corn production in the Southern U.S.A. Clim Change 16:53–82.

Cooter EJ, Eder BK, LeDuc SK, Truppi L (1993) Climate change models and forest research. J For 91:38–43.

Cooter EJ, Dhakhwa GB (1996) A solar radiation model for use in biological applications in the South and Southeastern U.S. Ag For Meteorol 78:31–51.

Dale VH, Rauscher HM (1994) Assessing impacts of climate change on forests: the state of biological modeling. Clim Change 28:65–90.

Folland CK, Karl TR, Nicholls N, Nyenzi BS, Parker DE, Vinnikov KYA (1992) Observed climate variability and change. In Houghton JT, Callander BA, Varney SK (Eds) *Climate Change 1992: The supplementary report to the IPCC scientific assessment*. University Press, Cambridge, 137–170.

Fouquart Y, Bonnel B, Ramaswamy V (1991) Intercomparing shortwave radiation codes for climate studies. J Geophys Rsch 96(D5):8955–8968.

Friend AD (1996) Parameterization of a global daily weather generator for terrestrial ecosystem and biogeochemical modelling. Ecol Modelling.

Gaffen DJ, Rosen RD, Salstein DA, Boyle JS (1997) Evaluation of tropospheric water vapor simulations from the atmospheric model intercomparison project. J Clim (10)7:1648–1661.

Gates WL, Mitchell JFB, Boer GJ, Cubasch U, Meleshko VP (1992) Climate modeling, climate prediction and model validation. In Houghton JT, Callander BA, Varney SK (Eds) *Climate change 1992, the supplementary report to the IPCC scientific assessment*. University Press, Cambridge, 97–134.

Gates WL, Henderson-Sellers A, Boer GJ, Folland CK, Kitoh A, McAvaney GJ, Semazzi F, Smith N, Weaver AJ, Zeng, Q C (1996) Climate models—evaluation. In Houghton JT, Meira Filho LG, Callander BA, Harris N, Kattenberg A, Maskell K (Eds) *Climate change 1995, the science of climate change (IPCC WGI)*. Cambridge University Press, Cambridge, UK, 233–284.

Giorgi F, Mearns LO (1991) Approaches to the simulation of regional climate change: a review. Rev Geophys 29:191–216.

Gyalistras D, von Storch H, Fischlin A, Beniston M (1994) Linking GCM-simulated climatic changes to ecosystem models: case studies of statistical down-scaling in the Alps. Clim Res 4(3):167–189.

Hansen J, Russell G, Rind D, Stone P, Lacis A, Lebedeff S, Ruedy R, Travis L (1983) Efficient three-dimensional global models for climate studies: models I and II. Mon Weather Rev 3:609–662.

Hodges T, French V, LeDuc S (1985) Estimating solar radiation for plant simulation models. AgRISTARS Technical Report JSC-20239; YM-15–00403.

Hodges T, Botner D, Sakamoto C, Hays-Haug J (1987) Using the CERES-Maize model to estimate production for the U.S. cornbelt. Ag For Meteorol 40:293–303.

IPCC (1990) Climate change: the IPCC scientific assessment, Houghton JT, Jenkins GJ, Ephraums JJ (Eds) Cambridge University Press, Cambridge, UK.

IPCC (1996) Summary for policymakers and technical summary of the working group I report, Cambridge University Press, Cambridge, UK.

Karl TR, Jones PD, Knight RW, Kukla G, Plummer N, Razuvayev V, Gallo KP, Lindseay J, Charlston RJ, Peterson TC (1993) Asymmetric trends of daily maximum and minimum temperatures. Bull Am Meteor Soc 74:1007–1023.

Kattenberg A, Giorgi F, Grassl H, Meehl GA, Mitchell JFB, Stouffer RJ, Tokioka T, Weaver AJ, Wigley TML (1996) Climate models—projections of future climate. In Houghton JT, Meira Filho LG, Callander BA, Harris N, Kattenberg A, Maskell K (Eds) *Climate change 1995, the science of climate change (IPCC WGI)*. Cambridge University Press, Cambridge, UK, 289–357.

Koss WJ, Owenby JR, Steurer PM, Ezell DS (1988) Freeze/frost data. United States Department of Commerce National Oceanic and Atmospheric Administration National Climatic Center Climatography of the U.S. No. 20, Supplement No. 1.

Manabe S, Wetherald RT (1987) Large-scale changes in soil wetness induced by an increase in carbon dioxide. J Atmos Sci 44:1211–1235.
Mearns LO, Schneider SH, Thompson SL, Daniel LR (1989) Analysis of climate variability in general circulation models: comparisons with observations and changes in variability in $2 \times CO_2$ experiments. In Smith JB, Tirpak DA (Eds) *The potential effects of global climate change on the United States: Appendix I—Variability.* U.S. Environmental Protection agency, Office of Policy, Planning, and Evaluation, Washington, DC, 1–59.
Mearns LO, Rosenzweig C, Goldberg R (1996) The effect of changes in daily and interannual climatic variability on CERES-wheat: a sensitivity study. Clim Change 32:257–292.
Mitchell JFB, Manabe S, Meleshko V, Tokioka T (1990) Equilibrium climate change and its implications for the future. In Houghton JT, Jenkins GJ, Ephraums JJ (Eds) *Climate change, the IPCC scientific assessment.* University Press, Cambridge, 134–172.
Murray FW (1967) On the computation of saturation vapor pressure. J Appl Meteorol 6:203–204.
NREL (1992) User's Manual, National Solar Radiation Data Base (1961–1990). National Renewable Energy Laboratory, Golden, Colorado. 93.
Randall DA, Cess RD, Blanchet JP, Boer GJ, Dazlich DA, Del Genio AD, Deque M, Dymnikov V, Galin V, Ghan SJ, Lacis AA, Le Treut H, Li Z-X, Liang X-Z, McAvaney BJ, Meleshko VP, Mitchell JFB, Morcrette J-J, Potter GL, Rikus L, Roechner E, Royer JF, Schlese U, Sheinin DA, Slingo J, Sokolov AP, Taylor KE, Washington WM, Wetherald RT, Yagai I, Zhang M-H (1992) Intercomparison and interpretation of surface energy fluxes in atmospheric general circulation models. J Geophy Res 97(D4):3711–3724.
Richman MB and Montroy DL (1996) Nonlinearities in the signal between el niño/la niña events and North American precipitation and temperature. In Preprints of the 13th conference on Probability and Statistics in the Atmospheric Sciences, February 21–23, 1996, San Francisco, CA, 90–97.
Richman MB, Lamb PJ (1985) Climatic pattern analysis of three- and seven-day summer rainfall in the central United States: some methodological considerations and a regionalization. J Clim Appl Meteorol 24:1325–1343.
Robinson PJ, Finkelstein PL (1991) The development of impact-oriented climate scenarios. Bull Am Meteorol Soc 72:481–489.
Running SW, Nemani RR, Hungerford RD (1987) Extrapolation of synoptic meteorological data in mountainous terrain, and its use for simulation forest evapotranspiration. Can J For Res 17:472–483.
Schlesinger ME, Zhao Z C (1989) Seasonal climate changes induced by doubled CO_2 as simulated by the OSU Atmospheric GCM/Mixed-layer Ocean Model. J Clim 2:459–495.
Smith JB, Tirpak DA (Eds) (1990) *The Potential Effects of Global Climate Change on the United States.* Hemisphere, New York.
von Storch H, Zorita E, Cubasch U (1993) Downscaling of global climate change estimates to regional scales: an application to Iberian rainfall in wintertime. J Clim 6:1161–1171.
Wilson CA, Mitchell JFB (1987) A doubled CO_2 climate sensitivity experiment with a global climate model including a simple ocean. J Geophys Res 92(D11):13,315–13343.

Appendix

This appendix contains 1° latitude × longitude maps Figures 2.2–2.11 of historical and double CO_2 scenarios for the Southern Global Change Program study area. Plotting categories have been selected to facilitate comparison *across models* for

(*text continued on p. 54*)

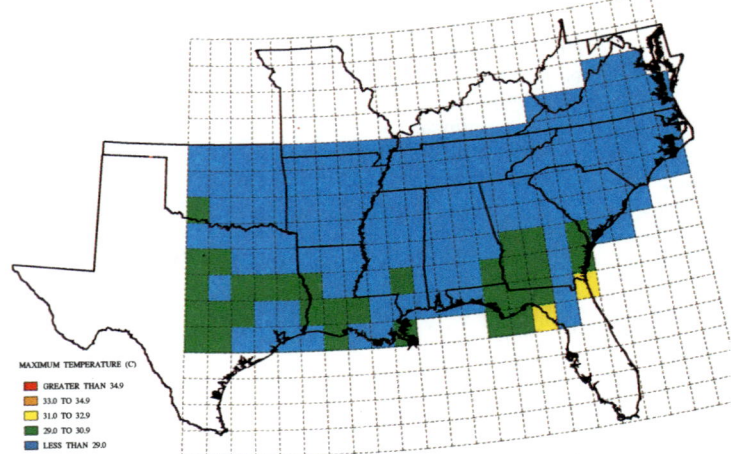

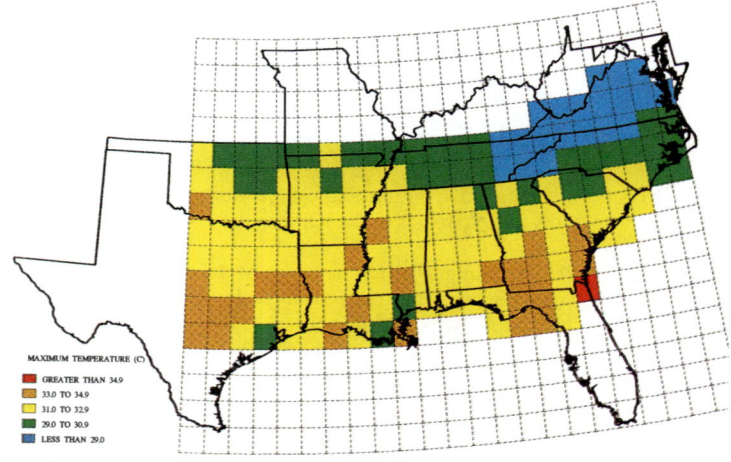

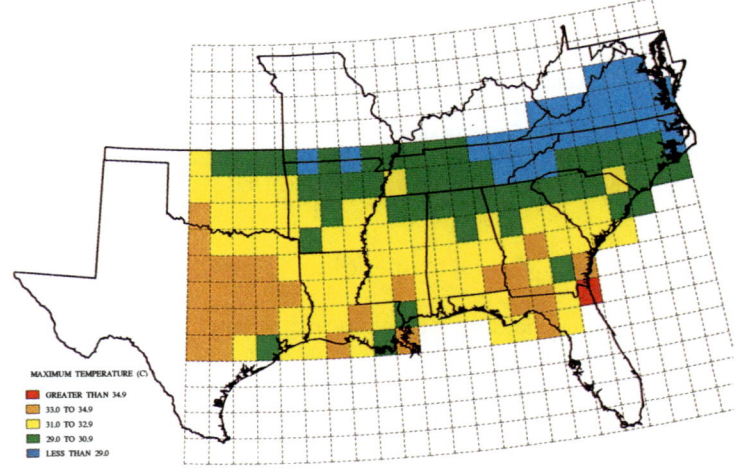

2. Circulation Models for the Southern United States 35

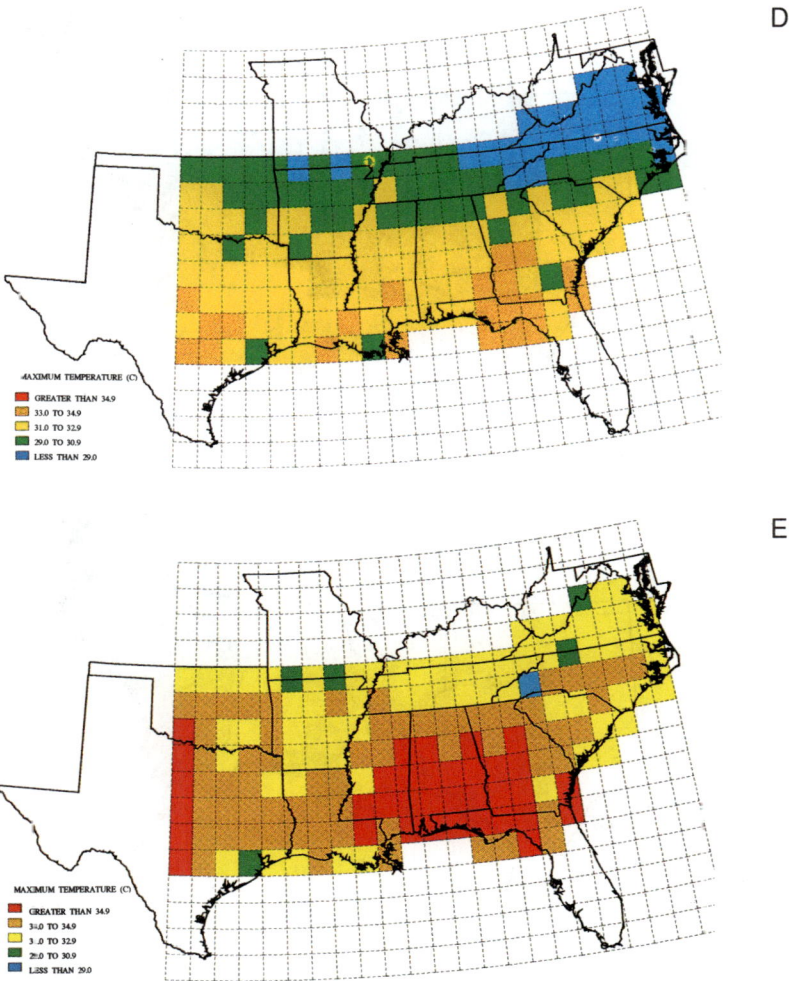

Figure 2.2. (A) Historical early growing season maximum temperature. (B) GISS early growing season maximum temperature. (C) GFDL early growing season maximum temperature. (D) OSU early growing season maximum temperature. (E) UKMO early growing season maximum temperature.

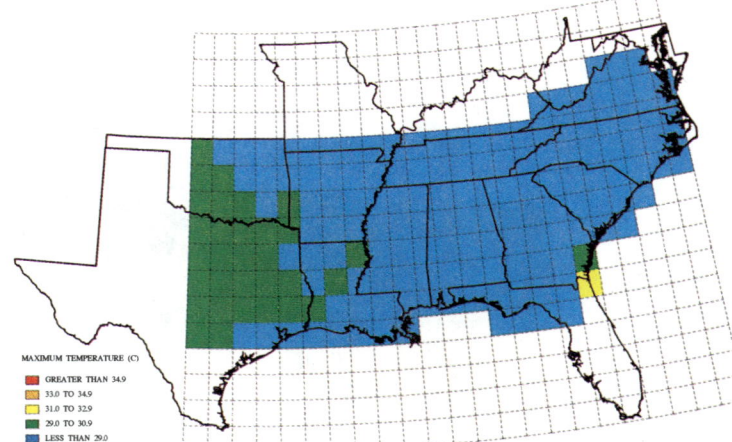
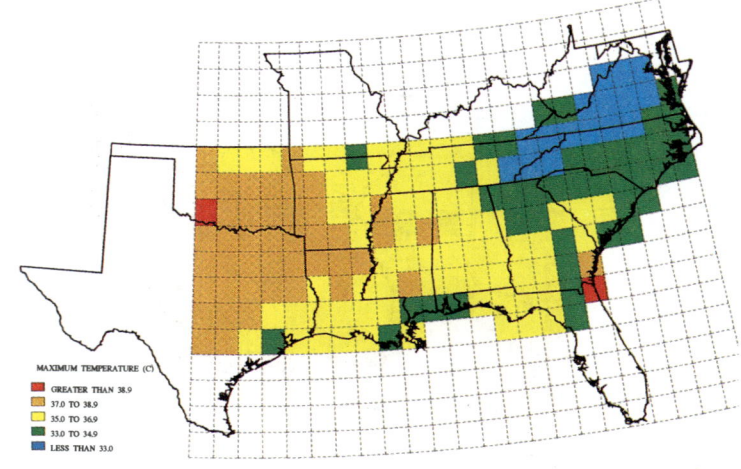
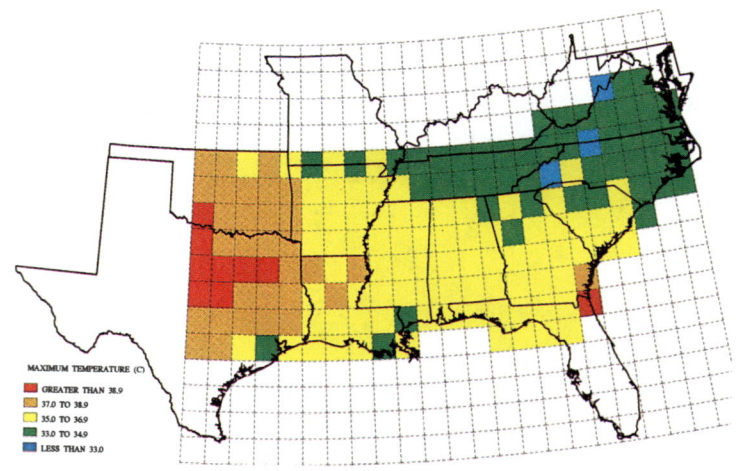

2. Circulation Models for the Southern United States

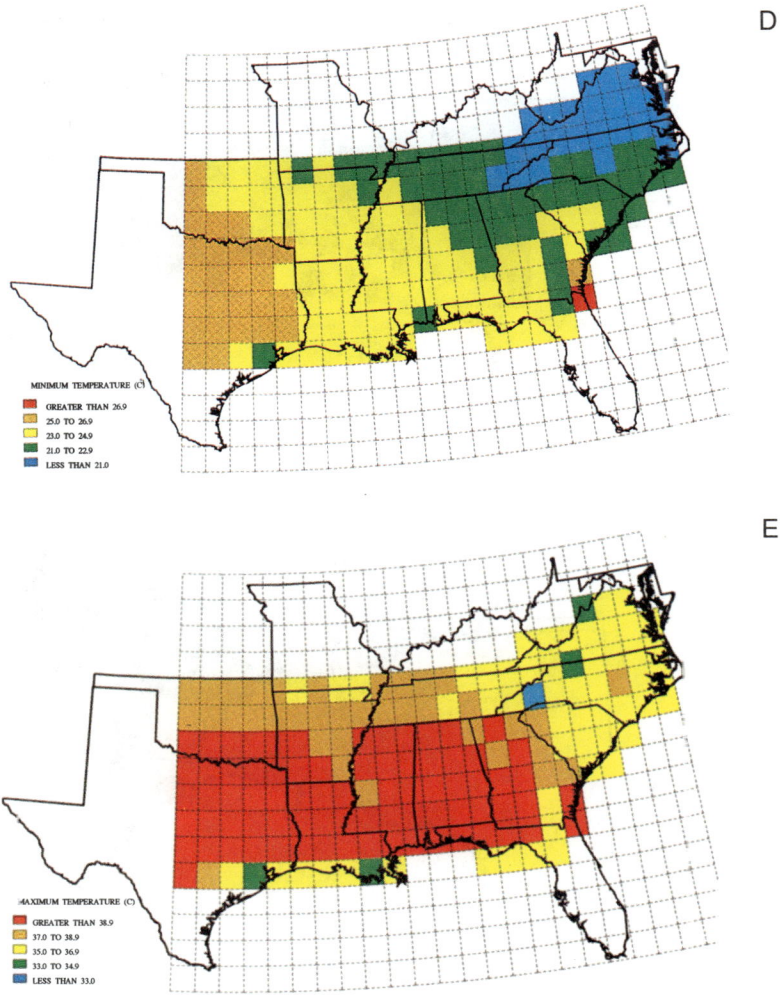

Figure 2.3. (A) Historical late growing season maximum temperature. (B) GISS late growing season maximum temperature. (C) GFDL late growing season maximum temperature. (D) OSU late growing season maximum temperature. (E) UKMO late growing season maximum temperature.

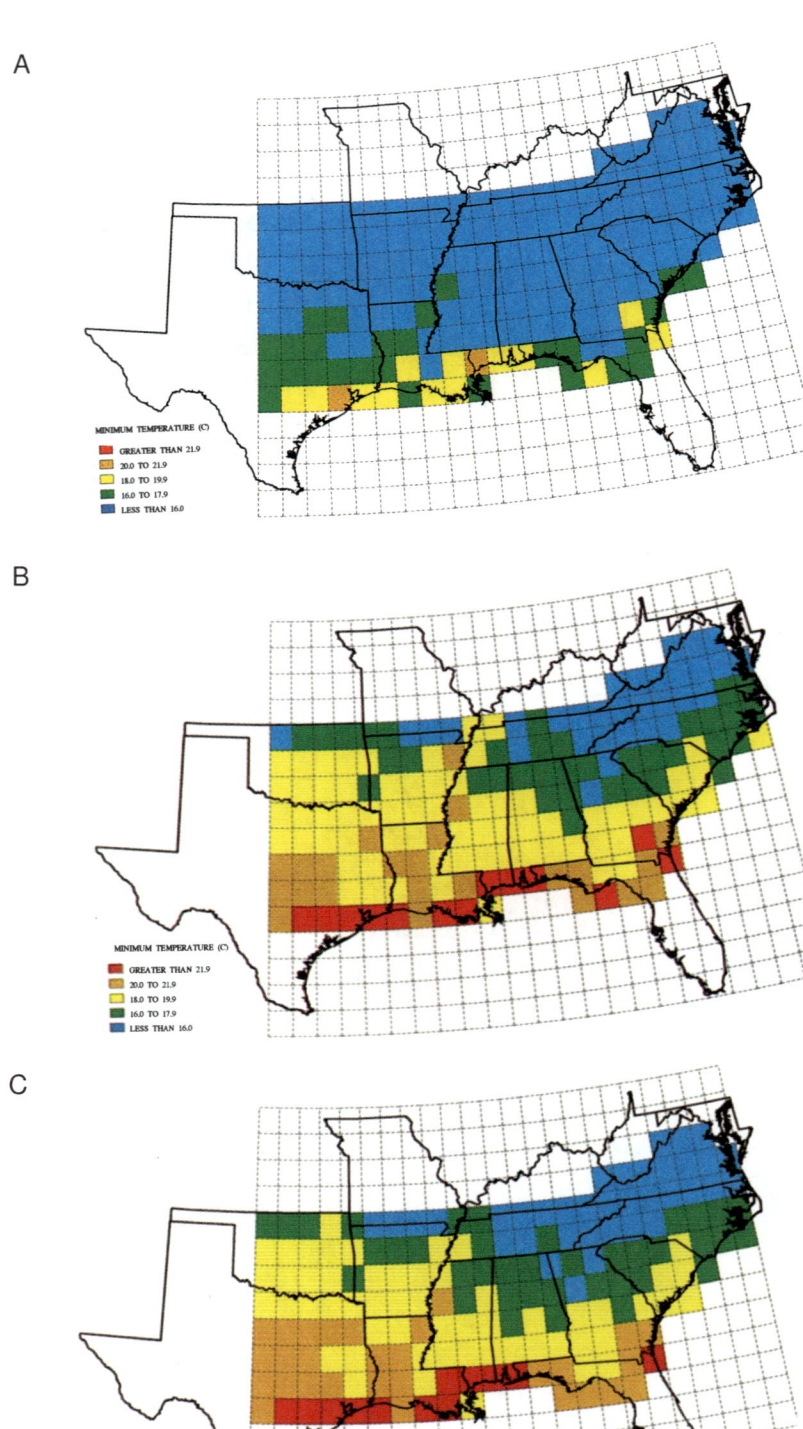

2. Circulation Models for the Southern United States   39

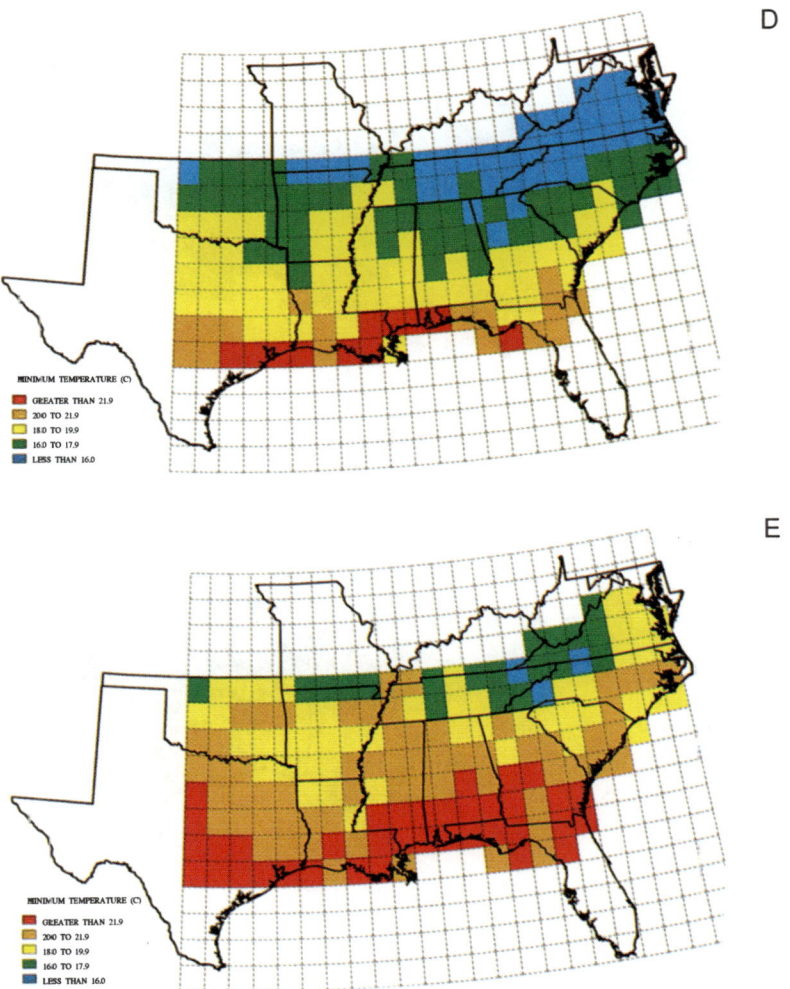

Figure 2.4. (A) Historical early growing season minimum temperature. (B) GISS early growing season minimum temperature. (C) GFDL early growing season minimum temperature. (D) OSU early growing season minimum temperature. (E) UKMO early growing season minimum temperature.

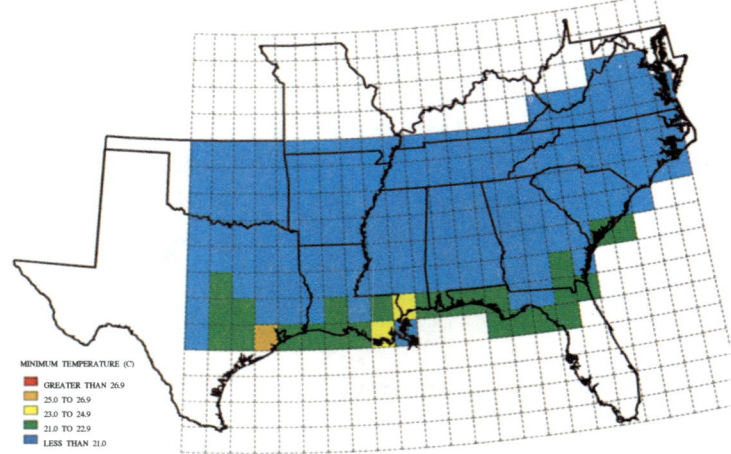

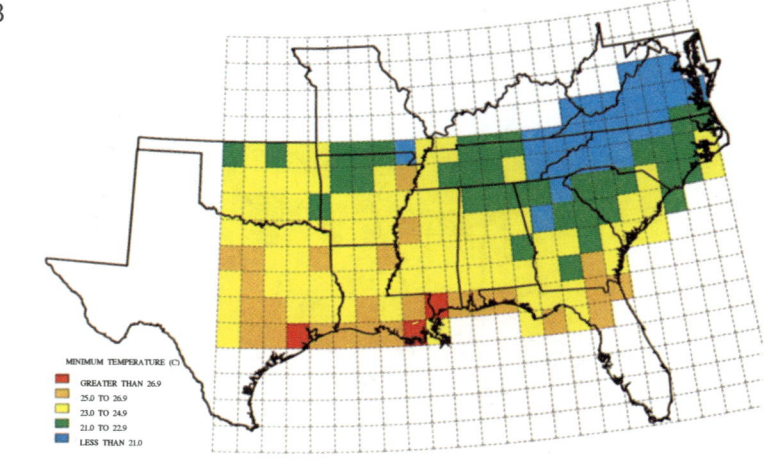

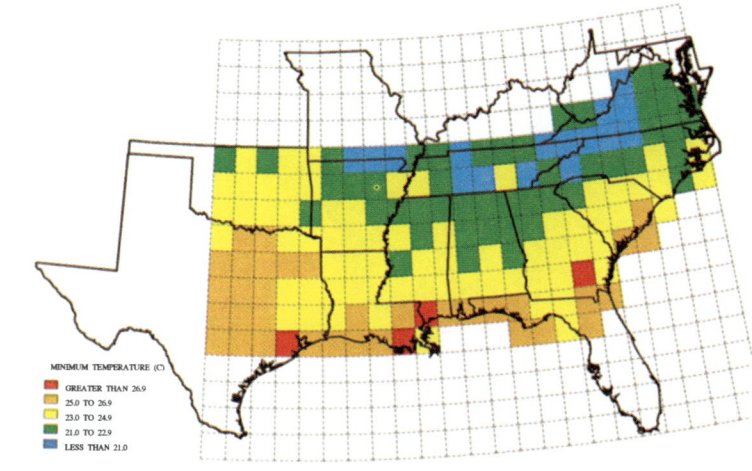

2. Circulation Models for the Southern United States   41

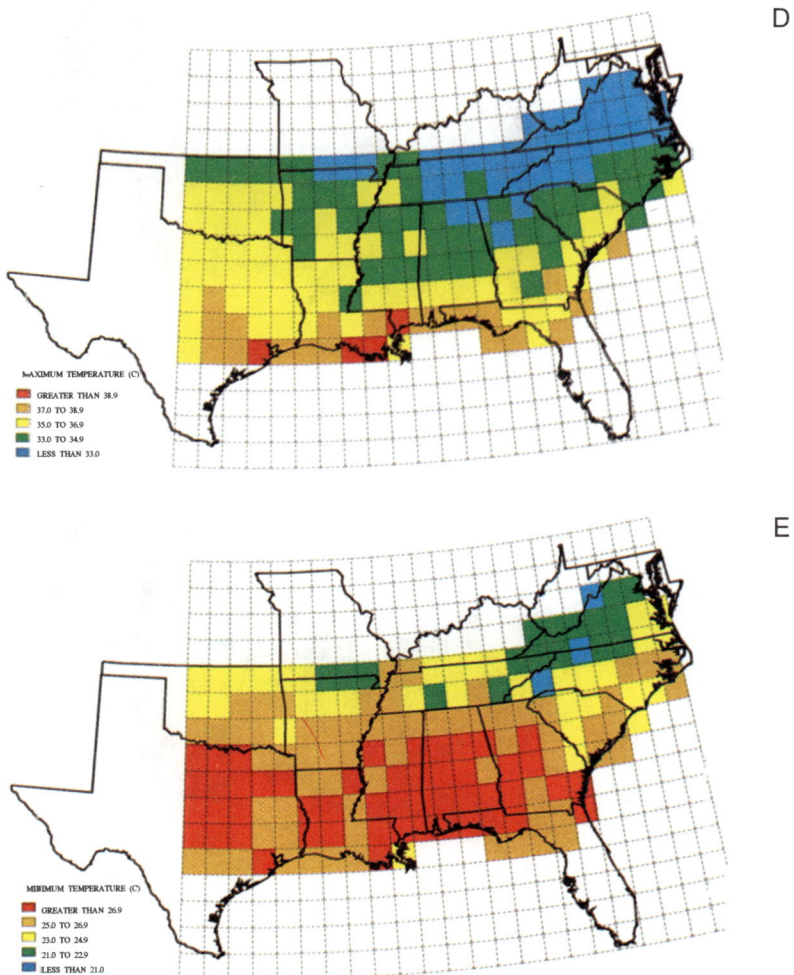

Figure 2.5. (A) Historical late growing season minimum temperature. (B) GISS late growing season minimum temperature. (C) GFDL late growing season minimum temperature. (D) OSU late growing season minimum temperature. (E) UKMO late growing season minimum temperature.

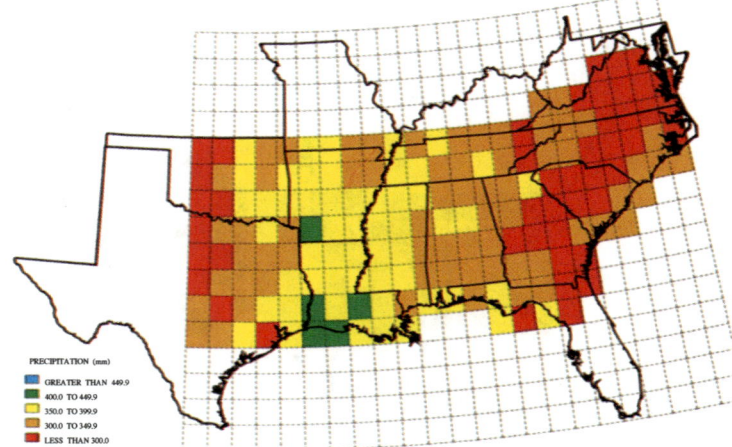
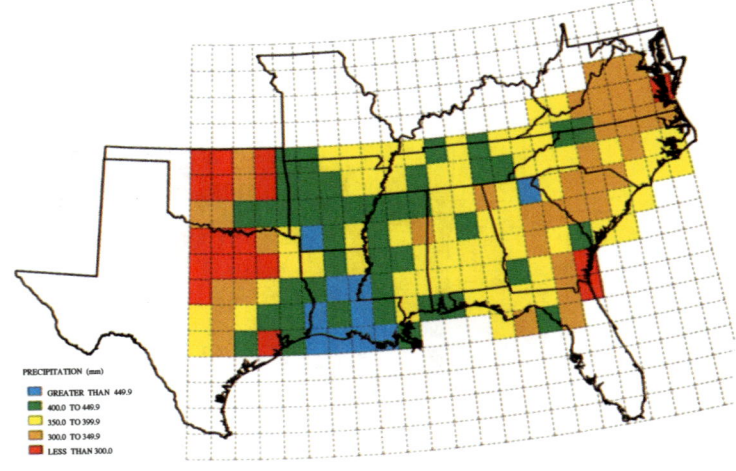
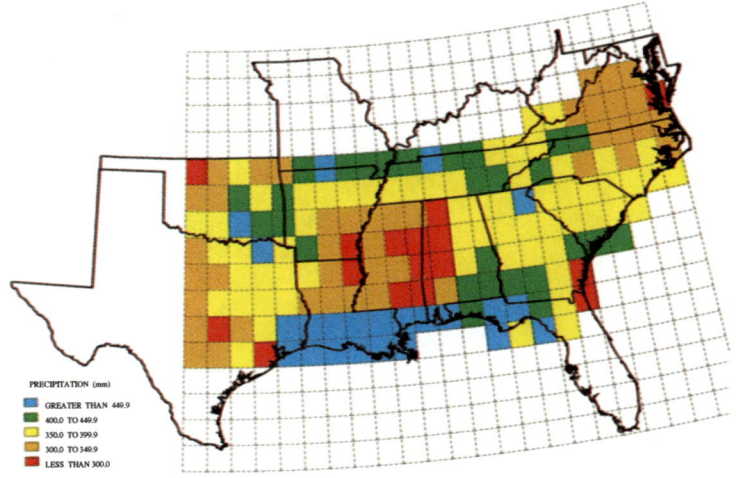

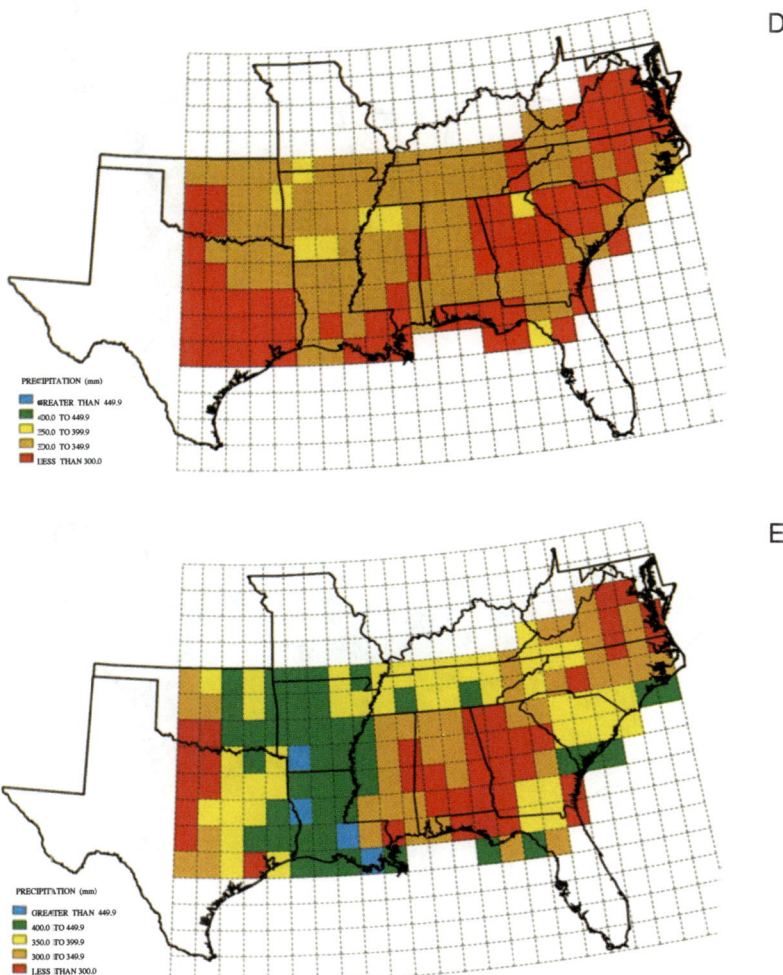

Figure 2.6. (A) Historical early growing season precipitation. (B) GISS early growing season precipitation. (C) GFDL early growing season precipitation. (D) OSU early growing season precipitation. (E) UKMO early growing season precipitation.

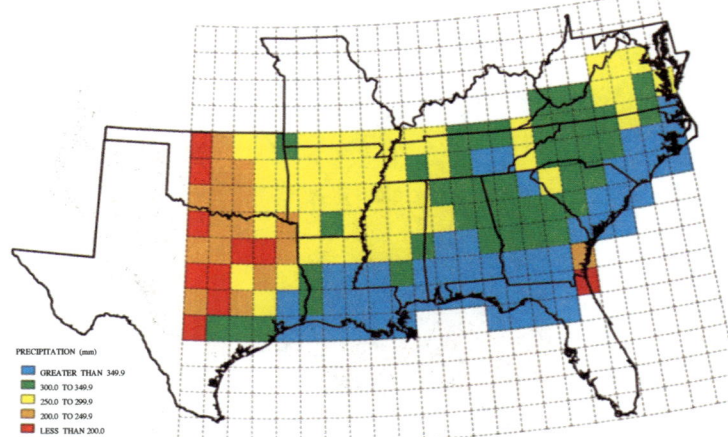

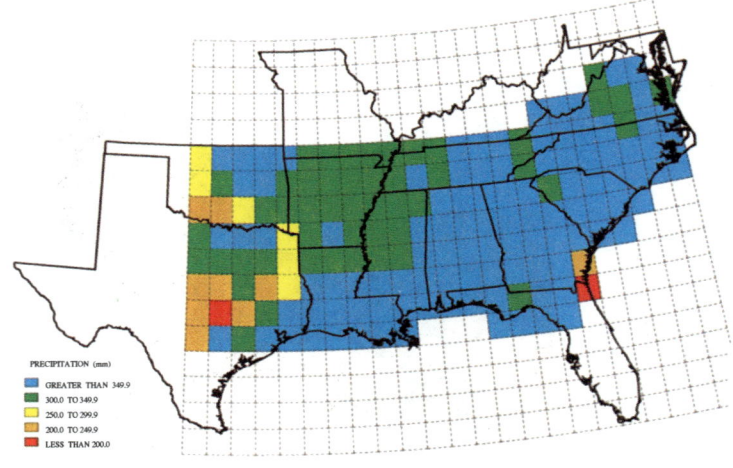

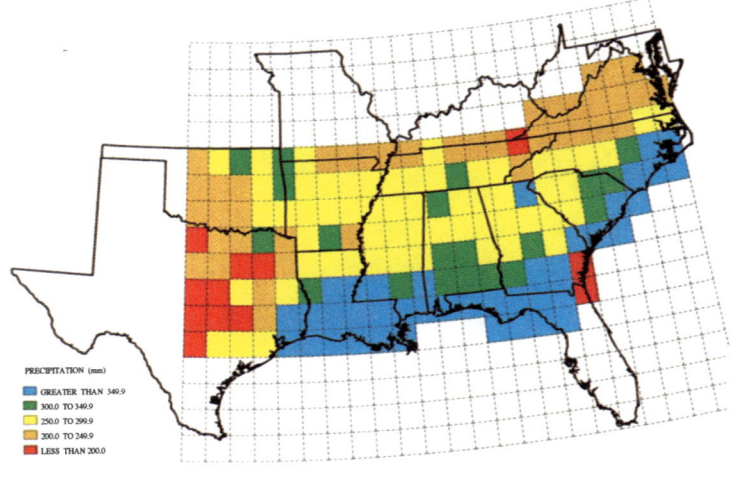

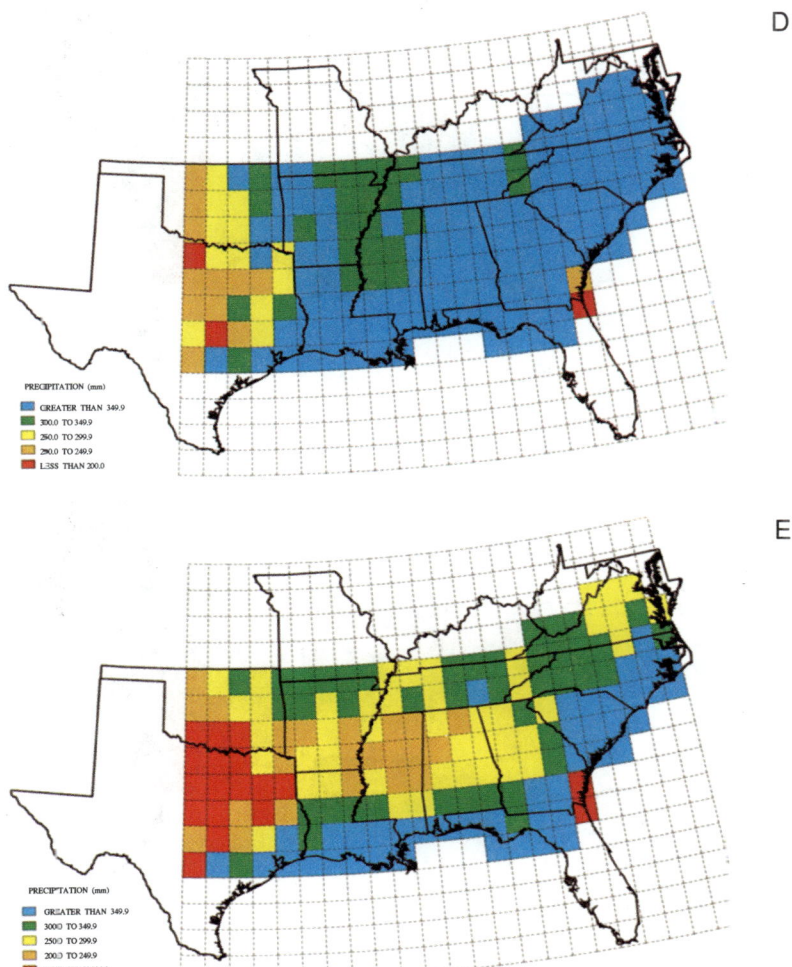

Figure 2.7. (A) Historical late growing season precipitation. (B) GISS late growing season precipitation. (C) GFDL late growing season precipitation. (D) OSU late growing season precipitation. (E) UKMO late growing season precipitation.

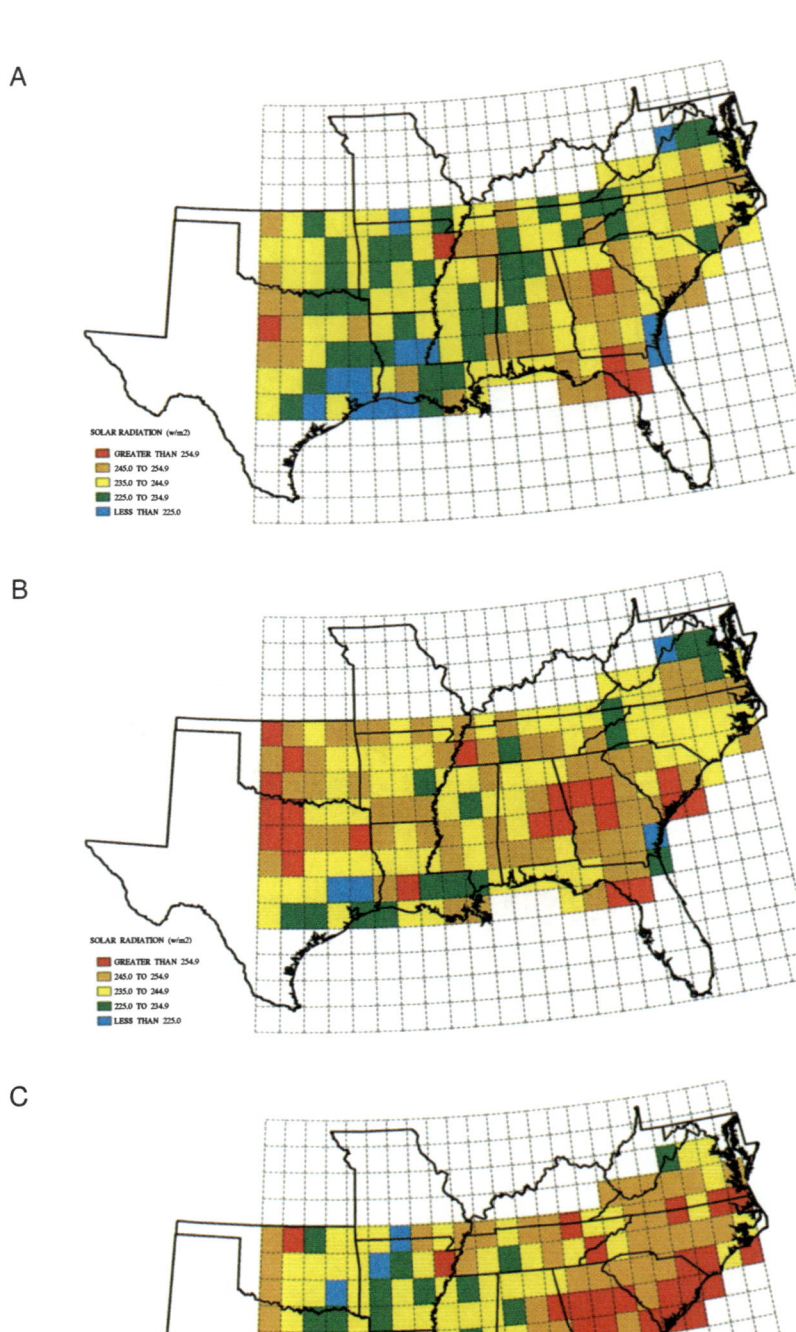

2. Circulation Models for the Southern United States   47

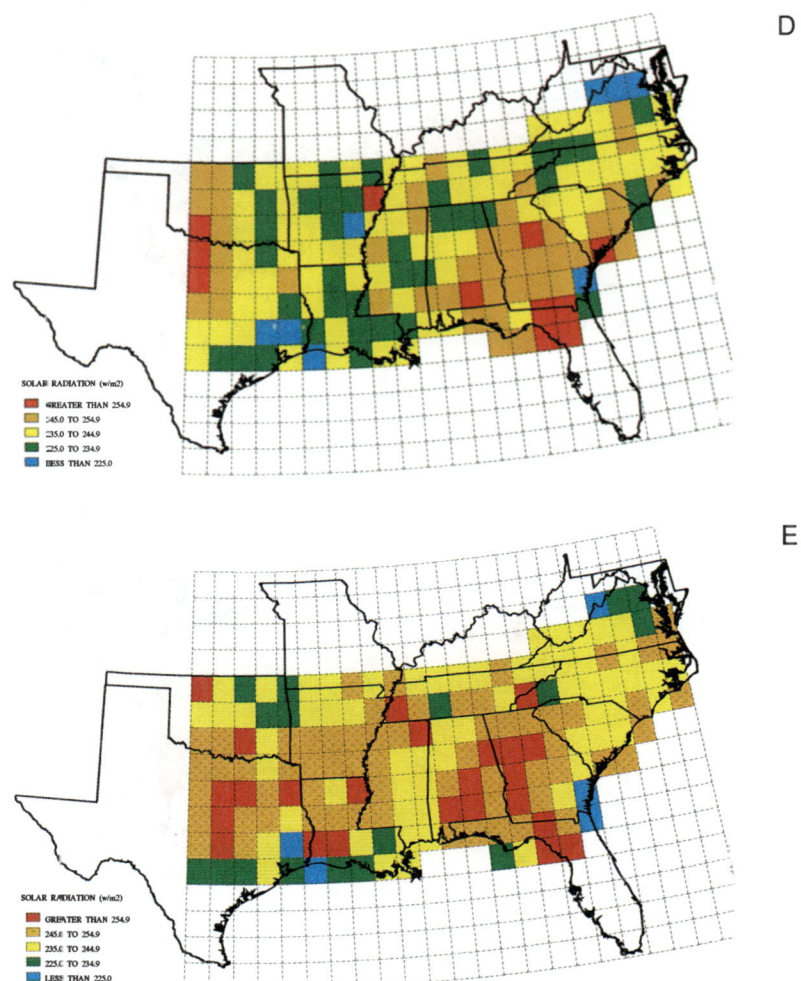

Figure 2.8. (A) Historical early growing season solar radiation. (B) GISS early growing season solar radiation. (C) GFDL early growing season solar radiation. (D) OSU early growing season solar radiation. (E) UKMO early growing season solar radiation.

A

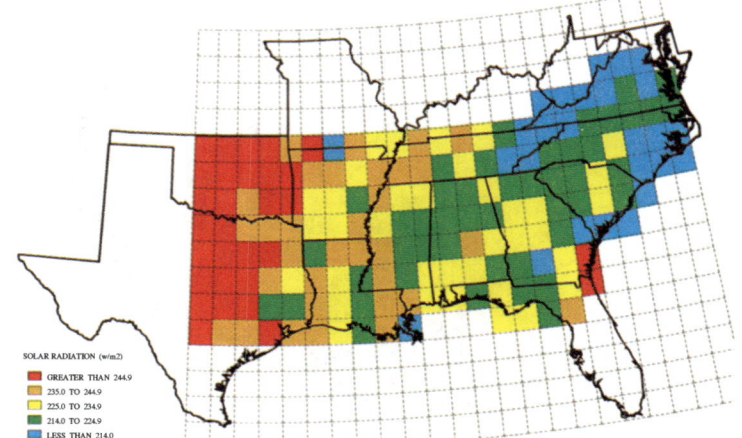

B

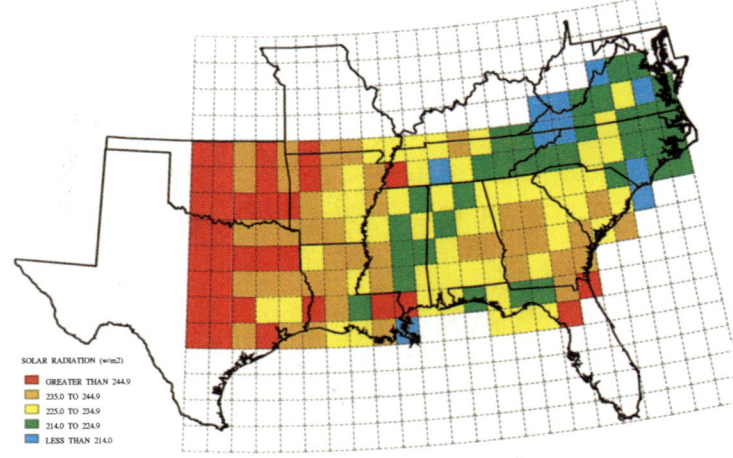

C

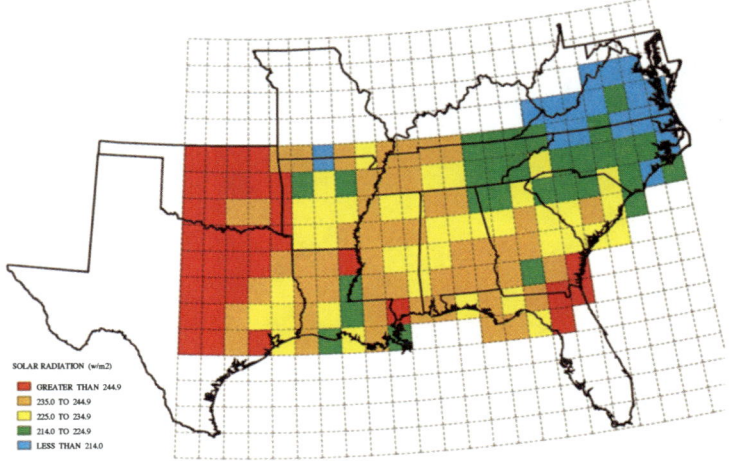

2. Circulation Models for the Southern United States 49

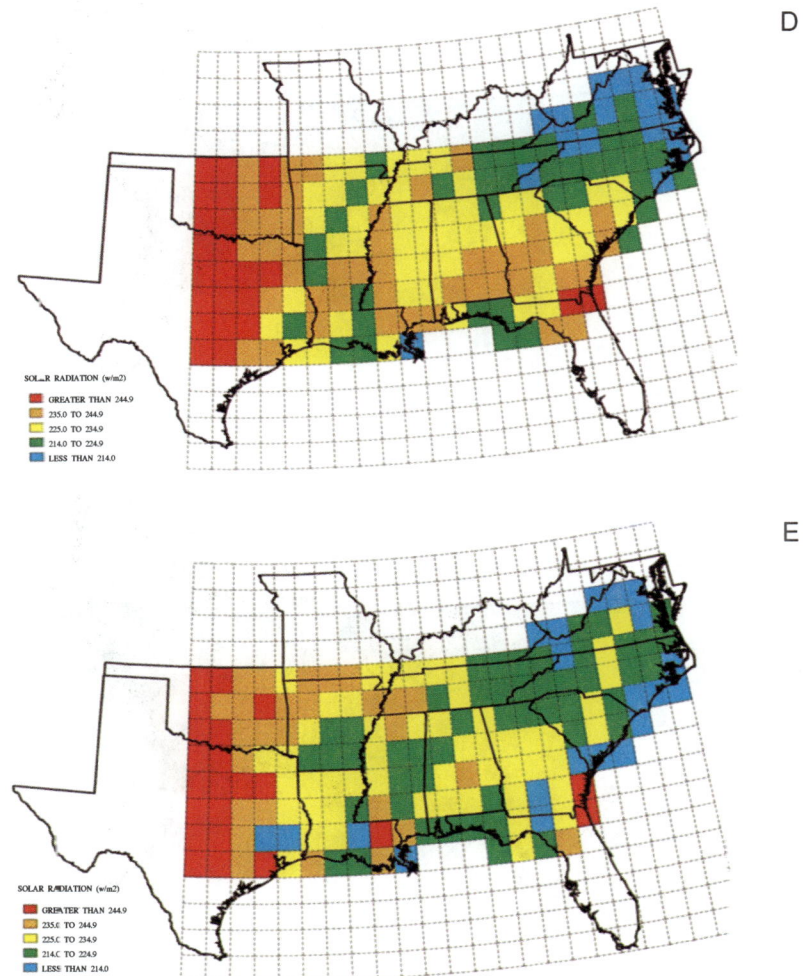

Figure 2.9. (A) Historical late growing season solar radiation. (B) GISS late growing season solar radiation. (C) GFDL late growing season solar radiation. (D) OSU late growing season solar radiation. (E) UKMO late growing season solar radiation.

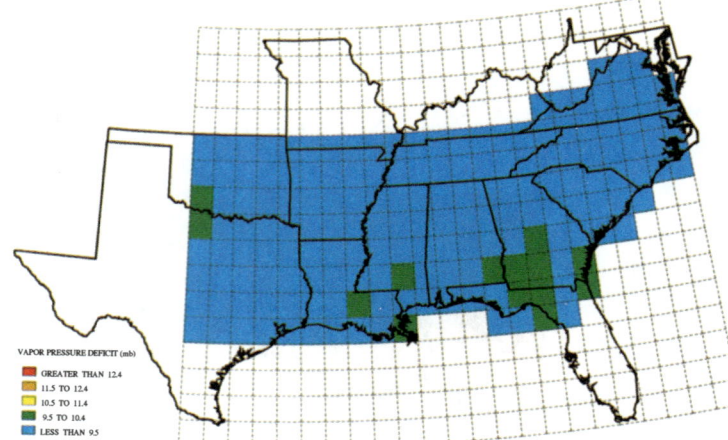
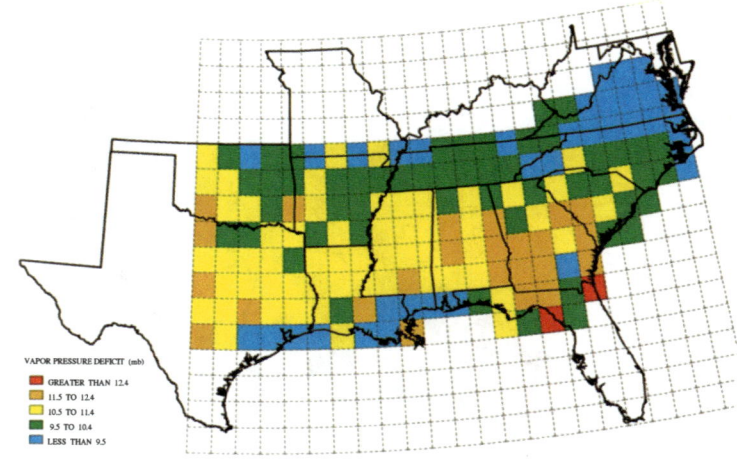
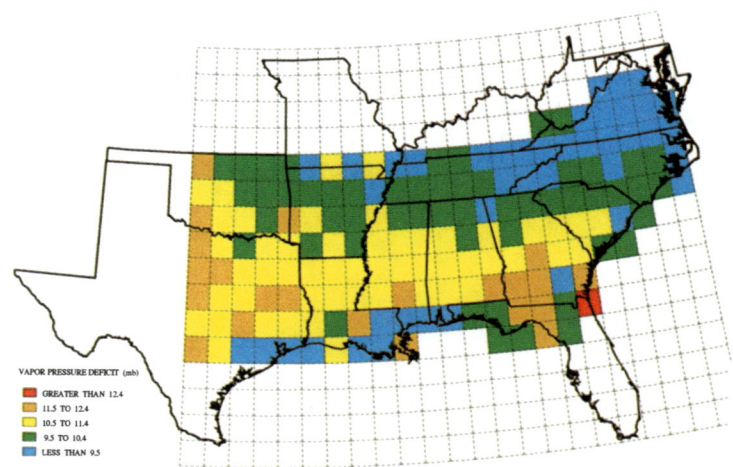

2. Circulation Models for the Southern United States   51

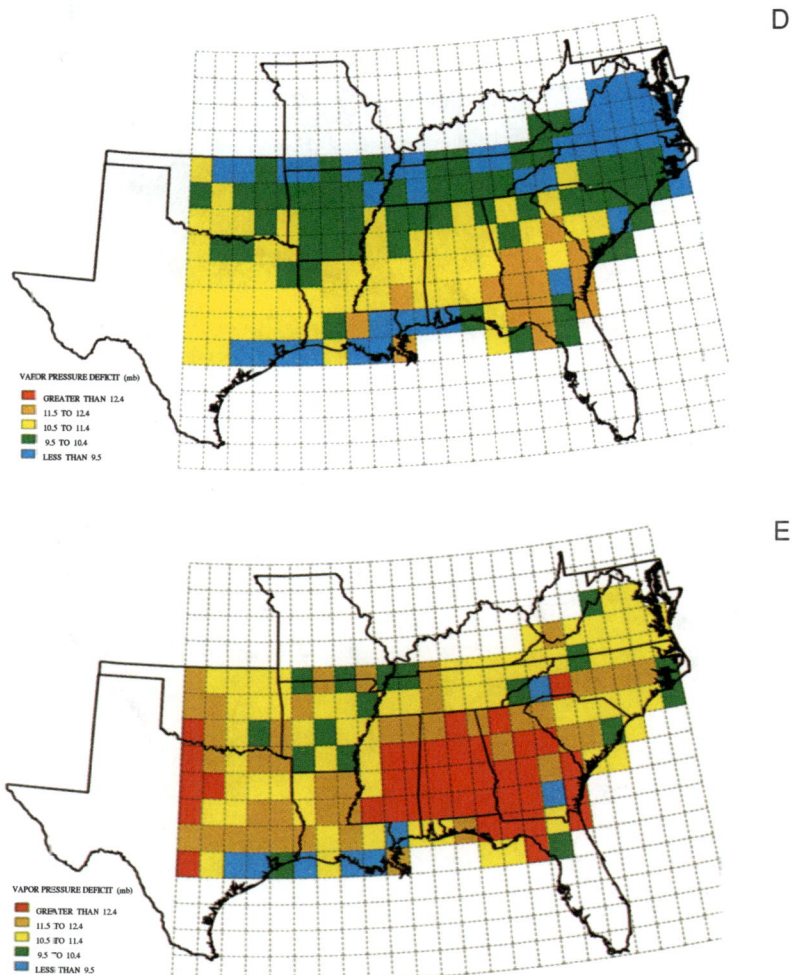

Figure 2.10. (A) Historical early growing season vapor pressure deficit. (B) GISS early growing season vapor pressure deficit. (C) GFDL early growing season vapor pressure deficit. (D) OSU early growing season vapor pressure deficit. (E) UKMO early growing season vapor pressure deficit.

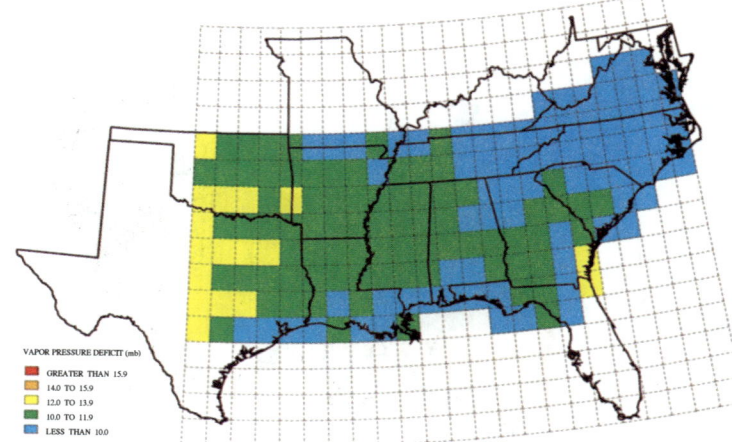

A

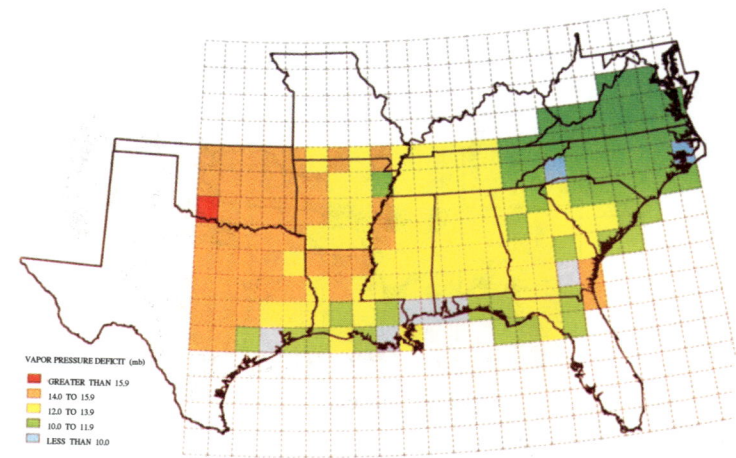

B

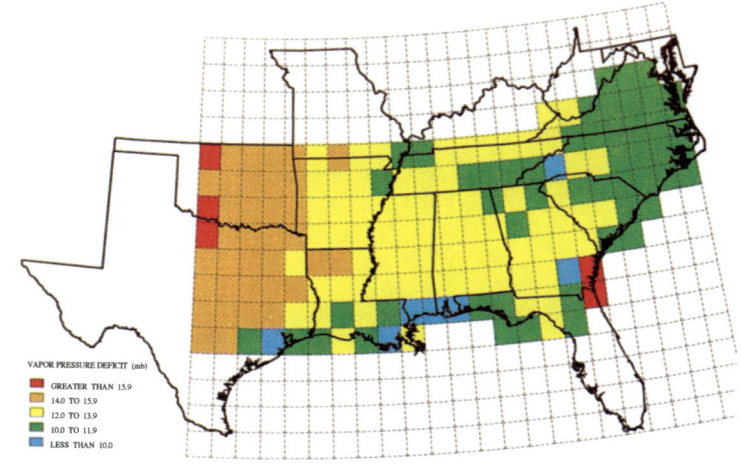

C

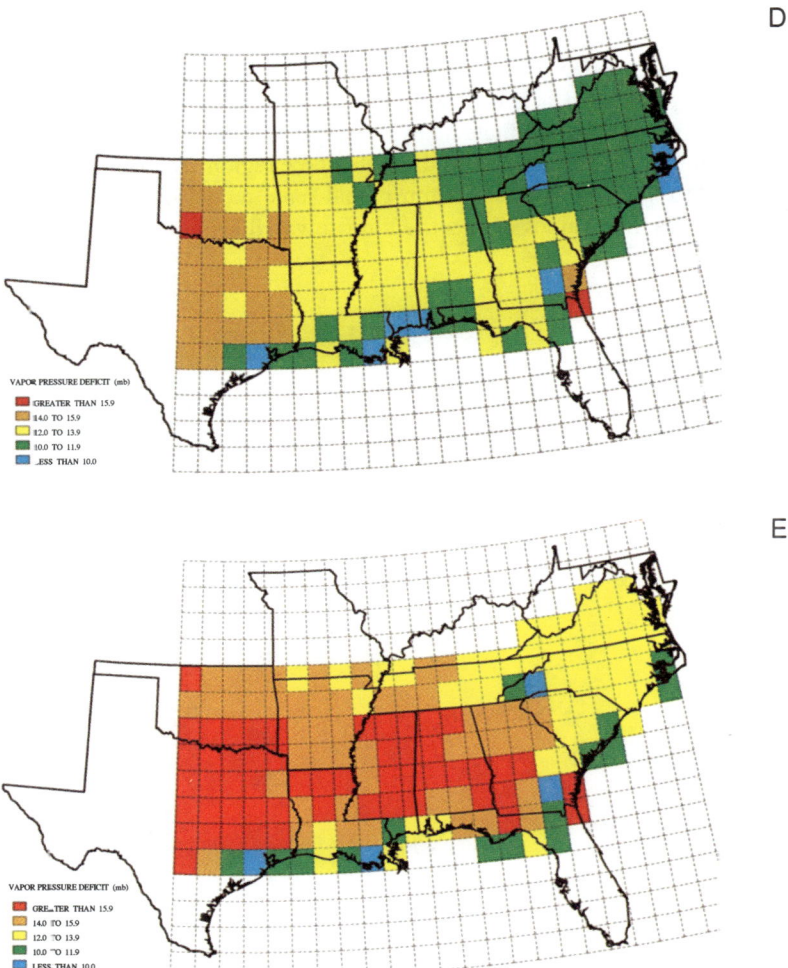

Figure 2.11. (A) Historical late growing season vapor pressure deficit. (B) GISS late growing season vapor pressure deficit. (C) GFDL late growing season vapor pressure deficit. (D) OSU late growing season vapor pressure deficit. (E) UKMO late growing season vapor pressure deficit.

the specified time period. They are not necessarily appropriate for comparison *across time periods*. In most cases, the number of plotting categories and interior plotting category size for a climate variable has been maintained, but category limits may be shifted between time periods. Plotting categories rather than isolines are presented to facilitate analysis of changes in the magnitude and spatial distribution of climate conditions throughout the region. The uncertainty inherent in making regional statements from global-scale model output discourages reliance on absolute values of specific climate variables.

3. Developing Policy-Relevant Global Climate Change Research

J. Christopher Bernabo

Effective communication between researchers and decision makers is a crucial ingredient for successfully addressing society's pressing environmental concerns. The increase in policy makers' demands for research that is relevant to solving societal issues highlights the communication gap between the technical and policy communities. The gap, largely caused by lack of mutual understanding, results in flawed and inadequate communication that hinders decision making and confuses the public.

This chapter examines the reasons for this communication gap and describes the significance of recent efforts to develop more fruitful science-policy dialogues on the issue of global climate change. First, the post-Cold War shift in government priorities for research funding is described; then the underlying relationship between science and policy is explored to identify key sources of ongoing communication inadequacies. This chapter then explains the importance of defining policy-relevant science questions that research can address. Finally, a project is described involving the elicitation of decision makers' information needs.

Policy-Relevant Research

Fifty years after World War II, the major political, social, and economic changes sweeping the globe are causing an historic shift in the emphasis of research funded by governments. In the United States, national security was a major societal justification for massive public funding of the natural sciences and engineering research. The end of the Cold War military competition has caused a wide spread reevaluation of funding priorities for science (Carnegie Commission, 1992a, 1992b).

Furthermore, the public's faith in science as an unquestioned source of increasing material living standards has been shaken by the emergence of many technologically induced environmental problems (Committee on Science, Space, and Technology, 1992). With economic constraints to growth and global competition rapidly increasing, there are greater demands to direct government-funded science and engineering toward solving pressing societal problems.

The emerging post-Cold War rationale for government funding of research has five priority factors:

- Emphasizing science that provides societal benefits;
- Linking research programs to the needs of decision makers;
- Providing economic development and competitive advantages;
- Developing partnerships with diverse stakeholders; and
- Leveraging international research activities and programs.

None of these factors are new, but the increased emphasis upon them in guiding research investments is a major development for science in the post-Cold War period. This greater attention to investment return and the societal relevance of research will require enhanced efforts to improve the communication between scientists, decision makers, and the public.

Relationship of Science to Policy

"Science has the first word about everything and the last word about nothing," Victor Hugo observed. The truth of this is inherent in the relative roles that both "objective" scientific information and "subjective" human values inevitably play in decision making. Environmental policies are developed by interpreting and applying technical information in light of the needs and human values of society (Figure 3.1). Viable policies must not only be technically sound but also socially, politically, and economically acceptable.

Science alone cannot provide answers to policy makers' ultimate questions because science necessarily is silent on the human values that underlie the decisions societies make. The scientific method itself is designed to screen out the value preferences and biases of the subjective human beings who conduct research. Technical information is useful in identifying issues, developing options, providing understanding, and evaluating consequences for policy actions. But in the end, human values must be applied to determine what is "good" policy for a given society on a specific issue.

Take the example of nuclear energy: is promoting it a good or bad policy? On the surface this appears to be a scientifically answerable question, yet nations with access to the same technical information have made different choices about the

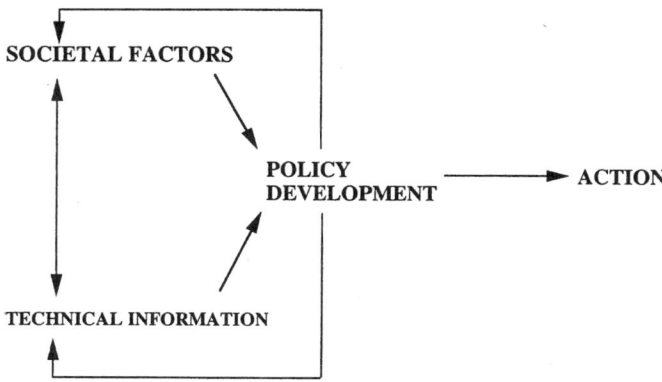

Figure 3.1. The relationship of science and human values in policy development, showing primary interactions and feedback.

best policy for their societies. Indeed, there are Nobel Laureates who staunchly argue opposite sides of the case because the question ultimately involves human values. Science can only approximate the risks and benefits, but a subjective value judgement must be applied to decide what ratio between the two is acceptable to a given individual or society (Bernabo, 1986).

Many of the difficulties scientists and policy makers face in communicating and working together arise from differences in their professional cultures (Table 3.1). Both the scientific and decision making communities experience frustration over the paradoxical relationship between information development and policy development. The public and policy makers often perceive that science is more effective at identifying uncertain problems than it is at providing certain solutions. On the other hand, the technical community becomes frustrated by the perceived inability of policy makers to grasp the "facts" and take what they personally judge is the "logical" action.

Applying technical information to decision making is a fundamentally different type of activity from discovering new knowledge. Alvin Weinberg coined the term trans-science to describe the process of using technical information in making decisions that inherently transcend the bounds of science (Weinberg, 1972). He points out that facts alone are not sufficient even for weighing the benefits and

Table 3.1. Contrasting Professional Cultures of Scientists and Policy Makers

Science	Policy
Objective facts	Subjective values
Proof	Beliefs
Rational	Emotional
Measurements	Perceptions
Incremental progress	Deadlines and crises

costs in policy issues, because subjective values must be applied in choosing what facts to use and how. Harvey Brooks concludes that, "the facts that are selected and the way they are presented to the public may have a greater political impact than the facts themselves" (Brooks, 1975).

Scientific Uncertainty and Policy Decisions

Environmental policy debates typically involve discussion of uncertainties and how much certainty is "enough" to justify a proposed action. The question of how much information is adequate for a given policy always involves a value judgment and cannot be answered by scientific research alone. There is no objective point in science that defines enough certainty for policy.

Research can only quantify the uncertainty in the science, and even that is often done with great difficult. Policy debates involve not only scientific but many other types of inherent uncertainties. Policy decisions must consider uncertainties about such matters as the significance of facts, the perceptions of the issue (opinion polls), the economic and social viability of the proposed solutions, and the actual versus intended consequences of the action.

The degree of scientific consensus is just one part of the information needed for decision making. Brooks cautions that we "should be careful not to expect that scientific consensus should be a necessary condition for policy consensus, an expectation to which scientists tend to be too prone" (Brooks, 1984). For instance, we might have no fiscal policies if action required consensus on economic predictions. The policy makers' roles include making subjective judgments about which information should be acted on and how much certainty is enough for decision making.

The degree of certainty that is adequate for policy can be viewed as an equation balancing scientific uncertainty and political uncertainty. Two general principles apply to environmental issues:

1. The greater the societal consensus on an issue, the less scientific certainty required for action.
2. The higher the societal costs of a policy, the greater the scientific certainty required for action.

The inverse of these principles also is true. They imply that enough certainty in the science is always defined relative to the political certainty in the issue. Therefore, enough scientific certainty in the policy process is a dynamic factor, not a static end point from research.

Two examples illustrate these principles. The United States and Canada fully shared scientific information on acid deposition; they had joint monitoring programs and the same degree of technical certainty on the issue. Nonetheless, lower scientific certainty was required to justify policy action by Canada because there was much higher political certainty than in the United States. More than 90% of Canadians believed that acid rain was a serious problem, while there was no such consensus in the United States. Canadians saw a threat to their major industries—timber, fisheries, and tourism—from the potential damages. In the United States, pollution control costs were instead perceived to be a threat to industry and jobs. In essence, all details of politics aside, the reason for the national differences in the thresholds of scientific certainty required for action was simple and predictable.

In the United States, chlorofluorocarbons (CFCs) were banned as spray can propellants back in 1978. At that time no ozone hole had appeared, and scientific certainty about the issue was lower than about acid rain in 1980 or global climate change in 1996. The threshold of scientific certainty was low for the initial CFC ban because there was political consensus that the risks of skin cancer were not judged to be worth the benefits of protecting a few jobs. Further, banning CFCs from all other uses awaited higher scientific certainty because of the greater societal costs involved.

Policy-Relevant Science Questions

For scientists to more effectively assist in the development of policy, their research needs to be focused on the questions of greatest value to decision makers. Examining past experiences in applying science to address environmental issues helps illustrate the importance of defining the policy-relevant science questions to guide research.

Policy relevancy is determined by the specific needs of the information user (policy maker), not the interests of the information producer (researcher). Unfortunately, the questions investigated by curiosity-driven science are often different than those required to provide the most policy-relevant information. This occurs because decision makers only require the information that can materially assist their specific deliberations, whereas scientists seek greater fundamental understanding of their subjects. Other mismatches exist because of the different values attached to information in the research and decision-making realms, and because policy makers often need information that cuts across fields of research.

There are three general ways to define policy-relevant research questions:

1. *Educated guesses.* This has been the traditional means whereby scientists who study an issue presume to formulate what questions they deem relevant to decision makers. Although quick, this investigator-initiated approach fails to examine the needs of the real decision makers. Without the benefit of decision makers' input, curiosity-driven questions tend to dominate these scientists' "best guess" agendas.
2. *Multistakeholder dialogues.* This approach involves systematically eliciting the information needs of decision makers in the various stakeholder groups for the issue. Interviews and meetings are utilized to determine what the information users' need. Then, scientists are involved in examining and responding to these needs in a facilitated process that ensures results reflecting the best input from both communities. This process can be accomplished over several months and builds direct dialogue between the participants, helping bridge the

science-policy communication gap. A limitation of this method is that it does not allow distinguishing what information participants say they need from what they may use in practice.
3. *Social science research.* This is the most intensive approach and goes beyond eliciting the expressed needs of decision makers to studying their actual behavior in applying information. It involves carefully-designed research and field studies observing the behavior of subjects involved in decision making. This approach provides valuable insights into the use of technical information in policy development. This scholarly work requires extended periods, usually years, during which the policy-relevant questions may shift. Moreover, it does not necessarily build understanding between the science and policy communities.

Whereas the educated guess approach has typically been used, a mixture of the multistakeholder dialogues and social science research is most effective. The dialogues facilitate timely development of broad policy-relevant science questions and build mutual understanding as a basis for consensus among the participants. This approach directly enhances the effectiveness of linking science and policy. The longer-range and more intensive social studies of decision makers' and scientists' behaviors help provide deeper understanding for designing more effective communication. Interactions between these two types of approaches is valuable in assisting each to reach its goal.

Congress' Role in Defining a Climate Change Research Agenda

Today's quest for more policy-relevant climate information and analyses began at least 18 years ago when Congress began hearings on the relationship between climate change and its impact on human activities. By 1978, under the leadership of Congressman George Brown, legislation was enacted to create a pioneering interagency National Climate Program (NCP). That program was explicitly charged with conducting research and assessment activities to provide input to decision makers and its applied mission was to be carried out in "consultation with current and potential users" of climate information.

As defined in the National Climate Program Act (P.L. 95–367), the first element of the program was "assessments of the effects of climate on the natural environment, agricultural production, energy supply and demand, land and water resources, transportation, human health, and national security." The program also was directed to evaluate the social and economic consequences of such changes. This visionary act is still law, but the program is now a ghost, existing only in name.

During its active years in the late 1970s and early 1980s, the National Climate Program's budget grew from about $54 million to more than $300 million per year. Many worthwhile goals were accomplished, such as establishment of the experimental climate forecast centers. However, the program never developed into the user-friendly, policy-relevant applied effort called for by the act.

In the early 1980s, there was a decline in policy interest in climate change, as the acid rain issue dominated decision makers' attention. With the lack of policy attention to the climate change issue by the Reagan administration and Congress, the NCP continued to do research but lost the unique assessment mission that was central to its mandate. Eventually, when the hot summers of the late 1980s put climate change back on the front policy burner, the program had atrophied to the point that a new science-driven effort sprang up to fill the void, leading to the current U.S. Global Change Research Program (USGCRP).

Had the original NCP been successful, we might today be reviewing the results of almost 20 years of progressively more useful policy-relevant assessments to guide U.S. decision making. Unfortunately for the nation, this is not the case. The USGCRP in effect replaced the NCP, but it was born in a different political environment with a much narrower mission. When the USGCRP took form in 1988, the Bush administration's position was to fund more fundamental research to determine if climate change was a real issue worthy of further assessment. The USGCRP budget grew to more than $1.3 billion annually and was the centerpiece of the Bush climate policy. With the administration's position of taking no major actions until more certain predictions of future climate could be made, research on impacts and responses was a low priority. Comprehensive policy-relevant assessments also were not deemed necessary until policy makers judged the science to be certain enough to warrant formulating specific policy options.

Ironically, when the USGCRP was codified in 1990, the Global Change Research Act (P.L. 101–606) instructed the program to "consider" the research of the defunct relic of the original NCP. The 1990 act requires the USGCRP to submit a ten-year research plan to Congress and produce periodic policy-relevant assessments, including outputs providing "usable information on which to base policy decisions." The plan that Congress called for has never been submitted nor have such comprehensive U.S. national assessments ever yet been conducted. The 1990 act did not specify in what time frame useful outputs were expected, but it did borrow the 1978 climate act language for "consultation with . . . information users" to ensure greater relevancy.

During the past eight years, the USGCRP has played a major role in the United Nations Intergovernmental Panel on Climate Change (IPCC) and contributed tremendously to the science assessments and other activities of that group. Nonetheless, these highly political international activities are not a substitute for a strong national assessment capability. Once we move beyond consensus on the state of the science, U.S. decision makers (and those of other nations) need credible analysis and evaluation of potential policy options and the specific consequences for the country. To implement a global agreement, each country will have to conduct internal policy-relevant assessments on how to comply most effectively.

The growing imbalance between federal research on the climate system and other areas such as effects and responses was increasingly recognized by 1992, largely as a consequence of international policy developments related to the U.N.

Conference on Environment and Development (UNCED). Under the Committee on Earth and Environmental Sciences (CEES), a short-lived working group on Mitigation and Adaptation Research Strategies (MARS) was created. MARS attempted to plan for the required research in the areas intentionally excluded from the USGCRP, but the effort foundered in a policy environment where studying impacts and solutions sounded too close to admitting the need for possible action. By late 1992 with the election approaching, the USGCRP structure was reorganized to include an "assessment group," but its role, outputs, and budget were undefined. To date, the USGCRP has only included method development work, shunning the real task of conducting actual integrated climate assessments for the United States.

Identifying U.S. Decision Makers' Climate Information Needs

By 1990, a number of United States research organizations became concerned that the government-sponsored U.S. Global Change Research Program may not provide an adequate basis for the inevitable information demands of future policy development. They decided that a first step in moving toward a policy-relevant research agenda was to determine generally what information decision makers needed, and they launched the "Joint Climate Project to Address Decision Makers' Uncertainties" (Bernabo and Eglinton, 1992). This unique private–federal partnership was sponsored by the Electric Power Research Institute (EPRI), the U.S. Environmental Protection Agency (EPA), the U.S.D.A. Forest Service (USFS), and the U.S. Departments of Energy (DOE), Agriculture (USDA), and Interior (DOI). The project was designed and conducted by Science & Policy Associates, Inc. The Joint Climate Project established a multistakeholder dialogue to help identify some major questions United States decision makers had about global climate change and then had scientists determine what research and time frames would be required to address those questions.

Focusing on the Needs of Decision Makers

The Joint Climate Project identified policy-relevant research using two interactive phases: United States decision makers first defined their information needs; then scientists gave feedback on these needs and determined the research required to address the policy-relevant questions. During the first phase of the project, the needs of the users of climate information were identified through interviews, workshops, and focus groups involving national-level decision makers. These individuals included dozens of United States government and private sector officials, ranging from working-level experts to members of Congress, Administration officials, and industry CEOs. They were invited to participate in the project on the basis of their active roles in climate change policy and their diverse perspectives, from federal regulators and resource managers, to industrial representatives and environmental groups. The interactive process lasted six months and resulted in a consensus set of policy-relevant general questions for researchers to address.

Then, leading experts in climate-related fields were convened at a workshop to discuss the specific questions developed by the decision makers. The scientists were chosen for their activities in research or in the synthesis of research results. They represented a broad range of expertise, including climate system modeling and monitoring, managed and unmanaged ecosystems, energy and technology, as well as economics and social sciences. The workshop participants examined the research needed to address the questions and expectations for providing better information over the next two, five, and ten years, and beyond.

Findings of the Joint Climate Project

The consensus-identifying approach of this project yielded several key findings that reflect the general concerns of decision makers and the responses of the research community. In discussions with these two communities, several common themes emerged for enhancing communication and increasing the value of research results.

The Concerns of Decision Makers

• *International perspectives drive policy.* The participating decision makers identified several general principles that define policy-relevant questions for research. The project was conducted during the year before the 1992 United Nations Conference on Environment and Development (UNCED). Talks were well underway to craft a Framework Convention on Climate Change. Therefore, many government policy makers focused on these and other ongoing international negotiations and conferences. The officials specifically asked for information to support follow-up actions to UNCED and preparations for future events. For their part, nongovernment decision makers expressed concern with the possible regulatory implications of proposed actions.

• *Climate change impacts and human responses are key to decision making.* Aside from pressing international policy issues, decision making is driven by concerns about the potential impacts of changing climate at the regional level, rather than predictions of changing global mean values of climate variables. Specifically, input is needed from the economic, social, and ecological sciences on the potential regional impacts of climate change and the consequences of possible response strategies. Any response to the threat of climate change must be measured against what is at stake. Therefore, more information is needed on the ecosystems, regions, and human populations that are most at risk from potential climate changes, even if atmospheric research is still unable to provide reliable predictions of the specific changes that will drive effects.

• *Implications of uncertainties need clarification.* Researchers need to clarify the sources and implications of policy-relevant scientific uncertainties and estimate time frames for reducing them. Many uncertainties, although scientifically profound, may be relatively insignificant for developing policies. Likewise there are questions that, although they would not be a primary focus for researchers, are of the greatest value to decision makers. The role of sulfate aerosols in determining the patterns of regional climate change is an example. Modelers are just now grappling with this issue, and yet this regulated emission long has been a priority policy concern. There is a need to define better which uncertainties are most important for policy development and resource management, and to articulate the practical implications of these uncertainties for decision makers.

• *Certainty is not a prerequisite for action.* During the project, several decision makers stressed that the resolution of all scientific uncertainties is not a prerequisite for policy action. Decisions are regularly made in the face of significant uncertainty. Decision makers will apply their constituents' values to determine how much certainty they judge is enough to take political action. Although the rhetoric of policy debates often includes assertions that we must have scientific "proof" before acting, in fact the level of certainty required for decision making typically is much less stringent than for purely scientific purposes.

The Response of Researchers

In the next phase of the project, a diverse group of United States experts in climate-related fields were convened to examine how research could best address the questions posed by decision makers. Specifically, the scientists examined what types of research are needed to reduce the uncertainties in the policy-relevant questions and estimated the time frames for possible results.

• *Timely results.* Some of the key questions decision makers have about climate change can be addressed within a short-time frame on the basis of analysis and interpretation of currently available scientific information. Although more complete scientific understanding of climate change may be decades away, much of the information needed to begin addressing decision makers' questions can be provided within two to five years. This could include a comprehensive evaluation of indicators of global climate change, a preliminary vulnerability analysis for systems and regions most sensitive to climate change, and an assessment of the sources and levels of greenhouse gas emissions for use in identifying potential mitigation and adaptation options.

• *Parallel approach to climate and human responses research.* Scientists need not wait for accurate climate predictions before beginning their research on potential impacts and response options. It is neither necessary nor practical for research to progress sequentially from the climate system, to the impacts, and then to the potential human responses in order to provide useful results for decision makers. Much can be done to improve the understanding of impacts without waiting for accurate regional climate predictions. For example, integrated regional and multi-sectoral models—using climate, ecological, demographic, economic, and social data collected at the regional level—can provide essential information on potential climate responses, the vulnerability and adaptability of key systems, the extreme ranges of change, and the impacts of climate change on the global marketplace.

- *Greater emphasis on impacts and human responses research.* Information on climate change impacts and response strategies has the greatest potential for assisting decision makers, yet these fields are the least researched. Many of the key questions identified by decision makers involve a significant amount of new socioeconomic, behavioral, and ecological research. However, only modest increases in funding for these disciplines would be necessary to achieve useful information for policy within a few years. Social science and economic research, in particular, receive a small percentage of federal funding, but are critical for making decisions about climate change.
- *Integrated assessments and case studies.* Integrated assessments of the causal linkages from emissions through impacts and human responses would help structure information for effective use in decision making. Such assessments would incorporate natural and physical sciences, economics, and social factors, including technological change and adaptation. In addition, a coordinated examination of case studies of regional climate variability is needed, based on historically documented events that show how societies have responded to past climatic variations. This information would provide valuable insights on how to treat future events.
- *Expect the unexpected.* Multidisciplinary research on potential surprises is also important, given its potentially serious implications for decision making (i.e., climate change could be much worse than anticipated, or it could be insignificant). Decision makers and scientists should frequently reexamine research on potential surprises, given that scientific progress is incremental and new information may become available. Based on this information, contingency plans could be developed to prepare for unforeseen events.
- *International perspective.* Because of the global dimensions of the issue, an international perspective for research is essential. Although decision makers may be most concerned with regional and local consequences, developing world issues (such as population and economic development, as well as the pace, quality, and sustainability of development) will be critical. Assessing the ability of the international community to implement mitigation and adaptation measures is important for evaluating the effectiveness of response strategies on the climate system.

The project asked researchers to identify the potential types of information that research could provide to address decision makers' concerns in two, five, and ten years. The participants provided educated estimates of the potentially available information for time frames of interest to decision makers. These estimates were developed without regard to financial or other resource constraints. Furthermore, the researchers suggested what research could do, and not what currently planned efforts will do.

Lessons in Communication

Discussions during the Joint Climate Project with representatives of both communities provided ample evidence that decision makers and researchers are uncomfortable with the present situation. Both are anxious to develop and sustain a productive dialogue. Both would like to increase the effectiveness of the research community in the decision-making process. Both agree that a two-way bridge must be developed to span the communications gap between the two communities.

But to truly close this gap, to construct a bridge between the two communities, will take more than wistful expressions and lofty pronouncements. There is no substitute for sustained effort and innovative institutional arrangements. The decision makers and researchers who participated in the project agreed that greater attention must be paid to the development of systemic communications processes. In particular, both sides need to recognize the following points.

- *Not an either/or decision.* Decision makers' choices are not simply between pursuing research or implementing response strategies. Rather, the challenge is to define the appropriate levels of each over time. Researchers need to provide a broad array of information to address the complex and interacting decisions on global climate change. Decision makers, for their part, need to recognize the long-time scales involved in research and, thus, the importance of continuity of funding and program goals.
- *Global climate change in a relative risk context.* Prediction of changes in mean global temperatures does not give an adequate picture of the societal risk that can be related to everyday experiences. The risk of global climate change needs to be compared to the risks of other economic, social, and environmental issues. Because the public tends to respond to perceived crises, assigning relative risk would help decision makers distinguish between verifiable serious threats and possibly misplaced public concern. Given that risk is a function of both the probability and the magnitude of the expected consequences, better data on possible impacts are critical to better estimates of societal risk.
- *Urgent need for education.* A concerted effort is needed to educate decision makers on the facts and uncertainties of global climate change. Because public concern is often the impetus for formulating policy, scientists need to communicate technical information to the public more effectively and more frequently. In addition, scientists need to learn more about the decision-making process and the types of information most useful for policy. Frequent, two-way communication between decision makers and researchers is essential if research is to play an effective role in the decision-making process.
- *Research does not always provide the answer.* Decision makers should understand that additional research can increase the amount of uncertainty in some areas. Researchers should inquire about how much certainty decision makers require to take a specific action. To this end, uncertainties that are not relevant to decision making should be identified early in the process. Decision makers and researchers should also seek ways to manage continuing uncertainties. For example, building resilient institutions would provide a flexible response to any future changes in climate, albeit at potentially significant costs. Contingency plans allow decision makers to prepare for possible climate outcomes through R&D on response technologies, without needing to deploy them.
- *Develop an ongoing assessment process for research.* To improve communication and better inform decision makers, research efforts should include an iterative

assessment process. These assessments not only help to identify the relevant questions, but also serve to structure the research results and, thus, facilitate clearer communication between the two communities. Furthermore, the assessment process provides valuable input to the planning of policy-relevant research.

Project Significance

The Joint Climate Project represents an initial step in determining how researchers can assist U.S. decision makers over the coming years and decades, thereby helping to bridge the communication gap between these two communities. A more frequent and systematic two-way dialog will be needed between decision makers and researchers in order for research to inform the decision-making process. Discussions with decision makers and researchers during the project revealed that both communities are very interested in developing and sustaining a productive dialogue. Both would like to increase the effectiveness of the research community in the decision-making process.

Following the successful dialogue established by the Joint Climate Project, other similar efforts were initiated for climate change in The Netherlands and for biodiversity in the United States (Bernabo, 1989). These types of dialogues also need to be supplemented by more in-depth social science studies to elicit greater understanding of the behavior of decision makers in applying science. A better mutual understanding of the professional cultures of researchers and decision makers is required to enhance the effectiveness of linking science to policy.

The results of the Joint Climate Project indicate that federal research efforts only partially address the current information needs of decision makers. Clearly, impacts and responses are high priorities for policy, yet research on climate system dynamics has received nearly all of USGCRP's past attention and funding. Outputs from impacts and responses research are required to conduct integrated assessments relevant to policy development, but the information is lacking. Furthermore, to be most effective, research on the climate system, impacts, and responses must be specifically designed and managed to support integrated assessments (Bernabo, 1993). The science-driven outputs of typical research cannot simply be adapted for direct use in policy-relevant assessments.

The United States needs a viable new interagency effort that ensures that assessment-oriented research is coordinated and synthesized into policy-relevant products that address the near- and mid-term questions of decision makers. Such assessments must integrate the information from many disciplines—including the social and economic sciences—and then communicate the results in terms most useful to decision makers. An open review process of the methods, assumptions, and results by both the producers and users of the information is crucial to maintain the necessary scientific and political credibility for the assessments.

Lessons from Other Policy-Relevant Assessment Programs

There are several lessons that can be drawn from existing and past policy-relevant assessment programs, particularly the Dutch National Research Program on

Global Air Pollution and Climate Change (NRP) and the U.S. National Acid Precipitation Assessment Program (NAPAP). Both programs have years of experience directly applicable to developing a U.S. assessment-oriented climate effort.

By design, the Dutch National Research Program (NRP) includes climate-related research that supports policy development. The broader fundamental climate research is carried out separately in other efforts not directly linked to assessment goals. The Dutch assessment program also extensively leverages international and other sources for research to provide needed inputs. There are five interrelated components to the Dutch NRP effort:

1. Climate system—atmosphere/ocean/land systems
2. Causes—greenhouse gas emissions
3. Effects—natural and human system impacts
4. Solutions—technical, regulatory, market, and behavioral options to control emissions
5. Assessments—end-to-end evaluation of policy options using integrated assessment models (i.e., IMAGE) and other assessment tools

The Dutch are attempting the difficult but necessary task of organizing and focusing the components of the research to feed the central assessment activities that will increasingly drive the program. They have conducted a project based on the Joint Climate Project approach to better define decision makers' questions and policy options for use in the assessment effort. The Dutch National Research Program is a model the United States should closely consider (Bernabo and Smythe, 1992).

An example of a successful interagency effort to relate science to policy in the United States was NAPAP. NAPAP produced the only existing series of policy-relevant environmental assessments. The triumphs and failures of this program contain lessons not yet properly applied in the climate area. Although the real significance of NAPAP is sometimes misunderstood, the program is the benchmark by which other such efforts are being measured. Some policymakers felt the program was too much science, whereas some researchers felt it was too "soft" and responsive to policy needs. Perhaps these opposite reactions mean a balance was achieved in these circumstances.

In any case, claims that the program was unsuccessful (Rubin et al., 1992), stem from the false assumption that enactment of the Clean Air Act Amendments of 1990, before NAPAP's final assessment, meant the program had no effect on policy. To the contrary, no astute observer expected the legislative process to be paced by NAPAP's time table and indeed outputs of the program had profound influences on every stage of the policy development, directly and indirectly (Regens, 1993). The assessment program's results (1) expanded the policy debate beyond sulfur to nitrogen and ozone, (2) reduced concern for crop damages, (3) established the size and location of the aquatic resources at risk, (4) defined the uncertainties about forest damages, (5) established the sources of emissions, and (6) analyzed various control strategies under consideration. Each of these and other outputs of research and assessments played a role in shaping the resulting national policy.

Of course, NAPAP was not without major problems and deficiencies that accompany such pioneering efforts. The struggle for relevancy was constantly waged and interagency turf battles reduced the program's ultimate effectiveness. The most debilitating single problem resulted with the first published assessment in 1987, when the long-standing open process of review and consensus-building was abridged by the second director and his own policy judgments influenced the interpretation of findings in a way not supported by objective analyses. This transgression severely damaged the credibility of the program. It took a new director and several years for perceptions about the program's objectivity to be restored. The lesson is that in conducting assessments, *process* is as important as results for credibility.

The program-wide reviews NAPAP employed to examine the relevancy of research to its assessment goals were a crucial tool for integrating the efforts of diverse agencies and disciplines. The program reached out to all the stakeholders for comments and input, and although that was a slow and cumbersome process, the dividends were many. The data bases, technical information, and assessments of NAPAP still continue to be widely used today. The current USGCRP spends $500 million every three months, equal to the entire ten years of NAPAP expenditures. When the policy-relevant-output-per-dollar ratio is examined by future decision makers, climate programs will have some explaining to do. NAPAP was not perfect, but to date it has come closest to reaching the goals of useful policy relevancy.

Another lesson of both NAPAP and the Dutch NRP is that policy-relevant assessments require a dedicated staff of specialists. Assessments are not effectively conducted by the usual scientific disciplinary experts. Without the assessment-oriented professionals, the gap between research results and decision makers' needs will remain wide because of the cultural and motivational differences in the professions. Science ultimately seeks understanding, but policy guides actions by applying human values to current knowledge. A core team is essential to effective synthesis, integration, and communication of the assessments significance.

Assessments must be targeted to specific audiences to pose and address the most useful questions. They serve as an effective vehicle for communicating the significance of complex information from diverse multidisciplinary sources. Most importantly, assessments must be an integral driver of the supporting research agenda, and not something one retrofits at the end of scientific research. The joint development of research agendas and priorities, incorporating the needs of decision makers and the input of researchers, will facilitate a more successful translation of science into policy-relevant information.

Conclusions

Developing policy-relevant research requires the involvement of both scientists and decision makers in framing the appropriate questions. Policy users of the research results must articulate their information needs and consult scientists on the feasibility of research providing meaningful answers. Scientists can examine those initial requirements to determine the limitations and strengths of investigation and to meet them within available budgets and time frames. An ongoing mutual learning process between the users and producers of the information is required to refine and then periodically update the policy-relevant research questions as both science and policy evolve.

References

Bernabo JC (1986), Science and policy: notes from a former Congressional Fellow. In Proceedings of the American Geophysical Union, EOS 7:82.

Bernabo JC (1989) Global change research for decision making. Statement before the U.S. House of Representatives Committee on Science, Space and Technology, July 27, 1989.

Bernabo JC (1993) Climate research relevant to policymakers' needs: lessons learned. Statement before the U.S. House of Representatives Committee on Science, Space and Technology, May 19, 1993.

Bernabo JC, Eglinton P (1992) Final Report of the Joint Climate Project to address decision makers' uncertainties. Electric Power Research Institute Technical Document No. TR-100772, Palo Alto, CA.

Bernabo JC, Smythe K (1992) Evaluation of the technical emphasis, policy relevance, and management performance of the Dutch National Research Program on Global Air Pollution and Climate Change. Science and Policy Associates, Inc. and Holland Consulting Group, Bilthoven, NL.

Brooks H (1975) Expertise and politics: problems and tensions. In Proceedings of the American Philosophical Society 119:257.

Brooks H (1984) The resolution of technically intensive public policy disputes. Science and Human Values 9.

Carnegie Commission (1992a) Enabling the future: linking science and technology to societal goals., Washington, DC.

Carnegie Commission (1992b) Environmental research and development: strengthening the federal infrastructure. Washington, DC.

Committee on Science, Space, and Technology (1992) Report of the task force on the health of research. U.S. House of Representatives 102nd Congress, Washington, DC.

Regens JL (1993) Acid deposition. In Uman MF (Ed) Keeping pace with science and engineering: case studies in environmental regulation. National Academy Press.

Rubin ES, Lave LB, Morgan MG (1991) Keeping climate research relevant. Issues in Science and Technology, Winter 1991–1992.

Weinberg AM (1972), Science and trans-science. Minerva 10:207.

Section 2. Global Change Impacts on Tree Physiology and Growth

4. Influence of Drought Stress on the Response of Shortleaf Pine to Ozone

Richard B. Flagler, John C. Brissette, and James P. Barnett

The gaseous composition of the atmosphere has changed significantly during the past century as the result of anthropogenic activity. Changes in the physical and chemical climate of the southern United States may have significant detrimental effects on the forest tree species that grow in this extensive and complex region. Although the consequences of these changes for forest ecosystems are manifold, the most significant outcomes may be 1) chronic exposure of ecologically and economically important species to elevated levels of phytotoxic pollutants, and 2) global and regional scale changes in precipitation regimes. Knowledge of the combined impacts of the climatic and air pollutant stresses on forests and forest species is essential to ensuring their productivity and sustainability.

The impacts of phytotoxic air pollutants on forest health have drawn global attention for the past three decades (Chappelka and Chevone, 1992; Cowling, 1989; Fox and Mickler, 1996; Olson et al., 1992; Nilsson and Duinker, 1987; Schulze et al., 1989). Ozone is the most phytotoxic of the regional air pollutants in North America (Krupa and Kickert, 1989; National Academy of Sciences, 1977; Wang and Schaap, 1988; U.S. Environmental Protection Agency, 1986) and there have been reports of forest injury and growth effects in virtually every forest type in the United States. Western forests have been the subject of several studies in which growth, injury, and ecosystem processes have been found to be impaired (Miller, 1983; Olson et al., 1992; Peterson and Arbaugh, 1988). Both coniferous and hardwood forests in the northeastern and eastern United States have been affected (Benoit et al., 1982; Berry and Hepting, 1963; Eager and Adams, 1992;

Jensen and Dochinger, 1989; Scott et al., 1984). Similarly, there are now many reports of air pollutant effects on forests and forest tree species in the southern United States (Flagler, 1992; Flagler and Chappelka, 1996; Fox and Mickler, 1996; Sheffield and Cost, 1987; Zahner et al., 1989).

Water is the primary factor that limits agricultural production in the United States and other parts of the world (Boyer, 1982); water deficit is the factor that most limits growth of forest trees in the southern United States. Water deficit decreases growth and alters carbon allocation in trees, generally favoring allocation to the roots and decreasing allocation to foliage (Schulze, 1986; Teskey and Hinckley, 1986). Water deficit also reduces stomatal conductance (Kramer and Kozlowski, 1979) and negatively affects photosynthesis and other gas exchange responses (Brix, 1962; Seiler and Johnson, 1985).

As a result of global change, predictions for a warmer, drier environment for many areas of the United States suggest that water will become increasingly more limited (Manabe and Wetherald, 1986; Schneider, 1989). In many areas of the southern United States, where drought is already a frequent occurrence, models suggest that summer soil moisture will decrease by approximately 20 to 40% in response to a doubling of atmospheric CO_2 (Manabe and Wetherald, 1986). Thus, combined ozone and drought stress is perhaps already common in some areas of the southern United States, and may become an even greater problem if general circulation model predictions are accurate. In addition to these problems, the predicted effects of global climate change may increase the concentrations of ozone in the troposphere (Krupa and Kickert, 1989) leading to possibly greater impacts on forests and tree species.

Although there are now several reports in the literature concerning the effects of ozone and water-stress interactions on crop plants, there are still relatively few reports about the effects on tree species. The research reported in this chapter is intended to provide some of the necessary information to assess the impacts of combined ozone and water-deficit stress on shortleaf pine (*Pinus echinata* Mill.), a major southern pine species.

Two studies, conducted at the same site in east Texas, have addressed the effects of drought stress on the ozone exposure-response relationships for shortleaf pine. The study site is located 31° 30′ N latitude, 94° 46′ W longitude, and lies within the contiguous range of shortleaf and loblolly pine (*P. taeda* L.). The 10 ha research site is surrounded by a mature shortleaf and loblolly pine forest. The mean annual maximum/minimum temperatures are 24.2 °/11.2 °C; the site receives a mean annual precipitation of 115.6 cm, with snowfall being rare.

Container-Grown Seedling Study

This study was conducted for a single growing season, April through November 1992. Seedlings were grown in containers so that access to the roots would be possible without disturbing adjoining seedlings. The specific objectives of this study were to determine 1) the functional relationship between growth and physiological response of shortleaf pine seedlings and ozone exposure, 2) if soil-water deficit affects the ozone exposure-response functions, and 3) if responses differ between seedlings from a "woods-run" collection and genetically improved populations.

Specific Methods

Seedlings were grown at the Alexandria Forestry Center (Pineville, LA) until significant secondary needles had emerged. In January 1992, these seedlings were transplanted into 8-l plastic rootrainer pots containing washed, fritted clay, a growth medium that has excellent dry-down characteristics for water-deficit studies (van Bavel et al., 1978). Each container was 41 cm tall with a square, 16-cm opening that tapered to a 10-cm square base. Seedlings were grown in a greenhouse until April, when they were moved to open-top chambers. They were fertilized with a slow-release, complete fertilizer.

The experimental design was a split-plot, within a completely randomized design with two complete replications. The main plots were five ozone treatments. Sub-plot treatments were a factorial combination of three soil-water levels and the three seed selections. The study utilized ten exposure chambers. Twelve pots of each selection/water-regime combination were placed in each exposure chamber, resulting in twenty-four plants per treatment combination for a total of 1080 plants in the study.

Ozone treatments were applied in 3-m diameter open-top chambers (Heagle et al., 1973) fitted with rain exclusion caps. Exposures to ozone began on April 21, 1992 and were ended November 25, 1992. The ozone levels used in the study were charcoal-filtered air (CF), nonfiltered air (NF), and 1.5×, 2.0×, and 2.5× the ambient ozone at the study site. Ozone additions were applied 12 h day^{-1} (0800 to 2000 CST, [Central Standard Time]), as a proportion of ambient ozone. Fans were operated 18 h day^{-1} (0600 to 0000 CST). A corona discharge ozone generator was used to generate ozone from oxygen, which was dispensed to each chamber via calibrated flowmeters. Ozone concentrations in each chamber were monitored on a time-share basis with ozone-specific, UV-photometric instruments three times each hour. Each ozone monitor was calibrated with a UV-photometric transfer standard and that ozone data were recorded on a data acquisition computer. The exposure statistic used for data analyses was the 12-h cumulative ozone exposure in ppm-h.

The three water regimes were designed to provide mild, moderate, and severe plant water deficits. Plants were watered when volumetric water content of the medium reached 32%, 22%, and 18%. The water potentials of the medium represented by these percentages are ca -0.11 MPa, -0.63 MPa, and -0.86 MPa soil-water potential (ψ_{soil}), respectively. Moisture content of the medium was checked daily using a time-domain reflectometry soil-moisture meter, which gives a direct reading of volumetric water content for the medium. A moisture-release curve for the medium was also generated to estimate the water potential of the medium.

Tree height and root collar diameter were measured on a monthly basis. At the end of the study, the trees were harvested for foliage area and biomass determinations. The foliage from each tree was removed and was measured for foliage area with a leaf area meter. The root systems from a subsample of six trees per treatment combination were separated into coarse roots (> 1mm diameter) and fine roots (< 1mm diameter). All plant material was dried to a constant weight at 65 °C and then weighed.

Net photosynthesis (A), stomatal conductance (g), transpiration (E), and instantaneous water-use efficiency (A/E) were measured monthly with a LI-6200 portable photosynthesis system equipped with a 0.25-l cuvette. Measurements were taken on two fascicles per tree and four trees per treatment combination.

Using a pressure chamber, plant-water potential (ψ_{plant}) was measured concurrently with gas exchange. Determinations of ψ_{plant} were also made predawn and midday on several occasions prior to and following rewatering to obtain estimates of maximum stress and subsequent recovery. Measurements were made on one fascicle from each of four trees per treatment combination.

Measurements of root system water flux (L_R) were made to assess root function according to the methods of Brissette and Chambers (1992). Six trees per treatment combination were measured as described by Flagler et al. (1993).

Data were analyzed by analysis of variance (ANOVA). Regression analyses were used only when significant linear or curvilinear responses were evident from the ANOVAs. When significant responses were found, the regression model was used to estimate the response for the ambient ozone levels occurring at the study site. The percentage differences reported are relative to the subambient, CF treatment. Statistical differences are significant at the $p \leq 0.05$ level, unless otherwise indicated.

Results and Discussion

The peak 1-h average ambient ozone concentration was 94 ppb and occurred on June 5. The highest 1-h average ozone concentration in the 2.5× treatment was 236 ppb. The 12-h cumulative ozone exposures for the CF, NF, 1.5×, 2.0×, and 2.5× treatments were 17, 87, 148, 172, and 208 ppm·h, respectively. The actual seasonal averages associated with the three soil moisture treatments were 29.1% (-0.22 MPa ψ_{soil}), 25.6% (-0.42 MPA ψ_{soil}), and 22.5% (-0.61 MPa ψ_{soil}) for the mild, moderate, and severe stress levels, respectively (Flagler et al., 1993).

At the end of the study, ozone had not affected height and diameter growth, but there was a pronounced impact of water stress on these variables. There were also differences among the three seed sources, with trees from the woods-run selection being slightly larger than the other two selections. Compared to the well-watered treatment, soil-water deficit resulted in reductions in diameter growth of 22.0% and 39.8% for the moderate and severe stress levels, respectively. These results were expected and have been reported for many tree species (Schulze, 1986; Teskey and Hinckley, 1986).

Seedling biomass was affected by both ozone and water regimes, with no

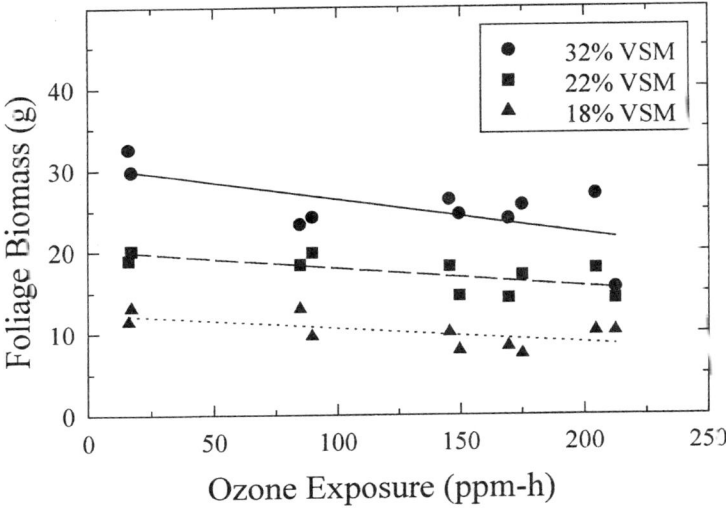

Figure 4.1. Ozone exposure-foliage biomass response functions for shortleaf pine seedlings grown under three soil water treatments ranging from 32% to 18% volumetric soil water for one growing season.

significant differences among the seed sources. Water deficit had a much greater impact than ozone treatments on biomass, affecting all biomass fractions and leaf surface area. Compared to the mild treatment, decreases in the moderate stress treatment were 38.9% for total weight, 31.3% for foliage weight, 44.4% for stem weight, 40.5% for root weight, and 33.9% for leaf surface area. In the severe water-stress treatment, decreases were 66.2% (total weight), 59.6% (foliage weight), 70.0% (stem weight), 68.5% (root weight), and 62.6% (leaf surface area). Because of the importance of water to tree growth (Schulze, 1986; Teskey and Hinckley, 1986), these results were also not unexpected.

Ozone decreased total dry weight of seedlings in a few of the treatment combinations, but the most consistent effects were on the foliage. This result was surprising, because ozone is known to affect biomass of many tree species, including shortleaf pine (Flagler et al., 1992a; 1992b; Pye, 1988). Ozone decreased foliage biomass and leaf surface area for all three seed sources. At ambient ozone levels, foliage biomass and foliage area were decreased by 12.7% and 12.3%, respectively. This is consistent with reports in the literature indicating that foliage is a sensitive indicator of ozone damage (Allen et al, 1992; Chappelka and Chevone, 1992; Flagler et al., 1992a; 1992b; Flagler and Chappelka, 1996) There was no interaction of ozone and water regime (Figure 4.1); however, there were differences among the seed sources in their response to ozone (Figure 4.2). One selection, S3PE9, was responsible for the significant ozone response and decreased linearly in foliage dry weight as ozone concentration increased. The woods-run selection showed a slight negative response to ozone (significant at

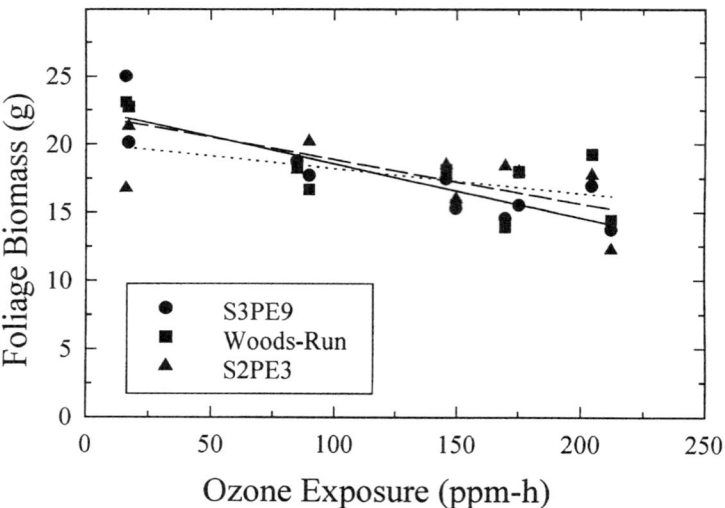

Figure 4.2. Ozone exposure-foliage biomass response functions for three shortleaf pine selections after one growing season.

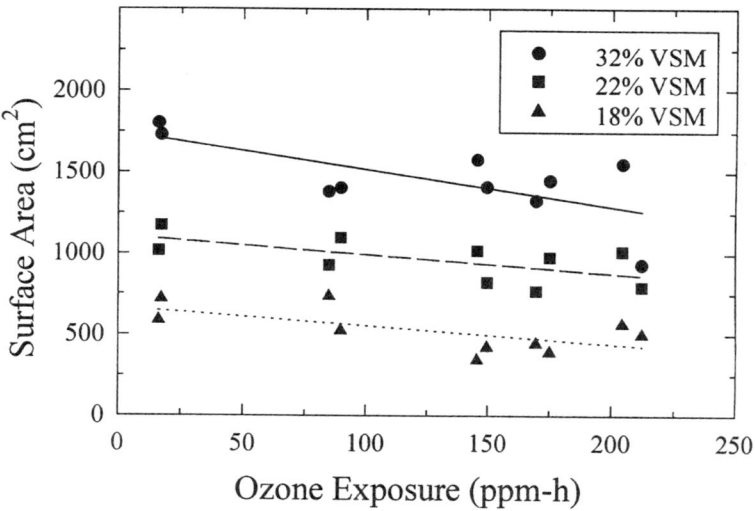

Figure 4.3. Ozone exposure-foliage surface area response functions for shortleaf pine seedlings grown under three soil water treatments ranging from 32% to 18% volumetric soil water one growing season.

4. Interaction of Drought Stress and Ozone

p < 0.10) and selection S2PE3 showed no response to ozone. Leaf surface area also decreased as ozone concentration increased with no interaction of ozone and water regimes (Figure 4.3). This decrease was the result of the response of selection S3PE9 (Figure 4.4). Decreases in pine seedling leaf biomass or surface area because of ozone exposure usually result from premature senescence (Allen et al., 1992; Flagler et al., 1992b; Flagler and Chappelka, 1996). Coarse root biomass decreased linearly as ozone concentration was increased in S2PE3, the only selection affected.

Seedling gas exchange was decreased significantly by soil-water deficit, with effects on A and g evident throughout the study. Water deficit is known to affect gas exchange by causing stomatal closure (Brix, 1962; Kramer and Kozlowski, 1979; Seiler and Johnson, 1985). There were no consistent differences among the three selections over the duration of the study. Compared to the mild stress treatment, during the October (final) measurement period, A decreased 13.0% and 48.1%; g decreased 27.6% and 57.7%; E decreased 27.1% and 56.5%; and A/E increased 15.2% and 23.7% for the moderate and severe stress levels, respectively.

Ozone effects became evident only during the October measurement period. This is not unusual because the seedlings need to accumulate a sufficient internal ozone dose to impair the photosynthetic process (Sasek and Flagler, 1996; Sasek and Richardson, 1992). Ozone did not affect either E or A/E. The woods-run selection was affected by ozone only in the severe water regime; and although rates were only 50% of those in the mild stress treatments, there was still sufficient

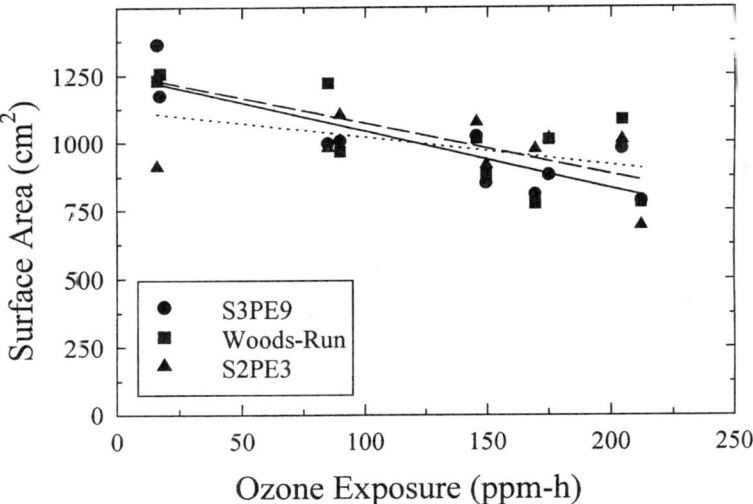

Figure 4.4. Ozone exposure-foliage surface area response functions for three shortleaf pine selections after one growing season.

ozone uptake to elicit a response. A and g were not affected in selection S3PE9, but were decreased in selection S2PE3. The response was strongest in the mild and moderate regimes; however, there was a decrease in A with increasing ozone in the severely water-stressed seedlings indicating that gas exchange was continuing, even though the seedlings were very water limited (Figure 4.5). Decreases in A because of ambient ozone were 15.1% and 17.3% for the mild and moderate water-stress regimes, respectively. Although water deficit may afford limited protection from ozone by stomatal closure, this protection is incomplete at best. Ponderosa pine (*P. ponderosa* Laws.) was afforded some partial protection from ozone via this mechanism (Beyers et al., 1992) and exhibited less foliar injury (Temple et al., 1992). This has also been observed in other pine species (Dobson et al., 1990).

Predawn ψ_{plant} measurements give an indication of a seedling's potential to recover after a dry-down cycle. Predawn ψ_{plant} after a drying cycle was influenced by soil-moisture deficit, but not directly by ozone treatments. At the end of a water-stress cycle, trees in the mild water-stress regime recovered to -0.38 MPa, trees in the moderate regime recovered to -0.70 MPa, and trees in the severe regime recovered to only -0.92 MPa. These differences may be a result of osmotic adjustment in the more severely water-stressed seedlings (Kramer and Kozlowski, 1979; Teskey and Hinckley, 1986). Ozone and moisture regimes interacted significantly such that in the severe soil moisture regime, recovery of the seedlings was less at the higher ozone levels (Table 4.1). While this interaction was apparent across all three selections, the selections did not vary significantly in

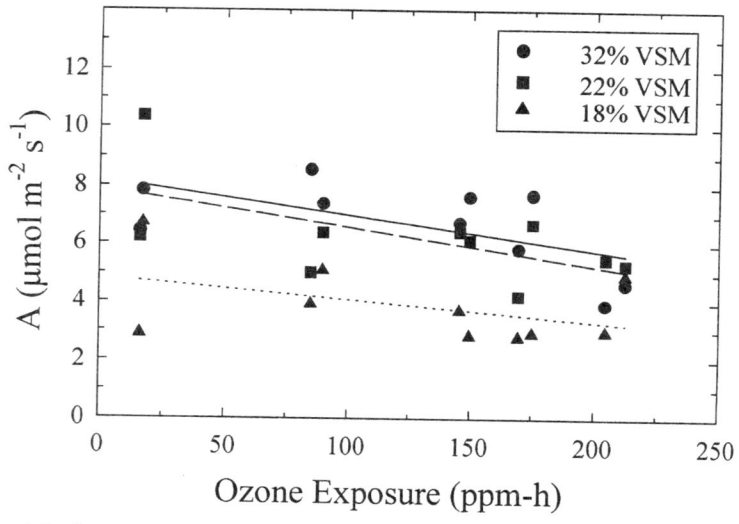

Figure 4.5. Ozone exposure-response functions for net photosynthesis of shortleaf pine selection S2PE3 for each of three soil water regimes after six months of exposure.

Table 4.1. Predawn and Midday Plant Water Potentials for Three Selections of Shortleaf Pine Exposed to Five Levels of Ozone and Three Soil Moisture Regimes[1]

	Selection											
	S3PE9			Woods-Run			S2PE3			Average		
	\multicolumn{12}{c}{Soil Moisture Treatment}											
Ozone	32%	22%	18%	32%	22%	18%	32%	22%	18%	32%	22%	18%
	\multicolumn{12}{c}{Predawn Plant Water Potential (MPa)}											
CF	−0.37	−0.67	−0.78	−0.37	−0.63	−0.77	−0.40	−0.62	−0.77	−0.38	−0.64	−0.77
NF	−0.37	−0.76	−0.81	−0.39	−0.87	−0.79	−0.35	−0.82	−0.83	−0.37	−0.82	−0.81
1.5×	−0.36	−0.57	−1.07	−0.36	−0.64	−1.02	−0.36	−0.68	−1.02	−0.36	−0.63	−1.04
2.0×	−0.44	−0.66	−1.04	−0.45	−0.84	−1.03	−0.41	−0.83	−1.21	−0.43	−0.78	−1.09
2.5×	−0.36	−0.59	−1.18	−0.38	−0.58	−0.94	−0.37	−0.67	−0.98	−0.37	−0.61	−1.04
Mean	−0.38	−0.65	−0.96	−0.39	−0.71	−0.88	−0.38	−0.72	−0.92	−0.38	−0.70	−0.92
	\multicolumn{12}{c}{Midday Plant Water Potential (MPa)}											
CF	−1.15	−1.45	−1.60	−1.18	−1.48	−1.60	−0.99	−1.42	−1.73	−1.11	−1.45	−1.64
NF	−1.25	−1.39	−1.54	−1.33	−1.55	−1.76	−1.31	−1.46	−1.71	−1.30	−1.47	−1.67
1.5×	−1.09	−1.32	−1.55	−1.18	−1.48	−1.54	−1.19	−1.55	−1.76	−1.15	−1.45	−1.62
2.0×	−1.14	−1.42	−1.59	−1.23	−1.48	−1.63	−1.00	−1.54	−1.90	−1.12	−1.48	−1.71
2.5×	−1.12	−1.39	−1.62	−1.25	−1.49	−1.69	−1.12	−1.56	−1.59	−1.16	−1.48	−1.63
Mean	−1.15	−1.40	−1.58	−1.23	−1.50	−1.64	−1.12	−1.50	−1.74	−1.17	−1.47	−1.65

[1] Data were taken at the end of respective dry down cycles, just prior to rewatering, when water stress was at its maximum (from Flagler et al., 1993)

their predawn ψ_{plant} values. A possible explanation for this interaction is that ozone affects plant membranes by making them "leaky." With the low ψ_{plant} levels in the severe regime, membrane damage may have been sufficient to upset internal water balance and prevent full recovery.

By midday, ψ_{plant} had decreased in all treatment combinations (Table 4.1). Significant effects occurred because of water regimes, but no ozone effects and no ozone x moisture-regime interactions were evident. There were no differences among the three selections. Mean ψ_{plant} for the soil-moisture regimes was -1.17 MPa, -1.47 MPa, and -1.65 MPa for the mild, moderate, and severe regimes, respectively.

Root-system water-flux was measured to assess the integrity and function of the root system and because it is a good indicator of new root growth (Brissette and Chambers, 1992). Root-system water-flux was greatly affected by soil-moisture regimes, and to a lesser extent by ozone. There were no significant differences in L_R among the three seed sources. Values of L_R decreased as water stress increased. Across seed sources, decreases averaged 60.3% and 77.6%, for the moderate and severe water regimes, respectively, relative to the mild regime. These decreases were attributed to smaller root systems in the more severely water-stressed seedlings. Prior to this study, a decrease in L_R caused by soil-water deficit had also been demonstrated with shortleaf pine (Brissette and Chambers, 1992).

Ozone had a tendency to increase L_R in all three seed sources, especially in the 2.5× ozone treatment. However, only the woods-run selection was statistically significant. In this selection, L_R increased linearly ($p = 0.07$) as ozone concentration increased (Table 4.2). This relationship was primarily the result of the effects in the mild soil-moisture regime. Ozone has been reported to decrease allocation of resources to roots (Cooley and Manning, 1987), which is consistent with the finding that coarse root biomass decreased in one selection in response to ozone. The increase in L_R with increasing ozone in the woods-run selection may be the result of senescent roots making the root system more permeable. The response was not the result of more roots, as there was no effect of ozone on the root system in this selection. Qualitatively, the root systems in the higher ozone treatments

Table 4.2. Root System Water Flux for the Woods-Run Seed Source as Influenced by Five Levels of Ozone and Three Soil Moisture Regimes[1]

Ozone Treatment	Soil Moisture Treatment			Mean
	32%	22%	18%	
CF	0.071	0.122	0.184	0.126
NF	0.245	0.330	0.197	0.257
1.5×	0.307	0.285	0.309	0.300
2.0×	0.381	0.357	0.289	0.342
2.5×	0.506	0.413	0.231	0.383
Mean	0.302	0.301	0.242	0.282

[1] Data are expressed on a fine root biomass basis (mg H_2O s^{-1} g^{-1} dwt).

appeared less healthy (fewer new growing tips, darker brown coloration) than the roots in the CF and NF treatments.

Summary

After one growing season, shortleaf pine seedlings exposed to a range of ozone concentrations and levels of soil-water deficit exhibited several negative impacts. Soil-water deficit was clearly the factor that most limited both the growth and physiology of the seedlings. The most significant impact of ozone treatments was on foliage biomass and area; these decreased through the mechanism of premature senescence. Ozone effects on foliage were evident at the ambient levels of ozone present at the study site. There were the significant ozone exposure-response relationships that occurred in the moderate and sometimes in the severe soil-water deficit treatments, indicating that under the conditions of this study, ozone uptake continued at detrimental doses in shortleaf pine even when the water availability is severely limited.

In Situ Large Chamber Study

This study was conducted over four growing seasons beginning in late 1991 and continuing through November 1994. The trees in this study were planted directly into the soil at the study site so that they would grow under as natural conditions as possible. The two specific objectives of this study were to determine 1) the effects of soil-water deficit on the response of shortleaf pine seedlings/saplings to a range of ozone exposures, and 2) the comparative responses of two half-sibling families with known ozone susceptibility with a naturally regenerated population of shortleaf pine.

Specific Methods

The trees used in this study were grown as described in the prior section on the container-grown study but were transplanted directly into the soil at the study site in November 1991. They were allowed to acclimate to their environment for a period of four months. After being transplanted, the seedlings were not fertilized and received only natural rainfall until treatment initiation began.

The experimental design was a completely randomized split-plot. The main plots were a factorial combination of four ozone treatments applied in large (4.6-m diameter) open-top chambers (Heagle et al., 1989) and three levels of simulated rain. The subplots contained the three shortleaf pine selections. There were two complete replications and the study utilized twenty-four chambers. The area inside each chamber was divided into three sections. Twenty-four seedlings of each selection were randomly assigned to one of the sections, for a total of 1728 trees in this study.

The four ozone treatments were CF, NF, 1.7×, and 2.5× the ambient ozone at the study site. Ozone dispensing and monitoring was as described in the previous study. The three rain treatments were 100%, 80%, and 60% of the thirty-year average rainfall (TYAR) at the study site. Rain treatments were applied via an overhead solid cone nozzle operated at constant pressure to deliver approximately 2-cm H_2O h^{-1}. The treatments were applied so that average weekly rainfall totals were achieved in the 100% TYAR treatment. All treatments began in March 1992 and were continued until November 1994.

Tree height and diameter were measured on a monthly basis. At the end of each growing season (nominally November), trees were harvested for biomass and foliage area determinations. The foliage from each tree was removed and was measured for projected surface area with a leaf area meter. The plant material was dried to a constant weight at 65° C and then weighed.

A, g, E, and A/E were measured on a monthly basis with a LI-6200 portable photosynthesis system equipped with a 0.25-l cuvette. Measurements were taken on two fascicles per tree and two trees per treatment combination. Plant-water potential was also measured concurrently with gas exchange measurements using a pressure chamber.

Results and Discussion

Ambient ozone exposures varied among the three growing seasons of exposure. Cumulative ambient 12-h exposures were 136 ppm·h, 183 ppm·h, and 154 ppm·h, for 1992 to 1994, respectively. Total cumulative exposures for the four treatments at the end of the study were 183 ppm·h, 363 ppm·h, 740 ppm·h, and 1056 ppm·h, for the CF, NF, 1.7×, and 2.5× treatments, respectively. In 1992, ozone levels were relatively low compared with previous years (1988 to 1994) (Flagler et al., 1994). This gave the plants a relatively "even start," with few ozone effects at the beginning of the second growing season. In 1993, cumulative ozone exposure was relatively high, but there were only a few hours greater that 100 μL L^{-1}. The peak 1-h average was 125 μL L^{-1}. 1994 was a more typical year in east Texas, although the distribution of moderate and high concentrations was somewhat unusual. The peak concentration of 102 μL L^{-1} occurred early in May, when the saplings were actively growing. This one particular peak resulted in significant foliar symptom development.

Soil-moisture treatments varied with the natural distribution of rain for the study site, therefore average values of ψ_{soil} are of little value. The 100% TYAR treatment separated well from the other two treatments. The main difference between the 80% and 60% TYAR treatments was that the 60% TYAR treatment typically went to lower soil-water potentials during periods between rain events. The lowest ψ_{soil} values occurred during August 1993 after a prolonged hot and dry period. Average ψ_{soil} minima during this period were −1.40 MPa, −1.45 MPa, and −1.53 MPa for the 100%, 80%, and 60% TYAR treatments, respectively. Conversely, March 1994, was very mild; average ψ_{soil} minima during this period were −0.47 MPa, −0.61 MPa, and −0.65 MPa for the 100%, 80%, and 60% TYAR treatments, respectively. Despite the nearly three-fold range of ψ_{soil} between these two months, the values for ψ_{plant} were much closer together:

−1.62 MPa and −1.32 MPa for the 60% TYAR treatment in August 1993, and March 1994, respectively. These two contrasting time periods will be used to highlight some of the physiological effects on the trees.

Both A and g decreased linearly as ozone exposure increased during most sample periods during 1993 and 1994. In August 1993, mean A (across all treatments) was 3.6 μmol m^{-2} sec^{-1}, which was almost half of the 6.7 μmol m^{-2} sec^{-1} rate in March 1994. Mean g was similarly affected with mean rates of 0.089 mol m^{-2} sec^{-1} and 0.167 mol m^{-2} sec^{-1} in August 1993 and March 1994, respectively. During both months, there were decreasing linear trends in A and g because of ozone, but only the March data were statistically significant. The trends observed and the rates of A and g are similar to data reported earlier for shortleaf pine (Flagler et al., 1994). Ozone had little effect on ψ_{plant} in either month.

The rain treatments did not have a significant effect on A in either August 1993 or March 1994, although there were decreasing trends as water stress increased. There were similar trends for g, with the March 1994 data being statistically significant. Significant decreases occurred in ψ_{plant} with increasing water stress in both months. Significant differences were not found among the three selections for any of these responses and there were no significant interactions among treatments.

At the final harvest, growth and biomass responses were affected significantly by ozone, and not by the moisture-stress treatments (although trees were generally smallest in the 60% TYAR treatment). The response of the three shortleaf pine selections was very similar, and not statistically significant. The response of height growth to ozone was quadratic in nature, while diameter growth, the biomass responses, and foliage surface area responded with linear decreases as ozone exposure increased. These types of responses have been observed in prior studies of shortleaf and other pine species (Flagler et al., 1992a; 1992b).

Regression models were used to predict the changes in growth and biomass that are occurring at present levels of ambient ozone exposures in the southern United States (Table 4.3). Data from all treatments were used in the model to estimate the ozone response in a broad sense. The models used the theoretical value of one-half the present level of ozone exposure as a surrogate for clean air for the region. The

Table 4.3. Predicted Values for Response Variables Exposed to 0.5× Ambient Ozone for Three Growing Seasons and Their Percentage Change at Ambient and Above-Ambient Ozone Exposures

Response	Predicted Value at 0.5× Ambient	Percentage Change at		
		1.0×	1.25×	1.5×
Height	3.27 m	+ 1.2	+ 0.3	− 1.6
Diameter	5.99 cm	− 5.1	− 7.7	− 10.3
Total Dwt.	2671.6 g	− 15.1	− 22.7	− 30.2
Woody Dwt.	1713.5 g	− 12.4	− 18.6	− 24.7
Foliage Dwt.	958.3 g	− 20.0	− 30.0	− 40.0
Foliage Area	5.37 m^2	− 18.3	− 27.4	− 36.6

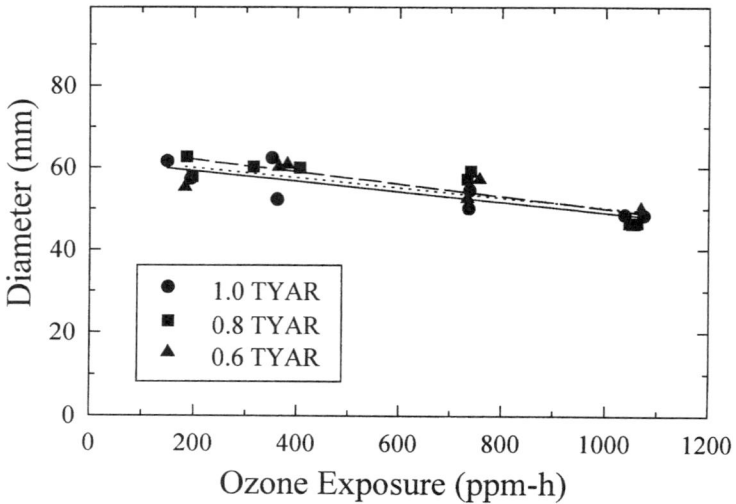

Figure 4.6. Ozone exposure-response functions for diameter growth of shortleaf pine exposed for three growing seasons under three simulated rain regimes.

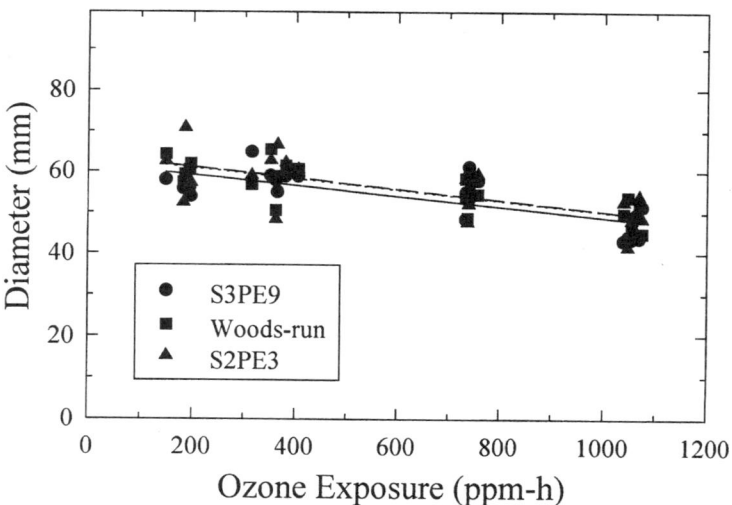

Figure 4.7. Ozone exposure-response functions for total above-ground biomass of shortleaf pine exposed for three growing seasons under three simulated rain regimes.

responses ranged from no negative effects at present ambient ozone exposures for height growth, to a decrease of 20% for foliage biomass, the most affected response. When the percentage changes are evaluated at ozone exposures of 1.5× the present ozone levels, all responses were negative (Table 4.3).

When the effects of ozone on diameter growth were evaluated for each rain treatment separately, there were significant linear decreases that occurred for all three treatments, but there were no differences among the three responses (Figure 4.6). The total biomass response across rain treatments was somewhat similar, although more pronounced (Figure 4.7). All the responses were linear and negative, but the response was more pronounced in the 60% TYAR treatment than in the other two treatments. This means that for diameter growth and total biomass at the end of three growing seasons of exposure to ozone, a 40% decrease in the amount of water the trees received did not alter the response to ozone. This is counter to some of the literature for other pines (Beyers et al., 1992; Temple et al., 1992) in which drought was found to provide some protection against ozone injury to foliage by reducing the stomatal conductances. But although conductances were decreased in this study, the long-term translation of decreased carbon uptake into reduced growth and biomass does not appear certain.

Similarly, when diameter growth and total biomass were evaluated for each selection, all responses were significant, with linear decreases in diameter or biomass as ozone exposure increased; with no differences among the three selections (Figures 4.8 and 4.9). Because the two half-sibling families had been determined at an earlier date to be susceptible to ozone (Flagler et al., 1992a; 1992b), and there were no differences between the half-sibling families and the woods-run selection, this leads to the conclusion that there are many ozone-susceptible genotypes among the naturally regenerated populations in east Texas, and possibly throughout the southern United States. Although there are wide variation in the response of southern pine genotypes to ozone (Flagler and Chappelka, 1996; Taylor, 1995), the results of this study lead to the question of finding the quantity of the extensive areas of shortleaf pine that are susceptible to ozone.

Summary

In this study, ozone was by far the factor that most impacted growth and biomass of shortleaf pine. Although there were some effects resulting from the water regimes and also some differences among the three genetic selections used in the study, there were very few interactions among treatments. For all response variables, except height growth, present ambient ozone exposures caused detrimental effects. There were significant decreases in diameter growth, biomass, and foliage surface area. Foliage biomass was affected most, with a 20% decrease predicted when compared to half of the present ambient ozone.

Soil-water deficit had some effects on growth and biomass, but did not ameliorate the responses to ozone. This is an important finding as it contradicts the crops literature. Also, because some predictions for climatic change in the southern United States predict increased frequency and severity of drought, our data do not

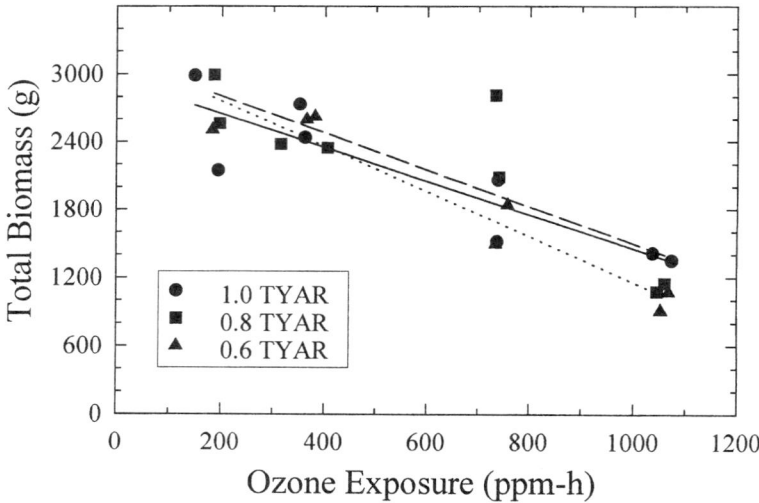

Figure 4.8. Ozone exposure-response functions for diameter growth of three shortleaf pine selections exposed for three growing seasons.

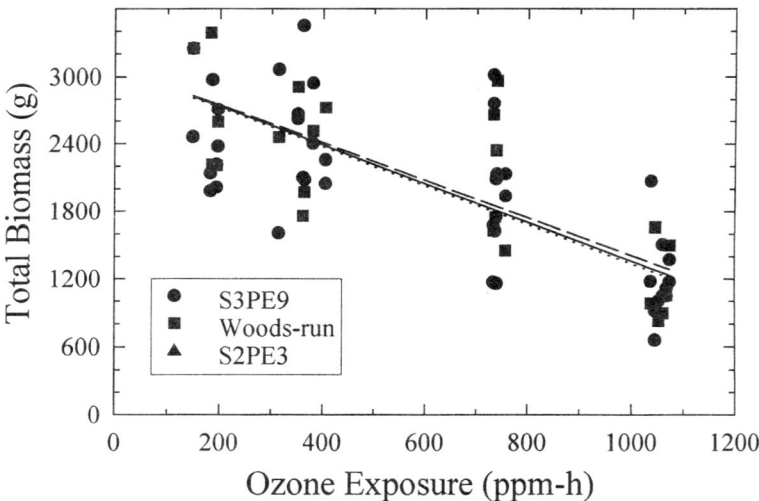

Figure 4.9. Ozone exposure-response functions for total above-ground biomass of three shortleaf pine selections exposed for three growing seasons.

indicate that ozone stress will be less. If this is coupled with predictions for increased ozone concentrations in the region resulting from increased temperatures, then the effects of ozone in the southern United States may become even more pronounced.

The three genetic selections used in this study all responded similarly and negatively to ozone. The two half-sibling families used, S2PE3 and S3PE9, were determined to be ozone susceptible in some of our earlier work. The fact that the woods-run selection was found to be as ozone susceptible (and, in some cases, more susceptible) as these two selections, indicates that there are some very ozone-susceptible genotypes among the naturally regenerated populations.

Summary

Both studies provide valuable data that will help in understanding the response of shortleaf pine to ozone and drought stress. Between the two studies, there were many similarities in the responses to ozone. In both studies, ozone had the greatest effect on the foliage, with decreases in foliage biomass resulting from premature senescence of foliage, even after only one growing season. Net photosynthesis was decreased in both studies after a threshold exposure had been exceeded. This combination of foliage loss and decreased photosynthesis leads to decreases in biomass. This was evident in the longer study, and most certainly would have occurred in the shorter study, had it been continued for a second year. There were some differences in response to ozone among the three selections between the two studies, but in general, all responded negatively to ozone. This result indicates little difference between the two half-sibling families determined at a piror date to be affected by ozone and the genetically unimproved material collected in the woods. The modeling results from the longer study indicate that present ambient levels of ozone in the southern United States are sufficient to cause decreased growth and biomass of shortleaf pine.

The response to drought was quite different between the two studies. However, the differences in methods between the two studies provide the reasons for the respective results. The container-grown seedlings had a limited volume of growth medium to exploit, while the field-grown trees had a relatively large soil volume to utilize. Consequently, as dry-down occurred, plant-water potentials reached lower values in the container-grown pines compared to the field-grown seedlings/saplings. Dry-down also occurred over shorter time spans in the container-grown seedlings, compared to the field-study trees. This combination of factors resulted in more severe effects of water deficit in the container-grown seedlings.

Water deficit did not interact with ozone to affect the ozone exposure-response relationships in the container-grown shortleaf pine study. Biomass and photosynthesis were affected to some extent in all three water regimes. In the field study, the ozone exposure-response relationship for diameter growth was not affected by water regime; however, the relationship for total biomass was affected. In this case, the water-limited trees responded more to the increased ozone exposure than the trees that received the TYAR.

Within the context of these studies, if the climate of the southern United States changes such that temperatures increase and drought becomes more prevalent, it is reasonable to expect that the growth of shortleaf pine will decrease. Such climatic conditions as these may lead to increased concentrations of ozone in the troposphere, which will affect growth; the increased severity or incidence of drought will not affect shortleaf pine response to ozone, and may exacerbate responses.

References

Allen HL, Stow TK, Chappelka AH, Kress LW, Teskey RO (1992) Ozone impacts on foliage dynamics of loblolly pine. In Flagler RB (Ed) *The response of southern commercial forests to air pollution.* Air & Waste Management Assoc., Pittsburgh. 149–162.

Benoit LF, Skelly JM, Moore LD (1982) Radial growth reductions of *Pinus strobus* L. correlated with foliar ozone sensitivity as an indicator of ozone-induced losses in eastern forests. Can J For Res 12:673–678.

Berry CR, Hepting GH (1963) Ozone, a possible cause of white pine emergence tipburn. Phytopathology 53:552–557.

Beyers JL, Riechers GH, Temple, PJ (1992) Effects of long-term ozone exposure and drought on the photosynthetic capacity of ponderosa pine (*Pinus ponderosa* Laws.). New Phytol 122:81–90.

Boyer JS (1982) Plant productivity and environment. Science 218: 443–448.

Brissette JC, Chambers JL (1992) Leaf water status and root system water flux of shortleaf pine (*Pinus echinata* Mill.) seedlings in relation to new root growth after transplanting. Tree Physiol 11:289–303.

Brix H (1962) The effect of water stress on the rates of photosynthesis and respiration in tomato plants and loblolly pine seedlings. Physiol Plant 15:10–20.

Chappelka AH, Chevone BI (1992) Tree Responses to Ozone. In Lefohn AS (Ed) *Surface level ozone exposures and their effects on vegetation.* Lewis Publishers, Chelsea, MI, 271–324.

Cooley DR, Manning WJ (1987) The impact of ozone on assimilate partitioning in plants: A review. Environ Pollut 47:95–113.

Cowling EB (1989) Recent changes in chemical climate and related effects on forests in North America and Europe. Ambio 18:167–171.

Dobson MC, Taylor G, Freer-Smith PH (1990) The control of ozone uptake by *Picea abies* (L.) Karst. and *P. sitchensis* (Bong.) Carr. during drought and interacting effects on shoot water relations. New Phytol 116:465–474.

Eager C, Adams MB (Eds) (1992) *Ecology and decline of red spruce in the eastern United States.* Springer-Verlag, NY.

Elsik CG, Flagler RB, Boutton TW (1992) Effects of ozone and water deficit on growth and physiology of *Pinus taeda* and *Pinus echinata*. In Flagler RB (Ed) *The response of southern commercial forests to air pollution.* Air & Waste Management Assoc., Pittsburgh, PA, 225–245.

Flagler RB (Ed) (1992) *The response of southern commercial forests to air pollution.* Air & Waste Management Assoc., Pittsburgh.

Flagler RB, Brissette JC, Elsik CG, Isbell VR, Lock JE (1993) Response of shortleaf pine to ozone and drought stress. *Proc Air & Waste Management Association,* 93-TA-43.04, Air & Waste Management Association, Pittsburgh, PA.

Flagler RB, Chappelka AH (1996) Growth response of southern pines to acidic deposition and ozone. In Fox S, Mickler RA (Eds) *Impact of air pollutants on southern pine forests.* Springer-Verlag, NY, 388–424.

Flagler RB, Lock JE, Elsik CG (1994) Leaf-level and whole-plant gas exchange characteristics of shortleaf pine exposed to ozone and simulated acid rain. Tree Physiol 14:361–374.

Flagler RB, Lock JE, Toups BG (1992a) Growth and gas exchange characteristics of shortleaf pine exposed to ozone and acid rain. *Proc. 9th World Clean Air Congress*, IUA92–5, Air & Waste Management Association, Pittsburgh, PA.

Flagler RB, Spruill SE, Chappelka AH, Dean TJ, Kress LW, Reardon JC (1992b) Growth of three southern pine species as affected by acid rain and ozone: A combined analysis. In Flagler RB (Ed) *The response of southern commercial forests to air pollution.* Air & Waste Management Assoc., Pittsburgh, PA, 207–224.

Fox S, Mickler RA (Eds) *Impact of air pollutants on southern pine forests.* Springer-Verlag, NY.

Heagle AS, Body DE, Heck WW (1973) An open-top field chamber to assess the impact of air pollution on plants. J Environ Qual 2:365–368.

Heagle AS, Philbeck RB, Ferrell RE, Heck WW (1989) Design and performance of a large field exposure chamber to effects of air quality on plants. J Environ Qual 18:361–367.

Jensen KF, Dochinger LS (1989) Response of eastern hardwood species to ozone, sulfur dioxide and acid precipitation. J Air Pollut Control Assoc 39:852–855.

Kramer PJ, Kozlowski TT (1979) *Physiology of woody plants*, Academic Press, NY.

Krupa SV, Kickert RN (1989) The greenhouse effect: Impacts of Ultraviolet-B (UV-B) radiation, carbon dioxide (CO_2), and ozone (O_3) on vegetation. Environ Pollut 61:263–393.

Manabe S, Wetherald RT (1986) Reduction in summer soil wetness induced by an increase in atmospheric carbon dioxide. Science 232:626–628.

Miller PR (1983) *Air pollution and productivity of the forest.* In Davis DD, Miller AA, Dochinger LS (Eds) Izaak Walton League, Arlington, VA, 161–197.

National Academy of Sciences (1977) *Ozone and other photochemical oxidants.* Washington, D.C.

Nilsson S, Duinker P (1987) The extent of forest decline in Europe. Environment 29 (9):4–31.

Olson RK, Binkley D, Böhm M (Eds) (1992) *The response of western forests to air pollution.* Springer-Verlag, NY.

Peterson DL, Arbaugh MJ (1988) An evaluation of the effects of ozone injury on radial growth of Ponderosa pine (*Pinus ponderosa*) in the southern Sierra Nevada. J Air Pollut Control Assoc 38:921–927.

Pye JM (1988) Impact of ozone on the growth and yield of trees: A review. J Environ Qual 17:347–360.

Sasek TW, Flagler RB (1996) Physiological and biochemical effects of air pollutants on southern pines. In Fox S and Mickler RA (Eds) *Impact of air pollutants on southern pine forests.* Springer-Verlag, NY, 424–463.

Sasek TW, Richardson CJ (1992) The dose-response approach for characterizing the effects of near-ambient ozone concentrations on photosynthesis. In Flagler RB (Ed) *The response of southern commercial forests to air pollution.* Air & Waste Management Assoc., Pittsburgh, PA, 257–271.

Schneider SH (1989) The greenhouse effect: science and policy. Science 243:771–781.

Schulze E-D (1986) Whole-plant responses to drought. Aus J Pl Physiol 13:127–141.

Schulze E-D, Lange OL, Oren R (Eds) (1989) *Forest decline and air pollution: A study of spruce (Picea abies) on acid soils.* Springer-Verlag, NY.

Scott JT, Siccama TG, Johnson AH (1984) Decline of red spruce in the Adirondacks, New York. Bull Torrey Bot Club 111:438–444.

Seiler JR, Johnson JD (1985) Photosynthesis and transpiration of loblolly pine seedlings as influenced by moisture stress conditioning. For Sci 31:742–749.

Sheffield RM, Cost ND (1987) Behind the decline. J For 85:29.

Taylor GE (1994) Role of genotype in the response of loblolly pine to tropospheric ozone: Effects at the whole-tree, stand, and regional level. J Environ Qual 23:63–82.

Temple PJ, Riechers GH, Miller PR (1992) Foliar injury responses of ponderosa pine seedlings to ozone, wet and dry acidic deposition, and drought. Environ Exp Bot 32:101–113.

Teskey RO, Hinckley TM (1986) Moisture: Effects of water stress on trees. In Hennessey TC, Dougherty PM, Kossuth SV (Eds) *Stress physiology and forest productivity*, Martinus Nijhoff, The Netherlands, 9–33.

US Environmental Protection Agency (1986) *Air quality criteria for ozone and other photochemical oxidants.* EPA-600/8–84/020cF. Environmental Criteria and Assessment Office, Research Triangle Park, NC.

van Bavel CH, Lascano R, Wilson DR (1978) Water relations of fritted clay. Soil Sci Soc Am J 42:657–659.

Wang D, Schaap W (1988) Air pollution impacts on plants: current research challenges. ISI Atlas of Science: Animal and Plant Sciences 1988: 33–39.

Zahner R, Saucier JR, Myers RK (1989) Tree-ring model interprets growth decline in natural stands of loblolly pine in the southeastern United States. Can J For Res 19:612.

5. Effects of Elevated Carbon Dioxide on the Growth and Physiology of Loblolly Pine

Makonnen Alemayehu, Douglas R. Hileman,
Gobena Huluka, and Prosanto K. Biswas

The level of carbon dioxide (CO_2) in the atmosphere has been increasing at an annual rate of approximately 1.5 μmol mol^{-1} (Keeling et al., 1989). The present atmospheric CO_2 is expected to double by the end of the next century. Carbon dioxide is one of the 'greenhouse gases' that is believed to cause gradual warming of the earth's atmosphere with consequent changes in natural ecosystems. Carbon dioxide is also one of the substrates of photosynthesis. Therefore, it may be reasonable to assume that the rise in atmospheric CO_2 will be advantageous to plant metabolism. Short-term studies, conducted mainly on annual crops, indicate that exposure of plants to elevated CO_2 improved rates of growth and photosynthesis and total dry matter production with varying effects under different environmental conditions (Kramer, 1981; Strain and Sionit, 1982; Kimball, 1983; Dahlman et al., 1985; Sionit et al., 1985; Strain and Cure, 1985; Strain, 1987; Rogers and Dahlman, 1993). In general, C_3 species respond more to elevated CO_2 than C_4 species.

Loblolly pine (*Pinus taeda* L.) is an economically important perennial C_3 forest species. According to Lieth (1975), forests cover about one-third of the global land area and play a dominant role as sources and sinks for terrestrial carbon. However, relatively few studies have been reported on the effects of elevated CO_2 on woody plants compared to crop plants. This may be attributed to the logistical difficulty of conducting CO_2-enrichment studies (e.g., large size and perennial habit). Therefore, ontogenetic CO_2 enrichment studies of forest trees remain as the only alternative to providing information on the effects of global climate change on forest ecosystems.

The limited information available on performance of woody plants under elevated CO_2 was obtained from studies that, for the most part, were conducted on container-grown plants. Kramer and Sionit (1987) concluded that 1) dry weight, stem diameter, and height of seedlings generally increase when CO_2 concentrations is in the range of 400 to 700 ppm, and 2) seedlings of various species react differently to increasing CO_2 levels, making the study of each individual species a necessity. In some of the studies, contradictory results have been reported. For example, O'Neill et al., (1987), Kramer and Sionit (1987), and Thomas et al., (1994) have reported that tree seedlings benefit from increasing CO_2 under non-limiting growth conditions. However, Larigauderie et al., (1994) studied performance of loblolly pine seedlings subjected to different levels of CO_2 and nitrogen (N) treatments and reported comparable growth stimulation of elevated CO_2 under both high and low N treatments.

In a similar study, Samuelson and Seiler (1993) grew red spruce (*Picea rubens* Sarg.) seedlings under two levels of CO_2, nutrients, and water for five months. They reported greater stem diameter, plant height, branching density, total biomass, and mean relative growth rate of plants grown under elevated CO_2 with either the presence or absence of water, in addition to nutrient limitations. The authors concluded that even if nutrient and moisture availabilities are limited, seedling establishment under natural environments may be enhanced by the rise of atmospheric CO_2. Conversely, studies conducted on Alaskan tundra species under CO_2 enrichment and different levels of nutrients indicated no increase in productivity during three years of exposure to CO_2 enrichment (Oechel and Riechers, 1986; Tissue and Oechel, 1987). The objective of this study was to evaluate morphological growth, physiological performance, and biomass production of loblolly pine seedlings planted directly into the ground and exposed to different CO_2 levels under natural field conditions.

Materials and Methods

The experiment was conducted at Tuskegee University's George Washington Carver Agricultural Experiment Station in Tuskegee, Alabama. The soil series was Norfolk sandy loam (kaolinitic, thermic, Typic Paleudult).

The CO_2 treatments were open-field plot (ambient), ambient $+ 0$ μmol mol^{-1}, ambient $+ 150$ μmol mol^{-1}, and ambient $+ 300$ μmol mol^{-1} CO_2. Liquid CO_2 was stored at the site in a fourteen-ton receiver from which CO_2 gas was delivered through a custom-made dispensing manifold to field chambers. Six open-top field chambers (Rogers et al., 1982) with 45° frustum were used to expose plants to the CO_2 treatments. Each chamber was 3.0 m in diameter and 2.4 m in height and were made of structural aluminum frames covered with clear polyvinyl chloride plastic. Carbon dioxide concentrations in all chambers and open-field plots were monitored twenty-four hours a day for the entire period of the study. Carbon dioxide sampling lines were run from each chamber and open-field plot to a sampling manifold located in the building adjacent to the field. Air samples were drawn in continuously through these lines by a pump attached to the sampling manifold. In the sampling manifold, each sample first passed through an adjustable flowmeter and then through a three-way solenoid valve. A computer was used to activate the solenoids sequentially and divert the samples to an infrared gas analyzer (IRGA). Outputs from the IRGA were recorded continuously by a computer and on a strip-chart recorder. The IRGA were calibrated on a daily basis using a series of high-pressure tanks of known CO_2 concentrations. After this, the CO_2 values of each of the field chambers were checked and any necessary adjustments were made to the flowmeters on the dispensing manifold. Application of CO_2 began on June 1, 1993 and continued until the time of final harvest (November 30, 1994).

Loblolly pine tree seedlings of uniform size, which were originated from seeds collected from an orchard at the Atlantic coastal plain of southeastern Georgia, were obtained from a commercial forest seed company and planted on May 20, 1993. The seedlings were planted in the soil in rows with a spacing of 38 cm between plants. A total of 32 seedlings were planted in each of the eight plots.

Determinations of growth were made on randomly selected plants. Measurements of plant height and stem diameter, in addition to countings of branching densities and growth-flush frequencies were made on a monthly basis beginning at the time of planting. Plant height was defined as distance from the ground line to tip of terminal bud on the main stem. Diameter measurements were made using Vernier calipers at a marked point at the base of the main stem. Aboveground biomass growth was determined using plant samples harvested at twelve and eighteen months after planting (June 15 and December 2, 1994, respectively). Five plants were randomly selected during the first harvest and then seven for second harvest. All harvested plants were separated into needles and stems; their fresh and dry weights were then determined. The dry matter content of the samples was determined after drying at 60 °C for seven or more days until constant weight was reached.

Gas exchange measurements (photosynthesis, transpiration, and stomatal conductance) were made on pairs of consecutive days of every month starting a month after planting (July 1993). The measurements were made with a LI-6200 portable photosynthesis system and one-quarter liter cuvette on five randomly selected plants in each of the eight plots and two mature fascicles per plant. The fascicles were clipped from the main stem, placed in the cuvette, and measured within one minute of clipping (Ginn et al., 1991).

A randomized complete block design (RCBD) with two replications was used. The data were analyzed by the analysis of variance method (ANOVA) using a microcomputer-based statistical package.

Results and Discussion

The monthly growth measurements of loblolly pine seedlings subjected to different CO_2 treatments indicated that plants with significantly greater plant heights and stem diameters were produced under the highest level of CO_2 enrichment ($+300$ μmol mol^{-1}), starting from three and four months after planting, respectively (Figure 5.1). Differences between plants grown in the highest and the lowest level CO_2 treatments, especially the $+300$ μmol mol^{-1} CO_2 and the

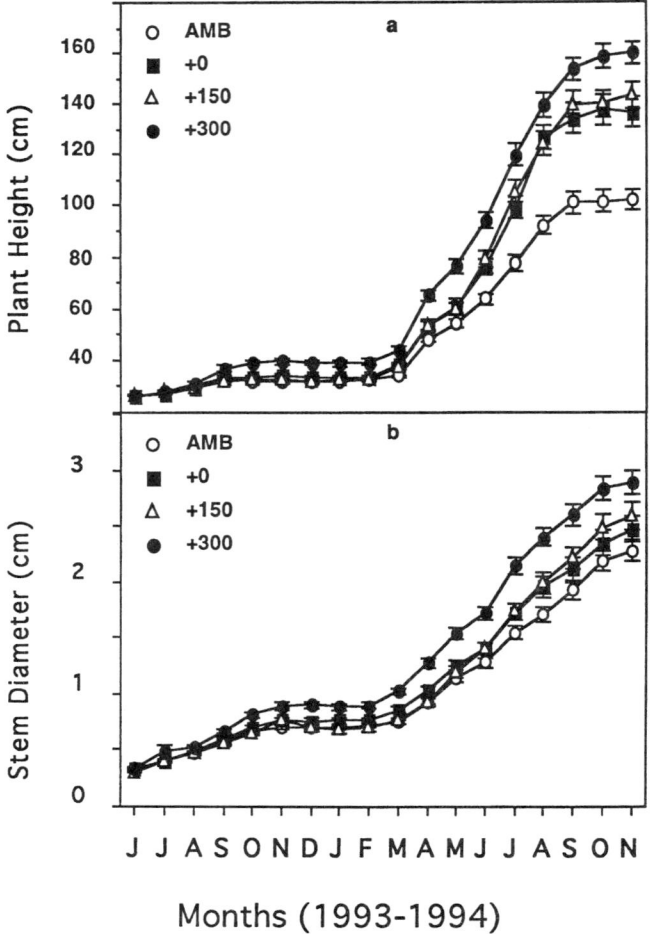

Figure 5.1. Monthly measurements of plant height (a) and stem diameter (b) of loblolly pine seedlings subjected to different CO_2 treatments. Vertical bars represent ± the standard error of the mean.

ambient plots, were magnified from the beginning of spring (March, 1994). In both plant height and stem diameter, the measurements of plants grown in the ambient plots were significantly smaller than those grown in the +0 and +150 μmol mol^{-1} CO_2 treatments. There was no significant difference in plants subjected to the latter two CO_2 treatments. The reduced growth of loblolly pine seedlings in the ambient plots may also have been partly the result of a pine-tip moth (*Rhyacionia spp.*) infestation that was observed about a month after planting. Pine-tip moth did not attack plants inside the open-top chambers, possibly because of the protection offered by the plastic coverings.

5. Effects of Elevated Carbon Dioxide

Similar to plant height and stem diameter, the overall number of flushes and branches of plants grown in the +300 μmol mol^{-1} CO_2 were significantly higher (p ≤ 0.01) than at the lower CO_2 treatments (Figure 5.2). Ratios of plant height/ number of flushes were 24.1, 27.1, 26.9, and 27.6 for plants grown under the ambient, +0, +150 and +300 μmol mol^{-1} CO_2, respectively. This indicates that regardless of the CO_2 level, chambered plots had similar and higher growth rates than plants grown in the ambient plots.

The average of monthly gas exchange measurements indicated that net photosynthesis increased significantly (p ≤ 0.001) and proportionally with increasing

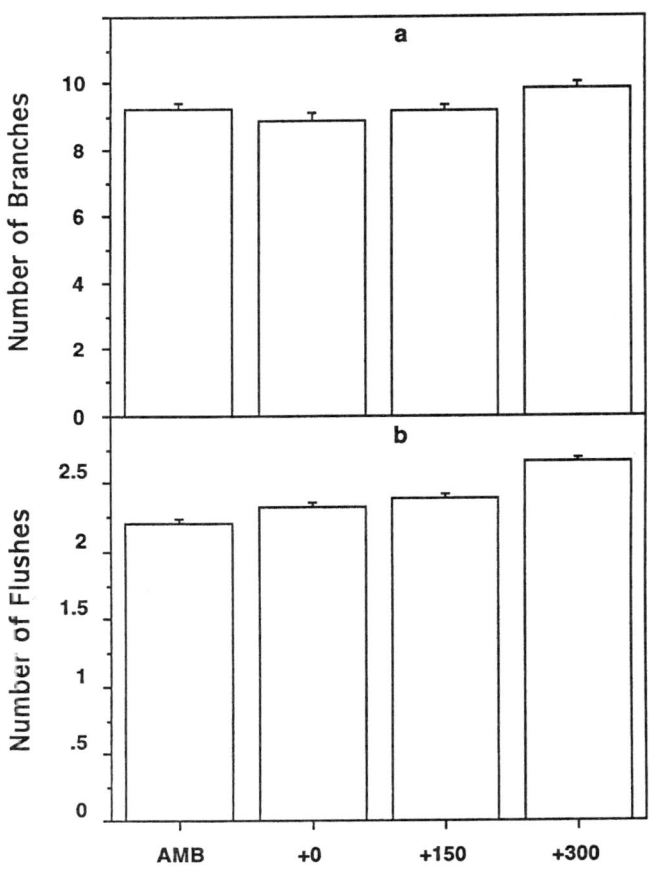

Figure 5.2. The overall number of branches (a) and number of flushes (b) of loblolly pine seedlings subjected to different CO_2 treatments. Vertical bars represent ± the standard error of the mean.

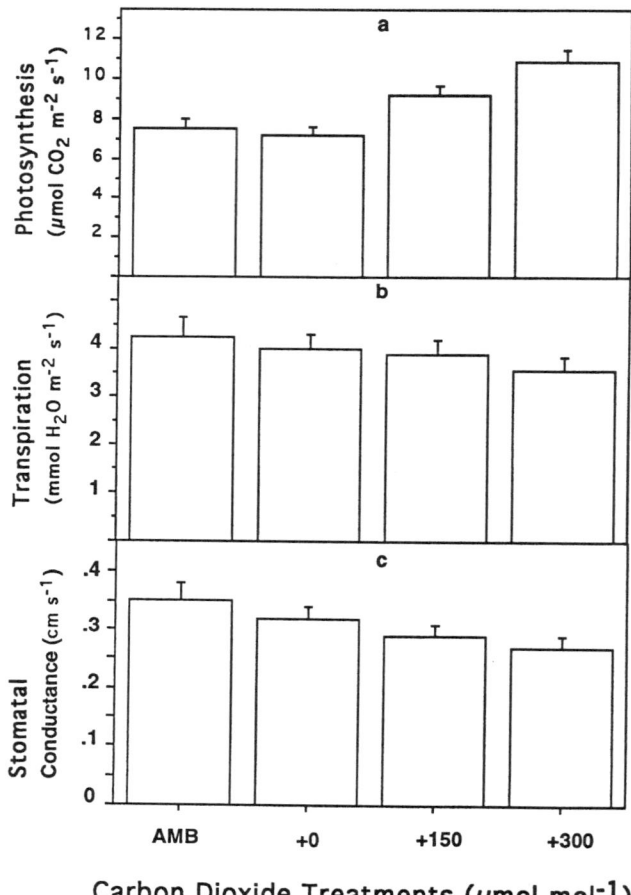

Figure 5.3. Average of monthly net photosynthetic rate (a) transpiration rate (b) and stomatal conductance (c) of loblolly pine seedlings subjected to different CO_2 treatments. Vertical bars represent ± the standard error of the mean.

levels of CO_2 treatments (Figure 5.3a). However, increases in photosynthetic rates did not always translate into greater growth rates and dry matter production, especially when plants grown under the +150 and +0 μmol mol^{-1} CO_2 treatments were compared. In a similar study, Tissue et al., (1996) reported nonsignificant differences in biomass production between loblolly pine plants treated with 15 Pa CO_2 and ambient CO_2. Dutton et al. (1988), however, reported better correlation between plant growth and photosynthetic rate using whole plant net CO_2 exchange as opposed to measurements made on a few leaves or needles per plant.

Carbon dioxide-enrichment reduced transpiration rates and stomatal conductances in proportion to increasing CO_2 concentration (Figure 5.3b, 5.3c). Ratios of

5. Effects of Elevated Carbon Dioxide

rates of photosynthesis/transpiration were 1.7, 1.8, 2.4, and 3.1 for plants subjected to the ambient, +0, +150, and +300 μmol mol^{-1} CO_2 treatments, respectively. These results suggest higher water use efficiency (WUE) of plants exposed to higher CO_2 levels. Increases in the ratios of carbon gain to water loss at elevated CO_2 have also been reported in similar studies with different species (Baker et al., 1990; Morison, 1985; Sionit et al., 1984).

Needle dry weight, stem dry weight, and total biomass of plants grown under the +300 μmol mol^{-1} CO_2 were significantly greater ($p \leq 0.001$) than those grown under the lower CO_2 level (Figure 5.4). This result is in agreement with

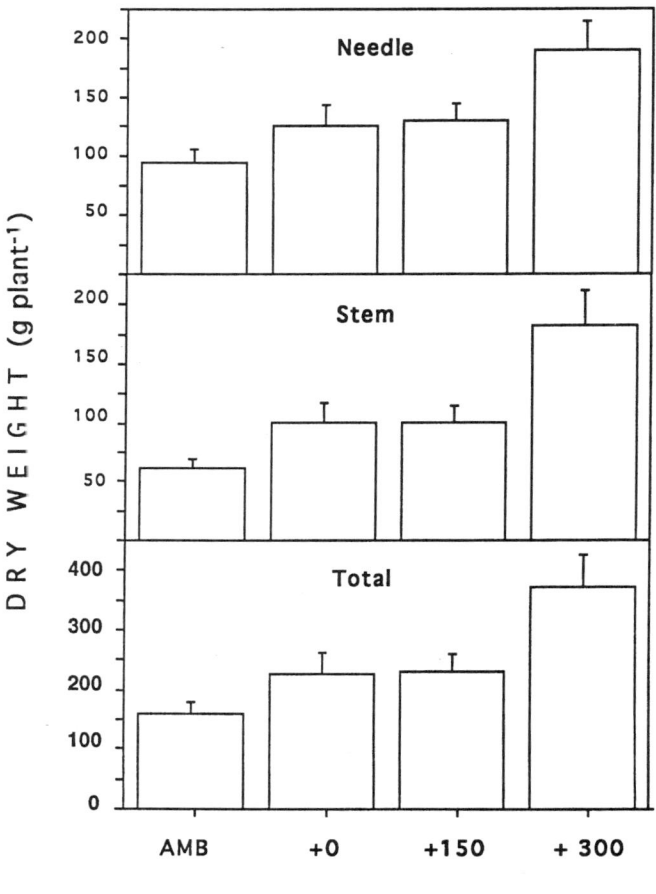

Figure 5.4. Needle dry weight (a) stem dry weight (b) and total dry weight (c) of loblolly pine seedlings subjected to different CO_2 treatments. Vertical bar represent ± the standard error of the mean.

other studies that have shown increased productivity of various tree seedlings subjected to increased CO_2 environments (O'Neill et al., 1987; Kramer and Sionit, 1987; Samuelson and Seiler, 1993; Thomas et al., 1994; Larigauderie et al., 1994; Tissue et al., 1996). Plants that had higher biomass production were also those that had higher growth measurements.

Ratios of needle dry weight (NDW) to stem dry weight (SDW) were 1.6, 1.4, 1.5, and 1.2 for the ambient, $+0$, $+150$, and $+300$ μmol mol^{-1} CO_2 treatments. The decreased NDW/SDW ratio for the highest CO_2 treatment may imply that plants grown under elevated CO_2 accumulate more photosynthetic carbon in the woody plant tissues than those grown in lower-level CO_2 treatments.

Summary

The results suggest that plant establishment, growth, and the subsequent economic wood yield of loblolly pine may be enhanced by elevated atmospheric CO_2. This is in conformity with many studies that have reported increment of biomass and economic yield of C_3 species under increasing CO_2 (Kramer, 1981; Strain and Sionit, 1982; Kimball, 1983; Dahlman et al., 1985; Sionit et al., 1985; Strain and Cure, 1985; Strain, 1987; Kramer and Sionit, 1987; O'Neill et al., 1987; Rogers and Dahlman, 1993; and Ceulemans and Mousseau, 1994). However, most of these reports are based on short-term investigations and predicting long-term effects of the rise of atmospheric CO_2 on loblolly pine forests may be difficult. Because of photosynthetic 'acclimation', reduction in both photosynthetic enhancements (Gundeerson and Wullshlegeger, 1994; Sage, 1994) and biomass production (Tissue et al., 1996) in response to elevated CO_2 over time have also recently been reported.

References

Baker JT, Allen LH Jr, Boote KJ, Jones P, Jones JW (1990) Rice photosynthesis and evapotranspiration in subambient, ambient, and super ambient carbon dioxide concentrations. Agron J 82:834–840.

Ceulemans R, Mousseau M (1994) Effects of elevated atmospheric CO_2 on woody plants. New Phytol 127:425–446.

Dahlman RC, Strain BR, Rogers HH (1985) Research on the response of vegetation to elevated atmospheric carbon dioxide. J Environ Qual 14:1–8.

Dutton RG, Jiao J, Tsujita MJ, Grodzinski B (1988) Whole plant CO_2 exchange measurements for nondestructive estimation of growth. Plant Physiol 86:355–358.

Ginn SE, Seiler JR, Cazell BH, Kreh RE (1991) Physiological and growth responses of eight-year-old loblolly pine stands to thinning. For Sci 37:1030–1040.

Keeling CD, Bacastow RB, Carter AF, Piper SC Whorf TP, Heimann M, Mook WG, Roelffzen H (1989). A three dimentional model of atmospheric CO_2 transport based on observed winds. I. Analysis of observational data. Ameriacan Geophysical Union Monograph 55:165–234.

Kimball B (1983) Carbon dioxide and agricultural yield; an assemblage and analysis of 430 prior observations. Agron J 75:779–788.

Kramer P (1981) Carbon dioxide concentration, photosynthesis, and dry matter production. Bioscience 31:29–33.

Kramer P, Sionit N (1987) Effects of increasing CO_2 concentration on the physiology and growth of forest trees. In Shands WE, Hoffman JS (Eds) *The greenhouse effect, climate change, and US Forests.* The Conservation Foundation, Washington, DC, 219–246.

Larigauderie A, Reynolds JF, Strain BR (1994) Root response to CO_2 enrichment and nitrogen supply in loblolly pine. Plant and Soil 165:21–32.

Lieth H (1975) Primary productivity of major vegetative units of the world. In Lieth H, Whittaker RH (Eds) *Primary productivity of the biosphere.* Springer-Verlag, NY.

Morison JL (1985) Sensitivity of stomata and water use efficiency to high CO_2. Plant, Cell and Environ 8:467–474.

Oechel WC, Riechers GH (1986) Impacts of increasing CO_2 on natural vegetation, particularly the tundra. In Rosenzweig C, Dickinson D (Eds) *Climate-Vegetation Interactions:* Proc of a Workshop 27–29 Jan 1986. 36–42.

O'Neill E, Luxmoore R, Norby R (1987) Elevated atmospheric CO_2 effects on the seedling growth, nutrient uptake, and rhizosphere bacterial populations of *Liriodendron tulipifera* L. Plant Soil 104:3–11.

Rogers HH, Dahlman RC (1993) Crop responses to CO_2 enrichment. Vegetatio 104/105:117–131.

Rogers HH, Heck WW, Heagle AS (1982) A field technique for the study of plant responses to elevated carbon dioxide concentrations. J Air Pollut Control Assoc 33:42–44.

Samuelson LJ, Seiler JR (1993) Interactive role of elevated CO_2, nutrient limitations and water stress in the growth responses of red spruce seedlings. For Sci 39:348–358.

Sionit N, Strain B, Hellmers H, Riechers G, Jaeger C (1985) Long-term atmospheric CO_2 enrichment affects the growth and development of *Liquidambar styraciflua* and *Pinus teada* seedlings. Can J For Res 15:468–471.

Sionit N, Rogers HH, Bingham GE, Strain BR (1984) Photosynthesis and stomatal conductance with CO_2-enrichment of container- and field-grown soybeans. Agron J 76:447–451.

Strain BR (1987) Direct effects of increasing atmospheric CO_2 on plants and ecosystems. Trends Ecol Evol 2:18–21.

Strain BR, Cure JD (1985) Direct effects of increasing carbon dioxide on vegetation. DOE/ER-0238. Office of Energy Research, U. S. Dept. of Energy. Washington, DC.

Strain BR, Sionit N (1982) Direct effects of carbon dioxide on plants: a bibliography. Department of Botany, Duke University, Durham, NC.

Thomas RB, Lewis JD, Strain BR (1994) Effects of leaf nutrient status on photosynthetic capacity in loblolly pine (*Pinus taeda* L.) seedlings grown in elevated atmospheric CO_2. Tree Physiol 14:947–960.

Tissue DT, Oechel WC (1987) Response of *Eriophorum vaginatum* to elevated CO_2 and temperature in the Alaskan tussock tundra. Ecology 68:401–410.

Tissue DT, Thomas RB, Strain BR (1996) Growth and photosynthesis of loblolly pine (*Pinus taeda* L.) after exposure to elevated CO_2 for 19 months in the field. Tree Physiol 16:49–59.

6. Environmental Stresses and Reproductive Biology of Loblolly Pine (*Pinus taeda* L.) and Flowering Dogwood (*Cornus florida* L.)

Kristina F. Connor, Timothy C. Prewitt, Franklin T. Bonner, William W. Elam, and Robert C. Parker

We have long recognized that natural climatic shifts have influenced the development of plant and animal life on earth. These slow temperature fluctuations have resulted in either the extinction or the evolution of various species. However, human activities in the last century have so altered the chemical composition of the atmosphere that it is hypothesized that the process of climatic change has become more rapid. The resultant phenomenon, global warming, could greatly alter the existing vegetation regions on the planet.

The overall goal of most global change research with trees is to measure the impact of these environmental shifts and increased concentrations of the greenhouse gases—carbon dioxide and ozone—on individual species and ecosystems. Most of the research concerns the effects of pollutants and climatic change on seedlings and saplings. This study, however, examines changes occurring in young, yet reproductively active trees and focuses on one of the most sensitive phases of plant development, the reproductive cycle.

Some major environmental factors that are expected to change because of global warming are light intensity and quality, temperature, and moisture. These, in turn, are major factors affecting the various stages of flowering and fruiting in plants (Kramer and Kozlowski, 1979), namely floral bud initiation, flowering/pollen shed, pollination, gametophyte development, fertilization, and embryo/seed development. The effects of these climatic factors vary, depending on the stage of flower or fruit development. For example

1. Light. A high light intensity may increase the number of floral buds in *Tectonia* and *Pinus* (Nanda, 1962; Sarvas, 1962, at the same time favoring female over male flowers in *Acer, Juglans,* and a number of *Pinus* species (Hibbs and Fischer, 1979; Matthews, 1963; Giertych, 1977; Ryugo et al., 1980, 1985). Light quality also affects *Pinus* pollen germination (Dhawan and Malik, 1981); germination is decreased by white light and enhanced by red light.
2. Temperature. High temperatures in late summer and fall favor the formation of floral buds over vegetative buds in various temperate tree species (Matthews, 1963; Jackson and Sweet, 1972; Menzel, 1983; Owens and Blake, 1985; Southwick and Davenport, 1986). These same temperatures in the following spring, however, may dry the pollination drop present in some species, and thus reduce pollination. Pollen tube growth is slowed or stopped by both high and low temperatures (Mellenthin et al., 1972; Griggs and Iwakiri, 1975; Sedgley, 1977; Sedgley and Annells, 1981; Staudt 1982; Sedgley and Grant, 1983).
3. Moisture. A heavy rain can adversely affect pollen dispersal in wind-pollinated species (Kramer and Kozlowski, 1979), yet can enhance fruit enlargement for cherries, grapes, peaches, pecans, and apples (Uriu and Magness, 1967; Goode and Ingram, 1971). La Bastide and Van Vredenburch (1970), Eriksson et al. (1975) and Fober (1976), Lindgren et al. (1977), Menzel (1983), Southwick and Davenport (1986), and Burgess (1972) all report that moisture stress may enhance floral initiation and seed-cone bud formation in tropical species, although Rehfeldt et al. (1971) assert that any moisture deficit after cone initiation has an adverse effect. Finally, high relative humidity reduces electrolyte leakage and enhances pollen germination (Hoekstra and van der Wal, 1988; Yates and Sparks, 1989).

All of these three factors can be affected by air pollution, and now some studies have reported that acidity (Van Ryn et al. 1985), air pollutants CO_2, SO_2, and NO_2 (Wolters and Martens, 1987), and ultraviolet radiation (Pfahler, 1981) can affect in vitro pollen germination and tube growth.

Materials and Methods

The loblolly pine study area was on the campus of Mississippi State University. The seventeen-year-old open-grown trees were from a North Mississippi loblolly pine source and ranged in height from 7.3 to 10.0 m. The site had a mean annual maximum temperature of 23.3 °C, a mean annual minimum temperature of 11.2 °C, and a mean annual precipitation of 136.2 cm. The soil underlying the site was a Kipling silty clay loam; such soils are strongly acidic with slow permeability and, generally, available water capacity is high.

The flowering dogwood study area was located in a natural stand on the Noxubee National Wildlife Refuge, approximately 25 km south of Starkville, MS. The trees ranged in age from thirty to fifty-two years and 4.3 to 18 m in height. Mean annual temperature and precipitation differed little from that of the pine site. The underlying soil was a Smithdale sandy loam that was strongly acidic and moderately permeable; available water capacity was usually moderate.

Trees and branches were observed in the year preceding the study and only those that produced fruit or cones were used. Seven trees were selected at each site. Branch chambers constructed on the design of Teskey et al. (1991) were used to enclose four test branches on each tree, and two branches per tree were used as ambient controls. The chambers were 1.48 m long and 0.58 m in diameter and were equipped with zippers to allow easy access to the branch. In the first year of the two year study (1993) the effects of temperature on loblolly pine flowering and fruiting were examined; in the second year (1994) 2× ambient CO_2 was added as a treatment to half of the chambered loblolly pine branches, and the temperature treatments on flowering dogwood began. The CO_2 treatment was added to the dogwood site in late summer of 1994. During the stage when temperature was the only treatment, centrally placed thermisters monitored the conditions of the following three treatments on each tree:

1. Ambient temperature on a chamberless branch.
2. Air temperature on a branch inside an unheated chamber (the reference temperature).
3. Air temperature on a branch inside a chamber heated 2 °C above the reference temperature.

Each of these three treatments was replicated on each test tree (six test branches per tree).

The addition of the CO_2 treatment in the second year yielded the following combinations in the four branch chambers on each tree:

1. 2× ambient CO_2, reference temperature (U^+ branches)
2. 2× ambient CO_2, elevated (2 °C) temperature (H^+ branches)
3. ambient CO_2, elevated (2 °C) temperature (H^- branches)
4. ambient CO_2, reference temperature (U^- branches)

Treatment branches were guyed to prevent damage in high winds. Branch chambers were attached at the top and bottom to a platform situated near each test tree as recommended by Teskey et al. (1991). Each chamber had its own one-half horse power industrial model blower mounted on the platform slightly below the level of the chamber. Air was blown through a plastic plenum, 16.5 cm in diameter and of varying length, into the chamber. Plenum diameters were individually adjusted with heavy twine or metal clamps to circulate roughly ten air changes per minute. Air flow was measured manually at installation with a Davis® Turbo-Meter mounted on a metal shaft and confirmed periodically throughout the study.

Omega® series 900 thermisters were placed in a monitored water bath to determine their accuracy prior to being placed on the branches. These temperature probes were found to be accurate to within ± 0.2 °C over a connecting cable distance of 30 m. Thermisters were placed in the center of each chamber and protected from the direct sun by an opaque plastic funnel. A PC computer, a

Keithly Metrabyte® Series 500 data logger and monitoring system, and a program developed by a cooperator in the project were used to monitor both temperature and CO_2. The program 1) activated the 750-watt open element heating units in the $+2\,°C$ chambers if the temperature fell below the reference temperature of $+1.7\,°C$, and 2) shut the heaters off if the temperature rose above the reference temperature of $+2.1\,°C$. Temperatures were monitored by the computer in each chamber at the rate of thirty-five measurements per minute; average temperatures were automatically recorded every fifteen minutes throughout the day.

Large, refrigerated storage tanks at each site were the source for the CO_2. High-density polyethylene tubing delivered CO_2 from the tanks to the branch chambers. Flow from the tank to the chambers was controlled by a series of rotometers (one per chamber). A LI-6252 gas analyzer was used to measure CO_2 concentrations every hour; these readings were calibrated against readings taken every two weeks with a separate line that bypassed the normal intake system. The analyzer was calibrated once each week.

Throughout the study, measurements were made on the relative humidity and solar radiation levels both inside and outside the chambers to determine if any seasonal variations existed. Relative humidity was measured with a Jenway® Relative Humidity Temperature Meter (model 5075) and solar radiation with a YSI-Kettering® radiometer (model 65).

Pollen and seeds were collected by hand; pine cones and flowering dogwood drupes were placed in paper bags identified by a code specifying both tree and treatment. Similarly labelled plastic petri dishes were used for pollen collection, which, in the pine trees, began on the third day of pollen shed. Pollen on the test branch was dusted into the labelled petri dish and thoroughly mixed. This pollen mixture was used for all tests of pollen moisture content, viability, and differential scanning calorimeter (DSC) studies. Flowering dogwood is an entomophilous species, however, which it proved to be very difficult to remove the sticky pollen from the anthers. This problem was resolved by removing entire anthers with the adhering pollen for testing, drying and storage.

All of the seeds produced by the test branches were collected. Loblolly pine cones were picked when color and specific gravity tests on cones on adjacent branches indicated ripeness. Flowering dogwood fruit were picked when they changed color (from green to red). Seeds were logged into a sample record as soon as they arrived at the laboratory.

The pine cones were placed in paper sacks in the laboratory under ambient conditions until they had completely opened. Seeds were separated with hand screens, dewinged by hand, and floated to remove empties. The pine seeds were placed in small, clear plastic bags containing water and soaked overnight. The water was then removed, and the the bags sealed and placed in cold storage ($4\,°C$) for twenty-eight days prior to germination. This is a standard pregermination treatment for loblolly pine. Any extra cleaned seeds were air dried to 10% moisture content, placed in labelled containers, and stored at $4\,°C$. The flowering dogwood drupes were depulped in a laboratory blender or by hand, spread to dry, and then stratified for germination tests. The seeds were placed in the same type of bags containing a damp paper towel and stored at 4 °C for at least ninety days as a pregermination treatment. Any extra seeds were air-dried to 10% moisture content and stored at 4 °C in labelled containers.

Upon arrival at the laboratory, some of the freshly collected pollen mixture was immediately tested for viability and moisture content. Pollen quality was evaluated by germination tests and the DSC, a thermal analyzer that measured the energy associated with melting lipids and water in the pollen grains. For moisture-content determination, pollen samples were weighed to the nearest 0.0001 g on an electronic balance, dried at 104 °C in a convection oven for two hours, cooled at room temperature for one minute and reweighed. Excess pollen was dried, tested again for viability, and stored at -20 °C for future use. If the pollen did not need to be dried (i.e. if moisture content was already 10% or less), the viability was not retested prior to storage at -20 °C.

Fresh pollen samples collected from each treated or control loblolly pine branch were germinated for seventy-two hours in a Brewbaker–Kwack solution (Brewbaker and Kwack 1963) containing 10% sucrose (Sowa et al. 1991). Pollen samples from flowering dogwood were germinated for seventy-two hours in a Brewbaker–Kwack solution containing 20% sucrose. There were four replicate vials for each sample. Vials were placed in a shaking water bath set at 25 °C to allow both temperature control and aeration while pollen grains were germinating. Germination counts were made on 400 pollen grains per vial for each sample collected from the tree. A pollen grain was considered germinated when the pollen tube length was at least two times the diameter of the grain.

DSC studies were performed with a Perkin Elmer® DSC-7 to determine the effects of temperature and CO_2 on pollen desiccation and chilling sensitivity. Scans were made on whole pollen grains to find if shifts in peaks or glass formations were detectable.

Seed quality was evaluated by germination tests and seed weight. For seed weight, two 25-seed samples from each treatment branch were weighed to the nearest 0.0001 g. These samples were then used in the standard germination test. Seeds germinated on moist blotters for twenty-eight days in Stults® water-curtain germinators, on a cycle of eight hours of light at 30 °C and sixteen hours of dark at 20 °C (Association of Official Seed Analysts, 1993). Temperature and relative humidity of the germinators was recorded on a hygrothermograph. Temperature was checked daily and adjusted if necessary. Light during the 30 °C cycle was provided by cool-white fluorescent bulbs at levels of at least 500 lux and not more than 1000 lux. Germination was counted three times weekly, and all ungerminated seeds were cut at the end of the test to identify dormant, empty, and dead seeds. On count days, trays were moved up one shelf space and rotated 180° to minimize any effect of temperature stratification or light inequity in the germinator. Germination values were expressed as percentages.

Germination tests were conducted according to the official rules of the Associa-

Table 6.1. Data Quality Objectives for Biological Variables*

Variable	Units	Technique	Range	Quantitative Limits	Precision (CV)	Accuracy (%)	Completeness (%)
Vegetative/reproductive buds	**	visual count	n/a	0–10	15	—	99
female/male buds	**	visual count	n/a	0–5	15	—	99
Pollen moisture	%	electronic balance	0–200	5–50	5	95	99
pollen germination	%	visual count	n/a	0–100	15	—	99
Lipid/water peaks[1]	deg. C	Calorimeter	−196–100	−100–40	5	95	99
seed weights[2]	g	electronic balance	0–200	.5–1.5	10	99	80
seed per cone	#	visual count	n/a	5–50	5	—	99
seed germination	%	visual count	n/a	0–100	5	—	99

*Variables were measured annually.
**These values are reported as proportions.
[1] These measurements were taken only on the first year's pine pollen; because differences were not found, they were discontinued. The sticky dogwood pollen could not be used for such tests.
[2] Pine seed weights were not taken in 1993.

tion of Official Seed Analysts, 1993 revision. Under these rules, a "normal" seedling was a "seedling possessing those essential structures that are indicative of its ability to produce a plant under favorable conditions."

Counts of male, female, and vegetative buds were made every spring for two years. Ratios of female/male buds and vegetative/flowering buds were determined for each treatment branch. Date and time of pollen shed and cone/fruit harvest was also recorded. Statistical comparisons between years were not made, because each year's treatment may affect flowering in subsequent years. Quality control information is presented in Table 6.1.

Results and Conclusions

Loblolly pine. The comparative time at which pollen shed began was very predictable over the three-year period of the pine study. Pollen shed started three to eleven days earlier on heated (H) chambered branches than on ambient branches and one to eight days earlier in H than in unheated (U) chambers. Pollen shed on the H branches stretched over a seven to twenty-one day period; on the U branches the pollen shed period was five to thirteen days; on ambient branches the pollen shed period was nine to fourteen days. This suggests a very strong influence of heat sums on time and duration of pollen shed (Boyer 1973, 1978) in loblolly pine. The effects of CO_2 on the date of pollen shed were negligible, and the beginning of pollen shed varied little on CO_2 branches within a treatment.

A higher number of male strobili were found on ambient branches than on those in chambers in all three years of the study (Table 6.2). Although there were no

Table 6.2. 1993 to 1995 Male and Female Strobili Counts and Cones Harvested at the Loblolly Pine Site

Characteristic[1]	Year	Chambers				Ambient
		H^+	H^-	U^+	U^-	
# Female strobili	1993	—	36	—	31	96
	1994	1	0	0	0	4
	1995	5	0	0	0	38
# Male strobili	1993	—	2182	—	2118	2646
	1994	83	7[2]	91	14[3]	348
	1995	1577	947[3]	1992	1714	5569
# Male strobili/cluster	1993	—	223	—	205	227
	1994	16	2	11	2	57
	1995	136	84	160	146	458
Total # cones harvested	1993	—	42	—	63	42
	1994	3	24	5	14	55
	1995	—	—	1	—	2
Average cone length (cm)	1993	—	8.7	—	8.8	9.1
	1994	10.8[2]	9.4[3]	9.1[3]	8.9	8.9
	1995	—	—	8.5	—	8.6

[1] Only the temperature treatment was in effect in 1993.
[2] Significantly different from ambient ($p = 0.05$).
[3] Significantly different from H^+ ($p = 0.05$).

significant differences between treatment branches and controls in 1993, this was not the case in 1994 and 1995. In both of these years, branches in heated chambers with ambient CO_2 (H^-) had significantly fewer male strobili than did ambient branches. A significant difference also existed between unheated branches with ambient CO_2 (U^-) and ambient branches in 1994. This latter distinction is questionable, however, because the trend wasn't repeated in 1995. Also, male strobili were so scarce on all branches in 1994 that there was great difficulty in finding them amongst the foliage. This drop in numbers makes strobili formation data from that year, and hence the differences among the treatments observed during that time period, of questionable value. The reduction in male strobili in 1994 was not considered the result of either temperature or CO_2-treatment because the numbers on ambient branches were also greatly reduced, and also because the strobili numbers rebounded in 1995. However, there was more than a twenty-fold reduction in strobili in chambers compared to an eight-fold reduction on ambient branches, which may indicate a chamber effect. Although it has been reported that hot, dry weather during the period of bud formation favors floral buds over those that are vegetative (Owens and Blake 1985), the very favorable conditions in 1993 produced very few male strobili; and in 1994, which had triple the normal amount of the rainfall and lower than average temperatures, yielded a twenty-fold increase in male strobili. The effect of the poor reproductive bud formation in 1994 is reflected in the bud ratio data shown in Table 6.3. A large number of male strobili

Table 6.3. Bud Ratios from the Pine and Dogwood Sites

Site	Treatment	Year	Vegetative/Reproductive	Female/Male
Pine	Heated	1993	0.21	0.016
		1994	6.36[1]	0.011
		1995	0.25	0.002
	Unheated	1993	0.17	0.016
		1994	5.26[1]	0.000
		1995	0.19	0.000
	Ambient	1993	0.18	0.031
		1994	0.55	0.011
		1995	0.16	0.007
Dogwood	Heated	1994	0.17	
		1995	0.46	
	Unheated	1994	0.15	
		1995	0.46	
	Ambient	1994	0.19	
		1995	0.70	

[1] Significantly different from ambient ($p = 0.05$).

Table 6.4. Pollen Germination for Three Years of Loblolly Pine and One Year of Flowering Dogwood

Treatment	Pine			Dogwood
	1993	1994	1995[1]	1994
Heated	85.2	2.2	53.7	
Heated + CO_2		14.1	59.0	0.1
Unheated	54.3	0.4	30.7	
Unheated + CO_2		2.7	65.9	0.6
Ambient	75.1	1.2	28.2	0.4

[1] These are preliminary figures; pollen has been germinated but counts are not yet complete.

being formed normally results in a low vegetative/reproductive bud ratio, as indicated by the 1993 and 1995 figures. The poor showing in 1994 resulted in significant differences between chambered and ambient vegetative/reproductive bud ratios. The female/male bud ratios did not differ significantly from year to year.

Pollen moisture at shedding averaged 7.0% on all branches tested in 1993 and 6.5% in 1995. None of the 1994 pollen was used for moisture determinations because pollen was scarce and germination tests were considered a higher priority. Pollen germination varied from year to year but generally was unaffected by chamber, temperature, or CO_2 (Table 6.4). Even though germination of 1995 pollen from ambient branches was lower than that from treated branches, it could not be assumed that this was a treatment effect because there was great variability in germination from ambient branches and the differences were not significant. The pollen from the few strobili present in 1994 was of very poor quality, and germination was negligible in all treatments. Pollen counts for 1995 indicate that pine pollen germination appears to have rebounded. DSC scans of pine pollen from the various treatment branches indicated no change in peak melting and enthalpy values for either lipids or water in the pollen. These results support the conclusion that pollen quality was not altered by treatment.

Female strobili in the H chambers began expanding eight to nine days prior to those in U chambers and twelve to eighteen days before those on ambient branches. This growth period, encompassing stages 3 through 5 (Bramlett and O'Gwynn 1980), lasted from six to sixteen days. Although female strobili formation appeared to be adversely affected by the presence of the chamber, significant differences did not exist between treatment and ambient branches (Table 6.2). The few female strobili that formed in chambers in 1994 aborted after pollen shed; however, the few number that formed on all fourteen of the ambient branches that year were not enough to make significant differences among the treatments. An interesting abnormality was observed in 1995 on Tree 6 in a H⁻ and a U⁻ chamber. Several bisexual cones were found on the terminal and first order laterals of these branches; the lower half was male, the upper half formed a female conelet. Pollen was shed from the male portion of the strobili, and, after the bottom half dried, the female conelets were shed along with their male counterparts. This phenomenon was not observed on any other branches of Tree 6, nor on any of the other test trees, nor on any other trees in the stand. In prior studies, this condition has been seen in *Pinus* by Lanner (1966) and in *Cupressus* by Lev-Yadun (1992) but not in loblolly pine. Lev-Yadun suggested that the abnormalities that were found in *Cupressus sempervirens* were the result of changes in the normal hormonal balance of the tree.

The 1994 cones harvested from H^+ branches were significantly larger than those that developed on ambient, H^-, and U^+ branches (Table 6.2). However, they also averaged significantly fewer full seeds/cones than those from ambient and H^- branches, as did those from H^+ vs H^- branches in 1993 (Table 6.5). Unfortunately, the absence of cones from heated chambers in 1995 and the few number of cones on unheated branches prevented the determination that this characteristic was definitely the result of the applied treatments.

The percent of full seeds in the 1994 cones from heated chambers dropped from the 1993 average, stayed about the same in cones from unheated chambers, and also dropped in ambient branch cones (Table 6.5). This variability may have been because of the reduced amount of pollen produced that year. Germination of full seed was over 90% regardless of treatment.

Flowering dogwood. The average number of flowers per inflorescence, which varied little from year to year or among treatments, was 20.5 (Table 6.6). To date,

Table 6.5. 1993 to 1995 Seed Data from the Loblolly Pine Site

Characteristic[1]	Year	Chambers				Ambient
		H^+	H^-	U^+	U^-	
Total # seeds	1993	—	4779	—	7387	4612
	1994	65	782	243	749	3067
	1995	—	—	19	—	178
Average # seeds/cone	1993	—	113.8	—	117.3	109.8
	1994	21.7	24.1	48.6	53.5	56.8
	1995	—	—	19.0	—	89.0
Total # full seeds	1993	—	2661	—	4805	3166
	1994	16	151	136	439	1328
	1995	—	—	13	—	83
Average # full seeds/cone	1993	—	63.4	—	76.3[2]	75.1
	1994	5.3[2]	6.3[3]	27.2	31.4	24.6
	1995	—	—	13.0	—	41.5
% Germination	1993	—	94.4	—	95.1	98.3
	1994	93.8	95.4	100.0	97.9	99.1
	1995	—	—	38.5	—	36.1
Seed weight (mg)	1994	28.1	24.7	30.9	25.9	20.1
	1995	—	—	19.8	—	21.6

[1] Only the temperature treatment was in effect in 1993; 1995 cone data not statistically analyzed because the sample size was too small.
[2] Significantly different from ambient (p = 0.05).
[3] Significantly different from H^+ (p = 0.05).

6. Environmental Stresses and Reproductive Biology

Table 6.6. 1994 to 1995 Flower, Fruit, and Seed Data from the Flowering Dogwood Site

Characteristic[1]	Year	Chambers				Ambient
		H+	H−	U+	U−	
# Inflorescences	1994	—	515[1]	—	425	258
	1995	443	377	209	201	383
# Flowers	1994	—	10626	—	8962	5341
	1995	8120	9064	4067	4131	7359
# Flowers/Inflor.	1994	—	20.6	—	21.1	20.7
	1995	18.3	24.0	19.5	20.6	19.2
# Fruits	1994	—	1212	—	1009	938
	1995	57	77	37	83	98
# Fruits/Cluster	1994	—	3.9	—	3.1[1,2]	3.9
	1995	1.3[1]	1.4[1]	1.2[1]	1.9[1]	2.5
Average fruit wt. (g)	1994	—	0.55[1]	—	0.48[1,2]	0.32
	1995	0.41[1]	0.48	0.37	0.36	0.29
Average fruit diam. (mm)[3]	1994	—	8.63[1]	—	8.14[1,2]	7.01
	1995	7.94[1]	8.41[1]	7.47[1]	7.59[1]	7.04
Average seed length (mm)	1994	—	9.12[1]	—	8.86[1,2]	8.08
	1995	7.91	7.88	7.94	8.19	7.90
Average seed diam. (mm)	1994	—	5.09[1]	—	4.89[1,2]	4.57
	1995	4.85	4.94[1]	4.88	4.86	4.70
% Germination	1994	—	39.4	—	50.5[1]	31.5

[1] Significantly different from ambient (p = 0.05).
[2] Significantly different from H (p = 0.05)
[3] All possible comparisons of 1995 fruit diameters are significantly different from each other except U+ vs U−.

the only significant difference in flower production was between the number of inflorescences produced on H vs ambient branches in 1994.

There was no difference in the number of fruits formed per cluster on H and ambient branches in 1994 (Table 6.6); however, there were significantly fewer fruits per cluster on unheated branches than on either the heated or ambient branches. An interesting fact to note, however, is that a steady progression in both fruit size (weight and diameter) and seed size (length and diameter) was found, with the largest occurring on H branches and the smallest on ambient branches. This may be a result of the extension of the growing season in the heated and unheated chambers. Flowers opened ten days sooner and fruits were collected seven days later from chambered branches than from non-chambered branches. Germination differed little between H and ambient branches, but seeds from U branches had an inexplicably higher percentage of germination than those on ambient branches.

The number of fruits per cluster was significantly fewer on all chambered vs ambient branches in 1995 (Table 6.6). A difference also existed between U+ and U− branches but not between H+ and H− branches, suggesting that the addition of CO_2 was not a consistently significant factor in increased fruit production. Again, fruits from the chambered branches were significantly larger in diameter;

however, only those from H+ branches were also significantly heavier. In 1995, increases in fruit size did not result in increased seed size. The larger fruits did not produce larger seeds even though flowers opened five days sooner and fruits were harvested seven days later on chambered branches. The question of whether or not the extended growing season was responsible for the increase in fruit size in 1995 is complicated by the reduction in number of fruit per cluster. The size increase may simply be the result of reallocation of carbohydrate and nutrient resources to a fewer number of developing fruit rather than to the extension of the growing season by the temperature enhancement.

Dogwood pollen was first collectable from heated, and then from the unheated and ambient branches. Within a treatment, collection occurred over a four to seventeen day period. Collection of dogwood pollen proved to be difficult; in 1994, collection of the pollen was attempted without removing any of the flower structures. In 1995, whole anthers were collected and used in germination and moisture tests. Average moisture was 12%, a relatively low figure when considering that anther tissue was present. However, it is believed that moisture content was greatly reduced by the length of time occurring between collection and time of arrival in the laboratory. Dogwood pollen germination was very low in 1994 (Table 6.4), and, despite changes to the growing medium, the low counts were repeated in 1995. Sucrose in the Brewbaker-Kwack (1963) medium had been reduced to 15% for 1995 dogwood pollen germinations in the hope that this might correct the problem. Our inability to isolate significant quantities of dogwood pollen in either 1994 or 1995 precluded the planned DSC experiment. The DSC tests were set up to examine moisture and lipids in pure pollen and could not be done with the anther tissue present.

Results to date indicate that temperature plays a role in reproductive structure formation and in both fruit and cone development. Although the results in pine are somewhat overshadowed by possible chamber effects and by an overall reduction in female reproductive structures, there are still marked changes in both male and female strobili phenology and in cone size. The chambers did not have a deleterious effect on flower bud formation in dogwood. Inflorescence and flower counts were always higher on chambered branches than on ambient branches, flowering phenology was noticeably changed, and fruits were larger. At this time, the effect of CO_2 on the reproductive biology of both species appears negligible.

References

Association of Official Seed Analysts (1993) Rules for testing seeds. J Seed Tech 16(3):1–113.
Boyer WD (1973) Air temperature, heat sums, and pollen shedding phenology of longleaf pine. Ecology 54(2):420–426.
Boyer WD (1978) Heat accumulation: an easy way to anticipate the flowering of southern pines. J For 76(1):20–23.
Bramlett DL, O'Gwynn CH (1980) Recognizing developmental stages in southern pine flowers: the key to controlled pollination. USDA FS Gen Tech Rept SE—18.
Brewbaker JL, Kwack BH (1963) The essential role of calcium ion in pollen germination and pollen tube growth. Amer J Bot 50(9):859–865.

Burgess PF (1972) Studies on the regeneration of the hill forest of the Malay peninsula. Malay Forest 35:103–123.

Dhawan AK, Malik CP (1981) Effect of growth regulators and light on pollen germination and pollen tube growth in *Pinus roxburghii* Sarg. Ann Bot 47:239–248.

Eriksson G. Lindgren K, Werner M (1975) Granensblomningsbiologi ett hinder för en rationell skogstradsfordling? Sveriges Skogsliardsforb. Tidskr. 73:413–426.

Fober H (1976) Relation between climatic factors and Scots pine (*Pinus sylvestris* L.) cone crops in Poland. Aboretum Kornickie 21:367–374.

Giertych M (1977) Role of light and temperature in flowering of forest trees. In: *Third World Consultation on Forest Tree Breeding, FAO/IUFRO. Vol 2*, CSIRO, Canberra, Australia, 1014–1021.

Goode JE, Ingram J (1971) The effect of irrigation on the growth, cropping, and nutrition of Cox's Orange Pippin apple trees. J Hort Sci 46:195–208.

Griggs WH, Iwakiri BT (1975) Pollen tube growth in almond flowers. Calif Agric 29 (July):4–7.

Hibbs DE, Fischer BC (1979) Sexual and vegetative reproduction of striped maple (*Acer pensylvanicum* L.). Bull Torr Bot Club 106:222–227.

Hoekstra FA, van der Wal EW (1988) Initial moisture content and temperature of imbibition determine extent of imbibitional injury in pollen. J Plt Physiol 133:257–262.

Jackson DI, Sweet GB (1972) Flower initiation in temperate woody plants. Hort Abstr 42:9–24.

Kramer PJ, Kozlowski TT (1979) *Physiology of Woody Plants*. Academic Press, Inc., Orlando.

La Bastide, JGA, Van Vredenburch CLH (1970) The influence of weather conditions on the seed production of some forest trees in the Netherlands. Meded. Bosbouwproefstation No. 102:1–12.

Lanner RM (1966) Phenology and growth habits of pines in Hawaii. United States Department of Agriculture Forest Service Res. Paper No. PSW—29, 25.

Lev-Yadun S (1992) Abnormal cones in *Cupressus sempervirens*. Aliso 13(2):391–394.

Lindgren K, Edberg I, Eriksson G (1977) External factors influencing female flowering in *Picea abies* (L.). Karst. Stud For Suec 142:53.

Matthews JD (1963) Factors affecting the production of seed by forest trees. For Abs 24(1):i–xiii.

Mellenthin WM, Wang CY, Wang SY (1972) Influence of temperature on pollen tube growth and initial fruit development in "d'Anjou" pear. HortSci 7:557–559.

Menzel CM (1983) The control of floral initiation in lychee: a review. Sci Hort 21:201–215.

Nanda KK (1962) Some observations on growth, branching behavior and flowering of Teak (*Tectonia grandis* L.F.) in relation to light. Indian Forester 88:207–218.

Owens JN, Blake MD (1985) Forest tree seed production. Information Report PI-X-53, Petawawa National Forestry Institute, Chalk River, Ontario, Canada.

Pfahler PL (1981) *In vitro* germination characteristics of maize pollen to detect biological activity of environmental pollutants. Env Health Persp 37:125–132.

Rehfeldt GE, Stage AE, Bingham RT (1971) Strobili development in western white pine: periodicity, prediction, and association with weather. For Sci 17:454–461.

Ryugo K, Bartolini G, Carson RM, Ramos DE (1985) Relationship between catkin development and cropping in the Persian walnut "Ser." Hort Sci 20:1094–1096.

Ryugo K, Marangoni B, Ramos DE (1980) Light intensity and fruiting-effects on carbohydrate contents, spur development and return bloom of "Hartley" walnut. J Amer Soc Hort Sci 105:223–227.

Sarvas R (1962) Investigations on the flowering and seed crop of *Pinus sylvestris*. Inst For Fenn 53:1–198.

Sedgley M (1977) The effect of temperature on floral behaviour, pollen tube growth and fruit set in the avocado. J Hort Sci 52:135–141.

Sedgley M, Annells CM (1981) Flowering and fruit-set response to temperature in the avocado cultivar "Hass." Sci Hort 14:27–33.

Sedgley M, Grant WJR (1983) Effect of low temperature during flowering on floral cycle and pollen tube growth in nine avocado cultivars. Sci Hort 18:207–213.

Sedgley M, Griffith AR (1989) *Sexual Reproduction of Tree Crops.* Academic Press, Inc., San Diego.

Southwick SM, Davenport TL (1986) Characterization of water stress and low temperature effects on flower induction in citrus. Plant Physiol 81:26–29.

Sowa, S, Connor KF, and Towill LE. (1991) Temperature changes in lipid and protein structure measured by Fourier Transform infrared spectrophotometry. Plant Sci 78:1–9.

Staudt G (1982) Pollenkeimung und Pollenschlauchwachstum *in vivo* bei *Vitis* und die Abhängigkeit von der Temperatur. Vitis 21:205–216.

Teskey RO, Dougherty PM, Wiselogel AE (1991) Design and performance of branch chambers suitable for long-term ozone fumigation of foliage in large trees. J Environ Qual 20(3):591–595.

Uriu K, Magness JR (1967) Deciduous tree fruits and nuts. In: Hagan RM, Haise HR, Edminster TW (Eds) *Irrigation of Agricultural Lands,* Monograph 11, American Society of Agronomy, Madison, WI.

Van Ryn DM, Jacobson JS, Lassoie JP (1985) Effects of acidity on *in vitro* germination and tube elongation in four hardwood species. Can J For Res 16:397–400.

Wolters JHB, Martens MJM (1987) Effects of air pollutants on pollen. Bot Rev 53(3):372–410.

Yates IE, Sparks D (1989) Hydration and temperature influence *in vitro* germination of pecan pollen. J Amer Soc Hort Sci 114(4):599–605.

7. Interactions of Elevated Carbon Dioxide, Nutrient Status, and Water Stress on Physiological Processes and Competitive Interactions Among Three Forest Tree Species

John W. Groninger, John R. Seiler, Alexander L. Friend, Paul C. Berrang, and Shepard M. Zedaker

Loblolly pine-hardwood forest systems cover vast areas throughout the southern United States, and contribute profoundly to the economic and ecological stability of the region. Concern has been voiced that increased atmospheric carbon dioxide (CO_2) concentrations could ultimately produce changes in forest composition or increase the cost associated with maintaining forests of desired species composition. In 1994, forest landowners in the South incurred site-preparation and crop-release costs in excess of $50 million to establish and maintain desired stand composition (Dubois et al., 1995). Forest managers, and corporate and regional planners would benefit from understanding the nature of these potential changes in forest composition in order to adjust silvicultural strategies, cost, and resource flow expectations for these forests during the next century.

Much of the present scientifically based interest in clarifying the potential effects of elevated CO_2 on the South's pine-hardwood systems can be attributed to a series of studies addressing the relative response to CO_2 of two widely cooccurring southern tree species, which are 1) loblolly pine (*Pinus taeda* L.), and 2) sweetgum (*Liquidambar styraciflua* L.). These studies found larger growth increases in sweetgum relative to loblolly pine in response to increased atmospheric CO_2 concentration (Tolley and Strain, 1984a; Sionit et al., 1985). Furthermore, sweetgum continued to increase growth into the drought period in contrast to loblolly pine (Tolley and Strain 1984b). These results suggest that sweetgum might become a stronger competitor to loblolly pine and may ultimately occupy sites presently dominated by loblolly pine. This research, based on the response of seedlings or saplings grown as individuals, was unable to account for the effects of direct competition on the growth of these species. Several studies across a range of plant communities have shown that seedlings grown in simulated communities will differ in their relative response to elevated CO_2 compared to individually grown seedlings (Bazzaz and McConnaughay, 1992).

The Importance of Competition

Competitive interactions ultimately determine tree-stand composition in forests throughout much of the region. Competitive success between cooccurring species, and the resultant vegetative composition, is largely driven by the differing ability of species to capture resources under varying levels of resource availability (Grime, 1979; Tilman, 1982). Uptake of a limited resource is an important consideration for a given individual species for two reasons which are 1) the resource can be converted to growth, further improving resource uptake ability, and 2) uptake of a resource precludes its use by a competitor. Competitive relationships in managed forests are manipulated to maximize the growth of the crop species through silvicultural treatments including site preparation, cleaning, and thinning.

Researchers wishing to understand the influence of competition on such plant communities as forest stands must create experimental conditions that account for this phenomenon. The availability of such resources as light, water, and nutrients can also influence community response to increased CO_2 concentrations.

The importance of growing plants in competitive mixture to determine CO_2 enrichment response of communities has been demonstrated in a simulated community of annual plants (Williams et al., 1988). These researchers found that two of six species from a serpentine grass community showed a significant growth response to elevated CO_2 levels when grown individually, although no species responded when grown in mixture. In another study of annual plants, response to elevated levels of CO_2 was found to be more strongly dependent upon light and nutrient availability in individuals relative to communities (Zangerl and Bazzaz, 1984). In two simulated temperate forest communities, atmospheric CO_2 concentration did not significantly affect the total biomass growth of either community, but did affect the proportion of total biomass accumulated by certain species (Williams et al., 1986).

Our approach to addressing the question of the effects of elevated CO_2 levels on future forest composition involved the use of miniature forest stands grown under varying levels of resource availability. This chapter summarizes our experience with using miniature stands in global change research and the implications they may have for future southern forests developing in an elevated CO_2 environment.

Experimental Approach

Miniature stands have been used in the effort to accelerate interspecific competition dynamics between forest tree species under varying levels of resource availability (Williams et al., 1986; Reekie and Bazzaz, 1989; Fredericksen et al., 1993). In contrast to seedlings grown noncompetitively, those grown in competition have the opportunity to exploit resources at the expense of neighboring plants, which is similar to those processes associated with early stand development in forests. As in forest stands, miniature stands also provide a patchy environment in which root proliferation and leaf deployment is affected by the presence of neighboring vegetation.

Our approach involved growing from seed monoculture and mixed stands of loblolly pine along with commonly associated hardwood competitors under several levels of resource availability. Both studies were conducted near Macon, Georgia in continuously stirred tank reactors (CSTRs) in a greenhouse with supplemental lighting. Methodology is explained in greater detail elsewhere (Groninger et al., 1995, 1996; Jifon et al., 1995). To provide similar chemical and physical conditions encountered in forests of the region, common native soils where these species cooccur were used as a growth medium. Both monocultures and mixed stands employed a 2.5 cm² square spacing. Containers were 30 cm in diameter and 37 cm deep. Two buffer rows surrounded a five by five block of seedlings used for biomass and whole canopy photosynthesis measurements. The close spacing, pot size, and buffers caused seedling size to be limited by competition rather than container size as typically occurs in greenhouse studies.

The first experiment used loblolly pine and sweetgum monocultures and mixtures grown under ambient and ambient + 400 ppm CO_2 concentrations, well-watered and one-half watered (drought) conditions, and high (400 kg N ha^{-1}) and low (20 kg N ha^{-1}) supplemental nitrogen (N) additions (Groninger et al., 1995, 1996; Jifon et al., 1995). The second experiment used monocultures and mixtures of loblolly pine and red maple (*Acer rubrum* L.) conducted with similar CO_2 and water treatments as the first experiment (Groninger, 1995). No nutrition treatment was imposed in this latter experiment. In both studies, mixed stands consisted of alternating loblolly pine with the hardwood competitor, which resulted in a 50:50 mixture of the two species.

Both experiments used accelerated growing seasons to compress two growing seasons into one calendar year (Samuelson and Seiler, 1994; Groninger et al., 1995). This procedure involved growing seedling from seed for several weeks and then inducing dormancy through shortened photoperiods and reduced air temperatures. Dormancy was then broken by several weeks of cold storage followed by resumption of sixteen-hour photoperiods and warmer temperatures at the onset of a second growing season. The use of accelerated growing seasons in combination with native soils and suboptimal watering regimes produced seedlings morphologically similar to those found in field conditions. One shortcoming of this approach, however, was the unavailability of light during the cold storage period preventing the loblolly pines from photosynthesizing during that time.

Fine root development was monitored by handsorting roots from soil cores collected with a soil punch (Jifon et al., 1995). Photosynthesis was measured on individual seedlings from each stand using a LI-6200 gas exchange system (Groninger et al., 1996). Whole seedling stand photosynthesis was measured at the conclusion of the experiment using a specially constructed open gas exchange

Table 7.1. Data Quality of Biological Response Variables

Variable[1]	Mean Error (%)
Seedling diameter	2
Seedling height	3
Sapling diameter	4
Sapling height	1
Individual seedling photosynthesis measurement leas mass	<1
Root length density	2
Foliar N content	1
Seedling stand biomass	

[1] Measurements are expressed in terms of average deviation of remeasurement values from actual data.

system on a five by five array of seedlings in the middle of the miniature stand in each pot. Following gas exchange measurements, seedling stands were harvested and separated into leaf, stem, and root components for each species, and then oven dried and weighed. Quality assurance data are summarized in Table 7.1.

Summary and Synthesis

Photosynthesis

Studies evaluating the effects of elevated CO_2 levels on net photosynthesis usually rely on measurements of canopies or single leaves from seedlings grown as individuals. These studies provide useful information regarding physiological responses at the leaf level but poorly represent the environmental conditions found within forest canopies. Under closed-crown conditions, leaves ranging in age from emergent to two years in loblolly pine, and from emergent to several months in hardwoods are found growing under levels of light availability ranging from full sunlight to below-light compensation point. In using miniature stands, our hope was to move toward better representing the range of age and light conditions found throughout forest canopies while still maintaining controlled experimental conditions typically associated with CO_2-enrichment studies.

Leaf photosynthetic rates have consistently been reported to increase under increased CO_2 concentrations in previous studies with loblolly pine and sweetgum (Fetcher et al., 1988; Tissue et al., 1993; Griffin et al., 1993; Tolley and Strain 1985; Tschaplinski et al., 1995). In miniature stands, both whole stands and foliage from individual seedlings taken from these stands responded to elevated CO_2 levels with increased photosynthetic rates. As expected, whole stands had overall decreased leaf mass-based photosynthetic rates than individually measured seedlings, which probably reflects the suboptimal light conditions in lower canopy leaves (Groninger et al., 1996; Groninger, 1995). Also, foliage from isolated individual seedlings was proportionally more responsive to elevated levels of CO_2 than whole seedling stands in sweetgum and red maple monocultures.

These results suggest that hardwood canopy response to elevated CO_2 levels may be decreased by less responsive lower canopy leaves (Groninger et al., 1996). Loblolly pine produced ambiguous results in the relative response of individual seedlings vs stands; the first study found no differences between the relative strength of individuals and stand responses although the second study suggested that individuals were more responsive than stands. The ambiguity of these results, however, may have been driven by differences in age, genetics, soil, water status, or seedling size.

Previous studies have reported elevated levels of CO_2 and the availability of N to have an interactive effect on photosynthetic response. For example, loblolly pine carbon exchange rate (CER) increases under elevated CO_2 levels were found to be dependent on nitrogen availability (Griffin et al., 1993; Tissue et al., 1993). In our work, N availability in the loblolly pine–sweetgum study did not affect photosynthetic responsiveness to CO_2 concentration, which suggests that this nutrient was available in greater amounts as compared to the aforementioned studies (Groninger et al., 1996). Additionally, a nitrogen treatment was not included in the loblolly pine–red maple study.

The compensatory effects of elevated CO_2 levels for low water availability have been widely discussed (Chaves and Pereira, 1992; Idso and Idso, 1994). Elevated levels of CO_2 allowed plants to maintain higher photosynthetic rates under drought conditions (Tolley and Strain, 1985; Lenham, 1994; Townend, 1993). This response has been attributed to higher instantaneous water use efficiency (WUE) and decreased stomatal conductance (Conroy et al., 1986; Jarvis, 1989; Bowes, 1993).

In our studies, a variety of responses to CO_2 concentration and water availability occurred. Water availability affected individually measured loblolly pine seedling responses to elevated levels of CO_2 on a foliar leaf-mass basis in one of the two CSTR studies (Groninger et al., 1996; Groninger, 1995). Well-watered and droughted loblolly pine seedling stands did not differ in their response to elevated CO_2 levels when measured on a whole stand foliar leaf-mass basis in either of these studies.

Under the drought conditions used in our studies, elevated levels of CO_2 compensated for low water-induced inhibition of photosynthesis (Groninger et al., 1996; Groninger, 1995). Loblolly pine, sweetgum, and red maple seedlings, whether measured as individuals or as whole seedling stand monoculture canopies, had photosynthetic rates under elevated CO_2 drought conditions that were equal to or higher than stands grown under present-day CO_2 concentrations and well-watered conditions. Red maple monoculture stands kept at drought-level showed the strongest compensatory response. Elevated CO_2 stands of red maple monoculture kept at drought-level had more than double the photosynthetic rates of well-watered, present-day CO_2 concentration stands. It is important to note, however, that photosynthetic rates are very poorly correlated with biomass production under both greenhouse and forest conditions.

Because resource depletion and microenvironmental conditions differ between forest stand types, the photosynthetic response of a tree to ambient CO_2 concentration is apt to depend on the species composition of its neighbors. Individually measured red maples grown in mixed stands with loblolly pine were more responsive to increased CO_2 availability than those grown in monoculture. Greater light availability in mixed stands resulting from leaf architectural differences between loblolly pine and red maple may have contributed to this difference (Fredericksen et al., 1993. Measuring the contribution of each species to stand-level photosynthetic response was impossible because of logistical constraints associated with the small scale of these experiments.

In several container-grown tree studies, the photosynthetic responsiveness of foliage grown in elevated CO_2 levels has been observed to decrease in magnitude over time, a process known as acclimation (El Kohen and Mousseau 1994: Samuelson and Seiler, 1994). Physical restriction of roots grown in containers has been suggested as one explanation for this phenomenon (Arp, 1991; Thomas and Strain, 1991). Loblolly pine, sweetgum and red maple all exhibited at least a trend toward acclimation despite the absence of root restriction (Groninger et al., 1996; Groninger, 1995). Acclimation needs to be examined in stands subjected to free air fumigation to determine whether this response occurs under field conditions as resources become limited. Because forest ecosystem-level CO_2-enrichment experiments have not yet been conducted for a sufficient duration, it is not known whether acclimation is an artifact of greenhouse experimental conditions or if photosynthetic potential per unit leaf mass will decrease over time in future forest ecosystems. It might be speculated that as stand closure occurs and other environmental factors begin to limit growth, photosynthetic acclimation will occur.

Changes in carbon fluxes from forests are apt to occur in a world with elevated CO_2 levels. An understanding of stand-level photosynthetic responses will aid in the development of regional and global carbon budgets, which may be useful in the establishment of environmental policy. Studies geared toward elucidating the nature of forest tree response to elevated levels of CO_2 have suggested increasing photosynthetic rates per unit leaf and increased total leaf area in young trees to be probable responses. The magnitude and duration of these responses under forest conditions remains uncertain, however.

In our studies, photosynthesis was expressed on a unit soil-area basis in order to integrate photosynthetic and foliar-mass response to elevated CO_2 levels across entire seedling stands. In this way, miniature stand data could be used to generate hypotheses for research in forest stand-level research. An expression of photosynthesis based on soil area indicates that loblolly pine monoculture seedling stands in an environment with elevated CO_2 levels have higher photosynthetic rates when grown under well-watered vs drought conditions, which is a response not detected when photosynthesis is expressed on a foliar-mass basis (Groninger et al. 1996; Groninger 1995). Similarly, an expression of photosynthesis based on soil area in mixed loblolly pine–red maple stands and monoculture red maple seedling stands suggests that the compensatory effect of elevated CO_2 levels for well-watered conditions found in leaf-mass expressions of photosynthesis would be overstated relative to measurements expressed in a soil-surface area basis. This pattern is not consistent for all stand type combinations. In mixed stands of loblolly pine and sweetgum, the compensatory effect of elevated CO_2 levels of was greater in expressions of photosynthesis based on soil area. It is important to note, however, that elevated CO_2 levels still show a compensatory effect for drought conditions in all stand types tested in both leaf-mass and soil-area expressions of photosynthesis. More detailed studies of canopy-level photosynthetic responses to atmospheric CO_2 concentrations are needed to explain these responses.

Biomass

Differences between loblolly pine and sweetgum biomass response to elevated CO_2 levels have been widely reported. Across a variety of resource availability levels, individually grown sweetgum seedlings and saplings showed more positive growth responses to elevated levels of CO_2 than did loblolly pine (Tolley and Strain 1984a; Sionit et al., 1985). Similarly, our results found that across all treatments sweetgum had a 33% mean biomass increase compared to 14% in loblolly pine grown under elevated CO_2 levels across water, N, and stand-type treatments (Groninger et al., 1995). It should be noted that in our study loblolly pine has total biomass growth responses exceeding 40% of the growth reported in other studies (Griffin et al., 1993; Tschaplinski et al., 1993; Groninger, 1995)

Given the dependence of growth on the availability of other resources, CO_2 effects have been measured under a range of experimental conditions, including differing availabilities of water and nitrogen. Drought stress continued to reduce growth when compared to well-watered conditions under elevated CO_2 in loblolly pine (Tschaplinski et al., 1993) and sweetgum (Tolley and Strain 1984b; Tschaplinski et al., 1995). However, seedlings grown under elevated CO_2 levels and in drought conditions were the same size or larger than those grown under present-day CO_2 concentrations in these prior studies, which reflects the compensatory effect of elevated CO_2 levels for water occasionally observed with the photosynthetic response.

In our miniature stand studies, the main effects of water stress were the significant reduction of biomass growth across all other treatments in loblolly pine, sweetgum, and red maple (Figures 7.1 and 7.2). When analyzing total seedling stand biomass for each species, when grown in mixture, the compensatory effect of CO_2 for water was not prevalent. Only the loblolly pine grown in mixture with sweetgum under elevated levels of CO_2 and low water accumulated biomass equal to or greater than well-watered, low CO_2 level stands. A study of similar design growing these species but at a wider spacing showed similar results (Burdick, 1996).

Belowground Growth and Biomass Accumulation

Tree species respond to their environment through differential accumulation of carbon to particular organs, presumably to increase the uptake of growth-limiting resources. Typically, plants grown under elevated CO_2 concentrations exhibit increased root/shoot ratio (Acock and Allen, 1985; Enoch and Zieslin, 1988). Extremely variable results have also been observed in loblolly pine and sweetgum

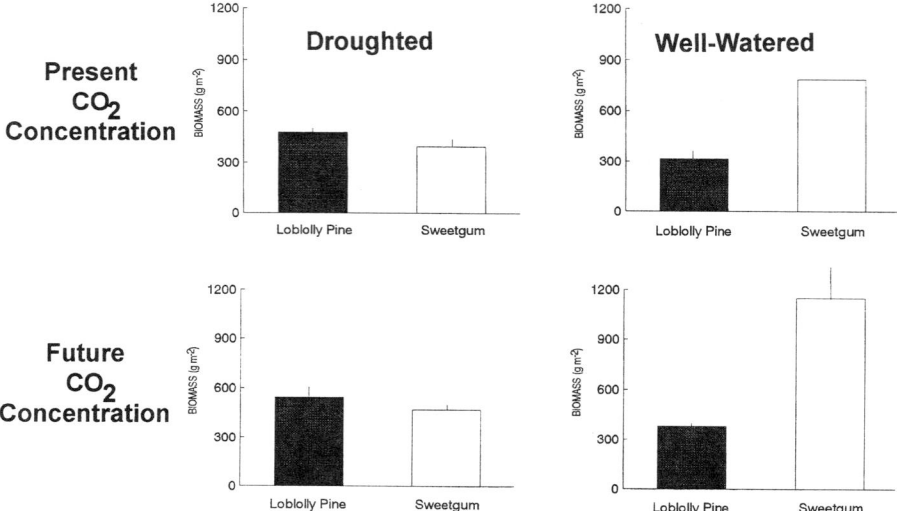

Figure 7.1. Total biomass of loblolly pine and sweetgum grown together in miniature stands under factorial combinations of ambient CO_2 concentration and water availability.

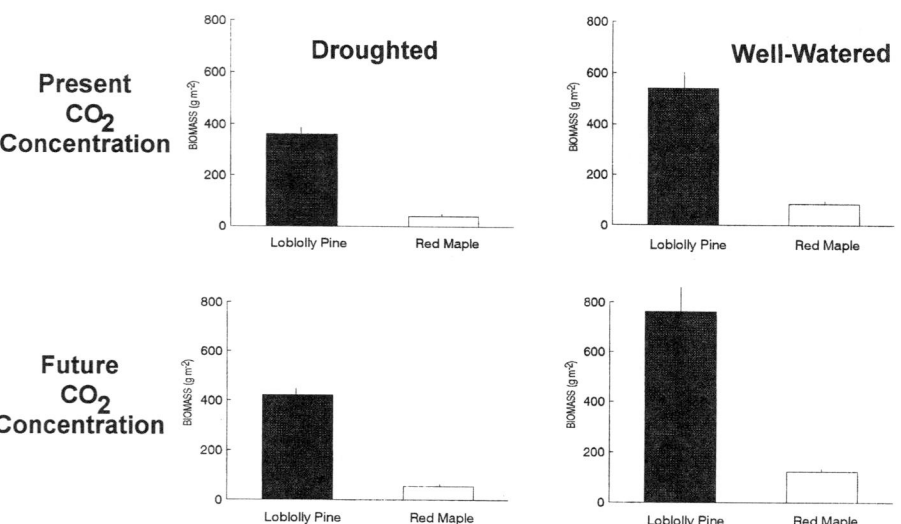

Figure 7.2. Total biomass of loblolly pine and red maple grown together in miniature stands under factorial combinations of ambient CO_2 concentration and water availability.

in which the root/shoot ratio has increased (Rogers et al., 1983), has decreased (Tolley and Strain, 1984b), and has remained unchanged (Sionit et al., 1985; Tschaplinski et al., 1993, 1995). Our data suggest that both loblolly pine and sweetgum root/shoot ratio are insensitive to elevated CO_2 levels alone or in combination with water, N, or growth in monoculture or mixture. This suggests that elevated CO_2 levels will not impact below ground competitive dynamics. Total root/shoot biomass ratio in sweetgum reflected typical responses to water and nitrogen availability with higher values at the lower levels of availability for each resource (Groninger et al., 1995).

Root length density (RLD), defined here as the cumulative length of roots less than 0.5 mm in diameter per unit soil volume, is a more direct measure of the plant's ability to absorb belowground resources than total biomass; this measurement provided some interesting insights into how atmospheric CO_2 might alter the performance of loblolly pine and sweetgum (Jifon et al., 1995). In mixed stands, under conditions of increased water and N availability, sweetgum RLD is considerably greater than that of pine (Jifon et al., 1995). However, this response is present under both ambient and elevated levels of CO_2, which suggests that high RLD under conditions of increased resource availability may at least partially account for the presently observed dominance of sweetgum over loblolly pine in resource-rich sites. If incremental increases in RLD occur in an elevated CO_2 environment, and resource availability levels less extreme than those tested here do occur, competitive dynamics between the two species might be affected on sites of intermediate quality. Nevertheless, related studies of loblolly-sweetgum mixtures in open-top chambers have found no increases in N uptake by sweetgum grown under elevated CO_2 levels, despite a doubling of sweetgum RLD under low to high CO_2, and a high level of available N (Friend et al., 1995). Neither pine RLD nor pine N uptake were affected by CO_2 in this study. Evidently, more detailed information is needed on how CO_2 influences plant N acquisition before the potential influence of CO_2 on belowground competition can be elucidated.

As was true for total biomass, RLD of sweetgum grown in monoculture generally had a stronger positive response to elevated levels of CO_2 than in mixture. One exception, however, was the very strong positive CO_2 response observed for RLD in mixtures under the factorial combination of increased N and water. In monoculture, RLD under decreased water regimes was more responsive to CO_2 than under increased water regimes. This resulted in a CO_2-induced increase in the proportion of absorbing roots under these conditions, which may translate into an advantage for sweetgum on drier sites but only in the absence of pine.

Red maple seedlings responded to elevated CO_2 levels with higher leaf/root allocation. Low stomatal conductance values under elevated levels of CO_2 may have permitted red maple to maintain a higher ratio of water-using to water-absorbing tissues under these conditions. Whether this plastic response constitutes an adaptation is not clear. Assessing the long-term competitive success of red maple is impossible in short-term studies because the observed ability of this species to persist in the understory and attain a dominant canopy position following several years of suppression.

Given the variability of biomass accumulation responses reported in the literature, results from our studies should be considered as additional information in a complex plant-environment interrelationship, which suggests that other environmental and cultural factors may be important to determining how CO_2 will affect biomass accumulation patterns.

Competition

Competitive relationships observed in our studies reflected those occurring in many southeastern forests. When both loblolly pine and sweetgum are present, loblolly pine tends to be more dominant under drought conditions although sweetgum has a distinct advantage under increased levels of water availability (Figure 7.1). As for red maple, loblolly pine was dominant under both levels of water availability, but this relationship was slightly stronger under drier conditions (Figure 7.2). In these studies, CO_2 concentration did not affect total biomass response of any species to the availability of other resources. These results suggest that these species may continue to behave according to present-day silvical characteristics under increased CO_2 concentrations. The stronger influence of water on the competitive status of loblolly pine with sweetgum compared to red maple reflects the greater sensitivity to water availability observed in sweetgum compared to red maple.

Even though competitive relationships did not change between loblolly pine and its hardwood competitors, the proportion of hardwoods on a total seedling biomass basis was slightly greater in all cases under elevated CO_2 conditions. It is significant to note, however, that this increase in hardwood proportion under elevated CO_2 levels was never associated with a decrease in total pine biomass. Studies of greater duration are needed to determine whether this trend could eventually result in either a competitive advantage for hardwoods over pines or continued increases in stand productivity for both loblolly pine and hardwoods.

The results of these greenhouse-based studies reflect competitive outcomes under a certain set of resource availabilities, species compositions, and environmental conditions that may be found in present and future forests. However, given spatial and temporal constraints associated with the miniature stand methodology, a corresponding stand age cannot be determined for our miniature stands. This may or may not be an important factor concerning the validity of these studies. In managed forests of the region, the actual outcomes between hardwood and pine competitors are often determined very early in the life of the stand, with harvest occurring long before the loblolly pine overstory breaks up and gives way to the hardwood mid- and understory.

The role hardwoods may play in early stand development also depends on the presence of herbaceous vegetation. In mixed loblolly pine–red maple stands on the Virginia Piedmont, loblolly pine continues to dominate red maple seven years after establishment from seedlings both in the presence and absence of herbaceous cover. Over time, however, red maple appears to become less suppressed in the absence of herbaceous cover. In contrast, under herbaceous cover, loblolly pine continues to become increasingly dominant over red maple.

Within the context of a miniature stand study, loblolly pine was becoming increasingly dominant over red maple under both CO_2 concentrations. Unfortunately, the duration of this study was insufficient to determine whether this would continue to occur.

Summary

Given the many limitations associated with drawing inferences from greenhouse studies of short duration, this research represents the closest approximation of forest competition dynamics in loblolly pine-hardwood systems to date. The extent that miniature stands represent competitive processes in forests has yet to be determined. With this uncertainty in mind, the following inferences may be drawn from our results concerning the characteristics of southern pine-hardwood forests under future high CO_2 environments:

- Loblolly pine, sweetgum, and red maple will have more rapid growth, at least in the earliest stages of stand development. If sustained, this would accelerate stand development, potentially reducing ages of thinnings and rotation length.
- Sustained growth could increase present site index values if any other growth-limiting factor does not come into play.
- Increasing CO_2 concentrations alone would not affect species dominance. Loblolly pine would continue to dominate dry sites and sweetgum would continue to be dominant on wet sites. Although the proportion of hardwoods would increase on both dry and wet sites, this change would not come at the expense of loblolly pine biomass.
- The nature or intensity of site preparation would not change in managed loblolly pine-dominated forests. Because most hardwood competitors are stump sprouts, however, research should be conducted to determine whether coppiced hardwoods are more responsive to atmospheric CO_2 concentration than seedling- or seed-originated hardwoods.

References

Acock B, Allen LH (1985) Crop responses to elevated carbon dioxide concentrations. In Strain, BR, Cure, JD (Eds) *Direct Effects of Increasing Carbon Dioxide on Vegetation.* DOE/ER-0238. U.S. Department of Energy, Washington DC.

Arp WJ (1991) Effects of source-sink relations on photosynthetic acclimation to elevated CO_2. Plant Cell Environ 14:1003–1006.

Bazzaz FA, McConnaughay KDM (1992) Plant-plant interactions in elevated CO_2 environments. Aust J Botany 40:547–563.

Bowes G (1993) Facing the inevitable: plants and increasing atmospheric CO_2. Ann Rev Plant Phys Plant Mol Bio 44:309–332.

Burdick TE (1996) Season effects of elevated carbon dioxide, competition and water stress on gas exchange and growth of loblolly pine and sweetgum grown in open-top chambers. Unpublished MS thesis, Virginia Polytechnic Institute and State University, Blacksburg, VA.

Chaves MM, Pereira JS (1992) Water stress, CO2 and climate change. J Exp Bot 43:1131–1139.

Conroy JP, Smillie RM, Kuppers M, Bevege DI, Barlow ES (1986) Chlorophyll a fluorescence, and photosynthetic growth responses of *Pinus radiata* to P deficiencies, drought stress and high CO_2. Plant Phys 81:423–429.

Dubois MR, McNabb K, Straka TJ, Watson WF (1995) Costs and cost trends for forestry practices in the South. For Farm 54(3):10–17.

El Kohen A, Mousseau M (1994) Interactive effects of elevated CO_2 and mineral nutrition on growth and CO_2 exchange of sweet chestnut seedlings (*Castanea sativa*). Tree Phys 14:679–690.

Enoch HZ, Zieslin N (1988) Growth and development of plants in response to carbon dioxide concentrations. Appl Agr Res 3:248–256.

Fetcher N, Jaeger CH, Strain BR, Sionit N (1988) Long-term elevation of atmospheric CO_2 concentration and the carbon exchange rates of saplings of *Pinus taeda* L. and *Liquidambar styraciflua* L. Tree Phys 4:255–262.

Fredericksen TS, Zedaker SM, Seiler JR (1993) Interference interactions in simulated pine-hardwood seedling stands. For Sci 39:383–395.

Friend AL, Jifon JL, Berrang PC, Seiler JR (1995) CO_2 and water stress alter nitrogen acquisition in *Pinus taeda-Liquidambar styraciflua* seedling mixtures. Bull Ecol Soc Amer (supplement) 76(2):86.

Griffin KL, Thomas RB, Strain BR (1993) Effects of nitrogen supply and elevated carbon dioxide on construction cost in leaves of *Pinus taeda* (L.) seedlings. Oecologia 95:575–580.

Grime JP (1979) Plant Strategies and Vegetation Processes. John Wiley and Sons, Chichester, UK.

Groninger JW (1995) Stand dynamics and gas exchange in loblolly pine and hardwood seedling stands: impact of elevated carbon dioxide, water stress and nutrient status. Unpublished PhD dissertation, Virginia Polytechnic Institute and State University, Blacksburg, VA.

Groninger JW, Seiler JR, Zedaker SM, Berrang PB (1995) Effects of elevated carbon dioxide, water stress and nitrogen level on competitive interactions of simulated loblolly pine and sweetgum stands. Can J For Res 25:1077–1083.

Groninger JW, Seiler JR, Zedaker SM, Berrang PB (1996) Photosynthetic response of loblolly pine and sweetgum seedling stands to elevated carbon dioxide, water stress and nitrogen level. Can J For Res 26:95–102.

Idso KE, Idso, SB (1994) Plant responses to atmospheric CO_2 enrichment in the face of environmental constraints: a review of the past 10 years' research. Agr For Meteor 69:153–203.

Jarvis PG (1989) Atmospheric carbon dioxide and forests. Phil Trans Roy Soc Lon B 324:369–392.

Jifon JL, Friend AL, Berrang PC (1995) Species mixture and soil-resource availability affect the root growth response of tree seedlings to elevated atmospheric CO_2. Can J For Res 25:824–832.

Lenham PJ (1994) Influences of elevated atmospheric CO_2 and water stress on photosynthesis and fluorescence of loblolly pine, red maple and sweetgum. Unpublished MS thesis, Virginia Polytechnic Institute and State University, Blacksburg, VA.

Reekie EG, Bazzaz FA (1989) Competition and patterns of resource use among seedlings of five tropical trees grown at ambient and elevated CO_2. Oecologia 79:212–222.

Rogers HH, Bingham GE, Cure JD, Smith JM, Surano KA (1983) Responses of selected plant species to elevated carbon dioxide in the field. J Environ Qual 12:569–574.

Samuelson LJ, Seiler JR (1994) Red spruce seedling gas exchange in response to elevated CO_2, water stress, and soil fertility treatments. Can J For Res 24:954–959.

Sionit N, Strain BR, Hellmers H, Reichers GH, Jaeger CH (1985) Long-term atmospheric

CO$_2$ enrichment affects the growth and development of *Liquidambar styraciflua* and *Pinus taeda* seedlings. Can J For Res 15:468–471.

Thomas RB, Strain BR (1991) Root restriction as a factor in photosynthetic acclimation of cotton seedlings grown in elevated carbon dioxide. Plant Physiol 96:627–634.

Tilman D (1982) *Resource competition and community structure*. Princeton University Press. Princeton, NJ.

Tissue DT, Thomas RB, Strain, BR (1993) Long-term effects of elevated CO$_2$ and nutrients on photosynthesis and rubisco in loblolly pine seedlings. Plant Cell Environ 16:859–865.

Tolley LC, Strain BR (1984a) Effects of CO$_2$ enrichment and water stress on growth of *Liquidambar styraciflua* and *Pinus taeda* seedlings. Can J Bot 62:2135–2139.

Tolley LC, Strain BR (1984b) Effects of CO$_2$ enrichment on growth of *Liquidambar styraciflua* and *Pinus taeda* seedlings under different irradiance levels. Can J For Res 14:343–350.

Tolley LC, Strain BR (1985) Effects of CO$_2$ enrichment and water stress on gas exchange of *Liquidambar styraciflua* and *Pinus taeda* seedlings grown under different irradiance levels. Oecologia 65:166–172.

Townend J (1993) Effects of elevated carbon dioxide and drought on the growth and physiology of clonal sitka spruce plants (*Picea sitchensis* (Bong.) Carr.). Tree Physiol 13:389–394.

Tschaplinsk TJ, Norby RJ, Wullschleger SD (1993) Response of loblolly pine seedlings to elevated CO$_2$ and fluctuating water supply. Tree Physiol 13:283–296.

Tschaplinsk TJ, Stewart DB, Hanson PJ, Norby RB (1995) Interactions between drought and elevated CO$_2$ on growth and gas exchange of seedlings of three deciduous tree species. New Phytol 129:63–71.

Williams WE, Garbutt K, Bazzaz FA, Vitousek PM (1986) The response of plants to elevated CO$_2$-IV. Two deciduous-forest tree communities. Oecologia 69:454–459.

Williams WE, Garbutt K, Bazzaz FA (1988) The response of plants to elevated CO$_2$-V. Performance of an assemblage of serpentine grassland herbs. Environ Exper Bot 28:123–130.

Zangerl AR, Bazzaz, FA (1984) The response of plants to elevated CO$_2$-II. Competitive interactions among annual plants under varying light and nutrients. Oecologia 62:412–417.

8. Effects of Elevated Carbon Dioxide Levels and Air Temperature on Carbon Assimilation of Loblolly Pine

Robert O. Teskey

Changes in the atmospheric carbon dioxide (CO_2) concentration directly affects photosynthesis, which, in turn, indirectly affects all other plant processes. Carbon dioxide is the substrate used in photosynthesis for the creation of sugars needed by the plant for all anabolic and catabolic activities. For trees, the present-day ambient CO_2 concentration (C_a) is not at saturation with respect to photosynthesis, therefore an increase in C_a can result, at least initially, in an increase in the rate of net carbon assimilation (A_n) (Ceulemans and Mousseau, 1994). For the most part, this effect is caused by a reduction in photorespiration. The higher CO_2 concentration causes an increase in the frequency of carboxylation reactions in relation to oxygenation reactions catalyzed by the enzyme rubisco in the initial step of the dark reactions, or Calvin cycle (Webber et al., 1994). An increased concentration of CO_2 in the atmosphere also speeds the rate of diffusion of CO_2 into the mesophyll. But when conditions are less than optimum, a complex variety of interacting factors may significantly reduce the potential for carbon gain in elevated CO_2 concentrations. Nutrient limitations, drought, and excessive temperatures are among the common stresses in the field that may reduce the increase in A_n expected under increased CO_2 concentrations (Conroy et al., 1990, Tschaplinski et al., 1993; El Kohen and Mousseau, 1994; Guehl et al., 1994; Lewis et al., 1994; Ziska and Bunce, 1994). The relationship between A_n, elevated CO_2 levels, and environmental stress is a critical issue because this relationship will affect ecosystem productivity, the magnitude of the carbon sink in the ecosystems of the world, and the global carbon cycle (Gifford 1994). Because forests are the most extensive vegetation type in the world, occupying over one-third of the land surface of the earth, it is vital to know how forest species will respond to the increase in atmospheric CO_2 that the world is experiencing, and to the potential changes in climates that are apt to accompany it.

Two key components of the issue of forest and tree response to CO_2 are 1) the magnitude of the increase in A_n under field conditions, and 2) whether the increase will be sustained over long periods of time. To understand these components, the relative limitation on the rate of A_n imposed by CO_2 levels must be compared to that of other such external and internal regulating factors as irradiance, temperature, humidity, nutrition, and available carbohydrates. In this study both of these components were addressed. An assessment of the environmental conditions caused by differences in the magnitude of the CO_2 effect on assimilation was made by measuring the diurnal and seasonal changes in A_n in treatments of ambient and elevated CO_2 levels under field conditions. The question of whether the CO_2 response could be sustained over time was examined by periodically measuring the gas exchange of the foliage from the time it had recently expanded to when it was near senescence.

An additional aspect of the study was an investigation of the effect of both increased CO_2 concentration and elevated air temperature on A_n and transpiration (E). Along with a doubling of the atmospheric CO_2 concentration, an increase in air temperature of approximately 2 °C has also been predicted (Bloomfield, 1992). Although there is great uncertainty associated with present estimates of future temperature changes, CO_2 is not expected to increase without an accompanying increase in air temperature, and therefore both CO_2 concentration and air temperature were manipulated in this experiment. Long (1991) examined the possible effects on A_n that might result from an increase in both CO_2 and air temperature, using the photosynthesis model of Farquhar et al. (1980). His analysis indicated that the effects of CO_2 on rates of photosynthesis would be significantly modified by air temperature. The model predicts that the temperature optimum for photosynthesis will shift to higher temperatures under elevated CO_2 concentrations and that there will be a relatively greater increase in A_n in elevated CO_2 concentrations at higher air temperatures than at lower temperatures. The present study was designed, in part, to determine if these sorts of interactions between CO_2 and temperature would be apparent under field conditions, where there are many factors simultaneously limiting rates of assimilation.

Materials and Methods

The study was conducted in a twenty-one-year-old loblolly pine (*Pinus taeda* L.) plantation, at the Whitehall Forest near Athens, GA. This area is in the Piedmont region of the southern United States and has a warm temperate climate with an average annual precipitation of 1257 mm, evenly distributed throughout the year. The average annual temperature is 16.5 °C and the region has hot, humid summers and mild winters. Mean summer and winter temperatures are 25.4 and 7.2 °C,

8. Effects of Elevated Carbon Dioxide Levels and Air Temperature

respectively. The soil at the study site is in the Pacolet series (USDA Soil Conservation Service), which is an eroded sandy loam soil with little A horizon, and a depth of approximately 0.9 m. This soil is classified as having moderate to low nutrient availability, which is reflected in the site index of the stand which, at 16.5 m at age twenty-five, is low for the region (Bailey et al., 1985). Although no fertilizer amendments were added to the site, it was irrigated to maintain soil water potentials above -0.05 MPa, as indicated by tensiometers placed at depths of 0.15 and 0.30 m near each study tree.

Six trees were selected within the stand for study. Towers and walkways were constructed around the crowns of these trees to provide platforms for mounting branch chambers. Six chambers were used in each tree so that all treatments could be replicated within a tree. The construction and performance of these chambers has been previously described (Teskey et al., 1991). Briefly, they consisted of an cylindrical aluminum frame (0.5 by 1.5 m) covered by a clear polyvinylchloride (PVC) plastic film. Each chamber had a separate blower to supply an adequate air flow through the chamber. The volume within the chamber was exchanged ten times per minute to minimize differences between outside and inside chamber conditions.

Environmental conditions inside each chamber were characterized by daily mean air temperature (calculated from 360 measurements per hour using a 0.8 mm diameter copper constantan thermocouple mounted in a ventilated shield), mean daily CO_2 concentration (calculated from measurements made twice hourly using an infrared gas analyzer (LI-6252) and computer controlled sampling system) and mean daily irradiance (calculated from ten measurements per hour using photodiodes. Three photodiodes were placed horizontally within each chamber at midchamber height. They were individually calibrated against a quantum sensor (LI-190SA). Air temperature, CO_2 concentration, and irradiance were measured outside the chambers in the center of the stand in the upper-middle canopy, using the same sensors, placed at approximately the same height as the chambers were mounted. The frequency of measurement was the same as within the chambers.

Usually, the air temperatures within the ambient temperature treatment chambers were slightly higher than the outside ambient temperature. Mean daily air temperatures within the chambers differed from the outside ambient air temperature by 0.2 °C or less, on 61% of the 914 days of the experiment. Ambient chamber air temperatures were within 0.5 °C of outside temperatures on 86% of the days. Daily mean temperature deviations did not exceed 1.6 °C.

At all times, the CO_2 concentration within the ambient chambers was within 10% of the outside ambient concentration; 88% of the time, it was within 5% of the ambient CO_2 concentration. The yearly mean CO_2 concentration at this site was 351 $\mu mol\ mol^{-1}$. On average, irradiance inside the chambers was reduced by 12% resulting from the use of the PVC film and chamber orientation. This value varied with solar angle and azimuth and also increased as the PVC film weathered to a maximum reduction in irradiance of 17% at the end of the calendar year. To minimize any effects caused by discoloration, the PVC film on the chambers was changed each year during the dormant season.

Carbon dioxide and air temperature treatments were imposed within the chambers. The three levels of CO_2 that were used were 1) ambient, 2) ambient + 165 μmol mol^{-1}, and 3) ambient + 330 μmol mol^{-1}. However, only results for ambient and ambient + 330 μmol mol^{-1} treatments will be reported in this chapter. In all respects, the middle CO_2 treatment responded in both an intermediate and linear manner between the ambient and ambient + 330 μmol mol^{-1} treatments, and will not be discussed here.

The air temperature treatments were ambient and ambient + 2 °C. These treatments were applied continuously for three growing seasons beginning in April of 1992 and extending through September of 1994. The air was heated using nichrome wire placed in a housing between the blower and the chamber. The temperature was controlled using a computer monitoring and switching system. The thermocouples in each chamber were monitored every 10 seconds. The mean difference in temperature of all ambient temperature chambers (not outside ambient) and all elevated temperature chambers defined the treatment. If the mean difference was less than 2.0 °C, the heaters were switched on, if the mean difference exceeded 2.2 °C, the heaters were turned off. The high frequency of sampling and switching was necessary to compensate for the relatively rapid heating and cooling effects caused by clouds. The elevated temperature was maintained in the chambers within ± 0.2 °C of the 2.0 °C elevated temperature treatment for 83% of the experiment, and within ± 0.4 °C for 87% of the experiment.

The ambient + 330 μmol mol^{-1} CO_2 treatment was imposed using a single mass flow controller connected to flowmeters (one per chamber to provide individual adjustment), which dispensed a constant flow of CO_2 into the air stream of each chamber where it was mixed with the ambient air before entering the chamber. The CO_2 concentration was monitored hourly in each chamber using an infrared gas analyzer (LI-6252) and a computer controlled sampling system. Because the ambient CO_2 concentration changed diurnally and seasonally, the ambient (outside) CO_2 concentration was compared daily with each elevated CO_2 chamber to ensure that there was an acceptable difference (defined as ± 10% of 330 μmol mol^{-1} CO_2 addition) between them. If not, the flow into any aberrant chamber was adjusted manually. The performance of the chambers for the period of the experiment indicates that the target value ± 10% was maintained on 77% of all days, and ± 20% of the target value was maintained for 94% of the experiment.

Gas Exchange Measurements

Diurnal measurements of water vapor and CO_2 exchange were made using an infrared gas analyzer (IRGA) and an automated sampling system. During the first year, an ADC-LCA2 IRGA was used. In the second and third years, this unit was replaced by a LI-6262 IRGA, which allowed the measurement of both CO_2 and water vapor concentrations in the air stream thereby allowing both A_n and E to be calculated. On a single day, all of the chambers in one tree (one replicate of the treatments) were measured twice-hourly from dawn to dusk. This was repeated until measurements had been made in all six trees. Four narrow (10 mm) cylindrical cuvettes, made of propafilm-C, completely enclosed individual needles of a single age class in each chamber. To minimize heating, air was pulled continuously through these cuvettes at a flow rate of 0.5 l min^{-1}. An ambient air reference line was also placed in each chamber near the cuvettes. The cuvettes were lightweight and small to allow the needles to remain in their original positions on the shoots during measurements without disturbing adjacent foliage.

In the piedmont region, loblolly pine foliage typically emerges in April, finishes elongation in late June, and abscission occurs in October of the following year, for a total life span of about eighteen months. During each year, older trees usually have two foliage growth flushes; a large, first flush occurs in April and a much smaller second flush takes place in June or July. In this study, measurements were made only on the first flush foliage, which represents the largest proportion of the foliage in the crown. During 1992, summer measurements were restricted to the current-year foliage because the foliage from the year prior had not been exposed to the elevated CO_2 levels or temperature treatments during its first growing season. During the summers of 1993 and 1994, foliage from the year prior and the fully expanded foliage from that year were measured. In the autumn and spring measurements, only the current-year foliage was measured because these measurements were made after abscission of the prior year's foliage in the autumn, and before leaf emergence in the spring. Total leaf surface area was determined by measuring the length and radius of each needle, and calculating and summing the areas of the outer and inner external surfaces. Additionally, gas exchange values presented in this chapter are based on the total surface area.

For analysis of each seasonal period, daily means were calculated for each treatment and the values from all six trees were compared using two-way analysis of variance (ANOVA) (SigmaStat). When significant treatment differences were indicated, they were compared using the Student-Newman-Kuels pairwise multiple comparison procedure. Hourly gas exchange values were also calculated and summed for comparisons of total daily net assimilation and transpiration, and for calculation of mean daily water use efficiency (W_e).

Results

From the diurnal pattern of gas exchange, differences in A_n could be clearly distinguished between the ambient and elevated CO_2 treatments (Figure 8.1). Within a CO_2 treatment, the differences between the ambient and elevated temperature treatments were much smaller. There appeared to be no effects on rates of transpiration, which could be attributed to either the CO_2 or temperature treatments.

Consistent with the responses shown in Figure 8.1, during all measurement periods there were highly significant differences between the mean daily A_n in ambient and elevated CO_2 treatments (Table 8.1). On average, over the three years, A_n was 2.2 times greater in the elevated CO_2 treatment, compared with the ambient treatment (Figure 8.2). The mean daily ratio of A_n in elevated CO_2 concentrations

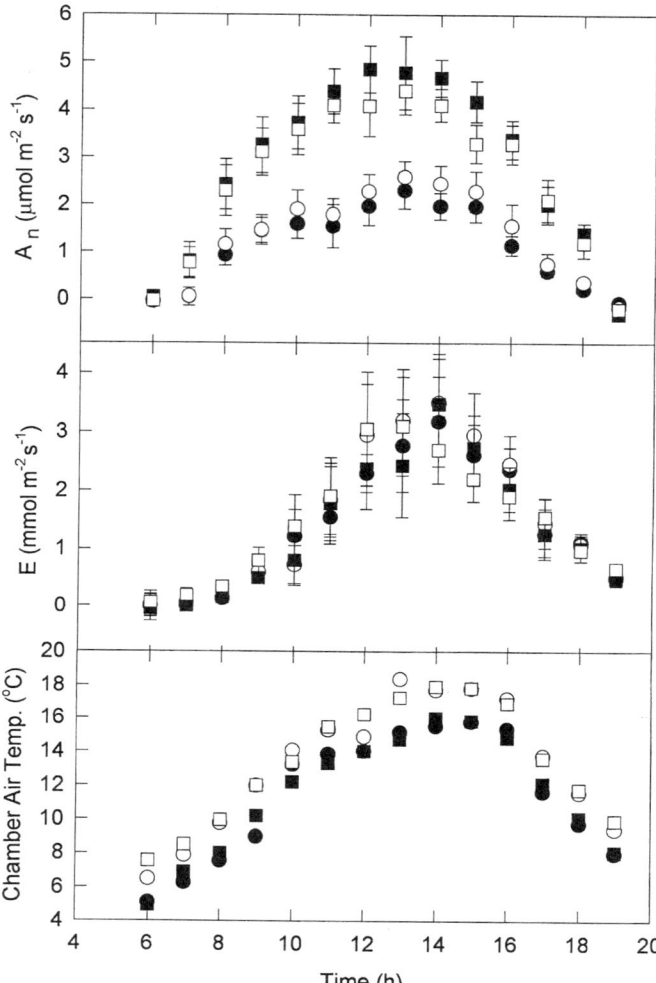

Figure 8.1. Mean hourly net assimilation (A_n), transpiration (E), and chamber air temperature for the autumn, 1993 measurement period. Treatments shown are 1) ambient CO_2 and ambient temperature (●), 2) ambient CO_2 and elevated temperature (○), 3) elevated CO_2 and ambient temperature (■), and 4) elevated CO_2 and elevated temperature (□). Each point represents the mean of the six replicates of each treatment, each replicate measured on a separate day over an eight-day period. The error bars represent ± 1 standard error of the mean.

relative to that in the ambient CO_2 concentrations ($A_{elevated}/A_{ambient}$) varied from 2.7 to 1.5 over the seven measurement periods. The magnitude of the ratio did not decrease with foliage age or length of exposure to elevated CO_2 concentrations because for both the 1992 and 1993 foliage age classes, $A_{elevated}/A_{ambient}$ was either equal to or higher than the second growing season—compared

8 Effects of Elevated Carbon Dioxide Levels and Air Temperature

Table 8.1. Significance Levels (p values) of the Two-Way Analysis of Variance on Daily Mean Net Assimilation

Year, season, and foliage age class	Source of Variation		
	CO_2	Temperature	$CO_2 \times$ Temperature
1992 Summer (1992 age class)	<0.001	0.971	0.002
1993 Spring (1992 age class)	<0.001	0.362	<0.001
1993 Summer (1992 age class)	<0.001	0.594	0.875
1993 Summer (1993 age class)	<0.001	0.016	0.007
1993 Autumn (1993 age class)	<0.001	0.775	0.199
1994 Summer (1993 age class)	<0.001	0.210	0.895
1994 Summer (1994 age class)	<0.001	0.744	0.036

to that achieved in the first growing season. Both the lowest and highest $A_{elevated}/A_{ambient}$ were observed in the summer of 1993, in current-year and one-year-old foliage, respectively, under similar temperature conditions.

In four of the seven measurement periods there were significant interactions between the CO_2 and temperature treatments (Table 8.1). For measurements made on current-year foliage in the summers of 1992, 1993, and 1994, and in the spring

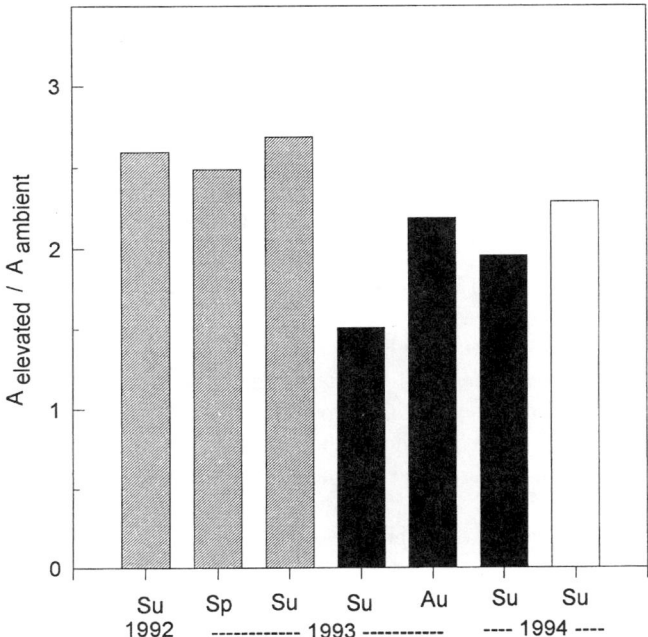

Figure 8.2. Comparison of the ratio of A_n achieved in the elevated CO_2 treatments with that achieved in the ambient CO_2 treatments. Each set of bars represents a foliage age class. Foliage that emerged in the 1992 growing season is shown as the shaded bars. Foliage that emerged in 1993 and 1994 is represented by the solid black and white bars, respectively.

1993 on one-year-old foliage, elevated air temperature reduced A_n in the elevated CO_2 treatment, while increasing it in the ambient CO_2 treatment (Table 8.2). During these four periods, the mean ambient air temperatures ranged from 26.0 to 29.6 °C (Table 8.3). The same pattern of interaction between CO_2 and temperature was also evident in the autumn measurements, but was not statistically significant. Mean air temperatures were much lower in autumn (11.6 °C in ambient and 13.2 °C in elevated treatment) relative to the spring and summer measurement periods (Table 8.3) and may have contributed to the lack of statistical significance. In summer measurements of older foliage, that is, the 1992 age class sampled in summer 1993, and the 1993 age class sampled in summer 1994, there was no CO_2 × temperature interaction. Additionally, only during the summer 1993 measurement period was there a significant difference in A_n resulting from the +2 °C air temperature treatment (Table 8.1). In the current-year foliage in summer 1993, A_n was lower in the elevated CO_2-elevated temperature treatment compared with the elevated CO_2-ambient temperature treatment (Table 8.2). At the same time, in the ambient CO_2 treatments, the effect of temperature on A_n was not statistically significant. In one-year-old foliage no effect of air temperature on A_n could be detected.

In contrast to the effects of CO_2 and air temperature treatments on A_n, these treatments did not have a significant effect on transpiration (E) (Table 8.4). Transpiration rates, averaged across all four treatments, varied with changing weather conditions, from 2.7 mmol m^{-2} s^{-1} (autumn, 1993) to 9.6 mmol m^{-2} s^{-1} (summer, 1993), but all foliage in all treatments achieved similar rates of E, even though the rates of A_n in the elevated CO_2 treatments were about twice that in the ambient CO_2 treatments.

To estimate W_e, the total daily net assimilation (A_d) and total daily transpiration (E_d) were calculated based on the period of time in which A_n was positive in each sample day (Tables 8.5 and 8.6). This period extended from approximately twelve hours in autumn to fifteen hours in summer. The values of A_d and E_d were consistent with those for the instantaneous rates of A_n and E. For the ambient CO_2 treatments, A_d ranged from 76 to 214 mmol m^{-2} d^{-1}, whereas in the elevated CO_2 treatments the range was 183 to 501 mmol m^{-2} d^{-1} (Table 8.5). In contrast, there was much less difference in E_d between CO_2 treatments (Table 8.6). Total daily transpiration was occasionally either higher or lower in the ambient CO_2 treatment relative to the elevated CO_s treatment. From these values it was not surprising that W_e was always higher in the elevated CO_2 treatments (Figure 8.3). For individual measurement periods, the ratio of W_e in the elevated and ambient treatments ranged from 1.5 to 2.8. On average, across all measurement periods and age classes, W_e was 2.2 times greater in the elevated CO_2 treatment compared with the ambient treatment.

Discussion

Increasing the atmospheric CO_2 concentration by 330 μmol mol^{-1} caused an increase in A_n of 116%, averaged across all measurement periods and temperature

Table 8.2. Mean Daily Rates of Net Assimilation ($\mu mol\ m^{-2}\ s^{-1}$) for Ambient and Elevated CO_2 and Temperature Treatments

	Treatments			
	Ambient CO_2		CO_2 + 330 $\mu mol\ mol^{-1}$	
Year, season, and foliage age class	Ambient Temp.	Ambient Temp. + 2 °C	Ambient Temp.	Ambient Temp. + 2 °C
1992 Summer (1992 age class)	3.46 a	4.25 a	9.93 b	9.13 b
1993 Spring (1992 age class)	3.27 a	4.01 a	9.69 b	9.20 b
1993 Summer (1992 age class)	3.71 a	3.87 a	10.08 b	10.16 b
1993 Summer (1993 age class)	5.01 a	5.14 a	8.29 b	6.27 a
1993 Autumn (1993 age class)	3.48 a	4.07 a	8.16 b	7.79 b
1994 Summer (1993 age class)	3.64 a	4.21 a	7.33 b	7.79 b
1994 Summer (1994 age class)	4.80 a	5.64 a	11.81 b	11.19 b

Different letters in a row represent statistically significant differences at p = 0.05.

Table 8.3. Daily Mean, Maximum, and Minimum Air Temperatures Calculated from Within-Chamber Measurements on the Days That Gas Exchange Was Measured

	Treatments			
	Ambient CO_2		CO_2 + 330 μmol mol^{-1}	
Year, season, and foliage age class	Ambient Temp.	Ambient Temp. + 2 °C	Ambient Temp.	Ambient Temp. + 2 °C
1992 Summer (1992 age class)	26.6 (23.0, 28.7)	28.2 (25.2, 30.6)	26.5 (22.8, 28.5)	28.4 (25.6, 30.7)
1993 Spring (1992 age class)	26.0 (16.7, 29.4)	27.7 (18.3, 31.5)	26.2 (17.1, 29.3)	27.5 (17.3, 31.5)
1993 Summer (1992 age class)	29.6 (23.5, 32.9)	31.2 (25.0, 34.8)	29.6 (23.4, 33.4)	31.3 (25.2, 34.9)
1993 Summer (1993 age class)	28.4 (20.9, 32.2)	30.5 (21.8, 35.3)	28.4 (20.7, 32.5)	30.4 (22.0, 35.6)
1993 Autumn (1993 age class)	11.6 (6.3, 16.4)	13.2 (6.0, 18.4)	11.7 (5.5, 17.5)	13.2 (7.6, 17.9)
1994 Summer (1993 age class)	26.9 (21.1, 30.5)	27.7 (19.6, 31.7)	26.7 (19.9, 30.6)	27.7 (19.9, 31.9)
1994 Summer (1994 age class)	27.6 (20.1, 31.4)	28.3 (19.9, 32.9)	27.6 (20.2, 31.2)	28.2 (18.9, 33.1)

[1] Mean (min., max.) Chamber Air Temperature (°C)

Table 8.4. Mean Rates of Transpiration (E) (mmol m^{-2} s^{-1}) for Ambient and Elevated CO$_2$ and Temperature Treatments

Year, season, and foliage age class	Treatments			
	Ambient CO$_2$		CO$_2$ + 330 μmol mol^{-1}	
	Ambient Temp.	Ambient Temp. + 2 °C	Ambient Temp.	Ambient Temp. + 2 °C
1992 Summer (1992 age class)	—	—	—	—
1993 Spring (1992 age class)	8.98	8.79	9.53	10.11
1993 Summer (1992 age class)	8.23	9.48	10.12	10.48
1993 Summer (1993 age class)	6.28	9.66	6.90	7.30
1993 Autumn (1993 age class)	2.78	2.84	2.64	2.70
1994 Summer (1993 age class)	6.69	6.08	6.92	7.12
1994 Summer (1994 age class)	6.54	7.27	7.16	7.98

Table 8.5. Mean Daily Total Assimilation (mmol m^{-2} day^{-1}) for Ambient and Elevated CO_2 and Temperature Treatments

	Treatments			
	Ambient CO_2		CO_2 + 330 μmol mol^{-1}	
Year, season, and foliage age class	Ambient Temp.	Ambient Temp. + 2 °C	Ambient Temp.	Ambient Temp. + 2 °C
1992 Summer (1992 age class)	174.6	214.1	500.7	460.3
1993 Spring (1992 age class)	135.7	113.8	387.8	379.7
1993 Summer (1992 age class)	145.6	145.6	406.2	410.2
1993 Summer (1993 age class)	170.0	184.4	283.1	212.0
1993 Autumn (1993 age class)	76.1	94.1	189.7	182.6
1994 Summer (1993 age class)	131.2	169.4	282.7	309.4
1994 Summer (1994 age class)	160.2	199.4	455.6	454.3

Table 8.6. Mean Daily Total Transpiration (E_{tot}) (mmol m^{-2} day^{-1}) for Ambient and Elevated CO_2 and Temperature Treatments, calculated from twice-hourly measurements made throughout daylight hours. Values shown in this table represent the grand mean of the daily E_{tot} (n = 6), each measured on a separate day.

	Treatments			
	Ambient CO_2		CO_2 + 330 µmol mol^{-1}	
Year, season, and foliage age class	Ambient Temp.	Ambient Temp. + 2 °C	Ambient Temp.	Ambient Temp. + 2 °C
1992 Summer (1992 age class)	—	—	—	—
1993 Spring (1992 age class)	387.8	379.7	411.5	436.6
1993 Summer (1992 age class)	414.6	477.8	509.9	528.1
1993 Summer (1993 age class)	316.7	486.8	347.8	367.8
1993 Autumn (1993 age class)	100.0	102.1	95.2	8.63
1994 Summer (1993 age class)	325.1	295.5	336.3	346.0
1994 Summer (1994 age class)	317.8	353.3	348.0	387.8

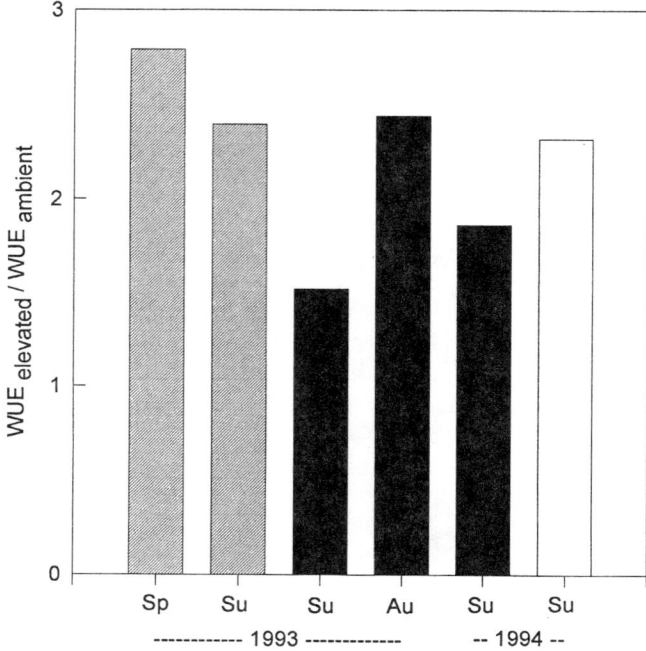

Figure 8.3. The ratio of the mean daily water use efficiency (mol CO_2 mol H_2O^{-1}) of the elevated CO_2 treatments compared with that of the ambient CO_2 treatments ($WUE_{elevated}/WUE_{ambient}$). As in Figure 8.2, the X axis represents the season and year the foliage was measured (Sp, spring; Su, summer; Au, autumn). Each set of bars represents a foliage age class. Foliage that emerged in the 1992 growing season is shown as the shaded bars. Foliage that emerged in 1993 and 1994 is represented by the solid black and white bars, respectively.

treatments. The rates of A_n and the percentage of increase in A_n resulting from elevated CO_2 levels reported here agree closely with spot measurements made on this site using a commercially available portable photosynthesis system (Liu and Teskey, 1995) as well as with measurements made in these treatments using a steady state gas exchange system (Teskey, 1995). The percentage of increase in net assimilation is also similar to that reported for other tree species including *Pinus radiata* (Conroy et al., 1990), *P. eldarica* (Garcia et al., 1994), *P. sylvestris* (Wang et al., 1995), *Picea sitchensis* (Barton et al., 1993), *Liriodendron tulipifera* and *Quercus alba* (Gunderson et al., 1993), *Fagus sylvatica* (El Kohen et al., 1993), *Populus tremuliodes* (Sharkey et al., 1991), and *Alnus rubra* (Arnone and Gordon, 1990)) but is higher than that reported in some other studies using other tree species (see review by Ceulemans and Mousseau, 1994). All field studies have reported a sustained increase in carbon assimilation in response to elevated CO_2 levels (cf Gunderson et al., 1993), as have most of the recent controlled environment studies that have adequately supplied water and nutrients (Idso and

Idso, 1994). The large increase in carbon assimilation in loblolly pine caused by increasing C_a by 330 μmol mol^{-1} was not seasonal, and did not depend on the age of the foliage. Although rates of A_n changed with environmental conditions, the ratio $A_{elevated}/A_{ambient}$ remained relatively constant and did not appear to be particularly correlated with air temperature. The best example of this is the comparison of the response in autumn 1993 (when the mean ambient air temperature was c 11 °C) with the summer measurement periods (when the mean ambient air temperatures ranged from approximately 26 to 30 °C). All of these periods have nearly equal $A_{elevated}/A_{ambient}$ ratios. This lack of response to temperature is in contrast with that reported for some other species (e.g., *Lolium perenne*, Nijs et al., 1992) in which the ratios increased with increased temperatures. For loblolly pine, the ratio did decrease in one measurement period, however. During the summer of 1993, in current-year foliage, A_n was significantly lower in the elevated CO_2-elevated temperature treatment, compared with the elevated CO_2-ambient temperature treatment. The specific reason for this response was not identified, but it occurred on days when air temperatures exceeded 35 °C in the elevated temperature treatment. But high temperatures alone cannot explain this response, because older foliage was not affected the same way under similar temperature conditions. It may be that the current-year foliage was not sufficiently acclimated to high temperatures or associated stresses.

There was a significant interaction between air temperature and CO_2 in many of the measurement periods but the effect was opposite of what would be expected based on the Farquhar et al. (1980) model of photosynthesis, which predicts that the enhancement of photosynthesis in elevated C_a will be greater when air temperature increases (Kirschbaum, 1994; Long, 1991). For example, Long (1991) estimated from model simulations that at 10 °C the maximum rate of assimilation would increase 20% when C_a increased from 350 to 650 μmol mol^{-1}. In contrast, the increase over the same range of C_a would be 105% at 35 °C. Our data are in agreement with the estimates of the effects of elevated CO_2 levels in warm conditions, but not with those made for cool conditions because the relative effect of elevated C_a was nearly identical at 11 °C and 30+ °C. Additionally, the data indicated that higher air temperatures had a slight negative effect on carbon assimilation in elevated C_a rather than a stimulating effect. Averaged across all measurement periods, increasing air temperature by 2 °C caused an increase in A_n of 14% in ambient CO_2 treatments and a decrease of 6% in ambient +330 μmol mol^{-1} treatments. In all cases, however, the effect of increasing C_a by 330 μmol mol^{-1} was much greater than the effect caused by increasing air temperature by 2 °C. Wang et al. (1995) found a very similar effect of air temperature on rates of net assimilation under elevated C_a for *Pinus sylvestris*. Callaway et al. (1994) also reported an interaction between CO_2 and temperature in *P. ponderosa*. But in that study, air temperature had a negative effect on A_n at both ambient and elevated C_a. Taken together, the results from *P. taeda, P. sylvestris,* and *P. ponderosa* indicate that a stimulation in A_n resulting from a positive interaction between air temperature and elevated CO_2 concentrations should not be expected, or at least assumed, in pines.

Transpiration rates in loblolly pine did not change significantly between ambient and ambient $+330$ µmol mol^{-1} CO_2 treatments, indicating that the stomata were not sensitive to CO_2. There appears to be a wide range of stomatal responsiveness to elevated CO_2 levels among tree species (Eamus and Jarvis, 1989; Bunce, 1992). Transpiration and stomatal conductance were 18% lower in elevated CO_2 in *Fagus sylvatica* (Overdieck and Forstreuter, 1994). Lower stomatal conductances were also reported in elevated C_a for *Picea sitchensis* seedlings (Townend, 1993), but were found to be higher in elevated CO_2 for *P. sitchensis* trees (Barton et al., 1993). But even without a change in transpiration rate there was a large increase in W_e in elevated CO_2 concentrations in loblolly pine, similar to that reported in other tree species (Townend, 1993; Overdieck and Forstreuter, 1994; Barton et al., 1993; Ceulemans and Mousseau, 1994).

Summary

In conclusion, on a site with an eroded soil of moderate fertility and a relatively low site index, the foliage of loblolly pine trees in elevated CO_2 concentrations was able to sustain high rates of carbon assimilation and to have high water use efficiencies over three growing seasons. Elevating air temperature by 2 °C had much less effect and must be considered a less important modifier of rates of carbon assimilation in comparison to the CO_2 concentration. However, air temperature may still be important in altering the total carbon available for tree growth through its effect on woody tissue respiration rates (see Chapter 18). But if the CO_2 and temperature scenarios used in this experiment are accurate, and the trees remain free to grow so that carbon assimilation is not restricted by feedback inhibition, it appears from these results that carbon assimilation of loblolly pine will be greatly enhanced on moderate-to-low fertility sites as a result of elevated CO_2 concentrations.

References

Arnone JA, Gordon JC (1990) Effect of nodulation, nitrogen fixation and CO_2 enrichment on the physiology, growth and dry mass allocation of seedings of *Alnus rubra* Bong. New Phytol 116:55–66.

Bailey RL, Grider GE, Rheney JW, Pienaar LV (1985) Stand structure and yields for site-prepared loblolly pine plantations in the piedmont and upper coastal plain of Alabama, Georgia and South Carolina. The Univ GA, Agric Exp Sta Bull 328.

Barton CVM, Lee HSJ, Jarvis PG (1993) A branch bag and CO_2 control system for long-term CO_2 enrichment of mature Sitka spruce (*Picea sitchensis* (Bong,) Carr.): Technical report. Plant Cell Environ 16:1139–1148

Bloomfield P (1992) Trends in global temperature. Clim Change 21:1–16.

Bunce JA (1992) Stomatal conductance, photosynthesis and respiration of temperate deciduous tree seedlings grown outdoors at an elevated concentration of carbon dioxide. Plant Cell Environ 15:541–549.

Callaway RM, DeLucia EH, Thomas EM, Schlesinger WH (1994) Compensatory responses of CO_2 exchange and biomass allocation and their effects on the relative growth rate of ponderosa pine in different CO_2 and temperature regimes. Oecologia 98:159–166.

Ceulemans R, Mousseau M (1994) Tansley review No. 71. Effects of elevated atmospheric CO_2 on woody plants. New Phytol 127:425–446.

Conroy JP, Milham PJ, Mazur M, Barlow EWR (1990) Influence of phosphorus deficiency on the growth response of four families of *Pinus radiata* seedlings to CO_2 enriched atmospheres. For Ecol Manage 30:175–188.

Eamus D, Jarvis PG (1989) The direct effects of increases in the global atmospheric CO_2 concentration on natural and commercial temperate trees and forests. Adv Ecol Res 19:1–55

El Kohen A, Mousseau M (1994) Interactive effects of elevated CO_2 and mineral nutrition on growth and CO_2 exchange of sweet chestnut seedlings (*Castanea sativa*). Tree Physiol 14:679–690.

El Kohen A, Venet L, Mousseau M (1993) Growth and photosynthesis of two deciduous forest tree species exposed to elevated carbon dioxide. Func Ecol 7:480–486.

Farquhar GD, von Caemmerer S, Berry JA (1980) A biochemical model of photosynthetic CO_2 assimilation in leaves of C_3 species. Planta 149:78–90.

Garcia RL, Idso SB, Wall GW, Kimball BA (1994) Changes in net photosynthesis and growth of *Pinus eldarica* seedlings in response to atmospheric CO_2 enrichment. Plant Cell Environ 17:971–978.

Gifford RM (1994) The global carbon cycle: A viewpoint on the missing sink. Aust J Plant Physiol 21:1–15.

Gunderson CA, Norby RJ, Wullschleger SD (1993) Foliar gas exchange responses of two deciduous hardwoods during 3 years of growth in elevated CO_2: no loss of photosynthetic enhancement. Plant Cell Environ 16:797–807.

Guehl JM, Picon C, Aussenac G, Gross P (1994) Interactive effects of elevated CO_2 and soil drought on growth and transpiration efficiency and its determinants in two European forest tree species. Tree Physiol 14:707–724.

Idso KE, Idso SB (1994) Plant responses to atmospheric CO_2 enrichment in the face of environmental constraints: a review of the past 10 years' research. Agric For Met 69:153–203.

Kirschbaum MUF (1994) The sensitivity of C_3 photosynthesis to increasing CO_2 concentration: a theoretical analysis of its dependence on temperature and background CO_2 concentration. Plant Cell Environ 17:747–754.

Lewis JD, Griffin KL, Thomas RB, Strain BR (1994) Phosphorus supply affects the photosynthetic capacity of loblolly pine grown in elevated carbon dioxide. Tree Physiol 14:1229–1244.

Liu S, Teskey RO (1995) Responses of foliar gas exchange to long-term elevated CO_2 concentrations in mature loblolly pine trees. Tree Physiol 15:351–359.

Long SP (1991) Modification of the response of photosynthetic productivity to rising temperature by atmospheric CO_2 concentrations: Has its importance been underestimated? Plant Cell Environ 14:729–739.

Nijs I, Impens I, Van Hecke P (1992) Diurnal changes in the response of canopy photosynthetic rate to elevated CO_2 in a coupled temperature-light environment. Photosyn Res 32:121–130.

Overdieck D, Forstreuter M (1994) Evapotranspiration of beech stands and transpiration of beech leaves subject to atmospheric CO_2 enrichment. Tree Physiol 14:997–1003.

Sharkey TD, Loreto F, Delwiche CF (1991) High carbon dioxide and sun/shade effects on isoprene emission from oak and aspen tree leaves. Plant Cell Environ 14:333–338.

Teskey RO (1995) A field study of the effects of elevated CO_2 on carbon assimilation, stomatal conductance and leaf and branch growth of *Pinus taeda* trees. Plant Cell Environ 18:565–573.

Teskey RO, Dougherty PM, Wiselogel AE (1991) Design and performance of branch chambers suitable for long-term fumigation of foliage in large trees. J Environ Qual 20:591–595.

Townend J (1993) Effects of elevated carbon dioxide and drought on the growth and physiology of clonal Sitka spruce plants (*Picea sitchensis* (Bong.) Carr.). Tree Physiol 13:389–399.

Tschaplinski TJ, Norby RJ, Wullschleger SD (1993) Responses of loblolly pine seedlings to elevated CO_2 and fluctuating water supply. Tree Physiol 13:283–296.

Wang K, Kellomäki S, Laitinen K (1995) Effects of needle age, long-term temperature and CO_2 treatments on the photosynthesis of Scots pine. Tree Physiol 15:211–218.

Webber AN, Nie G-Y, Long SP (1994) Acclimation of photosynthetic proteins to rising atmospheric CO_2. Photosyn Res 39:413–425.

Ziska L, Bunce JA (1994) Increasing growth temperature reduces the stimulatory effect of elevated CO_2 on photosynthesis or biomass in two perennial species. Physiol Plant 91:183–190.

9. An Investigation of the Impacts of Elevated Carbon Dioxide, Irrigation, and Fertilization on the Physiology and Growth of Loblolly Pine

Phillip M. Dougherty, H. Lee Allen, Lance W. Kress, Ramesh Murthy, Chris A. Maier, Timothy J. Albaugh, and D. Arthur Sampson

Southern pine forests that are dominated by loblolly pine (*Pinus taeda* L.) are the most intensively managed forests in the United States. They provide more than 50% of the total softwood being harvested annually in the United States and represent the first or second most economically important agricultural crops in nine of the twelve southeastern states (U.S. Department Agriculture Forest Service, 1988). Thus, any changes in environmental conditions that will alter productivity of these forests will have important ecological, economical, and sociological consequences. Over the past several decades, the environment of southeastern forests has been changing. Increases in acidic deposition (SO_4 and NO_x), nitrogen inputs (Husar, 1986), atmospheric CO_2 concentration (Conway et al., 1988; Keeling et al., 1989), and tropospheric ozone have all been documented to parallel the increase in population since the beginning of the industrial revolution. Climate change has also been predicted for the southeastern United States for the future. Each of these atmospheric and climatic elements that are being altered by human activities has the potential to affect productivity of southern pine forests. Nutrient availability, water availability, atmospheric CO_2 concentration, and temperature are presently the principal factors that are limiting the productivity of southern pine forests. Thus, it is extremely important that we understand how changes in these factors will interact to affect physiological processes of forest stands. This chapter summarizes our results to date concerning the effects of CO_2, irrigation, and fertilization on budbreak and photosynthesis and the effects of irrigation and fertilization on leaf area, volume growth, biomass production, and biomass partitioning. In addition to these results, our efforts to scale up the effects of elevated CO_2 levels to stand-level estimates of gross primary production (GPP) and net primary productivition (NPP) using an adaptation of the ecophysiological model BIOMASS are reviewed.

Methods and Materials

Site Description

The research site is located in the Sandhills physiographic province in Scotland County, NC at 34.55 °N and 79.30 °W on lands owned by Bowater, Inc. Annual air temperature averages 17 °C. Air temperatures in the summer (June to September) and winter (December to March) average 26 °C and 9 °C, respectively. Average annual rainfall is 1210 mm, with periods of drought often occurring in late summer and early autumn.

The soil is a deep sand belonging to the Wakulla series (sandy, siliceous, thermic psammentic hapludult). Soil drainage is classified as excessively well-drained and available water-holding capacity in the upper two meters of the soil profile is 18 to 20 cm. Pretreatment foliar nitrogen (N) concentration in the dormant season (December to February) was 0.98%. This concentration represents an extreme in N deficiency for loblolly pine growing in the southeastern United States, being well below the critical concentration established for obtaining an economical response to fertilization (Allen, 1987) and the optimum foliar nitrogen concentration of 1.4% for loblolly pine.

After harvesting a 65-year-old stand of longleaf pine (*Pinus palustris* Mill), the site was machine planted in March of 1985 at a spacing of 2.4 × 2.4 m with a ten half-sibling family mix of North Carolina piedmont seedlings. During 1992, the stand was thinned to a density of 1260 trees per hectare to standardize the number of trees in each treatment plot. After thinning, average height was 3.4 m and average bole diameter at 1.3 m height was 4.6 cm. Associated understory vegetation (hardwoods and grasses) was controlled with repeated applications of 1.5% glyphosphate (by volume) in water during the 1992 and 1993 growing seasons.

The experimental design was a 2 × 2 × 2 factorial combination of fertilization, irrigation, and CO_2 treatments imposed as a split-plot. The four main plot treatments included 1) control (C), 2) irrigation only (I), 3) fertilization only (F), 4) irrigation and fertilization (IF). These main plot treatments were randomly assigned to one of four treatment plots in each of four blocks. Each treatment plot was 50 × 50 m with an interior measurement plot of 30 × 30 m.

Fertilization treatments consisted of an initial application of nitrogen (N: 200 kg ha^{-1}), phosphorous (P: 50 kg ha^{-1}), and potassium (K: 100 kg ha^{-1}) in March of 1992. This was followed by an application of calcium (Ca: 120 kg ha^{-1}), magnesium (Mg: 50 kg ha^{-1}) and boron (B: 1.5 kg ha^{-1}) in the April to June period of 1992. Nitrogen, P, K, and Mg (23, 20, 19 and 0.16 kg ha^{-1}, respectively) were foliar applied in April of 1993, followed by K, S, and Mg applications (82, 107 and 50 kg ha^{-1}) in June to August of 1993. Additional applications of N were made in August of 1993 (50 kg ha^{-1}) and in March of 1994 (100 kg ha^{-1}).

Irrigation was initiated in May of 1993 and continued until November of 1995. Available soil-water percentage was calculated as

(observed available soil water/maximum total available soil water)*100.

The target of the irrigation treatment was to irrigate to field capacity when 40% of the maximum total available soil water was depleted from the upper 50 cm of the soil profile. Water was applied to treatment plots by a head-to-head sprinkler system with nozzles located below the canopy. Soil water of the irrigated plots was assessed every two days using the time-domain reflectometry (TDR) technique during the growing season to determine the need for irrigation. Additionally, soil water content at depths of 10, 25, and 50 cm were measured every two weeks on all plots.

Subplot CO_2 treatments consisted of exposing midcrown branches to ambient, ambient + 175, and ambient + 350 μmol mol^{-1} CO_2. The CO_2 treatments were randomly assigned to three branches of the 1989 or 1990 whorl of a single tree in each plot (sixteen trees total). CO_2 exposures were accomplished using the branch chamber technology developed by Teskey et al. (1991). Exposures ran for twenty-four hours per day starting in March of 1993 and continuing until January of 1995. Additional details of the exposure protocol are presented in Murthy et al. (1996).

Results

Budbreak

Budbreak date of the sixteen midcrown branches subjected to the three CO_2 concentrations and four fertilization and irrigation treatments was assessed in 1993, 1994, and 1995. Budbreak date ranged from early March to late March. Budbreak date was not affected by CO_2, but occurred earlier (up to two weeks) on trees receiving irrigation or fertilization.

Photosynthesis Rates

Detailed assessments by Murthy et al. (1996) over a fifteen-month period indicated that light-saturated net photosynthesis (Amax) was strongly affected by foliar age, as well as CO_2, irrigation, and fertilization. Foliage exhibited its peak Amax in July of its first growing season, which was the youngest age at which assessments were made (Figure 9.1A). Photosynthetic potential remained high throughout the first summer but declined rapidly in the fall and winter period to a minimum level in January. This decline in fall and winter Amax was highly correlated with a decline in maximum needle conductance to water vapor (Gmax; Figure 9.1B) and soil temperature. In the fall and winter period, Gmax was only 42% of the summer period Gmax values, even though soil-moisture supply was

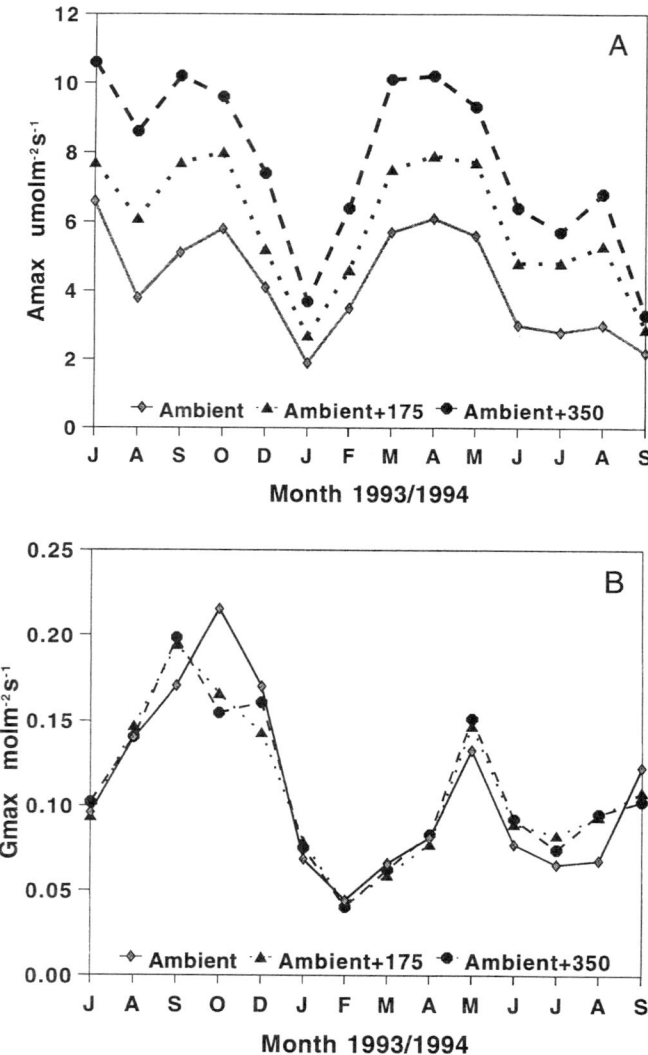

Figure 9.1. Trends in average monthly light-saturated net photosynthesis (Amax) (A); and stomatal conductance (Gmax) (B) for loblolly pine foliage initiated in 1993 and exposed to three concentrations of CO_2. Exposures began in March of 1993 and continued throughout the study period.

high and vapor-pressure deficits were low during the fall and winter period (Murthy et al., 1996). In the next spring (March to May), the Amax of mature needles increased to the peak level observed in the year prior. After May, Amax rates again declined, and by September, just prior to senescence, Amax rates were only 35% of those values observed in April. The decline in Amax during the

9. Impacts of Elevated Carbon Dioxide, Irrigation, and Fertilization   153

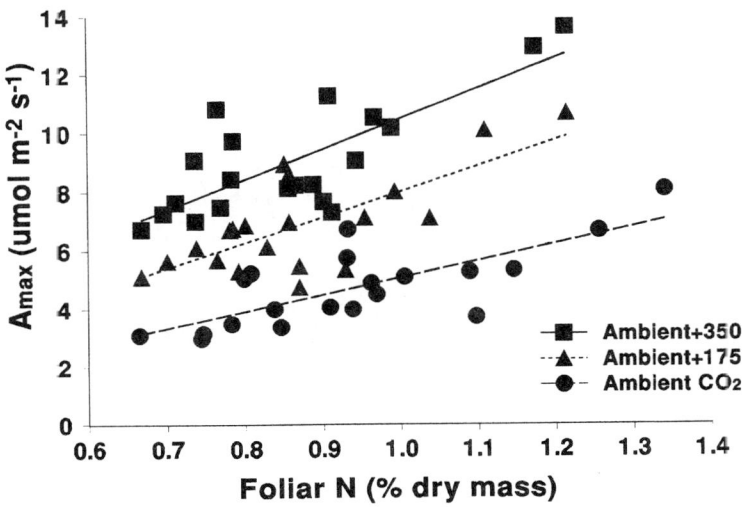

Figure 9.2. Relationships between of light-saturated net photosynthesis (Amax) and foliage nitrogen concentration (N) for loblolly pine foliage grown at three concentrations of CO_2.

summer of the foliage's second year was highly correlated (r = 0.8) with a decline in foliar N concentration (Murthy, 1995; Figure 9.2). However, it must be noted that Gmax also showed a similar decline during this period.

Increasing the CO_2 concentration by 177 μmol mol^{-1} or 350 μmol mol^{-1} over the ambient CO_2 concentration significantly increased Amax for the entire exposure period (Figure 9.3). Over the life of the 1993 foliage Amax increases averaged 40% and 86% for the +175 and +350 treatments, respectively. Similar increases in Amax were observed by Liu and Teskey (1995) for 21-year-old loblolly pine. No reduction of Amax was observed based on A-Ci curves developed for both current year (C) and the year prior (C+1) foliage, which is also in agreement with Liu and Teskey (1995). Our research and that of Liu and Teskey (1995) also indicated that increased CO_2 did not significantly reduce Amax through stomatal closure. In contrast, however, Ellsworth et al. (1995), using the Free-Air Carbon dioxide Enrichment (FACE) exposure system, found that stomatal closure was directly reduced by approximately 5% at elevated levels of CO_2 and that sap-flux density reduction was less than or equal to 7%. They also observed no long-term (80 days) adjustments in stomatal conductance to elevated CO_2 levels. Irrigation significantly (α = .05) increased Amax during drought periods (four months in 1993 and one month in 1994) (Figure 9.3B). Water stress may have reduced carbon gain during other periods of the year; however, because Amax was assessed only in the early morning hours, which is the period of the lowest possible water stress, such effects would not have been detected.

Fertilization significantly increased Amax by an average of over 20% during

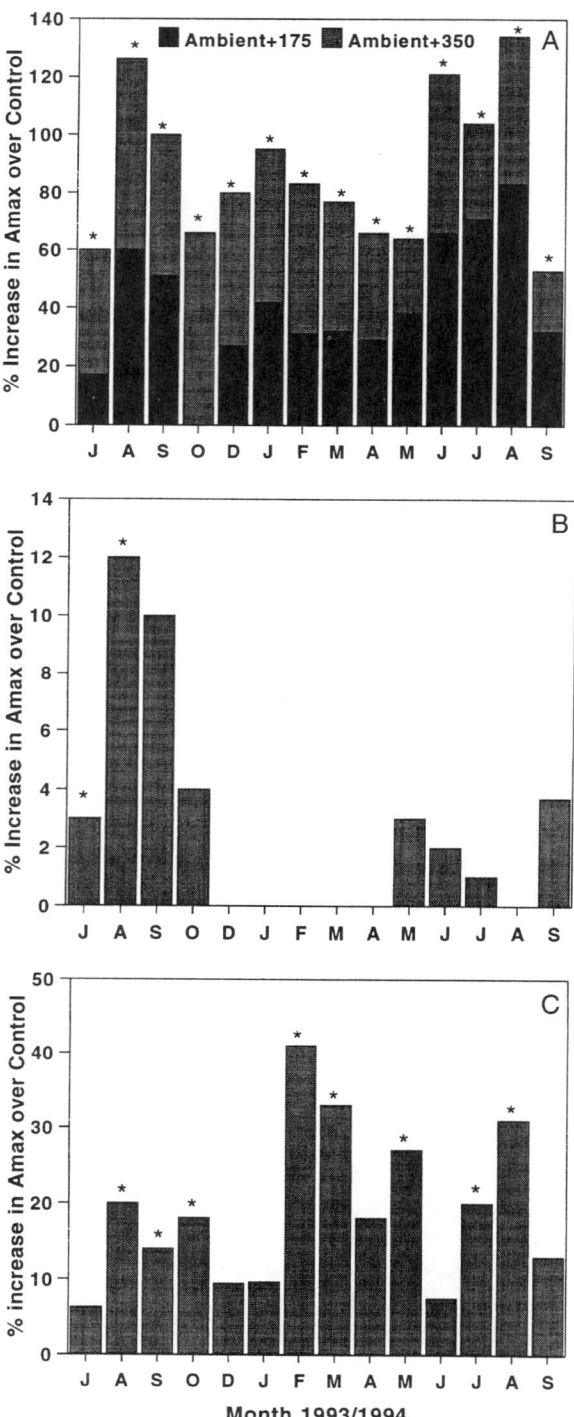

Figure 9.3. Average percent increase in light-saturated net photosynthesis of loblolly pine foliage resulting from elevated CO_2 (A); irrigation (B); and fertilization (C). The percent increase in Amax is statistically significant at the p = .05 level if it is designated with an *.

the study period (Figure 9.3C). This relative increase in Amax was similar to the relative increase observed in foliage N concentration resulting from the fertilization treatment. Because of the strong positive effects of fertilization on both Amax and leaf area (Allen et al., 1996), potential canopy carbon gain could be dramatically increased. An evaluation of the effects of fertilization and irrigation on both shoot-leaf area and Amax demonstrated that carbon gain in midcrown shoots in irrigated and fertilized plots would be as much as 70% greater than that of shoots in control plots. Our results and those of Liu and Teskey (1995) indicated that 1) the large response in Amax to a doubling of atmospheric CO_2 was much larger than the response observed to the dramatic improvement in site resources induced by the irrigation or fertilization treatments, and 2) the increase in Amax to CO_2 occurred throughout the life span of the foliage and was not dependent on the supply of water or nutrients.

Woody Tissue Maintenance Respiration

Dormant seasons (November through February) maintenance respiration rates per unit size (i.e., surface area, $\mu mol\ CO_2\ m^{-2}\ s^{-1}$; sapwood volume, $\mu mol\ CO_2\ m^{-3}\ s^{-1}$; sapwood dry weight, $nmol\ CO_2\ g^{-1}s^{-1}$), varied with tissue size, but was constant with respect to tissue nitrogen content ($\mu mol\ CO_2\ mole\ N^{-1}\ s^{-1}$) for bole and branch tissue (Maier et al., 1997). Cambium temperature accounted for 61 and 77% of the variation in bole and branch maintenance respiration, respectively (Figure 9.4). Irrigation and fertilization did not significantly alter the Q_{10} values. However, fertilization did increase the basal respiration rate (maintenance respiration at 0 °C; Figure 9.4). Improved nutrition more than doubled bole basal respiration rates and increased branch basal respiration by 38%. Basal respiration rates were linearly correlated with tissue nitrogen content. These results suggest adjusting for tissue-N status can improve those estimates of carbon losses from woody tissue substantially.

Leaf Area and Stemwood Production

Both peak leaf area and annual stemwood volume growth on C plots increased for the first three years after study initiation because of increased site occupancy. Peak leaf area increased from 0.65 to 1.25 $m^2\ m^{-2}$ and annual volume growth from 4.6 to 7.8 $m^3\ ha^{-1}\ yr^{-1}$. Irrigation did not significantly affect peak leaf area in any of the years after irrigation was initiated. However, irrigation did significantly increase annual volume growth by almost 30% in 1993 and 1995, the two years with extended droughts. In contrast, peak leaf area and annual stemwood volume growth were dramatically increased in each of the four years (1992 to 1995) since fertilization was initiated (Allen et al., 1996; Figure 9.5). Nutrient additions increased peak leaf area by 54%, 65%, 82%, and 100% as compared to nonfertilized plots in 1992, 1993, 1994, and 1995, respectively (Figure 9.5A). Treatment effects on annual stemwood volume growth followed a similar trend with increases of 52%, 109%, 120%, and 152% on fertilized as compared to nonfertilized plots (Figure 9.5B).

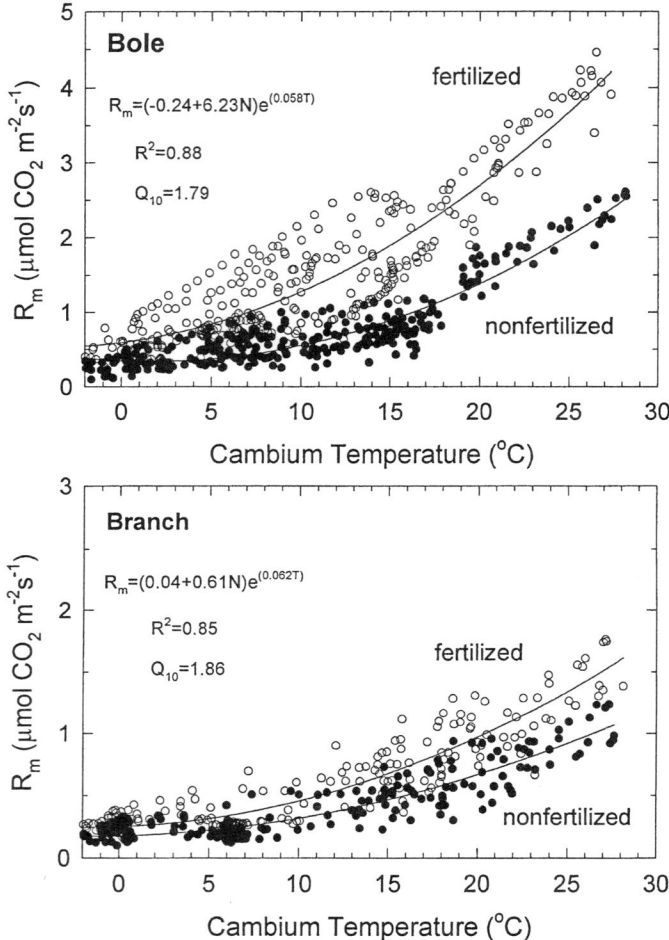

Figure 9.4. Relationships between maintenance respiration and cambium temperature for boles and branches of loblolly pine trees that are either fertilized or not fertilized.

Biomass Production and Partitioning

Detailed assessments of aboveground and belowground biomass production were undertaken in 1993 (Mignano, 1995). Both irrigation and fertilization significantly increased foliage, branch, stemwood, coarse root, and total biomass production and reduced fine root production (Figure 9.6), although irrigation had a much smaller effect on production than fertilization. As a result of the opposite effects of increasing resource availability on fine root production as compared to production of other biomass components, the partitioning of the total 1993 bio-

9. Impacts of Elevated Carbon Dioxide, Irrigation, and Fertilization 157

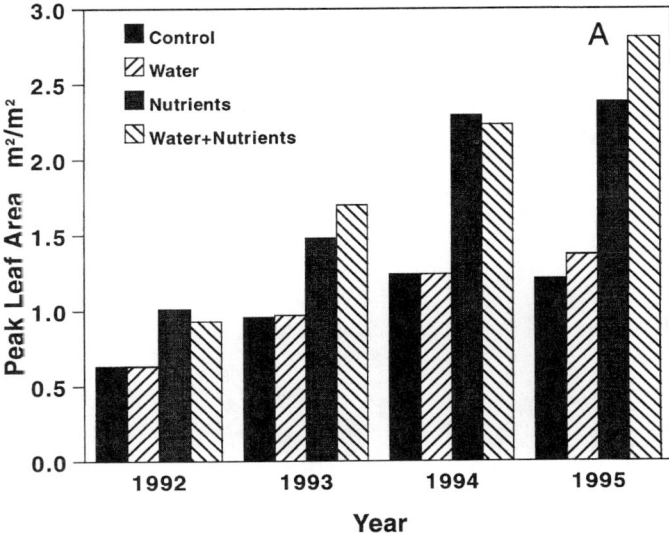

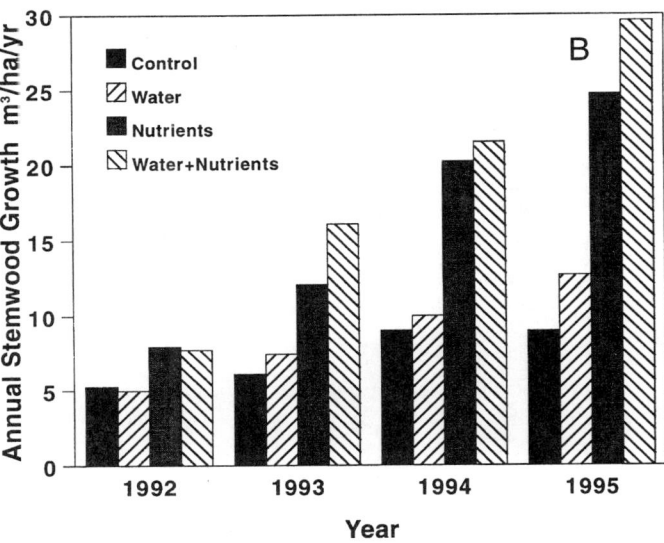

Figure 9.5. Effects of water and nutrient availability on peak leaf area (A) and annual stemwood volume (B) growth for loblolly pine for each of the four years after treatment initiation.

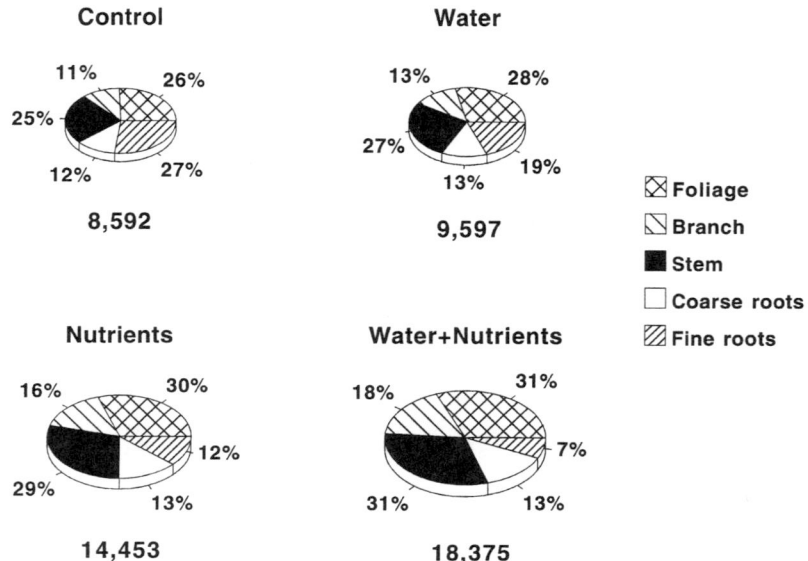

Figure 9.6. Effects of water and nutrient availability on the partitioning of loblolly pine biomass production (kg ha^{-1}) in 1993.

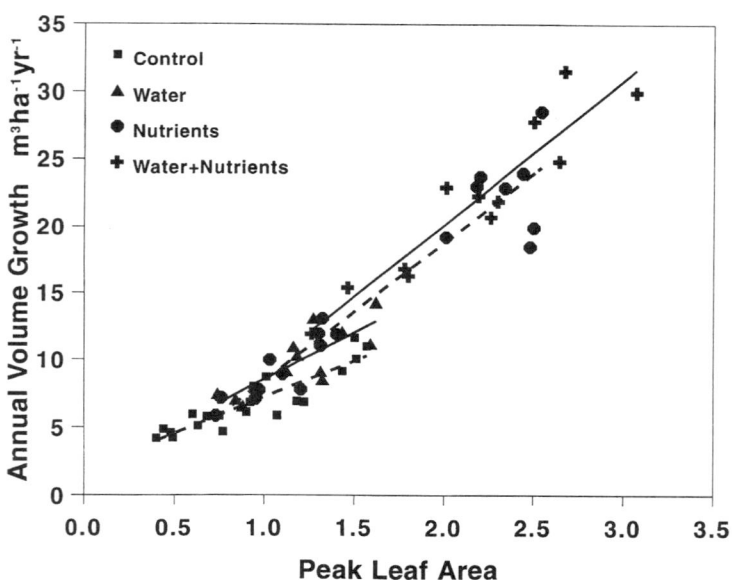

Figure 9.7. Relationships between annual stemwood volume growth and peak leaf area from 1992 through 1994 for the four water and nutrient availability treatments.

mass production to fine roots decreased from 27% on C plots to 7% on IF plots (Figure 9.6). However, the relative partitioning of biomass production for the aboveground components (foliage, branch, stem) was not affected by treatment.

When data from the four treatments were combined for the three years, the simple linear relationship between annual stemwood growth and peak leaf area was positive and highly significant (Figure 9.7). In fact, 93% of the variation in annual volume growth was accounted for by the variation in peak leaf area. An additional 2% of the variation in volume growth was accounted for by the significant positive effect that improved water and nutrient availability had on growth efficiency (the slope parameter). Growth efficiency increased from 7.2 m^3 ha^{-1} per unit of leaf area on C plots to 9.2 m^3 ha^{-1} per unit of leaf area on IF plots, representing a 28% increase. Growth efficiency was increased by 20% and 8%, respectively, by the addition of nutrients and water. Because no whole tree exposures to CO_2 were made within the fertilized and irrigated treatments, no assessment of the combined influence of these factors on leaf area and productivity could be made directly. However, the responses of leaf area to the fertilization and irrigation treatments and the responses of Amax and Gmax to elevated carbon dioxide levels were incorporated into BIOMASS, a process model, in order to estimate the combined effects of changes in water, nutrient, and carbon dioxide supplies.

Estimates of Stand-Level Carbon Gain Under Doubled Atmospheric Carbon Dioxide

Although doubling atmospheric CO_2 concentration increased light-saturated net carbon gain by 86% over a foliage cohort's life (Murthy et al., 1996), whole canopy carbon gain is expected to be much less because much of a stand's foliage is not exposed to light-saturated conditions. To address this issue, an adaptation of the BIOMASS 13.0 model, developed and parameterized for loblolly pine stands of the southeastern United States (Sampson and Dougherty, Chapter 21) was used to estimate the effects that a doubling of CO_2 would have on GPP for control and fertilized plots. Model estimates indicated that GPP would be increased by only 28% and 25%, respectively, for the C and F plots (Figure 9.8) indicating a substantial amount of the stand's foliage is below light saturation. Clearly, accounting for light extinction within the canopy when estimating stand-level responses to changes in photosynthesis rates caused by increases in the level of CO_2 or other resources will be very important. No attempt was made to adjust light extinction coefficients in the model for any effects that CO_2 or nutrients may have had on foliage density and distribution and, therefore, light extinction.

The impact of a doubling in CO_2 on NPP depends on both the effects of increased CO_2 on carbon gain and respiration. Model estimates indicated that doubling CO_2 would increase NPP by 48% on C plots and 49% on F plots (Figure 9.9). A greater increase in NPP as compared to GPP would be realized with

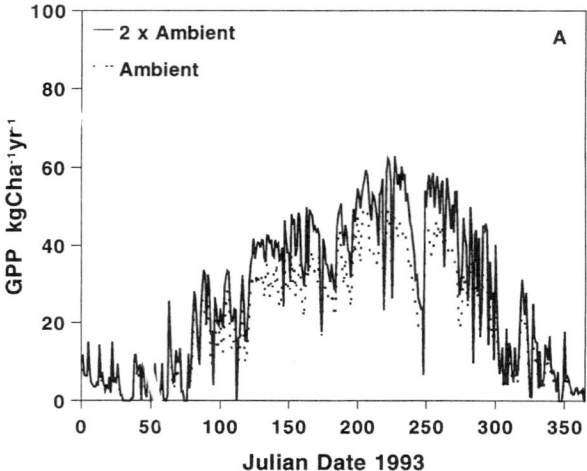

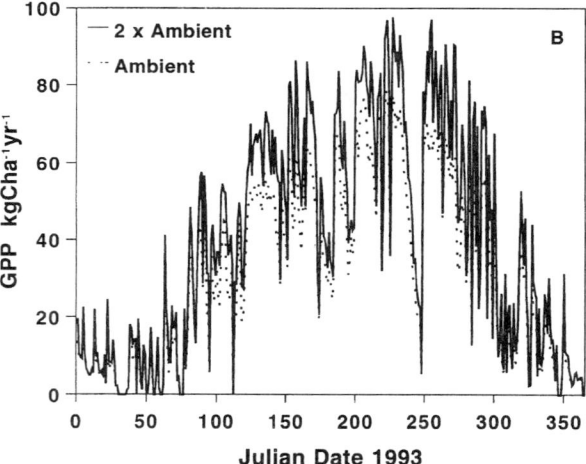

Figure 9.8. Daily trends in gross primary productivity (GPP) predicted for control (A); and fertilized (B) plots under ambient and 2× ambient concentrations of CO_2, using BIOMASS 13.0 parameterized using physiological responses and stand structural properties.

increasing CO_2 because most of the absolute gain in GPP will be allocated to NPP. Absolute respiration costs, particularly maintenance respiration, would not be expected to increase much with increased CO_2; with loblolly pine there is evidence that dark respiration is decreased with enhanced atmospheric CO_2 concentration (Teskey, 1995).

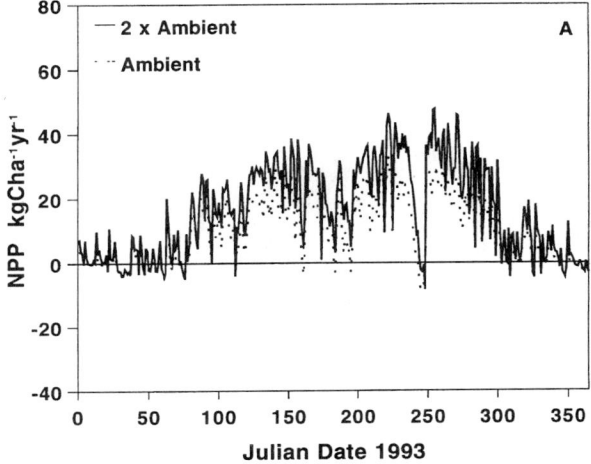

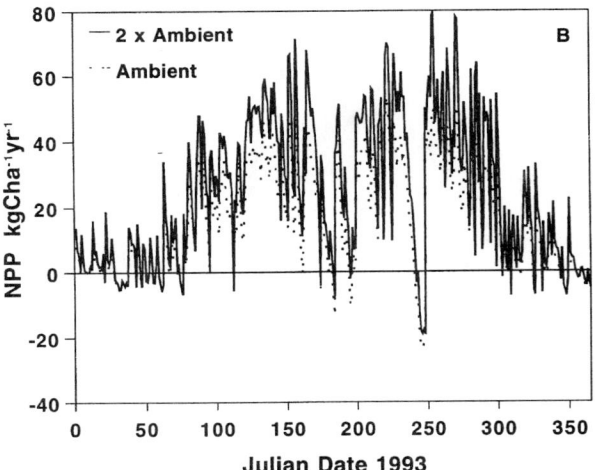

Figure 9.9. Daily trends in net primary productivity (NPP) predicted for control (A); and fertilized (B) plots under ambient and 2× ambient concentrations of CO_2 using BIOMASS 13.0 parameterized using physiological responses and stand structural properties.

Discussion

Amax Responses to CO_2

The results indicate that increased atmospheric CO_2 levels have the potential to almost double photosynthesis under light-saturated conditions (Figure 9.1). This increase in Amax was obtained whether the trees were growing in the C, I, F, or IF

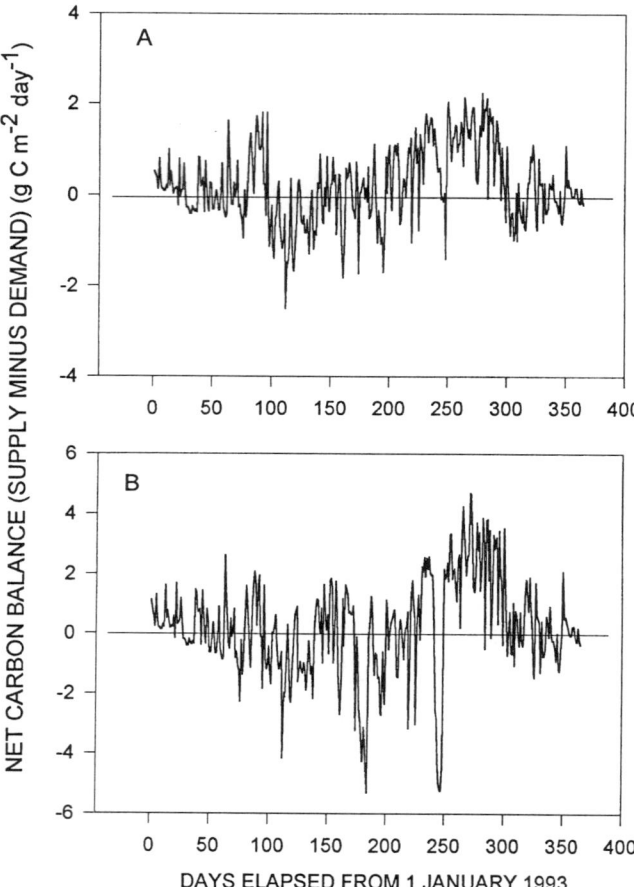

Figure 9.10. Net carbon balance (supply minus demand) for control (A); and fertilized (B) plots for nine-year-old loblolly pine stands. Carbon supply represents net carbon assimilate simulated using BIOMASS version 14.0. Carbon demand was estimated from empirical data.

treatment, which suggests that the carbohydrate production capacity of loblolly pine growing on all site classes will probably be improved with elevated levels of CO_2. The response in Amax was maintained throughout the life of the foliage and during all seasons (Figure 9.1). This has important implications for such coniferous species as loblolly pine living in temperate ecosystems where significant winter season photosynthesis does occur (Strain et al., 1976; Murthy et al., 1996). Winter photosynthesis of loblolly pine appears to play a vital role in determining growth and survival of loblolly pine. Seasonal carbon budgets developed for our site (Figures 9.9 and 9.10) and by Horne (1993) for branches indicate that stored carbon is depleted during summer to meet growth and maintenance respiration requirements. The fall and winter appear to serve as an important period for storage recharge. Based on our results, winter carbon gain should increase with elevated atmospheric CO_2 levels. It is even more probable if winter temperatures also increase as predicted because present-day winter temperatures are below the optimum for loblolly pine for much of the winter period (Strain et al., 1976). Also, decreasing leaf area index (LAI) resulting from approximately 50% of a stand's maximum leaf area being shed in the fall and winter period should increase light penetration into the lower canopy and differentially increase the response of Amax of lower canopy foliage to elevated CO_2 levels.

There are uncertainties about extrapolating the measured responses to increased levels of CO_2 on isolated branches to the whole tree or stand-level responses. Two major concerns about using CO_2-response data obtained from branch chamber studies are 1) the possible lack of "feed back inhibition" and thus the overestimation of assimilation (A) when only branches are exposed to elevated levels of CO_2, and 2) scaling branch-level responses observed under optimum light and temperature conditions to higher spatial levels.

Feedback Inhibition

Concerns about "feedback inhibition" occur when isolated branches are exposed to elevated CO_2 levels and the rest of the tree crown is not. These concerns have both a classical biochemical end-product feedback inhibition component and a "system" feedback inhibition component. If the carbohydrate production capacity of all branches in a tree's crown were enhanced as a result of exposure to elevated CO_2 concentration, there would be more probability that carbohydrate sinks and transport pathways might become "saturated" than if only a single branch within a tree crown was exposed to elevated levels of CO_2. Thus, end-product feedback inhibition might occur under whole tree exposure conditions and not under branch-chamber conditions. In this case, the expected Amax responses observed with branch chambers would be higher than those obtained from whole tree exposure studies. Interestingly, the Amax response to increased levels of CO_2 (65%) obtained by Ellsworth et al. (1995) for loblolly pine trees using the FACE system was greater than the 40% increase that we obtained or the 53% that Liu and Teskey (1995) obtained with branch chambers. However, it must be remembered that the study by Ellsworth et al. (1995) was only conducted for 80 days.

Concerns about the validity of using branch chambers to estimate Amax responses to enhanced CO_2 concentrations are based on the assumption that feedback inhibition occurs in loblolly pine trees; however, no evidence for feedback inhibition in loblolly pine trees has ever been reported. An assumption has also been made that Amax rates are being controlled by sink strength (growth, maintenance, storage, and so forth). Both evidence for feedback inhibition (Tissue et al., 1993; Thomas et al., 1994) and sink-activity limitations (Tissue et al., 1996) on A have been reported for loblolly pine seedlings. That the percentage increases in Amax to elevated CO_2 levels were similar across the range of fertilization and irrigation treatments, which resulted in extreme differences in growth and maintenance sink strengths (Figure 9.6), in addition to the lack of evidence of reduction in any of the main effect treatments do not provide evidence of feedback inhibition in loblolly pine trees. The apparent lack of feedback inhibition, even in trees that are extremely deficient in N, should not be surprising. In temperate ecosystems, the largest carbohydrate sink for trees is maintenance respiration. Our modeling results suggest a net depletion of storage reserves for much of the year even for very N-limited conditions (Figure 9.10), largely because of the high maintenance respiration cost. The maintenance respiration sink is expected to increase in the future as temperature increases. Based on our respiration-temperature response functions (Figure 9.4), we estimate that a 4 °C increase would increase maintenance respiration by 26% annually (Maier et al., 1997). Thus, in temperate ecosystems where maintenance respiration is the major sink for carbohydrates for trees (Kinerson, 1977), concerns about feedback inhibition may not be a large issue.

Concerns regarding "system-level feedback inhibition" are based on the assumption that with increased carbohydrate production, the demand for such resources as nutrients will be so increased that the site's capacity to supply those needed resources to assimilate the additional carbohydrates produced under doubled CO_2 conditions will be exceeded. Thus, the potential for increased growth resulting from greater carbohydrate production capacity would not be realized because of resource limitations. It is possible, however, that as carbohydrate production capacity of foliage is increased, carbohydrates could be allocated away from foliage, thus "freeing up" or reducing the use of nutrients for foliage construction. More of the available annual nutrient pool could then be allocated to woody tissue production without changing the annual utilization of nutrients. That is, with increasing levels of CO_2, leaf area could be less and GPP greater than today (i.e., growth efficiency would increase). While evidence for direct feedback inhibition of the photosynthesis does not exist for loblolly pine trees, evidence is provided for reduced photosynthetic capacity resulting from insufficient nutrient supply. Amax decreased linearly with decreasing foliar nitrogen concentration (Figure 9.3) and peak leaf area was also much less with low nutrient availability (Figure 9.5). Although growth and maintenance respiration sinks were reduced under low nutrient conditions, there was no indication that the internal carbohydrate status of any tissue component was higher on the C treatment than for the F or IF treatments. Thus, there is little probability of carbohydrate saturation in trees growing under poor or good resource conditions. This is consistent with the observed increases in Amax to enhanced CO_2 concentrations for all treatments, although the absolute increases in Amax rates were different.

Scaling

The second major concern with branch-chamber studies is how to scale the results obtained at the branch level in relation to higher spatial levels. Crown architecture, leaf area, leaf area distribution, and tree spacing all affect light attenuation through the canopy and will have a great influence on the amount of the light saturated "CO_2-fertilization effect" realized on a whole crown or canopy basis. In our simulations, substantially reduced CO_2 responses were predicted for GPP and

NPP than were found for Amax assuming no change in quantum efficiency with increasing CO_2 levels. If quantum efficiency does increase, our estimates of whole canopy gain in response to a doubling of the levels of CO_2 will be low. This aspect is presently under investigation.

The scaling issue is extremely important for branch studies but will be equally important for whole tree exposures and even for the FACE system in which physiological responses are measured at the branch or leaf level and then have to be scaled to the whole tree or stand level. Additionally, our research was conducted only at one stage of stand development. Changes in crown attributes with age, site quality, stand density, and such cultural treatments as thinning and fertilization should also be considered before physiological responses to CO_2 can be accurately scaled to the stand level across the region (Baldwin et al., 1997, see Chapter 17). Our research places some bounds, however, on the range of response that nutrients and water can have on leaf area and growth.

Water and Nutrient Availability Effects

Increased nutrient availability had strong positive effects on both the leaf area and growth efficiency components of production. Similar to prior reports, the effect of improved nutrition on leaf area was found to be greater than its effect on growth efficiency (51% and 28%, respectively). Because of the multiplicative effect of these two components, production was dramatically increased with improved nutrition. Apparently, the gains in growth efficiency with increased nutrient availability have resulted in part from proportionately greater partitioning of biomass production from fine roots to aboveground components. In addition to these factors, the increased rates of Amax with improved nutrition could also have contributed to the increase in growth efficiency.

One of the most surprising findings was the dramatic positive response to nutrition additions on what is typically thought to be a site principally limited by water. Clearly, stand-volume growth at the site is limited primarily by nutrients and secondarily by water. The imposition of an optimum nutrition treatment, rather than N alone, was probably responsible for the strong response to nutrient addition. This finding brings into question our present-day prescription technology that would rank such sites as this poor candidates for fertilization. Apparently, the addition of water alone did not elicit a leaf area response because there was not sufficient nutrients to produce more leaves. Similarly, Gillespie et al. (1994) reported that nutrient limitations—rather than light limitations—were responsible for low leaf area production in lower and mild crown positions of loblolly pine. The potential for a "complete" nutrition treatment to greatly enhance growth on what have been considered droughty sites is intriguing. However, we also have to wonder what the risks are in increasing leaf area and reducing fine roots on such sites.

Although the positive effects of improved resource availability on aboveground vs belowground partitioning of biomass production have been observed in many seedling studies, such effects have been postulated, but rarely tested at the stand level (Cannell, 1985; Gower et al., 1992; Haynes and Gower, 1995; Linder, 1987).

The absolute and proportional decrease in fine root production is a dramatic illustration of loblolly pine's plastic response to environmental stress. Are other species as plastic? It is also interesting to note that although the absolute and relative partitioning of biomass to foliage and stem production increased with improved resource availability, the ratio of stem production to foliage production remained unchanged. Apparently, the major effects of resource availability on partitioning resulted from aboveground vs belowground tradeoffs, rather than tradeoffs among aboveground components.

Nutrition has major effects on leaf area, biomass allocation, and yield; also modest effects on such physiological processes as respiration and net photosynthesis rates. In contrast, CO_2 concentration appears to have a major effect on Amax and a yet-to-be-determined effect on leaf area. In one study in which leaf area responses were assessed, Liu and Teskey (1995) reported a 30% increase in leaf area in the ambient +350, as compared to the ambient CO_2 treatment in which Amax was increased by 111%. No interaction between nutrition and CO_2 occurred for physiological processes. Thus, the effects of nutrition and CO_2 appear additive. At the stand level, however, it is expected that a significant interaction of nutrition and CO_2 on stand GPP does exist because GPP is the product of leaf area and average canopy net photosynthesis rates.

Summary

The results from our study clearly show that estimating GPP, NPP, and yield of loblolly pine forests under doubled atmospheric CO_2 concentrations will require consideration of the expected changes in forest nutrient status and the "CO_2 fertilization effect." Just as CO_2 is increasing, it must also be recognized that human-influenced N inputs into forest ecosystems are also increasing as population increases (Husar, 1986; Vitousek, 1994). Vitousek (1994) reported that human-influenced nitrogen additions now exceed those that are fixed naturally. Additionally, Vitousek pointed out that the changes in atmospheric CO_2 concentration and N inputs are more important components of anthropogenic global change than changes in climate. The combined effects of all components of global change must be considered in concert to evaluate the future productivity of our forests. Integrated studies that consider the interactive effects of site resources on carbon gain and allocation continue to be essential for providing the data necessary for developing an understanding of how loblolly pine forest will respond to future atmospheric and climatic conditions.

References

Allen HL (1987) Forest fertilizers. J For 87(2):37–45.
Allen HL, Albaugh, T and Dougherty TM (1995) The influence of nutrient and water availability on leaf area and productivity: A case study with Loblolly Pine (*Pinus taeda* L.) in the Southeastern U.S. Proceeding's 25th Congress of the Brazilian Soil Science Soc, Federal University of Vicosa, Minais Gerais, Brazil.

Cannell MGR (1985) Dry matter partitioning in tree crops. In Cannell MGR and Jackson JE (Eds) *Attributes of trees as crop plants*. 160–193.

Conway TJ, Tans P, Waterman LS, Thoning KW, Masarie KA, Gammon RM (1988) Atmospheric carbon dioxide measurements in the remote global troposphere. Tellus 40:81–115.

Ellsworth DS, Oren R, Huang C, Phillips N, and Hendrey GR (1995) Leaf and canopy responses to elevated CO_2 in a pine forest under free-air CO_2 enrichment. Oecologia 104:139–146.

Gillespie AR, Allen HL, and Vose JM (1994) Amount and vertical distribution of foliage of young loblolly pine trees as affected by canopy position and silviculture treatment. Can J For Res 24:1337–1344.

Gower ST, Vogt KA, and Grier CC (1992) Carbon dynamics of Rocky Mountain Douglas-fir: Influence of water and nutrient availability. Ecol Monogr 62:43–66.

Haynes BE and Gower ST (1995) Belowground carbon allocation in unfertilized and fertilized red pine plantations in northern Wisconsin. Tree Physiol 15:317–325.

Horne AM (1993) The effects of shade on growth and source-sink relations in branches of loblolly pine (*Pinus taeda* L.). PhD dissertation, Dept For, Yale Univ.

Husar RB (1986) Emissions of sulfur dioxide and nitrogen oxides and trends for eastern North America. In Gibson J (Ed) *Acid deposition long-term trends*. National Academy Press, Washington DC. 48–92.

Kinerson RS, Ralston CW, and Wells CG (1977) Carbon cycling in a loblolly pine plantation. Oecologia 29:1–10.

Linder S (1987) Responses to water and nutrients in coniferous ecosystems. In Schulze ED and Wolfer HZ (Eds) *Potentials and Limitations of Ecosystem Analysis*. Springer-Verlag, NY. 180–202.

Liu S and Teskey RO (1995) Responses of foliar gas exchange to long-term elevated CO_2 concentrations in mature loblolly pine trees. Tree Physiol 15:351–359.

Maier CA, Zarnoch SJ and Dougherty PM (1998) Modeling the effects of temperature and tissue nitrogen on dormant season stem and branch maintenance respiration in a young loblolly pine (*Pinus taeda* L.) plantation. Tree Physiol 18:11–20.

Mignano J (1995) The effects of water and nutrient availability on root biomass, necromass and production in a nine-year-old Loblolly pine plantation. MS thesis, Dept For, NC State Univ.

Murthy R (1995) Effects of CO_2, nutrients, and water on the physiology of Loblolly pine. PhD dissertation, Dept For, NC State Univ.

Murthy R, Dougherty PM, Zarnoch SJ, and Allen HL (1996) Effects of carbon dioxide, fertilization, and irrigation on photosynthetic capacity of loblolly pine trees. Tree Physiol 16:537–546.

Strain BR, Higginbotham KO, and Mulroy JC (1976) Temperature preconditioning and photosynthetic capacity of *Pinus taeda* L. Photosynthetica 10(1):47–53.

Teskey RO (1995) A field study of the effects of elevated CO_2 on carbon assimilation, stomatal conductance and leaf branch growth of *Pinus taeda* trees. Plant Cell Environ. 18:565–573.

Teskey RO, Dougherty PM, and Wiselogel AE (1991) Design and performance of branch chambers suitable for longterm ozone fumigation of foliage of large trees. J Environ Qual 20(3):591–595.

Tissue DT, Thomas RB, and Strain BR (1993) Long-term effects of elevated CO_2 and nutrients on photosynthesis and rubisco in loblolly pine seedlings. Plant Cell Environ 16:859–865.

Tissue DT, Thomas RB, and Strain BR (1996) Growth and photosynthesis of loblolly pine (*Pinus taeda*) after exposure to elevated CO_2 for 19 months in the field. Tree Physiol 16:49–59.

Thomas RB, Lewis JD, and Strain BR (1994) Effects of leaf nutrient status on photosynthetic capacity in loblolly pine (*Pinus taeda* L.) seedlings grown in elevated atmospheric CO_2. Tree Physiol 14:947–960.

Vitousek PM (1994) Beyond global warming: Ecology and global change. Ecol 75:1861–1876.

U.S. Department of Agriculture Forest Service (1988) The south's fourth forest: Alternatives for the future. Forest Resources Report No 24. Washington DC.

10. Effects of Elevated Carbon Dioxide, Water, and Nutrients on Photosynthesis, Stomatal Conductance, and Total Chlorophyll Content of Young Loblolly Pine (*Pinus taeda L.*) Trees

Thomas C. Hennessey and Venkatesh K. Harinath

Global atmospheric carbon dioxide concentration, presently at about 350 $\mu l\, l^{-1}$, is expected to continue to increase in the future (Lindzen, 1993) and may double by the end of the next century (Gates, 1983; Keeling et al., 1989; Houghton and Woodwell, 1989). Higher levels of carbon dioxide may increase the growth rate of trees and the productivity of forests (Teskey, 1995). At the present-day, ambient carbon dioxide concentration and under optimal conditions, the photosynthesis of plants is limited by the supply of carbon dioxide (Arp, 1991). Numerous studies have shown increased plant growth in elevated levels of carbon dioxide (Higginbotham et al., 1983, 1985; Stewart and Hoddinott, 1993; Gunderson et al., 1993), but these studies have used potted seedlings, optimum levels of resources (including water and nutrients), and short-term exposure to higher carbon dioxide concentrations. Because seedlings differ from older trees both physiologically and morphologically (Cregg et al., 1989), it is unknown how much carbon gain in trees will be affected by long-term exposures to elevated carbon dioxide levels under field conditions in which water and nutrient availability may limit growth.

Prior work with mature trees has shown increased carbon assimilation in elevated carbon dioxide (Teskey, 1995; Liu and Teskey, 1995; Barton et al., 1993), but in these experiments water and nutrient availability were not manipulated. The availability of nutrients has been shown to alter the potential carbon gain in woody plants at increased concentrations of carbon dioxide (Conroy et al., 1986; Tissue et al., 1993). However, even under nutrient deficient conditions, accelerated tree growth occurred in response to carbon dioxide enrichment (Gunderson et al.,

1993; Norby and O'Neill, 1989; Norby et al., 1986). Drought has also been shown to reduce expected gains in photosynthesis as a result of elevated carbon dioxide (Guehl et al., 1994; Tschaplinski et al., 1993), although Tolley and Strain (1985) found similar rates of photosynthesis in loblolly pine seedlings grown in enriched levels of carbon dioxide for both well-watered and drought treatments.

Because forests are distributed extensively over the world, fluxes of carbon dioxide and water vapor from trees and forests are major components that must be quantified if reliable models are to be developed to predict the effects of increased carbon dioxide on global climate. The role of terrestrial ecosystems as a sink for rising concentrations of carbon dioxide is unknown (Tans et al., 1990).

Our research was conducted in southeastern Oklahoma, which represents the northwestern edge of the natural range of loblolly pine (*Pinus taeda* L.) in the United States. The climate is typically hotter and dryer than that found across most of the range of loblolly pine. Miller et al. (1987) have suggested the western edge of the loblolly pine range is a critical area in which climate change may produce detrimental effects on productivity leading to reductions in the distribution of loblolly pine forests. The objective of this study was to determine the extent that nitrogen and water availability alter branch carbon and water exchange of loblolly pine trees in response to elevated carbon dioxide.

Materials and Methods

Site Description

The study site is located in southeastern Oklahoma near the town of Antlers. The average annual precipitation is 120 cm, with drought periods commonly during the summer months and early fall. The average annual temperature is 17 °C. Mean daytime summer and winter temperatures are 33.6 °C and 13.3 °C, respectively. The total number of growing days is 200.

The soil at the study site is deep, loamy, fine sand belonging to the Glenpool series, which is described as sandy, siliceous, thermic psammentic paleudalf. The soil has poor water-holding capacity and low nutrient availability, which, as reflected in the site index of a nearby stand (14.9 m at age 25), is low for the region (Woods et al., 1988).

The site was planted in 1990 with a mixture of unimproved Arkansas–Oklahoma families of loblolly pine (*Pinus taeda* L.). During 1994, at age four, the average height was 2.1 m and the average ground line diameter was 6.2 cm. The density of the stand was 498 trees per hectare. The understory herbaceous vegetation was controlled using a mixture of Roundup® (glyphosate) and Oust® (sulfometuron methyl), and hardwoods were controlled using Garlon® (triclopyr) as a basal spray.

Study Design and Layout

The study design was a 2 × 2 factorial split-plot with a combination of irrigation and fertilization treatments. The study included two levels of irrigation and two levels of fertilization. The four main plot treatment combinations were 1) control (C—no irrigation and no fertilization), 2) irrigated (I—irrigation only), 3) fertilized (F—fertilization only), and 4) fertilized and irrigated (IF—both irrigation and fertilization). The treatment combinations were established as a randomized, complete block design. The four treatment combinations were assigned at random to the four treatment plots in one block and replicated in the other three blocks. The treatment plots were 50 × 50 m and the measurement plots within the treatment plots were 30 × 30 m. Fertilizer was applied in April at the rate of 200 kg/ha of nitrogen An additional application of fertilizer, based on foliar nutrient analysis, was made in August, and consisted of 200 kg/ha of nitrogen, 50 kg/ha of phosphorus, 100 kg/ha of potassium, 120 kg/ha of calcium, 50 kg/ha of magnesium, and 1.5 kg/ha of boron.

Irrigation was initiated in July. Irrigated plots were watered with a sprinkler irrigation system to maintain soil-water potentials above -0.05 MPa, as indicated by tensiometers placed near each tree. The subplot treatments were three levels of carbon dioxide, which were 1) ambient CO_2 (350 $\mu l\, l^{-1}$), 2) ambient CO_2 + 175 $\mu l\, l^{-1}$ (525 $\mu l\, l^{-1}$), and 3) ambient CO_2 + 350 $\mu l\, l^{-1}$ (700 $\mu l\, l^{-1}$).

A single tree was selected from each of the sixteen treatment plots and the carbon dioxide treatments were individually assigned at random to three branches on each tree. The carbon dioxide fumigation began in April. The branches were exposed to the different levels of carbon dioxide for twenty-four hours per day throughout the study period using branch-chamber technology. Branch chambers have been shown to satisfactorily allow the fumigation of tissue of mature trees while maintaining adequate control of the microenvironment within the chamber and thereby permitting treatment effects to be distinguished (Teskey et al., 1991). Thus, long-term manipulative studies on large trees can be easily conducted using branch chambers. The chambers consisted of a cylindrical aluminum frame (0.5 × 1.5 m) covered by a clear polyvinyl plastic film. Air flow through each chamber was supplied by a blower that provided ten air exchanges per minute to minimize heat gain within the chamber. The distribution and sampling of carbon dioxide to each of the chambers was accomplished by using a computer-based control system. A data logger (Keithley 500A, Keithley Inc.) was connected to the computer. This data logger controlled the opening and closing of the solenoid valves that directed the sample air coming from the branch chambers sequentially to the infrared gas analyzer (LI-6262). The infrared gas analyzer measured the carbon dioxide and water vapor concentration in the sample air from each branch chamber. All of the forty-eight branch chambers were sampled within thirty minutes. Elevated concentrations of carbon dioxide were dispensed to selected branch chambers using a mass-flow controller connected to flow meters (one per chamber). Blowers mixed the known amount of carbon dioxide with ambient air and this mixture was then delivered to each branch chamber.

The light intensity and temperature were measured inside each branch chamber. The photosynthetic photon flux density was measured using photodiodes (G1118, Hamamatsu Corp.). The light sensors, located above and below the sample branch, were individually calibrated against a quantum sensor (LI-190SA). The temperature was measured using a 0.8 mm diameter copper-constantan thermocouple. The output from each of these sensors was measured and averaged over each hour and stored in data loggers (CR-7, Campbell Scientific Inc.). An on-site weather station measured the ambient weather conditions.

Physiological Measurements

Photosynthesis, stomatal conductance to water vapor, and needle total chlorophyll content were determined once a month at the first flush of the current-year foliage (1994) on branches within each chamber. A portable infrared gas analyzer (PS-301, CID Inc.) was used for obtaining photosynthesis and stomatal conductance measurements. Photosynthesis and stomatal conductance of needles were measured under saturated photosynthetic photon flux density. A source of artificial light (CI-301LA) was used during measurements. The light-saturated rate of photosynthesis (P_{max}) is an index of photosynthetic capacity; the stomatal conductance can be defined as the maximum stomatal conductance to water vapor (G_{max}). The P_{max} and G_{max} measurements were obtained at the carbon dioxide concentrations similar to the fumigation concentrations. Using a standard carbon dioxide gas, the PS-301 was calibrated before each measurement day. The P_{max} and G_{max} measurements were obtained for the months of July, August, and September. During September, measurements were obtained between 0700 and 1230 Central Standard Time. During July and August, measurements were obtained between 0100 and 0800 Central Standard Time. The latter times of measurement were chosen to reduce the temperature and water stress. Prior to the gas exchange measurements made in July and August, the needles were exposed to a source of artificial light for a period of about ninety minutes using tungsten-halogen lamps (Osram Corp.) placed in the branch chambers. This was done to ensure that the needles would more quickly reach equilibrium with the high light intensity (> 1200 μmol m^{-2}s^{-1}) provided during gas exchange measurements. Three fascicles (nine needles) were enclosed in the leaf chamber during gas exchange measurements. The equation used to determine the total needle surface area was

$$A \text{ (cm}^2) = 2RFL (N + \pi).$$

In this equation, R is the average radius of the fascicles, F is the number of fascicles, L is the total fascicle or average fascicle length, and N is the number of needles per fascicle (Bingham, 1983). The radius was measured using a magnifying glass. After the gas exchange measurements were completed, the needles were harvested and, using the acetone method, the chlorophyll content was determined (Arnon, 1949).

Statistical Analysis

The main effects of carbon dioxide, water, and nutrient fluxes on the P_{max}, the G_{max} and the total chlorophyll content were analyzed using the standard split-plot analysis. The SAS® PROC GLM procedure was used for analysis with Fischer's least significant difference (LSD) being utilized for separation of the means of the dependent variables. All the analyses were interpreted at the P = 0.05 probability level. The data sets collected in July, August, and September were combined during analysis because the results of the analyses had a similar pattern during each month.

Results

Environment

The amount of precipitation received during the year was greater than the thirty-year average for the region, and precipitation received during the months of July and August was much greater than the thirty-year average for the region (Figure 10.1). Pan evaporation exceeded precipitation in the months of June, August, and September (Figure 10.2). Within the treatment chambers, air temperatures were only slightly higher than ambient temperatures. Mean daily air temperatures within the chambers differed from the outside temperature by an average of 1.4 °C over the year (Figure 10.3). Irradiance inside the chamber was reduced by the plastic film. During the week of June 26 to July 2, 1994, the average photosynthetic photon flux density above and below the branch was about 70% and 37%, respectively, of the ambient average photosynthetic photon flux density (Figure 10.4). The performance of the chambers indicates that target values of carbon dioxide concentration were well maintained during the 1994 growing season (Figure 10.5).

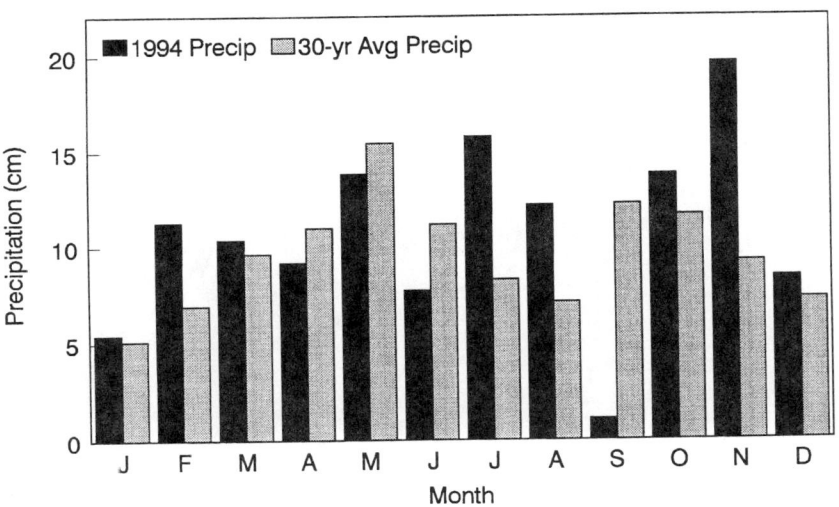

Figure 10.1. Comparison of monthly precipitation at the Antlers, OK, study site in 1994 and thirty-year average monthly precipitation.

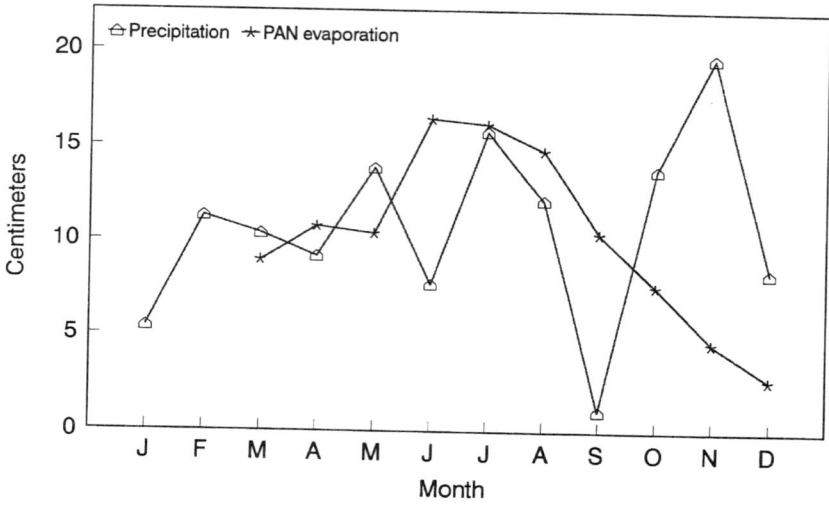

Figure 10.2. Mean monthly precipitation and pan evaporation at the Antlers, OK, study site in 1994.

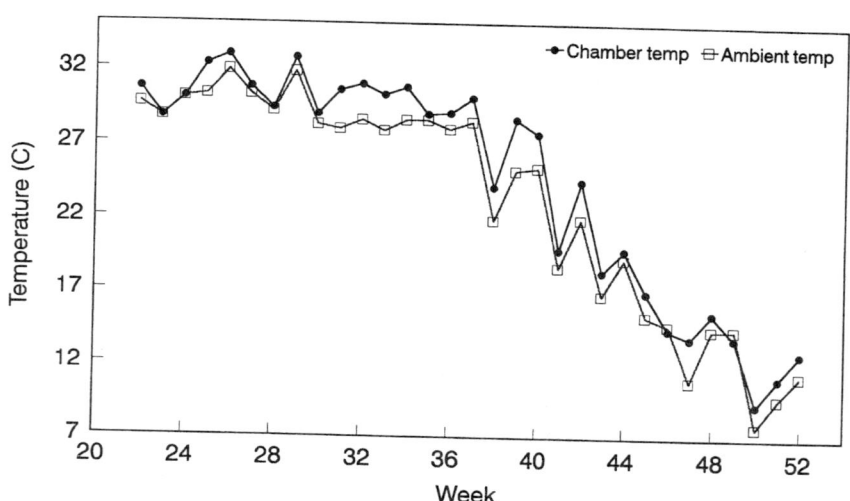

Figure 10.3. Comparison of mean ambient weekly daytime temperature with those inside of the branch chambers in 1994. Chamber temperature is the mean of forty-eight branch chambers.

10. Effects of Elevated Carbon Dioxide, Water, and Nutrients 175

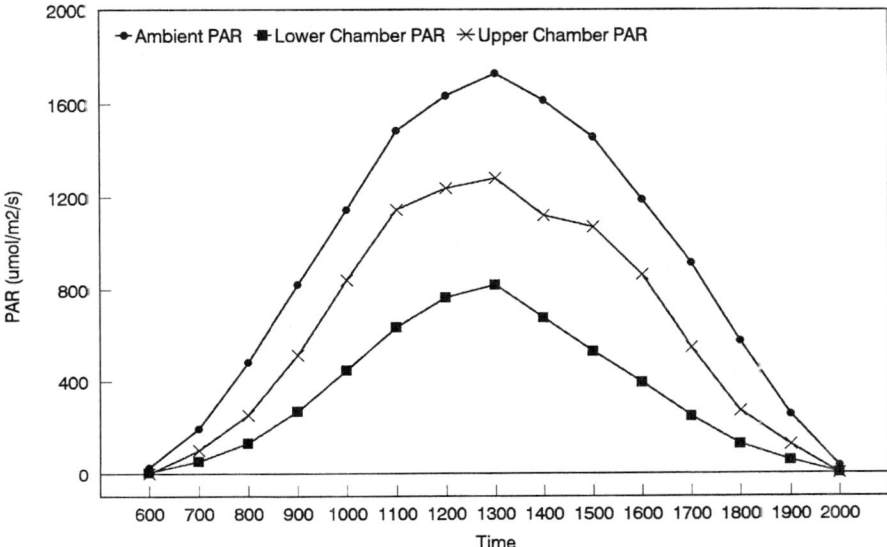

Figure 10.4. Comparison of hourly ambient photosynthetically active radiation (PAR) with PAR inside the chamber as measured above and below the sample branch. Data were collected between June 26 and July 2, 1994.

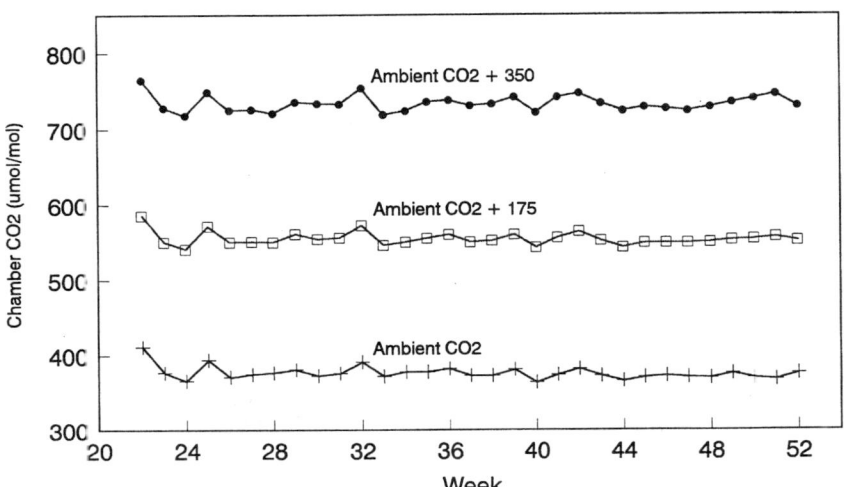

Figure 10.5. Trends in carbon dioxide concentration within branch chambers. Elevated carbon dioxide was applied as fixed additions to ambient concentrations. Each point represents the mean of sixteen branch chambers.

Light-Saturated Rate of Photosynthesis

The results from the split-plot analysis indicated that the branches growing under elevated carbon dioxide concentration had a substantially higher P_{max} compared to the branches growing under ambient carbon dioxide concentration (Table 10.1; Figure 10.6). Irrigation and fertilization did not have any effect on the P_{max} of the 700 $\mu l\ l^{-1}$ treated branches was higher than the P_{max} of 525 and 350 $\mu l\ l^{-1}$ treated branches. In the C, I, and F plots the responses were similar that is, P_{max} of the 350 $\mu l\ l^{-1}$ treated branches was significantly lower than the 525 and 700 $\mu l\ l^{-1}$ treated branches and the P_{max} of the 525 $\mu l\ l^{-1}$ treated branches was not significantly different from the 700 $\mu l\ l^{-1}$ treated branches. However, for the IF plots, the P_{max} of the 350, 525, and 700 $\mu l\ l^{-1}$ treated branches were varied significantly from each other (Figure 10.6). When averaged across the main plot treatments, the P_{max} was 3.1, 5.2, and 5.9 $\mu mol\ m^{-2} s^{-1}$ for the 350 $\mu l\ l^{-1}$, 525 $\mu l\ l^{-1}$ and 700 $\mu l\ l^{-1}$ treatments, respectively. In other words, the P_{max} was about 67% and 90% greater for the 525 $\mu l\ l^{-1}$ and 700 $\mu l\ l^{-1}$ treated branches, respectively, compared to the 350 $\mu l\ l^{-1}$ treated branches. There were no significant two-way or three-way interactions for irrigation, fertilization, and carbon dioxide (Table 10.1).

Maximum Stomatal Conductance to Water Vapor

The results from the split-plot analysis indicated that the G_{max} was significantly affected by carbon dioxide concentration. Irrigation and fertilization did not have any effect on the G_{max} (Table 10.1). In general, the branches growing under 525 $\mu l\ l^{-1}$ carbon dioxide concentration had a substantially higher G_{max} compared to the 350 $\mu l\ l^{-1}$ and 700 $\mu l\ l^{-1}$ carbon dioxide concentration (Figure 10.7). The G_{max} of the branches treated with 700 $\mu l\ l^{-1}$ carbon dioxide in the FI plots was significantly lower than the branches treated with either the 350 or the 525 $\mu l\ l^{-1}$ carbon dioxide concentration. When averaged across the main plot treatments, the G_{max} was 79.0, 96.0, and 72.0 $mmol\ m^{-2} s^{-1}$ for the 350 $\mu l\ l^{-1}$, 525 $\mu l\ l^{-1}$, and 700 $\mu l\ l^{-1}$ treated branches, respectively. The G_{max} at 525 $\mu l\ l^{-1}$ was significantly different from the 350 and 700 $\mu l\ l^{-1}$ carbon dioxide treated branches. The G_{max} did not vary between the 350 $\mu l\ l^{-1}$ and 700 $\mu l\ l^{-1}$ carbon dioxide treated branches and there were no significant two-way or three-way interactions for irrigation, fertilization and carbon dioxide (Table 10.1).

Total Chlorophyll Content

The results from the split-plot analysis indicated that total chlorophyll content was significantly affected by carbon dioxide concentration. Irrigation and fertilization did not have any effect on the total chlorophyll content (Table 10.1). In general, the total chlorophyll content decreased as the carbon dioxide concentration was increased. In the I plot, the total chlorophyll content of the branches grown at 350 $\mu l\ l^{-1}$ was significantly different from the branches grown at 700 $\mu l\ l^{-1}$ (Figure 10.8). When averaged across the main plot treatments, the total chlorophyll

10. Effects of Elevated Carbon Dioxide, Water, and Nutrients

Table 10.1. P-Values Obtained from the Split-Plot Analysis of Net Photosynthesis, Maximum Stomatal Conductance and Total Chlorophyll Content for the Data Collected in the Months of July, August and September. The Data Collected Was Combined for Analysis.

	Source of Variation		
	Net Photosynthesis	Stomatal Conductance	Total Chlorophyll
Irrigation	0.6213	0.5817	0.9931
Fertilizer	0.5202	0.7458	0.0667
Irrigation × Fertilizer	0.7747	0.8776	0.4029
CO_2	0.0001	0.0033	0.0048
Irrigation × CO_2	0.8837	0.6612	0.4547
Fertilizer × CO_2	0.4088	0.7581	0.2838
Irrigation × Fertilizer × CO_2	0.9594	0.4457	0.7790

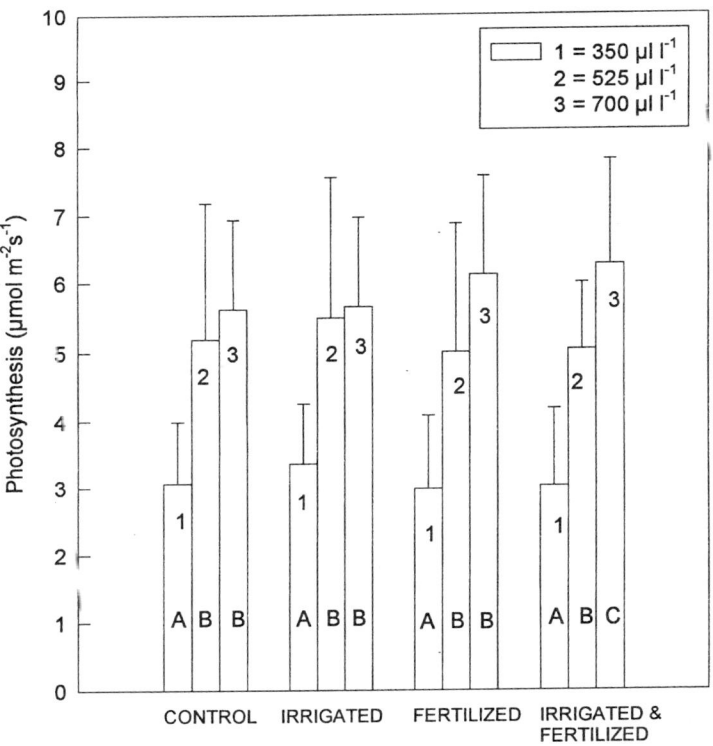

Figure 10.6. Light-saturated rate of photosynthesis (P_{max}) at 350, 525, and 750 $\mu l\ l^{-1}$ carbon dioxide concentration for C, I, F, and IF treatments. Each bar represents the mean of data collected in July, August, and September. Nitrogen varies from 3 to 4. Error bars are the standard errors of the mean. Letters A, B, and C indicate statistical significance. Similar letters on the bar graph (within each treatment) indicate they are not statistically different from each other.

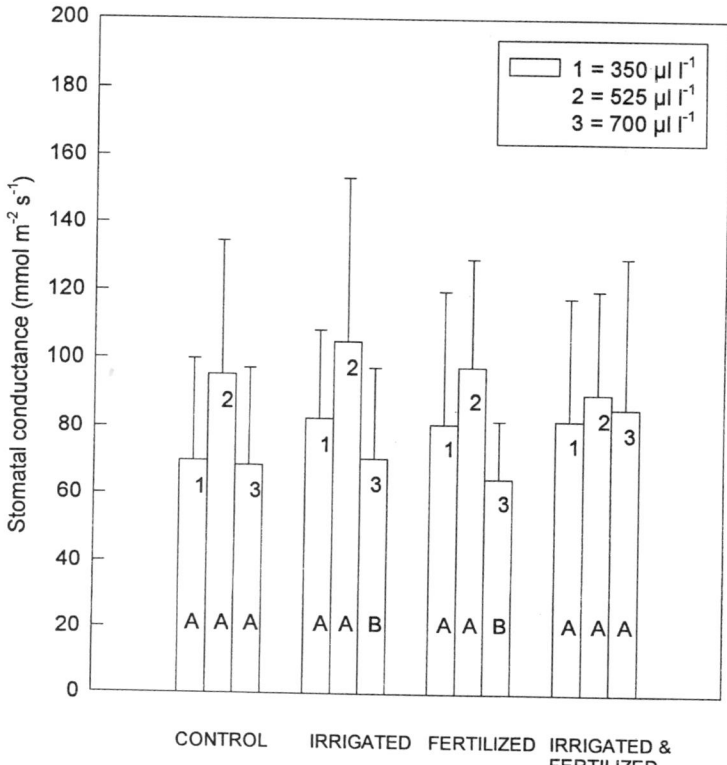

Figure 10.7. Maximum stomatal conductance to water vapor (G_{max}) at 350, 525, and 750 µl l^{-1} carbon dioxide concentration for C, I, F, and IF treatments. Each bar represents the mean of data collected in July, August, and September. Nitrogen varies from 3 to 4. Error bars are the standard errors of the mean. Letters A, B, and C indicate statistical significance. Similar letters on the bar graph (within each treatment) indicate they are not statistically different from each other.

content was 349.8, 324.3, and 304.4 mg/m² for the 350 µl l^{-1}, 525 µl l^{-1}, and 700 µl l^{-1} treated branches, respectively. The total chlorophyll content at 350 µl l^{-1} was significantly different from the 525 and 700 µl l^{-1} carbon dioxide treated branches. The total chlorophyll content did not vary between the 525 µl l^{-1} and 700 µl l^{-1} carbon dioxide treated branches and there were no significant two-way or three-way interactions for irrigation, fertilization, and carbon dioxide (Table 10.1).

Discussion

The P_{max} was increased from 67% to 90% on exposure to elevated carbon dioxide concentrations. The percentage increases in the P_{max} are within the range of

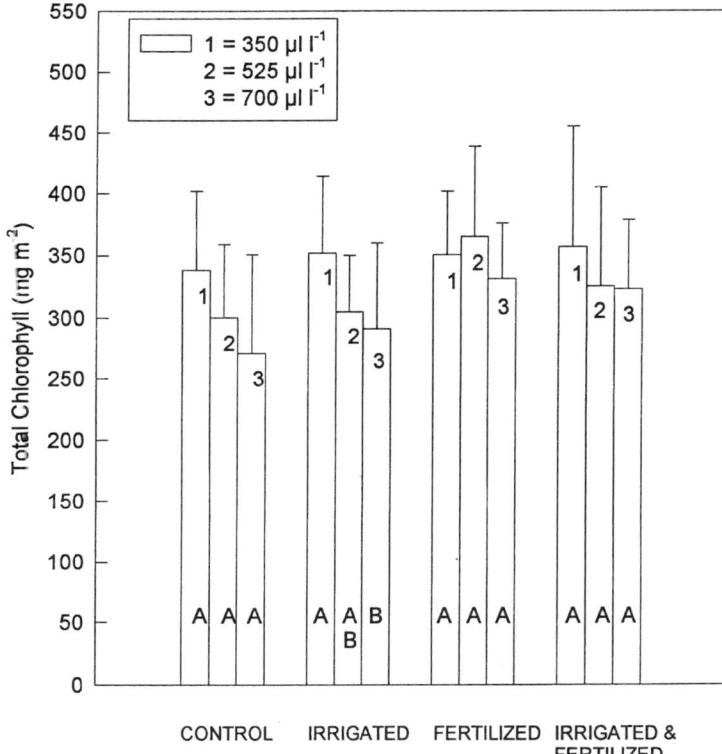

Figure 10.8. Total chlorophyll content in needles fumigated with 350, 525, and 750 μl l^{-1} carbon dioxide concentration for C, I, F, and IF treatments. Each bar represents the mean of data collected in July, August, and September. N varies from 3 to 4. Error bars are the standard errors of the mean. Letters A, B, and C indicate statistical significance. Similar letters on the bar graph (within each treatment) indicate they are not statistically different from each other.

values that have been obtained in other studies on loblolly pine (Teskey, 1995; Liu and Teskey, 1995; Murthy et al., 1995). Studies on other tree species such as *Pinus sylvestris* L. (Wang et al., 1995), *Pinus eldarica* (Garcia et al., 1994), *Castanea sativa* (El Kohen and Mousseau, 1994), three Australian tree species (Idso and Kimball, 1993), *Castanea sativa* (Mousseau, 1993), *Betula pendula* Roth (Evans et al., 1993), *Populus grandidentata* Michx. (Curtis and Teeri, 1992), *Abies fraseri* Poir. (Samuelson and Seiler, 1992), *Pinus radiata* D. Don (Conroy et al., 1988), and *Picea glauca* (Moench) Voss (Higginbotham, 1983) have reported similar types of responses. Kimball et al. (1993), Rogers and Dahlman (1993), Poorter (1993), and Drake (1992) have reported similar types of responses in crop and range species. The highest values of P_{max} obtained in this study were somewhat lower than those found for mature loblolly pine by Murthy et al. (1995), who reported a P_{max} of 9.2 μmol m^{-2}s^{-1} at 350 μl l^{-1} carbon dioxide concentration. However, their P_{max} values were obtained from selected families of loblolly pine, whereas our P_{max} values were collected from a mixture of nonimproved sources. Additionally, Boltz et al. (1986) have shown seed-source variation in P_{max} among diverse sources of loblolly pine.

Irrigation and fertilization did not have any effect on the P_{max}. This is in contrast to the results obtained in other studies (El Kohen and Mousseau, 1994; Conroy, 1992; Conroy et al., 1990) in which nutrient availability was shown to increase P_{max} under elevated carbon dioxide concentrations. Although preliminary tests of the soil at our site showed low levels of nitrogen and other nutrients, monthly sampling of foliage indicated growing season levels of nitrogen ranging from 1.3% (C) to 1.8% (IF). Because a value of 1.1% foliar nitrogen is considered to be adequate in loblolly pine (Allen, 1987), it is not surprising that a main treatment effect was not found during the first year of this study. Irrigation did not have a significant effect on the average P_{max}. This was probably the result of the excessive precipitation received during July and August, which negated the C treatment. Additionally, during 1994, unlike a more typical year in Oklahoma, pan evaporation did not greatly exceed precipitation over the entire growing season (Cregg et al., 1988). Geuhl et al. (1994), Townend (1993), Miao et al. (1992), and Tolley and Strain (1984) reported that the availability of water affected the P_{max} under elevated carbon dioxide concentrations. There have also been a number of studies reporting that neither water nor nutrients affect P_{max} under conditions of elevated carbon dioxide concentrations (Murthy et al., 1995; Teskey, 1995; Liu and Teskey, 1995; Conroy et al., 1988; Norby et al., 1986; and Tolley and Strain, 1985).

The G_{max} was affected by elevated carbon dioxide concentration but not by the main plot treatments of either irrigation or fertilization. The G_{max} 525 μl l^{-1} varied significantly from the 350 and 700 μl l^{-1} carbon dioxide treated branches, whereas the G_{max} did not vary between the 350 μl l^{-1} and 700 μl l^{-1} carbon dioxide treated branches. Murthy et al. (1995), Teskey (1995), and Liu and Teskey (1995) have reported that both the maximum stomatal conductance and stomatal sensitivity in loblolly pine were not affected by different levels of carbon dioxide concentration. These results differ from those obtained in other studies (Tyree and Alexander, 1993; Townend, 1993; Eamus et al., 1993; Samuelson and Seiler, 1992; Eamus and Jarvis, 1989; Hollinger, 1987; Cure and Acock, 1986) who reported that the G_{max} decreased in response ast the levels of carbon dioxide concentration were increased, indicating that the stomates do respond to increasing levels of carbon dioxide concentration.

In general, the total chlorophyll content decreased as the carbon dioxide concentration was increased. This was similar to the results obtained in other studies (Tissue et al., 1995; El Kohen and Mousseau, 1994; Wilkins et al., 1994; Cui and Nobel, 1994; Lee et al., 1993; Evans et al., 1993; Drake, 1992). The decrease in chlorophyll content may be the result of accumulation of carbohydrates and distortion of chloroplasts (Reining, 1994; Wilkins et al., 1994). In contrast to these studies, Eamus et al. (1993) reported that the chlorophyll content was not affected by elevated carbon dioxide concentration.

Summary

In conclusion, current-year, first-flush foliage of branches grown in 525 $\mu l\, l^{-1}$ and 700 $\mu l\, l^{-1}$ of carbon dioxide had much greater rates of P_{max} compared to the P_{max} of foliage grown in 350 $\mu l\, l^{-1}$ carbon dioxide. These findings are similar to other long-term field studies with loblolly pine (Teskey, 1995; Murthy, 1995). Elevated carbon dioxide concentration was also significantly affected the G_{max}, however, higher rates were only found at the 525 $\mu l\, l^{-1}$ carbon dioxide concentration. Generally the total chlorophyll content decreased as the carbon dioxide concentration was increased.

The data presented here represent first-year responses to the carbon dioxide and cultural treatments. This experiment will continue to determine whether increased maximum net photosynthetic rate resulting from elevated carbon dioxide will persist over the life of the foliage and over an anticipated greater range of moisture and nutrient availability than existed during the first year of the study. In addition to this determination, evidence will also be collected to test for the possibility of downward acclimation of photosynthesis by foliage exposed to long-term elevated carbon dioxide concentrations. Detailed phenology measurements of branches and whole trees are expected to further the knowledge of how loblolly pine trees growing at the edge of the natural range respond to variations in carbon dioxide concentration, water, and nutrient supply.

If the enhanced rates of maximum potential carbon assimilation observed after one growing season are found to be sustained over the life of the foliage, as has been reported by others, the results then will suggest that an increase in forest productivity may be expected with global increases in atmospheric carbon dioxide concentrations. Large increases in pine growth rates, both in plantations and natural ecosystems, have several implications for forest management. These include lower planting densities, earlier thinning regimes, and shorter rotation lengths. However, future climate changes that cause increased temperatures and decreased precipitation might limit the potential gain in forest productivity in response to elevated atmospheric carbon dioxide concentrations.

References

Allen HL (1987) Forest fertilizers. J For 2:37–46.
Arp WJ (1991) Effects of source-sink relations on photosynthetic acclimation to elevated carbon dioxide. Plant Cell Environ 14:869–875.
Barton CVM, Lee HSJ, Jarvis PG (1993) A branch bag and carbon dioxide control system for long-term carbon dioxide enrichment of mature Sitka spruce (*Picea sitchensis* (Bong.) Carr.). Plant Cell Environ 16:1139–1148.
Bingham GE (1983) Leaf area measurement of pine needles. In *LI-1600 Steady State Porometer Instruction Manual*. Li-Cor, Inc. Publication No. 8107–01R2, Lincoln, NE.
Boltz BA, Bongarten BC, Teskey RO (1986) Seasonal patterns of net photosynthesis of loblolly pine from diverse origins. Can J For Res 16:1063–1068.
Conroy JP (1992) Influence of elevated atmospheric carbon dioxide concentrations on plant nutrition. Aust J Bot 40:445–456.

Conroy JP, Barlow EWR, Bevege DI (1986) Response of *Pinus radiata* seedlings to carbon dioxide enrichment at different levels of water and phosphorus: growth, morphology and anatomy. Ann Bot 57:165–177.

Conroy JP, Kuppers M, Kuppers B, Virgona J, Barlow EWR (1988) The influence of carbon dioxide enrichment, phosphorus deficiency and water stress on the growth, conductance and water use of *Pinus radiata* D. Don. Plant Cell Environ 11:91–98.

Conroy JP, Milham PJ, Mazur M, Barlow EWR (1990) Growth, dry weight partitioning and wood properties of *Pinus radiata* D. Don after 2 years of carbon dioxide enrichment. Plant Cell Environ 13:329–337.

Cregg BM, Dougherty PM, Hennessey TC (1988) Growth and wood quality of young loblolly pine trees in relation to stand density and climatic factors. Can J For Res 18:851–858.

Cregg BM, Halpin JE, Dougherty PM, Teskey RO (1989) Comparitive physiology and morphology of seedlings and mature forest trees. In Noble RD, Martin JL, and Jensen KF (Eds) *Air Pollution Effects on Vegetation*. USDA Forest Service, Broomall, PA, 111–118.

Cui M, Nobel PS (1994) Gas exchange and growth responses to elevated carbon dioxide and light levels in the CAM species *Opuntia ficus-indica*. Plant Cell Environ 17:935–944.

Cure JD, Acock B (1986) Crop responses to carbon dioxide doubling: A literature survey. Agric For Meteor 38:127–145.

Curtis PS, Teeri JA (1992) Seasonal responses of leaf gas exchange to elevated carbon dioxide in *Populus grandidentata*. Can J For Res 22:1320–1325.

Drake BG (1992) A field study of the effects of elevated carbon dioxide on ecosystem processes in a Chesapeake Bay wetland. Aust J Bot 40:579–595.

Eamus D, Berryman CA, Duff GA (1993) Assimilation, stomatal conductance, specific leaf area and chlorophyll responses to elevated carbon dioxide of *Maranthes corymbosa*, a tropical monsoon rain forest species. Aust J Plant Physiol 20:741–755.

Eamus D, Jarvis PG (1989) The direct effects of increase in the global atmospheric carbon dioxide concentration on natural and commercial temperate trees and forests. Adv Ecol Res 19:1–55.

El Kohen A, Mousseau M (1994) Interactive effects of elevated carbon dioxide and mineral nutrition on growth and carbon dioxide exchange of sweet chestnut seedlings (*Castanea sativa*). Tree Physiol 14:679–690.

Evans L, Pettersson R, Lee HSJ, Jarvis PG (1993) Effects of elevated carbon dioxide on birch (*Betula pendula*). Vegetatio 104/105:452–453.

Garcia RL, Idso SB, Kimball BA (1994) Net photosynthesis as a function of carbon dioxide concentration in pine trees grown at ambient and elevated carbon dioxide. Environ Exp Bot 34:337–341.

Gates DM (1983) An overview. In Lemon ER (Ed), *Carbon dioxide and plants: The response of plants to rising levels of atmospheric carbon dioxide*. Westview Press, Boulder, CO, 7–20.

Guehl JM, Picon C, Aussenac G, Gross P (1994) Interactive effects of elevated carbon dioxide and soil drought on growth and transpiration efficiency and its determinants in two European forest tree species. Tree Physiol 14:707–724.

Gunderson CA, Norby RJ, Wullschleger SD (1993) Foliar gas exchange of two deciduous hardwoods during 3 years of growth in elevated carbon dioxide: No loss of photosynthetic enhancement. Plant Cell Environ 16:797–807.

Higginbotham KO (1983) Growth of white spruce (*Picea glauca* [Moench.]Voss) in elevated carbon dioxide environments. Agric For Bull 6(3):31–33.

Higginbotham KO, Mayo JM, Hirondelle SL, Krystofiak DK (1985) Physiological ecology of lodgepole pine (*Pinus contorta*) in an enriched carbon dioxide environment. Can J For Res 15:417–421.

Hollinger DY (1987) Gas exchange and dry matter allocation responses to elevation of atmospheric carbon dioxide concentration in seedlings of three tree species. Tree Physiol 3:193–202.

Houghton RA, Woodwell GM (1989) Global climatic change. Sci Amer 260:36–44.

Idso SB, Kimball BA (1993) Effects of atmospheric carbon dioxide enrichment on net photosynthesis and dark respiration rates of three Australian tree species. J Plant Physiol 141:166–171.

Keeling CD, Bacastow RB, Carter AF, Piper SC, Whorf TP, Heimann M, Mook WG, Roeloffzen H (1989) A three dimensional model of atmospheric carbon dioxide transport based on observed winds. I. Analysis of observational data. In Peterson DH (Ed) *Aspects of climate variability in Pacific and the Western Americas*. Geophys Monogr 55:165–235.

Kimball BA, Mauney JR, Nakayama FS, Idso SB (1993) Effects of elevated carbon dioxide and climate variables on plants. J Soil Wat Cons Jan–Feb:9–14.

Lee HSJ, Muray M, Evans L, Pettersson R, Leith I, Barton CVN, Jarvis PG (1993) Effects of elevated carbon dioxide on Sitka spruce seedlings. Vegetatio 104/105:458–459.

Lindzen R (1993) Absence of scientific basis. Res Explor 9:191–200.

Liu S, Teskey RO (1995) Responses of foliar gas exchange to long-term elevated carbon dioxide concentrations in mature loblolly pine trees. Tree Physiol 15:351–359.

Miao SL, Wayne PM, Bazzaz FA (1992) Elevated carbon dioxide differentially alters the responses of coocurring birch and maple seedlings to a moisture gradient. Oecolgia 90:300–304.

Miller WF, Dougherty PM, Switzer GL (1987) Effect of rising carbon dioxide and potential change on loblolly pine distribution, growth, survival and productivity. In Shands WE, Hoffman JS (Eds) *The greenhouse effect, climate change, and U.S. forests*. The Conservation Foundation, Washington DC. 157–188.

Mousseau M (1993) Effects of elevated carbon dioxide on growth, photosynthesis and respiration of sweet chestnut (*Castanea sativa* Mill.). Vegetatio 104/105:413–419.

Murthy R, Dougherty PM, Zarnoch SJ, Allen HL (1995) Effects of carbon dioxide, fertilization and irrigation on photosynthetic capacity of loblolly pine trees. Tree Physiol 16:537–546.

Norby RJ, O'Neill EG (1989) Growth dynamics and water use of seedlings of *Quercus alba* L. in enriched atmospheres. New Phytol 111:491–500.

Norby RJ, O'Neill EG, Luxmoore RJ (1986) Effects of atmospheric carbon dioxide on the growth and mineral nutrition of *Quercus alba* seedlings in nutrient-poor soil. Plant Physiol 82:83–89.

Poorter H (1993) Interspecific variation in the growth response of plants to an elevated ambient carbon dioxide concentration. Vegetatio 104/105:77–97.

Reining E (1994) Acclimation of C_3 photosynthesis to elevated carbon dioxide: Hypotheses and experimental evidence. Photosyn 30(4):519–525.

Rogers HH, Dahlman RC (1993) Crop responses to carbon dioxide enrichment. Vegetatio 104/105:117–131.

Samuelson LJ, Seiler JR (1992) Fraser fir seedlings gas exchange and growth in response to elevated carbon dioxide. Environ Exp Bot 32:351–356.

SAS (1988) SAS procedures guide, SAS Institute Inc., Cary, NC.

Stewart JD, Hoddinott J (1993) Photosynthetic acclimation to elevated atmospheric carbon dioxide and UV irradiation in *Pinus banksiana*. Physiol Plant 88:493–500.

Tans PP, Ting IY, Takakhashi T (1990) Observational constraints on the global atmospheric carbon dioxide budget. Science 247:1431–1438.

Teskey RO (1995) A field study of the effects of elevated carbon dioxide on carbon assimilation, stomatal conductance and leaf and branch growth of *Pinus taeda* trees. Plant Cell Environ 18:565–573.

Teskey RO, Dougherty PM, Wiselogel AE (1991) Design and performance of branch chambers suitable for long-term ozone fumigation of foliage in large trees. J Environ Qual 20:591–595.

Tissue DT, Griffin KL, Thomas RB, Strain BR (1995) Effects of low and elevated carbon dioxide on C_3 and C_4 annuals. II. Photosynthesis and leaf biochemistry. Oecologia 101:21–28.

Tissue DT, Thomas RB, Strain BR (1993) Long-term effects of elevated carbon dioxide and nutrients on photosynthesis and rubisco in loblolly pine seedlings. Plant Cell Environ 16:859–865.

Tolley LC, Strain BR (1985) Effects of carbon dioxide enrichment and water stress on gas exchange of *Liquidambar styraciflua* and *Pinus taeda* seedlings grown under different irradiance levels. Oecologia 65:166–172.

Tolley LC, Strain BR (1984) Effects of carbon dioxide enrichment and water stress on growth of *Liquidambar styraciflua* and *Pinus taeda* seedlings. Can J Bot 62:2135–2139.

Townend J (1993) Effects of elevated carbon dioxide and drought on the growth and physiology of clonal Sitka spruce plants (*Picea sitchensis* (Bong.) Carr.). Tree Physiol 13:389–399.

Tschaplinski TJ, Norby RJ, Wullschleger SD (1993) Responses of loblolly pine seedlings to elevated carbon dioxide and fluctuating water supply. Tree Physiol 13:283–296.

Tyree MT, Alexander JD (1993) Plant water relations and the effects of elevated carbon dioxide: a review and suggestions for future research. Vegetatio 104/105:47–62.

Wang K, Kellomaki S, Laitinen K (1995) Effects of needle age, long-term temperature and carbon dioxide treatments on the photosynthesis of Scots pine. Tree Physiol 15:211–218.

Wilkins D, Van Oosten JJ, Besford RT (1994) Effects of elevated carbon dioxide on growth and chloroplast proteins in *Prunus avium*. Tree Physiol 14:769–779.

Woods ED, Wittwer RF, Dougherty PM, Crockett JJ, Tauer CG (1988) Influence of site factors on growth of loblolly pine and shortleaf pine in Oklahoma. Res. Rep. P-900, Oklahoma Agricultural Experiment Station.

11. Ecophysiological Response of Managed Loblolly Pine to Changes in Stand Environment

Mary A. Sword, Jim L. Chambers, Dennis A. Gravatt,
James D. Haywood, and James P. Barnett

Anticipated shifts in our global climate may expose southern pine ecosystems to such environmental stimuli as elevated carbon dioxide and water and nutrient deficiencies (Hansen et al., 1988; Kirschbaum et al., 1990; Peters, 1990). Global climate change may also increase the degree of stress to which trees are presently exposed (Kirschbaum et al., 1990; Peters, 1990). For example, the western extent of loblolly pine (*Pinus taeda* L.), now dictated by moisture availability for seedling establishment, is predicted to shift eastward with temperature and precipitation changes that may occur with global climate change (Miller et al., 1987).

Forest management practices that most effectively respond to global climate change can be identified only if we understand how the physiological and growth processes of trees in forest stands are affected by climate-mediated shifts in such essential resources as light, carbon dioxide, water, and mineral nutrients. The interdependence of above ground and root system processes, and the interactions between these processes and the availability of site resources, emphasize the need to simultaneously study above ground, root-system, and soil responses to environmental change. Tree responses to silvicultural manipulation should also be intensively evaluated so that cultural scenarios that maintain stand productivity and ecosystem integrity can be implemented in the event of global climate change.

Unfortunately, the physiology and growth of large trees in forest stands is poorly understood. The study of trees larger than seedlings has long been avoided because of the expense and logistical problems encountered when accessing large tree crowns on a continuous basis. Because microenvironmental variation within the canopy of a forest stand strongly influences crown physiology, the study of crown processes requires intensive measurement vertically within the forest canopy. Also, most physiological studies conducted on large trees have relied on information derived from only one to five trees, and have excluded observations of root system dynamics.

Materials and Methods

Study Establishment

The study is located on the Palustris Experimental Forest in Rapides Parish, LA, on a Beauregard silt loam soil (fine-silty, siliceous, thermic, plinthaquic paleudult) with a one to three percent slope (Kerr et al., 1980). In 1981, container-grown loblolly pine were planted (1.83 × 1.83 m). Percent survival and diameters at breast height (DBH), measured during September of 1987, indicated that initial survival (92%) and tree size were evenly distributed.

During April of 1988, twelve treatment plots, thirteen rows of thirteen trees each, were established (Haywood, 1994). Blocking was not considered necessary because initial stand productivity was homogenous. Thinning and fertilization treatments were randomly assigned to the plots in a two by two factorial design with three replications. Levels of thinning were either maintenance of the original stocking (2,732 trees ha^{-1}) or removal of every other row of trees and every other tree in residual rows during November of 1988 (721 tree ha^{-1}). Levels of fertilization were either no fertilization or broadcast application of 747 kg ha^{-1} diammonium phosphate (150 kg ha^{-1} P and 135 kg ha^{-1} N) during April of 1989. The fertilization rate was based on recommendations for loblolly pine grown on the inherently nutrient-poor soil in this study (Kerr et al., 1980; Shoulders and Tiarks, 1983).

Tree Growth, Canopy Environment, and Crown Physiology

Height and DBH measurements of the interior twelve trees on each plot were repeated quarterly through 1993 (Haywood, 1994). Outside-bark stem volume per tree was calculated (Baldwin and Feduccia, 1987), and stem volume per hectare was determined.

Two replications were chosen as blocks for intensive measurement of the stand environment and tree physiology. Blocks were identified based on the influence of topography on soil drainage with one block appearing more poorly drained than the other. Free-standing, steel, radio towers that supported wooden walkways in the lower and upper one-third of the canopy were installed to access at least eight dominant or codominant trees on the interior of each plot.

Canopy environmental measurements were limited to one block that was chosen based on the proximity of plots relative to each other. At three north-facing and three south-facing locations in both the upper and lower one-third of measurement plot canopies air temperature and photosynthetic photon flux density (PPFD) were monitored using custom-designed sensor units that were wired to one data acquisition system per plot (Model 576, Keithley/Metrabyte/Asyst/DAC Inc.). Sample branches were randomly selected from those that were logistically available, which resulted in measurement of the microclimate of two to four trees per plot and canopy level.

Sensor units consisted of two sensor housings, 50 cm apart, that were attached to polyvinyl chloride pipe (1.5 cm diameter). Sensor units were affixed to the towers and positioned adjacent to sample branches. Each sensor housing contained one shielded, solid-state temperature sensor (AD-592C, Analog Devices), and two photodiodes (BS500B, Sharp Electronics Corp.) which were positioned on opposite sides of the sensor housing. Sensors were wired so that mean branch temperature and mean branch PPFD were recorded. Sensor units were reassigned to different sample trees and branches, and PPFD sensors were calibrated twice per year against a recently calibrated quantum sensor.

Using walkways, crown physiological processes were monitored in the upper and lower one-third of the canopies of two blocks. Once during 1992 and four times during 1993, three upper and lower crown branches from each of three trees were randomly chosen from logistically available branches. Time and sampling procedure constraints prevented measurement of all treatments on the same day. Thus, data were collected over a two-day period with each fertilization treatment being measured in one day. One set of measurements was collected from both blocks in the morning and a second set of measurements was collected in the afternoon. The plot measurement sequence in the afternoon was the reverse of that in the morning.

Net photosynthesis measurements were conducted on two to three fascicles (six to nine needles) from the mature foliage on south-facing, terminal, or adjacent lateral shoots. In situ net needle carbon dioxide (CO_2) uptake rate was quantified under ambient conditions with a portable photosynthesis system (LI-6200, Li-Cor Inc.), and expressed as the mean of data collected at each of three branches per canopy level and treatment.

Branch carbon exchange index (BCEI), was calculated to express the net amount of carbon assimilated by the most recently mature internode. To develop this index, the mean fascicle length and projected needle surface area (PNSA) of ten fascicles on each of three randomly selected terminal shoots per treatment and canopy level of one block were quantified once during 1993. Linear regression equations that predicted PNSA per fascicle from fascicle length were developed for each treatment and canopy level. Equations for each treatment had significantly different slopes. Therefore, separate equations were used to predict PNSA of fascicles used for physiological measurements. After completion of physiological measurements 1) the mean fascicle length of foliage used for the physiological measurements was quantified, 2) PNSA was predicted using the appropriate equation, and 3) BCEI was calculated by multiplying the total number of fascicles on the shoot by the predicted PNSA, and then multiplying by the mean rate of net CO_2 uptake per unit of projected needle surface area.

Soil Environment and Root System Growth

On the two blocks used for physiological measurements, vertical Plexiglass® rhizotrons were installed at three interior locations per plot. Rhizotron locations were randomly chosen from those that had a stable microtopography and were associated with a dominant or codominant tree that was not adjacent to a missing or dead tree, or a tree that was heavily infected with *Cronartium fusiforme* Hedge & Hunt (fusiform rust). At each location, one longitudinal side of an excavated area (35 × 20 × 80 cm) was patched with a mortar prepared with sieved soil from the site. Sheet metal screws were placed at 10 cm intervals around the periphery and down the center of Plexiglass® sheets (0.3 × 35.4 × 76 cm) to secure rhizotrons onto the soil face. Rhizotrons were insulated with styrofoam between root observations.

Soil temperature was measured with solid-state temperature sensors (AD592C, Analog Devices), embedded in epoxy resin inside 5 cm pieces of stainless steel tubing that were insulated with waterproof electrical sealant. Insulated sensors were inserted at 5, 15, and 30 cm through ports in rhizotrons. Sensors were also installed at two randomly chosen locations, independent of rhizotrons, for a total of five series of soil temperature sensors per measurement plot. Soil temperatures were measured at ten-day intervals beginning in June of 1992 and continuing through the study.

In all replications, one set of stationary time-domain reflectometry sensors was installed vertically at plot centers to quantify volumetric soil-water content in 0 to 20 and 20 to 40 cm depths of the soil. Measurements were taken at ten- to fourteen-day intervals beginning in June of 1992 and continuing through the study. Volumetric soil-water content was also measured with sensors inserted horizontally at 5, 15, and 30 cm through ports in one rhizotron per plot of one block. These measurements, taken at six-hour intervals, began in May of 1993 and continued through the study. Climate data were recorded in an open field 25 m from the study with an electronic weather station (Omnidata International, Inc.).

At ten-day intervals beginning in April of 1993, the long lateral new root length observed in rhizotrons was traced with permanent marker onto heavy-duty acetate sheets (21.6 × 30 cm) attached to left and right sides of the plexiglass. Observations were recorded cumulatively. After each measurement date, acetate sheets were photocopied, and a computer image file of each photocopy was created using a desktop scanner. The length of the lines contained in each image file was quantified using GSROOT software (PP Systems Inc.). Net lateral root elongation occurring in the 0 to 30 cm depth of rhizotrons was calculated by subtraction. After each measurement period, the number of new roots (≥ 0.5 cm) initiated in the 0 to 30 cm depth was also quantified.

On June 17 (day 168), August 29 (day 241), and December 6, 1993 (day 340), visible lateral roots were traced and their lengths were quantified as described. These lengths, expressed as a fraction of the length of lateral roots that had accumulated since April 13, 1993, described lateral root persistence. When roots were traced for lateral root persistence measurements, ectomycorrhizal and non-mycorrhizal roots were differentiated by marker color. Roots were considered ectomycorrhizal by the presence of one or more ectomycorrhizae or swollen short roots. Ectomycorrhizal colonization was expressed as the percentage of lateral root length that appeared ectomycorrhizal.

Data Analyses

During December of 1993, tree height, DBH, and stem volume, as well as stem volume per hectare, were subjected to analyses of covariance using a completely random, two by two factorial experimental design with three replications (Haywood, 1994). The factors were two levels each of fertilization and thinning. Covariates were initial height, DBH, stem volume, and stem volume per hectare measured in early March of 1989 before height-growth began.

For each fertilization treatment, branch air temperature and PPFD were subjected to analyses of variance using a repeated measures split-split-plot design with two replications. Whole plots were level of thinning and subplots were canopy level (upper or lower crown). Net needle CO_2 uptake rate and BCEI were analyzed by fertilization treatment using a repeated measures, split-split-split-plot design with four replications. Whole plots were level of thinning (thinned or not thinned), and subplots were canopy level and time of day (morning or afternoon). Analyses of variance were conducted by measurement date on soil temperature (5, 15, 30 cm) and water content (0 to 20, 20 to 40 cm) using a randomized complete block design with two blocks, and a completely random design with three replications, respectively. Significant treatment effects were noted if trends were consistent over time. Net root elongation and root initiation were analyzed using a randomized complete block, split-plot in time design with two blocks. Two levels each of thinning and fertilization were the whole plot treatments, and time was the subplot treatment. Lateral root persistence and ectomycorrhizal colonization were analyzed by measurement date using a randomized complete block design with two blocks. Main and interaction effects were considered significant at $p \leq 0.05$ unless otherwise noted.

Results

Tree Growth, Canopy Environment, and Crown Physiology

Analyses of covariance indicated that thinning significantly increased tree DBH and volume (Table 11.1). Fertilization significantly increased tree height, DBH and volume, as well as stand volume. Nearly significant interactions between thinning and fertilization were observed in tree DBH ($p = 0.0559$) and volume ($p = 0.0571$), with fertilization causing a greater increase in these variables on the thinned plots than on plots that were not thinned.

On plots that were not fertilized, lower canopy PPFD was significantly greater on thinned plots when compared to those that were not thinned (Figure 11.1).

Table 11.1. Growth and Yield of Loblolly Pine Before, and Four Years After Fertilization With Nitrogen and Phosphorus, and Manipulation of Stand Density With Row Thinning. Analyses of Covariance Were Conducted on Data Collected in the Thirteenth Growing Season. For Each Variable, the Covariant Was the Same Variable at the Start of the Study in March of 1989

Treatment	Trees Surviving in 1993 (number ha^{-1})	March 1989, 9th growing season				December 1993, 13th growing season			
		Height (m)	DBH (cm)	Tree Volume (m^3)	Total Volume (m^3 ha^{-1})	Height (m)	DBH (cm)	Tree Volume (m^3)	Total Volume (m^3 ha^{-1})
Not Thinned, Not Fertilized	2,783	9.1	11.2	0.0488	135.8	13.1	13.2	0.1009	280.6
Not Thinned, Fertilized	2,600	9.1	11.4	0.0514	133.7	14.1	13.9	0.1220	317.0
Thinned, Not Fertilized	731	8.9	10.9	0.0483	35.4	12.4	16.5	0.1446	105.8
Thinned, Fertilized	711	8.6	10.6	0.0420	29.8	13.4	18.4	0.1897	134.9

	Analysis of Variance			Analyses of Covariance				
	df	MS			df	MS			
						Height (m)	DBH (cm)	Tree Volume (m^3)	Total Volume (m^3 ha^{-1})
Thinning (T)	1	11,647,311.6			1	0.6701	34.414	0.0082	2255.84
Fertilizer (F)	1	30,947.3			1	3.2858	5.916	0.0037	2784.22
T × F	1	19,806.3			1	0.0222	1.757	0.0009	31.34
Error mean square	8	15,783.1			7	0.1258	0.3356	0.00017	169.09
Covariant					1	0.4691	0.8000	0.0005	15.95
	(Probability > F − value)				(Probability > F − value)				
Thinning (T)	0.0001					0.0543	0.0001	0.0002	0.0082
Fertilizer (F)	0.1990					0.0014	0.0040	0.0022	0.0048
T × F	0.2951					0.6866	0.0559	0.0571	0.6798

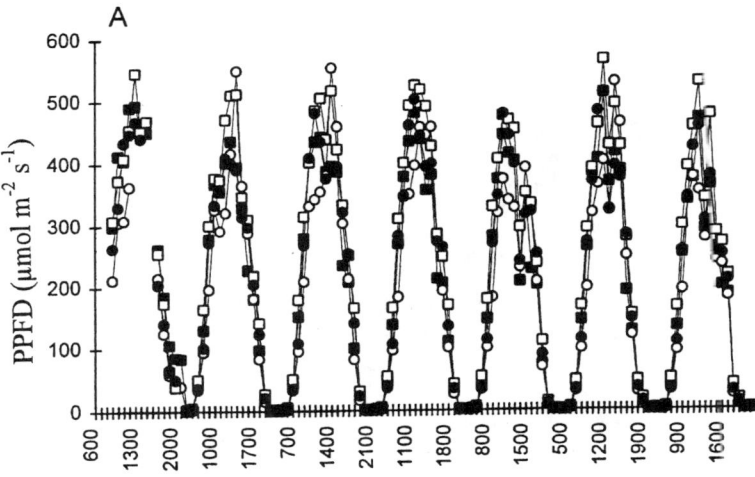

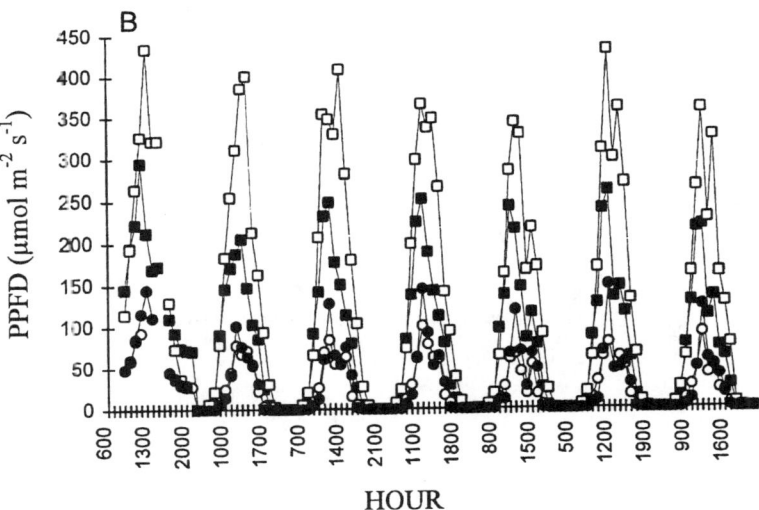

Figure 11.1. Typical photosynthetic photon flux density (PPFD) (μmol m^{-2} s^{-1}) in the upper (A) and lower (B) one-third of the canopy between 0600 and 2300 hours during one week in August of 1993. Data are hourly means of four 15-minute measurements collected at six branch locations on plots that were not thinned and either were fertilized (●) or were not fertilized (○), and plots that were thinned and either were fertilized (■) or were not fertilized (□).

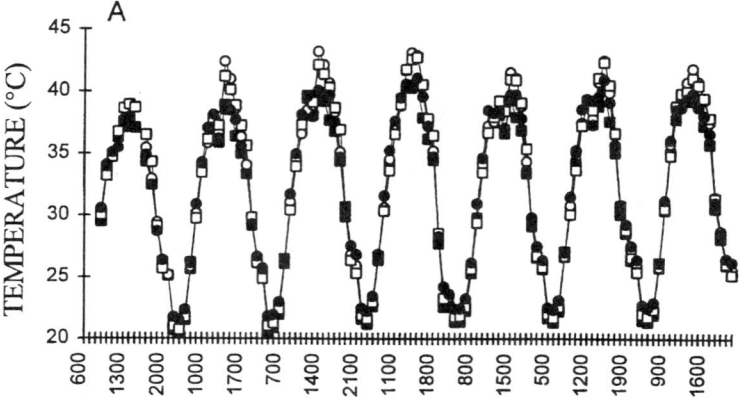

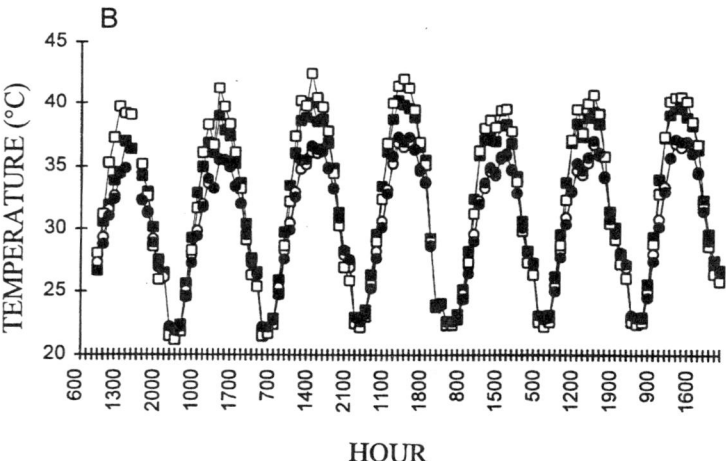

Figure 11.2. Typical branch temperature (°C) in the upper (A) and lower (B) one-third of the canopy between 0600 and 2300 hours during one week in August of 1993. Data are hourly means of four 15-minute measurements collected at six branch locations on plots that were not thinned and either were fertilized (●) or were not fertilized (○), and plots that were thinned and either were fertilized (■) or were not fertilized (□).

Although a similar trend was found, thinning did not significantly increase PPFD in the lower canopy of fertilized plots. Branch temperature in the lower canopy of both fertilization treatments was significantly increased by thinning with midday branch temperature differences of 2 to 5 °C on sunny days (Figure 11.2). Upper canopy PPFD and branch temperature were not affected by thinning in either fertilization treatment.

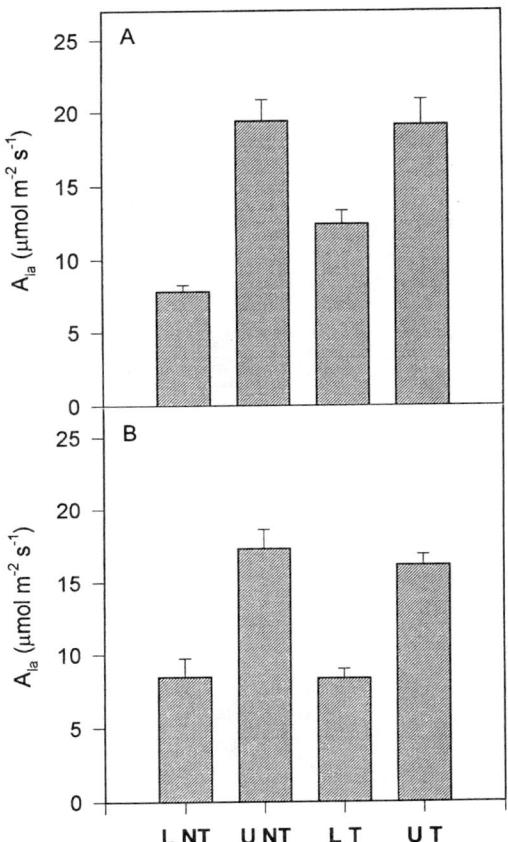

Figure 11.3. Typical net needle CO_2 uptake (A_{la}) (μmol m^{-2} s^{-1}) on plots that either were fertilized (A) and were not fertilized (B). Data are the mean of three measurements collected in the upper (U) and lower (L) one-third of the canopy on plots that were either thinned (T) or were not thinned (NT). Vertical bars indicate the standard error of the mean.

In both fertilization treatments, the rate of needle net photosynthesis was significantly higher in the upper crown than in the lower crown (Figure 11.3). Thinning did not affect needle net photosynthesis in either fertilization treatment. Although experimental design limitations prevented direct statistical comparison, needle net photosynthesis appeared slightly higher on plots that were fertilized when compared to plots that were not fertilized.

Upper crown BCEI was significantly higher than that of the lower crown (Figure 11.4). In both fertilization treatments, BCEI was significantly higher in the lower crown of the thinned plots when compared to the lower crown of plots that were not thinned. Thinning did not significantly affect BCEI in the upper crown of either fertilization treatment. Although experimental design limitations prevented statistical comparison, BCEI in both thinning treatments and in both

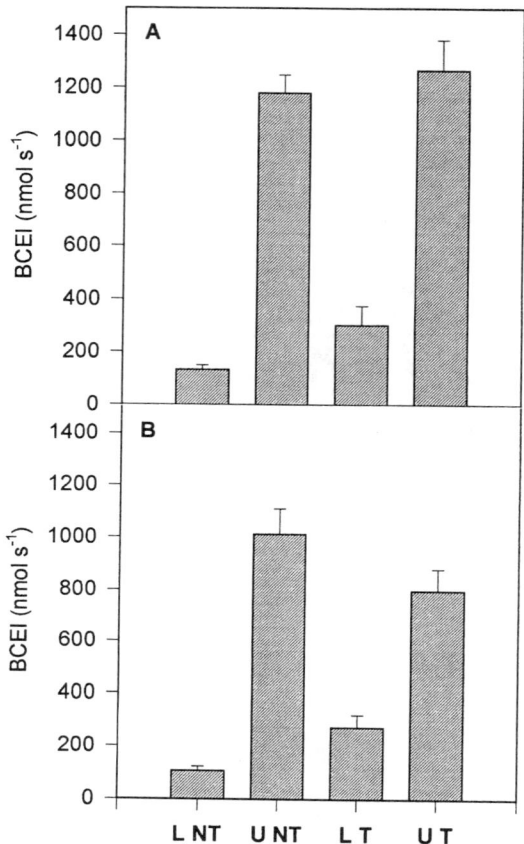

Figure 11.4. Typical branch carbon exchange index (BCEI) (nmol s^{-1}) on plots that either were fertilized (A) or were not fertilized (B). Data are the mean of three measurements collected in the upper (U) and lower (L) one-third of the canopy on plots that either were thinned (T) or were not thinned (NT). Vertical bars indicate the standard error of the mean.

crown levels appeared to be higher in response to fertilization. This effect was most pronounced in the upper crown of the thinned plots in which BCEI was 68% higher four years after fertilization.

Soil Environment and Root System Growth

During 1992 and 1993, 146 and 138 cm of precipitation were received, respectively. Approximately 57% of the precipitation in both years occurred during March through September. In June through September, the volumetric soil-water content at depths of 0 to 20 and 20 to 40 cm averaged 27.3 and 24.5%, respectively during 1992, and 26.5 and 23.6%, respectively during 1993. Between October of 1992 and May of 1993, the volumetric soil-water content at depths of 0 to 20 and

20 to 40 cm averaged 35.4 and 30.5%. Between late June and September of 1992, the volumetric soil-water content at the 0 to 20 cm depth was significantly greater on the thinned plots when compared to plots that were not thinned. Otherwise, thinning and fertilization treatments did not significantly affect volumetric soil water content.

Volumetric soil-water content, measured biweekly, decreased 32 and 27% at depths of 0 to 20 and 20 to 40 cm, respectively, between June and September of 1993. Daily measurements allowed greater resolution of soil-water content trends. Lack of precipitation between late May and mid-June, and between mid-July and early August were associated with 59 and 70% reductions, respectively, in soil-water content measured at 15 cm (Figure 11.5). A similar trend was observed at depths of 5 and 30 cm.

Soil temperature at the 15 cm depth was significantly higher (0.7 °C) on the thinned plots during the growing season; in winter, the soil temperature was significantly higher (0.6 °C) on plots that were not thinned (Figure 11.6). A similar response was observed at depths of 5 and 30 cm. Throughout the year, soil temperatures were significantly lower (0.4 °C) at depths of 15 and 30 cm on plots that were fertilized than on plots that were not fertilized.

Lateral root elongation and initiation in the 0 to 30 cm depth of the soil exhibited a multimodal pattern between April of 1993 and March of 1994 with the greatest growth in spring and early summer (Figures 11.7 and 11.8). During peak root growth, lateral root elongation exhibited a bimodal pattern. Root elongation and initiation declined 92 and 87%, respectively, in midsummer and continued at a reduced rate through fall and winter with several short periods of accelerated growth.

The thinning operation, which had been conducted during the fall of 1988, significantly increased lateral root elongation and initiation in the 0 to 30 cm depth of the soil during 1993 (Table 11.2). Fertilization did not significantly affect lateral root elongation or initiation. However, lateral root elongation was significantly affected by an interaction between time, thinning, and fertilization. Analyses of variance by day indicated that on the thinned plots, lateral root elongation was greater in response to fertilization on five of eighteen measurement days between June and January (Figure 11.7). Significant treatment effects during peak root growth corresponded to lower coefficients of variation and more stable soil environmental conditions relative to seasonal trends (Figure 11.9).

Of the lateral root length that accumulated in rhizotrons during 1993, approximately 99, 79, and 48% persisted and were visually observed on June 17 (day 168), August 29 (day 241), and December 6, 1993 (day 340), respectively. Losses were attributed to both mortality and obstruction of roots from view. Thinning did not significantly affect the persistence of lateral roots in rhizotrons, but fertilization significantly increased their persistence between April 13 and August 29, 1993 (Figure 11.10). The percentage of lateral root length with visible signs of ectomycorrhizal colonization increased from approximately 29 to 45% between June 17 (day 168) and December 6, 1993 (day 340), but was not significantly affected by thinning or fertilization.

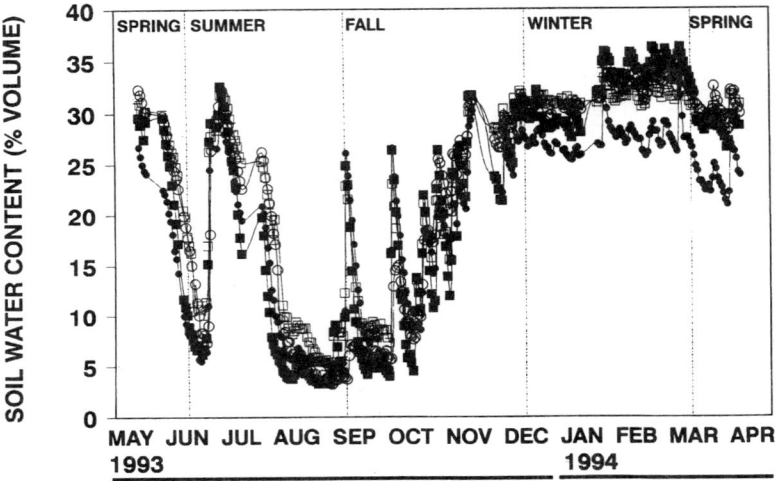

Figure 11.5. Daily percent volumetric soil-water content at 15 cm between May of 1993 and March of 1994. Data are the mean of four measurements collected at six-hour intervals at one location in each plot of one replication (not thinned and not fertilized (○); not thinned and fertilized (●); thinned and not fertilized (□); thinned and fertilized (■)).

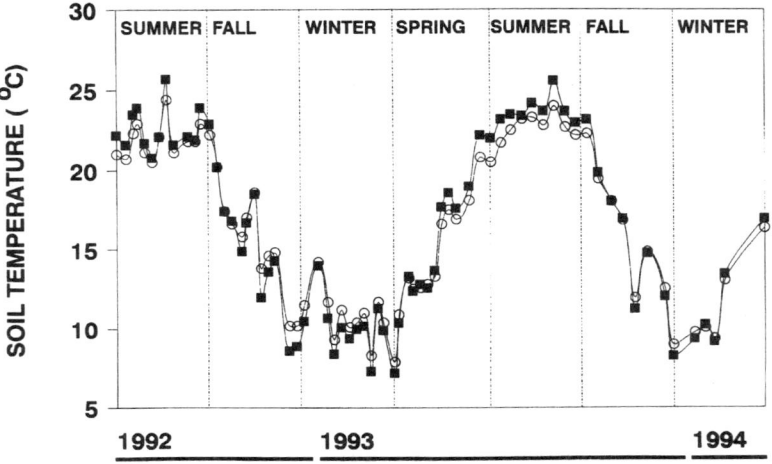

Figure 11.6. Soil temperature (°C) at 15 cm between June 1992 and March 1994 on plots that either were thinned (■), or were not thinned (○). Data are the mean of two replications of five measurements that were collected at approximately two-week intervals.

11. Ecophysiological Response of Managed Loblolly Pine   197

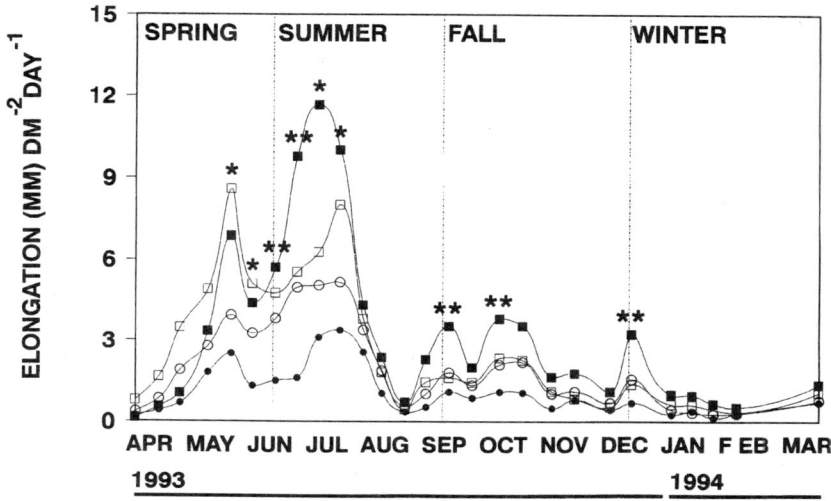

Figure 11.7. Seasonal rate of long lateral root elongation (mm dm^{-2} day^{-1}) in rhizotrons between April of 1993 and March of 1994. Data are the mean of two replications of six measurements (not thinned and not fertilized (○); not thinned and fertilized (●); thinned and not fertilized (□); thinned and fertilized (■)). The thinning effect was significant on dates noted with "*", and both thinning and T × F ($P \leq 0.07$) effects were significant on dates noted with "**".

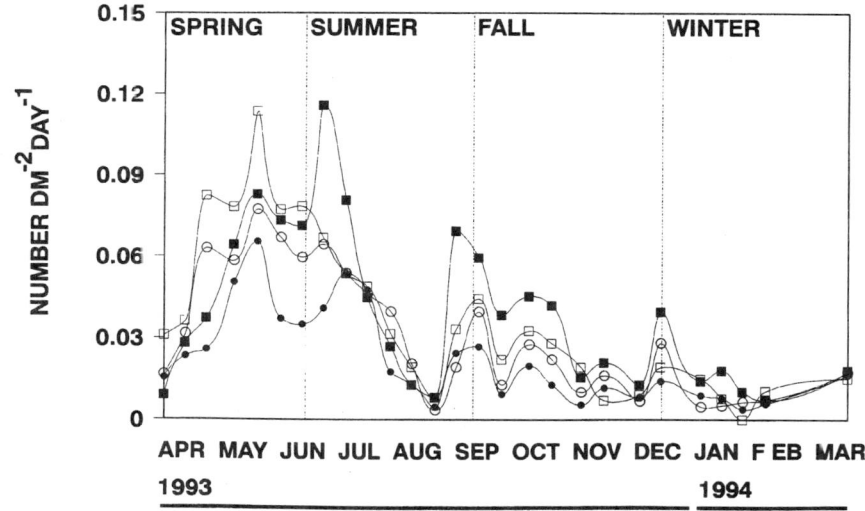

Figure 11.8. Seasonal rate of long lateral root initiation (number dm^{-2} day^{-1}) in rhizotrons between April of 1993 and March of 1994. Data are the mean of two replications of six measurements (not thinned and not fertilized (○); not thinned and fertilized (●); thinned and not fertilized (□); thinned and fertilized (■)).

Table 11.2. Analyses of Variance of Long Lateral Root Elongation (mm dm^{-2} day^{-1}), and Initiation (number dm^{-2} day^{-1}) at the 0 to 30 cm Depth of the Soil During April 1993 through March 1994

Source of Variation	Long Lateral Root Elongation			Long Lateral Root Initiation		
	df	MS	Probability > F – value	df	MS	Probability > F – value
Block (B)	1	22.5764	0.1502	1	0.00065	0.2193
Thinning (T)	1	110.4932	0.0238	1	0.00739	0.0138
Fertilization (F)	1	0.7444	0.7500	1	0.00034	0.3457
T × F	1	30.4516	0.1117	1	0.00159	0.0945
B × T × F (error a)	3	6.1057		3	0.00027	
Time	26	29.3242	0.0001	26	0.00428	0.0001
Time × B (error b)	26	0.8691		26	0.00016	
Time × T	26	4.4926	0.0001	26	0.00024	0.2645
Time × F	26	1.0868	0.1688	26	0.00030	0.0767
Time × T × F	26	2.2345	0.0003	26	0.00019	0.5060
Time × T × F × B (error c)	78	0.8173		78	0.00020	

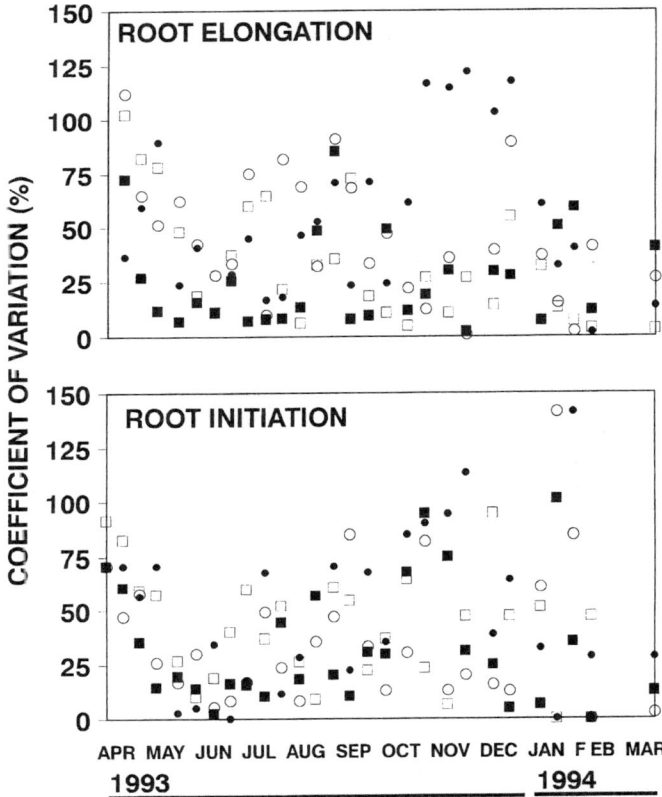

Figure 11.9. Coefficients of variation of root elongation (mm dm^{-2}) and initiation (number dm^{-1}) observed between April of 1993 and March 1994. Data are coefficients of variation associated with the mean of two replications.

Discussion

Thinning significantly increased both aboveground productivity and root growth. Significant increases in aboveground growth were consistently observed in response to fertilization, and on 25% of the measurement days between May and October, root growth on the thinned plots was significantly stimulated by fertilization. This is in contrast to the results of others who have found an inverse relationship between fine root production and such variables as site quality and fertilization (Comeau and Kimmins, 1989; Gower et al., 1992; Haynes and Gower, 1995; Keyes and Grier, 1981; Santantonio and Santantonio, 1987; Vogt et al., 1983; Vogt et al., 1987). The difference between our observations and those reported elsewhere demonstrate the complexity of growth responses to the forest stand environment.

Thinning and fertilization reduce the competition among trees for such es-

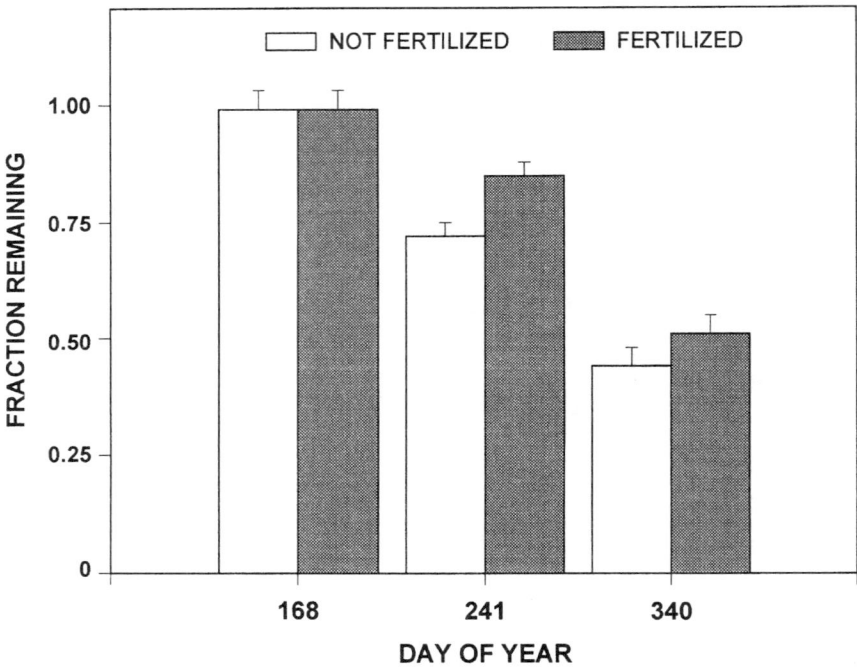

Figure 11.10. Fraction of the long lateral root length on plots that were and were not fertilized, that persisted in rhizotrons when compared to that accumulated between April 13, 1993 and each of three dates (June 17, day 168; August 29, day 241; December 6, day 340). Vertical bars indicate the standard error of the mean.

sential resources as light, water, and mineral nutrients (Kozlowski et al., 1991). By age thirteen, the plots that were not thinned were overstocked and had live-crown ratios less than 36%. Smith (1986) suggested that stands with live-crown ratios below 30 to 40% may not be vigorous enough to respond positively to thinning. In our study, light competition and reduced rates of photosynthate production in the canopies of plots that were not thinned may have limited the availability of carbohydrates for root metabolism, and therefore, root growth.

Beauregard silt loam soils are generally low in available phosphorus, and loblolly pine has responded well to the amounts of nitrogen and phosphorus applied in this study (Shoulders and Tiarks, 1983; Tiarks, 1982). Typically, nitrogen fertilization favors aboveground rather than root growth and phosphorus amendments stimulate root growth (Marschner, 1986). Four years after fertilization, insufficient phosphorus on plots that were not fertilized may have reduced root growth. Stimulation of root growth may be one mechanism by which phosphorus fertilization has eliminated this inhibition and has increased the growth of loblolly pine on some coastal plain soils (Haywood and Burton, 1990; Pritchett and Gooding, 1975). Our results suggest that conclusions about carbon partitioning to the root system in response to silvicultural treatments cannot be made without information on the environmental and physiological status of a stand.

Anticipated shifts in the earth's climate include significantly higher global temperatures (Hansen et al., 1988; Peters, 1990), and a simultaneous reduction in precipitation (Hansen et al., 1988). Furthermore, elevated atmospheric temperature and carbon dioxide, together with reduced precipitation, may alter the chemistry of organic matter in soil (Johnson, 1995; Van Cleve and Powers, 1995). Because the cycling of mineral nutrients through forest ecosystems is strongly influenced by the quantity and quality of organic matter deposited in the soil (McColl and Gressel, 1995), mineral nutrient availability may also be affected by global climate change.

Carbon partitioning among plant organs is the principal mechanism by which plants respond to seasonal patterns of water, mineral nutrient, and carbohydrate availabilities (Dickson, 1989). For example, water and mineral nutrient limitations generally result in a shift in biomass accumulation from the crown and bole to the root system (Gower et al., 1992; Santantonio and Santantonio, 1987; Vogt et al., 1983); under water and mineral nutrient sufficiency, but light limitation, however, crown growth is favored (Waring, 1991). To use these relationships to manipulate forest responses to climate change, interactions between physiological and growth processes within whole trees in a stand environment must be understood. With this information, it may be possible to use silvicultural treatments to manipulate both resource availabilities and internal tree physiological relationships in an effort to compensate for potential effects of global climate change on forests.

Older woody roots and newly elongated roots provide both water and mineral nutrients for tree growth (Eissenstat and Van Rees, 1994; MacFall et al., 1991; Van Rees and Comerford, 1990). However, the absorption of water and mineral nutrients in environments where these resources are limited may primarily rely on the growth of new roots (Eissenstat and Van Rees, 1994). Furthermore, calcium uptake primarily occurs apoplastically through the immature endodermis of newly elongated roots (Russell and Clarkson 1976). In our study, 63 and 62% of the lateral root elongation and initiation, respectively between April of 1993 and March of 1994, occurred from May to July, 1993. Because new roots that were produced during this twelve-week period not only supplied resources for tree growth, but also provided a foundation from which additional new roots grew, it is probable that environmental and physiological variables during this period affected root function throughout the year.

A 36% reduction in the rate of root elongation, which was observed in late spring of 1993, was associated with rapid expansion of new shoots and fascicles. Root growth is a weaker photosynthate sink than branch growth, and a complementary pattern of shoot and root growth is typical of such species with recurrent shoot growth as loblolly pine (Dickson, 1989; Dickson, 1991). The drop in root elongation that we observed could have been a response to reduced carbohydrate availability for root metabolism. However, a similar simultaneous reduction in new root initiation was not observed in late spring, which suggests that carbohydrates were not limited for root growth. Reduced root elongation with no effect on root initiation was also associated with twenty-two days of drought in June that caused a 59% decrease in soil-water content. Similarly, others have reported that water deficits reduce lateral root elongation but have minimal effects on root initiation (Hipps et al., 1995; Teskey and Hinckley, 1981). These results suggest that the extension of new loblolly pine roots into the soil during peak root growth may be strongly influenced by fluctuations in soil moisture.

Four years after thinning and fertilization, temperatures of the soil and lower crown were elevated in response to thinning but were reduced by fertilization. Temperature responses appeared to be related to changes in shading caused by branches and foliage in the canopy that were either removed with the thinning operation or produced in response to fertilization. Only a small increase in soil-water content in response to thinning was observed, and no significant soil-water response to fertilization. Lack of a stronger effect on soil-water content may have been caused by reequilibration of the foliage distribution in the canopy, and similarity of transpiration potentials that evolved after application of silvicultural treatments (Cregg et al., 1990).

Gower et al. (1993) found that new foliage was predominantly produced in the mid- and lower canopy of a ponderosa pine (*Pinus ponderosa* Dougl. ex P. Laws.) stand, but was produced in the upper and mid-canopy of a similarly stocked red pine (*Pinus resinosa* Ait.) stand. This phenomenon was attributed to the open nature of the ponderosa pine crown compared to that of red pine. In our study, the upper canopy received significantly more light than the lower canopy, and PPFD in the upper canopy was unaffected by silvicultural treatment. In the lower canopy, however, we found that loblolly pine responded to a thinning-induced increase in PPFD by increasing foliage production. Similar to the observations of Vose (1988), we also found that on the thinned plots, fertilization increased loblolly pine foliage production in the lower canopy. However, as the lower canopy became more dense, PPFD was reduced. Comparable relationships have been presented between fertilization, canopy leaf areas, and light levels in pine forests (Gower et al., 1993; Vose and Allen, 1988).

Changes in the canopy environment with crown depth were positively related to rates of photosynthesis. Similar to the results of Teskey et al. (1994), rates of photosynthesis within upper and lower portions of the canopy were only slightly affected by silvicultural treatment. However, fertilization increased fascicle production in both the upper and lower canopies and thinning increased fascicle production in the lower canopy. As a result, when rates of photosynthesis were expanded to express branch level carbon exchange, it was apparent that thinning and fertilization greatly increased carbon fixation in the crown. Clearly, an understanding of crown processes that control forest productivity requires the simultaneous measurement of environmental, physiological, and growth variables vertically within the forest canopy.

Our results present the potential for manipulation of root growth using silvicultural tools that alter photosynthate production in tree crowns. Starch accumulates in root parenchyma cells in fall and winter and is a primary carbohydrate source for root growth early in the growing season (Ericsson and Persson, 1980; Ford and Deans, 1977; Gholz and Cropper, 1991). In addition to early growth, a fall peak in root activity has been reported (Dickson, 1991; Grier et al., 1981; Santantorio and Santantonio, 1987). However, the consistent occurrence of a second peak in root growth is strongly dependent, in part, on environmental conditions earlier in the growing season (Dickson, 1991). A larger amount of carbon fixation in the canopy may stimulate root growth throughout the growing season by increasing the amount of photosynthate translocated to, and the amount of starch stored in, the root system.

In our study, canopy environmental responses to silvicultural treatments were isolated in the lower portion of the crown. It may be possible to manipulate carbon fixation in the lower canopy alone to affect the amount of photosynthate translocated to the root system. In a young tree, photosynthate produced by the most recently matured foliage is translocated acropetally to be metabolized by developing internodes and fascicles, whereas, that which is produced by older foliage is partitioned basipetally to support root growth (Dickson, 1991; Dickson, 1989; Watson and Casper, 1984). A similar pattern of carbon partitioning has been found within the branches of *Populus* trees (Dickson, 1986). Dickson (1986) suggested that the terminal shoot and lateral branches in the crowns of young *Populus* trees contributed more photosynthate for root growth than did the lateral branches of the lower crown. However, they also stated that the magnitude of this relationship may change with light levels in the canopy. Research should be conducted to determine if maintenance and stimulation of a positive carbon balance in the lower canopy of a managed forest stand contributes significantly to the amount of photosynthate allocated for root growth.

In addition to root elongation and initiation, the rate of fine root anatomical development also may be influenced by resource availability. Accelerated transformation of primary roots into secondary roots could reduce the susceptibility of new roots to such environmental extremes as water deficit. This would result in reduced mortality, and lead to a larger infrastructure from which new roots could initiate as conditions became favorable. In our study, the persistence of lateral roots that grew in rhizotrons between April and August was greater in plots that were fertilized than in those that were not fertilized. Because of dissimilarity between root environments in undisturbed soil and in rhizotrons, we cannot make inferences about fine root mortality based on the persistence of lateral roots in rhizotrons. However, increased lateral root persistence in response to fertilization may be attributed to the developmental status that roots achieved in response to mineral nutrient availability.

This relationship is apparent in the data presented by Gower et al. (1993), in which biomass partitioning by Douglas fir (*Pseudotsuga menziesii* (Mirb.) Franco) was quantified in response to five levels of soil amendment. Although biomass partitioned to the root system was greater on control than on amended plots, a larger proportion of the total small ($> 2 \leq 5$ mm diameter) plus very fine (≤ 2 mm) root biomass occurred in the small, rather than the very fine category, on the amended plots. The reverse was true on the control plots.

We found that under present-day climate conditions, the potential exists for water limitations to reduce root growth. If global climate change significantly alters the pattern or amount of precipitation received by southern pine forests, the growth of tree roots could be inhibited. In the event of global climate change, factors that may strongly influence root system function are the amount of root growth that occurs before the onset of water deficits, and the rate at which primary roots undergo secondary development. Our results provide important insight on how stand-management practices could be used to manipulate these factors.

Summary

Many of the potential negative impacts of global climate change on southern pine forests will manifest through reduced availabilities of soil resources. We found that during the period of peak root growth, the elongation of new loblolly pine roots was closely related to the rate at which soil moisture declined. This level of responsiveness was also noted by Vogt et al. (1993) who emphasized the sensitivity of fine roots to environmental change, in part, because of their direct contact with the soil.

The ability of tree root systems to procure key resources in a changing climate may be limited. Thus, aggressive forest management strategies that maintain optimum stand conditions will be needed as climate change occurs. Such silvicultural treatments as thinning and fertilization are effective through their influence on stand environment and resource variables, many of which are similar to those expected to shift with global climate change. We found that silvicultural manipulation of stand density influenced the vertical distribution of light in the canopy. The intensity and distribution of light in the canopy, in addition to the branch carbon exchange indices suggest that carbon fixation at the canopy level was stimulated by manipulation of stand density. Elevated root growth observed in response to thinning may have been caused by an increase in the amount of photosynthate translocated from either the entire crown or the lower crown alone. This suggests that such silvicultural tools as thinning and fertilization offer an opportunity to manipulate relationships between light, water, and mineral nutrient availabilities in forest stands, and carbohydrate partitioning within trees to buffer the possible negative effects or enhance the potential positive effects of global climate change on southern pine forests.

References

Baldwin VC, Feduccia DP (1987) Loblolly pine growth and yield prediction for managed west gulf plantations. US Dept Agric For Ser South For Exper Sta Gen Tech Rep SO-236.

Comeau PG, Kimmins JP (1989) Above- and below-ground biomass production of lodgepole pine on sites with differing soil moisture regimes. Can J For Res 19:447–454.

Cregg BM, Hennessey TC, Dougherty PM (1990) Water relations of loblolly pine trees in southeastern Oklahoma following precommercial thinning. Can J For Res 20:1508–1513.

Dickson RE (1986) Carbon fixation and distribution in young *Populus* trees. In Fujimori T, Whitehead D (Eds) *Crown and canopy structure in relation to productivity.* Forest and Forest Products Research Institute, Ibaraki, Japan.

Dickson RE (1989) Carbon and nitrogen allocation in trees. In Dreyer E, Aussenac G, Bonnett-Masimbert M, Dizengremel P, Favre JM, Garrec JP, Le Tacon F, Martin F (Eds) *Forest tree physiology.* Ann Sci For 46 (suppl), Elsevier and Institut National de la Recherche Agronomique, Paris, France.

Dickson RE (1991) Assimilate distribution and storage. In Raghavendra, AS (Ed) *Physiology of trees.* John Wiley and Sons, NY.

Eissenstat DM, Van Rees KCJ (1994) The growth and function of pine roots. Ecol Bull 43:76–91.

Ericsson A, Persson H (1980) Seasonal changes in starch reserves and growth of fine roots of 20-year-old scots pine. In Persson T (Ed) *Structure and function of northern coniferous forests—An ecosystem study.* Ecol Bull, Stockholm, Sweden, 32:239–250.

Ford ED, Deans JD (1977) Growth of a sitka spruce plantation: Spatial distribution and seasonal fluctuation of lengths, weights and carbohydrate concentrations of fine roots. Plant and Soil 47:463–485.

Gholz HL, Cropper WP Jr (1991) Carbohydrate dynamics in mature *Pinus elliottii* var. *elliottii* trees. Can J For Res 21:1742–1747.

Gower ST, Haynes BE, Fassnacht KS, Running SW, Hunt ER Jr (1993) Influence of fertilization on the allometric relations for two pines in contrasting environments. Can J For Res 23:1704–1711.

Gower ST, Vogt KA, Grier CC (1992) Carbon dynamics of Rocky Mountain Douglas-fir: Influence of water and nutrient availability. Ecological monographs 62:43–65.

Grier CC, Vogt KA, Keyes MR, Edmonds RL (1981) Biomass distribution and above- and below-ground production in young and mature *Abies amabilis* zone ecosystems of the Washington Cascades. Can J For Res 11:155–167.

Hansen J, Fung I, Lacis A, Rind D, Lebedeff S, Ruedy R, Russel G (1988) Global climate changes as forecast by the Goddard Institute for Space Studies three-dimensional model. J Geophys Res 93(D8):9341–9364.

Haynes BE, Gower ST (1995) Belowground carbon allocation in unfertilized and fertilized red pine plantations in northern Wisconsin. Tree Physiol 15:317–325.

Haywood JD (1994) Seasonal and cumulative loblolly pine development under two stand density and fertility levels through four growing seasons. US Dept Agric For Ser South For Exper Sta Res Paper SO-283.

Haywood JD, Burton JD (1990) Phosphorus fertilizer, soil and site preparation influence loblolly pine productivity. New For 3:275–287.

Hipps NA, Pages L, Huguet JG, Serra V (1995) Influence of controlled water supply on shoot and root development of young peach trees. Tree Physiol 15:95–103.

Johnson DW (1995) Role of carbon in the cycling of other nutrients in forested ecosystems. In McFee WW, Kelly JM (Eds) *Carbon forms and functions in forest soils.* Soil Science Society of America Incorporated, Madison, WI.

Kerr A Jr, Griffis BJ, Powell JW, Edwards JP, Venson RL, Long JK, Kilpatrick WW (1980) Soil survey of Rapides Parish Louisiana. US Dept Agric Soil Conserv Ser For Serv in cooperation with LA St Univ, LA Agric Exper Sta.

Keyes MR, Grier CC (1981) Above- and below-ground net production in 40-year-old Douglas-fir stands on low and high productivity sites. Can J For Res 11:599–605.

Kirschbaum MUF, Alvarez A, Cannell MGR, Cruz RV, Fischlin A, Galinski W, Odera JA, Xu D (1990) The impacts of climate change on forest ecosystems. In Houghton JT, Jenkins GJ, Ephraums, JJ (Eds) *Climate change—The IPCC (Intergovernmental Panel on Climate Change) Scientific Assessment.* Cambridge University Press, Cambridge.

Kozlowski TT, Kramer PJ, Pallardy SG (1991) *The physiological ecology of woody plants.* Academic Press, San Diego, CA.

MacFall JS, Johnson GA, Kramer PJ (1991) Comparative water uptake by roots of different ages in seedlings of loblolly pine (*Pinus taeda* L.). New Phytol 119:551–560.

Marschner H (1986) *Mineral nutrition of higher plants.* Academic Press, London.

McColl JG, Gressel N (1995) Forest soil organic matter: Characterization and modern methods of analysis. In McFee WW, Kelly JM (Eds) *Carbon forms and functions in forest soils.* Soil Science Society of America Incorporated, Madison, WI.

Miller WF, Dougherty PM, Switzer GL (1987) Effect of rising carbon dioxide and potential climate change on loblolly pine distribution, growth, survival and productivity. In Shands W E, Hoffman JS (Eds) *The greenhouse effect, climate change, and U.S. Forests.* Conservation Foundation, Washington, DC.

Peters RL (1990) Effects of global warming on forests. For Ecol Manage 35:13–33.

Pritchett WL, Gooding JW (1975) Fertilizer recommendations for pines in the southeastern coastal plain of the United States. Univ Fl Agric Exp Sta Bull 774.

Russell RS, Clarkson DT (1976) Ion transport in root systems. In Sunderland N (Ed) *Perspectives in experimental biology, volume 2, botany.* Pergamon Press, Oxford.

Santantonio D, Santantonio E (1987) Effects of thinning on production and mortality of fine roots in a *Pinus radiata* plantation on a fertile site in New Zealand. Can J For Res 17:919–928.

Shoulders E, Tiarks AE (1983) A continuous function design for fertilizer rate trials in young pine plantations. In Jones EP Jr (Ed) *Proceedings of the second biennial southern silvicultural research conference.* USDA For Ser, South For Exper Sta, Ashville, NC.

Smith DM (1986) *The practice of silviculture.* John Wiley and Sons, New York.

Teskey RO, Gholz HL, Cropper WP Jr (1994) Influence of climate and fertilization on net photosynthesis of mature slash pine. Tree Physiol 14:1215–1227.

Teskey RO, Hinckley TM (1981) Influence of temperature and water potential on root growth of white oak. Physiol Plant 52:363–369.

Tiarks AE (1982) Phosphorus sorption curves for evaluating phosphorus requirements of loblolly pine (*Pinus taeda*). Commun Soil Sci Plant Anal 13:619–631.

Van Cleve K, Powers RF (1995) Soil carbon, soil formation, and ecosystem development. In McFee WW, Kelly JM (Eds) *Carbon forms and functions in forest soils.* Soil Science Society of America Incorporated, Madison, WI.

Van Rees KCJ, Comerford, NB (1990) The role of woody roots of slash pine seedlings in water and potassium absorption. Can J For Res 20:1183–1191.

Vogt KA, Moore EE, Vogt DJ, Redlin MJ, Edmonds RL (1983) Conifer fine root and mycorrhizal root biomass within the forest floors of Douglas-fir stands of different ages and site productivities. Can J For Res 13:429–437.

Vogt KA, Publicover DA, Bloomfield J, Perez JM, Vogt DJ, Silver WL (1993) Belowground responses as indicators of environmental change. Environ Exp Bot 33:189–205.

Vogt KA, Vogt DJ, Moore EE, Fatuga BA, Redlin MR, Edmonds RL (1987) Conifer and angiosperm fine-root biomass in relation to stand age and site productivity in Douglas-fir forests. J Ecol 75:857–870.

Vose JM (1988) Patterns of leaf area distribution within crowns of nitrogen-and phosphorus-fertilized loblolly pine trees. For Sci 34:564–573.

Vose JM, Allen HL (1988) Leaf area, stemwood growth, and nutrition relationships in loblolly pine. For Sci 34:547–563.

Waring RH (1991) Responses of evergreen trees to multiple stresses. In Mooney HA, Winner WE, Pell EJ (Eds) *Response of plants to multiple stresses.* Academic Press, San Diego, CA.

Watson MA, Casper BB (1984) Morphogenetic constraints on patterns of carbon distribution in plants. Ann Rev Ecol Syst 15:233–258.

12. Dynamic Responses of Mature Forest Trees to Changes in Physical and Chemical Climate

Samuel B. McLaughlin and Daryl J. Downing

The annual variations in stem growth expressed in a typical tree ring-width series, as well as the distribution and composition of biomass measurable in forest stands, reflect the time-integrated capacity of trees to respond to changes in physical, chemical, and biological environment. The environmental changes along a temporal spectrum range from very rapid shifts in either photosynthetically active radiation or temperature to intermediate scale shifts in air quality or water supply rate, to such slower changes as shifts in competition from adjoining vegetation or changes in soil-nutrient supply. Trees respond to fluctuations in their chemical and physical environments on a temporal scale that can range from seconds to years and on spatial scales that can range from molecular to morphological changes in canopy or root architecture. Understanding the critical dynamics, both in rate and magnitude, of environmental change and biological response is particularly important when evaluating potential responses of such perennial species as forest trees to future levels and fluctuations in climate variables.

This chapter reports on exploratory studies that have examined responses of mature loblolly pine trees to changes in both physical climate (temperature, rainfall, and soil moisture supply), and chemical climate (tropospheric ozone) on scales that range from days to years. The rationale for these studies was based on the lack of data on responses of mature pine trees to ozone and environmental variables in combinations that were 1) known to affect forest-growth processes (Fox and Mickler, 1996), and 2) anticipated to provide increasingly stressful environments for future forest growth (Joyce et al., 1990; Chameides et al., 1994).

In our studies, the nature of individual tree responses to fluctuations in the present-day ambient environment has been explored to better understand and project probable future responses, as well as the health of pine forests to the interactive influences of increasing ozone in combination with a potentially hotter, drier climate. We considered ozone an important component of this study because of its regional distribution at levels that are phytotoxic to plants and because it has been implicated in the present health of southeastern pine forests.

The health of southeastern pine forests has been a subject of great interest and substantial controversy because significant growth declines were detected in unmanaged pine stands by U.S. Forest Service surveys in the mid-1980s (Sheffield et al., 1985; Sheffield and Cost, 1987). Modeling studies focusing on the possible role of stand competition and climate on observed changes found that most of the growth reduction occurring in recent decades was not explained by natural factors (Zahner et al., 1989; Bechtold et al., 1991). Air pollution, particularly ozone, has been suggested as a possible causative factor, based on regional growth patterns of large trees and the known sensitivity of tree seedlings to ozone exposure (Zahner et al., 1989; McLaughlin, 1985; Shriner et al., 1990; McLaughlin, 1994). Dissenting opinions on the significance and possible causes of observed growth trends, however, have been based on the suggested instability of growth trends of some individual trees (Van Deusen, 1992), in addition to the mensurational techniques (Zeide, 1992) used to establish predecline reference growth rates.

Controlled studies of responses of pine tree seedlings and saplings to ozone in the field chambers have provided many valuable insights into the sensitivity of southern pines to regional ozone pollution (Fox and Mickler, 1996). Major concerns with most controlled studies to date, however, have been the significant uncertainties involved in extrapolating data derived from seedling and sapling exposures to elevated ozone in chambers to expected responses of mature trees under ambient ozone levels in the field. Principal points of concern include 1) differences in metabolism between juvenile and mature trees, 2) the use of ozone-exposure levels significantly above present-day ambient exposure regimes, and 3) chamber-induced modifications in plant growing environment, which may modify plant responses.

We have evaluated ozone impacts on mature forest trees by measuring the influence of fluctuations in ambient levels of ozone and other environmental variables on intra- and interannual changes in stem growth of individual trees. Five years of field research were conducted in which repeated measures and statistical analyses of stem expansion patterns were used to identify growth responses of mature trees to fluctuations in ambient levels of ozone and other environmental variables in an undisturbed forest environment. These results were reported in two previous publications (McLaughlin and Downing, 1995; 1996). In this chapter, we will describe only the basic elements of the dendrometer study design and focus on the relationship of our findings to other research on both mechanistic and regional scales. Because moisture status plays a dominant role in growth of loblolly pine (Teskey et al., 1987; Dougherty, 1996) and is projected to be an important component of climate change (Joyce et al., 1990), a major focus in this chapter is the influence of ozone levels on tree responses to moisture stress.

Methods

Growth Pattern Analysis

For analysis of seasonal growth patterns, we developed a spring-tension dendrometer band system that was sensitive enough to allow us to precisely measure changes in stem circumference at time steps of ≤ 1 day (McLaughlin and Downing, 1995). The analytical approach involved statistical evaluation of the influences of changes in a suite of environmental variables on changes in rates of stem expansion within and between years. Environmental variables included daily rainfall, temperature, and ozone exposure, and weekly changes in moisture stress index (MSI), an integrating index of weekly scale soil moisture availability.

In this research, we have evaluated factors affecting the rate of stem expansion measured over short-term intervals (typically three- to seven-day intervals) as an indicator of the relative importance of these factors on seasonal growth. Short-term measurements integrate the growth-related processes of cambial cell growth with shifts of internal water status of the tree during periods of fluctuating water supply (Lassoi et al., 1985). Stem shrinkage occurs as water is moved from the stem to the canopy during periods of increasing water stress (Kramer and Kozlowski, 1979; Lassoi et al., 1985). Because cell growth is the most sensitive physiological process to changes in internal water status (Bradford and Hsaio, 1982), the existence of water potentials sufficiently low enough to cause stem shrinkage produces growth-limiting conditions. Both reduced leaf-water potentials and low stem-growth rates are significantly correlated with stem contraction (Lassoi et al., 1985). With adequate water supply during the active growing season, rehydration and growth-dominated processes lead to stem expansion. Thus, the seasonal patterns of expansion and contraction are associated with changes in growth rates that are closely tied to whole tree water relations. The frequency and duration of such limitations will be integrated at the whole tree level to influence other such processes as photosynthesis, assimilate transport, and growth (see Chapter 23).

To characterize the dynamics of seasonal growth patterns, short-term changes in stem circumference of twenty-eight to thirty-four mature trees located in the National Environmental Research Park in Oak Ridge, TN, were initially evaluated over five growing seasons (May to October, 1988 to 1992). Bands were measured manually with an electronic caliper approximately twice weekly during 1988, 1989, 1991, and 1992, and once every one to two weeks during 1990. Measurement precision on duplicate measurements was typically + or − 0.02 mm. Band measurements were initiated on twenty-eight trees in 1988 and six trees along the forest edge were added in 1989. Measurement frequency was reduced to once per

Table 12.1. Characteristics of Loblolly Pine Stands Used in a Study of Factors Influencing Seasonal Dynamics of Growth

				Stand Characteristics			
Stand	Topography	Trees measured	Fertility[1]	Basal Area $m^2\ ha^{-1}$	% Soil moisture 1992[2]	Mean Diameter cm (std)	Diameter growth range[3] cm stem^{-1} yr^{-1}
A	Upland	8	Low	28.1	14–24	30.6 (1.4)	0.15–0.84
B	Upland (unthinned)	8	Low	40.9	18–26	34.8 (6.1)	0.39–0.78
C	Terrace	6	High	29.2	32–38	33.5 (3.4)	0.62–1.12
D	Bottom	6	High	32.9	24–33	37.4 (2.8)	0.73–1.16
E	Bottom-edge	6	High	22.1	27–34	47.2 (6.8)	0.53–1.18

[1] Soil analyses representing A horizon soils of low vs high fertility sites were total nitrogen (0.089% vs 0.133%), calcium (632 vs 3034 mg kg^{-1}), and P (10.05 vs 11.10 mg kg^{-1}), respectively.
[2] Data represent ranges in soil moisture by volume occurring between May 29, 1992 and October 6, 1992. Approximately one measurement per week was made by time-domain reflectometry. Data are averages of two to three measurements per stand per date across the twenty-three measurement dates.
[3] Average diameter growth differences across five years.

Table 12.2. Variations in Rainfall, Temperature, Ozone Exposure Dose, and Moisture Stress Index Contributed to the Significant Variations in Stem Growth During the First Five Years of This Study (McLaughlin and Downing, 1996)

	Mean environmental conditions[1]				Tree growth (mm)[2]	
Year	Rainfall (mm)	Temperature (°C)	MSI	Ozone (ppb·h)[3]	Wet	Dry
1988	4.18	31.1	−0.12 (−1.62)	471	11.54	6.05
1989	7.32	27.7	1.17 (0.19)	165	18.75	13.72
1990	5.69	28.9	0.15 (−0.63)	438	14.89	11.50
1991	5.62	29.2	0.25 (−0.04)	375	13.68	7.51
1992	3.04	26.9	0.53 (−0.02)	325	16.65	11.88

[1] Environmental data include averages of daily rainfall (mm), daytime (0900–2000) temperature (7 h max), and weekly MSI with the seasonal minimum in parenthesis, and the average daily ozone dosage included as a time × concentration sum based on hours with ≥ 40 ppb ozone levels (ppb·h). Data were calculated over a May to September interval.
[2] Mean annual circumference growth is shown separately for twelve trees in wetter, better-quality sites (see stands C and D, Table 12.1) and the sixteen trees on drier sites (stands A and B). The number of measurements per year was 37, 23, 12, 31, and 35 during the 1988 to 1992 interval. Trees from stand F were not included in the five-year modeling.
[3] Ozone data are average daily exposure for the actual time period over which growth was analyzed during each year.

month during 1993 to characterize general annual trends (McLaughlin and Downing, 1996).

Summary data on site characteristics and annual variations in environmental variables measured during the first five years of the study are provided in Tables 12.1 and 12.2, respectively. The trees sampled were from five stand classifications, two of which were on a drier, less fertile upland site (stands A and B, respectively.) The three remaining stands were on more fertile bottomland sites, two from the forest interior (C and D) and one (F) along the forest edge. Average tree diameter for stands A and B was 32.7 cm; for stands C and D the average was 35.4 cm. Both stands D and F were on an old alluvial flood plain within 25 and 10 m, respectively, of a small stream. Sampled trees were selected at approximately 30 m intervals from line transects within each stand. All selected trees were canopy dominants, averaging approximately seventy years in age. Stands A, C, D, and F had undergone two commercial thinnings, the last taking place around 1980. Stand B was left unthinned at the last cut as a part of a nutrient cycling study, and had approximately 50% more basal area and density than the other four stands.

Rainfall data were collected from stands A and D at each growth-measurement date, but were used only to contrast the sites. Rainfall was 16% and 11% higher at Stand D compared to Stand A during 1991 and 1992, respectively. Daily patterns and standard deviations were very similar. Daily rainfall and hourly temperature data collected from the nearby Walker Branch Watershed (2 to 4 km from the study sites), which was provided by the National Oceanic and Atmospheric Administration, supplied the needed temporal resolution for growth analyses. Hourly ozone data collected from a rural station approximately 30 km from the study site were provided by the U.S. Environmental Protection Agency (EPA). A comparison of hourly ozone data collected adjacent to Stand F with the EPA data during thirty-one days in 1992 gave nearly identical means (53 ppb, research site; 55 ppb, EPA), standard deviations, and ranges.

Weekly values of a MSI derived from multiple stations in eastern Tennessee were obtained from the National Weather Service. The MSI used in these studies is a weekly derivation of the Palmer Drought Severity Index (Palmer, 1965), a monthly indicator that has proven very useful in studies of growth responses of southern pines to changing moisture status (Zahner et al., 1989). Moisture supply index is a calculated index of the balance between water supply, water storage in soil, and water demand resulting from evaporation and runoff. Values between $+1$ and -1 on this log scale index are considered within the normal range. Negative values of MSI indicate that evaporative demand has exceeded supply, and intermittent index values less than $-2.0, -3.0,$ or -4.0 are indicative of moderate, severe, and extreme drought, respectively.

Soil-moisture data were collected at soil depths of 15 cm and 30 cm from two to three locations in each stand using time-domain reflectometery (TDR) (Topp, 1987). Ranges in soil moisture across the five stands during the 1992 growing season are included in Table 12.1. These indicated substantial separation of soil-water content between the stands over the two measurement years (1991 and 1992) when TDR measurements were made.

Both correlation and regression analyses were used to evaluate seasonal growth response to climate variables (McLaughlin and Downing, 1996). These analyses were directed toward determining those environmental variables that most significantly influenced short-term fluctuations in stem circumference increment and to explore interactions among these factors within and across yearly time scales. Tests of the direct and interactive effects of environmental variables on growth patterns of each tree during each year were performed initially to evaluate the strength and consistency of influences of these variables over time. Thus, a separate model of seasonal growth responses was initially derived for each tree for each year. For this chapter, we emphasize the differences between individual trees and groups of trees on sites differing in soil-moisture content, as well as differences between years with widely varying rainfall amounts.

Initial screening analyses were performed at the individual tree level by first removing the linear time-based trends from both seasonal growth curves and climatic data and then evaluating the strength and consistency of partial correlation coefficients (Johnson and Wichern, 1988). Variations in both environment and growth variables around these trends were compared for twenty-seven variables describing ozone-exposure dose, rainfall, and temperature. Details on variables examined including three and seven-day lagged responses can be found in McLaughlin and Downing (1995). Following the screening of variables and response times across trees, sites, and years, a weekly scale multiple-regression model of stem growth was developed around the entire 1988 to 1992 data set. The purpose was to develop an empirical model to quantify influences of interacting environmental variables on seasonal stem-growth dynamics. Annual growth rates were then evaluated in relationship to annual ozone-exposure dose across six years to quantify the influences of prior selected indicator variables on annual stem growth over time (McLaughlin and Downing, 1996).

Results

There were significant intra- and interannual variations in both physical and chemical measures of climate during the five years of this study, as shown in Table 12.2. Rainfall, ozone exposure dose, and temperature (seven-hour maximum) varied by approximate factors of 2×, 3×, and 3 °C across the initial five years of the study providing an excellent range of conditions to test the sensitivity of growth to fluctuating climatic conditions.

Examination of the seasonal growth curves of six individual trees representing a wide range of growth rates during 1988 and 1992 (Figure 12.1) reflects the very high intra- and interannual variability. The two years included 1988, a warm, dry year with high ozone levels, and 1992, the coolest year, with relatively low ozone levels and moderate rainfall. Although individual annual circumference growth rates varied by a factor of 4 to 5 between trees in the same year, mean annual growth rates differed by approximately a factor of 3 in some cases for some individual trees for the same two years. The relative differences in growth rates among trees in the same stand showed marked shifts between years, which indicates that substantial differences in sensitivity to short-term climatic shifts can occur among trees from year to year even within the same stand.

Slowdown in stem expansion, sometimes accompanied by stem shrinkage, was typically associated with periods of higher ozone exposure throughout the five-year study. Sometimes these ozone episodes were also associated with periods of high temperature or drought, but the periodicity of episodes of highest ozone, temperature, and moisture stress was often quite varied, as shown in Figure 12.2. Variations in associations among environmental variables that developed over shorter time scales provided a basis of testing the timing and consistency of associations among variables influencing growth. During 1988, for example, high ozone exposure accompanied moderate drought during the spring and early summer. Although still responsive to short-term fluctuations in environmental variables, trees on the drier upland sites (stands A and B) experienced stem shrinkage during this unfavorable period while those trees on wetter sites grew steadily, albeit slowly (Figure 12.2). Differences between faster and slower growing trees developed at the earliest date on the poor-quality site and as ozone exposure increased significantly. At this stage, the weekly MSI was still at very modest levels. Differences were at the maximum approximately two weeks before the most negative values of a MSI indicated that the drought had peaked. Differences lessened with decreasing ozone around day 190, even though the MSI was at its most negative value.

Our analyses of the direct and interactive influences of ozone, rainfall, air temperature, and soil moisture on stem expansion rates included correlation

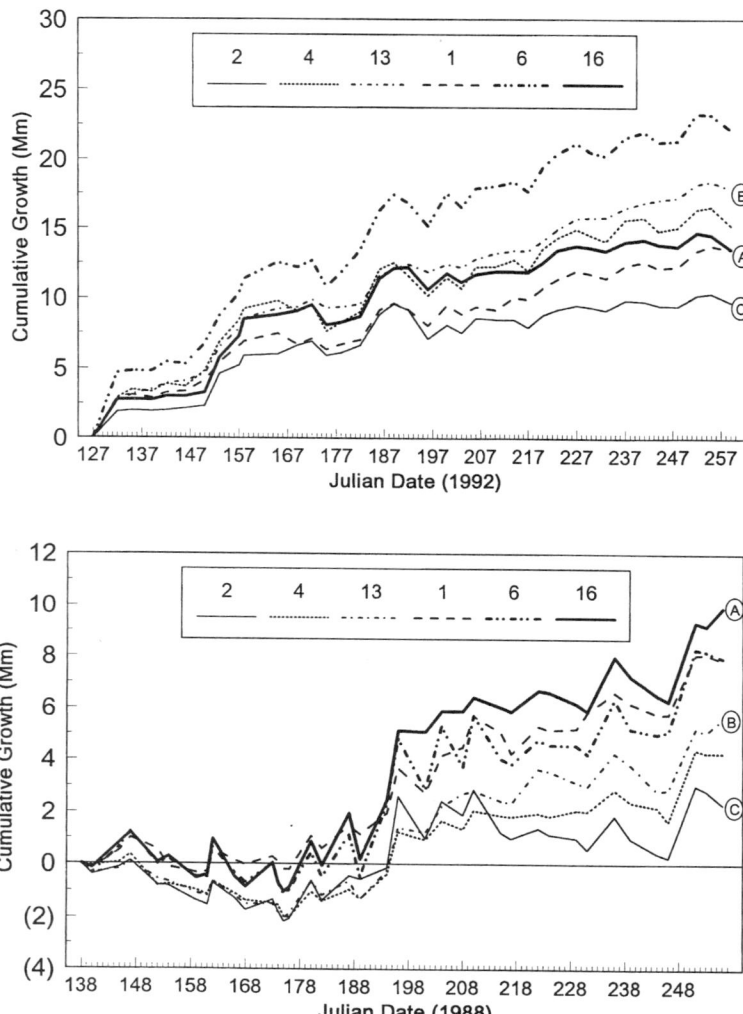

Figure 12.1. Contrasts in the seasonal dynamics of stem growth of six trees from a relative dry, infertile site (see Table 12.1) during two years that differed significantly in temperature, rainfall, and ozone exposure (see Table 12.2). Mean diameter was 31.5 cm and ranged from 28.6 cm for tree 13 to 35.3 cm for Tree 18. Note similarities in the short-term patterns of change among trees and differences in trends with time. Note also the shift in relative growth of Trees 16, 13, and 2 (indicated by A, B, and C) as growth differed markedly between the two years.

12. Responses of Mature Forest Trees to Climate Changes 215

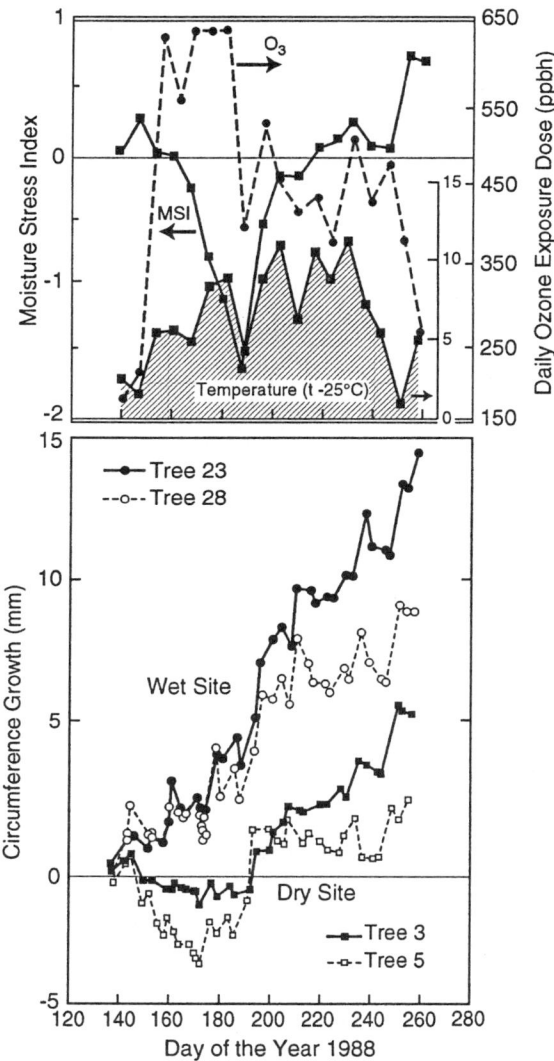

Figure 12.2. Comparison plots of the differences in the seasonal dynamics in circumference growth of the fastest and slowest growing trees were examined to evaluate the timing and cumulative significance of growth differences of trees in poor- and better-quality sites during 1988. Note that both the episodic departures in growth rates between tree categories early in the season were followed by chronic differences in growth rate that developed over time. Overlays of the weekly averages of three-day average ozone exposure dose ($\geq$ 40 ppb), maximum seven-hour temperatures, and MSI are indicated for comparison with emerging trends. Also note the differences in the timing of periods of highest and lowest values for environmental variables. Diameters for Trees 3, 5, 23, and 28 are 27.3 cm, 29.9 cm, 33cm, and 32.9 cm, respectively.

Table 12.3. Numbers of Individual Trees for Which Weekly Growth Rates Were Significantly Influenced by Indices of Chemical and Physical Climate. Multiple Regression was Used to Evaluate twenty-eight Trees in 1988 and thirty-four Trees Thereafter for Wetter Sites (W) and Drier Sites (D)

Variable	Significant responses (W/D) by year[1]				
	1988	1989	1990	1991	1992
Moisture supply	2/2	13/13	7/5	1/0	2/2
Rain	4/4	0	0	17/10	14/12
Temperature	4/2	13/11	8/10	16/16	18/16
Ozone	7/7	0	6/3	0	0
Ozone * MSI	5/13	0	0/4	0	0
Ozone + temp	0/9	0	0/1	4/1	0
Direct ozone	14	0	9	0	0
Total ozone[2]	27	0	11	5	0

[1] Responses are designated as the number of trees for wet sites (total population: twelve trees in 1988, eighteen thereafter)/dry sites (sixteen trees). A $P \leq 0.05$ significance level was used.
[2] Includes interactions.

analyses followed by stepwise regression analyses to test for significant influences of those variables. Initial correlation analyses were used to select the appropriate variables and time intervals for subsequent regressions analyses. These analyses indicated that ozone was a significant factor influencing short-term stem expansion rates across a wide range of conditions which included 1989—the year with the lowest ozone level. Multiple regression analyses were then used to evaluate the direct and interactive effects of all combinations of the dominant variables identified in the previous analyses.

Significant ozone effects on growth were detected ($p \leq 0.05$) by multiple-regression analysis of five or more trees during three of the five years. The significance of ozone impacts was directly proportional to the levels of ambient ozone that occurred, ranging from twenty-seven of twenty-eight trees in 1988, the year with highest ozone levels, to 0 in both 1989 and 1992, which were the two years with lowest ozone levels. In Table 12.3, both direct and interactive effects of regression variables on weekly growth rates of trees from the drier (A and B) and wetter (C, D, and F) stands are compared. The five years of data on weekly responses of stem growth to environmental variables were then examined collectively to generalize responses across years (McLaughlin and Downing, 1995). Multiyear environmental influences were coupled with annual growth trends determined for each year to develop the empirical model in which:

$$\text{Growth} = S \times \text{Time} + 0.0036 \times O_2 \times \text{Moist} + 0.0380 \times \text{Rain} + 0.0515 \times \text{Temp} - 0.0000533 \times O_2 \times \text{Temp}$$

in which:

Growth = The cumulative stem circumference determined at various times throughout each specific year.

12. Responses of Mature Forest Trees to Climate Changes   217

S =   Slope of the annual growth trend by year. The values ($+$ or $-$ S_e) for 1988 to 1992 were 0.0529 (0.005), 0.117 (0.005), 0.104 (0.004), 0.098 (0.004), and 0.143 (0.004), respectively.

Time =   Time from the beginning of the annual measurement cycle (approximately mid-May) in days. Time was reinitialized to zero at the beginning of each measurement year.

O_2 =   Average of three-day mean ozone exposure for all hours at or above 40 ppb.

Moist =   MSI, calculated weekly. The O_2 × Moist coefficient had a S_e of 0.0006.

Temp =   Average three-day air temperature during the hours 0900 to 0800. The O_2 × Temp coefficient had a S_e of 0.000026.

Rain =   Average three-day rainfall in mm d^{-1}. The Rain coefficient had a S_e of 0.014.

Diagnostic tests indicated that multicolinearity among variables included in this model was very low (Belsley et al., 1980). In addition to these indications, tests of autocorrelation (Durbin and Watson, 1951) of variables detected no significant autocorrelations for any variable within any annual series. Regression terms included were significant at the 0.05 level and the model form was generally consistent in structure for all the twenty-eight trees examined in the five-year data set. For example, ozone × MSI and rainfall were selected as the two most important variables in twenty-seven of the twenty-eight trees examined.

The model was constructed to include large sources of low frequency variance, by including variation across years and the phenological, time-dependent accumulation of growth within years. Thus, the high statistical significance of the model ($F_{8,82} = 906$, $p = 0.0001$) was not a meaningful test of model performance, but rather of model construction. High frequency variance that described the weekly scale responses to ozone and climate was a rather small component of the total variance in the model (Table 12.4), but was isolated from the larger, low frequency variance by stepwise regression. Approximately 51% of the high frequency variability, which remained after time trends were removed by stepwise regression, was explained by the environmental terms. Interactions between ozone and moisture stress accounted for 58% of this total, and rain, temperature, and temperature × ozone interactions accounted for 25%, 12%, and 5% respectively.

The form of the model, which includes interaction terms for ozone and MSI and ozone and temperature, but no individual terms for ozone or MSI, warrants comment. First, the absence of direct effects in a model developed across trees representing widely varying site conditions (basal area, soil moisture, and fertility) does not detract from documentation of measurements of direct effects of both MSI and ozone on individual trees from individual years as demonstrated in Table 12.4. Second, regression models with significant interactions terms that do not appear as significant individual terms can be used as valid prediction models (Draper and Smith, 1981). We consider the ozone × MSI and ozone × temperature

Table 12.4. Analysis of Variance of Terms Included in a Multiyear Empirical Model of Environmental Influences on Annual Stem Circumference Growth Rate

Source	Degrees of freedom	Sum of squares	Mean square	F-value	P-value
Regression	9	8769	974	945	0.0001
Annual slopes[1]	5	8672	1734.5	1684	0.0001
Ozone × SMI	1	54.4	54.4	53	0.0001
Rainfall	1	29.9	29.9	29	0.0001
Temperature	1	9.3	9.3	9	0.0021
Ozone × temperature[2]	1	3.3	3.3	3.2	0.0768
Error	82	84.5	1.03		

[1] The sum of squares reported here is the sequential sum of squares resulting from that term being added to the model that includes all terms above it. Thus, the sum of squares resulting from ozone × MSI is the increase in sum of squares resulting from regression by adding the ozone × MSI term to the model which includes annual slopes.

[2] Ozone + temperature was significant at the $P \leq 0.05$ level for most comparisons involving subsets of the data and was included in the general model.

terms in this empirical model to reflect the amplification of ozone effects and the effects of low moisture supply when either supply or demand for water is enhanced by low MSI or high evaporative demand, respectively.

We have used the empirical model as an exploratory tool to predict the influence of ozone and ozone × climate interactions on growth during the 1988 to 1992 analysis interval, as well as to isolate and estimate the contributions of ozone under two levels of soil-moisture stress (McLaughlin and Downing, 1995). The predictions involved utilizing mean values of empirical environmental data from the study interval as model input terms. Estimates of the contributions of combinations of environmental terms encountered during each study year on total growth were made using the empirical model. Combined annual influences of ozone and its interactions with the ambient environment on the high frequency components of growth ranged from a maximum of 13% in 1988 to almost 0% in 1989 and 1992 (Figure 12.3A). Because these terms express the combined effects of ozone interactions with MSI and temperature, the effects of annual variations in ozone were intrinsically evaluated by holding other environmental terms constant while inputing actual ozone levels occurring during each year. To estimate annual variations in effects attributable to ozone, 1989 was used as the "control year" against which the effects of relatively higher ozone levels for each other year were measured. Two model predictions were run for each year to evaluate the role of MSI in modifying ozone effects. The first involved a "no-water-stress" regime with MSI held constant at 0 across years. The second utilized a moderate (MSI = −1) stress level. Predicted growth losses resulting from ozone ranged from near 1% (1992) to a maximum of approximately 7% (1988) under the low-moisture stress regime (average = −2.8%). Increasing MSI from 0 to −1 amplified predicted ozone effects by a factor of approximately 5 with a range of −4.6 to −39% and a mean response across five years of −13.9%, thus emphasizing the potential

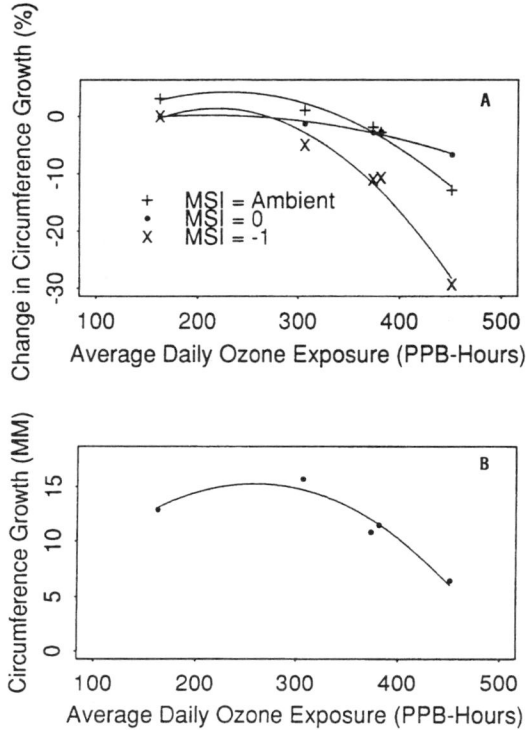

Figure 12.3. (A) Estimates of the combined high frequency effects of ozone and climate under ambient conditions (+), and the specific effects of ozone alone on total seasonal growth were made using an empirical model defined by five years of seasonal growth data. Combined influences of ozone and climate ranged from a maximum of −13% in 1988 to near 0 in 1989 and 1992. Estimates of ozone-specific effects on growth under two levels of soil moisture stress (MSI = 0, and MSI = −1) emphasize the importance of moisture availability on the level of ozone effects on growth. (B) The annual growth trends identified in the above empirical model reflect longer-term influences of both physical and chemical climate on growth patterns.

importance of ozone-water stress interactions on stem growth (Figure 12.3, after McLaughlin and Downing, 1995).

The empirical model and annual outputs represented in Figure 12.3 were derived from averaging responses across four stands, thereby including variability in soil-moisture availability from drier and wetter sites and differences in sensitivity to environmental variables determined on a tree-by-tree basis. In Table 12.5 we compare the effects of model input parameters on seasonal growth in 1988 for all twenty-eight trees, the trees on the dry site only, and the eight trees most consistently sensitive to ozone. The latter included trees for which direct effects of ozone were significant in 1990, all of which had significant direct or

Table 12.5. Predicted Effects of Environmental Influences on Loblolly Pine Growth over a 110-Day Period in 1988 Based on a Five-Year Empirical Model. All Terms Are Expressed as a Percent Contribution to Total Circumference Growth for Trees in Each Category. Ambient Environmental Conditions for 1988 Were Inputs to the Model

Variable	All trees[1]	Dry sites[2]	O_3 sensitive[3]
		% of total growth	
Time[4]	+88.7	+86.2	+89.6
Rain	+3	+5.1	2.5
Temperature	+21.6	+32.4	+19.8
$O_3 \times$ MSI	−3	−5.8	−2.5
$O_3 \times$ temp	−10.1	−17.8	−9.5
Total growth (mm)	6.59	3.82	7.61
Ozone effects (%)[5]	−13.1	−23.6	−12

[1] Twenty-eight trees (stand A − D).
[2] Sixteen trees (stand A + B).
[3] Eight trees (3 from dry sites and 5 from wetter sites).
[4] Time expresses the influence of the low frequency, linear trend of seasonal growth rates. Variations in other variables explain episodic deviations from that linear trend and which ultimately influence the linear trend.
[5] Summed influences of ozone × MSI and ozone × temperature effects.

interactive effects in 1988. Three of these were from drier sites and five were from wetter sites. Table 12.5 emphasizes the greater relative effects of ozone on drier sites. Trees on which ozone effects were detected (including individuals from both wet and dry sites) did not have a higher average predicted ozone effect. This points out the reality that statistical detection as tabulated here relates to both variance and consistency of effects of responding trees to environmental variables, and may not reliably predict the magnitude of effects. The relative magnitude of effects of input terms on total growth over the 110-day projected growth period should be noted, in addition to the temperature effects that were positive individually, but were negative when combined with higher ozone.

In addition to the high frequency components of growth, there were substantial differences in the linear growth trends identified by the empirical model for each of the years, 1988 to 1992 (Figure 12.3B). While year-to-year variability in trends was well-related to ozone as noted in Figure 12.3B, we cannot quantify the individual contributions of ozone and other environmental variables to the annual variations in these linear trends. Rather, we infer that variables most significantly influencing weekly growth rates are also important contributors to these trends.

As an exploratory test of the influence of ozone and other climatic variables on the longer-term trends, annual growth rates for each year were evaluated separately in a linear regression involving circumference growth, MSI, and ozone exposure dose (McLaughlin and Downing, 1996). Additional terms for both temperature (twenty-four hour mean) and ozone dose (ppb hours ≥ 80 ppb) were included in these analyses. Additionally, annual growth and environmental data were included from 1993, an extremely dry year in which seasonally averaged MSI (−1.57) was comparable to the driest weekly interval in 1988. Results

Table 12.6. Relative Influence of Ozone, Temperature, and Soil Moisture Stress on Annual (May to September) Circumference Growth of Loblolly Pine on Wet and Dry Sites over Six Years (after McLaughlin and Downing, 1996)

Wetter sites			Drier sites		
Variable	R^2	P	Variable	R^2	P
$O_3 > 80$ ppb	0.71	≤ 0.025	24 hour temp	0.75	≤ 0.05
$O_3 > 40$ ppb	0.66	≤ 0.05	Max 7 hour temp	0.40	≤ 0.10
24 hour temp	0.59	≤ 0.05	$O_3 > 40$ ppb	0.32	NS
MSI	0.44	≤ 0.10	$O_3 > 80$ ppb	0.36	NS
Max 7 hour temp	0.22	NS	MSI	0.17	NS

[1] Analysis were performed as single variable linear regressions to estimate and contrast relative influences of individual variables on stem growth from mid-May to mid-September.

shown in Table 12.6, which differentiates responses between wetter and drier sites, indicate a relatively higher level of statistical significance of ozone on the wetter sites, while temperature appeared more significant on drier sites. The implications of differences between these results and those in Table 12.5 are discussed in the next section.

Discussion

This exploratory study has examined both intra- and interannual variations in stem-growth rates of mature loblolly pine trees at localized sites in Tennessee, which represent widely varying soil-moisture supply, fertility, and stand-stocking levels. Significant variations in rainfall, temperature, and ambient ozone levels over the five-year study interval have allowed an examinination of the relative levels of influence of ozone and other predictor environmental variables on the patterns of growth observed. An important premise of this work has been that factors strongly and consistently influencing growth over successive three- to seven-day intervals will be important contributors to seasonal growth rates.

Collectively our data indicate that ambient levels of tropospheric ozone occurring in the southeastern United States can act individually and interactively with climate to significantly influence stem expansion and stem growth of mature loblolly pine trees. These effects were found to be highly variable between trees, and both between and within years and are expected to be related to repeated occurrences of levels above a biological response threshold. The rapidity of changes in stem expansion following short-term increases in ozone exposure noted in this study suggests that ozone is altering whole tree water balance and that the ozone-induced increases in water stress result in reduced rates of stem growth. We will evaluate the evidence that ozone is a contributor to such changes at two levels, which are 1) historical patterns of growth of mature trees in forest stands evaluated in dendroecological studies, and 2) mechanistic studies in the field and laboratory where varying effects of physiological processes, including both water relations and ozone effects on regulation of water relations, have been examined.

The importance of changes in physical climate on the annual stem growth of loblolly (Zahner et al., 1989) and shortleaf pine (Grissino-Mayer and Butler, 1993) have been evaluated through dendroclimatic analyses. However, the importance of chemical climate has only been inferred based on the inability of physical climate (Grissino-Mayer and Butler, 1993) or climate and competition (Zahner et al., 1989) to explain decreasing growth rates.

Grissino-Mayer and Butler (1993) specifically analyzed the inadequacy of their empirically based dendroclimatic model to predict growth during the last twenty-three years of the 1910 to 1986 analysis interval. They state ". . . climate variables do not adequately model growth beginning in 1963 as the residuals from the climate/growth model show increased variance over previous periods. This change in pine growth rates since 1963 must therefore be due to nonclimatic factors."

Two larger-scale phenomena that are plausible explanations of a shift in model performance are 1) a shift in competition as suggested by Van Deusen (1992), or 2) a change in chemical climate, tropospheric ozone. Van Deusen (1992) observed the growth decline found by others, but suggested that the dynamics of change represented a sequence of growth increase and decrease that made competition a plausible explanation for the growth slowdown. However, it should be noted that evaluating levels of stand-stocking and physical climate—specifically drought—in explaining growth patterns of southern pines (Zahner et al., 1989) is vitally important. By contrast, the influence of drought and annual rainfall patterns, as well as stand competition was included in the empirically based modelling of regional pine growth by Zahner et al. (1989). The latter researchers found that changing levels of competition and climate explained only about one-third of the 50% decrease in annual radial growth rate observed during the thirty-five-year period between 1949 and 1984. Both strong increases in growth in the early 1960s followed by decreases around 1980, which were closely related to Palmer Drought Severity Index, were identified in the growth series by Zahner et al. (1989).

Without considering annual levels of drought, the significant reduction in rainfall between the mid- and late 1970s makes any attempt problematic when annual growth patterns during this time are interpreted or shifts in competitive status are attributed (Van Deusen, 1992). Zahner et al. (1989) concluded that the nonclimatic factors, other than competition, had operated to ameliorate tree-growth responses to favorable climate during the 1970s and had sharply accelerated the decline in growth during the drier years (1980s) that followed.

The importance and rapidity of physical and chemical climate effects is well demonstrated in the present study by the 50% increase in growth rate noted from 1988 to 1989. The same approximate increase occurred in thinned and unthinned stands that differed by approximately 50% in stand density and basal area (Table 12.1). Our estimate of a 39% decrease in growth as a result of ozone effects under a mild water-stress regime based on the empirical model, is also similar to the

Zahner et al. (1989) estimates of a 38% decrease in annual growth resulting from factors other than physical climate in the drier 1980s.

Several other studies have suggested that tree-growth reduction or visible symptom expression are amplified by combinations of ambient ozone and drought. These include Ponderosa pine (Miller et al., 1989) and Douglas fir (Peterson et al., 1995) in California, and silver fir in the Jura Mountains of France (Bert, 1993). Additionally, controlled studies with field-grown soybean (Heggestadt et al., 1985), have documented the capacity of ambient ozone plus soil-moisture stress (SMS) to significantly amplify (5×) the effects of either ozone alone (-5%) or SMS alone (-4%) on yield at a Maryland site.

Physiological studies on the water relations of loblolly pine, controlled studies on ozone effects on water relations of loblolly pine seedlings and saplings, and a variety of studies on ozone effects on stomatal regulation of water loss provide a valuable basis for evaluating our experimental results. Although SMS is very important to the growth and development of loblolly pine (Dougherty, 1996 ; Teskey et al., 1987), the species has an considerable capacity to adjust physiologically and morphologically to high levels of moisture stress over time. This may take the form of increased water use efficiency in foliar gas exchange (Seiler and Johnson, 1985) or an increased proportion of root to aboveground growth (Bongarten and Teskey, 1987). The root:shoot balance may also be increased by reduced foliage production or reduced retention, which is a typical response of loblolly pine to drought (Dougherty, 1996), and which is also produced by chronic exposure to ambient ozone (Kress et al., 1992). The capacity of loblolly to adjust physiologically and morphologically to water stress explains its capacity to maintain similar water status in the canopy after thinning (Cregg et al., 1990). What appears to be critical to loblolly pine and other species' responses to water deficits is the rate of water loss during periods of maximum demand (Moehring and Ralston, 1967; Lassoi et al., 1985). Thus, Moehring and Ralston (1967) using dendrobands, found that diameter growth of loblolly pine was slowed when SMS depletion was rapid (that is, the demand was high), regardless of soil-moisture availability. This finding is particularly important in the present study because there is considerable evidence to indicate that ozone at near-ambient levels increases water demand in plants, including loblolly pine.

A variety of European studies of coniferous species have provided strong evidence that ozone can uncouple normal stomatal regulation of water loss (Keller and Hasler, 1984; Skarby et al., 1987; Maier-Maerker and Koch, 1992; Wallin and Skarby, 1992), which results in reduced water use efficiency and leads to decreased protection from drought. A particularly interesting aspect of the loss of stomatal regulation has been a failure of stomates to close in response to SMS and increased stomatal aperture at night (Wallin and Skarby, 1992).

Two studies with southeastern pine species support similar sensitivity of pines to combinations of ozone exposure and drought. Increased rates of transpirational water loss have been reported on loblolly pine seedlings exposed to ozone (100 ppb) and drought (Lee et al., 1990). Drought appeared to predispose these seedlings to ozone-induced changes in foliar gas exchange. Although photosynthesis

(P_n) increased by approximately 33%, transpiration increased 56% in response to ozone (0.10 µl l^{-1}), which was followed by a subsequent drought and rehydration treatment. Even though P_n was increased by O^3 at full hydration, P_n declined more rapidly in response to a subsequent drought in O^3-treated seedlings.

In working with three-year-old shortleaf pine in open-top chambers, Flagler et al. (see Chapter 4) found that drought and ozone did not cause stomatal closure and did not provide protection against further ozone uptake as suggested by other controlled studies with container-grown plants (Tingey and Hogsett, 1985; Temple et al., 1992). Yield data suggested some amplification of growth effects of ozone at highest SMS levels. Of particular interest in this study, however, was evidence that ambient-level ozone coupled with intermediate SMS levels amplified predawn plant-water stress and that the effect was typically greater than that occurring with ozone levels that exceeded the ambient level by up to 2.5×. All ozone treatments in this study exerted their influence on daily water balance by reducing predawn water potentials, not by reducing midday water potential. Thus, ozone reduced the recovery of water potential during the evening hours in all treatments.

In the present study, we used an empirical model to contrast effects of ozone across sites that differed widely in soil-water availability. Although the sites did not vary strongly in the number of trees that responded significantly to ozone and other variables, (Table 12.3), the magnitude of predicted effects on growth by the model during a high ozone year, and the relative importance of ozone and other variables was notably different between sites (Table 12.5).

We consider the importance of temperature as a predictor of annual growth rates on the dry sites (Table 12.6) to be a reflection of temperature-driven increases in evaporative demand and moisture stress, and not a direct effect of temperature on growth as such. Loblolly pine has the capability to adjust its photosynthetic rate to higher temperatures (Strain et al., 1976) and temperature effects on photosynthesis within the range experienced in this study are expected to be largely positive at this latitude (McNulty et al., 1996). Temperature had a positive effect on growth in our empirical model (Table 12.5). Negative temperature effects were only apparent in the empirical model when elevated temperature was combined with higher ozone. In Table 12.6 in which single variable regressions are used to examine influences of environmental variables at an annual scale, temperature had a negative regression coefficient as 1988 combined high ozone, high temperature, high SMS, and low annual growth. For loblolly pine growing at a similar latitude, McNulty et al. (1996), indicate that the effects of an approximate 1 °C increase in growing season temperature above the average levels experienced in 1988 (22.9 °C) would produce only a 2% decrease in net primary productivity. They found the major effects of temperature on loblolly pine were in water use. In this capacity, increasing temperature would be expected to amplify ozone effects on water stress by increasing water demand.

Our studies indicate that ozone and SMS interacted significantly to influence growth of mature loblolly pine trees. Despite the greater significance of annual ozone dose than MSI in explaining variations in annual growth (Table 12.6), it is clear that these two variables will often be inversely correlated in nature. Under our study conditions, we found this to be a time-dependent correlation. Weekly correlations among these variables were not significant within any single year, however correlations were highly significant when weekly measurements across five years were combined ($R^2 = 0.46$; $p < 0.01$). Annual average MSI data were very also highly correlated with chronic ozone exposure to > 40 ppb ($R^2 = 0.88$; $p < 0.01$). Thus, with annual scale dendroclimatic analyses in regions with significant ozone exposure, there is a high probability that growth responses that are associated with drought or temperature will be influenced by ozone and drought-ozone interactions. Under such conditions it will take either a long-time series of concurrent annual scale measurements of both ozone and climate variations or high frequency intraannual measurements to separate causal from associative influences.

The application of a repeated-measures approach involving serial measurements with dendrobands represents a sensitive tool for addressing the role of multiple environmental stresses on tree growth. Although our application of this technique in a multiple-regression approach to evaluating pollutant effects is a new approach, the concepts of multivariate analysis of intraannual tree growth patterns involving the use of dendrographs were introduced long ago (Fritts 1960; 1962). Our empirical model, derived from the in situ data on large trees, indicates that ozone can strongly influence tree growth particularly when acting in combination with high temperatures and periods of water stress. Failure to consider these interactions may lead to significant errors in identifying and estimating the effects of present-day and future physical climate and air pollution on tree growth.

The level of ozone at which such interactions occur is an important policy issue. In this study we have attempted to define an ozone-exposure threshold for adverse effects on growth based on evaluating the relative strengths of many different indices of exposure. A major issue in the process of quantifying the phytotoxicity of chronic ozone exposure has been the relative importance of episodic high level exposures and chronic lower levels of ozone in the overall exposure history (Hogsett et al., 1988; Lefohn and Foley, 1993). Our results suggest that ozone-exposure dose at and above a threshold level is a valuable index for evaluating growth responses of forest trees in the field. We did not exhaustively test various concentration-weighing functions in these analyses, but did find that the three-day average of ozone exposure hours at and above 40 ppb (D340) (McLaughlin and Downing, 1995) provided somewhat stronger relationships to growth than the concentration weighted function (W126) with a similar threshold as described by Lefohn et al. (1992).

The exposure-response threshold level for mature loblolly pine identified in this study agrees rather well with the response threshold of 45 ppb (twelve hour mean) suggested by Taylor (1994) based on studies of controlled exposures with seedlings. For sensitive individual trees, an even lower threshold (25 ppb) was indicated, a value only 30% above the natural background level of 19 ppb (Taylor, 1994). Recent analyses of crop-growth responses in the United States (Krupa et al., 1994) and a variety of studies in Europe (Legge et al., 1995) also support the importance of chronic exposure to low ozone levels (≤ 50 ppb) as significant to adverse impacts on plant growth. Under such low ozone levels, the importance of altered moisture stress may play a more significant role in limiting growth than reductions in photosynthesis (see Chapter 23). As ozone levels increase, the relative importance of responses involving foliage production and retention (Kress et al., 1992), carbohydrate production and utilization (Richardson et al., 1992) can be expected to play a greater role.

Summary

These results document growth inhibitions of mature forest trees that are significantly influenced by their exposure to ambient levels of ozone. The negative effects of ozone were exerted most strongly under SMS (when MSI becomes negative) and through negative interactions with temperature. These measurements on large trees represent a first step in documenting ozone influences on both water relations and growth processes at the whole tree level. Although documenting the specific mechanisms involved was not the focus of this research, there is ample evidence that disruption of foliar water use efficiency linked to increased transpiration and reduced stomatal control of water loss is an important component of tree responses to ambient ozone.

Our measurements and analyses have focused primarily on relationships between moderate levels of ozone exposure and the seasonal dynamics of growth of mature trees. These responses and correlations of annual differences in growth rate with ozone and other climatic variables support a significant role of ozone in longer-term growth patterns as well. Both short- and longer-term analyses indicate that ozone and climate, including high temperatures and SMS, interact to influence tree growth. These results are supported by inferential evidence from both controlled ozone exposure studies (Chappelka and Freer-Smith, 1995) and from modeling studies that have indicated that factors other than stand dynamics and climate have contributed significantly to periodic decreases in radial growth of some natural stands of southern pines during recent decades.

This study has focused on a single commercially important species, within one region. Our sampling scheme has broadened the applicability of these results by inclusion of widely varying site factors. Fortuitous and highly variable climatic conditions were also encountered. However, we suggest that similar analyses of loblolly and other species in other regions would be desirable in developing more refined predictive capabilities for assessing regional influences of ozone on forest growth.

Our data do not suggest that there will be a precipitous and steady decline in forest growth across the region in the near future, but rather that ozone will amplify responses to a variable climate, and to the natural stresses that affect forest growth. Recent surveys (1986 to 1993) of annual forest growth in the Piedmont region (Brown, 1993) and the coastal plain of South Carolina (Koontz and Sheffield, 1993) show an increase in average annual growth of softwoods over this most recent analysis period. These data do not relate to the presence or absence of an ozone effect, nor necessarily to the growth rates of individual trees. Rather, they reflect significant structural changes in the stands that occurred over this time interval. Analysis of the data reveals that the increases in net annual growth were driven by significant increases in plantation establishment and the inclusion of smaller size classes in the more recent sample (Brown, 1993; Koontz and Sheffield, 1993). Climate and competition will always significantly influence growth of southern pines, but we suggest that those influences could be amplified or ameliorated (Zahner et al., 1989) during time periods with higher ambient ozone levels.

The ambient environmental conditions during the five years of this study encompassed annual shifts in temperature and precipitation that have been predicted as more normal in the decades ahead (Joyce et al., 1990). Additionally, our measure of biologically significant ozone-exposure dose varied by a factor of approximately threefold (1988 vs 1989) over the course of this study. Predictions of approximately threefold increases in tropospheric ozone during the next three decades have been based on increases in NO_x emissions in large agroindustrial regions of the world (Chameides et al., 1994).

Ozone is already an important part of the present-day chemical climate. Possible increases in temperature as predicted by global scale models, particularly coupled with reduced rainfall, and increased NO_x emissions will probably increase the ecological and economic importance of tropospheric ozone. Increased rainfall or lower temperatures, on the other hand, could reduce ozone impacts. Because of the ozone-climate interactions demonstrated in these data, we suggest that stronger consideration of such interactions will be necessary in predicting the effects of both atmospheric pollution and future climate change on forest growth and forest ecosystem function.

References

Bechtold WA, Ruark GA, Lloyd FT (1991) Changing stand structure and regional growth reductions in Georgia's natural pine stands. For Sci 37:703–717.

Belsley D A, Kuh E, Welsch RE (1980) *Regression diagnostics*. John Wiley and Sons, New York.

Bert GD (1993) Impact of ecological factors, climatic stresses, and pollution growth and health of silver fir (*Abies alba* Mill.) in the Jura Mountains: An ecological and dendrochronological study. Acta Oecol 14(2):229–246.

Bongarten BC, Teskey RO (1987) Dry weight partitioning and its relationship to productivity in loblolly pine seedlings from seven sources. For Sci 33(2):255–267.

Bradford KJ, Hsiao TC (1982) Physiological responses to moderate water stress. In Pirson A and Zimmerman MH (Eds) *Encyclopedia of Plant Physiology, Vol. 12B*. Springer-Verlag, Berlin.

Brown, M (1993) Forest statistics for the Piedmont of South Carolina. U.S. For Ser, Southeast For Exper Sta, Res Bull SE 138.

Chameides WL, Kasibhatla PS, Yienger J, Levy, H II (1994) Growth of continental-scale metro-agro-plexes regional ozone pollution, and world food production. Science 264:74–77.

Chappelka AH, Freer-Smith PH (1995) Predisposition of trees by air pollutants to low temperatures and moisture stress. Environ Pollut 105:105–117.

Cregg BM, Hennessey TC, Dougherty PM (1990) Water relations of loblolly pine trees in southeastern Oklahoma following precommercial thinning. Can J For Res 20:1508–1513.

Dougherty PM (1996) Response of loblolly pine to moisture and nutrient stress. In Fox S and Mickler RA (Eds), *Impact of air pollutants on southern pine forests.* Springer-Verlag, NY.

Draper NR, Smith H (1981) *Applied regression analysis,* 2nd ed., John Wiley & Sons.

Durbin J, Watson GS (1951) Testing for serial correlation in least square regression. II. Biometrika 38:159–178.

Fox S, Mickler RA (1996) *Impact of air pollutatns on southern pine forests.* Springer-Verlag, New York.

Fritts HC (1960) Multiple regression analysis of radial growth in individual trees. For Sci 6(4):334–349.

Fritts HC (1962) The relevance of dendrographic studies to tree ring research. Tree Ring Bull 24:9–11.

Grissino-Mayer HD, Butler DR (1993) Effects of climate on growth of shortleaf pine (*Pinus echinata* Mill.) in northern Georgia: A dendroclimatic study. Southeast Geog 1:65–81.

Heggestad HE, Gish TJ, Lee EH, Bennett JH, Douglas LW (1985) Interaction of soil moisture stress and ambient ozone on growth and yields of soybeans. Phytopath 75:472–477.

Hogsett WE, Tingey DT, Lee EH (1988) Exposure indices: Concepts for development and evaluation of their use. In Heck WW, Taylor OC, Tingey DT (Eds), *Assessment of crop loss from air pollutants.* Elsevier Applied Science Publishing, London.

Johnson RA, Wichern DW (1988) Applied multivariate statistical analysis, 2nd ed, Washington, DC.

Joyce LA, Fosberg MA, Comanor JM (1990) Climate change and America's forests. USDA Gen Tech Rept RM-187.

Keller T, Hasler R (1984) The influence of fall fumigation on stomatal behavior of spruce and fir. Oecologia 64:284–286.

Koontz BJ, Sheffield RM (1993) Forest statistics for the southern coastal plain of South Carolina, 1993. U.S. For Ser, Southeast For Exper Sta, Res Bull SE 140.

Kramer PJ, Kozlowski TT (1979) *Physiology of woody plants.* Academic Press.

Kress LW, Allen HL, Mudano JE, Stow TK (1992) Impact of ozone on loblolly pine seedling foliage production and retention. Environ Tox Chem 11:1115–1128.

Krupa SV, Nosal M, Legge AH (1994) Ambient ozone and crop loss: Establishing a cause-effect relationship. Environ Poll 83:269–276.

Lassoi JP (1985) Stem dimensional fluctuations in Douglas-fir in different crown classes. For Sci 25:132–144.

Lee WS, Chevone BI, Seiler JR (1990) Growth and gas exchange of loblolly pine seedlings as influenced by drought and air pollutants. Water Air and Soil Poll 51:105–116.

Lefohn AS, Foley JK (1993) Establishing relevant ozone standards to protect vegetation and human health: Exposure/dose-response considerations. Air & Waste 43:106–112.

Lefohn AS, Shadwick DS, Somerville MC, Chappelka AH, Graeme B, Lockaby B, Meldahl RS (1992) The characterization and comparison of ozone exposure indices used in assessing the response of loblolly pine to ozone. Atm Env 26:287–298.

Legge AH, Grunhage L, Nosal M, Jager H-L, Krupa SV (1995) Ambient ozone and adverse crop response: an evaluation of North American and European data as they relate to exposure indices and critical levels. Angew Bot 69:192–205.

Maier-Maercker U, Koch W (1992) The effect of air pollution on the mechanism of stomatal control. Trees 7:12–25.

McLaughlin SB (1985) Effects of air pollution on forests: A critical review. J Air Pollut Contr Assoc 35:516–534.

McLaughlin SB (1994) Forest declines: Some perspectives on linking processes and patterns. In Alscher RG, Welburn AR (Eds), *Plant responses to the gaseous environment.* Chapman and Hall. New York.

McLaughlin SB, Downing DJ (1995) Interactive effects of ambient ozone and climate measured on growth of mature forest trees. Nature 374:252–254.

McLaughlin SB, Downing DJ (1996) Interactive effects of ambient ozone and climate measured on growth of mature loblolly pine trees. Can J For Res 26:670–681.

McNulty SG, Vose JM, Swank WT (1996) Potential climate change effects on loblolly pine forest productivity and water yield across the southern United States. Ambio 25:449–453.

Miller PR, McBride JR, Schilling SL, Gomez AP (1989) Trend of ozone damage to conifer forests between 1974 and 1988 in the San Bernadino mountains of southern California. In Olson RK and Lefohn AS (Eds), *Effects of air pollution on western forests.* APCA Transactions Series No. 16. Air and Waste Management Association, Pittsburgh.

Moehring DM, Ralston CW (1967) Diameter growth of loblolly pine related to available soil moisture and rate of soil moisture loss. Soil Sci Soc Amer Proc 31:560–564.

Palmer WC (1965) *Meteorological drought,* U.S. Weather Bureau, Washington, DC.

Peterson DL, Silsbee DG, Poth M, Arbaugh J, Biles FE (1995) Growth responses of Douglas fir (*Pseudotsuga macrocarpa*) to long term ozone exposure in southern California. J Air Waste Manage Assoc 45:36–45.

Richardson CT, Sasek TW, Fendick EA (1992) Implications of physiological responses to air pollution for forest decline in the southeastern USA. Env Toxic Chem. 11:1105.

Seiler JR, Johnson JD (1985) Photosynthesis and transpiration of loblolly pine seedlings as influenced by moisture-stress conditioning. For Sci 31:742–749.

Sheffield RM, Cost ND, Bechtold JP, McClure JP (1985) Pine growth reductions in the southeast. USDA Bull SE-83.

Sheffield RM, Cost ND (1987) Behind the decline. J For 87:29–33.

Shriner DS, Heck WW, McLaughlin SB, Johnson DW, Peterson CE (1990) Responses of vegetation to atmospheric deposition and air pollution. In Irving PM (Ed) *Acidic Deposition: State of Science and Technology, Vol. III Terrestrial, Materials, Health and Visibility Effects. National Acid Precipitation Assessment Program.* Washington, DC.

Skarby L, Troeng E, Bostrom C-A (1987) Ozone uptake and effects on transpiration, net photosynthesis, and dark respiration in Scots pine. For Sci 33:801–808.

Strain BR, Higginbotham KO, Mulroy JC (1976) Temperature preconditioning and photosynthetic capacity of *Pinus taeda.* L Photosyn 10:47–53.

Taylor GE (1994) Role of genotype in the response of loblolly pine to tropospheric ozone: Effects at the whole-tree, stand, and regional level. J Environ Qual 23:63–82.

Temple PJ, Riechers GH, Miller PR (1992) Foliar injury responses of ponderosa pine seedlings to ozone, wet and dry acidic deposition, and drought. Environ Exp Bot 32:107–113.

Teskey RO, Bongarten BC, Cregg BM, Dougherty PM, Hennessey TC (1987) Physiology and genetics of tree growth response to moisture and temperature stress: an examination of the characteristics of loblolly pine (*Pinus taeda* L.). Tree Physiol 3:41–61.

Tingey DT, Hogsett WE (1985) Water stress reduces ozone injury via a stomatal mechanism. Plant Physiol 77:944–947.

Topp GC (1987) The application of time-domain reflectometry (TDR) to soil water content measurement. Proceedings of International Conference on Measurement of Soil and Plant Water Status: in Commemoration of the Centennial of Utah State University. July 6–10, 1987. Utah State University, Logan, Utah.

Van Deusen PC (1992) Growth trends and stand dynamics in natural loblolly pine in the southeastern United States. Can J For Res 21:660.

Wallin G, Skarby L (1992) The influence of ozone on the stomatal and non-stomatal limitation of photosynthesis in Norway spruce, Picea abies (L.) Karst, exposed to soil moisture deficit. Trees 6:128–136.

Zahner R, Saucier JR, Meyers RK (1989) Tree-ring model interprets growth decline in natural stands of loblolly pine in the southeastern United States. Can J For Res 19:612–621.

Zeide B (1992) Reevaluation of forest inventory data from loblolly pine stands in the Georgia Piedmont and mountain areas. In Flagler RB (Ed), *The response of southern commercial forests to air pollution.* Air and Waste Manage Assoc, Pittsburgh.

13. Productivity of Natural Stands of Longleaf Pine in Relation to Competition and Climatic Factors

Ralph S. Meldahl, John S. Kush, Jyoti N. Rayamajhi, and Robert M. Farrar, Jr.

Prior to the arrival of settlers to the United States, natural communities dominated by longleaf pine (*Pinus palustris* Mill.) and maintained by periodic fire occurred throughout most of the southern Atlantic and Gulf coastal plains. These communities once covered an estimated twenty-four to thirty-six million hectare (ha) or two-thirds of the area in the Southeast (Vance, 1895; Chapman, 1932). The range of longleaf pine covers a broad arc along the coastal plain and portions of the Piedmont region from southern Virginia, south to central Florida, westward to eastern Texas, and extends further inland in the Cumberland Plateau and Ridge and Valley physiographic provinces in Alabama and Georgia. Dissimilar to the other southern pines, longleaf pine tolerates a wide variety of habitats. It is found growing on dry mountain slopes and ridges in Alabama and northwest Georgia, to the low, wet flatwoods, as well as the excessively drained sandhills found along the coast and fall line. Chapman (1932) commented that longleaf pine covered more acreage than any other North American ecosystem dominated by a single tree species.

Exploitation of longleaf pine-dominated forests has led to a steady decline of its acreage. By 1935, only 8.3 million ha were left, declining to 5 million ha by 1955. Today, estimates indicate that less than 1.3 million ha of longleaf pine exist. Noss (1989) noted that longleaf pine comprised 40.4% of the southern coastal plain in presettlement time. Today, that number has decreased to 0.7%. In comparison, wetlands have received much more attention from conservationists and preservationists but still comprise some 25 to 30% of their original area. Longleaf pine has virtually disappeared while everyone was looking but no one was seeing.

Longleaf pine is a medium-sized to large tree, 24 to 36 m tall and 60 to 80 cm in diameter (maximum 46 m tall and 122 cm diameter at breast height (DBH)). Its growth pattern is generally a long clear bole, with a small, open crown that has dense tufts of long needles at the ends of the branchlets. The needles are usually persistent until the end of the second growing season. Maturity is reached in 100 to 150 years, with some trees known to live in excess of 400 years. Seeds are borne at two- to four-year intervals but large crops considered to be adequate for natural regeneration are less frequent.

Longleaf pine stands cover some 1.3 million ha in the southern United States of which 1.18 million ha (91%) support natural stands and contain 94% of the species' growing stock volume. These natural stands are a very important source of high-value wood products, provide unique multiple-use benefits, maintain biological diversity, and supply necessary habitat for certain rare or endangered species. The significant loss of longleaf pine acreage did not change its regional distribution. Because of its broad geographic range and wide variety of habitats (climate niche), longleaf pine should be well-suited to adjust to possible changes in climate.

During 1964, the U.S. Forest Service established a regional longleaf pine growth study (RLGS) in a variety of natural, even-aged, longleaf stands in the mid-South. Research utilizing this existing longleaf pine database containing tree and stand-growth data spanning the past twenty-five years has been expanded to relate the productivity of natural longleaf pine stands to intraspecific competition and climatic factors.

Methods

Regional Longleaf Pine Growth Study

From 1964 to 1967, the U.S. Forest Service established a RLGS in the Gulf States (Kush et al., 1986). The original objective of the study was to obtain a database for the development of growth and yield predictions for naturally regenerated, even-aged, longleaf pine stands. Plots were installed to cover a range of ages, densities, and site qualities. The RLGS now consists of 284 permanent measurement plots located in central and southern Alabama, southern Mississippi, southwest Georgia, and northern Florida (Table 13.1).

The RLGS database is based on a rectangular distribution of cells formed by five stand-age classes ranging from twenty to one hundred years, five site-index classes ranging from 15 to 27 m at fifty years, and five density classes ranging from 7 to 35 m^2 ha^{-1}. At establishment, plots are assigned a target basal-area class of 7, 14, 21, 28, or 35 m^2 ha^{-1}. The trees in the plots are left unthinned to grow into that class if they are initially below the target basal area. In subsequent

Table 13.1. Name of Location, County of Location, Latitude and Longitude, Number of Plots, Range of Age, Site and Basal Area for the Regional Longleaf Pine Growth Study

Location	County, State	Latitude/ Longitude	# of Plots	Age (yrs)	Site Index (m)	Basal Area (m² ha⁻¹)
Apalachicola National Forest	Leon, FL	30.19/84.26	32	27–86	13–20	7–24
Blackwater River State Forest	Santa Rosa/Okaloosa, FL	30.52/86.52	24	58–92	17–26	7–35
Champion International Corp.						
Milton	Santa Rosa, FL	30.40/86.45	6	45–47	22–25	7–25
Hart Tract	Okaloosa, FL	30.53/86.29	2	51	23–24	14–22
Conecuh National Forest	Conecuh, AL	31.01/86.41	1	62	23	28
Cyrene Turpentine Company	Decatur, GA	30.86/84.38	6	47–81	26–27	7–34
Desoto National Forest	Perry, MS	30.55/88.53	12	17–24	17–24	7–24
Eglin Air Force Base	Okaloosa, FL	30.29/86.38	5	102–110	16–19	14–29
Escambia Experimental Forest	Escambia, AL	31.01/87.04	133	18–106	18–26	5–34
Gulf States Paper Company	Shelby, AL	33.10/86.32	8	38–40	17–19	7–35
Homochitto National Forest	Franklin, MS	31.20/91.10	5	74–81	20	8–35
International Paper Company						
Jack Springs	Escambia, AL	31.07/87.37	3	48–51	23–26	7–33
Southlands Exp. Forest	Decatur, GA	30.48/84.42	6	69–99	21–27	7–34
Kaul Trustee	Tuscaloosa, AL	33.02/87.29	3	79–81	19–23	34–35
Kimberly-Clark	Coosa, AL	32.57/86.21	6	68–98	17–23	17–35
Maschmayer Tract	Escambia, AL	31.07/87.15	11	42–46	24–26	7–34
Mobile County School Board	Mobile, AL	30.52/88.11	1	99	22	8
Resource Management Service	Covington, AL	31.01/86.22	8	68–73	25–27	7–21
Talladega National Forest						
Oakmulgee District	Bibb	32.52/87.06	4	68–75	15–16	7–29
	Tuscaloosa, AL	33.02/87.77	1	79–80	15–17	7–28
Talladega District	Talladega, AL	33.28/85.51	4	98–100	16–17	7–28

remeasurements, if the plot basal area has grown 1.7 m² ha⁻¹ or more beyond the target basal-area class, the plot is thinned back to the earlier assigned target. The thinnings are generally of low intensity and from below.

Within this distribution are three time replications of the youngest age class. These replications are located in the Escambia Experimental Forest in Brewton, AL, in Escambia County (Kush et al., 1986). As a part of the RLGS, plots in the youngest age class were first established in 1964 and new sets of plots have been added in this age class every ten years. Plots are located to achieve similar initial site qualities and ages, and are thinned to their target basal areas. Net (measurement) plots are circular and 0.08 ha (thirteen net plots are 0.04 ha) in size surrounded by a similar and like-treated 10-m wide isolation strip. Plots are inventoried, and treated as needed, every five years. The measurements are made during the dormant season (October to March) and it takes three years to complete a full remeasurement of all plots.

Each tree on the net plot with a DBH > 1.27 cm is numbered by progressive azimuth from magnetic north and its azimuth and distance from the plot center are recorded. At every remeasurement, each tree has its DBH recorded to the nearest 0.25 cm, and crown class and utility pole class and length determined. A systematic subsample of trees from each 2.54 cm DBH class has been permanently selected and measured for 1) height to the live-crown base, 2) total height, and, 3) if the tree is dominant or codominant, for age from seed.

Climatic data from weather stations located within thirty-five miles of every plot in the RLGS were obtained from the National Climatic Data Center (NCDC). These data included monthly minimum temperature, monthly maximum temperature, monthly total precipitation, and monthly Palmer's Drought Severity Index (PDSI) (Palmer, 1965).

Results

Needle Fall, Biomass, and Aboveground Net Productivity

The data and modeling efforts described in the following section are based on the analyses to date, using the independent variables that were readily available. Major variables include the plot values of 1) basal area ha⁻¹ (BApha in m² ha⁻¹), 2) trees/ha (Tpha), 3) site index (SI in m), 4) age (ring count at 1.2m), 5) latitude (Lat), and 6) longitude (Long). Selected logs, squares, and interactions were also considered in the analyses. Soil characteristics and climate variables will be added in future analyses. All of the models reflect the specific plots within the RLGS, their present-day and past management practices, and their environment. Almost all of the plots have received some management activity (thinning, fire, and so forth) and none of the plots were designed to test the limiting characteristics of longleaf pine stands (e.g., maximum BApha).

To this point, models have been developed to explore and identify predictive relationships. Models based more on biological relationships will be developed after the inclusion of soils and climate variables.

Needle Production

Tree growth is limited by photosynthetic efficiency and the amount of photosynthetic surface present. The quantity of foliage required to achieve a given amount of growth is probably a good index of photosynthetic efficiency. Most longleaf pine needles are shed during the late summer and fall of their second year. Presumably, the needles produced and shed each year are in balance and annual deposition is equivalent to production.

Leaf-litter traps for needle fall were set up on 205 of the 284 plots during the summer of 1992 to cover the range of age, density, and site classes. Three to four 0.8 m^2 litter traps were established on these plots at a random distance from the plot center. Only needles were collected from each litter trap on a four- to six-week basis. They were oven-dried at 70 °C for seventy-two hours and then weighed.

Needle fall from the 183 plots with two full years of data were used in this analysis. Table 13.2 presents the amount of needle fall collected by location and year. Figure 13.1 illustrates the differences in monthly needle collection, especially in the pattern or timing of the needle fall. The year 1993 showed an increase of needle fall from April through June with a small decrease in July before peaking in October. By contrast, there was a small drop in April of 1994 followed by a gradual increase that peaked in November. Boyer and Fahnestock (1966) and Boyer (1968) found small variations in total litter fall over three years, although they observed that the weight of needles deposited in one year may be two to three times that of another year. Litter deposition had a predictable seasonal pattern with 46 to 48% of total yearly fall occurring from September to November. A study from August 1932 to August 1933 showed a heavy litter fall began in the latter part of July reaching a peak in October and declining abruptly during November and December (Heyward 1933). Very little litter fell until May with a second peak in June. The amount of foliage dropped in May to July exceeded that for September to November. Heyward (1933) attributed the heavy fall during the summer months to a dry spring.

Average annual needle fall increased as basal area and site index increased. However, there was no clear relationship with age. Results from the RLGS tend to concur with the four-year study in a mature longleaf pine stand by Gresham (1982). Results revealed relatively high needle fall during the summer months of May to August and again in October and November. He found a positive relationship between needle fall and basal area but found no clear relationship with stand age. Although basal area integrates both trees/ha and size of trees, stand age does not reflect density. Therefore, two stands with similar densities but of different ages are expected to have similar needle-fall rates.

The correlations between needle weights and various independent variables are given in Table 13.3 (by year and by the average of two years). In general, needle weight was positively and significantly correlated with all variables evaluated except age, which was negatively correlated. There was little change in the correlations from year to year.

Table 13.2. Needle Fall, Specific Leaf Area, and Projected Leaf Area Index for Locations Where Needle Fall was Monitored

	ANF	BRSF	TNF1	TNFT	CTC	DNF	EEF
Needles (kg/ha)							
1993 mean	1590	2760	3020	3889	3273	4400	3733
range	675–2853	1085–5120	1817–4053	2567–4970	1911–5199	2391–6376	408–6254
1994 mean	2247	3147	3666	3428	3216	5039	4395
range	1245–3446	1417–14996	2446–4620	1423–5060	1795–4898	3076–6406	918–7587
Specific Leaf Area (cm²/gm)							
1993 Autumn mean	20.30	16.37	18.18	17.54	18.24	18.75	17.75
range	17.2–23.0	15.1–17.8	17.6–18.7	16.7–17.9	16.3–19.3	16.2–21.8	13.4–21.8
Winter mean	21.01	17.71	18.43	20.82	19.80	20.55	18.60
range	17.6–25.3	16.1–23.5	18.0–18.9	19.5–22.5	18.4–22.8	16.7–22.9	14.6–22.0
1994 Spring mean	19.14	17.58	16.87	17.15	19.19	17.91	18.61
range	16.8–20.3	16.6–18.9	16.0–19.2	15.8–18.3	18.3–20.6	16.6–20.1	12.8–23.9
Summer mean	19.54	17.59	17.63	18.31	18.52	19.34	19.8
range	17.2–22.4	16.0–18.7	17.3–18.3	17.6–19.	17.2–19.4	17.6–21.3	17.3–25.8
Autumn mean	21.06	18.58	19.5	18.38	19.43	21.25	19.41
range	18.6–22.6	16.5–19.9	18.6–20.7	17.8–19.0	18.5–20.4	18.8–22.9	17.0–22.4
Winter mean	19.90	18.31	20.01	17.44	19.37	19.68	15.89
range	17.2–22.8	16.8–20.1	18.6–21.6	15.8–18.6	18.5–20.0	17.6–21.0	12.0–22.2
Projected Leaf Area Index (m²/m²)							
mean	0.54	0.59	0.62	0.68	0.62	0.90	0.75
range	0.21–0.91	0.22–1.05	0.39–0.80	0.37–0.92	0.37–0.96	0.54–1.35	0.18–1.40

ANF - Apalachicola National Forest
BRSF - Blackwater River State Forest
TNF1 - Talladega National Forest - Oakmulgee District, Bibb County
TNFT - Talladega National Forest - Talladega District, Talladega County
CTC - Cyrene Turpentine Company
DNF - Desoto National Forest
EEF - Escambia Experimental Forest

Table 13.2. (Continued)

		EAFB	GSP	HNF	MST	SLEF	TNF2
Needles (kg/ha)							
	1993 mean	4027	3563	3384	4134	3336	3261
	range	3498–4397	2118–5166	1871–4945	2837–5701	1836–4795	1503–4825
	1994 mean	3406	3643	3882	4918	3029	3207
	range	2913–4214	2443–4955	1975–5164	3339–6849	1905–4359	1385–4312
Specific Leaf Area (cm²/gm)							
1993	Autumn mean	15.59	18.72	18.13	16.66	18.55	17.91
	range	14.5–16.8	16.8–20.4	17.6–19.2	15.4–17.9	17.5–19.6	16.7–19.0
	Winter mean	17.59	20.10	19.11	18.00	19.42	20.08
	range	16.4–19.8	16.4–23.5	17.4–20.7	17.0–19.6	18.0–21.3	18.2–23.6
1994	Spring mean	17.42	18.66	18.03	18.63	19.33	17.26
	range	16.6–18.9	17.8–20.2	16.1–19.0	16.5–20.0	17.8–20.8	16.5–18.3
	Summer mean	17.21	18.32	19.21	19.30	18.74	17.83
	range	16.9–17.7	16.8–19.2	16.7–21.3	18.3–20.5	17.9–20.3	16.6–19.4
	Autumn mean	18.20	20.18	21.11	18.81	19.22	18.87
	range	17.5–18.8	19.5–20.8	18.7–23.1	16.9–20.3	17.8–20.4	18.3–20.0
	Winter mean	17.72	19.87	18.85	16.49	18.65	19.12
	range	16.3–19.1	18.8–20.8	18.0–20.5	15.4–17.7	18.3–19.0	18.8–19.8
Projected Leaf Area Index (m²/m²)							
	mean	0.66	0.71	0.69	0.82	0.60	0.60
	range	0.59–0.74	0.45–0.97	0.36–0.91	0.57–1.11	0.35–0.85	0.27–0.86

EAFB - Eglin Air Force Base
GSP - Gulf States Paper
HNF - Homochitto National Forest
MST - Maschmayer Tract
SLEF - Southlands Experimental Forest
TNF2 - Talladega National Forest - Oakmulgee District, Tuscaloosa County

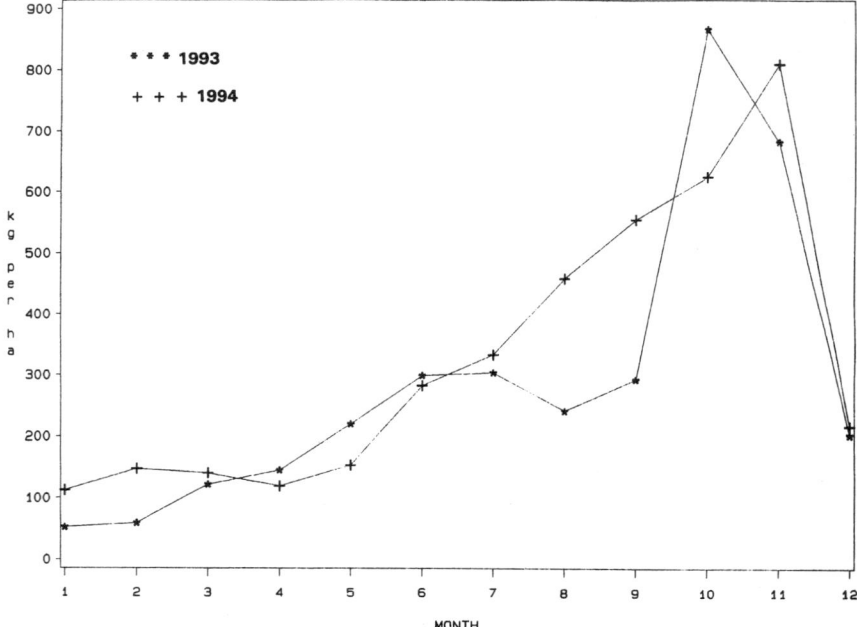

Figure 13.1. Monthly litter (dry weight) for 1993 (*) and 1994 (+).

Modeling efforts suggest that BApha, SI, Age^2, Age × SI, and year (as a class variable) were the most significant predictor variables ($R^2 = 79\%$). The addition of other variables (BApha^2, Lat, Long, and BApha × SI) significantly decreased residual sums of squares and made improvements to overall residual plots ($R^2 = 83\%$). The inclusion of specific independent variables made little difference in model-fit statistics (for example, LogSI vs SI or BApha vs LogBApha). Analyses of average needle fall (across years) showed similarly robust relationships ($R^2 = 85\%$).

Biomass

Aboveground standing biomass was estimated by applying the equations developed by Taras and Clark (1977) for longleaf pine to the RLGS data set (no harvesting was done as part of this project for biomass estimation). Dry weights for wood and bark were calculated using individual tree data from the RLGS, summed for a plot, and expanded to a per ha basis. Average annual needle data from actual plot measurements and those developed from a model based on the data were then added to the calculated wood and bark values to estimate total standing biomass. Aboveground standing biomass increased as site index and basal area increased. Biomass increased through age sixty levelling off in the eighty to one

Table 13.3. Correlations Between Dependent Variables and Independent Variables

Independent Variable	Needle Weight 1993	Needle Weight 1994	Average Needle Weight	Standing Biomass	Net Primary Productivity	Specific Leaf Area	Average Specific Leaf Area	Projected Leaf Area Index
	N = 183	N = 183	N = 183	N = 282	N = 265	N = 1089	N = 183	N = 183
BApha	0.76524	0.73637	0.76729	0.79543	0.57077	−0.04798	−0.04577	−0.04521
	0.0001	0.0001	0.0001	0.0001	0.0001	0.1135	0.5384	0.5434
LogBApha	0.77952	0.75374	0.78354	0.75775	0.59271	−0.04673	−0.04969	−0.05064
	0.0001	0.0001	0.0001	0.0001	0.0001	0.1233	0.5042	0.4960
BApha^2	0.71648	0.68459	0.71580	0.79255	0.52308	−0.04799	−0.04479	−0.04294
	0.0001	0.0001	0.0001	0.0001	0.0001	0.1135	0.5471	0.5639
Tpha	0.41737	0.48563	0.46324	−0.23782	0.54590	0.26613	0.54090	0.53142
	0.0001	0.0001	0.0001	0.0001	0.0001	0.0001	0.0001	0.0001
LogTpha	0.52585	0.64180	0.59959	−0.06742	0.65818	0.29630	0.59495	0.58973
	0.0001	0.0001	0.0001	0.2592	0.0001	0.0001	0.0001	0.0001
Tpha^2	0.34728	0.38117	0.37324	−0.17462	0.45940	0.17213	0.36483	0.35710
	0.0001	0.0001	0.0001	0.0033	0.0001	0.0001	0.0001	0.0001
SI	0.30147	0.33246	0.32485	0.38193	0.29312	−0.16868	−0.25038	−0.25402
	0.0001	0.0001	0.0001	0.0001	0.0001	0.0001	0.0006	0.0005
LogSI	0.32525	0.35516	0.34859	0.37711	0.31645	−0.18042	−0.26172	−0.26529
	0.0001	0.0001	0.0001	0.0001	0.0001	0.0001	0.0003	0.0003
SI^2	0.27698	0.30691	0.29923	0.38581	0.26780	−0.15681	−0.24009	−0.24375
	0.0001	0.0001	0.0001	0.0001	0.0001	0.0001	0.0011	0.0009

(*Continued*)

Table 13.3. (Continued)

	Needle Weight 1993	Needle Weight 1994	Average Needle Weight	Standing Biomass	Net Primary Productivity	Specific Leaf Area	Average Specific Leaf Area	Projected Leaf Area Index
Age	−0.18758	−0.37086	−0.28924	0.45672	−0.52831	−0.31713	−0.63451	−0.63122
	0.0110	0.0001	0.0001	0.0001	0.0001	0.0001	0.0001	0.0001
Age^2	−0.15605	−0.34239	−0.25886	0.39717	−0.48477	−0.29568	−0.58233	−0.57974
	0.0349	0.0001	0.0004	0.0001	0.0001	0.0001	0.0001	0.0001
Age × SI	−0.13574	−0.29054	−0.22129	0.54467	−0.44151	−0.33895	−0.65756	−0.65621
	0.0669	0.0001	0.0026	0.0001	0.0001	0.0001	0.0001	0.0001
Lat	0.14312	0.05238	0.09804	0.20100	0.11306	−0.01815	−0.5485	−0.05426
	0.0533	0.4813	0.1867	0.0007	0.0661	0.5497	0.4609	0.4657
Lat^2	0.13866	0.04623	0.09258	0.19821	0.10765	−0.01595	−0.05306	−0.05240
	0.0612	0.5343	0.2126	0.0008	0.0802	0.5991	0.4756	0.4811
Long	0.32179	0.38356	0.36202	0.11865	0.38357	−0.09816	−0.11172	−0.11333
	0.0001	0.0001	0.0001	0.0465	0.0001	0.0012	0.1321	0.1266
Long^2	0.31798	0.37958	0.35803	0.11696	0.38068	−0.09469	−0.10726	−0.10883
	0.0001	0.0001	0.0001	0.0497	0.0001	0.0018	0.1484	0.1425
BApha × SI	0.76699	0.75556	0.77837	0.84391	0.59970	−0.08956	−0.10881	−0.10899
	0.0001	0.0001	0.0001	0.0001	0.0001	0.0031	0.1426	0.1419
Tpha × SI	0.46864	0.52888	0.51140	−0.20142	0.60774	0.23539	0.49243	0.48296
	0.0001	0.0001	0.0001	0.0007	0.0001	0.0001	0.0001	0.0001
BApha × Age	0.34000	0.20998	0.27848	0.87996	−0.00521	−0.25276	−0.47744	−0.47386
	0.0001	0.0043	0.0001	0.0001	0.9328	0.0001	0.0001	0.0001
Lat × Long	0.27132	0.22998	0.25548	0.21132	0.26697	−0.06207	−0.09771	−0.09803
	0.0002	0.0017	0.0005	0.0004	0.0001	0.0406	0.1882	0.1867

Pearson Correlation Coefficients/Prob > |R| under Ho: Rho = 0

13. Productivity of Natural Stands of Longleaf Pine

Table 13.4. Total Aboveground Standing Biomass (Longleaf Pine Only) and Aboveground Net Primary Productivity for Plot Locations in the RLGS (Mean with Range in Parentheses)

	Biomass (Mt ha^{-1})	NPP (Mt ha^{-1} yr^{-1})
Apalachicola National Forest	55.64 (16.4–98.3)	4.31 (1.7–7.6)
Blackwater River State Forest	147.01 (52.9–275.4)	5.76 (2.4–8.8)
Conecuh National Forest	202.14 —	8.29 —
Cyrene	166.76 (59.4–270.9)	6.51 (4.0–9.2)
Desoto National Forest	61.24 (35.6–81.7)	11.95 (8.6–15.0)
Eglin Air Force Base	140.92 (100.5–201.9)	4.62 (4.1–5.0)
Escambia Experimental Forest	115.2 (34.7–303.3)	8.97 (1.4–17.7)
Gulf States	113.56 (45.1–161.1)	8.37 (4.6–11.1)
Hart Tract	131.37 (45.1–161.1)	6.97 (6.2–7.7)
Homochitto National Forest	157.42 (59.7–219.6)	6.75 (3.3–9.2)
Jack Springs	162.20 (51.3–222.2)	8.47 (4.1–11.2)
Kaul Trustee	283.51 (251.1–337.9)	9.46 (8.2–10.7)
Kimberly-Clark	201.73 (139.2–303.9)	6.86 (5.1–8.4)
Maschmayer Tract	153.56 (54.5–276.3)	10.11 (5.3–16.0)
Milton	136.40 (44.8–191.0)	7.38 (3.2–10.0)
Mobile County	62.92 —	2.82 —
Resource Management	142.71 (57.0–205.4)	6.04 (2.8–7.7)
Southlands Experimental Forest	173.95 (65.6–194.8)	5.46 (2.6–7.7)
Talladega National Forest (Bibb)	113.21 (61.6–148.8)	6.35 (3.7–8.6)
Talladega National Forest (Talladega)	117.45 (47.6–189.0)	5.61 (2.6–7.9)
Talladega National Forest (Tuscaloosa)	123.87 (56.7–194.8)	5.55 (3.0–7.9)

hundred-year-old age classes. Table 13.4 presents standing biomass (metric tons(Mt ha^{-1}) for plot locations in the RLGS.

Biomass estimates were significantly and positively correlated with almost all of the variables evaluated. The exception was Tpha, which had significant negative correlations with biomass. Correlations and their significance levels are given in Table 13.3.

A wide range of independent variables can be used to predict standing biomass estimates. For most combinations of variables R^2 values are above 95%. This suggests a reduction in mean square error (MSE) may be a better indicator of model performance. The best models in terms of R^2 and MSE included several interaction terms. Analyses of residual plots indicated that models with few interactions tended to display disturbing patterns.

Aboveground Net Primary Productivity

Aboveground net primary productivity (NPP) using only the longleaf pine component of the ecosystem was calculated from the most recent remeasurement of the RLGS, the time period from 1984 to 1987 and 1989 to 1991. Wood and bark estimates were calculated using Taras and Clark (1977) for each remeasurement. Wood and bark biomass change on each plot was obtained by taking the difference between the two remeasurements and dividing by the time period. Average annual

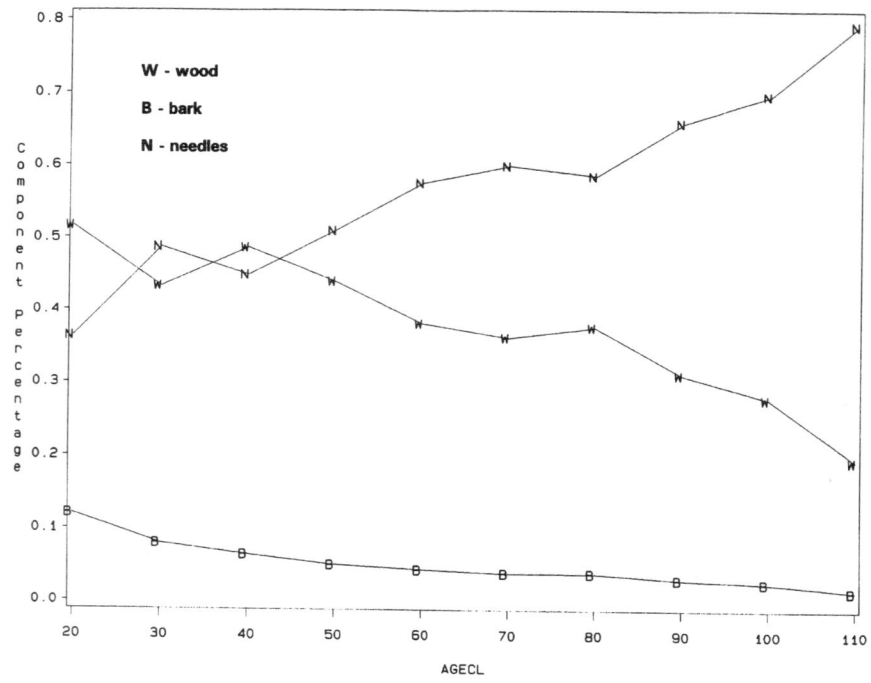

Figure 13.2. Percentage of annual aboveground longleaf pine biomass production by age class (B = bark, N = needles, W = wood).

needle production for each plot was added to wood and bark production values for the estimate of NPP (Mt ha^{-1} yr^{-1}). Results for RLGS locations are shown in Table 13.4. Figure 13.2 shows the percentage of NPP for the components. As age increases, wood and bark decrease with needles accounting for an greater portion of total productivity.

The correlations between NPP and the independent variables tend to follow the same trends as the other dependent variables. Net primary production is positively and significantly correlated with all variables evaluated except age, which was negatively correlated (Table 13.3).

The most important predictors of NPP tend to be BApha (or some transformations of it), SI, and age (R^2 = 85%). The second group of important predictor variables included Age × SI, Lat × Long, Lat, and Lat^2 (R^2 = 92%). The inclusion of other variables decreases MSE and improves overall residual plots.

While modeling NPP, two trends became apparent. The first was that the Apalachicola National Forest plots had a lower NPP for a given Tpha than the other locations. These plots are subjected to severe prescribed fires on a three-year cycle. These burns have been observed to scorch all foliage on an entire plot. Boyer (1987) noted that even low intensity fires will significantly reduce both diameter and height growth when compared to unburned stands on some sites.

Through age thirty-three, unburned longleaf pine stands produced volume yield 24% higher than stands burned by low-intensity fires at two-year intervals (Boyer, 1994). Another item of interest was that model-fit statistics were more sensitive to the specific variables in the model than were observed in the other modeling efforts in this study.

Specific Leaf Area

Specific leaf area (SLA) measurements were made on a subsample of litter traps for ten needle collections from August 1993 to January 1995. Ten fascicles were randomly selected from one trap per plot before drying. The fascicles were run through a leaf area meter to obtain a projected one-sided measure. The fascicles were dried and then weighed.

SLA values were highly variable. Table 13.2 presents SLA ($cm^2 g^{-1}$) values by year and growing season for locations that had litter traps. Correlations between average SLA (across years and seasons) and selected independent variables are given in Table 13.3. Usually, SLA estimates are significantly and negatively correlated with most variables except Tpha (and its transformations), which have significant, positive correlations.

Modeling SLA with stand-level variables and a Year × Season interaction explains approximately 32% of the variability. The inclusion of significant location variables increases the R^2 to about 38%. Although the percentage of variation explained is relatively low, many variables are significant predictors.

As expected, using average SLA (calculated by averaging SLA across year and season) greatly reduces the observed variation and correlations with several variables decreased in significance level (most notably BApha and Long). Using only stand-level independent variables resulted in an $R^2 = 65.2\%$; using only age and location variables, $R^2 = 65.9\%$; using SI, age, and location variables, $R^2 = 72.6\%$; and using stand-level and location variables, $R^2 = 74.7\%$. Although the variables included in these models varied considerably and all were highly significant, none of these models suffered from serious trends in the residuals.

Projected Leaf Area Index

Projected leaf area index (PLAI) was calculated by multiplying the average SLA by the average annual needle fall for each plot. Using average annual needle fall provides a conservative estimate of PLAI because longleaf pine retain needles for an average of two years. Table 13.2 presents the mean and range of PLAI for each location.

Table 13.3 gives correlations and significance levels with many other variables. PLAI is significantly correlated with Tpha (+), SI (−), Age (−), and their forms. Plant leaf area index can be predicted with a number of combinations of independent variables. Using only common stand-level variables (Tpha, SI, and Age) explains 64.1% of the observed variation in the data. Site index and age had an $R^2 = 61.8\%$; SI, age, and location, had an $R^2 = 72.1\%$. The combination of several stand-level variables and locations accounts for only slightly more variation, $R^2 =$ 72.9%. None of the models selected indicated any serious patterns in the residuals.

Impacts of Climate on Tree and Stand-Growth Prediction Models

Linear and nonlinear regression have been widely used in the measurement of biological relationships. Empirical fits to portray this relationship are often used (Clutter, 1963) and in some cases these fits explain, to a reasonable extent, the biological relationships (Somers and Farrar, 1991). When a biological relationship is represented using a linear or a nonlinear regression, it may be necessary to investigate whether that relationship remains stable for a long period of time. When a model is used to project growth and yield for a long projection period (usually fifteen years or more), it is essential to investigate whether that relationship is valid for such projections. This question can be answered statistically by testing the parameters of the model at different time periods. Furthermore, the results of the tests could also be used to monitor changes in productivity. Models are developed to predict the average responses to a given set of observed conditions. If conditions (ranges of stand characteristics and other uncontrolled variables) change over time, it becomes necessary to modify the model to reflect these changes and to provide accurate estimates.

The study of the impacts of climate on tree and stand-growth prediction models was conducted in two stages. First, changes in productivity over time were evaluated. Second, the inclusion of climatic variables in stand level and individual tree growth models was investigated to reduce prediction errors.

Data Quality Assessment

Field measurements followed the guidelines of Zedaker and Nicholas (1990). A minimum of five trees were remeasured for DBH, crown class, height to base of the live crown, and total height. On plots with more than fifty trees, one tree was remeasured for every ten trees. This provided at least a 10% remeasurement of the data. Additionally, measurements from the prior inventory were in the field at the time of plot remeasurement. As measurements were taken, they were compared to the prior inventory. If a discrepancy appeared between measurements, the variable in question was remeasured. As an example, if a tree did not gain in diameter or height growth, or actually lost growth, that tree was measured again to verify the discrepancy. Laboratory measurements followed the guidelines of Robarge and Fernandez (1986). A minimum of 10% of all soil samples subjected to laboratory analyses were duplicated. Data quality values are given in Table 13.5.

Stability of Model Parameters

Regional longleaf pine growth study time replication plots were used to fit stand-level models to evaluate parameter stability over time. To establish a change over time (possibly resulting from climate), the effects resulting from other such fac-

Table 13.5. Data Quality Assessment for Field and Laboratory Measurements

Variable	Data Quality Objectives					
	Target			Actual		
	Accuracy	Precision	Completeness	Accuracy	Precision	Completeness
DBH	95%	10%	99%	99%	< 1%	100%
Total Height	95%	20%	99%	99%	< 1%	100%
Crown Height	95%	10%	99%	98%	< 2%	100%
Crown Class	n/a	n/a	99%	n/a	n/a	100%
Age	n/a	n/a	99%	n/a	n/a	100%
Aspect	90%	20%	99%	97%	< 3%	100%
Slope	90%	20%	99%	97%	< 3%	100%
Basal Area	n/a	n/a	99%	n/a	n/a	100%
Site Index	n/a	n/a	99%	n/a	n/a	100%
Soil Horizon Depth	n/a	n/a	99%	n/a	n/a	100%
Sand	90%	20%	99%	90%	< 10%	100%
Silt	90%	20%	99%	90%	< 10%	100%
Clay	90%	20%	99%	90%	< 10%	100%
Litter	90%	20%	99%	98%	< 2%	100%
Surface Soil Bulk Density	90%	20%	99%	97%	< 3%	100%

tors as stand dynamics should be isolated properly. The controlled nature of the time replication (timerep) plots already isolates most of the effects induced by stand characteristics. The location of plots near to each other and with similar soil types have isolated effects that are the result of site quality. Still, there are small variations in age, trees/ha, and site quality within and between sets of timerep plots. For the purpose of further isolating these differences, a subset of plots termed as a "band" that are similar in such measured stand characteristics as site quality or index, trees/ha, and age have been selected from the three sets of timerep plots to make this comparison. The band represents a series of non-overlapping measurements for stands of similar ages (timerep I was from 1964 to 1973; timerep II was from 1974 to 1984; and timerep III was from 1985 to 1990).

The timerep band includes ages from 15 to 25 years. There were a few high density plots in the complete timerep III data set, but to reduce the variation between timereps resulting from density, plots with trees ha^{-1} greater in number than 7,240 were eliminated from the band data.

One other important factor to be considered was thinning. Quicke et al. (1994), in an investigation of including a thinning variable in an individual tree model, found that there was no consistent pattern of under- or overprediction in plots of residuals against actual basal area removed in the thinning process. The inclusion of a thinning variable was also investigated as part of this study and found to be insignificant. This is attributed to the thinnings being of low intensity and from below.

Selection of a suitable prediction equation using stand characteristics (SI, density, and age) was necessary so that it could be used in determining whether there were any changes in productivity among the timereps. The following basal area projection model was chosen as a candidate yield projection model to be used to identify any changes across timereps. The model was generalized by Clutter and Jones (1980) and is flexible to predict yield in most circumstances.

The general basal area projection model form is

$$BA_2 = BA_1^{\left(\frac{A_1}{A_2}\right)^{\beta_0}} e^{\beta_1 \left[1 - \left(\frac{A_1}{A_2}\right)^{\beta_0}\right]} \tag{1}$$

in which

BA_2 = Projected basal area (m^2 ha^{-1}) at age A_2
BA_1 = Basal area (m^2 ha^{-1}) at age A_1
β_i's are the coefficients to be estimated, $i = 0, 1$.

A generalization of this equation that included SI as a dependent variable has also been considered. The SI coefficient was not significant when fitted with the band data. We attribute this result to having a limited range of SI in the timerep band data set. Parameters in equation (1) were estimated using nonlinear estimation techniques.

A likelihood ratio (LR) criterion was selected to test the stability of the parameters estimated. The parameters of the fitted equations were tested simultaneously,

13. Productivity of Natural Stands of Longleaf Pine

as well as separately for various combinations by using this LR test. The LR is a general large sample test, the estimation of which is based on maximum likelihood methods. In essence, the LR test involves fitting a model under a null hypothesis that restricts its parameters and compares this with its unrestricted form.

To test for an overall significant difference among the three timereps for model (1), the following hypothesis was tested

$$H_O: \beta_{01} = \beta_{02} = \beta_{03} = \beta_0 \qquad H_A: \beta_{01} \neq \beta_{02} \neq \beta_{03} \neq \beta_0$$
$$\beta_{11} = \beta_{12} = \beta_{13} = \beta_1 \qquad \beta_{11} \neq \beta_{12} \neq \beta_{13} \neq \beta_1$$

To make this test, the model parameters were decomposed into β_{01}, β_{02}, β_{03}, and β_{11}, β_{12}, β_{13} (for example, β_{01} is the β_0 coefficient for timerep I). The test indicated a strong overall significant difference ($P \leq 0.0001$). Subsequently, the individual parameters were tested by the following hypotheses

(a) $H_O: \beta_{01} = \beta_{02} = \beta_{03} = \beta_0$ \qquad (b) $H_O: \beta_{11} = \beta_{12} = \beta_{13} = \beta_1$
$\qquad \beta_{11} \neq \beta_{12} \neq \beta_{13} \neq \beta_1$ \qquad\qquad $\beta_{01} \neq \beta_{02} \neq \beta_{03} \neq \beta_0$
$\qquad H_A: \beta_{01} \neq \beta_{02} \neq \beta_{03} \neq \beta_0$ \qquad $H_A: \beta_{11} \neq \beta_{12} \neq \beta_{13} \neq \beta_1$
$\qquad \beta_{11} \neq \beta_{12} \neq \beta_{13} \neq \beta_1$ \qquad\qquad $\beta_{01} \neq \beta_{02} \neq \beta_{03} \neq \beta_0$

The significance of hypothesis (a) ($P = 0.0308$) and the lower significance of hypothesis (b) ($P = 0.0508$) identified β_0 as the parameter that is most unstable over time and may be the best candidate for modification to reflect climate.

Further analyses found that timerep II and III coefficients (β_{02} and β_{03}) could be combined and the timerep I coefficient (β_{01}) was significantly different from II and III. The significant variation could be attributed to a change in climate because the other major variables that cause changes are minimized in the data set or included in the model. Details of these hypothesis tests and those repeated for other models can be found in Rayamajhi (1996). The general result was that for a number of dependent variables and model forms the model parameters were unstable over time.

Inclusion of Climatic Variables in Tree and Stand Models

To identify variables related to the observed changes over time, several models were investigated for the possible inclusion of a climatic variable using the full RLGS data set. The models evaluated included 1) a basal-area projection model (equation (1), $N = 905$), 2) a basal-area increment model ($N = 905$), and 3) an individual tree model ($N = 71955$).

The model form for the basal area increment model is (Rayamajhi, 1996)

$$BAI = \beta_0 + \beta_1 \left(\frac{BA_1}{A_1}\right) + \beta_2 \left(\frac{N_1}{A_1}\right) + \beta_3 \left(\frac{SI}{A_1}\right) \qquad (2)$$

in which

BAI = Basal area increment (m^2 ha^{-1} yr^{-1})

N_1 = Number of trees ha^{-1} at Age$_1$
BA_1 = Basal area (m^2 ha^{-1}) at Age$_1$
SI = Site Index (m) at base age 50
$\beta_0, \beta_1, \beta_2, \beta_3$ are coefficients to be estimated.

The individual tree-growth model was developed by Quicke et al. (1994) and has the form

$$bai = a_0 e^{a_1 BA^{0.5}} e^{b_0 BAL} e^{(c_1(1 - e^{c_2 DBH}) - c_0)A} \qquad (3)$$

in which

bai = Individual tree annual basal area increment (cm^2 tree^{-1})
BA = Stand basal area (m^2 ha^{-1})
BAL = Sum of the basal areas of all the trees larger than the subject tree (cm^2 tree^{-1})
A = Mean age of dominant and codominant trees
DBH = individual tree diameter at breast height (cm) outside bark
$a_0, a_1, b_0, c_0, c_1, c_2$ are coefficients to be estimated.

Because the RLGS plot remeasurement periods were not annual, it was necessary to average weather variables over the remeasurement period. Weather variables included monthly minimum and maximum temperatures, monthly average temperatures, and monthly total precipitation. In addition to these variables, many researchers have used climatic indices derived from raw climatic variables in order to investigate climatic impacts (Meyer et al., 1993; Jordan and Lockaby, 1990). One of the most widely used meteorological drought indices is the PDSI (Palmer, 1965). It consists of monthly values to indicate the severity of dry or wet periods of weather, and was developed to provide a standard indicator of relative importance of prolonged dry or wet periods for agricultural systems at a specific site or region. Based on this, monthly PDSI was added to the analysis.

Correlation analyses were performed with the residuals from equation 1 and the various common climatic variables using the full RLGS data set (N = 905). Several climatic variables for different months were found to be significantly correlated with the residuals (Table 13.6).

Based on the relationships between the model parameters and various climatic variables, each model was modified to assess the inclusion of a climate variable or index. The modified form of the basal area projection model is:

$$BA_2 = BA_1 \left(\frac{A_1}{A_2}\right)^{\kappa_0 + \kappa_1 f(x)_i} e^{\beta_1 \left[1 - \left(\frac{A_1}{A_2}\right)^{\kappa_0 + \kappa_1 f(x)_i}\right]} \qquad (4)$$

in which

$f(x)_i$ = Climatic variable is included in each of the modified models.

Similarly, for the basal area increment model [equation (2)], the form of the model including an index for climate is

Table 13.6. Pearson Correlation Coefficients / Prob > |R| Under H_0: $\rho = 0$/N = 905 Between Residuals from the Basal-Area Projection Model [Equation (1)] and the Climatic Variables

Climatic Variable	JAN	FEB	MAR	APR	MAY	JUN	JUL	AUG	SEP	OCT	NOV	DEC
Precipitation	.0317 .3412	0.295 .3752	−.0573 .0848	.0528 .1123	.0638 .0549	−.1417 .0001	−.0863 .0094	−.1138 .0006	−.0035 .9160	.1355 .0001	.1355 .0001	.2344 .0001
Maximum temperature	−.0608 .0674	−.1570 .0001	−.1142 .0006	−.0541 .1038	−.0822 .0134	−.0974 .0034	−.0917 .0057	−.1427 .0001	−.1978 .0001	−.1043 .0017	−.1863 .0001	−.2013 .0001
Minimum temperature	−.1271 .0001	−.1853 .0001	−.2380 .0001	−.1253 .0002	−.2719 .0001	−.2297 .0001	−.2692 .0001	−.2684 .0001	−.3435 .0001	−.1296 .0001	−.1821 .0001	−.1346 .0001
PDSI	−.1637 .0001	−.1801 .0001	−.1825 .0001	−.0821 .0135	−.0391 .2401	−.0262 .4315	−.0198 .5527	−.0394 .2366	−.0154 .6445	.0336 .3131	.1020 .0021	.1355 .0001

The model form used is $BA_2 = BA_1 \left(\frac{A_1}{A_2}\right)^{B_0} e^{B_1 \left[1 - \left(\frac{A_1}{A_2}\right)^{B_0}\right]}$

$$BAI = \beta_0 + \beta_1\left(\frac{BA_1}{A_1}\right) + \beta_2\left(\frac{N_1}{A_1}\right) + \beta_3\left(\frac{SI}{A_1}\right) + \kappa_1 f(x)_i \quad (5)$$

The modified form of the individual tree growth model [equation (3)] is

$$bai = a_0 e^{(k_0 + k_1 f(x)_i)BA^{0.5}} e^{b_0 BAL} e^{(c_1(1 - e^{c_2 DBH}) - c_0)A} \quad (6)$$

For a more detailed discussion of this procedure see Rayamajhi (1996).

In this study, numerous functions, $[f(x)_i]$, were considered and evaluated based on correlations of residuals from the original equations (1 to 3) and the climatic variables using the full RLGS data set. Combinations of climatic variables for various variables and months were also evaluated. Combinations providing the least mean square error from a large number of combinations of the climatic variables were explored and the following indices were considered

$f(x)_1$ = Sum of September, October, and November precipitation divided by the sum of September, October, and November maximum temperature.
$f(x)_2$ = Average September, October, and November precipitation.
$f(x)_3$ = Average September, October, and November maximum temperature.
$f(x)_4$ = Average September, October, and November minimum temperature.

For the purposes of this study, the analyses were primarily based on observed relationships among the different variables rather than trying to determine or explain the cause and effect.

The means, standard deviations, range, and correlations with residuals from equation 1 and the indices $f(x)_1, f(x)_2, f(x)_3,$ and $f(x)_4$ are

Index	Mean	Std. Dev.	Range	Correlations with residuals	Prob > |R|
$f(x)_1$	4.1 mm/°C	0.8678	2.4–6.1	0.11165	0.0008
$f(x)_2$	10.6 cm	2.2354	6.4–16.0	0.12122	0.0003
$f(x)_3$	26.2 °C	0.7559	23.1–27.6	−0.19996	0.0001
$f(x)_4$	11.9 °C	1.1369	9.6–14.7	−0.26326	0.0001

Table 13.7 shows the parameter estimates and the related statistics for the modified basal-area projection model [equation (4)], basal area increment model [equation (5)], and the individual tree growth model [equation (6)] with and without climatic index $f(x)_1$. The inclusion of $f(x)_1$, decreased the MSE by 9.4% (2.2761 to 2.06279) in the basal-area projection models [equations (1) and (4)]; by 1.7% (0.08797 to 0.08644) in the basal-area increment models [equations (2) and (5)]; and by 6.4% (13.3315 to 12.4719) in the individual tree models [equations (3) and (6)]. The inclusion of $f(x)_1$ was significant in all cases.

Numerous climatic variables and indices resulted in significant reductions in MSE for each model evaluated. The climatic indices $f(x)_2, f(x)_3,$ and $f(x)_4$ reduced the MSE from 2.2761 to 2.1160, 2.1652, and 2.0849 for the modified basal-area projection model [equation (4)], respectively. For the modified basal area increment model [equation (5)], however, the reduction was from 0.08797 to 0.08732, 0.08277, and 0.08552, respectively. Work will continue to refine our models and to identify superior climate indices.

Table 13.7. Parameter Estimates and Related Statistics for the Basal-Area Projection, Basal-Area Increment, and Individual Tree-Growth Models With and Without Climate Variable, $f(x)_i$

Parameter	Estimate	Standard error*	Mean square error	Estimate	Standard error*	Mean square error
		Basal-area projection model				
Full RLGS (N = 905)	Equation (1)			Equation (4)		
β_0	1.48170	.03993	2.27610	−0.28622	.18219	2.06279
β_1	3.80872	.03333		0.21313	.02354	
κ				3.82435		
		Basal-area increment model				
Full RLGS (N = 905)	Equation (2)			Equation (5)		
β_0	0.00141	.02402	.08797	−0.20263	.05503	.08644
β_1	0.13174	.04388		0.12703	.04352	
β_2	0.00231	.00017		0.00215	.00018	
β_3	0.56696	.02855		0.59218	.02896	
κ				0.48069	.11687	
		Individual tree-growth model				
Full RLGS (N = 71955)	Equation (3)			Equation (6)		
a_0	74.36910	.72763	13.33150	75.91090	.65398	12.47190
a_1	−0.18732	.00194		−0.19202	.00194	
b_0	−0.01733	.00044		−0.01803	.00046	
c_0	0.30303	.00176		0.30517	.00010	
c_1	0.29641	.00170		0.29882	.00002	
c_2	−0.14082	.00064		−0.14076	.00067	
κ				−0.00359	.00083	

* Asymptotic standard error in case of basal-area projection models [equations (1) and (4)] and individual tree-growth model [equations (3) and (6)].

Summary

There is a paucity of pertinent information in the literature on longleaf pine biomass, NPP, specific leaf area, and so forth. The only truly comparative values are for needle fall from Heyward (1933), Boyer (1968), Boyer and Fahnestock (1966), Wiegert and Monk (1972), and Gresham (1982). Results from all these studies indicate a very low period of needle fall from January to March, which increases for a few months in the spring and early summer, and peaking in October to November. Additionally, all studies exhibited tremendous variation in year-to-year needle fall indicating that weather and site play major roles in this process.

As expected, stand-level variables can be used to explain the majority of the variation in most of the dependent variables examined. The predictive power and significance of several other variables (year, season, Lat and Long) were stronger than expected. We associate year and season directly to climate variables that have not yet been included in the analyses. We believe the significant location variables (Lat, Long, and their transformations) are the most probable surrogates for regional long-term climate and soil characteristics. The inclusion of climate and soils variables in future work will either support or reject this assumption. Several other items also deserve mention. First is the range and significance of the simple correlations between dependent and potential independent variables. For example, average needle weight is significantly correlated with all of the variables except Lat and Lat^2. Second, we were surprised at the very high variability in the specific leaf area values and the importance of year and season. If other species have similar patterns, this may indicate an area in which simulation (projection) models need to focus additional attention. Third, if the patterns of needle fall are strongly influenced by climate, it could easily be concluded that any long-term changes in climate will affect needle retention and ultimately will be reflected in tree growth and mortality.

An overview of an investigation into the possibility that growth-model coefficients change over time and the significance and impacts of climate variables in growth-prediction models has been presented in this chapter. This was accomplished using the RLGS data that represents one of the longest continually measured growth studies in the southeastern United States.

The selection of a particular model or model form did not change the fact that model coefficients have changed significantly over the last twenty-five years. It was also shown that numerous climate variables or indices could be used to improve the prediction of growth models. These factors have strong implications for long-term projections and the inclusion of climate variables in models that will be used to the predict the responses of possible changes in climate.

As a means of quantifying the impacts of short-term changes in climate on stand growth, we predicted basal-area increment for each of our plots and re-measurement periods using equation (5) and adjusting the climatic variable in 5% increments from $+20\%$ to -20% (that is, the model coefficients were fixed and the values of the climate variable were modified). The sensitivity to changes in

Table 13.8. Sensitivity of Average Basal-Area Increment Predictions to Changes in Climate Indices Using the Basal-Area Increment Model [equation (5)]

Percent change (%) in climate variables	Average Predicted Basal-Area Increment (m² ha⁻¹ yr⁻¹) at Time 2				
	$f(x)_{11}$	$f(x)_{12}$	$f(x)_2$	$f(x)_3$	$f(x)_4$
−0.20	.5242	.6131	.5370	1.0674	.6765
−0.15	.5341	.5985	.5437	.9415	.6483
−0.10	.5339	.5856	.5503	.8156	.6201
−0.05	.5538	.5741	.5570	.6896	.5959
0.00	.5637	.5637	.5637	.5637	.5637
+0.05	.5736	.5543	.5703	.4378	.5355
+0.10	.5834	.5457	.5770	.3118	.5073
+0.15	.5933	.5379	.5836	.1859	.4791
+0.20	.6032	.5308	.5903	.0600	.4509

$f(x)_{11}$ = An index in which maximum temperature is held constant and precipitation changes
$f(x)_{12}$ = An index in which precipitation is held constant and maximum temperature changes
$f(x)_2$ = Precipitation index
$f(x)_3$ = Maximum temperature index
$f(x)_4$ = Minimum temperature index

climate were evaluated for each of the four functions of climate ($f(x)$). The results of these runs are given in Table 13.8.

Average BAI ha⁻¹ yr⁻¹ predictions changed by approximately 0.7% to 2.8% for each 5% change in the climate variables $f(x)_{11}, f(x)_{12}, f(x)_2$, and $f(x)_4$, and up to 12.6% for $f(x)_3$. Although this is an evaluation of a short-term response, it does provide an indication of magnitude and direction of changes in productivity as climate changes. Again, the use of this analysis must be viewed with caution. We are violating the interdependency of the regression coefficients by varying one of the independent variables and holding all others constant.

This study has shown highly significant seasonal and yearly fluctuations in litter fall and LAI. These observed short-term fluctuations drive periodic growth, which we have shown changes significantly over time. The magnitude and direction of the change will be dependent upon actual weather patterns, but our analyses of short-term responses indicate that the change may be quite severe. If our findings are generalized to other southern pine species, it is imperative that developers and users of stand and tree-growth projection models seriously consider the potential errors they are making when conducting long-term projections using models without climate variables. The parameter estimates will be unstable and will lead to inefficient and biased estimates.

The long-term nature of this overall study and the general findings should provide considerable initial insight for future studies of longleaf pine dynamics. If there are a long-term changes in climate conditions, longleaf pine, with its broad geographic distribution and ability to occupy a variety of sites, may be the best-adapted of the southern pines for these conditions. With concerns over increasing carbon dioxide emissions and the need to retain carbon on site, longleaf pine is the best suited of the southern pines for longer rotations. Not only will it provide a more valuable product, if that is a concern, but these ecosystems are among the most species-rich outside the tropics.

References

Boyer WD (1968) Foliage weight and stem growth of longleaf pine. USDA For Ser SO Res Note 86, 2.

Boyer WD (1987) Volume growth loss: A hidden cost of periodic prescribed burning in longleaf pine? South J Appl For 11(3):154–157.

Boyer, WD (1994) Eighteen years of seasonal burning in longleaf pine: Effects on overstory growth. In *Proceedings of the 12th International Conference on Fire and Forest Meteorology*. Society of American Foresters, Bethesda, MD.

Boyer WD, Fahnestock GR (1966) Litter in longleaf pine stands thinned to prescribed densities. USDA For Serv SO Res Note 31, 4.

Chapman HH (1932) Is the longleaf type a climax? Ecol 13:328–334.

Clutter JL (1963) Compatible growth and yield models for loblolly pine. For Sci 9:354–371.

Clutter JL, Jones EP Jr (1980) Prediction of growth after thinning of old-field slash pine plantations. USDA For Serv SE Res Paper 217, 14.

Gresham CA (1982) Litterfall patterns in mature loblolly and longleaf pine stands in coastal South Carolina. For Sci 26(2):223–231.

Heyward FD (1933) Monthly trend of needle fall in longleaf pine in northern Florida for period August, 1932 to August, 1933. Nav Stores Rev 43(34):12.

Jordan DN, Lockaby BG (1990) Time series modelling of relationships between climate and long-term radial growth of loblolly pine. Can J For Res 20:738–742.

Kush JS, Meldahl RS, Dwyer SP, Farrar RM Jr (1986) Naturally regeneratedlongleaf pine growth and yield research. In Phillips DR (Ed) *Proceedings of the Fourth Biennial Southern Silvicultural Research Conference*. USDA For Serv SE Gen Tech Rep 42:343–344.

Meyer SJ, Hubbard KG, Wilhite DA (1993) A crop specific drought index for corn. II. Application in drought monitoring and assessment. Agron J 85:396–399.

Noss RF (1989) Longleaf pine and wiregrass: Keystone components of an endangered ecosystem. Nat Areas J 9(4):211–213.

Palmer WC (1965) Meteorological drought. US Dept Comm Weath Bur Res Paper No. 45, 58.

Quicke HE, Meldahl RS, Kush JS (1994) Basal area growth of individual trees: A model derived from a regional longleaf pine growth study. For Sci 40(3):528–542.

Rayamajhi, JN (1996) *Productivity of Natural Stands of Longleaf Pine in Relation to Climatic Factors*. PhD dissertation, Auburn University.

Robarge WP, Fernandez I (1986) Quality assurance methods manual for laboratory analytical techniques. US Environ Prot Agen, For Resp Prog, Environ Res Lab, Corvallis, OR.

Somers GL, Farrar RM Jr (1991) Biomathematical growth equations for natural longleaf pine stands. For Sci 37(1):227–244.

Taras MA, Clark A III (1977) Aboveground biomass of longleaf pine in anatural sawtimber stand in southern Alabama. USDA For Serv SE Res Paper 162, 32.

Vance LJ (1895) The future of the longleaf pine belt. Gard and For 8:278–279.

Wiegert RG, Monk CD (1972) Litter production and energy accumulation in three plantations of longleaf pine (*Pinus palustris* Mill.). Ecol 53(5):949–953.

Zedaker SM, Nicholas NS (1990) Quality assurance methods manual for siteclassification and field measurements. United States EnvironmentalProtection Agency, Forest Response Program, Environmental Research Laboratory, Corvallis, OR.

14. The Impacts of Acidic Deposition and Global Change on High Elevation Southern Appalachian Spruce-Fir Forests

Samuel B. McLaughlin, J. Devereux Joslin, Wayne Robarge, April Stone, Rupert Wimmer, and Stan D. Wullschleger

The present distribution of high elevation spruce-fir forests in the southern Appalachian Mountains is the result of a retreat of red spruce (*Picea rubens.* Sarg.), Fraser fir (*Abies fraseri* Poir.), and associated species during the last post-glacial era to the coolest and most moist locations in this mountain range (White and Cogbill, 1992). The original range of red spruce in this region is estimated to have been from 450 km^2 to 2000 km^2. That range became more restricted, to approximately 300 km^2, and discontinuous within the region as the climate warmed to its present level. Thus, this forest type represents what is perhaps the most sensitive forest system in the eastern United States to climate change, particularly climate warming.

Red spruce in the southern Appalachians is presently confined to approximately 30,500 ha in the cooler and more moist high elevations, above the 1200 m level, approximately. Over 75% of that forest type occurs in the Great Smoky Mountains in Tennessee and North Carolina (Dull et al., 1988). These forests have historically been less disturbed by logging, and to date, red spruce mortality has been low compared to rates at high elevations in the northern Appalachians (Peart et al., 1992). In contrast, the mortality of a companion species, the Fraser fir, has been extremely high, and over 90% of the Fraser fir > 12.8 cm in diameter was found to be dead in the Smoky Mountains in a 1988 survey (Dull et al., 1988). Mortality is commonly ascribed to the Balsam woolly adelgid (*Adelges picea*), an introduced insect pathogen that has been active in the southern Appalachians since about 1956.

The southern Appalachian red spruce forest, because it has been less disturbed in the past and now is undergoing perhaps the first stages of a more serious decline offers very interesting opportunities to gain additional understanding of the mechanisms underlying forest responses to combinations of natural and anthropogenic stresses. There has now been an active research program in the southeastern United States for ten years, which has provided many valuable insights into both the patterns of responses of large trees, as well as the processes by which those patterns have been altered (McLaughlin and Kohut, 1992). We will review that research briefly here as a basis of 1) contrasting conditions of the northern and southern components of the spruce-fir forest and 2) providing a framework for evaluating the interactive effects of present and future changes in chemical and physical climate on these systems.

A Summary of Recent Research Findings

Linkages between observed spatial and temporal patterns of forest-system response and biological and chemical processes have played an important role in analyses of the health of spruce-fir forests throughout its northeastern range in the United States (Eagar and Adams, 1992; Johnson et al., 1992). Several of these patterns apparent in the southern Appalachians are highlighted in Figure 14.1, which describes changes in tree growth in relationship to regional atmospheric emissions and wood chemistry of high elevation red spruce.

Mature Tree Growth Patterns

A slowdown in radial growth of red spruce trees has been a significant early indicator of alterations in forest health in both the northern and southern Appalachians. In this chapter, we use this slowdown as an indicator of a "decline" in health of subject stands. Such declines may ultimately lead to either an increased probability of mortality or recovery depending on the causes and changes in interacting environmental influences. Cook and Zedaker (1992) provided a comprehensive review of dendroecological studies and the issues involved in evaluating the relative roles of climate, competition, and atmospheric pollution in observed growth declines. These conclusions, as well as those of McLaughlin et al. (1987) reached in an earlier study, indicate that the decline began in the late 1950s to early 1960s in the northeastern United States and was delayed five to ten years in the southeastern United States.

Neither unusual climate nor stand competition appear to have played a major role in initiating the stress, although interacting roles, particularly for climate, cannot be excluded (McLaughlin et al., 1987). Most examined stands have been much less than fully stocked and a changing growth response to climate, and not climate change, has characterized the most recent decline period. An important differentiating feature of the southern growth decline relative to patterns observed in the northeastern United States is its confinement to higher elevations and the absence of alarmingly high levels of spruce mortality, which have typically

14. High Elevation Southern Appalachian Spruce-Fir Forests   257

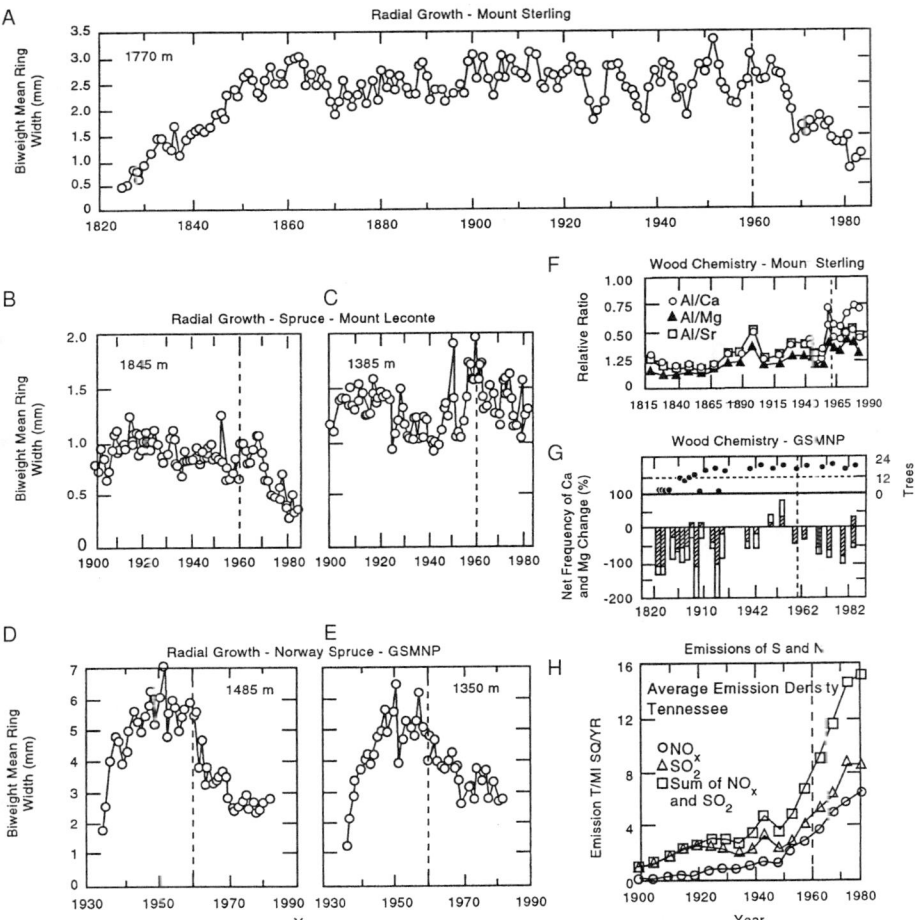

Figure 14.1. Composite chronology of red spruce growth, tissue chemistry and atmospheric emissions: southern Appalachians. (A) Mean ring-width chronology of fifteen canopy-dominant trees near summit (Adams et al., 1990); (B–C) high- and low-elevation ring width chronologies, 15 trees each (McLaughlin et al., 1987); (D–E) ring-width chronologies from two mid-elevation forty-year-old Norway spruce stands collected within 10 km of Mt. LeConte, North Carolina, samples (Adams et al., 1990); (F) aluminum/base cation ratios (Bondietti et al., 1989); (G) wood chemistry: cation trend (Bondietti et al., 1990); (H) regionally averaged emissions of SO_2 and NO_x in tons/mi_2/y in south and southwest quadrants within a 900 km radius of the Great Smoky Mountains of eastern Tennessee. Dashed vertical line on each graph indicates the year 1960 as a common reference point (after Johnson et al., 1992).

occurred in northern spruce-fir forests (Peart et al., 1992). Deterioration of canopies of red spruce was observed to have increased markedly over a 1984 to 1988 study interval in the southern Appalachians (Peart et al., 1992).

An apparent increased sensitivity of northern forests to damage by low winter temperatures has been an important early visual symptom of the decreasing health of northern spruce forests (Johnson et al., 1988). Increased sensitivity of present foliage to low temperatures has now been tied to exposure of branches to acidity in ambient-mist exposures (Vann et al., 1992). A much less severe, but cumulative, winter discoloration of needles has been found in southern red spruce (Andersen et al., 1991). Reduced cold tolerance of red spruce foliage as a consequence of exposure of ambient cloud chemistry at Whitetop Mountain has also been detected (Thornton et al., 1994), however, significant foliage loss from winter injury has not yet been detected in the southern Appalachians.

Soil Solution Chemistry

The soils of the southern spruce-fir forest are naturally acidic, poorly buffered, high in aluminum, and presently nitrogen-saturated (Johnson et al., 1991; Joslin et al., 1992). Under these conditions, the addition of strong sulfate (SO_4) and nitrate (NO_3) anions in acidic deposition provide the stimulus to mobilize aluminum in soil solutions (Reuss and Johnson, 1986). Soil-solution monitoring in the Great Smoky Mountain National Park (GSMNP) has documented the close relationship between strong anion inputs and peaks in soil-solution aluminum (Johnson et al., 1991). Aluminum levels in soil solutions periodically reach levels known to be directly toxic to roots (Joslin and Wolfe, 1992). However, the greater significance of the high aluminum levels in these soils is their interference with uptake of calcium. Levels of calcium to aluminum (Ca/Al) in soil solutions at high elevations are frequently in the range at which interference with the uptake of calcium and alteration of root function occurs (Johnson et al., 1991; Joslin and Wolfe 1992).

Atmospheric Deposition

Atmospheric deposition of SO_4 and NO_3 at a high elevation Smoky Mountain site was found to be the greatest of thirteen interregional sites examined in the recently completed Integrated Forest Study (Johnson and Lindberg, 1992). Deposition of S (36 kg ha^{-1} yr^{-1}) and N (31 kg ha^{-1} yr^{-1}) within the region increases with elevation and is heavily influenced by the frequent exposure of these forests to acidic clouds (Lindberg and Lovett, 1992). At nearby Mount Mitchell, exposure of spruce canopies to acidic fogs was found to occur on approximately 70% of the days and 35% of the time (Saxena and Lin, 1990). Over 50% of the exposures were at pH levels of ≤ 3.5, and in two successive years, cloud events with pH values < 3.0 occurred 5 and 30% of the time. Direct foliar N uptake levels for high elevation red spruce have been estimated at 5 kg N ha^{-1} yr^{-1}, or approximately 60% of the total annual dry deposition (Tjoelker et al., 1992). The minimum recorded pH level was 2.7 (Saxena and Lin, 1990). The high wet and dry deposition rates of N to these forests are a source of concern for both vegetation and stream-water quality. The high elevation spruce-fir soils are presently considered nitrogen-saturated, with the high levels of nitrate deposition from the atmosphere being transmitted with associated cations, principally aluminum and calcium, to associated streams (Nodvin et al., 1996).

Temporal Changes in Emissions and Tree Chemistry

In the absence of long-term records of atmospheric deposition, the two indicators that have been used to evaluate temporal changes in exposure of southern Appalachian forests to acidic deposition are 1) historical changes in emissions, and 2) shifts in tree-ring chemistry. Historical changes in emission density of the S and N precursors of acidic deposition upwind of the Smoky Mountains are shown in Figure 14.1. The source area includes emission density in the predominant upwind quadrants (SE and SW) within a 900-km radius, an index that described regional wet deposition of SO_4 and NO_3 well in earlier analyses (McLaughlin et al., 1984). The emission density pattern of S and N shows the strong upswing in regional emission patterns in the late 1950s and early 1960s.

Tree-ring chemistry patterns provide a means of describing changes in the chemical environment experienced by the tree as a function of shifts in soil solution chemistry (Bondietti and McLaughlin, 1992). Analysis of shifting patterns of Ca/Al in red spruce wood from elevated sites in the GSMNP (Figure 14.1) have shown that increases in availability of aluminum relative to calcium occurred in the late 1950s in parallel with changes in regional emission density of the strong anions known to mobilize aluminum (Bondietti et al., 1989, 1990).

Alterations in Physiology of Saplings and Trees

McLaughlin and Kohut (1992) have reviewed a series of published field and laboratory studies to evaluate physiological aspects of several hypotheses for changes in the vigor of spruce and fir during recent decades (Eagar and Adams, 1992). They concluded, based on growth and tree chemistry patterns observed in mature trees and physiological studies of sapling and seedling trees, that increased levels of foliar aluminum decreased the levels of foliar calcium. They also concluded that the shifts in carbon metabolism associated with increasing exposure to acid deposition, and reduced soil Ca/Al levels appeared to be the most strongly supported hypothesis for reduced vigor of red spruce at high elevation sites. Alteration in carbon metabolism was mostly a function of increased dark respiration rates and was significantly increased by decreasing foliar calcium and high soil aluminum (McLaughlin et al., 1991). Foliar calcium has also been decreased through leaching by the acidity in ambient cloud exposures (Joslin et al., 1988; Thornton et al., 1994).

The decreased ratio of photosynthesis to dark respiration (P/R ratio) observed in foliage is expected to lead to reduced carbohydrate availability and reduced growth rates of trees at high elevation sites. Confirmation of the role of acidic deposition in reducing P/R ratios, reducing growth, and altering root distribution of red spruce seedlings was provided in controlled greenhouse studies using ambient range mist and rain chemistry, as well as exposure frequency and native soil from the red spruce zone of the GSMNP (McLaughlin et al., 1993). Reduced growth of fine roots in deeper soil horizons was also noted across an increasing gradient in exposure to acidic fogs at Whitetop Mountain in Virginia (Joslin and Wolfe, 1992).

Additional evidence of the role of calcium in reducing growth and altering P/R ratios of sapling trees in the field has been provided by calcium fertilization studies (Van Miegroet et al., 1993). Large trees have also been shown to respond to calcium fertilization with both increased foliar calcium and increased growth of shoots in the canopy (Joslin and Wolfe, 1994).

Two important associative hypotheses for causality should be noted here as well. These include significant roles for ozone or nitrogen deposition as stressors. With respect to ozone, experiments with red spruce seedlings at a high elevation site in southwest Virginia, (Thornton et al., 1994) found minimal effects of ambient ozone on seedling-response parameters, whereas ambient cloud exposure adversely affected several indicators of response, including reduced foliar calcium and magnesium, increased dark respiration, and reduced cold tolerance by 3 to 5 °C. Seedling biomass was not reduced by either ozone or ambient cloud exposure in this study. Similarly, Laurence et al., (1989) found no effects on red spruce saplings at a low elevation site in New York with ozone levels up to twice the ambient levels. It should be noted that simulated rainfall down to pH 3.1 also had no measurable effect in these experiments, which involved sapling trees grown in containers of soil from a low elevation site. Thus, under the conditions evaluated to date, evidence for significant adverse effects of ambient levels of ozone is not strongly supportive. A second atmospheric oxidant, hydrogen peroxide, which occurs as a wet and dry deposited pollutant, has also been determined to have very low potential to affect red spruce physiology, based on controlled exposure studies (Hanson and McLaughlin, 1989).

In contrast, nitrogen deposition is clearly a significant contributor to the chemistry of clouds and rain, soil solutions, and streams of high elevation sites (Lovett, 1992; Johnson and Lindberg, 1992; Nodvin et al., 1996). The role of input of nitrogen oxides in influencing soil-solution chemistry, which in turn influences foliar chemistry by increasing aluminum, is well established (Johnson et al., 1991; McLaughlin et al., 1991), however the direct effects of nitrogen deposition on canopy physiology is less clear. Red spruce has the biochemical capability to utilize atmospheric nitrogen oxides (Norby et al., 1989) and field studies on Clingman's Dome have documented a seasonal trend in nitrate reductase activity in red spruce foliage with a late summer maxima (Tjoelker et al., 1992). Estimated potential nitrogen assimilation capacity of the foliage from atmospheric sources was about 5 kg N ha^{-1} yr^{-1} in the latter study (Tjoelker et al., 1992), compared to estimates of canopy uptake from soil of 3 kg N ha^{-1} g^{-1} (Johnson et al., 1991).

Presently, in spite of nitrogen-saturated soils and direct atmospheric uptake of nitrogen, red spruce at high elevations have relatively low levels of foliar nitrogen (approximately 1.0%, Johnson et al., 1991). This and the results of fertilizer trials in which added nitrogen reduced foliar growth and did not increase foliar nitrogen of mature trees (Joslin and Wolfe, 1994) suggest that nitrogen deposition is playing an important role in high elevation red spruce physiology, but primarily in the context of contributing strong anions, mobilizing aluminum, and reducing cation uptake. There are many other possible dimensions to the nitrogen deposition issue involving shifts in the balance of roots and shoots induced by foliar feeding (McLaughlin, 1985), and the chemical and physiological consequences of changing proportions of uptake of ammonia and nitrate from soil solutions. The latter have not been explored physiologically in the United States, but appear to have played an important part in some aspects of physiological symptoms noted on Norway spruce in Europe (Schulze, 1989).

The series of studies that have been briefly described provides a strong argument for the significance of acidic deposition in altering nutrient availability, physiological function, and growth of red spruce in the southern Appalachian Mountains. In this capacity, acid deposition appears to act as a significant modifier of natural stresses including naturally high acidity, high aluminum levels in soil, and high hydrologic fluxes. A primary mechanism by which the system is altered appears to be a disruption of the availability and uptake of calcium by atmospheric input of strong anions (McLaughlin and Kohut, 1992).

Calcium is an extremely important plant nutrient because it is required for formation and function of membranes and cell walls, serves as a enzyme cofactor, and regulates many physiological functions at the cell and plant level (Bangerth, 1979). Because calcium is not stored in the plant, but must be supplied from the soil uptake when it is needed, calcium is placed in a position to be limiting to growth under conditions in which plant supply may be altered by chemical interference in soil or by foliar leaching. Both of these conditions are being influenced by ambient acidic deposition at high elevation sites in the southern Appalachians.

Responses of Red Spruce to Physical Climate

Past Studies

Dendroclimatic analyses of red spruce growth patterns in relationship to climate have been an important part of our understanding of the extent to which the present-day decline in growth rates and the condition of mature red spruce is an expected result of shifts in climate (McLaughlin et al., 1987; Cook, 1988). A comparison of northern and southern red spruce stands with respect to the influences of both climatic and nonclimatic influences (Figure 14.2) provides a useful perspective in contrasting northern and southern red spruce growth patterns and for evaluating the effects of physical climate on observed growth rates (McLaughlin et al., 1987). These analyses, which used a climate growth model developed over the interval from 1900 to 1940 to predict growth over the two successive twenty-year periods, demonstrated that beginning around 1960 growth began to slow to rates below those that were predictable from past climate. The degree of slowdown was typically greater as well as being five to ten years earlier

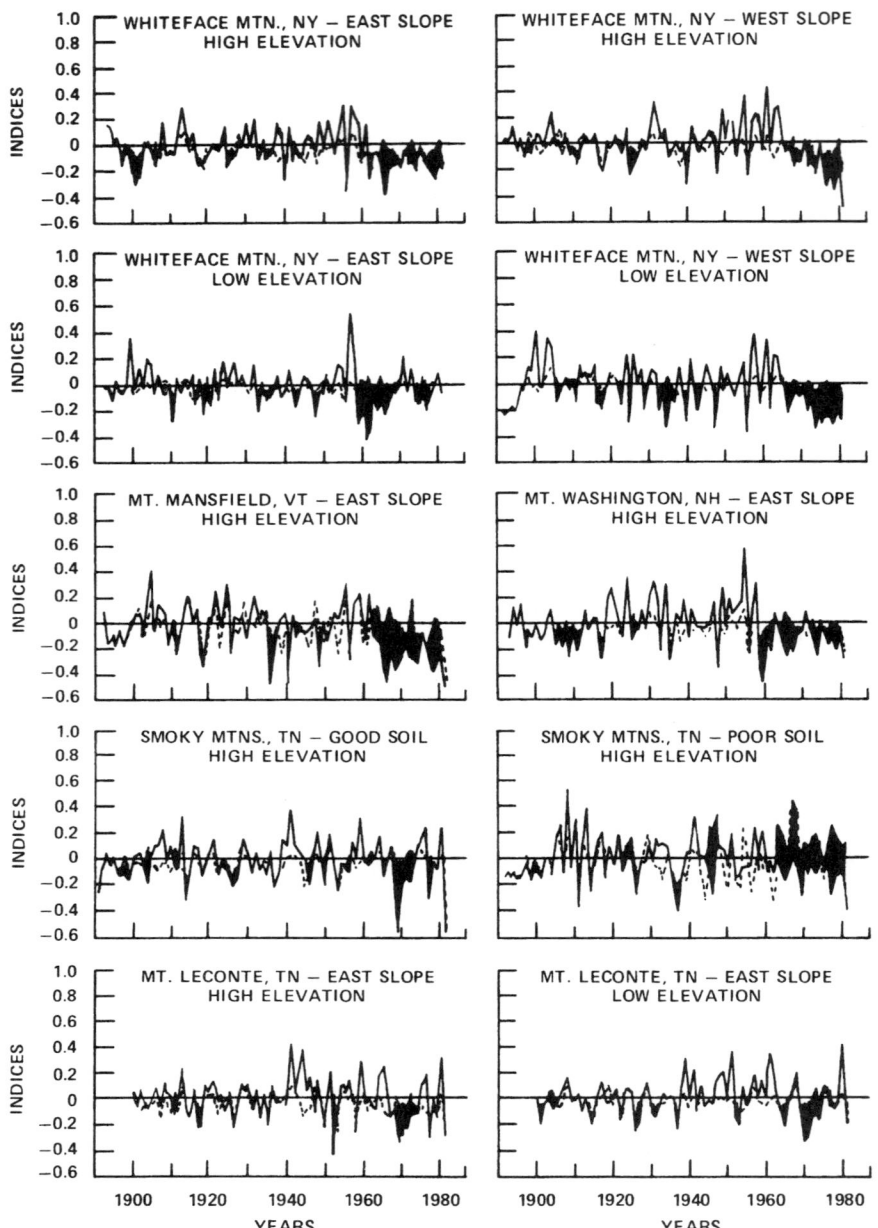

Figure 14.2. Comparison between predicted (dashed line) and actual standardized ring widths (solid line) of red spruce growth in response to climate for ten sites representing both northern and southern red spruce sites. Periods when predictions based on a model developed over the 1900 to 1940 interval exceeded actual growth are darkened (after McLaughlin et al., 1987).

in northern stands than in those from the southern Appalachians. Climate also explained a larger amount of the baseline high frequency variance in the southern stands (26% to 58%, mean = 40%) than in those from the North (6% to 54%, mean = 22%).

Analyses of Present Red Spruce Climate-Growth Relationships

Because the previous analyses were based on data collected during the 1982 to 1983 time interval, during the summer of 1995 we resampled a collection of forty-two dominant and old-grown red spruce trees at elevations of 1524 to 1829 m. These trees were associated with permanent plots in the Clingman's Dome area in the GSMNP and were not logged prior to this date. We have examined those growth trends as a basis of 1) characterizing the growth trends of the past decade and 2) evaluating seasonal climate-growth relationships and their possible shifts during this century.

Dendrochronological Methods

We have used correlation-function analysis as an interpretive guide by examining the series of correlation coefficients between the ring-width chronologies and each of several sequential seasonalized climatic variables (Blasing et al., 1984). First, we assessed the overall pattern of climate-growth relationships using present- and prior-year monthly climatic data. Second, we split up the last ninety-five years into seven thirty-five-year segments and shifted them sequentially in ten-year steps. This approach provided an easy way to evaluate seasonal patterns as they have shifted over the years.

Tree-Ring Data

We extracted seventy-five cores from forty-five mature and dominant red spruce trees from several locations over an elevational range between 1524 and 1829 m on Clingman's Dome. Because cores will also be used in a related dendrochemistry study, we surfaced them using a regular microtome with disposable blades rather than using sandpaper. After surfacing, the cores were dried and mounted in nonpermanent core holders. We preexamined all cores and used the extreme-ring match-mismatch method (Phipps, 1985) to aid in crossdating. Cores were measured with the VELMEX system at the Tennessee Valley Authority in Norris, Tennessee, and at the University of Agriculture in Vienna, Austria. Crossdating quality control was done using the COFECHA computer program (Holmes, 1983) to evaluate correlation between ring-width chronologies for individual trees within each site. Ring-width sequences exhibiting a high degree of intercorrelation were assumed to be dated correctly. For climate analysis, we used latest National Climatic Data Center-divisional climate data for eastern Tennessee covering a span of 100 years. With the ARSTAN (Cook, 1985) program, we detrended all measured series by fitting appropriate smoothing splines with a 50% cutoff to remove the low frequency variation in our chronologies (Cook and

Table 14.1. Statistical Characterizations of the Three Elevational Standard Red Spruce Chronologies at Clingman's Dome

	Low elevation 1524–1676 meters	Mid elevation 1676–1829 meters	High elevation >1829 meters
No. of trees	17	15	13
No. of radii	28	24	23
Standard deviation	0.1719	0.1655	0.1626
Mean sensitivity	0.1343	0.1250	0.1255
Skewness	0.4904	−0.0509	0.2803
Kurtosis	1.6487	0.2505	0.0247
First order autocorrelation	0.468	0.492	0.513
Mean interseries correlation	0.262	0.320	0.339
Mean correlation between trees	0.255	0.304	0.333
Mean correlation within-trees	0.561	0.677	0.463
Signal-to-noise-ratio	5.804	5.678	5.495
Variance of first eigenvector	30.68%	35.91%	38.43%

Peters, 1981), and then obtained dimensionless indices by dividing the actual ring measurement by the spline value. The ARSTAN program averaged these indices by year from all series to develop a standard index chronology accompanied with some descriptive statistics for the index chronology. Descriptive statistics are given in Table 14.1, which characterizes the investigated tree-ring chronologies taken at three elevational ranges at Clingman's Dome.

Trees grown at higher elevations generally show reduced sensitivity but higher first-order autocorrelation. High elevation plots typically exhibit lower portions of high frequency variance but, in contrast, have increased low frequency signals in their series (Fritts, 1976).

Results

Growth curves in Figure 14.3 represent red spruce growth across three elevational ranges on Clingman's Dome. All elevational curves were considerably variable, particularly the highest elevation site, showing periods of noticeably slow growth in the past twenty-five years. It should be noted that air pollution is not a new stress to the Smoky Mountains as emissions from the Copper Hill smelter significantly influenced the region from approximately 1850 to 1910 (McLaughlin et al., 1994). Only the highest elevation chronology from Clingman's Dome shows evidence of a significant slowdown during this time. Annual ring widths of the two highest elevation sites began a period of decline starting in the 1950s and 1960s that continued over an approximate thirty year interval to a minima around 1984.

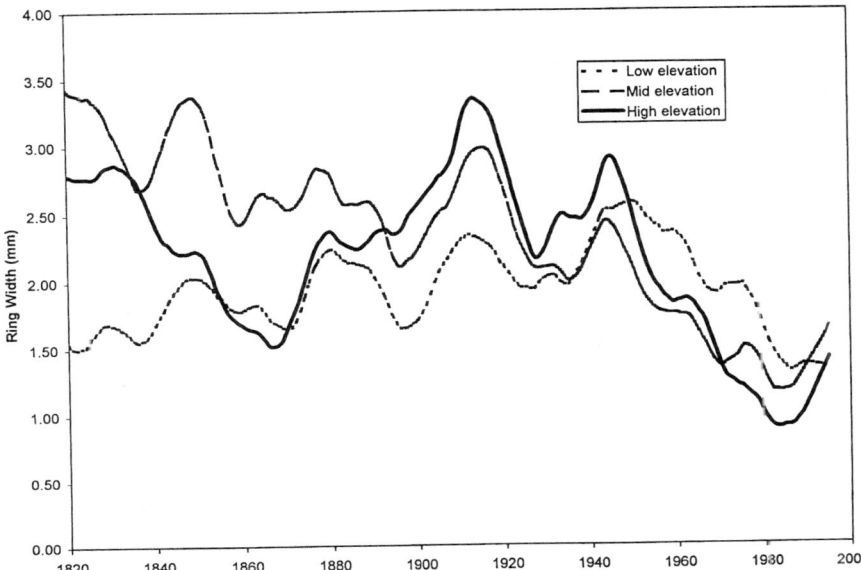

Figure 14.3. A 175-year chronology of annual ring widths from forty-five red spruce trees (seventy-five cores) extracted from three elevations (5000–> 6000 ft) in the Great Smoky Mountains National Park, North Carolina and Tennessee. (See Table 14.1 for sources and statistics on cores and growth data.)

From that year onward, all three curves show a generally increasing pattern, perhaps related to more favorable climatic conditions. Approximately 85% of all measured trees included in this period showed this growth improvement. In spite of this most recent increase, however, present-day red spruce growth at these higher elevation sites, but not lower elevation sites, is occurring at average levels that are approximately half those of the first half of this century. This has occurring despite the fact that upper elevation sites are not fully stocked (McLaughlin et al., 1997).

Overall Seasonal Patterns and Temporal Shifts in Climate Signal

To evaluate seasonal patterns and possible shifts in response to climate, we calcualted correlation functions for seven consecutive and overlapping thirty-five-year periods that were shifted by ten years at each step. The seven periods are 1) 1900 to 1935, 2) 1910 to 1945, 3) 1920 to 1955, 4) 1930 to 1965, 5) 1940 to 1975, 6) 1950 to 1985, and 7) 1960 to 1995. These growth portions were correlated with three months of seasonalized climate data that started in June of the prior year and were shifted month-by-month until October of the present year. In total, seventeen months were used in this approach.

For temperature, red spruce radial growth of both elevational ranges can be significantly and negatively influenced by higher late summer and fall tempera-

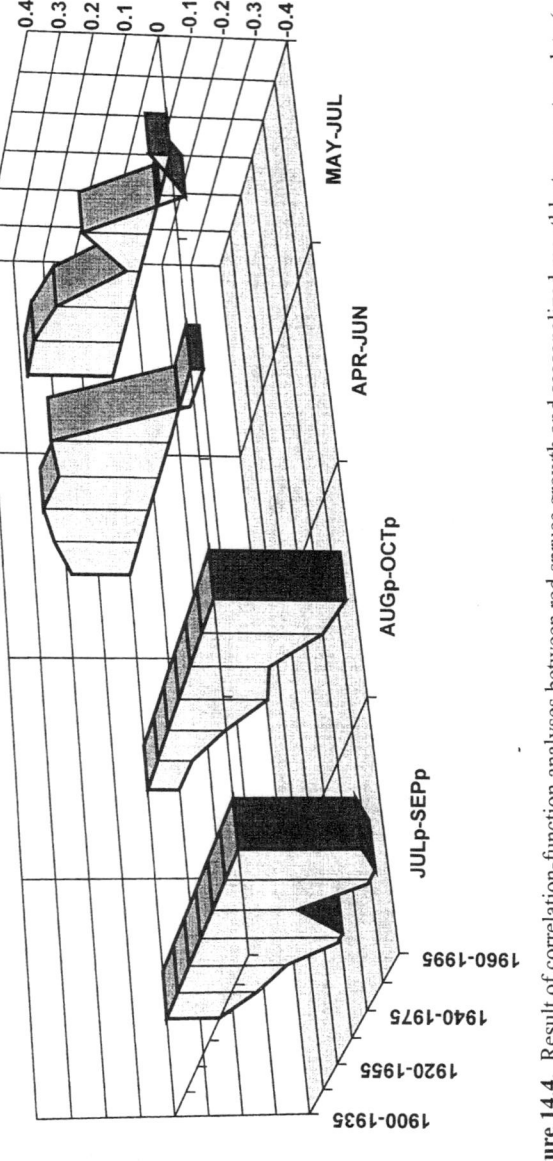

Figure 14.4. Result of correlation-function analyses between red spruce growth and seasonalized monthly temperature data (mid and high elevation) including prior (p) July to present August for seven thirty-five-year subperiods between 1900 to 1995. Only trends with at least one significant ($p < 0.05$) subperiod are shown.

ture of the prior year (July through September, Figure 14.4). Although higher temperatures during the present growing season have historically been a positive influence on growth, in more recent years (1960 to 1995) this influence has shifted from positive to near neutral or negative (Figure 14.4).

In contrast, higher July to September precipitation of the preceding year has had a consistently positive effect on growth for trees from both elevation ranges (Figure 14.5). Present year summer rainfall patterns have shown a strong negative shift in influence during this century, particularly from 1950 to 1995. In Table 14.2, contrasts in correlation coefficients are presented for both higher and lower elevation chronologies for three of the thirty-five-year comparison intervals and three climatic influence periods. These data emphasize the strength of the influence of prior late growing season conditions on present year growth, reflecting the importance of carbohydrate reserves on growth potential during the following year (McLaughlin et al., 1991). They also suggest that more subtle shifts have occurred that have resulted in both increasing growing season temperatures and rainfall amounts, which tend to be less favorable for red spruce growth in the last three decades than during the first half of this century.

Thus, the relative importance of climate variables has apparently shifted over time and has especially affected mid- and high-elevation populations represented by these cores. Such shifts may provide an indication of changes in tree-growth processes, soil-nutrient cycles, or both, as we will discuss in later sections.

Red Spruce Physiology and Responses to Climate

Red spruce has experienced significant shifts in several aspects of its growing environment during the past three decades. These are expected to be important relative to its past and projected future responses to changes in both physical and chemical climate. Those changes include significant increases in emissions of strong acid precursors of acidic deposition (Figure 14.1) and significant changes in both forest structure and forest micrometeorology.

Based on the series of studies on tree physiology and forest-nutrient cycles described in the first half of this chapter, we concluded that the primary effect of acidic deposition has been to reduce growth through both limitations on nutrient supply, principally calcium, and through reduced availability of carbon because of increased rates of dark respiration. Both of these processes would also be strongly influenced by change. They would also influence responses to climate change. Nutrient supply via soil biogeochemical cycles is heavily influenced by temperature, which affects release of nutrients through its influence on the rate of decomposition of litter. This may also accelerate soil acidification through increasing levels of nitrate in soil solutions (see Chapter 31). Nutrient fluxes of mineralized or leachable nutrient pools are also influenced by higher rainfall amounts, which can accelerate nutrient leaching (Jenny, 1980), even in the absence of strong anion inputs.

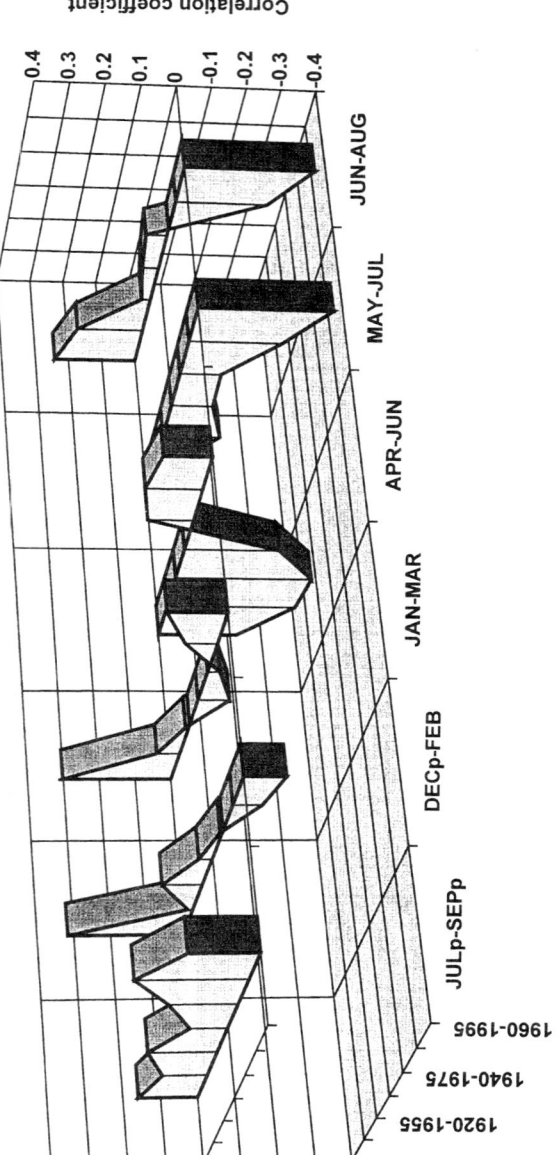

Figure 14.5. Result of correlation function analyses between red spruce growth and seasonalized monthly precipitation data including previous (p) July to current August for seven 35-year subperiods between 1900–1995. Only trends with al least one significant (p<0.05) subperiod are shown.

Table 14.2. Seasonal Months as Used in the Climate Change Scenario

Season	Months
Winter	December to February
Spring	March to May
Summer	June to August
Fall	September to November

Temperature also has a very significant effect on rates of dark respiration, with a typical elevation in respiration of approximately 100% for every 10 °C increase in temperature. This relationship, known as the respiratory quotient (or Q_{10}), is normally assumed to be $Q_{10} = 2$ for most plants. Thus, even a 1 °C increase in growing season temperature may elevate respiratory losses by 10%. Conversely, greenhouse studies on the effects of elevated CO_2 levels on seedling gas exchange and growth indicate that both red spruce and Fraser fir could respond favorably to a doubling of atmospheric CO_2 (Samuelson and Seiler, 1992, 1994). Thus, elevated CO_2 levels and higher temperatures might ultimately have antagonistic effects, although the significance of compensatory effects are conjectural at this stage and would probably depend on the relative rates of increase in temperature and CO_2.

Carbon Allocation

The significance of a small change in respiration rate to forest-growth rates may greatly exceed the absolute magnitude of the rate change, because net primary productivity (NPP) may be influenced much more than gross primary productivity (GPP). Kira (1975) estimates that aboveground NPP ranges from only about 21 to 68% of GPP for world forests. Thus, a loss of 10% of GPP could possibly be translated into a 14 to 50% loss in NPP if there were no compensatory adjustments in the NPP/GPP ratio associated with increasing temperature.

Field studies over two seasons in the spruce-fir forests of the southern Appalachians indicate that there is an approximate 30% increase in dark respiration associated with exposure of foliage to the higher levels of acidic deposition at the high elevation sites (McLaughlin et al., 1990, 1991). An increase in respiratory costs of this magnitude could explain the approximate 30% slowdown that occurred in parallel with elevations in acidic deposition around 1960 (Figure 14.1).

Although these studies focused on elevational- and nutrient-dependent shifts in dark respiration rates, a significant additional inference from this work related to the potential importance of late season carbon gain on total seasonal carbon allocation (McLaughlin et al., 1991). Used as an indicator of physiological health of foliage, P/R was found to be approximately two to four times higher during measurements in October of 1988 than any mid season measurements during 1987 or 1988. This P/R shift during the fall was a consequence of cooler temperatures reducing respiration more than photosynthesis, and led McLaughlin et al., (1991)

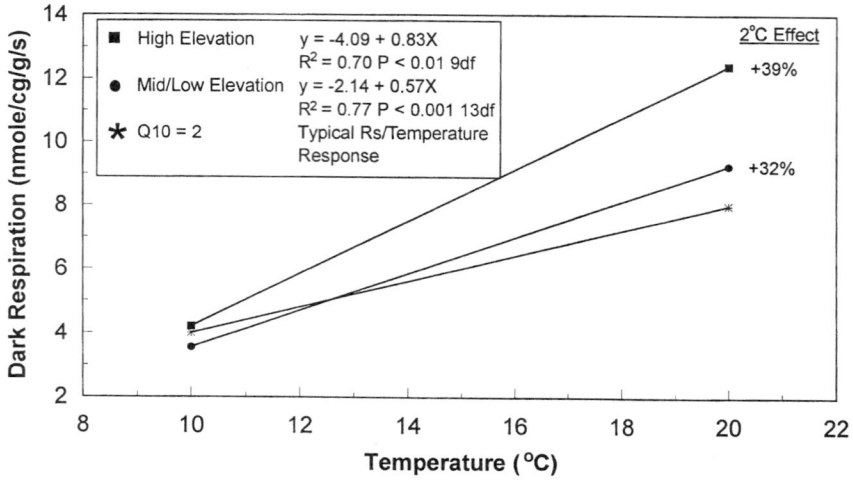

Figure 14.6. Effects of increasing temperature on dark respiration rates of red spruce foliage from high and mid elevation sites in the Smoky Mountains studied by McLaughlin et al., 1990. The effects of a 2 °C increase at upper and lower temperature extremes are noted for comparison.

to speculate that late autumns with delayed cold damage to photosynthetic systems could be very important to increasing carbohydrate storage for the subsequent growth year.

To quantify the role of possible changes in temperature on changes in dark respiration of red spruce, we have reexamined the data of McLaughlin et al., (1990, 1991) to define Q_{10} values for both high and low elevation red spruce foliage. The data shown in Figure 14.6 indicate that Q_{10} values are temperature dependent and generally higher for the high elevation sites than for lower elevations. Over the temperature range of 15 to 25 °C, for example, the Q_{10} for the highest elevation sites was approximately 50% higher (2.97) than that of the lower- and mid-elevation sites (1.85). Thus, an increase in temperatures during the summer would be more damaging potentially to the carbohydrate storage of trees at higher elevation sites than to those at lower elevations.

During the fall and early winter, in contrast, warmer temperatures would probably have a positive effect in delaying the "dormancy" of photosynthesis and extending the late season period of high carbohydrate fixation efficiency. Both of these processes would be expected to become relatively more important over time if basal rates of respiration were increased and carbon reserves were reduced by acid deposition or other stresses that limit growth. Such a pattern is suggested by the importance of low temperatures in both the prior summer and fall, and the relatively less positive effects of increasing temperature during the present summer in more recent decades (Figure 14.4).

Nutrient Flux

In addition to the effects of increasing temperature on litter decomposition, and rainfall on nutrient fluxes, the high levels of mortality of Fraser fir have released significant quantities of organic matter and associated nutrients to the remaining forests. The input of additional foliar nutrients, particularly the calcium and magnesium, should be expected to provide a fertilization effect as this material is mineralized and nutrients are released into soils in which calcium and possibly magnesium are limiting growth.

Acidic deposition is now recognized to have both long- and short-term impacts on nutrient fluxes and as such is an important part of the chemical climate of spruce-fir forest in the southern Appalachians. Longer-term impacts will probably be associated with the alteration of nutrient pools and cycles, which, coupled with naturally acidic geological substrates and high hydrological fluxes has led to very low levels of base saturation in soils (Johnson and Lindberg 1992; Joslin et al., 1992). Such changes have been slow to develop and are apt be slow to respond to remediation.

Foliar nutrient fluxes associated with acidic deposition (Joslin et al., 1988; McLaughlin et al., 1993) however, are a shorter-term response and can be expected to be reduced relatively rapidly if the chemical climate is changed by reduction of emissions of strong anions. The relative importance and reversibility of these processes and the implications of nutrient deficiency in the spruce-fir forest is focus of several lines of research.

Present-Day Research Emphasis in the Southern Spruce-Fir Ecosystem

Calcium is limiting growth and physiological function in red spruce and has led us to explore additional implications of ecosystem level shifts in nutrient cycles for the spruce-fir ecosystem. The two primary areas of emphasis have evolved are 1) changes in structural integrity of wood and branches, and 2) potential alterations in susceptibility of Fraser fir to lethal attack by the balsam wooly adelgid. A brief description of the rationale for these lines of research and some preliminary findings are detailed in the next sections.

Wood Structural Changes

Increasing levels of canopy thinning and deterioration have been the most obvious visual sign of reduced vigor in the southern Appalachian red spruce forest (Peart et al., 1992). Although damage from ice storms has played an obvious role in some of the canopy loss occurring in the Black Mountains of North Carolina (Nicholas and Zedaker, 1989), thinning has also occurred in the GSMNP without an obvious increase in such episodes, which are a natural part of the high elevation climatology. One possible explanation for increasing canopy deterioration is enhanced sensitivity of branches to winter ice damage and wind because of structural weakening of wood. The rational for such a hypothesis is that calcium is known to play an important role in wood formation. Calcium is an important crosslinking molecule in the chemical bonding associated with lignin formation (Westermark, 1982, Eklund, 1991). It is also important in formation of such noncellulosic cell wall polysaccharides as pectins (Eklund and Eliasson, 1990). Because soil aluminum has played a role in low foliar calcium levels, we anticipate that similar decreases may have occurred in the fine branches of the upper canopy.

Investigations into possible structural changes in wood are proceeding by using a combination of techniques to examine calcium and lignin distribution and structural attributes in wood (Wimmer and McLaughlin, 1996). Energy dispersive X-ray spectroscopy, coupled with transmission electron microscopy has been used to measure calcium and lignin distribution at various locations in early wood and late wood of spruce tracheids. These measurements have been combined with microtechniques used for testing material hardness. Results indicate that such techniques, though labor intensive, can be applied to tissues removed from increment cores to evaluate such properties as wood hardness and plasticity (Wimmer and McLaughlin, 1996). An inventory of red spruce and Fraser fir cores collected from elevations ranging from 1200 to 2000 m on two mountains in GSMNP is being examined to focus measurements across an appropriate range of conditions to efficiently test the hypothesis that acid deposition has structurally weakened the forest by reducing calcium availability.

Fraser Fir Ecophysiology and Adelgid Damage

The damage to mature Fraser fir in the southern Appalachians by the wooly adelgid has been extensive over the past fifteen years (Dull et al., 1988). Although heavy infestation noted on many dying trees is unquestionable evidence that the adelgid plays a major role in killing these trees, it is also important to consider the role of predisposing factors in susceptibility of stressed forests to pathogens (Manion, 1981).

Several bits of evidence warrant examination in evaluating factors that influence resistance or susceptibility of Fraser fir to adelgid-induced mortality. Some fir trees have a resistance to this introduced pathogen and can survive attack. Within the southern Appalachians, the most resistant population has been at Mount Rogers, in southwestern Virginia, where mortality has been very light and evidence of resistance pockets in the bark indicates that resistant mechanisms do exist (Eagar, 1984). As the northernmost extension of the range of Fraser fir, the apparent resistance at Mount Rogers could be the result of genetic differentiation within the Fraser fir population. However, there is now increasing evidence of attack in the Mount Rogers population primarily at low elevation sites, (Dull et al., 1988) which suggests that site-related factors may influence sensitivity.

Significant temporal and spatial variability in susceptibility of the closely related Balsam fir (*Abies balsameri*) has been evidenced across its eastern range since the adelgid was introduced around 1900 (Timmel, 1986). Mortality of this species has been significant in the Canadian maritime provinces, particularly in the mid-1980s. Tree vigor, bark characteristics, and the formation of compression wood are apparently related to resistance to the adelgid (Eagar, 1984; Timmel, 1986).

Several seemingly converging lines of inference have led us to examine the role of acid deposition on calcium supply as an additional resistance factor. First, there is dendroecological evidence that Fraser fir in GSMNP began a growth decline around 1960 in parallel to that experienced by red spruce (McLaughlin et al., 1984). Second, recent physiological measurements on Fraser fir indicate a decline in P/R ratio with increasing elevation (and acid deposition exposure) parallel to that documented for red spruce. Third, preliminary contrasts between Mount Rogers and Clingman's Dome indicate much higher foliar and soil calcium and lower aluminum levels at the high elevation Mount Rogers site than found for more sensitive populations at Mount Rogers low elevation or either high or low elevation GSMNP sites. There was also no significant P/R gradient with increasing elevation at Mount Rogers. Finally, the known role of calcium in disease resistance, including wound repair (Bangerth et al., 1979) and formation of lignin, which is a major constituent of compression wood, make calcium a probable modifier of resistance of Fraser fir to fatal adelgid infestation.

Our present-day investigations of Fraser fir ecophysiology in the southern Appalachians include 1) contrasts in deposition, soil solution chemistry, and tree growth among sites differing in levels of acid deposition or cation nutrient status, 2) fertilizer studies to examine the role of nutrition on physiology and growth, and 3) wood structural analysis to explore the relationships among wood chemistry and structural properties that may provide defense against adelgid attack or stress from wind and ice.

Summary

The spruce-fir forests of the southern Appalachian Mountains are geographically, topographically, and ecologically positioned to serve as a sensitive indicator of changes in both chemical and physical climate. Chemical changes include the increased dry and wet deposition of strong anions, sulfate and nitrate, in rain and cloudwater. Physical changes of primary interest include the potential effects of increasing temperatures, which can adversely affect both plant respiration and soil-chemical changes. A series of ecophysiology studies indicate that calcium, which is reduced in availability by acidic deposition, is an important modifier of the physiology, growth, and function of high elevation ecosystems. Inferential lines of evidence point to, but have not yet established, a much broader role for calcium in regulating the resistance of spruce and fir to a wide range of stresses in the naturally stressful high elevation environment. High levels of nitrogen deposition to this nitrogen-saturated system are of continuing interest and concern relative to both forest and stream chemistry. Understanding the nature of these interactions will probably play an important role in our capacity to understand and perhaps manage the future health of these systems in a changing environment. Climatic analysis of historical growth patterns of red spruce indicate that warmer temperatures in both the preceding fall and present summer negatively affect annual diameter growth. These findings appear to corroborate ecophysiology studies that have examined carbon allocation in red spruce. Physiological measurements indicate that rates of dark respiration have a strong influence on net carbon assimilation rates of red spruce and that seasonal patterns of carbon assimilation are very sensitive to temperature. Present-day levels of acidic deposition appear to have increased dark respiration rates of red spruce by an amount expected from a 2 to 3 °C increase in temperature. Red spruce and Fraser fir at higher elevations are apt to be more sensitive to predicted increases in global temperatures because of these combined effects. Conversely, increasing atmospheric levels of CO_2 in the longer term may help reduce stress from carbohydrate limitations of these forests.

References

Anderson CP, McLaughlin SB, Roy WK (1991). Foliar injury symptoms and pigment concentrations in red spruce saplings in the southern Appalachians. Can J For Res 21:1119–1123.

Bangerth F (1979). Calcium related disorders in plants. Ann Rev Phytopathol 17:97–122.

Blasing TC, AM Solomon, DN Duvick (1984). Response functions revisited. Tree Ring Bull 44:1–15.

Bondietti EA, Baes CF III, McLaughlin SB (1989). Radial trends in cation ratios in tree rings as indicators of the impacts of atmospheric deposition on forests. Can J For Res 19:586–594.

Bondietti EA, McLaughlin SB (1992). Evidence of historical influences of acidic deposition on wood and soil chemistry. In DW Johnson, Lindberg SE, (Eds), *Atmospheric deposition and forest nutrient cycling*. Springer-Verlag, New York.

Bondietti EA, Momoshima N, Shortle WC, Smith KT (1990). A historical perspective on changes in divalent cation availability to red spruce in relationship to acidic deposition. Can J For Res 20:1850–1858.

Cook ER (1985). A time series approach to tree ring standardization. PhD dissertation, Univ Ariz, Tucson.

Cook ER (1988). A tree ring analysis of red spruce in the southern Appalachian Mountains. In Van Drusen P (Ed), *Analysis of Great Smoky Mountain red spruce tree ring data*. USFS General Technical Report SO-69.

Cook ER, Zedaker SM (1992). The dendroecology of red spruce decline. In Eagar C, Adams B (Eds), *Ecology and decline of red spruce in the eastern United States*. Springer-Verlag, New York.

Cook ER, Peters K (1981). A conceptual linear aggregate model for tree rings. In Cook ER, Karirukstis LA (Eds), *Methods for dendrochronology: applications in the environmental science*. International Institute for Applied System Analysis, Kluwer Academic Publisher, Boston.

Dull CW, Ward JE, Brown HD, Ryan GW, Clerke WH, Uhler RJ (1988). Evaluation of spruce and fir mortality in the southern Appalachian mountains. USDA For Ser, Prot Rep R8-PR 13, Atlanta.

Eagar, C (1984). Review of the biology and ecology of the balsam woolly aphid in southern Appalachian spruce-fir forests. In White, PS (Ed), *The southern Appalachian spruce-fir ecosystem: Its biology and threats.* USDI Nat Park Ser, South Reg Off, Res/Res Manage Rep SER-71, Atlanta.

Eagar C, and Adams B (1992). *Ecology and decline of red spruce in the eastern United States.* Springer-Verlag, New York.

Eklund L (1991). Regulation of cell wall formation in Norway spruce (*Picea abies*). Doctoral dissertation, Dept Bot, Stockholm Univ.

Eklund L. Eliasson L (1990). Effects of calcium ion concentration on cell wall synthesis. J Exp Bot 41:863–867.

Fritts HC (1976). *Tree rings and climate.* Academic Press, New York.

Hanson PJ, McLaughlin SB (1989). Growth, photosynthesis, and chlorophyll concentrations of red spruce seedlings treated with mist containing hydrogen peroxide. J Environ Qual 18:499–503.

Holmes RL (1983). Computer-assisted quality control in tree-ring dating and measurement. Tree Ring Bull 43:69–75.

Jenny H (1980). *The soil resource: origin and behavior.* Springer-Verlag, New York.

Johnson DW, Van Miegroet H, Lindberg SE, Todd DE, Harrison RB (1991). Nutrient cycling in red spruce forests of the Great Smoky Mountains. Can J For Res 21 769–787.

Johnson AH, Cook ER, Siccama TG (1988). Climate and red spruce growth and decline in the northern Appalachians. Proc Natl Acad Sci USA 85:5369–5373.

Johnson AH, McLaughlin SB, Adams MB, Cook ER, DeHayes DH, Eagar C, Fernandez IJ, Johnson DW, Kohut RJ, Mohnen VA, Nicholas NS, Peart DR, Schier GA, White PS (1992). Synthesis and conclusions from epidemiological and mechanistic studies of red spruce decline. In Eagar, C, Adams MB (Eds), *Ecology and decline of red spruce in the eastern United States.* Springer-Verlag, Ecological Studies 96, New York.

Johnson DW, Lindberg SE (1992). *Atmospheric deposition and forest nutrient cycling.* Springer-Verlag, New York.

Joslin JD, Kelly JM, Van Miegroet H (1992). Soil chemistry and nutrition of North American spruce-fir stands: Evidence for recent change. J Environ Qual 21:12–30.

Joslin JD, McDuffie CM, Brewer PF (1988). Acidic cloudwater and cation loss from red spruce foliage. Water Air Soil Pollut 39:355–363.

Joslin JD, Wolfe MH (1992). Red spruce soil solution chemistry and root distribution across a cloudwater deposition gradient. Can J For Res 22(6):893–904.

Joslin JD, Wolfe MH (1994). Foliar deficiencies of mature southern Appalachian red spruce determined from fertilizer trials. Soil Sci Soc Am J. 58:1572–1579.

Kira T (1975). Primary productivity of world forests. In Cooper JP (Ed), *Photosynthesis and productivity in different environments.* Cambridge University Press, New York.

Laurence JA, Kohut RJ, Amundson RG (1989). Response of red spruce seedlings exposed to zone and simulated acidic precipitation in the field. Arch Environ Contam Toxicol 18:285–290.

Lindberg SE, Lovett GM (1992). Deposition and forest canopy interactions of airborne sulfur: results from the Integrated Forest Study. Atmos Environ 26:1477–1492.

Lovett GM (1992). Atmospheric deposition and canopy interactions of nitrogen. In Johnson DW, Lindberg SE (Eds), *Atmospheric deposition and forest nutrient cycling.* Springer-Verlag, New York.

Manion PD (1981). *Tree Disease Concepts.* Prentice-Hall, Englewood Cliffs, NJ.

McLaughlin SB (1985). Effects of air pollution on forests: a critical review. J Air Pollut Cont Assoc 35:512–534.

McLaughlin SB, Blasing TJ, Duvick DN (1984). Summary of a two year study of forest responses to anthropogenic stress. Draft synthesis report (FORAST). Oak Ridge National Laboratory, Oak Ridge, Tennessee.

McLaughlin SB, Downing DJ, Blasing TJ, Cook ER, and Adams HS (1987). An analysis of climate and competition as contributors to decline of red spruce in high elevation Appalachian forests of the eastern United States. Oecologia 72:487–501.

McLaughlin SB, Andersen CP, Hanson PJ, Tjoelker MG, Roy WK (1991). Increased dark respiration and calcium deficiency of red spruce in relation to acidic deposition at high-elevation southern Appalachian Mountain sites. Can J For Res 21:1234–1244.

McLaughlin SB, Andersen CP, Edwards NT, Roy WK, Layton PA (1990). Seasonal patterns of photosynthesis and respiration of red spruce saplings from two elevations in declining southern Appalachian stands. Can J For Res 20:485–495.

McLaughlin SB, Kohut R (1992). The effects of atmospheric deposition on carbon allocation and associated physiological processes in red spruce. In Eagar C, Adams B (Eds), *Ecology and decline of red spruce in the eastern United States*. Springer-Verlag, New York.

McLaughlin SB, Tjoelker MG, Roy WK (1993). Acid deposition alters red spruce physiology: laboratory studies support field observations. Can J For Res 23:380–386.

McLaughlin SB, Blasing TJ, Duvick DN (1984). Summary of a two year study of forest responses to anthropogenic stress. Draft synthesis report. Oak Ridge National Laboratory, Oak Ridge, Tennessee.

McLaughlin SB, Blasing TJ, Downing DJ (1994). Two hundred year variation of southern red spruce radial growth as estimated by spectral analysis: Comment. Can J For Res 24:2299–2304.

Nicholas NS, Zedaker SM (1989). Ice damage in spruce-fir forests of the Black Mountains, North Carolina. Can J For Res 19:1487–1491.

Nodvin SC, Van Miegroet H, Lindberg SE, Nicholas NS, Johnson DW (1995). Acidic deposition, ecosystem processes, and nitrogen saturation in a high elevation southern Appalachian watershed. Water Air Soil Poll 85(3):1647–1652.

Norby RJ, Weerasuriya Y, Hanson PJ (1989). Induction of nitrate reductase activity in red spruce needles by NO_2 and HNO_3 vapor. Can J For Res 19:889–896.

Peart DR, Nicholas NS, Zedaker SM, Miller-Weeks MM, Siccama TG (1992). Condition and recent trends in high-elevation red spruce populations. In Eagar C, Adams B (Eds), *Ecology and decline of red spruce in the eastern United States*. Springer-Verlag, New York.

Phipps RL (1985). Collecting, preparing, crossdating, and measuring tree increment cores. U.S. Geological Survey, Water-resources investigations report 85–4148.

Reuss JO, Johnson DW (1986). *Acid deposition and the acidification of streams and waters*. Springer-Verlag, New York.

Samuelson LJ, Seiler JR (1994). Red spruce seedling response to elevated CO_2, water stress, and soil fertility treatments. Can J For Res 24:954–959.

Samuelson LJ, Seiler JR (1992). Fraser fir seedling gas exchange and growth in response to elevated CO_2. Environ Exper Bot 32:351–356.

Saxena VK, Lin NH (1990). Cloud chemistry measurements and estimates of acidic deposition on an above cloudbase coniferous forest. Atmos Environ 24A(2):329–352.

Schulze E-D (1989). Air pollution and forest decline in a spruce (*Picea abies*) forest. Science 244:777–783.

Thornton FC, Joslin JD, Pier PA, Neufeld H, Seiler JR, Hutcherson JD (1994). Cloudwater and ozone effects upon high elevation red spruce: A summary of results from Whitetop Mountain, Virginia. J Enviorn Qual 23:1158–1167.

Timmel TE (1986). Compression wood induced in firs by balsam woolly aphid (*Adelges picea*). In Timell TE (Ed), *Compression wood, in gymnosperms. Vol 3*. Springer-Verlag, Berlin, Germany.

Tjoelker MG, McLaughlin SB, DiCosty RJ, Lindberg SE, Norby RJ (1992). Seasonal variation in nitrate reductase activity in needles of high-elevation red spruce trees. Can J For Res 22:375–380.

Van Miegroet H, Johnson DW, Todd DE (1993). Foliar response of red spruce saplings to fertilization with calcium and magnesium in the Great Smoky Mountains National Park. Can J For Res 23:89–95.

Vann DR, Strimback GR, Johnson AH (1992). Effects of air borne chemicals on freezing resistance of red spruce foliage. For Ecol Manag 51(1–3):69–80.

Westermark U (1982). Calcium promoted phenolic groups by superoxide radical—a possible lignification reaction in wood. Wood Science Techno 16:71–78.

White PS, Cogbill CV (1992). Spruce-fir forests of eastern North America. In Eagar C, Adams B (Eds), *Ecology and decline of red spruce in the eastern United States.* Springer-Verlag, New York.

Wimmer R, McLaughlin SB (1996). Possible relationships between chemistry and mechanical properties in the microstructure of red spruce xylem. In Dean JS, Meko DM, Sweetnam TW (Eds), *Tree tings, environment, and humanity.* Radiocarbon 1996.

Zedaker SM, Nicholas NS, Eagar C (1989). Assessment of forest decline in the Southern Appalachian spruce-fir forest, USA. In Bucher JB and Bucher-Wallin I (Eds), *Air pollution and forest decline.* Proceedings of the 14th International Meeting for Specialists in Air Pollution Effects on Forest Ecosystem. IUFRO P2.05, Interlaken, Switzerland. October 208, 1988. Birmensdorf.

15. The Influences of Global Change on Tree Physiology and Growth

Robert O. Teskey, Phillip M. Dougherty, and Robert A. Mickler

This chapter is a synthesis of the findings of those projects in the Southern Global Change Program (SGCP) that measured the effects of environment on physiological and growth processes. This synthesis will not attempt to be exhaustive or encyclopedic, instead it will focus on key findings from the program, building conclusions from these results whenever possible. In addition to these conclusions, we will discuss the similarities and differences in results among the projects in the program, as well as with the general body of knowledge of global change effects on trees.

In many instances, the primary influence of such global change factors as carbon dioxide (CO_2), temperature, precipitation and ozone are on physiological activities of plants. Carbon dioxide is the substrate for photosynthesis, ozone enters plants through the stomata and affects cellular processes within the leaf, temperature affects the rate and timing of most physiological processes, and water is a fundamental component of the cytoplasm of cells, which allow them to function and grow. The physiological processes at the cellular and metabolic level are numerous and varied, but can be usefully grouped into the processes involved in carbon gain (photosynthesis and leaf development), carbon loss (respiration, shedding of plant parts), and carbon allocation (growth of roots, stems, and leaves). The rate and magnitude of physiological activities ultimately determine the productivity of forests, although the scale at which the processes operate are far removed from from actual measures or indices of productivity, such as yearly biomass accretion. While individual physiological processes themselves cannot simply be directly related to productivity, changes in their rates and magnitude are important indicators of the direction and magnitude of forest growth in response to such perturbations as global change factors. The experimental procedures used in this program provided detailed and extensive measures of photosynthesis, respiration, and growth under a variety of experimental conditions and have helped us understand the potential of climate change to affect tree growth and forest productivity.

The issue that was central to many of the studies in the SGCP was the direct effect of elevated CO_2 on physiological processes. This issue was studied in five separate locations in four field experiments located in North Carolina (Dougherty et al., Chapter 9), Georgia (Teskey, Chapter 8), Alabama (Alemayehu et al., Chapter 5), Oklahoma (Hennessey and Harinath, Chapter 10), and in a greenhouse experiment in Georgia (Groninger et al., Chapter 7). In all of these experiments the principle species examined was loblolly pine (*Pinus taeda* L.). The greenhouse experiment also included sweetgum (*Liquidambar styraciflua* L.), and the Mississippi study included flowering dogwood (*Cornus florida* L.), but neither hardwood species can be considered a major component of the projects in the SGCP. The emphasis of the SGCP on the response of loblolly pine to climate change is reflected in our synthesis.

There was clear consistency among these studies on one important point which is that net photosynthesis was enhanced substantially in elevated CO_2 concentrations. When the ambient CO_2 concentration was doubled, rates of net photosynthesis in loblolly pine increased by 50% to 130%, depending on the study and the treatment conditions. The lowest increase was reported in an open-top chamber study using loblolly pine seedlings (Alemayehu et al., Chapter 5), although much larger percentage increases (80 to 130%) were reported in the other studies (Dougherty et al., Chapter 9; Groninger et al., Chapter 7; Hennessey and Harinath Chapter 10; Teskey, Chapter 8; Teskey 1995; Murthy et al., 1996). This increase was consistent for seedlings, saplings, and trees. With the exception of the Alemayehu et al. study (Chapter 5), the average increase in net photosynthesis under twice-ambient CO_2 concentration, relative to the rates in the present-day ambient concentration, was about 90 to 100%, irrespective of growing conditions or age of the trees. This corresponds to a linear response of net photosynthesis to CO_2 concentration in the range of 350 to 700 μmol mol^{-1} CO_2 (Teskey, 1995). In this range, a change in CO_2 concentration of 10 μmol mol^{-1} produces an approximate 3% change in the rate of net photosynthesis in loblolly pine and, more important, indicates that the process is more limited by the availability of the substrate (i.e., CO_2), than by nutrients or other environmental factors.

The response of net photosynthesis to CO_2 concentration in loblolly pine is consistent with the findings of others. For example, Ellsworth et al. (1995) reported that over the course of three days in August, net photosynthesis was found to be significantly enhanced in elevated CO_2 (550 μmol mol^{-1}) in a free air CO_2 exposure study in a loblolly pine stand in North Carolina. Additionally, they reported that rates under controlled conditions, measured in situ, were 63% greater in the 550 μmol mol^{-1} CO_2 treatment. This finding corresponds almost exactly to the estimate of a 3% increase in net photosynthesis per 10 μmol mol^{-1} change in CO_2 concentration that came from measurements in the branch chamber studies in mature trees (e.g., Teskey, 1995; Liu and Teskey, 1995; Dougherty et al., Chapter 9; Murthy et al., 1996). Similarly, in one- and two-year-old loblolly pine seedlings, rates in twice-ambient CO_2 concentrations more than doubled compared to the rates achieved in ambient CO_2 concentrations at most times of the year (except in winter) in an open-top chamber study (Tissue et al., 1996). In growth-chamber studies, loblolly pine also has exhibited similar increases in net photosynthesis in elevated CO_2 conditions (Fetcher et al., 1988).

Studies on other pine species generally have reported similar responses. Large increases in rates of net photosynthesis have been reported in *Pinus sylvestris* L. trees when placed in elevated CO_2 concentrations (Wang et al., 1995), as was also the case in *Pinus eldarica* L. (Garcia et al. 1994) and *Pinus radiata* D. Don (Conroy et al. 1990). Exceptions to this finding have occurred in studies in which photosynthetic compensation, or down regulation, has been reported in container studies using ponderosa pine seedlings (*Pinus ponderosa* Laws.) (Callaway et al., 1994; Grulke et al., 1993). But reports of down regulation in pines seem to be the exception rather than the norm.

For other tree species, the effect of elevated CO_2 concentrations on net photosynthesis is almost always positive, but the magnitude of the response appears to be quite variable, depending on species and growth conditions. Responses in the range found for loblolly pine and other pine species have been reported in such deciduous hardwood species as *Fagus sylvatica* L. (El Kohen et al., 1993); *Liriodendron tulipifera* L. (Norby and O'Neill 1991); *Quercus alba* L. (Gunderson et al., 1993); *Populus tremuloides* (Michx.) (Sharkey et al., 1991); *Populus* hybrids (Ceulemans et al., 1993); *Quercus prinus* L. (Bunce, 1994); as well as other conifers, e.g., *Picea sitchensis* (Bong.) Carr. (Townend, 1993). Lower percentage increases have also been reported, even in the same species, depending on growth conditions (Conroy et al., 1990; Norby and O'Neill, 1991; Tschaplinski et al., 1995; Koike et al., 1996), and can change seasonally (Curtis et al., 1996; Vivin et al., 1995) (see species comparisons in the review by Ceulemans and Mousseau (1994) for the general range of response in trees).

In addition to the direct effect of CO_2 concentration on net photosynthesis, the CO_2 concentration also appeared to directly affect dark respiration. Rates of dark respiration in ambient and elevated CO_2 concentrations were only measured in two studies, and both reported reductions in the rate of respiration when foliage was exposed to higher concentrations of CO_2 (Teskey, 1995). The amount of reduction, approximately 20%, is significant, but the cause of this effect remains unknown, and the findings are controversial. Decreases in dark respiration have been reported in nonwoody species (Reuveni and Gale, 1985; Bunce and Caulfield, 1991; Thomas et al., 1993; Ziska and Bunce, 1994), as well as in tree species (Wullschleger and Norby, 1992; Mousseau, 1993; Idso and Kimball, 1992) including ponderosa pine (Griffin et al., 1996), but the response has not always been detected (see review by Bunce, 1994). The response seems to be reversible when the foliage is exposed to lower CO_2 concentrations, and several mechanisms have been proposed but none have been firmly established (Bunce 1994). The apparent reduction in dark respiration appears to be further evidence that carbon gain will be enhanced in elevated CO_2 conditions, but the relative contribution to overall carbon balance will be less than the stimulatory effect of CO_2 on rates of net photosynthesis because rates of dark respiration are generally only about 10 to 20% that of net photosynthesis.

Elevated CO_2 concentrations greatly increased net photosynthesis and decreased foliar respiration, and therefore it is not surprising that measures of growth increased in elevated CO_2 concentration. Large increases were reported in branch length (Dougherty et al., Chapter 9; Teskey, 1995), stem dry weight (Alemayehu et al., Chapter 5), total plant biomass (Griffin et al., 1993, Larigauderie et al., 1994), and root growth (Jifon et al., 1995). Leaf area also increased (Teskey, 1995, Chapter 8; Alemayehu et al., Chapter 5), but the growth of individual needles remained fixed. Needle length remained the same in ambient and elevated CO_2 treatments, and the specific leaf area of individual needles was only reduced slightly, and usually not significantly, in elevated CO_2 concentrations (Sword et al., Chapter 11; Murthy et al., 1996). A large decrease in specific leaf area would be indicative of an increase in leaf weight, in this case probably resulting from a buildup of carbohydrates in the leaf. As this did not happen, it suggests that loblolly pine was able to effectively export carbohydrates from the foliage when exposed to elevated CO_2.

The direct effects of elevated CO_2 on physiological factors affecting productivity, including net photosynthesis, respiration and leaf area, were large and this leads to the expectation that increased growth should be expected in elevated CO_2 environments. But an important issue examined by the SGCP, and other studies involving loblolly pine, was whether less-than-optimum environmental conditions would diminish carbon gain in elevated CO_2 concentrations. This issue was addressed in the studies of Dougherty et al., (Chapter 9), Hennessey and Harinath (Chapter 10), Teskey (Chapter 8), and Thomas et al., (1994). The site factors of nutrition and water (Dougherty et al., Chapter 9; Murthy et al., 1996; Hennessey and Harinath, Chapter 10), nutrition alone (Thomas et al., 1994), and elevated temperature (Conner et al., Chapter 6; Teskey, Chapter 8), were studied in factorial experiments that lead to some substantive findings about the effects of elevated CO_2 concentration on loblolly pine. Consistently, the effect of elevated CO_2 on physiological processes was found to be primarily additive rather than interactive. This is a remarkable because it means that growth in the field on sites of high-growth potential, as well as sites of low-growth potential, will probably benefit from elevated atmospheric CO_2 concentrations. Important examples of this effect were the maximum rates of net photosynthesis in fertilized and unfertilized trees receiving elevated or ambient CO_2 treatments (Murthy, 1996; Dougherty et al., Chapter 9). At all levels of leaf nitrogen the same relative photosynthetic response to CO_2 concentration was found. This is in contrast to the reports of interactions between nutrient availability and the response of tree species to CO_2 enrichment in terms of growth, leaf area development, and photosynthetic rates (Conroy et al., 1990; El Kohen and Mousseau, 1994) The additive response was not found in seedling studies in which photosynthetic down regulation was induced in elevated CO_2 treatments placing the plants in a very low nutrient regime (Tissue et al., 1993; Thomas et al., 1994). The plants that exhibited down regulation were severely nutrient deficient, with foliar nitrogen contents of approximately 0.5 to 0.7%. This level of deficiency is extreme, and is improbable that it will ever occur in the field where typical nitrogen concentrations on nutrient-poor sites are 1.07% (Van Lear et al., 1984) to 0.90% (Valentine and Allen, 1989). Additionally, the sandhills site in North Carolina where the Dougherty et al., (Chapter 9) study took place is considered very nutrient poor, yet under the twice-ambient CO_2 treatment, the net photosynthetic rates in the control trees were almost double that of the ambient CO_2 treatment, yet the mean foliar nitrogen concentration was 0.88%. Therefore, although it is possible to induce photosynthesis down regulation under conditions of severe nutrient stress, it is not probable to be a significant factor in the field.

In the SGCP the effect of elevated air temperature was examined directly by manipulative temperature treatments in only two studies. The results from these studies provide some indication of the magnitude of the effect we can expect from air temperature on carbon gain (Teskey, Chapter 8) and reproductive effort (Conner et al., Chapter 6). The effect of a 2 °C increase in air temperature on net photosynthesis was small, almost always less than a 10% increase or decrease in rates of photosynthesis. In comparison, the effect of elevated CO_2 was very much larger. At twice-ambient CO_2 concentrations, net assimilation was over 100% greater than that at ambient CO_2 concentrations. Branch growth and leaf area development were slightly lower in the higher air temperature treatment, but again the effect was much smaller than the effect of elevated CO_2 concentration. There was no apparent effect of the 2 °C elevation in air temperature on the timing of budburst or the duration of the growing season, but elevated air temperature caused pollen release and the initiation of female strobili development to occur earlier in loblolly pine trees (Conner et al., Chapter 6). Flowering dogwood pollen release was also slightly accelerated in the heated chambers.

Air temperature is expected to increase as the concentration of greenhouse gases increases in the atmosphere. Assuming that +2 °C is a reasonable estimate for the increase in air temperature in the southeastern United States, it can be concluded that a temperature rise of this magnitude will have little negative impact on carbon gain or phenological growth patterns in loblolly pine, particularly because it will occur under conditions of much higher CO_2 concentrations, which appeared to substantially increase carbon gain. This conclusion must be made with caution, however, because whole plant responses to temperature were not addressed in any of these studies. For example, the effect of an increase in air temperature on woody tissue respiration was not measured, nor was the effect of air temperature on evapotranspiration and the development of diurnal and seasonal water stress.

The importance of air temperature as an interactive factor modifying the effect of elevated CO_2 concentrations has been investigated using the Farquhar et al., (1980) photosynthesis model. These simulations have lead to the conclusion that there will be a strong interaction between air temperature and CO_2 concentration on the rates of assimilation by the foliage of C_3 plants (Kirschbaum, 1994; Long, 1991; McMurtrie and Wang, 1993). This interaction was not clearly demonstrated in between-season comparisons of assimilation or in comparisons of temperature treatments. An explanation for the lack of differences between the temperature treatments may be that the small difference in temperatures between treatments (2 °C) produced relatively small differences in physiological responses, and so were not easily detected under field conditions. The lack of a clear CO_2 x temperature interaction in seasonal comparisons is more difficult to reconcile with theory. According to the Farquhar et al., (1980) model, under cool conditions the relative difference in assimilation between ambient and elevated CO_2 treatments should have been much smaller than in warm conditions. However, this was not apparent under field conditions. This discrepancy was probably caused by the control on assimilation exerted by other factors that change simultaneously with season, such as foliar nutrient concentrations, vapor pressure deficit, irridiance, growth rates, and respiration rates. The difference between theory and measurement in this case serves as a reminder that the response of loblolly pine to climate change will involve the complete complex of internal and external factors that contribute to the growth and productivity of this species and that they may change seasonally in relative importance to carbon assimilation and use.

In addition to changes in temperature, precipitation may also change, but the magnitude and direction of change are uncertain (Cooter et al., 1993). Water availability and water use will remain important factors contributing to the actual level of productivity of forests in the region, so understanding potential changes in water use is important (Teskey et al., 1987). It has been demonstrated that drought can significantly decrease the stimulatory effect of CO_2 enrichment in tree species (Guehl et al., 1994; Tschaplinski et al., 1995), yet it has also been shown that CO_2 enrichment can decrease whole plant water use (Overdieck and Forstreuter, 1994). The effect of water availability was studied by Dougherty et al., (Chapter 9) by using irrigated and unirrigated treatments, and indications of water use under elevated CO_2 concentrations were provided in a number of the studies that measured stomatal conductance to water vapor and leaf area development. Irrigation increased annual volume growth by 30% in drought years, and had less effect in years with average or above precipitation. Although a 30% increase in growth is appreciable, it was far less than the 108% average yearly increase over four years in fertilized plots, compared with growth in the unfertilized plots. Because the soil at this site is a deep sand, these results illustrate the relative importance of nutrients in determining productivity in the region, even on apparently dry sites. The effect of elevated CO_2 on stomatal conductance was not consistent among studies involving loblolly pine which, when considering the consistency in photosynthetic measurements, is an interesting result. Most of the studies found no significant changes in stomatal conductance to water vapor for foliage measured in both ambient and elevated CO_2 concentrations, including all of the studies using trees and saplings (Murthy et al., 1996; Sword et al., Chapter 11, Teskey, Chapter 8) and the greenhouse seedling study (Groninger et al., 1995). But the studies in open-top chambers reported a reduction in stomatal conductance in the range of 15% (Thomas et al., 1994) to 30% (Alemayehu et al., Chapter 5). The differences in the results among the studies suggest that the stomata of loblolly pine trees may be relatively insensitive to changes in CO_2 concentrations, at least in the 350 to 700 μmol mol^{-1} range, but also may indicate that the stage of development and the growth conditions may alter the sensitivity of the stomata to CO_2 concentration. In work with loblolly pine seedlings under controlled conditions, we have also found reductions in stomatal conductance in the range of 10 to 30% yet have been unable to detect them in trees. The inability to detect differences in stomatal conductance in field studies also may be partly the result of such changing environmental conditions as light intensity and the slow response of stomata to these changes (Whitehead and Teskey, 1995). But overall, because it could not be readily detected in the field, if the stomata actually do react to CO_2, it is probably a small response. This is important because it affects the water use by the trees, and indicates that the critical factor determining water use will be leaf area (Teskey and Sheriff, 1996).

A positive response of carbon gain to elevated CO_2 concentrations lasting up to three years has been demonstrated in these studies of loblolly pine. A key question still remaining is whether the effect will last for extensive periods of time and result in a dramatic long-term stimulation of growth. The declining response over time has been called acclimation, or down regulation, to elevated CO_2 and has been discussed and examined many times (e.g., Gunderson and Wullschelger, 1994; Curtis, 1996). It occurs when plants are grown in pots that are too small for their size (Sage, 1994; Thomas and Strain, 1991; Samuelson and Seiler, 1992; Arp and Drake, 1991) and in conditions of low nutrient availability (Tissue et al., 1993). With few exceptions, photosynthetic acclimation has not been reported in field studies (Arp and Drake, 1991; Idso and Kimball, 1992; Gunderson et al., 1993). The answer to whether long-term increases in productivity are possible in an enriched CO_2 environment appears to be determined by whether or not the plants are free to grow. A species such as loblolly pine has a great potential for utilizing CO_2. It has a long growing season, including multiple flushes of foliage growth, and it is not a high nutrient demanding species. Its growth characteristics allow it to incorporate carbohydrates into growing sinks for long periods during the year when photosynthetic rates are at there highest and the potential for the accumulation of carbohydrates is greatest. But the growth potential may be constrained by the availability of nutrients for growth. In managed situations this probably will not to pose a severe problem because fertilization is a common practice in the region (Colbert and Allen, 1993). In unmanaged situations it is unclear whether nutrients will become less available because of changes in litter quality, decomposition rates, and increased immobilization of nutrients (Rastetter et al., 1991) or whether the increased energy supplied to the ecosystem through increased productivity will result in increased nutrient availability (Gifford, 1994). Evidence of a positive feedback between increased carbon allocation to roots and increased nitrogen availability was reported for *Populus grandidentata* (Michx.) growing in elevated CO_2 concentrations and low-soil nutrient conditions

(Zak et al., 1993). This issue still remains unresolved but is critical to determining the long-term growth potential that could result from increased atmospheric CO_2 concentrations.

The concentrations of gases other than CO_2 are also rising in the atmosphere. Among those known to be capable of modifying plant physiological processes, ozone is considered one of the most significant to forests in the southeastern United States. The effects of ozone on loblolly pine growth and physiological processes have been studied in the region for over ten years. In the SGCP this emphasis was continued with two studies, one in loblolly pine (McLaughlin and Downing, Chapter 12) and one in shortleaf pine (*Pinus echinata* Mill.) (Flagler et al., Chapter 4). The McLaughlin et al. study was an uncontrolled correlative study of diameter growth patterns on five sites in eastern Tennessee, and the Flagler et al. study was a controlled exposure study of seedings in open-top chambers. In both studies, significant physiological effects from ambient ozone concentrations were reported. Flagler et al. found that ambient ozone concentrations reduced foliage biomass and leaf area compared to a charcoal-filtered control. Additionally, there was no interaction between ozone and soil-water availability, and therefore ozone uptake continued even when the plants were under water stress. The McLaughlin et al. study also reported a significant effect of ozone on growth. In that study, air temperature, rainfall, soil moisture, and ozone concentrations were statistically correlated with diameter growth using linear regression. Their analysis indicated that the present-day ozone concentrations in that region were correlated with reductions in growth. Both studies have provided indications that ambient levels of ozone may be capable of reducing growth of pines in the region. Although the findings are not conclusive, they add to the existing body of information that has clearly demonstrated that concentrations of ozone that are higher than the present-day ambient concentration are detrimental to the growth and physiological activity of southern pines, and have indicated that there is a potential for present-day ambient concentrations to be detrimental. However, the balance between ozone and CO_2 concentrations will be important in determining the growth response of pines in the region. For example, if ozone concentrations rise as rapidly as CO_2 concentrations and both double in the next 100 years, then tropospheric ozone could significantly reduce the beneficial effects of elevated CO_2 on pine growth and productivity. Alternatively, if ozone concentrations remain near their present-day levels, a doubling of the atmospheric CO_2 concentration will overcome the detrimental effects of ozone.

This program has demonstrated that elevated CO_2 concentrations have positive effects on the physiological processes related to growth and productivity of loblolly pine. Perhaps the most important findings from these studies in the program were 1) the lack of acclimation to elevated CO_2, and 2) the additive nature of the physiological responses to CO_2 enrichment. Of course, the environment greatly alters the potential and actual productivity that can be achieved on a given site, and it is important to recognize that although the relative response to elevated CO_2 concentrations was very similar under favorable and unfavorable growth conditions, under elevated CO_2 conditions the actual overall increase in growth and productivity on poor sites will not be as great as on good sites. Yet the findings indicate that pine forests on a wide variety of sites across the region will respond positively if the CO_2 concentration in the atmosphere continues to rise. This conclusion is supported by many studies that have shown that the relative growth-enhancing effect of elevated CO_2 concentration is often as great, or greater, in resource limited environments as in resource rich environments (Idso and Idso, 1994).

References

Arp WJ, Drake BG (1991) Increased photosynthetic capacity of *Scripus olneyi* after 4 years of exposure to elevated CO_2. Plant Cell Environ 14:1003–1006.

Bunce JA (1992) Stomatal conductance, photosynthesis and respiration of temperate deciduous tree seedlings grown outdoors at elevated concentration of carbon dioxide. Plant Cell Environ 15:541–549.

Bunce JA (1994) Responses of respiration to increasing atmospheric carbon dioxide concentrations. Physiol Plant 90:427–430.

Bunce JA, Caufield F (1991) Reduced respiratory carbon dioxide efflux during growth at elevated carbon dioxide in three herbaceous perennial species. Ann Bot 67:325–330.

Callaway RM, DeLucia EH, Thomas EM, Schlesinger WH (1994) Compensatory responses of CO_2 exchange and their effects on the relative growth rate of ponderosa pine in different CO_2 and temperature regimes. Oecologia 98:159–166.

Ceulmans R, Prez-Leroux A, Impens I (1993) The direct impact of rising CO_2 and temperature on plants, with particular reference to trees. In Symoens JJ, Devos P, Rammeloo J, Verstaeten C (Eds), *Biological indicatos of global change*. Royal Academy of Overseas Sciences Publish. Brussels, Belgium.

Ceulemans R, Mousseau M (1994) Tansley Review No. 71. Effects of elevated atmospheric CO_2 on woody plants. New Phytol 127:425–446.

Colbert SR, Allen HL (1996) Factors contributing to variability in loblolly pine foliar nutrient concentrations. So J App For 20:45–52.

Conroy JP, Milham PJ, Mazur M, Barlow EWR (1990) Influence of phosphorus deficiency on the growth response of four families of *Pinus radiata* seedlings to CO_2 enriched atmospheres. For Ecol Manag 30:175–188.

Cooter EJ, Eder BK, LeDuc SK, Truppi L (1993) General circulation model output for forest climate change research and applications. USDA For Ser South For Exper Sta Gen Tech Rep SE-85.

Curtis PS (1996) A meta-analysis of leaf gas exchange and nitrogen in trees grown under elevated carbon dioxide. Plant Cell Environ 19:127–137.

Curtis PS, Vogel CS, Pregitzer KS, Zak DR, Terri JA (1995) Interacting effects of soil fertility and atmospheric CO_2 on leaf area growth and carbon gain physiology in *Populus* x *euramericana* (Dode) Guinier. New Phytol 129:253–263.

El Kohen A, Mousseau M (1994) Interactive effects of elevated CO_2 and mineral nutrition on growth and CO_2 exchange of sweet chestnut seedlings (*Castanea sativa*). Tree Physiol 14:679–690.

El Kohen A, Venet L, Mousseau M (1993) Growth and photosynthesis of two deciduous forest tree species exposed to elevated carbon dioxide. Func Ecol 7:480–486.

Ellsworth DS, Oren R, Huang C, Phillips N, Hendry GR (1995) Leaf and canopy responses to elevated CO_2 in a pine forest under free-air CO_2 enrichment. Oecologia 104:139–146.

Farquhar GD, von Caemmerer S, Berry JA (1980) A biochemical model of photosynthetic CO_2 assimilation in leaves of C_3 species. Planta 149:78–90.

Fetcher N, Jaeger CH, Strain BR, Sionit N (1988) Long-term elevation of atmospheric CO_2 concentration and the carbon exchange rates of saplings of *Pinus taeda* L. and *Liquidambar styraciflua* L. Tree Physiol 4:255–262.

Garcia RL, Idso SB, Wall GW, Kimball BA (1994) Changes in net photosynthesis and growth of *Pinus eldarica* seedlings in response to atmospheric CO_2 enrichment. Plant Cell Environ 17:971–978.

Gifford RM (1994) The global carbon cycle: A viewpoint on the missing sink. Aust J Plant Physiol 21:1–15.

Griffin KL, Ball JT, Strain BR (1996) Direct and indirect effects of elevated CO_2 on whole-shoot respiration in ponderosa pine seedlings. Tree Physiology 16:33–41.

Griffin KL, Thomas RB, Strain BR (1993) Effects of nitrogen supply and elevated carbon dioxide on construction cost in leaves of *Pinus taeda* L. seedlings. Oecologia 95:575–580.

Groninger JW, Seiler JR, Zedaker SM, Berrang PC (1995) Effects of elevated CO_2, water stress, and nitrogen level on competitive interactions of simulated loblolly pine and sweetgum stands. Can J For Res 25:1077–1083.

Grulke NE, Hom JL, Roberts SW (1993) Physiological adjustment of two full-sib families of ponderosa pine to elevated CO_2. Tree Physiol 12:391–401.

Guehl JM, Picon C, Aussenac G, Gross P (1994) Interactive effects of elevated CO_2 and soil drought on growth and transpiration efficiency and its determinants in two European forest tree species. Tree Physiol 14:707–724.

Gunderson CA, Norby RJ, Wullschleger SD (1993) Foliar gas exchange responses of two deciduous hardwoods during 3 years of growth in elevated CO_2: no loss of photosynthetic enhancement. Plant Cell Envir 16:797–807.

Gunderson CA, Wullschleger SD (1994) Photosynthetic acclimation in trees to rising CO_2: A broader perspective. Photosyn Res 39:369–388.

Idso KE, Idso SB (1994) Plant responses to atmospheric CO_2 enrichment in the face of environmental constraints: A review of the past 10 years' research. Agric For Meteor 69:153–203

Idso S, Kimball BA (1992) Effects of atmospheric CO_2 enrichment on photosynthesis, respiration, and growth of sour orange trees. Plant Physiol 99:341–343.

Jifon JL, Fiend AL, Berrang PC (1995) Species mixture and soil resource availability affect the root growth response of tree seedlings to elevated CO_2. Can J For Res 25(5):824–832.

Kirschbaum MUF (1994) The sensitivity of C_3 photosynthesis to increasing CO_2 concentration: a theoretical analysis of its dependence on temperature and background CO_2 concentration. Plant Cell Environ 17:747–754.

Koike T, Lei TT, Maximov TC, Tabuchi R, Takahashi T, Ivanov BI (1996) Comparison of the photosynthetic capacity of Siberian and Japanese birch seedlings grown in elevated CO_2 and temperature. Tree Physiol 16:381–385

Larigauderie A, Reynolds JF, Strain BR (1994) Root response to CO_2 enrichment and nitrogen supply in loblolly pine. Plant and Soil 165:21–32.

Liu S, Teskey RO (1995) Responses of foliar gas exchange to long-term elevated CO_2 concentrations in mature loblolly pine trees. Tree Physiol 15:351–359.

Long SP (1991) Modification of the response of photosynthetic productivity to rising temperature by atmospheric CO_2 concentrations: Has its importance been underestimated? Plant Cell Environ 14:729–739.

McMurtrie RE, Wang YP (1993) Mathematical models of the photosynthetic response of tree stands to rising CO_2 concentrations and temperatures. Plant Cell Environ 16:1–13.

Mousseau M (1993) Effects of elevated CO_2 on growth, photosynthesis and respiration of sweet chestnut (*Castanea sativa* Mill.). Vegetatio 104/105:413–419.

Murthy R, Dougherty PM, Zarnoch SJ, Allen HL (1996) Effects of carbon dioxide, fertilization, and irrigation on photosynthetic capacity of loblolly pine trees. Tree Physiol 16:537–546.

Norby RJ, O'Neill EG (1991) Leaf area compensation and nutrient interactions in CO_2-enriched seedlings of yellow-poplar (*Liriodendron tulipifera* L.). New Phytol 117:515–528

Overdieck D, Forstreuter M (1994) Evapotranspiration of beech stands and transpiration of beech leaves subject to atmospheric CO_2 enrichment. Tree Physiol 14:997–1003.

Rastetter EB, Ryan MG, Shaver GR, Nadelhoffer JM, Hobbie KL, Aber JD (1991) A general model describing the responses of the C and N cycles in terrestrial ecosystems to changes in CO_2, climate and N deposition. Tree Physiol 9:101–126.

Reuveni J, Gale J (1985) The effect of high levels of carbon dioxide on dark respiration and growth of plants. Plant Cell Environ 8:623–628.

Sage RF (1994) Acclimation of photosynthesis to increasing atmospheric CO_2: The gas exchange perspective. Photosyn Res 39:351–368

Samuelson LJ, Seiler JR (1992) Interactive role of elevated CO_2, nutrient limitations, and water stress in the growth responses of red spruce seedlings. For Science 39:348–358.

Sharkey TD, Loreto F, Delwiche CF (1991) High carbon dioxide and sun/shade effects on isoprene emission from oak and aspen tree leaves. Plant Cell Environ 14:333–338.

Teskey RO (1995) A field study of the effects of elevated CO_2 on carbon assimilation, stomatal conductance and leaf and branch growth of *Pinus taeda* trees. Plant Cell Environ 18:565–573.

Teskey RO, Bongarten BC, Cregg BM, Dougherty PM, Hennessey TC (1987). Physiology and genetics of tree growth response to moisture and temperature stress: An examination of the characteristics of loblolly pine (*Pinus taeda* L.). Tree Physiol 3: 41–62.

Teskey RO, Sheriff DW (1996) Water use by *Pinus radiata* trees in a plantation. Tree Physiol 16:273–279.

Thomas RB, Lewis JD, Strain BR (1994) Effects of leaf nutrient status on photosynthetic capacity in loblolly pine (*Pinus taeda*, L.) seedlings growth in elevated atmospheric CO_2. Tree Physiol 14:947–960.

Thomas RB, Reid CD, Ybema R, Strain BR (1993) Growth and maintenance components of leaf respiration of cotton grown in elevated carbon dioxide partial pressure. Plant Cell Environ 16:539–546.

Thomas RB, Strain BR (1991) Root restriction as a factor in photosynthetic acclimation of cotton seedlings grown in elevated carbon dioxide. Plant Physiol 96:627–634.

Tissue DT, Thomas RB, Strain BR (1993) Long-term effects of elevated CO_2 and nutrients on photosynthesis and rubisco in loblolly pine seedlings. Plant Cell Environ 16:859–865.

Tissue DT, Thomas RB, Strain BR (1996) Growth and photosynthesis of loblolly pine (*Pinus taeda*) after exposure to elevated CO_2 for 19 months in the field. Tree Physiol 16:49–59.

Townend J (1993) Effects of elevated carbon dioxide and drought on the growth and physiology of clonal Sitka spruce plants (*Picea sitchensis* (Bong.) Carr.). Tree Physiol 13:389–399.

Tschaplinski TJ, Stewart DB, Hanson PJ, Norby RJ (1995) Interactions between drought and elevated CO_2 on growth and gas exchange of seedlings of three deciduous tree species. New Phytol 129:63–71.

Valentine DW, Allen HL (1990) Foliar responses to fertilization identify nutrient limitation in loblolly pine. Can J For Res 20:144–151.

Van Lear DH, Waide JB, Teuke MJ (1984) Biomass and nutrient content of a 41-year old loblolly pine (*Pinus taeda* L.) plantation on a poor site in South Carolina. For Science 30:395–404.

Vivin PM, Gross P, Aussenac G, Guehl JM (1995) Whole-plant CO_2 exchange, carbon partitioning and growth in *Quercus robur* seedlings exposed to elevated CO_2. Plant Physiol Biochem 33:201–211.

Wang K, Kellomäki S, Laitinen K (1995) Effects of needle age, long-term temperature and CO_2 treatments on the photosynthesis of Scots pine. Tree Physiol 15:211–218.

Whitehead D, Teskey RO (1995) Dynamic response of stomata to changing irridiance in loblolly pine (*Pinus taeda* L.). Tree Physiol 15:245–251.

Wullschleger SD, Norby RJ (1992) Respiratory cost of leaf growth and maintenance in white oak saplings exposed to atmospheric CO_2 enrichment. Can J For Res 22:1717–1721.

Zak DR, Pregitzer KS, Curtis PS, Teeri JA, Fogel R, Ranlettt DL (1993) Elevated atmospheric CO_2 and feedback between carbon and nitrogen cycles. Plant and Soil 151:105–117.

Ziska LH, Bunce JA (1994) Direct and indirect inhibition of single leaf respiration by elevated CO_2 concentrations: Interaction with temperature. Physiol Plant 90:130–138.

Section 3. Modeling the Biophysical Effects of Global Change

16. Modeling Nutrient Uptake as a Component of Loblolly Pine Response to Environmental Stress

J. Michael Kelly and Ruth D. Yanai

The ability of plants to acquire nutrients and fix carbon depends on the below ground processes associated with soil-nutrient supply, uptake kinetics, and root-surface area, combined with the physiological processes that capture, fix, and redistribute carbon from the aboveground portions of the plant to the belowground parts. Plants are exposed to multiple environmental stresses that act both individually and collectively to limit plant growth, by reducing rates of photosynthesis, growth, and carbon storage. Although the relative importance of various stresses depends on the site, for southern forests, available water and nutrient supplies in addition to tropospheric ozone are generally the factors of greatest concern (McLaughlin, 1985). Any stress that directly or indirectly impairs the ability of the plant to fix and store carbon can exacerbate nutrient and water stress, because root growth, the development of mycorrhizal associations, and active uptake of nutrients all depend on carbon supply.

Unfortunately for predictive purposes, in real world situations the various combinations of multiple stresses can be either competitive or offsetting. Although single-factor studies can clearly define the impact of an individual stress, the much-needed experiments with multiple stress factors are frequently more costly and time-consuming to conduct, difficult to design, and more problematic to interpret. Consequently, it is difficult to determine experimentally those processes that probably will control overall plant behavior under various circumstances. Mechanistic models provide a means to circumvent some of these limitations and investigate the effects of combinations of stresses. To be effective as well as accurate, nutrient-uptake models must be combined with detailed whole plant carbon allocation models so that interactions between the two models can occur and the results can describe verifiable changes in key parameters.

Model Descriptions

We employed two nutrient-uptake models, the Barber-Cushman model (Barber, 1984) as modified for the personal computer by Oates and Barber (1987) and a steady-state model (Yanai, 1994). The two models are similar in approach and share many assumptions with their predecessors (Nye and Spiers, 1964; Nye and Marriot, 1969; Claassen and Barber, 1976; Cushman, 1979; Barber and Cushman, 1981). They simulate uptake by the average absorbing root in the average soil; there is no consideration of the geometry of the root system or of differences in root properties with age or morphology. The root is essentially a uniform, linear sink and the amount of soil surrounding the root is defined by the average distance to the next root. In both models, nutrients move toward the root by both mass flow (the movement of solution to the root to support the transpiration stream) and by diffusion along the concentration gradient created by active uptake at the root surface. Uptake at the root surface is described by Michaelis-Menten kinetics. Neither model considers mycorrhizal association, except as it affects the values of parameters in the model, nor root modification of the rhizosphere, except for nutrient depletion.

Both modeling approaches have been extensively tested in a number of applications through comparison of model predictions to observed responses. Although this is not a fail-safe process, experience has shown that opportunities for error propagation through these modeling approaches lie more with the development of the data used to provide initial values than with the concepts or codes within the models. Consequently, a portion of the discussion in this chpater is devoted to an analysis of how best to develop key initial values.

The major difference between the models is that the Barber-Cushman model simulates uptake of a growing root system over a period of time, but without time-varying input. The rate of root growth is one of the values input to the model; that is, it must be specified in advance. Similarly, there are no inputs of nutrients to the soil system during the duration of a model run except as defined by the ability of the solid phase (C_s) to maintain solution-phase concentration (C_l) through buffering. We applied this model with considerable success in a simulation of loblolly pine seedlings for one growing season (Kelly et al., 1992). However, the Barber-Cushman model as presently configured could not be effectively linked to a plant simulator because it does not accept time-varying input to root growth and soil conditions. We also wanted to allow such feedback as the effect of nutrient uptake on root growth and the effect of litter quality on nutrient supply. Therefore, in addition to using the Barber-Cushman model, we developed a new model to allow a more dynamic simulation of nutrient uptake (Yanai, 1994). This model calculates the steady-state solution for uptake at any point in time and can be invoked repeatedly to simulate uptake over time under changing plant and soil conditions. Dissimilar to prior steady-state models (Baldwin et al., 1973; Nye and Tinker, 1977), which ignored the effect of root growth into unexplored soil, this new model includes the nutrient extracted from soil in the process of forming a depletion zone. At the other extreme, the Barber-Cushman model assumes that all roots start the simulation in unexplored soil, which is not realistic for the uptake of immobile nutrients by mature plants.

Results and Discussion

Factors Controlling Nutrient Uptake

Many factors interact in determining nutrient uptake by plants. In the uptake models we used, these factors are represented as parameters, which can be divided into those related to the plant and those describing the supply of nutrient from soil. Factors related to the plant can be further divided into those describing uptake kinetics (defined per unit surface of root) and those describing the development of the root system, which determine the absorbing length and surface area.

Uptake Kinetics

The three processes that combine to determine the movement of nutrients to the root surface are (1) nutrient uptake at the root, which tends to create a concentration gradient in the vicinity of the root, (2) flow of water to the root to support the transpiration stream, and (3) diffusion in response to the concentration gradient created by active uptake and solution flow. As soon as nutrients are delivered to the root surface, uptake in the natural environment will reflect the combined influences of varying degrees of active and passive uptake. Both the Barber-Cushman (Barber, 1984) and Yanai (1994) models use Michaelis-Menton parameters to describe nutrient uptake as a function of nutrient concentration at the root surface.

To use the Michaelis-Menton approach, it is necessary to conduct solution studies to find values of parameters representing (1) the maximum rate of nutrient influx (I_{max}) at high solution concentrations, (2) the nutrient concentration in solution (k_m) at which influx is one-half of I_{max}, and (3) the concentration in solution below which influx ceases (C_{min}) (Barber, 1984). Typical values for these three parameters by nutrient for loblolly pine are presented in Table 16.1; these values are consistent with values reported by others (Van Rees et al., 1990a; Williams and Yanai, 1996). Experimental work and modeling efforts revealed some important factors that need to be considered when attempting to model nutrient uptake.

First, as illustrated by the sensitivity analysis depicted in Figure 16.1, I_{max} is the most influential of the uptake kinetics parameters for the situation modeled and therefore must be chosen very carefully. In this case, high solution concentrations (C_1) relative to I_{max} account for the dominance of I_{max} (Figure 16.2a). At very low

Table 16.1. Root Uptake Kinetics Variables Used in the Barber-Cushman Model to Predict Mg, K, and P Uptake

Variable	Units	Mg	K	P
I_{max}	$\mu mol\ cm^{-2}\ s^{-1}$	1.29E−7	1.40E−6	2.68E−7
k_m	$\mu mol\ cm^{-3}$	9.83E−3	3.0E−2	1.60E−2
C_{min}	$\mu mol\ cm^{-3}$	0.001	0.001	0.0006

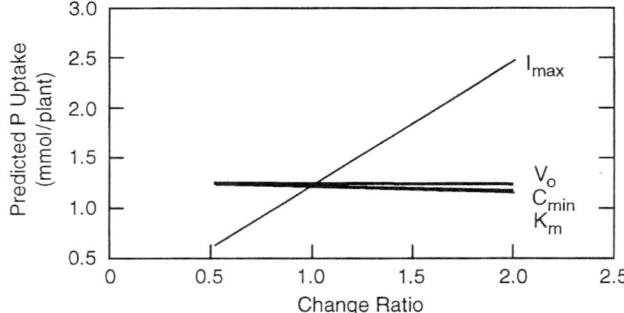

Figure 16.1. Sensitivity analysis of predicted phosphorus uptake in response to changing the maximum influx rate (I_{max}), solution-nutrient concentration at 0.5 I_{max} (k_m), solution concentration at which influx is zero (C_{min}), and water-uptake rate (v_o). Each parameter was varied individually by the indicated ratio while all other parameters were held constant. Figure redrawn from Kelly et al. (1992).

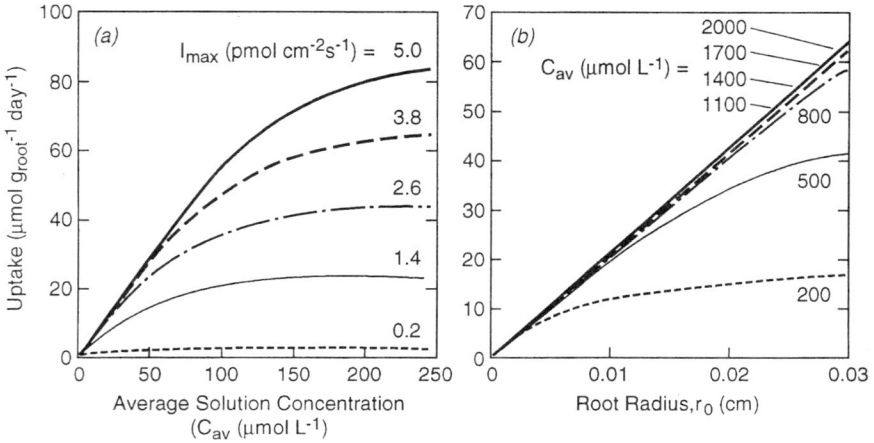

Figure 16.2. Two dimensional sensitivity analyses of calculated uptake (U_{est}) as (a) a function of the average concentration in soil solution (C_{av}) for varying values of the Michaelis-Menton uptake parameter I_{max}; and (b) a function of root radius (r_o) for various values of C_{av} with I_{max} = 4 E−12 mol cm^{-2} s^{-1} and D = 5 E−12 cm^{-2} s^{-1}. Other variables were held constant. Figure redrawn from Yanai (1994).

concentrations, uptake at the maximum rate may not occur or is relatively short-lived and therefore has only limited effect on uptake. Uptake will increase linearly with increasing C_l until the uptake approaches I_{max}, which it cannot exceed. The rate of approach of the I_{max} limitation with increasing C_l depends on the value of k_m (Yanai, 1994). When influx rates are close to I_{max}, k_m is not important in defining uptake (Kelly et al., 1992; Yanai, 1994). The C_{min} value, similar to k_m, often is not important to estimate nutrient uptake, because it is much lower than simulated uptake rates. Very low values of C_{min}, such as those of loblolly seedlings for NH_4-N (Kelly et al., 1995a), mean that nutrient uptake can continue even when nutrient depletion is severe.

The I_{max} value is both the most important and the most problematic of the uptake kinetics parameters to define for perennial plants. Kelly and Barber (1991) noted differences in the magnesium I_{max} value of at least an order of magnitude when they compared values for 365-d and 180-d seedlings. Similarly, Kelly et al. (1995a) found that I_{max} values can vary between loblolly pine families, as well as possibly differing across the growing season. The latter possibility is supported in part by an earlier observation (Kelly and Barber, 1991) in which seedlings that were not experiencing a shoot-growth flush exhibited a lower I_{max} than would have been observed if the experiments were performed during a growth flush. Although there are circumstances when this value will be less critical (for example, conditions of very low nutrient availability), finding appropriate methods for measuring I_{max} and mechanisms to describe its dynamic nature are key to future progress in nutrient uptake modeling.

Root Length and Surface Area

In addition to the parameters that define uptake kinetics, the parameters that define root length and surface area are extremely important in determining rates of nutrient uptake. In the steady-state models, these parameters are root radius and root length; root-growth rate is a factor in calculating nutrients acquired in the formation of depletion zones. In the Barber-Cushman model, the parameters are root radius, initial root length, and the rate of increase of root length. Both models calculate uptake as the product of root-surface area and the simulated uptake rate per unit area. For this reason, uptake might be expected to be proportional to root-surface area, and a number of one-dimensional sensitivity analyses (Nye and Tinker, 1977; Barber, 1984; Kelly et al., 1992) support this relationship (Figure 16.3). This relationship holds as long as uptake is proportional to the absorbing surface area, even if variation in surface area is the result of root radius (Figure 16.2b). On the other hand, when uptake is controlled by soil supply, increases in root radius offer little improvement in nutrient uptake. Uptake kinetics are not a limiting factor in this situation as the roots take up all the solute that arrives at the root surface; the limiting factor is the rate of delivery of solute to the root surface. (The parameters controlling the supply of nutrients by the soil will be discussed in the next section.) When the rate of delivery of solute to the roots is a limiting factor, the root behaves approximately as a linear sink. Uptake under these condi-

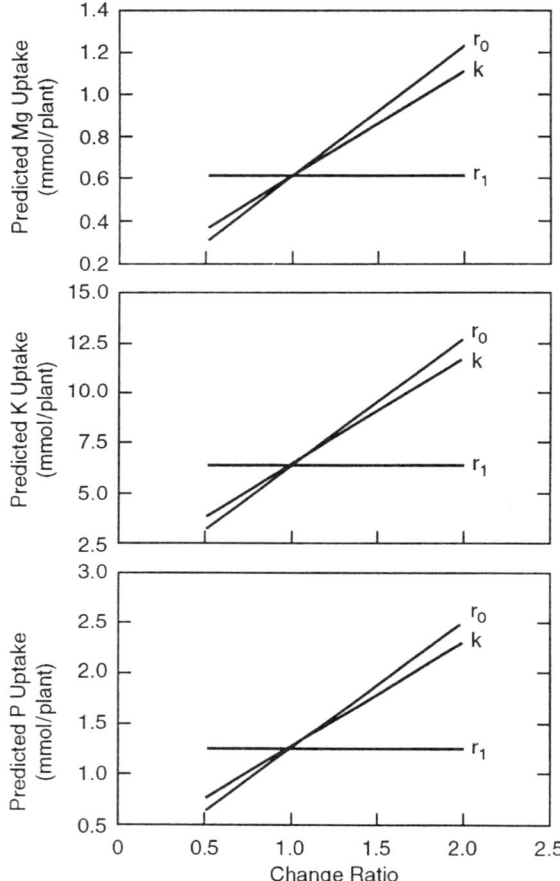

Figure 16.3. Sensitivity analysis of predicted Mg, K, and P uptake in response to changing the initial root length (L_o), root-growth rate (k), mean root radius (r_o), and half-distance between root axes (r_1). Each parameter was varied individually by the indicated ratio while all other parameters were held constant. Figure redrawn from Kelly et al. (1992).

tions will be more dependent on root length than on root-surface area (Yanai, 1994; Williams and Yanai, 1996).

Increased root length will result in increased nutrient uptake, but as root density increases, incremental additions of root length bring diminished returns to the plant. In the models, root-length density is represented by the interroot distance, r_1, which describes the average radius of the zone of influence of the root. In the Barber-Cushman model, this distance is constant; the plant is assumed to occupy a proportionately larger soil volume as the root length increases. In the steady-state models, the interroot distance is calculated at each timestep.

The separate treatment of new root growth in the steady-state model made it possible to assess the importance of the solute obtained in the formation of depletion zones. This contribution to uptake is most important for immobile nutrients and rapidly growing root systems (Yanai, 1994).

Soil Supply

Defining soil-supply parameters for modeling purposes is less problematic than defining plant parameters. The solution concentration (C_l) is often the most influential of the soil-supply parameters (Figure 16.4). As illustrated by the ammonium data plotted in Figure 16.5, the C_1 value not only varied among the four fertility treatments depicted, but also changed substantially from the initial sampling in early May through the final sampling in October. Although one of the weaknesses

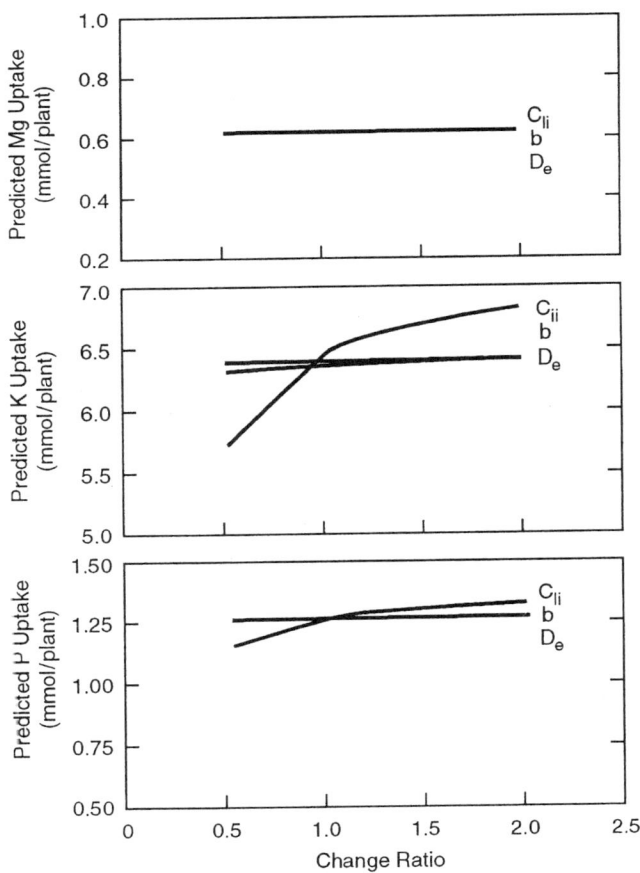

Figure 16.4. Sensitivity analysis of predicted Mg, K, and P uptake in response to changing the initial soil solution concentration (C_{li}), the diffusion coefficient (D_e), and buffer power (b). Each parameter was varied individually by the indicated ratio while all others parameters were held constant. Figure redrawn from Kelly et al. (1992).

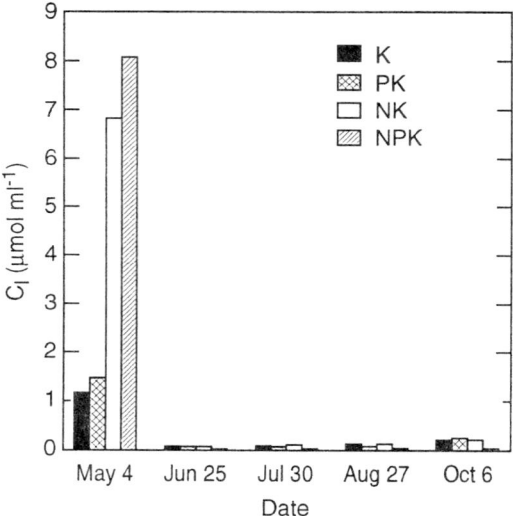

Figure 16.5. Mean equilibrium solution-phase concentrations (C_l) of ammonium by harvest date and fertility regime in soil collected from pots in which a single loblolly pine seedling was growing. Figure redrawn from Kelly et al. (1995a).

of the Barber-Cushman approach (Barber, 1984) is the lack of a mechanism in the model to allow resupply of nutrient except through transfer from the solid to the liquid phase, the actual soil-solution concentration data plotted in Figure 16.5 indicate that little or no resupply occurred during the growing season. However, had these data been collected under field conditions, it is more probable that resupply would have been observed.

Measured solid-phase values (C_s) also exhibit variation across the growing season as illustrated by the potassium values plotted in Figure 16.6. As might be anticipated, the values generally decline as the growing season progresses, reflecting a transfer to the solution phase in response to plant uptake or leaching loss (Kelly et al., 1995a). This relationship between the solution and solid phases is represented in the model through the buffer power (b), which is roughly C_s/C_l (Van Rees et al., 1990b). Although Kelly et al. (1992) found uptake to be relatively insensitive to changes in the b value using a single-factor sensitivity analysis, Yanai (1994), who worked with essentially the same data set, found the b value to be somewhat more influential in her multifactor approach to sensitivity.

Modeling Nutrient Uptake and Supply

The Barber-Cushman approach, although successful at simulating growth over a growing season, could not be linked to a plant simulator because it, as with similar

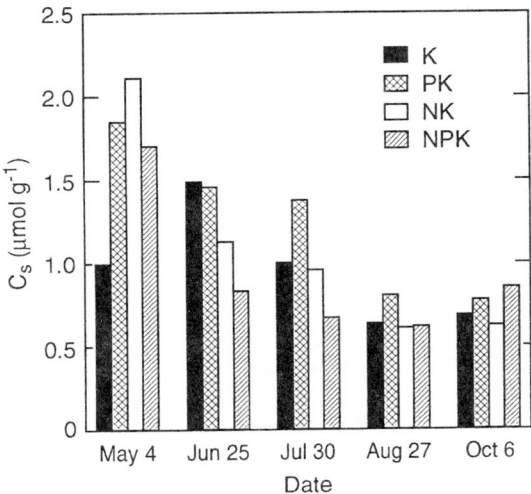

Figure 16.6. Mean concentrations of potassium in the solid phase (C_s) by harvest date and fertility regime in soils collected from pots in which a single loblolly pine seedling was growing. Figure redrawn from Kelly et al. (1995a).

numerical models, cannot accept time-varying input. As an alternative, the iterative steady-state approach (Nye and Tinker, 1977) provides a simple method for calculating nutrient uptake that is more appropriate to plants that have long-lived roots and multiple periods of root growth. Our version (Yanai, 1994) is an improvement over prior steady-state uptake models (Baldwin et al., 1973; Nye and Tinker, 1977) because the nutrient extracted from soil in the process of forming a depletion zone is included in the uptake calculation. Additionally, the inclusion of Michaelis-Menton kinetics is an improvement if non-linear uptake kinetics are required.

In the case of loblolly pine seedlings, the calculation of uptake by established roots was most sensitive to root length and soil-solution concentration (Yanai, 1994). The amount of uptake provided by the formation of a depletion zone by growing roots was most sensitive to root density, solution concentration, and the effective diffusion coefficient. These results, however, are dependent on the situation studied because model sensitivity to one parameter is dependent on the values of other parameters (Williams and Yanai, 1996).

Because the factors limiting uptake vary with environmental conditions and plant status, a model of solute uptake that considers only one or two limiting factors, such as root mass and soil-solution concentration, will not be applicable under a wide range of conditions. Such multifactor models, as those discussed here are therefore better suited to assess plant response to environmental stress. For example, the Barber-Cushman model can simulate nutrient uptake for a single growing season (Van Rees et al., 1990a; Kelly et al., 1992). The steady-state model of nutrient uptake (Yanai 1994) has been incorporated in the plant model

TREGRO (Weinstein et al., 1992), allowing feedbacks between the plant and soil that influence estimates of nutrient uptake and plant growth.

Nutrient Availability and Plant Response to Stress

The effects of multiple environmental stresses can be difficult to predict because the combined effects of individual stresses are often not additive (Van Heerdeen and Yanai, 1995). For example, Temple et al. (1993), as well as Runeckles and Chevone (1992), found that drought stressed plants responded less to ozone than did well-watered plants. Conversely, trees weakened by ozone can be more susceptible to damage from other stresses (Hain, 1987; Davidson et al., 1988; Edwards et al., 1990). Understanding the impact of nutrient limitation in combination with other stresses on the plant is essential to predicting the response of vegetation to air pollutants and other environmental changes.

Experimental work with loblolly pine (Kelly et al., 1993) and northern red oak (Kelly et al., 1995b) showed that ozone reduced root growth, presumably as a result of carbon allocation to foliar repair. Under nutrient limitation, decreased allocation of carbon to the root system could exacerbate a nutrient stress. Conversely, if nutrient limitation results in a reduced carbon supply, a plant could become more susceptible to damage from ozone exposure as a result of an insufficient supply of carbon to compensate for damage (Pell et al., 1994). Simulation models provide a means of assessing the possible interactions of nutrient limitation with other environmental factors.

Summary

Mechanistic models of nutrient uptake are useful tools in refining our understanding of the chemical, physical, and biological processes that control plant nutrition. Prior work with woody species has raised important questions on how best to derive model input values given that many of these values change substantially over a growing season. For example, models using fixed-root morphology should use weighted seasonal average values rather than values describing initial conditions. Alternatively, variable growth rates across the growing season based on actual observation can be used in an iterative steady-state model. Decisions on these two options will be influenced by the intent of the modeling exercise and the data available. Changes in soil-supply parameters should also be taken into consideration; again, seasonal variation must be examined even if time-varying input is not used. Equally important to reasonable model representations is the recognition that the age and growth stage of the plant can influence the kinetics of nutrient uptake and carbon allocation to roots.

A model of solute uptake that accepts root growth, water uptake, and soil-solution concentration as time-varying input is required to interactively link plant and soil processes. The advantage of the steady-state approach to solute uptake over more exact numerical solutions lies in the independence of the mathematical solution to prior conditions. Uptake thus calculated can accommodate unpredictable changes in root growth and mortality, root density, water-uptake rates, and such sources and sinks of nutrients as decomposition and leaching. This level of flexibility is required in simulating plant growth for multiple seasons in a dynamic soil environment. Prior steady-state models were modified to include nonlinear uptake kinetics and the contributions of new root growth to uptake.

References

Baldwin, JP, Nye PH, Tinker PB (1973) Uptake of solutes by multiple root systems from soil. IV. A model to calculate the uptake by a developing root system or root hair system of solutes with concentration variable diffusion coefficients. Plant Soil 38:621–635.

Barber, SA (1984) *Soil nutrient bioavailability: A mechanistic approach*, John Wiley and Sons, New York.

Barber, SA, Cushman JH (1981) Nitrogen uptake model for agronomic crops. In Iskandar IK (Ed) *Modeling waste water renovation-land treatment*, Wiley-Interscience, New York.

Claassen N, Barber SA (1976) Simulation model for nutrient uptake from soil by a growing plant root system. Agron J 68:961–964.

Cushman, JH (1979) An analytical solution to solute transport near root surfaces for low initial concentration: I. Equation development. Soil Sci Soc Am J 43:1087–1092

Davidson, AW, Barnes JD, Renner CJ (1988) Interactions between air pollutants and cold stress. In Schulte-Hostede S, Darrall NM, Blank LW, Wellburn AR (Eds) *Air pollution and plant metabolism*, Elsevier Applied Science, London.

Edwards, GS, Pier PA, Kelly JM (1990) Influence of ozone and soil magnesium status on the cold hardiness of loblolly pine (*Pinus taeda* L.) seedlings. New Phytol 115:157–164.

Hain, FP (1987) Interactions of insects, trees, and air pollutants. Tree Physiol 3:93–102.

Kelly JM, Barber SA (1991) Magnesium uptake kinetics in loblolly pine seedlings. Plant Soil 134:227–232.

Kelly JM, Barber SA, Edwards GS (1992) Modeling magnesium, phosphorus, and potassium uptake by loblolly pine seedlings using a Barber-Cushman approach. Plant Soil 139:209–218.

Kelly JM Taylor GE, Edwards NT, Adams MB, Friend AL (1993) Growth, physiology, and nutrition of loblolly pine seedlings stressed by ozone and acidic precipitation:A summary of the ROPIS-South project. Water Air Soil Pollut 69:363–391.

Kelly JM, Chappelka AH, Lockaby BG (1995a) Measured and estimated parameters for a model of nutrient uptake by trees. New Zeal J For Sci 24:213–225.

Kelly JM, Sameulson L, Edwards G, Hanson P, Kelting D, Mays A, Wullschleger S (1995b) Are seedlings reasonable surrogates for trees? An analysis of ozone impacts on *Quercus rubra*. Water Air Soil Pollut 85:1317–1324.

McLaughlin SB (1985) Effects of air pollution on forests: A critical review. J Air Pollut Contr Assoc 35:516–534.

Nye PH, Spiers JA (1964) Simultaneous diffusion and mass flow to plant roots. In Trans Int Congr Soil Sci 8th Bucharest. 31 Aug–9 Sept 1964. Rompresfilatelia, Bucharest.

Nye PH, Marriott FHC (1969) A theoretical study of the distribution of substance around roots resulting from simultaneous diffusion and mass flow. Plant Soil 30:459–472.

Nye PH, Tinker PB (1977) *Solute movement in the soil-root system*, Blackwell Scientific Publications, Oxford, England.

Oates K, Barber SA (1987) Nutrient uptake: A microcomputer program to predict nutrient absorption from soil by roots. J Agron Educ 16:65–68.

Pell EJ, Temple PJ, Friend AL, Mooney HA, Winner WE (1994) Compensation as a plant response to ozone and associated stresses: An analysis of ROPIS experiments. J Environ Qual 23:429–436.

Runeckles VC, Chevone BI (1992) Crop responses to ozone. In Lefohn AS (Ed) *Surface level ozone exposures and their effects on vegetation.* CRC Press, Boca Raton, FL.

Temple PJ, Riechers RH, Miller PR, Lennox RW (1993) Growth responses of ponderosa pine to long term exposure to ozone, wet and dry acidic deposition, and drought. Can J For Res 23:59–66.

Van Heerden K, Yanai R (1995) Effects of stresses on forest growth in models applied to the Solling spruce site. Ecological Modeling 83:273–282.

Van Rees KCJ, Comerford NB, McFee WW (1990a) Modeling potassium uptake by slash pine seedlings from low-potassium-supplying soils. Soil Sci Soc Am J 54:1505–1507.

Van Rees KCJ, Comerford NB, Rao PSC (1990b) Defining soil buffer power: Implications for ion diffusion and nutrient uptake modeling. Soil Sci Soc Am J 54:1505–1507.

Weinstein DA, Beloin RM, Yanai RD, Zollweg CG (1992) The response of plants to interacting stresses: TREGRO version 1.74. Description and parameter requirements. EPRI TR-101061, Electric Power Research Institute, Palo Alto, CA.

Williams M, Yanai R (1996) Multi-dimensional sensitivity analysis and ecological implications of a nutrient uptake model. Plant Soil 180:311–324.

Yanai RD (1994) A steady-state model of nutrient uptake accounting for newly grown roots. Soil Sci Soc Am J 58:1562–1571.

17. A Linked Model for Simulating Stand Development and Growth Processes of Loblolly Pine

V. Clark Baldwin, Jr., Phillip M. Dougherty, and
Harold E. Burkhart

Linking models of different scales (e.g., process, tree-stand-ecosystem) is essential for furthering our understanding of stand, climatic, and edaphic effects on tree growth and forest productivity. Moreover, linking existing models that differ in scale and levels of resolution quickly identifies knowledge gaps in information required to scale from one level to another, indentifies future research needs to fill these information gaps, and provides a test of the present state of modeling sciences for creating model systems for predicting responses to natural and human-based disturbances.

Today there is a need to assess how such interacting stressors as air pollutants, (e.g., carbon dioxide (CO_2), ozone, acid rain), and climate change will affect forest health (forest function), tree growth, and stand productivity. Standard growth and yield models usually operate at an annual or longer resolution level and assume environmental conditions remain constant over time. As such, they are of limited use for assessing the effects of changes in weather and atmospheric chemistry on forest functioning or growth. However, growth and yield models are our best tools for predicting tree morphology and stand structural characteristics as a function of stand density, age, and site quality. Conversely, process models operate at hourly or daily levels and are suitable for estimating the effects of annual regimes of air pollution or weather on individual tree functioning (net photosynthesis, respiration, allocation, growth), but require that tree and stand characteristics be provided as model inputs (Landsberg, 1986). Because of the large input data requirements, hourly time steps, and inability to account for morphological changes with age and competition, most present-day process models are not suitable for predicting forest functioning or growth over multiple years.

By linking a growth and yield model and a process model, it should be possible to first utilize the output of the growth and yield model (i.e., predicted morphological and structural characteristics of trees grown at a given stand density, on a given site, for a given length of time) as direct inputs into a process model. Next, the process model could be used to assess the expected impact of environmental changes on tree functioning and growth for a one- to three-year period. The resulting information could then be fed back into the growth and yield model to update future predictions through modification of, for instance, the site index (SI) function. Thus, the procedure would be to 1) use the growth and yield model to grow a stand to a given age and describe the stand and structural characteristics of the constituent trees, 2) use the tree-structure descriptions and the process model to assess forest function at that age, 3) feed back the resulting information to the growth and yield model to adjust its prediction equations, and 4) repeat steps 1 to 3 until the end of the rotation.

Applications

The distant-dependent individual tree growth and yield model, PTAEDA2 (Burkhart et al., 1987), and the biological process model, MAESTRO (Wang and Jarvis, 1990a) parameterized for loblolly pine (Jarvis et al., 1991), were selected for this linkage experiment. Preliminary results of the linkage were described in Baldwin et al. (1993). The initial linked model was successfully developed through computer program changes and modeling using both existing data and some newly developed equations. Outputs from PTAEDA2 were used as driver variables for MAESTRO. Since the time of the initial development, by combining additional new equations and procedures developed from data collected specifically to fit the selected models, the linked model has been developed to its present-day stage.

Summary of the Component Prediction Systems

The growth and yield prediction system, PTAEDA2 (Burkhart et al., 1987) can be used to either simulate a plantation of loblolly pine from the time of planting through a desired rotation or to accept data from an existing stand and project that stand through desired time periods. When simulating a plantation from the time of planting, the model employs two main subsystems—the first generating an initial precompetitive stand, and the second developing the growth and dynamics of that stand.

When applying PTAEDA2 to simulate plantations from time of planting, SI at base age twenty-five is specified and trees are assigned x and y coordinate locations in an 'n' by 'n' dimensional simulation plot. The juvenile stand is then advanced to age eight at which time intraspecific competition is assumed to begin. At that point, predicted juvenile mortality is assigned at random. Individual tree dimensions are then generated for the residual stand. The diameter distribution is generated from a two-parameter Weilbull distribution; the parameters of the distribution are estimated as functions of plantation age (eight years), number of trees surviving (from a stand-level survival function), and average height of the dominant and codominant trees at age eight (from a SI equation). Total tree height and crown ratio are predicted for every tree. After assigning dimensions to each tree, the competition effect of neighboring trees is calculated for every tree; this competition index takes into account both the size of and the distance to neighboring trees.

After generation of the precompetitive stand, competition is evaluated, and simulated trees are grown individually on an annual basis. In general, the growth in height and diameter is assumed to follow some theoretical growth potential. An adjustment or reduction factor is applied to this potential increment based on a tree's competitive status (as measured by the competition index) and photosynthetic potential (as expressed by the crown ratio), and a random component is then added, representing microsite and genetic variability. The probability that a tree remains alive in a given year is assumed to be a function of its competition index value and crown ratio. Survival probability is calculated for each live tree every year and is used to determine annual mortality.

The net carbon gain of an array of trees in a stand is estimated by the MAESTRO-system. A description of the model, including all environmental and tree characteristics required to operate it, is found in Wang and Jarvis (1990a). Two key input requirements are 1) the positions of all individual trees in the stand as specified by their x, y, and z coordinates, and 2) individual descriptions of each tree by the crown radii in the x and y directions, crown length, height to the crown base, and the total area of leaves within the tree crown.

The positions of leaves in both the vertical and radial directions are defined by functions describing the leaf area density distribution. The slope of the ground in the x and y directions and the orientation of the x axis are also specified. The time scale for MAESTRO is in hours, and the spatial scale is up to 120 subvolume grid points within each tree crown. For every tree, an estimate is made of the radiation at a selected number of grid points within the crown, which takes into account both within-tree and between-tree light penetration. Foliage density in each of the selected crown grids within the tree crown is estimated and foliage is classified with respect to age, position, and attendant physical and physiological properties.

First, MAESTRO calculates the radiation absorbed by the leaves, the CO_2 and water vapor exchanges between the leaves, and the ambient air for each of the selected grids. After integrating these factors to the crown level, MAESTRO then outputs daily amounts of 1) radiation absorbed, 2) photosynthesis minus leaf respiration. 3) respiration amounts for leaves, branches, the bole, and course and fine roots, and 4) transpiration of the defined target tree. Multiple runs of MAESTRO that designate different target trees can then be performed, and the output values calculated to acquire stand-level predictions.

The Linkage Process

Limitations to making the PTAEDA2–MAESTRO linkage are the same as those expected for any analogous model linkages, that is, most tree attributes needed to describe a tree for input into MAESTRO were not standard outputs of PTAEDA2. In addition to diameter at breast height (DBH), total height, crown ratio or crown length, and stem weight or volume, MAESTRO required crown width, crown shape or crown volume, the horizontal and vertical distributions of the foliage biomass or density (leaf area per unit volume of crown), as well as branch and bole surface area or weight.

The user was required by MAESTRO to input actual values of these key stand and tree structural variables, all of which vary interdependently according to age, stand density, and site quality. Thus, simulations over several years were prohibitive. There were only three options for describing crown shape, therefore, the same crown shape and horizontal and vertical foliage distributions were assumed for all trees in the stand. For an individual tree-process model, it is essential that an accurate description of the crown shape, physical characteristics (i.e., weight, surface area), and distribution of its components be provided. Therefore, in this linkage project it was deemed essential to not only accomplish the linkage but to also improve upon estimation of the key tree-structure characteristics that would optimize the functions of MAESTRO to produce accurate estimates of crown assimilation and net carbon gain.

Because foliage density within a crown of loblolly pine is largely a function of nutrient and water availability (Dougherty et al., 1990; Vose and Allen, 1988), it is not possible to predict foliage density using only growth and yield information. Models for predicting total stand leaf biomass and leaf area duration for all age classes of foliage as a function of climate, available water, and stand characteristics are being developed for loblolly pine (Dougherty et al., 1990; Dougherty et al., 1995). Furthermore, because it is not known if a changing environment (for example elevated levels of CO_2 in the atmosphere) will change crown allometry, for the present stage of development it is assumed that crown-shape characteristics will not change as a result of those factors.

The stand measurements and modeling needed for the initial linkage were reported in Baldwin et al., (1993). Recent improvements deemed essential by the authors included 1) development of a tree-specific crown-shape model consisting of a second degree polynomial function constrained to equal zero at the crown tip for the outer shape, in addition to a straight-line inner-shape function to describe the cone-shape inner defoliated area of a loblolly pine crown (Baldwin et al., 1995; Baldwin and Peterson, 1997), 2) improved prediction equations for both foliage (old and new) and branch-weight, 3) improved prediction equations for both vertical and horizontal foliage distribution (old and new foliage) that predict foliage weight at its actual vertical location in the vertical plane, and 4) improved surface-area distribution equations for new and old foliage, branch, and surface area that have the same properties as the improved prediction equations (Baldwin et al., 1996). The branch surface-area equations allow MAESTRO the option of predicting woody mass respiration by surface-area relationships rather than by weight. Details of the data collected and research procedures (field, laboratory), methods (data analysis and modeling), and final results pertaining to these improved prediction equations and other factors required for the one-way linkage of the models have also been reported in the publications cited.

The effects of stand- and tree-structure changes on trends of tree functions over time (from ten to forty years) were tested with the PTAEDA2–MAESTRO linked models for a selected codominant-dominant tree (CDtree) for each of four treatments. To follow the CDtrees through an entire simulated rotation, it was necessary to select each CDtree from the survivors at the end of the rotation for each treatment. The four trees, of about average DBH at forty years, were traced back through time to age ten. Their characteristics, and the stand characteristics for each treatment, at ages ten, twenty, thirty, and forty years, are reported for the following treatments:

1. A loblolly pine plantation planted at 1683 trees hectare (ha) (2.4 × 2.4 m) spacing on site index (base age twenty-five years) land of 16.8 m, and unthinned through age forty;
2. A loblolly pine plantation planted at 1683 trees/ha (2.4 × 2.4 m) spacing on site index (base age twenty-five years) land of 16.8 m, thinned selectively from below at ages fifteen and twenty-five to a residual basal area of 16.1 m^2/ha;
3. Same as (1), except a higher SI of 21.3 m;
4. Same as (2), except a higher SI of 21.3 m.

The physiological parameters for both aboveground and belowground loblolly pine tissues were based on laboratory and field measurements of loblolly pines in Athens, GA (Teskey et al., 1986) and the similar southern pine, *Pinus elliottii* (Gholz et al., 1986; Cropper and Gholz, 1991). The climatic data, used for the MAESTRO portion of these simulations, were 1988 hourly climate data from Athens, GA. In that year, mean monthly air temperatures ranged from 6.2 to 27.5 °C, and soil moisture was never a limiting factor.

Results

Combining the allometric functions with the growth and yield model PTAEDA2 and then coupling PTAEDA2 with MAESTRO permitted us the opportunity to examine the effects of SI and thinning on the trends in tree-structure components and their effects on carbon gain and carbon loss. In this study, we evaluated the trends in crown volume, crown length and width, leaf area amount, leaf area distribution, mean crown leaf area density, aboveground biomass, branch and bole weight and surface area, and the ratio of foliage to structural mass or surface area. The results of the model simulations for the branch, foliage, and entire crown (e.g., volume, biomass, surface area) of each CDtree for each treatment and age are summarized in Table 17.1. Corresponding and additional predicted stand

Table 17.1. Aboveground Example of Loblolly Pine Tree Values from Simulations by the PTAEDA2-MAESTRO System from Ages Ten to Forty Years

Treatment	Age	DBH (cm)	Height (m)	Foliated crown depth (m)	Crown radius (m)	Height of max. crown radius (m)	Crown volume (m³)	New foliage biomass (kg)	Old foliage biomass (kg)	Branch biomass (kg)	Bole biomass (kg)	Total woody biomass (kg)	New foliage surface area (m²)	Old foliage surface area (m²)	Branch surface area (m²)	Bole surface area (m²)	Total woody surface area (m²)	Leaf area index
Medium spacing, low site, unthinned	10	12.3	8.0	4.7	1.3	3.9	13.7	2.66	2.41	4.5	16.7	21.2	26.58	24.08	4.2	1.7	5.9	3.7
	20	18.4	15.1	5.4	1.6	11.3	25.1	4.19	3.38	9.8	78.1	87.9	41.87	33.77	6.8	4.7	11.5	3.0
	30	21.7	19.5	5.6	1.9	15.6	37.0	5.18	3.78	12.9	150.8	163.7	51.76	37.77	8.0	7.2	15.2	2.4
	40	23.1	21.7	5.7	2.3	17.7	53.1	5.70	3.85	13.8	204.8	218.6	56.96	38.47	8.2	8.5	16.7	1.8
Medium spacing, low site, thinned at ages 15 and 25	10	15.1	8.6	5.4	1.5	4.4	22.0	4.10	3.52	8.4	26.9	35.3	40.97	35.17	7.1	2.2	9.3	3.5
	20	22.3	15.9	6.4	2.0	11.6	43.2	6.42	4.99	18.4	120.1	138.5	64.15	49.86	11.6	6.0	17.6	2.6
	30	26.6	20.8	6.7	2.3	16.4	63.6	8.01	5.69	25.1	239.7	264.8	80.04	56.86	14.0	9.4	23.4	2.2
	40	28.8	23.3	6.9	2.7	18.7	88.0	8.98	5.94	28.1	339.1	367.2	89.73	59.36	14.9	11.4	26.3	1.7
Medium spacing, high site, unthinned	10	14.5	10.2	5.5	1.3	6.1	17.4	3.11	3.03	7.1	30.5	37.6	31.08	30.28	5.8	2.5	8.3	3.5
	20	20.2	18.6	5.8	1.5	14.9	23.8	4.38	3.74	12.4	120.3	132.7	43.77	37.37	7.9	6.4	14.3	3.4
	30	22.7	22.9	5.7	1.8	19.3	30.3	4.99	3.82	13.9	198.9	212.8	49.86	38.17	8.1	8.8	16.9	2.9
	40	25.1	26.7	5.8	2.1	23.0	42.5	5.71	4.08	16.4	306.2	322.6	57.06	40.77	9.0	11.4	20.4	2.3
Medium spacing, high site, thinned at ages 15 and 25	10	18.0	11.2	6.5	1.6	6.7	30.4	4.89	4.56	13.9	51.4	65.3	48.86	45.57	10.1	3.4	13.5	3.1
	20	26.6	20.9	7.4	2.0	16.4	53.3	7.63	6.35	29.8	233.5	263.3	76.24	63.45	16.1	9.4	25.5	2.6
	30	30.4	26.1	7.4	2.3	22.7	68.7	9.00	6.75	35.7	406.0	441.7	89.93	67.45	17.7	13.5	30.2	2.3
	40	33.5	30.2	7.5	2.7	25.6	90.5	10.26	7.20	41.9	614.6	656.5	102.52	71.95	19.4	17.2	36.6	1.9

17. A Linked Model for Simulating Stand Development and Growth 311

characteristics for each treatment and age are found in Table 17.2. Various graphical techniques are used to illustrate the trends observed.

Predicted Crown Volume Trends

The trends predicted for crown volume of the CDtree in each of the four SI-thinning combinations considered are illustrated in Figure 17.1a, b. Crown volume increased with age in a linear manner for all SI-thinning combinations. Thinning permitted crown volume to increase at more than twice the rate of the CDtree in the unthinned regime. In the thinned regime, (Figure 17.1a) crown volume of the CDtree growing on high SI land was slightly greater than the CDtree on low SI land through age twenty. At age thirty and forty, the CDtrees in the thinned stand had about the same crown volume for both SI conditions.

In the unthinned stands, crown volume of the CDtree was also initially greater for the high SI condition than for the low SI regime (Figure 17.1b). However, from age twenty to age forty, crown volume was predicted to be larger for the low SI CDtree than for the high SI CDtree. These results were predicted because the high SI tree had only slightly greater crown length (Figure 17.1d), although the low SI tree had greater crown width (Table 17.1). The pattern of foliated crown depth was consistent in both thinned and unthinned stands; although, as would be expected, the length was greater in thinned stands (Figure 17.1 c and d).

Both thinned and unthinned trees develop a portion of the inner crown that is not foliated (Figure 17.2). The percentage of the crown volume that is unfoliated is actually very low; it ranges from about 4% of the total crown volume at age ten to about 3% at age forty. Also, the volume of crown that does not contain foliage is mostly in the lower half of the tree crown (Figure 17.2). The difference in crown shapes and inner defoliated volumes results in the leaf area being displayed in different manners for trees in the thinned and low and high SI combinations.

Trends in Total Leaf Area

The trends in total amount of leaf area per CDtree determined for the four growing regimes also varied. For CDtrees in thinned stands, leaf area increased in a curvilinear relationship with age from ten to forty years (Figure 17.3). Leaf area of the high SI CDtree in the thinned regime was 15% (age forty) to 25% (age ten) greater than on the low SI CDtree. Because of the thinnings at ages fifteen and twenty-five, leaf area index (LAI) in thinned stands was predicted to decline from age ten until age forty (Figure 17.4).

The trend in total leaf area for high and low SI CDtrees in the unthinned regime also increased in a curvilinear manner, but at a much slower rate, especially beginning at age twenty (Figure 17.3). Only slight differences in leaf area were predicted for these trees. When leaf area was projected to the stand level, LAI was found to increase until age twenty and then declined slowly until age 40 (Figure 17.4). It is probable that LAI actually peaked earlier than age twenty years, but no simulations were made between ages ten and twenty. The decline in LAI occurred at a more rapid rate for the high SI CDtrees than the low CDtrees with the largest difference in LAI between highland and low SI trees being at the intermediate ages.

Table 17.2. Aboveground Loblolly Pine Stand Structure Values from Simulation of the PTAEDA2-MAESTRO System from Ages Ten to Forty Years

Treatment	Age	Quadratic mean DBH (cm)	Mean stand height (m)	Mean dominant height (m)	Number of trees surviving (trees/ha)	Basal Area (m²/ha)	Bole volume (m³/ha)	Foliage biomass (t/ha)	Branch biomass (t/ha)	Bole biomass (t/ha)	Woody biomass (t/ha)	Leaf Area (m²/ha)	Branch surface area (m²/ha)	Bole surface area (m²/ha)	Woody surface Area (m²/ha)
Medium spacing, low/and site, unthinned	10	11.9	7.8	8.4	1337	14.8	60.7	4.06	6.68	21.72	28.40	45691	5365	2082	7447
	20	17.2	13.8	14.7	1223	28.3	203.6	5.96	11.78	80.25	92.03	63308	7729	4892	12621
	30	20.0	17.4	18.4	1030	32.5	291.3	5.98	12.47	122.85	135.32	61792	7562	6050	13612
	40	22.2	19.9	20.8	796	30.7	314.0	5.22	11.30	142.00	153.51	53025	6512	5931	12443
Medium spacing, low/and site, thinned at ages 15 and 25	10	11.9	7.8	8.4	1337	14.8	60.7	4.06	6.68	21.72	28.40	45691	5365	2082	7447
	20	20.5	15.2	14.7	628	20.7	164.9	4.47	9.93	60.77	70.69	47362	6119	3215	9335
	30	26.0	19.5	18.4	329	17.4	179.2	3.41	.52	70.54	79.06	35142	4688	2838	7526
	40	28.4	22.3	20.8	294	18.6	228.9	3.53	9.17	96.64	105.82	35718	4786	3297	8083
Medium spacing, high/and site, unthinned	10	13.4	9.5	10.2	1371	19.4	96.5	5.17	9.36	35.40	44.76	57619	7010	2941	9951
	20	18.9	17.0	18.6	1221	34.4	304.0	6.89	14.61	123.55	138.16	72251	9007	6608	15614
	30	22.0	21.4	23.5	1001	38.0	420.6	6.61	14.56	183.07	197.63	67397	8356	7971	15328
	40	24.4	24.6	26.9	761	35.6	448.2	5.71	12.96	209.21	222.18	57197	7096	7735	14831
Medium spacing, high/and site, thinned at ages 15 and 25	10	13.4	9.5	10.2	1371	19.4	96.5	5.17	9.36	35.40	44.76	57619	7010	2941	9951
	20	24.3	19.1	18.6	445	20.7	207.0	4.35	10.91	78.52	89.42	45435	6144	3457	9600
	30	29.8	24.5	23.5	245	17.1	225.1	3.27	9.13	91.35	100.47	33245	4607	3085	7692
	40	32.8	28.1	26.9	215	18.2	258.8	3.02	8.73	112.83	121.56	30134	4182	3236	7418

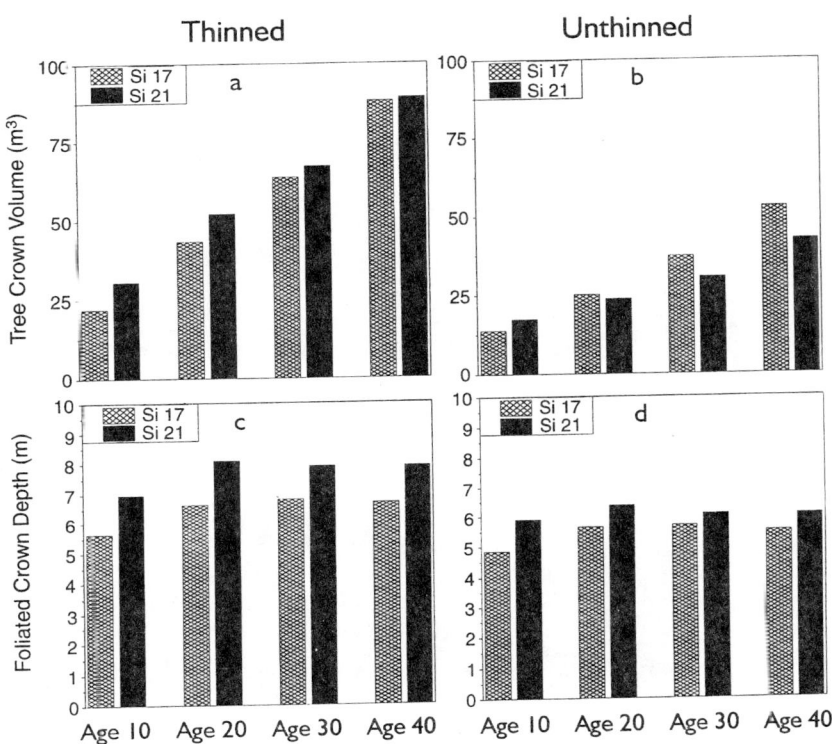

Figure 17.1. Predicted trends in crown volume and foliated crown length of an example codominant-dominant loblolly pine tree in stands of two site indices either thinned or unthinned.

Leaf Area Density

In general, because, high SI stands carry more leaf area than trees in low SI regimes (even though crown volume may be similar), leaf area density (LAD) has to be greater in high SI trees. The trends in CDtree LAD relative to crown volume are illustrated in Figure 17.5. Trees in thinned stands maintain a higher LAD than in unthinned stands at a given crown volume. High SI trees do have a slightly greater LAD than low SI trees, but the effect of stand density on mean tree LAD is greater than high or low SI effects on LAD at a specified crown volume.

For all SI-thinning combinations, CDtree LAD decreased with age (Table 17.1). The rate of decrease was greater in the unthinned stands. The distribution of CDtree LAD with age also changed and was different for thinned vs unthinned stands (Figure 17.6). For a given thinning regime, LAD decreased curvilinearly with age at all relative crown height positions. In the thinned regime, there were

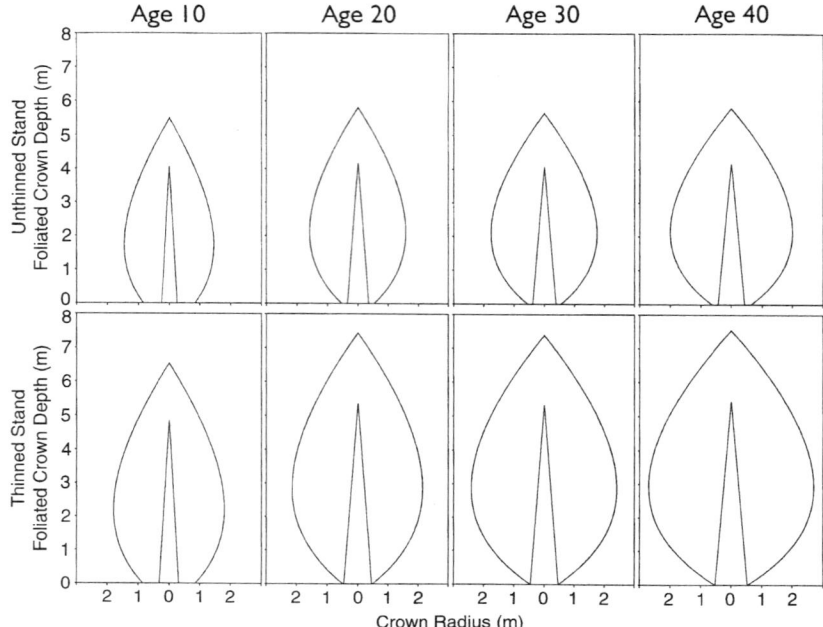

Figure 17.2. Predicted changes in crown shape over time of an example codominant-dominant loblolly pine tree in either a thinned or unthinned stand of site index equalling 16.8 m.

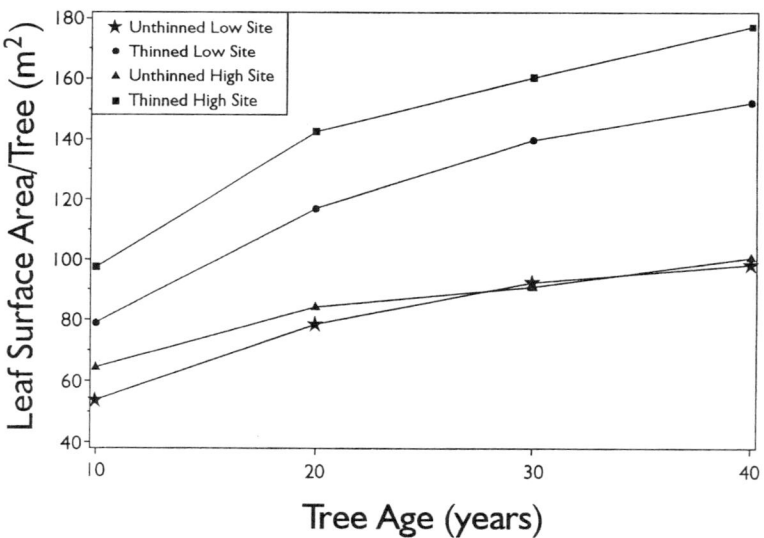

Figure 17.3. Predicted trends in leaf area through time of an example codominant-dominant loblolly pine tree in stands of two site indices either thinned or unthinned.

17. A Linked Model for Simulating Stand Development and Growth   315

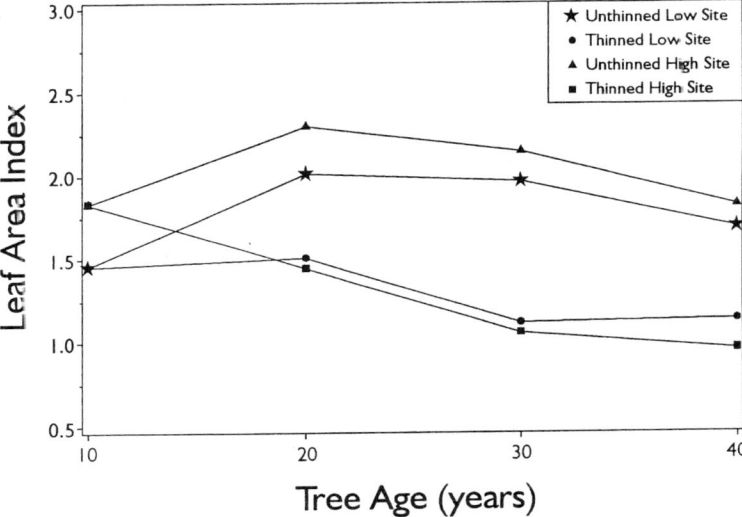

Figure 17.4. Predicted trends in stand leaf area index through time in loblolly pine stands of two site indices either thinned or unthinned.

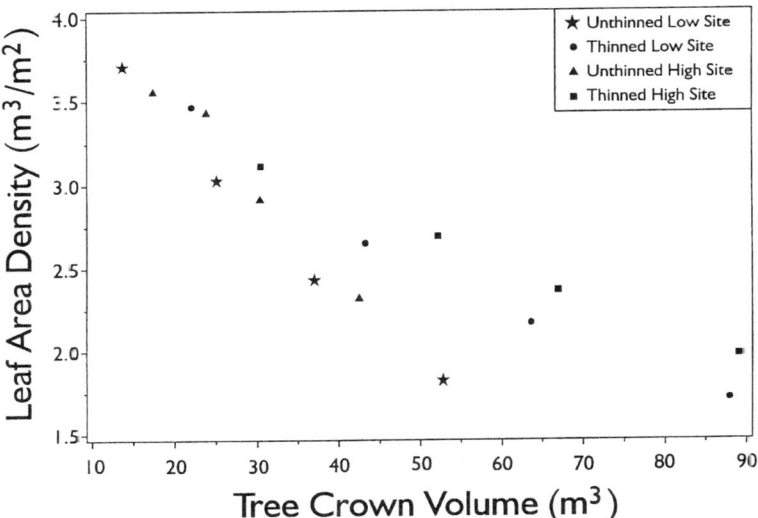

Figure 17.5. Predicted leaf area density compared to predicted tree crown volume at ages ten, twenty, thirty, and forty for an example comdominant-dominant loblolly pine tree in stands of two site indices either thinned or unthinned.

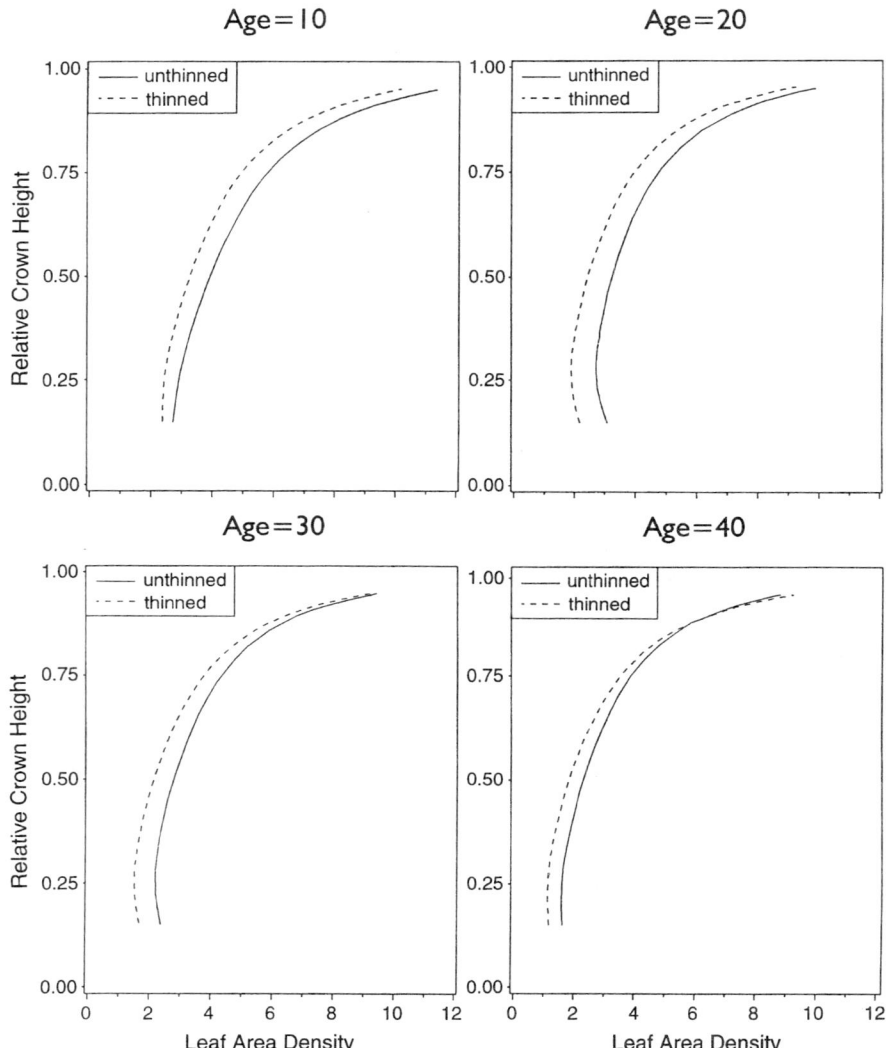

Figure 17.6. Predicted leaf area density for various relative crown heights at ages 10, 20, 30, and 40 of an example codominant-dominant loblolly pine tree in thinned or unthinned stands of site index equalling 21.3 m.

only small changes in LAD in the upper-crown after age ten; LAD in the upper-crown was approximately 10.3 m^2/m^3 at age ten and at age forty was still near 9.3 m^2/m^3. Near the lower one-quarter of the crown, LAD decreased from approximately 2.3 m^2/m^3 to about 1 m^2/m^3.

In the unthinned regime, LAD in the CDtree upper-crown changed more dramatically than in the thinned regime. Upper-crown LAD decreased from 11.5

m²/m³ at age ten (Figure 17.6) to 8.8 m²/m³ at age forty. However, LAD at the lower one-quarter relative crown height position decreased less than in the thinned regime. LAD at this crown position decreased from 2.8 m²/m³ to 1.7 m²/m³.

Trends in Branch and Bole Biomass and Surface Area

The range in branch biomass contribution to the total biomass (Figure 17.7a, b) across all four SI-thinned CDtree combinations considered was 19 to 24% at age ten. By age forty, the range in the percentage of the total biomass made up of branches was only 5 to 8%. This relationship indicates that on a biomass basis, branches are a minor component. However, this is not true on a surface area basis.

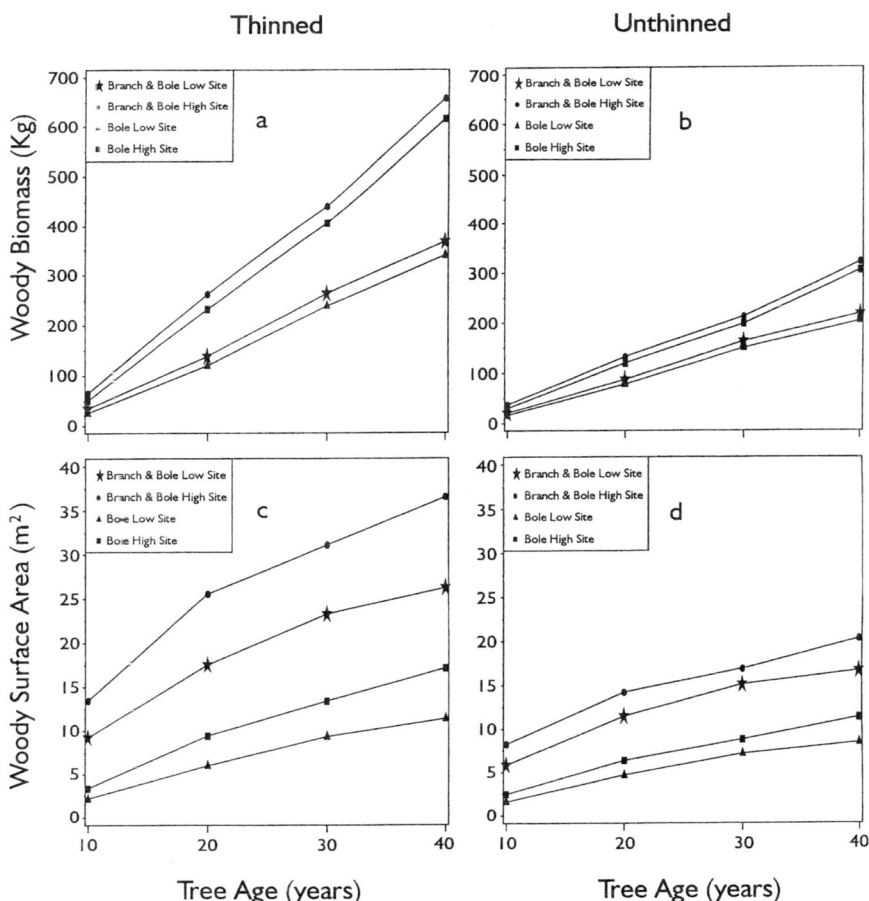

Figure 17.7. Comparison of predicted woody tissue weight and predicted surface area over time of an example codominant-dominant loblolly pine tree in stands of two site indices either thinned or unthinned.

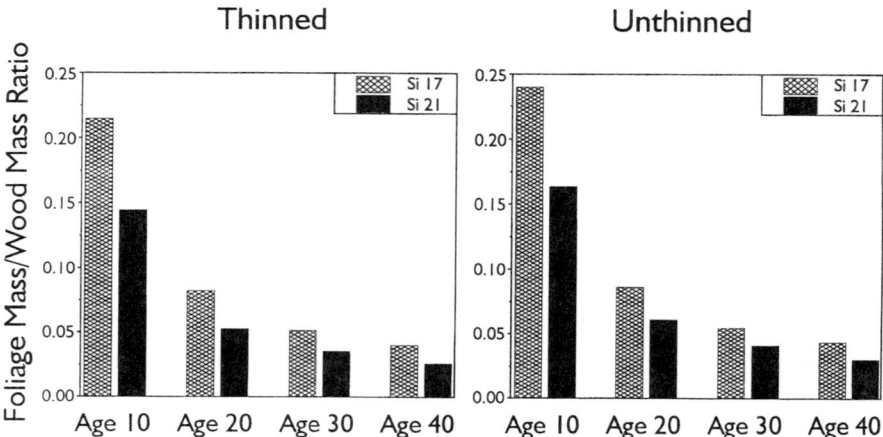

Figure 17.8. Trends over time of the ratio of predicted foliage biomass to predicted woody tissue biomass of an example codominant-dominant loblolly pine tree in stands of two site indices either thinned or unthinned.

The trends in CDtrees branch and bole surface areas are illustrated in Figure 17.7d, c. In the thinned regime at age ten, bole surface area constituted about 24 to 35% of the total CDtree aboveground woody tissue surface area (branch and bole) irrespective of SI class. By age forty, bole surface area was 43% (low SI) to 70% (high SI) of the total CDtree aboveground woody tissue surface area. Similar trends were predicted for the unthinned regime. However, bole surface area at all ages constituted a slightly greater percentage of the CDtree total surface area in unthinned stands. At age ten, the contribution of bole surface area to the total CDtree aboveground surface area was 4 to 6% greater in the unthinned than in the thinned regime and by age forty the difference was 8 to 9%. These predictions indicate a greater amount of branch material is developed and retained in the thinned regime.

Ratio of Photosynthetic Tissue to Woody Tissue

One of the greater differences between seedlings and trees is the ratio of foliage biomass to woody tissue biomass. At the seedling stage, this ratio is near 1.0. As can be seen in our simulations, this ratio changes rapidly with age (Figure 17.8). By age ten, the ratio of foliage biomass to woody tissue biomass is between .2 and .25 for the lower SI and approximately .15 for the high SI CDtrees. The rapid drop in the ratio of foliage biomass to woody biomass continued to age twenty and then decreased slowly thereafter. By age forty, this ratio was less than 5% in all cases.

Trends in the Ratio of Foliage Surface Area and Woody Tissue Surface Area

Similar trends to those observed for foliage weight to woody tissue biomass were also apparent for the trends in foliage surface area to woody surface area ratio

17. A Linked Model for Simulating Stand Development and Growth 319

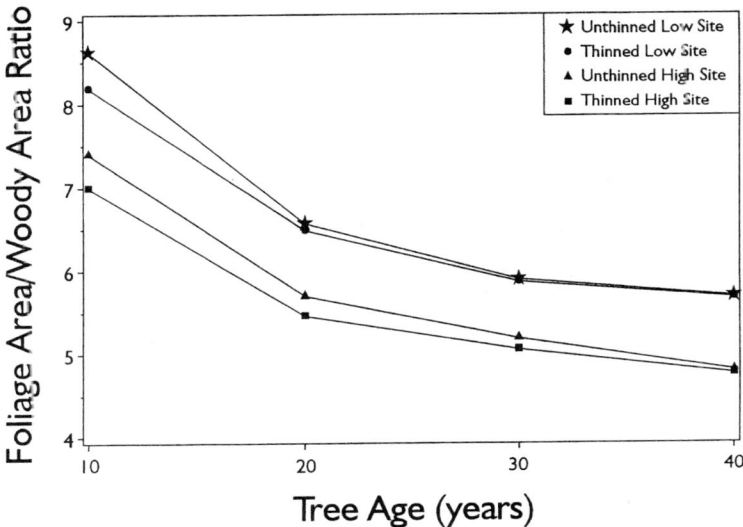

Figure 17.9. Trends over time of the ratio of predicted foliage surface area to predicted woody tissue surface area of an example codominant-dominant loblolly pine tree in stands of two site indices either thinned or unthinned.

(Figure 17.9). High SI CDtrees had slightly lower ratios than low SI CDtrees. One surprising difference in the surface area vs weight ratios is that the CDtrees in the unthinned stands appeared to maintain a greater "equilibrium" foliage surface area to woody surface area from age thirty to forty than the CDtrees in the thinned stand with unthinned trees having a ratio of only 4.8. This was apparently related to the larger size and perhaps greater number of developed branches on the thinned trees than on the unthinned trees.

Trends in Canopy Carbon Gain and Maintenance Respiration

The effects of the structural changes in trees related to age, SI, and thinning, on crown assimilation, total tree maintenance respiration, and net annual carbon gain (crown carbon gain minus total tree respiration) are shown in Figure 17.10a–f. Carbon gain increased at all ages for the CDtrees in the thinned scenario. Carbon gain of the CDtree in the high SI thinned regime had 40% greater carbon gain at age ten and 47% greater carbon gain at age forty than the low SI CDtree in the thinned regime. This gain occurred even though CDtree crown volumes in the thinned regime were actually greater for the low SI regime. However, total surface leaf area was 24% greater at age ten and 17% greater at age forty in the high SI vs low SI regime. Thus, LAD was greater in the thinned CDtree crowns (Figure 17.5) than in low SI CDtrees.

Crown assimilation in the unthinned regime increased with age for both the high and low SI CDtrees. Differences in crown carbon gain for high and low SI CDtrees in the unthinned regime were small (Figure 17.10b). This small differ-

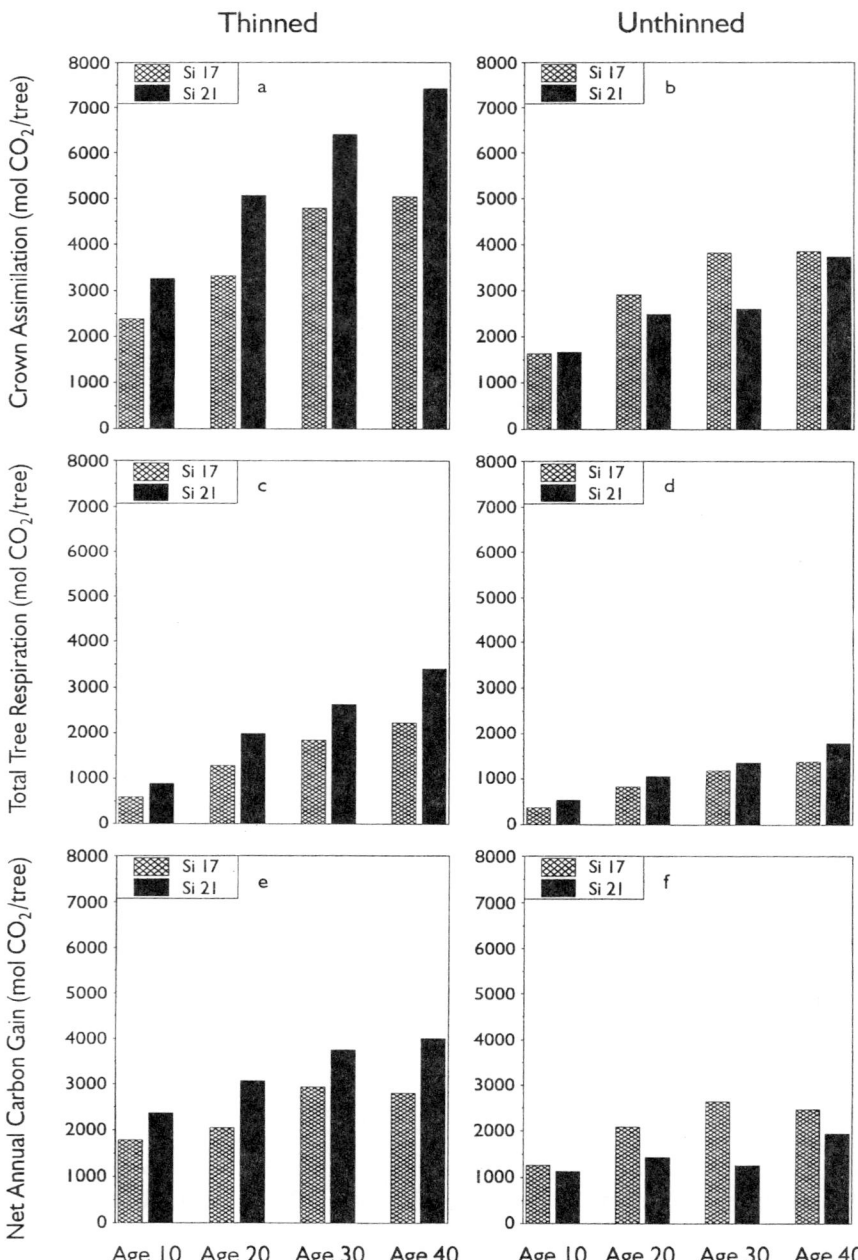

Figure 17.10. Predicted trends through time of total crown carbon assimilation, total tree maintenance respiration, and net annual carbon gain of an example codominant-dominant loblolly pine tree in stands of two site indices either thinned or unthinned.

ence occurred because both the total leaf area per tree (Figure 17.3) and LAD (Figure 17.5) for unthinned trees was about the same, irrespective of SI class. Thus, the reduction in crown carbon gain as a result of not thinning was much greater for the CDtree in the high SI regime than in the low SI regime (contrast Figure 17.10a–c). Crown carbon gain per CDtree was about 50 to 57% more in the thinned regime than in the unthinned regime for the SI-50 CDtree, and 47 to 100% more in the thinned SI-70 CDtree than in the unthinned SI-70 CDtree. In fact, the thinned SI-50 CDtree was estimated to have greater crown assimilation than the unthinned SI-70 CDtree.

Trends in Total Tree Maintenance Respiration

Annual total CDtree maintenance respiration in the thinned regime increased almost linearly with age. The thinned SI-70 CDtree at all ages had higher respiration than the thinned SI-50 CDtree (Figure 17.7c, d). Total tree respiration in the unthinned plots was much lower than that in the thinned plots. The difference in respiration between the high SI CDtree and the low SI CDtree was also less than that determined for the similar trees in the thinned regime. At age ten, the total tree respiration for the CDtree in the SI-50 thinned regime was estimated to be 45% greater than the CDtree in the SI-50 unthinned regime. At age forty, this difference had increased to 63%. At age 10, total tree respiration of the thinned SI-70 CDtree was predicted to be 70% more than the unthinned SI-70 CDtree. This difference at age forty was estimated to be 91%. Thinning was clearly the major factor that influenced individual tree respiration cost.

Trends in Tree Net Annual Carbon Gain

The major factor that affected the trend in annual carbon gain was whether the stand was thinned or not. In the thinned scenario, the SI-70 CDtree maintained a more favorable annual net carbon gain than the SI-50 CDtree. In the unthinned regime, annual net carbon gain of the individual CDtrees in the high or low SI classes was not consistently different. This trend apparently resulted because at each age, foliage surface area of unthinned high and low SI trees were about the same (Figure 17.3) and the woody biomass component was predicted to be less for the low SI CDtree than for the high SI CDtree (Figure 17.7d). This combination of structural characteristics resulted in net carbon gain of the low SI CDtrees being about equal or slightly greater at most ages evaluated. In MAESTRO, the unit rates of photosynthesis and respiration do not vary by SI class.

Trends in Mean Annual Increment

To estimate mean annual increment (MAI) for the CDtrees in the four SI-thinning combinations, PTAEDA2 was used. In the thinned regime, SI had a large positive effect on MAI at all ages evaluated (Figure 17.11). CDtree MAI increased in the thinned scenario from age ten to age forty but only at a slow rate after age twenty. This result is important because crown volume, which is one of the major factors

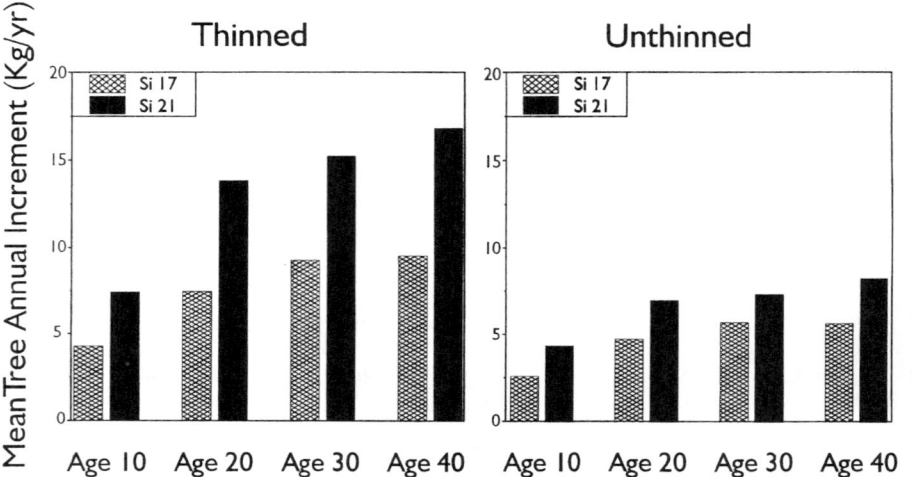

Figure 17.11. Trends of predicted mean annual woody biomass increment for an example codominant-dominant loblolly pine tree in stands of two site indices either thinned or unthinned

that determines an individual tree's growth potential in PTAEDA2, increased linearly with age and was not very different for the high and low SI categories.

A similar trend in CDtree MAI was observed for the unthinned scenario as for the thinned scenario. However, MAI was about one-half of what is was in the thinned regime. Even though low SI CDtree crown volume was actually larger in the unthinned regime than for high SI CDtree, PTAEDA2 predicted that MAI would be greater for the high SI CDtree (Figure 17.11).

Relationship of Crown Annual Net Carbon Gain and Yield

If annual net carbon gain is allocated in a fixed ratio to stemwood production there should be a linear relationship between yield estimates made with PTAEDA2 and net carbon gain estimates made with MAESTRO. As shown in Figure 17.12, the relationship between yield and annual net carbon gain was linear. However, there was considerable variation in the relationship. This result suggests that a fixed allocation coefficient may not be correct.

Discussion and Conclusions

The objectives of this chapter were to show the value of linking a growth and yield model (PTAEDA2) with a process model (MAESTRO) to better understand the effects of tree and stand structure changes over time on tree and stand functioning. This objective was accomplished in two steps. First, by using stand-density man-

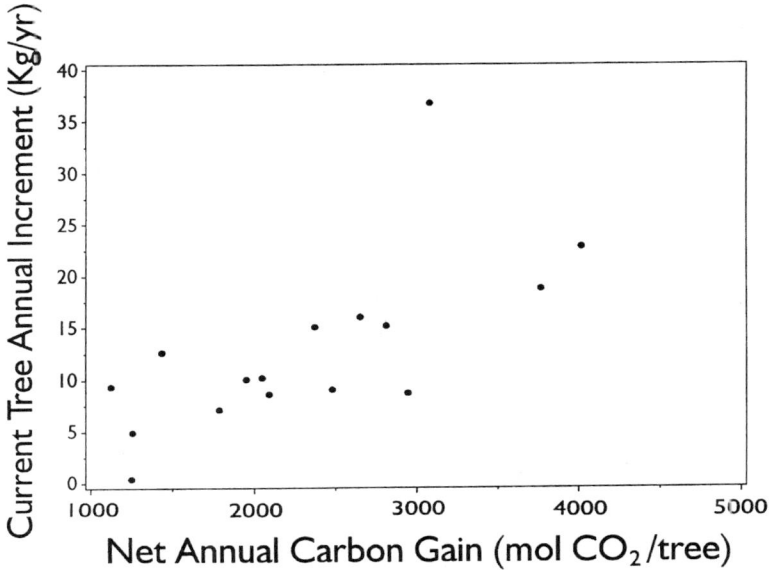

Figure 17.12. Predicted present-day annual woody biomass yield compared to predicted net annual carbon gain at ages ten, twenty, thirty, and forty for an example codominant-dominant loblolly pine tree in stands of two site indices either thinned or unthinned.

agement, the effects of stand structure changes on individual tree structure were predicted using PTAEDA2. Second, the effects of the resulting tree structure changes on various tree functions or physiological processes were presented. The results with a minimum of discussion all were presented in prior sections; several of these results are further discussed in this section.

This analysis showed that age, SI, and thinning affect many crown features that are important for predicting tree physiological functioning. Crown features altered by age, SI, and thinning include 1) total leaf area per tree, 2) mean crown leaf area density, 3) leaf area density distribution, 4) crown length, and 5) the nonfoliated fraction of the crown volume. In a prior study, Baldwin et al., (1993) demonstrated that vertical and horizontal leaf area distributions are affected both by age and stand density. Although some of these crown-structure changes may only have minor effects on net carbon gain, collectively the effect on light interception may be substantial. Cropper et al., (1996) evaluated potential effects of changing loblolly pine foliage distribution model from a Beta function to a truncated Weibull function that reflects the changes in the distribution parameters as a result of stand development (Baldwin et al., 1997). The impacts on annual net carbon assimilation varied by tree size—small trees exhibited a 9% reduction in carbon assimilation relative to that estimated with the Beta function. Larger trees exhibited a 30% increase in annual carbon assimilation with the new foliage distribution function relative to the Beta function. These results therefore demonstrate that having accurate descriptions of crown-structure characteristics is important.

Wang and Jarvis (1990b) used MAESTRO to evaluate the importance of various crown-structure properties of a Sitka spruce (*Picea sitchensis* (Bong.) Carr.) tree crown with respect to the absorption of photosynthetically active radiation, photosynthesis, and transpiration. Their results suggested that total tree leaf area and LAD are the crown-structure characteristics that have the greatest effect on light absorption. In the future, it is probable that accurate descriptions of crown structure with respect to potential carbon gain will be even more important, because under elevated levels of atmospheric CO_2 the potential for increased carbon assimilation will be greatly enhanced. However, the amount of enhancement appears to be closely related to the amount of photosynthetically active radiation that the foliage receives. Thus, more accurate descriptions of light distribution within the crowns of trees will be needed.

In temperate ecosystems, it is equally important to accurately describe tree-structure properties that affect maintenance respiration cost because maintenance respiration can consume more than 60% of gross primary productivity in pines (Ryan et al., 1994). This study clearly indicates that large differences in woody tissue maintenance respiration are apt to result if respiration is expressed as a function of woody surface area rather than as a function of woody biomass. The latter relationship increases in a linear fashion with age, whereas woody tissue surface area increases in a curvilinear manner (Figure 17.7a–c). Thinning also appears to substantially alter the ratio of foliage surface area to woody tissue surface area (Figure 17.9).

Maintenance respiration as a function of woody tissue surface area, foliage surface area, and root biomass is predicted by MAESTRO, but this system does not account for foliage, branch, and bole component variations in tissue nitrogen concentration, or that respiration rates may vary by two-fold depending on the nitrogen concentration of the tissue (Dougherty et al., 1996). Additionally, it is probable that all tissue components in high SI trees would have a higher nitrogen concentration than those of trees in a low SI stand. However, the combined effects of SI and thinning on woody tissue and photosynthetic tissue amounts and distribution ultimately determines the potential maintenance respiration cost, and carbon gain, and therefore, the net annual carbon gain and growth potential.

One important conclusion from this study is the overriding effect of thinning in determining annual net carbon gain. Thinnings were much more influential in the determination of individual tree annual net carbon gain than SI. This result was especially noticeable at the older ages evaluated, and should be considered seriously by those agencies responsible for maintaining "healthy forests."

It has been shown that as a research tool, the linked PTAEDA2–MAESTRO system allows those microeffects on stand structure and function to be considered that would have been impossible to predict using stand-process models. The ability of the linked model to predict changes in characteristics of individual trees was the desired objective of this system, even though hundreds of simulations and summarizations would be required to obtain stand averages. Several crown characteristics of loblolly pine needed as inputs to MAESTRO can now be predicted from PTAEDA2 for trees grown under a wide range of conditions. These changes replace the need for user input of several measures into MAESTRO and the linked model may now be used to assess the impacts of ozone, weather, or CO_2 regimes on carbon gain, respiratory demand, carbon storage, and growth trends. Some of these applications are demonstrated in the next chapter.

References

Baldwin VC Jr, Burkhart HE, Dougherty PM, Teskey RO (1993) Using a growth and yield model (PTAEDA2) as a driver for a biological process model (MAESTRO). Research Paper SO-276, US DOA, For Ser, South For Exper Sta. New Orleans, LA.

Baldwin VC Jr, Peterson KD (1997) Predicting the crown shape of loblolly pine trees. Can J For Res. 27:102–107.

Baldwin VC Jr, Peterson KD, Burkhart HE, Amateis RL, Dougherty PM (1997) Equation for estimating loblolly pine branch and foliage weight and surface area distributions. Can J For Res 27:918–927.

Baldwin VC Jr, Peterson KD, Simmons DR (1995) A program to describe the crown shape of loblolly pine trees. Compiler 13:17–19.

Burkhart HE, Farrar KD, Amateis RL, Daniels RF (1987) Simulation of individual tree growth and stand development in loblolly pine plantations on cutover, site-prepared areas. FWS-1-87, Virginia Polytechnic Institute and State University, School of Forestry and Wildlife Resources, Blacksburg, VA.

Cropper WP Jr, Gholz HL (1991) In situ needle and fine root respiration in mature slash pine (*Pinus elliottii*) trees. Can J For Res 21:1589–1595.

Dougherty PM, Hennessey TC, Zarnoch SJ, Stenberg PT, Holeman RT, Wittwer RF (1995) Effects of stand development and weather on monthly leaf biomass dynamics of a loblolly pine (*Pinus taeda* L.) stand. For Ecol Manage 72:213–227.

Dougherty PM, Oker-Blom P, Hennessey TC, Witter RE, Teskey RO (1990) An approach to modelling the effects of climate and phenology on the leaf biomass dynamics of a loblolly pine stand. Silv Carel 15:133–143.

Gholz HL, Vogel SA, Cropper WP Jr (1986) Organic matter dynamics of fine roots in plantations of slash pine (*Pinus elliottii*) in north Florida. Can J For Res 16:529–538.

Jarvis PG, Barton CVM, Dougherty PM, Teskey RO, Massheder JM (1991) MAESTRO. In Irving PM (Ed) *Acid deposition: state of science and technology. Vol. 3. Terrestrial, materials, health, and visibility effects.* National Acid Precipitation Assessment Program, Government Printing Office, Washington, DC.

Landsberg JJ (1986) *Physiological ecology of forest production.* Academic Press, London.

Ryan MG, Linder S, Vose JM, Hubbard RM (1994) Dark respiration of pines. In Gholz HL, Linder S, McMurtrie RE (eds) *Environmental constraints on the structure and productivity of pine forest ecosystems: a comparative analysis.* Ecol Bull (Copenhagen) 43:50–63.

Teskey RO, Fites JA, Samuelson LJ, Bongarten BC (1986) Stomatal and nonstomatal limitations to net photosynthesis in *Pinus taeda* L. under different environmental conditions. Tree Physiol 2:131–142.

Vose JM, Allen HL (1988) Leaf area, stemwood growth, and nutrition relationships in loblolly pine. For Sci 34:547–563.

Wang YP, Jarvis PG (1990a) Description and validation of an array model-MAESTRO. Agri & For Meteorol 51:257–280.

Wang YP, Jarvis PG (1990b) Influence of crown structural properties on PAR absorption, photosynthesis, and transpiration in Sitka spruce: application of a model (MAESTRO). Tree Physiol 7:297–316.

i

18. MAESTRO Simulations of the Response of Loblolly Pine to Elevated Temperatures and Carbon Dioxide

Wendell P. Cropper, Jr., Kelly Peterson, and Robert O. Teskey

An important tool in assessing the sensitivity of forests to global change is the simulation model of tree physiology. This tool must be used in conjunction with laboratory and field experiments. Population, ecosystem, and landscape-level models and analyses are also necessary to fully evaluate potential sensitivities to climate change. Physiological simulation models provide a unique method of exploring the complex nonlinear response surface of photosynthesis, respiration, transpiration, carbon allocation, and growth in trees. In principle this information could be obtained from controlled factorial experiments, however, a purely experimental approach would be extremely costly and time-consuming to implement for all of the species and regions of interest.

To answer questions of physiological responses to elevated temperature and carbon dioxide (CO_2) it is necessary to simulate a short time step (typically an hour) and include a large amount of site-specific information. The necessary information can include hourly meteorological data (e.g., soil and air temperature, relative humidity, photosynthetically active radiation, wind speed, and so forth), characterization of the site's soil, biomass or surface area estimates of tree components, and topography. A disadvantage to this approach is the requirement of many computer resources to obtain answers in reasonable time intervals, and the difficulty of determining initial conditions and model parameters for each site and species chosen for analysis.

The MAESTRO model, described by Wang and Jarvis (1990a, 1990b) and Wang et al. (1990, 1991), is the physiological model we have chosen to help evaluate the sensitivity of loblolly pine (*Pinus taeda* L.) to elevated temperature and CO_2. Light distribution within the canopy is a critical issue for scaling up from the leaf-level physiological processes to the stand-level (Grace, 1990; Stenberg et al., 1994). The three-dimensional position of each tree and tree crown in a stand is required by MAESTRO. For physiological simulations, an individual target tree is chosen and the light penetration to and within the crown is calculated for seventy-two points within the target tree. Assuming that seventy-two points adequately represent the variability within a tree crown, scaling up to the whole tree level merely requires summing the CO_2 and water exchanges for each point. We used this basic scheme for simulating loblolly pine crown physiology, but a new crown shape and foliage density distribution model was substituted (Baldwin et al., see Chapter 17) for the original formulation.

Methods

For MAESTRO to be a useful tool in assessing the potential responses of loblolly pine to global climate change, a number of modifications were made in the code provided by the University of Edinburgh. These changes included simulating maintenance respiration for needles, branches, stem, coarse roots, and fine roots, parameterizing MAESTRO photosynthesis for loblolly pine, and modifying the crown shape and foliage distribution to more accurately represent the distribution of foliage within the loblolly canopy (Baldwin et al., see Chapter 17). Modifications to simulated loblolly crown structure and allometric relationships were based on destructive field sampling of individual trees and periodic measurements of stand characteristics (Baldwin et al., see Chapter 17). Physiological parameters for aboveground loblolly pine tissues were based on laboratory and field measurements of loblolly pines near Athens, GA (Teskey et al., 1986) and root respiration rates and fine root biomass were based on measurements of the similar southern pine, *Pinus elliottii* (Gholz et al., 1986; Cropper and Gholz, 1991).

The standard MAESTRO model uses Beta distributions to distribute foliage vertically and horizontally within the crown (Wang et al., 1990). The Beta distributions provide continuous foliage density distributions based on six empirically determined parameters. Separate parameters can describe the spatial distribution of two foliage age-classes. The standard crown shape in MAESTRO is based on a half-ellipsoid envelope. A new linked model system, PTAEDA2–MAESTRO (Baldwin et al., 1993) has been developed to provide an appropriate framework for scaling between tree, stand, and regional levels. For our simulations, the stand was generated by the modified PTAEDA2 growth and yield model (Burkhart et al., 1987) portion of the linked system. The simulated loblolly stand had a mean tree diameter at breast height (DBH) of 17.8 cm (8.4 to 27.5 cm), a mean tree height of 15.2 m (10.2 to 19.0 m), and a mean initial tree projected leaf area index (LAI) of 15.2 m^2 (3.3 to 43.5 m^2). Changes to the MAESTRO model include the use of tree-specific right-truncated Weibull distributions to describe

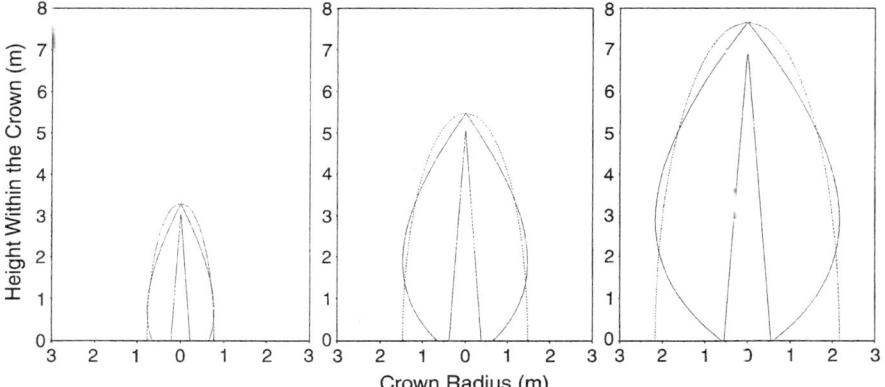

Figure 18.1. Simulated loblolly pine crown shapes for original and modified versions of MAESTRO. The new crown shape is inside the envelope of the original for the small, midsize and large target trees.

the foliage density distribution, and tree-specific equations that describe the inner and outer crown shape (Baldwin et al., see Chapter 17).

We chose three loblolly pines (9.98 cm DBH, 17.63 cm DBH, and 25.66 cm DBH) to represent approximately the natural variability within the simulated twenty-year-old loblolly pine stand. The three trees ranged in height from 12.0 to 18.6 m, and initial projected leaf areas ranged from 4.7 to 36.2 m². The physiology simulations by MAESTRO are based on a specific target tree; in this study each of the three identified loblolly pines was used in a common location in the stand as a target tree. Using this procedure results in the same neighbor trees for each of the target trees, however the incident light reaching the target trees still differs because of the variation in the height and position of the target tree crown. The original and modified crown shapes (Figure 18.1) differ on a size-specific basis. As a test of the significance of these changes each target tree was compared for both crown shape and foliage distribution schemes. There was no consistent pattern of differences for photosynthesis and transpiration between the old and new crown schemes (Table 18.1). Using the new scheme, annual assimilation ranged from 18% greater in the midsize tree to 22% smaller for the small tree, and the old and new crown schemes were nearly identical in the large tree. Because the new scheme is based on more complete canopy and stand data, it was chosen for all subsequent analyses.

To examine the physiological response surface of loblolly pine to climate change and elevated CO_2, we simulated the following six scenarios for each of the three representative loblolly pine trees:

1. Base case. 1988 hourly climate data from Athens, GA. Mean monthly air temperature for 1988 ranged from 6.2 to 27.5 °C (Figure 18.2). Base CO_2 was set at a daily minimum of 330 ppm at 04:00 and a daily maximum of 390 ppm at 16:00.

Table 18.1. Comparison of Old and New Crown Shape Foliage Distribution Schemes Used in MAESTRO

	annual assimilation (mol CO_2 tree^{-1})	annual transpiration (mol H_2O tree^{-1})
Small tree, old scheme	359	160,534
Small tree, new scheme	279	139,609
Midsize tree, old scheme	1,081	623,397
Midsize tree, new scheme	1,277	650,776
Large tree, old scheme	4,566	1,956,200
Large tree, new scheme	4,501	1,884,451

2. Plus 2 degrees: Air and soil temperatures were uniformly increased each hour by 2 °C.
3. Plus 4 degrees: Air and soil temperatures were uniformly increased each hour by 4 °C.
4. Doubled CO_2: The daily minimum was set at 660 ppm and the daily maximum was set at 780 ppm.

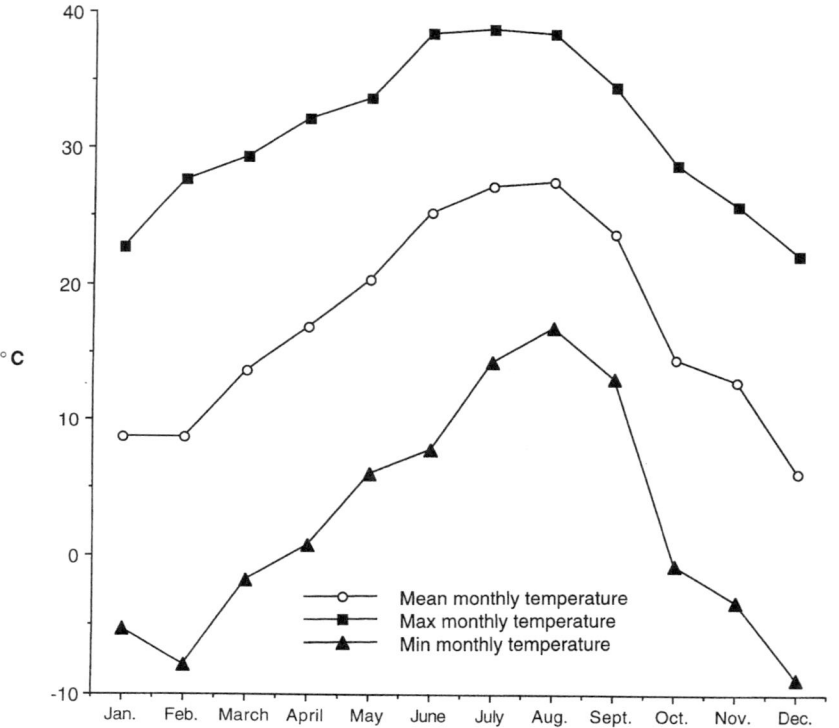

Figure 18.2. Mean, maximum, and minimum monthly air temperatures from hourly climate data recorded in Athens, GA, during 1988.

5. Plus 2 degrees and doubled CO_2.
6. Plus 4 degrees and doubled CO_2.

The response variables monitored for these scenarios included seasonal and annual photosynthesis, total maintenance respiration, and total transpiration for each target tree. Additionally, hourly assimilation, respiration, and transpiration were examined on both the June 22 and December 22 simulations.

Results and Discussion

Simulated MAESTRO loblolly assimilation responded strongly to all of the imposed global change scenarios. Hourly assimilation on June 22 (Figure 18.3) increased substantially above the base scenario for doubled CO_2. This phenomenon has also been observed in many other experimental exposures of plant species to elevated CO_2 (Rogers et al., 1983; Acock and Allen, 1985; Strain 1985; Breen et al., 1986). Increased assimilation in response to elevated CO_2 has also been observed by Teskey (1995) in twenty-one-year-old loblolly pines at the Athens, GA site. These mature pines exhibited no evidence of acclimation to an ambient $+330$ $\mu mol\ mol^{-1}$ CO_2 branch exposure (Teskey, 1995). Increased assimilation rates associated with elevated CO_2 were also evident on December 22 (Figure 18.4), but the magnitude of increase was lower than during the summer.

With a uniform temperature increase of 4 °C, doubled CO_2 was no longer sufficient to increase the simulated net crown CO_2 uptake above the base level on June 22 (Figure 18.3). We attribute this reduction primarily to the effects of increased temperature on simulated respiration. From 14:00 to 18:00 on June 22, temperatures were between 36 and 38 °C. These temperatures are well above the optimum of 29 °C for mesophyll conductance, but were responsible for only a 9% reduction in assimilation rates. Increasing temperatures by 4 °C caused an additional decline of assimilation of 14%, but the same temperature increase caused a 32% increase in leaf maintenance respiration.

In the December 22 simulation, the 4 °C temperature increase had only a small effect on simulated assimilation (Figure 18.4). During the peak period of assimilation on December 22 (hours 12–14), air temperature ranged from 18.3 to 19.9 °C. With a 4 °C increase from 19 to 23 °C, assimilation increases by 10% as a result of a closer approach to the optimum temperature of 29 °C for mesophyll conductance. The same temperature increase of 4 °C causes an increase of 32% in leaf respiration, but the net result on December 22 was only a small decrease in net crown carbon gain resulting from due to the relative magnitude of assimilation and leaf respiration under cool conditions.

Although the present version of MAESTRO does not include effects of CO_2 on stomatal conductance, transpiration is sensitive to temperature alterations. Teskey (1995) found no effect of elevated CO_2 on stomatal conductance, but did observe a transpiration response. The effect of temperature on stomatal conductance in MAESTRO is simulated as a 0 to 1 multiplier with decreasing stomatal conduc-

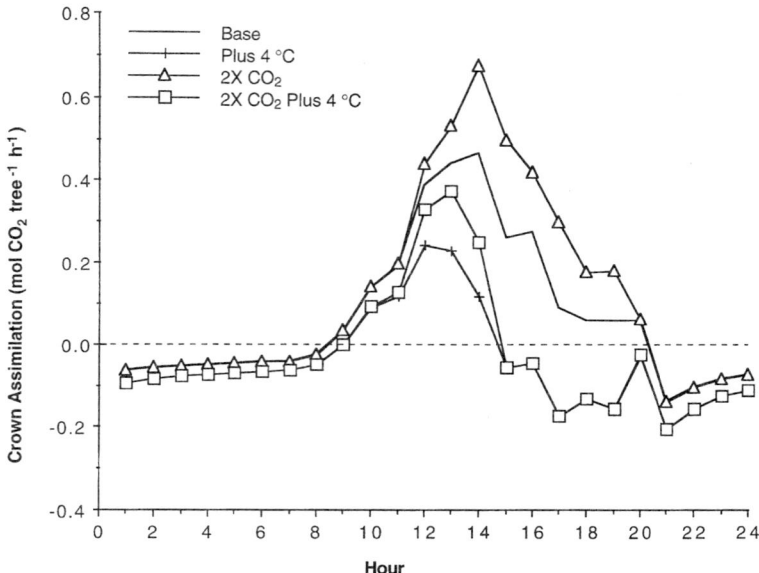

Figure 18.3. Simulated net crown CO_2 uptake for June 22, 1988 meteorology data.

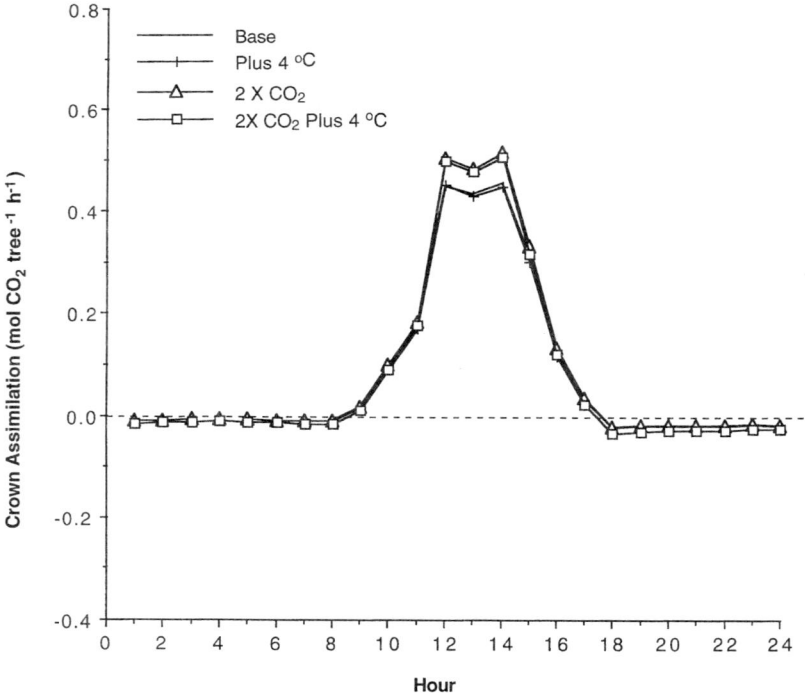

Figure 18.4. Simulated net crown CO_2 uptake for December 22, 1988 meterology data.

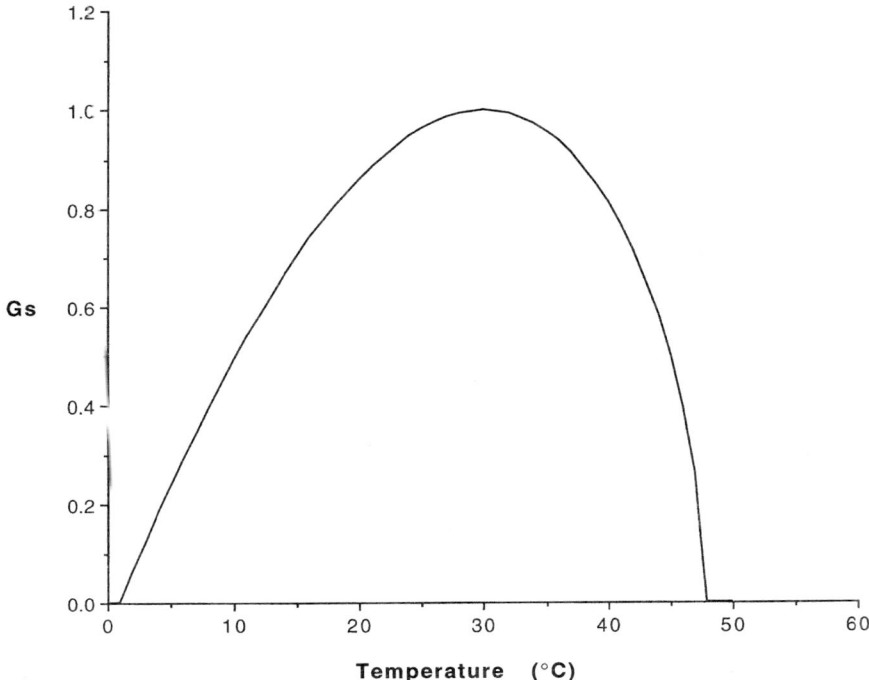

Figure 18.5. MAESTRO temperature response function for stomatal conductance.

tance occurring for temperatures both above or below an optimum temperature (Figure 18.5), but the primary control of stomatal conductance in the MAESTRO system is the incident light.

On December 22, the simulated target tree transpiration was increased by elevating the ambient temperature (Figure 18.6). December 22, air temperatures ranged from 9.8 °C at 06:00 to a high of 19.9 °C at 14:00. This entire temperature range was below the 30 °C optimum temperature for simulated stomatal conductance. Transpiration decreased substantially on June 22 with the 4 °C temperature increase (Figure 18.7) because the simulated vapor pressure deficit under conditions of low relative humidity and elevated temperature completely shut down stomatal conductance from 15:00 to 19:00.

The seasonal patterns of assimilation indicate that temperature has a variable effect. The highest seasonal net tree carbon gain was during the spring (days 81 to 172). In the winter (days 1 to 80 and 356 to 366), assimilation is somewhat limited by low temperatures and reduced day length. Under these conditions, simulated winter assimilation increases with a 4 °C temperature elevation. During the spring (days 81 to 172) and summer (days 173 to 266) seasons, elevated temperatures cause a net decrease in assimilation (Figure 18.8). Maintenance respiration is more than 1.75 times greater during the summer than any other season. Doubled

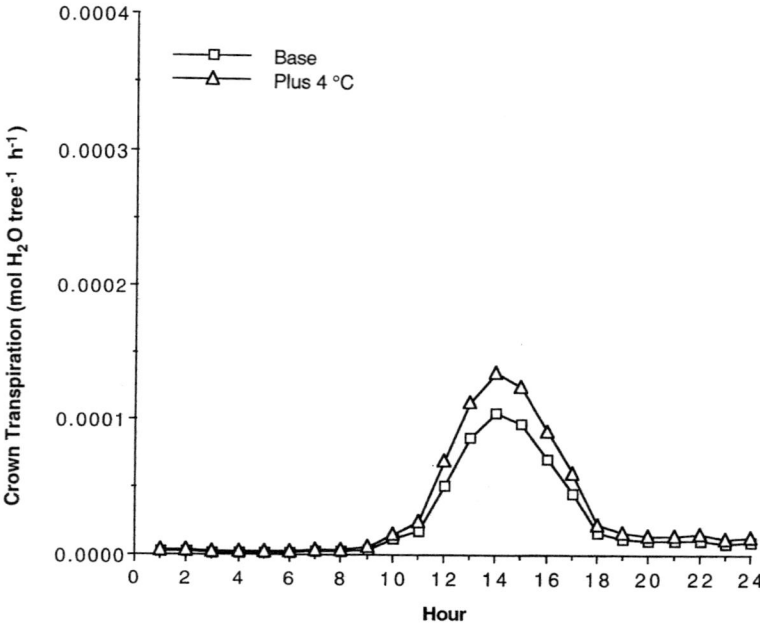

Figure 18.6. Simulated crown transpiration for December 22, 1988 meteorology data.

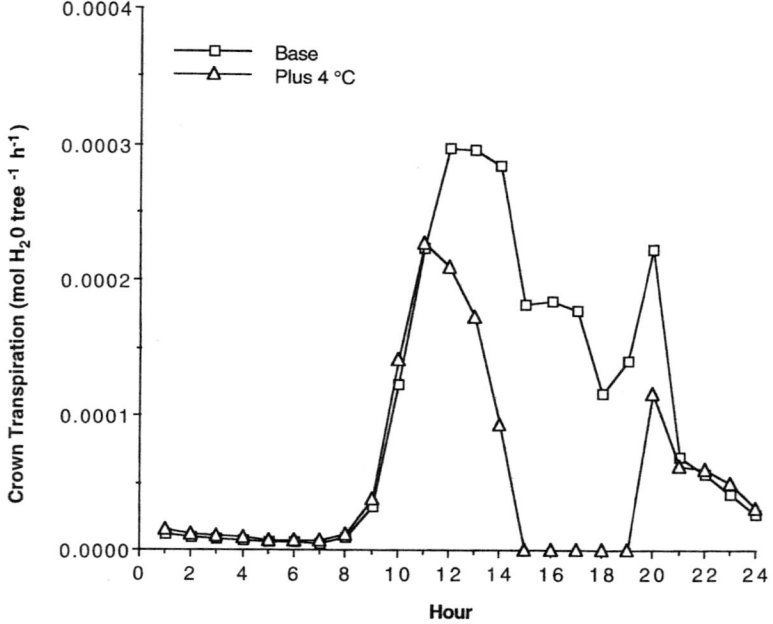

Figure 18.7. Simulated crown transpiration for June 22, 1988 meteorology data.

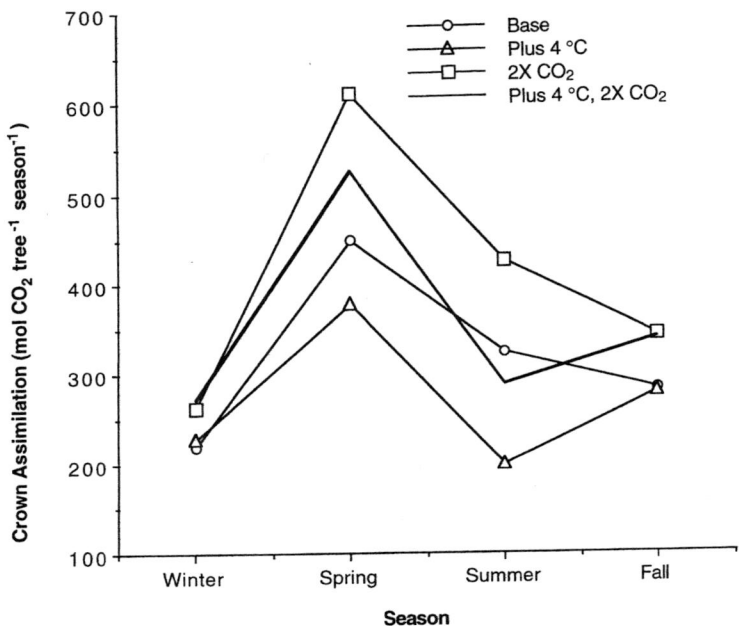

Figure 18.8. Seasonal total net crown CO_2 uptake for the midsize target tree.

atmospheric CO_2 increases assimilation substantially in all seasons. Transpiration is increased during the fall and winter seasons by the 4 °C increase, but decreases during the warmer seasons (Figure 18.9).

Temperature is extremely important in the simulation of maintenance respiration in loblolly pine tissues. In the MAESTRO system, respiration is represented by an exponential response function. Total target-tree respiration (Figure 18.10) is always larger with an imposed temperature increase. The effect is not uniform among seasons, with the warmest seasons exhibiting the largest increases in tissue-maintenance respiration.

Carbon available for loblolly pine growth, reproduction, and production of secondary chemicals (plant defense compounds and so forth) can be approximated as the difference between net assimilation and total maintenance respiration (Table 18.2). Increased temperature scenarios resulted in a negative annual carbon balance for only the small loblolly pine tree with a 4 °C increase. There were individual days and seasons with net carbon deficits for the larger trees, and this could be important during sensitive phenological events. Elevated temperatures did significantly reduce carbon available for growth; the small tree would be able to sustain a maximum of 0.37 kilogram carbon of annual growth under the plus 4 °C and the doubled CO_2 scenario. With a foliage turnover rate of 50% per year in loblolly pine, it is improbable that the small tree could survive under these circumstances. In these simulations, the same physiological parameters were used

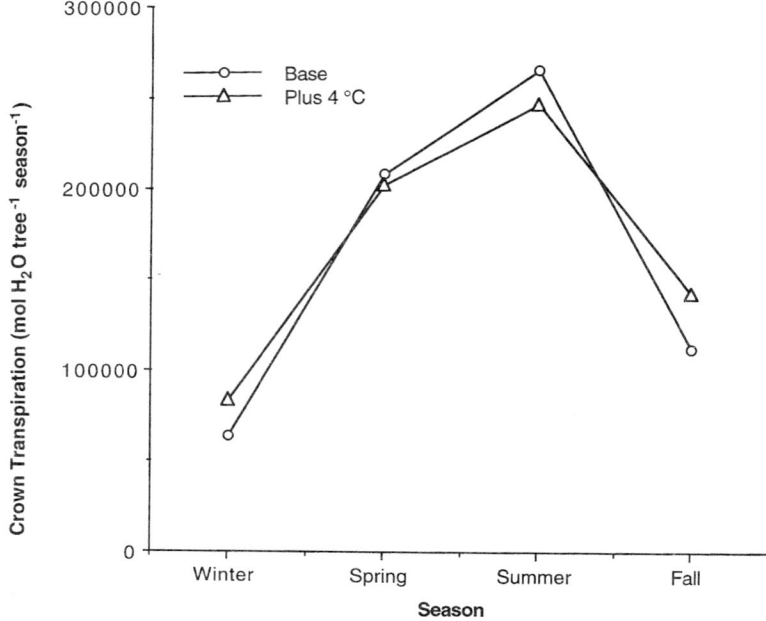

Figure 18.9. Seasonal total crown transpiration for the midsize target tree.

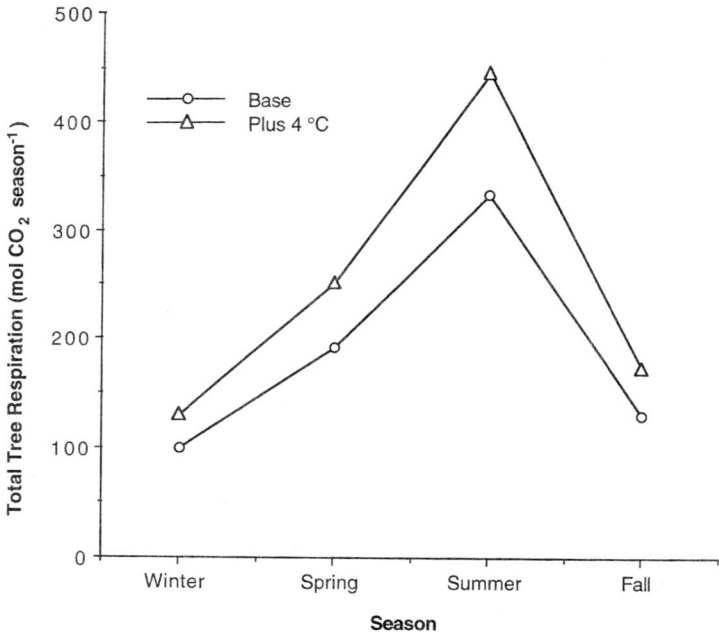

Figure 18.10. Seasonal total maintenance respiration for the midsize target tree.

Table 18.2. Net Annual Carbon Gain[1] for Ambient and Elevated Temperature and CO_2 Scenarios

Simulations	Small tree	Midsize tree	Large tree
Base	67	523	2,700
+2 °C	19	328	2,199
+4 °C	−40	84	1,565
2× CO_2	142	891	4,106
+2 °C, 2× CO_2	93	688	3,570
+4 °C, 2× CO_2	31	427	2,866

[1] Net annual carbon gain equals assimilation-total maintenance respiration, mol CO_2 per tree.

for all three tree sizes, however, small tree foliage may be composed primarily of "shade leaves" with low respiration and water turnover (Larcher, 1980). There may also be differences in assimilation rate between large and small plantation trees similar to the reduction associated with aging in western pines (Yoder et al., 1994), but it should be noted that mortality of suppressed pines is not unusual in southern plantations.

Increased atmospheric CO_2 would result in significant gains in the net carbon available for growth. The negative effects of elevated temperatures on warm season assimilation and the increase in maintenance respiration in response to those elevated temperatures were compensated completely by the doubled CO_2 scenario for the large tree. Doubled CO_2 resulted in an additional growth potential of approximately 4.2 kg carbon for the midsize tree with temperature increases of 2 or 4 °C.

Comparisons of forest carbon budgets over a wide latitudinal range support the importance of maintenance respiration as an increasing fraction of net assimilation in warmer climates (Ryan et al., 1995). Consideration of temperature alone, however, may overestimate the projected effects of climate change on net available carbon. Elevated CO_2 levels may decrease the respiration rates of trees, although the mechanism of this response is not well understood (Ryan, 1991; Wullschleger et al., 1994). This phenomenon has been observed in loblolly pine foliage when exposed to elevated levels of CO_2 (Teskey, 1995). These data indicate a linear (r^2 = 0.99) decrease of foliage maintenance respiration with ambient CO_2 ranging from 348 to 1,025 ppm. Using this relationship, we calculated the effect of a linear decrease in maintenance respiration on the annual carbon balance (Table 18.3). With this assumption, doubled CO_2 resulted in increased net available carbon for all of the trees, even with a 4 °C temperature increase. For the large tree, suppression of respiration resulted in an additional 3.78 kg carbon in the plus 4 °C scenario, available for growth, storage, or the production of defense chemicals. The total annual available carbon pool for the doubled CO_2 and plus 4 °C scenario was 38.2 kg for the large tree.

Although these results do not indicate major changes in loblolly pine productivity in the southeastern United States, a number of potentially significant factors are not considered in the version of MAESTRO used in this study. Soil-nutrient

Table 18.3. Net Annual Carbon Gain[1]

Simulations	Small tree	Midsize tree	Large tree
Base	67	523	2,700
+2 °C	19	328	2,199
+4 °C	−40	84	1,565
2× CO_2	170	990	4,344
+2 °C, 2× CO_2	126	803	3,844
+4 °C, 2× CO_2	68	558	3,181

[1] Assimilation-total maintenance respiration, mol CO_2 per tree. A 13.2% CO_2 induced depression of respiration is assumed.

status, plant-nutrient uptake, and the effects of nutrient limitation on physiology and growth are not presently included in the MAESTRO model. In addition to these factors, water limitation is not assumed to affect loblolly pine physiology (parameters were estimated from an irrigated site). Increased temperatures could greatly enhance the effects of water stress on loblolly pines during the summer because high temperatures increase evaporation rates, and may lead to such non-linear effects as stomatal closure (Figure 18.7). These, and other potential limitations could greatly alter the responses of photosynthesis, maintenance respiration, transpiration, and ultimately loblolly pine allocation, storage, and growth patterns. Additional research is needed to describe the interaction of these multiple limiting factors on the range of sites that loblolly pine is found in the southeast, but the linked PTAEDA2–MAESTRO model system is a potentially powerful tool for analyzing regional loblolly pine responses to global change.

References

Acock B, Allen LH Jr. (1985) Crop responses to elevated carbon dioxide concentrations. In Strain BR, Cure JD (Eds) *Direct effects of increasing carbon dioxide on vegetation.* US DOE, Washington, DC.

Baldwin VC Jr., Burkhart HE, Dougherty PM, and Teskey RO (1993) Using a growth and yield model (PTAEDA2) as a driver for a biological process model (MAESTRO). Research Paper SO-276, USDA, For Ser, South For Exper Sta New Orleans, LA.

Breen PJ, Hesketh JD, Peters DB (1986) Field measurements of leaf photosynthesis of C3 and C4 species under high irradiance and enriched CO_2. Photosyn 20:281–285.

Burkhart HE, Farrar KD, Amateis RL, Daniels RF (1987) Simulation of individual tree growth and stand development in loblolly pine plantations on cutover, site-prepared areas. FWS-1–87. VA Poly Inst State Univ, Sch For Wild Res, Blacksburg, VA.

Cropper WP Jr, Gholz HL (1991) In situ needle and fine root respiration in mature slash pine (*Pinus elliottii*) trees. Can J For Res 21:1589–1595.

Gholz HL, Vogel SA, Cropper WP Jr (1986) Organic matter dynamics of fine roots in plantations of slash pine (*Pinus elliottii*) in north Florida. Can J For Res 16:529–538.

Grace J (1990) Modeling the interception of solar radiant energy and net photosynthesis. In Dixon RK, Meldahl RS, Ruark GA, Warren WG (Eds). *Process modeling of forest growth responses to environmental stress.* Timber Press, Portland, OR.

Larcher W (1980) *Physiological plant ecology.* Springer-Verlag, Berlin.

Rogers HH, Bingham GE, Cure JD, Smith JM, Surano KA (1983) Responses of selected plant species to elevated carbon dioxide in the field. J Environ Qual 12:569–574.

Ryan MG (1991) Effects of climate change on plant respiration. Ecol Appl 1:157–167.

Ryan MG, Gower ST, Hubbard RM, Waring RH, Gholz HL, Cropper WP Jr. Running SW (1995) Woody tissue maintenance respiration of four conifers in contrasting climates. Oecclogia 101:133–140.

Stenberg P, Kuuluvainen T, Kellomaki S, Grace JC, Jokela EJ, and Gholz HL (1994) Crown structure, light interception and productivity of pine trees and stands. Ecol Bull 43:20–34.

Strain BR (1985) Physiological and ecological controls on carbon sequestering in terrestrial ecosystems. Biogeochem 1:219–232.

Teskey RO (1995) A field study of the effects of elevated CO_2 on carbon assimilation, stomatal conductance and leaf and branch growth of *Pinus taeda* trees. Plant Cell Environ 18:565–573.

Teskey RO, Fites JA, Samuelson LJ, Bongarten BC (1986) Stomatal and nonstomatal limitations to net photosynthesis in *Pinus taeda* L. under different environmental conditions. Tree Physiol 2:131–142.

Wang YP, Jarvis PG (1990a) Description and validation of an array model-MAESTRO. Agri For Met 51:257–280.

Wang YP, Jarvis PG (1990b) Influence of crown structural properties on PAR absorption, photosynthesis, and transpiration in Sitka spruce: application of a model (MAESTRO). Tree Physiol 7:297–316.

Wang YP, Jarvis PG, Benson ML (1990) Two-dimensional needle area density distribution within crowns of *Pinus radiata*. For Ecol Manage 32:217–237.

Wang YP, Jarvis PG, Taylor CMA (1991) PAR absorption and its relation to above-ground dry matter production of sitka spruce. J Appl Ecol 28:547–560.

Wullscheger SD, Ziska LH, Bunce JA (1994) Respiratory responses of higher plants to atmospheric CO_2 enrichment. Physiol Plant 90:221–229.

Yoder BJ, Ryan MG, Waring RH, Schoettle AW, Kaufmann MR (1994) Evidence of reduced photosynthetic rates in old trees. For Sci 40:513–527.

19. Projections of Growth of Loblolly Pine Stands Under Elevated Temperatures and Carbon Dioxide

Harry T. Valentine, Timothy G. Gregoire, Harold E. Burkhart, and David Y. Hollinger

Over the past 200 years, the concentration of carbon dioxide (CO_2) in the atmosphere has increased from about 280 ppm (Neftel et al., 1991) to 360 ppm. An eventual doubling of the present-day ambient concentration along with increases in atmospheric concentrations of other greenhouse gasses are expected. Predictions that these higher concentrations will cause alterations in climates in many regions of the world have been widely disseminated by atmospheric physicists and others (e.g., Houghton et al., 1990). Tree physiologists have indicated that an increase in CO_2, by itself, may foster faster growth and more efficient use of water by trees. Conversely, if the rise in CO_2 is accompanied by an altered climate, gains that might otherwise accrue from CO_2 fertilization could be either partially or entirely negated. Plant geographers have predicted changes in forest types in given regions under various climatic change scenerios, and concern has been voiced that if change is too rapid, some species will not be able to migrate fast enough to remain with those environmental conditions to which they are adapted.

The rise in atmospheric CO_2, together with scientific uncertainty, has fueled much concern and speculation about the future growth of forests. In an effort to reduce uncertainty, the Southern Gobal Change Program (SGCP) was chartered to 1) determine those processes in forested ecosystems that are sensitive to physical and chemical changes in the atmosphere, 2) evaluate how future physical and chemical changes influence the structure, function, and productivity of forest and related ecosystems, and to what extent forest ecosystems will change in response to atmospheric changes, and 3) evaluate what the implications are for forest management and how forest management activities must be altered to sustain forest productivity, health, and diversity.

The objectives of our work under the SGCP were to 1) develop a carbon-balance model of stand growth for loblolly pine that includes the effects of climatic variables in the parametrizations of key physiological rates, and 2) analyze the sensitivity of growth projections of the model to climatic variation. The carbon-balance model, called Pipestem, has been derived, calibrated for loblolly pine, and described in detail elsewhere (Valentine et al., 1997). In this chapter, we report some projections of the effects of elevated CO_2 concentrations, with and without alterations of climatic norms.

Our analyses are limited to planted stands of loblolly pine between the ages of three and fifty. We begin at age three because the success and growth of a plantation normally can not be predicted with an acceptable degree of certainty or precision prior to that age. We end at age fifty because 1) most loblolly pine plantations are harvested and reestablished before then, and 2) the growth of a plantation of that age, or older, normally is slow, nil, or even negative.

The Model

First, a brief overview of how Pipestem works. The mathematical model consists of a set of differential equations and additional ancillary functions. The differential equations completely describe the dynamic properties of a single model stand. The state variables that characterize the model stand ordinarily are initialized with measurements from a real stand that has aged three or more years since planting. Simultaneous numerical integration of the differential equations provide values of the state variables at any subsequent point in time, and these values serve as predictions for the real stand.

The rate of growth of the model stand is defined as the difference between the rates of production and loss of dry matter. Dry matter is measured in units of carbon (C) and the rate of growth per unit land area is measured in units of C/hectare (ha)/year. The rate of production of dry matter is defined as the difference between the rate of production of C substrate and the rate of consumption of C substrate through maintenance and constructive respiration. These metabolic rates also have dimensions of units of C/ha/year.

Dry matter is divided into foliar, feeder-root, and woody components. Carbon allocation rules based on pipe-model theory (Valentine 1988) are used to divide dry matter production into new leaf, fine root, woody root, and woody stem tissue in proper proportions for functional balance. Losses of dry matter result from the turnover of foliage and feeder roots, the death and selfpruning of branches associated with crown rise, and the death of trees associated with selfthinning. The latter two processes are driven by the production of dry matter.

To account for year-to-year differences in environmental conditions, specific rates of production and consumption of C substrate by the model stand are adjusted each year. Calculation of the adjustment factors involves the integration, over the yearly intervals, of steady-state physiological models. The integrations yield specific annual gas exchange rates that reflect the within-year environmental conditions. These specific annual rates are divided by baseline or average annual rates to furnish the requisite adjustment factors. Maintenance respiration per unit of live dry matter over a year is modeled as function of the ambient temperature record. The production of substrate per unit of foliar dry matter is calculated with the steady-state carbon-flux model from MAESTRO (Jarvis et al., 1990; Horne 1993). This carbon-flux model is driven by ambient temperature, photosynthetically active photon-flux density, vapor pressure deficit, and predawn xylem-water potential. The carbon-flux model is solved on a quarter-hour time step. Other carbon-flux models could substitute, however, the one chosen was already calibrated for loblolly pine.

In addition to information about the carbon balance and the production and loss of dry matter, Pipestem also furnishes total basal area (m^2/ha), average tree height (m), average height to the base of a crown (m), and tree density (1/ha) on an annual time step. Estimates or measurements of these four variables (ordinarily obtained from a real stand of age three or older) suffice to initialize the Pipestem model. By constrast, the steady-state physiological models require time streams of values of each of the driving variables for the period of interest. Values of driving variables on short time steps (quarter-hour, half-hour, or hour) can be calculated, under reasonable assumptions, from daily, weekly, or monthly averages.

Carbon Dioxide Scenerios

Presently, the ambient CO_2 concentration is increasing by approximately 1.6 ppm per year (Conway et al., 1991). Over a thirty- to fifty-year rotation, this amounts to a total increase of 48 to 80 ppm (i.e., 14 to 23%). The changing ambient CO_2 concentration eventually may cause climatic warming and changes in the amounts and patterns of precipitation in the present-day range of loblolly pine. Will these changes be detrimental to loblolly pine? To address this question the carbon-flux model of Jarvis et al., (1990) was coupled with Pipestem to project how transient increases in the ambient CO_2 concentration in the near term, and a constant concentration of 700 ppm in the long term, will affect the growth of loblolly pine plantations.

We calibrated the Pipestem model with data from spacing trials initiated in 1983 by the Loblolly Pine Growth and Yield Research Cooperative (Amateis et al., 1988). Three replicates of sixteen experimental plots of loblolly pines were planted in each of four locations, viz., Buckingham, Halifax, and Middlesex Counties in Virginia and Northampton County in North Carolina. Each of the sixteen plots per replicate contained forty-nine trees at planting and were remeasured annually through 1992. Variables measured on the trees of each plot included basal diameter, total height, height to the base of the live crown, and—beginning the fourth year after planting—diameter at breast height (DBH) (1.37 m). The measurements from the third year after planting, together with stem count/ha, were converted into initial values of the state variables. Measurements taken in later years were used to precisely calibrate three of the twenty-two parameters of the model namely 1) the annual production of C substrate per unit of foliar dry matter, 2) the specific rate of maintenance respiration of woody dry matter, and 3) a carbon-allocation parameter.

We obtained average monthly values of maximum and minimum daily temperatures from weather stations closest to the locations of the spacing trial for the years 1949 through 1992. We also obtained values of the Palmer drought severity index (Palmer, 1965) for the same span of years. These data were used to calculate time streams of values of the driving variables of the steady-state carbon-flux model on a quarter-hour time step. The temperature stream was calculated from the monthly average maximums and mimimums under the assumptions that 1) each day of a given month was identical, and 2) the temperature trace over the course of a day was sinusoidal with a two-hour lag from solar noon to the time of maximum daily temperature. A second driving variable, predawn xylem-water potential (ψ_χ), measured in MPa, was assumed to relate to the monthly Palmer drought severity index (I_ρ) thus:

$$\psi_\chi = \min(-0.4, -0.4 + I_\rho/5).$$

We calculated vapor pressure deficit with Buck's (1981) vapor-density formula under two assumptions, which are 1) the air is saturated with water vapor when the temperature is at its predawn mimimum, and 2) the vapor pressure does not change over the course of a day. Photosynthetic photon-flux density, the fourth and final driving variable, was calculated from standard formulae (e.g., Campbell, 1977, equations 5.6–5.11) assuming a constant atmospheric transmissivity of 0.6.

Loblolly pines manifest photosynthesis and maintenance respiration throughout the year, but meristematic growth (at least that which is observed aboveground) and accompanying constructive respiration begin in the spring and—in the vicinity of the spacing trial—end in September. Thus, carbon fixed in the fall and winter may contribute to meristematic growth in the spring and summer. Accordingly, we defined a physiological year to run from October 1 to September 30.

We solved the steady-state carbon-flux model with our streams of driving variables that ran for October 1, 1949 until September 30, 1992 on the quarter-hour time step. The carbon-flux model estimated the rates of both assimilation and dark respiration (μmol CO_2 (m^2 projected leaf area)$^{-1}$ s^{-1}). Empirical models (Kinerson et al., 1974, equations 3 and 4) were used to approximate the foliar dry matter present in a loblolly pine stand on any given day of the year relative to the amount present on May 1. With these latter estimates, the carbon-flux solutions were adjusted for the amount of foliage present in each quarter-hour (900 seconds) interval. Estimates of assimilation and dark respiration were accumulated over all the quarter-hour intervals of each physiological year and then averaged to establish baseline annual rates. Ambient CO_2 concentration was fixed at 348 ppm for these estimations.

Near-Term Transient Effects

Three of the four spacing trials established in 1983 were successful. The fourth, in Middlesex County, VA initially failed but was successfully reestablished in 1984. The ambient CO_2 concentration was approximately 343.2 ppm when the plantations were established and this concentration has been increasing by approximately 1.6 ppm per year since that time (Conway et al., 1991). We assumed that 1) this rate of increase will continue over the span of a rotation, and 2) climatic norms will not differ from those evident from 1949 to 1992. Initial values of the state variables of the Pipestem model were calculated with measurements obtained from the spacing trial after the third growing season (late fall of 1985 in three locations, 1986 in the fourth).

Baseline projections were run from the start of fourth to the end of the fiftieth year after planting for each of the spacing trial plots. Ambient CO_2 concentration was fixed at 348 ppm. The rates of C substrate production and maintenance respiration of live dry matter were adjusted for year-to-year variation in weather using output from the steady-state carbon-flux model. The weather data used to calculate the carbon-flux driving variables for the first seven years of the projection were those which actually obtained when the stands were in their fourth through tenth physiological years. After that, earlier data were recycled so weather for the eighth physiological year of the projection was concocted from the last three months of 1992 followed by the first nine months of 1949. The next physiological year's weather came from 1949 and 1950, and so forth.

To project responses to the increasing CO_2 concentration, an identical set of runs was performed—with one exception. We increased CO_2 by 1.6 ppm per year from the initial value of 348 ppm.

The results were similar for all plots, regardless of location or initial tree spacing. The constant annual increase of 1.6 ppm in ambient CO_2 concentration is predicted to yield larger trees in terms of average height, average diameter, and total woody dry matter (Figure 19.1). Selfthinning is predicted to accelerate in plantations, but because the surviving trees are predicted to have larger diameters, stand basal area is predicted to differ very little from that forecasted for a constant ambient CO_2 concentration. Average crown lengths are projected to be longer with the increasing levels of CO_2. In year fifty of the projections, the predicted 23% increase in ambient CO_2 concentration will result in a 12% increase the annual rate of photosynthesis and a similar increase in standing crop of woody dry matter.

Long-Term Effects

That ambient CO_2 will continue to increase in our lifetime is indisputable but the eventual equilibrium concentration and the accompanying climate are uncertain. One prevailing notion is that the preindustrial CO_2 concentration eventually will double at the very least, and the climate in many regions will become warmer (Houghton et al., 1990). Carter et al., (1992) suggested scenerios with CO_2 concentrations at twice the 1990 level.

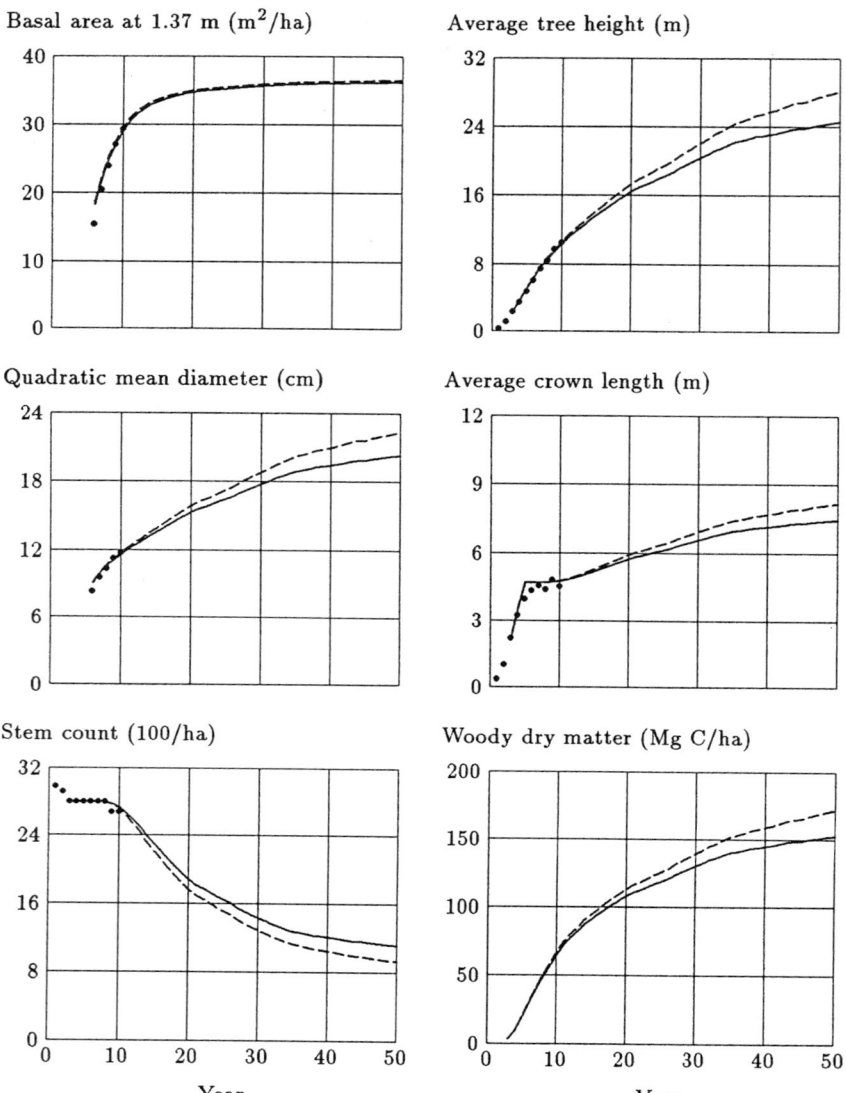

Figure 19.1. Projections for a loblolly pine stand planted with 1.83 m × 1.83 m spacing in Buckingham County, VA in 1983. The solid lines signify a constant ambient CO_2 concentration of 348 ppm. The dashed lines signify an increase of 1.6 ppm per year. The dots are measurements.

To understand how loblolly pine stands will grow in a warmer, higher CO_2 climate, it was assumed that the ambient CO_2 concentration eventually will equilibrate at 700 ppm. We posited three very simple climatic change scenerios 1) no change from present-day climatic norms, 2) an increase in average daily temperature of 2 °C, and 3) an increase in average daily temperature of 4 °C. For the last two scenerios it was assumed that solar radiation and precipitation patterns and amounts would not vary from present-day norms. Our procedures for initializing the Pipestem model and streaming the driving variables through the carbon-flux model were identical to those of the prior analysis.

As with the prior analysis, results were similar for all spacing plots, regardless of location or initial tree spacing, therefore it suffices to look at the results for a single stand (Figure 19.2). At an ambient CO_2 concentration of 700 ppm, trees are predicted to grow more quickly and larger than at 348 ppm. The addition of warmer temperatures are predicted to partially negate some of the gains resulting from the CO_2 fertilization. The partial negation is the result of an increase in the consumption of C substrate through maintenance respiration. The instantaneous specific rate of maintenance respiration roughly doubles with each 10 °C increment in ambient temperature. But even with an addition to the daily average temperature of 4 °C, the early growth rates and final yields of future loblolly plantations are expected to exceed those presently in the ground.

Model projections obtained with our time stream of weather data and assumed atmospheric CO_2 concentrations of either 280, 348, 550, or 700 ppm are compared in Figure 19.3. These projections suggest that there has already been an increase in the rate of stand growth over the last 200 years associated with the increase in atmospheric CO_2 concentration from the preindustrial 280 ppm to the recent 348 ppm. With climatic norms held constant, the projected woody yield of the stand grown under a concentration of 348 ppm is 21% greater than the same stand grown under a concentration of 280 ppm. Moreover, an eventual CO_2 concentration of 700 ppm would be 150% greater than preindustrial level of 280 ppm, and woody yields 100% greater under 700 ppm than under 280 ppm would be expected.

All of Pipestem's projections of increased woody dry matter are achieved with no increase in foliar or feeder-root dry matter per unit land area and, therefore, no real increase occurs in the respiratory consumption of carbon by foliage or feeder roots. On the average, a doubling of CO_2 tends to increase the carbon-flux model's estimate of annual gross photosynthesis per unit foliar dry matter by about 50%. Because carbon allocation to foliar and feeder-root dry matter remains more or less constant, the 50% elevation in carbon fixation may increase the amount of carbon available for woody production by 100% or more, and therefore, large increases in woody yield with CO_2 fertilization are predicted.

Discussion

It is important to bear in mind that these projections with the carbon-flux and Pipestem models are extrapolations with unknown degrees of error. The carbon-

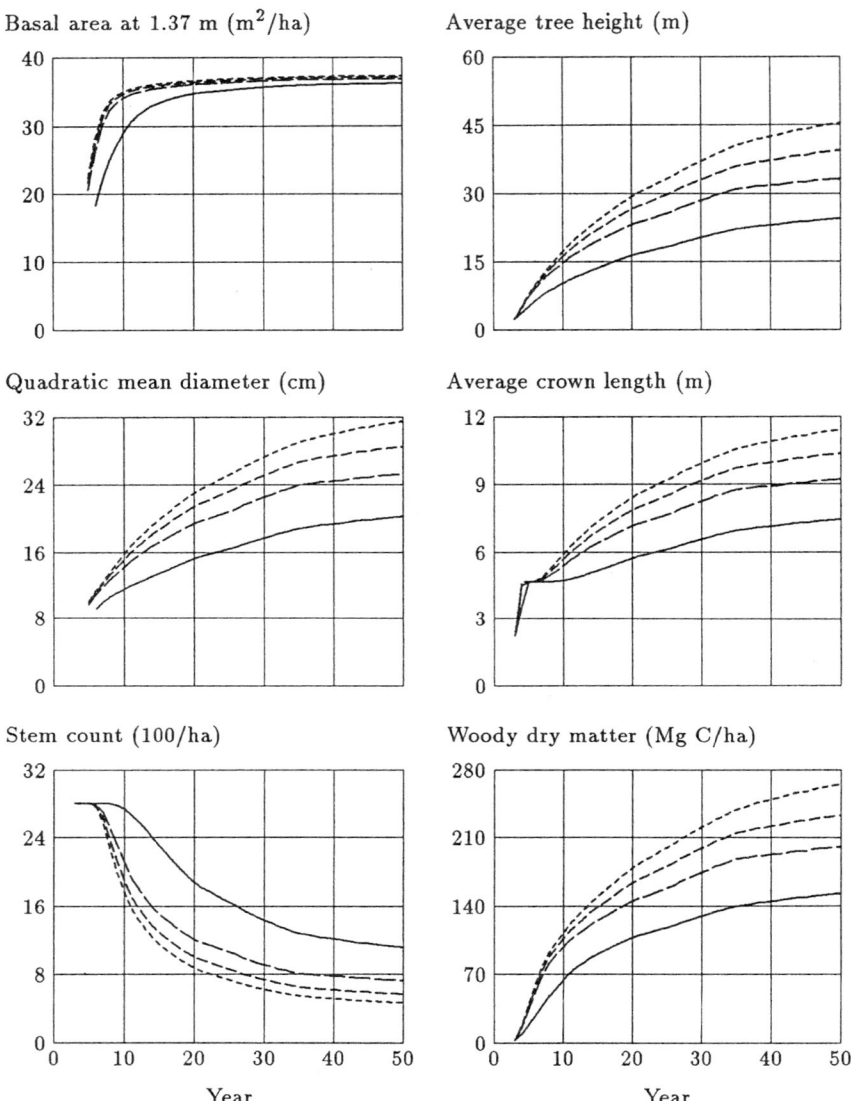

Figure 19.2. Projections for a loblolly pine stand in a constant ambient CO_2 concentration of 700 ppm. Short dashes signify no change from the present-day climatic norms for Buckingham County, VA. Medium dashes or long dashes, respectively, signify 2° or 4°C increases in both average minimum and maximum daily temperatures. Solid lines signify a constant ambient CO_2 concentration of 348 ppm.

19. Projections of Growth Under Elevated Temperatures and Carbon Dioxide   349

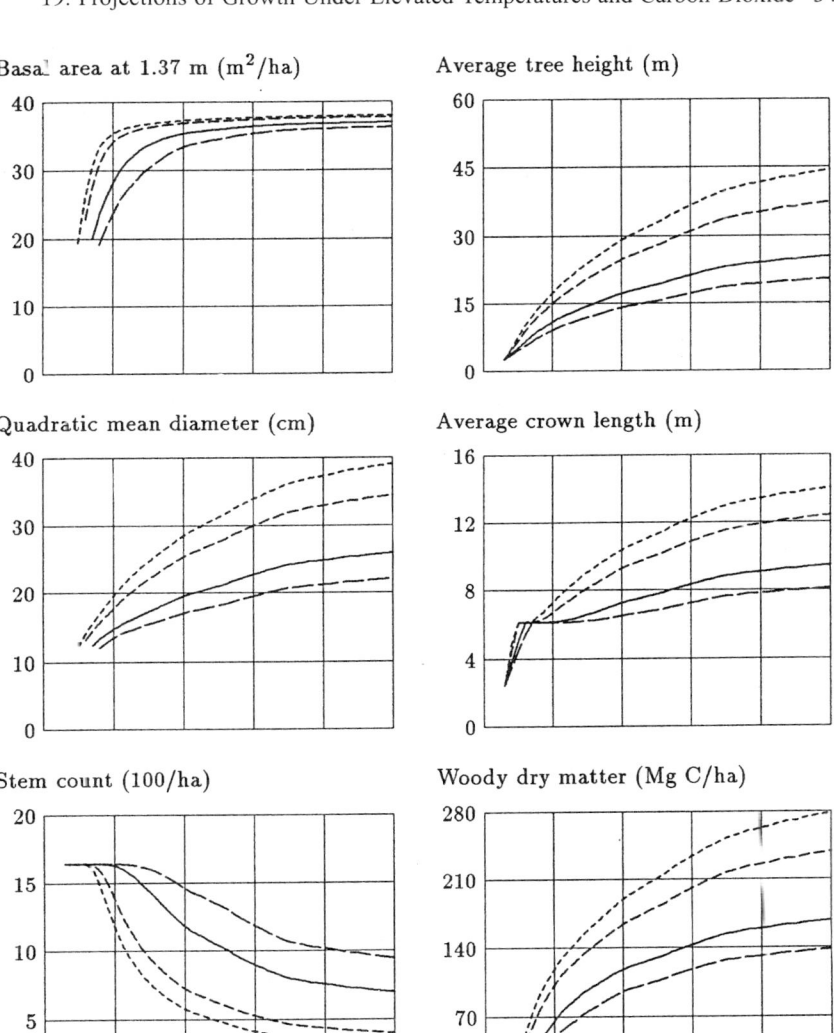

Figure 19.3. Projections for a loblolly pine stand planted with 2.44 m × 2.44 m spacing. Long, medium, or short dashes, respectively, signify atmospheric CO_2 concentrations of 280, 550, or 700 ppm. Solid lines signify 348 ppm. All projections assume present-day climatic norms for Buckingham County, VA.

flux model, which furnishes the photosynthetic response to ambient CO_2 concentration and the four driving variables, was primarily fitted to experimental data from immature trees. It was assumed, for our projections, that mature forest trees will continue to respond as immature trees growing in experimental situations.

Jarvis (1989) found that the carbon-flux submodel, parametrized for *Picea sitchensis*, overestimated carbon production for seedlings grown in elevated levels of CO_2. This was because seedlings in the increased-CO_2 treatment had a lower photosynthetic capacity than seedlings in the control treatment. There is evidence (e.g., Tolley and Strain 1984a,b) that the photosynthetic capacity of container-grown loblolly pine seedlings also may decline with prolonged exposure to elevated CO_2 levels. Many plant species show this response, which may be related to a decline in the content or activation state of the photosynthetic enzyme RuBP-carboxylase/oxygenase (rubisco) (Bowes, 1991). These changes may be associated with feedback inhibition caused by either inadequate translocation or carbon sinks. Plants growing in natural conditions, however, appear to maintain a stronger carbon sink and are less susceptible to this decline in photosynthetic capacity (e.g., Arp, 1991). In recent work, Liu and Teskey (1995) found that mature, field-grown loblolly pines did not exhibit downward acclimation of leaf-level photosynthesis following long-term exposure to elevated levels of CO_2.

Norby et al., (1992) found that the photosynthetic capacity of *Liriodendron tulipifera* trees grown at elevated CO_2 levels in open-top chambers did not decline and yet the growth of both control and elevated-CO_2 trees did not differ. In this case, additional carbon appeared to be going to feeder-root turnover, which would imply a need for modification of the feeder-root longevity parameter of Pipestem. Further long-term whole-tree studies such as those planned at the Duke Forest (e.g., Culotta 1995) may help resolve these and other questions. In the meantime, it is best to use or disseminate our projections with caution.

It also is important to bear in mind that factors other than the driving variables of the carbon-flux model affect the rates of photosynthesis and growth of mature forest trees. Our projections apply to stands that are carbon limited. Supplies of other nutrients, especially nitrogen, may regulate metabolic rates. Such stress agents as ozone, insects, or pathogens—singly, or in combination—may adversely affect loblolly pine stands. Other subtle direct or indirect effects of increasing CO_2 levels may also affect stand growth. These include the positive effects of increased water-use efficiency (e.g., Eamus and Jarvis, 1989) and decreased respiration (e.g., Wullschleger and Norby, 1992).

Uncertainty notwithstanding, our projections suggest more rapid growth by loblolly pine stands in the future. Highest yields will be achieved if warming does not eventuate. But even 4 °C warming will not entirely negate the positive effects of CO_2 fertilization; future yields are predicted to exceed current yields on given sites. Pipestem's projections of increased woody dry matter are achieved with no increase in foliar or fine root dry matter per unit land area. This suggests that the additional woody dry matter can be produced with little or no nitrogen added to loblolly pine sites.

Our projections regarding the near-term transient increases in the ambient CO_2

concentration should be of immediate interest to forest managers. An increase in the ambient CO_2 concentration of 1.6 ppm per year is predicted to result in increased growth in loblolly pine plantations in the next thirty years, assuming the climate does not warm from present-day norms. Thus, if extant yield tables or models are accurately predicting present-day yields, then those same tables or models will probably underpredict yields in the next rotation.

References

Amateis RL, Burkhart HE, Zedaker SM (1988) Experimental design and early analyses for a set of loblolly pine spacing trials. In Ek AR, Shifley SR, Burk TE (Eds) *Forest Growth Modelling and Prediction, Vol 2*. USDA For Ser, Northcen For Exper Sta, General Technical Report NC-120.

Arp WJ (1991) Effects of source-sink relations on photosynthetic acclimation to elevated CO_2. Plant Cell Environ 14:869-875.

Bowes G (1991) Growth at elevated CO_2: photosynthetic responses mediated through rubisco. Plant Cell Environ 14:795-806.

Buck AL (1981) New equations for computing vapor pressure and enhancement factor. J Appl Meteor 20:1527-1534.

Campbell GS (1977) *An Introduction to Environmental Biophysics*. Springer-Verlag, New York.

Conway TJ, Tans PP, Waterman LS (1991) Atmospheric CO_2-modern record, Key Biscayne. In Boden TA, Sepanski RJ, Stoss FW (Eds) *Trends '91: A Compendium of Data on Global Change*. Oak Ridge National Laboratory, Oak Ridge, TN.

Carter TR, Parry ML, Nishioka S, Harasawa H (1992) *Preliminary Guidelines for Assessing Impacts of Climate Change*. IPCC, The Environmental Change Unit, Oxford, UK.

Culotta, E (1995) Will plants benefit from high CO_2? Science 268:654-656.

Eamus D, Jarvis PG (1989) The direct effects of increase in the global atmospheric CO_2 concentration on natural and commercial temperate trees and forests. Adv Ecol Res 19:1-55.

Horne AL (1993) The effects of shade on growth and carbon allocation in branches of loblolly pine (*Pinus taeda* L.). PhD Dissertation, Yale Univ, New Haven, CT.

Houghton JT, Jenkins GJ, Ephraums JJ (Eds) (1990) *Climate Change: The IPCC Scientific Assessment*. Cambridge University Press, New York.

Jarvis PG (1989) Atmospheric carbon dioxide and forests. Phil Trans Roy Soc B 324:369-392.

Jarvis PG, Barton CVM, Dougherty PM, Teskey RO, Massheder JM (1990) MAESTRO. In *Development and Use of Tree and Forest Response Models*. State-of-Science/Technology Report 17, National Acid Precipitation Assessment Program, Washington, DC.

Kinerson RS. Higginbotham KO, Chapman RC (1974) The dynamics of foliage distribution within a forest canopy. J Appl Ecol 11:347-353.

Liu S, Teskey RO (1995) Responses of foliar gas exchange to long-term elevated CO_2 concentrations in mature loblolly pine trees. Tree Physiol 15:351-359.

Neftel A, Friedli H, Moor E, Lötscher H, Oeschger H, Siegenthaler U, Stauffer B (1991) Atmospheric CO_2-historical record from ice cores, Siple Station. In Boden TA, Sepanski RJ, Stoss FW (Eds) *Trends '91: A Compendium of Data on Global Change*. Oak Ridge National Laboratory, Oak Ridge, TN.

Norby RJ, Gunderson CA, Wullschleger SD, O'Neil EG, McCracken MK (1992) Productivity and compensatory responses of yellow poplar trees in elevated CO_2. Nature 357:322-324.

Palmer, W.C. (1965) Meteorological drought. US Weather Bureau, Res Pap No 45.

Tolley LC, Strain BR (1984a) Effects of CO_2 enrichment on growth of *Liquidambar styraciflua* and *Pinus taeda* seedlings under different irradiance levels. Can J For Res 14:343–350.

Tolley LC, Strain BR (1984b) Effects of CO_2 enrichment and water stress on growth of *Liquidambar styraciflua* and *Pinus taeda* seedlings. Can J Bot 62:2135–2139.

Valentine HT (1988) A carbon-balance model of stand growth: a derivation employing pipe-model theory and the self-thinning rule. Ann Bot 62:389–396.

Valentine HT, Gregoire TG, Burkhart HE, Hollinger DY (1997). A stand-level model of carbon allocation and growth, calibrated for loblolly pine. Can J For Res 27:817–830.

Wullschleger SD, Norby RJ (1992) Respiratory cost of leaf growth and maintenance in white oak saplings exposed to atmospheric CO_2 enrichment. Can J For Res 22:1717–1721.

20. Modeling the Potential Sensitivity of Slash Pine Stem Growth to Increasing Temperature and Carbon Dioxide

Wendell P. Cropper, Jr.

There is a widely recognized possibility that the earth's climate may change at an unprecedented rate during the next century. General circulation models (GCMs) of the atmosphere predict that the global annual average temperature will increase 2 to 5 °C in response to increased levels of greenhouse gases. This is the equivalent to a doubling of atmospheric carbon dioxide (CO_2) concentrations from preindustrial levels (Schneider et al., 1991). Although there are many uncertainties concerning the magnitude and pattern of regional climate changes (Dickinson, 1989), the time-scale of forestry management requires serious consideration of possible future effects now.

If GCMs adequately describe the present-day regional climates, it may be possible to use the GCM outputs as a guide for developing specific scenarios for regional or local impact assessments (Ackerman and Cropper, 1988). Even when this approach is not feasible, biological simulation models can be used to assess the sensitivity of ecosystems to assumed scenarios of climate change. It would be very expensive and impractical to assess the potential impacts of climate change and increasing levels of atmospheric CO_2 through experimental programs on all tree species of interest. This is particularly true if responses of trees beyond the seedling stage are of interest. Under these circumstances, mechanistic simulation models provide a valuable tool.

In this chapter, a simulation model of Florida slash pine plantation carbon dynamics (SPM) (Cropper and Gholz 1993a, 1993b, 1994) is used to evaluate the sensitivity of coastal plain plantations to both increasing temperatures and elevated levels of atmospheric CO_2. Planted slash pines have largely replaced natural pine forests in north Florida, and represent a commercially important species in the southeastern United States coastal plain. Dissimilar to annual agricultural crops, the appropriate time frame for considering the management implications of climate change in slash pine plantations is eighteen to forty years for one rotation.

Model Description and Methods

The SPM pine state variables are new foliage, second-year foliage, branches, stems, coarse roots, fine roots in the litter layer, and fine roots in the mineral soil. The canopy is divided into nine vertical layers with the vertical distribution of relative gap frequency simulated to produce a more realistic light environment than a standard Beer-Lambert formulation. Additional state variables include needle litter, dead fine roots, and soil-organic matter. The processes simulated are net canopy assimilation, maintenance respiration, growth respiration, carbon partitioning (growth), soil CO_2 evolution, decomposition, and mortality. Assimilation and maintenance respiration are simulated with an hourly time step and all other processes are on a daily time step. A detailed description of the model can be found in Cropper and Gholz (1993a).

Because reliable hourly meteorological data are not widely available for multiple years on slash pine sites, the SPM was modified to accept daily maximum and minimum air temperature and daily total incoming solar radiation. Hourly temperature was approximated using the following equation (Landsberg, 1986):

$$T(h) = Tmean + \frac{Tmax - Tmin}{2} \cdot \frac{\cos(2 \cdot \pi \cdot (h - hmax))}{24} \quad (1)$$

in which T(h) is the air temperature (°C) at hour h; Tmax is the daily maximum air temperature (°C); Tmin is the daily minimum air temperature (°C); Tmean is the mean of Tmin and Tmax; hmax is the hour when the maximum temperature occurs. A similar scheme (but with night photosynthetically active radiation (PAR) values set to zero) was used to simulate hourly PAR at the top of the canopy. Comparing this approximation to measured hourly meteorological data indicated that both temperature and light were overestimated. The daily data cannot reflect such transient events as the passage of clouds over the sun. In terms of simulated annual pine growth (Figure 20.1), the daily approximation was only a small overestimate, in part because the errors of overestimating temperature (increased maintenance respiration) were canceled by errors of overestimating light (increased assimilation).

Assumed climate change scenarios were applied to twenty-three years of historical climate data from north Florida (primarily from Gainesville, FL, 29° 44' N, 82° 9' W). The simulated temperature change was based on the United Kingdom Meteorological Office (UKMO) GCM of the atmosphere. It was assumed that solar radiation was not changed from the historical data. The UKMO temperature

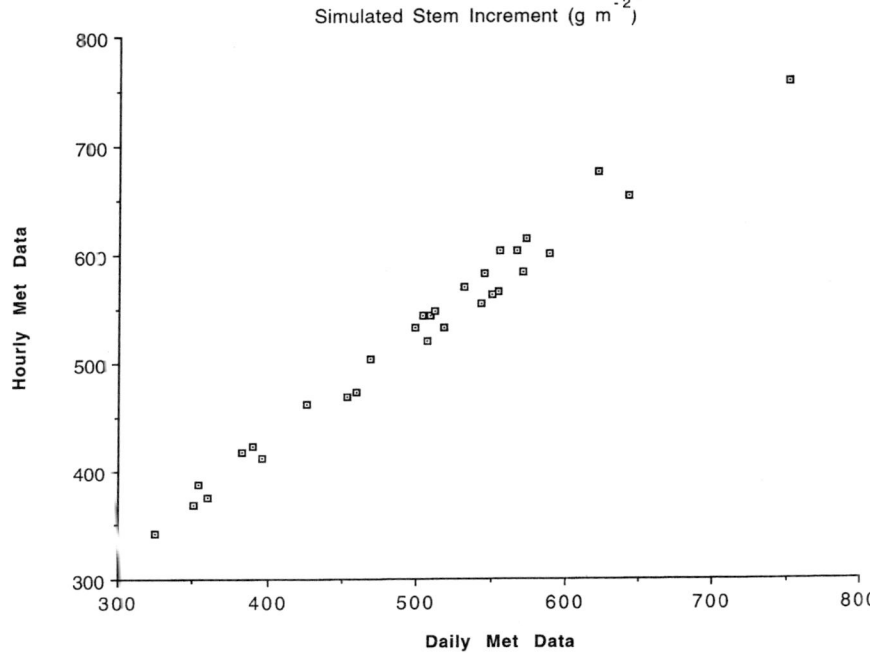

Figure 20.1. Simulated annual stem growth using hourly meteorological data or daily maximum and minimum temperatures and daily total incoming solar radiation.

results matched Gainesville long-term means adequately in the spring and summer, but underestimated the seasonality of the midpeninsula site (Table 20.1). This scenario was used as a guide to the potential sensitivity of slash pine to temperature increases, not as a prediction of future Gainesville temperatures.

To help characterize slash pine responses to increasing temperatures and ele-

Table 20.1. Monthly Mean Climate for Gainesville Florida

Month	Base historical mean	Base UKMO	Increase 2× CO_2 (UKMO)
1	12.9	22.2	4.41
2	13.7	21.9	4.47
3	17.0	21.1	5.19
4	20.1	21.0	5.17
5	23.4	21.9	5.12
6	26.6	24.1	5.02
7	27.4	26.6	5.02
8	27.4	27.6	5.11
9	26.1	27.1	4.69
10	22.0	25.3	4.16
11	17.4	23.8	3.96
12	14.3	22.9	4.27

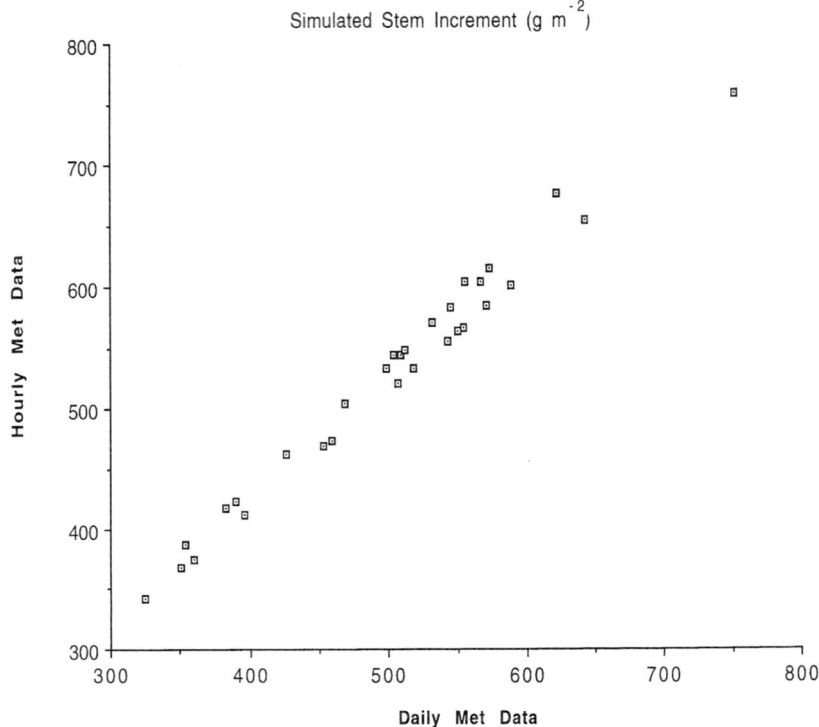

Figure 20.2. Location of ten FIA slash pine sites selected for simulation.

vated levels of CO_2 within the southeastern coastal plain, ten locations for additional analysis were chosen, ranging from latitudes of 28 to 33°N and longitudes of 82 to 89° W (Figure 20.2). Sites were chosen from the U.S. Department of Agriculture Forest Service Forest Inventory Assessment (FIA) slash pine plots. These FIA plots ranged in basal area from 18.25 m² to 31.72 m². Initial aboveground biomass for these plots was estimated by matching basal areas to previously measured plots in Florida (Gholz et al., 1991). Long-term daily climate for each of the ten plots was derived from the Vegetation/Ecosystem Model Analysis Project (VEMAP) database 0.5 degree latitude/longitude grid for the conterminous United States (Kittel et al., 1995).

As a measure of sensitivity of daily stem growth to elevated temperatures, the following equation was used (Luckyanov et al., 1995):

$$M = \frac{(\Sigma b_t \cdot p_t)^2}{(\Sigma b^2)_t \cdot (\Sigma p^2)_t} \quad (2)$$

in which b_t is the daily simulated stem growth (g m⁻²) on day t with the baseline unperturbed temperature; p_t is the daily simulated stem growth with an imposed

20. Modeling the Potential Sensitivity of Slash Pine Stem Growth

temperature increase. Values of M can range from 0 to 1 (no sensitivity), with the M = 0.95 threshold used to identify potentially significant changes in system response. This technique is particularly useful for identifying nonlinearities that are reflected in departures from the principle of superposition (Patten, 1975).

The slash pine assimilation was simulated with the following equation:

$$\text{ASSIM} = \frac{\alpha_p \cdot \text{PAR} \cdot k_c \cdot \text{Ci}}{\alpha_p \cdot \text{PAR} + k_c \cdot \text{Ci}} \quad (3)$$

in which PAR is the photon-flux density (μmol m^{-2} s^{-1}) of photosynthetically active radiation, α_p is the quantum efficiency (mol/mol), k_c is the mesophyll conductance (μmol m^{-2} s^{-1} μbar^{-1}), and Ci is the internal leaf CO_2 concentration (μbar). Additionally, simulation of the effects of elevated levels of CO_2 on slash pine photosynthesis is complicated by a lack of data.

I know of no published measurements of mature slash pine assimilation or transpiration under conditions of elevated CO_2, but data from another southern pine species, loblolly pine (*Pinus taeda* L.), are available (Tolley and Strain, 1984a, 1984b, 1985; Fetcher et al., 1988). Although the assimilation rate of loblolly pine seedlings generally increases with greater ambient CO_2 concentrations, the magnitude of observed assimilation increase varies from insignificant to a two-fold increase. Because the internal leaf concentration (Ci) increases linearly with ambient CO_2 concentration in C3 plants (Polley et al., 1993), the slash pine response to elevated levels of atmospheric CO_2 was simulated by increasing the Ci by values ranging from 25 to 100 ppm. A Ci increase of 100 ppm approximates the increase observed by Fetcher et al., (1988) in loblolly seedlings exposed to an ambient CO_2 concentration of 500 ppm when compared to control trees grown at 350 ppm CO_2.

Although slash pine assimilation has not been observed to decrease at temperatures above 30 °C (Teskey et al., 1994), conifer photosynthesis generally declines above optimum temperatures between 15 and 30 °C (Teskey et al., 1995). Sensitivity of assimilation to increasing temperature was simulated with a modified form of a temperature response function developed by O'Neill et al., (1972). With this function, slash pine assimilation is not decreased below the optimum temperature (Topt), but decreases monotonically between Topt and Tmax (Figure 20.3). Because no evidence of actual declines of slash pine assimilation at high temperatures was available, Topt is assumed to be 37 °C (the highest observed value in three years of hourly data collection) and Tmax is 50 °C. For sensitivity analysis, Topt values of 35 and 33 °C were selected.

Maintenance respiration is simulated using the following equation:

$$\text{MRESpi} = \text{BR}(i) \cdot X_i \cdot Q10^{\left(\frac{\text{TEMP} - 20}{10}\right)} \quad (4)$$

in which MRESPi is the maintenance respiration rate (g CO_2 g^{-1} tissue hr^{-1}) of pine component i; Q10 is the rate of change of respiration per 10 degree change in temperature; TEMP is air temperature (degrees C); Xi is the tissue mass (g m^{-2});

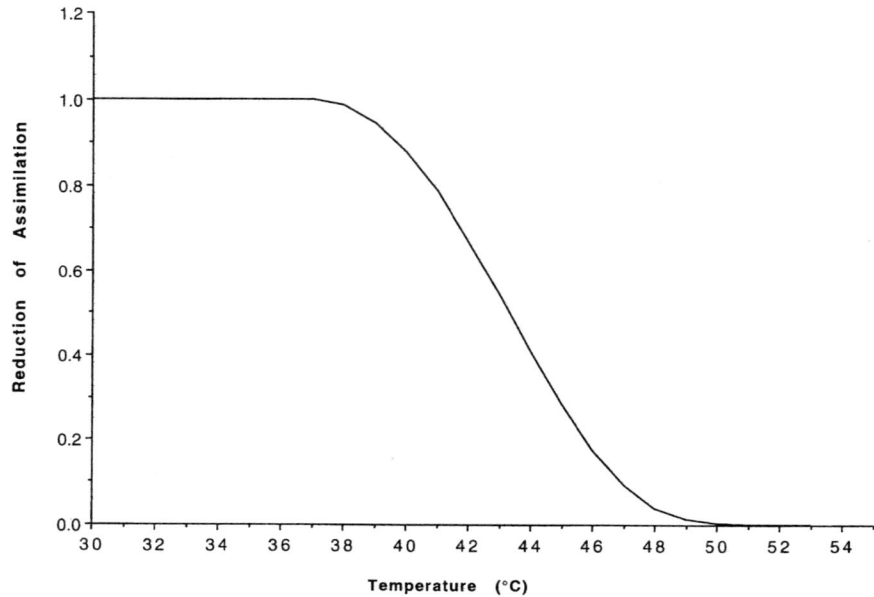

Figure 20.3. Response function for assimilation to increasing temperature. Topt is 37 °C.

and BR(i) is the respiration rate at 20 °C. The maintenance respiration rate of slash pine tissues approximately doubles with a 10 °C increase in temperature (Cropper and Gholz, 1991).

Results and Discussion

Global change associated with increasing levels of atmospheric CO_2 might affect the temperature, precipitation, relative humidity, and wind regimes experienced by most terrestrial plants. Increases in Florida's mean temperature by 2 °C, coupled with an annual precipitation change of ± 200 mm could shift Holdridge life zones 300 to 500 km north (Harris and Cropper, 1992).

Unfortunately, GCMs generally are not reliable prediction tools below the scale of continents (Schneider, 1993). Changes in regional mean precipitation, variability, and seasonal distribution are particularly difficult to accurately predict. Global average precipitation is expected to increase, but regional decreases are also expected (Schneider, 1993). Although it is clear that large changes in precipitation could eliminate the slash pine silviculture on Florida flatwoods sites, in this chapter it is assumed that the site-water balance remains within the historical variability.

Because slash pine assimilation and stomatal conductance is relatively insensitive to vapor pressure deficit and air temperature when mature trees are growing in wet Florida flatwoods sites (Teskey et al., 1994), the sensitivity analysis was limited to the effects of elevated levels of CO_2 on assimilation. Similarly, as a first-order sensitivity analysis, only the effects of increased temperature on main-

20. Modeling the Potential Sensitivity of Slash Pine Stem Growth   359

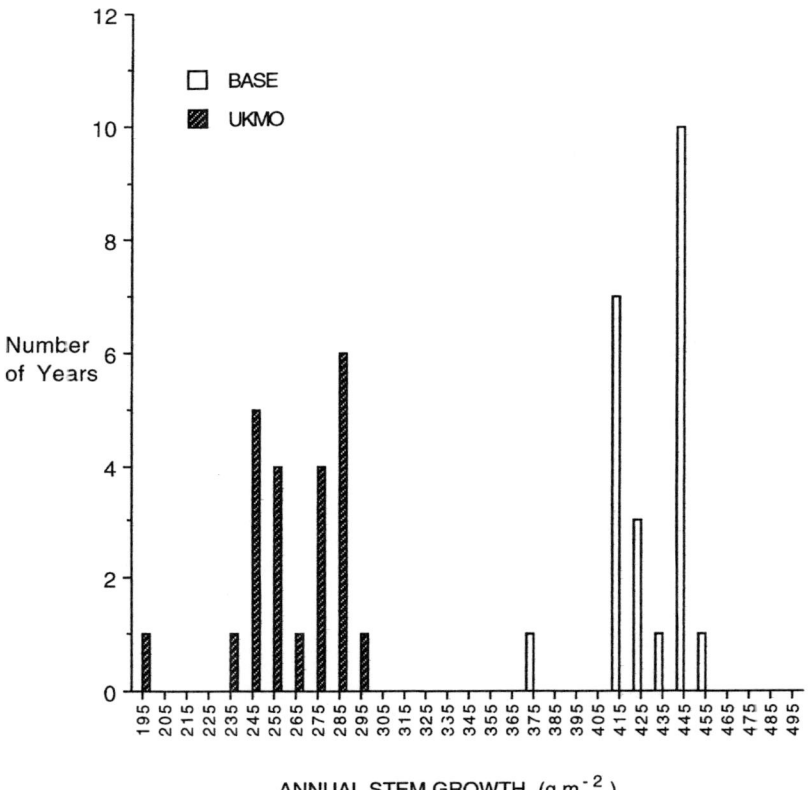

Figure 20.4. Frequency of simulated annual stem growth values with twenty-five years of daily historical weather data (base) and with the UKMO climate change scenario imposed on the historical data.

tenance respiration and assimilation rate are considered. Predicting the actual response of CO_2 exchange rates to climate change is complicated by potential interactions with nutrient availability, water stress, and effects of increased atmospheric CO_2 levels (Amthor, 1989; Ryan, 1991).

Historical weather data exhibit substantial variability, even without additional forcing from greenhouse CO_2. It is necessary to consider this variability when evaluating the potential sensitivity of biological systems to climate change. Threshold responses to extremes may be more important than responses to changes in means (Harris and Cropper, 1992). Twenty-three years of historical weather data were used in this research to evaluate the sensitivity of slash pine growth to increased temperatures.

Simulated mean annual stem growth was reduced from 430 g m^{-2} (historical weather data) to 262 g m^{-2} when the UKMO climate change scenario was applied to the same input weather data. All of the UKMO scenario years produced substantially smaller stem growth than any of the base climate years (Figure 20.4).

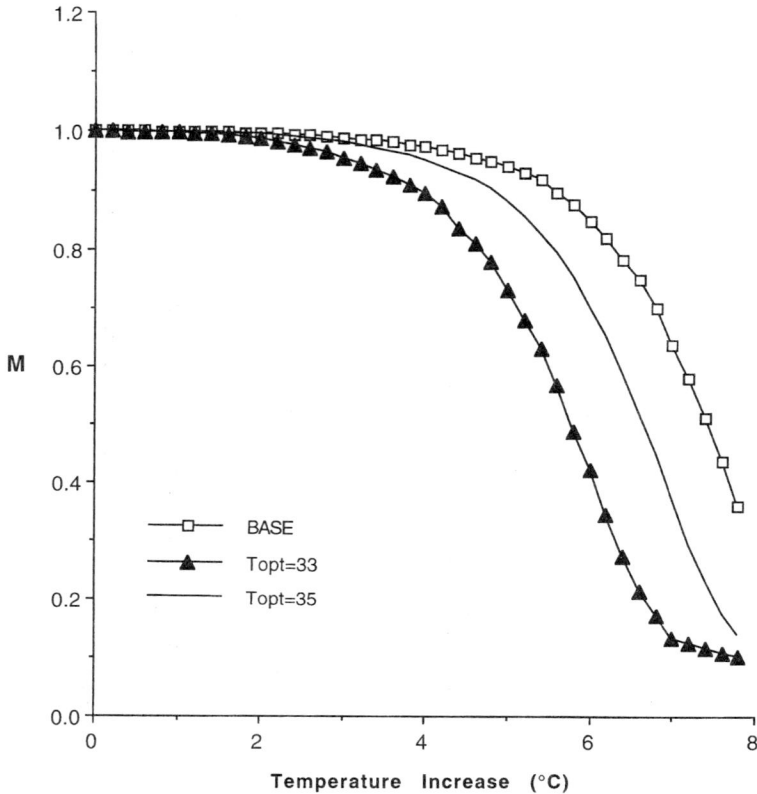

Figure 20.5. The sensitivity of simulated daily stem growth to mean temperature increases for different values of Topt. Base Topt is 37 °C.

The range of simulated annual growth values increased slightly from 83 g m^{-2} (372–455 g m^{-2}) in the base climate runs to 97 g m^{-2} (193–290 g m^{-2}) in the UKMO scenario. With the lower limit of the labile carbon pool set at 90 g m^{-2}, the range increased to 145 g m^{-2}. This increased variability implies less predictable slash pine growth rates, as well as lower growth rates. Clearly, present-day slash pine yield tables would be of little value under these circumstances.

As a test of the sensitivity of stem growth to the Topt for assimilation, the temperatures were elevated in steps of 0.2 °C from the base to a total increase of 8 °C. There was little sensitivity to temperature elevation for the base Topt (37 °C) or Topt values of 35 or 33 °C below a temperature elevation of 3 °C (Figure 20.5). Values of M decreased sharply for temperature elevations above 3 °C, indicating substantial sensitivity to these higher temperatures. As expected, the highest sensitivity to increasing temperature was associated with the lowest Topt (33 °C).

In a prior simulation study, Cropper and Gholz (1993a) identified the lower permissible limit of the simulated labile carbon (C) pool as potentially important

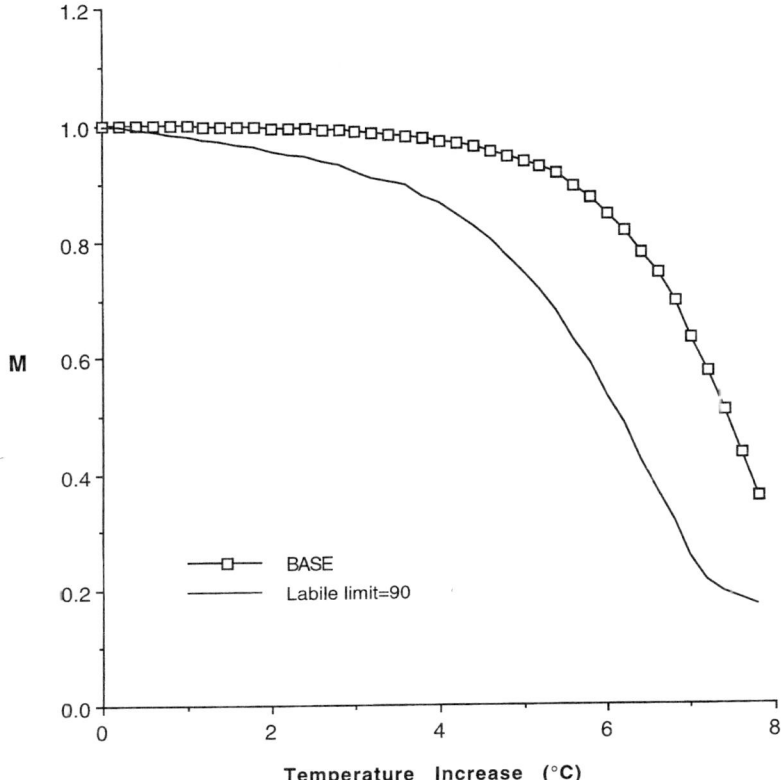

Figure 20.6. The sensitivity of simulated daily stem growth to mean temperature increases for different values of the lower limit for the labile C pool.

to the response of slash pines to temperature increase. When daily maintenance respiration is greater than daily net canopy assimilation, SPM simulated growth can only occur by drawing from the labile C pool (starch and soluble sugars). Field observations (Gholz and Cropper, 1991) indicate that the tissue sugar concentrations are relatively constant with the total labile C pool always at or above 90 g C m^{-2} (Cropper and Gholz, 1993a). Predictions of the effect of elevated temperatures on stem growth are very sensitive to the simulated labile C pool lower limit (Figure 20.6). Even the normal (and arbitrary) SPM lower limit of 35 g C m^{-2} can be reached with a UKMO based scenario. This may partially explain the increased variance of stem growth observed in the twenty-three-year simulations (Figure 20.4). If the lower labile C pool limit is raised to 90 g C m^{-2}, carbon allocation to wood is stopped sooner and stem growth is reduced. The magnitude of sensitivity to the lower limit of the labile C pool is similar to that of a Topt of 33 °C (Figure 20.7), however, the lower limit is associated with greater sensitivity at smaller temperature increases, whereas Topt is more important for temperature increases above 4.5 °C.

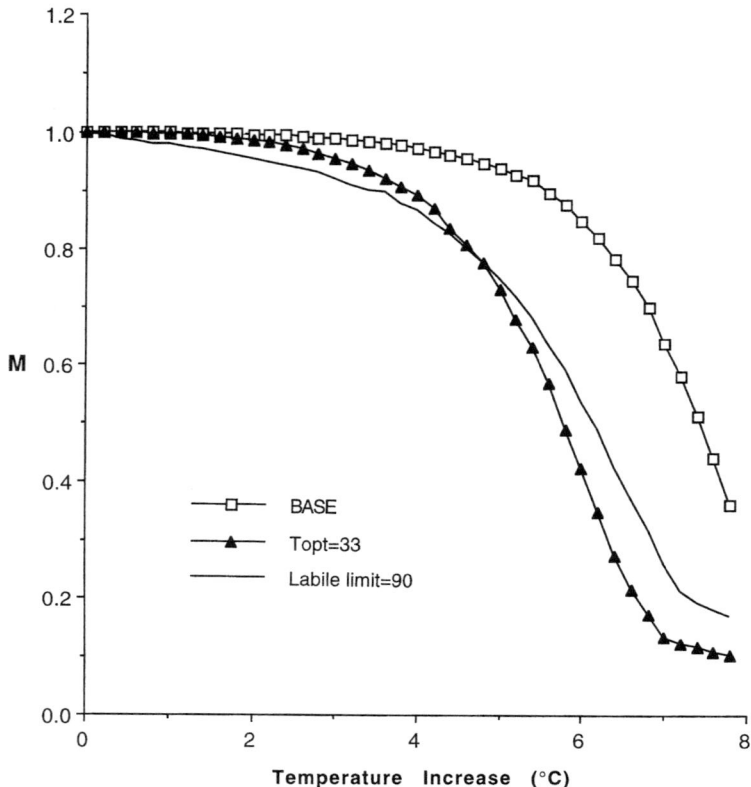

Figure 20.7. The sensitivity of simulated daily stem growth to mean temperature increases for Topt = 33 °C and the lower limit of the labile C pool = 90 g C m^{-2}.

Although there are substantial uncertainties concerning the magnitude of future temperature increases, there is little doubt that the atmospheric CO_2 concentration has risen steadily since the onset of widespread industrialization from about 275 ppm to the present-day level of approximately 360 ppm (MacDonald, 1990). With no increase in the atmospheric CO_2 concentration (and therefore the Ci), slash pine stem growth would be significantly reduced at temperature elevations similar to the UKMO scenario. If mature slash pines were as responsive as loblolly pine seedlings, Ci would increase by approximately 100 ppm when grown in an ambient concentration of 500 ppm (Fetcher et al., 1988). An elevation of this magnitude would increase stem growth above the present-day level (Figure 20.8), even for uniform temperature increases of 7 °C.

The pattern of SPM simulated response to both elevated temperature and levels of CO_2 among the ten FIA plots (Figure 20.9) was similar to the response of the north Florida research site. A uniform temperature increase of 4 °C reduced predicted annual stem growth from 574.3 ± 28.7 g m^{-2} (mean ± standard error)

20. Modeling the Potential Sensitivity of Slash Pine Stem Growth   363

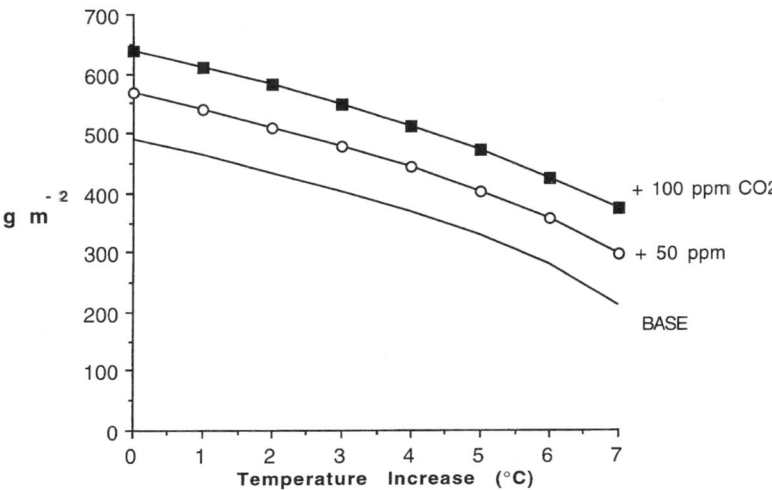

Figure 20.8. Simulated annual slash pine stem growth as a response to increased temperature and leaf internal CO_2 concentrations.

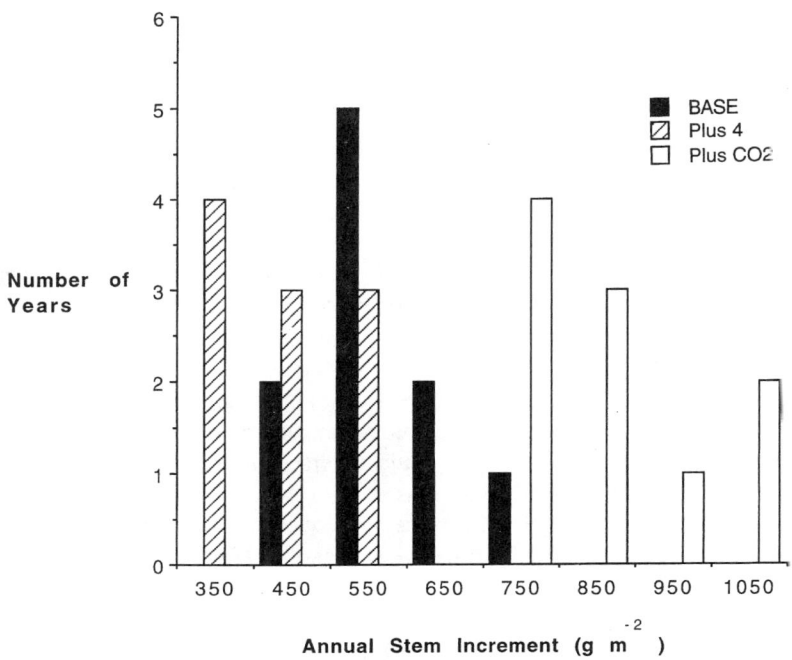

Figure 20.9. Frequency of simulated annual slash pine stem increment for ten FIA plots in the southeast coastal plain.

to 445.8 ± 25.8 g m^{-2}. Stem growth reduction ranged from 20 to 27% less than the base climate levels. Elevation of C_i by 100 ppm was responsible for an increase in simulated annual stem increment to 862.6 ± 39.5 m^{-2}. The combined effect of increasing temperature and CO_2 was very consistent across the ten plots, with net increases in stem growth ranging from 26 to 27%. More variation would be expected if a range of historical weather data was used to drive the model in place of the single VEMAP-derived daily weather series for each site.

There are major uncertainties concerning the degree of response of mature slash pine trees to CO_2 enrichment under field conditions. Florida slash pines stands are typically found on sandy, poorly drained, nutrient-poor soils. Phosphorous limitation is common on these sites (Gholz et al., 1985), and phosphorus fertilization at time of planting can increase the site index by as much as 4.5 m at twenty-five years (Jokela et al., 1991). The primary initial response of slash pine to fertilization is a large increase in leaf area (Gholz et al., 1991; Cropper and Gholz, 1994), but there is little evidence of changes in respiration rate (Cropper and Gholz, 1991) or assimilation rate (Teskey et al., 1994).

There has been little study of the effects of phosphorous limitation on the response of trees to elevated CO_2 levels, but Conroy et al., (1986, 1988) demonstrated that phosphorous limitation can significantly limit photosynthesis and growth gains in *Pinus radiata* D. Don seedlings. In a future of increased temperatures and elevated levels of atmospheric CO_2, widespread fertilization might be necessary to maintain or increase present-day growth rates.

Conclusions

Forest management requires a long-term perspective because of the relatively long life span of trees. During the next 100 years, we can expect rapid changes in climate associated with increasing levels of atmospheric CO_2. We should also recognize that changes in the frequency of storms, fires, and pathogen outbreaks could be significantly altered. Unfortunately, predictions of these potentially crucial changes cannot be made with any degree of certainty.

The uncertainties involved in predicting future climates complicate management decisions, but significant biological uncertainties also exist. Simulations by SPM indicate that increased temperatures would significantly reduce growth as a result of the greater carbon costs of maintenance respiration. Increased assimilation as a response to elevated CO_2 could reverse this trend if slash pines are responsive under typical field conditions.

Simulation models can only reflect the consequences of the assumptions built into them. Until additional experimental studies that are conducted under field conditions and include mature trees clarify how appropriate those assumptions are, it can only be concluded that slash pine stands are sensitive to projected climate changes. Because of this conclusion, plantation managers should begin to consider the economic consequences of increased fertilization requirements and to also consider alternative tree species or land-use strategies.

References

Ackerman TP, and Cropper WP Jr (1988) On scaling global climate projections to local biological assessments. Environ 30(5):31–34.

Amthor J.S(1989) *Respiration and crop productivity.* Springer-Verlag, New York.

Conroy JP, Smillie RM, Kuppers B, Bevage DI, Barlow EW(1986) Chlorophyll a fluorescence and photosynthetic and growth responses of *Pinus radiata* to phosphorous deficiency, drought stress, and high CO_2. Plant Physiol 81:423–429.

Conroy JP, Kuppers B, Virgona J, Barlow EWR(1988) The influence of CO_2 enrichment, phosphorous deficiency and water stress on growth, conductance and water use of *Pinus radiata* D. Don. Plant Cell Environ 11:91–98.

Cropper WP Jr, Gholz HL(1991) In situ needle and fine root respiration in mature slash pine trees. Can J For Res 21:1589–1595.

Cropper WP Jr, Gholz HL(1993a) Simulation of the carbon dynamics of a Florida slash pine plantation. Ecol Model 66:213–249.

Cropper WP Jr, Gholz HL(1993b) Constructing a seasonal carbon balance for a forest ecosystem. Clim Res 3:7–12.

Cropper WP Jr, Gholz HL(1994) Evaluating potential response mechanisms of a forest stand to fertilization and night temperature: A case study using *Pinus elliottii*. Ecol Bull 43:154–160.

Dickinson RE(1989) Uncertainties of estimates of climatic change: A review. Clim Change 15:5–13.

Fetcher N, Jaeger CH, Strain BR, Sionit N(1988) Long-term elevation of atmospheric CO_2 concentration and the carbon exchange rates of saplings of *Pinus taeda* L. and *Liquidambar styraciflua* L. Tree Physiol 4:255–262.

Gholz HL, Fisher RF, Prichett WL(1985) Nutrient dynamics in slash pine plantation ecosystems. Ecol 66:647–659.

Gholz HL, Vogel SA, Cropper WP Jr, McKelvey K, Ewel KC, Teskey RO, Curran PJ(1991) Dynamics of canopy structure and light interception in *Pinus elliottii* stands of north Florida. Ecol Mon 61:33–51.

Gholz HL, Cropper WP Jr(1991) Carbohydrate dynamics in mature *Pinus elliottii* var *elliottii* trees. Can J For Res 21:1742.

Harris LD, Cropper WP Jr(1992) Between the devil and the deep blue sea: Implications of climate change for Florida's fauna. In Peters RL, Lovejoy TE (Eds). *Global warming and biological diversity.* Yale University Press, New Haven, CT.

Jokela EJ, Allen HL, Mcfee WW(1991) Fertilization of southern pines at establishment. In Duryea ML, Dougherty PM (Eds). *Forest regeneration manual.* Kluwer Academic Publishers. Dordrecht.

Kittel TGF, Rosenbloom NA, Painter TH, Schimel DS, VEMAP Modeling Participants (1995) The VEMAP integrated database for modeling United States ecosystem/vegetation sensitivity for climate change. J Biogeogr 22:857–862.

Landsberg JJ(1986) *Physiological ecology of forest production.* Academic Press, New York.

Luckyanov NK, Cropper WP Jr, Harwell MA(1995) State Analyses of ecological models: Model reaction to parameter change. Ecol Model 82(1):99–104.

MacDonald GJ(1990) Global climate change. In MacDonald GJ, Sertorio L (Eds). *Global climate and ecosystem change.* Plenum Press, New York.

O'Neill RV, Goldstein RA, Shugart HH, Mankin JB(1972) Terrestrial ecosystem energy model. US IBP Eastern Deciduous Forest Biome Memo Report Number 72–19. Oak Ridge National Laboratory, Oak Ridge, TN.

Patten BC(1975) Ecosystem linearization: An evolutionary design problem. In Levin SA (Ed) *Ecosystem analysis and prediction.* Society for Industrial and Applied Mathematics, Philadelphia, PA.

Polley HW, Johnson HB, Marino BD, Mayeux HS(. 1993) Increase in C3 plant wateruse efficiency and biomass over glacial to present CO_2 concentrations. Nature 361:61–64.

Ryan MG (1991) Effects of climate change on plant respiration. Ecol App 1:157–167.

Schneider SH (1993) Scenarios of global warming. In Kareiva PM, Kingsolver JG, Huey RB (Eds) *Biotic interactions and global change.* Sinauer Associates, Inc., Sunderland, MA.

Schneider SH, Mearns L, Gleik PH (1991) Climate-change scenarios for impact assessment. In Peters RL, Lovejoy TE (Eds) *Global warming and biological diversity.* Yale University Press, New Haven, CT.

Teskey RO, Gholz HL, Cropper WP Jr (1994) Influence of climate and nutrient availability on net photosynthesis of mature slash pine. Tree Physiol 14:1215–1227.

Teskey RO, Sheriff DW, Hollinger DY, Thomas RB (1995) External and internal factors regulating photosynthesis. In Smith WK, Hinckley TM (Eds) *Resource physiology of conifers: Acquisition, allocation, and utilization.* Academic Press, San Diego.

Tolley LC, Strain BR (1984a) Effects of CO_2 enrichment on growth of *Liquidambar styraciflua* and *Pinus taeda* seedlings under different irradiance levels. Can J For Res 14:343–350.

Tolley LC, Strain BR (1984b) Effects of CO_2 enrichment and water stress on growth of *Liquidambar styraciflua* and *Pinus taeda* seedlings. Can J Bot 6:2135–2139.

Tolley LC, Strain BR (1985) Effects of CO_2 enrichment and water stress on gas exchange of *Liquidambar styraciflua* and *Pinus taeda* seedlings grown under different irradiance levels. Oecologia 65:166–172.

21. An Index for Assessing Climate Change and Elevated Carbon Dioxide Effects on Loblolly Pine Productivity

David Arthur Sampson, H. Lee Allen, and Philip M. Dougherty

Loblolly pine (*Pinus taeda* L.) forests represent the major forest type in the southern United States. The loblolly pine region extends from Delaware and central Maryland south to central Florida and west to eastern Oklahoma and Texas (Fowells, 1965). The wide range of loblolly pine largely results from its rapid growth and its successful adaptation to many varieties of soil types and environmental conditions. These and other factors have made loblolly pine an important commercial species in the region. However, although loblolly pine occurs on a many types of sites, its commercial value, as measured by net primary productivity (NPP), varies tremendously and is strongly determined by variability in the local climate and stand and site conditions (McNulty et al., 1997). Uncertainty regarding potential changes in climate as a result of increasing atmospheric carbon dioxide (CO_2) concentration has caused concern for the future commercial viability of loblolly pine forests.

General circulation models (GCMs) project that atmospheric CO_2 concentrations will double near the middle of the next century, with significant concomitant changes in climate predicted (Houghton et al., 1990). These GCMs predict that temperatures may increase by 4 °C, to as much as 8 °C or more, and that precipitation may increase or decrease under a doubling of CO_2 for the geographic range presently occupied by the loblolly pine forest type (Cooter et al., 1993). Process models sensitive to climate and CO_2 concentration on carbon gain, loss, and allocation can be used to assess the potential impact of potential global climate change on productivity and yield of loblolly forests.

Such process models as the Forest-Biogeochemical Cycles Model (FOREST-BGC) (Running and Coughlan, 1988), the Canadian Climate–Vegetation Model (CCVM) (Lenihan and Neilson, 1993), and the Terrestrial Ecosystem Model (TEM) (Raich et al., 1991) have been used to assess ecosystem structure, function, and productivity. For example, CCVM has been used to examine potential climate change effects on forest species distribution (Lenihan and Neilson, 1995; Smith et al., 1992); FOREST-BGC and TEM have been used to assess present and potential climate change effects on net primary production (NPP) (Raich et al., 1991; Running and Nemani, 1991; Melillo et al., 1993). However, these simulations have generally been conducted at continental to global scales, with broad generalizations required for species distributions and site parameterization. TEM has been used to access the present and potential change in NPP for temperate mixed forests that geographically correspond to the loblolly pine forest type of the southern United States (Melillo et al., 1993; McGuire et al., 1992). Because TEM was not species-specific to loblolly pine, such important driver variables as leaf area index (LAI), which varies two-fold yearly in loblolly pine, may not be adequately characterized in the NPP simulations. Stand-level process models that incorporate species-specific physiological controls over NPP may provide information unavailable in more generic, larger scale models (Malanson, 1993).

Methods

Model Overview

A modified version of the process model BIOMASS version 12.0 was used that had been adapted for loblolly pine stands of southern United States. Originally developed for Monterey pine (*Pinus radiata* D. Don) stands, BIOMASS used data from the Biology of Forest Growth experiments conducted in Australia (McMurtrie and Landsberg, 1992; Benson et al., 1992). There are strong similarities between Monterey and loblolly pine with respect to physiological controls and response to nutrient amendments that made the modification feasible. Additionally, and perhaps most notably, both species exhibit a strong seasonal change in LAI. However, some changes to BIOMASS version 12.0 were necessary to adapt it for use in loblolly pine because BIOMASS version 12.0 partitions daily net available carbon on a monthly cycle, with the balance of negative and positive carbon days met prior to monthly allocation to growth. This approach bypassed the need for such processes as labile carbon storage and removal from storage. Additionally, growth was distributed throughout the twelve month calendar year based on fixed carbon partitioning coefficients. Loblolly pine, in comparison, exhibits discrete seasonal aboveground growth periods that last from seven to ten months, with monthly estimates of production thought to be too coarse in resolution for examining loblolly pine carbon budgets. Moreover, preliminary hypotheses suggested that there may be periods in which combined growth and maintenance costs depend, in part, on stored carbohydrates to meet daily carbon demands. Carbon storage as such was not present in BIOMASS version 12.0.

BIOMASS version 13.0 represents a different approach to modelling forest productivity and for making stand-level estimates of NPP. BIOMASS version 13.0 differs from other models (e.g., BGC and TEM) because it uses empirical estimates of LAI directly in the model and it contains dynamic, daily, time step carbon partitioning and labile carbon storage algorithms that balance on a seasonal schedule. The daily budgeting of carbon enables a more suitable accounting of the seasonal pattern in carbon supply and use. Therefore, a more tightly coupled analyses of the role of woody tissue biomass accretion and its associated construction respiration (Rc) and maintenance respiration (Rm) costs on yearly carbon dynamics of loblolly pine forests could be made. Tissue Rm represents a fundamentally important cost in the daily and seasonal carbon budget of loblolly pine stands (Kinerson 1975, 1977). Present estimates of annual Rm costs are 50% or more of the annual gross primary productivity (GPP) (Kinerson, 1975). Because Rm increases exponentially with temperature, and because ambient temperatures are predicted to increase under a doubling of CO_2, accurate woody tissue biomass accretion and accurate representation of the yearly dynamics in LAI are of particular importance to modelling forest production in a 2× CO_2 environment.

Daily partitioning of carbon required the inclusion of component relative growth rates (RGR), and labile carbon storage. Daily RGR for foliage, stem, and branch components are approximated from the first-derivative solution of a closed-formed logistic equation fit to normalized empirical observations of bimonthly tissue growth phenology. Photoperiod and a soil-heat index initiates the tissue growth phenologies. Labile carbon storage and removal follows a fixed hierarchy in which: foliage > root > branches > stems, with the maximum allowable storage determined from present standing mass of each component. Although the dynamics of the labile carbon pool estimates are modelled, and therefore are not based on empirical data, the seasonal trend in the total pool and the magnitude in each pool at any given time, are comparable with published data.

Carbon partitioning may include 1) allocation to storage, 2) removal from storage, 3) partitioning to component tissues for growth, 4) allocation to a "transparent" belowground carbon pool, or 5) any combination thereof. The partitioning coefficients are estimated from stand age, live-crown length, and foliar nitrogen concentration based on empirical data. The belowground pool serves as a surrogate to coarse and fine root production, Rm, Rc, and turnover. Transparency of the belowground pool, as used here, implies no interaction of carbon with other carbon sources or sinks following allocation to this pool. A lack of information on coarse and fine root production, standing mass, and mortality, especially on a site-to-site basis, continues to be a common weakness in modelling carbon budgets of forest systems. Our simulations used a constant fine root biomass with initial and ending fine root mass equivalent to initial standing foliage mass. Limited fine root data for slash pine (*Pinus elliottii* Engelm.) (Gholz et al., 1986), and loblolly pine (Mignano, 1995) supports this approximation.

Model Parameterization and Validation

BIOMASS version 13.0 was parameterized using twelve loblolly pine stands from across the southeastern United States representing a broad range in LAI, climatic regimes, soils, and nutritional status. Leaf area index was estimated at a prior time using needle litterfall techniques (Vose and Allen, 1988). Monthly estimates of LAI were determined from a program that models foliage cohort accretion and senescence. We validated BIOMASS version 13.0 using twelve test stands also from a range in stand structure, soil types, and climatic regimes. The performance of BIOMASS as a stand model has been thoroughly reviewed (McMurtrie and Wang, 1993; McMurtrie et al., 1992; McMurtrie and Landsberg, 1992).

Stemwood Yield Index

A yield index was developed to evaluate the relationship between loblolly pine stemwood yield and simulated NPP. We used empirically measured annual yield data for thirty-eight intermediate-age loblolly pine stands from across the Southeast and the corresponding simulated NPP for each stand (Table 21.1). The empirical data were from a series of long-term regional studies established to examine production response in midrotation loblolly pine stands to fertilization (NCSFNC, 1993). Annual or biannual measurements of diameter and height growth, and litterfall permitted fairly accurate estimates of stemwood production using published equation (Shelton et al., 1984) and LAI using litterfall techniques (Vose and Allen, 1988). Direct measures of LAI for each of the thirty-eight stands minimized uncertainty in the NPP estimates. Needle litterfall estimates of LAI are very expensive and labor intensive, and require two years of data collection for a one-year estimate. As such, needle litterfall estimates for the thirty-eight stands were only available for the 1988 growth year. Therefore, the yield index represents a one-year evaluation of the stemwood yield/NPP relationship.

Table 21.1. Stand Characteristics of the Thirty-Eight Intermediate-Age Loblolly Pine Stands Used in the Development of the Stemwood Yield Index

Stand Attribute	Units	Mean* and (range)
Initial standing mass in:		
Foliage and fine roots	Mg Biomass ha^{-1}	2.70(1.92–3.72)
Stems	Mg Biomass ha^{-1}	54.5(32.1–90.9)
Branches	Mg Biomass ha^{-1}	8.9(6.3–14.0)
Stand density	Trees ha^{-1}	1387 (864–1870)
Average tree diameter	cm	15.4(12.4–20.0)
Peak LAI (projected)	m^2 m^{-2}	2.4(1.8–3.15)
Canopy height	m	12.2(9.2–17.0)
Age	Years	14.8(11–18)
Available soil-water in first profile	mm	53.1(35.1–76.4)
Available soil-water in total profile	mm	177(117–203)
Foliar nitrogen	%	1.18(1.04–1.91)

[1] Foliar nitrogen concentration represents the median value for the thirty stand data set.

Simple linear regression was used to evaluate the "goodness-of-fit" in the relationship between stemwood yield and simulated NPP. The slope parameter in the regression model provided an estimate of the average proportion of NPP allocated to stemwood production. Climate data was used from meteorological stations closest to the location of each stand for the stemwood yield/NPP simulations. These data were acquired from the National Climatic Data Center (NCDC).

Response to Climate Change Index

Overview

Historical climate data and climate projections from four general circulation models were used for a doubling of atmospheric CO_2 concentrations to simulate NPP for a matrix of $1° \times 1°$ rectangular grids across the southeastern United States. We used standardized stand-structure and soil-site parameters because grid-level information required to run the model was unavailable. Therefore, these simulations represent an index of NPP response to a $2\times CO_2$ environment rather than an estimate of present and projected future NPP in the loblolly forest type.

A sensitivity analysis determined that the NPP projections were more sensitive to stand and site conditions assumed, as well as to the interannual variability in climate, than to adjacent $1° \times 1°$ rectangular grid differences in projected climate. Therefore our analyses were restricted to a subset of the total number of grids available.

Geographic Considerations

Eighteen $1° \times 1°$ grid cells were chosen in a systematic matrix that represents the latitudinal and longitudinal extent of loblolly pine plantations across the Southeast. These $1° \times 1°$ cells cover the topographic, edaphic, and climatological gradients that exist across the loblolly pine forest type. Specifically, cells were chosen to represent lower, middle, and upper coastal and piedmont regions of the loblolly pine forest type.

Climate Considerations

The eighteen-cell subset is from a daily precipitation database initially consisting of 766 cooperative observation locations east of the Rocky Mountains in the United States and Canada developed by Richman and Lamb (1985, 1987). Station locations were selected for completeness of record and their locations relative to the $1° \times 1°$ rectangular grid. The precipitation database is more fully described in Richman and Lamb (1985, 1987). These studies summarize initial quality control procedures, as well as documenting research analyses of precipitation event totals and distributions that contributed to quality assurance for the database. Similar procedures have been followed in the development of a comparable daily maximum and minimum temperature database for the period 1949 to 1993.

Widely available GCM output is usually not suitable for model application on the temporal and spatial scales proposed for our study, therefore climate scenarios must be constructed. Such scenarios should offer possible sets of future climate conditions, should be internally consistent and based on sound scientific reasoning, but have no specific probability of occurrence. Several general approaches to simulating smaller-scale or regional climate change scenarios are discussed in Giorgi and Mearns (1991), and Carter et al., (1992).

One approach that has been used in many national and regional assessments was employed in Smith and Tirpak (1990) and Rosenzweig et al., (1993). It consists of modifying an existing historical time-series variable by adding the difference between GCM estimated values for $2\times CO_2$ and GCM estimates for present $1\times CO_2$, or multiplying the original value by the ratio of $2\times CO_2$ to $1\times CO_2$. In general, the GCM difference is applied to temperature variables with the ratio approach then applied to daily precipitation values. The ratio approach, as it has been applied to daily precipitation has, in particular, come under repeated criticism as producing physically unjustified scenarios (no change in storm frequency and proportionately larger increases in heavy rain events). Recent studies of regional and national climatic time series suggest that this may not be unreasonable given that a similar trend may already be present in the recent climate record (Karl et al., 1995).

Output from four GCMs was obtained from the National Center for Atmospheric Research (NCAR) archives. They are the Geophysical Fluid Dynamics Laboratory of the National Oceanic and Atmospheric Administration-NOAA (GFDL), Goddard Institute for Space Studies, National Aeronautics and Space Administration-NASA (GISS), Oregon State University (OSU), and United Kingdom Meteorological Office (UKMO) models (Table 21.2). The grid-scale of each of these models is different, therefore, to facilitate later analysis, the long-term mean monthly temperature and precipitation differences and ratios discussed earlier have been interpolated to a $1° \times 1°$ rectangular grid common with the historical Richman-Lamb data base. Our initial scenario, which assumes that equal day and nighttime warming occur as estimated by the selected GCMs, was applied to the historical forty-year time series of daily maximum and minimum temperature and precipitation data.

We used the forty-year historical data record to conduct base NPP simulations for each of the eighteen $1° \times 1°$ grid cells examined using BIOMASS version 13.0. Then the corresponding forty-year future climate projections were used for a "$2\times CO_2$ climate" for each grid cell for the four GCMs to predict future NPP. Daily minimum and maximum temperatures, in addition to daily precipitation were the climatic elements required as inputs into the model from these data sets. Such important physiological driver variables as shortwave radiation, vapor-pressure deficit, and absolute humidity are estimated in the model using published algorithms present in BIOMASS version 12.0 (McMurtrie and Landsberg, 1992).

For the sensitivity analysis, a "typical" year of weather observations was selected from a cell of typical climate. The typical climate cell was determined by graphical analyses of the eighteen cell climate data and the forty-year historical

Table 21.2. Attributes of the Four GCMs Used in This Study[1]

Model	GFDL	GISS	OSU	UKMO
Date of output generation	1988 (Q-flux)	1982 (Q-flux)	1984–1985	1986
Numerical solution technique	Spectral (R15)	Finite difference	Finite difference	Finite difference
Horizontal resolution (latitude × longitude)	4.5° × 7.5°	7.8° × 10.0°	4.0° × 5.0°	5.0° × 7.5°
Vertical resolution	9	9	2	11
Surface characterization	Uniform	Fractional grid	Uniform	Uniform
Convective parameterization	Moist adiabatic	Penetrating convection	Penetrating convection	Penetrating convection
Initial CO_2 concentration (ppm)	300	315	326	320

[1] Adapted from Cooter et al. (1993).

climate record. These analyses produced a temperature-precipitation domain for the loblolly pine forest type. The typical climate cell was chosen as that grid cell whose forty-year climate observations fell within the center of this domain. This cell corresponds to latitude 34.5° N and longitude 88.5° W, the approximate center of the southern forest region. Average climate could not be used because, even though average daily temperatures can be computed, average daily precipitation has little meaning. Additionally, the averaging process does not take into account natural patterns of autocorrelation and multicorrelation among temperature and precipitation observations. Such patterns can have important implications for productivity simulations.

The typical year was determined in the followng manner. First, annual means of temperature and precipitation were computed for each of the forty-years. Only those years that fell within the central one-half of the distribution were retained. Next, the process was repeated for seasonal means of this subset of years. The goal was to select a year whose annual and seasonal means/totals fell within the central portion of the sample distribution. For the observation set representing the one degree grid cell with centroid located at 32.5° N latitude and 88.5° W longitude, the year that best met these criterion was 1964. Thus, the typical year is represented by observations recorded in 1964.

Stand Structure/Site Considerations

BIOMASS is a stand-level process model. We used a standardized stand-input data set to enable examination of the main effects of climate and increased CO_2 on NPP for the loblolly pine region. Therefore simulations were conducted for two defined stand structure and site conditions, which were 1) low LAI and poor soil-water-holding capacity, and 2) high LAI and good soil-water-holding capacity conditions (Table 21.3). Both stands were similar with respect to initial standing biomass in stems and branches, tree height, canopy depth, stand density, and component tissue nitrogen concentrations (except foliage).

The two stand and site conditions provided an estimate of the potential minimum and maximum response for loblolly pine attainable given the restrictions set by the stand structure used. Consequently, we were able to generate productivity projections even though we lacked cell-specific stand and site data for all required driving variables. There are presently no data sets available to conduct site-specific simulations at the regional level using this model. Our approach permits an evaluation of the direct effects of climate and CO_2 concentration on NPP and NPP response; local and regional differences in stand structure and site conditions strongly influence NPP simulations. As such, simulations represent an index of change rather than an estimate of present-day productivity and potential future production response to a doubling of atmospheric CO_2.

There were several assumptions necessary for this approach. It was assumed that both stand structure and site conditions exist, or could exist, for each cell examined. Nitrogen availability and uptake predominantly determines the maximum LAI obtainable in most loblolly stands. The occurrence and relative

Table 21.3. Characteristics of the Model Loblolly Pine Stand Used in the Simulations for the Low LAI, Poor Soil-Water-Holding Capacity and the High LAI, Good Soil-Water-Holding Capacity Conditions Used in These Analyses

		Stand/Site Condition	
Parameter	Units	Low LAI, poor soil-water-holding capacity	High LAI, good soil-water-holding capacity
Initial standing mass	Mg Biomass ha^{-1}		
Foliage		1.65	3.72
Stems		56.3	56.3
Branches		9.03	9.03
Roots		1.65	3.72
Projected LAI	m^2 m^{-2}		
Minimum		0.78	1.60
Maximum		1.58	3.15
Foliar N concentration	%	1.08	1.54
Average tree height	m	12.3	12.3
Soil profile depth	m	0.3	2.0
Available water storage in total profile	mm	56	374

distribution of low and high LAI stands in any cell would depend on N availability, the soil-water relations, and forest management practices (operational fertilization, thinning, and so forth) specific to each cell. Additionally, the occurrence of good and poor soil-water-holding capacity situations may or may not be found in all cells examined.

Simulation Outputs

Sensitivity to ambient and twice-ambient CO_2 concentrations required the use of the Farquhar et al., (1980) biochemical photosynthesis model found in BIO-MASS. We assumed that the nitrogen effects incorporated into this algorithm were suitable for loblolly pine simulations.

Annual NPP was simulated for each of the eighteeen grid cells for the forty-year historical climate record and the four GCM modified forty-year future climate projections. The annual predictions were used to calculate contemporary and future average annual NPP for each cell as:

NPP contemporary or future = (i = 1,40) [Σ Annual NPP/40]

and the average percent change in annual NPP, or NPP response, as:

Average NPP response = ((NPP future − NPP contemporary) / NPP contemporary) × 100

Although we present average NPP estimates from across the Southeast, actual estimates of NPP for these regions would depend on the average contemporary stand and site conditions that exist in each cell. Therefore, we focus on the NPP response for the forty-year GCM climate projections and consider these predictions to be an index of change rather than an absolute change. Additionally, the NPP response data from the eighteen grids were smoothed using a cell-search algorithm (Anonymous, 1995). This enabled a region-wide presentation of the NPP and NPP response simulations.

Sensitivity Analyses of Environmental Variables

Interpretation of the simulation output was aided by conducting a sensitivity analysis of the driving environmental variables predicted to change in a 2× CO_2 environment. The three primary environmental variables examined here were CO_2 concentration, temperature, and precipitation. Such important effects as cloud cover and the accompanying changes in solar radiation were reflected in the temperature data.

Stand structure and site condition were held constant as we modified CO_2 concentration, temperature, and precipitation separately. We used the typical climate year for the typical climate cell for the sensitivity comparisons. First, we used ambient (350 ppm)CO_2, 1.5× ambient, and 2× ambient CO_2 concentrations to simulate NPP using the two defined stand and site conditions to examine the CO_2 affect in the model. Second, we increased daily minimum and maximum temperatures by 3 and 6 °C to examine the effect of a constant temperature increase on NPP response. Finally, for days exhibiting precipitation in the typical climate year, we increased and decreased precipitation by 25 and 75% of one standard deviation of the forty-year average precipitation for that day to examine the sensitivity of precipitation in the model in relation to the defined stand structure and soil-water-holding capacity conditions.

Results

Empirical Yield Index

Simulations of NPP for the index year ranged from 4.16 to 10.14 Mg C ha^{-1} year^{-1} for intermediate-age loblolly pine stands from across the southern United States (Figure 21.1). The corresponding empirical estimates of stemwood production ranged from 2.35 to 5.30 Mg C ha^{-1} year^{-1}, and averaged 3.31 Mg C ha^{-1} year^{-1}. Approximately 47% of NPP was used in stemwood production (Figure 21.1). Proportional carbon allocated to stemwood increment, however, ranged from roughly 33 to 63% of NPP for the stands examined here, with 53% of the variation in stemwood production explained by NPP.

Regional Trends in Contemporary Net Primary Production

Contemporary net primary productivity ranged from 2.6 to over 8.0 Mg C ha^{-1} year^{-1} for both stand conditions across all regions examined. High LAI, good soil-water-holding capacity conditions resulted in 1.5 to 2 times the productivity

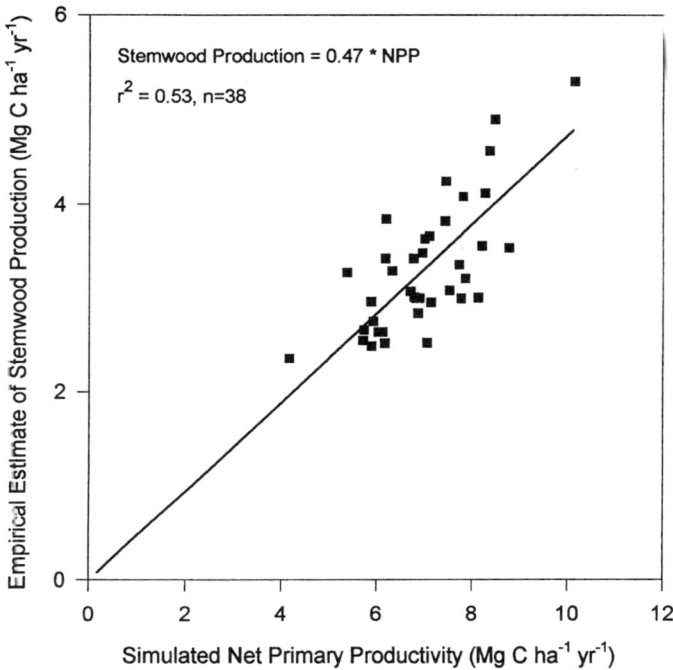

Figure 21.1. The relationship between simulated NPP from the process model BIOMASS version 13.0 and empirical estimates of stemwood production. Data are from thirty-eight intermediate-age loblolly pine stands from across the southeastern United States.

of low LAI, poor soil-water-holding capacity simulations (Figure 21.2a, b). These results indicate that, for stands with similar attributes to those used in this study, regions along the east coast of North and South Carolina have the highest NPP potential while the westernmost regions in eastern Oklahoma and Texas exhibit the lowest NPP potential.

Regional trends in contemporary NPP between the two stand/site conditions were found, with NPP varying by latitude and longitude, and proximity to the Atlantic and Gulf coasts. In general, NPP was high along both coasts, with decreased productivity observed toward interior regions (Figure 21.2a, b). Additionally, NPP generally decreased with increased longitude; Western and north Western regions had the lowest productivity.

Mean daily growing season precipitation and temperature varied considerably across the loblolly pine forest type. Precipitation varied two-fold, ranging from 2.5 mm rainfall per day (day^{-1}) in Virginia and eastern Texas to 5 mm rainfall day^{-1} along coastal Louisiana (Figure 21.2a). Mean daily growing season temperature increased by 50% from Virginia to Florida, ranging from roughly 16.5 to 24.5 °C (Figure 21.2b). Precipitation and temperature both generally increased with decreasing latitude.

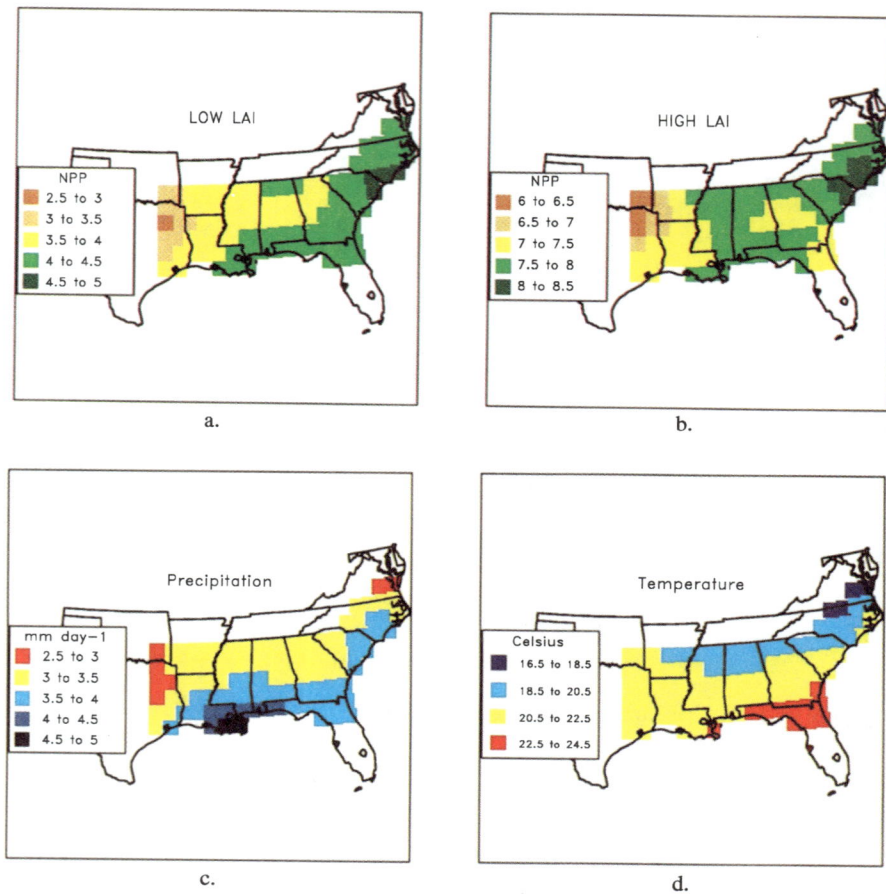

Figure 21.2. Simulations of loblolly pine contemporary NPP for the southeastern United States for two stand and site conditions and the corresponding climate. Data represent an estimate of forty-year average (1949 to 1988) NPP for a) low LAI, poor soil-water-holding capacity simulations, and b) high LAI, good water-holding capacity simulations. Climatic variables are c) mean daily growing season precipitation, and d) mean daily growing season temperature.

The regional trends in contemporary NPP were coupled to the temperature and precipitation patterns. For example, moderately high to high productivity regions corresponded to moderate to high precipitation and moderate to low temperatures (Figures 21.2a–d). This was especially true for low LAI, poor soil-water-holding capacity simulations (Figures 21.2a, c). The low productivity areas for both low and high LAI, and the accompanying soil-water conditions were associated with very low growing season precipitation in conjunction with moderately high growing season temperatures.

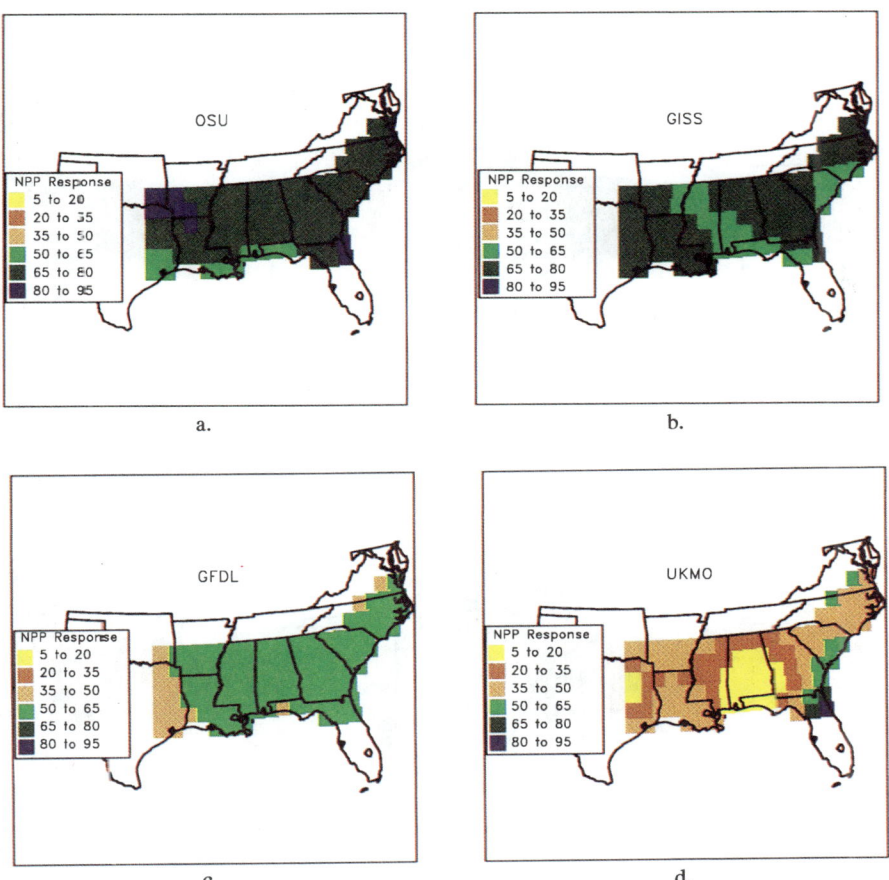

Figure 21.3. Simulations of NPP response (future NPP − present/present * 100) to a doubling of atmospheric CO_2 concentration and the associated climate change. Data represent NPP response for low LAI, poor soil-water-holding capacity conditions for the a) Oregon State University (OSU), b) Goddard Institute for Space Studies (GISS), c) the Geophysical Fluid Dynamics Laboratory (GFDL), and d) United Kingdom Meteorological Office (UKMO) general circulation models.

Regional Trends in Simulated Net Primary Production Response to Future Climate

Simulations of expected NPP response to climate change across the range of loblolly pine depended more on the GCM climate scenario used and less on the stand/site condition assumed. Overall, simulations suggest increased productivity in a 2× CO_2 environment throughout most of the loblolly pine forest type (Figures 21.3a–d and 21.4a–d).

The magnitude and range in the NPP response varied according to the GCM climate projection used and stand structure and water-holding capacity used.

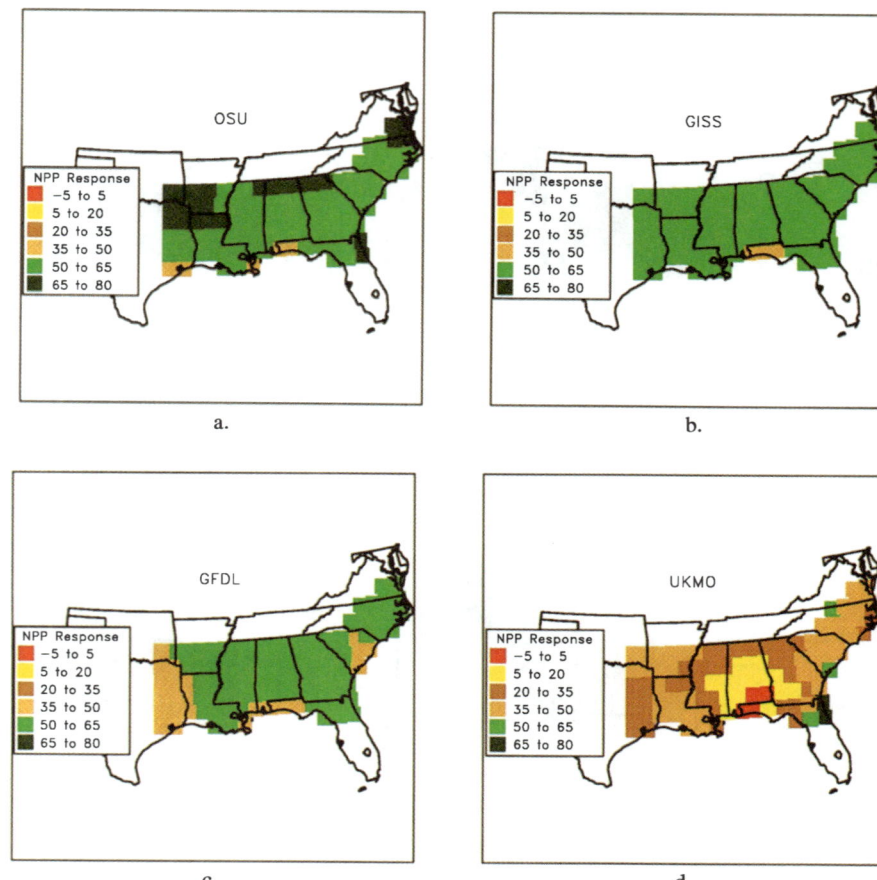

Figure 21.4. Simulations of NPP response (future NPP − present/present * 100) to a doubling of atmospheric CO_2 concentration and the associated climate change. Data represent NPP response for high LAI, good soil-water-holding capacity conditions for the a) Oregon State University (OSU), b) Goddard Institute for Space Studies (GISS), c) the Geophysical Fluid Dynamics Laboratory (GFDL), and d) United Kingdom Meteorological Office (UKMO) general circulation models.

Simulations using the OSU climate projections had the greatest regionally average NPP response, with a 65 to 80% increase in NPP projected for most of the loblolly pine forest type for high LAI, good soil-water-holding capacity conditions (Figure 21.4a). Although gulf coastal regions had a somewhat lower NPP response, northwestern regions had an even greater response for this GCM (80 to 95%). The GISS climate projections also indicated 65 to 80% increases in NPP for much of the loblolly pine range, however there were areas along both coasts where 50 to 65% increases were projected for similar stand/site conditions (Figure 21.4b).

There was a 50 to 65% increase in NPP response for the GFDL climate projections for high LAI, good soil-water-holding capacity simulations throughout most of the loblolly pine forest type (Figure 21.4c). A somewhat reduced response was observed for western regions along eastern Texas and Oklahoma. In stark contrast to the other GCMs, NPP response for the UKMO climate projections ranged from 5 to 95% for the high LAI, good soil-water-holding capacity simulations, with strong regional patterns observed (Figure 21.4d). Low NPP response was projected for most of Alabama, western Georgia, and eastern Texas. However, comparable NPP response estimates were found for Atlantic coastal areas of South Carolina, Georgia, and Florida as observed in the other GCMs (Figure 21.4a, b, c).

The range in NPP response, and the regional trends, for low LAI, poor soil-water-holding capacity conditions were generally similar to those for high LAI, good water-holding capacity simulation for each GCM, although the magnitude of the response was reduced (Figure 21.3a, b, c, d). Additionally, NPP response was similar for the OSU, GISS, and the GFDL simulations for the low LAI, poor water-holding capacity conditions throughout much of the loblolly pine range (Figure 21.3a, b, c). As such, stand structure and water-holding capacity had little affect on NPP response for the GFDL climate projections. Also in contrast to the other GCMs, these stand and site conditions resulted in a negative NPP response for the UKMO simulations for the western Florida panhandle and south-central Alabama (Figure 21.3d). The western panhandle of Florida had a lower NPP response when compared to other regions for all four GCMs.

These simulations indicate that the interregional average magnitude of the NPP response to the climate projections was inversely correlated to the CO_2 sensitivity of the GCM examined. However, the range in the forty-year average response, and the absolute change in the NPP response predicted, was directly correlated. The models may thus be ordinated from greatest forty-year mean regional NPP response to least response as: OSU > GISS > GFDL > UKMO.

Environmental Variable Sensitivity

Elevated CO_2 had a greater effect on NPP response than temperature or precipitation in the sensitivity comparison. Both stand and site conditions had a similar response to increased ambient CO_2 and increased temperature. Specifically, NPP increased by roughly 40 and 70% for 1.5 and 2× ambient CO_2, respectively, for both stand and site conditions (Figure 21.5a). The slightly higher response for low LAI, poor soil-water-holding capacity conditions was attributed to increased belowground allocation. Increasing temperatures by 3 and 6 °C decreased NPP by approximately 20 and 40%, respectively. However, the two stand/site conditions diverged in their response to addition or subtraction of precipitation. Increasing or decreasing daily precipitation by 25 and 75% of one standard deviation of the forty-year mean for those days that it rained had no affect in high LAI, good water-holding capacity simulations. Additionally, increasing precipitation had little effect for low LAI, poor water-holding capacity conditions (Figure 21.5b).

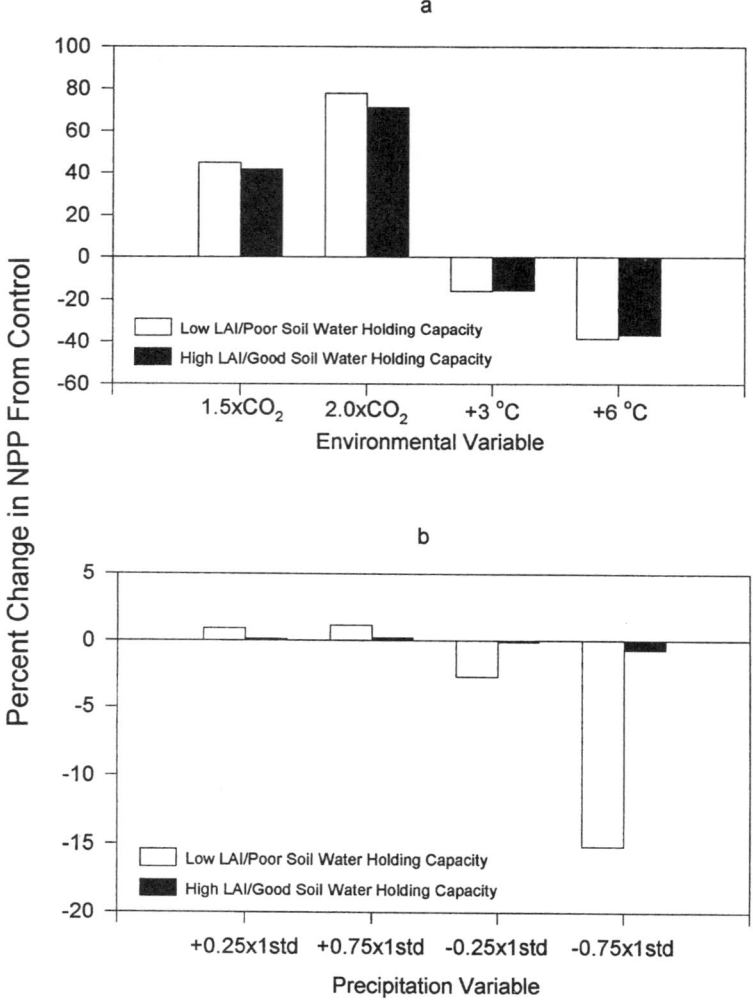

Figure 21.5. A sensitivity comparison of the effect the environmental variables projected to change in a 2 × CO_2 environment on NPP response in BIOMASS version 13.0. The comparisons are for a) two levels of ambient CO_2 concentration and two levels of increased temperature, and b) two levels of altered precipitation. Daily precipitation for those days that it rained in a typical climate year was increased and decreased by 25 and 75% of one standard deviation of the forty-year daily mean.

However, decreasing precipitation decreased NPP response by 3 and 15% for the 25 and 75% of one standard deviation, respectively, for low LAI, poor water-holding capacity simulations.

Discussion

The response index developed here suggests that loblolly pine NPP will increase in a 2× CO_2 environment. The overall increase can largely be attributed to CO_2 fertilization. However, much regional variation in the NPP response can be expected, with the potential magnitude of absolute changes in productivity set by the LAI present, but moderated by the soil-water-holding capacity, and determined by the local climate projected for the region.

Predicted changes in NPP may not directly correlate with changes in yield (merchantable stem component). If climate or site resources shift carbon allocation among the various tissue components, then the amount of yield expected at a given NPP level would vary (Allen et al., 1995). However, the range in the relationship between yield and predicted NPP response presently provides the best estimate of the range in stemwood production response expected for intermediate-age loblolly pine stands (Figure 21.1). Thus, the yield index is proportional to the NPP response index, with corresponding projected changes in stemwood production proportional to the regional changes in NPP predicted.

No direct comparison of NPP or NPP response to an "instantaneous" doubling of CO_2 from other process model predictions for loblolly pine could be made. However, TEM (Melillo et al., 1993) and (McGuire et al., 1992) has been used to predict NPP and NPP response to a 2× CO_2 climate for temperate mixed forests of the southern United States. Assuming that NPP simulations for these forests are representative of loblolly pine forests, our NPP estimates were very comparable with TEM (Table 21.4). Indeed, the similarity in the contemporary NPP estimates between results presented here and TEM suggests that our two stand structure/site conditions may have represented the broad natural variability in loblolly pine productivity attainable. Our contemporary NPP estimates are generally within those typically found for southern forests (Table 21.4).

Our NPP response estimates were also generally similar to those predicted by TEM (Melillo et al., 1993) for the OSU, GISS, and GFDL GCMs (Table 21.5). Noticeably different, however, are both the range in NPP response predicted, as well as the minimum response observed, with TEM simulations indicating negative NPP response for two of the three GCMs examined here. Our estimates are slightly more conservative than TEM, and no direct examination of the differences observed here can be evaluated. However, the two models do differ in their structure and time step (Raich et al., 1991; McMurtrie and Landsberg 1992). Notwithstanding, the similarity in the NPP estimates for the contemporary climate suggest that either 1) the climate scenarios used may differ, or 2) the processes affecting NPP associated with a 2× CO_2 environment may differ between the two models. The climate projections may have varied sufficiently enough to explain,

Table 21.4. A Comparison of Contemporary NPP for Loblolly Pine Forests from This Study With Those from Various Sources in the Literature

Comparison source	Aboveground NPP plus fine root production[8] (Mg C ha^{-1} yr^{-1})
Simulated regional forty-year average from this study	3.13–8.22
[1]Simulated mixed temperate forests of the southern United States using TEM	2.31–10.66
[2]Simulated mixed temperate forests of the southern United States using TEM	3.81–10.20
[3]Simulated using PnET–IIS for the southern United States	2.0–8.0
[4]Loblolly pine estimates from various sources	1.4–7.0
[5]Seven-year-old *P. elliottii* stand from Florida	2.43
[6]Twenty-seven-year-old *P. elliottii* stand from Florida	8.23
[7]Sixteen-year-old *P. taeda* stand from North Carolina	9.83

[1] Melillo et al. (1993).
[2] McGuire et al. (1992).
[3] McNulty et al. (1997).
[4] Teskey et al. (1987). Converted to carbon units using 0.5 × biomass estimate.
[5] Vogt, K. (1991).
[6] Kinerson (1977).
[7] Includes shrub and herb production.
[8] <=4 mm.

in part, these differences. We used a larger grid resolution and a separate final database than the TEM simulations.

Regional differences in the NPP response to a 2× CO_2 climate in this study are directly explained by the relative and absolute difference, and combination of differences, in the projected environmental variables. The GCM NPP response hierarchy for the OSU, GISS, and GFDL models can be causally linked to the ambient CO_2 concentration that each GCM assumes and, therefore, the 2× ambient CO_2 concentration used in the NPP response simulations (Table 21.2). The dominant affect of CO_2 on NPP in the model, in comparison to the generally moderate temperature and precipitation affect (Figure 21.4), largely explains the hierarchy for these three GCMs. Although temperature and precipitation patterns

Table 21.5. A Comparison of NPP Response for the Southeastern United States to Climate Projected for a Doubling of Atmospheric CO_2 Concentration as Predicted by Three General Circulation Models. The NPP Response Data from This Study Using BIOMASS Version 13.0 Were Compared to Response Data Generated from the TEM

GCM[1]	Range of NPP response (% Change in NPP)	
	TEM	BIOMASS version 13.0
OSU	−5 to +50	+51 to +84
GISS	+5 to +100	+47 to +75
GFDL	−5 to +50	+43 to +60

[1] OSU = Oregon State University, GISS = Goddard Institute for Space Studies, and GFDL = Geophysical Fluid Dynamics Laboratory.

projected by the GCMs vary both by region and season (Cooter et al., 1993), sufficient commonality between the GFDL, GISS, and OSU model projections, and their minimum overall projected change, especially in comparison to the UKMO GCM (Cooter et al., 1993), deemphasizes their importance in relation to the CO_2 influence on carbon gain. Using a GCM average ambient CO_2 concentration for the contemporary climate probably also influenced the GCM affect. This would have underestimated the response for the GFDL model and overestimated the response for the OSU model; separation in the NPP predictions between the models would be further reduced.

The dramatic increases in temperature (5 to 45%), and the increases and decreases in precipitation ($+30$ to -20%) predicted by the UKMO GCM (data not shown) were reflected in the UKMO NPP response matrix, with broad separation in the regional response patterns directly explained by the absolute change predicted for temperature and precipitation combinations (Figure 21.3). The NPP response ordination of the four GCMs found here was observed when examining terrestrial aboveground carbon storage for global simulations of major ecosystem complexes (Smith et al., 1992).

The NPP response simulations to the four climate scenarios represent response to an instantaneous doubling of ambient CO_2 concentration with no acclimation of the photosynthetic machinery considered. Although CO_2 fertilization affect has often been shown for C3 and C4 species, questions remain whether positive growth response will be maintained in a CO_2-enriched environment (Bazzaz and Fajer, 1992; Conroy et al., 1990; Acock and Allen, 1985). Empirical estimates for an eleven-year-old loblolly pine stand indicate a 50 to 100% increase in maximum photosynthesis (AMAX) associated with $2\times CO_2$ concentrations for two nutrient and water regimes (Murthy et al., 1996). At present, no downward trend in AMAX, or the associated elevated growth rates, have been observed under short-term, chronic, elevated levels of CO_2 for eleven-year-old loblolly pine trees (Murthy et al., 1996), or long-term exposure for field-grown mature, twenty-one-year-old loblolly pine trees (Liu and Teskey, 1995; Teskey 1995). A second concern in CO_2 exposure studies involves the effect of chronic exposure on stomatal conductance and, therefore, water use efficiency (Reynolds et al., 1992). A significant reduction in stomatal conductance for a number of C3 and C4 species has been observed (Thomas and Strain, 1991; Morison, 1985). However, present evidence suggests no changes in stomatal conductance in $1.5\times$ and $2\times CO_2$ treatments for eleven-year-old loblolly pine trees in branch chamber experiments (Murthy et al., 1996), or mature loblolly pine trees (Liu and Teskey, 1995).

Nutrient limitations, and in particular nitrogen, affect the absolute photosynthetic response of plants to elevated levels of ambient CO_2. Increased foliar nitrogen concentration results in increased AMAX, with nitrogen limitations to photosynthesis, at time, potentially more important than CO_2 concentration (Griffin et al., 1993). Dilution of nitrogen in foliage from elevated CO_2 levels would reduce litter quality, ultimately resulting in reduced nitrogen mineralization. This would have a negative impact on carbon gains from CO_2 fertilization. Comins and McMurtrie (1993), using a generic analytical ecosystem model, project initial increases (27%) in productivity that decline to below present-day levels because of nitrogen limitations to photosynthesis. At present, we are unable to examine this phenomena with our model. However, nitrogen feedbacks in loblolly pine may ultimately be more responsible for changes in the foliage dynamics.

Loblolly pine generally carries only two foliage cohorts, with present-year foliage production strongly sensitive to soil nitrogen availability (NCSFNC, 1991). As such, nitrogen limitations in loblolly pine forests restrict the maximum LAI attainable, provided water is not limiting (NCSFNC, 1991). These simulations demonstrate that LAI strongly influences both the absolute response, as well as the relative response to changes in temperatures and precipitation. Year-to-year feedbacks, or NPP associated with developing stands are not considered in this model, and would have important implications on NPP and NPP response to climate change. For example, decreased precipitation in one year would result in lower leaf area production in the following year (Dougherty et al., 1995; Hennessey et al., 1992), resulting in reduced NPP and a different response to climate projected for that year (Figures 21.3, 21.4). Decreased nitrogen availability would decrease LAI, and therefore reduce NPP, but may increase the NPP response (Figures 21.3, 21.4). Increased NPP response for low LAI, poor water-holding capacity conditions when compared to high LAI, good water-holding capacity conditions for the same region is the result of lower Rm costs associated with the lower foliage mass. Additionally, differences in stand structure would result in differential NPP responses when, on a relative basis, a stand with higher standing mass in structural, nonphotosynthetic tissue would have greater maintenance respiration costs and, therefore, lower NPP than a stand of similar LAI with lower-standing mass (Kinerson, 1975).

The index of potential NPP response to an instantaneous doubling of CO_2, and the associated climate projections to this CO_2 increase, provides one estimate of the climate-CO_2 effect for fixed stand and site conditions for loblolly pine forests. Although we may have generally encompassed the natural variability in loblolly pine NPP, actual productivity would, of course, depend on the local stand structure and site conditions present. Regional estimates of NPP on a site-specific basis using this model would require data sets not presently available. Efforts to couple database requirements for process models at the spatial and temporal scales needed for assessments continue to be of primary importance in modeling climate change effects on forest systems (Dale and Rauscher, 1994). Because of the importance of nutrition and leaf area on loblolly pine productivity, a planned revision of our model to incorporate resource-driven functions for foliage phenology, production, and litterfall would enable broader application of the model. However, site-determined estimates of the resource base at the level of the simulation desired would still be required.

Conclusions

Empirical estimates of stemwood production for one year ranged from 2.35 to 5.30 Mg C ha^{-1} year^{-1} with 47% NPP used in stemwood production. Carbon allocation to stemwood production ranged from roughly 33 to 63% of NPP, with NPP explaining 53% of the variation in yield. Regional estimates of forty-year average NPP for the southern United States ranged from approximately 3.0 to 8.2 Mg C ha^{-1} year^{-1}, and were comparable to those predicted by the TEM for temperate mixed forests of the Southeast.

The NPP response to the OSU, GISS, GFDL, and UKMO climate projections varied regionally, and depended on the GCM scenario, the stand structure and water-holding capacity used, and the region examined. Variation in the CO_2 sensitivity of the four GCMs resulted in predictions for the forty-year mean NPP response of -1 to $+94\%$ NPP under a doubling of atmospheric CO_2. The interregional average NPP response to the climate projections was inversely correlated to the CO_2 sensitivity of the GCM examined. However, the absolute range in the forty-year regional response, and the magnitude of the NPP predictions, were directly correlated. The GCM ranking for regional average NPP response from greatest to least response was: OSU > GISS > GFDL > UKMO. Differences in the NPP projections are attributed to the precipitation and temperature projections from the models, with CO_2 fertilization largely explaining the increased production observed. The process model sensitivity to CO_2 more than offsets declines in NPP resulting from increased temperatures or decreased precipitation for three of the four GCMs examined. Negative production response was observed for the UKMO climate scenarios, and was attributed to the extreme increases in temperature and decreases in precipitation predicted by the UKMO model for that region.

References

Acock B, Allen HL (1985) Crop responses to elevated carbon dioxide. In Strain BR, Cure JD (Eds) *Direct effects of increasing carbon dioxide on vegetation*. ER-0238. USDE, Dep Environ, Washington, DC.

Anonymous (1995) TriMetrix, Inc. Seattle, WA.

Bazzaz FA, Fajer ED (1992) Plant life in a CO_2-rich world. Sci Amer 266(1):68–74.

Benson ML, Landsberg JJ, Borough CJ (1992) The biology of forest growth experiment: An introduction. For Ecol Manage 52:1–16.

Carter TR, Parry ML, Nishioka S, Harasawa H (1992) Preliminary guidelines for assessing impact of climate change. Environ Change Unit and Cent for Global Environ Res, Oxford, UK.

Comins HN, McMurtrie RE (1993) Long-term response of nutrient-limited forests to CO_2 enrichment; Equilibrium behavior of plant-soil models. Ecol Appl 3:661–681.

Conroy JP, Milham M, Mazur M, Barlow EWR (1990) Growth, dry weight partitioning and wood properties of *Pinus radiata* D. Don after 2 years of CO_2 enrichment. Plant Cell Environ 13:329–337.

Cooter EJ, Eder BK, LeDuc SK, Truppi L (1993) General circulation model output for forest climate change research and applications. USDA For Ser Gen Tech Rep SE-85.

Dougherty PM, Hennessey TC, Zarnoch SJ, Stenberg PT, Holeman RT, Wittwer RF (1995) Effects od stand development and weather on monthly leaf biomass dynamics of a loblolly pine (*Pinus taeda* L.) stand. For Ecol Manage 72:213–227.

Farquhar GD, von Caemmerer, Berry JA (1980) A biochemical model of photosynthetic CO_2 assimilation in leaves of C3 species. Planta 149:78–90.

Fowells HA (1965) Loblolly pine. In *Silvics of forest trees of the United States*. USDA For Ser Agric Handbook 271.

Gholz HL, Hendry LC, Cropper WP Jr (1986) Organic matter dynamics of fine roots in plantations of slash pine (*Pinus elliottii*) in north Florida. Can J For Res 16:529–538.

Giorgi F, Mearns LO (1991) Approaches to the simulation of regional climate change: A review. Rev Geophys 29(2):191–216.

Griffin KL, Thomas RB, Strain BR (1993) Effects of nitrogen supply and elevated carbon dioxide on construction cost in leaves of *Pinus taeda* (L.) seedlings. Oecologia 95:575–580.

Hennessey TC, Dougherty PM, Cregg BM, Wittwer RF (1992) Annual variation in needlefall of a loblolly pine stand in relation to climate and stand density. For Ecol Manage 51:329–338.

Houghton JT, Jenkins GJ, Ephramus JJ (1990) *Climate Change: The IPCC Scientific Assessment Report*. Cambridge University Press, Cambridge, England.

Karl TR, Knight RW, Easterling DR, Quayle RG (1995) Trends in US climate during the twentieth century. Conseq 1:2–12.

Kinerson RS (1975) Relationships between plant surface area and respiration in loblolly pine. J Appl Ecol 12:965–971.

Kinerson RS, Ralston W, Wells CG (1977) Carbon cycling in a loblolly pine plantation. Oecologia 29:1–10.

Lenihan JM, Neilson RP (1995) Canadian vegetation sensitivity to projected climatic change at three organizational levels. Clim Change 30:27–56.

Liu S, Teskey RO (1995) Responses of foliar gas exchange to long-term elevated CO_2 concentrations in mature loblolly pine trees. Tree Physiol 15:351–359.

Malanson GP (1993) Comment on modelling ecological response to climatic change. Clim Change 23: 95–109.

McGuire AD, Melillo JM, Joyce LA, Kicklighter DW, Grace AL, Moore B III, Vorosmarty CJ (1992) Interactions between carbon and nitrogen dynamics in estimating net primary productivity for potential vegetation in North America. Global Biogeochem Cycles 6:101–124.

McMurtrie RE, Landsberg JJ (1992) Using a simulation model to evaluate the effects of water and nutrients on the growth and carbon partitioning of *Pinus radiata*. For Ecol Manage 52:243–260.

McMurtrie RE, Leuning R, Thompson WA, Wheeler AM (1992) A model of the canopy photosynthesis and water use incorporating a mechanistic formulation of leaf CO_2 exchange. For Ecol Manage 52:261–278.

McMurtrie RE, Wang Ying-ping (1993) Mathematical models of the photosynthetic response of tree stands to rising CO_2 concentrations and temperatures. Plant Cell Environ 16, 1–13.

McNulty SG, Vose JM, Swank WT (1997) Scaling predicted pine forest hydrology and productivity across the southern United States. In Quattrochi D, Goodchild J (Eds) *Scaling in remote sensing data and GIS*. Lewis Publishers. New York.

Melillo JM, McGuire AD, Kicklighter DW, Moore III B, Vorosmarty CJ, Schloss AL (1993) Global climate change and terrestrial net primary production. Nature 363:234–240.

Mignano J (1995) Effects of water and nutrient availability on root biomass, necromass, and production in a nine-year-old loblolly pine plantation. MS Thesis. Dep For, NC State Univ, Raleigh.

Morison JIL (1985) Sensitivity of stomata and water use efficiency to high CO_2. Plant Cell Environ 8:467–474.

Murthy R, Dougherty PM, Zarnoch SJ, Allen HL (1996) Effects of carbon dioxide, fertilization, and irrigation on photosynthetic capacity of loblolly pine trees. Tree Physiol 16:537–546.

NCSFNC (1993) Six-year growth responses of midrotation loblolly pine plantations to N and P fertilization. NCSFNC Report 31. Coll For Res. NC State Univ, Raleigh.

NCSFNC (1991) Leaf area variation in midrotation loblolly pine plantations. NCSFNC Research note 6. Coll For Res. NC State Univ, Raleigh.

Raich JW, Rastetter EB, Melillo JM, Kicklighter DW, Steudler PA, Peterson BJ (1991) Potential net primary productivity in South America: Application of a global model. Ecol Appl 1:399–429.

Reynolds JF, Hilbert DW, Chen J, Harely PC, Kemp PR, Leadley PW (1992) Modelling the response of plants and ecosystems to elevated CO_2 and climate change. TR-054. USDE. Dep Environ, Washington, DC.

Richman MB, Lamb PJ (1985) Climatic pattern analysis of three- and seven-day summer rainfall in the central United States: Some methodological considerations and a regionalization. J Clim App Met 24:1325–1343.

Richman MB, Lamb PJ (1987) Pattern analysis of growing season precipitation in southern Canada. Atmosphere-Ocean. 25(2):135–158.

Rosenzweig C, Parry ML, Fischer G, Frohberg K (1993) Climate change and world food supply. Environmental Change Unit. Research report 3. University of Oxford, England.

Running SW, Nemani RR (1991) Regional hydrologic and carbon balance responses of forests resulting from potential climate change. Clim Change 19:349–368.

Running SW, Coughlan JC (1988) A general model of forest ecosystem processes for regional applications I. Hydrologic Balance, Canopy Gas Exchange and Primary Production Processes. Ecol Mod 42:125–154.

Shelton MG, Nelson LE, Switzer GL (1984) The weight, volume, and nutrient status of plantation-grown loblolly pine trees. Technical Bulletin 121. Mississippi Agricultural and Forestry Experiment Station, MS State Univ, MS.

Smith TM, Leemans R, Shugart HH (1992) Sensitivity of terrestrial carbon storage to CO_2-induced climate change: comparison of four scenarios based on general circulation models. Clim Change 21:347–366.

Smith JB, Tirpak DA (Eds) 1990. The potential effects of global climate change on the United States. Hemisphere Pubulic Corporation, New York.

Teskey RO (1995) A field study of the effects of elevated CO_2 on carbon assimilation, stomatal conductance and leaf and branch growth of Pinus taeda trees. Plant, Cell, and Environment 18:565–573.

Teskey RO, Bongarten BC, Cregg BM, Dougherty PM, Hennessey TC (1987) Physiology and genetics of tree growth response to moisture and temperature stress: An examination of the characteristics of loblolly pine (*Pinus taeda* L.). Tree Physiol 3:41–61.

Thomas RB, Strain BR (1991) Root restriction as a factor in photosynthetic acclimation of cotton seedlings grown in elevated carbon dioxide. Plant Physiol 96:627–634.

Vogt K (1991) Carbon budgets of temperate forest ecosystems. Tree Physiol 9:69–86.

Vose J, Allen HL (1988) Leaf area, stemwood growth, and nutrition relationships in loblolly pine. For Sci 34:547–563.

22. Predictions and Projections of Pine Productivity and Hydrology in Response to Climate Change Across the Southern United States

Steven G. McNulty, James M. Vose, and Wayne T. Swank

The southeastern United States is one of the most rapidly growing human population regions in continental United States, and as the population increases, the demand for commercial, industrial, and residential water will also increase (USWRC, 1978). Forest species type, stand age, and the climate all influence the amount of water use and yield from these areas (Swank et al., 1988). Because forests cover approximately 55% of the southern United States land area (Flather et al., 1989), changes in water use by forests could significantly change water yields and potentially lead to water shortages within the region. Hence, estimates of future water supply from forested areas are needed and this will require a model that can accurately predict potential change in forest wateruse at the regional scale.

In addition to water resources, an accurate estimate of future loblolly pine forest productivity is essential to the development of a management plan to provide enough timber to meet consumer demand. At present, it is uncertain if the southern forests will be able to maintain (or increase) present-day levels of productivity. For example, Zahner et al., (1988) recorded a decrease in radial growth of loblolly pine (*Pinus taeda*) during the years from 1949 to 1984 in Piedmont stands.

During the next century, substantial changes are expected to occur in a variety of environmental variables including temperature and precipitation (Melillo et al., 1989; Mitchell et al., 1989). The magnitudes of these changes are expected to vary both temporally and spatially, and they may have profound effect on forest productivity (Melillo et al., 1993) and wateruse. Although some of these changes may directly affect the physiology of trees, others may increase fire, insect damage, and flooding. Thus, environmental changes and stresses have the potential to alter not only the function of forest ecosystems but also the structure and composition of forests.

Models of forest response to environmental change will be useful tools to help manage our nation's forest resources into the next century. We will need detailed plant physiology models that operate at small spatial and temporal scales to integrate our mechanistic understanding of forest responses to environmental changes at a detailed level. In a dramatically changing environment, we will also need to manage forest resources at regional and national scales both over decades and over centuries. This will require forest ecosystem models that operate at larger spatial and temporal scales. These large-scale models must be realistically demanding in both computational capacity and the information needed to initialize and run the models. As an example of this type of large-scale model, PnET–IIS is a regional scale model developed to predict hydrology and productivity across a range of climate scenarios (McNulty et al., 1994; McNulty et al., 1996a). The objective of this chapter is to validate the use of a regional scale process-based wateruse and productivity model (PnET–IIS) using historic data, and to then predict how climate change could affect pine forest wateruse and productivity across the southern United States.

Model Structure

A derivation of the PnET–II model developed by Aber et al., (1995) to predict forest hydrology and productivity in the northeastern United States (McNulty et al., 1994; McNulty et al., 1996b, 1997), PnET–IIS utilizes site-specific soil-water-holding capacity (SWHC), four monthly climate parameters (i.e., minimum and maximum air temperature, total precipitation, and solar radiation) and species-specific process coefficients to predict evapotranspiration (ET), water drainage and net primary productivity (NPP) from the stand level ($<$ one hectare (ha)) to a $0.5° \times 0.5°$ grid cell resolution (approximately 50×75 km) across the southern United States (Aber et al., 1992, 1995; McNulty et al., 1994, 1996a, 1997). The model calculated the maximum amount of leaf area that could be supported on a site based on the soil, the climate, and the parameters specified for the vegetative type. Leaf area is a major component in calculating NPP and water use. The model that we used, PnET–IIS, assumed that all stands were fully stocked and that leaf area was equal to the maximum amount of foliage that could be supported as a result of soil and climate limitations. Predicted NPP was defined as total gross photosynthesis minus growth and maintenance respiration for leaf, wood and root compartments. The respiration was calculated by PnET–IIS as a function of the present and prior month's minimum and maximum air temperature. The optimum temperature for net photosynthesis varied from 23 to 27 °C, and the maximum air temperature for gross photosynthesis ranged from 30 to 43 °C (Strain et al., 1976). As air temperature became elevated beyond the optimum photosynthetic temperature, the respiration rate increased and gross photosynthesis either increased slightly or decreased, and therefore, proportionally less net carbon per unit leaf area was fixed (Daniel et al., 1979; Kramer, 1980). Total gross photosynthesis was a function of both gross photosynthesis per unit leaf area and leaf area. Changes in water availability and plant-water demand placed limitations on the amount of leaf area produced, hence, as vapor-pressure deficit and air temperature increased above optimum levels, leaf area and total gross photosynthesis decreased.

Annual transpiration was calculated from a maximum potential transpiration that was modified by plant-water demand (a function of gross photosynthesis and water use efficiency). In the model, water interception loss was a function of leaf area and total precipitation, and ET was equal to transpiration plus interception loss. Drainage was calculated as water in excess of ET and SWHC. In PnET–IIS, plant-water demand depended on monthly precipitation and the amount of water stored in the soil profile. If precipitation inputs exceeded plant-water demand, the soil was first recharged to the SWHC and if water was still available, water was output as drainage. Monthly drainage values were summed to provide an estimation of annual water outflows.

Climate, Vegetation, and Soil Input Data

The model PnET–IIS required site-specific soils and climate data, as well as species-specific vegetation information. To predict monthly loblolly pine growth and wateruse, climate data from 1951 to 1984 were used as model inputs. The 900 + cooperative climate station point databases were interpolated on a 0.5° × 0.5° grid across the southern United States (Marx, 1988). The gridded databases of minimum and maximum air temperature, relative humidity, and precipitation were compiled into a single database and were used to calculate average monthly solar radiation (Nikolov and Zeller, 1992). Solar radiation values were then combined with monthly maximum and minimum air temperatures, and total monthly precipitation as input for PnET–IIS.

No site-specific vegetation indices were required to run PnET–IIS. Instead, loblolly pine-specific vegetation coefficients were used (Table 22.1). These coefficients were largely derived from the published literature (Aber and Federer, 1992; Aber et al., 1995; McNulty et al., 1994, 1996b).

Soil-water holding capacity was the only soil parameter needed to run PnET–IIS. The data were derived from a geographic information systems (GIS)-based soils atlas compiled by the Soil Conservation Service (SCS) (Marx, 1988). In developing a coverage of average SWHC, soils unsuitable for growing loblolly pines were excluded from the data set. If all SWHC were averaged across a grid cell, very low and high SWHC areas would have been averaged within the same grid cell to produce a cell with a pseudoaverage SWHC that appeared suitable for pine growth. To eliminate this source of input error, we used forest inventory and analysis (FIA) data, which consisted of stand volume, growth, and species composition information remeasured at more than 21,000 permanent plots across the

Table 22.1. PnET–IIS Default Coefficients[1] Used for Model Predictions and Parameter Coefficients Used in Sensitivity Analysis

Parameter name	Parameter abbreviation	Model default value	Sensitivity analysis values
Light extinction coefficient	k	0.5	0.4, 0.5*, 0.6
Foliar retention time (years)		2.0	
Leaf specific weight (g)		9.0	
NetPsnMaxA (slope)		2.4	
NetPsnMaxB (intercept)		0	
Light half saturation (J m^2 sec^{-1})	HS	70	60, 70*, 80
Vapor deficit efficiency constant	VPDK	0.03	0, 0.03*, 0.05
Base leaf respiration fraction		0.10	
Water use efficiency constant	WUE C	10.9	10, 10.9*, 12.0
Canopy evaporation fraction		0.15	
Soil-water release constant	F	0.04	0.03, 0.04*, 0.05
Maximum air temperature for photosynthesis (°C)	TMAX	variable*	35, 45
Optimum air temperature for photosynthesis (°C)	TOPT	variable*	17, 23
Change in historic air temperature (°C)	DTEMP	0	+2, −2
Change in historic precipitation (% difference)	DPPT	0	+10, −10

[1] Default coefficients are listed with an *.

southern United States. A database that contained plot locations of loblolly pine FIA plots across the southern United States was selected. A GIS was used to layer regional scale map of SWHC over FIA plot locations of loblolly pine. The pine stands were located on FIA plots and SWHC ranged from 3.8 to 15.8 cm H$_2$O for soil depths of 102 cm (McNulty et al., 1994).

Using the selected range of SWHC where loblolly pine grow, the 0.5° × 0.5° grid cell was placed over the region and a weighted average of all remaining SWHC polygons within each grid cell was computed. This GIS database formed the basis for the soils input to the PnET–IIS model.

Model Validation

Model validation is often overlooked in large geographic scales. Because models designed for use in large spatial scales are based on numerous assumptions about forest structures and functions as soil-water storage and stand stocking, for a specific forest stand, one or more of the assumptions may be inaccurate. Depending on the degree and type, inaccurate assumptions may result in erroneous model predictions of wateruse and productivity for any particular site. Therefore, regional scale models should not be expected to accurately predict annual wateruse and productivity for all sites and all years. However, the model should generally correlate with site wateruse and productivity, across numerous sites occupying a wide geographic range. If general relationships are not found between predicted

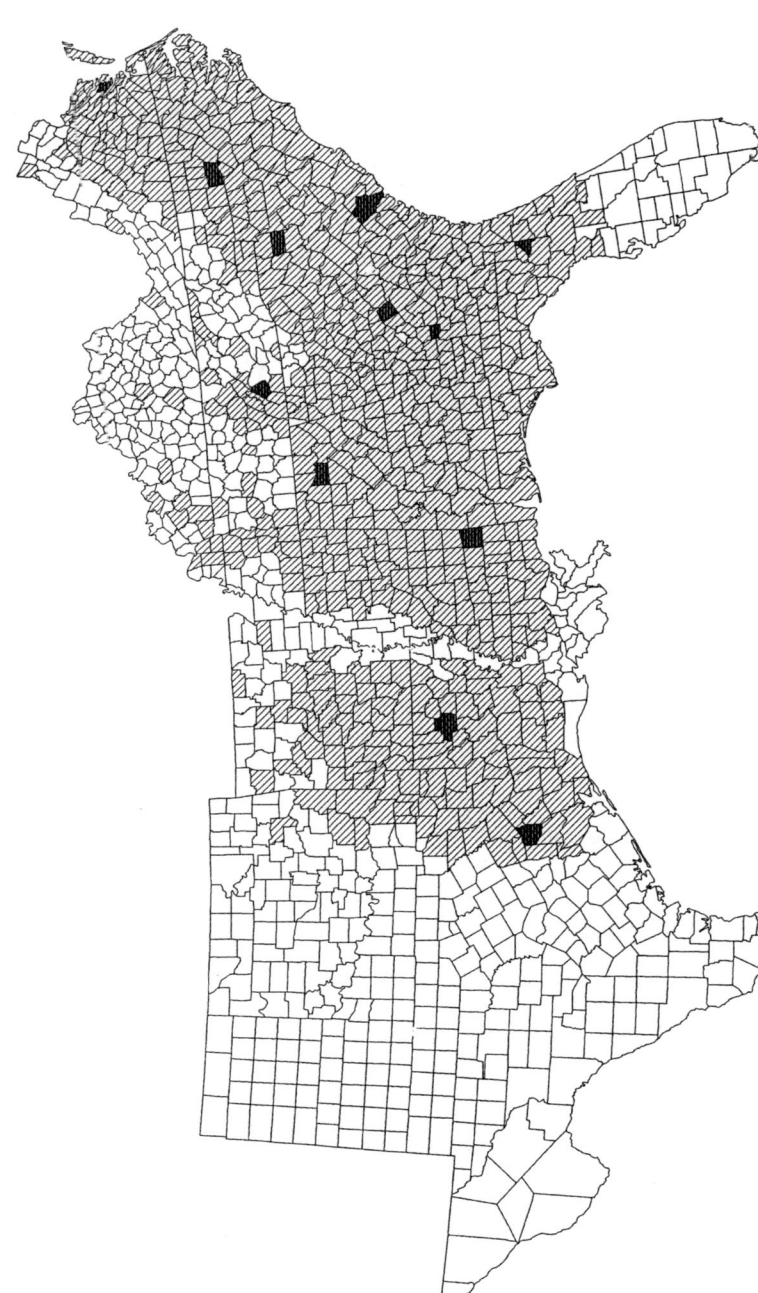

Figure 22.1. Site locations of the twelve pine sites sampled (in black), overlaid on range map of loblolly pine (in grey).

and measured wateruse and productivity across sites, the model logic is flawed or the model has too many incorrect assumptions to produce accurate predictions.

Net Primary Production Validation

Predicted productivity (t ha^{-1} yr^{-1}) was compared with measured basal area growth (cm^2 tree^{-1} yr^{-1}) for twelve loblolly pine stands located across the southern United States (Figure 22.1). These sites represented a wide range of climate and soil conditions (Table 22.2), as well as meeting the following selection criteria: 1) stands were fully stocked at the time of sampling; 2) more than 95% of the stand basal area contained loblolly pine; 3) the site had not been thinned, burned, fertilized, or heavily damaged by insect or disease; 4) all sites were on relatively level terrain (< 10% slope). PnET–IIS was run on each of the twelve sites using climate data from 1951 to 1990.

Forest Growth Measurements

Two tree core samples were collected 1.4 m above the forest floor (diameter at breastheight (DBH)) from each of twenty trees per site. The selected trees were randomly located within the plot, but represented the dominant or codominant size class. The first core was selected at a random azimuth, and the second core was extracted at 90° to the first core. Cores were returned to the laboratory, mounted, and sanded prior to ring-width measurements. The cores were the crossdated, and ring width was measured using a Model 3 increment measurer (Fred C. Henson Co.), which has an accuracy of 0.01 mm. All cores were measured twice and if the difference in measured annual ring width was > 10% between the two readings, the core was measured a third time and an average of the three measurements was used. Annual basal growth (cm^2) was calculated as $\pi \times$ (tree radius) of the present year ring area $- \pi \times$ (tree radius) of the prior year ring area.

Predicted Forest Growth by PnET–IIS

The years of record (YOR) that basal area growth could be compared to predicted NPP varied between sites, because plantation establishment times and rates of canopy closure differed (Table 22.2). The shortest record of basal area growth (eight years) was from the site at Chester County, SC, and the longest (twenty-eight years) was from the site at Wayne County, MS. Predicted average annual NPP was greatest on both the sites at Colleton County, SC and Wayne County, MS, and smallest at the site in Wilkinson County, GA (Table 22.2). The Colleton County, SC site had the largest average annual basal area growth, and the Dooly County, GA site had the smallest average annual basal area growth (Table 22.2). PnET–IIS predicted NPP ranged from 2 to 18 t biomass ha^{-1} yr^{-1}, with an average annual value of 11.3 t biomass ha^{-1} yr^{-1}, and average annual basal area growth was highly correlated ($r^2 = 0.66$, $P < 0.001$, n = 12) with average annual predicted NPP (Figure 22.2). Predicted growth rates were within the general range of forest growth measured by others. Teskey et al., (1987) measured a range of

Table 22.2. Climatic Data for Twelve Measured Loblolly Pine Sites[1]

SITE	YOR[1]	Lat.[2] (°)	Growing season avg. solar radiation[3] (j m^{-1} sec^{-1})	Annual avg. temp. (°C)	Annual avg. PPT (cm H$_2$O)	GIS SWHC[4] (cm H$_2$O)[5]	Predicted NPP (t biomass ha^{-1} yr^{-1})	Avg. basal area growth[6] (cm^2 yr^{-1})[7]
Bradford, FL	21	30.0	465 (10)	20.2 (0.2)	130 (3)	6	11.4 (0.9)	11.4 (0.5)
Bienville, LA	15	32.3	446 (5)	18.0 (0.2)	149 (9)	14	10.2 (0.7)	11.3 (0.9)
Chatham, NC	14	35.6	414 (6)	15.0 (0.2)	117 (6)	14	10.8 (0.9)	11.8 (0.5)
Chester, SC	8	34.8	432 (10)	16.2 (0.3)	120 (6)	13	10.3 (1.3)	13.1 (0.4)
Colleton, SC	11	32.9	436 (6)	18.7 (0.2)	121 (4)	13	12.8 (1.2)	21.1 (1.3)
Gloucester, VA	13	37.5	400 (7)	15.0 (0.3)	117 (7)	12	10.8 (0.9)	11.7 (0.7)
Dooly, GA	11	32.1	466 (8)	20.2 (0.3)	108 (4)	14	9.4 (0.8)	8.7 (0.8)
McMinn, TN	14	35.5	417 (9)	14.8 (0.2)	134 (9)	9	9.3 (0.8)	11.3 (0.7)
Morgan, AL	10	34.5	424 (9)	15.0 (0.2)	136 (5)	14	11.9 (0.9)	13.6 (0.4)
Walker, TX	12	31.0	477 (8)	19.5 (0.2)	114 (7)	11	9.9 (0.8)	14.2 (1.6)
Wayne, MS	28	31.6	432 (4)	18.2 (0.2)	149 (6)	16	13.1 (0.5)	17.4 (0.7)
Wilkinson, GA	14	32.8	449 (6)	17.9 (0.2)	111 (4)	13	9.1 (0.8)	11.4 (0.6)

[1] YOR = years of record since canopy closure.
[2] Lat. = latitude.
[3] Growing Seas. Avg. Solar Radiation = growing season solar radiation.
[4] GIS SWHC = site soil-water-holding capacity derived from a Soil Conservation Service map of the soils on each site.
[5] Per 102 cm soil.
[6] Avg. basal area growth = average annual basal area growth for the measured trees on each site.
[7] Standard errors are included in ().

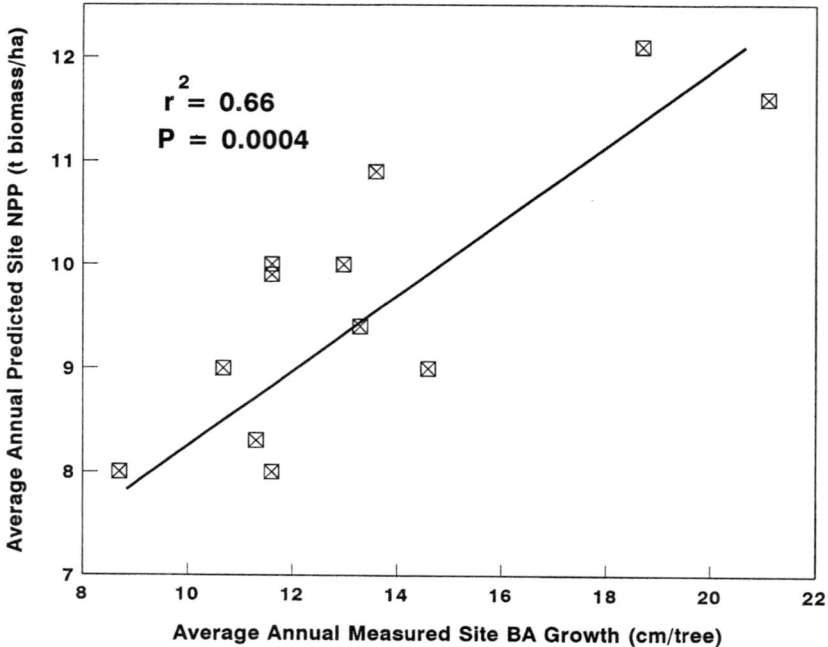

Figure 22.2. Average annual predicted NPP vs average annual measured basal area growth for all twelve loblolly pine sites.

aboveground NPP between 2 and 10 t dry matter ha^{-1} year^{-1} on loblolly pine sites. Other studies have estimated that belowground production equals approximately 40% of aboveground NPP (Nadelhoffer et al., 1985; Whittaker and Marks, 1975). Multiplying Teskey et al., (1987) measurements of aboveground NPP by 1.4, which represents the approximate 30% additional NPP that occurs belowground, yielded a measured range of total (aboveground and belowground) NPP between 2.8 and 14.0 t biomass ha^{-1} yr^{-1}, with most site NPP (aboveground only) > 8.5 t biomass ha^{-1} yr^{-1} (Teskey et al., 1987). The NPP predicted by PnET–IIS fell within this measured range of NPP.

Wateruse Validation

Although runoff is related to precipitation, numerous factors affect the amount of runoff from a basin. Nationally, 8% of all runoff is removed for industrial, commercial, and residential purposes (USGS, 1992). The other principle factor affecting runoff is vegetation. Approximately 50% of precipitation either evaporates from leaf surfaces or transpires through plant stomates (Swift et al., 1975). Species type, age, and morphology all influence ET and, therefore, the rates of runoff. Researchers have long used United States Geologic Survey (USGS) streamflow data for hydrologic modeling, but traditionally, the emphasis was on model calibrations (James, 1972; Dawdy et al., 1972; Magette et al., 1976). Basin streamflow data are useful in broad-scale modeling, calibration, and validation because measurements integrate ecosystem water input, movement, and usage. The USGS has more than 6,000 stream-gauging stations across the continental United States (USGS, 1992), some of which were used in model validation of regional drainage (McNulty et al., 1994). Average annual runoff data for the southern United States was calculated from gauge-station data from 1951 to 1980 (Moody et al., 1986). The 0.5° × 0.5° grid cell was placed over an isopleth map and a weighted average of mean cell runoff was calculated that was based on the area size and value of all isopleths within each cell. Historic rates of predicted average annual drainage varied widely across the region. Low rates of average annual precipitation and elevated annual air temperatures combined to give the eastern Texas and central Georgia the lowest measured rates of annual water drainage (Figure 22.3a). Conversely, cool temperatures and high rates of precipitation combined to make southern Appalachian mountains in western North Carolina the area of highest predicted drainage. PnET–IIS predicted that the lowest drainage would occur in eastern Texas and along the coastal plain, and the highest drainage would occur in the high elevation Appalachian Mountains in southwestern North Carolina and northeastern Georgia (Figure 22.3a). Predicted drainage corresponded with measured USGS annual runoff data collected from 1951 and 1980 ($r^2 = 0.64$, $P < 0.0001$, $n = 502$) (Figure 22.4). Measured average annual precipitation was not as well correlated with measured USGS average annual runoff ($r^2 = 0.42$, $P < 0.0001$, $n = 502$).

Climate Change Scenarios

After model-predicted wateruse and productivity estimates were validated against historic measurements, PnET–IIS was used to predict the potential influence of climate change on southern pine forest wateruse and productivity. Only changes in precipitation and air temperature were considered in the climate change scenarios. Other such potential atmospheric changes as carbon dioxide (CO_2), ozone (O_3), or sulfur oxides (SO_x), which may be important to future forest growth, are not addressed in this chapter.

Two climate change scenarios were developed using historic climate databases in conjunction with two GCMs. The Oregon State University (OSU) (Schlesinger and Zhao, 1989), and United Kingdom Meteorological Office (UKMO) (Mitchell, 1989), were selected because of their common application and range of climate change predictions. All of the GCMs predicted variation in monthly temperature and precipitation, based on a doubling of atmospheric CO_2 by the year 2050 (Cooter et al., 1993). The predicted monthly degree (°C) changes in air temperature and percent changes in precipiation by GCMs were projected onto a 0.5° × 0.5° grid across the southern United States. Because predictions of climate change by GCMs are static between years, historic monthly air temperature and precipitation data (1951 to 1984) were combined with predicted changes in air temperature

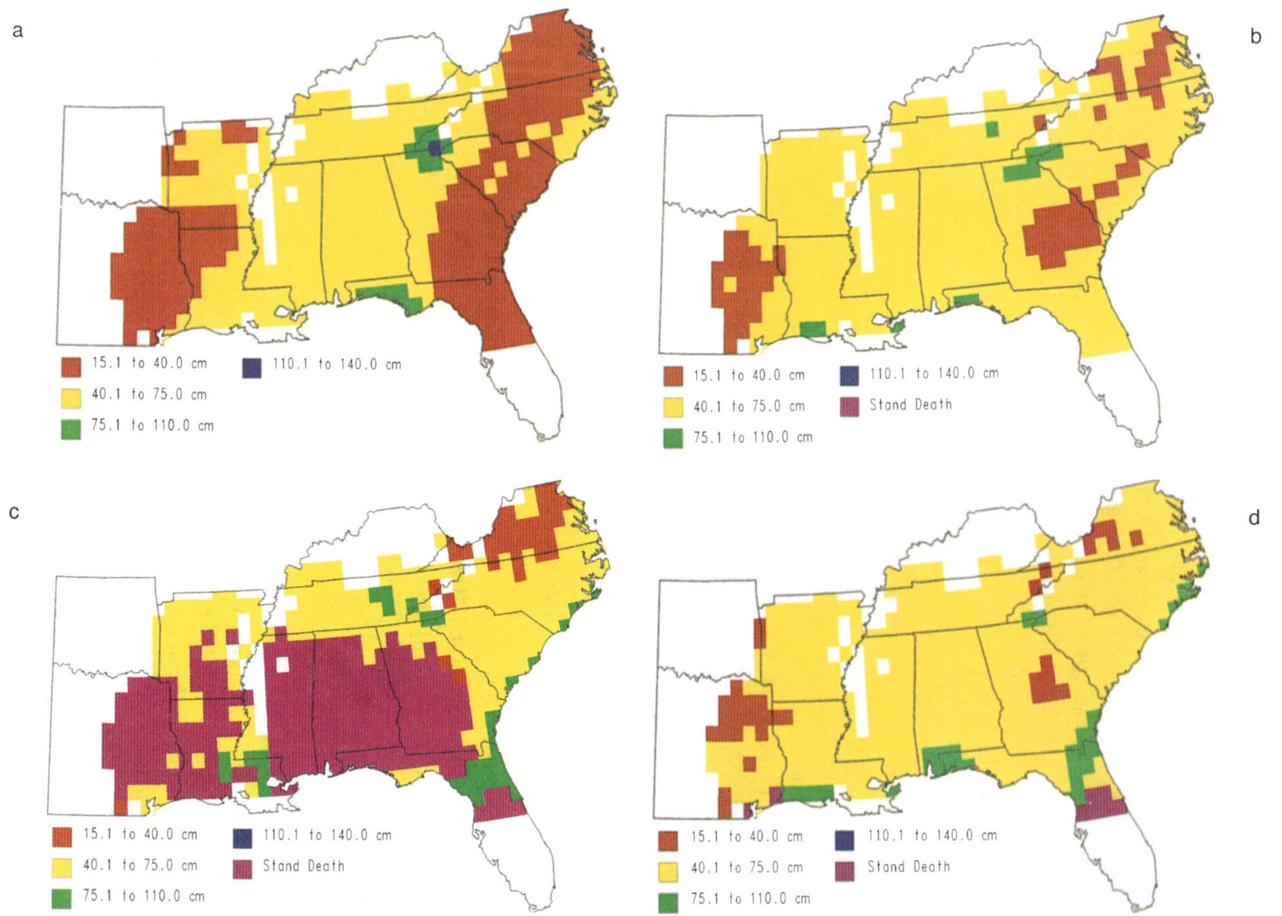

Figure 22.3. Measured USGS average annual historic (1951–1980) drainage (a); PnET–IIS predicted average annual water

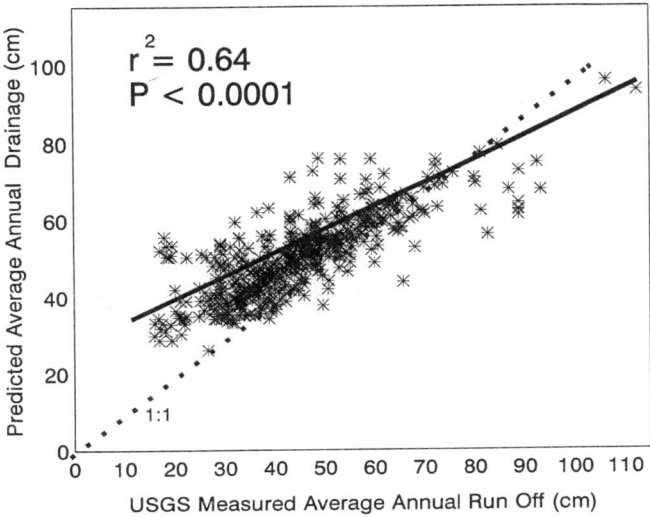

Figure 22.4. Measured USGS average annual historic (1951–1980) drainage v PnET–IIS predicted average annual drainage using historic (1951–1980) climate data on 0.5° × 0.5° grid cells across the southern United States.

and precipitation to provide thirty-five years of dynamic climate change scenario data.

The GCM-predicted precipitation and air temperature change estimates under a doubled CO_2 environment were very different. Across the southern United States, the OSU–GCM predicted a smaller increase (+3 °C) in average annual air temperature compared to the UKMO–GCM (+7 °C). The OSU–GCM also predicted above-average precipitation in the late summer and fall, as well as below historic average levels of precipitation in the late winter and spring. Finally, the OSU–GCM predicted that although the total annual precipitation would decrease in the central portion of the South and would increase along the Atlantic coast, generally across the region average annual precipitation would increase by 3% as compared to historic total annual precipitation. The UKMO–GCM predicted that regional precipitation would be greater than historic amounts during the spring and smaller during the summer and fall; average annual precipitation would decrease in the central and southwestern portion of the region and increase along the southern Atlantic coast, with the total annual precipitation decreasing by 1% region-wide compared to historic levels.

Results and Discussion

Predicted Net Primary Production

When PnET–IIS was run in conjunction with the two climate change scenarios, predicted NPP was reduced across most of the southern United States (except in some high-elevation, mountainous areas) but the severities of the reductions were dependent on the GCM applied. Using the OSU–GCM scenario, PnET–IIS predicted a 26% average reduction in growth across the southern United States, but the envirnomental conditions were not predicted to be severe enough to cause a large reduction in the pine range. PnET–IIS predicted that approximately 2% of the present-day loblolly pine range would be lost across the southern United States. if the OSU–GCM scenario occurred. Predicted NPP generally decreased across the region but would increase in the cooler, mountainous areas of the region (Figure 22.5a), and the model suggested that the range of loblolly pine could shift significantly northward if global climate change should occur. When the OSU–GCM scenario was applied to northern sections of the loblolly pine range, the climate in this area was very similar to historic climate in eastern Texas, and consequently, predicted NPP for the northern loblolly pine range was then similar to the eastern Texas predicted NPP under historic conditions. Future ecosystem research will need to account for shifts in species range resulting from climate change.

In the UKMO–GCM scenario, predicted NPP was reduced by 100% of historic NPP across most of the south-central and southwestern portions of the region including most of Florida, Georgia, Alabama, Mississippi, Louisiana, and Texas (Figure 22.5a), which suggested that the climate in these states would no longer be suitable for growing loblolly pine. The UKMO–GCM predicted less severe reduction in NPP for the northern and eastern portions of the region (Figure 22.5a), because these areas have historically cooler air temperatures and relatively high rates of precipitation. Across the region, average NPP was reduced by 46% and the range of loblolly pine was reduced by 42% when the UKMO–GCM scenario was applied to the model.

Climate Change Scenario Effects on Wateruse

Because drainage was equal to precipitation minus ET, and ET and NPP are a function of leaf area and temperature, the pattern of drainage is similar to NPP. Using the OSU–GCM, predicted drainage decreased in the central and north-central areas, and increased across the southern and eastern portions of the region. Compared to historic drainage, the OSU scenario average annual drainage increased by 6% across the region. The UKMO–GCM scenario caused a larger deviation in predicted ecosystem hydrology. In areas of mortality, predicted ET was zero, and drainage was equal to precipitation. The UKMO scenario predicted increased drainage throughout the region, except along the cooler Appalachian Mountains where drainage decreased. Compared to historic drainage, the UKMO scenario average annual drainage increased by 82% across the region, including areas of mortality where drainage equaled precipitation. If only areas where loblolly pine NPP $>$ 0 are included, drainage increased by only 42% as compared to historic levels.

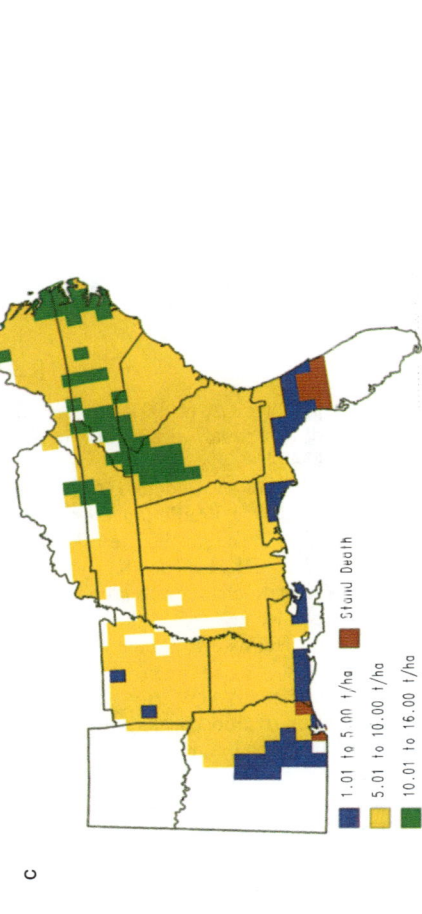

Figure 22.5. PnET–IIS predicted average annual NPP using historic (1951–1980) climate data (a); UKMO (b); and OSU (c) climate scenarios.

Conclusions

Depending on the climate scenario and site location, loblolly pine NPP could be significantly reduced and drainage could be increased across forested areas in the southern United States. Sites located in the warmest sections of the present range of loblolly pine are more susceptible to changes in productivity and wateruse than pine sites located in cooler areas. The model also suggested that the region is much more susceptible to changes in air temperature than changes in precipitation. Depending on the climate scenario, annual NPP could be reduced from 26% to 46%, and drainage could be increased from 6% to 82% across the region. These projections have serious potential socioeconomic implications for the southern United States. However, additional research is needed to assess the effects that other atmospheric changes (e.g., CO_2, O_3, NO_x, SO_x), weather changes (e.g., solar radiation), genetics, and species replacement may have on forest processes, before a complete assessment can be made of potential climate change effects on forest productivity and wateruse.

References

Aber JD, Federer CA (1992) A generalized, lumped-parameter model of photosynthesis, ET and net primary production in temperate and boreal forest ecosystems. Oecologia 92:463–474.

Aber JD, Ollinger SV, Federer CA, Reich PB, Goulden ML, Kicklighter DW, Melillo JM, R Lathrop (1995) Predicting the effects of climate change on water yield and forest production in the northeastern U.S. Climate Res 5:207–222.

Cooter EJ, Eder BK, LeDuc SK, and Truppi L (1993) General circulation model outputs for forest climate change research and application. USDA, For Ser, South For Exper Stat Gen Tech Rep SE-85, Asheville, NC.

Daniel TW, Helms JA, Baker FS (1979) *Principles of silviculture.* McGraw-Hill, New York.

Dawdy DR, Lichty RW, Bergmann JM (1972) A rainfall-runoff simulation model for estimation of flood peaks for small drainage basins. USGS Prof Pap 506-B.

Flather CH, Joyce LA, King RM (1989) Linking multiple resources analysis to land use and timber management: application and error considerations. USDA, For Ser, Pac Northw For Exper Stat, Gen Tech Rep PNW GTR-263.

James LD (1972) Hydrologic modeling, parameter estimation, and watershed characteristics. J Hydrol 17:283–291.

Kramer PJ (1980) Drought, stress and the origin of adaptations. In Turner NC, Kramer PJ (Eds) *Adaptation of plants to water and high temperature stress.* John Wiley & Sons, New York.

Magette WL, Shanholtz VO, Carr JC (1976) Estimating selected parameters for the Kentucky Watershed Model from watershed characteristics. Water Resour Res 12:472–480.

Marx DH (1988) Southern forest atlas project. In *The 81st annual meeting of the association dedicated to air pollution control and hazardous waste management.* APCA, Dallas, TX.

McNulty SG, Vose JM, Swank WT, Aber JD, Federer CA (1994) Regional scale forest ecosystem modeling: Data base development, model predictions and validation using a geographic information system. Clim Res 4:223–229.

McNulty SG, Vose JM, Swank WT (1997) Potential climate change affects on loblolly pine productivity and hydrology on four sites in the southern United States. Ambio: 25:449–453.

McNulty SG, Vose JM, Swank WT (1996b) Loblolly pine hydrology and productivity across the southern United States. For Ecol Manage: 86:241–251.

McNulty SG, Vose JM, Swank WT (1996c) Prediction and validation of forest ecosystem processes at multi-spatial scales using a GIS. In Quattrochi DA, Goodchild MF (Eds) *Scaling Of Remote Sensing Data For GIS*. Lewis Publishers, Chelsea, MI.

Melillo JM, McGuire AD, Kicklighter DW, Moore B III, Vorosmarty CJ, Schloss AL (1993) Global climate change and terrestrial net primary production. Nature 363:234–240.

Melillo JM, Steudler PA, Aber JD, Bowden RD (1989) Atmospheric decomposition and nutrient cycling In Andreae MO, Schimel (Eds) *Exchange of trace gases between terrestrial ecosystems and the atmosphere*. John Wiley & Sons, New York.

Mitchell JFB (1989) The greenhouse effect and climate change. Rev Geophys 27(1): 115–139.

Mitchell MJ, David MB, Harrison RB (1989) Sulfur dynamics of forest ecosystems In Howarth RW, Stewart JWB (Eds) *Sulfur Cycling in Terrestrial Ecosystems and Wetlands*. John Wiley & Sons New York.

Moody DW, Chase EB, Aronson DA (1986) National water summary 1985 hydrologic events and surface-water resources. USGS water-supply paper 2300. US Government Printing Office, Washington, DC.

Nadelhoffer K, Aber JD, Melillo JM (1985) Fine roots, net primary production, and soil nitrogen availability: A new hypothesis. Ecol 66:1377–1390.

Nikolov NT, Zeller KF (1992) A solar radiation algorithm for ecosystem dynamic models. Ecol Mcdel 61:149–168.

Schlesinger ME, Zhao ZC (1989) Seasonal climatic change introduced by doubled CO_2 as simulated by the OSU atmospheric GCM/mixed-layer ocean model. J Climate 2:429–495.

Strain BR, Higginbotham KO, Mulroy JC (1976) Temperature preconditioning and photosynthetic capacity of *Pinus taeda* L. Photosyn 10:47–53.

Swank WT, Swift LW Jr, Douglass JE (1988) Streamflow changes associated with forest cutting species conversions and natural disturbances. In Swank WT, Crossley DA Jr, (Eds) *Forest hydrology and ecology at Coweeta*. Springer-Verlag, New York.

Swift LW, Swank WT, Mankin JB, Luxmoore RJ, Goldstein RA (1975) Simulation of evapotranspiration and drainage from mature and clearcut deciduous forests and young pine plantation. Wat Resour Res 11:667–673.

Teskey RO, Bongarten BC, Cregg BM, Dougherty PM, Hennessey TC (1987) Physiology and genetics of tree growth response to moisture and temperature stress: an examination of the characteristics of loblolly pone (*Pinus taeda* L.). Tree Physiol 3:41–61.

United States Geological Survey (1992) Regional hydrology and the USGS stream gaging network. National Academy Press, Washington, DC, 1.

United States Water Resources Council (1978) The nation's water resources, 1975–2000. Volume 1: Summary. US Government Printing Office, Washington, DC, 1.

Whittaker RH, Marks PL (1975) Methods of assessing terrestrial productivity. In Lieth H, Whittaker RH (Eds) *Primary production of the biosphere*. Springer-Verlag, New York.

Zahner R, Saucier JR, Myers RK (1988) Tree-ring model interprets growth decline in natural stands of loblolly pine in the southeastern United States. Can J For Res 19:612–621.

23. Scaling Up Physiological Responses of Loblolly Pine to Ambient Ozone Exposure Under Natural Weather Variations

Robert J. Luxmoore, Scott M. Pearson, M. Lynn Tharp, and Samuel B. McLaughlin

The challenge of scaling up (i.e., of incorporating such small-scale information as leaf physiology into larger-scale processes of the canopy, stand, or ecosystem) has been the subject of considerable discussion in recent years (e.g., Ehleringer and Field, 1993; King 1993; Luxmoore et al., 1991; Rastetter et al., 1991; Reynolds et al., 1992). Two particular concerns in scaling up from physiological to landscape scales are 1) accounting for relevant processes that express the behavior of the soil-plant-atmosphere system at particular scales, and 2) the integration and transfer of relevant information from one scale to the next. One approach for scaling up involves simulation with models of differing scales and the transfer of information from the smaller-scale to larger-scale simulators. This linked modeling approach addresses scaling concerns by 1) explicitly modeling processes at the scale of their operation using available mechanistic understanding and appropriate data, and 2) by explicitly transferring integrated information from one scale to the next through a hierarchy of linked models.

There are two major consequences in extending the space and time domains of soil-plant-atmosphere processes. The first one is an increase in the variability of soil, plant, and weather variables, and the second is the possible gain of larger-scale phenomena not represented at the small-scale level (Luxmoore et al., 1991). These two scaling-up consequences are accounted for in this investigation through the use of uncertainty analysis (i.e., accounts for increasing variability) and hierarchical information transfer between three modeling scales (i.e., accounts for changing phenomena at different scales (Figure 23.1) (Luxmoore, 1992).

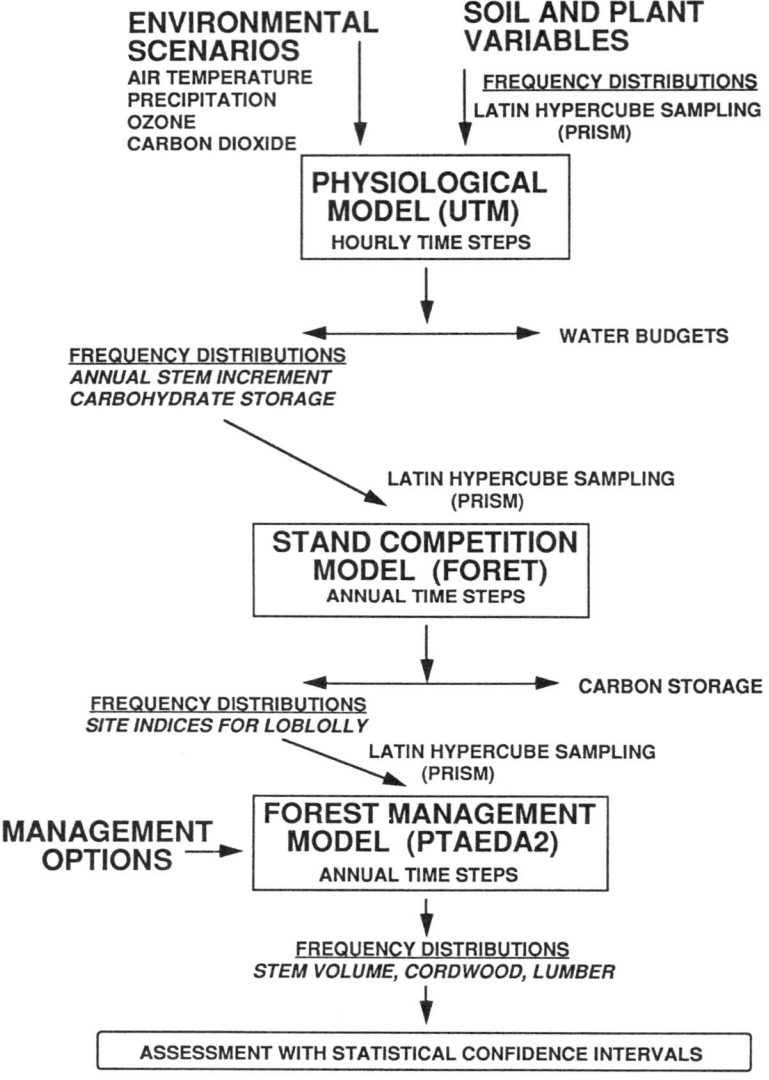

Figure 23.1. Linkage of a physiological growth model (UTM), a stand competition simulator (FORET), and a forest management model (PTAEDA2) for evaluation of long-term growth responses of loblolly pine to variation in ozone exposure and climate (adapted from Luxmoore, 1992).

We report the development and application of this scaling-up approach in an examination of ambient ozone effects on the growth of loblolly pine (*Pinus taeda* L.) plantations. McLaughlin and Downing (1995) recently reported decreases in stem-growth rate of mature loblolly pine trees during periods of high concentrations of tropospheric ozone in combination with water stress or elevated air temperature. We have used modeling to evaluate ambient ozone effects on the productivity of loblolly pine plantations under natural variation in weather conditions.

Modeling Methods

The component models used in this analysis account for tree physiological processes (Unified Transport Model), competition between individuals in a forest stand (FORET), and forest plantation management (PTAEDA2). The combination of these three models provides a comprehensive basis for scaling up (Figure 23.1). The steps taken in this scaling-up investigation are

1. Installation of the physiological, competition, and stand-management simulators on UNIX workstations,
2. Linkage of sensitivity analysis and uncertainty analysis with each simulator for propagation of frequency distributions of sensitive input variables through each simulator,
3. Incorporation of algorithms for ozone effects on physiological processes into the physiological model,
4. Application of the models to a field monitoring study of McLaughlin and Downing (1995) conducted on the Oak Ridge Reservation in eastern Tennessee from 1988 to 1992.

A particular feature of our scaling-up approach is the use of sensitivity and uncertainty analyses with each component of the modeling hierarchy. Sensitivity analysis identifies input variables with significant influence on model outputs. In uncertainty analysis, the identified sensitive input variables are described by frequency distributions representative of the heterogeneity of the soil-plant-atmosphere system under consideration. These frequency distributions capture the extended ranges of variables appropriate for scaling up. Propagation of these distributions through the models generates outputs in the form of frequency distributions. Some of these output distributions become input distributions to the next simulator of the modeling hierarchy. The final outputs from scaling up are frequency distributions of plantation growth and productivity that can be statistically compared for alternative ozone exposure scenarios. The three models, outlined in the next sections, have been developed over many years and have been tested in a range of applications.

Unified Transport Model—Carbon, Water, and Nutrient Coupled Processes Model

The Unified Transport Model (UTM) is a linkage of five component simulators of whole plant physiological processes conducted within a watershed framework of a hydrologic transport simulator (Luxmoore, 1989). The diurnal cycle of 1) soil-plant-water relations (PROSPER), 2) photosynthesis, respiration, translocation, growth, mortality, and litter decomposition (CERES), 3) nutrient uptake by diffusion and mass flow to roots (DIFMAS), 4) foliar uptake of gaseous (ozone) and particulate pollutants, in addition to chemical movement within the plant

(DRYADS), and 5) soil-chemical adsorption (CADIL) provide a comprehensive representation of coupled soil-plant-atmosphere processes at a consistent level of detail for both aboveground and belowground processes.

The UTM uses hourly time steps that reduce to fifteen-minute steps during precipitation events. Ozone uptake calculations apply a concentration gradient/ resistance equation using hourly tropospheric ozone concentration data and a zero ozone concentration within needles (Laisk et al., 1989). The boundary layer and stomatal resistances for ozone uptake are modified from the PROSPER values for water-vapor transport according to the ratio of diffusion coefficients for ozone and water vapor. Translocation is driven by sucrose gradients operating in a source (photosynthesis)—sink (growth, respiration) framework that results in dynamic carbon allocation responses to stress (Luxmoore, 1991).

The UTM simulates annual increments in stem growth and internal carbohydrate storage for a stand; these values are denoted as potential responses because competition between individuals is not considered. The UTM determines ozone effects on annual stem growth for variable weather conditions of the five-year field study of McLaughlin and Downing (1995). The ratio of these results obtained with and without ozone exposure provides a stem-growth multiplier (0 to 1 range) that is supplied to the stem-growth algorithm of FORET. This index is an integrated form of whole plant response to ozone exposure compatible with the modeling structure of FORET.

FORET—Forest Stand Growth and Competition Model

FORET simulates diameter and height growth, establishment, and mortality of trees using annual time steps to generate successional dynamics of mixed age-class forest stands of eastern North America (Shugart and West, 1977). The diameter of individual trees is simulated for a circular 1/12 hectare (ha) plot; diameter growth is incremented annually according to a function that produces 2/3 of a tree's diameter at breast height (DBH) at one-half of its age of longevity.

Stochastic variations in monthly air temperature and precipitation, derived from long-term climate records for the Oak Ridge area, are used in the calculation of evapotranspiration and soil-water storage. The model computes the monthly water budget to determine available soil-water and the number of days with low water availability. Stem growth is reduced according to a growth factor defining tree tolerance to water stress. Additional growth reduction factors are determined from total growing degree days (5.6 °C base), shading from the total leaf area of taller trees, and crowding from other trees on the plot. Direct temperature and precipitation effects on forest growth involve both increases and decreases in productivity, depending, in part, on whether soil-water becomes less or more limiting to growth. The optimum growth for a tree is reduced when the tree is exposed to a less-than-optimum climate, is shaded, or becomes crowded. The most limiting of these growth-reduction factors is chosen as the growth-reduction variable in any particular year. Recruitment and mortality of individual trees in a stand are stochastically determined by mammal browsing, sprouting characteristics, and expected longevity. FORET generates realized growth responses in a competitive environment from an average of twenty-five simulated plots, which is sufficient for averaging stochastic climate, recruitment, and mortality effects.

The code is applied as a single-species stand with 1334 stems/ha for simulation of loblolly pine plantations. A stem-growth multiplier from the UTM provides an integrated effect of ozone exposure; ozone concentration data are not directly used in the competition model. FORET generates site index by simulation of dominant and codominant tree heights at a tree age of twenty-five years. The site index results obtained with and without ozone exposure are used as input to the plantation management model.

PTAEDA2—Loblolly Pine Plantation Management Model

Using annual time steps, PTAEDA2 is a distance-dependent, individual tree growth and yield model for loblolly pine that simulates a plantation from the time of planting through a desired rotation (Burkhart et al., 1987). Trees are assigned coordinate locations in a stand and annual growth is determined as a function of tree size, site index, and competition from neighbors. Growth increments are adjusted stochastically for genetic and local site variability. Mortality is determined randomly by a variable based on competitive relationships and crown ratio. Subroutines for hardwood competition, stand thinning, and fertilization also determine tree and stand development. Options for varying individual tree spacing can also be used to mimic variability in machine and hand-planting operations. The mensuration data used to develop the empirical relationships in PTAEDA2 came from a wide range of loblolly stands distributed through the southern United States. (Burkhart et al., 1985; Burkhart, 1987). Site index simulations from FORET incorporate ambient ozone effects on loblolly pine under variable weather conditions. There are no specific inputs of climate or ozone concentration data to PTAEDA2. Site index is useful as an index of site quality because it is insensitive to variation in stand density and has been found to be a quantitative measure of site productivity. Simulations with PTAEDA2 may be used to evaluate alternative management adjustments (e.g., planting density, thinning, fertilization) to air pollution and climate impacts.

Sensitivity and Uncertainty Analysis

The PRISM code of Gardner et al., (1983) provides sensitivity and uncertainty analyses when linked with simulation models. We linked PRISM to the UTM, FORET, and PTAEDA2 to account for the effects of variability of model inputs. This variation is represented by frequency distributions derived from available data sources or by assuming representative distributions. These frequency distributions are used in PRISM to generate input data sets with an efficient Monte Carlo sampling procedure called Latin hypercube sampling. Briefly, all input frequency distributions for a given model are divided into N equal probability classes and these are sampled without replacement to generate N input data sets for the model. In this way, the complete input distributions are included in the input data sets. An N of 100 was used in all applications with the three models. Either random or correlated relationships between variables can be specified and incorporated in the input data sets generated; we used random relationships between all input variables for all simulations. Each model is sequentially run 100 times with the 100 input data sets and selected output variables are assembled into frequency distributions for propagation through the modeling hierarchy. A valuable feature of this scaling-up approach is that statistically based comparisons may be made between output distributions obtained for alternative modeling scenarios. In our applications, we compare the results obtained both with and without ambient ozone exposure.

Adaptation of the Unified Transport Model for Simulation of Ozone Effects

Simulations were conducted on a calendar year basis and were initiated for a mature loblolly plantation with a leaf area index (LAI) of 3, an aboveground phytomass of 118 Mg/ha, and a total phytomass of 147 Mg/ha. Several changes to the UTM were made to account for ozone effects on physiological processes. Cumulative ozone exposure during the growing season and daily ozone exposure values were determined for the daylight hours by summing hourly ozone concentration data. We arbitrarily included a cumulative ozone exposure of 40 ppm hours for winter exposure during the October to March period prior to the growing season (Sasek and Richardson, 1992). A 12- to 14.5-hour photoperiod, including hours 7:00 to 19:00, was used for determining daily ozone exposure during the growing season. The exposure units used in this report of ppm hours and ppb hours correspond on a volumetric basis to $\mu L\ L^{-1}$ hours and $\eta L\ L^{-1}$ hours, respectively.

Ozone Effects on Initiation of Leaf Senescence and Needle Retention

Cumulative ozone exposure was used to decrease the time at which foliar senescence was initiated. Guided by the results of Stow et al., (1992), we reduced the initiation of needle senescence in autumn by one day for each 33.3 ppm hours of cumulative ozone exposure in excess of 150 ppm hours. Also, enhanced needle senescence, resulting in reduced foliage overwintering into the next growing season, was represented by an increase in leaf mortality equivalent to 1 g/m² for each 20 ppm hours that cumulative ozone exposure exceeded 150 ppm hours.

Ozone Effects on Leaf Respiration

We chose an arbitrary function to increase foliar respiration with increase in ozone exposure. Needle respiration rate was made a linear function of cumulative ozone exposure, doubling at exposures of 60 ppm hours and tripling at 120 ppm hours. A threshold exposure value was not used in this algorithm.

Ozone Effects on Relative Photosynthesis

Richardson et al., (1992) reported the following exponential relationship between cumulative ozone exposure (in units of ppm hours) and the relative rate of photosynthesis:

$$\text{Relative photosynthesis} = \exp[-(\text{cumulative ozone exposure}/W)^L]$$

in which $W = 224$ and the exponent $L = 3.18$ are mean values determined from results for three loblolly genotypes. Relative photosynthesis varies over the interval (0, 1) with a relative response of 0.5 at an exposure of 200 ppm hours. At 100 ppm hours of ozone exposure, relative photosynthesis is 0.93. This function does not provide any representation of such short-term ozone impacts as may occur during summer air stagnation events. The relative photosynthesis response was applied to ozone directly as a modifier of photosynthesis calculated with the carbon dixoide (CO_2) gradient/resistance equation used in the UTM. Photosynthesis is reduced with buildup of cumulative ozone exposure; thus, ozone has increasing effects later in the growing season. Richardson et al., (1992) demonstrated that ozone effects on photosynthesis mainly impacted on biochemical mechanisms. Concentrations of chlorophyll and carotenoids were lowered with increasing cumulative ozone exposure. Sasek and Richardson (1989) and Richardson et al., (1990) hypothesized that ozone caused destruction of photosynthetic pigments.

Loss of Stomatal Regulation With Elevated Daily Ozone Exposure

McLaughlin and Downing (1995) demonstrated an association between buildup of ozone during air stagnation events and reduced stem growth, if accompanied by elevated temperatures or water stress. We have arbitrarily introduced stomatal modeling adjustments as a hypothesis that loss of stomatal function occurs on days with elevated ozone concentrations. Following calculation of daily ozone exposure above a daily threshold of 300 ppb hours, two variables determining foliar water relations are modified for the remainder of the day. The minimum stomatal resistance is divided by five and the plant-water potential associated with maximum stomatal resistance is reduced by a factor of two. On the next day, these variables are set back to the nonstress values until the next daily threshold limit is exceeded. McLaughlin and Downing (1995) have predicted a decrease in loblolly pine growth at daily ozone exposures above 300 ppb hours and Johnson et al., (1995) have shown that ozone exposure in slash pine (*Pinus elliottii* Engelm.) seedlings can lead to high needle conductance of the youngest needle age-class.

Our modifications to stomatal regulation result in lower stomatal resistance, reduced plant-water potential, enhanced transpiration, and enhanced soil-water utilization in simulations with ambient ozone exposure. Reduced stomatal resistance initially favors a small increase in photosynthesis, as well as a large increase in transpiration. Enhanced plant-water loss leads to low plant-water potentials and reduced tissue growth. Slower growth may eventually cause down-regulation of photosynthesis in the UTM modeling structure, driven by sucrose accumulation in foliage (Luxmoore, 1991).

Adaptation of FORET for Simulation of Loblolly Pine Plantations

Growth and yield information from Fowells (1965) was used to determine the species-specific input values for loblolly pine appropriate for plantation conditions. The temperature (degree day) control of tree growth was inactivated because loblolly pine is out of its normal range in the Oak Ridge, TN area and the algorithm excessively restricted tree growth. Recruitment was also inactivated because this is not relevant for a managed plantation. For simulation purposes, the loblolly plantation was established in FORET at a tree age of three years with a stem DBH of 1.27 cm. A planting density of 1334 stems/ha was established. The tree mortality algorithm allowed 3% mortality during years three to five. Simulations were conducted for twenty-five years and the heights of dominant and codominant trees were determined for a tree age of twenty-five years.

Scaling-Up Application

Data applicable to the field monitoring study of McLaughlin and Downing (1995; see Chapter 12), conducted in mature loblolly pine stands of the Oak Ridge Reservation, were used in this scaling-up study. Precipitation, meteorological, and ozone concentration data for the Oak Ridge area were assembled for UTM simulation of the 1988 to 1992 field-study period. Hourly ozone concentration data were obtained for the April to September periods of 1988 to 1992 from the Environmental Protection Agency (EPA) monitoring station in Knoxville, TN. Soil and plant variables were selected to represent the podzolic soil (Ultisol) and loblolly pine attributes of the field sites. Mean soil-hydraulic properties were taken from Luxmoore (1983), and soil heterogeneity was represented with a lognormal distribution of scaling factors (Sharma and Luxmoore, 1979) with a mean of 1.0 and a standard deviation of 0.2. Variability of sensitive plant and litter variables was represented with normal distributions, including the two variables (W, L) relating relative photosynthesis and cumulative ozone exposure.

Preliminary UTM simulations were conducted with 1987 meteorological conditions to remove the effects of initial conditions from predictions. Soil-water content values simulated for the end of 1987 were used as input for simulations with the 1988 data. This process was repeated for each of the years from 1988 to 1992 to incorporate carryover effects of stored soil water on water budget and plant-growth calculations.

The ratio of annual stem-sapwood growth plus carbohydrate storage, obtained with and without ozone (0 to 1 range), provided an integrated plant response to ambient ozone exposure. A frequency distribution of these values was generated from 100 UTM simulations. This distribution was transferred as a growth multiplier to the stem-growth algorithm of FORET. One hundred simulations with FORET generated a frequency distribution of site index values for ambient ozone exposure, which was transferred to PTAEDA2. A deterministic simulation (without heterogeneity) of site index was made with FORET for the case without ozone exposure using a stem-growth multiplier of unity. In our procedure, one variable (stem-growth multiplier) was transferred from the UTM to FORET and one variable (site index) was transferred from FORET to PTAEDA2 to propagate ozone effects from the physiological scale to the plantation-management scale.

Applications were made by PTAEDA2 with a tree density of 1537/ha hand-planted on a grid with 3 m between rows and 2.1 m between trees within a row. A simulated thinning cut was applied after seventeen growing seasons to obtain a basal area not exceeding 15 m^2/ha at this stage of development. A plantation timber harvest was made at thirty-five years. One hundred simulations were made with PTAEDA2 using frequency distributions of site index obtained with and without ozone exposure. Timber harvests were calculated with algorithms that estimated stem volume of trees > 2.6 cm DBH, cordwood yield of trees > 12.8 cm DBH, and marketable lumber obtained from trees > 20.5 cm DBH. Statistical t-tests were used to compare output frequency distributions (N = 100) obtained with the modeling hierarchy. The high number of degrees of freedom (200 − 2) in the t-tests favors determination of small, significant differences between means.

Results

Annual UTM simulations, conducted for the 1988 to 1992 period, were made with 1) mean values for input variables in deterministic applications (ie., without heterogeneity), and 2) with Latin hypercube sampling involving frequency distributions for sensitive input variables. Simulations were made with and without ambient ozone concentration data giving a set of ten simulations (five years × two ozone) for each of the deterministic and uncertainty approaches.

Deterministic Results from the Unified Transport Model

Cumulative ozone exposures for the 1988 to 1992 growing seasons showed a range of 70 to 102 ppm hours for the five growing seasons, and these were associated with simulated foliar uptake values from 0.9 to 1.5 g O_3/m^2 (land-area basis). We note that the first growing season cumulative ozone exposure was initiated at 40 ppm hours to account for winter exposure prior to the growing season.

In the water budget results, precipitation ranged from 1018 to 1799 mm/year and evapotranspiration increased with ozone exposure relative to the results obtained without ozone (Table 23.1). The evapotranspiration simulations obtained with ambient ozone exposure may be excessive because the stomatal resistance algorithm used in the UTM modeling of ozone effects was selected arbitrarily. Lower soil-water content and reduced drainage accompanied higher evapotranspiration (Table 23.1). Leaf area indices ranged between 3 and 8 in the simulations and were significantly reduced in association with the loss of stomatal

Table 23.1. Simulated Evapotranspiration, Drainage, and Soil-Water Storage Change

	Precipitation (mm)	Evapotranspiration (mm)	Drainage (mm)	Soil Storage Change (mm)
Year Without Ozone				
1988	1241	833	287	121
1989	1799	683	1142	−26
1990	1546	788	689	69
1991	1471	755	694	22
1992	1018	745	287	−14
Mean	1415.0	760.8	619.8	34.4
Standard Deviation	298.0	55.4	355.1	60.9
Year With Ozone				
1988	1241	989	14	238
1989	1799	742	844	213
1990	1546	939	515	92
1991	1471	892	622	−43
1992	1018	872	291	−145
Mean	1415.0	886.8	457.2	71.0
Standard Deviation	298.0	92.7	317.9	164.4

regulation during days with elevated ozone concentration, which led to water stress and inhibition of needle growth. Inclusion of ozone in the simulations resulted in a 19% decrease in LAI for 1988, a year with reduced spring rainfall and high ozone concentrations during the growing season (Figure 23.2).

Uncertainty Applications

Selected results are presented for each model of the modeling hierarchy. In most cases, the output distributions were normally distributed but there was a tendency for right-sided skewness (to higher values) in some instances.

Results from the UTM

Cumulative annual respiration (all woody tissues plus nighttime needle respiration) was lower in simulations without ozone than with ozone as shown in the frequency distribution results for 1989 (Figure 23.3). This was a consistent result for all five years with the mean annual respiration showing a range from 2394 to 2708 g/m^2 without ozone and from 2562 to 2987 g/m^2 with ozone.

Cumulative annual net photosynthesis was higher in 1990 simulations with ambient ozone exposure than without ozone (Figure 23.4). Similar results occurred in four of the five annual simulations (not in 1988 as a result of low LAI). The cumulative ozone exposures (maximum of 142 ppm hours) were sufficient to slightly reduce photosynthesis late in the growing season of some years. Mean annual photosynthate ranged from 3631 to 3936 g/m^2 without ozone and from 3567 to 4087 g/m^2 with ozone. Enhanced photosynthesis rates with ambient

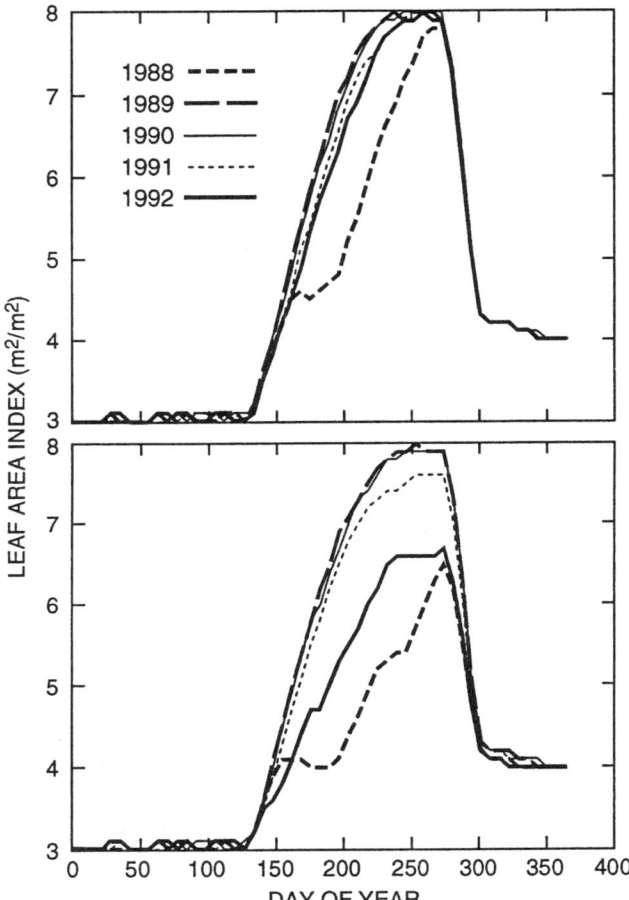

Figure 23.2. Leaf area indices for loblolly pine during five years (1988 to 1992) simulated with the UTM without ambient ozone (upper) and with ambient ozone exposure (lower).

ozone exposures were a consequence of 1) algorithms that allowed respiration to have a higher sensitivity to ozone exposure than photosynthesis, 2) the source-sink modeling framework of the UTM that favored enhanced photosynthesis when the sink for photosynthate increased (e.g., ozone-induced respiration), and 3) loss of stomatal control during days when daily ozone exposures exceeded 300 ppb hours (lower stomatal resistance favored photosynthesis).

Stem-sapwood plus carbohydrate storage was higher without ozone than with ozone, as shown in the results for 1991 (Figure 23.5). All simulations were initiated with a stem-sapwood of 1500 g/m^2. A t-test conducted with the means and standard deviations for the stem-sapwood plus storage frequency distributions showed ozone exposure caused a significant reduction (0.1% significance level)

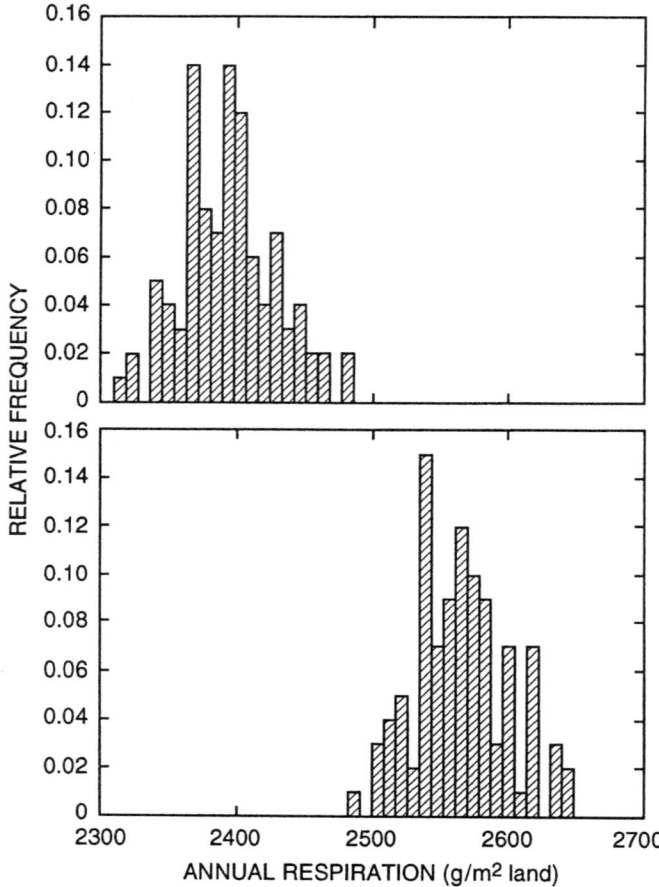

Figure 23.3. Relative frequency of annual respiration simulated for 1989 with the UTM using Latin hypercube sampling without ambient ozone (upper) and with ambient ozone exposure (lower).

in each of the five annual simulations. The mean annual stem-sapwood plus carbohydrate storage showed a range from 1709 to 1783 g/m² without ozone and from 1563 to 1725 g/m² with ozone during the years 1988 to 1992. Annual stem-sapwood plus storage was reduced on average by 91 g/m² with ambient ozone, equivalent to a mean reduction of 5.4% in sapwood mass. The stem-growth multiplier for ambient ozone effects had a mean of 0.946 and a standard deviation of 0.35. Without ambient ozone, the stem-growth multiplier was 1.0 and this was used in FORET as a deterministic simulation without heterogeneity.

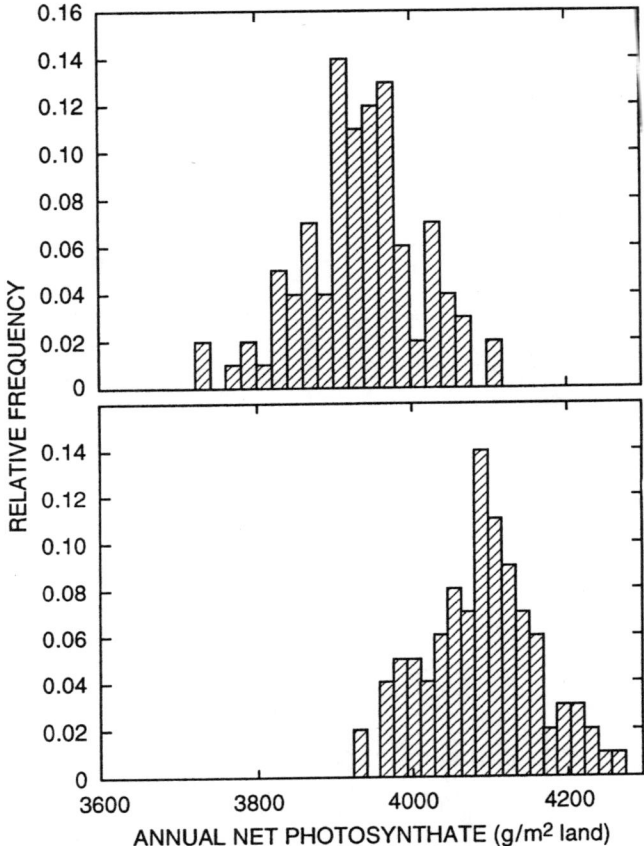

Figure 23.4. Relative frequency of annual photosynthate simulated for 1990 with the UTM using Latin hypercube sampling without ambient ozone (upper) and with ambient ozone exposure (lower).

The mean annual evapotranspiration for the five years was 732 mm (somewhat less than results for the deterministic case, Table 23.1) with a range from 655 to 808 mm in simulations without ozone. The loss of stomatal regulation on days with ozone exposures in excess of 300 ppb hours caused greater evapotranspiration for ambient ozone simulations as shown in the results for 1988 (Figure 23.6). Mean annual evapotranspiration ranged from 714 to 965 mm in the simulations with ozone. Volumetric soil-water content in the root zone was higher without ozone than with ozone, particularly during the summer, as shown in results for the end of August, 1992 (Figure 23.7).

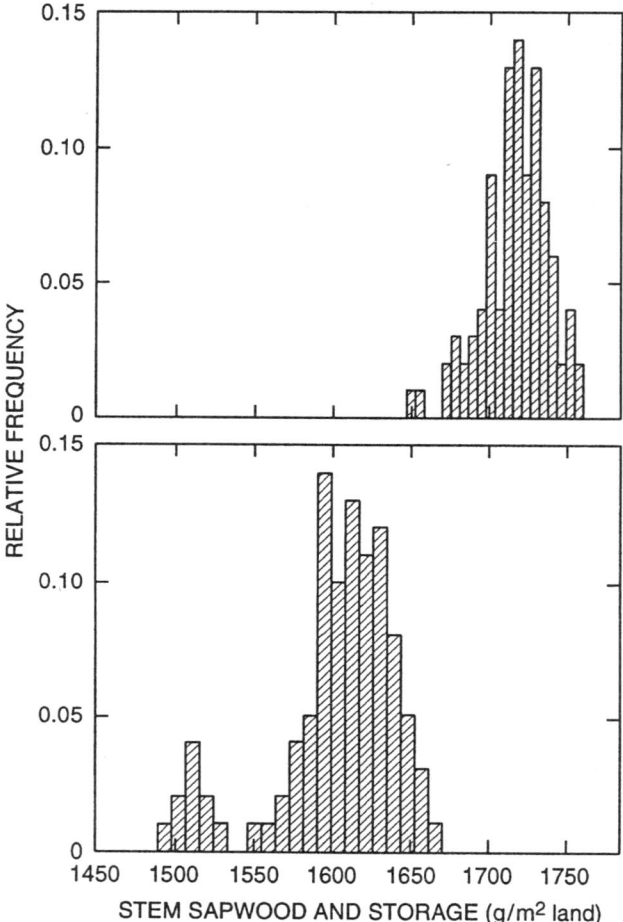

Figure 23.5. Relative frequency of stem-sapwood and storage simulated for 1991 with the UTM using Latin hypercube sampling without ambient ozone (upper) and with ambient ozone exposure (lower).

Results from FORET

Simulations of tree height from FORET for dominant and codominant individuals at age twenty-five years showed a mean of 19.0 m with a standard deviation of 0.35 m with ambient ozone exposure (Figure 23.8). This was a 5% decrease in site index in comparison to the deterministic result of 20.0 m obtained for site index without ozone. The site index for this latter case was assumed to have a frequency distribution with a standard deviation identical with that obtained for ozone exposure. Applying a standard deviation of 0.35 m for the simulation without ozone gave a t-test result suggesting a significant ($< 0.01\%$ significance level) reduction in site index with ambient ozone exposure.

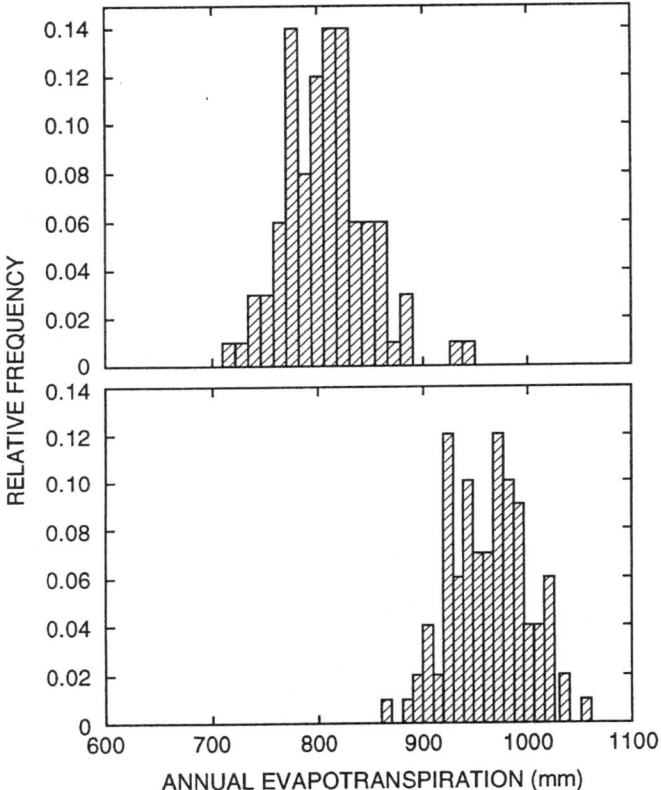

Figure 23.6. Relative frequency of annual evapotranspiration simulated for 1988 with the UTM using Latin hypercube sampling without ambient ozone (upper) and with ambient ozone exposure (lower).

Results from PTAEDA2

Simulation of two ozone exposure cases were conducted for loblolly pine plantations with the frequency distributions of site index developed from FORET. The PTAEDA2 results suggest that ambient ozone exposure impacts several attributes of trees harvested at the seventeen-year thinning harvest, and the rotation harvest at year thirty-five. Average DBH, average tree height, crown ratio, density, and total basal area of trees thinned at seventeen years were higher for the plantation simulated without ozone than with ozone exposure (Table 23.2). Thus, a greater volume of trees were harvested at seventeen years from the plantation with higher site index. In contrast, the density and total basal area of trees harvested at thirty-five years were higher with ozone exposure resulting in slightly higher stem volume and cordwood yield (one cord is a wood stack 8 × 4 × 4 feet) at the final harvest. However, a lower lumber volume (in units of board feet; 1 square foot × 1 inch thickness) was simulated with ambient ozone exposure resulting from a low density of trees with DBH > 20 cm. The t-test comparisons between ozone

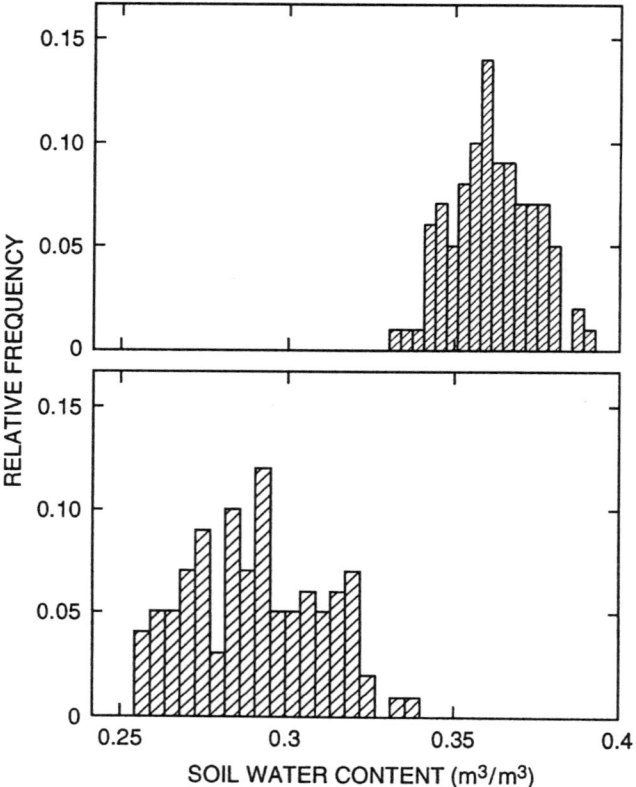

Figure 23.7. Relative frequency of soil-water content for the 0–35 cm depth at the end of August, 1992, simulated with the UTM using Latin hypercube sampling without ambient ozone (upper) and with ambient ozone exposure (lower).

exposure simulations showed many highly significant results for rather small mean differences (Table 23.2) resulting from the high number (198) of degrees of freedom.

The frequency distribution of lumber volume at the thirty-five-year harvest was significantly higher and exhibited less variability for the plantation without ozone than for the ambient ozone exposure simulation (Figure 23.9). The ambient ozone exposure case had a bimodal lumber volume distribution. The total lumber volume for the two harvests was 6% lower with ambient ozone exposure (Table 23.2) showing the possibility that a small annual effect (5.4%) at the physiological scale can lead to significant impacts on timber yield over a thirty-five-year plantation rotation.

Discussion

Investigations by Sheffield and Cost (1987) and Zahner et al., (1990) have suggested that growth declines in some southern pine stands over the last thirty years

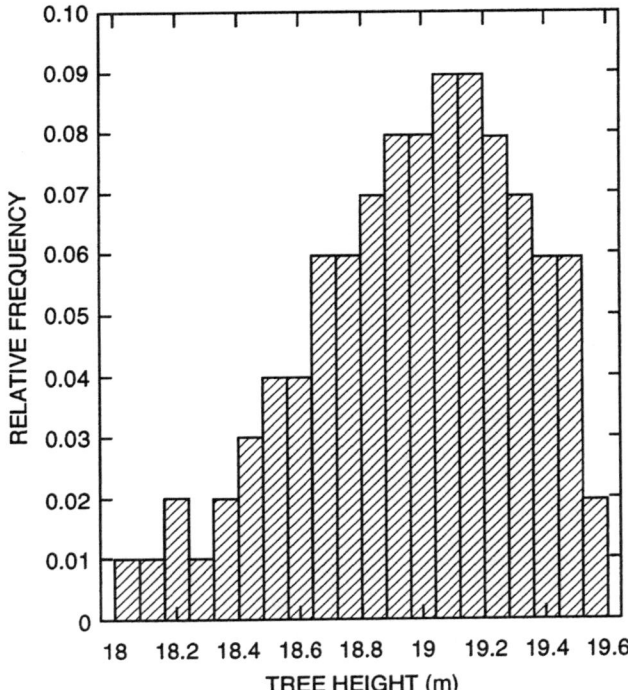

Figure 23.8. Relative frequency of site index (height of dominant and codominant trees at twenty-five) simulated with FORET using Latin hypercube sampling for ambient ozone conditions. This distribution has a mean of 19.0 m. For comparison, results obtained without ozone had a site index of 20.0 m.

are the result of a decline in air quality or some factor other than climate variation. Because much of the experimental information on tree responses to air pollution has been obtained for physiological processes, this scaling-up approach offers one means for evaluating ambient ozone effects on the productivity of loblolly pine plantations under variable meteorological conditions. Ozone effects were specifically represented in the UTM only. There were no specific ozone algorithms used in FORET or PTAEDA2. The scaling-up method resulted in many physiological variables being integrated into one input variable (stem-growth multiplier) for the competition model and one input variable (site index) for the plantation-management model.

The simulation results presented here are more of an example application of scaling up than a prediction of ambient ozone effects. We do not advocate quantitative interpretation of our modeling results at this stage of development because many assumptions have been made that require further examination and justification. These assumptions are hypotheses of tree physiological responses to ozone (eg., respiration, stomatal regulation) that were made because suitable data are not

Table 23.2. Plantation Attributes for Loblolly Pine

Plantation Attribute	Unit	No Ozone Mean	No Ozone Standard Deviation	Plus Ozone Mean	Plus Ozone Standard Deviation	t-test Prob > |t|
Seventeen-Year Thinning						
Average DBH	cm	15.0	0.12	14.4	0.21	< 0.001
Average height	m	13.9	0.15	13.3	0.21	< 0.001
Crown ratio	m/m	0.357	0.0011	0.359	0.008	< 0.001
Tree density	stems/ha	860	12.4	837	4.5	< 0.001
Total basal area	m²/ha	16.0	0.46	14.3	0.46	< 0.001
Stem volume	m³/ha	111.3	4.08	95.7	4.54	< 0.001
Cordwood yield	cord/ha	36.0	1.51	29.8	1.91	< 0.001
Lumber yield	board feet	46	43	0	0	< 0.001
Thirty-Five-Year Harvest						
Average DBH	cm	31.3	0.44	30.1	0.25	< 0.001
Average height	m	24.8	0.46	23.3	0.32	< 0.001
Crown ratio	m/m	0.317	0.0016	0.324	0.0016	< 0.001
Tree density	stems/ha	282.8	16.3	326.7	10.8	< 0.001
Total basal area	m²/ha	21.6	0.74	23.1	0.61	< 0.001
Stem volume	m³/ha	247.6	5.77	249.8	6.74	0.014
Cordwood yield	cord/ha	92.1	2.13	93.1	2.48	0.005
Lumber yield	board feet	10538	292	9911	380	< 0.001
Total Harvest						
Stem volume	m³/ha	358.8	5.72	345.5	7.7	< 0.001
Cordwood yield	cord/ha	128.1	2.12	122.8	2.82	< 0.001
Lumber yield	board feet	10584	317	9911	380	< 0.001

23. Scaling Up Physiological Responses to Ambient Ozone Exposure 425

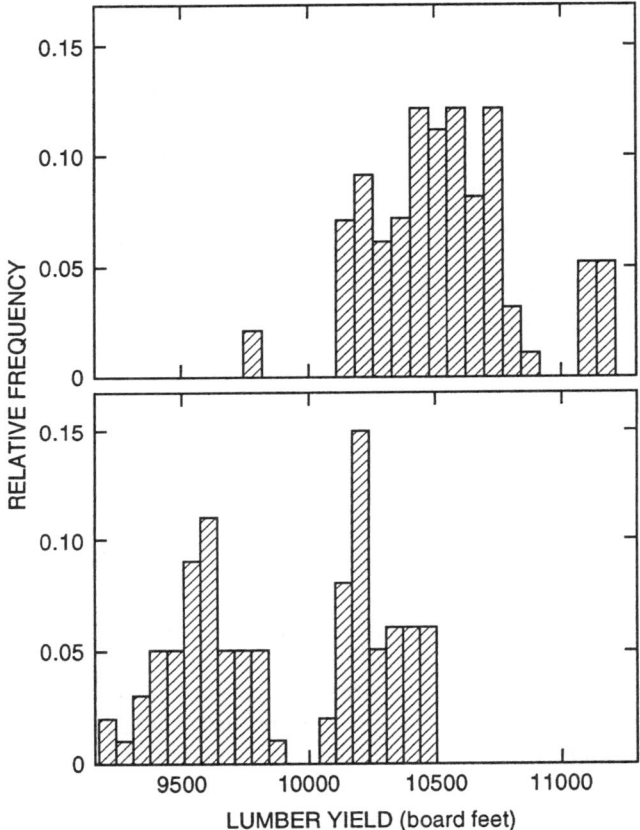

Figure 23.9. Relative frequency of lumber volume (board feet) harvested at thirty-five years in simulations with PTAEDA2 using Latin hypercube sampling without ambient ozone (upper) and with ambient ozone exposure (lower).

presently available for mature trees. Hanson et al., (1994) have shown that seedlings of red oak (*Quercus rubra*) are less sensitive to ozone exposure than older trees, highlighting the need for caution in the use of seedling data for forestry applications. Experimental investigation of the physiological responses of mature trees to ambient ozone are urgently needed.

The ambient ozone exposures examined in this study did not exceed 142 ppm hours. A considerable amount of physiological research has been conducted with ozone exposures of several hundreds of ppm hours. It is not clear that data from such high-exposure experiments will have ready application in regional assessments of ambient ozone impacts in the southern United States.

The present study suggests that small physiological responses to ambient ozone exposure can impact the longer time-scales of a plantation harvest cycle. This result contrasts with results from an earlier scaling-up study with the UTM and

FORET. Significant stem-growth responses of oak and hickory trees to elevated CO_2 obtained with the UTM were propagated to FORET as a stem-growth multiplier (Luxmoore et al., 1990). The FORET simulations showed an initial increase in forest growth but this did not persist beyond a few decades. Variation in forest phytomass, resulting from stochastic variation in tree mortality and ingrowth of new individuals, eventually masked the physiological growth-enhancement response to CO_2 enrichment. In a plantation setting, where mortality and establishment are largely controlled or modified by thinning practices, physiological responses should not be masked during a plantation harvest cycle. Thus, physiological effects of ambient ozone, elevated CO_2, and climate change could significantly impact plantation growth and productivity.

Summary

This modeling study evaluated several physiological processes of mature trees that may be sensitive to ambient ozone concentrations. Conclusions from physiological modeling and the simulated ozone effects on the productivity of loblolly pine plantations, along with possible management adjustments to ozone impacts, are summarized as:

1. Loss of stomatal control, hypothesized to occur on days with ozone concentration above 300 ppb hours, resulted in increased plant-water stress and reduced tree growth. Diurnal studies of ozone effects on needle physiology and growth of mature loblolly pine are needed for evaluation of such short-term ozone impacts as may occur during summer air stagnation events.
2. Increase in plant-water stress with ozone exposure may have a more significant impact on tree growth (cell expansion) at ambient concentrations than disruption of photosynthesis.
3. Elevated needle respiration induced by cumulative ozone exposure was significantly compensated for by enhanced photosynthesis in the source-sink framework of the physiological model. Species having a greater sensitivity of respiration to ambient ozone exposure than that of photosynthesis may produce a similar response.
4. Cumulative ozone exposure at ambient concentrations had negligible or sometimes very small direct effects on photosynthesis. These effects occurred late in the growing season. Ozone-exposure research with seedling and sapling trees at annual cumulative exposures above 200 ppm hours may have limited applications in regional assessments of ambient ozone impacts on forested lands.
5. Annual stem-sapwood growth was reduced by a mean of 5.4% at ambient ozone concentrations in whole plant physiological simulations for five years with differing weather conditions. Transfer of this effect to a competition model resulted in a 5% decrease in site index determined at a tree age of twenty-five years.
6. Small reductions in tree growth and site index predicted for ambient ozone exposures were associated with a 6% reduction in lumber volume simulated

for a plantation harvest at thirty-five years. Stem volume and cordwood yield from the thinning plus final harvest were slightly lower with ambient ozone exposure. Cumulative effects of small physiological disturbances from ambient ozone concentrations may lead to a significant decline in timber quality and small reductions in total yield.

7. A higher density (15.5% higher) of trees was harvested at thirty-five years in simulations with ambient ozone exposure than for the simulations without ozone. Plantation sites with significant ozone impacts may require additional thinning to obtain trees with lumber-grade timber quality.

References

Burkhart HE (1987) Data collection and modeling approaches for forest growth and yield prediction. In Chappell HN, Maguire DA (Eds) *Predicting forest growth and yield: Current issues, future prospects.* Contribution 58. Inst For Res, Univ WA, Seattle.

Burkhart HE, Cloeren DC, Amateis RL (1985) Yield relationships in unthinned loblolly pine plantations on cutover, site-prepared lands. South J Appl For 9:84–91.

Burkhart HE, Farrar KD, Amateis RL, Daniels RF (1987) Simulation of individual tree growth and stand development in loblolly pine plantations on cutover, site-prepared areas. FWS-1-87. VA Poly Inst State Univ, Sch For Wildlife Resour, Blacksburg.

Ehleringer J, Field C (Eds) (1993) *Scaling physiological processes: Leaf to globe.* Academic Press, San Diego.

Fowells HA (1965) Silvics of forest trees of the United States. Agr Handbook No. 271. US Dep Agr, Washington, DC.

Gardner RH, Röjder B, Berström U (1983) PRISM: A systematic method for determining the effect of parameter uncertainties on model predictions. Studsvik Energiteknik AB Report NW-83-555. Nykoping, Sweden.

Hanson PJ, Wullschleger SD, Samuelson LJ, Tabberer TA, Edwards GS (1994) Seasonal patterns of light-saturated photosynthesis and leaf conductance for mature and seedling *Quercus rubra* L. foliage: Differential sensitivity to ozone. Tree Physiol 14:1351–1366.

Johnson JD, Byers DP, Dean TJ (1995) Diurnal water relations and gas exchange of two slash pine (*Pinus elliottii* Engelm.) families exposed to chronic ozone levels and acidic rain. New Phytol 131:381–392.

King AW (1993) Considerations of scale and hierarchy. p. 19–45. In Woodley S, Francis G, Key J (Eds) *Ecological integrity and the management of ecosystems.* Lewis Publishers Inc., Chelsea, MI.

Laisk A, Kull O, Moldau H (1989) Ozone concentration in leaf intercellular air spaces is close to zero. Plant Physiol 90:1163–1167.

Luxmoore RJ (1983) Water budget of an eastern deciduous forest stand. Soil Sci Soc Am J 47:785–791.

Luxmoore RJ (1989) Modeling chemical transport uptake and effects in the soil-plant-litter system. In Johnson DW, Van Hook RI (Eds) *Biogeochemical cycling processes in Walker Branch watershed.* Springer-Verlag, New York.

Luxmoore RJ (1991) A source-sink framework for coupling water, carbon, and nutrient dynamics of vegetation. Tree Physiol. 9:267–280.

Luxmoore RJ (1992) An approach to scaling up physiological responses of forests to air pollutants. In Flagler R (Ed) *The response of southern commercial forests to air pollution.* Air & Waste Management Association, Pittsburgh, PA.

Luxmoore RJ, King AW, Tharp ML (1991) Approaches to scaling up physiologically based soil-plant models in space and time. Tree Physiol 9:281–292.

Luxmoore RJ, Tharp ML, West DC (1990) Simulating the physiological basis of tree ring responses to environmental changes. In Dixon RK, Meldahl RS, Ruark GA, Warren WG

(Eds) *Process modeling of forest growth responses to environmental stress.* Timber Press, Inc., Portland, OR.

McLaughlin SB, Downing DJ (1995) Interactive effects of ambient ozone and climate measured on growth of mature forest trees. Nature 374:252–254.

Rastetter EB, Ryan MG, Shaver GR, Melillo JM, Nadelhoffer KJ, Hobbie JE, Aber JD (1991) A general biogeochemical model describing the responses of the C and N cycles in terrestrial ecosystems to changes in CO_2, climate, and N deposition. Tree Physiol 9:101–126.

Reynolds JF, Hilbert DW, Chen J-I, Harley PC, Kemp PR, Leadley PW (1992) Modeling the response of plants and ecosystems to elevated CO_2 and climate change. DOE/ER-60490T-H1. Carbon Dioxide Research Program, USDE, Washington, DC.

Richardson CJ, Sasek T, Di Giulio RT (1990) Use of physiological markers for assessing air pollution stress in trees. In Wang W, Gorsuch JW, Lower WR (Eds) *Plants for toxicity assessment.* ASTM STP 1091. American Society for Testing and Materials, Philadelphia, PA.

Richardson CJ, Sasek T, Fendick EA, Kress LW (1992) Ozone exposure—response relationships for photosynthesis in genetic strains of loblolly pine seedlings. For Ecol Manage 51:163–173.

Sasek T, Richardson CJ (1989) Effects of chronic doses of ozone on loblolly pine: Photosynthetic characteristics in the third growing season. For Sci 35:745–755.

Sasek T, Richardson CJ (1992) The dose—response approach for characterizing the effects of near ambient ozone concentrations on photosynthesis. In Flagler R (Ed) *The response of southern commercial forests to air pollution.* Air and Waste Management Association, Pittsburgh, PA.

Sharma ML, Luxmoore RJ (1979) Soil spatial variability and its consequences on simulated water balance. Water Resour Res 15:1567–1573.

Sheffield RM, Cost ND (1987) Behind the decline. J For 85:29–33.

Shugart HH, West DC (1977) Development of an Appalachian deciduous forest succession model and its application to assessment of the impact of the chestnut blight. J Environ Manage 5:161–179.

Stow TK, Allen HL, Kress LW (1992) Ozone impacts on seasonal foliage dynamics of young loblolly pine. For Sci 38:102–119.

Zahner R, Saucier JR, Meyers RK (1990) Tree-ring model interprets growth decline in natural stands of loblolly pine in the southeastern United States. Can J For Res 19:612–621.

24. Integrating Research on Climate Change Effects on Loblolly Pine: A Probabilistic Regional Modeling Approach

James E. Smith, Peter B. Woodbury, David A. Weinstein, and John A. Laurence

Synthesizing probable effects of climate change on such large, complex ecological systems as forests is not readily achieved through experimental manipulation. Therefore, numerical models and assessments based on "expert opinion" are often the bases for projections of future climate effects. The essential goal of each of these two processes is the same: scientific data from short-term, small-scale experiments are projected to larger temporal and spatial scales. Inherent uncertainties and biases are often obscured—usually in proportion to the amount of scaling necessary. Despite potential biases of models, they are an attractive means of synthesizing diverse data because quantitative expressions of predictions are possible.

Forest productivity has been most accurately and precisely modeled through statistical methods resulting in such models as PTAEDA2 (Burkhart et al., 1987). However, these have limited generality and their application may be difficult as climate changes alter normal patterns of growth and ecosystem processes. Another approach entails use of bottom-up mechanistic models that balance reality and generality (Sharpe and Rykiel, 1991), which differ widely according to purpose, structure, and assumptions. These models include: 1) FORET (Solomon, 1986), which is based on estimates of species-population-biology parameters, 2) MAESTRO (Wang and Jarvis, 1990), which explicitly addresses spatial aspects of canopy structure, and 3) G'DAY (Comins and McMurtrie, 1993), which is largely based on resource balance. Analyses of model predictions, including estimates of biases and range of expected results, are possible for these large mechanistic model structures through sensitivity and uncertainty analyses (Dale et al., 1988; Luxmoore et al., 1991; Summers et al., 1993). However, influences of individual model components on final model prediction are usually obscured in complex model structures that contain a myriad of elements and interactions. Conceptually, it is difficult to use these models to determine the influence of each process on prediction uncertainty, and the ways the model could be improved by future research are not easy to identify.

The use of data from one level of ecological organization to interpret and predict events at a different level will necessarily involve increased uncertainty. Research summaries, compilations of expert opinion, and simulation models are all liable to introduce different uncertainties and biases in predicting forest growth. Our model is focused on quantitatively expressing a synthesis of these uncertainties; it employs a minimal mechanistic structure for a tractable representation of the essential components of stand productivity potentially affected by climate change. We do not presume an a priori list of model elements that must be included to correctly predict forest growth, but rather derive these from results of both experimental and modeling research. Thus, we are developing a mechanistic model with a reduced structure relative to the three types of detailed models mentioned. Reduction of model complexity is regularly a part of development for most large, mechanistic models, but it is usually done for ease of computation, and therefore preserves original model assumptions and purpose. In contrast, our minimal model structure is based on adding model components only as their importance is demonstrated.

We are developing a rudimentary model structure based on linking published experimental results to components of annual stand carbon (C) budget. Our model relies on simple generalizations about how experimental data relate to annual carbon gain, loss, or allocation in a loblolly pine individual or stand. Most researchers have some notion of how their work relates to larger issues, even if only in a simple quantitative sense. Thus, the model utilizes both 1) minimum, mechanistic structure, and 2) input of experienced researchers to reduce the amount of scaling required by any single step. A necessary part of the compromise of sacrificing mechanistic detail for tractability is an increase in uncertainty in remaining model components, however, we feel that this is acceptable for developing syntheses useful for policy and management.

Experimental research results are summarized to express effects of climate changes on components of annual carbon budget for a loblolly pine stand. These summaries, expressed in probabilistic terms, are then compiled in Monte Carlo simulations to produce probabilistic estimates of induced climate change effects on stand growth across the region. Results of stand-level simulations are aggregated to express predictions across the twelve-state range of loblolly pine (from Texas to Virginia). A geographic information system (GIS) is created to integrate data, extrapolate stand results to the region, and create regional summaries and maps. The model is intended to demonstrate our approach to synthesizing data in a form amenable to needs of managers and policy-makers, and the results we present are preliminary. Our overall goal is to create a model that identifies and quantifies sources of uncertainty in an estimate of climate change effects on annual growth of loblolly pine forests.

Overview of Modeling Approach

In our study, we model change in loblolly pine basal-area growth rate as affected by climate change. The focus of the model is on the uncertainty of model predictions, particularly in defining contributions to uncertainty at the stand level and then propagating them to a regional assessment. Growth of loblolly pine was selected as the endpoint because measures of productivity for this species figure prominently among Southern Global Change Program (SGCP) research projects (Figure 24.1). The effects of climate on basal-area growth as predicted through three distinct modeling steps are 1) summaries of research are formulated as stand-level model inputs, 2) simulations produce estimates of stand-level growth, and 3) stand-level results are propagated to form a regional estimate of climate change effects on loblolly pine forests. We have named this modeling system SOuthern FOrest CLimate Effects Synthesis, or SOFOCLES.

Objective

We are developing this model system to use both experimental and modeling data as inputs. However, for the example presented in this chapter, we used results of SGCP experimental research projects as inputs. Our objective was to synthesize inputs from experimental results into a probabilistic estimate of change in loblolly pine stand-growth rate for the region. Therefore, the model includes

1. Summaries of experiments. Results of SGCP experimental research projects (Figure 24.1) are represented by one or more relatively simple numerical functions that summarize results and relate them to stand growth.
2. Quantitative expressions of uncertainty. Definitions of uncertainties, in the form of probability density distributions, are defined for model inputs and, thus, affect outputs at both the stand and regional levels.
3. A flexible and tractable model structure. A model structure readily amenable to redefinition or addition of inputs allows biologists and analysts to debate and identify key influences on model behavior and predictions.

Model Organization

Stand-level simulations of a change in annual loblolly pine growth rate are central to our modeling system. The stand model is structured to estimate annual stand carbon budget. Model prediction of annual basal-area growth is based on estimates of gross carbon gain by the canopy, carbon lost to respiration, and allocation to stem growth. To modify components of the annual carbon budget, results of individual research projects are generalized to describe an effect on either

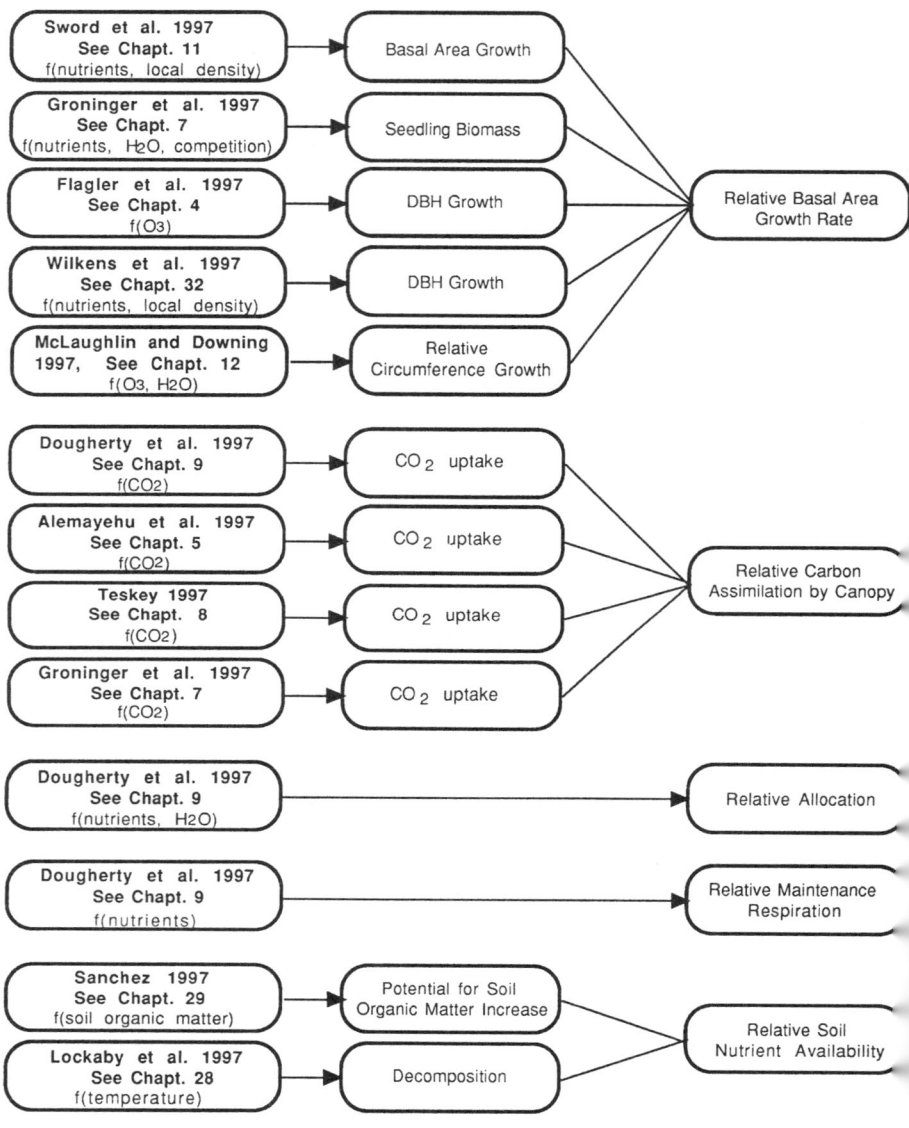

Figure 24.1. Southern Global Change Program experimental research projects used as input summary functions. Diagram indicates projects, experimental treatments (independent variables), dependent variables, and type of summary function formed.

basal-area growth or carbon gain, loss, or allocation. These probabilistic functions are incorporated in a Monte Carlo simulation of stand growth. A GIS facilitates use of regional data as input to a large number of such stand-level estimates across the region. The GIS is subsequently used as the principal means of presenting

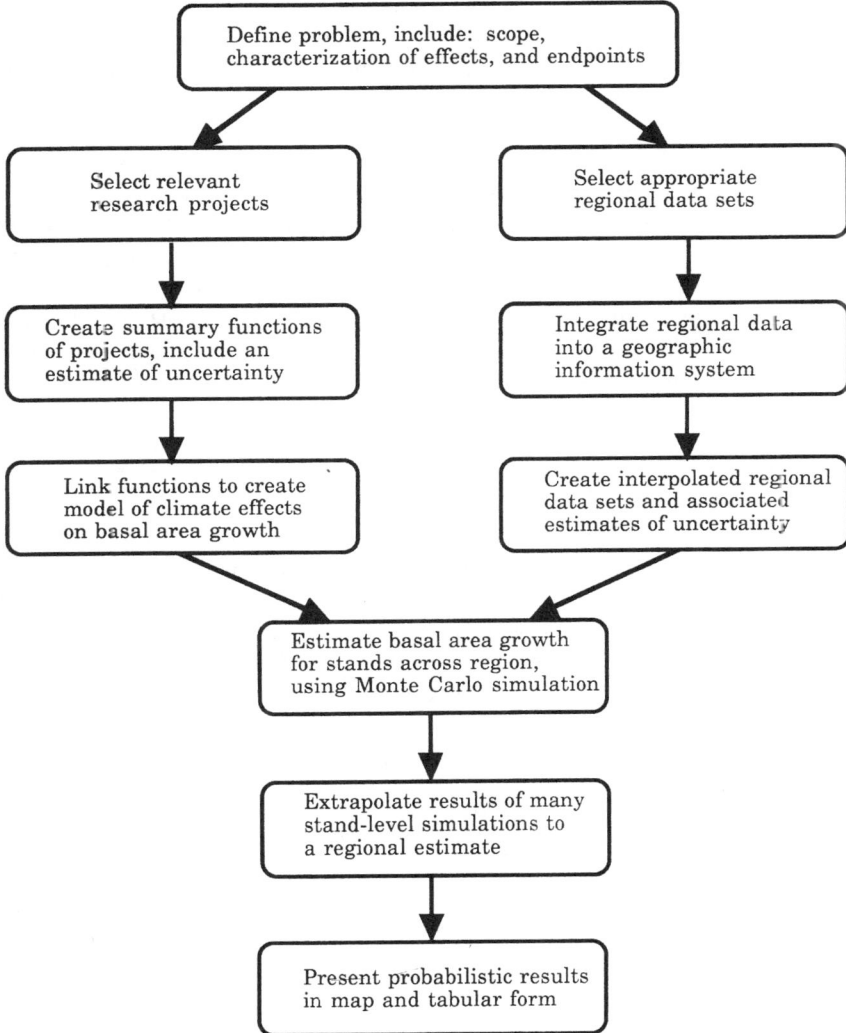

Figure 24.2. Steps taken in creating the regional assessment.

results through maps of probabilities of forest change for the twelve-state SGCP region. Figure 24.2 summarizes the general steps taken in the modeling process; these methods are described in further detail in the next section.

The term "uncertainty" is used here to describe the range of values possible for a variable. Uncertainty is quantitatively defined as a probability density distribution, that is, the relative possibility that a variable will take on a particular value along the defined range of workable values. Usually, statistical variability (i.e., error) of individual experiment summaries is only a minor part of overall variable uncertainty (Rowe, 1994). Uncertainty further increases, for example, as experimental results are pooled and extrapolated to express complex biological systems at the scales required for a regional assessment. Variable uncertainty can be represented as a single frequency distribution, even when derived from multiple sources.

The concept of a simple and flexible model structure is to allow any or all inputs, parameters, and mathematical relationships among variables to be defined in terms of probabilities, that is, with uncertainty. However, the only uncertainties included as inputs to this implementation of the model are the research summary functions defining changed growth in response to climate change. Present-day stand structure and climate, as well as future scenarios of climate are all model inputs defined without uncertainty. Obviously, treating these variables as certain produces somewhat unrealistic boundary conditions, but the present focus is on synthesis and scaling of experimental results.

Methods

The overall modeling process can be conveniently divided into two levels of organization—region and stand. The initial and final steps are at the regional level with the organization of region-wide data and presentation of results, respectively. Formation of research summary-functions and Monte Carlo simulations are at the stand level. Methods are presented as two sections.

Construction of Regional Analysis System

The analysis region consisted of the following twelve states: 1) Texas, 2) Oklahoma, 3) Louisiana, 4) Arkansas, 5) Mississippi, 6) Alabama, 7) Tennessee, 8) Florida, 9) Georgia, 10) South Carolina, 11) North Carolina, and 12) Virginia. The major steps required to construct the GIS and produce regional predictions are summarized in Table 24.1. Each of these steps is subsequently described in greater detail.

Table 24.1. Steps in SOFOCLES Regional Analysis

1. *Select data:* Collect available regional data on forest growth, distribution of forest type, climate, and soil.
2. *Create geographic information system:* Integrate all data into a single georeferenced database.
3. *Interpolate:* Create complete regional data sets by interpolation.
4. *Estimate uncertainty:* Create regional estimates of uncertainty for each type of data.
5. *Simulate climate effects:* Send data to the stand-level simulation model, which integrates research results with regional data to estimate growth rate.
6. *Produce regional analysis:* Extrapolate stand-model results to the region, calculate regional summary statistics, display summary maps.

Step 1. Select Data

Loblolly pine growth. Loblolly pine forest-growth data derived from the U.S.D.A. Forest Service forest inventory analysis (FIA) survey. Data were selected from survey plots (0.4 hectare (ha)) consisting primarily of loblolly pine that were largely undisturbed between consecutive surveys. Growth rates, stand density, and basal area were calculated for loblolly pine and other species. The most recent available data were selected for Alabama (1990), Arkansas (1988), Louisiana (1991), Mississippi (1987), Oklahoma (1993), Tennessee (1989), and Texas (1992). Data from the sixth survey were selected for Florida, Georgia, South Carolina, North Carolina, and Virginia. This selection regime resulted in 615 plots for the twelve-state region for use in the analysis described here.

Loblolly pine distribution. Forest type distribution data were obtained from a map derived from advanced very high resolution radiometer data, FIA surveys, thematic mapper data, and other sources. This map shows forest types on a 1 km grid cell size and was produced in support of the 1993 Resources Planning Act (RPA) update. For our investigation, we examined only grid cells classified as Loblolly–Shortleaf Pine. Methods for forest density mapping are described by Zhu (Zhu, 1994; Zhu and Evans, 1992); further details may be obtained from Forest Inventory and Analysis, Southern Research Station, USDA Forest Service, Asheville, NC.

Ozone exposure. Tropospheric ozone concentration data for May to September for 1988 to 1991 were obtained from the U.S. Environmental Protection Agency (EPA) laboratory in Corvallis, Oregon. These data are derived from the aerometric information and retrieval system and represent nearly continuous measurements of known accuracy. Ozone dose was expressed as sum06, that is, the sum of all hourly average concentrations greater than or equal to 0.06 $\mu L\ L^{-1}$ minus 0.06 $\mu L\ L^{-1}$. These values were adjusted for the few days on which data were not collected. For the analysis presented in this chapter, monitoring sites within the twelve-state region and within neighboring states were used.

Climate scenarios. The SGCP has supported an effort to produce an internally consistent climate database for climate change studies. This project, the Vegetation/Ecosystem Modeling and Analysis Project (VEMAP) is designed so that numerous models can be compared using a common set of climate scenarios and other data (Kittell et al., 1995). The VEMAP database contains single year steady-state climate scenarios based on the results of several general circulation models (GCMs). For our initial investigation, we selected the scenario based on the GCM of the United Kingdom Meteorological Office (UKMO) because it predicted the most extreme temperature change. Monthly data on precipitation, solar radiation, and maximum and minimum air temperature were obtained from the VEMAP database for the base case (no climate change) and for the UKMO scenario. All VEMAP data are calculated for a grid cell size of 0.5° by 0.5° latitude and longitude.

Step 2. Create Geographic Information System

All data were imported into a single database using Arc/Info software. Data were converted to a common geographic projection (Lambert equal area). This projection was selected to minimize the generation of errors in area extent of data features because the forest distribution data were already in this projection. Other data were either in a geographic projection or were point locations, and consequently there was little distortion or interpolation when converting them to another projection. For each type of data, a grid with a cell size of 1 km was created. Additional grids were created to represent the uncertainty of each of several data types, as described later. To supply data to the forest stand-growth model, data were aggregated into 30 km analysis cells. However, data are stored at the 1 km resolution in order that smaller analysis cells could be used in the future if they are required.

Steps 3 and 4. Interpolate and Estimate Uncertainty

Our modeling strategy requires that complete regional data sets with estimated uncertainties be generated for each type of data. Complete regional data sets, when required, were obtained by interpolation. Methods of interpolation are described in the following subsections for each data type. Part of the purpose of the SOFOCLES modeling system is to simultaneously examine uncertainties in both regional data sets and in estimates of the effects of climate on forest growth. However, for the example presented in this chapter, the focus is on uncertainty in climate effects on forests, and uncertainty in regional data is not examined. Subsequent analyses will incorporate estimates of uncertainty in regional data, as well as uncertainty in interpolating these data within the region.

Loblolly pine growth. Values between FIA plots were simply interpolated from the nearest plot. No attempt was made to use distance-weighted averages or other spatial methods because examination of the data did not suggest strong spatial autocorrelation in the growth rate of loblolly pine. An example of this interpolation procedure is shown for Alabama in Figure 24.3.

Loblolly pine distribution. Because these data provided complete coverage of the region, no interpolation was required. Although some classification error may be expected to occur with such a data set, we had no estimate of such error and did not include it in our analysis.

Ozone exposure. Ozone values were interpolated as the weighted average of values from all monitoring stations within 300 km of each analysis cell (Figure 24.4). Weighting decreased with the square of the distance from the station. This weighting scheme is a reasonable first approximation for the distance that ozone may travel from sources, and is appropriate because stations are generally located near population centers (LeFohn et al., 1987, 1988). An improved interpolation would take into account wind direction as well as distance; such a method is under development. For the model presented herein, interpolated ozone sum06 values

24. Integrating Research on Climate Change Effects 437

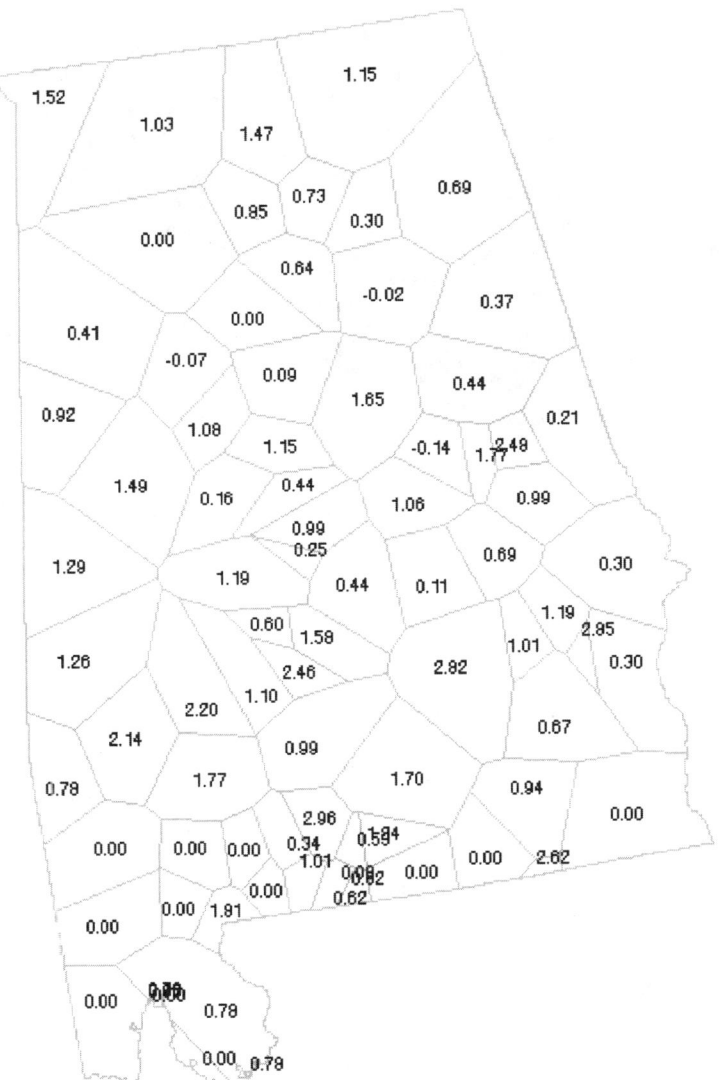

Figure 24.3. Interpolated loblolly growth data for Alabama (m^2 ha^{-1} yr^{-1}). The text shows the basal-area-growth rate of loblolly pine greater than 12.7 cm DBH at selected U.S.D.A. Forest Service survey plot locations. All stands within a polygon are assumed to have the same growth rate. A similar interpolation procedure was followed for the entire twelve-state region.

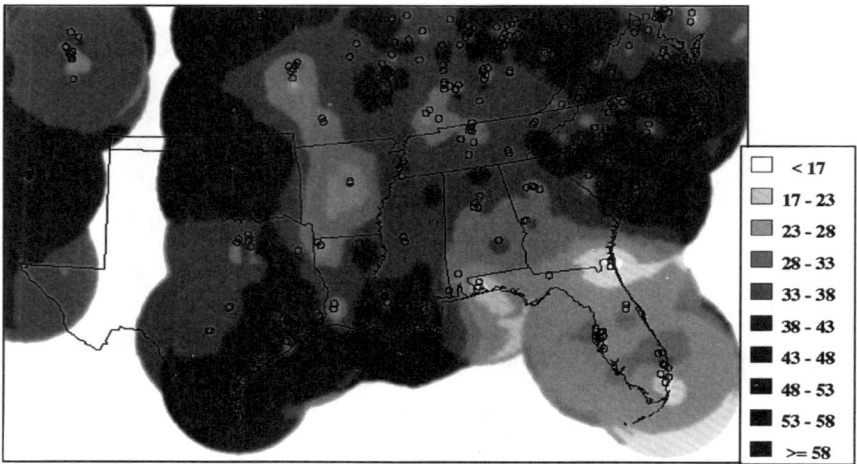

Figure 24.4. Interpolated ozone dose (sum06 May to Oct, average for 1988 to 1991, μL L^{-1} h). The open circles show the locations of monitoring stations. Interpolated values were calculated based on a weighted average of values from all stations within 300 km. Weighting decreased with the square of the distance from the station (see text for details).

for each year from 1988 to 1991 were averaged to obtain a value for present-day exposure. Future exposure was defined as a 50% increase.

Climate scenarios. No climate interpolation was required because these data provided complete coverage. Uncertainties in weather data exist as a result of interpolation and averaging error, yearly variations, and, in particular, uncertainties in predicting climatic change. However, the example presented in this chapter focuses on uncertainty in forest response to a specified climate change scenario. The VEMAP scenarios are based on a doubling of tropospheric carbon dioxide (CO_2) concentration. We used a value of 350 μL L^{-1} CO_2 for the base case (present) condition.

Step 5. Simulate Climate Effects

Regional data were extracted for 30 km grid cells throughout the twelve-state region in order to provide input data for the forest stand-growth model. This produced data sets for 942 separate forest-stand simulations.

Step 6. Produce Regional Analysis

The forest-stand model produces predictions on a 30 × 30 km square cell basis, as described above. These results are then applied to each of the 1 km² within the cell that contain growing loblolly pine stands. Hence, predictions are made for every square kilometer of loblolly forest within the analysis region. Based on the RPA map, there are 230,106 1-km grid cells of loblolly–shortleaf pine in the United

States, most of which lie within our twelve-state analysis region. However, predictions are made for only 203,227 of these 1-km cells because of missing data, location outside the study region, or a record of zero-growth rate for loblolly pine from the FIA data set.

Probabilistic Stand-Level Model

Summaries of experimental research and regional data sets were brought together in this portion of our model to produce the probabilistic estimates of stand growth. Initially, present-day growth rate, stand structure, and annual carbon budget were defined. Results of various SGCP research projects were then generalized to estimate effects of changed climate on a component of growth or carbon budget. Probabilistic outputs from these functions were then used in a Monte Carlo simulation to produce a probabilistic estimate of how basal-area-growth rate for the stand may change with climate.

Current Stand Growth

The starting point for the stand simulation model was an initial estimate of stand structure (species composition and size distribution) and present growth rate; for this, we used existing stands, as described by FIA data. We assumed that without climate change, future growth will be the same as has been measured over the past few decades. This expected rate of growth at present-day climate conditions was used as an input starting condition for each stand simulated. We divided the loblolly pine population in the stand into ten classes based on diameter size. The simulated responses for each modeled individual, "representative" tree was extrapolated to all trees of the stand in the same size class. Published respiration rates (Ryan et al., 1994) and allometric relationships (Nelson and Switzer, 1975; Gower et al., 1994) were used to decompose annual growth into probable components of annual carbon budget (that is, C gain, loss, and allocation) for the representative trees.

Project Summaries

Summary functions were created to generalize published research results as probability density distributions describing plausible changes in components of an individual tree's annual carbon budget. Conceptually, this means translating paired values of independent and dependent variables out of the context of the original experiment. The function should form links between 1) the experiment's independent variables and analogous values in regional data sets, and 2) the experiment's dependent variable and a component of annual carbon budget. As such, functions are generally formed as relatively simple linear or curvilinear relationships. However, they do not necessarily need to be based on statistical analyses. In some cases, the functions can be a best estimate, or a numerical expression of "expert opinion." The same translation—from experiment to stand model—is required for an estimate of uncertainty. The means of establishing simple links to scale-up results from the level of an experiment may sometimes be unclear. In these instances, correspondingly large increases in uncertainty may be more appropriate than nesting complex scaling models within the simple carbon-budget framework. Functions can either directly estimate relative change or estimate absolute effects on carbon budget at both present-day and future climate conditions. (Note that the ratio of the two is then the relative effect.)

Data from ten SGCP projects were included in this model (Figure 24.1). Measurements of responses to climate change variables ranged from gas exchange of leaves to stem growth in mature stands. Data sources included published research and preliminary analyses of experiments. Effects of all climate change variables were initially assumed to be additive unless indicated otherwise. Interactions among independent variables within a project were incorporated in that particular summary function. Project summaries used in this example represent only coarse approximations, and further refinements are possible through review and feedback of researchers.

The summary functions from SGCP projects (Figure 24.1) that we attempted to incorporate into the stand-level simulation fell into four categories, which are 1) carbon gain, expressed as stem growth or carbon assimilation, 2) carbon allocation, 3) carbon loss through maintenance respiration, and 4) ecosystem-level processes, such as the potential for increased soil nutrients. Sources of data are indicated in Figure 24.1; the unpublished data used in this figure were preliminary results provided by a member of the research project. For the example presented here, the use of additional data from the ecological literature was minimized. We will not elaborate the details of each summary function, but rather explain the general concept with the following examples.

Five projects were linked to changes in basal-area-growth rate. These were each initially reduced to an estimate of relative change in growth rate at the original measurement, scale, and experimental environment. The median expected value of relative change in growth was specified as the midpoint of a uniform distribution, which described expected effect on basal-area growth. One end of the uniform distribution was set to 1 (no relative change), and the other was twice the difference between 1 and the expected median value. Each project's results were used to produce one probability density distribution describing relative effect of the climate scenarios on basal-area growth for each climate variable for each representative tree. Allometric relationships and assumptions about effects of the climate variable on allocation and respiration were then applied to convert distributions to estimates of change in annual carbon assimilation required to affect the change in basal area.

An example of the process described in the prior paragraph is the summary-function of ozone response of loblolly pine growth described in McLaughlin and Downing (1995). Data were taken from Figure 3a of McLaughlin and Downing (1995) by developing regression equations to match the lines in their figure. Assuming that the modeled stand had a present-day ozone exposure of 280 nL L^{-1} h and the scenario for future exposure indicated a 1.5× increase, the future exposure could be 420 nL L^{-1} h. Their research suggested an interaction with water stress and had lines that we interpreted as "ambient" and "moderately stressed" conditions. We did not have a regional data set for determining a moisture-stress index, so we arbitrarily decided to use a decrease in estimated growing season soil-water content of 20% to distinguish among the two lines. Soil water was calculated at monthly intervals based on soil-water-holding capacity, precipitation, and potential evapotranspiration (Thornthwaite method, Rosenberg et al., 1983). Thus, a representative tree of 19 cm diameter at breast height (DBH), presently growing at 0.7 cm DBH per year, is expected to have an 8% decrease in basal-area growth in the specified future climate scenario in which growing season water content decreases by 18% at that site. The uniform distribution describing probable change in basal-area-growth rate extended from 0.84 to 1.0. Because the model did not include any effects of ozone on allocation or respiration, allometric relationships did not change for this example; thus, the summary-function describing the relative change in annual carbon assimilation had the same distribution, with a median value of 0.92.

Four SGCP projects were used to estimate effects of elevated CO_2 on annual canopy productivity (Figure 24.1). Each set of results represented different growth and measurement conditions but the summary process was the same for all projects in this example. An expected value for relative net CO_2 uptake was obtained from each project by dividing mean rate of uptake at 700 $\mu L\ L^{-1}$ by the rate at 350 $\mu L\ L^{-1}$. Because it is improbable that annual change in CO_2 uptake will be as great as instantaneous measurements, the median value was set at the upper limit for a triangular distribution summarizing that particular project. For example, data from Figure 2 in Teskey (1995) were used to set this limit at 2.0 for one summary-function. The minimum value for the distribution was set at 1 (no relative change), and the mode varied among representative trees according to estimated light-environment of individual representative tree crowns within the canopy.

Results from one project were used to estimate some influence on allocation and respiration (see Chapter 9). These functions were defined as uniform distributions, which followed procedures similar to those outlined. Effects of changed temperatures on maintenance respiration were defined by the same Q_{10} values (Ryan et al., 1994) used to define the carbon budget of present stand growth; the prediction was set as the midpoint of a uniform distribution (as described earlier for growth responses). The results of two additional SGCP projects were included in an attempt to develop means of incorporating ecosystem-level processes (Figure 24.1). We had difficulty linking these data with an effect on soil nutrients, therefore they were incorporated as a slight increase in uncertainty around soil-nutrient levels.

Stand-Level Model Process

Project summaries that produced estimates of change in annual carbon assimilation (described in the prior section) were pooled by sampling among distributions. The results were a single distribution, which described expected relative change in assimilation under the future climate scenario for each independent variable. New values for maintenance respiration, mostly in response to temperature, were then estimated. Expected maintenance respiration was subtracted from expected canopy productivity, and a portion of this carbon was allocated to stem growth and growth respiration. Finally, new basal area growth was determined by reference to allometric relationships.

Modeling Uncertainty

A probabilistic estimate of stand-growth rate is generated by selecting one value from the frequency distribution of each probabilistic variable for each iteration of the simulation. A Monte Carlo simulation involves repeating a large number of such model calculations. Each selection from a distribution is independent of those of other distributions. The cumulative result of this process is a probabilistic estimate of stand-growth rate defined by a frequency distribution of results from the Monte Carlo simulation.

Modeling software used for Monte Carlo simulations was Demos version 2.6b2 (Lumina Decision Systems). Stand-growth rate is estimated using a stratified sampling method known as Latin hypercube, which systematically samples from all parts of the distribution (Morgan and Henrion, 1990). This method has the advantage of requiring fewer repetitions of sampling before defining outer bounds and reaching a stable output distribution. Results presented here were run with sixty repetitions.

Results

These results are intended to demonstrate a methodology for conducting regional analyses based on available data. The forest-stand model presented in this chapter is a simple example, and does not incorporate all factors known to affect tree growth. Data summaries and model results should therefore be viewed as preliminary examples of methodology.

Stand-Level Growth

The stand-level simulation model produces distributions representing expected growth rate under the future climate scenario (for example, Figure 24.5). This model clearly predicted a reduction in loblolly forest-growth rate under the UKMO climate scenario of the VEMAP data set (Figures 24.5, 24.6). The distribution in Figure 24.5, representing probable basal-area growth rate for a selected stand (solid line), indicated that growth was most likely to be less than one-third of that stands present-day rate of 1.5 m^2 ha^{-1} yr^{-1}. The size and shape of the probability distribution was typical of all stand simulations presented on region-wide maps.

Sensitivity of the stand model to individual components of climate was examined by systematically altering such input values as 1) temperature, 2) carbon dioxide (CO_2), 3) precipitation, and 4) ozone (O_3) (Figure 24.5). Mean monthly

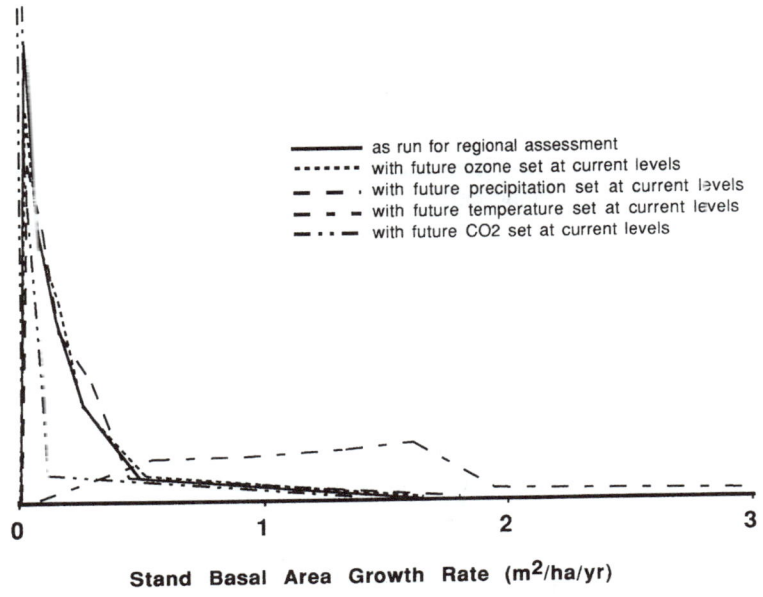

Figure 24.5. Climate influence on stand simulation model estimates of basal-area-growth rate. Frequency distributions for predicted growth from simulations run for a single, randomly selected stand (central Mississippi) while successively holding a single climate factor constant at present level. Present-day growth rate is 1.5 $m^2\ ha^{-1}\ yr^{-1}$.

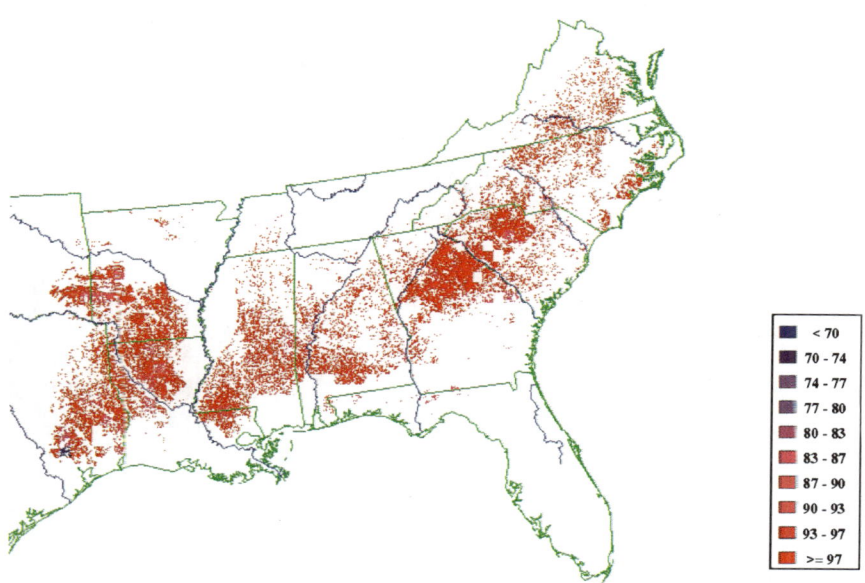

Figure 24.6. Chance of climate change decreasing loblolly growth. This map shows the predicted proability (%) that climate change will decrease loblolly growth. This prediction is based on a scenario derived from the UKMO GCM (see text for details).

temperature and atmospheric CO_2 concentration were the most influential climate variables in determining this model's estimate of changed growth rate. When elevated temperatures were included in model simulations, growth was greatly reduced (Figure 24.5). Inclusion of a doubling in atmospheric CO_2 concentration increased growth predictions but produced a smaller effect than temperature. Simulations with no change in level of either tropospheric ozone or annual precipitation had little effect on either the shape or location of the output distribution.

Influences of probabilistic variables on predictions of stand growth were determined by calculating correlations of such variables with model predictions as a part of the Monte Carlo simulation. Nonparametric correlations are calculated to avoid any need for imposing assumptions on frequency distributions, as required in a parametric correlation model (Morgan and Henrion, 1990). The degree of correlation is proportional to the influence of that variable. A probabilistic variable's influence in a model is a function of both its uncertainty and its functional relationship with model output. Annual maintenance respiration and carbon assimilation strongly influenced growth rate, with respiration having the greatest influence on the majority of stands simulated (Figure 24.7).

Respiration and canopy productivity are most strongly influenced by temperature and CO_2, respectively. The usefulness of the distinction between a variable's

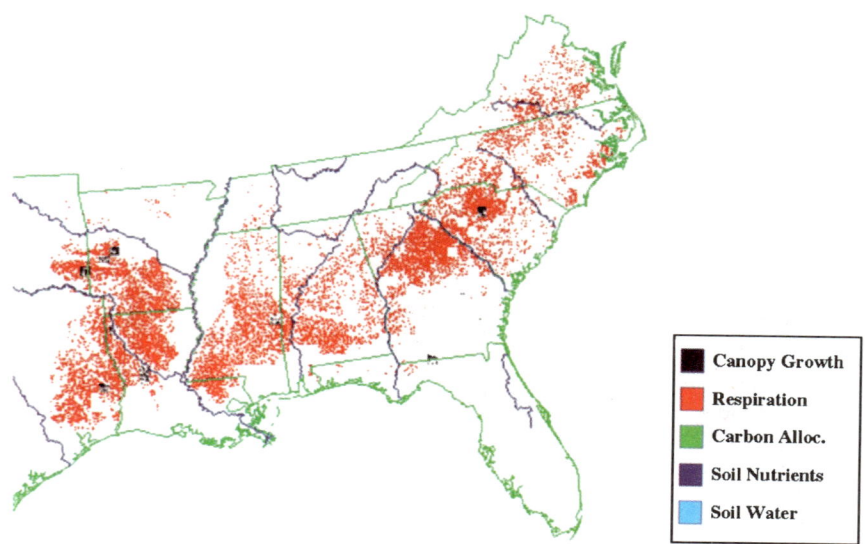

Figure 24.7. Most influential aspect of loblolly growth. This map shows that aspect of forest growth which the Monte Carlo forest-stand model most strongly influences the model prediction for each location. For example, for locations colored red, the most influential variable in the model is respiration; for locations colored black, the most influential variable is canopy growth. Although there are no locations where other variables are most influential, these other variables still may influence the model predictions.

uncertainty and its determinate relationship to model output can be illustrated by these two variables. Changing input values for either temperature or CO_2 affects both model output (Figure 24.5) and the relative importance of respiration and productivity in determining that output (e.g., Figure 24.7). However, simulations at only half the UKMO projection of temperature increase reduced the influence of respiration, which resulted in canopy productivity becoming most influential in the majority of plots (data not shown). In this case, the range of the frequency distribution used to define uncertainty in input values for temperature and CO_2 had little effect on model predictions or relative importance of respiration and canopy productivity. The range of uncertainties for future values of temperature and CO_2 became important in this model only as they became very large (i.e., as range of distribution approached $\pm$ 50% of median value).

Regional Growth

Results from stand-level simulations were extrapolated to a regional scale (Figures 24.6, 24.7, and 24.8; Table 24.2). Figure 24.6 shows the probability of a growth decrease; Figure 24.8 presents the magnitudes. These data show a high plausibility of decreased growth throughout the region, with a moderately high likelihood in some parts of Texas and Louisiana. The pattern of small squares containing dots of a single color is the result of a single prediction being made for all stands within a 30 km grid cell. These predictions are based on probability, hence there is no single value for any particular location. The frequency distributions from results of the Monte Carlo simulations can be represented by mapping selected percentiles of the output distribution. We use the 10th, 50th, and 90th percentile values of differences in growth rate to represent the range of predicted values. In all three cases, the effect of the climate scenario is to reduce loblolly basal-area growth throughout the region, although basal-area growth is decreased more in some areas than others. Values for each of the three predictions are summarized in Table 24.2. The substantial decrease in predicted growth shown in Figure 24.8 and in Table 24.2 is probably the result of the large increase in temperature (approximately 7° C) in the UKMO climate scenario.

Discussion

Our model is being developed to facilitate decision-making by forest managers and policy-makers. Ecological assessments require simple quantitative statements as inputs to summarize science; for example, expressions of relative probability of changes in forest productivity in response to climate. Therefore, SOFOCLES was constructed to summarize expectations, based on present biological understanding of forest response to climate. This is a complex subject, and it is implausible that it would be completely represented by a simple simulation model. Our model is based on the concept that almost all scientific research has relevance to larger issues, yet direct applications of findings involve an increase in uncertainty. Our

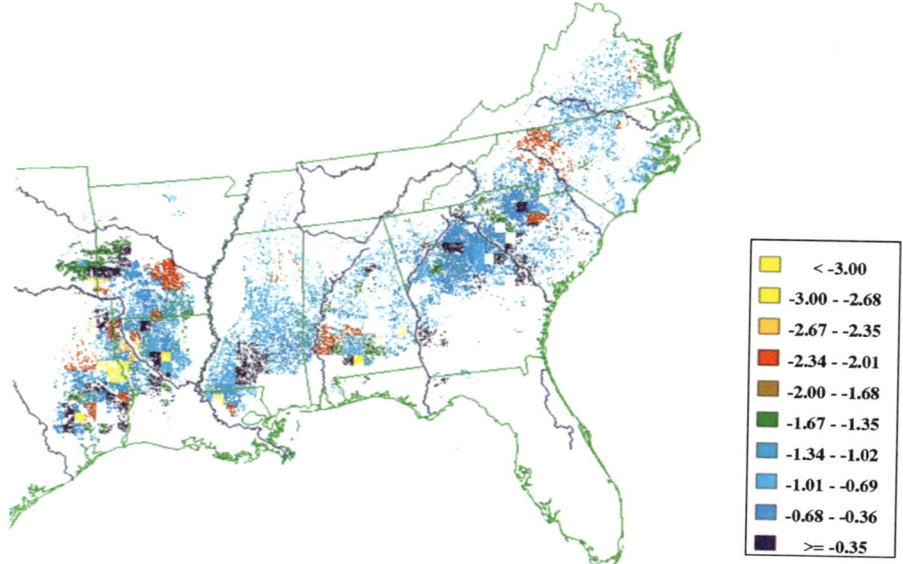

Figure 24.8. Most probable change in basal-area growth of loblolly pine as a result of a climate change scenario. This map shows the predicted change in basal-area growth (m² ha⁻¹ yr⁻¹) resulting from the UKMO-based climate change scenario. This map shows the most probable response (50th percentile) of the probability distribution of the prediction (see text for details).

goal is to provide a framework to help identify important processes controlling forest response at the stand level and then quantitatively project these results to a regional scale. This example is an illustration of the methodology we are developing to incorporate uncertainty into a data-synthesis amenable for use in ecological risk assessments. Uncertainty was addressed in three distinct steps, which are that 1) it was explicitly defined for summary-functions used to construct the stand-level model, 2) it was determined for model predictions through Monte Carlo simulations, and 3) the relationship between uncertainty in model input and predictions was analyzed. These results were then presented on a regional basis through the link to a GIS.

The extreme growth reduction indicated through these simulations is largely a

Table 24.2. Predicted Change in Loblolly Basal Area Growth Rate[1]

Worst case	−23
Most probable	−21
Best case	−14

[1] Basal-area-growth rate = 10^6 m⁻² yr⁻¹.

function of the rise in temperature (~7°C) projected by the UKMO climate change scenario selected for use with this example (Figures 24.5, 24.7). As such, these results are not intended as a projection of future events, but rather an illustration of a methodology. Results of any model are dependent on definitions of input data and model structure. The objective of this exercise is to explicitly address issues of influence and uncertainty in the results. Only after contrasting various model formulations and climate scenarios through an iterative review process will model results begin to approach predictions useful for decision-makers.

The homogeneity in predicted climate change effects across the region (Figure 24.8) was largely a function of the two dominant influences in this model—temperature and CO_2. These components of climate are represented by relatively smooth, continuous, spatial surfaces. In contrast, such resources as water and nutrients show more spatial heterogeneity resulting from their dependence on site-specific ecosystem processes and edaphic factors. Such site-specific data had little influence on projected change in growth rate. Although this may have the appearance of an important result, it may simply be a function of the model. We are working to improve our ability to estimate levels of soil resources. This includes mechanisms for incorporating ecosystem feedbacks on soil-nitrogen availability.

Present stand structure and growth were used as model inputs, but were not defined as variables with uncertainty. Similarly, the scenario of future climate was also defined without uncertainty. Although uncertainties in these variables exist, the focus of this effort was on how growth rate may be affected. The implicit assumption that future loblolly pine stands will be structured similarly to those under present-day climate is not a prerequisite for this model because any arbitrary stand structure and growth can be defined as a starting point. However, experience in risk assessment indicates that using well-defined scenarios is useful to easily interpreted assessment results.

Uncertainty and its effects on model predictions are both explicitly addressed in this tractable and flexible model. No particular assumptions or form of uncertainty need be imposed on model input. Uncertainty can arise from limitations in understanding of biology or can be the result of technical limitations of experiments. For our purposes, the origin of uncertainty is not so important as is an estimate of its magnitude. However, producing such estimates can be difficult because of the number of steps needed to link experimental results to stand-level effects. For example, a significant loss of realism is introduced by assuming that annual stand-canopy productivity is linearly proportional to instantaneous leaf-level carbon assimilation. Similar to scaling predicted values, scaling uncertainty in this example would be more straightforward with reference to such factors as light extinction through forest canopies, leaf response at limiting light levels, and the probability that other resource limitations will occur through the growing season. We hope to avoid the use of large, complex process models to scale-up research results to input functions. However, such simple scaling models as that of Kull and Jarvis (1995) may be useful in these instances. Nevertheless, large gaps in our understanding probably exist, and occasionally, the initial assignment of uncertainty may be little more than a "best" guess.

Uncertainties and relationships among parts of the model are liable to change continuously as research sheds new light on the system's biology. Flexibility in approach to model construction and definition of variables is an effort to include the best available information in the assessment. Simplifying a complex network of interrelationships, as exists in a forest ecosystem, can increase uncertainty as some components are deleted in favor of simplicity. It is incumbent on biologists to ensure that such decisions represent a summary of present biological understanding of the ecosystem. Obviously one scientist's reasonable summary may be another's gross oversimplification. Therefore, a simple starting structure allows the model to be open for scientific discussion and scrutiny. Implications of alternate interpretations, or sources, of data leading to additional complexities can be explored as they are suggested. This system allows for the identification of such disagreements, which can lead to iterative change in the model. This approach not only provides probabilitic summaries of research, but also provides a means of presenting issues of scientific debate to analysts who may not be forest biologists.

We do not suggest that a simple model is always the answer for analyzing a complex system, but we do propose that use of simplifying assumptions can focus an assessment on the most important issues in a complex system. One approach to assessing how climate change may affect forests is to simply summarize the results of a large body of research for policy-makers. However, such an approach does not guarantee a useful synthesis of research results and was heavily criticized when used in the National Acidic Precipitation Assessment Program (Loucks, 1992; Russell, 1992; Schindler, 1992). Such extensive research efforts will be most useful if attention is paid both to problem definition and to methods for summarizing and presenting results to policy-makers. The Ecological Risk Assessment (ERA) framework proposed by the U.S. Environmental Protection Agency (EPA) is a useful tool for these purposes (USEPA, 1992, 1994).

The ERA approach defines three major portions of the assessment process, which are 1) problem formulation, 2) analysis, and 3) risk characterization (USEPA, 1992). The analysis section is divided into exposure characterization and effects characterization. Most of SOFOCLES falls into the category of effects characterization, which determines potential effects of global climate change on forests in the South. However, we feel that the outlined ERA framework illustrates how our efforts fit into a larger effort to assess the potential risks of climate change for southern United States forests.

A summary of our approach restated in the ERA format is shown in Figure 24.9. Even though the focus of a risk assessment is normally on deleterious effects, it should be noted that global change effects may also have beneficial effects on forests. The stressors are defined as factors driving forest change. Selection of end points is a part of problem formulation; we selected aspects of loblolly pine forest growth as an end point because we felt that it would best serve to integrate the results of the SGCP research projects. Additionally, techniques for measuring growth are well developed and large databases of historical growth records are available.

The analysis portion of a risk assessment includes characterization of both exposure to a stressor and effects of the stressor on forests. We used scenarios

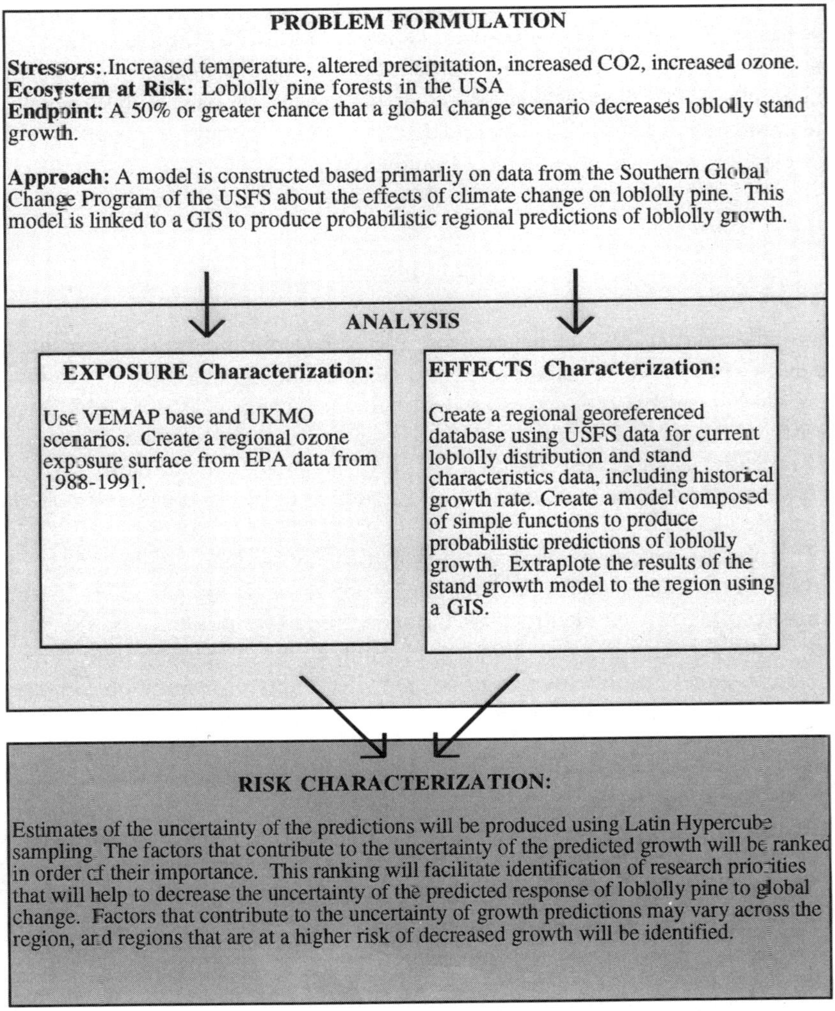

24. Integrating Research on Climate Change Effects

PROBLEM FORMULATION

Stressors: Increased temperature, altered precipitation, increased CO_2, increased ozone.
Ecosystem at Risk: Loblolly pine forests in the USA
Endpoint: A 50% or greater chance that a global change scenario decreases loblolly stand growth.

Approach: A model is constructed based primarliy on data from the Southern Global Change Program of the USFS about the effects of climate change on loblolly pine. This model is linked to a GIS to produce probabilistic regional predictions of loblolly growth.

ANALYSIS

EXPOSURE Characterization:

Use VEMAP base and UKMO scenarios. Create a regional ozone exposure surface from EPA data from 1988-1991.

EFFECTS Characterization:

Create a regional georeferenced database using USFS data for current loblolly distribution and stand characteristics data, including historical growth rate. Create a model composed of simple functions to produce probabilistic predictions of loblolly growth. Extraplote the results of the stand growth model to the region using a GIS.

RISK CHARACTERIZATION:

Estimates of the uncertainty of the predictions will be produced using Latin Hypercube sampling The factors that contribute to the uncertainty of the predicted growth will be ranked in order of their importance. This ranking will facilitate identification of research priorities that will help to decrease the uncertainty of the predicted response of loblolly pine to global change. Factors that contribute to the uncertainty of growth predictions may vary across the region, and regions that are at a higher risk of decreased growth will be identified.

Figure 24.9. SOFOCLES approach in ecological risk assessment context. This figure outlines key aspects of the SOFOCLES modeling approach in the form of the Ecological Risk Assessment Framework under development by the USEPA (see text for details).

developed as part of the VEMAP project. There are several single year, steady-state scenarios, including a base case and a number of future climate scenarios derived from the GCMs: CCC, GFDL, GISS, OSU, and UKMO. For this chapter, we compared the base case scenario to that derived from the UKMO GCM. These scenarios were specifically designed to be usable by different models of ecological effects, hence, a benefit of using them is the opportunity to compare our results with those of very different models.

The final step in the ERA framework, risk characterization, should be based on quantitative estimates of uncertainty about predictions. This is why we emphasized probabilistic analysis in our modeling effort. We also believe that model results should be presented in formats usable by decision-makers, such as maps and graphs, rather than technical summaries hundreds of pages long. Hence, we attempt to produce simple figures and data summaries that are understandable, but still capture scientific uncertainty of predictions.

Conclusions

Our model consists of three major steps, which are 1) formulation of experimental research summaries, 2) stand-level growth simulation, and 3) propagation of stand-level results to regional estimates of climate change effects on loblolly pine forests. The semi-independence of components allows for separate development and revision, as necessary. The SOFOCLES framework can therefore be used to iteratively improve each component of the model in order to use available data most effectively.

This model structure is unique in that it explicitly incorporates uncertainty surrounding model variables in determining predicted change in growth. Maintenance respiration was identified as the most important influence of the annual carbon budget. This suggests that any research aimed at better defining climate effects on either absolute rate or associated uncertainty of respiration can reduce prediction uncertainty of expected changes in growth.

References

Burkhart HE, Farrar KD, Amateis RL, Daniels RF (1987) Simulation of individual tree growth and stand development in loblolly pine plantations on cutover, site-prepared areas. FWS-1-87. VA Poly Inst State Univ, Sch For Wildlife Resour, Blacksburg, VA.

Comins HN, McMurtrie RE (1993) Long-term biotic response of nutrient-limited forest ecosystems to CO_2-enrichment: Equilibrium behavior of integrated plant-soil models. Ecol Appl 3:666–681.

Dale VH, Jager HI, Gardner RH, Rosen AE (1988) Using sensitivity and uncertainty analysis to improve predictions of broad-scale forest development. Ecol Model 42:165–178.

Gower ST, Gholz HL, Nakane K, Baldwin VC (1994) Production and carbon allocation of pine forests. Ecol Bull 43:115–135.

Groninger JW, Seiler JR, Zedaker SM, Berrang PC (1995) Effects of elevated CO_2, water stress, and nitrogen level on competitive interactions of simulated loblolly pine and sweetgum stands. Can J For Res 25:1077–1083.

Groninger JW, Seiler JR, Zedaker SM, Berrang PC (1996) Photosynthetic response of loblolly pine and sweetgum seedling stands to elevated carbon dioxide, water stress, and nitrogen level. Can J For Res 26:95–102.

Kittel TGF, Rosenbloom NA, Painter TH, Schimel DS, and VEMAP Modeling Participants (1995) The VEMAP integrated database for modeling United States ecosystem/vegetation sensitivity to climate change. J Biog 22:857–862.

Kull O, Jarvis PG (1995) The role of nitrogen in a simple scheme to scale up photosynthesis from leaf to canopy. Plant Cell Environ 18:1174–1182.

Lefohn AS, Knudsen HP, Logan JA, Simpson J, Bhumralkar C (1987) An evaluation of the kriging method to predict 7-h seasonal mean ozone concentrations for estimating crop losses. JAPCA 37:595–602.

Lefohn AS, Knudsen HP, McEvoy LR Jr (1988) The use of kriging to estimate monthly ozone exposure parameters for the southeastern USA. International Conference on Assessment of Crop Loss from Air Pollutants, Raleigh, NC, October 25–29, 1987. Environ Pollut 53:27–42.

Loucks O (1992) Forest response research in NAPAP: Potentially successful linkage of policy and science. Ecol Appl 2:117–123.

Luxmoore RJ, King AW, Tharp ML (1991) Approaches to scaling up physiologically based soil-plant models in space and time. Tree Physiol 9:281–292.

McLaughlin SB, Downing DJ (1995) Interactive effects of ambient ozone and climate measured on growth of mature forest trees. Nature 374:252–254.

Morgan MG, HeNrion M (1990) *Uncertainty: A guide to the treatment of uncertainty in quantitative policy and risk analysis.* Cambridge University Press, New York.

Nelson LE, Switzer GL (1975) Estimating weights of loblolly pine trees and their components in natural stands and plantations in central Mississippi. MS Agric For Exper Stat Tech Bull 73, MS State Univ, Mississippi State, MS.

Rosenberg NJ, Blad BL, Verma SB (1983) *Microclimate: The biological environment.* John Wiley and Sons, New York.

Rowe WD (1994) Understanding uncertainty. Risk Anal 5:743–750.

Russell M (1992) Lessons from NAPAP. Ecol Appl 2:107–110.

Ryan MG, Linder S, Vose JM, Hubbard RM (1994) Dark respiration of pines. Ecol Bull 43:50–63.

Schindler DW (1992) A view of NAPAP from north of the border. Ecol Appl 2:124–130.

Sharpe PJH, Rykiel EJ Jr (1991) Modelling integrated response of plants to multiple stresses. In Mooney HA, Winner WE, Pell EJ, (Eds) *Response of plants to multiple stresses.* Academic Press, San Diego, CA.

Solomon AS (1986) Transient responses of forests to CO_2-induced climate change: Simulation modeling experiments in eastern North America. Oecologia 68:567–579.

Summers JK, Wilson HT, Kou J (1993) A method for quantifying the prediction uncertainties associated with water quality models. Ecol Model 65:161–176.

Teskey RO (1995) A field study of the effects of elevated CO_2 on carbon assimilation, stomatal conductance and leaf and branch growth of *Pinus taeda* trees. Plant Cell Environ 18:565–573.

U.S. Environmental Protection Agency (1992) Framework for ecological risk assessment. EPA/630/R-92/001, Risk Assessment Forum Washington, DC.

U.S. Environmental Protection Agency (1994) Ecological risk assessment issue papers. EPA/630/R-94/009, Risk Assessment Forum Washington, DC.

Wang TP, Jarvis PG (1990) Description and validation of an array model—MAESTRO. Agric For Meteo 51:257–280.

Zhu Z (1994) Forest density mapping in the lower 48 states: a regression procedure. Gen tech rep SO-280. USDA For Ser, South For Exper Stat, New Orleans, LA.

Zhu Z, Evans DL (1992) Mapping midsouth forest distributions with AVHRR data J For 90:27–30.

25. Projected Impacts of Global Climate Change on Forests and Water Resources of the Southeastern United States

Jeffrey G. Borchers and Ronald P. Neilson

How will forest and water resources in the southeastern United States change over the next fifty to one hundred years? Resource managers looking this far into the future are faced with scenarios that include large ecological and economic impacts from global climate change. Although there is an emerging scientific consensus that anthropogenic global warming has been taking place (McCracken, 1995), the rate and magnitude of climate change and ecosystem responses cannot be predicted with any certainty. These and other uncertainties in the social, economic, and biophysical environments have engendered new, more ecosystem-oriented approaches to natural resources management (Christensen, 1996; Walters, 1986; Walters and Hulling, 1990). These new management paradigms seek to sustain the health of ecosystems (Borchers, 1996; Norton, 1992), for example, the productivity and diversity of aquatic and terrestrial ecosystems. In a rapidly changing climate however, sustaining ecosystem health becomes an even more elusive goal, a challenge that requires managers to be accountable for large uncertainties in the future.

Accounting for the complexity and uncertainty of global climate change impacts necessitates the use of simulation models in resource planning. For example, the implications of even subtle alterations in temperature and precipitation patterns in the southeastern United States are significant. Most, if not all, natural disturbances (e.g., insect or pathogen outbreaks, blowdowns, fire, sea-level rise) interact with climate in a complex fashion. For example, some climate scenarios suggest increased storm (hurricane) frequency and severity (Emanuel, 1987) that may alter forest structure. A warmer, drier regional climate may translate to more frequent droughts and their attending consequences (e.g., more frequent and severe fires, insect-pest epidemics). In the face of such complexity, resource managers can use model simulations in assessing the combined role of cross-scale linkages (e.g., landscape implications of local practices), complex interactions (e.g., feedback of vegetation changes to climate), and changing social and economic conditions (e.g., demand for wood products).

Our goal is to provide resource managers, policy-makers, and scientists with process-based simulations of southern regional terrestrial ecosystems at "equilibrium" with potential future climates as simulated under a doubling of present-day carbon dioxide (CO_2) radiative forcing levels. In this chapter, we use the biogeography model Mapped Atmosphere-Plant-Soil System (MAPSS) (Neilson, 1995) to predict several possible future equilibrium distributions of various types of forests and woodlands in the USDA Forest Service's southern region. These predictions include estimates of present and future runoff for the region. Ideally, simulations of vegetation responses to climate change in the southern region would incorporate many of the actual linkages of atmosphere, land, and oceans that underlie the uncertainties and complexities of climate change impacts. Although this goal may be realized in the future, full coupling among land, atmosphere, and ocean simulation models is not possible today.

Such present-day biogeographical and biogeochemical models as MAPSS use future climate scenarios obtained from general circulation models (GCMs). These GCM "snapshots" represent future climates at equilibrium with a doubled atmospheric CO_2 content ($2\times CO_2$) and do not portray the transient, that is, the time-dependent dynamics of atmospheric interactions with oceans and land. Ecological models driven by GCM climate scenarios are therefore generally restricted to simulating equilibrium conditions. (VEMAP, Members, 1995). Climate conditions are never constant, hence equilibrium simulations have no exact analog in nature. However, they are valuable in depicting theoretical equilibrium states toward which ecosystems might evolve under hypothetically stable, future climate scenarios. These states correspond roughly to estimates of potential natural "climax" vegetation (Küchler, 1964), a concept that has guided decision-making in forest management and silviculture for many years. Predicted changes in equilibrium or "climax" ecosystem states can be used to make broad inferences about changes in vegetation succession, disturbance, patch dynamics, and water resources.

Materials and Methods

The Mapped Atmosphere-Plant-Soil System Biogeography Model

In the past, biogeographers have frequently relied on descriptive and correlative methods for determining how the biotic and abiotic environments influence large-scale vegetation distribution (Emanuel et al., 1985). Occasionally, large-scale experiments have been designed to corroborate observations of plant-climate interactions (Neilson and Wullstein, 1983, 1986). With the advent of such models as MAPSS, it is now possible to produce process-based, regional-to global-scale simulations of vegetation distribution under present and future climate scenarios. The MAPSS model operates on the fundamental principle that an ecosystem will tend to maximize leaf area such that it can just be supported by the site's available soil moisture or energy (Woodward, 1987; Neilson, et al., 1989; Neilson 1993b; Neilson, 1995). The model calculates the potential natural vegetation (Küchler, 1964) that can be supported at any upland site in the world under a steady-state climate. The model iteratively calculates the leaf area index (LAI) of both woody lifeforms (trees or shrubs, but not both) and grass lifeforms in competition for both light and water, at the same time maintaining a site-water balance consistent with observed runoff (Neilson, 1995) and stomatal conductance. The conceptual framework for the approach is that vegetation distributions are, in general, constrained by either the availability of water in relation to transpirational demands or the availability of energy for growth (Neilson and Wullstein, 1983; Neilson et al., 1989; Woodward, 1987). In temperate latitudes, water is the primary constraint; at high latitudes energy is the primary constraint. The energy constraints on vegetation type and LAI are presently modeled in MAPSS using a growing degree day algorithm as a surrogate for net radiation (Botkin et al., 1972; Shugart, 1984).

The model is operated iteratively to calculate the LAI that can be supported at the site, given monthly inputs of precipitation, temperature, humidity and wind speed. Mixtures of grass and woody lifeforms are obtained in the model through a combination of light and water competition. Although water in the surface layer is apportioned to the two lifeforms in relation to their relative LAIs and stomatal conductances (i.e., canopy conductance), woody vegetation alone has access to deeper soil water. Total monthly runoff is estimated as the sum of two component, which are (1) fast-flow surface runoff, and (2) deeply percolated water in the bottom soil layer that is not transpired by grass or woody vegetation.

Neither biomes nor species are explicitly simulated in MAPSS. Rather, the model simulates the distribution of vegetation lifeforms (e.g., tree, shrub, grass), the dominant leaf form (e.g., broadleaf, needleleaf), leaf phenology (e.g., evergreen, deciduous), thermal tolerances, and vegetation density. These characteristics are then combined into a vegetation classification consistent with the biome level (Neilson, 1995)..The interacting effects of climate change and elevated CO_2 on carbon stocks are simulated indirectly using LAI. Higher concentrations of CO_2 can modulate vegetation responses to climate change by increasing water use efficiency (WUE), that is, carbon fixation per unit water transpired. The WUE effect often manifests as reduced stomatal conductance (Eamus, 1991). In MAPSS, increased WUE is imparted directly as a decrease in stomatal conductance that usually produces an increased LAI.

MAPSS Implementation for the Southern Region

The MAPSS model has been implemented at a prior date at a 10 km resolution over the continental United States and Mexico, and at a 0.5° resolution globally (Neilson, 1995; Neilson, 1993; Neilson and Marks, 1994). The model has been partially validated within the United States and globally with respect to simulated vegetation distribution, LAI, and runoff (Neilson, 1993a; Neilson 1995; Neilson and Marks, 1994). Predictions and analyses presented here for the southeastern United States are based on the USDA Forest Service's southern region and have been derived from a MAPSS implementation at 10 km resolution over the conterminous United States.

Scenarios of double CO_2 climatic change were derived from six GCM $2\times CO_2$ equilibrium simulations as supplied by the National Center for Atmospheric Research (NCAR). Outputs for present-day and $2\times CO_2$ climates were obtained from the following models: 1) Goddard Institute of Space Studies (GISS) (Hansen et al., 1988), 2) United Kingdom Meteorological Office (UKMO) (Mitchell and Warrilow, 1987); 3) Geophysical Fluid Dynamics Laboratory (GFDL) (Wetherald and Manabe, 1988); and 4) Oregon State University (OSU) (Schlesinger and Zhao, 1989). The GFDL versions R15 (approximately $4° \times 5°$ grid), R15 Q-flux, and R30 (ca. $2° \times 2.5°$ grid) were used. The Q-flux version of the GFDL model (GFDL-Q) includes a prescribed ocean-atmosphere coupling (Manabe et al., 1991). The coarse grid from each model was interpolated using a 4-point, inverse distance squared algorithm to a 10 km Albers grid in a raster-based geographic information system (GIS) (USA–CERL, 1993). The scenarios were applied as recommended and calculated by the NCAR data support section. Scenarios were constructed by applying ratios $((2\times CO_2)/(1\times CO_2))$ of all climate variables (except temperature) back to a baseline, long-term average monthly climate dataset (NOAA-EPA, 1993). Ratios were used to avoid negative numbers, but were not allowed to exceed a value of five in order to prevent unrealistic changes in areas having normally low rainfall. Temperature scenarios were calculated as a difference $((2\times CO_2) - (1\times CO_2))$ and applied to the baseline dataset.

The model was calibrated under present-day climate to a Küchler (1964) potential vegetation map. The simulations were then compared to a map of forest distribution (derived from satellite images) and forest area statistics contained in the 1992 Resource Planning Act (RPA) Update (Powell et al., 1993). (The RPA mandates that the USDA Forest Service periodically assess the status of forest resources in the United States.) The MAPSS model was then operated under several different $2\times CO_2$ climate scenarios produced by the various GCMs. Because MAPSS only simulates potential natural vegetation distribution, no direct incorporation of present-day land use can be derived, nor can future land use be predicted. Instead, a 1 km² resolution forest type groups map (derived from satellite images; see Powell et al., 1993; Zhu and Evans, 1992) was used to mask present-day nonforested areas from all subsequent maps and analyses. This "mask" was also applied to all potential future forest distributions, thereby imposing the assumption that the amount and distribution of nonforested areas (primarily agricultural) will not change. As of 1992, the total area of forest lands in the southern region had fluctuated only several percent in the preceding fifty years (Powell et al., 1993).

The MAPSS model is capable of simulating reductions in maximum leaf stomatal conductance as a function of increased CO_2 concentration. The net result is increased WUE (Eamus, 1991), but there is little agreement about the duration or significance of this "CO_2 effect" as a long-term, large-scale process outside of growth chambers (Eamus and Jarvis, 1989). For this reason, companion simulations with and without a WUE effect are also presented.

Classification and Analysis of Vegetation

The MAPSS model was explicitly calibrated to the Küchler potential vegetation map of the United States (Küchler, 1964). Its vegetation classification is based on lifeform, that is, such physiognomic properties as leaf form (e.g., broadleaf or needleleaf), leaf phenology (evergreen or deciduous), and stand structure (closed forest or open woodland/savanna). At the life form level, MAPSS classifications are process-based simulations. Further subclassifications are based on correlations between the distribution of lifeforms and such existing categories of vegetation as types, genera, and species.

A total of ten vegetation types were devised for this analysis of the southern region. The MAPSS model vegetation types were translated both geographically and logically into aggregations of forest type groups from the 1992 RPA Update (Powell et al., 1993) (Table 25.1). Of the twenty-two RPA forest type groups in the United States, twenty (excluding pinyon–juniper and chaparral types) were aggregated into nine types, four of which occur in the southern region. Two woodland types (i.e., open forest to savanna) derived from the oak–hickory forest and southeast (SE) mixed forest types are simulated in MAPSS when tree canopy cover falls below a critical LAI threshold and understory grass competition becomes significant. In addition to these six southern region vegetation types, it was necessary to create moist tropical forests and dry tropical forests from their MAPSS equivalents for several future climate scenarios. Finally, two vegetation types from outside the southern region, coastal moist forest and hardwood savanna were prominent in the future scenarios (see Results and Discussion pp. 460 to 474).

Large changes in forest and woodland structure (e.g., LAI, volume, basal area, density) are possible even with little variation in the land area covered by a vegetation type. In this analysis, a composite index was developed to depict such changes in structure. LAI (the area of leaves per unit area of ground) represents a coarse index of stem-basal area per hectare (ha), but is independent of changes in the area occupied by vegetation. Combining these two parameters, LAI and land surface area, yields total leaf area (TLA). Calculated as the product of average LAI and the total land area (hectares) occupied by a vegetation type, TLA provides an index of total volume, biomass, or density for vegetation types within the southern region. The relationships between TLA and the values they index (e.g., timber volume) are curvilinear, therefore TLA is provided only as a coarse indicator of the direction and magnitude of change in forest and woodland structure.

In this study, changes in vegetation land surface area, LAI, and TLA are geographically referenced to the southern region. Consequently, latitudinal shifts

Table 25.1. Cross-Classification Between Southern Region Forest and Woodland Types, RPA Forest Type Groups, and MAPSS Vegetation Classes

	Southern region forest and woodland types	1992 RPA forest type groups	Original MAPSS vegetation classes
Present Types	NE mixed conifers and hardwoods	White-red-jack pine Spruce-fir Aspen-birch	Forest mixed cool
	Maple-beech-birch	Maple-beech-birch	Forest hardwood cool
	Oak-hickory forest	Oak-hickory Elm-ash-cottonwood	Forest temperate deciduous
	SE mixed pines and hardwoods	Longleaf-slash pine Loblolly-shortleaf pine Oak-pine Oak-gum-cypress	Forest mixed warm (deciduous to evergreen, broadleaf to needleleaf)
	Oak-hickory woodland	Oak-hickory Elm-ash-cottonwood	Savanna temperate deciduous
	SE mixed woodland	Longleaf-slash pine Loblolly-shortleaf pine Oak-pine Oak-gum-cypress	Savanna mixed warm (DEB)
	Coastal moist forest	Hemlock-sitka spruce Redwood	Forest mixed warm (evergreen needle)
New Types	Hardwood savanna Moist tropical forest Dry tropical forest	Oak-pine	Tree savanna mixed warm (evergreen needle) Forest evergreen broadleaf tropical Forest seasonal tropical evergreen to deciduous

in the distribution of vegetation to or from areas outside the region appear as net losses or gains. Changes within such larger geographic regions as the eastern United States may appear less significant.

Quality of Data and Model Function

The MAPSS model and the simulations presented here have received extensive quality assurance and quality control, of which there are three aspects: 1) input data, including both actual climate and future scenarios, 2) MAPSS model structure and function; and 3) model output data. The MAPSS model was run on a gridded data set of mean monthly precipitation, temperature, vapor pressure, and wind speed that encompassed the lower forty-eight states. A full discussion of the methods used in development of the data layers is available from Marks (1990).

All of the data input layers were based on a 10-km resolution digital elevation model (DEM). The DEM was acquired from the National Oceanic and Atmospheric Administration (NOAA) National Geophysical Data Center as a 5-minute latitude longitude grid, and re-projected to Albers equal area using the image processing workbench (IPW) (Frew, 1990). Precipitation data were interpolated from meteorological station data using a DEM and the PRISM rainfall model (Daly et al., 1994). The PRISM model continually adjusts its frame of reference to accommodate local and regional variations in orthographically influenced precipitation regimes. A grid cell's precipitation is derived using an algorithm that employs precipitation-elevation relationships obtained from a local "window" of surrounding grid cells. The method has been compared with kriging, detrended kriging, and cokriging in a jackknife cross-validation exercise using data from the Willamette River Basin, Oregon (Phillips et al., 1992; Daly et al., 1994). The PRISM model exhibited lower overall bias and mean absolute error and has also been applied to northern Oregon and to the western United States (Daly et al., 1994). Detrended kriging and cokriging, however, could not be used in these regions, which had no overall relationship between elevation and precipitation. The PRISM model cross-validation bias and absolute error in northern Oregon increased a small to moderate amount compared to those in the Willamette River Basin; errors in the western United States showed little further increase.

Monthly average air temperatures were extracted from the historical climatology network database for the years 1948 to 1987 (Quinlan et al., 1987; Karl et al., 1990). Data from 1211 stations within the continental United States were used to construct long-term monthly averages for this forty year period. Topographic effects were removed from the measured air temperatures by converting them to their sea-level equivalent (i.e., potential temperature) prior to spatial interpolation. This temperature interpolation was performed using a simple linear inverse distance squared algorithm (Isaaks and Srivastava, 1989). Sea-level temperatures were then interpolated to the 10-km grid and converted back to air temperatures appropriate for DEM elevations.

Gridded vapor-pressure data were derived from interpolation of relative humidity. Input data consisted of dew point temperatures archived by the National

Climate Data Center (NCDC) (Spangler and Jenne, 1989), and from the worldwide airfield summaries database. Next, the dew point temperatures were converted to vapor pressures, then combined with the appropriate temperatures from the temperature data set to calculate temperature-corrected relative humidities. Relative humidities were then interpolated to a grid using the same linear inverse distance squared algorithm (Isaaks and Srivastava, 1989). Finally, vapor pressures were estimated from relative humidities using the gridded temperature data set.

Wind speeds were derived from a gridded US Department of Energy wind speed data set described in Elliot et al., (1987). Estimates were based on a combination of surface measurements and upper air data, and some attempt was made to account for topographic effects. Data were resampled from the original 1/3 × 1/4-degree latitude/longitude grid to the 10-km grid using the inverse distance squared approach (Isaaks and Srivastava, 1989). No attempt was made to account for the greater topographic detail in the 10-km DEM grid.

The six GCM scenarios used to derive model results in this chapter depict a range of temperature and precipitation regimes for the southern region. The OSU and UKMO models generally represent end points of a spectrum of temperature and precipitation changes relative to present climate (VEMAP Members, 1995). For example, in the southern region, the OSU model predicts an annual temperature increase of about 3.8 °C over the region; the UKMO increases are more variable, ranging from 5.5 °C to 8.0 °C. Similarly, the OSU model depicts a lower ratio of future to present annual precipitation for the region than the UKMO model. However, spatial variability of predicted annual precipitation are greater than with temperature. As a result, both increased and decreased annual precipitation are predicted for the region.

The MAPSS model output has been subjected to a rigorous intercomparison with other biogeography models. It is one of three global biogeography models participating in the Vegetation/Ecosystem Model Analysis Project (VEMAP), a core project within the International Geosphere-Biosphere Programme/Global Change and Terrestrial Ecosystems (IGBP/GCTE) program. VEMAP is an assessment of the impacts of climate change over the conterminous United States, as well as a formal model intercomparison and model linkage project. The first phase of VEMAP, now complete (VEMAP Members, 1995), compared three biogeography models (i.e., MAPSS, BIOME2, and DOLY) and three ecosystem productivity models (i.e., TEM, CENTURY, and BIOME-BGC) (Neilson, 1995; Woodward et al., 1995; Raich et al., 1991; Parton et al., 1988; Running and Hunt, 1993). The two classes of models were then linked (simple, one-way file transfer) to estimate the combined effects of climate change on vegetation distribution and productivity over the conterminous United States.

Results and Discussion

Existing Forests and Woodlands in the Southern Region

According to the 1992 RPA Update (Powell et al., 1993; Table 25.1), 85.7 million ha of the southern region's 216 million ha have been classified as forest land.

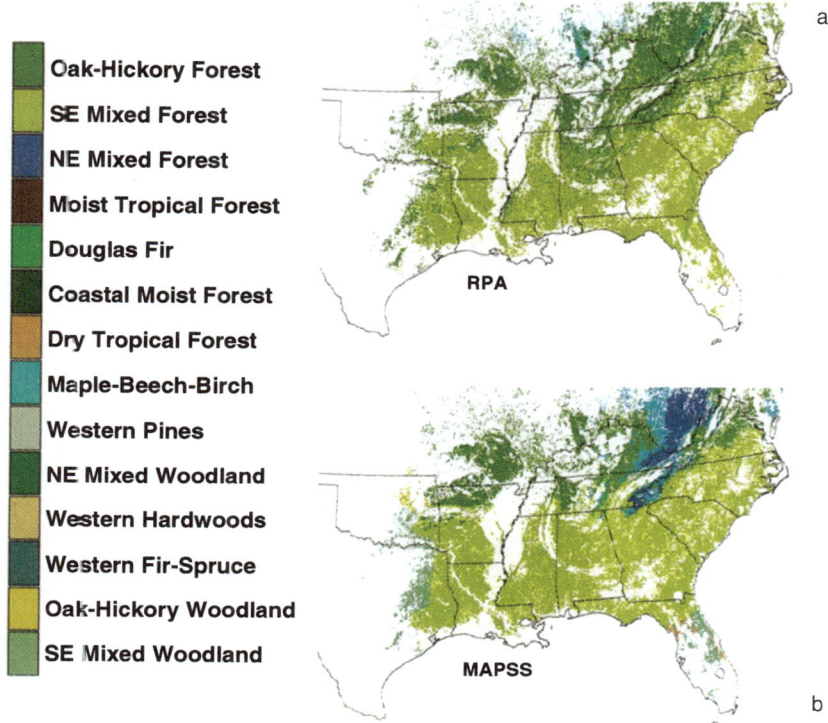

Figure 25.1. Distribution of forest and woodland types in the southern region based on (a) reclassified, remotely-sensed imagery (Powell et al., 1993), and (b) MAPSS simulations based on present climate conditions.

Under the present-day climate scenario (as masked by the RPA map), MAPSS classifies 84.3 million ha as forest (Figure 25.1a,b), only slightly less than the 85.1 million ha portrayed by the RPA map (Powell et al., 1993). Simulated LAI (all-sided) in forests ranged from a minimum of 6.6 in the oak–hickory forest type to a maximum of 15.0 in the northeast (NE) mixed forest type (Table 25.2). The average simulated LAI of woodland vegetation types, SE mixed woodland, oak–hickory woodland, and dry tropical forest, was 2.5, 3.1, and 2.2, respectively.

In general, MAPSS does quite well at simulating the correct spatial location and extent of the different types. However, MAPSS overestimates NE mixed forests along the Appalachian axis (Figure 25.1a,b; Table 25.3). The oak–hickory forest type is underestimated by MAPSS, in part, because the areal extent of the NE mixed forest, SE mixed forest, and maple–beech–birch types are overestimated. Additionally, both MAPSS and the RPA map obtained by remote sensing overestimate SE Mixed Forests in comparison to the survey statistics. The foregoing discrepancies are, in part, a function of comparing data at different spatial scales. For example, MAPSS simulates a single type for each 10 km pixel which, according to the RPA survey data, may encompass complex mixtures of pure and mixed

Table 25.2. Estimated Leaf Area Index by Forest and Woodland Type Under Present Climate and Six Future Climate Scenarios

	Present Climate	GFDL		GFDL-Q		GFDL-R30		GISS		OSU		UKMO	
		−WUE	+WUE	−WUE	+WUE	−WUE	+WUE	−WUE	+WUE	−WUE	+WUE	−WUE	+WUE
NE mixed forest	15.0	*		15.0	14.5			15.0	15.0	15.0	15.0		
Maple–beech–birch	11.4		9.2	11.9	11.3		11.0	12.8	13.9	11.2	13.3	0.0	10.0
Oak–hickory forest	6.6	4.0	5.8	5.2	6.6	5.1	7.5	7.7	7.0	7.0	8.6	5.4	7.1
SE mixed forest	7.4	4.2	4.8	4.8	5.9	4.6	5.7	5.7	8.2	5.0	7.4	4.1	5.0
Coastal moist forest	15.0		6.5		15.0	4.7	5.8	13.4	7.3	12.4	13.8		
Hardwood savanna		2.2	2.3			2.8	2.8	2.4					
Moist tropical forest			4.0		4.4		4.1		4.6		4.1		3.9
Dry tropical forest	2.2	2.1	2.9	2.3	3.2	2.3	2.9	2.4	2.2	2.2	3.0	2.0	2.7
Oak–hickory woodland	3.1	3.1	3.0	3.2	3.4	3.6							
SE mixed woodland	2.5	2.3	2.9	2.8	3.1	2.6	2.9	2.8	2.8	2.7	2.8	2.2	2.6

*Blank cell signifies forest type group not present.

Table 25.3. Land Area of Vegetation Types in Millions of Hectares

Vegetation Type	1992 RPA Update survey	1992 RPA Update forest type group map	MAPSS
NE mixed forest	0.278	0.236	1.87
Maple–beech–birch	0.440	0.375	3.18
Oak–hickory Forest Oak–hickory Woodland	32.3	25.5	12.6
SE mixed forest SE mixed woodland	47.7	59.0	66.2
Dry tropical forest	NA	NA	0.440
Other forest types Nonstocked Productive reserved	5.0	NA	NA
Total	85.7	85.1	84.3

[1] As depicted by 1) the 1992 RPA Update Survey statistics (Powell et al, 1992), 2) 1992 RPA Update forest type group map (based on satellite imagery from Powell et al., 1992), and 3) the MAPSS simulation based on present-day climate. (Also see Table 25.1 for explanations of forest types.)

vegetation classifications. Thus, areal estimates of mosaics of mixed and pure types scaled up to a 10-km resolution would be biased toward such "mixed" classifications as the SE mixed forest type. A similar overestimate for the SE mixed forest type may also exist in the 1-km resolution, forest type group map (Table 25.3).

Mapped Atmosphere-Plant Soil System Simulations Under Future Climate Scenarios

Vegetation Distribution

Under the six future climate scenarios, MAPSS generally depicts northward and elevational shifts for all forest types (Figure 25.2a,b). The largest projected decreases in forest and woodland area in the southern region occur under the GFDL and UKMO scenarios (no WUE effect), 43% and 42%, respectively (Figure 25.2a). In contrast, much less change in area (-18% to $+1\%$) was predicted with MAPSS simulating a WUE effect (Figure 25.2b), underscoring one of the most critical uncertainties in predicting ecosystem-level responses to climate change and elevated levels of CO_2.

The MAPSS model predicts that large areas of the SE mixed forest type will be converted to the SE mixed woodland type (no WUE effect) (Figure 25.2a). The extent of the conversion was greatly reduced by a WUE effect in MAPSS (Figure 25.2b). To some extent, projected declines in area of the SE mixed forest type are the result of a northerly migration out of the southern region, especially in the UKMO future climate scenario. Similarly, the greater than 90% losses in area of the oak–hickory forest type (Figure 25.3a,b) represent a prediction specific to the southern region. A somewhat different pattern, however, is evident when the

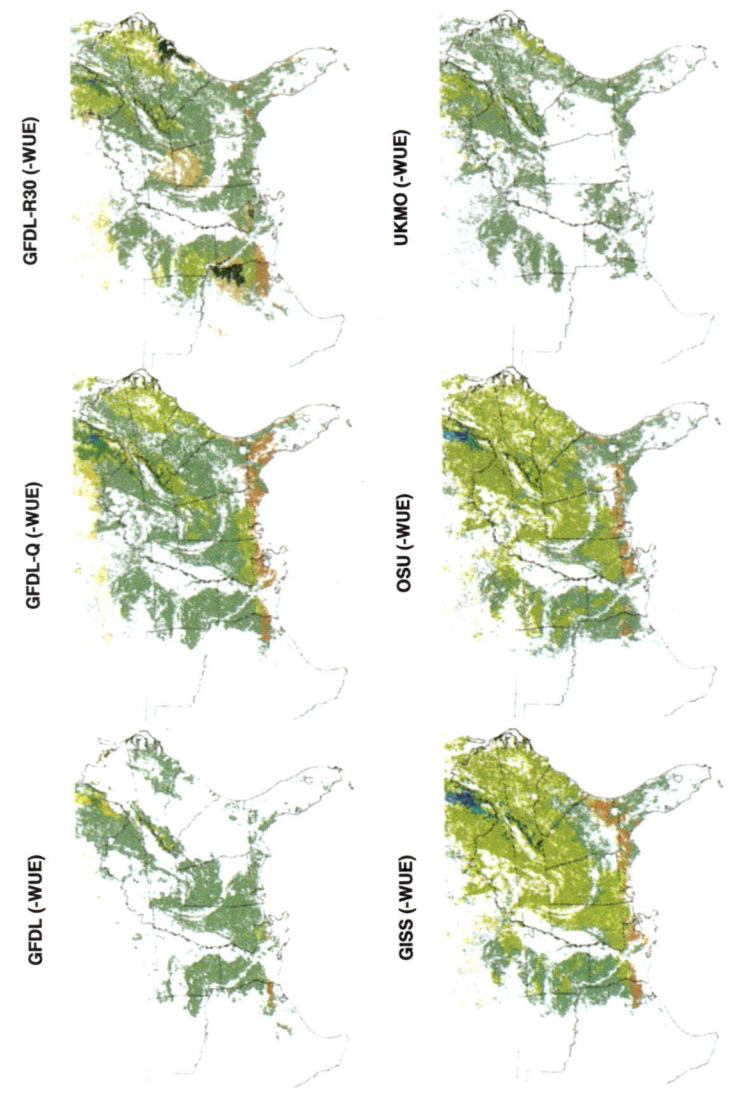

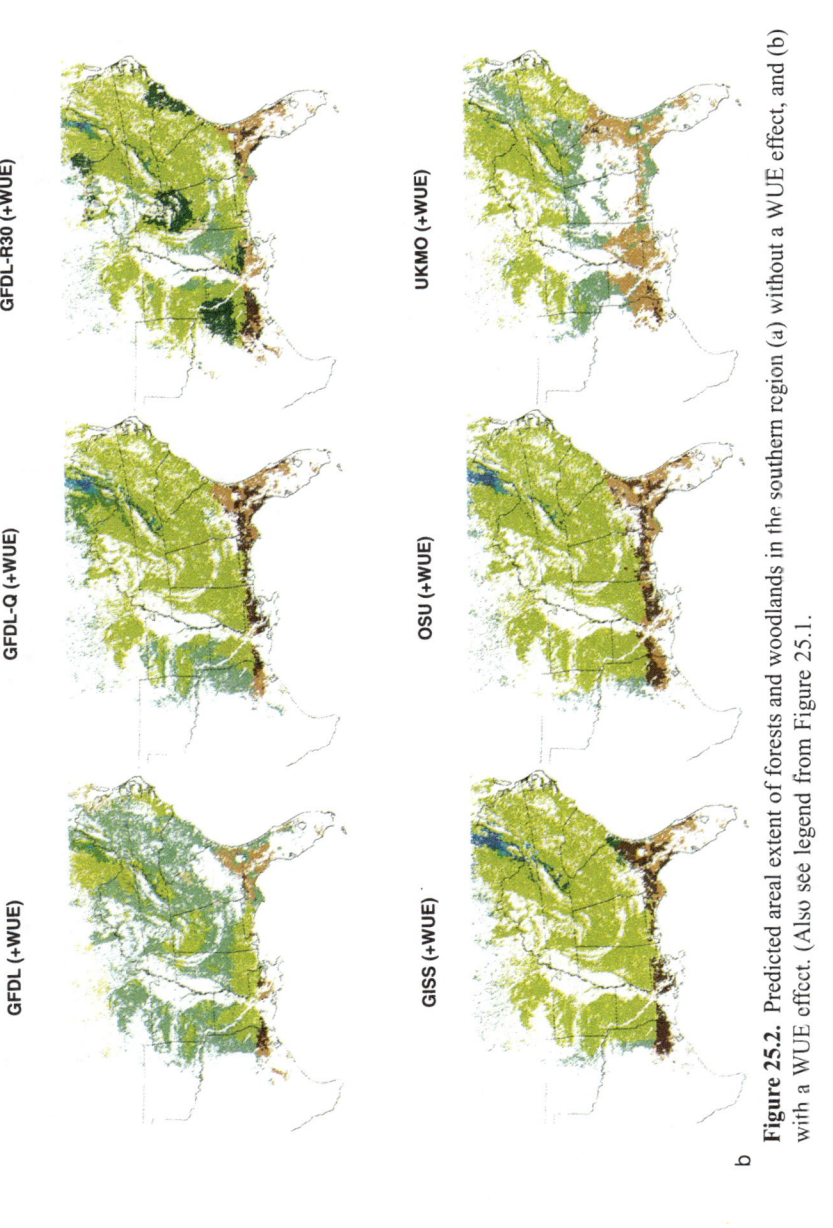

Figure 25.2. Predicted areal extent of forests and woodlands in the southern region (a) without a WUE effect, and (b) with a WUE effect. (Also see legend from Figure 25.1.)

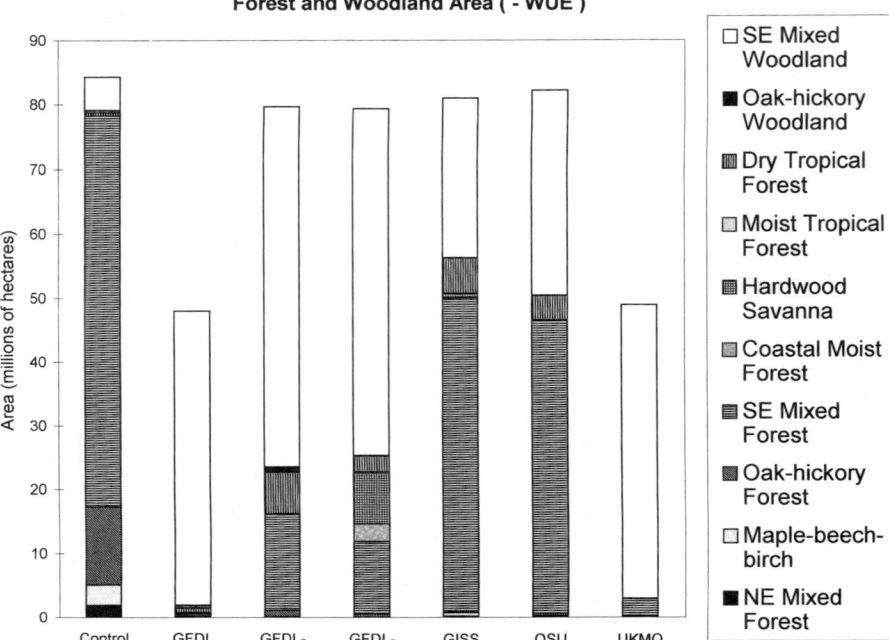
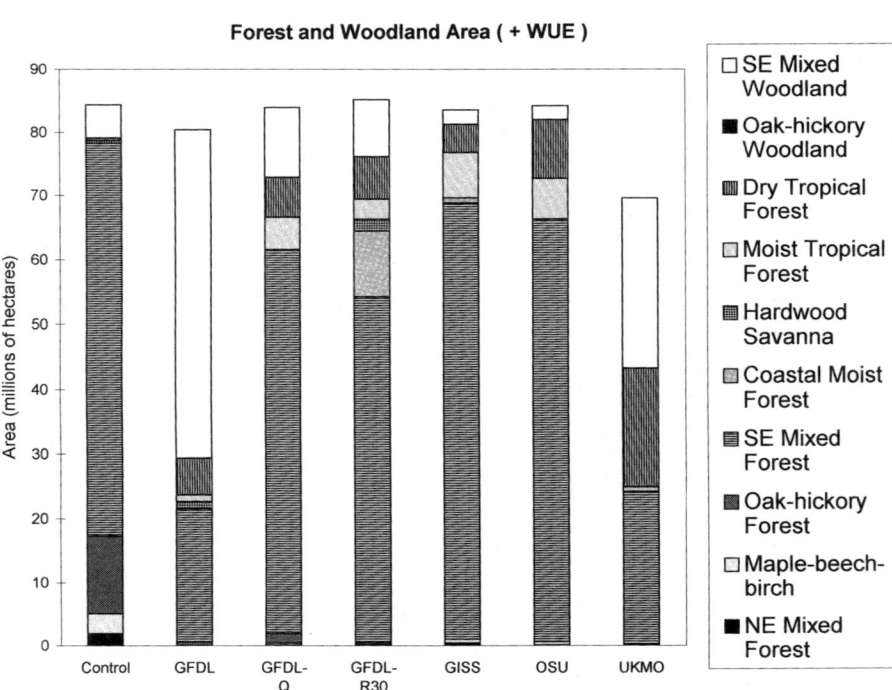

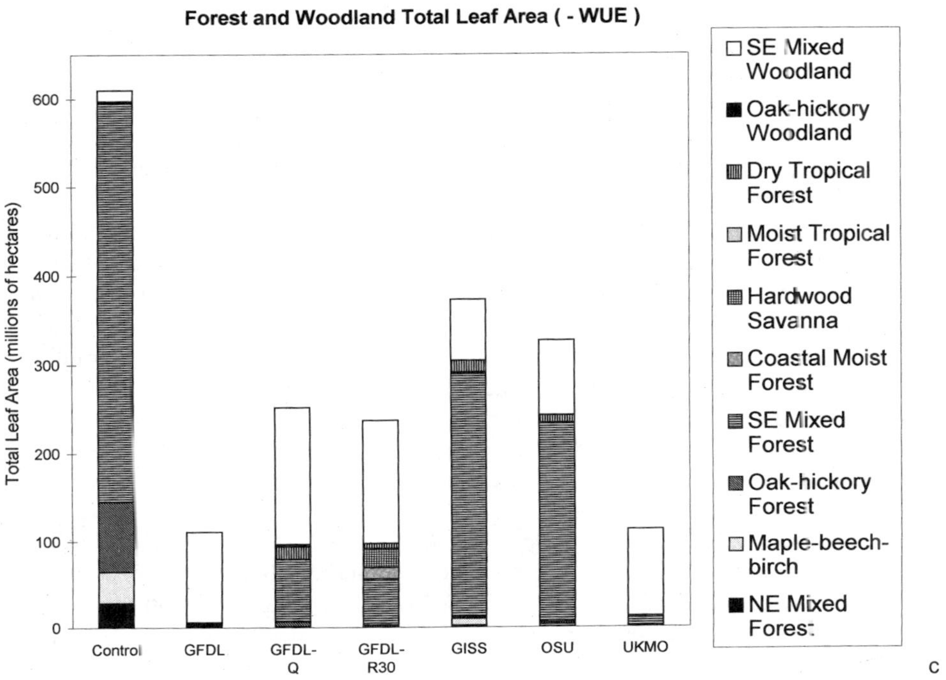

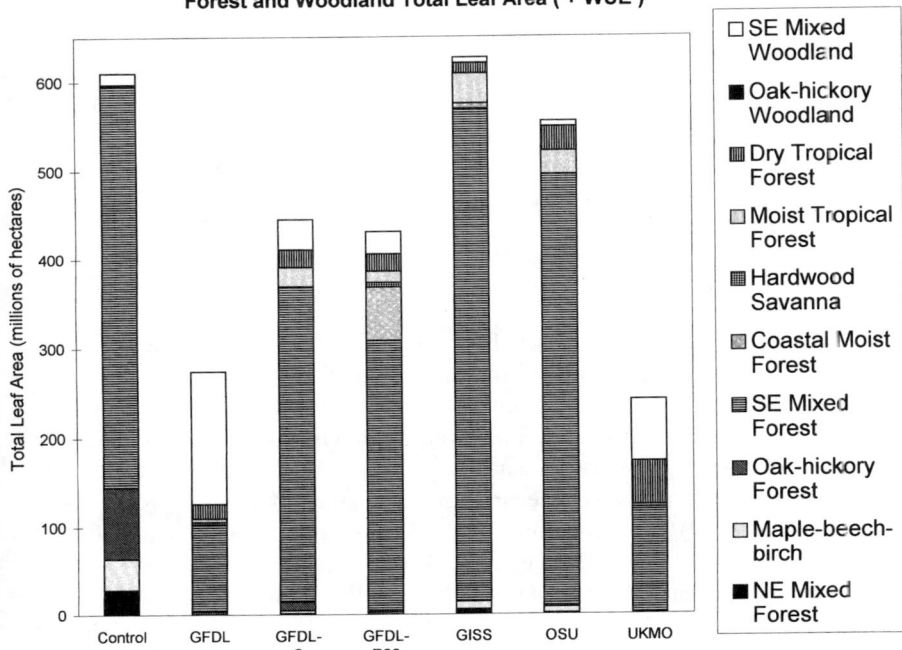

Figure 25.3. The MAPSS-based estimates of forest and woodland area (a) without a WUE effect, and (b) with a WUE effect; and forest and woodland total leaf area (c) without a WUE effect, and (d) with a WUE effect under present and future climate scenarios.

entire eastern United States is analyzed: Without a WUE effect, the area of the oak–hickory forest type decreased an average of 31% averaged across all climate scenarios (but increased 21% and 29% under the GISS and OSU scenarios, respectively). With a WUE effect, an even more modest increase in area for this vegetation type (6%) in the eastern United States resulted.

The emergence of the dry tropical forest type in the Gulf states represents a significant change of forest type, especially under particular future climate scenarios (Figure 25.2a,b). Generally, the WUE effect produced higher areal estimates for this type, but not under all scenarios. The moist tropical forest type was predicted to occupy significant areas of the Gulf states (0.71 million to 7.3 million ha), but only when WUE is implemented in MAPSS.

Based on an occasional dry summer month in the GFDL, GFDL-R30, and GISS scenarios, MAPSS predicted the occurrence of a hardwood savanna type in the southern region. The hardwood savanna type corresponds to the tree savanna mixed warm (evergreen needle) vegetation class in MAPSS and to the western hardwoods RPA forest type group (Table 25.1). The western hardwoods type is a fire climax, representing warm, broadleaf woodlands, a climatic climax of nearly closed needleleaf forests in the Willamette Valley, Oregon and other western locales. A probable southern region equivalent for the savanna mixed warm (evergreen needle) MAPSS class is the oak–pine forest type group, a constituent of the SE mixed forest (Table 25.1).

Leaf Area Index

Changes in LAI under all scenarios were not always related to changes in the areal extent of a vegetation type. For example, average LAI changes in the SE mixed forest type ranged from −44% to −23% for no WUE effect, and −36% to +11% with WUE (Table 25.2). Conversely, in scenarios where the coastal moist forest type was predicted to expand in area (mainly GFDL-R30 with a WUE effect), LAI generally decreased.

Total Leaf Area

The potential impact of decreased LAI on total timber volume, density, and biomass in southern region forests is depicted by large reductions in TLA (Figure 25.3c,d). Without a WUE effect, changes in the TLA (i.e., an index of total volume, basal area, and so forth) for all vegetation types ranged from −82% to −39% across the six future climate scenarios. A WUE effect mitigates this somewhat, but four of the six scenarios still depict significantly lower sums of TLA (Figure 25.3d) even when only slight decreases in the total area of forests and woodlands were predicted (Figure 25.3b).

Under the GISS and OSU scenarios, a WUE effect produced 23.3% and 8.1% increases, respectively, in the estimates for TLA of the SE mixed forest type. This increase represents the only predicted expansion of a native forest type in the southern region. The increase in TLA primarily reflects specific features of the GISS and OSU future climate scenarios (relatively small temperature increases combined with large precipitation increases) that underlie MAPSS predictions of increased areas of distribution and greater LAI values for this vegetation type. The SE mixed forest type shifts significantly northward from the southern region such that the eastern United States has an even larger increase in TLA, 54% and 39% for the two scenarios (with WUE effect). The possible expansion of nonindigenous species in the form of the dry tropical forest type occurs with or without a WUE effect under all but the UKMO scenario (Figure 25.2a,b). The TLA increased by one to two orders of magnitude (Figure 25.3c,d), but this change was mainly the result of areal expansion and not increased LAI.

Runoff

In prior studies, GCM scenarios have been used to drive hydrologic simulations of managed river systems. For example, Hains and Hains (1989) predicted a 14 to 27% decrease in runoff (i.e., meteoric water not evaporated, transpired, or stored) into the Appalachicola River in Florida. The relatively warmer and wetter GISS scenario used in their study yielded predictions of increased runoff. Conversely, MAPSS, predicted decreased runoff for the general area surrounding the Appalachicola River (Figure 25.4a,b). Such a discrepancy derives in large part from comparing a prediction that relies on panevaporation data with a prediction obtained using simulated linkages between soil hydrology and plant-canopy transpiration (Figure 25.4a,b).

Because MAPSS simulates an equilibrium between LAI and spatiotemporal patterns of water availability (Neilson, 1995), predicted changes in average annual runoff are closely related to vegetation changes. In the southern region, MAPSS depicts simultaneous increases and decreases in runoff, but for most future climate scenarios there appears to be a net decrease in runoff (Figure 25.4a,b). The GFDL series of scenarios generally produce less drastic declines in runoff, however, the greatest decreases in runoff occur in the UKMO scenario. In the GISS, OSU, and UKMO scenarios, MAPSS depicts a 100 to 500 mm decrease in annual runoff across broad areas of the southern region where averages are presently 200 to 600 mm. The WUE effect generally mitigates runoff decreases under all scenarios by reducing transpiration. Predictions of future precipitation using GCMs are generally less reliable than are estimates of temperature (Robcock et al., 1993), and therefore runoff data should be interpreted with caution. However, Neilson and Marks (1994) concluded that simulated regional changes in runoff tend to depict natural, regional sensitivities to global warming regardless of uncertainties associated with precipitation patterns. Thus, the consistent pattern of decreases for the southern region argues for cautious planning and management of water resources.

Model Uncertainties

The two most important MAPSS-related uncertainties contributing to variations in predicted vegetation patterns are (1) the direct effects of elevated levels of CO_2, and (2) mechanisms and degree of coupling between the canopy and the atmo-

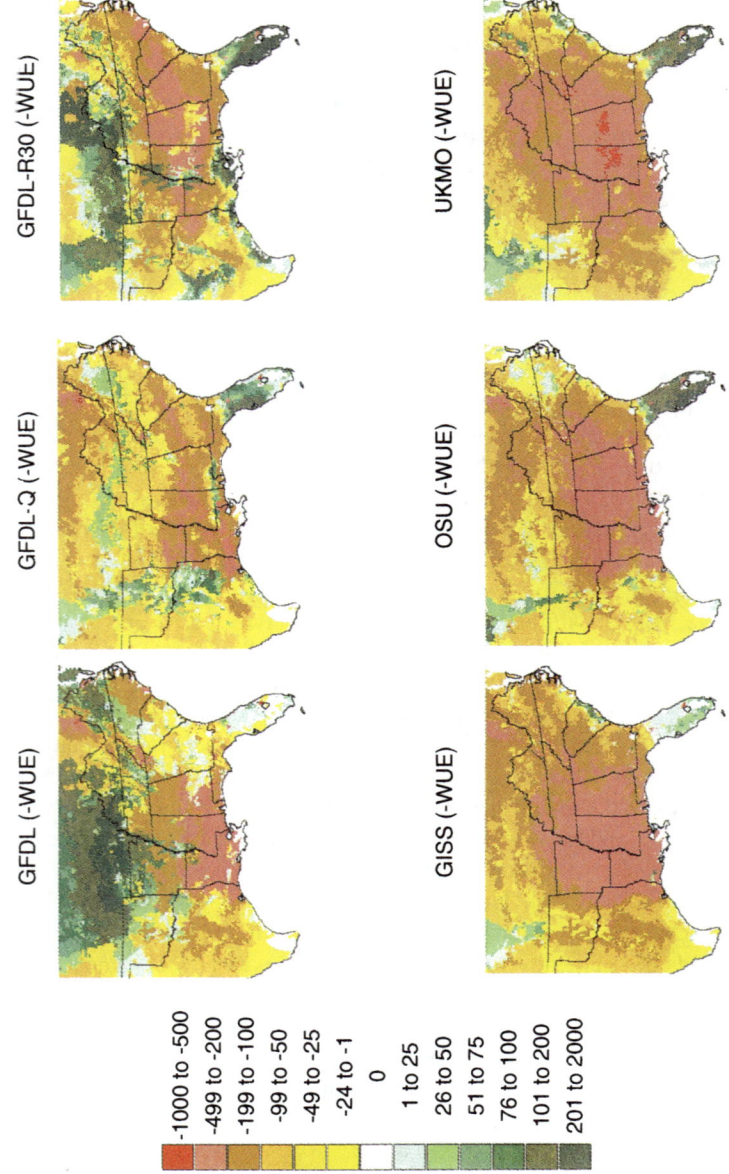

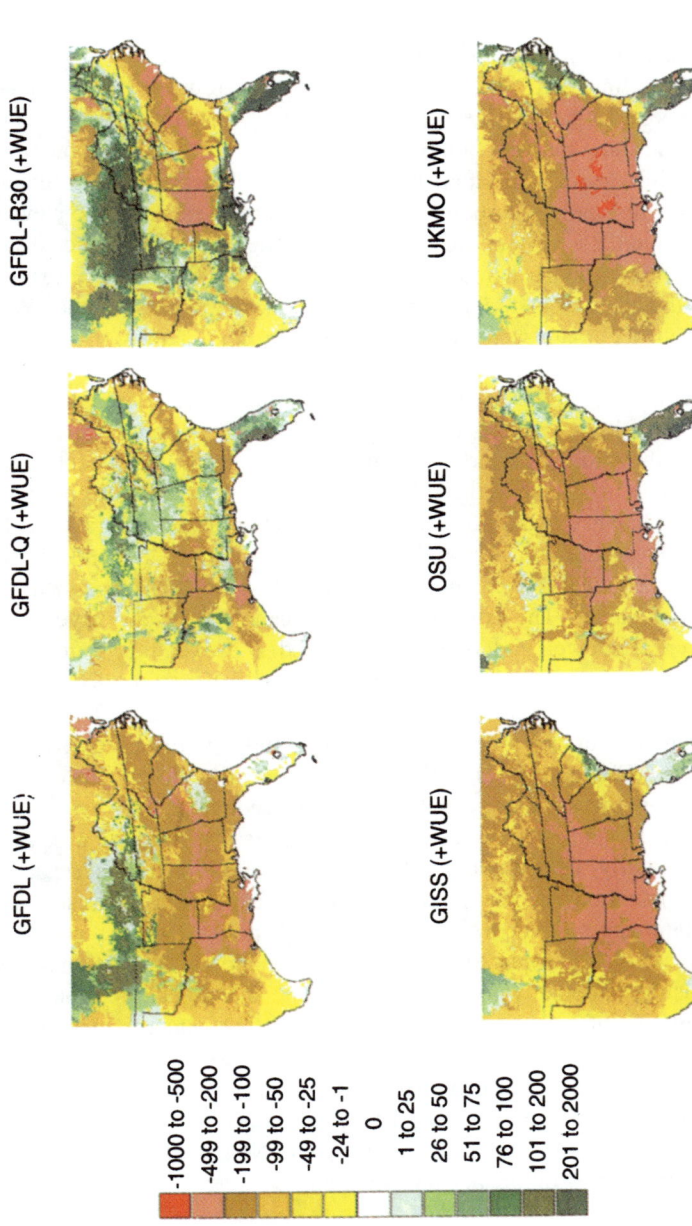

Figure 25.4. Predicted differences in average annual runoff for the southern region between present climate and future climate scenarios, (a) without a WUE effect, and (b) with a WUE.

sphere. The more "beneficial" scenarios of global warming occur generally when direct effects of elevated CO_2 levels are explicitly simulated. However, there are numerous uncertainties associated with how the CO_2-fertilization effect is modeled. In a recent comparison among biogeography and biogeochemistry simulation models (including MAPSS), this relationship was illustrated very clearly (VEMAP Members, 1995). Of the six models compared, none implemented direct CO_2 effects in the same manner. Some models simulated increased productivity, others depicted increased WUE, and some included both effects. Yet it remains unclear how to realistically incorporate these physiological responses to elevated levels of CO_2. Even less certain are CO_2 effects under the complex ecosystem constraints absent in growth-chamber studies. For MAPSS, the VEMAP exercise demonstrated that CO_2 sensitivity is probably too great, implying that WUE effects under future climate scenarios for the southern region are somewhat less than depicted.

Uncertainties related to the problem of canopy-atmosphere coupling are several. First, the quantitative aspects of the effects of canopy roughness are poorly understood, yet they dominate turbulent mixing processes between canopy and atmosphere. Hence, to quantify the relative importance of stomatal control on ecosystem-water balance, MAPSS must also accurately assess interactions between various canopy architectures and atmospheric conditions. Second, interactions include the boundary layer above canopies that gradually mixes with the upper atmospheric layers. The extent to which this canopy boundary layer mixes with upper layers is not well understood or measured. However, new syntheses of existing data indicate a relatively large degree of coupling between forest canopies and the atmosphere (Hollinger et al., 1994; Schulze et al., 1994). These canopy-atmosphere coupling processes may not be fully simulated until vegetation models and GCMs are dynamically coupled in a fully closed earth system model that better expresses energy and momentum feedbacks.

Finally, the fixed land-use map used in our study is a severe and unrealistic constraint, particularly with respect to 1) the role of natural disturbances, and 2) those future climate scenarios that portend significant expansion of forests. More deterministic land-use models are required that will portray various classes of land-use conversions (e.g., forest to agriculture as a function of population density and demand) as they interact with changes in climate and vegetation. The development and coupling of such models (e.g., Lee et al., 1992) into a framework that includes a biogeography modeling component represents an area of research critical to future integrated assessments of climate change impacts. Ideally, such a framework would contain models now being developed to simulate transient responses of ecosystems to climate change (King and Neilson, 1992), e.g., alterations in fire regimes, pest outbreaks, and so forth.

Implications and Limitations of the Equilibrium Approach to Modeling

The actual trajectory of ecosystems undergoing climatic stress cannot be inferred directly from studies based on equilibrium climate scenarios. The rate and probability of species migrating to new sites in the southern region and their eventual status depend on many factors not analyzed in this study. Nevertheless, the projected decline of the SE mixed forest type can be interpreted in light of what is known about its constituent species and their future distribution. Such forest type groups as the oak–gum–cypress and Loblolly–shortleaf pine (Table 25.1) may not fare as well in their new range. Miller et al., (1987) noted that a major southern commercial species, loblolly pine (*Pinus taeda* L.), presently occupies the better growing sites, that is, relatively deep and fertile soils. A migration toward the shallower, steeper, and rockier soils to the north and northeast probably would result in decreased productivity for the species. Less favorable soil conditions also may alter competitive dynamics, for example, a shift in forest composition toward shortleaf pine (*Pinus echinata* Mill.) (Lawson, 1990) in those areas where it occurs with loblolly pine. In our simulations, lower productivity resulting from changes in species composition or decreased site quality are represented as lower values of LAI and TLA in the SE mixed forest type. Hydroedaphic constraints on northward (i.e., inland) migration of the oak–gum–cypress forest type group may be even greater as this group is presently restricted to lowland areas. Most bald-cypress (*Taxodium distichum* (L.) Rich) stands (and related species) are situated at elevations that are less than 30 m above sea level, yet are salt-intolerant (Dicke and Toliver, 1990). A rise in sea level on the Atlantic and Gulf coasts may stress these and other lowland forests and woodlands in marginal habitats, for example, near estuaries.

It is well established that disturbance regimes significantly shape the patterns and processes of landscape and ecosystems (Turner et. al. 1987). Thus, the equilibrium scenarios presented in this chapter strongly imply a critical role for disturbance in determining the transient responses of southern region forests to climate change. Prior analyses have suggested that large areas of many forests will undergo precipitous rather than gradual responses to climate change and other related disturbances (King and Neilson, 1992; Neilson and King, 1992; Overpeck et al., 1990; Smith and Shugart, 1993; Neilson et al., 1994; Neilson, 1993a; Neilson, 1993b; Franklin et al., 1991). Most nonanthropogenic factors in the disturbance regimes of the southern region are influenced ultimately by climate, for example, fire, drought, blowdown, floods, and insect outbreaks. Human influences (e.g., forest harvest, prescribed burning) overlay these factors in a complex fashion and have created the forest landscapes visible today. Yet it is difficult to predict the spatial and temporal pattern of the disturbances that so greatly influence the dynamics of succession and the resulting age structure. For this reason, such equilibrium models as MAPSS perform best when simulating near steady-state condition, that is, forests and woodlands under a contemporary climate and large-scale disturbance regimes. Transient versions of MAPSS are being developed that will simulate ecosystem trajectories under dynamic climate change scenarios.

The equilibrium estimates of runoff for the southern region presented here also must be considered in light of potential transient dynamics of climate and ecosystems. As with vegetation, the significance of the MAPSS runoff estimates (particularly the large decreases) depends greatly on such extreme events as droughts, floods, and erosion. These events are difficult to predict because they are usually associated with extremes of climate. For example, even small changes in precipitation means can create large changes in the frequency of extremes, for example, droughts and floods (Waggoner and Revelle, 1990). The potential for ecological and economic damage may be greatest during such events, and our present inability to predict them with certainty suggests that more conservative short-term decisions in natural resources management are in order (Borchers, 1996). Such equilibrium scenarios as those provided by MAPSS can aid formulation of long-term management goals (i.e., over several rotations).

Conclusions

There is no clear agreement among the various scenarios of how southern region forests will respond to global warming. Because these simulations are based on a range of climate scenarios, they have a greater probability of bracketing actual forest responses in the future. This analysis suggests several relatively consistent patterns of change that appear almost inevitable. Forests that are presently constrained to cooler climates in the southeast (e.g., higher altitude forests) will probably be lost or reduced significantly in size. In particular, poorly developed soils at higher altitudes may constrain forest development. In warmer regions of the southeastern United States, forests may be particularly vulnerable to increased drought stress, particularly species in the SE mixed forest type. These closed forests could be replaced by woodlands, possibly composed of the same species.

The potentially catastrophic conclusions presented here are consistent with other studies of the potential impacts of global warming on United States forests (Franklin et al., 1991; Winjum and Neilson, 1990). The potential changes are of sufficient magnitude to warrant dramatic shifts in long-range forest-management policies, for example, regulations related to forest-regeneration practices. Although these results present a broad spectrum of possible futures, it is hoped that they will catalyze the formation of flexible, alternative policy options that can anticipate potential climate changes in the southern region of the United States.

References

Borchers JG (1996) A hierarchical context for sustaining ecosystem health. In Jaindl, RG, Quigley, TM (Eds) *Search for a solution: sustaining the land, people, and economy of the Blue Mountains.* American Forests Publications. Washington, DC.

Botkin DB, Janak JF, Wallis JR (1972) Some ecological consequences of a computer model of forest growth. J Ecol 60:849–872.

Christensen NL, Bartuska AM, Brown JH, Carpenter S, D'Antonio C, Francis R, Franklin JF, MacMahon JA, Noss RF, Parsons DJ, Peterson CH, Turner MG, Woodmansee RG (1996) The report of the ecological society of america committee on the scientific basis for ecosystem management. Ecol Appl. 6(3):665–691.

Daly C, Neilson RP, Phillips DL (1994) A statistical-topographic model for mapping climatological precipitation over mountainous terrain. J Appl Meteor 33:140–158.

Dicke SG, Toliver JR (1990) Growth and development of bald-cypress/water tupelo stands under continuous versus seasonal flooding. For Ecol Manag 33:523–530.

Eamus D (1991) The interaction of rising CO_2 and temperatures with water use efficiency. Plant Cell Environ 14:843–852.
Eamus D, Jarvis PG (1989) The direct effects of increase in the global atmospheric CO_2 concentration on natural and commercial temperate trees and forests. Adv Ecol Res 19:1–55.
Elliot, DL, CG Holladay, WR Barchet, HP Foote, and WF Sandusky (1987) Wind energy atlas. Solar Tech Info Prog, USDE. Washington, DC.
Emanuel KA (1987) The dependence of hurricane intensity on climate. Nature 326:483–485.
Emanuel WR, Shugart HH, Stevenson MP (1985) Climatic change and the broad-scale distribution of terrestrial ecosystem complexes. Clim Change 7:29–43.
Franklin JF, Swanson FJ, Harmon ME, Pery DA, Spies TA, Dale VH, McKee A, Ferrell WK, Means JE, Gregory SV, Lattin JD, Schowalter TD, Larsen D (1991) Effects of global climatic change on forests in northwestern North America. Northw Environ J 7:233–254.
Frew, JE 1990 The image processing workbench. PhD Thesis, Dep Geog, Univ CA, Santa Barbara.
Hansen J, Fung I, Lacis A, Rind D, Lebedeff S, Ruedy R (1988) Global climate changes as forecast by Goddard Institute for Space Studies three-dimensional model. J Geophys Res 93 9341–9364.
Hollinger DY, Kelliher FM, Schulze E-D, Kostner BMM (1994) Coupling of tree transpiration to atmospheric turbulence. Nature 371:60–62.
Isaaks, EH, Srivastava RM (1989) *Applied Geostatistics.* Oxford University Press, New York.
Karl TR, Williams CN, Quinlan FT, Boden TA (1990) United States Historical Climatology Network. Serial Temperature and Precipitation Data. USDE, Carb Diox Info Anal Cen, Oak Ridge, TN.
King GA, Neilson RP (1992) The transient response of vegetation to climate change: A potential source of CO_2 to the atmosphere. Water Air Soil Pollut 64:365–383.
Küchler, AW (1964) *Potential natural vegetation.* American Geographical Society, New York.
Lawson ER (1990) Shortleaf pine. In Burns RM, Honkala BH (Eds) *Silvics of North America. Volume 1, conifers.* USDA For Ser Agric Hand 654.
Lee RG, Flamm R, Turner MG, Bledsoe C, Chandler P, DeFerrari C, Gottfried R, Naiman RJ, Schumaker N, Wear D (1992) Integrating sustainability development and environmental vitality: A landscape ecology approach. In Naiman RJ (Ed) *Watershed management: Balancing sustainability and environmental change.* Springer-Verlag, New York.
Manabe S, Stouffer RJ, Spelman MJ, Bryan K (1991) Transient responses of a coupled ocean-atmpsheric model to gradual changes of atmospheric CO_2. Part I: Annual mean responses. J Climate 4:785–818.
Marks D (1990) A continental-scale simulation of potential evapotranspiration for historical and projected doubled CO2 climate conditions. In Gucinski H, Marks D, Turner D (Eds) *Biospheric feedbacks to climate change: the sensitivity of regional trace gas emissions,evapotranpsiration, and energy balance to vegetation redistribution—Status of ongoing research.* EPA report EPA/600/3–90/078. USEPA, Corvallis, OR.
McCracken M (1995) The evidence mounts up. Nature 376:645–646.
Miller WF, Dougherty PM, Switzer GL (1987) Effect of rising carbon dioxide and potential climate change on loblolly pine distribution, growth, survival, and productivity. In Shands WE, Hoffman JS (Eds) *The greenhouse effect, climate change, and U.S. forests.* Conservation Foundation, Washington, DC.
Mitchell JFB, Warrilow DA (1987) Summer dryness in northern mid latitudes due to increased CO_2. Nature 330(19):238–240.
Neilson RP (1993a) Vegetation redistribution: A possible biosphere source of CO_2 during climatic change. Water Air Soil Pollut 70:659–673.

Neilson RP (1993b) Transient ecotone response to climatic change: Some conceptual and modelling approaches. Ecol Appl 3:385–395.

Neilson RP (1995) A model for predicting continental-scale vegetation distribution and water balance. Ecol Appl 5:362–385.

Neilson RP, King GA (1992) Continental scale biome responses to climatic change. In McKenzie DH, Hyatt DE, McDonald VJ (Eds) *Ecological indicator volume 2*. Elsevier Science Publishers, London.

Neilson RP, King GA, DeVelice RL, Lenihan J, Marks D, Dolph J, Campbell W, Glick G (1989) Sensitivity of ecological landscapes to global climatic change, USEPA, EPA-600-3-89-073, NTIS-PB-90-120-072-AS, Washington, DC.

Neilson RP, King GA, Lenihan J (1994) Modeling forest response to climatic change: The potential for large emissions of carbon from dying forests. In Kanninen M (Ed) *Carbon balance of the world's ecosystems: Towards a global assessment*. Publications of the Academy of Finland, Painatuskeskus, Helsinki.

Neilson RP, Marks D (1994) A global perspective of regional vegetaion and hydrologic sensitivities from climatic change. J Veg Sci 5:715–730.

Neilson RP, Wullstein LH (1983) Biogeography of two southwest American oaks in relation to atmospheric dynamics. J Biogeogr 10:275–297.

Neilson RP, Wullstein LH (1986) Microhabitat affinities of Gambel Oak seedlings. Great Basin Natur 46:294–8.

NOAA-EPA Global Ecosystems Database Project (1993) Global ecosystems database version 1.1. (CD ROM) User's guide, documentation, reprints, and digital data. USDOC/NOAA National Geophysical Data Center, Boulder, CO.

Norton BG (1992) A new paradigm for environmental management. In Costanza R, Norton BG, Haskell BD (Eds) *Ecosystem health: New goals for environmental management*. Island Press, Washington DC.

Overpeck JT, Rind D, Goldberg R (1990) Climate-induced changes in forest disturbance and vegetation. Nature 343:51–53.

Parton WJ, Stewart JWB, Cole CV (1988) Dynamics of CNP and S in grassland soils: A model. Biogeochem 5:109–31.

Phillips, DL, Dolph J, Marks D (1992) A comparison of geostatistical procedures for spatial analysis of precipitation in mountainous terrain. Agric For Meteor 58:119–141.

Powell DS, Faulkner JL, Darr DR, Zhu Z, MacCleery DW (1993) Forest resources of the United States, 1992. USDA For Ser Gen Tech Rep RM-234.

Quinlan FT, Karl TR, Williams CN (1987) United States Historical Climatology Network. Serial Temperature and Precipitation Data. USDE, Carb Diox Info Anal Cen, Oak Ridge, TN.

Raich JW, Rastetter EB, Melillo JM, Kicklighter DW, Steudler PA, Peterson BJ, Grace AL, Moore BI, Vorosmarty CJ (1991) Potential net primary productivity in South America: Application of a global model. Ecol Appl 1:399–429.

Robock A, Turco RP, Harwell MA, Ackerman TP, Andressen R, Chang HS, Sivakumar MVK (1993) Use of general circulation model output in the creation of climate change scenarios for impact analysis. Clim Change 23:293–336.

Running SW, Hunt R (1993) Generalization of a forest ecosystem process model for other biomes, BIOME-BGC, and an application for global-scale models. In Ehleringer JR, Field C (Eds) *Scaling processes between leaf and landscape levels*. Academic Press, San Diego, CA.

Schlesinger ME, Zhao ZC (1989) Seasonal climatic change introduced by double CO_2 as simulated by the OSU atmospheric GCM/mixed-layer ocean model. J Climate 2:429–495.

Schulze E-D, Kelliher FM, Korner C, Lloyd J, Leuning R (1994) Relationships among maximum stomatal conductance, ecosystem surface conductance, carbon assimilation rate, and plant nitrogen nutrition: A global ecology scaling exercise. Annu Rev Ecol Syst 25:629–660.

Shugart, HH (1984) *A Theory of Forest Dynamics.* Springer-Verlag, New York.

Spangler WM, Jenne RL (1989) *World monthly surface station climatology (and associated datasets).* National Center for Atmospheric Research, Boulder, CO.

Smith TM, Shugart HH (1993) The transient response of terrestrial carbon storage to a perturbed climate. Nature 361:523–526.

Turner MG (1987) *Landscape Heterogeneity and Disturbance.* Springer-Verlag, New York.

USA-CERL (1993) *GRASS 4.1 User's Manual.* US Army Corps of Engineers, Champaign, IL.

VEMAP Members (1995) Vegetation/ecosystem modeling and analysis project: Comparing biogeography and biogeochemistry models in a continental-scale study of terrestrial ecosystem responses to climate change and CO_2 doubling. Glob Biogeochem Cycles 9:407–437.

Waggoner PE, Revelle RR (1990) Summary. In Waggoner PE (Ed) *Climate change and U.S. water resources.* John Wiley and Sons, New York.

Walters, CJ (1986) *Adaptive management of renewable resources.* MacMillan Publishing Co., New York.

Walters CJ, Holling, CS (1990) Large-scale management experiments and learning by doing. Ecol 71:2060–2068.

Wetherald RT, Manabe S (1988) Cloud feedback processes in a general circulation model. J Atmo Sci 45:1397–1415.

Winjum JK, Neilson RP (1990) Forests. Smith JB, Tirpak DA (Eds) *The potential effects of global climate change on the United States.* USEPA, EPA-230–05–89–050, Washington, DC.

Woodward, FI (1987) *Climate and plant distribution.* Cambridge University Press, London, England.

Woodward FI, Smith TM, Emanuel WR (1995) A global land primary productivity and phytogeography model. Glob Biogeochem Cycl 9:471–90.

Zhu Z, Evans DL (1992) Mapping midsouth forest distributions with AVHRR data. J For 90:27–30.

26. Summary of Simulated Forest Responses to Climate Change in the Southeastern United States

David A. Weinstein, Wendell P. Cropper, Jr., and Steven G. McNulty

During the next century, substantial changes are expected to occur involving such environmental variables as temperature, precipitation, cloudiness, atmospheric carbon dioxide (CO_2), tropospheric ozone, and atmospheric deposition of nutrients such as sulfur and nitrogen (Melillo et al., 1993; Mitchell et al., 1992). These changes, which are expected to vary temporally and spatially, may have profound effects on forest health, productivity, and distribution. Some of these changes may directly affect the physiology of trees, others may alter the susceptibility of trees to such disturbances as fire and flooding, and others may alter the establishment and competitive balance of forest communities. Thus, environmental changes and stresses have the potential to alter not only the function of forest ecosystems, but also the structure, composition, and distribution of forests.

Models of forest responses to environmental change will become tools that help us manage our nation's forest resources in the next century. We will need detailed plant physiology models that operate at small spatial and temporal scales to integrate our mechanistic understanding of forest responses to environmental changes at a detailed level. We will also require models of whole canopy and landscape processes to extrapolate this understanding to predictions of behavior of forests across the entire region and to evaluate whether such large-scale processes as hydrology, nutrient cycling, and competition will constrain and shape the behavior of these forests.

In the Southern Global Change Program (SGCP) a variety of models have been used to generate predictions of the effects of various aspects of global change. These models simulate dynamics at different scales of resolution in process and space, from photosynthetic light response in different portions of the canopies to aggregate carbon, nutrient, and water balance and biogeography across large landscapes. Nevertheless, each model was used to estimate the response of southern forest growth, and in the case of the larger-scale regional models, vegetation distribution, to scenarios of future climatic conditions. Consequently, taken as a set of models, their predictions estimate growth under increased CO_2 levels, temperature, tropospheric ozone, and decreased regional water availability. When these models have been used for predictions, and when they are under the same scenario of future conditions, their predictions can be compared and contrasted. This comparison either increases confidence in our understanding about the probable direction of these effects or draws attention to the importance of the differences in assumptions among the models.

The models in this section represent a wide variety of approaches to the problem of predicting the response of southern forests to climatic change. They differ in the way they calculate photosynthesis, and in the way that photosynthesis is utilized to build tree leaves, branches, stems, and roots. Several models use a mass-balance approach, predicting forest-stand growth by estimating how much carbon or water is available to construct plant organs or maintain plant community canopies. Other models use known correlations between overall forest productivity and climatic conditions without considering the details of carbon, energy, or water balance within those forests.

There are ecosystem processes that could be affected by altered climatic conditions that are not considered in any of the models reviewed here. For example, climate change could certainly alter the fire frequency in a region by drying out the dead organic matter and creating a flammable fuel-load environment. More fires, in turn, would undoubtedly reshape forests in many ways. A change in flooding frequency could modify the dynamics in forests that have become tuned to a particular rate for their normal cyclic processes, as well as having devastating effects on forests that rarely experience floods. Climatic change can also alter the ability of insects or pathogens to spread, and alter the susceptibility of trees to attack by these agents. We know much less about how to model these processes. Therefore, the models discussed here represent examinations of the better understood processes that are potentially affected by global change

This chapter summarizes the overall conclusions that can be drawn from comparing and contrasting these modeling results concerning forest productivity and distribution. Our goal is to summarize only the models used in the SGCP. We do not make any effort to justify why these models and not others were used, nor do we attempt to position these models in the entire field of models available elsewhere to simulate forest responses to climate change. Both of these would make fascinating exercises, but they are well beyond the scope here. Nevertheless, the models that were used are capable of simulating an excellent cross-section of processes and responses at each significant scale of forest dynamics. Thorough model descriptions can be found in the appropriate chapter elsewhere in this book;

26. Summary of Simulated Forest Responses to Climate Change

consequently, we discuss only the most relevant portions of the assumptions in these models.

Although many other aspects of forests could be altered, for example, structure, successional sequences, understory, and so forth, this review examines predictions of productivity and distribution only because they were simulated by a significant subset of the models. Through a comparison of the behavior of these models, we can evaluate the sensitivity of predictions to different assumptions about how forests grow. If two or more models, which were constructed from very different conceptual frameworks and were based on significantly different sets of assumptions, arrive at the same conclusions, it may suggest that the predictions are relatively insensitive to the exact method used to represent forest dynamics in a given model. That fact alone could greatly increase our confidence in the predictions being made.

Before we use models for policy and management planning, it is necessary to gain confidence in model predictions. It is not possible to validate predictions of responses to novel combinations of future climate, nutrition, pollution, and so forth, but we can attain greater insight into these predictions by comparing the results of independently developed forest-process models. Although we may gain confidence in predictions when different models simulate similar results, we must study the results of each model very carefully to determine whether models are agreeing for the wrong reasons. In many cases, comparisons among models does not bolster our confidence in any one model, as is the case in the recently reported VEMAP study (Schimel et al., 1997, Field et al., 1996). However, agreement in predictions often signals that independent groups agree that the same processes are responsible for driving the behavior of the forest system.

Each model has been demonstrated to be capable of accurately reproducing known dynamics in existing forest stands. However, those demonstrations do not provide an adequate test of whether they contain accurate representations of the mechanisms that will come into play in real forests after they are exposed to new climatic conditions. Each of these models, however, is the embodiment of a well considered and thoroughly studied set of concepts about how forests function, and therefore each provides a platform for logically extrapolating forest behavior under future scenarios of climate change.

Although there is considerable agreement among these models with respect to the general direction of response we are liable to see in forest growth under future climates, the greatest value of studying the similarities and differences among the models may not lie in their agreement. What we seek through the use of these models is an understanding of those mechanisms that will be of greatest importance in determining forest response in order that we may study them more thoroughly. No matter how thoroughly these models suggest we understand forest dynamics, scientists in this field know that our study of these dynamics is in its infancy. A comparison of the models and the differences in their predictions will help in an evaluation of our present understanding.

It is not easy to unravel the reasons for differing predictions among models; it is often no easier than understanding why models based on entirely different sets of relationships should produce the same results. Models of biological systems are, of necessity, complex interwoven structures with layers upon layers of interdependent relationships. The tools are not yet available to permit a sensitivity analysis of the importance of a given relationship to the response of a model to a change in a certain input such as temperature. Nevertheless, in the course of using each model to make predictions, the modelers gain an understanding of the key relationships governing the behavior they observed. We used these insights whenever possible to determine those relationships that might have contributed to the differences or similarities in predictions. The reader must keep in mind, however, that the major purpose of a discussion of these similarities and differences is to attempt to identify those points at which our conceptual framework regarding the mechanisms and relationships that drive forest dynamics needs to be improved through more extensive study. Better predictions can be developed by understanding what can be gained by including a given concept in a particular model and what is lost by ignoring a given relationship in another.

General Approaches to Forest-Growth Prediction in the Models

The models used to predict forest productivity under future climate scenarios in the SGCP (Table 26.1) represent a wide variety of approaches to making this projection. In Table 26.1, the models are arranged according to the scale at which they represent mechanisms contributing to the calculation of forest growth and distribution. At one end of the spectrum, MAESTRO (Wang and Jarvis, 1990) focuses on the availability of light throughout a canopy and the photosynthesis that can be achieved given that light regime. This model constructs a prediction of productivity from our knowledge of the most basic plant processes. On the other end of the spectrum, SOFOCLES, (Smith et al., Chapter 24) simulates regional forest growth without explicitly simulating biological mechanisms.

The models MAESTRO and SPM (Cropper and Gholtz, 1993) both focus on a rigorous calculation of photosynthesis for a small forest stand, but MAESTRO concentrates on accurate estimates of leaf-level functions (i.e., photosynthesis, transpiration, and respiration), and SPM focuses on the balance of the carbon budget within the forest. MAESTRO evaluates whether productivity will change based on the limitations to net photosynthesis, respiration, and transpiration; SPM considers whether photosynthesis will be limited by a lack of demand for the carbon products produced and made available for tissue growth. Similarly, in UTM (Dixon et al., 1978; Luxmoore 1989), winked to PTAEDA2, predictions focus on whether the carbon demand for tissue growth is sufficient to maintain the gradient of carbon within the trees. This gradient is necessary to keep carbon flowing from the leaves and prevent leaf-starch buildup which could downregulate photosynthesis.

These models may either underestimate or overestimate carbon demand by assuming that the allometry of carbon allocation throughout a tree will remain relatively constant under new environmental conditions. However, the demand

26. Summary of Simulated Forest Responses to Climate Change

Table 26.1. Summary of Climate Change Predictions of Models Used in the Southern Global Change Project Study

Model	Chapter Authors	Climatic Variable	Principal Conclusions
MAESTRO	Cropper et al.	Temperature ↗ CO₂ ↗	Increasing productivity
UPTAKE	Kelly and Yanai	Nutrient availability ↘	Decreasing productivity
SPM	Cropper	Temperature ↗ CO₂ ↗	Increasing productivity
PIPESTEM	Valentine et al.	Temperature ↗ CO₂ ↗	Increasing productivity
BIOMASS	Sampson et al.	Temperature ↗ Precipitation ↘ CO₂ ↗	Increasing productivity
PTAEDA2	Luxmoore et al.	Ozone ↗ Water stress ↗	Decreasing productivity
PnET–IIS	McNulty et al.	Temperature ↗ Precipitation ↘	Decreasing productivity Increased water yield
MAPSS	Borchers and Neilson	Temperature ↗ Precipitation ↘ CO₂ ↗	Decreased distribution Decreased leaf area Decreased wter yield
SOFOCLES	Smith et al.	Temperature ↗ Precipitation ↘ CO₂ ↗ Ozone ↗	Decreasing productivity

for carbon could be shifted dramatically if trees are able to develop alternative sinks for this carbon. For example, plants could be driven to allocate more carbon to roots to accelerate the uptake of nutrients to supply an increased aboveground demand for those nutrients. The focus of UPTAKE (Oates and Barber, 1987; Yanai, 1994) is to evaluate whether an increase in investment below ground would be of benefit to the plant, that is, whether nutrient uptake can be accelerated to keep up with increasing demands for nutrients coming from plants with abundant supplies of carbon available for structure building.

On the next-larger scale, the models focus less on specific trees or specific portions of trees and more on aggregate stand processes. In this group, PIPE-STEM (Valentine et al., Chapter 19) describes the growth of a stand in terms of production and loss of dry matter. The distinctions among types of models blur, of course, because PIPESTEM is similar to the earlier models mentioned in its physiological underpinning; dry matter production is proportional to photosynthesis minus construction and maintenance respiration. The PIPESTEM model divides dry matter production into new leaf, fine root, woody root, and woody stem tissue in proper proportions to maintain functional balance, using the pipe-theory assumption that trees build leaves to obtain carbon; leaves need stem and roots to supply them with support and resources. Forest growth is determined by the effect of environmental conditions on photosynthesis and respiration, with net photosynthesis determining the production of each type of tissue, added in proportion to maintain the functional relationship among them. Negative feedback between accumulated forest mass and crown rise and self-thinning rates cap the growth as forests age.

The BIOMASS model (McMurtrie and Wolf, 1983; McMurtrie, 1985; McMurtrie et al., 1992) also focuses on aggregate stand properties, assuming that trees are limited by the available photosynthetic fixation. Both BIOMASS and PIPESTEM assume that biomass will increase in a canopy if the leaves provide the carbon for building after the daily demands of maintenance, in the form of respiration, are accounted for. However, while PIPESTEM assumes that carbon allocation is constrained to maintain quantities of nonsupport and support tissues in functional balance, the daily carbon allocation in BIOMASS version 13.0 (Sampson et al., Chapter 21) is constrained only by observed relative growth rates for foliage, stem, and branch, and by the availability of carbon from a labile storage pool. Additionally, the size of the belowground sink for carbon is assumed to be equal to the leaf mass. A unique feature of BIOMASS is its accounting for a labile pool of reserve carbon that can be used to maintain growth rates on days when the balance between photosynthesis and respiration is negative.

The model, PTAEDA2, (Burkhart et al., 1987) focuses on the density dependence of forest-plantation growth. It, too, portrays forest growth as being heavily dependent on the crown ratio (a surrogate for the amount of leaf area) that can be generated. The principal limitation on the amount that can be created is the fixed architecture of canopies and how tightly they become packed within a forest. A competition index, based on historical tree growth and density relationships, is used to decrease the growth of trees disadvantaged by being too tightly packed into a stand. In the linkage with MAESTRO, in which information (but not carbon budgets) is passed between models (Cropper et al., Chapter 18), it is assumed that the stand structure will determine crown size, which, in turn, will determine net photosynthesis, which is used to build more structure. The predictions of this model are dependent on the assumption that climate change affects only photosynthesis.

The FORET model (Shugart and West, 1977), linked to PTAEDA2, is a gap-succession model operating at the stand level. The principal assumption in FORET is that stand growth and composition can be altered by changing the relative growth rates of individual trees. Also, different species are assumed to have different sensitivities to environmental changes. Stand dynamics are changed in this model when an alteration in environment is sufficient to cause a relative increase or decrease in the growth rate of competing species. In this model, the effect of environment on individual growth is reduced to a relative index of performance.

On the large scale, models are distinguished by the simplifications of biological processes, which have been made to predict the average behavior of stands on many different types of conditions over large landscapes. The PnET–IIS model (Aber and Federer, 1992), for example, uses only water-holding capacity, four monthly climate variables, and a relative leaf area potential to predict leaf area, and, in turn, water use and productivity. It is assumed that any plant adjustment to greater details of environment or other properties creates local variation in productivity but contributes little to altering regional production estimates. In tests, this model and its predecessor, PnET, have predicted known runoff rates from forested watersheds, and have proven that this assumption is reasonable for the scale on which the model operates.

A global biogeography model applied at a regional scale, MAPSS (Neilson, 1995) predicts the potential natural vegetation that can be supported at any upland site under a long-term, steady-state climate. The MAPSS model accomplishes this by calculating the available soil moisture and radiant input energy and, in turn, the maximum leaf area that water and energy input can support. It uses a set of biophysical rules that describe the leaf form, phenology, and behavioral patterns of plant types observed over large geographical regions. This model assumes that the distributions of vegetation, as represented by leaf area index (LAI) of both woody (shrub and tree, either deciduous or evergreen) and grass lifeforms, are constrained by either the availability of water in relation to transpirational demands or the availability of energy for growth. Elevated levels of CO_2 can affect vegetation responses to climate change through changes in carbon fixation and water use efficiency (WUE), (carbon atoms fixed per water molecule transpired), which is simulated as a reduction in stomatal conductance. This reduction results in increased LAI (carbon stocks) and usually a small decrease in transpiration per unit land area. Although the MAPSS approach has proven very successful in reproducing present patterns of vegetation, it may be unable to predict responses to rapid changes in climate that are historically unprecedented. This is particularly true if vegetation assemblages are able to deviate from the biophysical constraints presently in operation across major landscapes in ways we do not understand and therefore cannot anticipate.

The SOFOCLES model is an entirely different type of model that is built around simple representations of the results of experiments in which plants were subjected to altered climates or resource availability. The mechanisms in this model are relationships identified by these experiments. The emphasis in this model is to forecast whether definitive predictions can be made given the uncertainty (or error) surrounding the experimental results.

Scenarios Used for Predictions

Two factors that complicate the use of model output comparisons are 1) different data input sets and sources, and 2) combination of changing climate factors. In this section, nine models were used to predict changes in forest processes from the tree to regional scale. The input data necessary to run the models were derived from site measurements, empirical equations, literature reviews, and compiled regional or global data sets, including output from general circulation models (Table 25.2). Because they represent various processes, the models draw on somewhat different input data sources. It is not within the scope of this chapter to detail the sources of the databases for each model.

The models often differed in the variables of future climate they included in their simulations. Most models conducting simulations of increased CO_2 levels

Table 26.2. Sources of Input Data for the Nine Models Analyzed In This Chapter

Model	INPUT VARIABLE							
	Air temp	PPT	CO_2	Ozone	Climate Scenarios	Solar Radiation	Vegetation	Soil
BIOMASS	Cooter et al., 1993	Cooter et al., 1993	Cooter et al., 1993	none	Cooter et al., 1993	Cooter et al., 1993	NSCFNC 1991, NCSFNC 1993	NCSFNC 1991, NCSFNC 1993
MAESTRO	PTAEDA2	PTAEDA2	none	none	none	PTAEDA2	Wang and Jarvis, 1990a	Wang and Jarvis, 1990a
MAPSS	NOA-EPA data	NOAA-EPA 1993	NCAR data	none	NCAR GCM	NOAA-EPA data	Zhu and EVans, 1992	Kern, 1995
PIPESTEM	NCDC	NCDC	Conway, 1991	none	Static additive	Simulated Campbell, 1977	Amateis et al., 1988	Amateis et al., 1988
PnET–HS	Marx, 1988	Marx, 1988		none	EPA-GCM, static additive	simulated Nikolov et al.	FIA	Marx, 1988
PTAEDA2	Burkhart et al., 1987	Burkhart et al., 1987	none	none	none	Burkhart et al., 1987	Burkhart et al., 1987	Burkhart et al., 1987
SOFOCLES	Melillo et al., 1995	Melillo et al., 1995	Melillo et al., 1995	EPA	Melillo et al., 1995	Melillo et al., 1995	FIA	Melillo et al., 1995
SPM	Melillo et al., 1995	Melillo et al., 1995	Melillo et al., 1995	none	Melillo et al., 1995	Melillo et al., 1995	FIA	Melillo et al., 1995
UPTAKE	Kelly et al., 1992	Kelly et al., 1992	none	Kelly et al., 1992	none	none	Kelly et al., 1992	Kelly et al., 1992

compared growth under present CO_2 levels to that under a doubled CO_2 atmosphere (assumed to be close to a constant 700 ppm). Although most models simulated the effects of increased temperature, the magnitude of increases and their spatial distributions differed. The PTAEDA2/MAESTRO and SPM models used a 4 °C temperature rise; PnET–IIS used 3 °C and 7 °C. Such models as PnET–IIS, MAPSS, and SOFOCLES used global climate model data (GCM) (Table 26.2) with spatial variation in temperature change across the region, however, PIPESTEM ran scenarios of 2 °C and 4 °C temperature change above the long-term historical record from four sites in Virginia and North Carolina.

Multiple scenario model runs also complicated the interpretation of model outputs. Some models predicted the influence of multiple environmental factors (i.e., climate change, ozone, and increasing atmospheric CO_2 concentrations) on forest processes; others predicted the influence of a single or multiple climate factor on forest processes. Because of these two complicating factors, the range of predicted environmental change was large and only general statements (i.e., positive or negative) regarding the influence of each variable on ecosystem functions are possible.

Productivity Response to Increases in Temperature, Carbon Dioxide, and Ozone

In interpreting the general conclusions to be drawn from the aggregate results of the simulations, we attempted to ignore most model-specific behavior and, instead, looked toward the overall direction of forest response the models were suggesting. A given model may have separately tested the effects of rising levels of CO_2 and temperature. The modeler may have reported a positive influence of productivity in the first case and a productivity reduction in the second. We discuss each of these responses in the following sections, and we report the net effects of considering all climate change effects simultaneously.

Increasing Temperature

All models, with the exception of PTAEDA2, evaluated the effect of a climatic increase in temperature. Consequently, the effect of temperature is a logical starting point for the model comparison.

There was agreement among the models that increases in temperature will lead to decreases in productivity (Table 26.1). Even though many models did not examine temperature effects independently from other climatic changes, the strong negative response to temperature was apparent in almost all models. The degree of productivity decline depended heavily on the amount of temperature increase. Furthermore, at a given temperature increase the models varied widely in the amount of productivity decline predicted. However, even models as dissimilar as PnET–IIS and SPM produced very similar reductions in productivity (20 to 30%) under a 3 °C temperature increase.

The predominant mechanism in the various models that led to this decline was the effect of temperature on increasing the metabolic respiration of tissues. However, the effect of spatial variability in temperature across the southeastern landscape must be considered in applying this generality. For example, PnET–IIS predicted that where conditions were presently cool in northern sections of the southern United States, a 2 to 4 °C increase in temperature would increase net primary production (NPP) in stands of loblolly pine (*Pinus taeda* L.); where conditions were presently warm, increased temperatures would make loblolly pine existence marginal.

As will be discussed later, such models as BIOMASS, which considered temperature effects in combination with CO_2 increases and precipitation changes, found growth increases. The effects of temperature increases on respiration in BIOMASS were relatively insignificant in the broader context of the simultaneous effects of CO_2, precipitation, and temperature. Models in which the effects of both temperature and precipitation were integrated through the water and energy balance calculation (e.g., PnET–IIS and MAPSS), also predicted large negative effects. In PnET–IIS, changes in rainfall did not affect stand growth nearly as much as did temperature increases.

Because of the potentially large influence of temperature and respiration on these predictions, we must closely examine the different assumptions used in calculating respiration in different models. Among the relationships critical to these predictions are assumptions of the quantity of live woody tissue that must be supported by the existing leaf photosynthesis. Because the respiration rates of all tissues increases exponentially with increasing temperature, the ability of trees to withstand temperature increases depends on their continued ability to supply carbon to meet the needs of the woody tissues. For example, PTAEDA2 assumes that, in mature loblolly pine stands, leaves must support twenty times their own weight, while young stands must support only five times their weight (Baldwin et al., Chapter 17).

Variability in hourly temperatures or site fertility could also greatly affect the predictions of temperature effects. If a model uses a single annual average or twelve monthly averages to calculate relative respiration for the year (as do most of the large-scale models), it may underestimate the respiration during hot hours of hot days. Because respiration increases with temperature exponentially, the losses during these brief periods can amount to a significant percentage of the annual carbon budget, an amount which could cause an underestimate of total losses. In another source of possible error, the majority of these models ignore the effects of site fertility on respiration, yet Ryan (1991) showed that respiration can vary greatly with nitrogen content of tissue. Even though respiration is an intensively studied process, there is insufficient understanding of the effects of temperature on respiration of forest stands.

A number of models suggested that an important threshold exists near a 3 °C increase in average annual temperature at the regional scale. The SPM and PnETIIS models showed significant increases in effects as the annual temperature increases exceeded 3 °C; responses at 3 °C were slight to moderate, but, above this point, responses became severe.

Some models gave an indication that there is liable to be greater variability in growth among forest stands in future climates. This could be very significant because the key to the viability of a population in a given region is its ability to survive extreme years of poor conditions. Instead of all stands growing somewhat more poorly than present, it is possible that some stands would experience no growth decreases and others would be completely eliminated. This finding could result in segmented forests with large gaps on sites where conditions exceeded the threshold of viability for brief but sufficient periods of time. The MAPSS model predicted a wholesale conversion of vegetation types, from closed forest to woodlands and savannas because critical thresholds were exceeded.

Increasing Carbon Dioxide

All models that tested the effects of CO_2 alone showed growth increases. However, they diverged widely in their estimates of how much of the photosynthesis gain would be expressed in growth increases. Most models suggested that a doubling of CO_2 could double the photosynthesis rate, but the growth increases would probably be considerably less than double the present rates. The results from PTAEDA2/MAESTRO suggested that almost twice as much carbon could be available for growth in individual trees under a doubling of CO_2; SPM results indicated that the canopy average increase in growth is more likely to be 60 to 70%. The PIPESTEM and BIOMASS models similarly predicted increases of from 65 to 74% and 75 to 80%, respectively. In BIOMASS, the amount of increase was predicted to be highly dependent on the ability of the site soil to provide an ample supply of nutrients; PIPESTEM projected that increased rates of biomass accumulation could occur with no additional uptake of nitrogen through building woody material instead of foliar or fine root dry matter. The PTAEDA2/MAESTRO simulations also suggested that productivity increases would reach their potential only if fertilization treatments are begun.

It is interesting to compare these results with those of models using very different sets of assumptions to drive productivity. For example, PnET–IIS calculates a maximum leaf area that is sustainable at a location given the climate at that location (although nitrogen concentration in the leaves drives the amount of production from that leaf area). Because CO_2 is not one of the factors governing this maximum, PnET–IIS is incapable of directly producing an increase in leaf area under elevated levels of CO_2, but could indirectly increase leaf area under elevated CO_2 levels by changing WUE (permitting more leaf per unit of water available). A similar process and result is possible from MAPSS. However, over a large landscape there may be insufficient opportunities to capture more light in existing forest canopies to allow vegetation to take advantage of the higher photosynthesis rates physiologically possible under elevated CO_2 levels. The assumptions in a given model about those limitations in resources that constrain the possible behavior play an enormous role in determining whether or not the predictions will show a response.

Increasing Tropospheric Ozone Levels

The UTM/FORET/PTAEDA2 model and the PTAEDA2/MAESTRO model considered the effect of ozone on carbon gain. Both models concluded that, independent of any other environmental change, ambient ozone levels are causing a decrease in productivity and would expect larger decreases under higher levels of ozone anticipated in the future. The UTM/FORET/PTAEDA2 model simulated a direct impact of ambient ozone on stomatal regulation, which then translated to water stress for trees and stand-wide growth decline in combination with dry periods and elevated air temperature.

Combined Effects

Increases in CO_2 levels, temperature, and ozone are all liable to occur in concert with one another. Table 26.3 reports the predicted net-effect of these simultaneous changes on stand productivity, based on simulations using a temperature increase

Table 26.3. Processes Driving Model Behavior

Model	Prediction	Process Responsible
MAESTRO	Increasing productivity	Increased photosynthesis offsets increased respiration losses from temperature.
UPTAKE	Decreasing productivity	Reduced carbon fixation results in less root growth and consequently reduced uptake.
SPM	Increasing productivity	Increased photosynthesis, particularly with fertilization treatment, offsets increased respiration losses in carbon balance.
PIPESTEM	Increasing productivity	Increasing carbon assimilation drives increased leaf area, resulting in taller trees with longer crowns. Stand density decreases do not limit growth increases.
BIOMASS	Increasing productivity	Increased photosynthesis with CO_2 fertilization offsets respiration losses from elevated temperatures and growth losses from drought.
UTM/ FORET/ PTAEDA2	Decreasing productivity	Ozone causes loss of stomatal regulation and stem-growth decline, causing site index decline.
PnET–IIS	Decreasing productivity Increased water yield Species range change	Decreased leaf production causes decreased evapotranspiration in South; Increased leaf production causes increased evapotranspiration in North; higher temperatures cause high respiration and declines throughout the region.
MAPSS	Decreased distribution	Species tolerances to drought are exceeded.
SOFOCLES	Decreasing productivity	High respiration causes poor carbon balance.

26. Summary of Simulated Forest Responses to Climate Change

of 3 to 4 °C and a doubling of atmospheric CO_2 concentration. This table also reports the most probable reason for reported effects and the most probable process dominating model behavior in producing the overall response. Not all models simulated the response to increases in all variables or simulated an influence of temporal and spatial variability across the southeastern landscape.

Models that operate at the regional and greater scales, such as MAPSS and PnET–IIS, constructed predictions for each section of the heterogeneous southeastern region, using site-specific soil properties and temperature and precipitation predictions. Among all these sites, a variety of changes in leaf area (and, therefore, productivity) are predicted, with some sites showing declines because of large local temperature changes, some showing declines because of sensitive soil conditions, and some showing increases because of positive local climate changes. The results reported in Table 26.3 are an average of the local responses, and therefore reflect only an overall trend and not the underlying broad heterogeneity of site-specific responses.

The SPM model indicated that the positive response to CO_2 would outweigh the negative depression of growth caused by temperature-induced increases in respiration. The PIPESTEM model was in agreement with this result, indicating that the negative effects of a 4 °C rise would be overwhelmed by the positive growth response to an atmosphere of 700 ppm CO_2. This model predicts that the gain in growth will be half of what would be realized without temperature increases, but nevertheless represents a significant net positive response.

The BIOMASS model examined more extensively how temperature and CO_2 might interact over a wide variety of regional sites, and predicted that for almost all GCM scenarios the positive effects of CO_2 fertilization would outweigh the negative effects of increased temperature. Only in the GCM with the greatest predicted rise in temperature, 7 °C in the UKMO model, did CO_2 response fail to offset the decline in productivity caused by temperature-induced increased respiration costs. In this scenario, the two very nearly canceled each other, so that the UKMO model produced the least response in BIOMASS to climate change scenarios.

Analysis of simulations with the BIOMASS model also concluded that productivity would increase under probable future climates. In this model, increased photosynthesis with CO_2 fertilization offset increased respiration losses from temperature. The PTAEDA2/MAESTRO model also predicted increasing productivity with climate change for this reason. Additionally, the BIOMASS model predicted that photosynthesis increases would more than compensate for growth losses from drought, which, dissimilar to PTAEDA2/MAESTRO, it had considered.

Similarly, PIPESTEM predicted that assimilation would more than compensate for respiration losses, using an atmosphere of 700 ppm CO_2 and a temperature increase of 4 °C. Because the relationships in PIPESTEM focused on the significance of increased carbon assimilation for stand-structure development and because the model contains a density dependent feedback limitation to stand growth, it is significant that the model did not find the increase in productivity to be limited by the closing of the canopy. This may be the result the model's prediction that canopies would elongate into taller trees with longer crowns and more leaves. The resulting stand density decreases did not limit growth increases.

Although the SPM model concluded that the combination of a 7 °C temperature and CO_2 increases would cause a net decrease in growth, this result reflected the assumption that future CO_2 increase would be only 500 ppm. Given the reported rate of growth decline with 4 °C temperature increase (a loss of approximately 150 g/m^2) and growth increase at 500 ppm (an increase of 150 g/m^2), a linear extrapolation of the model results suggests a net growth increase at 4 °C and 700 ppm. Even at 4 °C an increase a growth increase was predicted.

The remaining models found the opposite result, a decrease in productivity. Each of the remaining models focuses on a particular aspect of forest dynamics with the capability of limiting growth responses to CO_2. Although the UPTAKE model did not make predictions of growth under climate change, it demonstrated that under moderate to low rates of soil-nutrient supply, nutrient uptake has limited ability to increase the supply of nutrients (Kelly and Yanai, Chapter 16). This means that a plant could have difficulty keeping up with an accelerated demand for nutrients generated in plants. Cropper et al., (Chapter 18) suggests that phosphorus availability limitations could prevent potential growth increases from being realized unless stands were fertilized. Low availability of nutrients could dilute nutrient concentrations of aboveground material, resulting in less carbon fixation and less construction of tissues to place newly fixed carbon.

None of the models considered in this section exhaustively evaluated whether nutrients are liable to become more available, either through increased decomposition rates caused by increased temperature and/or through increased atmospheric deposition. Also, none of the models considered whether nutrients are liable to become less available, through decreased decomposition caused by higher carbon to nitrogen ratios in decomposable plant material. No consideration was paid to the potential that increases in nitrogen deposition, a trend that will undoubtedly continue into the future and become a part of the altered climate, could lead to higher rates of availability and less need for root expansion and efficient uptake mechanisms. Alternatively, no models evaluated the probability that nitrogen will become less available as high carbon tissues become incorporated into the litter and soil-organic matter. With the exception of PnET–IIS, none of the photosynthesis routines in any of the physiologically based models consider the necessity of maintaining nitrogen levels in leaf tissues in order fix carbon at accelerated rates.

The UTM/FORET/PTAEDA2, SOFOCLES, and PTAEDA2/MAESTRO models assumed that an important part of the future climate will be increases in ozone exposure levels. These models conclude that present and future concentrations of ozone cause productivity reductions of a sufficient level to alter inter-tree competition and stand dynamics, but that these alterations will become dwarfed by reductions caused by the respiratory losses to elevated temperatures. However, a great number of local forest stands could be faced with a less favorable carbon balance than presently exists. Under these conditions, the negative effects of ozone could easily tip the scales from a positive to a negative effect.

In the PnET–IIS model, it was predicted that productivity will be decreasing because leaf production has a limited ability to increase and keep pace with high respiration losses caused by higher temperatures. Under moderate temperature increases, this model projects that forests in the warmest locations are presently at risk, with cooler sites predicted to have growth increases as photosynthesis gains outpace respiration losses. However, PnET–IIS may have underestimated potential photosynthesis increases in response to elevated levels of CO_2 by assuming that leaf area can be increased only through a more favorable water balance as a result of increased WUE.

When precipitation was predicted to decline, PnET–IIS simulated decreases in leaf area and growth, as the needs for evapotranspiration could not be met. This model projected that water availability and use would increase in environments that are presently cooler (and will be warmed moderately but not excessively in the future), decreasing runoff in these locations. However, if water availability did not increase as rapidly as the demand for water (to meet increased evapotranspirational needs) potential gains in leaf area and productivity would not occur. In locations that are presently warm, increased temperatures were predicted to cause leaf area reductions from excessive respiration losses, increasing water availability and runoff.

The MAPSS model predicted that species tolerances to drought will be exceeded. In MAPSS simulations vegetation increased leaf area until water availability became limiting. The MAPSS model simulated LAI (productivity) increases under elevated CO_2 levels if precipitation remained unchanged by assuming WUE reduces stomatal conductance, and hence, the water loss per unit leaf area. Furthermore, MAPSS results suggested that when water loss increased, because of increased temperature or decreased precipitation, sufficient leaf area could be lost to force replacement of the present vegetation by another type able to reach its maximum leaf area on that site. As a consequence, MAPSS predicted a great sensitivity of species ranges to the availability of water.

In MAPSS, stands in the southern portion of the region were projected to experience large decreases in density, and, in extreme cases, distribution, as elevated temperatures reduced water availability below critical thresholds. Ecosystems have historically replaced each other through time as conditions changed. MAPSS captured this behavior, predicting dominance by new vegetation types more suited to new levels of water availability. Because it predicted vegetation replacement, MAPSS produced some of the most drastic predictions of forest alteration. Although the changes seem out of the range of the other models, this may be because other models have failed to incorporate such thresholds. The biogeographic literature is rich with examples in which species ranges have been shown to be tightly correlated with temperature or water thresholds.

In a result similar to other models, SOFOCLES predicted productivity declines because of the overwhelming effect of temperature on respiration losses. However, SOFOCLES identified that gross carbon gain was usually the most important variable in determining the estimated change in growth rate. Where the gross carbon gain was highest, stands were predicted to show productivity increases. In some stands, the most important effect of temperature was on the rate of allocation to stem growth. The effect of temperature on maintenance respiration was never first in importance in determining growth response (it was usually third). This result reflects the relative certainty with which different plant processes are known. Respiration is much more predictable from a given climatic regime and therefore is less apt to be a major source of uncertainty in predictions.

Synthesis

Overall Conclusions

Will future climates produce increases in growth? The models generally suggest that if the only future deviation from the present climate and atmosphere were temperature increases, forests would show large declines in growth. A variety of models were in surprising agreement on the magnitude of growth decrease of approximately 20 to 30% with moderate temperature changes (3 °C). The uniformity of response to temperature alone is probably caused by the similarity of the equations predicting respiration cost in response to temperature in different models. A large body of experimental evidence supports a single form for this relationship, a doubling of respiration with each rise in temperature of 10 °C. Few other relationships in any of these models have as high a rate of change per unit change in environmental variable. Consequently, effects on respiration dominate the model predictions, and because the temperature-respiration relationships are similar, the models produce similar results. Although in agreement regarding the negative consequences of temperature increase, one model, MAPSS, predicts that temperature increases will have their greatest effect on plant-water balance. A decrease in the water available per unit demand could cause widespread change in the distributions of vegetation types.

Similarly, most of the models are in agreement concerning the magnitude of projected growth increases if CO_2 were the only climate variable to change. Here, too, there is general agreement in the literature concerning the form of the response of photosynthesis to CO_2 in plants not previously exposed to high CO_2 for extended periods. Most of the models contain a similar equation representing this relationship. Photosynthesis is a key driver for most of these models.

Although ozone is projected by UTM/FORET/PTAEDA2 and by SOFOCLES to cause significant growth reductions and to have an effect that is magnified by stand-level interactions, the models suggest that its importance will be dwarfed by temperature and CO_2 effects. These models suggest a synergistic interaction in which ozone works in combination with high evaporative conditions to decrease growth. However, there is an enormous absence of data on synergistic interactions. As a consequence, few relationships exist in any of the models that depend on the interaction of different environmental conditions. For example, a few physiological studies have demonstrated that drought conditions can reduce ozone injury (Temple et al., 1993, Beyers et al., 1992), but the UTM model alone considers the interaction of ozone with stomatal regulation. With the exception of WUE responses, mechanisms capable of producing such interactions tend to be missing from these models.

Climate change however, is liable to involve changes in CO_2, ozone, temperature, and precipitation—all occurring simultaneously. With the exception of the responses of WUE to the interaction of CO_2, temperature, and precipitation, the effects of climate appear to be additive throughout the simulations of the different models. The models predominately assumed that CO_2, ozone, temperature, and precipitation each would enhance or deplete the carbon balance of the trees in the forest stands independently. Simulated stands with high respiratory costs from elevated temperatures, therefore, were not less capable of using elevated CO_2 levels.

The majority of the simulators suggest that growth will increase as a result of the relative responses of the processes that we understand best, that is, photosynthesis and respiration, and those changes that have received the most experimental attention, increases in temperature and CO_2 levels. Some of the most complex models, however, suggest that such processes as nutrient limitation, drought response, ozone-induced loss of stomatal regulation, inter-tree competition, stand-closure dynamics, and temperature and water balance have the capacity to prevent growth increases. If a model included a representation of the functioning of one of these processes, the process usually played a major role in altering growth increases, suggesting that complex models may be needed to predict the interactions of growth and physiology processes in forest stands.

Additional Sources of Errors in Prediction

We must recognize that many poorly understood processes, either weakly interlinked in various models or not included at all, could play a major role in changing the predictions of these models. Considerable understanding of the environmental responses of photosynthesis to CO_2, and the responses of respiration to temperature exist. Consequently, these responses are represented in similar ways in the different models. As a result, the models produce similar predictions about the effects of climate change on photosynthesis and respiration. Some models begin to predict different results when they focus on different processes capable of constraining the influence of changes in photosynthesis and on forest productivity and distribution.

For example, a large amount of carbon is used annually by forests to replace fine roots that are lost during the course of a growing season. We do not know how much control plants have on the rate at which these are lost and replaced. In such models as BIOMASS and PIPESTEM it is assumed that the proportion of newly fixed carbon allocated to root development will remain the same under future conditions. The estimates of the amount of carbon that could be used differ by as much as an order of magnitude. Without sound estimates of root turnover, budgetary models of carbon balance in forests may be greatly in error in their estimate of total tree carbon demand. Overestimates of the demand could lead to a conclusion that excess carbon fixed under elevated CO_2 levels should be continually pumped belowground to meet the demand. If, on the other hand, belowground turnover, and therefore belowground demand, is and continues to be relatively small, photosynthesis may decline despite elevated levels of CO_2 because trees have no place to store additional carbon.

Many of the models inadequately treat the way in which physiological systems can be influenced by processes at higher scales. The UTM/FORET/PTAEDA2 linked modeling system examined the ways that effects caused by ozone can be altered by these higher-scale processes. In this case, effects on individual trees translated to effects on a whole stand property (site index), which influenced predicted regional volume increment. The UTM model suggested a 5.4% decline in annual stem growth, which caused the model of stand inter-tree competition, FORET, to produce a 5% decline in site index at age twenty-five. In turn, the decline in site index, when passed through the PTAEDA2 model, resulted in a 6% reduction in harvested lumber volume simulated for a thirty-five year rotation. This demonstrates that there is a significant potential for dynamics at one level of organization to be altered by processes at other scales.

Three of the models here, SPM, BIOMASS (the version by Sampson et al., Chapter 21), and UTM have storage pools that can be drawn on to meet the demands of growth when net photosynthate is insufficient. In each of these models, simulations highlighted the importance of this buffer to reducing the short-term sensitivity to climatic change. In the SPM model, growth was severely diminished at temperatures greater than 3 °C above the ambient level. This threshold may represent a point where the pool of labile carbon internal to the trees is essentially depleted by higher respiration rates.

Many of the forests in the southeastern United States are under intensive management at present. The degree to which these forests continue to be managed in the future may determine how responsive they will be to the changes discussed here. If the limitations imposed by new climatic conditions can be overcome, for example, with fertilizer or water additions or thinning of canopies, managed systems may show greater response to climate change than would be otherwise anticipated. A more realistic goal might be to maintain growth at its present rate with an infusion of energy into the growth and harvesting system. Of course, the management will not accomplish this task without an enormous input of funds.

Key Processes Influencing Predictions

Which processes have been identified as the most important in determining the direction of response to climate change? Two physiological processes, the response of photosynthesis to elevated CO_2 levels and the response of maintenance respiration to elevated temperature were the driving forces behind the predictions of the models focusing on the scale of individual stands and smaller. When the more indirect processes were examined, major changes were predicted to occur through decreasing the water availability. The results indicated it was improbable that the effects of climate change would increase water availability and alleviate drought in many locations. Several indirect processes were largely unexamined by these models, however. For example, the effect of temperature on increases in decomposition and release of nitrogen could increase nitrogen availability and, therefore, fertilize forests. Furthermore, nitrogen deposition from the atmosphere is increasing yearly, potentially improving the internal nitrogen supplies of stands. The result of nitrogen deposition was not considered.

Important features of forested ecosystems other than productivity and distribution were predicted to be vulnerable to modification by some models. For example, runoff from these ecosystems might be modified as a result of decreases in production. The PnET–IIS model concluded that large increases in temperature (of the order of 7 °C as predicted by the UKMO model) would cause large decreases in available water and, subsequently, LAI, decreased evapotranspiration, and increased runoff. However, in the simulations of MAPSS that similarly predicted a decrease in LAI, the thinning of the canopy allows accelerated growth of an understory which used up the remaining available water. As a consequence, runoff was not increased in these simulations. In fact, the assumptions driving MAPSS dictate that under water-limited conditions the vegetation will always use all the water available and runoff increases will never be possible.

Dissection and comparison of the models presented in this volume reveal critical areas in which more information is needed. Will the potentially positive response of photosynthesis rates decrease with prolonged exposure to elevated levels of CO_2? Some experimental evidence suggests they will, but none of these models incorporates such an adaptation. All of these models assume that there is little adaptive capacity built into the respiratory system. Respiration rates are assumed to keep their present rate of response to temperature.

Management Implications

Are there management techniques that could be employed to lessen the probability of the predicted changes occurring and increase the sustainability of these forests? First, because growth and yield tables are heavily relied on for proper forest management, and because the trajectory of growth in loblolly pine plantations is predicted to change by these models, it is essential that these tables be modified. It is difficult to identify exactly how the modification should be made based on these modeling results because the changes in climate will occur gradually (and forest growth will respond gradually to those changes), and the trees in these simulations were assumed to be grown under new climatic conditions during their entire lifespan. The exception was in the PIPESTEM simulations, where, with CO_2 fertilization projected to increase yields of recently planted stands by 5 to 10%, alteration of growth and yield tables may be dictated.

Secondly, the simulations presented here demonstrate how valuable forest models could be when employed as a regular part of an adaptive management strategy. Managers could anticipate possible changes in growth as dictated by the models, make alterations in management regimes, test these regimes in the models, and develop a better feel for the corrections necessary to produce the yields desirable given the climate-induced alterations in growth.

We would anticipate that process models will be relied on much more heavily to assist management. Rapid growth-pattern changes induced by a quickly changing climate will invalidate the assumption used in most management models today that forests will growth the same way in the future that they did in the past. Decisions need to be made concerning changes in species to be planted, with managers choosing species better adapted for high growth under new climates. The density of plantings may also need to be shifted dramatically to prevent water shortages from limiting productivity.

The analysis of this set of models and their results suggest that we can make some strong inference about growth trends under future climates. We are capable of making some logical extrapolations from our present knowledge base and those extrapolations will accurately reflect the relationships we now understand. However, these inferences, entirely based on a level of understanding about forest-climate interactions, need improvement. Nevertheless, application of these models has helped identify the information needed to create better simulators in the future.

References

Aber JD and Federer CA (1992) A generalized, lumped-parameter model of photosynthesis, ET and net primary production in temperate and boreal forest ecosystems. Oecologia 92:463.

Ameteis RL, Burkhart HE, Sedaker SM (1988) Experimental design and early analyses for a set of loblolly pine spacing trials. In Ek AR, Shifley SR, Burk TE (Eds) *Forest growth modeling and predictions, Vol 2*. USDA For Serv, Northcentral For Exper Stat, Gen Tech Rep NC-120.

Beyers JL, Riechers GH, Temple PJ (1992) Effects of long-term ozone exposure and drought on the photosynthetic capacity of ponderosa pine (*Pinus ponderosa* Laws). New Phytol 122(1):81–90.

Burkhart HE, Farrar KD, Amateis RL, Daniels RF (1987) Simulation of individual tree growth and stand development in loblolly pine plantations on cutover, site-prepared areas. FWS-1-87. VA Poly Inst State Univ, Sch For Wildlife Resour, Blacksburg, VA.

Campbell, GS (1977) *An Introduction to Environmental Biophysics*. Springer-Verlag, New York.

Conway TJ, Tans PP, Waterman LS (1991) Atmospheric CO_2-modern record, Key Biscayne. In Boden TA, Sepanski RJ, Stoss FW (Eds) *Trends '91: A Compendium of Data on Global Change*. Oak Ridge National Laboratory, Oak Ridge, Tennessee.

Cooter EJ, Elder BK, LeDuc SK, Truppi L (1993) General circulation model output for forest climate change research and applications. Gen Tech Rep SE-85. USDA For Ser, Southeast For Exper Stat, Asheveille, NC.

Cropper WP Jr, Gholz HL (1993) Simulation of the carbon dynamics of a Florida slash pine plantation. Ecol Model 66:213–249.

Dixon KR, Luxmoore RJ, Begovich CL (1978) CERES—A model of forest stand biomass dynamics for predicting trace contaminant, nutrient, and water effects. I. Model description. Ecol Model 5:17–38.

Forest Inventory and Analysis, Data Base Retrieval System. World Wide Web address http://www.srsfia.usfs.msstate.edu/scripts/ew.htm. Described in "Forest Service Resource Inventories: An Overview". Sept 1992. USDA Forest Service. Washington Of-

fice: Forest Inventory, Economics, and Recreation Research, W. Brad Smith, FIA WO-Staff.
Field CB, Ruimy A, Luo Y, Malmstrom CM, Randerson JT, Thompson MV (1996) VEMAP: Model shootout at the sub-continental corral. Trends in Ecology & Evolution 11(8):313–314.
Kelly M, Taylor GE, Edwards NT, Adams MB, Friend AL (1993) Growth, physiology, and nutrition of loblolly pine seedlings stressed by ozone and acidic precipitation. A summary of the ROPIS-South Project. Water Air Soil Pollut 69:363–391.
Kern JS (1995) Geographic patterns of soil water holding capacity in the contiguous United States. Soil Sci Soc Amer 59:1126–1133.
Luxmoore RJ (1989) Modeling chemical transport, uptake, and effects in the soil-plant-litter system. In Johnson DW Van Hook RI (Eds) *Biogeochemical cycling processes in Walker Branch watershed*. Springer-Verlag, New York.
Marx DH (1988) Southern forest atlas project. In *The 81st annual meeting of The Association Dedicated to Air Pollution Control and Hazardous Waste Management (APCA)*, Dallas, Texas, 1988.
McMurtrie RE (1985) Forest productivity in relation to carbon partitioning and nutrient cycling: A mathematical model. In Cannell MGR and Jackson JE (Eds) *Attributes of trees as crop plants*. Institute of Terrestrial Ecology, Abbots Ripton, Huntington, England.
McMurtrie RE, Wolf LJ (1983) Above and below ground growth of forest stands: A carbon budget model. Ann Bot 52:437–448.
McMurtrie RE, Leuning R, Thompson WA, Wheeler AM (1992) A model of the canopy photosynthesis and water use incorporating a mechanistic formulation of leaf CO_2 exchange. For Ecol Manage 52:261–278.
Melillo JM, Borchers J, Chaney J, Fisher H, Fox S, Haxeltine A, Janetos A, Kicklighter DW, Kittel TGF, McGuire AD, McKeown R, Neilson R, Nemani R, Ojima DS, Painter T, Pan Y, Parton WJ, Pierce L, Pitelka L, Prentice C, Rizzo B, Rosenbloom NA, Running S, Schimel DS, Sitch S, Smith T (1995) Vegetation/ecosystem modeling and analysis project: Comparing biogeography and biogeochemistry models in a continental-scale study of terrestrial ecosystem responses to climate change and CO_2 doubling. Global Biogeochem Cycl 9:407–437.
Melillo JM, McGuire AD, Kicklighter DW, Moore B III, Vorosmarty CJ, Schloss AL (1993) Global climate change and terrestrial net primary production. Nature 363:234–240.
Mitchell MJ, David MB, Harrison RB (1992) Sulfer dynamics of forest ecosystems. In Howarh RW, Stewart JWB, Ivanov MV (Eds) *Sulfur Cycling on the continents: Wetlands Terrestrial Ecosystems and Water Bodies*. John Wiley and Sons, New York.
National Climatic Data Center (NCDC). NOAA. World Wide Web adress http://www.ncdc.noaa.gov. Federal Building, 151 Patton Avenue, Asheville NC 28801-5001.
NOAA-EPA (1993) Global ecosystems database project. Global ecosystems database version 1.1. User's guide, documentation, reprints, and digital data. USDOC/NOAA National Geophysical Data Center, Boulder, CO.
NCSFNC (1991) Effects of site preparation, fertilization and weed control onthe growth and nutrition of loblolly pine. NCSFNC Report 26. Coll For Res. NC State Univ, Raleigh.
NCSFNC (1993) Six-year growth responses of mid-rotation loblolly pine plantation to N and P fertilization. NCSFNC Report 31. Coll For REs. NC State Univ, Raleigh.
Neilson RP (1995) A model for predicting continental scale vegetation distribution and water balance. Ecol Appl 5(2):362–385.
Nikolov NT, Zeller KF (1992) A solar radiation algorithm for ecosystem dynamic models. Ecol Modeling 61:149–168.
Oates K, and Barber SA 1987. Nutrient uptake: A microcomputer program to predict nutrient absorption from soil by roots. J Agron Educ 16:65–68.

Ryan MG (1991) Effects of climate change on plant respiration. Ecol Appl 1:157–167.

Schimel DS, Vemap-Participants, Braswell BH (1997) Continental scale variability in ecosystem processes: Model, data, and the role of disturbance. Ecological Monographs 67(2):251–271.

Shugart HH, West DC (1977) Development of an Appalachian deciduous forest succession model and its application to assessment of the impact of the chestnut blight. J Environ Manag 5:161–179.

Temple PJ, Riechers GH, Miller PR, Lennox RW (1993) Growth responses of ponderosa pine to long-term exposure to ozone, wet and dry acidic deposition, and drought. Can J For Res 23(1):59–66.

Wang TP and Jarvis PG (1990) Description and validation of an array model—MAESTRO. Agri For Meteor 51:257–280.

Yanai R 1994 A steady state model of nutrient uptake accounting for newly grown roots. Soil Sci Soc Am J 58:1562–1571.

Zhu Z, Evans DL (1992) Mapping the midsouth forest distribution with AVHRR data. J For 90:27–30.

Section 4. The Effects of Climate Change on Forest Soils

27. Simulated Effects of Atmospheric Deposition and Species Change on Nutrient Cycling in Loblolly Pine and Mixed Deciduous Forests

Dale W. Johnson, Richard B. Susfalk, and Wayne T. Swank

Forest soils of the South, similar to those in several other parts of the world, are undergoing changes that may have bearing upon the future health and productivity of southern forest ecosystems (Johnson et al., 1988, 1991; Binkley et al., 1989; Knoepp and Swank, 1994; Richter et al., 1994). In contrast to earlier indications that soil-nutrient pools were very large and therefore buffered from short-term changes, several studies both in the southeastern United States and elsewhere have shown that soils are changing on the time-scale of decades, and, in some cases, on a seasonal basis. Significant reductions in the pools of exchangeable base cations have been noted over periods of one to three decades in both deciduous (Johnson et al., 1988; Knoepp and Swank, 1994) and loblolly pine (Binkley et al., 1989; Richter et al., 1994) forests of the southeastern United States. Surface soil concentrations have also been shown to vary on a seasonal basis (Haines and Cleveland, 1981; Johnson et al., 1988). These changes have been attributed to sequestration of base cations (especially calcium (Ca)) in biomass and to leaching, the latter of which is accelerated by atmospheric deposition (Johnson and Todd, 1990; Knoepp and Swank, 1994; Richter et al., 1994).

There are some interesting potential interactions between tree uptake and leaching that can have profound effects upon the cation budgets of forests subject to both intensive management and acidic deposition (Johnson and Todd, 1987; Johnson et al., 1995). Cation exchange equations predict that depletion of one base cation on the exchanger by uptake (or any other process) will lead to a relative decrease in the leaching rate of that cation and an increase in the leaching of other cations. In essence, this interaction predicts that the base cation in greatest biological demand (often Ca) is conserved in the system because exchangeable pools are depleted and transferred into biomass, whereas cations in less demand (e.g., magnesium (Mg)) experience increased leaching rates. This kind of interaction has been observed in both field conditions (Johnson and Todd, 1987, 1990) and with simulation modeling (Johnson et al., 1995). Both field data and modeling suggest that, in deciduous forests, uptake generally has a larger effect upon soil Ca depletion than does leaching, even when the latter is accelerated by atmospheric sulfur (S) and nitrogen (N) deposition (Johnson and Todd, 1987, 1990; Johnson et al., 1995). In pine forests, Ca uptake rates are lower and leaching can be relatively more significant. Johnson and Todd (1987) estimated that both Ca^{2+} and Mg^{2+} leaching was ~ three times greater than vegetation accumulation for a loblolly pine stand in eastern Tennessee. Richter et al., (1994) calculated that leaching was somewhat greater than vegetation accumulation for both Ca and Mg over a twenty-eight-year period in a loblolly pine forest, but the sum of vegetation and forest floor accumulation were approximately equal to leaching.

In this chapter, we explore the interacting effects of atmospheric deposition and species change on soil changes through simulations with the nutrient cycling model (NuCM) (Liu et al., 1991; Johnson et al., 1993, 1995). The simulated effects of various atmospheric deposition or species change scenarios have been described for both sites at a prior time (Johnson et al., 1993, 1995); this chapter will summarize some of the effects noted ealier, along with standardizing the scenarios used and exploring the potential nutrient limitations at both sites over a long time period.

Sites and Methods

The Duke site is located in the Piedmont region at an elevation of 220 m about 60 km west of Durham, North Carolina. Mean annual temperature is 14.5 °C, and mean annual precipitation is 113 cm, with less than 2 cm as snow. Before acquisition of the land by Duke University in the 1960s, the land was used for farming various crops; no records were kept of soil treatments. Vegetation consisted of loblolly pine (*Pinus taeda* L.) planted by machine in 1966. The stand was thinned in 1984, and site index was 17.7 m at twenty-five years. Understory vegetation was negligible. Soils were of the Durham series, Typic hapludults derived from highly weathered felsic igneous rocks. The data used to calibrate the NuCM model was collected from 1985 to 1988.

The Coweeta site is located in the southern Appalachian mountains at an elevation of 725 m within the Coweeta Basin near Otto, North Carolina. Mean annual temperature is 12.6 °C, and mean annual precipitation is 180 cm, less than 10 cm of which typically falls as snow. Vegetation is a multistoried and uneven-aged mixture of *Quercus spp* (50%), *Oxydendrum arboreum* (11%), *Rhododendron maximum* (11%), *Acer rubrum* (10%), *Carya spp.* (6%), *Cornus florida*

(3%), and other, primarily deciduous species (9%). Soils are in the Fannin series, Typic Hapladults derived from gneiss.

NuCM Model Calibration

NuCM was developed as part of the Electric Power Research Institute's Integrated Forest Study (IFS; Liu et al., 1991; Johnson and Lindberg, 1991). The NuCM model links the soil-solution chemical components of the Integrated Lake-Watershed Acidification (ILWAS) model (Goldstein et al., 1984) with traditional conceptual models of forest-nutrient cycling on a stand level. The forested ecosystem is represented as a series of vegetation (i.e., foliage, bole, roots; overstory and understory) litter, and soil components. The soil includes multiple layers (up to ten), and each layer can have different physical and chemical characteristics. Tree-growth potential is defined by the user and is subject to reduction in the event that either nutrients or moisture become limiting.

Using mass-balance and transport formulations, the NuCM model tracks sixteen solution-phase components including the major cations and anions (analytical totals), acid-neutralizing capacity (ANC), an organic acid analog, and total monomeric aluminum (Liu et al., 1991). The concentrations of hydrogen ion, aluminum and carbonate species, and organic acid ligands and complexes are then calculated based upon the sixteen components. The model routes precipitation through the canopy and soil layers, and simulates evapotranspiration, deep seepage, and lateral flow. The movement of water through the system is simulated using the continuity equation, Darcy's equation for permeable media flow, and Manning's equation for free surface flow.

The processes that govern interactions among nutrient pools include decay, nitrification, anion adsorption, cation exchange, and mineral weathering. The NuCM model simulates the noncompetitive adsorption of sulfate, phosphate, and organic acid. Sulfate adsorption is simulated in NuCM using both linear and Langmuir (saturation), pH-dependent adsorption isotherms. Because sulfate is adsorbed noncompetitively, there is no direct mechanism by which additions of large amounts of phosphate will cause sulfate desorption. Phosphate adsorption in the model is represented by a linear isotherm.

Cation exchange is represented by the Gapon equation:

$$\frac{XM^{a+} (M^{b+})^{1/b}}{Xm^{b+} (M^{a+})^{1/a}} = Q$$

in which X equals exchange-phase equivalent fraction, ( ) equals soil-solution activity, M^{a+} equals cation of valence a, M^{b+} equals cation of valence b, and Q equals selectivity coefficient (constant).

Mineral weathering reactions are described in the model using rate expressions with dependencies on the mass of mineral present and solution-phase, hydrogen ion concentration taken to a fractional power. In these simulations, mineral weathering rates were set to low values appropriate for the Duke and Coweeta sites.

NuCM was calibrated for the Duke and Coweeta sites using data from the IFS (Johnson and Lindberg, 1991) according to the procedures outlined in the user's manual (Munsen et al., 1992). Details of the general calibration procedure have been described elsewhere (Liu et al., 1991; Johnson et al., 1993). Metrological and air quality data from input files for a given number of years are used to generate long-term inputs by repeating these intervals over the simulation period. In the Coweeta case, a one-year interval was repeated twenty times to simulate deposition to the site over a twenty-year time period; in the Duke case a two-year interval was repeated ten times for the same total time period.

Data for physiographic, vegetation, physical and chemical soil data (i.e., primary minerals, mineral dissolution rates, mineral stoichiometries, exchangeable cations, cation exchange capacity, adsorbed sulfate and phosphate, and anion adsorption isotherm parameters) were taken from the IFS data set (Johnson and Lindberg, 1991). At both the Duke and Coweeta sites, the soils have low stores of weatherable minerals, low weathering rates (April and Newton, 1991) and relatively high sulfate adsorption capacity (Harrison et al., 1989). Several model parameters (e.g., organic acid adsorption, snowmelt characteristics, fractions of leachable nutrients in litter) were left as in the original model formulation (Liu et al., 1991); these were of minor importance to our particular ecosystem and interests, and no better data were available.

After calibration, five atmospheric deposition scenarios were run, which were: 1) no change, 2) 200% N deposition, 3) 50% S deposition, 4) 50% base cation deposition, and 5) the combination of N, S, and base cation deposition. Prior to our study, both the 50% S deposition has been run for both sites and 200% N deposition scenarios has been run for the Duke site (Johnson et al., 1993, 1995). The 50% base cation deposition scenarios were added in this series of model runs because of recent indications that base cation deposition may be declining (Hedin et al., 1994).

The effects of harvesting and species change were simulated at both sites by substituting the Coweeta hardwood vegetation-nutrient concentrations at the Duke site and the Duke loblolly pine concentrations at the Coweeta site, and setting the stand age back to one in each case. This resulted in a simulated biomass reduction of approximately 95 to 97% in each site. These vegetation changes were used to approximate the effects of conifer to hardwood conversion (or vice versa) at each site while maintaining a constant sets of initial conditions for the purposes of comparisons. Similar simulations at the Duke site were run prior to our study, using average vegetation biomass and nutrient concentration data for a site near Oak Ridge, Tennessee (Johnson and Todd, 1987). Aside from changing the concentration data, changes were necessary in leaf area and litterfall to allow for the hardwood simulation. The changes in leaf area resulted in unrealistically large differences in evapotranspiration and soil-water flux between the hardwood and pine stands. We adjusted these differences to 20% greater in the pine than in the hardwood stands to match the observations at Coweeta where differences of this magnitude were observed between white pine and mixed deciduous stands (Swank and Douglass, 1977). Initial soil parameters, precipitation, and air concentrations were kept the same in both cases. The deposition and species-change simulations were run for twenty-year periods. Additionally, longer-term simulations were run at each site in order to determine those nutrients that became deficient. In the Coweeta case, this involved an eighty-year simulation; in the Duke case it involved a 198-year simulation.

Results and Discussion

Atmospheric Deposition Scenarios

As noted in prior studies, simulated changes in atmospheric deposition had little or no effect upon growth rates or nutrient pools at both sites. This is illustrated for N, Ca, and Mg in Figures 27.1 to 27.3. The largest effects occurred at the Duke site, where the 2X N scenario caused an 8 to 17% increase in vegetation, forest-floor mass, and nutrient content, and approximately 5% decreases in exchangeable Ca_2^+ and Mg^{2+}. The 0.5× base cation scenario caused only slight changes in simulated N distribution, but caused 5 to 10% reductions in exchangeable Ca_2^+ and Mg^{2+} pools. The 0.5× S scenario had the least effect on nutrient distribution at Duke. At the Coweeta site, the relative effects of each scenario followed the same general pattern as at Duke, but the overall effects were smaller.

As noted in prior studies (Johnson et al., 1993, 1995), the lack of changes in exchangeable cation pool sizes did not imply that no changes in soil exchangeable concentrations occurred. Simulated exchangeable cation concentrations varied on both a seasonal and decadal scale at both sites, and there were large effects of simulated atmospheric deposition in several instances. This is illustrated for exchangeable Ca_2^+ under the base case and the 0.5 deposition scenarios in Figures 27.4 and 27.5. The simulated soil changes were always not unidirectional, but sometimes showed increases followed by decreases (i.e., Coweeta BA horizon). If such short-term changes occur under field conditions, they could seriously confound efforts to monitor long-term changes in soils if not accounted for in some manner. Seasonal changes in exchangeable base cation concentrations have been noted by Haines and Cleveland (1981) and by Johnson et al. (1988) for both loblolly pine and mixed deciduous forest soils, respectively, but there is no data to either confirm or negate the decade-scale oscillations suggested by the NuCM simulations.

In both sites, there were substantial redistributions of exchangeable Ca_2^+ from surface to subsurface horizons under the base case (i.e., decreases in surface and increases in subsurface Ca_2^+ concentrations; Figures 27.4 and 27.5). This redistribution was much reduced in the 0.5 deposition scenario at each site (i.e., exchangeable Ca_2^+ remained higher in the surface horizons and did not increase as much in the subsurface horizons. The effects of this redistribution were most pronounced at the Duke site, where the net loss of Ca_2^+ from the Ap and E horizons in the base case at the Duke site was -4.88 kmol ha^{-1} over the twenty-year simulation, $+3.04$ kmol ha^{-1} of which accumulated in the Bt1 and Bt2

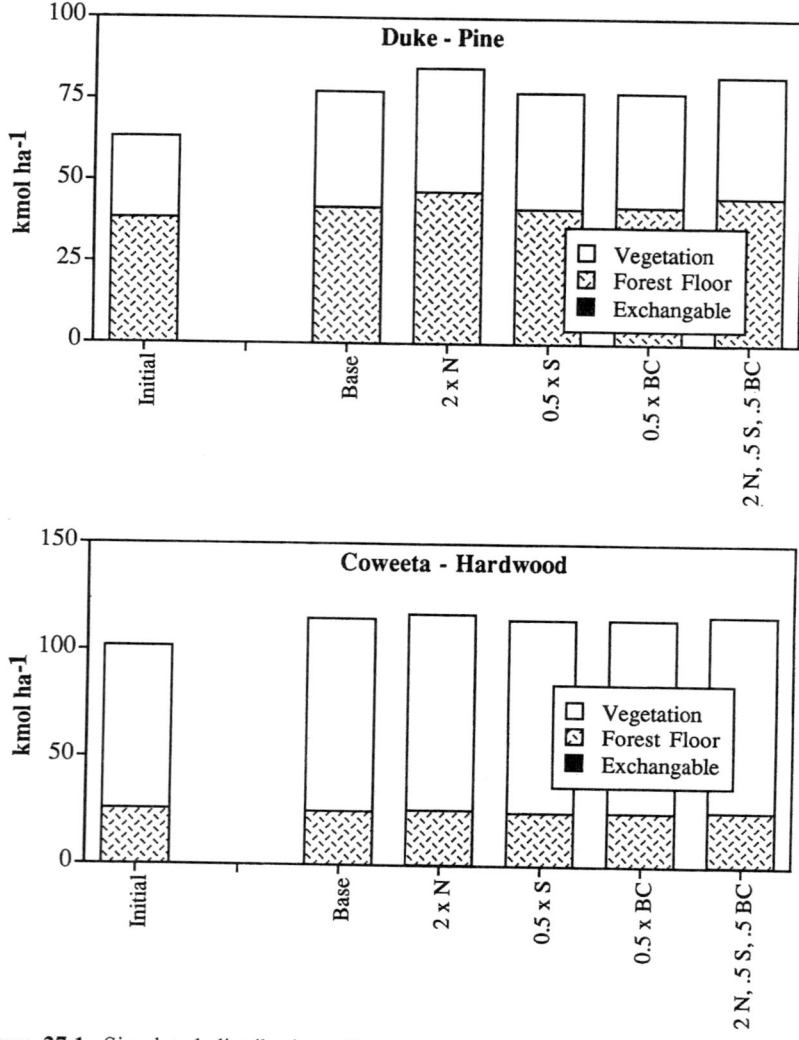

Figure 27.1. Simulated distribution of nitrogen at the Duke and Coweeta sites under various deposition scenarios. (Base equals no change in deposition; 2× N equals twice present N deposition; 0.5× S equals 50% of present S deposition; 0.5× BC equals 50% of present base cation deposition; and 2N, 0.5 S, and 0.5 BC equals combination of all three altered deposition scenarios).

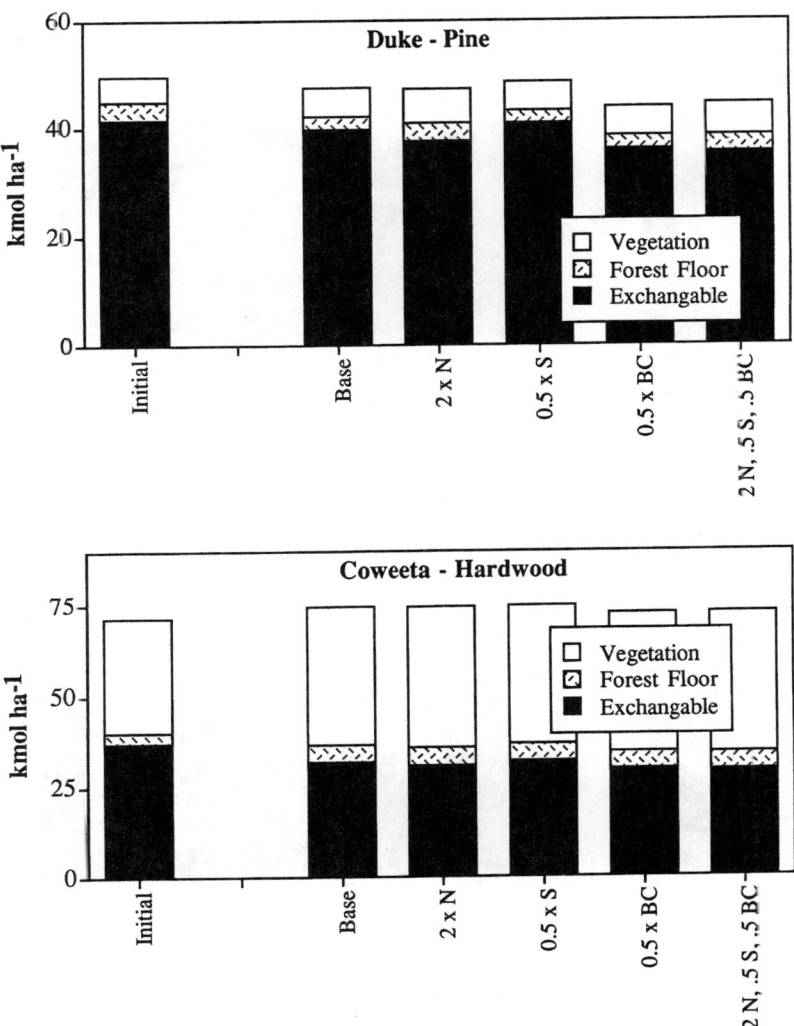

Figure 27.2. Simulated distribution of calcium at the Duke and Coweeta sites under various deposition scenarios. (See Figure 27.1 for legend).

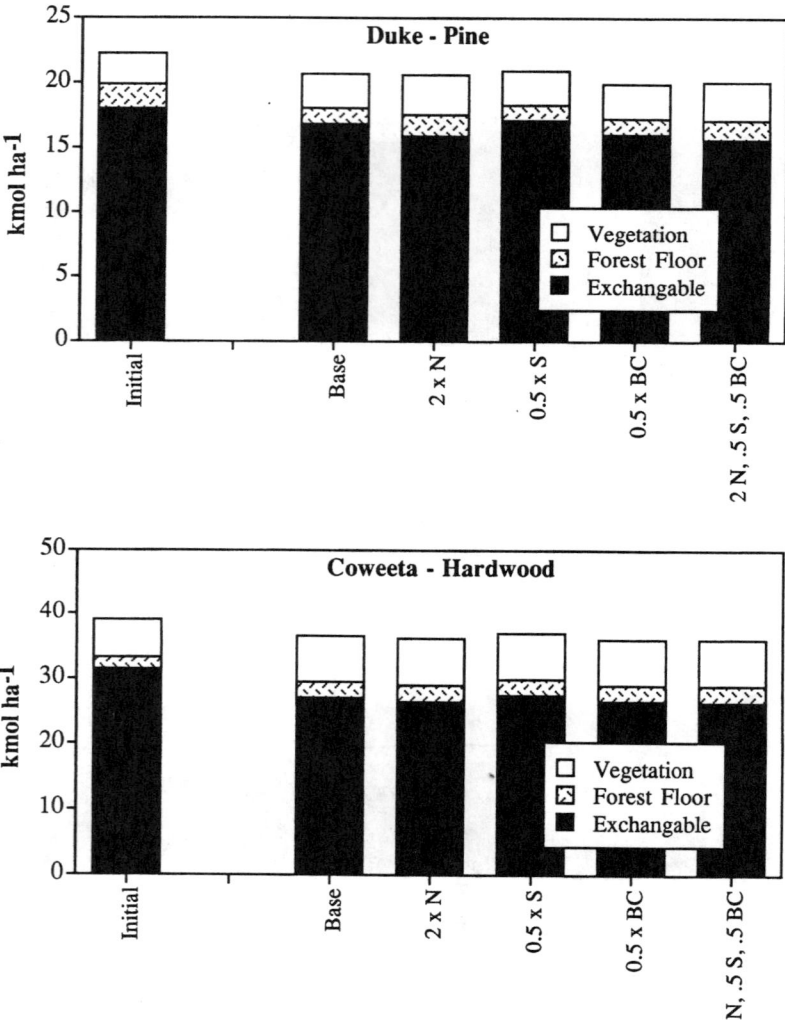

Figure 27.3. Simulated distribution of magnesium at the Duke and Coweeta sites under various deposition scenarios. (See Figure 27.1 for legend).

27. Simulated Effects of Atmospheric Deposition and Species Change

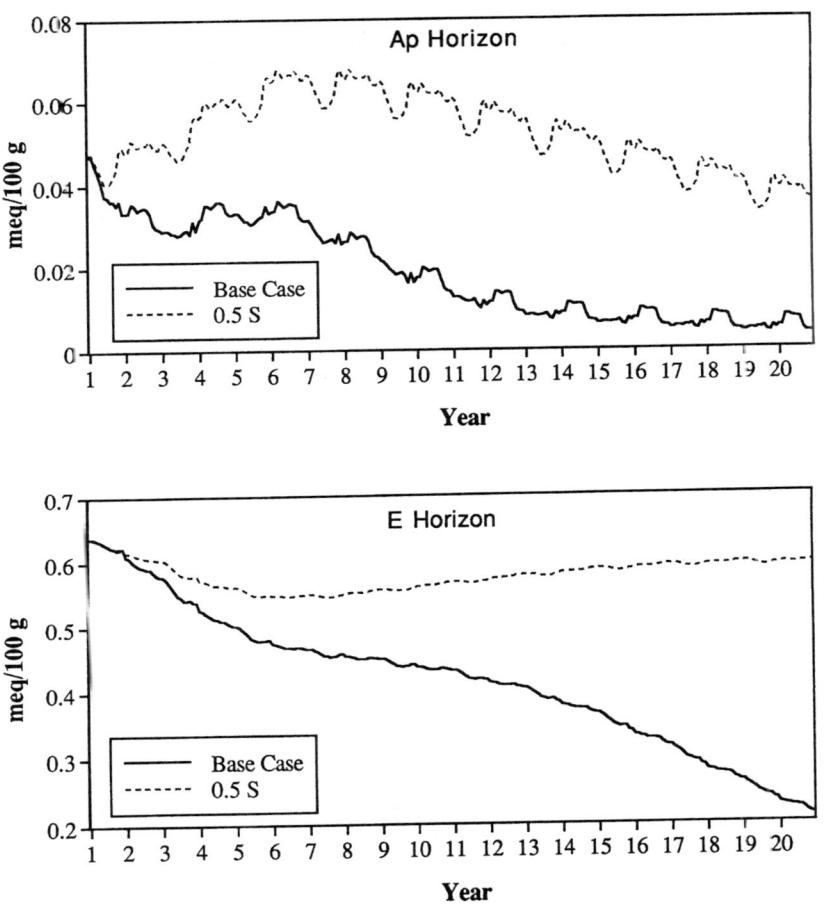

Figure 27.4a. Simulated exchangeable Ca2+ under the base case and 0.5× S deposition scenario at the Duke site (pine).

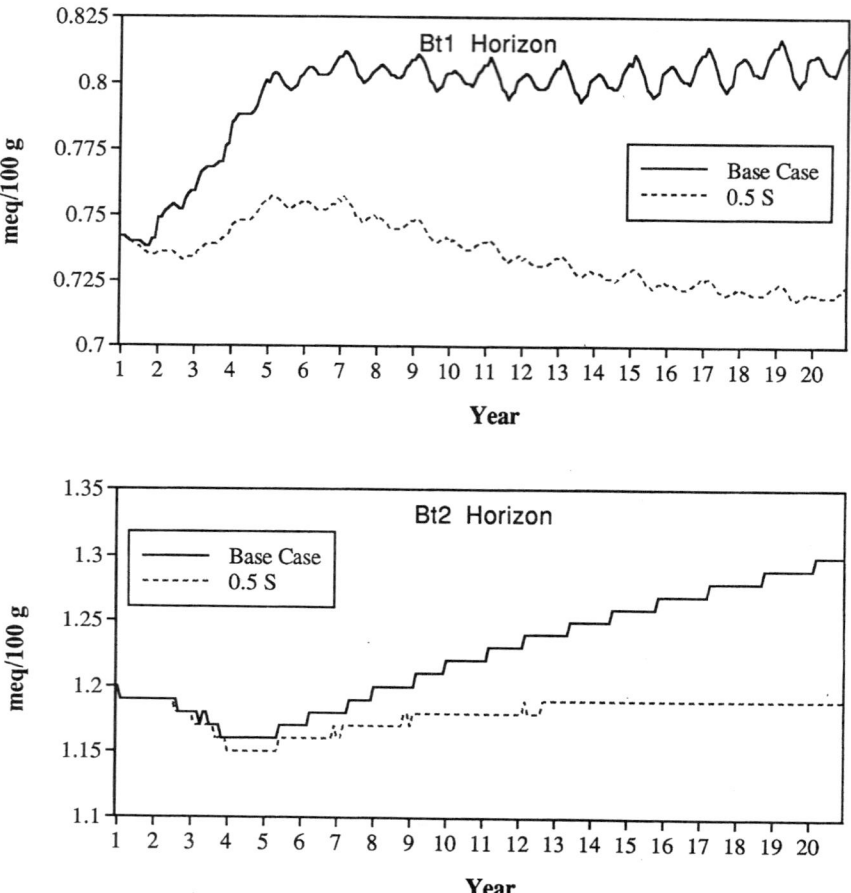

Figure 27.4b. Simulated exchangeable Ca2+ under the base case and 0.5× S deposition scenario at the Duke site (pine).

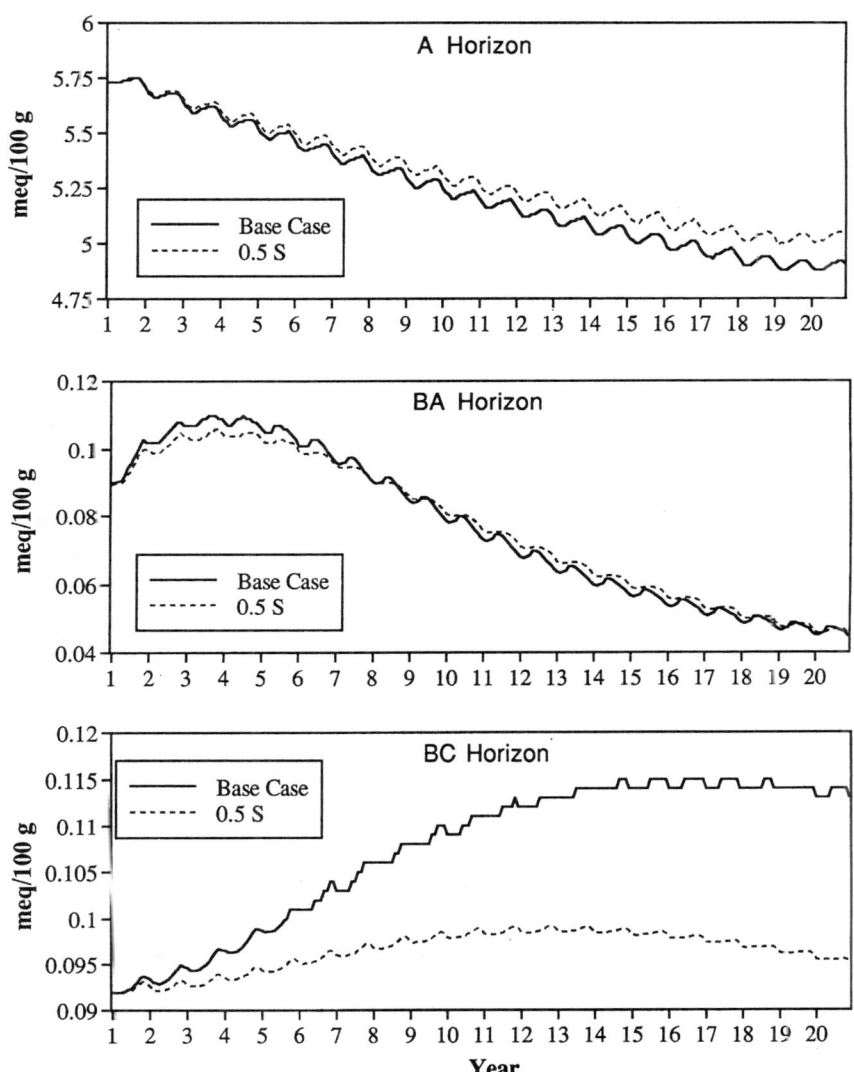

Figure 27.5. Simulated exchangeable Ca^{2+} under base case and $0.5\times$ S deposition scenario at the Coweeta site (hardwood).

horizons (Table 27.1). Total change in the exchangeable Ca_2^+ pool over this same period was only -1.84 kmol ha^{-1}. Under the 0.5× scenario, the net loss from the Ap and E horizons was reduced to -0.64 kmol ha^{-1}, $+0.54$ kmol ha^{-1} of which accumulated in the Bt1 and Bt2 horizons. The total change in the exchangeable Ca_2^+ pool with 0.5× was -1.18 kmol ha^{-1}.

Fluxes within the soil profile, and, therefore, exchangeable Ca_2^+ concentrations, changed substantially during the Duke simulations and were strongly affected by atmospheric deposition, even though total pool sizes were not affected. To a lesser extent the same was true at the Coweeta site (Table 27.1). Knoepp and Swank (1994) found large reductions in exchangeable Ca_2^+ and Mg_2^+ in soils in the top 20 cm (A and BA horizons) over a twenty-year period at Coweeta, as is also indicated by the simulations. There is no data to either verify or negate the prediction of increased exchangeable Ca_2^+ in the BC horizon.

In an effort to determine that nutrient besides N was potentially limiting to these systems, we ran simulations for sufficiently long periods such that deficiencies of one or more additional nutrients developed (i.e., growth rates dropped off and soil exchangeable pools approached zero). The base case and the combined deposition scenarios were run in these cases. The results of these long-term simulations indicated that phosphorus (P) was the nutrient projected to become most limiting (aside from N) over the long term. In both cases, growth rates in the combined deposition scenario increased slightly relative to those of the base case after twenty years (Figure 27.6). In the Coweeta simulation, there was a significant reduction in growth (i.e., a break in the biomass accumulation curve) at about year 15; the trend was smoother in the Duke simulations.

At the eighty-year mark, the Coweeta site had basically become N and P deficient, as soil available P levels dropped to approximately zero (Figure 27.7). At the Duke site, no nutrient other than N was deficient at eighty years, therefore the simulation was run to the maximum number possible (ninety-nine iterations of two years, or 198 years), and at that time, soil P dropped to near zero (Figure 27.7). Exchangeable pools of all other nutrients except N remained at levels sufficient to supply uptake rates.

Harvesting and Species Change

The effects of simulated harvest and species change were far greater than those of atmospheric deposition on ecosystem-nutrient pools at both sites. As noted in a prior study for the Duke site (Johnson et al., 1995), there were substantial redistributions of Ca from the exchangeable pool to vegetation in the hardwood stands compared to the pine stands at both sites (Figure 27.8). There were also substantial differences in Mg uptake by vegetation, but vegetation Mg pools were smaller and had little effect upon exchangeable Mg_2^+. As noted (Johnson et al., 1995), the changes in exchangeable Ca_2^+ caused changes in the Ca/Mg ratios on the exchange complex that in turn caused some nonintuitive changes in Ca_2^+ and Mg_2^+ leaching. Excluding all cations except Ca and Mg,

Table 27.1. Simulated Exchangeable Calcium (kmol ha^{-1}) at the Duke and Coweeta Sites in the Base Case and a 0.5× Sulfur Deposition Scenarios

		After 20 Years	
Horizon and depth (cm)	Initial	Base Case	0.5× S
Duke			
Ap (0–20)	0.72	0.56	0.54
E (20–35)	7.07	2.35	6.60
Bt1 (35–60)	14.84	16.28	14.46
Bt2 (60–80)	19.20	20.80	19.04
Total	41.83	39.99	40.64
Coweeta			
A (0–13)	34.64	29.68	30.47
BA (13–22)	0.52	0.26	0.26
BC (22–67)	2.48	3.05	2.57
Total	37.64	32.99	33.30

we can represent the cation-exchange complex with the Gapon equation for homovalent exchange:

$$Kg = \frac{X\ Ca_2^+\ (Mg_2^+)^{1/2}}{X\ Mg_2^+\ (Ca_2^+)^{1/2}}$$

in which X equals exchange phase equivalent fraction, ( ) equals soil solution activity, and Kg equals the Gapon selectivity coefficient.

If vegetative uptake decreases the concentration of Ca_2^+ on the soil exchanger, the exchange-phase equivalent fraction component ($X\ Ca_2^+ / X\ Mg_2^+$) will decrease. With the Gapon selectivity coefficient assumed to remain constant, this must result in an increase in the soil-solution activity component [$(Mg_2^+/Ca_2^+)^{1/2}$]. In essence, the uptake of Ca is balanced by the net release of Mg into the soil solution.

Therefore, the greater Ca consumption by hardwoods caused a reduction in Ca_2^+ leaching over time compared to the pine stands at both sites (Figures 27.9 and 27.10). This reduced Ca_2^+ leaching was offset in part by greater in Mg_2^+ leaching at both sites, but these increases were very small relative to the initial differences at the Coweeta site. The cause for this is the relatively low Ca_2^+ leaching rate at Coweeta compared to Mg_2^+ leaching (i.e., the percentage increase in Mg_2^+ leaching needed to offset the reduction in Ca_2^+ leaching at Coweeta was small). At the Duke site, the initial differences Mg_2^+ leaching rates approximately doubled by the end of the twenty-year simulation (Figure 27.10).

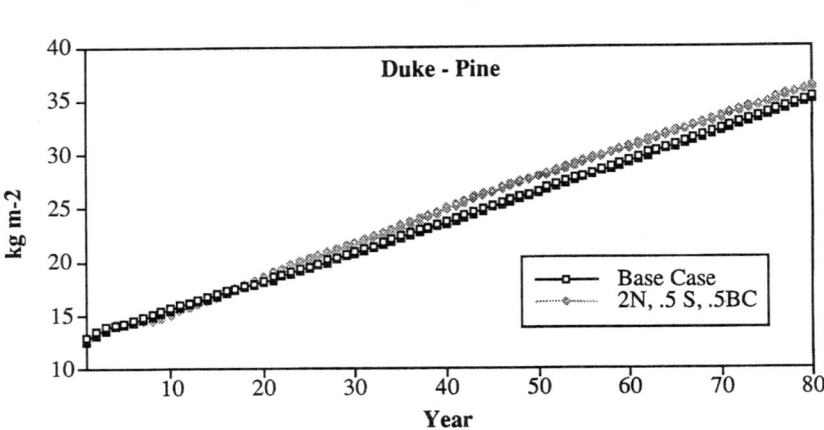

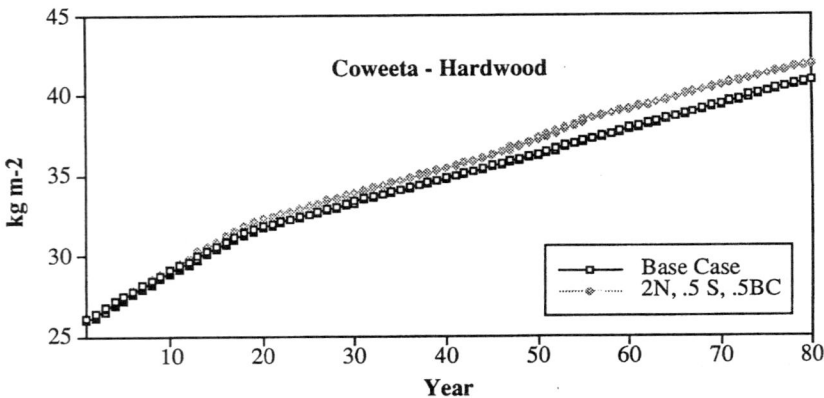

Figure 27.6. Simulated biomass accumulation under the base case and combined deposition scenarios at the Duke and Coweeta sites.

The reduction in exchangeable Ca_2^+ pools by the hardwoods also caused reductions in soil-solution pH (not shown), which resulted in the protonation of HCO_3^- and increased adsorption of SO_4^{2-}. Thus, over time, the leaching rates of ANC and SO_4^{2-} under hardwood vegetation were significantly reduced compared to the pine vegetation as the soils under the hardwoods became acidified (Figures 27.11 and 27.12). This pattern was most pronounced at the Duke site, where conversion to hardwoods also caused increased N leaching because of the greater rate of N cycling in the hardwood vegetation (Figure 27.12). No clear patterns of N leaching resulting from vegetation emerged at the Coweeta site (Figure 27.11).

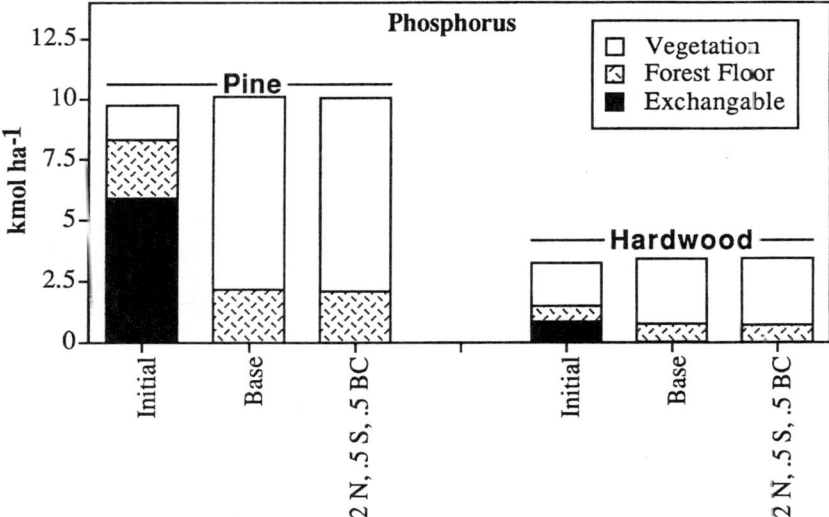

Figure 27.7. Simulated distribution of P in the Duke loblolly pine and Coweeta hardwood forests after 198 and eighty years, respectively.

Discussion

These simulation results suggest that soils from both Duke and Coweeta are sufficientlly buffered against nutrient depletion by atmospheric deposition. However, the simulations also suggest that soil exchangeable base concentrations can change greatly over a period of decades even when total pool sizes change little. Such interhorizon changes in the soil occurred in all scenarios, and were strongly affected by simulated changes in deposition.

The simulated changes in exchangeable cation concentrations are corroborated by field observations at Coweeta that showed a large reduction in exchangeable base cations in the surface 20 cm of Coweeta soils from 1970 to 1990 (Knoepp and Swank, 1994). These changes in the field were interpreted as a result of accumulation in vegetation and leaching (Knoepp and Swank, 1994). The simulations suggest that the changes in the surface horizons may have been only part of the story of decadal redistribution of nutrients within the soil profile; there may have been increases in exchangeable cations in deeper horizons, offsetting the apparent loss from the surface horizons.

The indications that P would become the most potentially limiting nutrient other than from N over the long term in the simulations were a consequence of two primary factors 1) low P deposition rates, and 2) low P weathering rates from primary minerals. It must be noted that the NuCM model includes only a fraction of the many interacting chemical processes that are known to control P availability in soils Specifically, NuCM includes only adsorption and a user-described

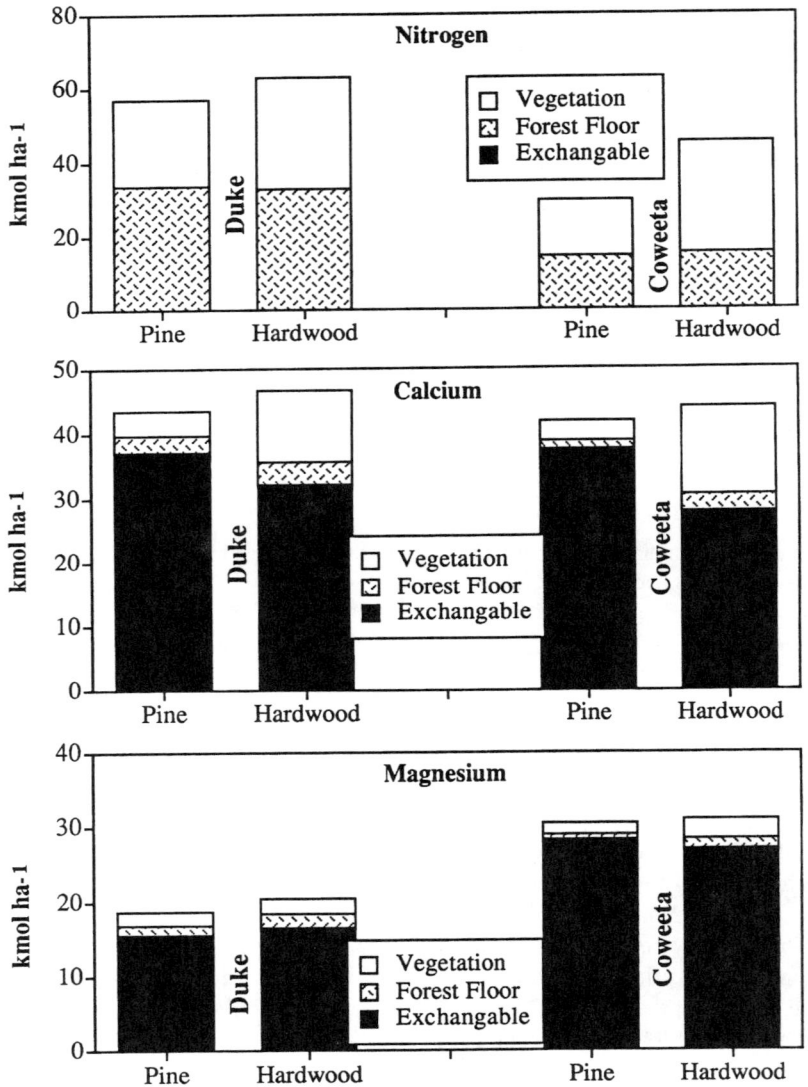

Figure 27.8. Simulated nutrient distribution of nitrogen, calcium, and magnesium in the Duke and Coweeta sites with loblolly pine and hardwood vegetation after twenty years.

P weathering rate, and does not include calcium, iron, and aluminum phosphate precipitation. NuCM predictions regarding long-term P status, therefore, must be viewed with caution and an appropriate level of skepticism. Nonetheless, the simulation results suggest the need for both further exploration of P dynamics with the NuCM model and further field studies both on soil-P dynamics and on long-term changes in soil-P status. With regard to long-term studies of soil-P change, it must be borne in mind that extractable P, as with exchangeable N, changes rapidly (Haines and Cleveland, 1981; Johnson and Todd, 1984, 1987).

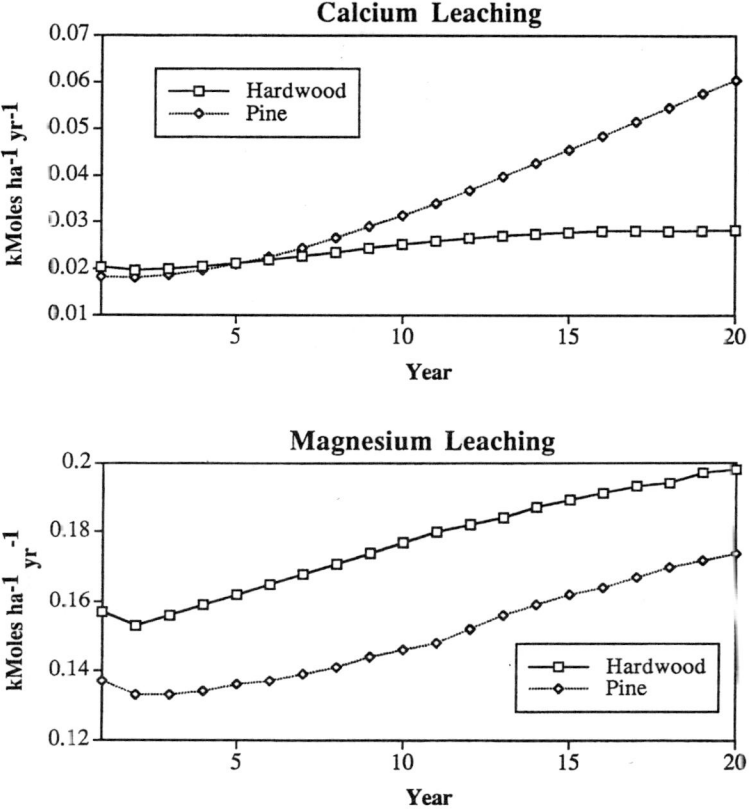

Figure 27.9. Simulated leaching of calcium and magnesium with loblolly pine and hardwood vegetation at the Coweeta site.

These rapid changes in extractable P, which are known to occur even on a seasonal basis, complicate efforts to detect long-term changes (Johnson et al., 1988).

One aspect of P cycling that has received little or no attention is the accurate measurement of atmospheric deposition. Problems with P adsorption to many surfaces require special techniques to obtain accurate P input measurements (e.g., acid washing of collection vessels). The long-term NuCM simulations are particularly sensitive to atmospheric P inputs, and any changes in P deposition estimates will have profound consequences for NuCM predictions for long-term change in P status.

These simulations, similar to those conducted for the Duke site at a prior date, indicate that Ca accumulation by hardwoods causes a chain of events resulting in the conservation of Ca in the system both by exchange for the less limiting Mg and by reduced total cation leaching. Specifically, reductions in exchangeable Ca_2^+ pools by uptake in the hardwood stands causes a shift to greater Mg_2^+ leaching because of cation exchange relationships and a reduction in total anion and cation leaching as a result of reduced base saturation, soil-solution pH, with consequent

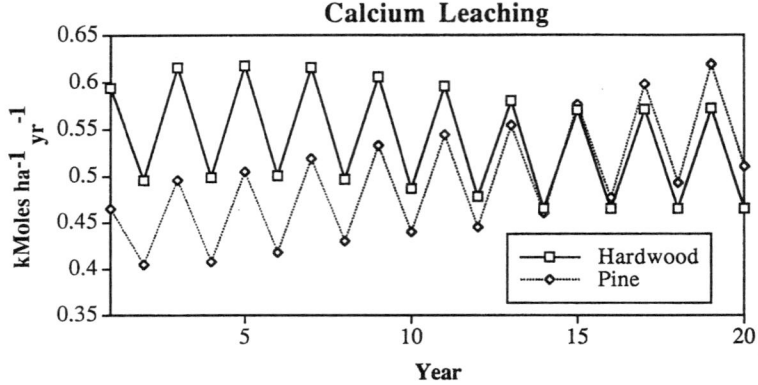

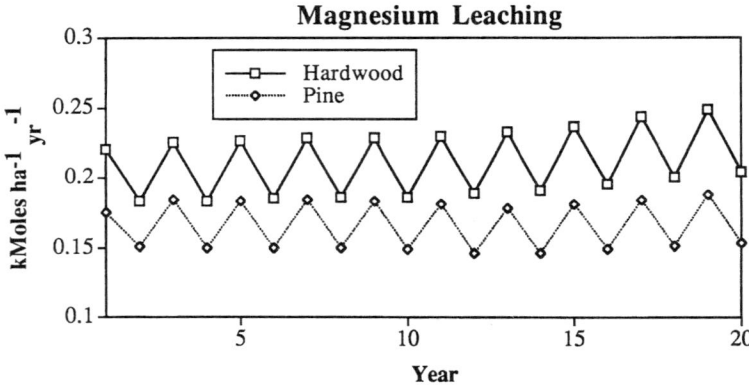

Figure 27.10. Simulated leaching of calcium and magnesium with loblolly pine and hardwood vegetation at the Duke site (The regular variations in leaching are the result of to the repitition of two-year interval meteorlogical and air quality data. See methods for details).

reductions in bicarbonate leaching and increased sulfate adsorption. Some of these interactions have been observed in prior field studies comparing nutrient cycling in mixed deciduous and loblolly pine stands (Johnson and Todd, 1987). Specifically, the reduction in soil exchangeable $Ca_2{}^+$ reserves and the shift from Ca to Mg, and the reduction in soil-solution pH and $HCO_3{}^-$ concentration have been observed (Table 27.2). A reduction in $SO_4{}^{2-}$ leaching was not observed, however; this may have been because of lower soil sulfate adsorption capacity in the Oak Ridge soils or to erroneous predictions by the model.

Summary

The NuCM simulations suggest that changes in atmospheric deposition can have major effects upon soil and soil-solution concentrations, but little effect upon

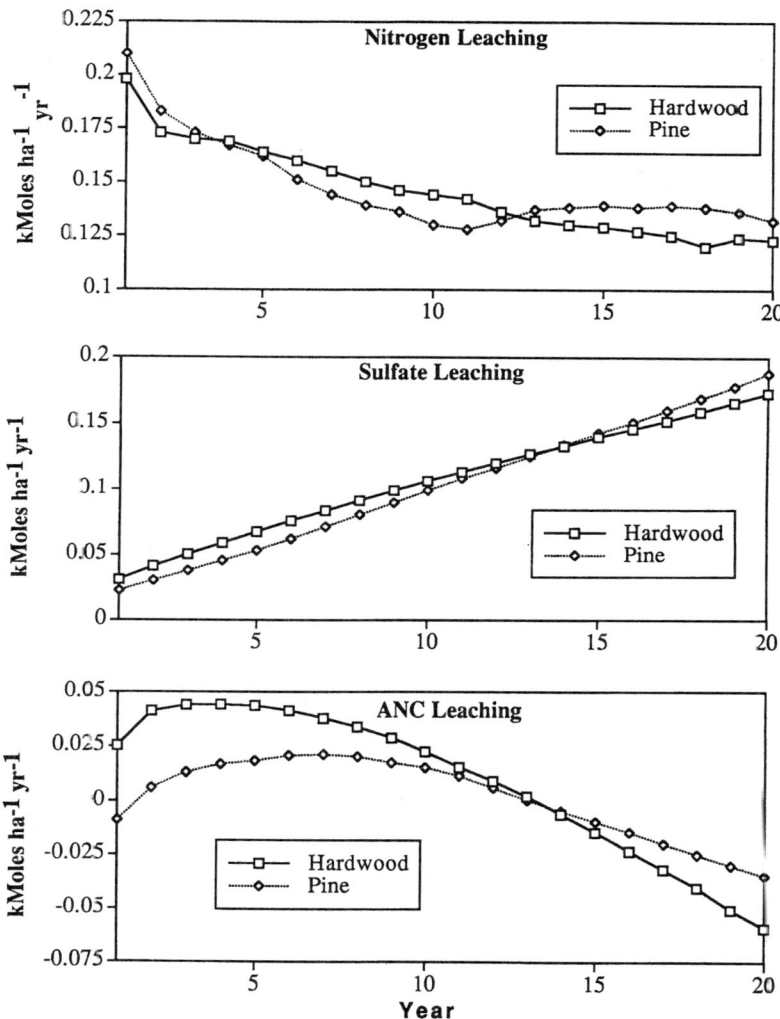

Figure 27.11. Simulated leaching of nitrogen, sulfate, and ANC with loblolly pine and hardwood vegetation at the Coweeta site.

growth or nutrient capital of the loblolly pine and hardwood stands simulated. This apparent contradiction is explained by marked changes in the distribution of nutrients in the soil profile that calculate sum up to only small changes in total pool sizes. The simulations suggest that soil-concentration change can occur over very short periods (even seasonally), and that deposition rates will strongly affect the nature of these changes.

Long-term simulations suggest that, other than N, the nutrient most probable to

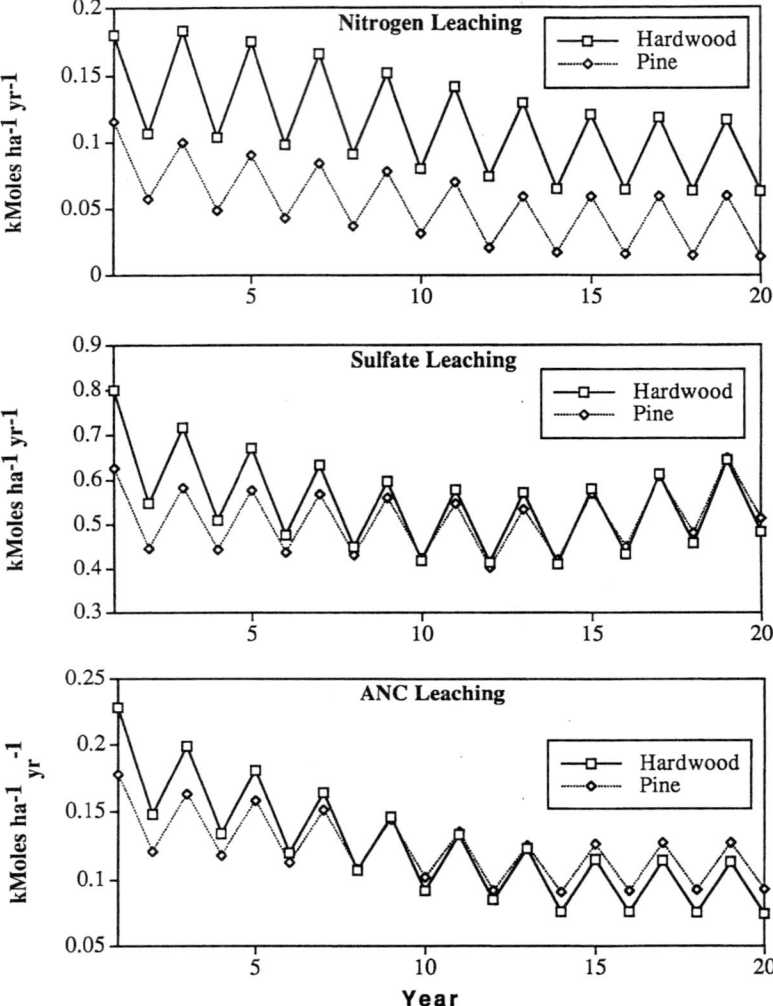

Figure 27.12. Simulated leaching of nitrogen, sulfate, and ANC with loblolly pine and hardwood vegetation at the Duke site. (The regular variations in leaching are the result of the repitition of two-year interval meteorlogical and air quality data. See methods for details).

become limiting with time in these forest ecosystems is P. However, NuCM lacks algorithms for many important soil-P cycling processes that may be relevant to long-term trends in soil P. The simulation output is very sensitive to atmospheric P deposition rates, pointing to the need for better field information on this input. These simulation results suggest a need for more field research and model development on long-term changes in both soil- and ecosystem-P status. In partic-

Table 27.2. Soil-Solution Concentrations ($\mu mol_c\ L^{-1}$) Beneath a Mixed Deciduous and a Loblolly Pine Stand Near Oak Ridge, Tennessee (after Johnson and Todd, 1987)

	Mixed deciduous	Loblolly pine
pH	4.9	5.4
Ca^{2+}	102 ± 23	207 ± 50*
Mg^{2+}	90 ± 27	63 ± 12*
HCO_3^-	32 ± 36	86 ± 76
SO_4^{2-}	228 ± 56	251 ± 63

ular, accurate measurements of P-deposition rates and incorporation of chemical and biological processes controlling soil-P availability are needed.

The simulations suggest that replacement of pines by hardwoods (or vice versa) causes large differences in Ca uptake and accumulation by vegetation. This, in turn, causes a chain of events (i.e., increased Mg leaching, reduced base saturation, reduced soil-solution pH, bicarbonate, and sulfate leaching) that significantly alters the fluxes and accumulation of most other nutrients in the system. Many of these changes have been corroborated in field studies of loblolly pine and mixed deciduous forests near Oak Ridge, Tennessee, but more information is needed to test the simulations for the Duke and Coweeta sites.

Finally, it should be noted that these simulations are presented in order to investigate the collective implications of our concepts of nutrient-cycling processes in forests. Because the NuCM model, as with all models, is only an approximation of the real world, it is inherently inaccurate at some level; it is both scientifically interesting and germane to policy to determine the level at which NuCM fails to reproduce the real world, and for those reasons, we continue to seek field data to compare to model output.

References

April R, Newton R (1991) Mineralogy and mineral weathering. In Johnson DW, Lindberg SE (Eds) *Atmospheric deposition and forest nutrient cycling: a synthesis of the integrated forest study.* Ecological Series 91, Springer-Verlag, New York.

Binkley D, Valentine D, Wells C, Valentine, U (1989) An empirical model of the factors contributing to 20-yr decrease in soil pH in an old-field plantation of loblolly pine. Biogeochem 8:39–54.

Goldstein RA, Gherini SA, Chen CW, Mok L, Hudson RJM (1984) Integrated acidification study (ILWAS): A mechanistic ecosystem analysis. Bull Phil Trans R Soc Lond 305:259–279.

Haines SG, and Cleveland G (1981) Seasonal variation in properties of five forest soils in southwest Georgia. Soil Sci Soc Amer J 45:139–143.

Harrison RB, Johnson DW, Todd DE (1989) Sulfate adsorption and desorption in a variety of forest soils. J Environ Qual 18:419–426.

Hedin LO, Granat L, Likens GE, Buishand TA, Galloway JN, Butler TJ, Rodhe H (1994) Steep declines in atmospheric base cations in regions of Europe and North America. Nature 367:351–354.

Johnson DW, Binkley D, Conklin P (1995) Simulated effects of atmospheric deposition, harvesting, and species change on nutrient cycling in a loblolly pine forest. For Ecol Managem 76:29–45.

Johnson DW, Cresser MS, Nilsson SI, Turner J, Ulrich B, Binkley D, Cole DW (1991) Soil changes in forest ecosystems: Evidence for and probable causes. Proc., Royal Society of Edinburgh 97B:81–116

Johnson DW, Henderson GS, Todd DE (1988) Changes in nutrient distribution in forests and soils of Walker Branch Watershed, Tennessee, over an eleven-year period. Biogeochem 5:275–293.

Johnson DW, Lindberg SE (Eds) (1991) *Atmospheric deposition and forest nutient cycling: a synthesis of the integrated forest study.* Ecological Series 91, Springer-Verlag, New York.

Johnson DW, Swank WT, Vose JM (1993) Simulated effects of atmospheric sulfur deposition on nutrient cycling in a mixed deciduous forest. Biogeochem 23:169–196.

Johnson DW, and Todd DE (1984) Effects of acid irrigation CO_2 evolution, extractable nitrogen, phosphorus and aluminum in a deciduous forest soil. Soil Sci Soc Am J 48:664–666.

Johnson DW, Todd DE (1987) Nutrient export by leaching and whole-tree harvesting in a loblolly pine and mixed oak forest. Plant Soil 102:99–109.

Johnson DW, Todd DE (1990) Nutrient cycling in forests of Walker Branch Watershed: Roles of uptake and leaching in causing soil change. J Environ Qual 19:97–104.

Knoepp JD, Swank WT (1994) Long-term soil chemistry changes in aggrading forest ecosystems. Soil Sci Soc Amer J 58:325–331.

Liu S, Munson R, Johnson DW, Gherini S, Summers K, Hudson R, Wilkinson K, Pitelka L (1991) Application of a nutrient cycling model (NuCM) to northern mixed hardwood and southern coniferous fores. Tree Physiol 9:173–182

Munsen RK, Liu S, Gherini SA, Johnson DW, Wilkinson KJ, Hudson RJM, White KS, Summers KV (1992) *NuCM Code Version 2.0: An IBM PC code for simulating nutrient cycling in forest ecosystems.* Tetra-Tech, Inc., Hadley, MA.

Richter DD, Johnson DW, Dai KH (1992) Cation exchange reactions in acid forested soils: Effects of atmospheric pollutant deposition. In Johnson DW, Lindberg SE (Eds) *Atmospheric deposition and forest nutrient cycling: a synthesis of the integrated forest study.* Ecological Series 91, Springer-Verlag, New York.

Richter DD, Markewitz D, Wells CG, Allen HL, April R, Heine PR, Urrego B (1994) Soil chemistry change during three decades in an old-field loblolly pine (*Pinus taeda* L.) Ecosystem Ecology 75:1463–1473.

Swank WT, Douglass JD (1977) Nutrient budgets for undisturbed and manipulated hardwood forest ecosystems in the mountains of North Carolina. In Correll DL (Ed) *Watershed research in eastern North America.* Vol. I. Smithsonian Institution, Edgewater, MD.

28. Influence of Microclimate on Short-Term Litter Decomposition in Loblolly Pine Ecosystems

B. Graeme Lockaby, Arthur H. Chappelka, Mary A. Sword, and Allan E. Tiarks

Recently, there has been much concern expressed over projected increases in global temperatures (Schneider, 1989), and the subsequent effects on terrestrial ecosystems (Woodward, 1992). Atmospheric levels of carbon dioxide and other trace greenhouse gases such as methane and chlorofluorocarbons are increasing at a dramatic rate (Mooney et al., 1987; Schneider, 1994). Increases in these gases and potential trapping of infrared radiation have led to predictions by general circulation climate models (GCMs) of global surface temperature increases from 1.5 to 4.5 °C over the next 50 to 100 years (Hansen et al., 1981; Bretherton et al., 1990). These models have also predicted shifts in the global water cycle, although the magnitude and direction of change is less certain.

Increases in trace gases and subsequent alterations in temperature and precipitation could result in species migration, and shifts in species competition, diversity, and productivity (Mooney et al., 1987; Woodward, 1992; Perry, 1994). Furthermore, changes in temperature and soil-moisture regimes may influence litter decomposition, nutrient cycling, and, subsequently, primary productivity of forest ecosystems (Field et al., 1992; Perry, 1994). In the laboratory, Taylor and Parkinson (1988) demonstrated that an increase in both temperature and moisture accelerated the decomposition of pine (*Pinus contorta* Loud. and *P. banksiana* Lamb.) and aspen (*Populus tremuloides* Michx.) litter. Between 2 and 26 °C, temperature generally had a greater effect than moisture. However, moisture became more important as temperature increased. In a similar study with Scots pine (*Pinus sylvestris* L.), Jansson and Berg (1985) reported that although 70% of the change in the rate of litter decomposition was explained by temperature, moisture explained 90% of this response. When temperature and moisture were considered simultaneously, they accounted for 95 to 99% of the variation in litter decomposition.

Klemmedson et al., (1985) found that stand condition influenced the rate of litter decomposition. Stands with fewer trees per hectare (ha) had higher surface temperatures and rates of litter decomposition than more heavily stocked stands. However, if moisture is limiting, this relationship may be reversed. For example, Carlyle and Ba Than (1988) found that at soil-water contents above 12.5%, carbon dioxide (CO_2) evolution from the soil and forest floor of an eighteen-year-old Monterey pine (*P. radiata* D. Don) stand was more closely related to soil temperature than soil water content. However, this relationship was reversed at soil water contents below 12.5%. Furthermore, De Santo et al., (1993) reported that moisture was a key rate-limiting factor in the early decomposition of litter from three species of pine.

Abbott and Crossley (1982) suggested that under moisture limitation, increased temperature may cause a decrease in populations of both macrofauna and microfauna, and a reduction in the rate of litter decomposition. Arthropod activities that cause the fragmentation of litter and subsequent release of nutrients are also very important in determining the rate of mineral-nutrient cycling in a stand (Swift et al., 1979). Because these macro-fauna are affected by changes in temperature and moisture, it is important to determine how modifications in climate alter their population dynamics.

Litter decomposition is also influenced by the quality of the decomposing substrate. Berg et al., (1993) found that if the lignin concentrations of Scots pine litter sources were similar, climate was the most important factor governing litter decomposition. However, if the lignin concentration of litter sources varied significantly, lignin concentration rather than climate most strongly influenced litter decomposition. Furthermore, in an experiment comparing four tree species grown in China and Wisconsin, Geng et al., (1993) reported that almost 50% of the variation in litter decomposition was explained by grouping species according to litter quality. Fogel and Cromack (1977) suggested that lignin content, rather than carbon/nitrogen ratio, was more closely related to the decomposition of Douglas fir [*Pseudotsuga menziesii* (Mirb.) Franco] litter. However, because microbial activity increases as the ratio of carbon to mineral nutrients increases in decomposing litter (Rai and Srivastava, 1982), both macronutrient and micronutrient concentrations are also important factors that control litter decomposition.

Several techniques are presently being used to investigate the influence of climatic variation on litter decomposition. However, none of these methods separate temperature and moisture effects. Buried electric-resistance wires have been used in several soil-warming experiments (Fernandez et al., 1993; Peterjohn et al., 1993; Van Cleve et al., 1990). This method permits precise control of soil temperature, but heats the soil from below (an unnataural process), which produces sharp temperature gradients and causes physical disruption to the soil (National Science Foundation, 1992). Another technique utilizes landscape cloth made of

28. Influence of Microclimate on Short-Term Litter Decomposition

fiber that is placed over the soil surface (Christiansen et al., 1993). The cloth modifies light quality and soil-moisture characteristics and, therefore, alters detritivore activity within the litter. Harte et al., (1995) used overhead heaters to warm the soil in a mountain meadow experiment in Colorado. This method maintained the integrity of the soil profile and provided a realistic global-warming scenario, but was expensive, labor intensive, and required an electric power source.

To understand and predict the relationship between climate and such forest processes as litter decomposition and nutrient cycling, such confounding variables as temperature and precipitation need to be separated experimentally. The goals of this chapter were to isolate and quantify the influence of climate on litter decomposition, and to evaluate the role of litter quality, and, therefore, site productivity on this relationship. Specifically, we determined the effects of climate variation on the litter decomposition rate, the dynamics of nitrogen (N), phosphorus (P), potassium (K), calcium (Ca) and magnesium (Mg), and the tannin concentrations (measure of litter quality) of two sources of loblolly pine (*P. taeda* L.) litter placed in the field for one year in two loblolly pine stands.

Materials and Methods

Study Locations

Study sites were established near Auburn, Alabama, and Pineville, Louisiana. The Auburn site is a thirteen-year-old loblolly pine plantation (1.8 × 1.8 m spacing), located in the Auburn University Forest in Lee County. The site index at this location is twenty-six (base age fifty), and the soil is a Cecil sandy loam (Typic kanhapludult) (McNutt, 1981) containing adequate P. The Pineville site is a twelve-year-old loblolly pine plantation (1.8 m × 1.8 m), located on the USDA Forest Service Palustris Experimental Forest in Rapides Parish. The site index at this location is twenty-seven (base age fifty) and the soil is a Beaureguard silt loam (Plinthaquic paleudult) that is low in N and P (Kerr et al., 1980). In spite of differential P availability at the two sites, productivity is similar. This is probably the result of greater rainfall amounts at the Louisiana site (Kerr et al., 1980).

Approach

Climate treatments, consisting of realistic variations in temperature and precipitation (Cooter et al., 1993), were imposed with rectangular plexiglass microcosms that raised the temperatures near the forest floor from 1 to 5 °C and permitted throughfall to be varied (Hornsby et al., 1995). Microcosm covers were placed over 1 m^2 areas that were prepared by digging a trench 6 cm wide and 15 cm deep, around wooden microcosm frames (120 × 60 × 2 cm). To prevent root growth and water infiltration into the microcosms, the inner side of each trench was lined with black polyethylene that was stapled to the frame; trenches were then backfilled.

Two types of microcosm covers were constructed of 0.65 cm nonphotoreactive plexiglass. Those associated with ambient temperature consisted of one sheet of plexiglass (120 × 60 cm), mounted 30 cm above the forest floor. Elevated temperatures were achieved using boxlike microcosm covers (120 × 60 × 30 cm). The appropriate temperature differential between ambient and elevated temperature treatments was maintained by adjusting the air passage through five holes (5 cm) on the longitudinal sides of the covers. This adjustment also allowed equilibration of relative humidity inside the microcosms to the outside conditions.

Copper-constant thermocouple sensors, linked to a data recorder (Campbell Scientific Inc.), were used to monitor the temperature of the forest floor in each microcosm. At each site, throughfall was collected with nine randomly located throughfall caches that consisted of polyethylene containers (114 l), each under a perforated plastic cloth (1.2 × 2.4 m) that was tied horizontally to adjacent trees. One plastic rain gauge was randomly placed at each replicate location. After a rainfall event, rain gauges were read, and throughfall volumes were calculated and applied manually to each microcosm with plastic watering cans.

The litterbag technique was used to determine rates of decomposition. Nylon mesh litterbags (30 × 15 cm), as described by Kelly and Beauchamp (1987) were preweighed then filled with approximately 10 g of air-dried, senescent loblolly pine foliage collected in autumn of 1992 at the study sites. On January 25, 1993, seven Pineville and seven Auburn litterbags were buried in random locations in the O_i horizon of all microcosms at each site. An additional litterbag was placed in each microcosm, and immediately after the study installation this litterbag was harvested and used to determine litter loss associated with handling during litterbag transport and study installation. One litterbag from each litter source (Auburn and Pineville) was harvested from all microcosms at both locations seven times between January 25, 1993 and February 25, 1994 (after 0.0, 0.3, 1.0, 2.0, 4.0, 9.0 and 13.0 months).

After collection, litter samples were lyophilized, ground (20-mesh), and analyzed for total-N by thermal conductivity. Phosphorus concentrations were determined colorimetrically, and K, Ca, and Mg concentrations were determined with atomic absorption spectrophotometry. Dry weight (ash-free basis) and nutrient concentration data, and initial litter characteristics (Table 28.1) were used to

Table 28.1. Concentration of Mineral Nutrients in Air-Dried Litter from Pineville and Auburn Sources Before Exposure to Climate Treatments

Nutrient	Source	
	Auburn	Pineville
Nitrogen (%)	0.51	0.62
Phosphorus (%)	0.08	0.04
Potassium (ppm)	1.98	0.73
Calcium (ppm)	3.04	2.72
Magnesium (ppm)	1.23	0.74

calculate the fractions of mineral nutrients remaining in the litter after each incubation.

Tannin concentrations were determined using sequential extraction followed by quantification of anthocyanidins formed in n-butanol-HCl (Tiarks et al. 1992). For this determination, 0.05 g of ground litter was extracted two times with 10 ml of methanol for one hour. Pellets were resuspended, extracted one time with 10 ml of acetone-water (1:1) (v/v), and washed one time with 2 ml of n-butanol. The acetone-water supernatant and the n-butanol wash were then combined. Condensed tannin concentrations were determined in both the methanol and the acetone-water-butanol extracts, as well as the residue after resuspension in n-butanol-HCl (95:5) (v/v) containing 12 mg per l of Fe^{++} (added as $FeSO_4$). Condensed tannin concentrations were determined spectrophotometrically using a standard curve generated from condensed tannin that was purified from loblolly pine foliage.

Litter Arthropod Collection

In October and December of 1993, soil arthropods were collected from the forest floor in microcosms at the Auburn site according to the method of Gist and Crossley (1975). Litter from the forest floor of each microcosm (1 m^2 area) was collected, brought back to the laboratory, and placed in large Teflon-coated funnels. Lamps were placed over the large opening of funnels, approximately 15 cm from the litter surface. Funnels were inserted into vials containing 70% ethanol. After five days, arthropods that were extracted into ethanol solutions were counted under a dissecting microscope.

Experimental Design and Statistical Analysis

The design was completely randomized with four replications per site. On April 13, 1993, two replications at the Pineville site were vandalized. As a result, research findings from this site are based on only two replications. The following climate treatments were studied:

T1P1 = Ambient temperature, ambient throughfall (covered with plexiglass but not enclosed on sides),
T2P1 = Increased temperature, ambient throughfall,
T2P2 = Increased temperature, throughfall decreased by 30%,
T2P3 = Increased temperature, throughfall increased by 30%.

Litterbag data were compared in terms of decay rate (k) over the thirteen-month study period and the percentage of original mass, N, P, K, Ca, and Mg remaining in addition to the concentration of tannins present after thirteen months of incubation (day 387). Decay-rate parameters for each treatment × replication combination were calculated using a single exponential nonlinear regression equation (SAS, 1985). For each site, decay rates, and litter variables at day 387 were compared using analyses of variance (ANOVA) with a statistical significance level of $P < 0.10$. Statistical comparisons were not made between the two locations because of the multiple environmental differences, which, out of necessity, existed between the two sites.

Litter arthropod data were analyzed using the distribution-free "randomization" method as described by Potvin and Roff (1993). This method is applicable to data that are not normally distributed. Using this method, pairwise comparisons were made among treatments.

Quality control and Quality assurance

Data were calculated in a careful, systematic manner. Samples were weighed on calibrated analytical balances checked with standard weights. Litter samples were randomly selected each collection period. For the arthropod collections, populations were counted and 10% of the samples were recounted. Regarding the laboratory analyses (N, P, K, Ca, Mg, and tannins) National Bureau of Standards (NBS) standard and duplicates were used (10% remeasurements). For tannins, a NBS standard was not available, but a standard check sample was used.

Temperature was measured using polyvinylchloride(PVC)-coated type T thermocouple wire and a recently calibrated data logger. The wires were calibrated by placing them in an ice bath to obtain an offset measurement, which was included in the program. Regarding throughfall application, the average of the four rain gauges was used to calculate the amount of rainwater applied to the microcosms [amount (liters) = rain (cm) × area / 1000]. The amount of water applied was measured with a graduated 2-l cylinder.

Results

The mean daily temperature of the forest floor was approximately 1 to 3 °C higher in the elevated temperature treatments when compared to the ambient temperature treatment (Figures 28.1a and 28.1b). Rainfall patterns varied between the two sites (Figure 28.2), and approximately 27 cm more throughfall was collected at the Pineville site than the Auburn site. The majority of this difference occurred from March to July.

Comparison of k parameters indicated that there were no significant differences between climate treatments or sources of litter at either location (Table 28.2). Values of k averaged 0.11 and 0.09 at the Auburn and Pineville sites, respectively. These values reflect turnover times of approximately ten years and are representative of litter decomposition rates in young loblolly pine forests (Lockaby et al., 1995).

At the Auburn site, the elevated temperature with reduced throughfall treatment (T2P2) exhibited significantly greater mass loss than the ambient climate treatment (T1P1) (Table 28.3). At Auburn, the sources of litter also differed with the Auburn source showing greater loss of mass. There were no significant differences between treatments or sources at the Pineville site.

Significant separation in the percentage of P remaining occurred between treat-

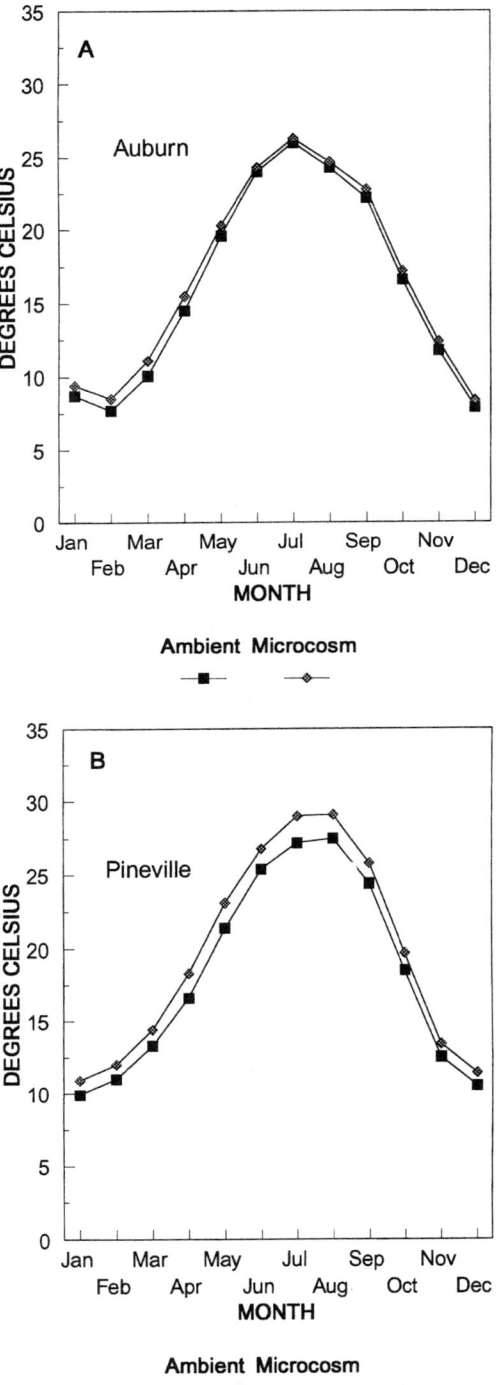

Figure 28.1. Mean daily temperature (°C) of the forest floor in response to four climate treatments at A. Auburn, Alabama, and (b) Pineville, Louisiana from January, 1993 to February, 1994.

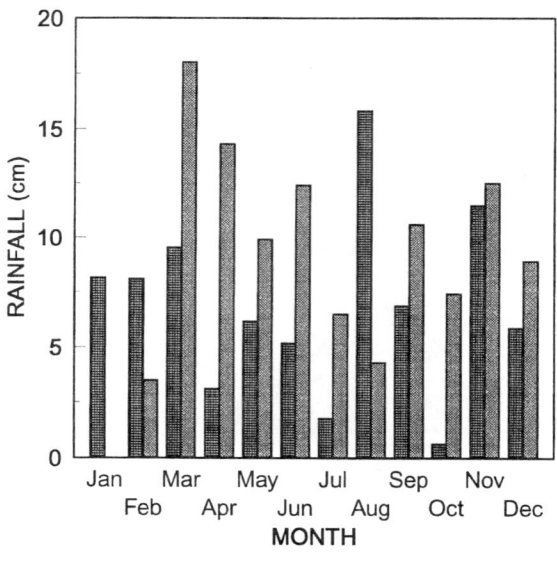

Figure 28.2. Monthly rainfall (cm) at Auburn, Alabama, and Pineville, Louisiana from January, 1993 to February, 1994.

Table 28.2. Decay Parameters[1] Compared Among Four Climate treatments and Litter Sources at Two Locations

Auburn Site			
Climate	k^2	Source	k^2
T1P1	0.11 a	Pineville	0.11 a
T2P1	0.12 a	Auburn	0.11 a
T2P2	0.11 a		
T2P3	0.11 a		
Pineville Site			
Climate	k^2	Source	k^2
T1P1	0.08 a	Pineville	0.09 a
T2P1	0.08 a	Auburn	0.08 a
T2P2	0.09 a		
T2P3	0.09 a		

[1] Decay parameters = k.
[2] Significant differences among climate and source are noted by different letters using Duncan's New Multiple Range Test ($P < 0.10$).

Table 28.3. Mean Percentage of Mass Remaining in Litterbags at Day 387 Compared Among Four Climate Treatments and Litter Sources at Two Locations

Auburn Site			
Climate	mass (%)[1]	Source	mass (%)[1]
T1P1	76 a	Pineville	75 a
T2P1	74 ab	Auburn	73 b
T2P2	73 b		
T2P3	74 ab		
Pineville Site			
Climate	mass (%)[1]	Source	mass (%)[1]
T1P1	76 a	Pineville	77 a
T2P1	76 a	Auburn	76 a
T2P2	77 a		
T2P3	76 a		

[1] Significant differences among four climate treatments and two sources are noted by different letters using Duncan's New Multiple Range Test ($P < 0.10$).

ments, which caused elevated temperature and either ambient (T2P1) or reduced (T2P2) throughfall at the Auburn site (Table 28.4). There were no significant differences in the percentage of P remaining between sources at either site or among climate treatments at the Pineville site. With the exception of the elevated temperature with ambient throughfall treatment (T2P1) at Auburn, the data reflected mineralization of P after a thirteen-month incubation period.

At both sites, the percentage of N remaining was the response variable most sensitive to climate and source treatments (Table 28.5). In general, climate and source comparisons at both sites reflected N immobilization. At Auburn, the elevated temperature with ambient throughfall treatment (T2P1) displayed a sig-

Table 28.4. Percent of Original P Remaining in Litterbags at Day 387 Compared Among Four Climate Treatments and Sources at Two Locations

Auburn Site			
Climate	% P[1]	Source	% P[1]
T1P1	93 ab	Pineville	94 a
T2P1	109 a	Auburn	98 a
T2P2	89 b		
T2P3	95 ab		
Pineville Site			
Climate	% P[1]	Source	% P[1]
T1P1	79 a	Pineville	89 a
T2P1	94 a	Auburn	89 a
T2P2	93 a		
T2P3	88 a		

[1] Significant differences among four climate treatments and source are noted by different letters using Duncan's New Multiple Range Test ($P < 0.10$).

Table 28.5. Percentage of Original N Remaining in Litterbags at Day 387 Compared Among Four Climate Treatments and Sources at Two Locations

Auburn Site			
Climate	% N[1]	Source	% N[1]
T1P1	109 b	Pineville	102 b
T2P1	121 a	Auburn	125 a
T2P2	112 ab		
T2P3	108 b		
Pineville Site			
Climate	% N[1]	Source	% N[1]
T1P1	108 b	Pineville	103 b
T2P1	117 ab	Auburn	145 a
T2P2	137 a		
T2P3	127 ab		

[1] Significant differences among four climate treatments and source are noted by different letters using Duncan's New Multiple Range Test ($P < 0.10$).

nificant increase in the precentage of N remaining when compared to both the ambient climate (T1P1), and the elevated temperature and throughfall (T2P3) treatments. Also at the Auburn site, the precentage of N remaining also differed by source, displaying a stronger tendency for N immobilization to occur in Auburn litter. At Pineville, the ambient climate (T1P1) and the elevated temperature with reduced throughfall (T2P2) treatments were significantly different, with more N immobilization occurring in response to treatment T2P2. At Auburn, litter sources at the Pineville site were significantly different with Auburn litter exhibiting greater N immobilization.

Climate treatment did not significantly affect the percentage of initial K, Ca, or

Table 28.6. Percentage of Original K Remaining in Litterbags at Day 387 Compared Among Treatments and Sources at Two Locations

Auburn Site			
Climate	% K[1]	Source	% K[1]
T1P1	55 a	Pineville	69 a
T2P1	59 a	Auburn	45 b
T2P2	61 a		
T2P3	56 a		
Pineville Site			
Climate	% K[1]	Source	% K[1]
T1P1	21 a	Pineville	44 a
T2P1	46 a	Auburn	18 a
T2P2	35 a		
T2P3	22 a		

[1] Significant differences among four climate treatments and source are noted by different letters using Duncan's New Multiple Range Test ($P < 0.10$).

Table 28.7. Percentage of Original Ca Remaining in Litterbags at Day 387 Compared Among Four Climate Treatments and Sources at Two Locations

Auburn Site			
Climate	% Ca[1]	Source	% Ca[1]
T1P1	93 a	Pineville	103 a
T2P1	95 a	Auburn	89 b
T2P2	97 a		
T2P3	101 a		
Pineville Site			
Climate	% Ca[1]	Source	% Ca[1]
T1P1	108 a	Pineville	111 a
T2P1	106 a	Auburn	101 a
T2P2	105 a		
T2P3	105 a		

[1] Significant differences among four climate treatments and source are noted by different letters using Duncan's New Multiple Range Test ($P < 0.10$).

Mg in decomposing litter after the thirteen-month incubation period (Tables 28.6 to 28.8). At Auburn, the percentage of initial K, Ca, and Mg remaining in the decomposing litter was significantly greater in the Pineville source when compared to that collected at the Auburn site. A similar response was observed at the Pineville site with the percentage of Mg remaining. Percentages of Ca and Mg in litter from Pineville reflect accumulations of these elements. Although statistical comparisons could not be made, the mean percentage of K remaining in litter after thirteen months was 46% less at the Pineville site when compared to the Auburn site.

At Auburn, the concentration of soluble and insoluble tannins together in the

Table 28.8. Percentage of Original Mg Remaining in Litterbags at Day 387 Compared Among Four Climate Treatments and Sources at Two Locations

Auburn Site			
Climate	% Mg[1]	Source	% Mg[1]
T1P1	101 a	Pineville	124 a
T2P1	104 a	Auburn	81 b
T2P2	106 a		
T2P3	102 a		
Pineville Site			
Climate	% Mg[1]	Source	% Mg[1]
T1P1	71 a	Pineville	100 a
T2P1	88 a	Auburn	61 b
T2P2	84 a		
T2P3	79 a		

[1] Significant differences among four climate treatments and source are noted by different letters using Duncan's New Multiple Range Test ($P < 0.10$).

Table 28.9. Concentration of Soluble, Nonsoluble, and Total Tannins in Litterbags at Day 387 Compared Among Four Climate Treatments at Two Locations

	Auburn Site			
	Source of Tannins, g/kg[1]			
Climate	Methanol-soluble	Acetone-soluble	Nonsoluble	Total
T1P1	5.46 a	5.04 a	18.26 a	28.76 a
T2P1	1.92 a	2.34 a	12.76 b	17.01 b
T2P2	2.16 a	2.39 a	13.89 b	18.43 b
T2P3	5.01 a	4.67 a	13.83 b	23.51 ab
	Pineville Site			
	Source of Tannins, g/kg[1]			
Climate	Methanol-soluble	Acetone-soluble	Nonsoluble	Total
T1P1	1.07 a	1.43 a	8.64 a	11.1 a
T2P1	1.66 a	2.15 a	9.34 a	13.1 a
T2P2	1.35 a	1.85 a	8.98 a	12.2 a
T2P3	1.49 a	1.51 a	8.49 a	11.5 a

[1] Significant differences among four climate treatments at each site are noted by different letters using Duncan's New Multiple Range Test ($P < 0.10$). At each location, litter sources (Auburn and Pineville) were combined.

decomposing litter was significantly greater in the ambient climate treatment (T1P1) when compared to 1) the climate treatments that caused an increase in temperature, 2) ambient treatments (T2P1), or 3) the reduced throughfall (T2P2) (Table 28.9). Climate did not significantly affect the tannin concentration of decomposing litter at the Pineville site. Sources of litter differed significantly in their concentration of tannins (Table 28.10). Again, this response was observed at the Auburn site only. Responses to climate and litter source were caused primarily by significant effects on recalcitrant, rather than soluble tannins. Although statisti-

Table 28.10. Concentration of Soluble, Nonsoluble, and Total Tannins in Litterbags at Day 387 Compared Between Two Sources at Two Locations

	Auburn Site			
	Source of Tannins, g/kg[1]			
Source	Methanol-soluble	Acetone-soluble	Nonsoluble	Total
Pineville	4.98 a	4.74 a	17.54 a	27.27 a
Auburn	2.15 a	2.34 a	11.55 b	16.04 b
	Pineville Site			
	Source of Tannins, g/kg[1]			
Source	Methanol-soluble	Acetone-soluble	Nonsoluble	Total
Pineville	1.59 a	1.98 a	10.60 a	14.10 a
Auburn	1.19 a	1.50 a	7.15 a	9.98 a

[1] Significant differences between two sources at each site are noted by different letters using Duncan's New Multiple Range Test ($P < 0.10$).

Table 28.11. Concentration of Arthropods[1] in the Litter of a Loblolly Pine Stand Located Near Auburn, Alabama, Compared Among Four Climate Treatments at Two Collection Periods

October				December			
T1P1	T2P1	T2P2	T2P3	T1P1	T2P1	T2P2	T2P3
0.26	0.34	1.31	1.00	0.56	0.79	0.26	0.35
P values[2]				P values[2]			
T1P1 vs T2P1 = 0.85				T1P1 vs T2P1 = 0.91			
T1P1 vs T2P2 = 0.03				T1P1 vs T2P2 = 0.06			
T1P1 vs T2P3 = 0.0001				T1P1 vs T2P3 = 0.31			
T2P1 vs T2P2 = 0.03				T2P1 vs T2P2 = 0.51			
T2P1 vs T2P3 = 0.06				T2P1 vs T2P3 = 0.79			
T2P2 vs T2P3 = 0.67				T2P2 vs T2P3 = 0.35			

[1] Concentration of arthropods = numbers/m^2.
[2] Based on the "randomization" method described by Potvin and Roff (1993).

cal comparisons could not be made, at the end of thirteen months, the tannin concentration of decomposing litter at the Pineville site was more than 40% less than that at the Auburn site.

In October, significantly more arthropods were present in litter exposed to treatments that caused elevated temperature and either reduced (T2P2) or elevated throughfall (T2P3), when compared to that of litter exposed to the ambient climate treatment (T1P1) (Table 28.11). Specifically, the concentration of arthropods was 24, 80, and 74% less in response to the ambient climate treatment (T1P1) when compared to the elevated temperature treatments with ambient, reduced, or elevated throughfall, respectively. This trend was not evident in December. Additionally, the concentration of arthropods in the October litter collection was significantly greater in the elevated temperature treatment with reduced (T2P2) and elevated (T2P3) throughfall when compared to that of the elevated temperature with ambient throughfall treatment (T2P1). In December, significantly more (115%) arthropods were found in litter exposed to an ambient climate (T1P1) than the elevated temperature with reduced throughfall treatment (T2P2). Although not significant, the concentration of arthropods in litter was generally less in December than in October. The majority of arthropods collected in this study were mites (qualitative observation).

The data quality objectives for the various parameters are shown in Table 28.12.

Table 28.12. Data Quality Objectives

Variable	Precision	Accuracy	Completeness
Litter weight	< ± 1%	< 2%	99%
Litter nutrients (N, P, Ca, K, Mg)	5% (cv)	< 5%	99%
Arthropod populations	± 12%	NA	99%
Litter quality, tannins, and so forth	15% (cv)	< 15%	99%
Temperature (microcosms)	± 1%	< 5%	99%

The arthropod counts and tannins were the most variable of the parameters measured. This is not surprising given the techniques in determining these values.

Discussion

Mass loss responded weakly to climate treatments, which may be attributable to the inherently slow decomposition associated with these ecosystems, combined with the short duration of our study. The single, significant response to climate that did occur was an increase in loss of mass caused by simultaneous elevation of temperature and reduction of precipitation at the Auburn site. Stimulation of litter decomposition in response to elevated temperature has been reported by others (Carlyle and Ba Than, 1988; De Santo et al., 1993). However, as moisture becomes limiting, litter decomposition is less responsive to elevated temperature and more dependent on moisture (De Santo et al., 1993). Thus, it is probable that the reduced throughfall treatment (i.e., rainfall) in our study was not limiting to litter decomposition.

In terms of source comparisons, the significantly greater mass loss of Auburn litter at the Auburn site may reflect exposure to microbial populations that have adapted through succession to the utilization of litter with a particular chemistry. Klemmedson et al., (1985) proposed that the indigenous litter in a stand with a specific structure may be characterized by a unique needle chemistry.

Greater mass loss of the Auburn litter may also be attributed to its litter quality when compared to the Pineville litter. Specifically, a lower concentration of tannins, a narrower N/P ratio, and a higher concentration of mineral nutrients indicate that the Auburn litter is a better microbial substrate when compared to the Pineville litter. Other studies have reported similar findings for lignin (Berg et al., 1993; Fogel and Cromack, 1977; Geng et al., 1993), mineral nutrients (Foster et al., 1980), and tannin (Hagerman and Robbins, 1987).

Although statistical comparisons could not be made, litter decomposition was more rapid at the Pineville site when compared to the Auburn site. The Pineville site received 49% more throughfall between March and July than the Auburn site, which may have accelerated litter decomposition at the Pineville site. This also may be one explanation for the more rapid loss of K from litter at Pineville when compared to that at Auburn. Early in our study, perhaps the climate at Pineville altered the litter chemistry or microflora so that a litter decomposition response to temperature at the end of the study was either absent or hidden.

Differential behavior between N and P in the decomposing litter of pine communities has been reported prior to our study (Klemmedson et al., 1985; Lockaby et al., 1995). The tendency for N and P to be immobilized and mineralized, respectively, during short-term (Phase I) decomposition is related to the availabilities of these elements in forest ecosystems. The soil at the Pineville site is inherently low in N and P; the Auburn site is low in N, but not P. Because a positive relationship exists between the C/N ratio of decomposing litter and N immobilization (Klemmedson et al., 1985), more N immobilization was expected in the Pineville litter in our study. However, we observed a stronger tendency for N immobilization in litter collected at the Auburn site when compared to that collected at the Pineville site. This response may be related to differences between the two sources in terms of initial N/P ratios (Auburn = 6.3, Pineville = 13.8). In addition to the N/P ratio, the higher tannin concentration of decomposing litter from Pineville may have affected the microbial processes driving N immobilization.

At the Auburn site, N immobilization was characteristic of the Auburn litter, but Ca and Mg accumulations occurred in the Pineville litter. An accumulation of Ca and Mg could be attributed to immobilization by decomposer organisms. However, dissimilarity between sources of litter exhibiting mineral-nutrient immobilization suggests that an artifact may be present in our data. Klemmedson et al., (1985) also reported an accumulation of Ca during early decomposition of pine litter and attributed this increase to dust and sediment contamination that was exacerbated by overland flow. In our study, throughfall applications often caused ponding in microcosms. As a result, our litterbags may have been similarly contaminated resulting in an apparent immobilization of both Ca and Mg. The observed source effects on Ca and Mg retention perhaps were caused by differences in needle morphology and integrity, and thus, a differential adherence of contaminants to the two sources of decomposing litter.

As in the case of mass, elevated temperature with reduced throughfall evoked the greatest response in elemental behavior. At the Auburn site, elevated temperature and reduced throughfall stimulated P mineralization. Furthermore, N immobilization, which is indicative of microbial activity, was maximized at Pineville under the same treatment, and at Auburn under elevated temperature with ambient throughfall. At the same time, the concentration of recalcitrant tannin in litter at the Auburn site was significantly greater in response to ambient temperature and throughfall when compared to the elevated temperature treatments. Additionally, in October, the concentration of arthropods in the forest floor was increased in two of the three elevated temperature treatments tested. These results indicate that elevated temperature may have accelerated the Phase I decomposition of juvenile loblolly pine litter.

Summary

Our study demonstrates the sensitivity of early litter decomposition processes to small shifts in the temperature of the forest floor. Litter, together with senescent roots and decaying macrofauna are the primary contributors to organic carbon in forest soils (Fisher, 1995). Moreover, organic components in the A horizon of a forest-soil account for up to 90% of the soil's cation exchange capacity, with the soil-organic matter containing nearly all mineralizable N and P (McColl and Gressel, 1995). As a result, the effects of temperature on macrofaunal activity in the forest floor, the dynamics of substrate quality, the mobilization of mineral nutrients, and the subsequent release of organic constituents into the soil solution may strongly affect the cycling of essential resources within forest ecosystems. However, the retention of K in the Pineville litter when compared to that in the Auburn litter, together with the occurrence of more tannins in Pineville litter, suggests that this positive effect of temperature on early litter decomposition may not be manifested on forest sites of poor foliage quality.

Throughfall treatments did not consistently affect litter decomposition in our study. However, variances were observed in litter decomposition occurring at the two sites in our study. These differences may be tied to precipitation patterns at these sites. Lack of an observed temperature effect on litter decomposition at Pineville may be associated with higher precipitation in spring and early summer relative to the Auburn site. Thus, positive responses of early litter decomposition on stand productivity may only be observed on sites that receive precipitation below a specific threshold.

The responsiveness of early litter decomposition to elevated temperatures suggests that the increases in temperature predicted to occur with global climate change may increase litter decomposition and thus, nutrient cycling in southern pine forests. Furthermore, such management practices as thinning could be used to regulate the rate of litter decomposition, and therefore, nutrient cycling in a forest stand (Klemmedson et al., 1985). To understand the potential effects of global change on nutrient cycling and whether we can operationally manipulate this process, further research is needed to study the possible precipitation and substrate quality constraints on litter decomposition that were observed in this study.

References

Abbott DT, Crossley, DA Jr (1982) Woody litter decomposition following clear-cutting. Ecol 63:35–42.
Berg B, McClaugherty C, Johansson MB (1993) Litter mass-loss rates in late stages of decomposition at some climatically and nutritionally different pine sites. Long-term decomposition in a Scots pine forest. VIII. Can J Bot 71:680–692.
Bretherton FP, Bryan K, Woods JD (1990) Time-dependent greenhouse-gas-induced climate change. In *Climate change: The IPCC scientific assessment.* Cambridge University Press, Cambridge, England.
Carlyle JC, Ba Than U (1988) Abiotic controls of soil respiration beneath an eighteen-year-old *Pinus radiata* stand in southeastern Australia. J Ecol 76:654–662.
Christiansen TA, Perry WB, Perry SA (1993) Effect of temperature and moisture on forest floor invertebrate dynamics, microbial biomass, and leaf litter decomposition in six hardwood forest catchments in West Virginia. In *Abstracts of joint meeting of northern and southern global climate change programs.* June 8–9, 1993, Washington, DC.
Cooter EL, Eder BK, LeDuc SK, Truppi L (1993) General circulation model output for forest climate change research and applications. USDA For Serv Southeast For Exp Stn Gen Tech Rep SE-85.
De Santo AV, Berg B, Rutigliano FA, Alfani A, Fioretto A (1993) Factors regulating early-stage decomposition of needle litters in five different coniferous forests. Soil Biol Biochem 25:1423–1433.
Fernandez IJ, Rustad LE, Briggs R, Simmons J (1993) The Howland integrated forest study (HIFS): Nutrient cycling and climate change effects research. In *Abstracts of joint*

meeting of the northern and southern global climate change programs. June 8–9, 1993, Washington, DC.

Field CB, Chapin FS, Matson PA, Mooney HA (1992) Responses of terrestrial ecosystems to the changing atmosphere: A resource-based approach. Ann Rev Ecol & System 23:201–236.

Fisher RF (1995) Soil organic matter: Clue or conundrum? In McFee WW, Kelly JM (Eds) *Carbon forms and functions in forest soils.* Soil Science Society of America, Inc., Madison, WI.

Fogel R, Cromack K Jr (1977) Effect of habitat and substrate quality on Douglas fir litter decomposition in western Oregon. Can J Bot 55:1632–1640.

Foster NW, Beauchamp EG, Corke CT (1980) Microbial activity in a *Pinus banksiana* Lamb. forest floor amended with nitrogen and carbon. Can J Soil Sci 60:199–209.

Geng X, Pastor J, Dewey B (1993) Decay and nitrogen dynamics of litter from disjunct, congeneric tree species in old-growth stands in northeastern China and Wisconsin. Can J Bot 71:693–699.

Gist CS, Crossley DA (1975) The litter arthropod community in a southern Appalachian hardwood forest: Numbers, biomass and mineral element content. Amer Mid Natur 93:107–122.

Hagerman AE, Robbins CT (1987) Implications of soluble tannin-protein complexes for tannin analysis and plant defense mechanisms. J Chem Ecol 3:1243–1259.

Hansen J, Johnson D, Lacis A, Lebedeff S, Lee P, Rind D, Russell G (1981) Climatic impact of increasing carbon dioxide. Science 213:957–966.

Harte J, Torn MS, Chang FR, Feifarek B, Kinzig AP, Shaw R, Shien K (1995) Global warming and soil microclimate: Results from a meddow-warming experiment. Ecol Appl 5 132–150.

Hornsby DC, Lockaby BG, Chappelka AH (1995) Influence of microclimate on decomposition in loblolly pine stands: A field microcosm approach. Can J For Res 25:1570–1577.

Jansson PE, Berg B (1985) Temporal variation of litter decomposition in relation to simulated soil climate. Long-term decomposition in a Scots pine forest. VIII. Can J Bot 63:1008–1016.

Kelly JM, Beauchamp JJ (1987) Mass loss and nutrient changes in decomposing upland oak and mesic mixed-hardwood leaf litter. Soil Sci Soc Amer J 51:1616–1622.

Kerr A Jr, Griffis BJ, Powell JW, Edwards JP, Venson RL, Long JK, Kilpatrick WW (1980) Soil survey of Rapides Parish Louisiana. US Dep Agri, Soil Conserv Ser For Ser in cooperation with LA State Univ, LA Agri Exp Stat.

Klemmedson JO, Meier CE, Campbell RE (1985) Needle decomposition and nutrient release in ponderosa pine ecosystems. For Sci 31:647–660.

Lockaby BG, Miller JM, Clawson RG (1995) Influences of community composition on biogeochemistry of loblolly pine systems. Amer Mid Nat 134:176–184.

McColl JG, Gressel N (1995) Forest soil organic matter: Characterization and modern methods of analysis. In McFee WW, Kelly JM (Eds) *Carbon forms and functions in forest soils.* Soil Science Society of America, Inc., Madison, WI.

McNutt RB (1981) Soil survey of Lee County, Alabama. US Dep Agri, Soil Conserv Ser in cooperation with the AL Agri Exp Stat and the AL Dep Agri Indus.

Mooney HA, Vitousek PM, Matson PA (1987) Exchange of materials between terrestrial ecosystems and the atmosphere. Science 238:926–932.

National Science Foundation (1992) Soil warming experiments in global change research: The report of a workshop held in Woods Hole MA, 27–28 Sept 1991. NSF Ecosystem Studies Prog., Washington, DC.

Perry DA (1994) *Forest Ecosystems.* Johns Hopkins University Press, Baltimore, MD.

Peterjohn W, Melillo JM, Bowles FP, Steudler PA (1993) Soil warming and trace gas fluxes: Experimental design and preliminary flux results. Oecologia 93:18–21.

Potvin C, Roff DA (1993) Distribution-free and robust statistical methods: Viable alternatives to parametric statistics. Ecol 74:1617–1628.

Rai B, Srivastava AK (1982) Microbial decomposition of leaf litter as influenced by fertilizers. Plant Soil 66:195–204.

SAS (1985) SAS User's Guide: Statistics. SAS Institute, Cary, NC.

Schneider SH (1989) The greenhouse effect: Science and policy. Science 243:771–781.

Schneider SH (1994) Detecting climatic change signals: Are there any "fingerprints?" Science 263:341–347.

Swift MJ, Heal VW, Anderson JM (1979) *Decomposition in terrestrial ecosystems.* University of California Press, Berkeley.

Taylor BR, Parkinson D (1988) Respiration and mass loss rates of aspen and pine leaf litter decomposing in laboratory microcosms. Can J Bot 66:1948–1959.

Tiarks AE, Meier CE, Flagler RB, Steynberg EC (1992) Sequential extraction of condensed tannins from pine litter at different stages of decomposition. In Hemingway RW, Laks PE (Eds) *Plant polyphenols.* Plenum Press, New York.

Van Cleve KW, Oechel WC, Hom JL (1990) Response of black spruce (*Picea mariana*) ecosystems to soil temperature modifications in interior Alaska. Can J For Res 20:1530–1535.

Woodward FI (1992) Predicting plant responses to global environmental change. New Phytol 122:239–251.

29. Soil Organic Matter and Soil Productivity: Searching for the Missing Link

Felipe G. Sanchez

Soil-organic matter (SOM) is a complex array of components including soil fauna and flora at different stages of decomposition (Berg et al., 1982). Its concentration in soils can vary from 0.5% in mineral soils to almost 100% in peat soils (Brady, 1974). Organic matter (OM) in the surface mineral soil is considered a major determinant of forest ecosystem productivity because it affects water retention, soil structure, and nutrient cycling (Powers et al., 1990; Paul 1991). Soil-organic matter is the major source of nitrogen available to plants and contains as much as 65% of the total soil phosphorus (Bauer and Black, 1994).

During decomposition, OM is broken down into various components including carbohydrates, amino acids, proteins, nucleic acids, lipids, tannins, lignin, and humus. Humus refers to the microbial resistant forms of SOM that remain after major portions of added plant and animal residues have decomposed. Its chemical nature has not been completely determined but it is believed to consist of various polymeric compounds, probably aromatic and aliphatic in composition (Schnitzer and Schulten, 1992). Humus has been found to affect the physical properties of soil (Elliot, 1986; Beare et al., 1994). The nonhumus components of SOM are crucial in nutrient cycling dynamics and are the primary source of food and energy for soil microorganisms (Cambardella and Elliot, 1992; Wander et al., 1994). As such, these components are the driving force for productivity.

The input into the OM pool in forest soils comes mainly from forest litter and fine root mortality, exudates, and sloughing (Vogt et al., 1986). Estimates of the net annual carbon (C) budget of a tree relegated to root growth and maintenance are disputed, but the range is from 20 to 65% (Persson, 1979). Global change could significantly alter the supplies of nutrients and water to the root systems, C-allocation patterns within and potentially between plants and, in this way, plants will be affected (Friend, 1988; Johnson, 1990; Wilson, 1988). Changes in the amount of carbon allocated to the root system will, in turn, affect the annual amount of fine root material entering the soil.

Nationally, efforts are underway to monitor forest ecosystem health and productivity. Soil parameters, including total SOM, account for many of the site quality indicators being collected. The recommended analytical procedure is to report SOM values in units of total elemental C (Nelson and Sommers, 1986). The feasibility of using total SOM as an indicator of soil productivity is problematic because of large seasonal fluctuations in its value. Haines and Cleveland (1981) showed that the seasonal variability of SOM concentration for some Georgia soils could be more than 100%. Nutrient turnover rates are generally more relevant to forest productivity than is the total amount of SOM (Cole and Rapp, 1981) and this is consistently reflected in equations for predicting nitrogen (N) fertilization responses and N uptake. Variation in the factors regulating decomposition (C/N ratio, moisture, temperature) are often more closely related to standard measures of productivity than is the total amount of SOM (Edmonds and Hsiang, 1987; Binkley and Hart, 1987). Total SOM on a site is, at best, a key to long-term productivity potential or carrying capacity. What may be more important to short-term (i.e., annual) productivity is the dynamic relationship between the labile fraction of the SOM pool and plant growth. It is this labile fraction that is most sensitive to environmental change and variations in this fraction will have the most immediate impact on forest ecosystem productivity (Ruark and Blake, 1991; Wander et al., 1994; Eswaran et al., 1995).

Various models describing C and N dynamics have been reported in the literature. Some describe SOM as a dynamic system spanning a continuum of characteristics and qualities (Bosatta and Ågren, 1985; Ågren and Bosatta, 1987; Janssen, 1984). Others fractionate SOM into pools based on C accessibility to microorganisms (Jenkinson and Rayner, 1977; Parton et al., 1988). Forest productivity may be sensitive to climate variations that result in alteration of SOM decomposition rates, particularly those associated with the highly labile fractions that direct nutrient cycling and therefore, productivity (Ruark and Blake, 1991; Wander et al., 1994). To date, direct measurement and characterization of this labile SOM fraction have been elusive.

Soil-Organic Matter Characteristics

Relating forest productivity to levels of total SOM ignores the separate contributions that different SOM components have in the soil. For example, it has been shown that the carbohydrate fraction of SOM significantly increases soil aggregate stability and water infiltration (Cheshire, 1979; Oades, 1984; Roberson et al., 1991). Sometimes these changes can be subtle as in situations in which changes in land-management practices have induced improvements in soil structure without changing total organic C or total carbohydrate content (Hamblin and Davies, 1977; Baldock et al., 1987). In these situations, it has been shown that not all the carbohydrate fractions contributed to the improved soil structure. Instead it was shown that only soil carbohydrates originating from microbial activity have a positive correlation with the improved soil structure (Roberson et al., 1995).

The labile fraction comprises plant and animal residues at various stages of decomposition. The size of this fraction is regulated by such external factors as litter input, microbe population, and climate (Bosatta and Ågren, 1985; Ågren and Bosatta, 1987). Its high variability throughout the year accounts for most of the seasonal variability of total soil C at a particular site. The annual dynamics of this fraction are hypothesized to positively correlate with SOM decomposition rates and, consequently, nutrient-cycling rates (Ruark and Blake, 1991; Wander et al., 1994). This fraction is believed to be directly affected by changes in C inputs and temperature fluxes induced by global change (Eswaran et al., 1995).

Past SOM Research

Past attempts to partition SOM into components that are biologically meaningful have been hindered by the various extraction methods used. The most prevalent method for separating SOM pools is with sodium hydroxide (NaOH) (Calderoni and Schnitzer, 1984). This procedure fractionates SOM into humic acids (HA), fulvic acids (FA), and humin. Humic acids are the SOM fractions that are only soluble in water at alkaline pH levels. Fulvic acids are soluble in water at both alkaline and acidic pHs but not at neutral pH. Humin is insoluble in water at all pH levels. This procedure, however, has come under much criticism in the literature. The primary concern is that NaOH can react with many organic compounds and may induce such sample modifications as polymerization, autoxidation, and hydrolysis (Stevenson, 1982). These chemical modifications create a series of compounds that are not in the soil and root environment and are not clear indicators of C-forms available in the soil. Another major concern is that the subdivision of SOM into HA, FA, and humin is arbitrary and bears no relationship to soil productivity. These criticisms have prompted development of other means of analyzing SOM that do not alter the chemical integrity in SOM; these results could then be related directly to relevant soil factors, particularly soil productivity.

Nondestructive methods have been used for the analysis of SOM. One such method is Fourier transform infrared spectroscopy (FTIR). This method has been used to obtain information on specific classes of compounds in the water soluble fraction (Sposito et al., 1976; Stevenson, 1982; Baes and Bloom, 1989). Unfortunately, the diversity of chemical functional groups present in SOM results in featureless spectra that provide little information on specific functional groups. However, some insight has been obtained by using the ratio of the absorption at two wavelengths, 465 nm and 665 nm, in the visible spectrum. This ratio, abbreviated as the ratio E_4/E_6, has been found to provide information on the chemical functional group character of SOM in solution. Chen et al., (1977) found that the E_4/E_6 ratio could be correlated to the molecular weight, the oxygen (O), C, and carboxylic acid (CO_2H) content, and the total acidity of the SOM. They found that small compounds with higher total acidity and higher CO_2H content had a larger E_4/E_6 ratio. Although the analysis is nondestructive and did not involve sample modification, this method only covers a small portion of the SOM pool (the water soluble fractions), and therefore, has apparent limited value in relating SOM and productivity.

One of the most promising techniques for SOM research has been nuclear magnetic resonance (NMR). Solid-state-cross-polarization–magic angle spinning ^{13}C NMR (CPMAS ^{13}C NMR) has been used to investigate decomposition processes. The CPMAS ^{13}C NMR spectra for decomposing litter showed that carbohydrates and proteins are rapidly degraded, lignin is slowly oxidized, and there is a relative increase of alkyl structures (Baldock et al., 1992; Baldock and Preston, 1995). Some oxidized lignin molecules are incorporated into humus, some are leached, and the rest are decomposed into stable subunits. The advantage of this technique is that the unsieved soil can be used in the analysis. Unfortunately, this procedure is limited by the SOM concentration in the soil and, hence, the abundance of the ^{13}C isotope. Naturally occurring ^{13}C represents only 1.1% of the total carbon concentration of a sample. This becomes particularly troublesome in the mineral soils of the southern United States where the carbon concentrations are usually low, typically ranging from 0.5% to 5%. Additionally, the presence of ferromagnetic species can degrade the homogeneity of the magnetic field and cause line broadening. In these situations, the soil must be treated to remove ferromagnetic species and concentrate the SOM. Even under "ideal" conditions (i.e., sufficient SOM concentration, no ferromagnetic species), substantial signal overlap in the spectrum occurs because of the complexity of the sample mixture. This makes it difficult to detect changes in organic functional groups resulting from microbial degradation or external factors (e.g., exposure to increased carbon dioxide (CO_2), temperature fluxes, moisture fluxes, and so forth). This technique only provides qualitative information about changes in SOM. It does not provide insight into how the changes in SOM occur or how the different SOM components relate to soil productivity.

More recently, supercritical fluid extraction (SFE) has emerged as a promising means of investigating SOM dynamics. The advantage of SFE over conventional extraction techniques is that it is an efficient method that is chemically inert (Richards and Campbell, 1991). However, this technique has been limited by the fact that the commonly used SFE solvents (i.e., CO_2, nitrous oxide (N_2O)) will only extract essentially nonpolar compounds. The range of the materials extracted can be extended by adding small amounts of a cosolvent that has specific capabilities (i.e., hydrogen bonding, dipole moment, and so forth) that the primary solvent does not. Binding of SOM in soils is primarily ionic, and thus limits the effectiveness of this technique for SOM research. Nevertheless, it has been successfully used to extract long-chain alkanes, alkenes, saturated and unsaturated fatty acids, alcohols, ketones, and alkyl esters from soils (Schulten and Schnitzer, 1991). In addition to CO_2 and N_2O, researchers have used n-pentane, ethanol, and water as solvents (Schnitzer et al., 1986; Schnitzer and Preston, 1987; Schnitzer et al., 1992) to extract varying amounts of aliphatic and aromatic species. These extractions were done to characterize SOM chemical composition and no attempt was done to relate this information to soil productivity.

Current Research

To date, no consistent quantitative correlation has been made between total SOM and forest productivity. This has prompted researchers to shift from investigating the relationship between total SOM and productivity, to devising means to fractionate SOM into biologically meaningful pools. Presently, efforts are directed at isolating the labile from the recalcitrant SOM components. Labile SOM fractions are theorized to be correlated to nutrient-cycling dynamics (Ruark and Blake, 1991; Wander et al., 1994). Because prior linkages have been made between nutrient cycling and productivity (Cole and Rapp, 1981), a link between labile SOM and nutrient-cycling dynamics would be useful in predicting potential productivity changes. This, in turn, could guide the development and use of land-management practices.

Several methods have been devised for the extraction and characterization of labile SOM while maintaining the chemical integrity of the extracted SOM. Complementary methods have been developed based on SOM particle size and density fractionation. These methods physically divide SOM into pools differing in structure and biological function (Christensen, 1992). Size fractionation studies have shown that SOM in the sand-size fractions (> 53 μm) are more labile than the SOM in the clay- and silt-size fractions (Tiessen and Stewart, 1983; Gregorich et al., 1988). Density fractionation methods have also shown that during humification, SOM increases in density through its associations with soil minerals. Thus, SOM can be density fractionated into a light fraction, which consists of mineral-free organic matter, and a heavy fraction, which is composed of SOM adsorbed on aggregate surfaces and sequestered within organo-mineral aggregates (Strickland and Sollins, 1987). Various methods have been developed that combine these complementary particle size and density fractionation methods (Strickland and Sollins, 1987; Cambardella and Elliot, 1992; Meijboom et al., 1995). Although the degeneracy of the fractionated material has been verified (Buyanovsky et al., 1994; Cambardella and Elliot, 1994), no clear correlation between these fractions and soil productivity has been achieved.

Dissolved organic matter (DOM) is another measurement designed to investigate the labile SOM fraction. The basic tenet of this metric is that the nutrients that are immediately available to plants and microbes are contained in the soil solution (Ellert and Gregorich, 1995). In contrast, solid OM is viewed as less available to plants and microbes. However, information on the relationship between DOM and

C-mineralization is inconclusive. Although some researchers have shown a strong correlation between DOM and C mineralization (Seto and Yanagiya, 1983; Davidson et al., 1987), others have found a weak or nonexistent correlation between the two (Linn and Doran, 1984; Walters and Joergensen, 1991; Cook and Allen, 1992). At present, no clear correlation between DOM and soil productivity has been made.

As described earlier, supercritical fluids (SCF) have been used for the extraction of organics from soils. In prior attempts to analyze SOM through SFE (Schnitzer et al., 1986; Schulten and Schnitzer, 1990; Capriel et al., 1990; Spiteller, 1985), only bulk extractions with no fractionation were achieved. Even so, Spiteller (1985) achieved four to five times better OM recoveries than with conventional NaOH extraction. Recent advances in instrumentation and the introduction of within-matrix derivatization techniques for the extraction of polar and ionic compounds (Miller et al., 1991) allow for considerable improvements in extracting SOM. Sanchez and Ruark (1995) developed a SOM extraction method that used supercritical Freon-22™ ($CHClF_2$) as a solvent. At a prior date, this solvent was shown to be superior to CO_2 and N_2O for the extraction of organic materials from soils (Hawthorne et al., 1992). The extract obtained with this method was composed of low molecular weight materials ($<$ 900 Daltons) that had significant polar character (Sanchez and Ruark, 1995; Sanchez and Bursey, 1996, in review). Additionally, the high quality and degeneracy of the extract were shown by its low C/N ratio (18) as compared with the unextracted soil (35) Sanchez and Ruark (1995) correlated the annual dynamics of the labile SOM fraction and the annual biomass production of two soils. As it is relatively new, this method will require extensive testing before definitive statements can be made concerning linkages between the SFE extract and soil productivity.

Comparisons Between Methods

Identifying changes in the factors that drive productivity is important to understanding and predicting the impacts of global change on forested ecosystems. Global change induced variations in the labile SOM may provide indications of concurrent or future changes in forest productivity. The question then arises as to which labile SOM pool measurement should be used. Although all three general fractionation methods (i.e., particle and density fractionation, DOM, and SFE) attempt to fractionate labile and recalcitrant SOM fractions, there are significant differences between the methods.

Particle size and density fractionation results in the removal of recognizable plant fragments. These fragments decompose quickly but have wide C/N ratios and immobilized N (Sollins et al., 1984; Christensen, 1992). Also, the percentage of the total SOM concentration contained in these fractions can be highly variable (3 to 50%) (Ellert and Gregorich, 1995).

The DOM fraction is decomposed material, with a low molecular weight range and high level of polarity. Inorganic carbon is also contained in this fraction. Dissolved organic matter represents only a very small percentage (0.04 to 0.39%) of the total C (Ellert and Gregorich, 1995). Attempts to correlate DOM to productivity have been both variable and conflicting.

Similarly to DOM fractionation, supercritical Freon-22™ removes decomposed material. The molecular weight range of the extract is low, as compared with the whole soil, but is probably larger than the DOM molecular weight range. In addition to the polar and ionic compounds in the DOM fraction, the materials extracted by SFE contain components that are not water soluble. In our laboratory we found that the extract could represent up to 60% of the total soil C (Sanchez and Ruark, 1995). Because DOM represents a very small percentage of the total SOM, most of the extract obtained by SFE is not water soluble. The low C/N ratio of the SFE extract suggests that these components are susceptible to microbial degradation. Although this procedure shows promise in separating labile and recalcitrant SOM fractions, it is a new technique that has not been tested in other laboratories.

It is important to note that none of the methods have demonstrated a strong correlation to soil productivity. The missing link has yet to be discovered. Nevertheless, investigations concentrating on the labile SOM fraction presently appear to be our best effort at finding this missing link.

Carbon Sequestration

In addition to regulating forest ecosystem productivity, SOM dynamics are critical to carbon sequestering. The environmental constraints on sequestering in a given soil system are not well understood. Evidence exists suggesting that there may be a negative feedback between SOM levels and carbon allocation to plant root systems that limit carbon storage in the soil (Ruark and Blake, 1991). Significant increases in the recalcitrant SOM fractions may suggest the potential for carbon sequestration. Recalcitrant SOM fractions are highly protected because of their associations with soil minerals and because of their chemical stability (via their presumed high aromatic and aliphatic character); therefore, there is little annual variation in either their size or identity. Because this fraction has been linked to the physical properties of the soil (Elliot, 1986; Beare et al., 1994), fluxes in the size or chemical identity of this fraction may result in significant long-term changes in the soil.

Individual trees establish a long-lived network of primary, secondary, and tertiary roots Transient crops of fine roots that turnover each year are attached to this main root system. As such, a concentration of labile SOM can be expected to build up adjacently to the large root system. A 38% increase in total soil C was found laterally within 30 mm of a permanent secondary root system; this increase was from a first rotation forty-year-old loblolly pine (*Pinus taeda* L.) stand grow-

Table 29.1. Soil-Carbon Level Within 3.0 cm Distances of Second-Order Laterals vs Levels in the 3.0 to 10.0 cm Distance Ranges

Distance (cm)	N	Mean† (%C)	Standard Error (%C)
0–3.0	32	0.50	0.03
3.0–10.0	64	0.37	0.01

†Means differ statistically at $P < 0.0001$ level

ing on an abandoned farm site at the Department of Energy's Savannah River Site (DOE–SRS) in Aiken, South Carolina (Table 29.1). This increase was at a depth of 10 cm in the Ap horizon, where prior agricultural practices probably left a homogeneous plow horizon. Ruark and Blake (1991) termed this area of fine root turnover the "reoccurring rhizosphere" and considered the increased SOM content to be the result of fine root activity over the length of the rotation. The large spatial variability in total SOM and N that occurs in first generation pine stands within the priorly homogenized Ap layer of abandoned farm land suggests that the variability was heightened by plant activity (Ruark and Zarnoch, 1992). Because OM input from the forest floor in these stands is predominantly in solution and from a single, dominant tree species, it should uniformly influence the mineral soil.

The SFE procedure developed by Sanchez and Ruark (1995) was used to investigate the carbon sequestration potential of soils in the southeastern United States. The study was conducted on three soil types that span an available moisture and texture gradient at the DOE–SRS. The soils were in the Lakeland (thermic, coated Typic quartzipsamment), Fuquay (loamy, siliceous, thermic Typic arenic plinthic paleudult), and Orangeburg (fine, loamy, siliceous, thermic Typic paleudult) series. A complete description of the analysis is presented elsewhere, but briefly, the existence of a reoccurring rhizosphere region for the Lakeland and Fuquay soils was confirmed. The increased SOM concentration in this area was greatest for first-order roots and declined going from secondary to tertiary roots. The carbon increase in the reoccurring rhizosphere region for the first order roots in the Lakeland soil was substantial (55% increase). Lakeland soils exhibited the largest detectable C buildup for each of the root orders. No reoccurring rhizosphere region was detected for the Orangeburg soil. It is possible that the finer texture and better water retention of the Orangeburg soil (sandy loam) discouraged fine root extension. If this is the case, it is possible that this soil has a more compact reoccurring rhizosphere region.

Supercritical fluid extraction of the soils from the reoccurring rhizosphere region for the three soil series showed that the additional carbon was soluble. This would suggest that there is not a significant amount of C sequestration in the recalcitrant SOM fractions occurring at these sites during the forty years that they have been established. However, spectral (^{13}C NMR) comparisons of the extracts obtained in and away from the reoccurring rhizosphere for the Lakeland and Fuquay soils showed subtle differences in the two regions. Extracts from the reoccurring rhizosphere region for each soil had a higher relative proportion of chemically stable compounds (e.g., aliphatics, aromatics) than did extracts away from this region. This may suggest that the potential exists for long-term carbon storage in these soils with continual carbon additions.

Potential Change Scenarios

Burning of fossil fuels and clearing of forests have resulted in an elevation of atmospheric CO_2 concentration by 25% since the industrial revolution began and it continues to rise at an approximate rate of 0.5% yr^{-1} (Houghton et al., 1990, 1992). The increased level of of CO_2 and other gases (i.e., methane, nitrous oxide, and others) may result in eventual surface warming because of the anticipated "greenhouse effect" (Rosenzweig, 1994). Several researchers have attempted to model the effects of increased CO_2 or surface temperature on forest ecosystems and have obtained significant variations in their results because of species differences, nutrient status, and air pollution inputs. However, there is a general trend for increasing litterfall, increasing decomposition rate, and decreasing SOM with increasing mean annual temperature (Johnson, 1995).

Most of the research on belowground responses has been conducted as experiments using container-grown seedlings (Norby et al., 1995). These research efforts have resulted in conflicting conclusions. For example, although several researchers have concluded that nutrient deficiency does not preclude growth responses to elevated CO_2 levels, other researchers have reached the opposite result (Norby et al., 1995). It is possible that the different conclusions obtained are the result of differences in tree species used. Despite the conclusions reached, it is a major leap to extrapolate any results obtained from container-grown seedlings to whole forest ecosystems.

There have been attempts, albeit on a small scale, at investigating whole ecosystem responses to elevated levels of CO_2. These experiments were conducted on the arctic tussock tundra (Oechel et al., 1991), the Chesapeake Bay salt marsh (Drake, 1992), and a tallgrass prairie (Owensby et al., 1993). Unfortunately, because of the economic, technical, and biological considerations, no attempt has been made to replicate these studies on intact forests. However, there are several ongoing studies looking at the aboveground and belowground responses of individual trees or small groups of trees to elevated CO_2 (Norby et al., 1995).

A very important observation obtained from the ongoing field studies has been an alteration of the C budget of trees. Norby et al., (1992, 1993) found that the fine root density of yellow poplar and white oak saplings was greater under elevated levels of CO_2. Because fine roots are major contributors to the SOM pool, this could signal an increase in C sequestration in the soil. The probability that this material is sequestered in the soil is lessened, however, by the observation that there is also an increase in microbial activity in response to increased levels of CO_2 and surface temperature (Johnson, 1995). Research on C sequestration in soils suggests that although the additional organic material entering the soil is labile, there is a shift in its chemistry that may allow for some stabilization in the soil.

Management Options

The challenge presently facing us is the management of our forests in order to minimize the effects of global change and perhaps even use it to our advantage. Considering that forests store 20 to 100 times more C per unit area than does cropland (Houghton et al., 1983), reforestation is an obvious avenue for C storage. Unfortunately, with the world's increasing demands for agriculture and wood products, implementing a reforestation strategy may be difficult to achieve on a global scale. We must be able to optimize the use of our forest ecosystems. Achieving maximum sustainable productivity from these forests may help in attenuating the rise in atmospheric CO_2 levels while lessening the need for deforestration.

Careful management of the forest soils is essential to achieving optimal sustainable productivity of forest ecosystems. Management systems comprising appropriate harvest, regeneration, and sivicultural options can be used to approach this goal. Extensive evidence exists that shows that extreme soil disturbance results in a loss of SOM, which subsequently contributes to the elevated atmospheric CO_2 concentration (Schlesinger, 1995). Stabilizing the SOM in forests can aid in obtaining optimum sustainable productivity without exacerbating the elevated atmospheric CO_2 problem. Although oxidation of SOM contributes to atmospheric CO_2, increases in recalcitrant SOM could provide a negative feedback to global warming (Schlesinger, 1995).

A potential method of stabilizing OM in forest soils is through application of municipal biosolids. Municipal biosolids have been demonstrated to be effective in enhancing forest productivity (Beckett et al., 1977; McCaslin and O'Connor, 1982). Also important is the evidence that biosolids can sometimes promote C and nutrient stabilization in soils (Harrison et al., 1995).

Agroforestry is another alternative that is gaining popularity when there is limited available farm land. Agroforestry involves growing trees in conjunction or succession with crops to change the SOM and nutrient dynamics to increase annual crop production. Strategies include using deep-rooting trees to cycle nutrients from below the crop root zone and using N-fixing trees to minimize N loss. Compared to soils, the C pool in crops is more responsive to changes in atmospheric CO_2 levels (Schlesinger, 1995). Agroforestry systems are a potential means of achieving a balance between SOM turnover and stabilization.

References

Ågren GI, Bosatta E (1987) Theoretical analysis of the long-term dynamics of carbon and nitrogen in soils. Ecol 68:1181–1189.
Baes AU, Bloom PR (1989) Diffuse reflectance and transmission Fourier transform infrared (DRIFT) spectroscopy of humic and fulvic acids. Soil Sci Soc Am J 53:695–700.

Baldock JA, Kay BD, Schnitzer M (1987) Influence of cropping treatments on the monosaccharide content of the hydrolysates of a soil and its aggregate fractions. Can J Soil Sci 67 489–499.

Baldock JA, Oades JM, Waters AG, Peng X, Vassallo AM, Wilson MA (1992) Aspects of the chemical structure of soil organic materials as revealed by solid-state ^{13}C NMR spectroscopy. Biogeochem 16:1–42.

Baldock JA, Preston CM (1995) Chemistry of carbon decomposition processes in forests as revealed by solid-state carbon-13 nuclear magnetic resonance. In McFee WW, Kelly JM (Eds) *Carbon forms and functions in forest soils.* SSSA, Madison, WI.

Bauer A, Black AL (1994) Quantification of the effect of soil organic matter content on soil productivity. Soil Sci Soc Am J 58:185–193.

Beare MH, Hendrix PF, Coleman D (1994) Water stable aggregates and organic matter fractions in conventional and no-tillage soils. Soil Sci Soc Am J 58:777–786.

Beckett PHT, Davis RD, Milward AR, Brindley P (1977) A comparison of the effect of different sewage biosolids on young barley. Plant Soil 48:129–141.

Berg B, Hannus K, Popoff T, Theander O (1982) Changes in organic-chemical components during litter decomposition. Long-term decomposition in a Scots pine forest. I. Can J Bot 60: 310–1319.

Binkley D, Hart SC (1989) The components of nitrogen availability assessments in forest soils. Adv Soil Sci 10:57–112.

Bosatta E, Ågren GI (1985) Theoretical analysis of decomposition of heterogeneous substrates. Soil Biol Biochem 17:601–610.

Brady NC (1974) *The nature and properties of soils.* 8th Ed. MacMillan Publishing Co. Inc., New York.

Buyanovsky GA, Aslam M, Wagner GH (1994) Carbon turnover in soil physical fractions. Soil Sci Soc Am J 58:1167–1173.

Calderoni G, Schnitzer M (1984) Effects of age on the chemical structure of Paleosol humic acids and fulvic acids. Geochim et Cosmochim Acta, 48:2045–2051.

Cambardella CA, Elliot ET (1992) Particulate soil organic matter changes across a grassland cultivation sequence. Soil Sci Soc Am J 56:777–783.

Cambardella CA, Elliot ET (1994) Carbon and nitrogen dynamics of soil organic matter fractions from cultivated grassland soils. Soil Sci Soc Am J 58:123–130.

Capriel P, Beck T, Borchert H, Harter P (1990) Relationship between soil aliphatic fraction extracted with supercritical hexane, soil microbial biomass and soil aggregate stability. Soil Sci Soc Am J 54:415–420.

Chen Y, Senesi N, Schnitzer M (1977) Information provided on humic substances by E_4/E_6 ratios. Soil Sci Soc Am J 41:352–358.

Chesire MV (1979) Origins and stability of soil polysaccharides. J Soil Sci 28:1–10.

Christensen BT (1992) Physical fractionation of soil and organic matter in primary size and density separates. Adv Soil Sci 20:1–90.

Cole DW, Rapp M (1981) Elemental Cycling. p 341–409. In Riechle DE (Ed) *Dynamic properties of forest ecosystems.* Cambridge University Press, Cambridge, England.

Cook BD, Allen DL (1992) Dissolved organic carbon in old field soils: Total amounts as a measure of available resources for soil mineralization. Soil Biol Biochem 24:585–594.

Davidson EA, Galloway LF, Strand MK (1987) Assessing available carbon: Comparison of techniques across selected forest soils. Commun Soil Sci Plant Anal 18:45–64.

Drake BG (1992) A field study of the effects of elevated CO_2 on ecosystem processes in Chesapeake Bay wetland. Aust J Bot 40:579–595.

Edmonds RL, Hsiang T (1987) Forest floor and soil influences on response of Douglas-fir to urea. Soil Sci. Soc. Am. 51:1332–1337.

Ellert BH, Gregorich EG (1995) Management-induced changes in the actively cycling fractions of soil organic matter. In McFee WW, Kelly JM (Eds) *Carbon forms and functions in forest soils.* SSSA, Madison, Wisconsin.

Elliot ET (1986) Aggregate structure and carbon, nitrogen and phosphorus in native and cultivated soils. Soil Sci Soc Am J 50:627–633.

Eswaran H, Van den Berg E, Reich P, Kimble J (1995) Global soil carbon resources. In Lal R, Kimble J, Levine E, Stewart BA (Eds) *Soils and global change.* CRC Press, Inc., Boca Raton, FL.

Friend AL (1988) Nitrogen stress and fine root growth of Douglas-fir. PhD dissertation, Univ WA–Seattle.

Gregorich EG, Kachanoski RG, Voroney RP (1988) Ultrasonic dispersion of aggregates: Distribution of organic matter in size fractions. Can J Soil Sci 68:395–403.

Haines SG, Cleveland G (1981) Seasonal variation in properties of five forest soils in southwest Georgia. Soil Sci Soc Am J 45:139–143.

Hamblin AP, Davies DB (1977) Influence of organic matter on the physical properties of some East Anglican soils of high silt content. J Soil Sci 28:11–22.

Harrison RB, Henry CL, Cole DW, Xue D (1995) Long-term changes in organic matter in soils receiving applications of municipal biosolids. In McFee WW, Kelly JM (Eds) *Carbon forms and functions in forest soils.* SSSA, Madison, WI.

Hawthorne SB, Langenfeld JJ, Miller DJ, Burford MD (1992) Comparison of supercritical $CHClF_2$, N_2O, and CO_2 for the extraction of polychlorinated biphenyls and polycyclic aromatic hydrocarbons. Anal Chem 64:1614–1622.

Houghton JT, Callander BA, Varney SK (Eds) (1992) Climate change 1992: The supplementary report to the IPCC scientific assessment. World Meteorological Organization and United Nations Environment Programme. Cambridge University Press, Cambridge, England.

Houghton JT, Jenkins GJ, Ephraums JJ (Ed) (1990) Climate change: The IPCC scientific assessment. World Meteorological Organization and United Nations Environment Programme. Cambridge University Press, Cambridge, England.

Houghton R, Hobbie J, Melillo J (1983) Changes in the carbon content of terrestrial biota and soils between 1860 and 1980: A net release of CO_2 to the atmosphere. Ecol Monogr 53:235–262.

Janssen BH (1984) A simple method for calculating decomposition and accumulation of "young" soil organic matter. Plant Soil 76:297–304.

Jenkinson DS, Rayner JH (1977) The turnover of soil organic matter in some of the Rothamsted classical experiments. Soil Sci 123:298–305.

Johnson JD (1990) Dry matter partitioning in loblolly and slash pine: Effects of fertilization and irrigation. For Ecol Manage 30:147–157.

Johnson DW (1995) Role of carbon in the cycling of other nutrients in forested ecosystems. In McFee WW, Kelly JM (Eds) *Carbon forms and functions in forest soils.* SSSA, Madison, WI.

Linn DM, Doran JW (1984) Effect of water-filled pore space on carbon dioxide and nitrous oxide production in tilled and nontilled soils. Soil Sci Soc Am J 48:1267–1272.

McCaslin BD, O'Connor GA (1982) Potential fertilizer value of gamma irradiated sewage biosolids on calcareous soils. NM Agric Exp Stat Bull 692.

Miller DJ, Hawthorne SB, Langenfeld JJ (1991) SFE with chemical derivatization for the recovery of polar and ionic analytes. In Proceedings of International Symposium on Supercritical Fluid Chromatography and Extraction, January 1991, Park City, Utah.

Meijboom FW, Hassink J, Van Noordwijk M (1995) Density fractionation of soil macroorganic matter using silica suspensions. Soil Biol Biochem 27:1109–1111.

Nelson DW, Sommers LE (1986) Total carbon, organic carbon, and organic matter. In Klute A (Ed) *Methods of soil analysis part II.* SSSA, Madison, WI.

Norby RJ, Gunderson CA, Wullschleger SD, O'Neill EG, McCracken MK (1992) Productivity and compensatory responses of yellow-poplar trees in elevated CO_2. Nature (London) 357:322–324.

Norby RJ, O'Neill EG, Wullschleger SD, Gunderson CA, Nietch CT (1993) Growth enhancement of *Quercus alba* saplings by CO_2 enrichment under field conditions. Bull Ecol Soc Am 74(Suppl.):375.

Norby RJ, O'Neill EG, Wullschleger SD (1995) Belowground responses to atmospheric carbon dioxide in forests. In McFee WW Kelly JM (Eds) *Carbon forms and functions in forest soils.* SSSA, Madison, WI.

Oades JM (1984) Soil organic matter and structural stability: Mechanisms and implications for management. Plant Soil 76:319–337.

Oechel WC, Riechers G, Lawrence WT, Prudhome TT, Grulke N, Hastings SJ (1991) Longterm *in situ* manipulation and measurement of CO_2 and temperature. Func Ecol 6:86–100.

Owensby CE, Coyne PI, Auen LM (1993) Nitrogen and phosphorus dynamics of a tall-grass prairie ecosystem exposed to elevated carbon dioxide. Plant Cell Environ 16:843–850.

Paul EA (1991) Decompositon of OM. In J. Lederburg (Ed) *Encyclopedia of microbiology.* Academic Press, San Diego, CA.

Parton WJ, Stewart JWB, Cole CV (1988) Dynamics of C, N, P and S in grassland soil: A model. Biogeochem 5:109–131.

Persson H (1979) Fine root production, mortality, and decomposition in forest ecosystems. Vegetatio 41:101–109.

Powers RF, Alban DH, Miller RE, Tiarks AE, Wells CG, Avers PE, Cline RG, Fitzgerald RO, Loftis NS (1990) Sustaining site productivity in North American Forest: Problems and prospects. In Gessel SP, Lacate DS, Weetman GF, Powers RF (Eds) *Sustaining productivity of forest soils.* 7th North American Forest Soils Conference, July 24–28, 1988, Vancouver, BC, Fac For, UBC, Canada.

Richards M. Campbell RM (1991) Comparison of supercritical fluid extraction, soxhlet, and sonication methods for the determination of priority pollutants in soil. LC·GC 9:358–364.

Roberson EB, Sarig S, Firestone MK (1991) Cover crop management of polysaccharide-mediated aggregation in an orchard soil. Soil Sci Soc Am J 55:734–739.

Roberson EB, Sarig S, Shennan C, Firestone MK (1995) Nutritional management of microbial polysaccharide production and aggregation in an agricultural soil. Soil Sci Soc Am J 59:1587–1594

Rosenzweig C (1994) Agriculture in a changing global environment. In *Soil and water science: key to understanding our global environment.* SSSA Special Publication 41. SSSA, Madison WI.

Ruark GA, Blake JI (1991) Conceptual stand model of plant carbon allocation with a feedback linkage to soil organic matter maintenance. In Dyck WJ, Mees CA (Eds) *Long-term field trails to assess environmental impacts of harvesting.* Forest Research Institute, Rotorua, New Zealand, FRI Bull 161.

Ruark GA, Zarnoch SJ (1992) Soil carbon, nitrogen, and fine root biomass sampling in a pine stand. Soil Sci Soc Amer J 56:1945–1951.

Sanchez FG, Ruark GA (1995) Fractionation of soil organic matter with supercritical freon. In McFee WW, Kelly JM (Eds) *Carbon forms and functions in forest soils.* SSSA, Madison, WI.

Schlesinger WH (1995) An overview of the carbon cycle. In Lal R, Kimble J, Levine E, Stewart BA (Ed) *Soils and global change.* CRC Press, Inc., Boca Raton, FL.

Schnitzer M, Hindle CA, Meglic M (1986) Supercritical gas extraction of alkanes and alkanoic acids from soils and humic materials. Soil Sci Soc Am J 50:913–919

Schnitzer M, Preston CM (1987) Supercritical gas extraction of soil with solvents of increasing polarities. Soil Sci Soc Am J 51:639–646.

Schnitzer M, Schulten HR (1992) The analysis of soil organic matter by pyrolysis-field ionization mass spectrometry. Soil Sci Soc Am J 56:1811–1817.

Schulten HR, Schnitzer M (1990) Aliphatics in soil organic matter in fine-clay fractions. Soil Sci Soc Am J 54:98–105.

Schulten HR, Schnitzer M (1991) Supercritical carbon dioxide extraction of long-chain aliphatics from two soils. Soil Sci Soc Am J 55:1603–1611.

Seto M, Yanagiya K (1983) Rate of CO_2 evolution from soil in relation to temperature and amount of dissolved organic carbon. Jpn J Ecol 33:199–205.

Sollins P, Spycher G, Glassman CA (1984) Net nitrogen mineralization from light- and heavy-fraction forest soil organic matter. Aust J Soil Res 30:195–207.

Spiteller M (1985) Extraction of soil organic matter by supercritical fluids. Org Geochem 8:111–113.

Sposito G, Holtzclaw KM, Baham J (1976) Analytical properties of the soluble, metal complexing fractions in sludge-soil mixtures: II. Comparative structural chemistry of fulvic acid. Soil Sci Soc Am J 40:691–697.

Stevenson FJ (1982) *Humus chemistry: Genesis, composition, reactions.* Wiley–Interscience, New York.

Strickland TC, Sollins P (1987) Improved method for separating light- and heavy-fraction forest soil organic matter. Soil Biol Biochem 16:31–37.

Tiessen H, Stewart JWB (1983) Particle size-fractions and their use in studies of soil organic matter. II. Cultivation effects on organic matter composition in size fractions. Soil Sci Soc Am J 47:509–514.

Vogt KA, Grier CC, Vogt DJ (1986) Production, turnover, and nutrient dynamics of above- and belowground detritus of world forests. In Macfadyen A, Ford ED (Eds) *Advances In ecological research.* Vol. 15. Academic Press, New York.

Walters V, Joergensen RG (1991) Microbial carbon turnover in Beech forest soils at different stages of acidification. Soil Biol Biochem 23:897–902.

Wander MM, Traina SJ, Stinner BR, Peters SE (1994) Organic and conventional management effects on biologically active soil organic matter pools. Soil Sci Soc Am J 58:1130–1139.

Wilson JB (1988) A review of evidence on the control of shoot:root ratio, in relation to models. Ann of Bot 61:433–449.

30. Effects of Soil Warming on Organic Matter Decomposition and Soil-Nitrogen Cycling in a High Elevation Red Spruce Stand

J. Devereux Joslin, Mark H. Wolfe, and Charles T. Garten

Of all the ecosystems located in the southeastern United States, probably the most sensitive to climatic warming is the spruce–fir ecosystem found in the higher elevations of the Appalachian Mountains. This ecosystem has experienced significant disturbances in recent decades, including extensive mortality of mature Fraser fir (*Abies fraseri* (Pursh) Poir.) resulting from an exotic disease carried by the balsam wooly adelgid (White and Cogbill, 1992) and reductions in growth rate (McLaughlin et al., 1987), and in crown condition of red spruce (*Picea rubens* Sarg.) (Peart et al., 1992). In addition to these recent disturbances, the high elevation spruce–fir ecosystem appears to be particularly vulnerable to the effects of warming of its belowground component. Spruce–fir soils typically have very high storage pools of organic matter, containing large pools of nitrogen (N) (Fernandez, 1992; Joslin et al., 1992). Among the world's ecosystems, subalpine conifer forests rank very high with respect to the amounts of carbon stored in soil-organic matter (Post et al., 1985). The combination of low mean annual temperature, high moisture, and plant litter that is resistant to decomposition because of high lignin and low N concentrations, results in slow rates of decomposition. In such systems, rates of organic matter production have historically exceeded decomposition rates, resulting in the buildup of soil-organic matter over time.

Table 30.1 summarizes the ranges in sizes of soil-organic matter pools and soil-nitrogen pools published for different forest types in the southeastern United States. In general, temperate zone low-elevation forests, regardless of species composition, have considerably smaller pool sizes of organic matter than high-

Table 30.1. Differences Betweenn Forest Types and Regions in the Size of Soil-Organic Matter Pools and Soil-Nitrogen Pools (Mg ha^{-1})

SITE	Organic Matter	Nitrogen
Low-elevation Pine (SE)[1]	—	1.4–5.2 (7)
Low-elevation Pine-hardwood (SE)	150–200 (1)	4–5.5 (8)
Low-elevation Hardwood (SE)	150–200 (1)	4–5.5 (8)
Low-elevation Spruce–fir (NE)	40–350 (2,3,4)	1–6 (2,3,4)
High-elevation Spruce–fir (SE)	250–600 (5,6)	10–15 (5,6)

[1] SE = Southeastern US; NE = Northeastern US (References: Cole and Rapp, 1981; Lang et al., 1981; Smith et al., 1986; Fernandez et al., 1993; Wolfe, 1967; Johnson et al., 1991; Van Miegroet et al., 1992; Johnson and Henderson, 1989).

elevation spruce–fir soils in the southern Appalachians. Because belowground pools of N are almost entirely bound up in soil-organic matter, subalpine conifer ecosystems also have much larger soil pools of N than do low-elevation temperate forests (Table 30.1; Post et al., 1985; Joslin et al., 1992). In the southern Appalachians, high-elevation, spruce–fir soil-nitrogen pools range from 10 to 15 metric tons per hectare, with about 20 % found in the forest floor (Wolfe, 1967; Johnson et al., 1991).

Although natural factors have resulted in the buildup of large, soil-organic pools of carbon (C) and nitrogen (N) in high-elevation spruce–fir ecosystems, recent anthropogenic alterations in atmospheric deposition chemistry, coupled with predicted increases in global temperatures may be reversing this buildup of organic matter. Over the last five decades, atmospheric deposition rates of nitrogen have continued to increase; this increase has been particularly dramatic in high-elevation cloud forests, where many spruce–fir stands are located in the southern Appalachians. In these forests, annual rates of N deposition have increased over the past fifty years from approximately less than 10 kg/hectare (ha)/yr to their present levels of from 20 to 45 kg/ha/yr (Lovett, 1992; Joslin and Wolfe, 1992; Joslin and Wolfe, 1994). These N deposition rates are among the highest in North America and are two- to three-fold higher than those in northeastern spruce–fir ecosystems (Lovett, 1992). Because mature spruce–fir forests exist in a climate with a short growing season and cool temperatures, tree-growth rates are relatively slow and have typically been incapable of incorporating these high inputs of N into new growth.

The fact that inputs of N have exceeded plant uptake of N in recent years has certainly contributed to an increase in C/N ratios in soil-organic matter in the southern Appalachians, where rates of nitrogen mineralization, nitrification and nitrate leaching are presently quite high (Strader et al., 1989; Sasser and Binkley, 1989). As noted, soil pools of nitrogen in the southern Appalachian spruce–fir forests are typically three times the size of northeastern low-elevation spruce–fir stands. Carbon to nitrogen ratios in the mineral soil are also approximately twice as large in the Northeast (27 to 30 vs 9 to 19; Joslin et al., 1992). Net nitrogen mineralization rates tend to be higher in southern Appalachian spruce–fir stands

(Van Miegroet et al., 1992; Joslin et al., 1992); nitrification rates and soil-solution nitrate concentrations are consistently higher in the southern Appalachians (Smithson, 1990; Johnson et al., 1991; Joslin and Wolfe, 1992; Joslin et al., 1992) than in northeastern spruce–fir stands (Fernandez and Rustad, 1990; Friedland et al., 1991).

Naturally, climatic warming is expected to increase rates of soil-organic matter decomposition. To date, studies that have manipulated temperatures have consistently shown such an increase (Lockaby et al. Chapter 28; Rustad et al., 1995; Mitchell et al., 1995; Joslin and Johnson, this volume). Depending on the magnitude and rapidity of this effect, soil warming in the C- and N-rich soils of the southern Appalachian spruce–fir forests could result in considerable release of carbon dioxide into the atmosphere. Of greater concern for the health of these particularly fragile ecosystems, is the potential rapid release of large stores of nitrogen resulting in both soil acidification, aluminum (Al) mobilization, and leaching of base cations in ecosystems that are already highly acidic (Van Miegroet and Cole, 1985; Joslin et al., 1992), sensitive to Al toxicity (Raynal et al., 1990; Cronan and Grigal, 1995), and suffering from deficiencies of base cations in some cases (McLaughlin et al., 1991; Joslin and Wolfe, 1994).

In a recently published paper, we explored the impact of soil warming in a high-elevation spruce–fir stand on Whitetop Mountain in southwestern Virginia, by comparing the soil-solution chemistry of the sun side to that of the shady side in a recently formed clearing in the forest (Joslin and Wolfe, 1993). The sun side of that clearing was found to be consistently warmer (on average, + 1.2 °C) than the shade side at a soil depth of 15 cm. Concurrent with this temperature difference was a consistent elevation of the nitrate concentration of the soil solution on the sun side. This increase in nitrate concentration of 71% also resulted in a 24% increase in Al soil-solution concentration and a 30% increase in Mg leaching.

The objective of this chapter was to examine the effects of soil warming on southern Appalachian spruce–fir forest organic horizons under more controlled conditions. It was also designed to examine more closely the component processes involved in the decomposition process and in the cycling of N within the belowground component of these systems. Not only was nitrate, Al, and base cation concentrations in soil solutions examined, but also rates of organic matter decomposition, nitrogen mineralization, and nitrification. Soil temperature was continually monitored at both the 5 cm and 15 cm depths.

Methods

Site Description

The experimental stand was located in a nearly pure stand of mixed-aged red spruce on a north-facing slope (1650 m elevation) near the summit of Whitetop Mountain in southwestern Virginia (81° 34' W, 36° 35' N). Soils are Typic or Pachic haplumbrepts, with O horizons 8 to 15 cm thick (Kelly and Mays, 1989; Joslin and Wolfe, 1992). Mean annual temperature is approximately 7 °C and mean annual precipitation is approximately 1300 mm. Atmospheric deposition chemistry, soil chemistry, foliar chemistry, and soil-solution chemistry has been described in detail elsewhere (Joslin and Wolfe, 1992; Joslin and Wolfe, 1994).

Preliminary Testing

Rather than using direct solar radiation incident at the soil surface as our source of temperature difference, as in the prior "sun vs shade" study, in this effort we chose to capture solar radiation in "minigreenhouses." These minigreenhouse enclosures were constructed so that approximately half of each enclosure protruded into a small, natural canopy gap. The solar radiation captured at this end of the enclosure heated the entire space of the minigreenhouse by convective transfer, including a more "natural" shaded end. Monitoring of soil temperatures, soil-solution collections, litter decomposition measurements, and soil-N mineralization and nitrification measurements were all conducted in the shaded end of each enclosure.

Preliminary testing of the effectiveness of various approaches to constructing minigreenhouses were tested. A minigreenhouse with a surface area of 9 m^2 was found capable of achieving an average increase in air temperature of nearly 2 °C over a growing season. Although smaller than the increase we had hoped to achieve, we were encouraged in these tests by a concomitant average increase in soil temperature at 5 cm depth of 1.7 °C. At this time, we felt that such a treatment would be sufficient to generate a measurable treatment effect upon soil-organic matter decomposition and nitrogen cycling.

Experimental Treatments

In May of 1994, we constructed three minigreenhouse enclosures within the mature red spruce stand described in a prior section. Approximately 3 m^2 of the surface area of each enclosure received direct sunlight during at least five hours daily. Minigreenhouses were constructed of 3.9 cm diameter polyvinylchloride (PVC) pipe, 0.6 to 1.5 m in height, 1.2 m in width and 5 to 6 m in length. These frames supported 0.15-mm thick PVC sheeting, which served as roof and sidewalls for the enclosures so that the sides and ends of the enclosures were completely closed off. Any small seedlings, ferns, or other plants growing within the enclosed space were left relatively undisturbed. At the shaded end of each enclosure, the plastic sheeting was punctured with approximately 250 5-mm diameter holes per m^2 to allow entry of rainfall through the roof. Within 15 m of each of the three minigreenhouses, control plots were established under canopy cover comparable to the shaded end of each minigreenhouse. Enclosures were maintained from May 15 to November 2, 1994, at which time they were disassembled in anticipation of winter snowfalls.

Into both the shaded portion of each of the three enclosures and each of the three unenclosed control plots were installed 1) sets of three tension lysimeters located immediately below the Oa horizon (humus layer), 2) sets of three soil bags for measuring nitrogen mineralization and nitrification, 3) sets of four litter decomposition bags, and 4) sets of three soil temperature probes located at 5 cm, two soil temperature probes at 15 cm, and one ambient air temperature sensor. Air temperatures were monitored continually at 5 minute intervals, and hourly averages recorded on a data logger (Li-Cor 1000), from May 24 to November 2. Soil thermistors similarly recorded hourly averages at 5 cm from June 8 to November 2 and at 15 cm from June 22 to November 2.

Soil Measurements

Tension lysimeter funnels (84 mm diameter polysulfonate funnels with 0.4 micrometer pore size polycarbonate filters) were installed immediately below the O horizon. These funnels were connected via Tygon tubing to 1-l polyethylene sample bottles buried outside the plots, and designed to maintain 10 kPa (0.1 bar) of tension via a hanging water column (Riekerk and Morris, 1983). Soil solutions were collected from sample bottles biweekly from June to October and maintained under refrigeration until they could be analyzed for pH (glass electrode), nitrate, sulfate, and chloride (ion chromatography), aluminum, base cations (Ca, Mg, Na, K, NH_4) and micronutrients (Zn, Cu, Fe, Mn) by inductively coupled plasma emission spectrophotometry.

To minimize its variability, forest floor material for decomposition bags was collected on site from a single location within the stand but below the plots. Undisturbed small blocks containing on average 200 g (fresh weight; 110 g ovendry weight) of the Oi and Oe horizons were cut as units out of the forest floor and placed intact in each litter decomposition bag, which were constructed of nylon stocking material. Prior to reinsertion on site on June 7, fresh weights were determined for each bag and moisture contents estimated gravimetrically from subsamples taken form each. The bags were inserted at each plot so that the fresh litter portion was flush with the level of the existing forest floor on the same day they were collected and weighed. Litter decomposition bags were exposed to the enclosure treatments for 163 days, from June 7 until November 16. They were removed intact on March 13, 1995, after an additional 116 days of autumn and winter. Bags were transported on ice to the laboratory, where they were ovendried and weighed to determine loss of mass over the entire period of 279 days. One time during the growing season, and once again at harvest, moss growing on bag surfaces was carefully removed with forceps from each bag.

Net nitrogen mineralization in the forest floor (O horizon) was measured by in situ buried bag incubations. Sample cores of the O horizon (10 cm deep) were taken with a bulb planter from the forest floor adjacent to each enclosure and each control plot. Four replicate samples from each plot were sealed in resealable plastic bags and placed in the O horizon. Companion sample cores were taken for each incubation sample to determine initial levels of extractable ammonium and nitrate in the O horizon and the fresh mass (FM) to dry mass (DM) conversion factor for each sample. Extractions with 2 molar potassium chloride were performed in the field and transported to the laboratory for analysis. The ratio of extraction solution to sample FM was maintained at approximately 10:1. Conversion factors for FM to DM were determined by drying forest floor samples for several days at 70 °C.

Mineralization bag incubations were begun on June 7 and recovered on August 2. Upon recovery, a sample of forest floor from each bag was extracted in the field with 2 molar KCl (approximately 5:1 ratio of solution to FM). Associated samples of forest floor were taken from each bag to determine FM/DM conversion factors. Both the initial and final extracts were filtered (Whatman #42) prior to chemical analysis. Ammonium and nitrate concentrations in the extracts were determined at our laboratory by colorimetric methods with a Bran Lubde TRAACS-800 autoanalyzer. Net N mineralization for each sample was calculated as the sum of the final extractable ammonium-N + nitrate-N concentrations minus the sum of the initial extractable ammonium-N plus nitrate-N concentrations. Net nitrification potential was calculated as the final concentration of extractable nitrate-N minus the initial nitrate-N concentration. All results are reported in μg N produced per g DM.

Results

Temperature Differential Created by Mini-Greenhouses

The three greenhouse enclosures, when compared to ambient plots, created an average ambient air temperature increase of 1.24 °C over the May to October treatment period ($p < 0.001$) (Table 30.2). The temperature gradients were consistent throughout the study period and varied little from month to month, although differences declined slightly in September to October (Figure 30.1). Examined on a diurnal cycle, temperature differences between enclosures and ambient plots were minimal (generally less than 3 °C) between 0 and 10:00 and 18:00 and 24:00 hours, but large differences (up to 10 °C) developed during midday on sunny days (Figure 30.2). Cloudy days, conversely, produced minimum differences between treatments.

The enclosures created significant ($p < 0.001$) increases in soil temperatures at depths of 5 cm and 15 cm, as well (Table 30.2; Figure 30.3). On average across all three enclosure and control pairs, the differences between treatments in soil temperature at 5 cm and 15 cm were 0.68 °C and 0.52 °C, respectively. Virtually no diurnal cycle was observable, however, at either depth.

Table 30.2. Treatment Differences in Air and Soil Temperature (Two Depths): Means and Standard Errors of Paired Differences in Daily Temperature Averages for All Three Pairs of Enclosures-Control Plots Over Entire Treatment Period

Air Temperature	(°C)	1.24 (0.05)	$p < 0.001$
Soil Temperature, 5 cm	(°C)	0.68 (0.02)	$p < 0.001$
Soil Temperature, 15 cm	(°C)	0.52 (0.02)	$p > 0.001$

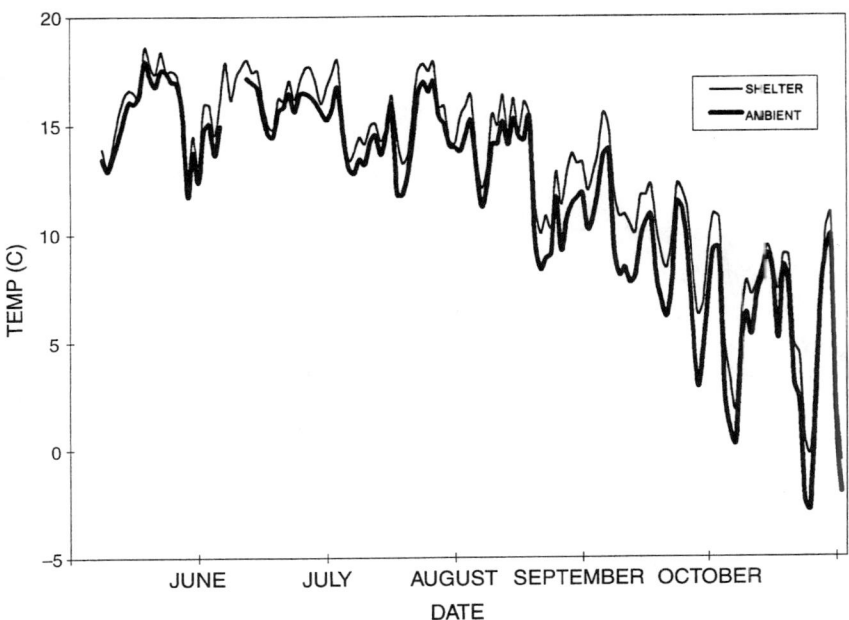

Figure 30.1. Daily mean air temperatures for the three shelters compared to daily mean ambient air temperatures in 1994.

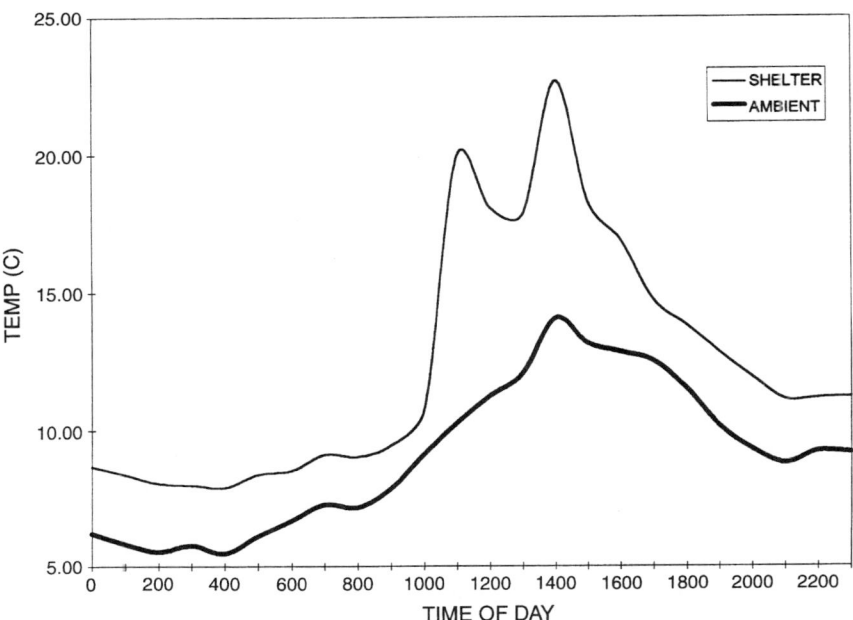

Figure 30.2. Hourly mean air temperatures for the three shelters compared to hourly mean ambient air temperatures on a sunny day in September of 1994.

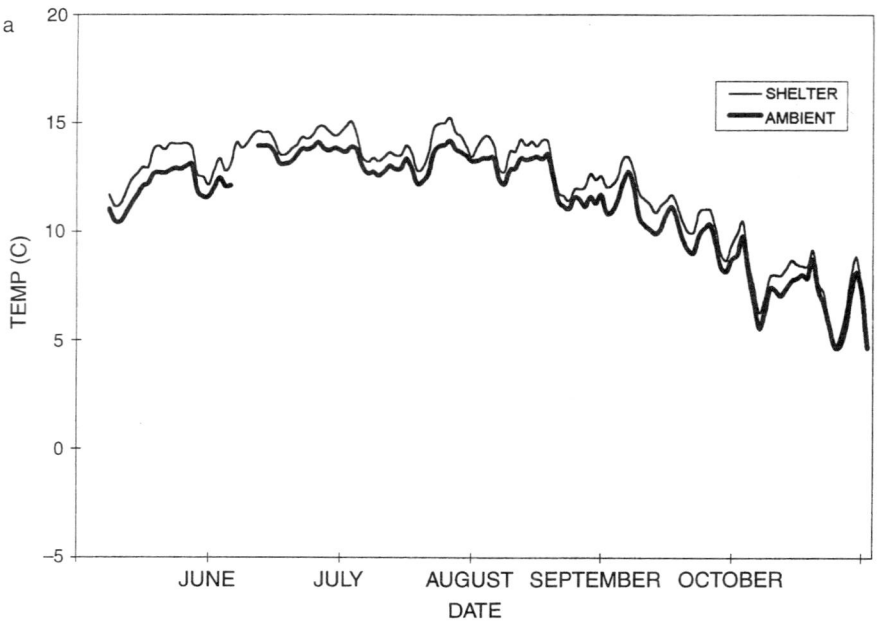

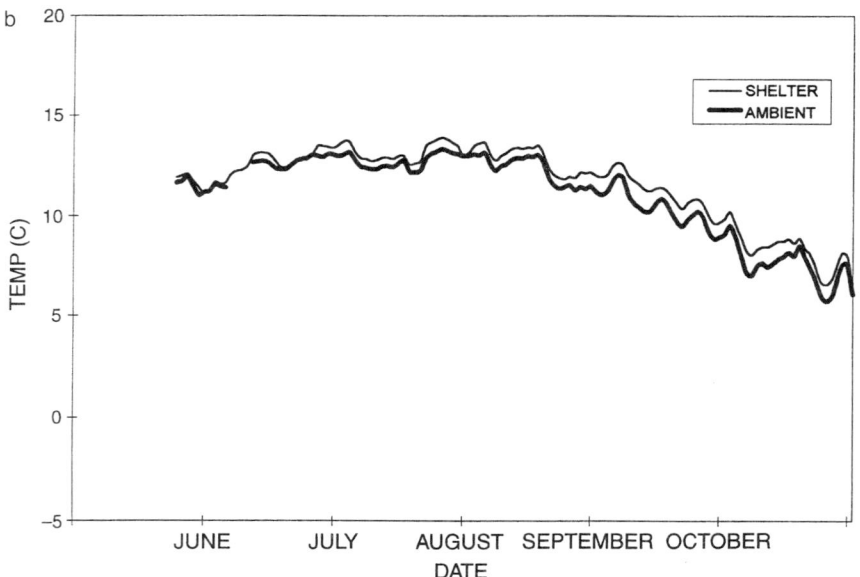

Figure 30.3. Daily mean forest floor temperatures at depths of 5 cm (a) and 15 cm (b) for the three shelters compared to the three ambient plots in 1994.

Table 30.3. Treatment Differences in Decomposition Rates (279 d), and Soil-Solution Concentrations (Volume-Weighted Means During Growing Season)

	AMBIENT	ENCLOSURES	
Decomposition rate			
Loss of mass (% per day)	0.127 (.005)[1]	0.125 (.005)[1]	ns[1]
Soil-solution Nitrate (mg/l)	3.24	2.78	ns
Soil-solution Al (mg/l)	1.36	1.67	ns
Soil-solution Ca (mg/l)	2.66	2.20	ns
Soil-solution Mg (mg/l)	0.55	0.68	ns

[1] Standard errors in parenthesis; ns = nonsignificant.

Litter Decomposition and Soil Solution Chemistry

No statistically significant differences were observed between treatments in the percentage of mass loss from the litter decomposition bags over the growing season. Actually, the percentage lost was almost identical in the two treatments (Table 30.3). Similarly, no significant differences between treatments in the volume-weighted elemental concentrations of soil solutions were observed (Table 30.3).

Nitrogen Mineralization and Nitrification

No statistically significant differences were observed between enclosures and controls (Figure 30.4). Treatment and control plots in each pair were nearly identical with respect to N mineralization. There was a trend for higher net nitrification in the enclosure treatment plots for two of the pairs, but nitrification was near zero in the third enclosure. Although there were no significant differences with treatment in any of the measures, both net N mineralization and nitrification different significantly ($p < 0.05$) with location of the treatment pair.

Discussion

One interesting result of this study is that soil temperatures at the shaded end of the minigreenhouses were raised an average of 0.68 °C at a depth of 5 cm and 0.52 °C at 15 cm as a result of raising the average air temperature in the enclosures by 1.24 C. Pretesting results indicated even greater warming of the forest floor at 5 cm with a comparable average increase in air temperature. These results occurred in soils in which the thick forest floor is generally considered a very poor conductor of heat. These findings indicate that small increases in air temperature will indeed be translated into significant impacts on forest-floor temperatures, even in well-insulated spruce–fir soils. As most recent field experiments to date that have manipulated soil temperature have applied heat directly below the surface (Rustad et al., 1995; Van Cleve et al., 1983; Peterjohn et al., 1994; Mitchell et al., 1995), this study supplies a rather unique set of information on the translation of air temperature increases to forest-floor temperature changes.

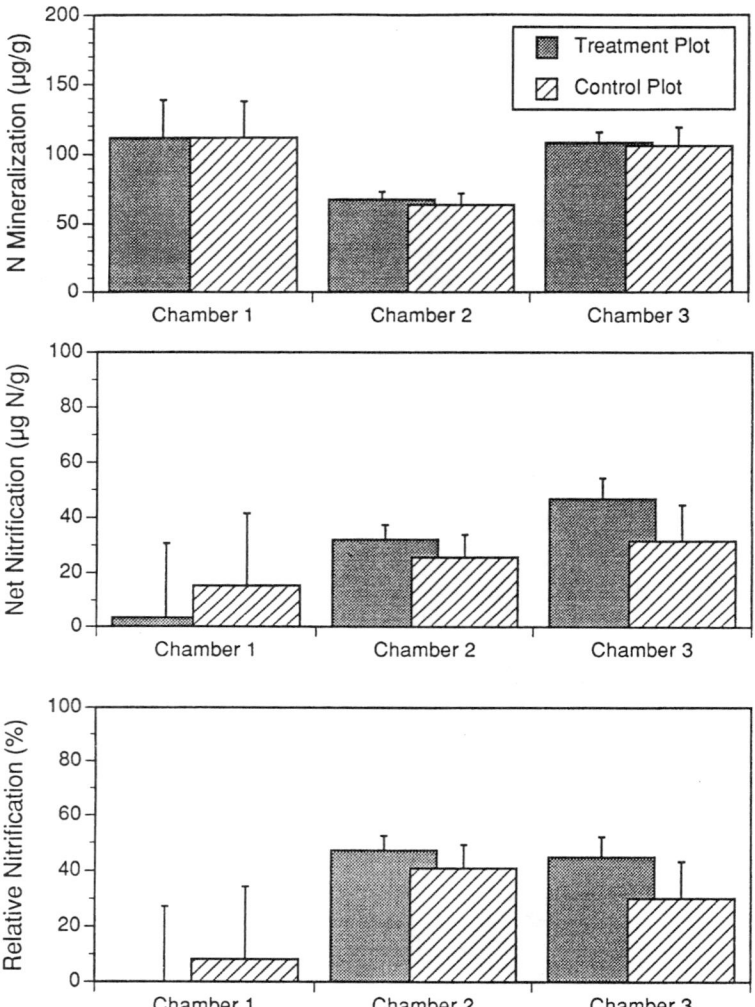

Figure 30.4. Mean (+/− SE) net N mineralization, net nitrification, and relative nitrification during eight-week period (6/7 to 8/2) within in situ incubations of forest floor samples. There were four replicates per plot.

The absence of any treatment effects on soil warming is certainly a departure from predictions. It is also a departure from prior results at this site (Joslin and Wolfe, 1993), as well as results from other soil warming studies in either spruce–fir (Rustad et al., 1995) or other ecosystems (Van Cleve et al., 1983; Peterjohn et al., 1994; Mitchell et al., 1995; Lockaby et al., Chapter 28), in which losses of organic matter or increases in nitrogen mineralization have been observed. One possible reason for the absence of any response to warming in this study was simply the size of the temperature differential we were able to achieve, which was somewhat smaller than that hoped for based upon preliminary work. Although average air temperature differences were only 1.24 °C, some acceleration of decomposition would still be expected, as the air temperature differential was somewhat comparable to that observed in Lockaby et al., (Chapter 28) (2.9 °C) and in Joslin and Wolfe (1993) (1.2 °C).

A second factor contributing to lack of significant differences was created by unexpectedly large variability within and between plots in soil-solution volumes collected, soil-solution chemistry, and soil-N availability. These parameters were considerably more variable than that experienced in prior studies at Whitetop Mountain (Joslin and Wolfe 1992, 1993), in spite of the fact that both the stand and the soils within the stand appeared to be very uniform. Large and statistically significant differences in N mineralization and nitrification rates were observed across distances of as little as 25 m between locations (Figure 30.4). Natural variability in soils was further intensified by variability within each chamber in the distribution of throughfall to the forest floor.

In this study, clearly no null hypotheses were rejected that would support hypotheses that soil warming will have dramatic impacts on high-elevation spruce–fir ecosystems. However, given the minimum temperature differential achieved in this study, the highly variable experimental conditions, the relative small sample sizes used to measure effects, and the short duration of the treatments, we feel that it is imprudent to draw any conclusions about the effects of soil warming on these ecosystems. The probability of a Type II error in which real significant differences would have been observed under more stringent treatments or more rigorous experimental conditions, remains relatively high.

References

Cole DW, Rapp M (1981) Elemental cycling in forest ecosystems. pp. 341–409. In Reichle DE (Ed) *Dynamic properties of forest ecosystems.* Cambridge University Press, London.

Cronan CS, Grigal DF (1995) Use of calcium/aluminum ratios as indicators of stress in forest ecosystems. J Environ Qual 24:209–226.

Fernandez IJ (1992) Characterization of eastern U.S. spruce–fir soils. In Eagar C, Adams MB (Eds) *Ecology and decline of Red Spruce in the eastern United States.* Springer-Verlag, New York.

Fernandez IJ, Rustad LE (1990) Soil response to S and N treatments in a northern New England low elevation coniferous forest. Water Air Soil Pollut 52:23–29.

Fernandez IJ, Rustad LE, Lawrence GB (1993) Estimating total soil mass, nutrient content, and trace metals in soils under a low-elevation spruce–fir forest. Can J Soil Sci 72:317–332.

Friedland AJ, Miller EK, Battles JJ, Thorne JF (1991) Nitrogen deposition, distribution and cycling in a subalpine spruce–fir forest in the Adirondacks, New York, U.S.A. Biogeochemistry, 14:31–55.

Johnson DW, Henderson GS (1989) Terrestrial nutrient cycling. In Reichle DE (Ed), *Dynamic properties of forest ecosystems.* Cambridge University Press, London.

Johnson DW, Van Miegroet H, Lindberg SE, Harrison RB, Todd DE (1991) Nutrient cycling in red spruce forests of the Great Smoky Mountains. Can J For Res 21:767–787.

Joslin JD, Kelly JM, Van Miegroet H (1992) Soil chemistry and nutrition of North American spruce–fir stands: evidence for recent change. J Environ Qual 21:12–30.

Joslin JD, Wolfe MH (1992) Red spruce soil solution chemistry and rot distribution across a cloud water deposition gradient. Can J For Res 22:893–904.

Joslin JD, Wolfe MH (1993) Temperature increase accelerates nitrate release from high-elevation red spruce soils. Can J For Res 23:756–759.

Joslin JD, Wolfe MH (1994) Foliar deficiencies of mature southern Appalachian red spruce determined from fertilizer trials. Soil Sci Soc Am J 58:1572–1579.

Kelly JM, Mays PA (1989) Root zone physical and chemical characteristics in southeastern spruce–fir stands. Soil Sci Soc Am J 53:1248–1255.

Lang GE, Cronan CS, Reiners WA (1981) Organic matter and major elements of the forest floors and soils in subalpine balsam fir forests. Can J For Res 11:388–399.

Lovett GM (1992) Atmospheric deposition and canopy interactions of nitrogen. In Johnson DW, Lindberg Se (Eds) *Atmospheric deposition and forest nutrient cycling*. Springer-Verlag, New York.

McLaughlin SB, Andersen CP, Hanson PJ, Tjoelker MJ, Roy WK (1991) Increased dark respiration and calcium deficiency of red spruce in relation to acidic deposition at high-elevation southern Appalachian mountain sites. Can J For Res 21:1234–1244.

McLaughlin SB, Downing DJ, Blasing TJ, Cook ER, Adams HS (1987) An analysis of climate and competition as contributors to decline of red spruce in high elevation Appalachian forests of the eastern United States. Oecologia 72:487–501.

Mitchell MJ, McHale PJ, Raynal DJ, Stehman SV, White EH, Driscoll CT, David MB, Bowles F (1995) Increasing soil temperature in a northern hardwood forest: Effects on elemental dynamics and primary productivity. Procedings 1995 Meeting of the Northern Global Change Program, Pittsburgh, PA, March 14–16, 1995. USDA For Ser General Technical Report NE-214, Northeast For Exp Stat, Radnor, PA.

Peart DR, Nicholas NS, Zedaker SM, Miller-Weeks MM, Siccama TG (1992) Condition and recent trends in high-elevation red spruce populations. pp. 125–191. In Eagar C, Adams MB (Eds) *Ecology and decline of Red Spruce in the eastern United States.* Springer-Verlag, New York.

Peterjohn WT, Melillo JM, Steudler PA, Newkirk KM, Bowles ST, Aber JD (1994) Responses of trace gas fluxes and N availability to experimentally elevated soil temperatures. Ecol Appl 4:617–625.

Post WM, Pastor J, Zinke PJ, Stangenberger AG (1985) Global patterns of soil nitrogen storage. Nature 317:613–616.

Raynal DJ, Joslin JD, Thornton FC, Schaedle M, Henderson GS (1990) Sensitivity of tree species to Al: III. red spruce and loblolly pine. J Environ Qual 19:180–187.

Riekerk H, Morris LA (1983) a constant-potential soil water sampler. Soil Sci Soc Am J 47:606–608.

Rustad LE, Fernandez IJ, Arnold S (1995) Experimental soil warming effects on C, N, and major element cycling a low elevation spruce–fir forest soil. Proceedings 1995 Meeting of the Northern Global Change PRogram, Pittsburg, PA, March 14–16, 1995. USDA Forest Service, General Technical Report NE-214, Northeast For Exper Sta, Radnor, PA.

Sasser CL, Binkley D (1989) Nitrogen mineralization in high-elevation of the Appalachians. II. Patterns with stand development in fir waves. Biogeochem 7:147–156.

Smith CT, McCormack ML Jr, Hornbeck JW, Martin CW (1986) Nutrient and biomass removals from red spruce-balsam fir whole tree harvest. Can J For Res 16:381–388.

Smithson (1990) Aluminum chemistry of forested soils subjected to acidic inputs. PhD dissertation. NC State Univ, Raleigh.

Strader RH, Binkley D, Wells CG (1989) Nitrogen mineralization in high elevation forests of the Appalachians. I. Regional patterns in southern spruce–fir forests. Biogeochem 7:131–145.

Van Cleve K, Oliver L, Schlentner R, Viereck LA, Dyrness CT (1983) Productivity and nutrient cycling in taiga forest ecosystems. Can J For Res 13:747–766.

Van Miegroet, H. and D.W. Cole. 1985. Acidification sources in red alder and Douglas fir soils—importance of nitrification. Soil Sci Soc Am J 49:1274–1279.

Van Miegroet H, Cole DW (1985) Acidification sources in red alder and Douglas fir soils—importance of nitrification. Soil Sci Soc Am J 49:1274–1279.

Van Miegroet H, Cole DW, Foster NW (1992) Nitrogen distribution and cycling. In Johnson DW, Lindberg SE (Eds), *Atmospheric deposition and forest nutrient cycling.* Springer-Verlag, New York.

White PS, Cogbill CV (1992) Spruce–fir forests of eastern North America. In Eagar C, Adams MB (Eds) *Ecology and decline of Red Spruce in the eastern United States.* Springer-Verlag, New York.

Wolfe JA (1967) Forest soil characteristics as influenced by vegetation and bedrock in the spruce–fir zone of the Great Smoky Mountains. PhD dissertation Univ TN, Knoxville.

31. Effects of Soil Warming, Atmospheric Deposition, and Elevated Carbon Dioxide on Forest Soils in the Southeastern United States

J. Devereux Joslin and Dale W. Johnson

Changes in the atmosphere have the potential to affect forest soils through a variety of pathways. Various impacts upon forest soils, in turn, may affect forest-nutrient cycles, and ultimately forest productivity. The primary atmospheric changes most probable to impact forest soils are 1) global warming, 2) atmospheric deposition, and 3) increasing atmospheric carbon dioxide (CO_2) concentrations. In addition to these impacts, changes in species composition may also impact upon forest-soil-nutrient cycles. After briefly reviewing the conclusions of the chapters in this section, the remainder of this chapter will be devoted to a review of relevant literature and discussion of the potential impacts of each of these types of changes.

Two papers presented in this section (Lockaby et al., Chapter 28, and Joslin et al., Chapter 30) experimentally examined the effects of warming on soil decomposition and nitrogen mineralization. Lockaby et al. noted small increases with warming in soil decomposition rates and increases in nitrogen (N) immobilization at two sites. Both types of responses were stronger at the Auburn site than at the Pineville site, probably because of differences in litter quality. The Joslin et al., study found no impacts of soil warming on these same processes in a high-elevation red spruce stand, despite having found increased nitrate production with soil warming in a prior study. The small temperature effect produced by the treatment, the short study period, and the large variability in soil chemistry all may have contributed to this result. The Sanchez study (Chapter 29) also focused upon decomposition processes, pointing out the primary importance of examining the dynamics of the labile soil-organic matter pool (as opposed to the more slowly decomposing pools) when looking at moisture and nutrient gradients. Finally, the modeling study by Johnson et al., (Chapter 27) pointed out that although atmospheric deposition may impact soil-solution concentrations and the distribution of nutrients in the soil profile, modeling indicated little immediate effect of deposition upon growth in loblolly pines and hardwood stands. In contrast, changing species composition from pine to hardwood, or vice versa, greatly impacted soil pools of base cations. The results of these chapters will be incorporated into the following of review of the three major topic areas (i.e., global warming, atmospheric deposition, and increasing levels of CO_2) relating to effects of atmospheric changes on forest soils.

Effects of Soil Warming on Forest Soils

Scope of This Review

Increases in temperature in the southeastern United States have the potential to affect the belowground portions of forest ecosystems through at least three major pathways, which are 1) decreases in soil moisture resulting from increased evapotranspiration, 2) increased rates of root respiration (and perhaps rates of root growth), and 3) increased rates of soil-organic matter decomposition and accompanying impacts upon nutrient availability, especially nitrogen. The reduced soil-moisture scenario (Pathway 1) will undoubtedly result if precipitation, solar radiation, humidity, and cloud cover remain constant while average temperature increases. Pathways 2 and 3 both will result in increased emissions of CO_2 to the atmosphere from the forest floor surface, but such increases will not necessarily result in reductions in belowground carbon (C) storage, as inputs of C to the belowground system could also increase.

The following discussion on soil warming effects will focus almost entirely on Pathway 3—the effects on soil-organic matter decomposition—in order to make the topic manageable. Pathway 1—decreases in soil moisture—is being addressed by a number of modeling efforts, and the ultimate prediction will be highly dependent upon predictions of other climatic variables (e.g., precipitation, and so forth). Pathway 2—increases in root respiration—has been directly studied by very few (e.g., Cropper and Gholz, 1991), but reliable predictions of various impacts could be accomplished given accurate data on root distribution, soil temperature, and soil moisture.

As we examine the potential effects of temperature on soil-organic matter decomposition, for the sake of simplicity, we will assume that soil moisture remains constant. The importance of this rather tenuous assumption should constantly be kept in mind throughout this discussion. The following discussion of soil warming impacts on soil-organic decomposition will focus on the four potential effects, which are 1) losses of C from belowground pools, that is, the forest floor and organic matter in the mineral soil[1], 2) redistribution of existing ecosystem nitrogen between living plant pools and below ground organic matter pools (including microorganisms), 3) losses of N from forest ecosystems; and 4) effects of changing nitrogen availability on forest productivity. Because the availability of soil N and soil moisture are the two most important variables limiting growth in most temperate zone forests, the focus in topics 2, 3, and 4 on N cycling appears appropriate. Much of the discussion of the impacts of changes in soil-organic matter decomposition rates therefore relates directly to changes in the availability of nitrogen for plant uptake, and to the ability of plants to respond to those changes in nitrogen availability.

Losses of Carbon from Belowground Pools

Both laboratory incubation studies of forest organic matter decomposition and field studies of relationships between temperature and decomposition have repeatedly verified the strong and consistent relationship between litter decomposition rates and soil temperature, whether the organic matter originated from aboveground or belowground plant parts (Edwards, 1975; McClaugherty et al., 1984; Moore, 1986; Joslin and Henderson, 1987; Ruark, 1993; MacDonald et al., 1995). Climatic gradient studies also indicate that decomposition rates, and the rate at which N is released in the process, are strongly correlated with temperature (Simmons et al., 1995). Virtually every field experiment that has examined the belowground effects of manipulations designed to increase soil temperature has reported an initial increase in the decomposition rate of soil-organic matter (Van Cleve et al., 1983; Peterjohn et al., 1994; Rustad et al., 1995; Mitchell et al., 1995; Lockaby et al., Chapter 28) or an increase in CO_2 emissions from the forest floor surface (Peterjohn et al., 1994; Rustad et al., 1995; Mitchell et al., 1995).

This initial increase in organic matter decomposition rate with a consistently maintained temperature increase might be expected to decline over time (as observed in Rustad et al., 1995 and Melillo et al., 1995) as a result of a gradual depletion of the so-called "labile" organic matter fraction (Sanchez, Chapter 29). After an initial rapid loss phase, the remaining increase in decomposition might be expected to occur largely in the slowly decomposing organic matter fraction (Schimel et al., 1994; Melillo et al., 1995). The third fraction that Schimel et al., (1994) discuss—the "resistant" or "stable" fraction—contributes little to changes in CO_2 emissions because of its recalcitrant nature. The nature of these three organic matter fractions—1) labile (rapid), 2) slow, and 3) stable (resistant)—have been discussed at length in the literature, but are believed to be related to a number of factors, including C/N ratios, lignin to N ratios, and the structural nature of the particular organic compounds.

---

[1] Note: The terms "soil-organic matter," "soil nitrogen," "belowground organic matter," and so forth will be used here to refer to the combination of the organic matter of both surface organic layers and the mineral soil.

That decomposition rates consistently increase with temperature in these studies is an important, if not unexpected, finding. Any attempt to apply a quantitative estimate of an actual amount of rate increase, however, should take into account the fact that 1) the amount of increase in decomposition rate may decline over time, 2) the amount of increase will vary considerably among forest ecosystems, just as the nature and amount of organic matter varies across them, and 3) any changes in soil moisture accompanying temperature changes will affect decomposition rates. A final caution concerning the use of CO_2 emission rates from the forest floor, is that these rates always reflect a combination of root respiration and organic matter decomposition, and rarely have the effects of temperature on these two separate processes been teased apart (Edwards, 1975; Hanson et al., 1993).

Redistribution of Nitrogen Between Soil Pools and Plant Pools

In agricultural ecosystems, it has been predicted that losses of C from belowground pools caused by temperature increases will largely result in net losses for the entire ecosystem (Bradbury and Powlson, 1994). In forest ecosystems, this may not necessarily be the case. Rather, in northern and temperate forest ecosystems, Pastor and Post (1988), Melillo et al., (1993, 1995), and Post et al., (1995) have all asserted that increased rates of organic matter decomposition probably will result in the increased availability of soil N, which, in turn will increase production of both above- and belowground plant matter. Because the C/N ratio of this living plant matter (typically 50:1) is higher than that of the soil-organic matter from which the nitrogen will have been released (typically, 15:1), the net effect would be an increased amount of C stored in the ecosystem (Post et al., 1995; Melillo et al., 1995). Of course, this effect will only continue until a new equilibrium is reached because soil-organic matter pools with high C/N ratios will gradually decline, as pools of the higher C/N ratio material increase.

The response hypothesized above by Pastor and Post (1988), Melillo et al., (1993), and Post et al., (1995) depends upon the following two factors being present 1) a soil-organic matter pool with a sufficiently high C/N ratio to allow a net release of N, rather than immobilization by microbial organisms, and 2) a forest that will respond to the release of N with an increase in net primary production. However, results of soil warming experiments to date indicate that temperate zone forest ecosystems vary greatly in their response to soil warming with respect to nitrogen as is shown in the following "cases":

Case I: In some cases, there is an increase in decomposition with warming and a trend towards a reduction in nitrogen released accompanying this increase (Lockaby et al., Chapter 28; southeastern United States loblolly pine). In this latter study, N in the decomposed organic matter appears to have been immobilized by the microbial population, which is feeding upon other organic matter with low C/N ratios.

Case II: In another case, there was an increase in decomposition with warming, but no significant change in the release of N (Rustad et al., 1995; northeastern United States, low-elevation spruce–fir). Again, this result suggests an ecosystem deficient in N and containing organic matter with high C/N ratios, consistent with supporting data from the site (Fernandez et al., 1993).

Case III: In a third study, an increase in soil warming resulted in an increase in both organic matter decomposition rates and net N mineralization rates, but no increase in nitrification or increase in nitrate levels in soil solution (Peterjohn et al., 1994; Melillo et al., 1995).

Case IV: In another study, resulting from a 9 °C temperature increase, there was an observed increase in net N mineralization, soil exchangeable ammonium, and soil-solution nitrogen, as well as increased foliar N and an increase in tree growth (Van Cleve et al., 1983).

Case V: In other studies, not only is there an increase in net N mineralization with increasing temperature, but there is also an increase in nitrate produced by nitrifying bacteria (Mitchell et al., 1995).

Case VI: In those systems in which nitrification is stimulated by soil warming, there is also the potential for increased nitrate leaching of base cations or aluminum mobilization, as exhibited in Joslin and Wolfe (1993).

The wide range in responses of these six forest ecosystem "cases" to soil warming with respect to N appears to be primarily related to soil differences in N mineralization potential. The six prior case studies represent a progression of availability in the belowground component, resulting in responses ranging from immobilization to increased N mineralization accompanied by elevated nitrification and nitrate leaching, and every step in between on this progression. The range in N availability is, in turn, related to both the size of the soil—N pool and to such substrate factors as the C/N ratio, and the C/lignin ratio, of the soil-organic fraction. MacDonald et al., (1995) even found that net N mineralization rate may be strongly dependent upon interactions between the nature of the organic substrate and temperature.

The soils of forest ecosystems studied in soil warming experiments and mentioned in the prior paragraph can be ordered with respect to mineralizable soil N (Figure 31.1). Those with the lowest N mineralization potential generally have high C/N ratios and are subject to nitrogen immobilization; those with the highest nitrogen mineralization potential have organic matter fractions with low C/N ratios, high nitrification potential, and high concentrations of nitrate in soil solutions.

Potential Forest Responses to Soil Warming

A second axis of importance with respect to ecosystem response to soil warming, is the potential for the existing vegetation to respond to increased N availability with increased growth. This characteristic affects not only the potential for redistribution of N between soil pools and plant pools, but also the potential for loss of N from the ecosystem through leaching or denitrification, and the potential for increases in forest productivity. Forest stands, communities, or ecosystems can be placed approximately on this axis based upon a number of factors, including 1) species composition, 2) stand age, 3) length of the growing season, 4) annual precipitation and other climatic variables, 5) physical and chemical characteristics of soil, and 6) the present N status of the vegetation. In short, many of the factors contributing to the traditional forestry concepts of "site," "site quality" or "site index" (Davis, 1966; Spurr and Barnes, 1973) are related to this potential for a forest stand to respond to increased nitrogen availability. The one major exception is, of course, current nitrogen status of the vegetation, which would normally might be a factor positively correlated with site quality, but in this case is negatively correlated with the potential to respond to additions of N.

These stands, as well as other forest types in North America, have been tentatively placed on the two axes in Figure 31.1. Although this figure is highly generalized, it clearly demonstrates the principles involved, fitting the results from existing research on the axes in a manner that is not only consistent with the experimental data, but compatible with intuition. A young, unfertilized southern pine stand (as studied in Lockaby et al., Chapter 28) would exemplify a stand with a high potential to respond to increased N availability with increased N uptake and increased growth. Such pine stands typically occur in climates with relatively long growing seasons and considerable rainfall; they usually consist of a fast-growing species that is capable of producing considerable new foliage every year, are at an

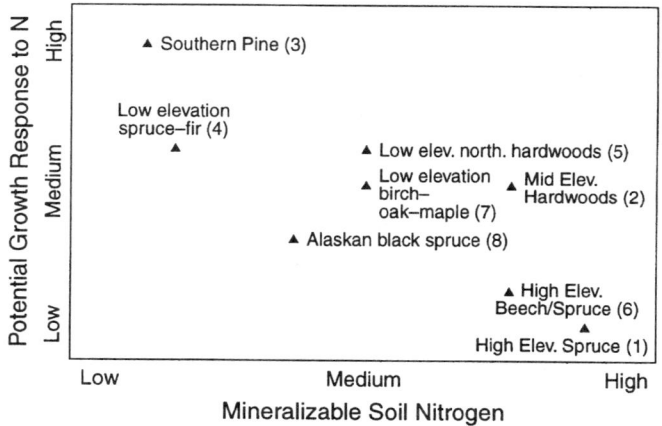

Figure 31.1. Apparent location of stands in which soil temperature has been manipulated, with respect to (X) net mineralizable soil nitrogen, and (Y) potential forest growth response. Key to Stands: (1) Joslin and Wolfe, 1993, high-elevation red spruce, Whitetop Mountain, VA, clearcut gradient; (2) Mitchell et al., 1995, midelevation northern hardwoods, Adirondack Mountains, NY, soil warming study; (3) Lockaby et al. Chapter 28, loblolly pine, Auburn, AL, soil warming study; (4) Rustad et al., 1995, low-elevation spruce–fir, Maine, soil warming study; (5) Simmons et al., 1995, low-elevation northern hardwoods, Maine, climate gradient; (6) Johnson and Lindberg, 1992, high-elevation beech–maple and spruce–fir, Smoky Mountains, TN, nutrient cycling study; (7) Peterjohn et al., 1994; Melillo et al., 1995, low-elevation birch–oak–maple, MA, soil warming study; (8) Van Cleve et al., 1983, black spruce, interior Alaskan taiga, soil warming study.

age when growth rates are near maximum, and frequently have foliage that is marginal to deficient in N (under unfertilized conditions). At the opposite extreme, a very mature red spruce stand growing at high elevations in the Appalachians (as studied by Joslin and Wolfe, 1993) would exemplify a stand with low potential to respond with a positive growth response to increased N availability, because such stands occur in climates with short growing seasons, are at a life stage during which growth is very slow, maintain needles for many years so that the N requirement for new needle production is at a minimum, and frequently have adequate foliar N nutrition under present N deposition rates in these portions of the United States.

These two extreme illustrations (with respect to the Y axis) also fall at the lower and upper ends of the mineralizable soil-N (X axis) in Figure 31.1. Other stands are depicted as falling intermediate to these two stands on both axes. Although there is a tendency for stands to fall along a line running from the upper left with a slope of negative one (as a result of the fact that foliar N status is a confounding factor affecting both variables) there a number of exceptions. For example, low-elevation red spruce stands (as studied in Rustad et al., 1995), have relatively low mineralizable soil N and are generally N-deficient, but rank intermediate in the ability to respond to N additions because of the short growing season and growth limitations of the species involved. In contrast, midelevation hardwoods in the Adirondacks (as studied by Mitchell et al., 1995) are also moderate in their ability to respond to such N releases, but apparently have relatively high mineralizable soil N.

Predicted Forest Response by Ecosystem Type

The relationship depicted in Figure 31.1 is not only useful for depicting where various forest ecosystems studied to date presently fall relative to each other, but it is also very useful, in a most general fashion, in predicting in what direction they are apt to respond to soil-organic matter changes as soil temperatures increase. Figure 31.2 depicts the predicted relative size of growth responses of stands falling in various positions on these axes. Growth responses are greatest toward the upper right, where stands have both the potential to respond to N additions, and where mineralizable soil-N pools are available to fertilize that growth if decomposition rates increase with temperature. Midelevation hardwoods in the southern Appalachians and the Northeast appear to be in the best position to respond positively to temperature increases because of moderately high N mineralization potential and forest site conditions and species composition favoring growth responses. Other hardwood stands in the low elevations of the Southeast might respond minimally because of more limited N availability in soil pools. Low-elevation coniferous stands, particularly in the Southeast, probably will not respond at all because of low available mineralizable soil N.

In contrast to all the above stands, high-elevation spruce–fir stands and high-elevation beech–maple stands (Johnson and Lindberg, 1992) of the southern Appalachians are apt to respond negatively to soil warming (Figure 31.2), because

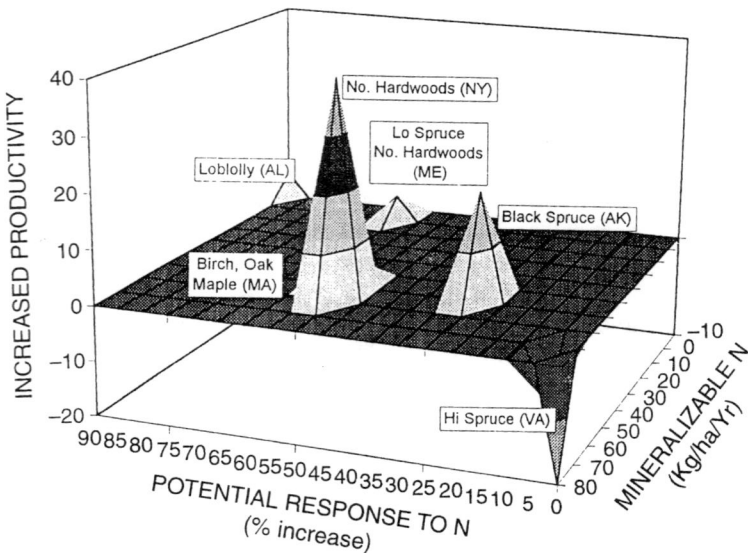

Figure 31.2. Predicted magnitude and direction of forest-growth responses to future soil warming based upon the location of a given forest type with respect to the two axes depicted in Figure 31.1.

additional N released from soil-organic matter will not only be of no value as a fertilizer, but because soil nitrate levels should elevate further thereby reducing the availability of base cations or increasing negative impacts of elevated aluminum (Al). Figure 31.3 depicts the potential for various stands to experience increased nitrate leaching with further soil warming. Because some of these stands are already "leaking" nitrate and leaching cations or mobilizing aluminum in the process, further acceleration of decomposition rates by warming their large organic matter pools should not benefit their forest health.

The prior review of the literature is intended to present the range of possible responses and to establish where forest types, which have been studied to date, fall within that range. The generalizations made in this review concerning the future responses of various forest types to soil warming are meant only as estimations of general trends within broad ecosystem categories. Much variability within those categories naturally exists, in N mineralization potential, as well as in the ability of forests to respond with increased growth. When specific information concerning a stand or a ecotype within a region is available, simulating the forest response with such computer models as the terrestrial ecosystem model (TEM) (Melillo et al., 1993), should provide more precise and reliable predictions. The wide range of stand responses clearly indicates, however, that results from such modeling efforts must not be extrapolated beyond the forests for which the input data used is appropriate.

31. Effects of Soil Warming, Atmospheric Deposition, and Elevated Dioxide   579

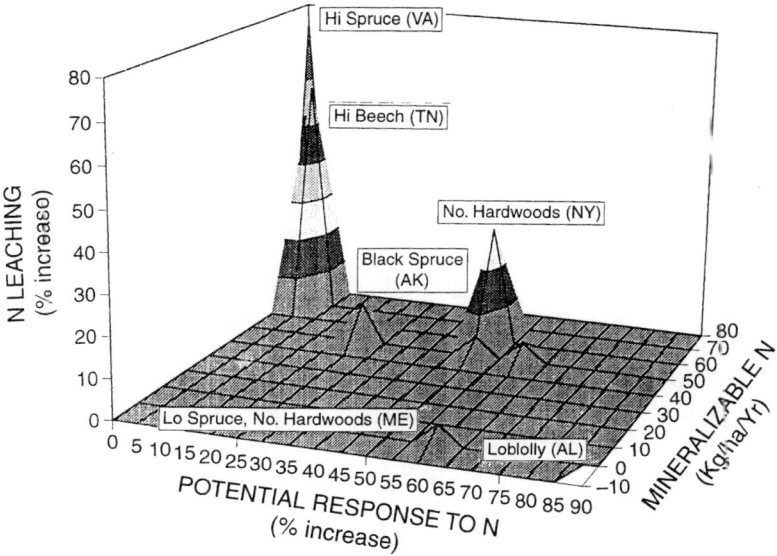

Figure 31.3. Potential for future nitrate leaching in response to soil warming based upon the location of a given forest type with respect to the two axes depicted in Figures 31.1 and 31.2.

Soil Changes: Effects of Atmospheric Deposition and Species Changes

Forest soils of the southeastern United States, as with those in several other parts of the world, are undergoing changes that may have bearing upon the future health and productivity of southern forest ecosystems (reviewed by Johnson et al., 1991). In contrast to earlier indications that soil-nutrient pools were very large and therefore buffered from short-term changes, several studies have shown that soils are changing on the time-scale of decades, and, in some cases, even on a seasonal basis. The first indication of the potential for short-term changes was the study by Haines and Cleveland (1981), which documented seasonal variations, in bulk density, pH, organic matter, extractable phosphorus (P), and exchangeable calcium (CA) and magnesium (Mg) in soils beneath both pine and pine–hardwood mixtures in Georgia. The authors attributed the changes to seasonal effects of tree nutrient uptake and litterfall return, and, in the case of organic matter, to contamination with root material. Johnson et al. (1988) found similar seasonal variations in extractable P, exchangeable Ca, Mg, and potassium (K) in the upper horizons of a mixed deciduous forest on Walker Branch Watershed, Tennessee. There was little or no seasonal variation in lower horizons, suggesting that the changes in the upper horizons were the result of seasonal effects of uptake and litterfall return.

Long-term trends in forest soils of the South have also been documented. Johnson et al (1988) found large decreases in exchangeable Ca and Mg over an eleven-year period in subsurface soils of Walker Branch, Tennessee. In a later study of nutrient budgets on selected plots, Johnson and Todd (1990) found that the decreases in Ca were mostly the result of uptake and sequestration in vegetation; the decreases in Mg, however, resulted from leaching. In that leaching was dominated by atmospherically derived SO_4^{2-}, the decreases in Mg were attributed to atmospheric deposition. Binkley et al. (1989) also noted large reductions in base cation pools over a twenty-five-year period in a loblolly pine plantation in South Carolina, which they attributed primarily to accumulation in the biomass of the rapidly growing plantations. More recently, Knoepp and Swank (1994) noted large decreases in exchangeable Ca and Mg over a twenty-year period in soils under both hardwood and white pine (*Pinus strobus* L.) at Coweeta, North Carolina. Nutrient budget analyses at this site also indicated that both uptake and leaching accounted for the observed decreases. Richter et al., (1994) reported on the changes in a South Carolina ultisol planted with loblolly pine over a period of three decades. They found substantial (from 50 to 400%) increases in acidity and decreases in exchangeable Ca and Mg over this period of time, but smaller changes in exchangeable K. Nutrient budget analyses revealed that the changes in Ca and Mg resulted in approximately equal measure from leaching and to sequestering in live biomass and forest floor of this developing stand. Of the leaching component, approximately half was accounted for by sulfate, thus, a maximum of 25% might be attributed to atmospheric sulfur (S) deposition.

There are some interesting potential interactions between tree uptake and leaching that can have profound effects upon the cation budgets of forests subject to both intensive management and acidic deposition (Johnson and Todd, 1987; Johnson et al., 1995). Cation-exchange equations predict that depletion of one base cation on the exchanger as a result of uptake (or any other process) will lead to a relative future decrease in the leaching rate of that cation and an increase in the leaching of other cations. In essence, this interaction predicts that the base cation in greatest demand (often Ca) is conserved in the system because exchangeable pools are depleted and transferred into biomass, whereas cations in less demand (e.g., Mg) experience increased leaching rates. This kind of interaction has been observed in both field conditions (Johnson and Todd, 1987, 1990) and with simulation modeling (Johnson et al., 1995; Johnson, et al., Chapter 27). Both field data and modeling suggest that uptake generally has a larger effect upon soil-Ca depletion than does leaching in deciduous forests, even when the latter is accelerated by atmospheric S and N deposition (Johnson and Todd, 1987, 1990; Binkley et al., 1989; Johnson et al., 1995). In coniferous forests, with lower Ca uptake or higher S and N deposition, leaching plays a larger role in Ca depletion (Johnson and Todd, 1987; Joslin et al., 1992; Richter et al., 1994). Species and stage of stand development are especially important with regard to the rate of Ca and other nutrient accumulation in vegetation. In contrast, leaching is often shown to be the major cause of soil Mg decline (Johnson and Todd, 1987, 1990; Richter et al., 1994).

Historically, forest ecosystems in the northern hemisphere have been N deficient, and hence there is reason to believe that increased N deposition has caused or will cause increased productivity and N uptake (Kauppi et al., 1992). In doing so, N deposition may also cause forest ecosystems to reach deficiency levels of the next most limiting nutrient; in central Europe, this appears to be Mg and K (Schulze et al., 1989). In southern forests, both field fertilization trials and modeling exercises suggest that the nutrient most probable to become limiting after N deficiencies are overcome is P; Mg and K deficiencies are relatively rare (NCSFNC, 1995). Indeed, some coastal plain forests are presently P deficient (Pritchett and Llewellyn, 1966; Pritchett and Comerford, 1982). Again, southern Appalachian high-elevation spruce–fir stands appear to be an exception in which Ca appears to limit growth, while N is being lost from the ecosystem (Van Miegroet et al., 1993; Joslin and Wolfe, 1994).

There is some indirect evidence of that elevated Al levels in soil solutions may be inhibiting base cation uptake in some red spruce forests growing on very acidic soils in the eastern United States. Shortle and Smith (1988) found a relationship between soil Al/Ca ratio and decline in the northeastern United States and hypothesize that the decline is the result of Al inhibition of Ca uptake. Bondietti et al. (1989) found an inverse correlation between woody Al concentration or Al/Ca ratio and ring width in red spruce tree cores from the Appalachians. Johnson et al., (1992) and Joslin and Wolfe (1992) observed soil-solution Al concentrations approaching red spruce toxicity thresholds and Ca/Al molar ratios of 0.3 to 0.5 in red spruce forests of the southern Appalachians. A recent review by Cronan and Grigal (1995) indicates that Ca/Al ratios in this range substantially increase the risk of reducing tree vitality in many species. Van Miegroet et al., (1993) and Joslin and Wolfe (1994) both found growth responses to Ca fertilization in high-elevation red spruce in the southern Appalachians where soil-solution Al and Al/Ca ratios are elevated. However, the connection between Al toxicity, potential cation deficiencies, and the dramatic decline of red spruce in the northeastern United States remains weak. No widespread mortality has occurred in the southern Appalachians where soil-solution Al levels are highest. In the Northeast, red spruce decline occurred on sites with abundant base cation supplies as well as sites with low base cation supplies. The red spruce decline in the Northeast appears to be chiefly attributable to winter injury, which has been exacerbated by exposure of foliage to acidic mist (Johnson et al., 1992).

Effects of Elevated Carbon Dioxide on Soils

Forest ecosystems throughout the world have experienced and will probably continue to experience significant increases in both N deposition and atmospheric CO_2 concentrations (Aber et al., 1989; Strain and Thomas, 1992). Since N is the most commonly limiting nutrient to forest growth in the northern hemisphere (Gessel et al., 1973; Aber et al., 1989; Johnson, 1992), it is important to understand the interactions between CO_2 and N in order to make reasonable forecasts of forest response to these nutrients and the potential for C sequestering as a result of such responses.

There are several ways in which CO_2 and N can interact so as to allow a growth response to CO_2 even under N limitation. Many studies have reported reduced tissue N concentrations under elevated CO_2 regimes, facilitating growth increases under suboptimum N conditions (Campagna and Margolis, 1989; Samuelson and Seiler, 1993; Johnson et al., 1994; Griffin et al., 1995). One explanation for the often observed decreases in foliar N concentration with increased CO_2 is that plants may produce lower concentrations of the enzymes of the photosynthetic carbon reduction (PCR) cycle, particularly the carboxylating enzyme ribulose-1,5-bisphosphate carboxylase/oxyenase (rubisco). If this response occurs under field conditions, it implies that the N deficiencies common to many forest ecosystems will not preclude a growth increase in response to CO_2.

Another factor that may allow N deficient forests to respond to CO_2 is increased uptake because of increased root growth. Several studies have shown that increased CO_2 causes greater carbohydrate allocation to roots and mycorrhizae, causing disproportionate growth increases in root growth and thereby facilitating greater soil exploration (Norby et al., 1986a, 1986b; Rogers et al., 1992; Walker et al., 1995). Some studies have suggested that stimulation of rhizosphere activity by elevated CO_2 can cause increased N availability in both tropical (Korner and Arnone, 1992) and temperate (Zak et al., 1993) broadleaf species.

In the case of P deficiency, no such convenient mechanism for the increased use efficiency exists. Several studies have shown that P deficiency virtually precludes growth responses to increased CO_2. Conroy et al. (1986, 1988, 1990a, 1990b) tested the responses of P deficient and water-stressed *Pinus radiata* D. Don to elevated CO_2 in a series of greenhouse studies using naturally P deficient soils. Deficiency of P severely limited growth at both elevated and present ambient CO_2. A growth increase at elevated CO_2 was seen at all watering levels with adequate P, but with deficient P the effect of CO_2 level on growth was eliminated in all but the most water-limited treatment. Deficiency of P was not overcome by increased root growth or mycorrhizal development. The reduced responses to both limited water supply and elevated CO_2 in the P deficient treatments was the result of lower photosynthetic rates, stomatal conductance and water use per unit area and significantly reduced leaf surface area. There were no effects of elevated CO_2 on tissue P concentration. Conroy et al., (1988) conclude that " . . . the growth of *P. radiata* will only be increased [by CO_2 enrichment] in areas where P nutrition is adequate."

Given the widespread occurrence of N and P deficiency in southern forest ecosystems, it must be concluded that the potential for a CO_2 "fertilization effect" in terrestrial ecosystems is strongly constrained by N and P availability and the effects of CO_2 on the cycling of these two nutrients. From the results to date on coniferous species, it appears that a long-term CO_2 fertilization response in N deficient forest ecosystems is possible only via changes in the N cycle, specifically, through an increase in N uptake. Given this premise, the question becomes,

31. Effects of Soil Warming, Atmospheric Deposition, and Elevated Dioxide

can N uptake be increased in mature, N deficient forests? As of this writing, this remains an open question.

Summary

The following summary attempts both to summarize succinctly this chapter's conclusions while simultaneously addressing three key questions that the Southern Global Change Program (SGCP) has posed. The first two questions are discussed together because they are intimately related. Productivity is emphasized over structure and function because of the lack of knowledge concerning the effects of atmospheric change via soils on the latter.

These first two questions are: 1) what processes in southern forest ecosystems are sensitive to physical and chemical changes in the environment? and 2) how will atmospheric change influence structure and function and productivity of forest ecosystems? As pointed out in the first section of this chapter, soil-organic matter decomposition rates are especially sensitive to soil warming, almost uniformly showing an increase, at least initially. These changing rates of decomposition, in turn, affect nutrient mineralization rates, though the direction and size of the effect varies greatly from one ecosystem to another, depending upon organic matter quality. Nitrogen mineralization and nitrification responses to these changes in decomposition rates also vary greatly across different ecosystems. Because nitrogen cycling can, in turn, greatly affect productivity and leaching processes. the direction of these impacts is similarly highly ecosystem-dependent (Figures 31.2, 31.3). In general, soil-mediated effects of warming would probably have little impact on low-elevation pine and hardwood forests of the Southeast because of the low level of mineralizable N in the organic matter. Conversely, N-saturated spruce–fir stands in the higher elevations would probably be negatively impacted by warming because of increased nitrate leaching.

Recent studies indicate that forest soils in the Southeast seem to be changing, with exchangeable base cations declining at some sites. A number of possible causes have been discussed in this chapter, and the situation bears monitoring. Calcium is especially susceptible to depletion resulting from tree uptake and subsequent harvest removal (especially in hardwood stands); Mg is particularly susceptible to leaching by strong acid anions (whether from atmospheric deposition or internal sources). Increased N deposition is probably having a fertilization effect on many southeastern forests, perhaps causing P to become the limiting nutrient in some stands. In contrast, some high-elevation forests appear to be experiencing base cation losses as a result of excess N deposition. Although some studies indicate increased tree-growth responses to elevated CO_2, it is highly probable that the widespread occurrence of N and P (and perhaps other nutrient) deficiencies in the Southeast will greatly limit any growth responses to CO_2.

The third question is: 3) how should forest management practices be altered to sustain southern forest productivity, health, and diversity? There are a limited number of ways that forest managers can respond to predicted effects of atmospheric change on forest soil processes. In many stands in the Southeast, it appears that effects will be at a minimum or will only become significant over long time spans. There are, however, some practices that can serve to anticipate changes in health, productivity, and diversity. Soil warming (through nitrate release), atmospheric deposition of strong acid anions, elevated CO_2, and the repeated harvesting of forests all can lead to depletion of base cation or micronutrient reserves through the processes of leaching and harvest removal. As the prior discussion points out, such effects will be highly site specific—dependent upon initial nutrient supplies, the degree of leaching potential, and the extent of harvest removal. One management practice that can certainly minimize such nutrient depletion is limiting the removal of nutrients from a site during harvest. Foliage, fine branches, and bark contain a high proportion of the total biomass pool of most nutrients, especially base cations. This is especially true of hardwood species. A second recommended practice is periodic monitoring of foliar nutrient levels in selected stands to discover trends towards incipient (or full-blown) nutrient deficiencies. Finally, fertilization with limiting nutrients (as determined by foliar or soil sampling) may be a practice to consider on such highly sensitive sites as high elevation spruce–fir stands, depleted sites, or intensively harvested sites.

References

Aber JD, Nadelhoffer KJ, Streudler P, Melillo JM (1989) Nitrogen saturation in northern forest ecosystems. Bioscience 39:378–386.

Binkley D, Valentine D, Wells D, Valentine U (1989) An empirical model of the factors contributing to 20-year decrease in soil pH in an old-field plantation of loblolly pine. Biogeochemistry 8:39–54.

Bondietti EA, Baes CF, McLaughlin SB (1989) Radial trends in cation ratios in tree rings as indicators of the impact of atmospheric deposition on forests. Can J For Res 19:586–594.

Bradbury NJ, Powlson DS (1994) The potential impact of global change on nitrogen dynamics In arable systems. In Rounsevell MDA, Loveland PJ (Eds) *Soil Responses to Climate Change* Springer-Verlag, Berlin.

Campagna MA, Margolis HA (1989) Influence of short-term atmospheric CO_2 enrichment on growth, allocation patters, and biochemistry of black spruce seedlings at different stages of development. Can J For Res 19:773–782.

Conroy JP, Barlow EWR, Bevege DI (1986) Response of *Pinus radiata* seedlings to carbon dioxide enrichment at different levels of water and phosphorus: Growth, morphology, and anatomy. Ann Bot 57:165–177.

Conroy JP, Kuppers M, Kuppers B, Virginia J, Barlow EWR (1988) The influence of CO_2 enrichment and water stress on the growth, conductance, and water use of *Pinus radiata* D. Don. Plant Cell Environ 11:91–98.

Conroy JP, Milham PJ, Bevage DI, Reed ML, Barlow EW (1990a) Influence of phosphorous deficiency on the growth response of four families of *Pinus radiata* seedlings to CO_2-enriched atmospheres. For Ecol Managem 30:175–188.

Conroy JP, Milham PJ, Bevege DI, Reed ML, Barlow EW (1990b) Increases In photosynthesis requirements for CO_2-enriched pine species. Plant Physiol 92:977–982.

Cronan CS, Grigal DF (1995) Use of calcium/aluminum ratios as indicators of stress In forest ecosystems. J Environ Qual 24:209–226.

Cropper WP Jr, Gholz HL (1991) In situ needle and fine root respiration in mature slash pine (*Pinus elliottii*) trees. Can J For Res 21:1589–1595.
Davis KP 1966 *Forest management: regulation and valuation*. McGraw-Hill, New York.
Edwards NT (1975) Effects of temperature and moisture on carbon dioxide evolution in a mixed deciduous forest floor. Soil Sci Soc Am Proc 39:361–365.
Fernandez IJ, Rustad LE, Lawrence GB (1993) Chemistry and distribution of nutrients and trace metals in soils under a low elevation spruce–fir forest. Can J Soil Sci 73:317–328.
Gessel SP, Cole DW, Steinbrenner EC (1973) Nitrogen balances in forest ecosystems of the Pacific Northwest. Soil Biol Biochem 5:19–34.
Griffin KL, Winner WE, Strain BR (1995) Growth and dry matter partitioning in loblolly and ponderosa pine seedlings in response to carbon and nitrogen availability. New Phytol 129:547–556.
Haines SG, Cleveland G (1981) Seasonal variation in properties of five forest soils in southwest Georgia. Soil Sci Soc Amer J 45:139–143.
Hanson PJ, Wullschleger SD, Bohlman SA, Todd DE (1993) Seasonal and topographic patterns of forest floor CO_2 efflux from an upland oak forest. Tree Physiol 13:1–15.
Johnson AH, McLaughlin SB, Adams MB, Cook ER, DeHayes DH, Eagar C, Fernandez IJ, Johnson DW, Kohut RJ, Mohnen VA, Nicholas NS, Peart DR, Schier GA, White PS (1992) Synthesis and conclusions from epidemiological and mechanistic studies of red spruce decline. In Eagar C, Adams MB (Eds) *Ecology and decline of red spruce in the eastern United States*. Ecological Studies Number 96. Springer-Verlag, New York.
Johnson DW (1992) Nitrogen retention in forest soils. J Environ Qual 21:1–12.
Johnson DW, Ball JT, Walker RF (1994) Effects of CO_2 and nitrogen on nutrient uptake in ponderosa pine seedlings. Plant Soil 168:144–153.
Johnson DW, Binkley D, Conklin P (1995) Simulated effects of atmospheric deposition, harvesting, and species change on nutrient cycling in a loblolly pine forest. For Ecol Manag 76:29–45.
Johnson DW, Henderson GS, Todd DE (1988) Changes in nutrient distribution in forests and soils of Walker Branch Watershed, Tennessee, over an eleven-year period. Biogeochem 5:275–293.
Johnson DW, Lindberg SE (1992) *Atmospheric deposition and nutrient cycling in forest ecosystems of the integrated forest study*. Springer-Verlag, New York.
Johnson DW, Todd DE (1987) Nutrient export by leaching and whole-tree harvesting in a loblolly pine and mixed oak forest. Plant Soil 102:99–109.
Johnson DW, Todd DE (1990) Nutrient cycling in forests of Walker Branch Watershed: Roles of uptake and leaching in causing soil change. J Environ Qual 19:97–104.
Johnson DW, Van Miegroet H, Lindberg SE, Harrison RB, Todd DE (1991) Nutrient cycling in red spruce forests of the Great Smoky Mountains. Can J For Res 21:769–787.
Joslin JD, Henderson GS (1987) Organic matter and nutrients associated with fine root turnover in a white oak stand. For Sci 33:330–346.
Joslin JD, Kelly JM, Van Miegroet (1992) Soil chemistry and nutrition of North American spruce–fir stands: evidence for recent change. J Environ Qual 21:12–30.
Joslin JD, Wolfe MH (1992) Red spruce soil solution chemistry and root distribution across a cloud water deposition gradient. Can J For Res 22:893–904.
Joslin JD, Wolfe MH (1993) Temperature increase accelerates nitrate release from high-elevation red spruce soils. Can J For Res 23:756–759.
Joslin JD, Wolfe MH (1994) Foliar deficiencies of mature southern Appalachian red spruce determined from fertilizer trials. Soil Sci Soc Am J 58:1572–1579.
Kauppi PE, Mielikainen K, Kuusela K (1992) Biomass and carbon budget of European forests, 1971 to 1990. Science 256:70–74.
Knoepp JD, Swank WT (1994) Long-term soil chemistry changes in aggrading forest ecosystems. J Soil Sci Soc Amer 58:325–331.
Korner C, Arnone JA (1992) Biomass and carbon dioxide in artificial tropical ecosystems. Science 257:1672–1675.

MacDonald NW, Zak DR, Pregitzer KS (1995) Temperature effects on kinetics of microbial respiration and net nitrogen and sulfur mineralization. Soil Sci Soc Am J 59:233–240.

McClaugherty CA, Aber JD, Melillo JM (1984) Decomposition dynamics of fine roots in forested ecosystems. Oikos 42:378–386.

Melillo JM, Kicklighter DW, McGuire AD, Peterjohn WT, Newkirk KM (1995) Global change and its effects on soil organic carbon stocks. In Zepp RG, Sonntag CH (Eds) *Role of nonliving organic matter in the earth's carbon cycle.* John Wiley and Sons, New York.

Melillo JM, McGuire AD, Kicklighter DW, Moore B III, Vorosmarty CJ, Schloss AL (1993) Global climate change and terrestrial net primary production. Nature 363:234–239.

Mitchell MJ, McHale PF, Raynal DJ, Stehman SV, White EH, Driscoll CT, David MB, Bowles F (1995) Increasing soil temperature in a northern hardwood forest: effects on elemental dynamics and primary productivity. In Abstracts of Meeting of the Northern Global Change Program, Pittsburgh, PA, March 14–16, 1995. USDA For Ser, Radnor, PA.

Moore AM (1986) Temperature and moisture dependence of decomposition rates of hardwood and coniferous leaf litter. Soil Biol Biochem 18:427–435.

NCSFNC (1995) North Carolina State forest nutrition cooperative—Twenty-fourth annual report. Dep For, Coll For Res, NC State Univ, Raleigh.

Norby RJ, O'Neill EG, Luxmoore RJ (1986a) Effects of atmospheric CO_2 enrichment on the growth and mineral nutrition of *Quercus alba* seedlings in nutrient-poor soil. Plant Physiol 82:83–89.

Norby RJ, Pastor J, Melillo JM (1986b) Carbon-nitrogen interactions in CO_2-enriched white oak: Physiological and long-term perspectives. Tree Physiol 2:233–241.

Pastor J, Post WM (1988) Response of northern forests to CO_2-induced climate change. Nature (London) 334:55–58.

Peterjohn WT, Melillo JM, Steudler PA, Newkirk KM, Bowles ST, Aber JD (1994) Responses of trace gas fluxes and N availability to experimentally elevated soil temperatures. Ecol Appl 4:617–625.

Post WM, Anderson DW, Dahmke A, Houghton RA, Huc A-Y, Lassiter R, Najjar RG, Neue H-U, Pedersen TF, Trumbore SE, Vaikmae R (1995) Group report: What is the role of non-living organic matter cycling on the global scale? In Zepp RG, Sonntag CH (Eds) *Role of nonliving organic matter in the earth's carbon cycle.* John Wiley and Sons, New York.

Pritchett WL, Comerford NB (1982) Long-term response to phosphorus fertilization on selected southeastern coastal plain soils. Soil Sci Soc Am J 46:640–644.

Pritchett WL, Llewellyn WR (1966) Response of slash pine (*Pinus elliottii* Engelm var) to phosphorus in sandy soils. Soil Sci Soc Am Proc 30:509–512.

Richter DD, Johnson DW, Dai KH (1992) Cation exchange reactions in acid forested soils: Effects of atmospheric pollutant deposition. In Johnson DW, Lindberg SE (Eds) *Atmospheric deposition and nutrient cycling in forest ecosystems of the integrated forest study.* Springer-Verlag, New York.

Richter DD, Markewitz D, WElls CG, Allen HL, April R, Heine PR, Urrego B (1994) Soil chemistry change during three decades in an old-field loblolly pine (*Pinus taeda* L.) Ecosystem. Ecology 75:1463–1473.

Rogers HH, Peterson CM, McCrimmon JN, Cure JD (1992) Response of plant roots to elevated atmospheric carbon dioxide. Plant Cell Environ 15:749–752.

Ruark GA (1993) Modeling soil temperature effects on in situ decomposition rates for fine roots of loblolly pine. For Sci 39:118–129.

Rustad LE, Fernandez IJ, Arnold S (1995) Experimental soil warming effects on C, N, and major element cycling in a low elevation spruce–fir forest soil. In Abstracts of Meeting of the Northern Global Change Program, Pittsburg, PA. USDA For Ser, Radnor, PA.

Samuelson LJ, Seiler JR (1993) Interactive role of elevated CO_2, nutrient limitations, and water stress in the growth responses of red spruce seedlings. Forest Sci 39:348–358.

Schimel DS, Braswell BH, Holland EA, McKeown R, Ojima DS, Painter TH, Parton WJ, Townsend AR (1994) Climatic, edaphic, and biotic controls over storage and turnover of carbon in soils. Global Biogeochem Cyc 8:279–293.

Schulze E-D (1989) Air pollution and forest decline in a spruce (*Picea abies*) forest. Science 244:776–783.

Shortle WC, Smith KT (1988) Aluminum-induced calcium deficiency syndrome in declining red spruce. Science 240:1017–1018.

Simmons JA, Fernandez IJ, Briggs RD (1995) Soil respiration and net N mineralization along a climate gradient in Maine. In Abstracts of Meeting of the Northern Global Change Program, Pittsburg, PA. USDA For Ser, Radnor, PA.

Spurr SH, Barnes BV (1973) *Forest ecology.* Ronald Press, New York.

Strain BR (1985) Physiological and ecological controls on carbon sequestering in terrestrial ecosystems. Biogeochem 1:219–232.

Strain BR, Thomas RB (1992) Field measurements of CO_2 enhancement and climate change in natural vegetation. Water Air Soil Pollut. 64:45–60.

Van Cleve K, Oliver L, Schlentner R, Viereck LA, Dyrness CT (1983) Productivity and nutrient cycling in taiga forest ecosystems. Can J For Res 13:747–766.

Van Miegroet H, Johnson DW, Todd DE (1993) Foliar responses of red spruce seedlings to fertilization with Ca and Mg in the Great Smoky Mountains National Park. Can J For Res 23:89–95.

Walker RF, Geisinger DR, Johnson DW, Ball JT (1995) Interactive effects of CO_2 enrichment and soil N on growth and ectomycorrhizal colonization of ponderosa pine seedlings. For Sci 41:491–500.

Zak DR, Pregitzer KS, Curtis PS, Teeri JA, Fogel R, Randlett DL (1993) Elevated atmospheric CO_2 and feedback between carbon and nitrogen cycles. Plant Soil 151:105–117.

Section 5. Disturbance Interactions With Global Change

32. Environmental Effects on Pine Tree Carbon Budgets and Resistance to Bark Beetles

Richard T. Wilkens, Matthew P. Ayres, Peter L. Lorio, Jr., and John D. Hodges

Pine trees are a dominant component of primary production in natural and managed ecosystems throughout the southeastern United States. Because of the economic importance of pines in the Southeast, the southern pine beetle (*Dendroctonus frontalis* Zimmerman, Coleoptera: Scolytidae) can cause losses in excess of $236 million per year by attacking and killing pine trees (Price et al., 1992), and is arguably the greatest source of natural disturbance in ecosystems of the southeast. Interactions between pine trees and bark beetles have become a focus of global change research because it has long been hypothesized that bark beetle outbreaks are linked to climatic patterns (Beal 1927; Beal, 1933; Berryman and Ferrel, 1988; Christiansen and Bakke, 1988; Craighead, 1925; Grégoire, 1988; Kalkstein, 1976; King, 1972; Kroll and Reeves, 1978; Michaels, 1984; Raffa, 1988; St. George, 1930; Wyman, 1924). This implies that climate change will probably alter the frequency and intensity of forest disturbance from pest outbreaks. However, the mechanisms by which climatic patterns impact bark beetle population dynamics have remained obscure (Martinat, 1987; Mattson, 1980; Reeve et al., 1995). The development of accurate, physiologically explicit models is an essential first step in assessing the ecological risks associated with global change (Ayres, 1993). The research reported in this chapter was designed to test and refine a model of environmental effects on tree carbon budgets that provides a promising tool for understanding and predicting the effects of global change on interactions between pine trees and the southern pine beetle (SPB) (Figure 32.1).

There are numerous pathways by which climatic change could affect trees and

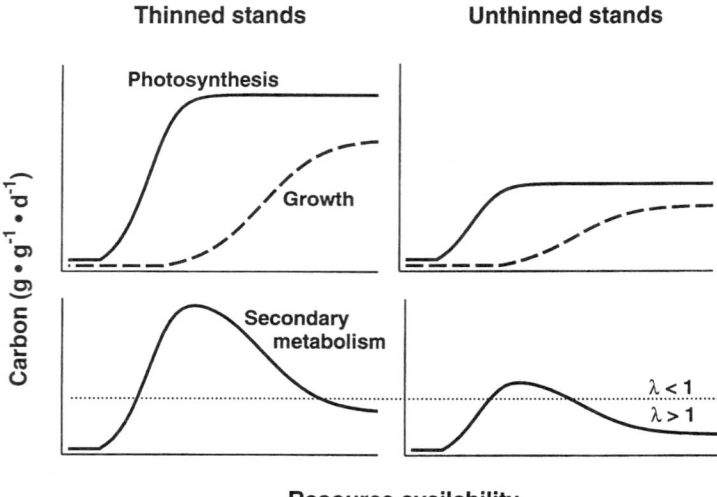

Figure 32.1. Hypothesized responses of loblolly pine in thinned and unthinned stands to changes in the availability of water and mineral nutrients. Moderate deficiencies of water or mineral nutrients are predicted to limit growth more than photosynthesis, and thus, carbon available for secondary metabolism (photosynthesis—growth) increases, oleoresin production rises, and attacking bark beetles suffer low reproductive success ($\lambda < 1$). Secondary metabolism is predicted to be low, and beetle reproductive success high ($\lambda > 1$), either at high resource availability (in which the majority of photosynthates go toward growth) or low resource availability (in which the total carbon budget is low). In dense, unthinned pine stands, the total carbon budget ($g \cdot g^{-1} \cdot d^{-1}$) is predicted to be less than in thinned stands, resulting in reduced secondary metabolism, and allowing reproductive success of attacking bark beetles across a broad range of resource availability.

influence pest resistance (Ayres, 1993). Precipitation patterns impact soil-water availability and thereby exert strong effects on trees. Any changes in precipitation patterns could impact the suitability of pine trees for bark beetles and influence beetle population dynamics. We have attempted to explain effects of soil moisture on bark beetles with a physiological model (Reeve et al., 1995) that is derived from principles of growth-differentiation balance in plants (Loomis, 1932; Loomis, 1953; Lorio, 1986; Wilkens et al., 1996a). There is a widely held perception that trees that are "vigorous" (a word that is often used interchangeably with "fast growing"), are also most resistant to insect attack. However, the model that we employed argues against this simple assumption, except under conditions of very low resources. The model predicts decreasing resistance when moving from conditions of moderate to high resources. In this model, moderate water deficiencies are predicted to limit growth more than they limit photosynthesis (Figure 32.1, upper) because growth is a function of cell division and expansion, which are strongly dependent upon water availability (Brown and Sommer, 1992).

As a consequence, trees experiencing moderate water deficits, compared to trees in well-watered conditions, are predicted to have a surplus of carbohydrates beyond that which can be invested in growth (the difference between photosynthesis and growth curves in Figure 32.1). Pine trees with relatively more surplus carbon are predicted to have higher rates of secondary metabolism (Figure 32.1, lower), produce more oleoresin (a mixture of monoterpenes and resin acids that impedes attacking bark beetles), and allow only limited reproductive success in attacking beetles. At low water availability, photosynthesis is predicted to drop, which should limit rates of secondary metabolism. Hence, our model predicts high reproductive success of attacking bark beetles (i.e., positive population growth, $\lambda > 1$) at either high water availability or low water availability.

Experimental and theoretical support for this general model of plant responses is accumulating (Ayres, 1993; Herms and Mattson, 1992; Wilkens et al., 1996a). As predicted, Lorio (1978) found that 72% of all infestations they observed were on moist sites that were closely associated with a high site index. Also as forecasted for loblolly pine, severe water stress can limit resin flow and promote beetle success (Lorio and Hodges, 1977; Lorio et al., 1995). More surprisingly (but also as predicted) moderate water stress can produce the opposite effect. Recently completed studies in central Louisiana (Reeve et al., 1995) showed that moderately water-stressed loblolly pine had markedly higher rates of secondary metabolism (but reduced growth) than either control or irrigated trees; the potential rate of increase in SPB was three to six times higher among those that attacked irrigated trees compared to moderately water—stressed trees. The magnitude of these effects indicates that even modest changes in precipitation patterns could have dramatic effects on bark beetle population dynamics. Furthermore, because the response of secondary metabolism appears to be nonlinear (Figure 32.1), the effect of changes in precipitation are not only dependent upon the direction of change (i.e., precipitation increases or decreases), but also on the extent of change and the initial conditions of the forest. For example, a 20% decrease in summer precipitation might lead to moderate water stress and limit bark beetle outbreaks, but a 40% decrease might lead to severe water stress and promote bark beetle outbreaks. Similarly, a decrease of 20% in summer precipitation might lead to moderate water stress and limit bark beetle outbreaks in Louisiana, but also might lead to severe water deficits and promote bark beetle outbreaks in east Texas where there is less average rainfall and moderate water deficits are presently the status quo.

Global change is also apt to produce changes in southern forests through alteration of mineral-nutrient availability (Anderson, 1991; Bassaz, 1990; Chapin, 1991; Pastor and Post, 1988). Mineralization rates and, therefore, the availability of mineral nutrients to plants are highly sensitive to temperature, moisture, and litter quality. In some forests, including large regions of the northeastern United States, soil-nutrient regimes are also changing as a result of the sustained deposition of airborne nitrogenous pollutants (Lovett and Kinsman, 1990; McNulty et al., 1990; Miller et al., 1993; Olliger et al., 1993). These pollutants are frequently deposited as ammonium and nitrate (Friedland et al., 1991), and hence, may be expected to produce the same effects on tree growth and physiology as fertilization. A review of published studies (Ayres, 1993), indicates that of all the abiotic factors associated with global change (temperature, carbon dioxide (CO_2), water, cloud cover, and nutrient availability), changes in nutrient regime may produce the largest effects on tree resistance to herbivores. Changes in nutrient availability can alter patterns of secondary metabolism in many plants with consequences for populations of insect pests (Ayres 1993, and references within). However, the effects of nutrient availability on oleoresin production in loblolly pine are not known. Before developing land-use strategies that will be appropriate under future climatic regimes, we must understand the physiological responses to existing nutrient regimes and the effects of these responses on the interaction between SPB and southern pines. We hypothesized that our model (summarized in Figure 32.1 can be generalized to predict responses to changing nutrient availability, as well as changing water availability. Based on Figure 32.1, we predicted that the effect of fertilizing pines would be to increase growth more than it increases photosynthesis, which would limit carbohydrates available for secondary metabolism, lead to reduced oleoresin flow, and increase the reproductive success of attacking beetles. This prediction runs counter to predictions that any silvicultural practices that increase tree growth will necessarily increase tree resistance to herbivores (Mason et al., 1992; Raffa, 1988; Waring, 1983).

We emphasize, however, that Figure 32.1 does not indicate that fertilization will always reduce tree resistance. Indeed, the model indicates that fertilization can increase secondary metabolism in plants that are in extremely stressed conditions (Figure 32.1). We make our prediction of decreased tree resistance in response to fertilization with the assumption that the site is not extremely nutrient deficient. This assumption is supported by the site index for our study area, which was approximately 88, and which is midrange for loblolly pine forests in Lousiana and Arkansas (Carmean et al., 1989). Our assumption is further supported by foliar nitrogen (N) and phosphorus (P) concentrations in the study site (Gravatt, 1994), which generally fall at or above established sufficiency levels (Allen, 1987).

The risk of bark beetle outbreaks might be mitigated by appropriate silvicultural practices. We hypothesized that thinning of pine plantations would tend to elevate total carbon budgets in loblolly pine (because it increases the crown size and reduces shading), and would, therefore, increase rates of secondary metabolism at any given level of resource availability (Figure 32.1). If this is true, thinned plantations would only allow reproductive success in attacking bark beetles ($\lambda > 1$) at extremely low or high resource availability, unthinned plantations would allow beetle success across most conditions of water or nutrient availability. This prediction is consistent with reports that SPB infestations tend to be less severe in pine forests with low stand density than in "overstocked" pine forests with high stand density (Brown et al., 1987; Coster and Searcy, 1981; Gara and Coster, 1968; Lorio, 1980; Lorio et al., 1982; Lorio and Sommers, 1981; Mason et al., 1985; Nebeker et al., 1985). Our research aimed to test the hypotheses represented in Figure 32.1.

Materials and Methods

The study was conducted on a 1.1 hectare (ha) plantation on Beauregard silt loam soil located in the Johnson Tract of the Palustris Experimental Forest in the Kisatchie National Forest of central Louisiana. During 1981, the site was planted with loblolly pine seedlings at a 1.8 × 1.8 m spacing (6 × 6 ft), yielding a rectangular grid of seventy rows and ninety columns containing 6,300 trees. Before the growing season of 1989, treatments were randomly assigned to eight plots (each plot containing thirteen rows of thirteen trees); two plots were left as controls, two were fertilized with diammonium phosphate, two were thinned, and two were thinned and fertilized. Thinned plots were reduced to 748 trees per ha from the original density of 2990 trees/ha (Haywood, 1994). In April of 1989, fertilizer was applied at the rate of 746 kg/ha, that is, 150 kg P and 134 kg N per ha. We selected experimental trees from each of these plots; nine from each thinned plot and eleven from each unthinned plot, for a total of 80 trees. Experimental trees in the unthinned plots were drawn equally from trees classified in May of 1993 as codominant (tall enough to share the upper canopy) or intermediate (being partly overgrown by neighboring trees). The thinned plots, because of the thinning treatment, contained only codominant trees. Immediately after treatments were applied in 1989, the basal areas were 28 to 29 m^2/ha for the unthinned plots and 5 to 7 m^2/ha for the thinned plots. In 1992 when our measurements began, the basal areas were 37 to 39 m^2/ha for the unthinned plots and 14 to 17 m^2/ha for the thinned plots (Haywood, 1994).

At intervals throughout the growing season in 1993 and 1994, we measured oleoresin flow, phloem thickness, photosynthesis, height growth, and cambial growth. On each sampling date, oleoresin flow was measured from standardized wounds (125 mm^2; 2/tree) to the xylem face (Dunn and Lorio, 1993). Phloem discs (125 mm^2), which were removed to produce the wounds, were returned to the laboratory, freeze-dried, and then weighed to yield estimates of phloem specific mass (mg/125mm^2), which is a correlate of thickness.

We measured carbon assimilation rates with a portable photosynthesis system (Li-6200, Li-Cor Corporation). On each sampling occasion, we measured two sets of two fascicles from each tree. The sampled needle tissue was freeze-dried, weighed, and subsequently analyzed for total nitrogen using a Carlo–Erba carbon/nitrogen analyzer. For each sampled tree, we recorded the average length and mass of fascicles (four fascicles per tree), and the number of fascicles per 10 cm of shoot. These data allowed us to express photosynthesis as 1) $\mu moles \cdot m^{-2} \cdot s^{-1}$; (2) $\mu moles \cdot g^{-1} \cdot s^{-1}$; (3) $\mu moles \cdot g\, N^{-1} \cdot s^{-1}$; (4) $\mu moles \cdot fascicle^{-1} \cdot s^{-1}$; and (5) $\mu moles \cdot 10\, cm\, shoot^{-1} \cdot s^{-1}$.

Cambial growth was measured every seven to nine days during the growing season from dendrometer bands permanently affixed to each experimental tree at 1.5 m aboveground. Height was measured twice per year, once in the first week of June and again in late fall after height growth had ceased for the year.

We tested for patterns in carbohydrate partitioning by tracking the fate of labeled carbon (^{14}C) assimilated by photosynthesizing needles as CO_2. In late summer of 1993 (September 2) and in spring of 1994 (May 12), two branches from each of three trees in each of four treatment groups were exposed to 200 μCuries (740k Bq) of labeled CO_2 for thirty minutes within a plastic bag (long enough to assimilate about 90% of the label). Five days later, the branches were harvested, and then separated into basal and distal regions (the distal region was within the bag during label uptake). Tissue for analysis included a mixture of the inner bark and recently formed xylem cells. Xylem tissue was obtained by carefully removing the phloem tissue, and then scraping the last-formed cells from the xylem. Extraction of the tissues was done in such a way that labeled material could be separated into one of two fractions, which were 1) hexane soluble (mainly monoterpenes and resin acids) and 2) structural compounds. Liquid scintillation counts were performed on each sample; one was performed on the hexane extract (representing the oleoresin fraction), and one on the residue (after oxidation in a carbon oxidizer). Resulting data allowed us to compare the proportion of assimilated carbon that was distributed to primary vs secondary metabolism.

Treatment effects were evaluated by analysis of variance (ANOVA) with a statistical model that tested for effects of 1) thinning (thinned and unthinned), 2) fertilization (fertilized and unfertilized), 3) crown class (codominant and intermediate), and 4) sampling date (varies by measurement). Data were analyzed as a partial factorial (Milliken and Johnson, 1984) because one treatment combination (i.e., thinned plot, intermediate crown class) did not exist. Measurements of resin flow, phloem thickness, photosynthesis, and needle morphology involved subsampling within trees, so the ANOVAs were modified to include nesting within trees. Preliminary analyses included treatment plots as a blocking factor, but the effects of blocking were always small and usually nonsignificant, so blocking was eliminated from the models for simplicity.

Results

Photosynthesis

Thinning elevated the rate of photosynthesis when considered on a per fascicle or per 10 cm shoot basis; however, other measures of photosynthesis were not affected (Table 32.1; Figure 32.2). Crown class had no discernible effect on photosynthesis for any scale measured (Table 32.1). Photosynthesis was consistently higher in fertilized trees, yielding significant differences for most contrasts (Table 32.1; Figure 32.2). However, on a per gram basis (A/g), fertilizer had no effect. Among unthinned intermediate trees, fertilization increased photosynthesis on a per mole N basis (A/mole N), but not the others (Table 32.1; Figure 32.2).

Needle Morphology and Chemistry

Nitrogen concentration was higher in unthinned trees (Table 32.2; Figure 32.3). Among the codominants, however, thinning did not affect nitrogen concentration

Table 32.1. Statistical Contrasts for the Effects of Thinning,[1] Fertilization, and Crown Class (Codominants vs Intermediates) on Carbon Assimilation Rates (A = μmoles CO_2/s) of *Pinus taeda*

Treatment contrast	F-statistics					
	A/m^2	A/g	A/fascicle	A/10 cm shoot	A/mole N
Thinned vs unthinned	3.58	0.16	9.34**[2]	9.17**	0.02
Thinned vs unthinned codominants	3.32	0.00	5.05*	3.98	0.06
Fertilized vs unfertilized	11.60**	2.12	20.87***	9.33**	21.32***
Fertilized vs unfertilized in thinned	6.03*	2.10	8.72**	7.14*	10.42**
Fertilized vs unfertilized in unthinned	6.50*	0.77	12.97***	4.14*	12.08**
Fertilized vs unfertilized in unthinned codominants	4.71*	1.41	13.78***	2.52	7.22*
Fertilized vs unfertilized in unthinned intermediates	2.26	0.02	2.50	1.74	5.31*
Codominants vs intermediates in unthinned	0.08	0.48	0.50	1.19	0.41
Fertilizer × crown class interaction in unthinned	0.06	0.43	1.41	0.00	0.02
Fertilizer × thinning interaction in codominants	0.01	0.00	0.95	0.23	0.04
Trees (Treatment)	6.50***	5.97***	9.41***	14.83***	

[1] Corresponds to data in Figure 32.2. Measurements included two replicates per tree (each based on a different pair of fascicles drawn from the same shoot) for all except A/mole N. All treatment contrasts used MS_{tree} (degrees of freedom = 48) as the F-test denominator except for Trees (Treatment), which used MS_{error} (degrees of freedom = 65).

[2] * $P < .05$; ** $P < .01$; *** $P < .001$

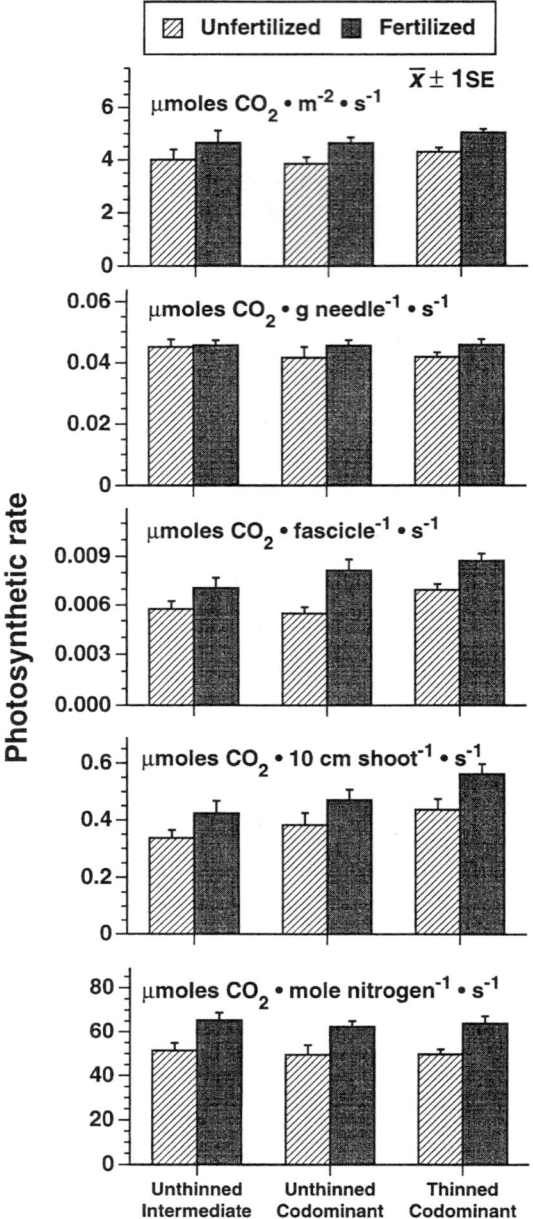

Figure 32.2. Photosynthetic rate of fertilized or unfertilized trees in thinned and unthinned plots (note: unthinned plots contain both codominant and intermediate trees and will be the case whenever unthinned and thinned trees are compared) as measured by several different methods. Starting from the top: on a per area basis, on a per gram basis, on a per fascicle basis, on a per shoot basis, and on a per unit of nitrogen basis.

Table 32.2. Statistical Contrasts for the Effects of Thinning, Fertilization, and Crown Class (Codominants vs Intermediates) on Needle Morphology and Chemistry of *Pinus taeda*[1]

Treatment contrast	F-statistics			
	Fascicle length	Fascicle mass	Fascicles / 10 cm	% Nitrogen
Thinned vs unthinned	10.54**	14.80***	0.18	4.18*
Thinned vs unthinned codominants	3.45	6.60*	0.06	1.37
Fertilized vs unfertilized	14.75***	14.80***	0.75	28.36***
Fertilized vs unfertilized in thinned	3.54	4.92*	0.17	10.20**
Fertilized vs unfertilized in unthinned	11.22**	10.08**	1.44	18.58***
Fertilized vs unfertilized in unthinned codominants	16.77***	8.68**	3.76	13.70***
Fertilized vs unfertilized in unthinned intermediates	0.86	2.79	0.00	6.56*
Codominants vs intermediates in unthinned	2.70	1.76	1.08	1.26
Fertilizer × crown class interaction in unthinned	3.74	0.39	1.70	0.09
Fertilizer × thinning interaction in codominants	3.81	0.73	3.08	0.34
Date				5.15*
Date × treatment[3]				0.73
Trees (treatment)				2.16**

[1] Corresponds to data in Figure 32.3. Nitrogen measurements included two dates (June 3, 1993 and July 5, 1995). All treatment contrasts used MS_{tree} as the *F*-test denominator (degrees of freedom = 48) except for Date, Date × Treatment, and Trees (Treatment), which used MS_{error} (degrees of freedom = 41).

[2] * $P < .05$; ** $P < .01$; *** $P < .001$

[3] The treatment classes included the six possible combinations of thinning, crown class, and fertilization (unthinned intermediate unfertilized, unthinned intermediate fertilized, unthinned codominant unfertilized, unthinned codominant fertilized, thinned codominant unfertilized, and thinned codominant fertilized).

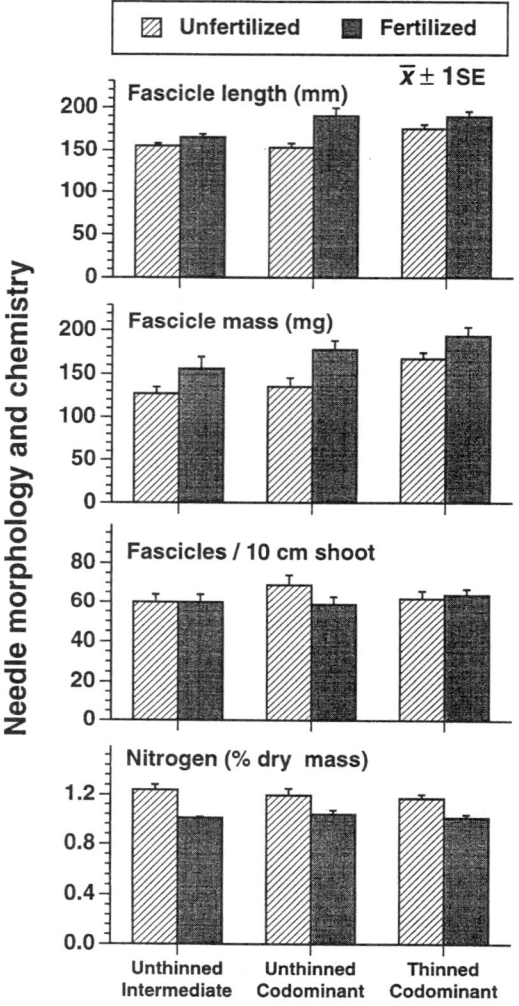

Figure 32.3. Needle morphology and chemistry of fertilized or unfertilized trees in thinned and unthinned plots.

(Table 32.2; Figure 32.3). Thinning resulted in greater fascicle length and mass (Table 32.2; Figure 32.3). Among the codominants, thinning increased fascicle mass but not fascicle length (Table 32.2; Figure 32.3). The number of fascicles per 10 cm was not affected by thinning or any other treatment (Table 32.2; Figure 32.3). Crown class had no discernible effect on needle morphology or chemistry (Table 32.2; Figure 32.3), but fertilization tended to increase the mass and length of the fascicles (Table 32.2; Figure 32.3). Fertilized trees had consistently lower nitrogen concentrations than the unfertilized trees (Table 32.2; Figure 32.3), which was also discovered by Gravatt (1994) at the same site in a different study. This result is consistent with a dilution effect. That is, fertilized trees had bigger needles than the unfertilized trees (mean ± standard error = 180 ± 7 vs 147 ± 6 mg per fascicle) but the same total amount of nitrogen per fascicle (mean ± standard error = 1.84 ± 0.09 vs. 1.76 ± 0.08 mg), and therefore, lower needle nitrogen concentrations (mean ± standard error = 1.022 ± 0.016 vs 1.200 ± 0.025).

Diameter Growth

Four years after treatments were applied, in 1988 trees in thinned plots had markedly larger trunks than those in unthinned plots (mean ± standard error = 18.01 ± 0.32 vs 15.45 ± 0.35 cm diameter, respectively). Thinned trees continued to show greater cambial growth than unthinned trees throughout the 1993 (not shown) and 1994 growing seasons (Table 32.3, Figure 32.4). Within the unthinned plots, trunk diameter in spring of 1993 was greater in codominant trees than intermediate trees (mean ± standard error = 15.45 ± 0.35 vs 11.76 ± 0.14 cm). Subsequent diameter growth in the unthinned plots continued to be greater for codominant trees than for intermediate trees throughout the 1993 and 1994 growing seasons (Table 32.3, Figure 32.4). Fertilizer had no effects on cambial growth (Table 32.3, Figure 32.4); cambial growth in all treatments was highly episodic. For example, trees in the thinned plot gained nearly 0.8 mm/week of cambial growth during three weeks in April and May of 1993, but barely grew at all ($<$ 0.2 mm/week) during the five weeks that separated these bouts of rapid growth (Figure 32.4). Periods of cambial growth were closely associated with short-term changes in soil moisture as each episode of cambial growth during 1993 occurred immediately following a significant rainfall event. The majority of cambial growth in all treatments occurred during the same brief episodes (Figure 32.4). Differences among treatments in cumulative cambial growth resulted largely from differences in the rate of growth during these episodes. Cambial growth patterns for 1993 (not shown) were nearly identical to those of 1994.

Height Growth

In 1992, four years after treatments were applied, unthinned trees were taller than thinned trees, codominant trees were significantly taller than intermediate trees, and fertilized trees were taller than unfertilized trees (Table 32.4; Figure 32.5). From 1993 to 1994, thinned trees grew more than unthinned trees in three of the four time-periods analyzed, thinned codominants grew more than unthinned codominants in one of the time-periods, and fertilized trees grew more in three of four time-periods (Table 32.4; Figure 32.6). Fertilizer increased height growth least in unthinned intermediate intermediate trees (Table 32.4; Figure 32.6).

Resin Flow

There were no significant differences in resin flow between thinned and unthinned trees (Table 32.5; Figures 32.7 and 32.8). Similarly, there were no differences in

Table 32.3. Statistical Contrasts for the Effects of Thinning, Fertilization, and Crown Class (Codominants vs Intermediates) on Trunk Diameter Growth in *Pinus taeda*

		F-statistics			
			Incremental diameter growth (mm)		
Treatment contrast	Diameter in 1992 (cm)	Spring 1993	Summer 1993	Spring 1994	Summer 1994
Thinned vs unthinned	145.04***[1]	256.86***	215.31***	178.22***	184.32***
Thinned vs unthinned codominants	34.55***	121.85***	98.73***	79.47***	78.19***
Fertilized vs unfertilized	0.08	0.69	1.32	0.14	0.01
Fertilized vs unfertilized in thinned	0.74	0.97	0.65	0.01	0.18
Fertilized vs unfertilized in unthinned	0.63	2.23	0.76	0.13	0.12
Fertilized vs unfertilized in unthinned codominants	0.20	2.78	1.06	0.20	0.24
Fertilized vs unfertilized in unthinned intermediates	0.45	0.22	0.05	0.01	0.00
Codominants vs intermediates in unthinned	56.86***	18.90***	18.33***	16.95***	20.99***
Fertilizer × crown class interaction in unthinned	0.03	0.66	0.30	0.06	0.11
Fertilizer × thinning interaction in codominants	0.79	3.68	0.09	0.08	0.42

[1] * $P < .05$; ** $P < .01$; *** $P < .001$

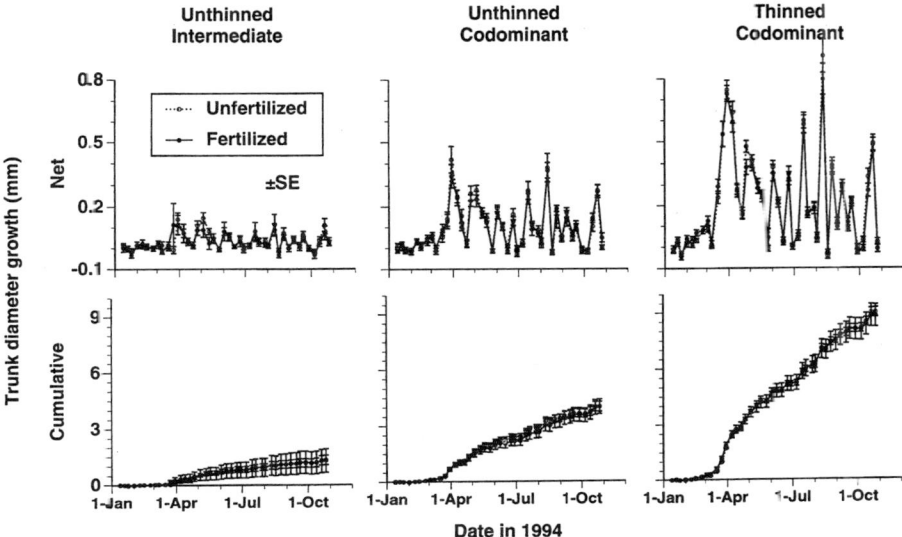

Figure 32.4. Diameter growth in 1994 of fertilized or unfertilized trees in thinned and unthinned plots presented in two different ways. Upper figure: The net growth, or incremental increase in growth after each seven to ten day period. Lower figure: The cumulative diameter growth over the growing season. Patterns of diameter growth for the 1993 growing season were nearly identical to the 1994 data shown here.

resin flow between codominant and intermediate crown classes (Table 32.5). In contrast, fertilizer generally reduced resin production (Table 32.5; Figures 32.7 and 32.8). However, there was a significant fertilizer × crown class interaction in the unthinned plots as fertilization affected resin flow in unthinned intermediates, but had no effect in the unthinned codominants (Table 32.5; Figures 32.7 and 32.8).

Phloem-Specific Mass

Phloem-specific mass was nearly twice as high in unthinned trees (Table 32.6; Figures 32.7 and 32.8). As the season progressed, there was a gradual decline in phloem-specific mass, especially in thinned codominants (Figures 32.7 and 32.8). Phloem-specific mass was greater in unthinned codominants than in unthinned intermediates, and fertilizer had no discernible effect on phloem specific mass (Table 32.6; Figures 32.7 and 32.8). All patterns were consistent across years. Phloem is a major storage site for starch and sugars in the tree, and is a primary food source for attacking beetles and their developing broods.

Labeled Carbon Patterns

Labeled carbon studies indicated strong seasonal shifts in partioning of carbon to growth vs secondary metabolism. In the spring, during a period of rapid growth of

Table 32.4. Statistical Contrasts for the Effects of Thinning, Fertilization, and Crown Class (Codominants vs Intermediates) on Height Growth in *Pinus taeda*

Treatment contrast	Height in 1992 (m)	F-statistics			
		\	Incremental height growth (cm)		
		Spring 1993	Summer 1993	Spring 1994	Summer 1994
Thinned vs unthinned	4.86*[1]	0.97	4.52*	26.49***	4.70*
Thinned vs unthinned codominants	31.15***	0.07	1.16	12.99***	0.67
Fertilized vs unfertilized	29.14***	6.89*	6.30*	16.65***	1.34
Fertilized vs unfertilized in thinned	11.19**	0.28	7.50**	9.62**	0.57
Fertilized vs unfertilized in unthinned	18.66***	7.30**	1.84	8.70**	0.82
Fertilized vs unfertilized in unthinned codominants	13.68***	6.66*	1.56	7.59**	5.49*
Fertilized vs unfertilized in unthinned intermediates	5.96*	1.62	0.47	2.09	0.97
Codominants vs intermediates in unthinned	43.39***	0.99	1.60	1.67	3.18
Fertilizer × crown class interaction in unthinned	0.62	0.74	0.14	0.74	5.44*
Fertilizer × thinning interaction in codominants	0.65	2.83	0.54	0.05	1.85

[1] * $P < .05$; ** $P < .01$; *** $P < .001$

32. Environmental Effects and Resistance to Bark Beetles 605

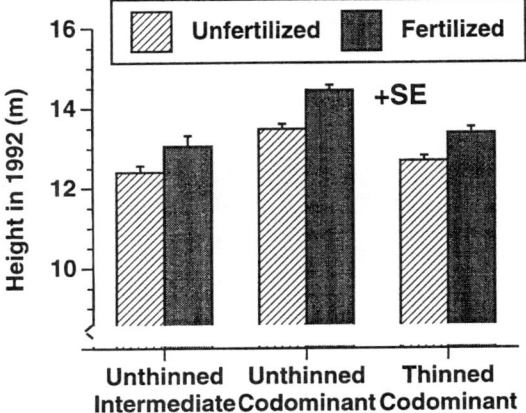

Figure 32.5. Height in 1992 of fertilized or unfertilized trees in thinned and unthinned plots. These heights were measured four years after treatments were applied and just before our intensive measurements began.

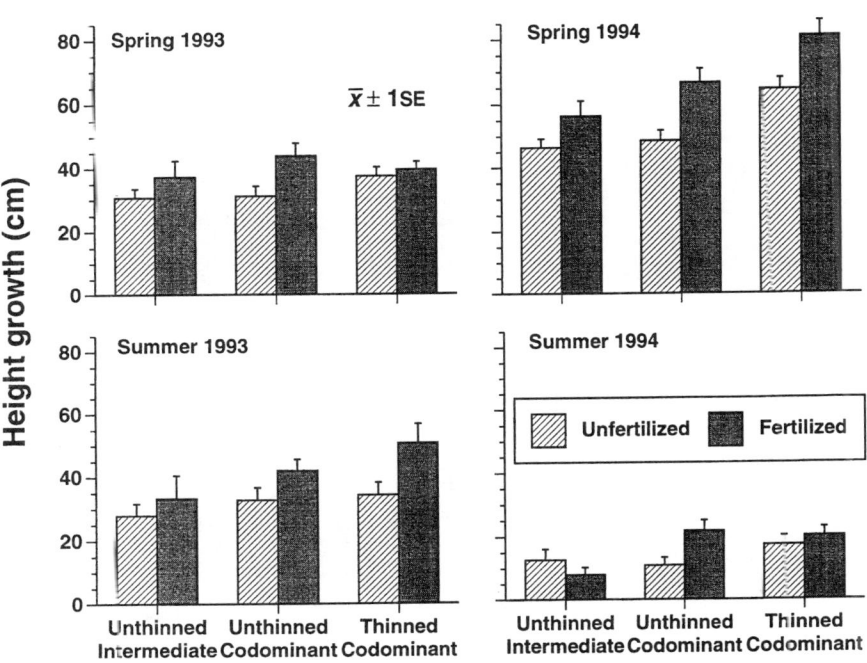

Figure 32.6. Height growth of fertilized or unfertilized trees in thinned and unthinned plots in the spring and fall of 1993 and 1994.

Table 32.5. Statistical Contrasts for the Effects of Thinning, Fertilization, and Crown Class (Codominants vs Intermediates) on Resin Flow in *Pinus taeda*[1]

Treatment contrast	d f	Error term	F	P
Thinned vs unthinned	1	MS_{tree}	1.63	0.21
Thinned vs unthinned codominants	1	MS_{tree}	1.33	0.25
Fertilized vs unfertilized	1	MS_{tree}	5.69	0.0196
Fertilized vs unfertilized in thinned	1	MS_{tree}	4.22	0.0436
Fertilized vs unfertilized in unthinned	1	MS_{tree}	2.53	0.12
Fertilized vs unfertilized in unthinned codominants	1	MS_{tree}	0.14	0.70
Fertilized vs unfertilized in unthinned intermediates	1	MS_{tree}	6.54	0.0126
Codominants vs intermediates in unthinned	1	MS_{tree}	0.02	0.89
Fertilizer × crown class interaction in unthinned	1	MS_{tree}	4.46	0.0381
Fertilizer × thinning interaction in codominants	1	MS_{tree}	2.50	0.11
Treatment[2]	5	MS_{tree}	73.85	0.0001
Date	14	MS_{error}	9.83	0.0001
Date × treatment	70	MS_{error}	1.42	0.0352
Tree(treatment)	74	MS_{error}	29.03	0.0001

[1] Data include nine dates in 1993 and six in 1994.
[2] The treatment classes included the six possible combinations of thinning, crown class, and fertilization (unthinned intermediate unfertilized, unthinned intermediate fertilized, unthinned codominant unfertilized, unthinned codominant fertilized, thinned codominant unfertilized, and thinned codominant fertilized).

32. Environmental Effects and Resistance to Bark Beetles 607

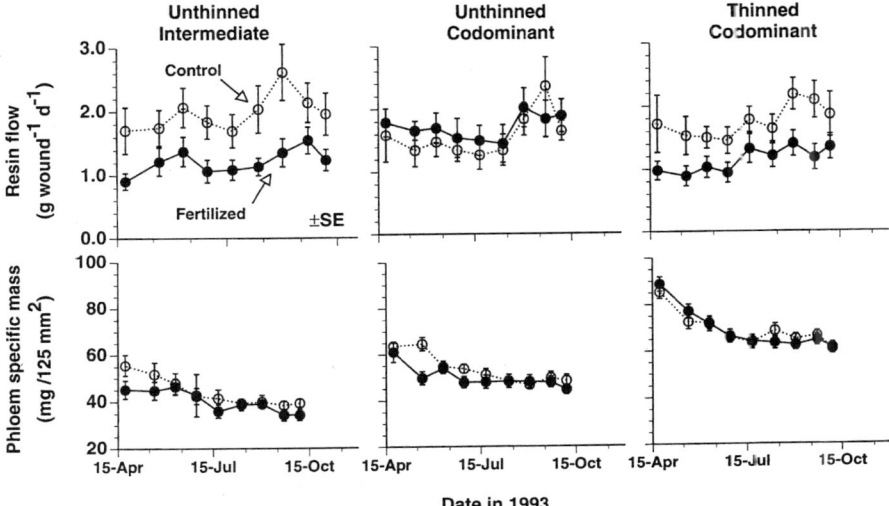

Figure 32.7. Resin flow and phloem-specific mass, in 1993, of fertilized or unfertilized trees in thinned and unthinned plots. The size of the wound made to measure resin flow and collect phloem samples was 125 mm².

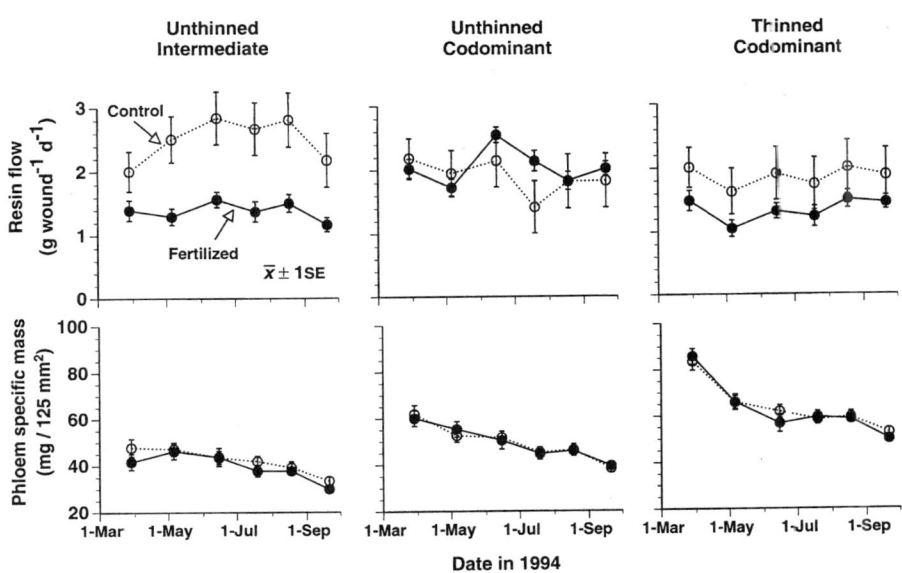

Figure 32.8. Resin flow and phloem-specific mass in 1994. See Figure 32.7 legend for details.

Table 32.6. Statistical Contrasts for the Effects of Thinning, Fertilization, and Crown Class (Codominants vs Intermediates) on Phloem Thickness (mg/125 mm^2) in *Pinus taeda*[1]

Treatment contrast	d f	Error term	F	P
Thinned vs unthinned	1	MS_{tree}	190.95	0.0001
Thinned vs unthinned codominants	1	MS_{tree}	83.01	0.0001
Fertilized vs unfertilized	1	MS_{tree}	2.18	0.14
Fertilized vs unfertilized in thinned	1	MS_{tree}	0.00	0.96
Fertilized vs unfertilized in unthinned	1	MS_{tree}	2.90	0.09
Fertilized vs unfertilized in unthinned codominants	1	MS_{tree}	0.91	0.34
Fertilized vs unfertilized in unthinned intermediates	1	MS_{tree}	2.06	0.16
Codominants vs intermediates in unthinned	1	MS_{tree}	21.05	0.0001
Fertilized × crown class interaction in unthinned	1	MS_{tree}	0.17	0.68
Fertilizer × thinning interaction in codominants	1	MS_{tree}	0.61	0.44
Treatment[2]	5	MS_{tree}	41.22	0.0001
Date	10	MS_{error}	58.27	0.0001
Date × treatment	50	MS_{error}	3.03	0.0001
Tree (treatment)	74	MS_{error}	6.56	0.0001

[1] Data include five dates in 1993 and six dates in 1994.
[2] The treatment classes included the six possible combinations of thinning, crown class, and fertilization (unthinned intermediate unfertilized, unthinned intermediate fertilized, unthinned codominant unfertilized, unthinned codominant fertilized, thinned codominant unfertilized, and thinned codominant fertilized).

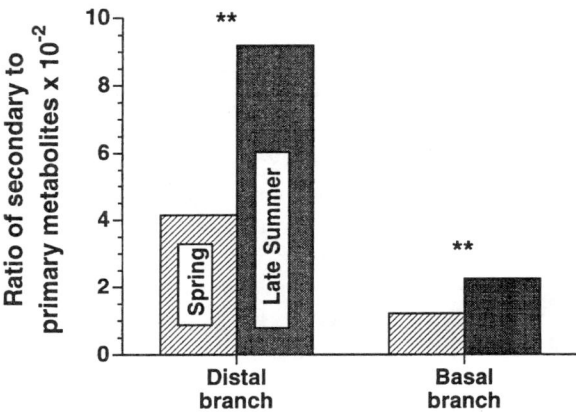

Figure 32.9. Ration of ^{14}C incorporated into secondary compounds relative to primary compounds (structure) in distal and basal branch portions in the spring (May 12 to 16, 1994) and in the late summer (September 2 to 7, 1993). Asterisks above bars indicate significant difference between the summer and spring measurements. (*$P = 0.01$).

shoots, the ratio of ^{14}C allocated to secondary compounds relative to structure was approximately 0.04 in the distal (inside the bag) part of branches and about 0.01 in the basal portion (Figure 32.9). Conversely, in late summer, during a very hot and dry period when shoot growth was negligible, the ratio of ^{14}C in secondary to primary metabolites was about 0.09 in the distal portion of branches and about 0.02 in the basal part (Figure 32.9).

Discussion

Patterns of Growth and Resin-Based Defenses

Figure 32.1 predicts that fertilization will ease nutrient limitations on tree growth, and result in lower allocation of carbon to resin-based defenses. As predicted, fertilized trees grew more than unfertilized trees. This was evident in height growth (Figures 32.5 and 32.6), fascicle growth (Figure 32.3), and root growth (Sword et al., Chapter 11), but not in cambial growth (Figure 32.4). Also as predicted fertilized trees had consistently lower resin flow than unfertilized trees. An exception to this pattern was that fertilization had no effect on resin flow in codominants found in unthinned plots (Figures 32.7 and 32.8). One explanation for this result is that fertilization increased photosynthesis and growth of codominants such that unfertilized and fertilized trees straddled the peak of the secondary metabolism curve in Figure 32.1.

A decrease in resin flow in response to fertilization is consistent with many studies that demonstrate a reduction in secondary defenses in response to fertilization (Bryant et al., 1987; Dudt and Shure, 1994; Larsson et al., 1986; Wilkens et al., 1996a; but see Honkanen, 1995 for exceptions). Pines in this study exhibited reductions in monoterpenes and resin acids in response to fertilization. Two other studies conducted at this site have also indicated a marked reduction in resin flow in response to fertilizer. The effects of fertilization ran counter to predictions that any silvicultural practices that increase tree growth and "vigor" will also increase tree resistance to herbivores (Raffa, 1988; Waring, 1983; Waring, et al., 1992). Our results indicate that increasing productivity using fertilizer may actually increase susceptibility to SPB damage through reduced resin flow. Resin flow has been shown to negatively affect SPB (Lorio et al., 1995; Reeve et al., 1995) and is widely considered to be the most important defense against bark beetles in general (Christiansen et al., 1987; Nebeker et al., 1993). Nevertheless, forestry practices are typically aimed at maximizing stand productivity; however, maximizing pine productivity through fertilization or by initiating plantations on highly productive sites may ultimately result in greater SPB damage. Indeed, Lorio (1978) found that 72% of all infestations observed were on moist productive sites. Our results, however, must be evaluated with the knowledge that this experiment was conducted on a single site of juvenile pines. More research is needed to evaluate the generality of these patterns. Nonetheless, these results suggest that it may be neccessary to balance the desire for rapid tree growth against concomitant decreases in resin flow, and increases in the risk of SPB damage.

Changes in soil fertility, resulting from changes in climate patterns, could decrease or increase the suitability of pines for SPB. The direction of this change will depend on whether global climate changes increase or decrease mineralization rates, and whether the affected population is under severe, moderate, or limited nutrient stress (Figure 32.1). Although it is difficult to predict how global change will influence soil-nutrient availability, the general pattern might be that of lower soil-nitrogen availability (Bazzaz, 1990). Our results suggest that this would tend to increase resin flow (and decrease success of attaching bark beetles) in pine populations experiencing moderate to little nutrient stress.

Soil moisture and, therefore, rainfall patterns also have strong effects on the physiology and resin production of loblolly pine (Lorio, 1986; Lorio et al., 1990). Water deficits lead to the development of latewood in which resin ducts are more plentiful (Lorio, 1993), and which leads to higher resin production (Lorio et al., 1990). In contrast, water surplus is associated with greater growth, but decreased resin flow (Lorio et al., 1990). Thus, fast-growing pines in well-watered sites may have increased susceptibility to SPB resulting from reduced resin flow. Because of this, foresters and planners must weigh the benefit of increased growth with the risk of decreased resistance to SPB. For instance, the placement of pine plantations on well-watered sites, such as bottomlands, may increase the risk of severe SPB problems.

The results of the ^{14}C study conducted in the spring and fall provide additional physiological evidence of shifts in carbon partitioning between primary and secondary metabolites associated with ontogeny and environmental conditions. These results are consistent with the principles of growth-differentiation balance (Loomis, 1932; Lorio, 1986), because rapid growth in the spring was associated with reduced carbon allocation to secondary metabolism (Figure 32.9). These represent the first data we know of that test the predictions of the growth-differentiation balance hypothesis using labeled carbon.

Figure 32.1 predicts that trees in thinned stands will have a larger carbon budget (g of fixed carbon $\cdot$ g^{-1} of plant tissue $\cdot$ d^{-1}), will invest relatively more carbon in resin-based defenses, and will have increased resistance to southern pine beetles. Increased secondary chemistry in response to light has also been shown in numerous studies (Coley, 1993; Dudt and Shure, 1994; Larsson et al., 1986; Shure and Wilson, 1993; Wilkens et al., 1996b; Wilkens et al., 1996a). However, our data did not support this hypothesis. Trees in thinned and unthinned plots plots had very similar resin flow (Figures 32.7 and 32.8). Three possible explanations for the similarity in resin flow between thinned and unthinned plots are identified in the next section.

Explanation 1

Resin production was similar between thinned and unthinned trees because thinned trees did not have a larger carbon budget on a per gram basis (g $\cdot$ g^{-1} $\cdot$ d^{-1}) than unthinned trees as predicted by Figure 32.1. On a per gram of needle-tissue basis, photosynthesis was similar between thinned and unthinned plots (Table 32.1; Figure 32.2). However, photosynthesis was higher in thinned trees on a per fascicle and per 10-cm shoot basis because of larger needles (Table 32.2; Figure 32.3). Visual assessment suggested that crowns were larger in thinned trees. Thus, it is probable that a larger crown containing more robust needles will produce more carbon per unit time (i.e., g $\cdot$ d^{-1}). We predicted, however, that the rate of carbon fixation would increase on a per gram of plant-tissue basis (i.e., g of fixed carbon $\cdot$ g^{-1} of plant tissue $\cdot$ day^{-1}), which includes the bole and the root system, in addition to the crown. There was much greater bole growth (Figure 32.4), and somewhat greater root growth in thinned plots (Sword et al., Chapter 11). Furthermore, faster-growing trees probably had greater respiration and maintenance costs (Kozlowski et al., 1991; Ryan 1991). If the increase in bole and root growth were commensurate with the increase in crown size, then the larger pool of fixed carbon in thinned trees would be partitioned between larger sinks, which would result in similar photosynthesis on a g $\cdot$ g^{-1} $\cdot$ d^{-1} basis between thinned and unthinned trees and, thus, similar resin production on a g $\cdot$ g^{-1} $\cdot$ d^{-1} basis. Although this explanation is plausible, it seems improbable that the response of trees to vastly improved light and nutrient conditions would lead to exactly proportional increases in crown, bole, and root systems. Empirical evidence repeatedly demonstrates that plants exposed to increased light have altered root/shoot ratios. Furthermore, Donner and Running (1986) found that the increased leaf hydration and light availability associated with thinned plots of lodgepole pine led to a 21% increase in photosynthesis. Thus, there are several reasons to believe that photosynthesis on a whole plant basis (g $\cdot$ g^{-1} $\cdot$ d^{-1}) was indeed higher in thinned trees, and we are inclined to reject explanation 1.

Explanation 2

Resin production was similar between thinned and unthinned trees because growth in unthinned trees was more limited by nutrients, water, and space than by carbon and, therefore, had a relative surplus of carbon to support secondary metabolism. Based on soil-moisture measurements, unthinned trees experienced more severe water shortages than the thinned trees, which is consistent with published results on the effect of thinning for loblolly pine (Basset, 1964; Zahner and Whitmore, 1960). If water and nutrient shortages in unthinned trees led to growth inhibition, the proportion of fixed carbon available for secondary metabolism would then increase (Figure 32.1). This shift to secondary metabolism might enable trees in unthinned plots to match the resin flow levels of the trees on thinned plots with larger carbon budgets.

Explanation 3

Resin production was similar between thinned and unthinned trees because thinned trees allocated proportionally more carbon to storage than thinned trees. Thinned trees had a dramatically higher phloem-specific mass, which indicates greater carbohydrate storage (Figures 32.7 and 32.8). Increased allocation to storage in thinned trees could explain why thinned and unthinned trees had similar levels of resin flow even though thinned trees had greater total carbon budgets.

Regardless of the physiological explanation, our results suggest that thinning does not affect tree physiology in a way that increases resistance to SPB attack. Thus, alternative explanations must be explored as to why SPB infestations tend to be less severe in pine forests with low stand density than in "overstocked" pine forests. One explanation is that attacking adult beetles are able to aggregate on host trees more efficiently in dense stands than in thinned stands (Gara and Coster, 1968; Johnson and Coster, 1978; Showalter and Turchin, 1993; Turchin, 1989; Turchin and Thoeny, 1993).

Summary

Our results indicate that increased nutrient availability leads to an increase in growth and a reduction in resin-based defenses. Because of this, we recommend that foresters and planners consider the effect of fertilization and site fertility on both tree growth and SPB risk. Unfortunately, increased tree growth may frequently be associated with increased SPB risk. The mechanism for the usefulness of thinning as a silvicultural technique to limit SPB infestations remain unclear.

If, as expected, global change alters precipitation patterns and soil-nutrient availability in the southern United States, patterns of growth and secondary metabilism in southern pines will also change, and this will probably produce changes in the spatial and temporal pattern of SPB infestations. Because SPB have such a large impact on southern forests, even modest changes in SPB activity can have significant economic and ecological repercussions. Our theoretical and empirical understanding of environmental effects on the physiology of southern pine remains imperfect but is growing. With this study, we now have some basis for predicting the direction and magnitude of changes in risk of SPB attack, anticipating how these patterns will vary from region-to-region or site-to-site, and suggesting appropriate changes in land-use strategies. Further testing and parameterization of such physiologically explicit models as depicted in Figure 32.1 should increase our ability to manage southern forests in a changing world.

References

Allen HL (1987) Forest fertilizers: Nutrient amendment, stand productivity, and environmental impact. J For 85:37–46.
Anderson JM (1991) The effects of climate change on decomposition processes in grassland and coniferous forests. Ecol Appl 1:326–347.
Ayres MP (1993) Plant defense, herbivory, and climate change. In Kareiva PM, Kingsolver JG, Huey RB (Eds) *Biotic interactions and global change.* Sinauer Associates Inc., Sunderland, MA.
Basset JR (1964) Diameter growth of loblolly pine trees as affected by soil moisture availability. US Forest Service Research Papers. SO–59.
Bazzaz FA (1990) The response of natural ecosystems to the rising global CO_2 levels. Ann Rev Ecol Syst 21:167–196.
Beal JA (1927) Weather as a factor in southern pine beetle control. J For 25:741–742.
Beal JA (1933) Temperature extremes as a factor in the ecology of the southern pine beetle. J For 31:329–336.
Berryman AA, Ferrel GT (1988) The fir engraver beetle in western states. In Berryman AA (Ed) *Dynamics of forest insect populations: patterns, causes, and implications.* Plenum Press, New York.
Brown CL, Sommer HE (1992) Shoot growth and histogenesis of trees possessing diverse patterns of shoot development. Am J Bot 79:335–346.
Brown MW, Nebeker TE, Honea CR (1987) Thinning increases loblolly pine vigor and resistance to bark beetles. South J For 11:28–31.
Bryant JP, Chapin FS III, Reichardt PB, Clausen TP (1987) Response of winter chemical defense in Alaska paper birch and green alder to manipulation of plant carbon/nutrient balance. Oecologia 72:510–514.
Carmean WH, Hahn JT, Jacobs RD (1989) Site index curves for forest tree species in the eastern United States. USDA, For Ser, Gen Tech Rep NC-128.
Chapin FS III (1991) Effects of multiple environmental stresses on nutrient availability and use. In Mooney HA, Winner WE, Pell EJ, Chu E (Eds) *Response of plants to multiple stresses.* Academic Press, San Diego.
Christiansen E, Bakke A (1988) The spruce bark beetle of Eurasia. In: Berryman AA (ed) *Dynamics of Forest Insect Populations: Patterns, Causes, and Implications.* Plenum Press, New York.
Christiansen E, Waring RH, Berryman AA (1987) Resistance of conifers to bark beetle attack: Searching for general relationships. For Ecol Man 22:89–106.
Coley PD (1993) Gap size and plant defenses. Tree 8:1–2.
Coster JE, Searcy JL (1981) Site, stand, and host characteristics of southern pine beetle infestations. USDA, Tech Bull Vol 1612.
Craighead FC (1925) Bark beetle epidemics and rainfall deficiency. J Econ Entomol 18:577–584.
Donner SL, Running SW (1986) Water stress response after thinning *Pinus contorta* stands in Montana. For Sci 32:614–625.

Dudt JF, Shure DJ (1994) The influence of light and nutrient on foliar phenolics and insect herbivory. Ecol 75:86–98.

Dunn JP, Lorio PL Jr (1993) Modified water regimes affect photosynthesis, xylem water potential, cambial growth, and resistance of juvenile *Pinus taeda* L. to *Dendroctonus frontalis* (Coleoptera: Scolytidae). Physiol Chem Ecol 22:948–957.

Friedland AJ, Miller EK, Battles JJ, Thorne JF (1991) Nitrogen deposition, distribution, and cycling in a subalpine spruce-fir forest in the Adirondacks, New York, USA. Biogeochem 14:31–55.

Gara RI, Coster JE (1968) Studies on the attack behavior of the southern pine beetle. III. Sequence of tree infestation within stands. Contrib Boyce Thompson Inst 24:77–85.

Gravatt DA (1994) Physiological variation in loblolly pines (*Pinus taeda* L.) as related to crown position and stand density. PhD dissertation, Louisiana State University, Baton Rouge, Louisiana.

Grégoire JC (1988) The greater European spruce beetle. In: Berryman AA (Ed) *Dynamics of forest insect populations: Patterns, causes, and implications.* Plemum Press, New York.

Haywood JD (1994) Seasonal and cumulative loblolly pine development under two stand density and fertility levels through four growing seasons. USDA, Res Paper, SO–283.

Herms DA, Mattson WJ (1992) The dilemma of plants: To grow or defend. Quart Rev Biol 67:283–335.

Honkanen T (1995) Plant defenses: The roles of intraplant regulation and resource availability. PhD dissertation, University of Turku, Turku Finland.

Johnson PC, Coster JE (1978) Probability of attack by southern pine beetle in relation to distance form an attractive host tree. For Sci 24:574–580.

Kalkstein LS (1976) Effects of climatic stress upon outbreaks of the southern pine beetle. Env Entomol 5:653–658.

King B (1972) Rainfall and epidemics of the southern pine beetle. Env Entomol 1:279–285.

Kozlowski TT, Kramer PJ, Pallardy SG (1991) *The physiological ecology of woody plants.* Academic Press, San Diego.

Kroll JC, Reeves HC (1978) A simple model for predicting annual numbers for southern pine beetle infestations in East Texas. S J Appl For 2:62–64.

Larsson S, Wiren A, Lundgren L, Ericsson T (1986) Effects of light and nutrient stress on leaf phenolic chemistry in *Salix dasyclados* and susceptibility to *Galerucella lineola* (Coleoptera). Oikos 47:205–210.

Loomis WE (1932) Growth-differentiation balance vs. carbohydrate nitrogen ratio. Am Soc Hort Sci 29:240–245.

Loomis WE (1953) Growth correlation. In Loomis WE (Ed) *Growth and differentiation in plants.* The Iowa State College Press, Ames.

Lorio PL Jr (1978) Developing stand risk classes for the southern pine beetle. USDA, For Ser Res Paper, SO–144.

Lorio PL Jr (1980) Loblolly pine stocking levels affect potential for southern pine beetle infestation. S J Appl For 4:162–165.

Lorio PL Jr (1986) Growth-differentiation balance: A basis for understanding southern pine beetle-tree interactions. For Ecol Man 14:259–273.

Lorio PL Jr (1993) Environmental stress and whole-tree physiology. In Showalter TD, Filip GM (Eds) *Beetle-pathogen interactions in conifer forests.* Academic Press, London.

Lorio PL Jr, Hodges JD (1977) Tree water status affects induced southern pine beetle attack and brood production. USDA For Ser Res Paper, SO–135.

Lorio PL Jr, Mason GN, Autry GL (1982) Stand risk rating for the southern pine beetle: Integrating pest management with forest management. J For 80:212–241.

Lorio PL Jr, Sommers RA (1981) Central Louisiana. In Coster JE, Searcy JL (Eds) *Site, stand, and host characteristics of southern pine beetle infestations.* USDA, Washington, DC.

Lorio PL Jr., Sommers RA, Blanche CA, Hodges JD, Nebeker TE (1990) Modeling pine resistance to bark beetles based on growth and differentiation balance principles. In Dixon RK, Meldahl RS, Ruark GA, Warren WG (Eds) *Process modeling of forest growth responses to environmental stress*. Timber Press, Portland, OR.

Lorio PL Jr, Stephen FM, Paine TD (1995) Environment and ontogeny modify loblolly pine response to induced acute water deficits and bark beetle attack. For Ecol Man 73:97–110.

Lovett GM, Kinsman JD (1990) Atmospheric pollutant deposition to high-elevation ecosystems. Atmos Environ 24A:2767–2786.

Martinat PJ (1987) The role of climatic variation and weather in forest insect outbreaks. In: Barbosa P, Schultz JC (Eds) *Insect outbreaks*. Academic Press, San Diego, CA.

Mason GN, Lorio PL Jr, Belanger RP, Nettleton WA (1985) Rating the susceptibility of stands to southern pine beetle attack. USDA, Agricultural Handbook, Washington, DC, Volume 645.

Mason RR, Wickman BE, Beckwith RC, Paul HG (1992) Thinning and nitrogen fertilization in a grand fir stand infested with western spruce budworm. Part I: Insect response. For Sci 38:235–251.

Mattson WJ Jr (1980) Herbivory in relation to plant nitrogen content. Ann Rev Syst Ecol 11:119–161.

McNulty SG, Aber JD, McLellan TM, Katt SM (1990) Nitrogen cycling in high elevation forests of the northeastern US in relation to nitrogen deposition. Ambio 19:38–40.

Michaels FJ (1984) Climate and the southern pine beetle in Atlantic Coastal and Piedmont regions. For Sci 30:143–156.

Miller EK, Friedland AJ, Arons EA, Mohnen VA, Battles JJ, Panek JA, Kadlecek J, Johnson AH (1993) Atmospheric deposition to forests along an elevational gradient at Whitface Mountain, NY, U.S.A. Atmos Environ 14:2121–2136.

Milliken GA, Johnson DE (1984) *Analysis of messy data volume I: Designed experiments*. Van Nostrand Reinhold Company, New York.

Nebeker TE, Hodges JD, Blanche CA (1993) Host response to bark beetle and pathogen colonization. In Schowalter TD, Filip GM (Eds) *Beetle-pathogen interactions in conifer forests*. Academic Press, New York.

Nebeker TE, Hodges JD, Karr BK, Moehring DM (1985) Thinning practices in southern pines—with pest management recommendations. USDA For Ser, Tech Bull, Vol. 1703.

Olliger SV, Aber JD, Lovett GM, Millham SE, Lathrop RG, Ellis JM (1993) A spatial model of atmospheric deposition for the northeastern US Ecol Appl 3:459–472.

Pastor J, Post WM (1988) Response of forests to CO_2-induced climate change. Nature 334:55–58.

Price TS, Dogget C, Pye JM, Holmes TP (1992) A history of southern pine beetle outbreaks in the southeastern United States. The Georgia Forestry Commission, Macon, Georgia.

Raffa K (1988) The mountain pine beetle in western North America. In: Berryman AA (Ed) *Dynamics of forest insect populations: Patterns, causes, and implications*. Plenum Press, New York.

Reeve JD, Ayres MP, Lorio PL Jr (1995) Host suitability, predation, and bark beetle population dynamics. In Cappuccino N, Price PW (Eds) *Population dynamics: New approaches and synthesis*. Academic Press, San Diego.

Ryan MG (1991) Effects of climate change on plant respiration. Ecol Appl 1:157–167.

Showalter TD, Turchin P (1993) Southern pine beetle infestation development: Interaction between pine and hardwood basal areas. For Sci 39:201–210.

Shure DJ, Wilson LA (1993) Patch-size effects on plant phenolics in successional openings of the southern Appalachians. Ecology 74:55–67.

St. George RA (1930) Drought-affected and injured trees attractive to bark beetles. J Econ Entomol 23:825–828.

Turchin P (1989) Population consequences of aggregative movement. J Anim Ecol 58:75–100.

Turchin P, Thoeny WT (1993) Quantifying dispersal of southern pine beetles with mark-recapture experiments and a diffusion model. Ecol Appl 3:187–198.

Waring RH (1983) Estimating forest growth and efficiency in relation to tree canopy area. Adv Ecol Res 13:327–354.

Waring RH, Savage T, Cromack K Jr, Rose C (1992) Thinning and nitrogen fertilization in a grand fir stand infested with western spruce budworm. Part IV: An ecosystem management perspective. For Sci 38:275–286.

Wilkens RT, Shea GO, Halbreich S, Stamp NE (1996b) Resource availability and the trichome defenses of tomato plants. Oecologia 106(2):181–191.

Wilkens RT, Spoerke JM, Stamp NE (1996a) Differential responses of growth and two soluble phenolics of tomato to resource availability. Ecol 77:247–258.

Wyman L (1924) Bark-beetle epidemics and rainfall deficiency. USDA For Serv Bull 8:2–3.

Zahner R, Whitmore FW (1960) Early growth of radically thinned loblolly pine. J For 58:628–634.

33. Predictions of Southern Pine Beetle Populations Using a Forest Ecosystem Model

Steven G. McNulty, Peter L. Lorio Jr., Matthew P. Ayres, and John D. Reeve

Dendroctonus frontalis Zimm. (southern pine beetle (SPB)) has caused over $900 million in damage to pines in the southern United States between 1960 and 1990 (Price et al., 1992). The damage of SPB to loblolly (*Pinus taeda* L.), shortleaf (*Pinus echinata* Mill.), and pitch (*Pinus rigida* Mill.) pine has long been established (Hopkins, 1899), however, extensive mapping of SPB infestations has only existed since 1960 (Price and Doggett, 1982). Early detection of SPB outbreak areas is essential to controlling population increases (Swain and Remion, 1981), but the range of SPB is large, SPB have six to eight generations per year, and there is inconsistency in the monitoring methods used to measure SPB populations across its range. Therefore, various models have been developed that attempt to predict SPB outbreak severity across the region (Hansen et al., 1973; Kalkstein, 1974; Michaels, 1984).

In addition to predicting SPB outbreak areas based on present-day climates, models could be useful for assessing the influence of future climate change on SPB populations. One unknown regarding climate change and forest ecosystems involves the effect that chronically warmer air temperature could have on insect populations. For example, a 1 °C increase in average annual air temperature caused a tripling of a principal insect herbivore that feeds on mountain birch (Herms, 1991). When combined with the potential for increased forest stress resulting from climate change, increased insect populations could seriously reduce forest productivity across the region.

Because the SPB has six to eight generations per year, it is impossible to accurately measure the population at one time across a large geographic area. Instead, we have chosen to predict how average annual SPB populations change given annual environmental change. We have used a process-based forest water use and productivity model called PnET–IIS to assess the potential impact of climate change on average SPB population and southern pine forest growth. The PnET–IIS model has been used before to predict regional scale drainage, soil-water stress (WATSTRS), and forest productivity under historic conditions and for a series of climate change scenarios (McNulty et al., 1994, 1996a, 1996b, 1996c). This research modifies formerly validated predictions by PnET–IIS of forest water use and productivity to estimate factors associated with the regulation of average annual SPB populations, given historic climate data and two climate change scenarios. This type of model could be used to alert forest managers to present areas of heavy SPB infestation or future areas that are susceptible to infestation given changing climate.

Factors Affecting Southern Pine Beetle Population

White (1974) hypothesized that insect performance is favored in stressed plants because consumed tissues have higher nitrogen concentrations, and Rhoades (1979) linked increased stress with reduced synthesis of defensive compounds. It has long been recognized that climatic effects, especially drought can weaken host trees and increase SPB populations (Wyman, 1924; Craighead, 1925; St. George, 1930). Although the relationships between SPB population, climate, and host tree vigor have been studied and refined during the past seventy years, experimental evidence often fails to correlate increased plant stress with increased insect performance (Larsson, 1989) as fast-growing trees may also be at risk to SPB infestation (Hopkins, 1899; Lorio et al., 1988; Lorio and Sommers, 1986; Price et al., 1992). Manipulations of loblolly pine have also shown that severe water deficits can enhance the success of attacking SPB (Lorio and Hodges, 1977; Lorio et al., 1990), but these conditions may exceed normal climatic variation (Reeve et al., 1995). These findings and 450 other studies were summarized by Waring and Cobb (1992) in a review of plant interactions between plants and herbivores, and they concluded that climatic stress could have a positive, negative, or negligible influence on the size of insect populations.

The apparent discrepancy in correlating plant health, environmental stress, and insect population size arises because the interactions are complex and researchers lack the adequate knowledge of the causes of plant stress that is essential for understanding the nature of plant and insect interactions. For example, pine trees in the southeastern United States grow very rapidly in the spring and early summer but by midsummer, tree growth typically becomes limited by soil-water deficits (Reeve et al., 1995). This environmental change correspond to an ontogenetic transition in the cambium from the production of earlywood to latewood. The latewood is dominated by vertical resin ducts that function in the synthesis and transport of oleoresin (a mixture of monoterpenes and resin acids that impedes attacking bark beetles but also plays a role in secondary attraction of SPB). Annual variation in climatic conditions influences the timing of seasonal water deficits and the timing of the transition from earlywood to latewood. Measurements of resin flow from standardized wounds to the face of the cambium indicate that spring and early summer is a period of low resin flow for loblolly pine. The transition to latewood formation is accompanied by an increase in resin flow, probably resulting from the increase in vertical resin ducts and because increased water stress would increase the production of secondary metabolites. A climate that protracts the period of earlywood production in southern pine (e.g., high precipitation in midsummer) favors SPB outbreaks by extending the time when trees are more easily colonized. To better predict the effects of SPB on forest function and changes in SPB populations size, improved physiologically based forest-process models are needed.

Modeling Southern Pine Beetle Populations

Many insect population models have attempted to correlate outbreaks of SPB with climatic conditions, including principle component analysis (Michaels, 1984; Michaels et al., 1986) but have had limited success (Hansen et al., 1973; Kalkstein, 1974). One reason for the lack of success in relating SPB activity to climate may be that other factors (e.g., oleoresin production, prior years' SPB population, growth and differentiation balance relationships) that are needed to predict SPB population changes are often not included in predictive models (Martinat, 1987). For example, Michaels et al., (1986) developed a model (SPBCMP) that incorporates climate and site-specific volume and growth data into predictions of future SPB population size. Although this model has many biological controls that have been cited as factors regulating SPB population size, the model makes many assumptions about forest processes that could be better simulated using a forest process model such as PnET–IIS. A better estimate of forest function would improve a model's ability to accurately predict changes in the average annual SPB population size.

Original PnET–IIS Model Structure and Predictive Outputs

To predict and project average annual SPB populations we used PnET–IIS (McNulty et al., 1994, 1996a, 1996b), a monthly time-step model, which is a derivation of the PnET–II model developed by Aber and Federer (1992) and Aber et al., (1995) for predicting forest water use and productivity. The PnET–IIS model has been modified to run for pine forests across the southern United States. The model uses site-specific, soil-water-holding capacity, four monthly climate parameters (i.e., minimum and maximum air temperature, total precipitation, and solar radiation) and species-specific process coefficients to predict evapotranspiration (ET), water drainage, soil water stress (WATSTRS), and net primary productivity (NPP) from the stand level ($<$ one hectare (ha)) to a $0.5° \times 0.5°$ grid cell resolution (approximately 50 × 75 km) across the southern United States (McNulty et al., 1994, 1996b).

Leaf area is a major component in calculating NPP, plant-water use and soil-water stress. The PnET–IIS model assumes that all stands are fully stocked and that leaf area is equal to the maximum amount of foliage that can be supported by soil, climate, and vegetative conditions. Predicted NPP is total gross photosynthesis minus growth respiration and maintenance respiration for leaf, wood, and root compartments. The model calculates respiration as a function of the present and prior month's minimum and maximum air temperature. The optimum temperature for loblolly pine net photosynthesis varies from 23 to 27 °C, and the maximum air temperature for gross photosynthesis ranges from 30 to 43 °C (Strain et al., 1976). As temperature increases beyond the optimum photosynthetic temperature, the respiration rate increases while gross photosynthesis increases slightly or decreases, so that a proportionally reduced amount of net carbon per unit leaf area is fixed (Kramer, 1980). Total gross photosynthesis is a function of gross photosynthesis per unit leaf area and total leaf area. Changes in water availability and plant-water demand place limitations on the amount of leaf area produced. When vapor pressure is deficient and air temperature increases, leaf area and total gross photosynthesis decrease.

Annual transpiration is calculated from a maximum potential transpiration that is modified by plant-water demand (a function of gross photosynthesis and water use efficiency (WUE)). Interception water loss is a function of leaf area and total precipitation; in closed canopy stands, interception water loss is approximately 15% of total precipitation. Evapotranspiration is equal to transpiration and interception loss, and drainage is calculated as water in excess of ET and soil-water-holding capacity (SWHC). Plant-water demand is dependent on monthly precipitation and water stored in the soil profile. If precipitation inputs exceed plant-water demand, the soil is first recharged to the SWHC, and if water is still available, it is output as drainage. Monthly drainage values are summed to provide an estimate of annual water outflow. Present-year growing season (April to September) soil water stress (WATSTRS0) equals $0.15 + (1 - $ (average growing season soil water/SWHC). Average growing-season soil water equals the sum of monthly calculated growing-season soil water that is $<$ SWHC, divided by the number of months in the growing season. Theoretically, soil water stress could range from 0.15 (no water stress) to 1.15 (maximum potential water stress); and has a minimum value 0.15, as opposed to 0, to reflect approximate percentage of hygroscopic water present in southern clay soils that cannot be removed by plant roots (Pritchett, 1979). Therefore, if no precipitation occurred during the growing season, soil water stress would be 1.15. Across the southern United States, average annual WATSTRS range between 0.15 and 0.50.

At a prior date, PnET–IIS model predictions of water use and productivity have been validated across the southern United States. Stream-gauge stations represent an integration of water yield across a region. Predictions of drainage across the southern United States by PnET–IIS were well correlated with average US Geological Survey measured discharge from 1951 to 1980 ($R^2 = 0.66$,

$P < 0.00001$, $n = 502$) (McNulty et al., 1996a). Predicted NPP was also well correlated with measured average annual basal area growth in twelve sites located from Texas to Virginia ($R^2 = 0.67$, $P = 0.005$, $n = 12$) (McNulty et al., 1996a).

Southern Pine Beetle Population Modeling Using PnET–IIS

Although many researchers have attempted to relate climatic factors with tree stress and SPB population change, most have had little success. However, oleoresin production has been linked with the failure rate of SPB to colonize host trees and, hence, is a partial controller of average annual SPB population size (Lorio, 1986; Reeve et al., 1995). Factors that regulate oleoresin production (i.e., soil water stress and photosynthesis) should, therefore, also partially affect SPB population size. In addition to tree stress, SPB population size is quasicyclic (Turchin et al., 1991), therefore the size of the prior year's SPB population is also an important factor in deciding the size of the present year's SPB population. These three factors (soil water stress, photosynthesis, and the prior year's SPB population) all influence the present year's SPB population and will be discussed in later sections.

The PnET–IIS is primarily a regional scale, forest-process model, and the model is not designed to predict some parameters that are important in forecasting SPB at the stand level. For example, PnET–IIS simulates stand growth from time zero to a point of canopy closure. Canopy closure and model equilabration requires approximately ten years of data. No provision is made for stand age or stand stocking because it would be impractical to input this information into the model on a regional scale. However, both stand stocking and forest age are important controllers of SPB infestation rates.

Soil Water Stress

Total oleoresin production is partially a function of plant-growth-differentiation balance, which is related to soil water stress. The principle of growth-differentiation balance asserts that tissue growth is negatively associated with tissue differentiation when the two processes compete for the same pool of carbohydrates (Loomis, 1932, 1953; Lorio, 1986, 1988; Herms and Mattson, 1992). Plants with a surplus of carbohydrates beyond that which can be invested in growth because of such limitations as soil water stress, invest proportionally more carbon in differentiation of such secondary compounds as oleoresin, which would be detrimental to such insect herbivores as the SPB. Conversely, plants limited in carbon supply invest less carbon into growth or differentiation processes. Plants with a high allocation to differentiation should have low concentrations of water and protein, but high concentrations of such secondary metabolites as oleoresin (Herms and Mattson, 1992).

Oleoresin production and growth-differentiation balance can be modeled using an inverse hyperbolic function of soil water stress (Reeve et al., 1995). In southern pines, although growth processes decline during moderate soil-water deficits, the production of secondary compounds tends to increase. When soil water stress is low, more of the photosynthate is devoted to tree growth compared to oleoresin production. Conversely, when soil water stress is very high (i.e., extreme drought) the production of photosynthates will be low because stomates remain closed and total photosynthate and oleoresin production remains low. Therefore, during periods of moderate water stress, pines may be less susceptible to SPB attack than during periods of either very low or high water stress. The PnET–IIS model uses an inverse hyperbolic function of soil water stress factor (WATSTRSFAC) to simulate the effect of soil water stress on oleoresin production (Figure 33.1). When soil water stress was either low or high, oleoresin production was low and WATSTRSFAC > 1.0; when soil water stress was moderate, oleoresin production was maximized and WATSTRSFAC ≤ 1.0. The relationship between oleoresin production and the probability of SPB colonization is expressed as the effect that present-year oleoresin production (WATSTRFAC0) would have on present-year

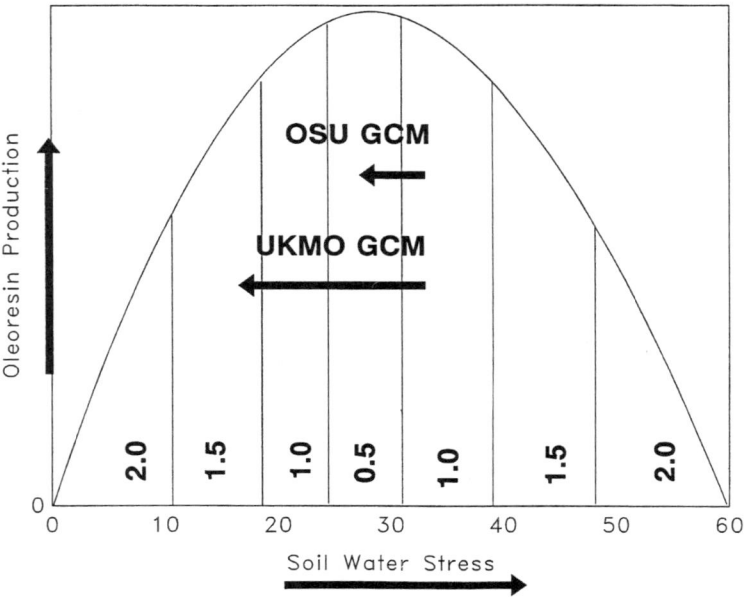

Figure 33.1. The influence of soil-water stress (WATSTRS) on oleoresin production is represented by an inverse hyperbolic function, and is expressed as a southern pine beetle population multiplication factor (WATSTRSFAC). Maximum oleoresin production and the maximum reduction in the WASTRSFAC (WASTRSFAC = 0.5) occurs under conditions of moderate soil water stress. Extreme water stress or extreme water excess reduces oleoresin production and increases the WASTRSFAC (WASTRSFAC > 1.0). Under historic conditions, the ecosystem predicted moderately severe water stress and WASTRSFAC = 1.0. Both the Oregon State University (OSU) and United Kingdom Meteorological Office (UKMO) caused a predicted decrease in regional soil water stress resulting from a reduction in total leaf area and a decrease in plant-water demand across the southern United States.

average SPB population size (SPBPOP0). The WATSTRFAC0 values > 1.0, $= 1.0$, and < 1.0 show that conditions are more favorable, equally favorable, or less favorable to future SPB populations, compared to the present-year average SPB population size.

Photosynthesis

The rate of plant photosynthesis also affects oleoresin production (Reeve et al., 1995). Although related to soil water stress, photosynthesis is also a function of air temperature and species type. Across the region, predicted average historic annual NPP is approximately 12 t ha^{-1} year^{-1} (McNulty et al., 1996b). If NPP were the only factor controlling oleoresin production, rapidly growing trees should have lower oleoresin production and higher incidence of SPB infestation compared to slower-growing trees. When rates of photosynthesis are low, little carbohydrate is available for growth or production of secondary compounds and when photosynthesis is high much more carbohydrate is available, but proportionally less of the carbohydrate may be allocated for secondary compounds (Reeve et al., 1995). Finally, when photosynthetic rates are moderate, much of the photosynthate that is being produced is devoted to such secondary compound production as oleoresin (Ayres, 1995). For this model, NPP will be used as an indicator of changes in photosynthesis and total photosynthate production. Similar to soil water stress, net primary productivity factor (NPPFAC) represents the inverse hyperbolic function between NPP, oleoresin production, and average annual SPB population size, which is related using changes in the SPB population size multiplication fraction (Figure 33.2). Present-year NPP > 10 t ha^{-1} year^{-1} or < 8 t ha^{-1} year^{-1} has a NPPFAC >1, while NPP < 10 t ha^{-1} year^{-1} and > 8 t ha^{-1} year^{-1} has a NPPFAC < 1.0.

Prior Year's Southern Pine Beetle Population Size

No matter how idyllic environmental and plant conditions are for the reproduction of SPB, outbreaks of SPB cannot occur without a pre-existing SPB population. Additionally, the larger the size of the present SPB population, the greater the potential to colonize new areas and thus increase the size of the future SPB population further. Conversely, however, under these same conditions, natural SPB enemy populations could also build up, and therefore reduce the SPB population. A relative measure of the prior year's average SPB population is the final equation factor used to calculate a relative measure of the present year's average SPB population. Population data can either be collected from the agencies that monitor SPB numbers or if multiple years of prediction are needed, prior years' predicted average annual SPB population data can be used. Although the insects may follow a cyclic pattern, other factors (e.g., climate, parasites, predators, or disease) could delay or accelerate the average annual population flux. This theory of density dependence is the basis of the Turchin et al., (1991) research on reasons for SPB population fluctuations. Using the average number of SPB spots in eastern Texas from 1955 to 1985, Turchin et al., (1991) found a significant lag

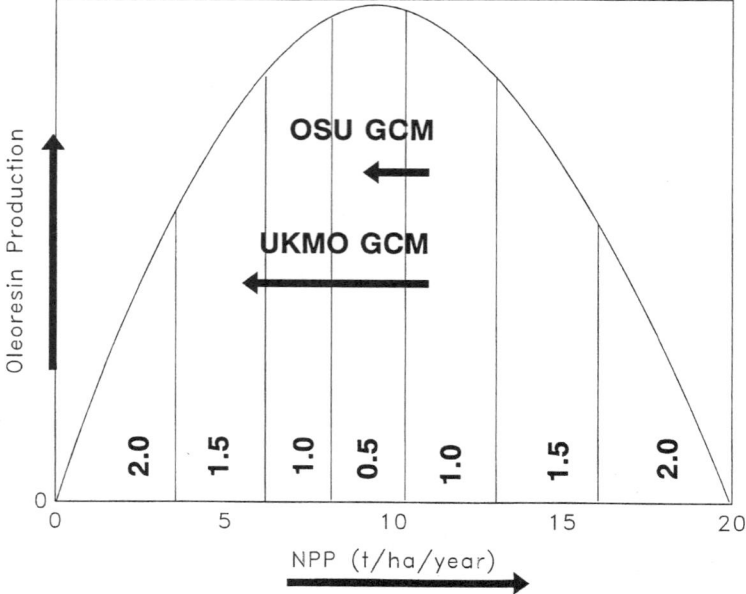

Figure 33.2. The influence of net primary productivity (NPP) on oleoresin production is represented by an inverse hyperbolic function, and is expressed as a SPB population growth multiplication factor (NPPFAC). Maximum oleoresin production and the maximum reduction in the SPB population factor (NPP = 0.5) occurs under conditions of moderately limited NPP. Extreme reductions or increases in NPP reduce oleoresin production and increases the SPB population factor (NPPFAC > 1.0). Under historic conditions, the ecosystem predicted moderately high rate of NPP and WASTRSFAC = 1.0. Both the Oregon State University (OSU) and United Kingdom Meteorological Office (UKMO) caused a decrease in NPP resulting from a reduction in total leaf area and decrease in net carbon fixed per unit of leaf area.

affect between the average present and prior year's SPB population size, and they did not detect any correlation between climate change and the rate of SPB population change. However, they did not examine the relationship between climate and vegetative response. Although Turchin et al., (1991) did not determine any cause for the cyclic nature of SPB populations, prior years' SPB population size were clearly related to the present year's population size.

Combining Factors in PnET–IIS

Predictions of average annual SPB populations were based on predicted changes in growing season WATSTRS, NPP (NPPFAC), and measured or predicted prior years' average SPB population size (SPBPOP). Although WATSTRS and NPP affect the oleoresin production that result from changes in total carbohydrate production and carbon partitioning patterns, and thus, the environmental conditions for SPB colonization and reproduction, the SPBPOP determines the starting condition for future changes in average annual SPB populations.

Turchin's theories on prior year's average SPB populations (SPBPOP1) controlling present-year average SPB population (SPBPOP0) were evenly weighted with the present-year's WATSTRSFAC0 and NPPFAC0 to produce a relative prediction of the present-year's average SPB population (SPBPOP0) as shown:

$$SPBPOP0 = WATSTRSFAC0 \times NPPFAC0 \times SPBPOP1 \tag{1}$$

Average southern pine beetle populations were also predicted using prior year's soil water stress (WATSTRSFAC1), NPP (NPPFAC1), and average annual SPB population data (SPBPOP1) as follows:

$$SPBPOP0 = WATSTRSFAC1 \times NPPFAC1 \times SPBPOP1 \tag{2}$$

Using the second-year's prior WATSTRS (WATSTRSFAC2), NPP (NPPFAC2), and average annual SPBPOP size (SPBPOP2), is shown as:

$$SPBPOP0 = WATSTRSFAC2 \times NPPFAC2 \times SPBPOP2 \tag{3}$$

Input Data

The PnET–IIS model uses constant, generalized, species-dependent process parameters (e.g., light extinction coefficient, and optimum temperature for gross photosynthesis) (Table 33.1) and site-specific soils and climate data. Soils series

Table 33.1. PnET–IIS Model Values

Parameter name	Parameter abbreviation	Model value
Light extinction coefficient	k	0.5
Foliar retention time (years)		2.0*
Leaf specific weight (g)		9.0*
NetPsnMaxA (slope)		2.4*
NetPsnMaxB (intercept)		0*
Light half saturation (J m² sec⁻¹)	HS	70
Vapor deficit efficiency constant	VPDK	0.05
Base leaf respiration fraction		0.10
Water use efficiency constant	WUE C	10.9
Canopy evaporation fraction		0.15
Soil-water release constant	F	0.04
Maximum air temperature for photosynthesis (°C)	TMAX	variable*
Optimum air temperature for photosynthesis (°C)	TOPT	variable*
Change in historic air temperature (°C)	DTEMP	0
Change in historic precipitation (% difference)	DPPT	0

* Values are derived specifically for loblolly pine. All other parameters are general vegetative values.

data were derived from a geographic information system (GIS)-based soils atlas compiled by the Soil Conservation Service (Marx, 1988). The soil series data was hand-digitized from a paper source at a scale between 1:500,000 and 1:1,500,000, depending on the state (Marx, 1988). Soil information associated with each series includes soil water holding capacity (SWHC) to a depth of 102 cm, which is the only site-specific soils data used in our simulations. All other soils parameter values were held constant across all sites and years (Table 33.1). The PnET–IIS model also requires four monthly climatic drivers, which are 1) minimum air temperature, 2) maximum air temperature, 3) precipitation, and 4) solar radiation. The Forest Health Atlas provided cooperator and first-order station data, which was originally obtained from the National Climatic Data Center (NCDC) (Marx, 1988). Cooperator station data include average minimum and maximum monthly air temperature and total monthly precipitation; first-order station records include relative humidity. Because these data had error rates between 5 and 40% (Marx, 1988), much of the data was removed before usage. After checking for accuracy, the database was interpolated on a $0.5° \times 0.5°$ grid across the southern United States (Marx, 1988). The gridded databases of minimum and maximum air temperature, relative humidity and precipitation were compiled into a single database and run through a program to calculate monthly solar radiation (Nikolov and Zeller, 1992) at a $0.5° \times 0.5°$ grid. Solar radiation values were then combined with average monthly maximum and minimum air temperature and total monthly precipitation and put into PnET–IIS.

A Case Study in Southern Pine Beetle Population Size Modeling

As an example of SPB population modeling, we predicted average annual SPB population changes for Walker County in Texas. This county was selected because this is an area with a variable SPB population that has been summarized at a prior date (Turchin et al., 1991). We obtained twelve years of climate, soils, and SPB population data[1] for this area, and then used PnET–IIS to predict how average SPB population size would change over time given changes in WATSTRS, NPP, and the prior year's average SPB population size.

Predicted changes in the average SPB population using the equation (1) were highly correlated with measured average SPB population ($R^2 = 0.76$, $P = 0.0004$, n = 12). If predictions of present average SPB populations (SPBPOP0) were based on the prior year soil water stress (WATSTRSFAC1), NPP (NPPFAC1) factors, and the prior year's average SPB population (SPBPOP1), without any knowledge of present year soil water stress or NPP (equation 2), the correlation is weaker but still significant ($R^2 = 0.47$, $P = 0.016$, n = 12). Environmental stress from the present and prior year are the controlling factors of present year average SPB populations, as no relationship ($P > 0.05$) was found with environmental stress

---

[1] SPBPOP data supplied by R.F. Billings.

from the two prior years (WATSTRS2, NPPFAC2, and SPBPOP2), and the present year's average SPB population (equation 3). The PnET–IIS model may predict SPB populations better than other SPB population models because climate and soils are linked to form a better measure of ecosystem function compared to using a coarse measure of such ecosystem stress as precipitation or days over 90 °F.

Projections of Climate Change Effects on Southern Pine Beetle Populations

In addition to predicting present-year average SPB populations, PnET–IIS could also be used to predict how climate change may affect future SPB populations. General circulation models (GCMs) estimate a 3 to 7 °C increase in annual air temperature by the year 2050 across the southern United States (Cooter et al., 1993). Although most of the major GCM scenarios predict a total annual precipitation change of < 10%, the timing of the precipitation could be markedly altered from historic patterns (Cooter et al., 1993). These changes could significantly influence the severity and frequency of SPB outbreaks. Such models as PnET–IIS could be used to predict changes in future SPB populations given present-day climate or future climate change scenarios.

Two climate change scenarios were developed using historic climate databases with two GCMs. The GCMs from both Oregon State University (OSU) (Schlesinger and Zhao, 1989), and United Kingdom Meteorological Office (UKMO) (Mitchell, 1989) were selected because of their common application and range of climate change predictions. All of the GCMs predict variation in monthly temperature and precipitation, based on a doubling of atmospheric carbon dioxide (CO_2) by 2050 (Cooter et al., 1993). Grid data from each of the two GCMs monthly climate change databases were added to historic (from 1950 to 1985) average monthly minimum and maximum air temperature or were multiplied by historic monthly precipitation to produce thirty-five years of climate change scenario data. The predictions by GCMs of precipitation and air temperature under a doubled CO_2 environment vary widely. Across the southern United States, the OSU GCM consistently predicted smaller increases (+ 3 °C) in average annual air temperature compared to the UKMO GCM (+ 7 °C). The OSU GCM predicted above-average precipitation in the late summer and fall, and below historic average levels of precipitation in the late winter and spring. The OSU GCM also predicted that total annual precipitation would decrease in the central portion of the South and increase along the Atlantic coast although region-wide, average annual precipitation would increase by 3% compared to historic total annual precipitation. The UKMO GCM predicted that regional precipitation would be greater than historic amounts during the spring and less during the summer and fall. Average annual precipitation would decrease in the central and southwestern portion of the region and increase along the southern Atlantic coast, and total annual precipitation would decrease by 1% region-wide when compared to historic levels.

Climate Change Scenario Effects on Net Primary Productivity

When PnET–IIS was run with the two climate change scenarios, predicted NPP was reduced across most of the southern United States (except in some high-elevation, mountainous areas) but the severity of the reductions were dependent on the GCM applied. Using the OSU GCM scenario, PnET–IIS predicted a reduction in growth across most of the southern United States, but these growth reductions were generally not severe enough to cause tree death or large reductions in species range. The PnET–IIS model predicted that approximately 2% of the present loblolly pine range would be lost across the southern United States if the OSU GCM scenario occurred, and that NPP would generally decrease across the region but could increase in the cooler mountain areas (Figure 33.3b) compared to historic rates on NPP (Figure 33.3a).

In the UKMO GCM scenario, predicted NPP would be reduced by 100% of historic NPP across most of the southcentral and southwestern portions of the region including most of Florida, Georgia, Alabama, Mississippi, Louisiana, and Texas (Figure 33.3c), suggesting that the climate in these states would no longer be suitable for growing loblolly pine. The UKMO GCM predicted less severe reductions in NPP for the northern and eastern portions of the region (Figure 33.3c), because areas had historically cooler air temperatures and high rates of precipitation. Average regional NPP was reduced by 46% and the range of loblolly pine was reduced by 42% when the UKMO GCM scenario was applied to the model (Figure 33.3c).

Climate Change Scenario Effects on Soil Water Stress

Because drainage is equal to precipitation—ET, and ET and NPP are functions of leaf area and temperature, the pattern of drainage is similar to NPP. Using the OSU GCM, predicted WATSTRS increased in the central and northcentral areas, and decreased across the southern and eastern portions of the region (Figure 33.4b), compared to historic WATSTRS (Figure 33.4a). The scenario from OSU also predicted that average annual WATSTRS would decrease by 22% across the region (Figure 33.4). Compared to the OSU GCM scenario, the UKMO GCM scenario caused a larger deviation in predicted ecosystem WATSTRS. In areas of mortality, predicted ET was zero, drainage was equal to precipitation, and WATSTRS was a function of precipitation. The UKMO scenario predicted decreased WATSTRS throughout the region, except along the cooler Appalachian Mountains where WATSTRS would increase (Figure 33.4c). Compared to historic WATSTRS, the UKMO scenario average annual WATSTRS decreased by 56% across the region. If only areas where loblolly pine NPP > 0 are included, WATSTRS decreased by only 27% compared to historic levels.

33. Predictions of Southern Pine Beetle Populations 629

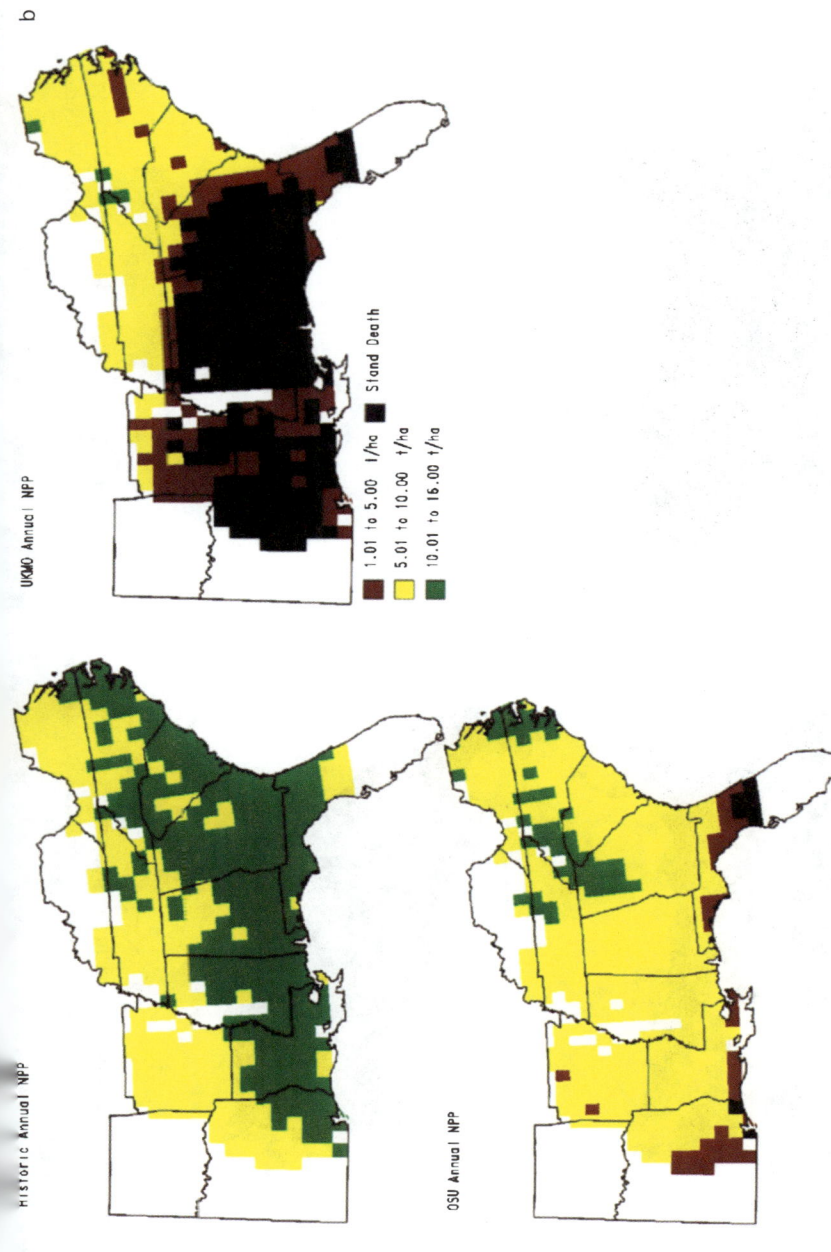

Figure 33.3. Regional changes in net primary productivity (NPP) across the southern United States for: historic (from 1951 to 1984) climate (a); the Oregon State University, general circulation model (GCM) scenario (b); and the United Kingdom Meteorological Office GCM scenario (c).

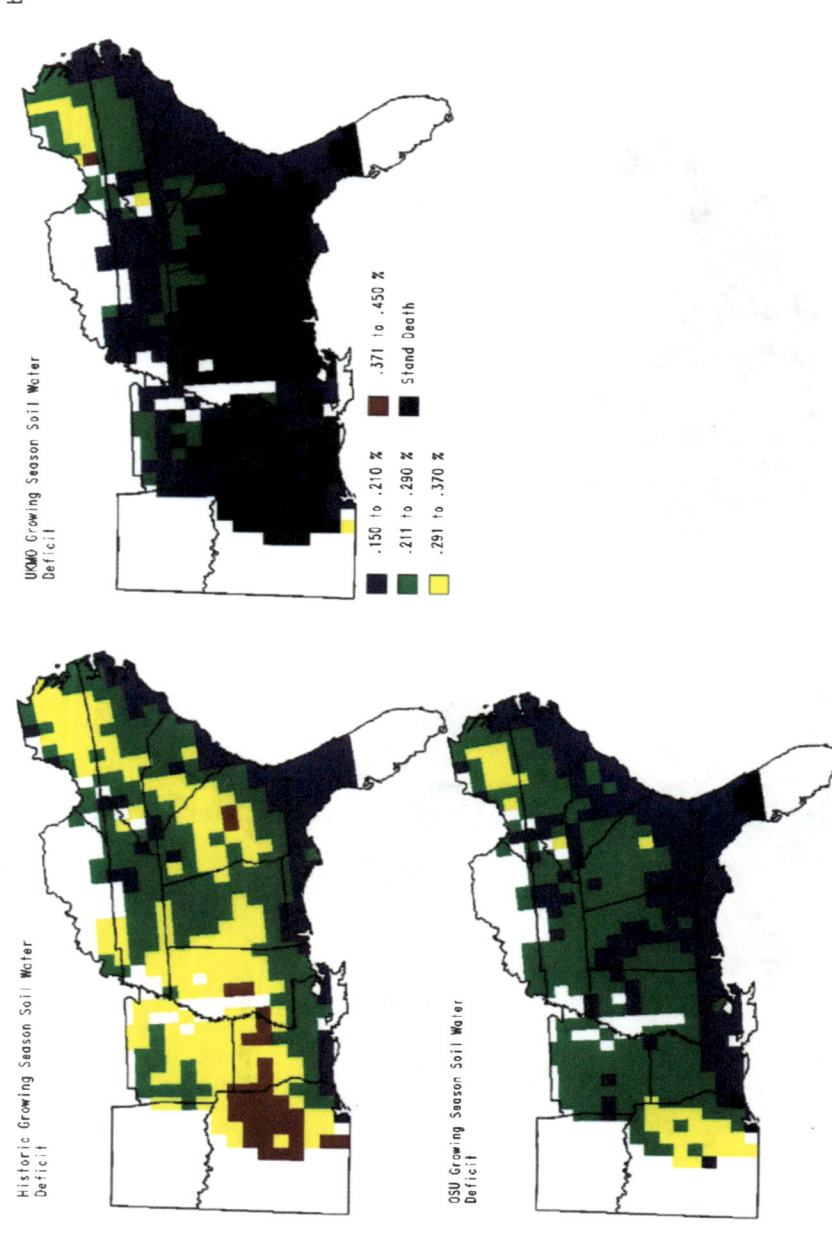

Figure 33.4. Regional changes in soil water stress (WATSTRS) across the southern United States for: historic (from 1951 to 1984) climate (a); the Oregon State University, general circulation model (GCM) scenario (b); and the United Kingdom Meteorological GCM scenario (c).

Effects of Changing Soil-Water Stress and Net Primary Productivity on Southern Pine Beetle Populations

Under conditions of climate change, average annual NPP and soil water stress will generally decrease across the region (Figures 33.1 and 33.2). With a reduction in NPP and ET, total plant-water demand will also decrease. As soil water stress decreases, pines across the south may become less water limited and more temperature limited. In the UKMO scenario, increased air temperatures reduced NPP and the plants become carbon limited. If carbon limitations become severe, oleoresin synthesis will probably also decline (Ayres, 1993) and SPB populations could rapidly increase. The OSU scenario projects less severe changes in climate and, therefore, less severe changes in forest ecosystem processes. Slight reductions in the NPP and soil water stress across the region under the OSU scenario could increase oleoresin production and therefore, cause a reduction in average annual SPB populations (Figures 33.1 and 33.2). Although the model predicts that within the region, certain areas could see NPP increases (e.g., Appalachian Mountains) or increased soil water stress (Figures 33.3 and 33.4), causing little or no change in average annual SPB populations, such areas as eastern Texas or central Florida could have large reductions in NPP and soil water stress and oleoresin production, leading to increases in the predicted average annual SPB population.

The size of these decreases will vary with year, and extreme event years will probably have a much greater impact on forest mortality than long-term averages. To better understand the potential effects of climate change on average annual SPB populations, individual years and locations need to be examined.

Model Refinement

The present version of the model only predicts SPB populations at an annual time step. Because PnET–IIS predicts water use and productivity at monthly intervals, future versions of the model could have increased sensitivity for predicting SPB populations if a finer temporal resolution were used. Reeve et al., (1995) noted that in the southeastern United States, pine trees grow very rapidly in the spring and early summer and vertical resin ducts are absent or rare (earlywood growth). By midsummer, tree growth generally slows resulting from soil-water deficits and the latewood formed typically contains many resin ducts. The timing of the transition from earlywood to latewood formation has significant impact on tree resistance to insect herbivores. Climatic patterns that protract the length of earlywood development favor SPB attack and, therefore, monthly predictions of WATSTRS and NPP may be better predictors of changing SPB populations compared to growing-season averages. Future SPB population models should attempt to incorporate monthly variation in forest processes into model predictions of SPB population change.

Conclusions

Initial efforts to correlate the present and prior year's soil water stress and NPP, and the prior year's average SPB population with present-year average annual SPB population have proven moderately successful. This success has lead to the attempt to predict how potential future climate change could affect future SPB populations. Depending on the climate scenario and site location, southern pine NPP and soil water stress could be significantly reduced across forested areas in the southern United States. Sites located in the warmest sections of the present pine range are more sensitive to changes in climate than are pine sites located in drier or cooler areas. These predictions also suggest that the region is much more susceptible to changes in air temperature than changes in precipitation. If our model predictions are correct, climate change would have serious potential socio-economic implications for the southern U.S. timber industry. However, additional research is needed to assess the effects that other atmospheric changes (e.g., CO_2, O_3, NO_x, SO_x), genetics, and species replacement may have on forest processes, before a complete assessment of climate change effects on forest productivity, water use and SPB populations can be made.

References

Aber JD, Federer CA 1992 A generalized, lumped-parameter model of photosynthesis, evapotranspiration and net primary production in temperate and boreal forest ecosystems. Oecologia, 92:463–474.

Aber JD, Ollinger SV, Federer CA, Reich PM, Goulden ML, Kicklighter DW, Melillo JM, Lathrop RG Jr, Ellis JM. (1995) Predicting the effects of climate change on water yield and forest production in the northeastern U.S. Climate Res 5:207–222.

Ayres MP (1993) Plant defense, herbivory, and climate change. In Kareiva PM, Kingsolver JG, Huey RB (Eds) *Biotic interactions and global change.* Sinauer Associates, Sunderland, MA.

Cooter EJ, Eder BK, LeDuc SK, Truppi L 1993 General circulation model output for forest climate change research and application. Gen Tech Rep SE-85. USDA, For Ser, Southeast For Exper Stat, Asheville, NC.

Craighead FC (1925) Bark beetle epidemics and rainfall deficiency. J Econ Entomol 18:577–584.

Hansen JB, Baker BH, Barry PJ (1973) Southern pine beetles on the Delaware Peninsula in 1971. J Ga Entomol Soc 8:3:157–164.

Herms DA (1991) Variation in the resource allocation patterns of paper birch: Evidence for physiological tradeoffs among growth, reproduction, and defense. PhD dissertation. MI State Univ, East Lansing.

Herms DA, Mattson WJ (1992) The dilemma of plants: To grow or defend. Q Rev Biol 67:283–335.

Hopkins AD (1899) Report on investigations to determine the cause of unhealthy conditions of the spruce and pine from 1880–1893. WV Agri Exper Stat Bull 56, 461.

Kalkstein LS (1974) The effect of climate upon outbreaks of the southern pine beetle. Publ Climatol 27:1–65.

Kramer PJ (1980) Drought, stress and the origin of adaptations. Adaptation of plants to water and high temperature stress. John Wiley and Sons, New York.

Larsson S (1989) Stressful times for the plant stress-insect performance hypothesis. Oikos 56:277–283.

Loomis WE (1932) Growth-differentiation balance vs. carbohydrate-nitrogen ratio. Proc Am Soc Hortic Sci 29:240–245.
Loomis WE (1953) Growth correlation. In Loomis WE (Ed). *Growth and differentiation in plants*. IA State Coll Press, Ames, Iowa.
Lorio PL Jr (1986) Growth-differentiation balance: A basis for understanding southern pine beetle-tree interactions. For Ecol Manage 14:259–273.
Lorio PL Jr (1988) Growth-differentiation balance relationships in pines affect their resistance to bark beetles (Coleoptera: Scolytidae). In Mattson WJ, Levieux J, Bernard-Dagan C (Eds). *Mechanisms of woody plant defenses against insects*. Springer-Verlag, New York.
Lorio PL Jr, Hodges JD (1977) Tree water status affects induced southern pine beetle attack and brood production. US For Serv Res Pap SO-135.
Lorio PL Jr, Sommers RA (1986) Evidence of competition for photosynthates between growth processes and oleoresin synthesis in *Pinus taeda* L. Tree Physiol 2:301–306.
Lorio PL Jr, Sommers RA, Blanche CA, Hodges JD, Nebeker TE (1990) Modeling pine resistance to bark beetles based on growth and differentiation balance principles. In Dixon RK, Meldahl RS, Ruark GA, Warren WG (Eds). *Process modeling of forest growth responses to environmental stress*. Timber Press, Portland OR.
Martinat PJ (1987) The role of climatic variation and weather in forest insect outbreaks. In Barbosa P, Schultz JC (Eds). *Insect outbreaks*. Academic Press, San Diego, CA.
Marx DH (1988) Southern forest atlas project. In *The 81st annual meeting of The Association Dedicated to Air Pollution Control and Hazardous Waste Management* (APCA), Dallas, Texas, 1988.
McNulty SG, Vose JM, Swank WT, Aber JD, Federer CA (1994) Landscape scale forest modeling: Data base development, model predictions and validation using a geographic information system. Clim Res 4:223–231.
McNulty SG, Vose JM, Swank WT (1997) Scaling predicted pine forest hydrology and productivity across the southern United States. In Quattrochi DA, Goodchild ME (Eds) *Scaling in remote sensing and GIS*. Lewis Publishers, Chelsea, MI.
McNulty SG, Vose JM, Swank WT (1996a) Loblolly pine hydrology and productivity across the Southern United States. For Ecol Manage 86:241–251.
McNulty SG, Vose JM, Swank WT (1996b) Potential climate change affects on loblolly pine productivity and hydrology across the southern United States. Ambio 25(7):449–453.
Michaels PJ (1984) Climate and the southern pine beetle in Atlantic coastal and piedmont regions. For Sci 30:143–156.
Michaels PJ, Sappington DE, Spengler PJ, Phillip J (1986) SPBCMP—A program to assess the likelihood of major changes in the distribution of southern pine beetle infestations. S J Appl For 10:158–161
Mitchell JFB (1989) The greenhouse effect and climate change. Rev Geophys 27(1):115–139.
Nikolov NT, Zeller KF (1992) A solar radiation algorithm for ecosystem dynamic models. Ecol Model 61:149–168.
Price PW (1991) The plant vigor hypothesis and herbivore attack. Oikos 62:244–251.
Price TS, Doggett C (1982) A history of southern pine beetle outbreaks in the southeastern United States. GA For Comm, Macon, GA.
Price TS, Doggett C, Pye JM, Holmes TP (1992) A history of southern pine beetle outbreaks in the southeastern United States. GA For Comm, Macon, GA.
Pritchett WL (1979) *Properties and management of forest soils*. John Wiley and Sons, New York.
Reeve JD, Ayres MP, Lorio PL Jr (1995) Host suitability, predation, and bark beetle population dynamics. In Cappuccino N, Price P (Eds) *Population dynamics*. Academic Publishers, New York.

Rhodes DF (1979) Evolution of plant chemical defense against herbivores. In Rosenthal GA, Janzen DH (Eds) *Herbivores: Their interaction with secondary plant metabolites.* Academic Press, New York.

St George RA (1930) Drought-affected and injured trees attractive to bark beetles. J Econ Entomol 59:955–957.

Schlesinger ME, Zhao ZC (1989) Seasonal climatic change introduced by doubled CO_2 as simulated by the OSU atmospheric GCM/mixed-layer ocean model. J Climate 2:429–495.

Strain BR, Higginbotham KO, Mulroy JC (1976) Temperature preconditioning and photosynthetic capacity of *Pinus taeda* L. Photosyn 10:47–53.

Swain KM, Remoin MC (1981) Direct control methods for the southern pine beetle. USDA Agric Handbook 575:15.

Turchin P, Lorio PL Jr, Taylor AD, Billings RF (1991) Why do populations of southern pine beetle (Coleoptera: Scolytidae) fluctuate? Environ Entom 20:401–409.

Waring GL, Cobb NS (1992) The impact of plant stress on herbivore population dynamics. In Bernays EA (Ed) *Plant-insect interactions.* CRC Press, Boca Raton, FL.

White TCR (1974) A hypothesis to explain outbreaks of looper catepillars, with special reference to populations of Selidosema suavis in a plantatiion of Pinus radiata in New Zealand. Oecologia, 16:279–301.

Wyman L (1924) Bark-beetle epidemics and rainfall deficiency. USDA For Ser Bull 8:2–3.

34. Soil Effects Mediate Interaction of Dogwood Anthracnose and Acidic Precipitation

Kerry O. Britton, Paul C. Berrang, and Erika Mavity

Dogwood anthracnose is a fungal disease caused by *Discula destructiva* Redlin. It was first reported in 1976 (Byther et al., 1979), and spread rapidly throughout the range of the Pacific dogwood (*Cornus nuttallii* Audubon) on the west coast. The disease was found in 1978 in New York, and swept through the eastern flowering dogwood (*Cornus florida* L.) population as far south as northern Alabama in just fifteen years (Redlin, 1991). This rapid spread led to speculation that the fungus may be exotic (Redlin, 1991), but *D. destructiva* has not been found elsewhere. Another theory is that such environmental factors as hard winters or air pollution, may have increased dogwood susceptibility (Hibben and Daughtrey, 1988; Hudler, 1985).

The relationship between acidic deposition and forest tree diseases has been the subject of much debate (Grzywacz and Wazny, 1973; Johnson and Taylor, 1989; Rehfuess, 1989; Skelly, 1989). Acidic deposition has both direct and indirect effects on plants and associated microorganisms (Campbell et al., 1988; Heagle, 1973; Killiam et al., 1983, Magan and McLeod, 1991; Musselman and McCool, 1989; Skelly, 1989). The sum of these effects can result in an increase or a decrease in disease severity. Acidic deposition inhibits many fungi, especially rusts and wood-decay organisms (Heagle, 1973; Magan and McLeod, 1991; Shafer et al., 1985; Shriner, 1978; Smith, 1990). Bruck and Shafer (1983) proposed the explanation that acidic deposition wounds host tissues, and induces generalized resistance responses effective against obligate parasites, which prefer a vigorous host. However, because wounding and stress increase the rate of senescence, acidic deposition could increase susceptibility to facultative parasites, which prefer a weakened host (Bruck and Shafer, 1983). In some plants, acidic rain increases leaf wettability, foliar water-holding capacity, foliar nutrient uptake, and the leaching of polar solutes (Evans, 1982; Lepp and Fairfax, 1976; Norby et al., 1986; Percy and Baker, 1988). Also, acidic fog affects water relations (Eamus et al., 1989; Mengel et al., 1989) in some plants. Such effects as these may have more impact on trees than on annual plants. Through the soil acidic rain can also indirectly produce cumulative effects. It can increase the concentration of aluminum and base cations in the soil solution and provide nitrogen (N), which is a limiting nutrient in many southeastern soils (Binkley et al., 1989).

Madden and Campbell (1987) noted that the effects of air pollutants on pathogen virulence and host resistance have rarely been studied. Most of the literature considers effects on spore germination and penetration or the ultimate effects on disease severity. The mechanisms by which acidic rain influences disease severity are largely unknown.

The potential for complexity when a disease and a pollutant interact is illustrated by the effects of acidic deposition on Scleroderris canker caused by *Gremmeniella abietina* on Norway spruce (*Picea abies* L. Karsten). *G. abietina* infection is usually latent in Norway spruce. But acidic precipitation enables the pathogen to produce disease symptoms by reducing competition and enhancing spore germination (Barkland et al., 1984). Competition is reduced when acidic precipitation affects epiphytes and endophytes (Barkland and Unestam, 1988). At the same time, acidic precipitation causes ion leakage that enhances spore germination of *G. abietina*. Intriguingly, Scleroderris canker severity on Scots (Barkland and Unestam, 1988) and red pines (Bragg and Manion, 1984), which are normally more susceptible than spruce, is not affected by acidic precipitation treatments.

In 1993, Anderson et al. (1993) reported that simulated acidic rain (SAR) increased the susceptibility of dogwoods to anthracnose under laboratory conditions. Britton et al. (1996) demonstrated that pretreatment with SAR also increased susceptibility to natural inoculation in the field. However, the mechanism for this effect is unknown.

Acid conditions do not directly favor the pathogen. In fact, *D. destructiva* is inhibited by highly acidic conditions in vitro. Conidial germination on agar was zero at pH 2.0, but not significantly affected from pH 3.0 to 5.6 (Britton, 1989). Mycelial growth was also reduced by acidic conditions (Britton, 1993).

Several studies have examined the effects of acidic precipitation on dogwood foliage. Haines et al. (1980) reported no visible damage on dogwoods treated with pH 2.0 SAR. However, scanning electron microscopy studies have since shown that SAR treatments eroded epicuticular wax and altered trichome morphology of dogwood leaves (Brown et al., 1994; Thornham et al., 1992). Furthermore, Willey and Hackney (1991) demonstrated that there was increased leaching of calcium and magnesium ions from dogwood leaves treated with SAR droplets. Plant leachates also often contain sugars and amino acids, although these have not been investigated in dogwood. Leakage of such foliar nutrients may either stimulate or inhibit pathogenic fungi (Blakeman, 1973). An excessive loss of nutrients through leaching could also stress the host. Stress from other sources, for example, drought, increases susceptibility to anthracnose (Gould and Peterson, 1994).

In 1992 and 1993, we established experiments to determine whether the observed increase in dogwood susceptibility to anthracnose was the result of aboveground factors (e.g., cuticular erosion) or belowground factors (e.g., altered nutrient availability). We designed other experiments to determine whether plants exposed to SAR were more stressed or more susceptible to drought than plants not exposed to SAR. The results reported here offer intriguing clues to the complex nature of this host–parasite interaction.

Materials and Methods

Soil vs Foliar Effects

To separate soil and foliar effects, we applied pH 2.5 and pH 5.5 SAR to the soil, to the foliage, or both. The experiment as described in the following sections was run twice in 1992 and twice in 1993. The experiments for each year were conducted simultaneously in order to complete ten pretreatment applications and expose the seedlings to natural inoculation early in June. We started in January with one-year-old flowering dogwood seedlings from the Georgia Forestry Commission nursery near Montezuma. In February, we planted them in 5-l containers in a mixture of equal parts by volume of peat moss, perlite, and topsoil, and placed them in an air-conditioned greenhouse maintained at 23 to 27 °C in Dry Branch, Georgia. We fertilized all the seedlings once two weeks after planting, with an excess volume of a dilute commercial fertilizer (15:30:15 at 100 ppm N).

After leaf emergence, we selected 240 seedlings for uniformity in size and vigor. We prepared SAR solutions, adjusted to pH 2.5 or pH 5.5 with a 0.65 molar (M) mixture of sulfuric and nitric acids in approximately the same ratio found in ambient rain (70 mequiv SO_4^{-2}:30 mequiv. NO_3^-) (Shafer et al., 1985).

We assigned sixty seedlings to each of the following four pretreatments. 1) soil/acid, which received pH 2.5 SAR on the growing medium surface and pH 5.5 SAR on the foliage, 2) foliage/acid, which received pH 2.5 SAR on the foliage and pH 5.5 SAR on the growing medium surface, 3) both/acid, with pH 2.5 SAR applied to both the growing-medium surface and foliage, and 4) none/acid, with pH 5.5 SAR applied to both the growing-medium surface and foliage. Before each rain event, we tied a plastic bag around the root collar of each seedling, covering the growing medium and pot. We applied the foliar pretreatments using a rain simulator (Chevone et al., 1984), and we removed the plastic bags between rain events. We applied treatments to the growing medium by pouring 300 ml of SAR of the appropriate pH directly on the soil surface. Additionally, we applied SAR of the appropriate pH as needed to maintain adequate levels of soil moisture for a total of about 5-l per container over the 10-week duration of the pretreatments.

After SAR pretreatment, we transported most of the dogwood seedlings to

Coweeta Hydrologic Laboratory in southwestern North Carolina and placed them under mature dogwoods that were naturally infected with *D. destructiva*. These seedlings received natural rainfall thereafter. A subgroup of forty-eight seedlings were kept at Dry Branch, Georgia for physiological studies.

In late June and September, we visually estimated the percent leaf area infected for each seedling. We used the means for the eight seedlings remaining in each pretreatment in each replication as data points (n = 6) in a regression analysis. An analysis of variance (ANOVA) was also performed, and the means were separated using Duncan's Multiple Range Test. Significant interaction between soil and foliar applications in 1993 necessitated comparison of simple effect means by Duncan's Multiple Range Test.

Plant Vigor and Drought Tolerance

At the end of the rain pretreatment period, we removed two seedlings from each pretreatment in each replication (forty-eight seedlings) and kept them in the greenhouse in Georgia for further study. We saturated the growing medium with deionized water, allowed it to drain, weighed it, and measured predawn xylem pressure potential (PXPP), midday net photosynthesis (P_{net}), and stomatal conductance (g_1). We used a pressure bomb (Soil Moisture Equipment Corp.) to measure PXPP, and a LI-6200 portable photosynthesis system (Li-Cor, Inc.) to measure P_{net} and g_1. Half the seedlings from each rain treatment continued to receive adequate amounts of deionized water; the other half received no additional water for the duration of this study. We repeated all measurements at approximately weekly intervals. We harvested the water-stressed seedlings when they stopped showing a positive rate of P_{net} and recorded fresh weights for leaves, stems, roots, and soil dry weight. We measured leaf areas with a Li-Cor leaf area meter (Li-Cor, Inc.) and we also recorded oven dry weights. We then mailed leaf samples and soil samples to the UGA Soil Testing Laboratory, Cooperative Extension Service in Athens, Georgia for nutrient analyses.

Results

Soil vs Foliar Effects

Application of pH 2.5 SAR directly on the growing medium increased disease severity in three of four experiments (Table 34.1). Applications of pH 2.5 SAR to the foliage alone did not increase disease in either year. However, in the second experiment in 1993, applications to both the foliage and the growing medium significantly increased disease over applications to the growing medium alone (Table 34.1).

Some abscission of uninfected leaves was noted in August of 1993. Earlier than

Table 34.1. Percentage of Leaf Area Infected with Dogwood Anthracnose Following Pretreatments with Either pH 2.5 or 5.5 Simulated Acidic Rain on the Soil and Foliage

Pretreatment with pH 2.5 SAR[1]	1992		1993	
	Experiment 1	Experiment 2	Experiment 1	Experiment 2
Soil only	5.7[2]	2.0[3]	1.0[2]	4.1[4]b
Soil and foliage	5.8	3.1	0.5	7.4 a
Foliage only	3.5	2.5	0.1	0.9 c
None	2.9	2.2	0.1	0.9 c

[1] Foliage or soil not treated with pH 2.5 simulated acidic rain (SAR) was treated with pH 5.5 SAR.
[2] Soil application treatments exhibited significant increase in disease (ANOVA main effect at $P < 0.01$). Foliage application main effect and soil × foliage application interactions were not significant.
[3] No significant differences in this experiment.
[4] Significant soil × foliage application interactions occurred. Simple effects tested by ANOVA. Means followed by the same letter do not differ significantly at $P \leq 0.05$ according to Duncan's Multiple Range Test.

expected for autumn leaf fall, this abscission was more common in Experiment 2, which was exposed to natural inoculum in a location receiving less light than experiment 1. The amount of leaf loss was small (2 vs 4%), but significantly less ($P = 0.05$) in seedlings receiving pH 2.5 SAR on the growing medium than in those receiving normal rain (pH 5.5).

Plant Vigor and Drought Tolerance

Well-watered seedlings that received pH 2.5 rain on the growing medium showed significantly higher rates of P_{net} and g_1 on most test dates than trees that received pH 5.5 rain on the medium, regardless of the pH of the foliar rain pretreatment. In water-stressed seedlings, those that received pH 2.5 rain on the growing medium initially showed higher rates of P_{net} and g_1 than those that received pH 5.5 rain on the medium. But after seventeen days, these rates reversed and water-stressed seedlings that received pH 2.5 rain on the medium showed lower rates of P_{net} and g_1 than those receiving pH 5.5 rain (Figures 34.1 and 34.2).

A steady decline of PXPP was observed in all water-stressed seedlings (Figure 34.3). Early in the study, the difference in PXPP between rain treatments was not significant, but after seventeen days without water, water-stressed trees that received pH 2.5 SAR on the growing medium showed significantly lower PXPP than seedlings that received pH 5.5 SAR on the medium (Figure 34.3). At the end of the study, leaf area and dry weight were significantly greater in plants that received pH 2.5 SAR only on the medium than in plants that received pH 2.5 SAR only on the leaves (Table 34.2). Stem weight followed the same trend, but the difference was not significant. The differences in root weights and shoot/root ratios were not significant.

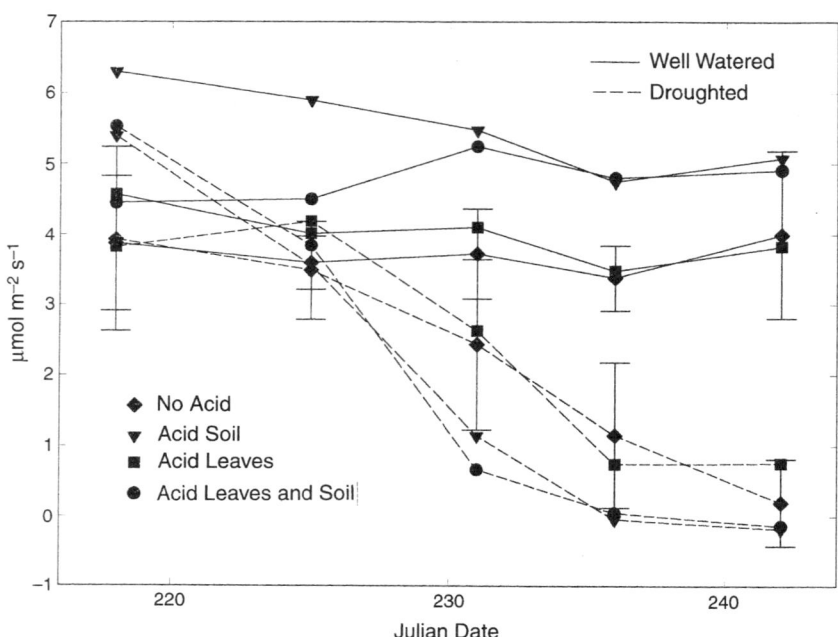

Figure 34.1. Net photosynthesis (P_{net}) of dogwood seedlings that were pretreated with either pH 2.5 or 5.5 simulated rain on foliage, soil, or both, on five measurement dates after two drought stress treatments were applied.

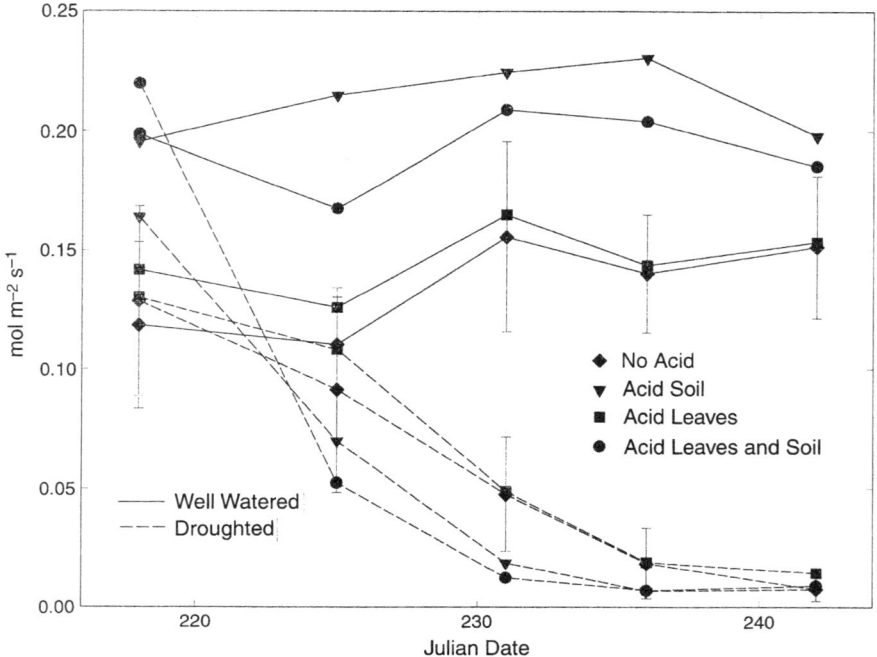

Figure 34.2. Stomatal conductance (g_l) of dogwood seedlings that were pretreated with either pH 2.5 or 5.5 simulated rain on foliage, soil, or both, on five mesurement dates after two drought stress treatments were applied.

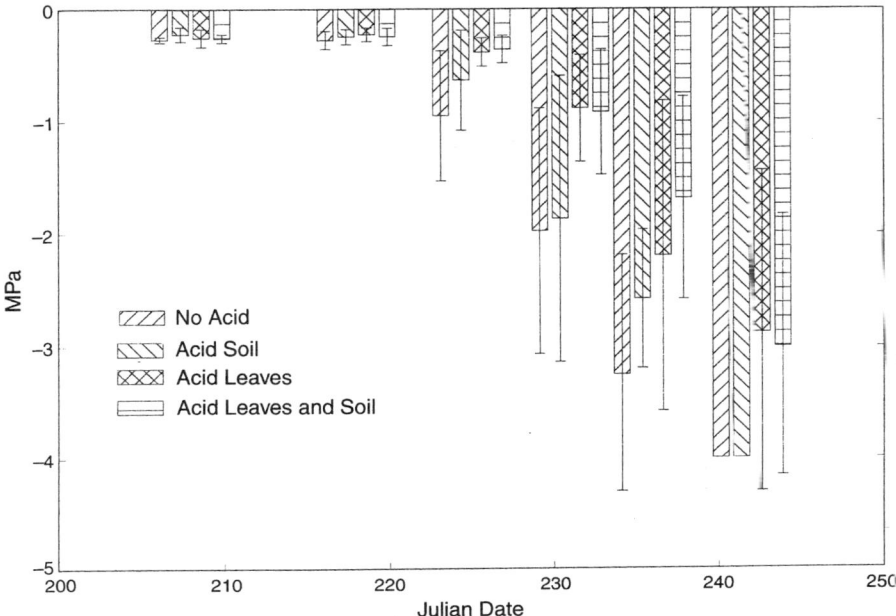

Figure 34.3. Predawn xylem pressure potential (PXPP) of dogwood seedlings that were pretreated with either pH 2.5 or 5.5 simulated rain on foliage, soil, or both, on five measurement dates after two drought stress treatments were applied.

Summary

Although other research has shown that acidic rain treatment erodes dogwood foliage (Brown et al., 1994; Thornham et al., 1992) and enhances nutrient leakage (Willey and Hackney, 1991), the studies reported here clearly demonstrate that the more important mechanisms for increased anthracnose severity are soil-mediated.

Acidic rain has the potential to affect a number of soil or root processes, but N fertilization appears to be the most probable explanation for the increased growth observed. All seedlings were lightly fertilized once during the study, but the SAR treatments applied to the medium provided more than twice this amount of N when the pH was 2.5 but they received almost no additional N when the pH was 5.5. Seedlings growing in the medium treated with pH 2.5 rain had significantly higher leaf N than seedlings growing in the pH 5.5 rain-treated medium (Table 34.2). Ludovici (1990) found SAR N increased growth in *Pinus taeda* L. (Hudler, 1985). Nitrogen fertilization would also explain the reduced leaf abscission observed in 1993, following soil applications of pH 2.5 rain. The tendency of high levels of soil N to delay leaf abscission has been recognized for many years (Kozlowski, 1971). Further studies are underway to determine whether N fertilization alone increases susceptibility to anthracnose. Neely (1986) found the

Table 34.2. Effects of Soil and Foliage Pretreatments with pH 2.5 or 5.5 Simulated Acid Rain on Date of Death, Biomass, Leaf Area, and Leaf Nitrogen Content and Shoot/Root Ratio of Drought-Stressed Trees

Pretreatment with pH 2.5 SAR[1]	Mean Julian date of death	Leaf biomass (gm)	Stem biomass (gm)	Root biomass (gm)	Leaf area (cm^2)	Leaf N content %	Shoot/Root Ratio
None	247 a[2]	5.7 ab	9.55 ab	22.2 a	780 b	0.75 b	0.67 a
Foliage only	244 ab	4.9 b	8.58 b	19.6 a	787 b	0.72 b	0.70 a
Soil only	237 bc	6.7 a	10.39 ab	22.2 a	1110 a	0.90 a	0.76 a
Soil & foliage	234 c	6.4 ab	11.13 a	23.2 a	1073 a	0.89 a	0.76 a

[1] Foliage or soil not treated with pH 2.5 SAR was treated with pH 5.5 SAR.
[2] Numbers within columns followed by the same letter are not significantly different ($P \leq 0.05$) according to Duncan's Multiple Range Test.

reverse in *Gnomonia leptostyla* infection of black walnut; in that host–pathogen system, the number of anthracnose lesions decreased with increasing leaf N content.

Because acidic deposition can affect the ability of plants to regulate internal moisture (Eamus et al., 1989; Mengel et al., 1989), increased susceptibility to drought could be another explanation for the effect of acidic rain on the intensity of dogwood anthracnose. Among water-stressed seedlings, those growing in media treated with pH 2.5 rain died ten days earlier than seedlings growing in media treated with pH 5.5 rain. Because these seedlings were larger, increased water usage may explain the faster mortality rate. The increase in leaf area associated with soil applications of pH 2.5 rain could also make the seedlings more susceptible to drought stress if leaf growth occurred at the expense of root growth. Although the trend in shoot/root ratio supports this hypothesis, differences were not significant with this small sample size. Norby et al. (1986) reported a similar increase in moisture stress resulting from acidic rain treatments that was associated with increases in growth rates of red spruce.

Factors that reduce dogwood vigor are believed to increase anthracnose severity (Gould and Peterson, 1994). This study suggests, however, that exposure of the soil to acidic rain actually increases vigor if the seedlings are well watered. Ludovici (1990) also found that soil applications of SAR increased root and shoot weight of *Pinus taeda* L. seedlings. An increase in photosynthesis following SAR treatments was also reported for *Phaseolus vulgaris,* but it was accompanied by a decrease in carbohydrate production and growth rate (Ferenbaugh, 1976). Ferenbaugh (1976) suggested that uncoupling of photophosphorylation explained this apparent anomaly. This possible explanation does not apply to dogwood because soil applications increased growth and photosynthesis in this study. Smith (1990) described a potential interaction of an intermediate dosage of air pollutant with temperate forest ecosystems. In this model interaction, individuals of a given species would be expected to undergo nutrient stress, decreased photosynthesis rate, decreased reproductive rate, and reduced vigor, thus becoming increasingly predisposed to disease and insect pests. The data presented here suggest that although well-watered plants can be more vigorous after receiving acidic rain treatments, water-stressed trees died faster following acid rain treatments. In nature, midsummer drought stress is common in dogwood, which is shallow-rooted. Therefore, it is important to remember that the host response to acidic precipitation interacts significantly with other climatic and site factors.

Most research on the interactions between acidic rain and dogwood anthracnose has focused on foliar mechanisms (Brown et al., 1994; Thornham et al., 1992). The results of these studies indicate that the belowground effects of acidic rain are more important. Acidic rain can change the availability of N and other nutrients (Binkley et al., 1989) and it may be that changes in nutrient composition or changes in carbohydrate levels make the leaves a better substrate for growth of the fungus. Perhaps an increase in succulence brought on by the addition of N in the SAR solution to the growing medium makes it easier for fungal hyphae to penetrate the leaf tissue. Unfortunately, too little is known about the effects of acidic rain on dogwood anthracnose to determine how important it is outside an experimental situation, or how to counteract its effects.

It seems improbable that acidic rain alone is responsible for the decimation of dogwood experienced in the Northeast and in the southern Appalachian *Cornus florida* L. population. Our studies do indicate, however, that acidic rain may play a role by increasing host vulnerability to drought stress, as well as increasing inherent susceptibility of dogwood to anthracnose.

The impact of anthracnose on dogwood populations varies by location (Langdon et al., 1993). At Catoctin Mountain National Park, dogwood populations have declined 94% in just ten years. Most surviving trees are growing in locations with partial exposure to direct sunlight. Understory trees, which receive only 2% of ambient photosynthetically active radiation (Chellemi and Britton, 1992) probably have few carbohydrate reserves to expend on refoliation after fungal attack. The microclimate of understory trees is also more favorable to infection than that of partially exposed trees (Chellemi and Britton, 1992). Temporary conditions favorable to severe infection, which might include acidic precipitation, may have profound direct and indirect effects on the forest understory in just a few years.

The long-term effects of species replacement in the understory could be numerous and significant. Dogwood fruit are high in fat, an important energy source for winter survival of migratory birds and other animals. The fruit makes up 25 to 50% of the diet of the evening grosbeak, and 5 to 10% of the diet of ruffed grouse, wild turkey, cardinals, robins, gray-checked and wood thrushes, and cedar waxwings (Halls, 1977; Martin et al., 1951). Dogwood leaves and twigs are high in calcium (DeGraff and Whitman, 1979), and are used by bear, beaver, rabbit, racoon, fox squirrel, chipmunks, deer, and moose. The decline of this essential nutrient may have long-term dietary consequences.

Dogwood also plays an important role in calcium cycling in forest soils. The foliage accumulates calcium, and leaf litter contains 2.0 to 3.5% calcium (dry weight) (Vimmerstedt, 1957). Thus, species replacement of the dogwood could have sweeping consequences for other vegetation.

References

Anderson RL, Berrang P, Knighten J, Lawton KA, Britton KO (1993) Pretreating dogwood seedlings with simulated acidic precipitation increases dogwood anthracnose symptoms in greenhouse-laboratory trials. Can J For Res 23:55–58.

Barkland P, Axelsson G, Unestam T (1984) *Gremmeniella abietina* in Norway spruce, latent infection, sudden outbreaks, acid rain, pre-disposition) In Manion PD (Ed) *Scleroderris canker of conifers*. Martinus Nijhoff/W. Junk, the Hague, Amsterdam.

Barkland P, Unestam T (1988) Infection experiments with *Gremmeniella abietina* on seedlings of Norway spruce and Scots pine. Eur J For Path 18:409–420.

Binkley D, Driscoll CT, Allen HL, Schoeneberger P, McAvoy D (1989) *Acidic deposition and forest soils*. Springer-Verlag, New York.

Blakeman JP (1973) The chemical environment of leaf surfaces with special reference to spore germination of pathogenic fungi. Pestic Sci 4:575–588.

Bragg RJ, Manion PD (1984) Evaluation of possible effects of acid rain on Scleroderris canker of red pine in New York. In Manion PD (Ed) *Scleroderris canker of conifers* Marinus Nijhoff/ W. Junk, The Hague, Amsterdam.

Britton KO (1989) Temperature, pH, and free water effects on in vitro germination of conidia of a *Discula* sp. isolated from dogwood anthracnose. Phytopath 79:1203.

Britton KO (1993) Anthracnose infection of dogwood seedlings exposed to natural inoculum in western North Carolina. Plant Dis 77:34–37.

Britton KO, Berrang P, Mavity E (1996) Effects of pretreatment with simulated acidic rain on the severity of dogwood anthracnose. Plant Dis 80:646–649.

Brown DA, Windham MT, Anderson RL, Trigiano RN (1994) Influence of simulated acid rain on the flowering dogwood (*Cornus florida* L.) leaf surface. Can J For Sci 24:1058–1062.

Bruck RI. Shafer SR (1983) Effects of acid precipitation on plant diseases. In Linthurst RA (Ed) *Direct and indirect effects of acidic deposition on vegetation*. Acid Precipitation Series No. 5., Butterworth Publishers, Boston.

Byther RS, Davidson RM Jr (1979) Dogwood anthracnose. Orn Northw News 3:20–21.

Campbell CL, Bruck RI, Sinn JP, Martin SB (1988) Influence of acidity level in simulated rain on disease progress in four plant pathosystems. Environ Pollut 53:219–234.

Chellemi DO, Britton KO (1992) Influence of canopy microclimate on incidence and severity of dogwood anthracnose. Can J Bot 70:1093–1096.

Chevone BI, Yang YS, Winner WE, Storks-Cotler I, Long SJ (1984) A rainfall simulator for laboratory use in acidic precipitation studies. J Air Pollut Control Assoc 31:355–359.

DeGraff RM, Whitman GM (1979) *Trees, shrubs and vines for attracting birds: A manual for the Northeast*. Univ MA Press, Amherst.

Eamus D, Leith I, Fowler D (1989) The influence of acid mist upon transpiration, shoot water potential and pressure-volume curves of red spruce seedlings. Ann Sci For 46:577–580.

Evans LS (1982) Biological effects of acidity in precipitation on vegetation: A review. Environ Exper Bot 22:155–169.

Ferenbaugh RW (1976) Effects of simulated acid rain on *Phaseolus vulgaris* L. (Fabaceae). Am J Bot 63:283–288.

Gould AB, Peterson JL (1994) The effect of moisture stress and sunlight on the severity of dogwood anthracnose in street trees. J Arboric 20:75–78.

Grzywacz A, Wazny J (1973) The impact of industrial air pollutants on the occurrence of several important pathogenic fungi of forest trees in Poland. Eur J For Path 3:129–141.

Haines B, Stefani M, Hendrix F (1980) Acid rain: Threshold of leaf damage in eight plant species from a southern Appalachian forest succession. Water Air Soil Pollut 14:403–407.

Halls LK (1977) Southern fruit-producing woody plants used by wildlife. USDA For. Serv. SE Forest Exper Stat Gen Tech Rep SO-16. Asheville, North Carolina.

Heagle AS (1973) Interactions between air pollutants and plant parasites. Ann Rev Phytopath 11:365–388.

Hibben CR, Daughtrey ML (1988) Dogwood anthracnose in the northeastern United States. Plant Dis 72:199–203.

Hudler GW (1985) Thinking out loud . . . Origins of dogwood lower branch dieback. New York State Arborists Shade Tree Notes 8(2):1–2.

Johnson DW, Taylor GE (1989) Role of air pollution in forest decline in eastern North America. Water Air Soil Pollut 48:21–43.

Killiam K, Firestone MK, McColl JG (1983) Acid rain and soil microbial activity: Effects and their mechanisms. J Environ Qual 12:133–137.

Kozlowski TT (1971) *Growth and development of trees,* Vols 1, 2. Academic Press, New York.

Langdon K, Parker C, Windham M, Powell S, Johnson K (1993) A preliminary hazard rating for dogwood anthracnose in the Southern Appalachians. In Results of the 1992 dogwood anthracnose impact assessment and pilot test in the southeastern United States. USDA For Serv South Reg Prot Rep R8-PR 24.

Lepp NW, Fairfax JAW (1976) The role of acid rain as a regulator of foliar nutrient uptake and loss. In Dickinson CH, Preece TF (Eds) *Microbiology of aerial plant surfaces*. Academic Press, London.

Ludovici KH (1990) Influence of different simulated rain chemistries on a hapludult and the root growth of *Pinus taeda*. MS thesis. NC State Univ, Raleigh.

Madden LV, Campbell CL (1987) Potential effects of air pollutants on epidemics of plant diseases. Agric Ecosys Environ 18:251–262.

Magan N, McLeod AR (1991) Effects of atmospheric pollutants on phyllosphere microbial communities. In Andrews JH, Hirano SS (Eds) *Microbial ecology of leaves*. Springer-Verlag, New York.

Martin AC, Zim HS, Nelson HL (1951) *American wildlife and plants*. McGraw Hill, New York.

Mengel K, Hogrebe AMR, Esche A (1989) Effect of acidic fog on needle surface and water relations of *Picea abies*. Physiol Plant 75:201–207.

Musselman RC, McCool PM (1989) Effect of acidic fog on productivity of celery and lettuce and impact on incidence and severity of diseases. Ann Appl Biol 114:559–565.

Neely D (1986) Total leaf nitrogen correlated with walnut anthracnose resistance. J Arboric 12:312–315.

Norby RJ, Taylor GE Jr, McLaughlin SB, Gunderson CA (1986) Drought sensitivity of red spruce seedlings affected by precipitation chemistry. In Tauer CG, Hennessey, TC (Eds) Proc No Amer For Biol Wkshp, Stillwater, OK.

Percy KE, Baker EA (1988) Effects of simulated acid rain on leaf wettability, rain retention and uptake of some inorganic ions. New Phytol 108:75–82.

Redlin SC (1991) *Discula destructiva* sp. nov., cause of dogwood anthracnose. Mycol 83:633–642.

Rehfuess KE (1989) Acidic deposition—Extent and impact on forest soils, nutrition, growth and disease phenomena in Central Europe: A review. Water Air Soil Pollut 48:1–20.

Shafer SR, Bruck RI, Heagle AS (1985) Influence of simulated acid rain on *Phytophthora cinnamomi* and Phytophthora root rot of blue lupine. Phytopath 75:996–1003.

Shafer SR, Grand LF, Bruck RI, Heagle AS (1985) Formation of ectomycorrhizae on *Pinus taeda* seedlings exposed to simulated acidic rain. Can J For Res 15:66–71.

Shriner DS (1978) Effects of simulated acidified rain on host-parasite interactions in plant diseases. Phytopath 68:213–218.

Skelly JM (1989) Forest decline versus tree decline—Pathological considerations. Environ Monit Assess 12:23–27.

Smith WH (1990) *Air pollution and forests: Interaction between air contaminants and forest ecosystems*. Springer-Verlag, New York.

Thornham KT, Stipes RJ, Grayson RL (1992) Effect of acid deposition on trichome morphology and dogwood anthracnose biology. VA J Sci 42:242.

Vimmerstedt JP (1957) Silvical characteristics of flowering dogwood. USDA For Serv, SE For Exper Stat Pap No 87.

Willey JD, Hackney JH (1991) Chemical interactions between acid rain and dogwood leaves. J of Elisha Mitchell Sci Soc 107:83–88.

35. Effects of Temperature and Drought Stress on Physiological Processes Associated With Oak Decline

Theodor D. Leininger

Oak decline is a term used to describe a sequence of events (decline syndrome) which is typically triggered by an abiotic stress and subsequently involves other biotic and abiotic factors that cause the progressive deterioration and eventual death of a tree. Decline diseases lack a single causal agent, and in that way are different from diseases caused by one pathogen or by a single abiotic injury. Decline and premature death of oaks in the oak-dominated eastern deciduous forests have been documented in at least twenty-six separate reports over the past 140 years (Ammon et al., 1989).

Drought-induced stress appears to trigger or contribute to the decline syndrome in many of these reports for the eastern United States (Beal, 1926; Hursh and Haasis, 1931; McIntyre and Schnur, 1936; True and Tyron, 1956; Gillespie, 1956; Fergus and Ibberson, 1956; Staley, 1965; Lewis, 1981; Tainter et al., 1983; Law and Gott, 1987; Maass, 1989; Tainter et al., 1990; Myers and Killingsworth, 1992). Under environmental conditions in which water availability may limit growth, abnormally high temperatures can alter normal energy flows and can increase both respiration and transpiration. Temperature and water availability are among the most critical abiotic conditions that must remain within certain ranges for optimum growth of any species, and oaks are no exception. Temperature can influence, growth and development, metabolism, carbon translocation, enzyme action, water potential, and transpiration (Salisbury and Ross, 1978). Water probably provides the strongest influence on productivity of vegetation in forest ecosystems than any other abiotic factor (Whittaker, 1975; Kozlowski, 1982).

Concentrations of such greenhouse gases as carbon dioxide (CO_2), methane, and oxides of nitrogen in the atmosphere are predicted to double in the next 100 years (Edmonds, et al., 1984; Friedli et al., 1986), thereby increasing the greenhouse effect and leading to an estimated increase in global mean temperature of 1.5 to 4.5 °C (National Academy of Sciences, 1983). In addition to these increases of gases, summertime precipitation is predicted to decrease from 5 to 10%, and wintertime precipitation is predicted to increase from 0 to 15% (Karl et al., 1991). However, opposite responses of greater precipitation, lower maximum temperatures, and higher minimum temperatures have also been predicted (Idso and Balling, 1992).

Under the warmer and drier climate change scenario, normal ecosystem functions and energy flows could be altered as individual trees and species adapt to the predicted changes. This adaptation will depend on a number of such factors as the magnitude and rate of climate change and the extent to which increased CO_2 can offset the potential growth-limiting effects of elevated temperature and drought (Keller, 1984). Woodman and Furiness (1988) predicted an increase in tree injury and death from insects and pathogens, acting as single agents, resulting from a warmer and drier climate. For instance, trees growing on suboptimum sites would be more susceptible to insect attacks (Miller et al., 1987). In general, pest problems would increase in stands containing less vigorous, stressed trees (Hedden, 1987) found in older, unmanaged, and denser forests. More than thirty years ago, Hepting (1963) described the potential for the incidence, severity, and northern ranges of diseases to increase should climate factors, which normally act as constraints against outbreaks, become altered. Climate modeling with regard to decline of evergreen oaks in the Mediterranean area predicts an increase in host range of the root pathogen *Phytophthora cinnamomi* Rands, in addition to increases in root disease severity and fungus survivability (Brasier and Scott, 1994). Tomlinson (1993) postulated that increased temperature and reduced rainfall led to nutrient deficiency, fine root mortality, and, eventually, crown dieback as nutrient cations were leached from soil by acids formed from mineralization and nitrification in excess of the tree's needs. An increased frequency of such stresses as drought and elevated temperature may trigger increasing numbers of decline events and therefore place a greater burden on present oak resources.

The southern United States produces 57% of the nation's hardwood lumber, 40% of its hardwood plywood, and 60% of its pulpwood, a large part of which is hardwood (Kronrad, 1993). Red and white oak lumber accounted for more than half the $3.5 billion value of hardwood lumber in the eastern United States in 1990. Within the next fifty years, the U.S. Forest Service predicts an 80% increase in hardwood harvesting at the same time growth is declining as a result of several factors including 1) growth decline in some species (oaks), 2) increased death, 3) low regeneration rates, and 4) declines in acreage (Kronrad, 1993). The most recent Forest Inventory Analysis (FIA) data indicate that of the 42 million hectares (ha) of hardwood forest type occurring in the twelve southern states, 44% are vulnerable to oak decline, and of this area, 6% of bottomland and 10% of upland types are affected by oak decline (Hoffard et al., 1995). Furthermore, volume losses resulting from oak death are about 1.4 times greater on affected areas compared to unaffected areas. Given that this important oak resource is declining in availability at a time of increasing demand while simultaneously facing the possibility of an increasingly stressful environment, it is imperative to understand the effects of such environmental stresses as a warmer average temperature and more frequent droughts on oak tree physiology and growth. Additional studies will also be needed to examine the effects of such stresses along with effects of insect attacks and disease occurrence.

Much of the bottomland oak resource, especially in the Lower Mississippi Alluvial Valley, has been removed for agriculture (Ford, 1994). The remaining bottomland hardwood forests, and the upland hardwood forests, will contribute greatly to the multiple resource needs of fiber, wildlife habitat, and recreation placed by society on the forests of the southern United States. Because decline diseases are triggered by stress, symptoms typically occur simultaneously over wide geographic regions and often affect a single species, or group of species. Therefore, it is essential to be able to predict how the bottomland and upland oak resources in the southern United States will respond to increased stress from elevated temperature and drought, and to determine whether oaks on better sites could become more susceptible to a decline-triggering stress.

During 1994, an experiment was conducted to examine the effects on seedling physiology and growth of three bottomland and three upland oak species from the combined stresses of elevated temperature and drought under a predicted 1.5 to 4.5 °C increase in average global temperature. The results of the first year are presented here and represent average responses of seedling oaks to the very low end of the predicted increases in global mean temperature and drought. Although studies of seedling physiology have limited application to mature trees and whole ecosystems, data from this experiment should be useful to process modelers interested in simulating effects of moderate temperature and drought stress on individual whole plants and in extrapolating the data to estimate effects at the ecosystem level.

Methods

Plant Culture

During May of 1993, seeds of the following four species in the red oak group were planted: 1) southern red (*Q. falcata* Michx.), 2) nuttall (*Q. nuttallii* E.J. Palmer), 3) willow (*Q. phellos* L.), and 4) scarlet (*Q. coccinea* Münchh.). Two species in the white oak group were also planted: 1) white (*Q. alba* L.), and 2) overcup (*Q. lyrata* Walter). All the seeds were planted in treepots (7.5 l; 41 × 15 × 15 cm) with fritted clay (van Bavel et al., 1978) as the potting medium. Acorns were collected in the fall of 1992 from the following sources: 1) nuttall, willow, and overcup from an area near Stoneville, Mississippi, southern red from near Redwood, Mississippi, 3) scarlet from near La Grange, Tennessee, and 4) white from near Cadiz, Kentucky. Seedlings were given 35 g of a controlled-release fertilizer containing nitrogen-phosphorus-potassium (N-P-K) (17–6-10) and other minor nutrients (Sierra Blend, Grace-Sierra Co.) at the start of each growing season. Seedlings were exposed to similar temperature and shading during 1993, and subjected to three soil moisture and three air temperature shading treatments during 1994.

Temperature Shading and Soil-Moisture Treatments

Three greenhouses provided three distinct summertime temperatures for three similar sets of plants by using a combination of evaporative cooling and shading that was needed to reduce heat loads inside the greenhouses. Evaporative cooler thermostats and shadecloths were adjusted to produce the treatments employed, that reflect differences from actual average monthly maximum temperatures during June to October of 1994 against a baseline average monthly maximum temperature of 1) 27.4 °C (Base), 2) a baseline + 0.7 °C (Base + 0.7), and 3) a baseline + 1.7 °C (Base + 1.7) (Figure 35.1). To produce these temperatures, greenhouse roof areas were covered year round with 50% shadecloth in the following three manners: 1) 100%-covered for the Base, 2) 40%-covered for the Base + 0.7, and 3) 25% covered for the Base + 1.7 treatments. Average year-round temperatures, from the time acorns were planted until seedlings were harvested, were 26.0 °C, 26.4 °C, and 26.8 °C, respectively, for the Base, Base + 0.7, and Base + 1.7 summertime treatments. During winter, temperatures were prevented from falling below 0 °C in all greenhouses.

Soil-moisture regimes were started in mid-July 1994 with adequate water (A), intermediate water (B), and drought (C) treatments maintained at or above, 27%

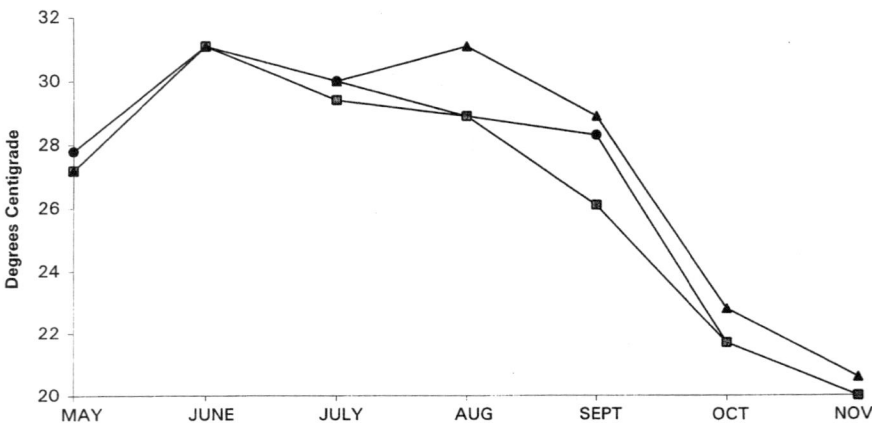

Figure 35.1. Average monthly maximum temperatures for the 1994 growing season; differences are described in the text. ■ = Base, ● = Base + 0.7 °C, ▲ = Base + 1.7 °C. Differences in temperatures of 0.7 °C and 1.7 °C above Base were acheived as averages from May to October.

(−.3 MPa), 22% (−.6 MPa), and 17% (−1.2 MPa) volumetric soil moisture (soil-moisture potential), respectively. Soil moisture was measured regularly by time-domain reflectometry (Soil Moisture Equipment Corp.), and seedlings were watered to field capacity with deionized water each time the predetermined minimum volumetric soil moisture was attained. On the warmest days, A seedlings were watered every 24 hours, and C seedlings were watered aprroximately every 36 to 48 hours.

Experimental Design and Biological Measurements

A randomized complete block design was used in which three replicates of each species × moisture regime combination were distributed in thirty-six positions on each of three benches per greenhouse. Each bench was a block with the first being closest to the evaporative cooler so that air in this block was presumably cooler and more humid than air in block three, which was farthest from the cooler and next to the exhaust fans. Half the number of seedlings per bench (eighteen) were harvested after the first year, and the remaining seedlings were harvested after the 1995 growing season. Analysis of variance (ANOVA), using SAS software was performed on values of net photosynthesis (P_{net}), stomatal conductance (g_s), transpiration rate (TR), leaf pigment concentrations (chlorophylls a and b, and carotenoids and xanthophylls), and predawn leaf xylem water potential (ψ) measured during the first, sixth, tenth, and fourteenth weeks (July to October) following the start of different soil-moisture regimes. Heights and diameters, biomass (leaves, stems, and roots), and foliar concentrations of calcium (Ca), magnesium (Mg), K, and P were measured at the end of the growing season. All variables were analyzed in a two-way analysis of variance (six species × three soil-moisture regimes) for different responses in species and soil-moisture treatments within each greenhouse.

Stomatal conductance and photosynthetic and transpiration rates were measured on fully expanded leaves with a LI-6250 portable photosynthesis system (LiCor, Inc.) and a 1–l cuvette. Measurements were made during the day with photosynthetic photon flux densities (PPFD; ± standard error) between 1100 and 1300 ± 2 to 5 $\mu mol\ m^{-2}\ s^{-1}$, mean leaf temperatures (± standard error) in the cuvette from 23 to 33 ± 0.1 to 0.2 °C depending on the month, and relative humidities (± standard error) in the cuvette from 61 to 83 ± 0.2 to 1.2%. Photosynthetic photon flux densities were generated by a quartz halogen projection lamp (GE–ESD, General Electric Co.) for consistency. Using a pressure chamber, predawn leaf-water potentials were measured on the same leaf of each seedling used to measure gas exchange nine to eighteen hours earlier. Chlorophyll, carotenoids, and xanthophylls were extracted in 10 ml of dimethyl sulfoxide (Hiscox and Israelstam, 1979) from two 0.78 cm^2 disks removed from the portion of each leaf that was in the cuvette for gas exchange measurements. Pigment abscrbencies were measured following extraction in the dark at 65 °C for fifteen to eighteen hours, and pigment concentrations were estimated according to the equations of Lichtenthaler and Wellburn (1983) for 100% acetone. For biomass determinations, leaves, stems, and roots of seedlings were dried to a constant

Table 35.1. Data Quality Objectives and Achievements for the Response Variables[1]

Response variable	Precision (%)		Completeness (%)	
Leaf xylem water potential	10	6.2	99	100
Net photosynthesis	10	4.6	99	99
Stomatal conductance	10	7.0	99	99
Transpiration	10	6.2	99	99
Chlorophyll a	10	5.3	99	98
Chlorophyll b	10	9.0	99	98
Carotenoids & xanthophylls	10	4.8	99	98
Leaf biomass	15	12.3	99	99
Stem biomass	15	16.2	99	99
Root biomass	15	10.6	99	99
Stem height	15	8.0	99	100
Stem diameter	10	4.9	99	100

[1] Precision was expressed as a coefficient of variation for the reduced data. Completeness was measured as the ratio of actual measurements recorded to the total number of possible measurements. The first value in each category was the objective, the second value was the achievement.

weight at 70 °C and then were weighed. Stem heights were measured to the nearest mm from 2.54 cm above the soil line to the tip of the dominant leader; stem diameters were measured to the nearest 0.01 mm at 2.54 cm above the soil line. Samples of dried leaf tissue were analyzed for Ca, Mg, and K concentrations using atomic absorption spectrophotometry, and for P concentrations using acid extraction and a colorimetric technique.

Data quality objectives of precision and completeness for measurement variables were met, or exceeded, in most cases (Table 35.1). Accuracy of data could not be measured for all variables but was ensured by calibrating all instruments according to manufacturers' specifications and by adhering to published standard operating procedures.

Results

Average ψ for the six species showed some differences (Base + 0.7), and general trends resulting from differences in soil-moisture treatments six weeks after treatments were in place (Table 35.2). Differences continued to be evident in Weeks 10 (Base + 1.7) and 14 (Base and Base + 0.7). Throughout the experiment, ψ tended to be more negative for the majority of seedlings in the warmer and drier air environment of block 3 in each greenhouse; nuttall oak tended to have greater negative ψ values (data not shown). The values of predawn ψ measured in this study are about seven to ten times less than those reported for sapling and mature *Q. alba* L. (Hinckley et al., 1978; Dougherty and Hinckley, 1981) under prolonged drought that caused soil-moisture potentials from -2.0 to -4.5 MPa. The drought stress treatment (-1.2 MPa) in the present study, although moderate in severity, could not be sustained in container-grown seedlings of this size for more

35. Effects of Temperature and Drought Stress Associated With Oak Decline   653

Table 35.2. Average Predawn Leaf Xylem Water Potentials (ψ) of Six Species of Two-Year-Old Oak Seedlings Exposed to Three Temperature Shade Treatments and Three Soil-Moisture Regimes[1]

Temp.	Moist.	Ψ (−MPa)			
		Wk. 1	Wk. 6	Wk. 10	Wk. 14
Base	A	.24 ± .01	.28 ± .02	.29 ± .02	.22^b ± .01
	B	.24 ± .02	.30 ± .02	.29 ± .02	.27^a ± .02
	C	.24 ± .01	.30 ± .02	.32 ± .02	.26^{ab} ± .02
Base + 0.7	A	.26 ± .02	.27^b ± .02	.27 ± .01	.18^b ± .01
	B	.22 ± .02	.32^a ± .01	.27 ± .01	.21^{ab} ± .01
	C	.23 ± .02	.33^a ± .01	.29 ± .02	.23^a ± .01
Base + 1.7	A	.26^a ± .02	.26 ± .02	.30^b ± .02	.22 ± .02
	B	.22^b ± .01	.29 ± .02	.31^b ± .02	.19 ± .01
	C	.27^a ± .01	.29 ± .02	.38^a ± .02	.22 ± .02

[1] Data are means of eighteen samples ± SE. Also, means with different letters between soil-moisture regimes, and within the same temperature treatment and time interval combinations, are different as determined by Duncan's Multiple Range Test (P = 0.05).

than forty-eight hours without inducing severe wilting and risking the possibility of branch dieback.

Mean P_{net} (Table 35.3) of all species without regard to soil-moisture regimes was greater in Base + 0.7 (8.2 ± 0.3 $\mu molm^{-2}s^{-1}$) and Base + 1.7 (7.8 ± 0.4 $\mu molm^{-2}s^{-1}$) temperatures than in the Base temperature (6.4 ± 0.3 $\mu molm^{-2}s^{-1}$). Furthermore, P_{net} was significantly greater (P = 0.05) in block 3 than in block 1 of each greenhouse at 6, 10, and 14 weeks (data not shown) probably because the air surrounding block 3 was warmer than that of block 1 because of its greater distance from the evaporative cooler pads. However, within temperature treatments, there was little relationship between mean P_{net} of all species and soil-moisture treatments. Mean P_{net} values averaged across temperature treatments were 7.3 ± 0.3 $\mu molm^{-2}s^{-1}$ for soil-moisture regime A, 7.6 ± 0.4 $\mu molm^{-2}s^{-1}$ for regime B, and 7.5 ± 0.4 $\mu molm^{-2}s^{-1}$ for regime C, again indicating little response to soil-moisture treatments. These same trends of an apparent greater response to warmer temperatures and little response to soil-moisture regime were evident for g_s and TR although individual analyses of g_s and TR responses to soil-moisture regimes within temperature regimes were generally inconclusive (Table 35.3).

Taken as a whole, these results suggest that exposing these oak seedlings to warmer air temperatures tended to increase rates of net photosynthesis and transpiration, and increase stomatal conductance. Seedlings of *Q. rubra* L. exposed to warmer temperatures and greater PPFD in an ozone-exposure study had greater stomatal conductances and greater rates of net photosynthesis than seedlings in a cooler environment with a lesser rate of PPFD (Samuelson, 1994). However, results of the present study are in contrast to reports in which prolonged drought conditions caused marked decreases in P_{net} of *Q. alba* (Hinckley et al., 1979; Dougherty and Hinckley, 1981), and in *Q. petraea* Matt. Liebl. prolonged drought

Table 35.3. Average Net Photosynthesis (P_{net}), Stomatal Conductance (g_s), and Transpiration Rates (TR) for Six Species of Two-Year-Old Oak Seedlings Exposed to Three Temperature and Shade Treatments and Three Soil Moisture Regimes[1]

Temp	Moist	P_{net} (μmolm^{-2}s^{-1})				g_s (molm^{-2}s^{-1})				TR (mmolm^{-2}s^{-1})			
		Wk. 1	Wk. 6	Wk. 10	Wk. 14	Wk. 1	Wk. 6	Wk. 10	Wk. 14	Wk. 1	Wk. 6	Wk. 10	Wk. 14
Base	A	7.1 ± .2	5.6 ± .3	5.3^b ± .3	6.7 ± .2	.21 ± .01	.12^b ± .01	.10^b ± .01	.17^b ± .01	2.9 ± .13	1.7 ± .12	1.3^b ± .10	1.1^b ± .04
	B	8.2 ± .4	6.1 ± .3	5.3^b ± .3	6.6 ± .3	.24 ± .01	.14^a ± .01	.11^{ab} ± .01	.18^b ± .01	3.1 ± .11	1.8 ± .12	1.4^{ab} ± .11	1.1^b ± .05
	C	8.0 ± .6	5.6 ± .3	6.0^a ± .3	7.0 ± .4	.28 ± .04	.12^b ± .01	.12^a ± .01	.22^a ± .02	3.3 ± .28	1.7 ± .11	1.6^a ± .10	1.3^a ± .09
Base + 0.7	A	7.9^b ± .4	8.3^a ± .3	7.9 ± .4	9.4 ± .4	.22 ± .02	.22^a ± .02	.15^{ab} ± .01	.28 ± .02	3.1 ± .21	3.5^a ± .20	2.4 ± .17	2.9^b ± .15
	B	8.9^a ± .3	8.4^a ± .3	7.5 ± .5	8.9 ± .4	.23 ± .01	.23^a ± .01	.17^a ± .01	.29 ± .02	3.3 ± .13	3.6^a ± .18	2.4 ± .18	3.5^a ± .21
	C	7.7^b ± .2	7.3^b ± .4	7.0 ± .4	8.7 ± .4	.24 ± .02	.17^b ± .01	.13^b ± .01	.30 ± .02	3.3 ± .18	2.8^b ± .21	2.2 ± .14	2.9^b ± .14
Base + 1.7	A	7.2^b ± .3	6.9^b ± .4	8.3^{ab} ± .3	6.8^{ab} ± .3	.23^b ± .01	.14^b ± .01	.17 ± .01	.16^b ± .01	2.2^b ± .11	2.5^b ± .23	2.1 ± .11	1.5 ± .09
	B	8.3^a ± .4	9.0^a ± .4	7.9^b ± .4	6.7^b ± .4	.34^a ± .02	.23^a ± .02	.20 ± .02	.20 ± .02	3.2^a ± .20	3.5^a ± .21	2.0 ± .12	1.7 ± .10
	C	8.9^a ± .3	7.5^b ± .3	8.8^a ± .3	7.4^a ± .4	.30^a ± .02	.17^b ± .02	.17 ± .01	.21^a ± .02	2.9^a ± .15	3.2^a ± .26	2.3 ± .13	1.7 ± .12

Data are means of eighteen samples ± SE. Also, means with different letters between soil-moisture regimes, and within the same temperature treatment and time interval combinations, are different as determined by Duncan's Multiple Range Test (P = 0.05)

Table 35.4. Average Heights and Diameters of Six Species of Two-Year-Old Oak Seedlings Exposed to Three Temperature and Shade Treatments, and Three Soil-Moisture Regimes[1]

Temp	Moisture	Height (cm)	Diameter (mm)
Base	A	156.6 ± 13.2	13.5 ± .7
	B	146.8 ± 13.6	12.8 ± .9
	C	147.8 ± 12.6	13.0 ± .8
Base + 0.7	A	139.1 ± 10.2	12.8 ± .5
	B	128.7 ± 10.6	12.3 ± .5
	C	138.6 ± 9.1	12.4 ± .5
Base + 1.7	A	127.7 ± 10.4	11.9 ± .5
	B	124.1 ± 9.7	11.3 ± .5
	C	142.0 ± 10.3	11.3 ± .6

[1] Data are means of thirty-six samples ± SE.

conditions caused marked decreases in P_{net} and g_s when predawn ψ dropped below -1.0 MPa (Epron and Dreyer, 1993). Similarly, P_{net} and g_s were reduced in *Q. robur* L., *Q. rubra,* and *Q. petraea* with predawn ψ as low as -3.5 MPa (Vivin et al., 1993).

The pattern of increasing carbon assimilation as temperature increased was not immediately evident in either the height and diameter data (Table 35.4) or the biomass data (Table 35.5) resulting primarily from a plant culture artifact during seedling establishment that resulted in larger seedlings in the Base temperature treatment than in Base + 0.7 and Base + 1.7 treatments. The effect of greater carbon uptake in warmer temperatures was evident as greater increases in average heights (Base = 124%, Base + 0.7 = 318%, Base + 1.7 = 283%) and average diameters (Base = 82%, Base + 0.7 = 184%, Base + 1.7 = 149%) expressed as percentages of pretreatment heights and diameters for all species across soil-moisture treatments. Drought-stressed seedlings exposed to the Base + 1.7 temperature treatment had greater leaf and total shoot biomass than seedlings in the intermediate or adequately watered treatments, although means separations for stem and root biomass between water regime treatments were inconclusive (Table 35.5). Root/shoot ratios indicate that in the Base + 0.7 and Base + 1.7 temperature treatments, carbon allocation favors roots at the expense of shoots. Within Base and Base + 0.7 treatments, root/shoot ratios tend to increase as soil moisture decreases. This allocation pattern would facilitate greater root growth for attaining more water under drought conditions. Seedlings in the A and B soil-moisture regimes exposed to the Base temperature had a near 1:1 carbon allocation balance between roots and shoots.

Comparisons of concentrations of chlorophylls a and b, and carotenoids and xanthophylls on a weight per area basis and a weight per weight basis between soil-moisture treatments at the various combinations of time and temperature treatments either did not differ or the comparisons were inconclusive (Table 35.6). White oaks tended to have the highest concentrations of all pigments on the basis of area; overcup oaks had the highest concentrations of chlorophylls a and b on a weight basis (data not shown). The effects of drought and elevated temperature on

Table 35.5. Average Dry Weights of Leaves, Stems, Shoots (Leaves and Stems), Roots, and Root/Shoot Ratios (R/S) of Six Species of Two-Year-Old Oak Seedlings Exposed to Three Temperatures and Shade Treatments, and Three Soil-Moisture Regimes[1]

Temp.	Moist.	Leaves (g)	Stems (g)	Shoots (g)	Roots (g)	R/S
Base	A	35.7 ± 4.5	91.4 ± 16.9	127.1 ± 20.7	133.1 ± 13.0	1.05
	B	32.9 ± 5.0	89.0 ± 18.2	121.9 ± 22.7	124.2 ± 21.5	1.02
	C	29.6 ± 4.1	81.2 ± 14.8	110.8 ± 18.3	125.7 ± 14.6	1.13
Base + 0.7	A	29.6 ± 3.1	62.5 ± 7.4	92.0 ± 10.1	116.2 ± 9.9	1.26
	B	27.2 ± 4.1	51.6 ± 10.7	78.7 ± 14.5	120.4 ± 14.2	1.53
	C	28.0 ± 2.8	54.0 ± 6.0	82.0 ± 8.3	113.4 ± 8.7	1.38
Base + 1.7	A	24.6[b] ± 2.7	55.0[ab] ± 7.7	77.8[b] ± 9.0	95.9[ab] ± 9.3	1.23
	B	21.3[b] ± 3.2	41.3[b] ± 6.1	62.5[b] ± 9.1	75.6[b] ± 8.7	1.21
	C	33.1[a] ± 2.6	65.5[a] ± 7.9	98.6[a] ± 8.8	112.1[a] ± 7.9	1.14

[1] Data are means of eighteen samples ± SE. Also, means separation ($P = 0.05$) between moisture regimes within temperature treatments are indicated by different letters as determined by Duncan's Multiple Range Test.

Table 35.6. Average Content of Chlorophyll A, Chlorophyll B, and Carotenoids and Xanthophylls, on a Weight/Weight and Weight/Area Basis, of Six Species of Two-Year-Old Oak Seedlings Exposed to Three Temperature and Shade Treatments and Three Soil-Moisture Regimes

Temp.	Moist	Chlorophyll a ($\mu g\ cm^{-2}$)				Chlorophyll b ($\mu g\ cm^{-2}$)				Car + Xan ($\mu g\ cm^{-2}$)			
		Wk. 1	Wk. 6	Wk. 10	Wk. 14	Wk. 1	Wk. 6	Wk. 10	Wk. 14	Wk. 1	Wk. 6	Wk. 10	Wk. 14
Base	A	48 ± 2	46 ± 2	45^{ab} ± 2	42 ± 2	27 ± 2	30 ± 2	26^{ab} ± 1	23 ± 2	8.7 ± .3	9.0 ± .3	9.9^{ab} ± .4	8.9 ± .4
	B	45 ± 2	46 ± 2	43^b ± 2	42 ± 2	26 ± 2	29 ± 2	24^b ± 1	23 ± 1	8.2 ± .2	9.0 ± .3	9.1^b ± .4	8.6 ± .6
	C	46 ± 2	45 ± 2	49^a ± 2	50 ± 7	27 ± 2	30 ± 2	27^a ± 2	37 ± 13	8.3 ± .3	8.6 ± .3	11^a ± .3	7.6 ± .2
Base + 0.7	A	41 ± 2	39 ± 2	41 ± 2	44 ± 2	29 ± 2	22 ± 2	21 ± 1	21 ± 2	7.7 ± .4	8.8 ± .4	9.3 ± .2	9.7^a ± .4
	B	41 ± 2	41 ± 2	37 ± 2	39 ± 3	29 ± 2	24 ± 2	20 ± 1	21 ± 2	7.8 ± .4	9.0 ± .3	9.6 ± .5	8.1^b ± .6
	C	40 ± 2	37 ± 2	39 ± 1	39 ± 2	27 ± 2	24 ± 2	21 ± 1	23 ± 2	8.2 ± .4	8.3 ± .4	10 ± .4	8.1^b ± .5
Base + 1.7	A	41 ± 3	40 ± 3	42 ± 2	42 ± 3	27 ± 2	29 ± 3	24 ± 2	25 ± 2	7.3 ± .5	8.2 ± .7	9.9 ± .5	8.4 ± .7
	B	43 ± 2	43 ± 2	44 ± 2	44 ± 3	26 ± 2	29 ± 2	24 ± 1	25 ± 2	7.5 ± .5	8.0 ± .5	9.9 ± .5	8.7 ± .6
	C	43 ± 2	41 ± 2	42 ± 2	44 ± 2	27 ± 2	28 ± 3	24 ± 2	25 ± 2	7.7 ± .4	8.1 ± .3	9.8 ± .5	8.8 ± .4

Temp.	Moist	Chlorophyll a ($\mu g\ mg^{-1}$)				Chlorophyll b ($\mu g\ mg^{-1}$)				Car + Xan ($\mu g\ mg^{-1}$)			
		Wk. 1	Wk. 6	Wk. 10	Wk. 14	Wk. 1	Wk. 6	Wk. 10	Wk. 14	Wk. 1	Wk. 6	Wk. 10	Wk. 14
Base	A	11.9 ± 1.8	10.7 ± .8	8.9 ± .7	7.6^{ab} ± .6	7.0 ± 1	7.1 ± .7	5.1 ± .5	4.3 ± .4	2.4 ± .3	2.3 ± .1	2.1 ± .1	1.7 ± .1
	B	11.5 ± 1.2	10.1 ± .4	8.3 ± .6	6.8^b ± .5	6.8 ± .9	6.4 ± .4	4.8 ± .5	3.8 ± .3	2.3 ± .2	2.2 ± .1	1.9 ± .2	1.5 ± .1
	C	12.4 ± 1.3	10.6 ± .6	8.9 ± .6	9.8^a ± 1.6	7.5 ± .9	7.0 ± .6	5.0 ± .4	7.4 ± 2.9	2.4 ± .2	2.2 ± .1	2.1 ± .1	1.5 ± .4
Base + 0.7	A	8.8 ± .7	8.3 ± .2	8.6 ± .5	7.6 ± .4	5.9 ± .6	4.7 ± .5	4.5 ± .3	4.0 ± .4	1.9 ± .2	2.0 ± .1	2.3 ± .1	1.8^a ± .1
	B	8.6 ± .6	8.4 ± .6	7.7 ± .4	6.5 ± .5	5.9 ± .6	5.0 ± .5	4.2 ± .3	3.6 ± .4	1.8 ± .1	2.0 ± .2	2.2 ± .1	1.5^b ± .1
	C	8.9 ± .6	8.3 ± .6	8.7 ± .5	7.0 ± .4	6.4 ± .6	5.3 ± .5	4.8 ± .4	3.8 ± .3	1.9 ± .1	2.0 ± .1	2.3 ± .1	1.6^{ab} ± .1
Base + 1.7	A	7.9 ± .6	9.0^b ± .7	9.9 ± .8	7.1 ± .8	5.3 ± .6	6.6 ± .7	5.8 ± .5	4.3 ± .5	1.5 ± .1	2.0 ± .2	2.5 ± .1	1.5 ± .2
	B	8.9 ± .6	10.7^a ± .7	11.2 ± .7	7.7 ± .5	5.7 ± 1.5	7.3 ± .8	6.4 ± .7	4.3 ± .3	1.7 ± .1	2.1 ± .1	2.8 ± .2	1.6 ± .1
	C	8.6 ± .6	9.0^b ± .5	9.7 ± .6	7.8 ± .4	5.6 ± .6	6.3 ± .6	5.5 ± .4	4.5 ± .3	1.7 ± .1	1.9 ± .1	2.5 ± .2	1.7 ± .1

[1] Data are means of eighteen samples ± SE. Also, means with different letters between soil moisture regimes, and within the same temperature treatment and time interval combinations, are different as determined by Duncan's multiple range test ($P = 0.05$).

pigment content are inconclusive and do not indicate any shift in resources to increase light-harvesting ability and thereby optimize the balance between carbon fixation and water lost through open stomata in plants exposed to drier soils or warmer air conditions within temperature treatments.

There were no differences in year-end foliar concentrations of K, Mg, Ca, and P between drought treatments in any of the temperature treatments (data not shown). At all three temperatures, white oak had greater foliar Ca concentrations ($P = 0.05$) than the other species (data not shown).

Discussion

On average, the oak seedlings in this experiment exhibited moderate increases in P_{net}, g_s, and TR, as well as corresponding increases in height and diameter growth as air temperature increased. However, P_{net}, g_s, TR, and height and diameter growth were all less in the Base + 1.7 temperature treatment compared to the Base + 0.7 treatment. Under these conditions, oak seedlings responded to increased temperature by increasing carbon uptake to a point at which greater respiration and general physiological stress began to have a damping effect on carbon uptake and growth.

There are no clear cut answers for predicting the climate of the future. If future regional climates produce warmer temperatures accompanied by short-term droughts, growth patterns similar to those in this present study might occur. In that climate scenario, oaks might benefit from warmer temperatures through increased carbon uptake and might exhibit some level of photosynthetic tolerance to drought thereby outperforming oaks in more mesic environments. This response could be detrimental to an oak species that becomes established on a site for which it is not well-suited in the long term. In this manner, an individual oak, a species of oak, or a group of oak species over a wide geographic region would be susceptible to decline initiated by a sudden and prolonged drought. Changes in soil-moisture availability, evapotranspiration, and the length of the growing season have been hypothesized with climate alterations of the magnitude predicted (Jones et al., 1994). It is clear that prolonged drought stress decreases gas exchange and net carbon assimilation in oaks (Hinckley et al., 1978; 1979), which if severe enough can reduce a tree's vigor and natural defenses. Severe drought can also cause embolisms in oak xylem vessels through cavitation (Tyree et al., 1992; Sperry and Sullivan, 1992; Tyree and Sperry, 1989). A large number of embolisms will limit growth by reducing water conductivity in the xylem (Schultze and Matthews, 1988). Drought stress, accompanied by elevated temperatures, can alter any number of normal physiological functions in oaks. Many reports relate oak decline to drought sites (relatively thin, rocky soils) on warmer, southern or western exposures to the extent that these conditions tend to define sites at high risk for increased oak mortality (Starkey et al., 1989). The probability of oak decline occurring on this type of site can be increased by an acute summer drought, a recent spring defoliation, or by the fact that oaks are physiologically mature.

Predicted changes in climate patterns over the next fifty years hold the potential to severely alter normal oak physiology and growth through increased drought and temperature stress in the southern United States. After a healthy tree has been stressed one or more times, its defense systems can become impaired making it vulnerable to attack by insects and diseases (Wargo and Haack, 1991), and the oak decline syndrome is fully expressed. Bottomland and upland oak resources are increasing in value and therefore it is critical to be able to predict how individual oak species, and groups of oak species, will respond to oak decline syndrome in a given region. Studies will continue to examine the many abiotic and biotic stresses that could be part of oak decline in the South.

References

Ammon V, Nebeker TE, Filer TH, McCracken FI, Solomon JD, Kennedy HE (1989) Oak decline. MS Agri For Exper Stat Tech Bull 161, MS State Univ, Mississippi State.

Beal JA (1926) Frost kills oaks. J For 24:949–950.

Brasier CM, Scott JK (1994) European oak declines and global warming: A theoretical assessment with special reference to the activity of *Phytophthora cinnamomi*. Bulletin OEPP/EPPO 24:221–232.

Dougherty PM, Hinckley TM (1981) The influence of a severe drought on net photosynthesis of white oak (*Quercus alba*). Can J Bot 59:335–341.

Edmonds JA, Reilly J, Trabalka JR, Reichle DE (1984) An analysis of possible future retention of fossil fuel CO_2. DOE OR/21400–1, USDO, Washington, DC.

Epron D, Dreyer E (1993) Photosynthesis of oak leaves under water stress: Maintenance of high photochemical efficiency of photosystem II and occurrence of non-uniform CO_2 assimilation. Tree Physiol 13:107–117.

Fergus CL, Ibberson JE (1956) An unexplained extensive dying of oak in Pennsylvania. Plant Dis Rep 40(8):748–749.

Ford VL (1994) Mississippi valley forest type. In Moorhead DJ, Coder KD (Eds) Southern hardwood management. Mgmt Bull R8-MB 67. USDA For Serv, South Reg, Coop Ext Serv, Athens, GA.

Friedli HL, Oescher H, Siegenthaler H, Stauffer U, Stauffer B (1986) Ice core record of the C^{13}/C^{12} ratio of atmospheric CO_2 in the past two centuries. Nature 324:237–238.

Gillespie WH (1956) Recent extensive mortality of scarlet oak in West Virginia. Plant Dis Rep 40(12):1121–1123.

Hedden R (1987) Impact of climate change on forest insect pests. In Meo M (Ed) Proceedings of symposium on climate change in the southern United States: Future impacts and present policy issues. Univ OK, May 22–29, 1987.

Hepting GH (1963) Climate and forest diseases. Ann Rev Phytopathology 1:31–50.

Hinckley TM, Aslin RG, Aubuchon RR, Metcalf CL, Roberts JE (1978) Leaf Conductance and photosynthesis in four species of the oak–hickory forest type. Forest Sci 24(1):73–84.

Hinckley TM, Dougherty PM, Lassoie JP, Roberts JE, Teskey RO (1979) A severe drought: Impact on tree growth, phenology, net photosynthesis, and water relations. Amer Mid Natur 102:307–316.

Hiscox JD, Israelstam GF (1979) A method for the extraction of chlorophyll from leaf tissue without maceration. Can J Bot 57:1332–1334.

Hoffard WH, Marx DH, Brown HD (1995) *The health of southern forests*. USDA For Serv, South Reg, Atlanta, GA.

Hursch CR, Haasis FW (1931) Effects of 1925 summer drought on southern Appalachian hardwoods. Ecol 12:380–386.

Idso SB, Balling RC Jr (1992) US temperature/precipitation relationships: Implications for future 'greenhouse' climates. Agric For Meteor 58:143–147.

Jones EA, Reed DD, Desander PV (1994) Ecological implications of projected climate change scenarios in forest ecosystems of central North America. Agric For Meteor 72:31–46.

Karl TR, Heim RH Jr, Quayle RG (1991) The greenhouse effect in central North America: If not now, when? Science 251:1058–1061.

Keller WE (1984) Rising CO_2: Impacts on climate explored at AAAS. Bioscience 34(8):475–476.

Kozlowski TT (1982) Water supply and tree growth. Part I. Water deficits. For Abstr 43(2):57–161.

Kronrad G (1993) Hardwoods as an investment for landowners. Tex For 1:6–7.

Law JR, Gott JD (1987) Oak mortality in the Missouri Ozarks. In Hay RL, Woods FW, DeSelm H (Eds) Proceedings of the sixth central hardwood forest conference. University Tennessee, Knoxville, TN.

Lewis R Jr (1981) *Hypoxylon* spp., *Ganoderma lucidum* and *Agrilus bilineatus* in association with drought related oak mortality in the South. Phytopath 71:890.

Lichtenthaler HK, Wellburn AR (1983) Determinations of total carotenoids and chlorophylls a and b of leaf extracts in different solvents. Biochem Soc Trans 60:591–592.

Maass D (1989) The 1988 drought: Some likely impacts on northern forests. North Log Timb Proc 5:18–40.

McIntyre AC, Schnur GL (1936) Effects of drought on oak forests. PA Agric Exper Stat Bull 325.

Miller WF, Dougherty PM, Switzer GL (1987) Effect of rising carbon dioxide and potential climate change on loblolly pine distribution, growth, survival, and productivity. In Shands WE, Hoffman JS (Eds) *The greenhouse effect, climate change, and U.S. forests.* The Conservation Foundation, Washington, DC.

Myers CC, Killingsworth PA (1992) Growth and mortality of black oak in southern Illinois. N J Appl For 9(1):33.

National Academy of Sciences (1983) Changing climate. Report of the carbon dioxide assessment committee. National Academy Press, Washington, DC.

Salisbury FB, Ross CW (1978) *Plant physiology*, 2nd Edition. Wadsworth Publishing Co., Inc., Belmont, CA.

Samuelson LJ (1994) The role of microclimate in determining the sensitivity of *Quercus rubra* L. to ozone. New Phytol 128:235–241.

Schultze HR, Matthews MA (1988) Resistance to water transport in shoots of *Vitis vinifera* L.: Relation to growth at low water potential. Plant Physiol 88:718–724.

Sperry JS, Sullivan JEM (1992) Xylem embolism in response to freeze-thaw cycles and water stress in ring-porous, diffuse-porous, and conifer species. Plant Physiol 100:605–613.

Staley JM (1965) Decline and mortality of red and scarlet oaks. For Sci 11(1):2–17.

Starkey DA, Oak SW, Ryan GW, Tainter FH, Redmond C, Brown HD (1989) Evaluation of oak decline areas in the South. USDA For Serv, South Reg, Atlanta, GA.

Tainter FH, Retzlaff WA, Starkey DA, Oak SW (1990) Decline of radial growth in red oaks is associated with short-term changes in climate. Eur J For Path 20:95–105.

Tainter FH, Williams TM, Cody JB (1983) Drought as a cause of oak decline and death on the South Carolina coast. Plant Dis 67:195–197.

Tomlinson GH (1993) A possible mechanism relating increased soil temperature to forest decline. Water Air Soil Pollut 66:365–380.

True RP, Tyron EH (1956) Oak stem cankers initiated in the drought year 1953. Phytopath 46:617–622.

Tyree MT, Alexander J, Machado J (1992) Loss of hydraulic conductivity due to water stress in intact juveniles of *Quercus rubra* and *Populus deltoides*. Tree Physiol 10:411–415.

Tyree MT, Sperry JS (1989) Vulnerability of xylem to cavitation and embolism. Ann Rev Plant Phys Mol Biol 40:19–38.

van Bavel CHM, Lascano R, Wilson DR (1978) Water relations of fritted clay. Soil Sci Soc Am J 42:657–659.

Vivin P, Aussenac G, Levy G (1993) Differences in drought resistance among 3 deciduous oak species grown in large boxes. Annales des Sciences Forestieres 50(3):221–233.

Wargo PM, Haack RA (1991) Understanding the physiology of dieback and decline disease and its management implications for oak. The Oak Resource in the Upper Midwest: Implications for Management, Conference Proceedings. June 3–6, 1991.

Whittaker RH (1975) *Communities and ecosystems,* Second edition. Macmillan, Inc., New York.

Woodman JN, Furiness CS (1988) Potential effects of climate change on U.S. forests: Case studies of California and the Southeast. US EPA, Off Pol, Plan Eval, Washington, DC.

36. Effects of Global Climate Change on Biodiversity in Forests of the Southern United States

Margaret S. Devall and Bernard R. Parresol

Climate has not been stable in the past. Fluctuations of pine (*Pinus*) pollen in a 50,000-year sequence from Lake Tulane in Florida indicate that major vegetation shifts occurred during the last glacial cycle. Phases of pollen dominated by pine (indicating a wet climate) were interspersed with periods with plentiful oak (*Quercus*), ragweed, and marsh elder (*Ambrosia* type) populations (Grimm et al., 1993). During the Holocene (i.e., the last 12,000 years), climate has fluctuated with periods of cooler, warmer, wetter or drier weather than at present. The greatest changes in climate probably occurred during deglaciation, approxiamtely 12,500 to 11,000 years ago. However in parts of the United States great shifts in plant distribution and composition occurred during the past 120 years, mainly resulting from anthropogenic factors (Miller and Wigand, 1994). From 1550 to 1850 a small ice age caused widespread starvation in Europe. Living things have been able to adapt to the warming since then, but widespread, rapid warming could be disastrous (Fajer and Bazzaz, 1992).

Integrated assessments of climate problems have been developed or are under way at various institutions, but none is completely satisfactory (Dowlatabadi and Morgan, 1993; Roberts, 1987). Nevertheless, most climate models predict that the continued buildup of carbon dioxide (CO_2) and other infrared absorbing greenhouse gases is apt to lead to average temperature increases of 1.5 to 5 °C, as well as changes in precipitation patterns during the next 50 to 75 years (Adams et al., 1990). Evidence from high alpine environments in the Alps suggests that global warming is already having a serious effect on alpine plants, which are being pushed upward toward higher ground. As a consequence, plant species and communities at the crests of mountains may be eliminated (Grabherr et al., 1994).

Lessons from the Past

Climatic variability is the driving exogenous factor affecting community structure. Through the study of ice cores and lake sediment cores along with knowledge of planetary shifts and the use of atmospheric general circulation models (GCMs), past climate and vegetation regimes have been examined and described (Davis, 1986; Kutzbach, 1987; Kutzbach and Guetter, 1986; Webb, 1986; Webb and Wigley, 1985; Webb et al., 1987). Knowledge of past trends in temperature, precipitation, radiation, and so forth, in addition to the consequent shifting of vegetation in terms of migration rates, distributional patterns, and so on, provide important clues to assessing present and future impacts of global climate change on the biodiversity of the southern United States.

The timing and direction of changes in the pollen record parallel the changing patterns of temperature, moisture, and radiation gradients. Webb et al., (1987) examined six pollen types, which are 1) sedge (Cyperaceae), 2) spruce (*Picea*), 3) northern pines (*Pinus*), 4) southern pines (*Pinus*), 5) oak (*Quercus*), and 6) prairie forbs (sum of sage (*Artemisia*), Compositae, and pigweed (*Chenopodiaceae-Amaranthaceae*)). Overpeck and Bartlein (1989) used these same six pollen records plus a seventh, birch (*Betula*). They plotted response surfaces showing the relationship between the percentages of the pollen types and mean July temperature, mean January temperature, and annual precipitation. They found, for example, that the replacement of oak in the southern United States by southern pines was related to the increase in January temperatures. The response surface for pine pollen showed that the abundance gradient for southern pines paralleled that for winter temperatures and not summer temperatures. This suggests that the southern pines increased in abundance as winter temperatures increased in the latter part of the Holocene. As summers cooled beginning 6000 years ago in response to decreased radiation, spruce increased in abundance and moved southward. The response surfaces for oak also indicated a sensitivity to summer temperature. Similarily, birch was sensitive to both summer and winter temperatures, decreasing in abundance with an increase in these temperature variables.

Schwartz (1991) provides insights about the speed and timing of the change and composition of forests over time from Holocene evidence. First, Holocene tree migrations proceeded at an average rate of 10 to 40 km per century, with a maximum rate of 200 km per century for spruce. Second, species-range changes often lagged in response to climate change. Third, the individual responses of species to climate change resulted in historical plant communities for which no present examples exist.

The temperature in North America apparently increased 2 to 3 °C several times during the Pleistocene. Sweetgum trees (*Liquidamber styraciflua* L.) grew as far north as southern Ontario (Wright, 1971). Osage oranges (*Maclura* sp.) and pawpaws (*Asimina* sp.) grew near Toronto, a good distance north of their present distribution; manatees (*Trichechus* sp.) occurred off the coast of New Jersey, and tapirs (*Tapirus* sp.) and peccaries (*Tayassu* sp.) lived in Pennsylvania (Dorf, 1976). In contrast, the full glacial vegetation of the Atlantic coastal plain was very different. A deposit with spruce cones in Louisiana (Brown, 1938) was carbon dated 7,240 years ago, and the boreal conifer forest probably extended to the south-central United States, stretching west from Georgia (Wright, 1971). However, Delcourt and Delcourt (1987) propose that there since the last full glacial period, has been continued vegetational stability and dynamic equilibrium in the Gulf coastal plain, in the southeastern evergreen forest composed predominantly of southern pines (*Pinus*), tupelo gum (*Nyssa*), cypress (*Taxodium distichum*), Atlantic white cedar (*Chamaecyparis thyoides*), and hickory (*Carya*).

Climate Change Effects

The southern United States has an extensive coast line and mountainous areas, features that influence the large-scale flow of the atmosphere, and help define the regional weather, demographics, and biodiversity. Regions that differ substantially in background climate should have different levels of sensitivity to climatic change (Neilson and Marks, 1994). In this section, we will examine the sweeping effects that could occur under different climate scenarios and look with some detail specifically at the effects on the southern United States, its species and populations, and natural processes.

To assess the potential impacts of climate change, Miller and Brock (1989) conducted a modeling study using the Weekly Scheduling Model of the Tennessee Valley Authority (TVA) to simulate reservoir levels, river flows, and hydropower generation for wet and dry scenarios[1], based on the runoff estimates from the Goddard Institute for Space Studies (GISS) doubled-CO_2 model run. Table 36.1 lists temperature results for a number of cities in the region from the GISS doubled-CO_2 scenario. Projected lake and reservoir levels have important implications for fish and wildlife populations, as well as recreation. Miller and Brock (1989) found that the wet scenario would largely eliminate present problems with low lake levels but that the dry scenario would make these problems the norm. Lower flows would reduce the dilution of municipal and industrial effluents discharged into the Tennessee river and its tributaries, thus the ability of streams to assimilate wastes would be reduced and water quality degraded. Water would remain at the bottom of reservoirs for a longer period of time, hence, the amount of dissolved oxygen would decline. This would directly harm fish, amphibians, aquatic invertebrates, and so forth. Although a drier climate would exacerbate many problems facing the TVA, a wetter climate would increase the risk of flooding and stream sediment loading. Under the wet scenario, Miller and Brock (1989) found that storage was inadequate at the tributary reservoirs, which could result in uncontrolled spillage over dams. The recent flooding along the upper Mississippi river in Iowa and elsewhere attest to the impacts of uncontrolled

---

[1] GISS estimates high runoff, Miller and Brock use the inverse of GISS as a dry scenario.

Table 36.1. The GISS Doubled-CO_2 Scenario: Frequency of Hot and Cold Days Given in Degrees Fahrenheit

	Number of winter days						Number of summer days					
	Daily low < 32		Daily high ≥ 70		Daily high < 80		Daily high > 90		Daily high ≥ 100			
Location	Hist	$2 \times CO_2$	Hist	$2 \times CO_2$	Hist	$2 \times CO_2$	Hist	$2 \times CO_2$	Hist	$2 \times CO_2$		
Atlanta, GA	38.3	20.5	4.2	13.6	10.0	2.2	17.1	53.3	0.6	4.2		
Birmingham, AL	35.5	8.1	7.1	30.7	4.5	0.4	34.1	72.5	1.5	10.7		
Charlotte, NC	42.1	23.8	3.4	9.9	11.9	3.7	23.1	56.5	0.1	5.9		
Jackson, MS	33.5	5.9	15.3	43.5	0.8	0.2	55.1	83.1	2.0	19.5		
Jacksonville, FL	9.3	1.7	34.6	49.6	2.3	0.3	46.4	81.3	0.6	14.1		
Memphis, TN	41.2	8.1	5.2	23.6	4.9	0.7	50.5	74.8	2.6	19.1		
Miami, FL	0.2	0.0	72.9	82.7	0.6	0.0	29.8	83.5	0.0	2.5		
Nashville, TN	42.5	15.4	0.3	8.6	60.4	33.7	10.5	20.2	0.3	3.5		
New Orleans, LA	14.9	3.5	24.9	39.5	0.9	0.1	55.4	84.9	0.3	13.5		

flooding, not only to human populations and property, but to plant and animal communities as well. In addition to the dangers of flooding, increased runoff would elevate the amount of suspended sediments, degrading the recreational quality of surface waters, filling reservoirs, raising water treatment costs for sediment removal by municipalities and industry, and disrupting fish spawning grounds.

To assess the possible impacts of climate change on southern forests, Urban and Shugart (1989) applied a forest simulation model to upland sites near Knoxville, Tennessee; Macon, Georgia; Florence, South Carolina; and Vicksburg, Mississippi. Their study considered the Oregon State University (OSU), Geophysical Fluid Dynamics Laboratory (GFDL), and GISS scenarios for doubled CO_2, as well as the GISS transient A scenario through the year 2060. These researchers used a modified version of FORET, a gap forest dynamics model originally developed by Shugart and West (1977) and they made a number of simplifying assumptions. For example, they assumed loblolly pine (*Pinus taeda* L.) could not tolerate more than 6,000 cooling degree days per year. Also, they did not consider the potentially beneficial effects of CO_2 fertilization on photosynthesis, improved water use efficiency, or leaf area.

Following World War II, substantial amounts of farmland have been removed from agriculture, and much of this land has reverted to forest. Unfortunately, the simulations by Urban and Shugart (1989) question the ability of southern forests to be regenerated from bare ground, particularly if the climate becomes drier and warmer—an important point, because seedlings and saplings may be more sensitive to climate change. For the Knoxville site, the dry GFDL scenario indicated that a forest could not be started from bare ground; the GISS and OSU doubled-CO_2 scenarios showed reductions in biomass of 10 to 25%. For the South Carolina site, only the GISS climate would support a forest, but at less than 50% of present productivity. The Georgia and Mississippi sites, based on the three climate scenarios, could not generate a forest from bare ground.

The GISS transient A analyses suggest that mature forests could die if the climate changes. The analyses indicate significant destruction would not occur before 2030, a lag effect, but all forests would be substantially affected by 2060. The Mississippi forest would be lost by 2040, and the South Carolina and Georgia sites by 2060. The Tennessee site, being much cooler, would remain somewhat healthy, but lose about 35% of its biomass. If forest decline and mortality truncate southern distributions of tree species, affected areas would then be susceptible to weed expansion and pest outbreaks, which can eliminate native species (USDA, 1971). For instance, Kudzu (*Pueraria lobata* Ohwi) and japanese honeysuckle (*Lonicera japonica* Thunb.), two exotic species, are predicted to expand northward approximately 500 km under a doubling of CO_2 (Sasek and Strain, 1990).

Animals

The most rapid responses to climate change occur among animals. Animal behaviors such as feeding and reproduction vary with climate and are apt to change considerably if global warming occurs. Such extremes in weather as severe storms and harsh winters can destroy food supplies and decimate local populations (Davis, 1986). An increase in local temperature can speed up insect metabolism, affecting local population density. Species could evolve the capacity for extra life cycles per year, with considerable economic impact in the case of crop pests. Warmer winter temperatures probably will expand the overwintering ranges of many insect pests (Smith and Tirpack, 1989). Drought stress on plants may indirectly alter the feeding and reproduction of insects (Rubenstein, 1992), for example, drought-stressed plants are more suitable for the growth and reproductive success of such insects as butterflies, moths, and grasshoppers (Mattson and Haick, 1987).

An increased CO_2 concentration may indirectly affect secondary plant metabolites in environments where water or nutrients are limiting. It may lead to reduced protein synthesis and to diversion of phenylalanine or tyrosine into phenolics. It is not clear whether this will result in decreased herbivory or if herbivores will eat more of plants that are less nutritious (Lambers, 1993).

Although the physiological responses of organisms to temperature are understood, ecological responses of populations are not, especially for terrestrial animals (Tracy, 1992). Examples from the southeastern United States can be useful in predicting responses of populations to climate change. Brisbin (1974) found that a cove of Par Pond at Savannah River Plant that received hot water from one of the nuclear reactors always had fewer numbers of species and less abundance of waterfowl than a second cove that was about 10 °F cooler.

Rapid changes in range are possible for birds, because they are mobile, and species that are present every year as visitors may begin to nest as soon as conditions become favorable. Other species may attempt to nest for several years after conditions have changed, and therefore may lag behind habitat and climate changes (Davis, 1986). Jarvinen and Ulfstrand (1980) suggest, however, that changes brought about by humans have had greater influence on birds than climate.

Ducks are dependent on coastal wetlands for breeding, food, or wintering grounds. Shorebirds are primarily dependent on sand beaches. Both groups are especially vulnerable to two indirect effects of global warming which are 1) drought, and 2) rise in sea level. Disruptions in relative timing of food availability and bird migration may occur as a result of climatic change at specific sites. Also, the onset of migration, especially for long-distance migrants, is not sensitive to local temperature, but the onset of spring activity by insect larvae and emerging adults in the birds' breeding sites is very sensitive to local temperature (Myers and Lester, 1992).

Nonmigrating bird species could face problems as a result of changes associated with global climate change, for example more frequent hurricanes. During 1987, Hurricane Hugo destroyed 70% of the nesting trees of the endangered red-cockaded woodpecker (*Picoides borealis*) in the Francis Marion National Forest in South Carolina, where the largest population of the woodpeckers was located.

The whooping crane, a migrating species, also underwent serious population declines as a result of hurricanes during this century.

Reptiles and amphibians are an important component of the food chain and constitute an amazing amount of biomass of the forest community. As an example, population densities of the salamander *Plethodon cinereus* in the Eastern deciduous forest have been recorded as high as 0.9 to 2.2 individuals/m^2 (Heatwole, 1962; Jaeger, 1980). Many herpetofauna require a heterogeneous habitat structure (Bennett et al., 1980). One major consequence of changing climate in the South is an 8% predicted increase in the occurrence of wildfires (Simand and Main, 1987), which will greatly reduce these habitat structural components with obvious negative impacts on herpetofauna diversity.

Whiting et al. (1987) found that increasing drought conditions impacted both winter and spring amphibians in the east Texas pine–hardwood ecosystem during three dry years by reducing breeding sites. Williams and Mullin (1987) obtained similar results in loblolly–shortleaf (*Pinus echinata* Mill.) and longleaf (*Pinus palustris* Mill.)–slash (*Pinus elliottii* Engelm) pine ecosystems in central Louisiana. In these ecosystems, the greatest herpetofauna diversity occurs in mature stands, which Urban and Shugart (1989) suggest could be lost in the South if climate changes. Pearson et al. (1987) noted that little breeding of anurans (toads and frogs) took place in the longleaf–slash pine ecosystem in southern Mississippi during a spring drought, except in bayheads (mesic-hydric hardwood habitats). A number of factors, including water, ground cover, and overstory composition, influence amphibian and reptile community composition and relative abundances.

Large animals are affected by genetic problems caused by small population size, because they usually have low population densities and large ranges. This will increase as rising sea level, habitat conversion, and change caused by climate continue. The Florida panther is already demonstrating genetic problems related to these factors. Historically it occurred throughout the southeastern United States, but human settlement has limited its distribution to the southern tip of Florida (Harris and Cropper, 1992).

Plants

Experiments have demonstrated that elevated levels of CO_2 increase photosynthesis and decrease stomatal conductance in crop plants, causing reduced transpiration rate per unit leaf area and overall increase in water use efficiency during the growing season. Nevertheless, the stomata of many tree species are unresponsive to CO_2 after long-term exposure (Hollinger, 1987). Loblolly pine foliage in an intact forest that grew for 50 to 80 days under elevated CO_2 levels showed no evidence of adjustment in stomatal conductance from foliage that developed under current ambient CO_2 (Ellsworth et al., 1995).

These benefits should offset predicted changes in precipitation and temperature to some extent, depending on the severity of the change and short-term differences. Plants are responsive to short-term changes in weather, and there can be important short-term differences with similar averages (Adams et al., 1990; Roberts, 1987).

Changes in CO_2 can alter the competititive abilities of plants, thus changing community composition. Plants possessing C_3 biochemistry (e.g., wheat, rice, and all trees species) grow better in CO_2-rich conditions than those species with C_4 biochemistry (e.g., corn, sugar cane and many dry land grasses). Even certain C_3 plant species grow better than others with increased levels of CO_2. Under competitive conditions, plants which are more responsive to elevated CO_2 conditions grow at the expense of less responsive plant species, coopting water, light, and nutrients (Fajer and Bazzaz, 1992).

Carbon dioxide concentration in the atmosphere has escalated from 280 ppm before the Industrial Revolution to 350 ppm (Keeling et al., 1982), and some scientists think that it will double before the end of the twenty-first century. For the eastern half of North America, models using a doubled-CO_2 scenario predict an average 4 to 6 °C increase in temperature, a 0 to 2 mm/day decrease in precipitation, and a loss of 1 to 2 cm of soil moisture during the growing season (Mitchell et al., 1990). Such predictions require trees to migrate an order of magnitude faster than they did during the Holocene to maintain populations within appropriate climatic parameters. Davis and Zabinski (1991) predicted potential future distribution for four tree species (beech (*Fagus grandifolia* Ehrh.), sugar maple (*Acer saccharum* L.), eastern hemlock (*Tsuga canadensis* (L.) Carr.), and yellow birch (*Betula alleghaniensis* Britton) under two models of CO_2 doubling. For all species, they predicted northward range shifts in excess of 500 km. Thus, predicted climate change for the next century may outstrip many species' ability to stay within suitable climatic ranges, with fragmented habitats further reducing their ability to migrate.

Modern communities of trees should not be considered highly evolved, tightly linked complexes of species. Tree species will have a wide range of responses and response times to climate change that result from differences in life spans, seed production and dispersal rates, vegetative and sexual propagation, genetic diversity, phenotypic plasticity, competition, and disturbance. Tree population changes may follow climatic shifts by decades or centuries (Brubaker, 1986). Historical records demonstrate that some species can expand rapidly as climate becomes less limiting. Species that are rare today have the potential to become common under a changed climate and vice versa (Brubaker, 1988). Species with shorter life spans may be able to adapt more quickly (Davis, 1986). In contrast, trees with long life spans can delay the movement of range boundaries. Adult trees can remain in the vegetation for hundreds of years after seeds can no longer become established as a result of climate change (Brubaker, 1986).

Shugart et al. (1980) carried out a simulation of a stand of beech and yellow poplar under gradual changes in growing degree day values (GDD) in order to investigate the effect of different tree sizes and successional positions on population responses. Beech is a tree that regenerates in gaps left by either species, while yellow poplar is a large tree that requires substantial gaps. Growing degree day values were increased or decreased equivalent to a 1/93 °C change per year in summer temperature during 1,500 years. At 3,800 GDD, only beech occurred in the plots; at 5,300 GDD only tulip poplar occurred. In both experiments, the stand composition shifted over a 200-year period. The authors attributed the differences in the results to the effect of gap size on species replacement. The lag in response is related to the amount of time that mature trees remain in the canopy. From this and other experiments, Shugart (1985) concludes that the time necessary for forest change depends on the characteristics of the species involved.

Climate has traditionally been considered the fundamental regulator of vegetation structure, composition, and productivity, but the studies of Johnson et al. (1993) support the hypothesis that CO_2 concentration is also important to vegetation. A large change occurred in the composition of twenty-six species of C_3 plants and seventeen species of C_4 plants grown from a native soil seed bank along a CO_2 gradient that was similar to the CO_2 increase of the last 150 years. Increasing CO_2 levels elicited a strong growth response for a number of C_3 species, but the C_4 plant increased in productivity as the CO_2 concentration decreased. It seems probable that changing CO_2 levels have had and will continue to have a significant influence on the control that climate exerts on vegetation productivity, species composition, and physiognomic structure.

Loblolly pine is limited on its southern border by moisture stress on seedlings. Miller et al., (1987) predict that the southern range of loblolly pine would move approximately 350 km to the north and northeast in response to global warming of 3 °C, based on its physiological requirements for moisture and temperature. Davis and Zablinski (1991) predict that with doubled CO_2 levels, beech would become rare or die out except at high elevations throughout the eastern United States, while a smaller, new habitat would open up in Ontario and Quebec. A scenario presented by Woodman and Furiness (1989) has the northerly range of the southern pine forests shifting several hundred kilometers into the regions presently occupied by mixed hardwood species.

Pitcher plant (*Sarracenia* sp.) bog communities are floristically highly diverse communities that occur across the southeastern mixed forests where sandy uplands are underlain by impermeable layers of clay. Eleuterius and Jones (1969) list 271 taxa, representing 134 genera and sixty-three families, occurring in southern Mississippi bogs. These bog communities are fragile and do not respond well to disturbance. Though the amount of acreage of such communities is small within the southeastern mixed forest, they are a major biological diversity resource. Folkerts (1982) speculates that no significant amount of this habitat will survive into the twenty-first century.

Wetlands

A 10 cm rise in sea level could cause tidal rivers to move inland as much as 1 km, and a 2 m rise could eliminate 80% of our coastal wetlands (Hoffman, 1987; Leatherman, 1987). Coastal wetlands would migrate inland if that were possible, but in many cases the way will be blocked by levees, highways, seawalls, housing, and other human-made structures. Inland wetlands are not expected to do much better (Myers, 1992). Wetlands are already degraded, fragmented, and dissipated largely because of draining, which will leave them vulnerable to drying out in a warmer atmosphere (Gopal et al, 1982).

Loss of forested barrier islands could cause problems for neotropical migrant birds, which already face problems in their breeding and wintering habitats in the eastern deciduous and coniferous forests. These islands are the first places to rest and find food that the birds encounter after the long migration north across the Gulf of Mexico or the Atlantic each spring. An increased rise in sea level would destroy the island forests, making the trip longer and more difficult for migrating birds.

In the southern United States, 95% of the rare plants may be vulnerable to extinction as a result of climatic warming (Schwartz, 1991); one-third of these are found in mountainous regions, and therefore they may find refuge by ascending in elevation. Many rare plants have characteristics that place species at risk, such as small populations, being habitat specialists, or being endemics with limited geographic ranges. Distributional data provided by Kral (1983) on 316 rare, threatened, and endangered plants of forests in the southeast United States indicate that 114 taxa (36%) are confined to areas spanning 100 km or less in latitude. Forty-one taxa (13%) have ranges that span greater than 500 km in latitude, and only 15 species (5%) have continuous distributions with no disjunctions of more than 100 km.

The native dune vegetation of the higher elevations of Florida, with fifty endemic plant and animal species (Christman, 1988, in Harris and Cropper, 1992) has survived the extremes of several ice ages, and could no doubt withstand global climate change, except that development in the area has restricted the options for adaptation (Harris and Cropper, 1992). Florida's lower-elevation plant communities are vulnerable to rising sea level. For example, the mangrove ecosystems of south Florida would be imperiled by coastal erosion, and the Everglades could be damaged by massive saltwater intrusion. As the sea level has risen, the narrow coastal habitat of several species of beach mice and the Cape Sable sparrow (*Ammospiza maritima mirabilis*) has migrated inland, but encroaching human populations have reduced and fragmented the habitat. Exotics are already a problem in south Florida, where moderate hydrological changes in the Everglades have allowed invasion of a number of exotic plant species, including *Casuarina,* peppertree (*Schinus*), *Hydrilla,* water hyacinth (*Eichornia),* and (*Melaleuca*) (Ewel, 1986; Myers, 1983).

Ground and Marine Diversity

Many people, when considering diversity, fail to consider what is happening on or below the ground, or in the water. To begin, the great majority of vascular plants have evolved to a dependence on mycorrhizae, or root-inhabiting fungi. Most woody plants require mycorrhizae to survive, and most herbaceous plants need them to thrive (Harley, 1969; Marks and Kozlowski, 1973; Trappe and Fogel,

1977). Mycorrhizae function as a mutualistic, symbiotic biotrophy between a fungus and a higher plant host, and are key links in belowground nutrient and energy cycling. The several thousand species of fungi believed to form mycorrhizae encompass great physiological diversity. In turn, these fungi are intimately linked to such small mammals as squirrels, rabbits, mice, and voles, also to insects and birds. These animals depend on the mycorrhizal fungi as a source of nutrition and are essential for the dispersal of spores (Fogel and Trappe, 1978; Maser et al., 1978; McMahon and Warner, 1984; Li et al., 1986; Malajczuk et al., 1987). This complex, tripartite relationship among plants, animals, and mycorrhizal fungi appears to be integral to the healthy functioning of ecosystems.

Assuming inocula are present, mycorrhiza formation depends on environmental factors, host physiology, and soil microorganisms. Alteration of any of these may influence the number of mycorrhizae that can be formed in a particular soil (Parke et al., 1983; Perry et al., 1987). Most studies find fewer mycorrhizae formed on disturbed than on undisturbed sites, for example, clear cut, burned, or eroded sites (Reeves et al., 1979, Harvey et al., 1980, Loree and Williams, 1984). Persistence of mycorrhizal fungal spores and hyphal fragments varies with climate and soil. To the extent that survival of hyphal fragments is related to their respiration rate, the period would be shorter in warmer climates (Perry et al., 1987). Dormant spores presumably survive for long periods; however, they can be lost through erosion or leaching, or can germinate prematurely from chemical secretions from nonhosts (Harley and Smith, 1983). With the myriad possible effects of climate change, from higher temperatures to altered rainfall patterns, increased fires, and shifting composition of species, both plant and animal, the complex interactions among plants, animals, and mycorrhizal fungi may be severely disrupted, with disastrous ecological effects.

The importance of snags and downed woody material in the forest has been recognized for some time (Davis et al., 1983; Maser and Trappe, 1984). Some eighty-five species of birds utilize snags for nesting; snags and fallen trees are also important to mammals, reptiles, amphibians, and invertebrates as breeding, roosting, and foraging sites. So too, the importance of driftwood (wood carried by water from the forest to the sea) is being increasingly recognized as a critically important source of habitat and food for the marine ecosystem, including the deep-sea floor (Maser and Sedell, 1994). Such human activities as stream cleaning, firewood cutting, and product-oriented forest management, have impacted the driftwood, snag, and fallen wood resources, and have had an overall negative impact on biodiversity. We anticipate that climate change, through increased fires and dieback of coastal forests, would exacerbate this situation, though in the short-run, the amount of woody material may increase (Maser, 1994). The southern United States has an extensive coast along the Atlantic Ocean and Gulf of Mexico with many estuaries, so a decrease or loss of the marine wood resource would have economic implications on the fisheries and shellfish industries, though this has been little recognized (Turner, 1977, 1981; Xavier et al., 1992, in Maser and Sedell, 1994).

Predictions

Global climate change projections for the Southeast vary considerably; some predict increased precipitation, which may compensate to some extent for increased temperature. Also, the models cover large areas and do not take into account such local features as mountains or islands, so the changes and species' response to them will not be nearly as even as the predictions imply. For the purposes of this discussion, we will assume that warming in the Southeast will average 3 °C, precipitation will decrease 25 cm/year (10 inches) and sea level will rise 10 cm during the next fifty to seventy-five years. We know how some plant and animal species respond to environmental and competitive changes, but it is difficult to predict how the flora and fauna of southern forests will respond to future environmental change, because we really do not know what controls the abundance and distribution of most species. Global warming will no doubt alter some southern species' distributions and change the composition of some communities in unexpected ways. Many of the relations of plants and animals will be changed if climate change occurs as projected (Fowells and Means, 1990). However, we can predict from the many climate scenarios and forest dynamics simulations that the diversity of plants and animals is apt to be reduced as a direct result of global warming, leading to an overall simplification of southern ecosystems. We suggest that the change will cause a domino effect in southern forests, with each ecosystem losing species as it moves north and causing the demise of other species in the ecosystem it replaces. (Although entire ecosystems do not move as a unit, there will be a general migration of species to the north.)

When plant species are introduced into continental areas, few are able to become established except in disturbed areas. Only a small fraction of fish introductions have been successful. Bird introductions into continental areas are usually failures. A few introductions have been highly successful; the European starling (*Sturnus vulgaris*) spread over the entire United States and much of Canada within sixty years. The chestnut blight (*Endothia parasitica*) spread throughout the southeastern United States, but within forty years it had caused the demise of the American chestnut (*Castanea dentata* (Marsh.) Borkh.), which made up 40% of the overstory of climax forests in the area. Most chestnuts were replaced by oaks, so the oak–chestnut forests are now oak or oak–hickory forests (Krebs, 1978).

As the species inhabiting pine forests move north into the very diverse (Braun, 1950) mixed hardwood forests, biological diversity will suffer. Although the trees will migrate, and foresters can plant the species of commercial importance in more favorable habitats, many of the plants and animals that are associated with southern forests may become threatened or extinct because they are unable to move as rapidly as the tree species that provide their habitat. Species with short seed-dispersal distances (many forest herbs) and low ability to colonize more favorable habitat will be especially at risk. The proximity of southern forests to the Atlantic Ocean and the Gulf of Mexico has important implications for the study area. The coast has been sinking for some time, and a rise in sea level associated with global climate change will increase the risk to wetlands and other species, and will increase the need for migration as species' former habitat is inundated.

Pioneer plant species (e.g., red cedar (*Juniperus virginiana* L.), the southern pines (*Pinus* spp.), and sweetgum occur in early successional stages. They exhibit characteristics that help them become established more quickly than competing vegetation, for example, large and frequent seed crops, efficient seed dispersal, adaptability to a wide variety of sites, and high juvenile-growth rates. These species are nomads, with each succeeding generation moving to new sites that have been disturbed or are vacant. Pioneer species probably can adapt to changing climatic conditions more readily than can species that occur in later successional stages.

With a decrease in precipitation, the water quality of southern streams will decrease, and lake levels will be lowered, with deleterious effects on aquatic animals and plants. Decreased precipitation will also increase the stress on forest species. Drier forests will lead to more frequent forest fires, with consequent alteration or destruction of forests. Loss of older forests will pose a particular problem for such species as the endangered red-cockaded woodpecker that require old-growth pine forests for nesting. Reptiles and amphibians that prefer mature stands also may be eliminated, because there will be no replacement forests for a long time.

Climate extremes are more important than averages. In the coming years, changes in the frequency of fires, hurricanes, and droughts may be more important to biodiversity than temperature change. Increased incidence and severity of fires with a hotter and drier climate could cause loblolly pine, for example, to become less common in the South as longleaf pine and other species favored by frequent fires become established over much of their formerly large ranges (Burns and Honkala, 1990). Increased frequency of hurricanes will also hasten the destruction of coastal plain forests and the plants and animals that inhabit them. The cypress–tupelo forests have evolved to withstand hurricanes better than other forest communities, but these are the forests that are most at risk from sea level rise.

Such large animals as the Florida panther and the black bear, which are already experiencing difficulties as a result of habitat alteration, will probably not survive. Climate change will probably play a major role in lessening herptofauna diversity, because all of the habitat characteristics that determine their community composition are dependent upon the age of the forest and the degree of disturbance to which it has been subjected. Although bird groups, for example, shorebirds and other coastal species (that may lose much of their habitat) and neotropical migrants (that are already threatened in their breeding and winter habitats) will be particularly impacted by climate change, other bird species that inhabit the more northern part of the study area may be better able to cope, and some of them will have potential areas of refuge available. Because of short life cycles, insects may be better able to cope with climate change than many other groups, but pest species may be more of a problem than they are at present. Climate warming will undoubtedly affect the flowering dates of many plant species, and insect pollinators may not be present when they are needed (Moore, 1995).

The effects of global climate change should be most severe in the coastal habitat, lessening to the North. A moderate rise in sea level, in addition to the rise that has been occurring for some time, will probably cause the destruction of many coastal wetlands. We agree with Myers (1992) that this may well be the worst wildlife-related disaster of the greenhouse effect in the United States because the wetlands are already very disturbed and migration will be difficult because of human development. Regeneration of cypress (*Taxodium distichum* (L.) Rich.) has been a problem for some time. The salinization of ground and soil water that results as sea level rises has caused a reduction of pine forests in the Florida keys. If the sea level continues to rise, the Keys and other low-lying island ecosystems will experience a decline in diversity as the species-rich upland communities are replaced by simpler mangrove communities (Ross and O'Brien, 1994). Loss of coastal forests will result in increased mortality for migrating birds and butterflies, as well as for the organisms that inhabit the forested wetlands. Inland wetlands, in contrast, will suffer from decreased precipitation and increased frequency of fire. The loss of most bog communities will be a serious blow to southern pine forests. Florida will especially suffer the effects of global climate change, because of its large coastal area, and because of human development. Exotic species will probably be more of a problem than they are now.

During the Cretaceous and the Pleistocene, the Interior Highlands of Arkansas served as a refuge for plants and animals (Dowling, 1956). This area may well serve as a refuge again if climate change disrupts plant and animal communities in the regions surrounding it. The Smoky Mountains are another large area that should provide shelter to numerous species during the disruptions caused by global climate change.

Many of the present goals and methods of conservation will not change as a result of global climate change, but conservationists already have much to do, and changing climate will reduce the time left in which to accomplish much of this work. Conservationists will try to ameliorate the effects of global climate change but conservation efforts may well be overwhelmed by the increased numbers of threatened and endangered species. Many less conspicuous species will probably not receive the help they are going to need.

Conclusion

These suggestions are for the southern United States, but many are appropriate for other parts of the country. A sensible way to begin preparing for climate change is to practice sound conservation measures now. Conservation plans should be flexible, to incorporate increased understanding of climate. Although there are many parks and preserves in the South, these protected areas are a small portion of the land, so any strategy to protect plant and animal species must consider public and especially private managed lands (public lands make up a small part of the acreage in the south). If we are concerned about conserving the region's biodiversity for the future, we must begin preparation now by practicing sound conservation and management, carrying out appropriate research, and using information about global climate change and biodiversity as it becomes available.

References

Adams RM, Rosenzweig C, Peart RM, Ritchie JT, McCarl BA, Glyer JD, Curry RB, Jones JW, Boote KJ, Allen, Jr LH (1990) Global change and US agriculture. Nature 345:219–224.

Bennett SH, Gibbons JW, Glanville J (1980) Terrestrial activity, abundance and diversity of amphibians in differently managed forest types. Am Mid Natur 103:412–416.

Braun EL (1950) *Deciduous forests of eastern North America*. Hafner Publishing Co., New York.

Brisbin IL (1974) Abundance and diversity of waterfowl inhabiting heated and unheated portions of a reactor cooling reservoir. In Gibbons JW, Sharitz RR (Eds) *Thermal ecology*. USAEC, Washington, DC.

Brown CA (1938) The flora of Pleistocene deposits in the Western Florida Parishes, West Feliciana Parish, and East Baton Rouge Parish, Louisiana. LA Dept of Conserv Geol Bull 12:59.

Brubaker LB (1986) Responses of tree populations to climatic change. Vegetatio 67:119–130.

Brubaker LB (1988) Vegetation history and anticipating future vegetation change. In Agee JK, Johnson DR (Eds) *Ecosystem Management for Parks and Wilderness*. University of Washington Press, Seattle.

Burns RM, Honkala BH (Tech coords) (1990) *Silvics of North America, Vol. 1. Conifers*. USDA For Serv, Agriculture Handbook 654, Washington, DC.

Christman S (1988) *Endemism and Florida's interior sand pine scrub*. Florida Game and Fresh Water Fish Commission, Final Report.

Davis MB (1986) Climatic instability, time lags, and community disequilibrium In Diamond J, Case TJ (Eds) *Community ecology*. Harper and Row, New York.

Davis JW, Goodwin GA, Ockenfels RA (Tech. coords) (1983) *Snag habitat management: Proceedings of the symposium*. USDA For Serv, Gen Tech Rep RM-99, Rocky Mtn For Exper Stat,.

Davis MB, Zabinski C (1991) Changes in geographical range resulting from greenhouse warming. Effects on biodiversity in forests. In Peters RL, Lovejoy TE (Eds) *Consequences of greenhouse warming to biodiversity*. Yale University Press, New Haven.

Delcourt PA, Delcourt HR (1987) *Long-term forest dynamics of the temperate zone*. Springer-Verlag, New York.

Dorf E (1976) Climatic changes of the past and present. In Ross CA (Ed) *Paleobiogeography: Benchmark papers in geology 31*. Dowden, Hutchinson and Ross, Stroudsburg, PA.

Dowlatabadi H, Morgan MG (1993) Integrated assessment of climate change. Science 259:1813, 1932.

Dowling HG (1956) Geographic relations of Ozarkian amphibians and reptiles. Southwestern Naturalist 1:174–189.

Eleuterius LN, Jones Jr, SB (1969) A floristic and ecological study of pitcher plant bogs in south Mississippi. Rhodora 71:29–34.

Ellsworth DS, Oren R, Huang C, Phillips N, Hendrey GR (1995) Leaf and canopy responses to elevated CO_2 in a pine forest under free-air CO_2 enrichment. Oecologia 104:139–146.

Ewel JJ (1986) Invasibility: lessons from south Florida. In Mooney HA, Drake JA (Eds) *Ecology of biological invasions of North America and Hawaii*. Springer-Verlag, New York.

Fajer ED, Bazzaz FA (1992) Is carbon dioxide a 'good' greenhouse gas? Global Envir Change, Dec:301–310.
Fogel RD, Trappe, JM (1978) Fungus consumption (mycophagy) by small animals. Northwest Sci 52:1–31.
Folkerts GW (1982) The Gulf Coast pitcher plant bogs. Am Sci 70:260–267.
Fowells HA, Means JE (1990) The tree and its environment. In Burns RM, Honkala BH (Tech coords) *Silvics of North America, vol 2. Hardwoods.* USDA For Serv, Agriculture Handbook 654, Washington, DC.
Gopal B, Turner RE, Wetzel RG (1982) *Wetlands ecology and management.* International Scientific Publishers, Jaipur, India.
Grabherr G, Gottfreid M, Pauli H (1994) Climate effect on mountain plants. Nature 369:448.
Grimm EC, Jacobson GL Jr, Watts WA, Hansen BCS, Maasch KA (1993) A 50,000-year record of climate oscillations from Florida and its temporal correlation with the Heinrich events. Sci 261:198–200.
Harris LD, Cropper WP Jr (1992) Between the devil and the deep blue sea: Implications of climate change for Florida's fauna. In Peters RL, Lovejoy TE (Eds) *Global warming and biological diversity.* Yale University Press, New Haven.
Harley JL (1969) *The biology of mycorrhizae.* Leonard Hill, London.
Harley JL, Smith SE (1983) *Mycorrhizal symbioses.* Academic Press, New York.
Harvey AE, Larsen MJ, Jurgensen MF (1980) Clearcut harvesting and ectomycorrhizae: Survival of activity on residual roots and influence on a bordering forest forest stand in western Montana. Can J For Res 10:300–303.
Heatwole H (1962) Environmental factors influencing the local distribution and abundance of the salamander *Plethodon cinereus.* Ecology 43:460–472.
Hollinger DY (1987) Gas exchange and dry matter allocation responses to elevation of atmospheric CO_2 concentrations in seedlings of three species. Tree Physiol 3:193–202.
Jaeger RG (1980) Microhabitats of a terrestrial forest salamander. Copeia 1980:265–268.
Jarvinen O, Ulfstrand S (1980) Species turnover of a continental bird fauna: Northern Europe, 1850–1970. Oecologia 46:186–195.
Johnson HB, Polley HW, Mayeux HS (1993) Increasing CO_2 and plant–plant interactions: Effects on natural vegetation. Vegetatio 104/105:157–170.
Kral R (1983) *A report on some rare, threatened, or endangered forest-related vascular plants of the South.* USDA, Atlanta, Tech Publ R8-TP-2.
Krebs CJ (1978) *Ecology: The experimental analysis of distribution and abundance.* Harper and Row, New York.
Kutzbach JE (1987) Model simulations of the climatic patterns during the deglaciation of North America. In Ruddiman WF, Wright Jr, HE (Eds) *North America and adjacent oceans during the last deglaciation.* Geological Society of America, Boulder, CO.
Kutzbach JE, Guetter PJ (1986) The influence of changing orbital parameters and surface boundary conditions on climate simulations for the past 18,000 years. J Atmos Sci 43:1726–1759.
Lambers H (1993) Rising CO_2, secondary plant metabolism, plant-herbivore interactions and litter decomposition: Theoretical considerations. Vegetatio 104/105:263–271.
Leatherman SP (1987) *Impact of the greenhouse effect on coastal environments: marshlands and low-lying population areas.* Laboratory for Coastal Research, University of Maryland, College Park.
Li CY, Maser C, Harlan F (1986) Initial survey of acetylene reduction and selected microorganisms in the feces of 19 species of mammals. Great Basin Natur 46:646–650.
Loree MAJ, Williams SE (1984) Vesicular-arbuscular mycorrhizae and severe land disturbance. In Williams SE, Allen MF (Eds) *VA mycorrhizae and reclamation of arid and semiarid lands.* WY Agric Exper Stat Sci, Report No SA1261.
Malajczuk N, Trappe JM, Molina R (1987) Interrelationships among some ectomycorrhizal trees, hypogeous fungi and small mammals: Western Australian and northwestern American parallels. Aust J Ecol 12:53–55.

Marks GC, Kozlowski TT (Eds) 1973. *Ectomycorrhizae—Their ecology and physiology.* Academic Press, New York.
Maser C (1994) *Sustainable forestry: Philosophy, science, and economics.* St. Lucie Press, Delray Beach, FL.
Maser C, Trappe JM, Nussbaum RA (1978) Fungal—small mammal interrelationships with emphasis on Oregon coniferous forests. Ecol 59:799–809.
Maser C, Trappe JM, (Tech eds) (1984) *The seen and unseen world of the fallen tree.* USDA For Serv, Pac NW For Exper Stat, Gen Tech Rep PNW-133, Portland, OR.
Maser C, Sedell JR (1994) *From the forest to the sea: the ecology of wood in streams, rivers, estuaries, and oceans.* St. Lucie Press, Delray Beach, FL.
Mattson WJ, Haick, RA (1987) The role of drought in outbreaks of plant-eating insects. Bioscience 37:110.
McMahon JA, Warner N (1984) Dispersal of mycorrhizal fungi: Processes and agents. In Williams SE, Allen MF (Eds) *VA mycorrhizae and reclamation of arid and semiarid lands.* WY Agric Exper Stat Sci, Report No SA1261.
Miller BA, Brock WG (1989) Potential impacts of climate change on the Tennessee Valley Authority reservoir system. In Smith JB, Tirpak DA (Eds) *The potential effects of global climate change on the United States—Report to congress. Volume A—Water Resources,* USEPA, Washington, DC.
Miller WF, Dougherty PM, Switzer GL (1987) Rising CO_2 and changing climate: major southern forest management implications. In Shands WE, Hoffman JS (Eds) *The greenhouse effect, climate change, and U.S. Forests.* Conservation Foundation, Washington, DC.
Miller RF, Wigand PE (1994) Holocene changes in semiarid pinyon-juniper woodlands. Bioscience 44:465–474.
Mitchell JFB, Manabe S, Meleshko V, Tokioka T (1990) Equilibrium climate change—and its implications for the future. In Houghton JT, Jenkins GJ, Ephraums JJ (Eds) *Climate change: The IPCC scientific assessment.* Cambridge University Press, Cambridge.
Moore PD (1995) Opening time by degrees. Nature 375:186–187.
Myers JP, Lester RT (1992) Double jeopardy for migrating animals: Multiple hits and resource asynchrony. In Peters RL, Lovejoy TE (Eds) *Global warming and biological diversity.* Yale University Press, New Haven.
Myers N (1992) Synergisms: joint effects of climate change and other forms of habitat destruction, In Peters RL, Lovejoy TE (Eds) *Global Warming and Biological Diversity.* Yale University Press, New Haven.
Myers RL (1983) Site susceptibility to invasion by the exotic tree *Melaleuca quinquenervia* in southern Florida. J Appl Ecol 20:645.
Neilson RP, Marks D (1994) A global perspective of regional vegetation and hydrologic sensitivities from climatic change. J Veg Sci 5:715–730.
Overpeck JT, Bartlein PJ (1989) Assessing the response of vegetation to future climate change: Ecological response surfaces and paleoecological model validation. In Smith JB, Tirpak DA (Eds) *The potential effects of global climate change on the United States—Report to congress. Volume D—Forests.* USEPA, Washington, DC.
Parke JL, Linderman RG, Trappe JM (1983) Effect of root zone temperature on ectomycorrhizal and vesicular-arbuscular mycorrhiza formation in disturbed and undisturbed forest soils of southwest Oregon. Can J For Res 13:657–665.
Pearson HA, Lohoefener RR, Wolfe JL (1987) Amphibians and reptiles on longleaf–slash pine forests in southern Mississippi. In Pearson HA, Smeins FE, Thill RE (Eds) *Ecological, physical, and socioeconomic relationships within southern national forests.* USDA For Serv, South For Exper Stat, Gen Tech Rep SO-68, New Orleans, LA.
Perry DA, Molina R, Amaranthus MP (1987) Mycorrhizae, mycorrhizospheres, and reforestation: Current knowledge and research needs. Can J For Res 17:929–940.
Reeves FB, Wagner D, Moorman T, Kiel J (1979) The role of endomycorrhizae in revegetation practices in the semi-arid west. I. A comparison of incidence of mycorrhizae in severely disturbed vs. natural environments. Am J Bot 66:6–13.

Roberts WO (1987) Time to prepare for global climatic change. In Shands WE, Hoffman JS (Eds) *The greenhouse effect, climate change, and U.S. forests*. Conservation Foundation, Washington, DC.

Ross MT, O'Brien JJ (1994) Sea-level rise and the reduction in pine forests in the Florida keys. Ecol Appl 491:144–156.

Rubenstein DI (1992). The greenhouse effect and changes in animal behavior: Effects on social structure and life-history strategies. In Peters RL, Lovejoy TE (Eds) *Global warming and biological diversity*. Yale University Press, New Haven.

Sasek TW, Strain BR (1990) Implications of atmospheric CO_2 enrichment and climatic change for the geographical distribution of two introduced vines in the USA. Clim Change 16:31–51.

Schwartz MW (1991) Potential effects of global climate change on the biodiversity of plants. For Chron 68:462–471.

Shugart HH (1985) *A theory of forest dynamics: Ecological implications of forest succession models*. Springer-Verlag, New York.

Shugart HH, Emanuel WR, West DC, DeAngelis DL (1980) Environment gradients in a simulation model of a beech-yellow poplar stand. Math Biosci 50:163–170.

Shugart HH, West DC (1977) Development of an Appalachian deciduous forest succession model and its application to assessment of the impact of the chestnut blight. J Envir Manage 5:161–179.

Simand AJ, Main WA (1987) Global climate change: The potential for changes in wildland fire activity in the southeast. In Meo M (Ed) *Proceedings of the symposium on climate change in the southern U.S.: Future impacts and present policy issues*. USEPA, New Orleans, LA.

Smith JB, Tirpak D (1989) *The potential effect of global change on the United States*. USEPA, Washington, DC.

Tracy CR (1992) Ecological responses of animals to climate. In Peters RL, Lovejoy TE (Eds) *Global warming and biological diversity*. Yale University Press, New Haven.

Trappe JM, Fogel RD (1977) Ecosystematic functions of mycorrhizae. In *The belowground ecosystem: a synthesis of plant-associated processes*. Range Sci Dep Sci Ser No 26, Colorado State University, Ft. Collins.

Turner RD (1977) Wood, mollusks, and deep-sea food chains. Bull Am Malacol Union 1976:13–19.

Turner RD (1981) "Wood islands" and "thermal vents" as centers for diverse communities in the deep sea. Soviet J Mar Biol 7:3–9 (translation of Biol Morya 7:3–10).

United States Department of Agriculture (1971) *Common weeds of the United States*. Dover Publications, New York.

Urban DL, Shugart HH (1989) Forest response to climatic change: A simulation study for southeastern forests. In Smith JB, Tirpak DA (Eds) *The potential effects of global climate change on the United States—Report to congress. Volume D—Forests*, USEPA, Washington, DC.

Webb T (1986) Is the vegetation in equilibrium with climate? How to interpret late-Quaternary pollen data. Vegetatio 67:75–91.

Webb T, Bartlein PJ, Kutzbach JE (1987) Climatic change in eastern North America during the past 18,000 years: Comparisons of pollen data with model results. In Ruddiman WF, Wright Jr, HE (Eds) *North America and adjacent oceans during the last deglaciation: The geology of North America*. Geological Society of America, Boulder, CO.

Webb T, Wigley TML (1985) What past climates can indicate about a warmer world. In McCracken MC, Luther FM (Eds) *The potential climatic effects of increasing carbon dioxide*. USDE, Rep DOE/ER-0237, Washington, DC.

Whiting RM, Fleet RR, Rakowitz VA (1987) Herpetofauna in loblolly–shortleaf pine stands of East Texas. In Pearson HA, Smeins FE, Thill RE (Eds) *Ecological, physical, and socioeconomic relationships within southern national forests*. USDA For Serv, South For Exper Stat, Gen Tech Rep SO-68, New Orleans, LA.

Williams KL, Mullin K (1987) Amphibians and reptiles of loblolly–shortleaf pine stands in central Louisiana. In Pearson HA, Smeins FE, Thill RE (Eds) *Ecological, physical, and socioeconomic relationships within southern national forests*. USDA For Serv, South For Exper Stat, Gen Tech Rep SO-68, New Orleans, LA.

Woodman JN, Furiness CS (1989) Potential effects of climate change on U.S. forests: Case studies of California and the Southeast. In Smith JB, Tirpack DA (Eds) *The potential effect of global change on the United States: Report to Congress. Volume D—Forests.* USEPA, Washington, DC.

Wright HE (1971) Late quaternary vegetational history of North America. In Turekian KK (Ed) *The late Cenozoic glacial ages.* Yale University Press, New Haven.

Xavier A, Delgado A, Fonteneau A, Gonzales C, Pilar F, Pallares P (1992) Logs and tunas in the eastern tropical Atlantic: A review of present knowledge and uncertainties. Paper given at the Inter-American tropical tuna commission, La Jolla, CA.

37. Regional Climate Change in the Southern United States: The Implications for Wildfire Occurrence

Warren E. Heilman, Brian E. Potter, and John I. Zerbe

Fires have always been an important factor in determining the composition of forests worldwide, but particularly in the southern United States. Wildfires were a common occurrence in American forests in the early twentieth century. Before 1930, wildfires typically accounted for the burning of eight to twenty million hectares (ha) in the United States each year. By the early 1940s, wildfires were still responsible for the annual burning of over eight million ha. Over 90% of the area burned during this time was on privately owned lands, primarily in the southern United States (Fedkiw, 1989). Between 1950 and 1980, the hectares burned by wildfires steadily decreased as the area receiving organized protection increased and the intensity of the protection efforts increased (Peterson, 1982). In recent years, the total area burned by wildfires in the United States has diminished to about one to two million ha per year (USDA Forest Service, 1992). Although the decrease in the number of hectares burned by wildfires across the United States has been significant over the last seventy years, the relative importance of wildfires in the southern United States in relation to other regions of the United States is significant. More hectares are burned by wildfires in the southern United States than in any other region of the country. A notable exception occurred in 1988 when the severe drought that affected many of the central and Rocky Mountain states contributed to the massive wildfires that occurred in the Rocky Mountain region. Between 1984 and 1990, wildfires in the southern region of the United States accounted for 20 to 43% of all hectares burned in the United States.

Wildfire occurrence is also much more common in the southern United States than in other parts of the country. The southern and northeastern regions are dominated by human-caused wildfires, and the southern region alone accounts for roughly 53% of all wildfires occurring in the United States, excluding Alaska. Even lightning-caused fires were more common in the southern region over the entire seven-year period between 1984 and 1990 than in any other region, excluding Alaska.

Regardless of the cause of wildfires in the southern region or any other region, the occurrence of severe wildfires depends to a large degree on the atmospheric conditions present before, during, and after the time of ignition. The prospect of future global or regional climate changes resulting from an increase in atmospheric carbon dioxide (CO_2) concentrations has raised concerns about how large-scale changes in the atmosphere associated with a changed climate will affect regional wildfire occurrence. Alteration of the large-scale mean thermal structure of the atmosphere resulting from increased CO_2 concentrations has the potential for affecting the dynamics of the atmosphere across the entire spectrum of scales that govern atmospheric processes. Inherent in these changes are interactions among the scales that could change, resulting in an alteration in the frequency of relatively short-term weather systems that enhance the probability of wildfire occurrence. Indeed, climatic variability and extreme weather events (e.g., drought, flood, extreme heat, fire-weather development) are much more important factors for wildfire occurrence than the effects of systematic climate change (e.g., long-term and large-scale temperature and precipitation trends) (Fosberg et al., 1993).

Given the importance of wildfires in the southern United States, this chapter describes some of the recent research efforts that have examined many of the critical atmospheric interactions with southern wildfire occurrence and the implications of regional climate change on the future development of fire-weather systems. The following sections provide an overview of present and recent research dealing with large-scale atmospheric circulation, temperature, and moisture patterns associated with severe wildfires in the region, useful atmospheric variables as indicators of severe fire potential in the region, soil-moisture effects on southern fire-weather development, El Niño/southern oscillation effects on fire occurrence in the southeastern United States, and potential changes in lightning-caused wildfires in the region.

Synoptic Circulation, Temperature, and Moisture Patterns

Atmospheric conditions play a critical role in affecting the severity of wildfires and the probability of their occurrence in the southern United States, as well as the rest of the country during specific times of the year. Figure 37.1 shows the distribution of severe wildland fires in the south-central and southeastern United States by month for years 1971 to 1984 and 1987 to 1991. Although severe wildland fires in the south-central United States typically occur during the months of March and April, severe fires in the southeastern United States generally occur

37. Climate Change: Implications for Wildfire Occurrence

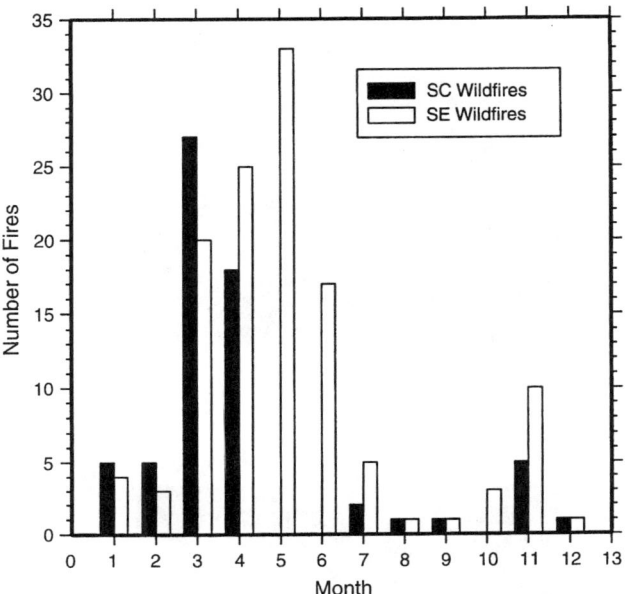

Figure 37.1. Number of wildfires burning more than 400 ha in the south-central United States (i.e., Texas, Oklahoma, Kansas, Missouri, Arkansas, Louisiana) and the southeastern United States (SE: Mississippi, Alabama, Georgia, Florida, South Carolina, North Carolina, Kentucky, Tennessee, Virginia) from 1971 to 1984 and 1987 to 1991.

between March and June. Secondary peaks in severe wildfire occurrence typically occur in the month of November. There are specific synoptic-scale circulation patterns in the middle troposphere during these months that tend to be associated with severe wildfires in the southern United States. These circulation patterns result in temperature and moisture distributions in the lower atmosphere that enhance the probability of severe fires occurring.

Heilman (1995) expanded upon the work of Schroeder et al., (1964) to examine the prevalent circulation patterns at the onset of past severe fires in six different regions of the United States. Using empirical-orthogonal-function analyses of the observed geopotential heights at the 500 mb pressure level in the atmosphere at the onset of severe wildfires in the United States, it was shown that severe wildfires in the south-central and southeastern United States are associated with three distinct mid-tropospheric circulation patterns. Examples of these patterns are shown in Figures 37.2a,b,c. Severe wildfires occurred on these days in the southeastern United States. The first pattern (Figure 37.2a) is described by a middle tropospheric ridge over the western half of the United States with an accompanying trough over the eastern United States that results in northwesterly to westerly flow over the southeastern United States. This circulation pattern tends to bring cool dry air from Canada into the southern and eastern states. The second pattern (Figure 37.2b) is described by a middle tropospheric ridge over the eastern half of the United States, with the western or central states dominated by a

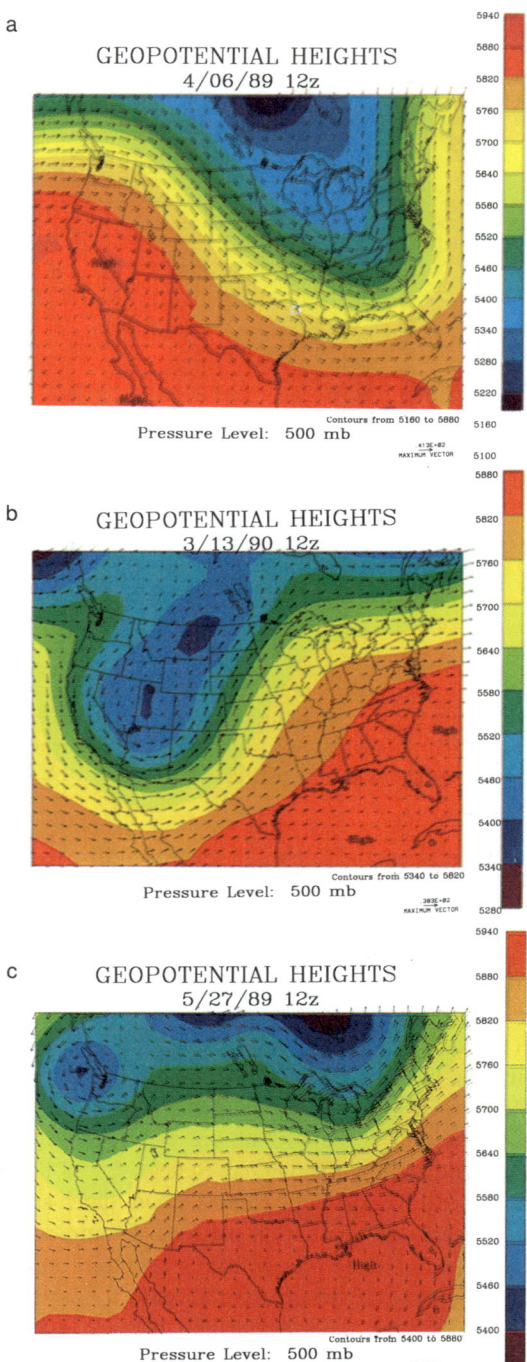

Figure 37.2. Recent examples of three 500 mb geopotential height (contours in meters) and circulation (vectors in m s^{-1}) patterns associated with severe wildfires in the southeastern United States that occurred on: (a) April 6, 1989, (b) March 13, 1990, and (c) May 27, 1989.

middle tropospheric trough. This pattern can be very conducive to the westward shift of the surface Bermuda high-pressure system, leading to hot and dry conditions over parts of the southeastern and northeastern United States. These first two patterns are equivalent to the positive and negative phases, respectively, of the Pacific/North American (PNA) teleconnection pattern (Wallace and Gutzler, 1981) that is responsible for much of the climate variability in the northern hemisphere during the autumn, winter, and spring months (Leathers et al., 1991). The third circulation pattern is shown in Figure 37.2c and is characterized by middle tropospheric zonal flow over the southern regions of the United States.

These circulation patterns that are associated with severe wildfires in the south-central and southeastern United States produce lower atmospheric temperature and moisture patterns that enhance the probability of fire occurrence if fuel conditions are adequate. Figures 37.3a,b,c and Figures 37.4a,b,c show the corresponding average lower atmospheric temperature and relative humidity anomalies over the United States when severe wildland fires occurred in the southeastern United States from 1971 to 1984 and 1987 to 1991. The first circulation pattern (Figure 37.2a) resulted in lower than normal 850 mb temperatures over the eastern half of the United States, with maximum temperature anomalies occurring over the Ohio River Valley. Average temperature anomalies at 850 mb over the southeastern United States ranged from -1 °C in Florida to about -5 °C in Kentucky (Figure 37.3a). The second circulation pattern (Figure 37.2b) resulted in warmer than normal temperatures at 850 mb over all the eastern United States. Average temperature anomalies reached about 5 °C over the northeastern United States, and decreased southward to about 1 °C over southern Florida (Figure 37.3b). The zonal circulation pattern over the southern United States (Figure 37.2c) resulted in positive 850 mb temperature anomalies over the southern half of the United States, with the southeastern United States experiencing average temperature anomalies of 1 to 2 °C (Figure 37.3c). For all three circulation patterns, lower-than-normal relative humidity values occurred over at least a portion of the southeastern United States (Figures 37.4a,b,c). The average lower atmospheric moisture anomalies associated with the second circulation pattern (Figures 37.2b and 37.4b) indicate that although this pattern tends to produce dry atmospheric conditions over most of the Atlantic coast states and eastern Great Lakes region, the states west of Alabama, Tennessee, and Kentucky typically experience higher-than-normal relative humidity values. It is not uncommon for significant portions of the south-central and southeastern United States to experience higher-than-normal relative humidity values under the second circulation pattern. The location of the Bermuda high-pressure system under this particular circulation pattern determines where the northward transport of Gulf moisture will occur and those southeastern and northeastern states that will be cut off from the supply of Gulf moisture.

The average lower atmospheric temperature and relative humidity anomalies occurring during severe wildland fire episodes in the south-central United States over the same time period for each of the three identified circulation patterns are shown in Figures 37.5a,b,c and Figures 37.6a,b,c, respectively. The temperature

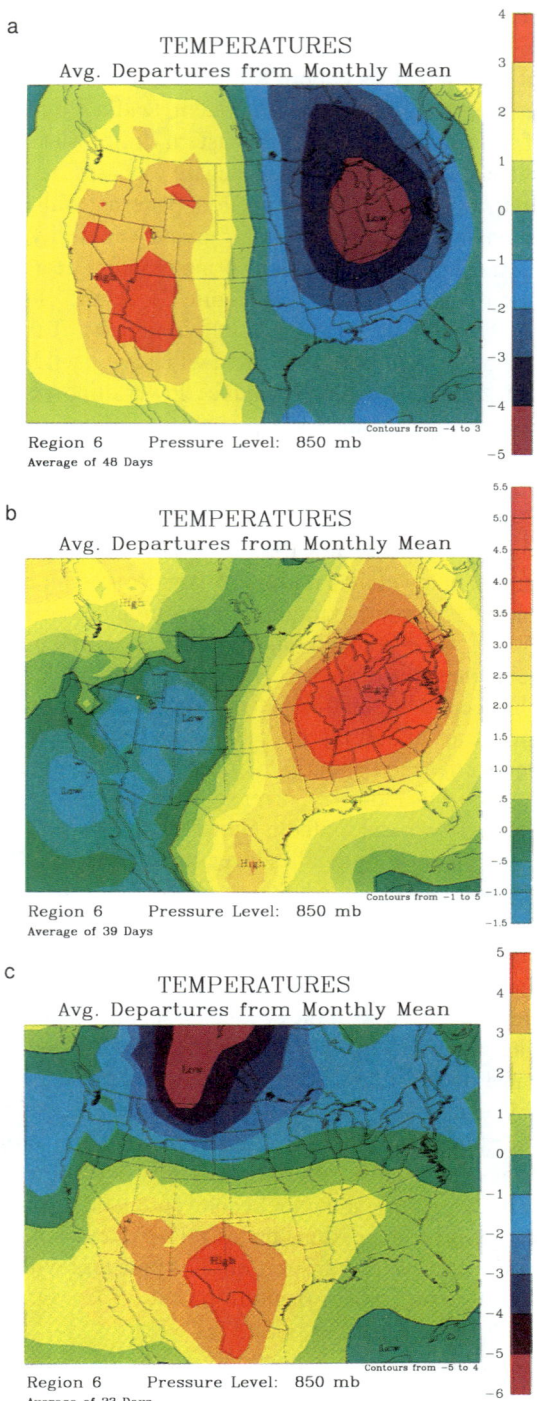

Figure 37.3. Average temperature anomalies (°K) at the 850 mb pressure level during past fire-weather episodes in the southeastern United States having circulation patterns similar to those shown in: (a) Figure 37.2a, (b) Figure 37.2b, and (c) Figure 37.2c.

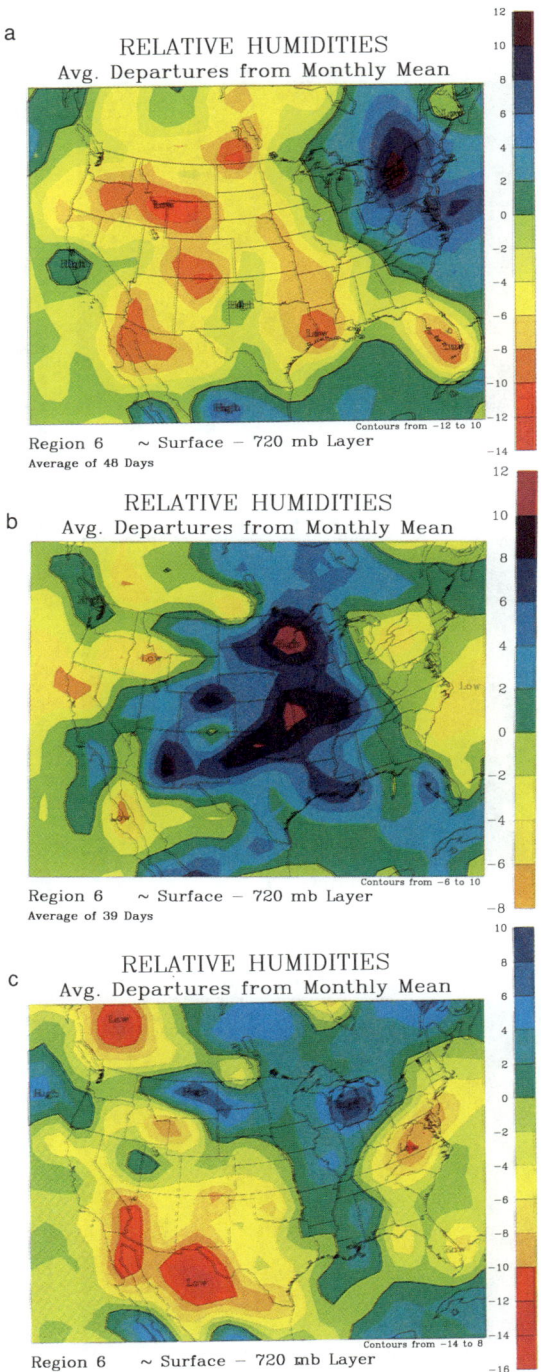

Figure 37.4. Average relative humidity anomalies (%) in the lower atmosphere during past fire-weather episodes in the southeastern United States having circulation patterns similar to those shown in: (a) Figure 37.2a, (b) Figure 37.2b, and (c) Figure 37.2c.

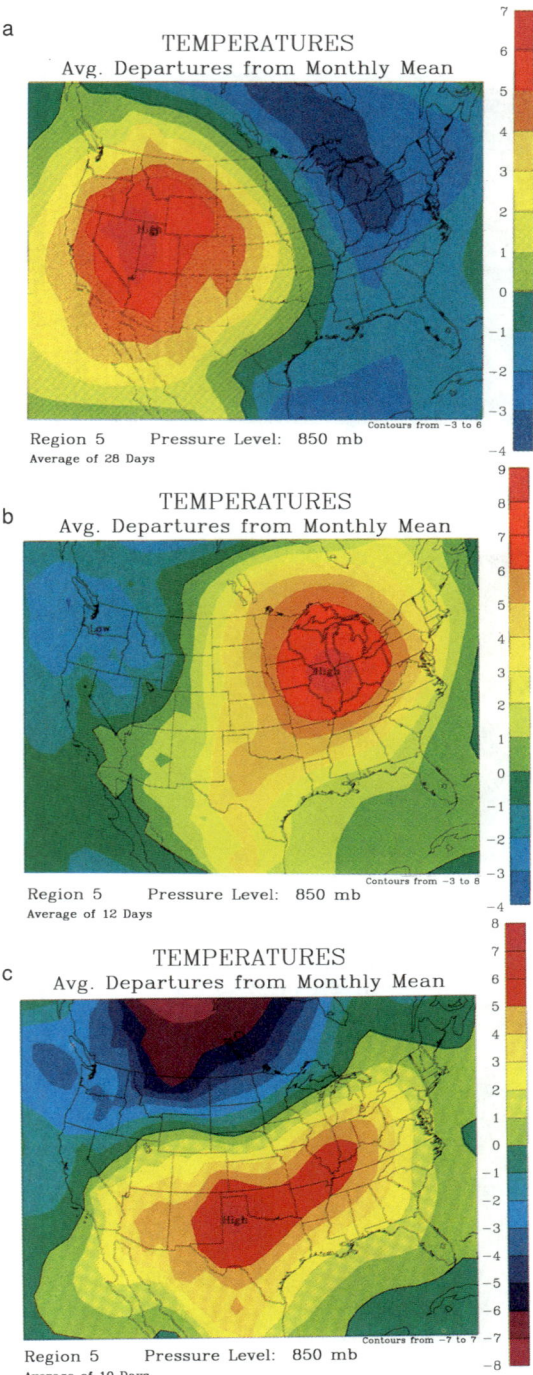

Figure 37.5. Average temperature anomalies (°K) at the 850 mb pressure level during past fire-weather episodes in the south-central United States having circulation patterns similar to those shown in: (a) Figure 37.2a, (b) Figure 37.2b, and (c) Figure 37.2c.

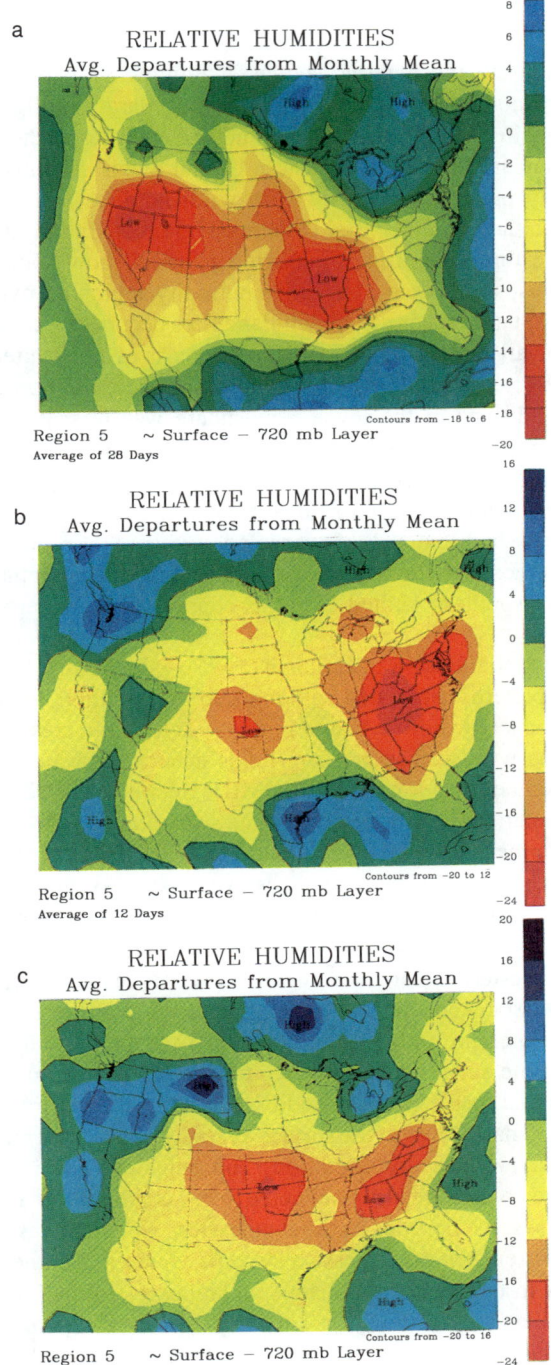

Figure 37.6. Average relative humidity anomalies (%) in the lower atmosphere during past fire-weather episodes in the south-central United States having circulation patterns similar to those shown in: (a) Figure 37.2a, (b) Figure 37.2b, and (c) Figure 37.2c.

patterns are similar to the patterns shown in Figures 37.3a,b,c, although the magnitudes of the computed anomalies are generally larger in Figures 37.5a,b,c because of the smaller sample size of wildland fire episodes in the south-central United States compared to the southeastern United States. The average relative humidity anomaly patterns associated with wildland fires in the southeastern (Figures 37.4a,b,c) and south-central (Figures 37.6a,b,c) United States and corresponding to the three identified circulation patterns show some differences, particularly in the spatial extent of negative humidity anomalies over the eastern half of the United States. When severe wildland fires occurred in the south-central United States for all three circulation patterns, most of the United States was characterized by drier-than-normal, lower atmospheric conditions. The spatial extent of dry conditions in the lower atmosphere was less pronounced when severe fires occurred in the southeastern U.S, particularly under the circulation patterns shown in Figures 37.2b,c. However, each circulation pattern associated with fires in the south-central and southeastern United States results in lower-than-normal relative humidity values over some portion of the regions. The significance of lower atmospheric moisture as an indicator of the potential for wildfire occurrence in the southern United States is supported by Potter (1995, 1996) as well.

In a study related to the work of Heilman (1995), Takle et al., (1994) examined surface-pressure patterns associated with reduced precipitation, high evaporation potential, and enhanced wildfire danger in the West Virginia area, and compared these observations with surface-pressure fields generated by the Canadian Climate Centre general circulation model (GCMII) (Boer et al., 1992; McFarlane et al., 1992) for simulations of the present climate and a doubled Co_2 climate. Using the eight synoptic meteorology patterns that tend to influence the eastern United States, as defined by Yarnal (1993), they observed that an extended high-pressure system situated over the eastern half of the United States, a high-pressure system situated off the eastern coast of the United States, and a high-pressure system centered over the western Great Lakes region are the three most common surface-pressure patterns associated with severe wildfires in the West Virginia area. Simulations of surface-pressure fields from the GCMII for the present climate and a doubled CO_2 climate suggested a tendency for the occurrence of more surface-pressure patterns that result in drier conditions in the northeastern United States under a changed climate. Most of the surface-pressure patterns identified by Takle et al., (1994) that produce dry conditions and enhanced fire potential in the northeastern United States are also conducive to dry conditions and increased fire potential in at least some portions of the southeastern United States. In particular, the increased occurrence of high-pressure systems centered over the Gulf coast of the United States in the changed climate simulations performed by Takle et al., (1994) suggests that the southeastern United States may experience more dry days. High-pressure systems located near the Gulf coast tend to block the transport of Gulf moisture northward over the eastern states. This surface-pressure pattern usually develops in response to mid-tropospheric circulation patterns similar to those shown in Figures 37.3b,c, which have been found to be associated with severe wildfires in the south-central and southeastern United States (Heilman, 1995). Thus, the potential for more fire-weather episodes in the south-central and southeastern United States would probably increase under this scenario.

Atmospheric Indicators of Severe Wildfire Potential

In addition to the studies that have identified those atmospheric synoptic patterns conducive to severe wildfire occurrence in the southern United States, other studies have focused on specific atmospheric variables that are indicative of severe wildfire occurrence. For example, such studies as Byram (1954), Davis (1969), and Brotak and Reifsnyder (1977) examined various atmospheric conditions at or near the earth's surface and their relationships to large wildfires, including near-surface wind speeds and profiles, frontal positions, and temperature profiles. The conditions generally considered as favorable to large wildfires are dry air and an unstable temperature profile (i.e., temperatures decreasing more than 1 °C for every 100 m increase in altitude). Brotak and Reifsnyder (1977) examined atmospheric profiles for fifty-two fires, more than half of which occurred in the southeastern United States. They concluded that a low-level wind jet often accompanied the occurrence of large wildland fires. They defined this jet as a wind maximum within 3,050 m of the surface where the wind speed is 2.2 m s^{-1} greater than the speed 305 m above or below. They did not examine conditions on nonfire days, hence there is no indication of whether such a jet is also found on days when no fire occurred.

In the most recent study of specific atmospheric variables associated with severe wildfires, Potter (1996) used analysis of variance (ANOVA) to examine how strongly atmospheric conditions differ on the days of large wildfires (over 400 ha) as compared to climatological conditions. Surface temperature, dewpoint depression, relative humidity, wind speed, wind shear, and stability were all considered. There was no indication that temperature at the surface, or stability or wind shear in the lowest 100 to 1000 m are any different when large wildfires occur than at any other time. Surface wind speed and relative humidity showed some tendency to differ from normal on large wildfire days, though not in Florida. Dewpoint depression showed the most significant difference and was most widespread in its ability to discriminate between normal conditions and large wildfire conditions. It was significant all along the southeastern Atlantic coast (from Virginia to Palm Beach, Florida), the region for which the most data were examined. The relationships between the findings of Brotak and Reifsnyder (1977) and Potter (1996) and the potential atmospheric conditions associated with a changed climate are difficult to determine. However, these studies do suggest that changes in the frequency or intensity of weather systems that produce low-level jets or that move dry air masses with large surface dewpoint depressions and low relative humidity values into the southern United States could affect the frequency of severe wildfires there.

Soil-Moisture Effects on Fire-Weather Development

Along with the specific large-scale circulation patterns that result in drier-than-normal conditions and enhanced wildfire probabilities in the south-central and southeastern United States, regional soil-moisture variations can also influence the development of fire-weather systems. Fast (1994) and Fast and Heilman (1996) examined the effects of soil-moisture deficits on the atmospheric mesoscale dynamics that influence fire severity. Using the regional atmospheric modeling system (RAMS) (Pielke et al., 1992), a three-dimensional nonhydrostatic mesoscale model, simulations of past fire-weather episodes in the southeastern United States were performed to examine the role of soil moisture and vegetation within and outside the southeastern United States in affecting fire-weather development in the region, as measured by the lower atmospheric severity index (LASI) (Haines, 1988).

Meteorological conditions were simulated during the period May 5 to 17, 1989 when Florida experienced numerous wildfires and moderate to severe drought conditions were observed in central and southern Florida and in the northern Great Plains. During this same period, moderate to severe wet conditions were observed over much of the northeastern United States and in the states north and west of Georgia in the southeastern United States. Circulations in the middle atmosphere during this period were very similar to one particular circulation pattern identified by Heilman (1995) as being conducive for severe wildfires in the southeastern United States (see Figure 37.2a). Sensitivity tests performed with the RAMS model indicated that the addition of soil moisture significantly affects the near-surface temperature and relative humidity fields, and affects cloud cover, precipitation, and wind speeds to a lesser extent. The simulations also indicated that the effects of soil moisture are most pronounced where the soil is sufficiently moist, but the advection of evaporated moisture from wet-soil regions can occasionally increase the relative humidity observed at locations downwind of the wet-soil regions. Daytime surface temperatures and relative humidity values are reduced and increased, respectively, when soil moisture is increased. Vegetation was found to moderate the transfer of water from the soil into the atmosphere. In regions where soil is wet, the presence of vegetation reduces the amount of water evaporated into the atmosphere, and where the soil is relatively dry, the presence of vegetation increases the amount of water evaporated compared to regions with little vegetation. Vegetation also reduces the speed of surface wind through the inhomogeneous roughness and plant lengths that increase the surface friction.

Fast (1994) and Fast and Heilman (1996) found that fire-weather development in the southeastern United States., as measured by the LASI, is affected by soil moisture and vegetation distributions that determine the degree of evapotranspiration in a region. Values of LASI were reduced in areas of significant evapotranspiration because the additional moisture added to the atmospheric boundary layer reduces the lower atmospheric dew point depression, thereby reducing the probability of severe wildfire occurrence. The implication of this phenomenon is that surface evapotranspiration in regions of moist soil or significant vegetation upwind of the south-central or southeastern United States have the potential for reducing the probability of severe wildland fire occurrence in the southern United States because of the advection of low-level moisture. Thus, although large-scale circulation patterns in the middle atmosphere may be conducive to severe wildland fires in the southern United States, regional soil moisture and surface evapotranspiration patterns within and outside the region must also be considered in assessing the potential for severe wildfires in the southern United States and elsewhere.

El Niño/Southern Oscillation Effects

Because severe wildfires are strongly linked to relatively short-term weather events that produce conditions favorable for their occurrence, any large-scale circulation changes in the atmosphere brought about by global climatic forcing factors have the potential for altering circulation patterns in the United States and modifying the normal frequency of weather events conducive to severe wildfires. One particular forcing factor that has a major influence on atmospheric circulation patterns over North America is the El Niño/southern oscillation (ENSO) phenomenon.

El Niño is a term that refers to the periodic extreme warming of surface ocean water in the eastern tropical Pacific Ocean (Wyrtki, 1979). This periodic warming has been shown to be associated with the southern oscillation, a periodic fluctuation in surface pressures observed in the western and eastern Pacific Ocean (Walker, 1928; Bjerknes, 1969). El Niño/southern oscillation events typically occur every three to eight years (Haston and Michaelsen, 1994), and they have been linked to changes in observed midlatitude circulation and weather patterns through atmospheric teleconnections. For example, Horel and Wallace (1981) showed that ENSO events are correlated with below-normal 700 mb geopotential heights in the northern Pacific and the southeastern United States, and above normal-heights over western Canada. Gray (1984a, 1984b) found that the seasonal number of hurricanes and hurricane days was negatively correlated with moderate to strong ENSO events. Ropelewski and Halpert (1986) used monthly precipitation and temperature data from 1875–1980 from sites throughout North America to show that ENSO events are associated with above normal precipitation and below-normal temperatures in parts of the southeastern United States starting in October of ENSO years and ending in March of the following years. Trenberth et al., (1988) identified tropical Pacific sea-surface temperature changes associated with the 1988 La Niña episode as the primary factor responsible for the atmospheric circulations over North America that caused the severe drought in the Great Plains region of the United States. Many other studies have also been conducted in recent years that indicate a relation between ENSO events and climatic and weather variability over the United States and North America.

The observed changes in atmospheric circulation and weather patterns associated with ENSO events, particularly trends in precipitation, suggest that ENSO events may influence the probability of severe wildfire occurrence and the number of hectares burned as a result of wildfires in certain regions of the United States. The most notable study examining ENSO/wildfire relationships was performed by Simard et al., (1985a, 1985b). They used fifty-seven years of fire activity data (1926 to 1982) along with El Niño occurrence and intensity level data (Quinn et al., 1978) to identify specific regions of the United States that tend to show changes in wildfire activity during ENSO episodes. It was found that annual fire activity in the southern region tends to decrease during El Niño years, the relation between fire activity and ENSO episodes in the north-central and eastern regions is weak, and that there is no evidence that ENSO episodes affect fire activity in the Pacific or Rocky Mountain regions.

Lightning

Large-scale circulation conditions in the atmosphere also play a role in the development of regional weather systems that produce lightning, a factor in wildfire occurrence throughout the country. Although large wildfires attributed to cloud-to-ground lightning strikes are much more common in the western United States., lightning is responsible for about 1,000 to 3,000 wildfires on Federal, State, and private lands in the southern United States each year (USDA Forest Service, 1992). Lightning-caused wildfires have been responsible for the burning of over 121,000 ha in some years in the southern United States. Changes in global and regional climates could result in increased frequency of weather patterns associated with drought conditions, leading to enhanced fuel dryness, and thunderstorms that generate cloud-to-ground lightning strikes, leading to fuel ignition. Price and Rind (1994) examined the impact of a changed climate resulting from an increase in atmospheric CO_2 concentration on lightning-caused wildfires in North America and in different regions of the United States, using the Goddard Institute for Space Studies (GISS) general circulation model (GCM) (Hansen et al., 1983). Their modeling study suggested that the number of lightning-caused wildfires in the entire United States could increase by approximately 44% over the present number. The total area burned in the United States was projected to increase by about 78%. For the south-central and southeastern regions of the United States, their modeling study suggested increases on the order of 40 to 50% in the number of wildfires caused by lightning each year. Contributing to these increases is the probable decrease in the average effective precipitation (defined as precipitation minus potential evapotranspiration) over the United States under a changed climate as a result of increased atmospheric CO_2 concentrations (Price and Rind, 1994).

Summary

An overview of present and recent research on atmospheric interactions with wildfire occurrence in the south-central and southeastern United States and the potential implications for fire occurrence in these regions under a changed climate have been presented. Because wildfire occurrence is very dependent on relatively short-term weather events along with fuel conditions, present research has partly focused on identifying the critical atmospheric circulation patterns that lead to enhanced wildfire activity in different regions of the United States., including the Southern Global Change Program (SCGP) study region. The results indicate that there are three specific circulation patterns in the middle atmosphere that are prevalent at the onset of severe wildfires in the south-central and southeastern United States. These circulation patterns produce drier-than-normal conditions over many parts of the south-central and southeastern United States during the spring and fall fire seasons. Examinations of the surface-pressure patterns associated with severe wildfire occurrence in the eastern United States and comparisons with general circulation model simulations under doubled CO_2 conditions tend to suggest the future occurrence of more surface pressure and atmospheric circulation patterns that produce drier conditions in the eastern and southeastern United States. Present-day research substantiates the importance of near-surface moisture (surface dew point depression values) as an indicator of enhanced wildfire activity in the southeastern United States.

Recent research has also focused on the additional effects of vegetation and soil-moisture variations on fire-weather development over the southeastern United States. Research results indicate the importance of soil moisture within and outside the southern United States in affecting fire-weather development in the region. Evaporation from wet-soil regions and the advection of moisture into regions that are much drier tend to reduce the potential for severe wildfire occurrence in the drier regions, as measured by the LASI. The presence of vegetation moderates the transfer of moisture to the atmosphere in wet- and dry-soil areas.

Other research has focused on lightning as a causative agent for wildfires. General circulation model simulations suggest that the south-central and southeastern United States will experience more lightning-caused wildfires under a changed climate resulting from increased atmospheric CO_2 concentrations.

The periodic fluctuations in sea-surface temperatures in the eastern Pacific Ocean, commonly referred to as El Niño episodes, have also been shown to have a major impact on wildfire occurrence in the southeastern United States. El Niño events tend to reduce wildfire activity in this region. However, it is uncertain what impact periodic El Niño events would have on circulation patterns over the United States and wildfire activity in the southern states under a changed climate because of increased atmospheric CO_2 concentrations.

Given the importance of wildfires in the south-central and southeastern United States, as reflected in the yearly totals of wildfire numbers and hectares burned, there is a need to better understand the potential ramifications of a changed climate and climate variability on wildfire activity in the region. The research results outlined in this chapter provide insight into some of the key atmospheric processes involved with wildfire occurrence in the southern states. They provide the foundation for new studies that are needed to further examine the relationship between these atmospheric processes relevant to large-scale climatic changes and the smaller-scale atmospheric dynamics that are most relevant to regional fire-weather development and wildfire occurrence in the southern states.

References

Bjerknes J (1969) Atmospheric teleconnections from the equatorial Pacific. Mon Wea Rev 97:163–172.
Boer GJ, McFarlane NA, Lazare M (1992) Greenhouse gas-induced climate change simulated with the CCC second-generation general circulation model. J Climate 5:1045–1077.
Brotak EA, Reifsnyder WE (1977) Predicting major wildfire occurrence. Fire Manage Notes 38:5–8.
Byram GM (1954) Atmospheric conditions related to blowup fires. USDA For Serv Stat Pap 35, 31 p.
Davis RT (1969) Atmospheric stability forecast and fire control. Fire Cont Notes 30:3–4,15.
Fast JD (1994) The effect of regional-scale soil-moisture deficits on mesoscale atmospheric dynamics that influence fire severity. WSRC-TR-94–0468, Westinghouse Savannah River Company, Aiken, SC.
Fast JD, Heilman WE (1996) The effect of regional-scale soil-moisture deficits on mesoscale atmospheric dynamics that influence fire severity. 22nd Conference on Agricultural and Forest Meteorology with Symposium on Fire and Forest Meteorology, American Meteorological Society.
Fedkiw J (1989) The evolving use and management of the nation's forests, grasslands, croplands, and related resources. RM-175, USDA For Serv, Fort Collins, CO.
Fosberg MA, Mearns LO, Price C (1993) Climate change-fire interactions at the global scale: Predictions and limitations of methods. In: Crutzen PJ, Goldammer JG (Eds) *Fire in the environment: The ecological, atmospheric, and climatic importance of vegetation fires.* John Wiley and Sons, New York.
Gray WM (1984a) Atlantic seasonal hurricane frequency. Part I: El Niño and 30 mb quasi-biennial oscillation influences. Mon Wea Rev 112:1649–1668.
Gray WM (1984b) Atlantic seasonal hurricane frequency. Part II: Forecasting its variability. Mon Wea Rev 112:1669–1683.
Haines DA (1988) A lower atmospheric severity index for wildland fires. Nat Wea Digest 13:23–27.
Hansen J, Russell G, Rind D, Stone P, Lacis A, Lebedeff S, Reudy R, Travis L (1983) Efficient three-dimensional global models for climate studies: Model I and II. Mon Wea Rev 111:609–662.
Haston L, Michaelsen J (1994) Long-term central coastal California precipitation variability and relationships to El Niño-Southern Oscillation. J Climate 7:1373–1387.
Heilman WE (1995) Synoptic circulation and temperature patterns during severe wildland fires. Ninth Conference on Applied Climatology, American Meteorological Society.
Horel JD, Wallace JM (1981) Planetary scale atmospheric phenomena associated with the southern oscillation. Mon Wea Rev 109:813–829.
Leathers DJ, Yarnal B, Palecki MA (1991) The Pacific/North American teleconnection pattern and United States climate. Part I: Regional temperature and precipitation associations. J Climate 4:517–528.
McFarlane NA, Boer GJ, Blanchet JP, Lazare M (1992) The Canadian Climate Centre second-generation general circulation model and its equilibrium climate. J Climate 5:1013–1044.
Peterson RM (1982) *An Analysis of the Timber Situation in the United States.* US Government Printing Office, Washington, DC.

Pielke RA, Cotton WR, Walko RL, Tremback CJ, Lyons WA, Grasso LD, Nicholls ME, Moran MD, Wesley DA, Lee TJ, Copeland JH (1992) A comprehensive meteorological modeling system—RAMS. Meteor Atmos Phys 49:69–91.

Potter BE (1995) Atmospheric stability, moisture, and winds as indicators of wildfire risk. Ninth Conference on Applied Climatology, American Meteorological Society.

Potter BE (1996) Atmospheric properties associated with large wildfires. Int J Wildland Fire, 6(2):71–76.

Price C, Rind D (1994) The impact of a $2 \times CO_2$ climate on lightning-caused fires. J Climate 7:1484–1494.

Quinn WH, Zopf DO, Short KS, Kuo Yang RTW (1978) Historical trends and statistics of the southern oscillation, El Niño, and Indonesian droughts. Fishery Bull 76:663–677.

Ropelewski CF, Halpert MS (1986) North American precipitation and temperature patterns associated with the El Niño/southern oscillation (ENSO). Mon Wea Rev 114:2352–2362.

Schroeder MJ, Glovinsky M, Hendricks VF, Hood FC, Hull MK, Jacobson HL, Kirkpatrick R, Krueger DW, Mallory LP, Oertel AG, Reese RH, Sergius LA, Syverson CE (1964) Synoptic weather types associated with critical fire weather. Pac Southwest For Range Exper Stat, Berkeley, CA.

Simard AJ, Haines DA, Main WA (1985a) El Niño and wildland fire: An exploratory study. Eighth Conference on Fire and Forest Meteorology, Society of American Foresters.

Simard AJ, Haines DA, Main WA (1985b) Relations between El Niño/southern oscillation anomalies and wildland fire activity in the United States. Agric For Meteor 36:93–104.

Takle ES, Bramer DJ, Heilman WE, Thompson MR (1994) A synoptic climatology for forest fires in the NE US and future implications from GCM simulations. Int J Wildland Fire 4:217–224.

Trenberth KE, Branstator GW, Arkin PA (1988) Origins of the 1988 North American drought. Science 242:1640–1645.

USDA Forest Service (1992) 1984–1990 Forest fire statistics. USDA For Serv, Washington, DC.

Walker GT (1928) World weather III. Mem Roy Meteor Soc 2:97–106.

Wallace JM, Gutzler DS (1981) Teleconnections in the geopotential height field during the Northern Hemisphere winter. Mon Wea Rev 109:784–812.

Wyrtki K (1979) El Niño. La Recherche 10:1212–1231.

Yarnal B (1993) *Synoptic climatology in environmental analysis: A primer.* Belhaven Press, London.

38. Detecting and Predicting Climatic Variation from Old-Growth Baldcypress

Gregory A. Reams and Paul C. Van Deusen

Tree-ring data can extend back in time for thousands of years allowing researchers to reconstruct certain environmental factors that have left an imprint or signal in the tree-ring record. Typically, these factors include reconstructions of annual precipitation or temperature for months or seasons to which a particular tree species is sensitive. Over the last several decades, scientists have used tree-ring records in novel ways to investigate the timing and extent of such natural phenomena as volcanoes (Baillie and Munro, 1988), earthquakes (Sheppard and Jacoby, 1987), El Niño/southern oscillation (Stahle and Cleaveland, 1993), fire (Swetnam 1993), carbon dioxide (CO_2) (Graybill and Idso, 1993), and synchronous landscape-level disturbances (Reams and Van Deusen, 1993) by recognizing the possibility that various signals may be recorded in the growth record of trees, depending on microsite characteristics, geographic location, and disturbance history (Fritts 1976).

Climate reconstruction from tree-ring data involves establishing a relationship between the tree-ring variable(s) and some measure of climate. The uniformitarian assumption (Fritts, 1976) is then called upon to allow for using this established relationship to reconstruct the climate variable during the period before climate measurements were available. Weather stations in the United States were rarely in place prior to approximately 1860, but many living trees provide data for centuries before that time. Thus, the motivation to use tree-ring derived variables to reconstruct climate is clear.

The usual procedure involves fitting a regression equation that uses climate as the dependent variable and tree-ring data as the independent variable over the period when climate has been recorded. This calibration equation (Fritts, 1976) is then used to predict the earlier, unknown climate data. Stahle and Cleaveland (1992) provide a recent example of this approach using baldcypress (*Taxodium distichum* (L.) Rich.). An alternative approach has been presented by Graumlich (1993), which uses response-surface methods to simultaneously reconstruct temperature and precipitation variables. However, neither of these studies presented confidence intervals around the reconstructions.

In this chapter, we present a new procedure for climate reconstruction that allows for the incorporation of prior knowledge about the climate variable, and that demonstrates the existence of a hurricane signal in the tree-ring data. The climate reconstruction method is applied to old-growth baldcypress data from Louisiana using the Palmer Drought Severity Index (PDSI) for the month of June. We show that the method is relatively easy to program and produces valid confidence intervals as a byproduct. The actual algorithm involved is the Gibbs sampler (Smith and Gelfand, 1992).

The identification of a hurricane signal in these data addresses the increased concern that climate change caused by increased emissions of greenhouse gases may lead to either intensification or greater frequency of extreme storms (Schmidt and von Storch, 1993). Emanuel (1987) found that, given August mean conditions over the tropical oceans with twice the present atmospheric CO_2 content, a general circulation model (GCM) predicts a 40 to 50% increase in the destructive potential of hurricanes. Jarrell and Elsberry (1994) suggested that global warming might increase the frequency of tropical cyclones because it would expand the area of oceans with temperatures above the 26 °C threshold required for the formation of tropical cyclones. Baldcypress is a uniquely valuable species for investigating the long-term frequency of hurricanes because baldcypress trees are long-lived, and accumulating evidence indicates they withstand hurricanes better than any other tree species in the southern United States (Putz and Sharitz, 1991; Sheffield and Thompson, 1992).

Baldcypress Habitat

Baldcypress is a deciduous conifer that grows on saturated and seasonally inundated soils of the southeastern and Gulf coastal plains. Inland, baldcypress grows along the many streams of the middle- and upper-coastal plains and northward through the Mississippi Valley. Humid, moist subhumid, and dry subhumid climatic types occur within the range of baldcypress. Baldcypress occurs most frequently on intermittently flooded sites, and therefore drainage may be more important than rainfall in determining site suitability. The growing season within the natural range of baldcypress increases from about 190 days in southern Illinois to nearly year-long in southern Florida.

More than 90% of the natural baldcypress stands are on flat topography or in slight depressions at elevations of less than 30 meters above sea level. Baldcypress sites are characterized by frequent, prolonged flooding. Floodwaters may be 3 meters deep or more and may flow at rates up to 6.5 kilometers per hour or may be stagnant. Normally the species is found on intermittently flooded and very poorly drained Spodosols, Ultisols, Inceptisols, Alfisols, and Entisols. The soil temperature regimes are classified as thermic to hyperthermic.

Cypress swamps and other forested wetlands that receive periodic nutrients subsided from floodwaters probably are some of the world's most productive ecosystems. The annual aboveground production of biomass in a baldcypress forest in Florida is 15,700 kg per hectare (ha). In comparison, terrestrial forest communities in the temperate region often produce 12,300 to 15,000 kg/ha annually. Stillwater forested wetlands do not receive nutrient subsidies from floodwaters, and they have production rates comparable to, or lower than, those of terrestrial forests.

Early estimates of the area of baldcypress forests in Louisiana range from 0.67 to 3.64 million ha (Conner and Toliver, 1990). Baldcypress was cut extensively from 1890 to 1925 when the last virgin stands were depleted. Recent estimates indicate that there are 0.14 million ha left in Louisiana (Conner and Toliver, 1990). Williston et al., (1980) estimates that there are between 1.2 and 2 million ha of cypress forest in the United States, with a total growing-stock volume of 155.7 million m^3. Over one-half of the cypress is in Florida and Louisiana, with Louisiana ranked first in timber quality of cypress. Baldcypress growing-stock increased from 15 million m^3 in 1954 to 41 million m^3 in 1984. At present, loblolly pine (*Pinus taeda* L.) and sweetgum (*Liquidambar styraciflua* L.) are the only species in Louisiana with timber volume greater than baldcypress (Rosson et al., 1988)

Methods

Sample

Old-growth baldcypress trees are scattered individually and within small stands throughout many river basins and estuaries in the southeastern United States. These old-growth remnants exist because the trees were considered uneconomical to harvest either because of defect or location. To locate our study sites and eliminate unnecessary field work, we utilized various remote sensing techniques including satellite imagery, NASA's Stennis Space Center calibrated airborne multispectral scanner (CAMS) data, and aerial photography. We began this study in the early summer of 1992 in the Tangipahoa River basin, Louisiana. During 1993 we expanded site selection to the Pearl River basin, Mississippi, and in 1994 began sampling old-growth baldcypress in the Atchafalaya River basin, Louisiana. The results presented here, however, will be restricted to the tree-ring material sampled from the Tangipahoa River and Pearl River basins.

After identifying candidate old-growth sites from remotely sensed data we visited the sites and collected our data using standard increment core sampling techniques with a few minor modifications that greatly increased the temporal depth of our tree cores. Increment cores were extracted from the base of each tree at heights between 60 and 137 cm aboveground, depending on the soundness of the main stem. In prior studies involving old-growth baldcypress, researchers sampled above the well-known root buttress (Stahle et al., 1985, 1988). Trees in our study site showed little to no buttressing, but exhibited heartrot (and therefore hollowness) that increased with sampling height. The heartrot is caused by a fungus, *Stereum taxodi,* and is prevalent in old-growth baldcypress. The fungus gains entrance in the crown and slowly works downward, frequently destroying heartwood to the base of the tree (Wilhite and Toliver, 1990). We often were able to extract more heartwood and thereby add hundreds of annual ring-widths by sampling as close to ground level as possible. Annual increments of each tree core were measured to within 0.001 mm and cross-dated using graphical and statistical procedures available in DYNACLIM (Van Deusen, 1990).

The main data quality concern in tree-ring studies is that the annual ring-widths be measured accurately and precisely, but even more important is that the ring-width be assigned to the correct calendar year. This process is known as cross-dating and is the cornerstone of dendrochronological studies. In dendrochronology, quality control takes place every time a tree core is measured because after the removal of a long-term trend (i.e., standardization), climate-sensitive tree species have the same standardized growth pattern across trees such that the actual data of any one ring of the pattern is the same among trees (Douglass, 1941). This is possible because similar environmental conditions have limited high frequency fluctuations in radial growth.

Bayesian Climate Model

We will begin by designating the tree-ring variable as y_t and the climate variable as x_t. The subscript indicates year t, in which $t = 1, \ldots, T$ is the period over which the tree-ring data are available. We will assume that the climate data are available over the period $t = t1, \ldots, T$ with $1 < t1 < T$. Note that the variable y_t could be ring-width, density or some other derived variable. When convenient, we refer to the entire time series of climate or tree data as X and Y, in which X includes the unknown part of the climate series unless otherwise specified.

Now assume that the probability of x_t given Y is a normal distribution:

$$L(X_t \mid Y) = N(f(Y), q\sigma_t^2) \tag{1a}$$

Basically, f(Y) could be any linear or nonlinear function of the tree data including a function in which y_t is a vector of multiple values from different chronologies or derived variables. We estimate σ_t^2 from the individual tree standardized data for each time period. The q-parameter is estimated from that part of the data in which both X and Y are known as described in the example application section following.

Next, we incorporate the prior information also as a normal distribution for x_t conditional on the rest of the climate data denoted as X_{-t}:

$$p(x_t \mid X_{-t}) = N(g(X_{-t}), p) \tag{1b}$$

The function, $g(X_{-t})$, could be a random walk, an autoregressive model of order r, AR(r), or any function of X_{-t} that sensibly describes the climate time series. The p-parameter is estimated from the known climate series as described later.

Finally, using Bayes rule and well-known results about normal probabilities and priors, we get the posterior distribution (Box and Tiao, 1973) for climate given the tree data:

$$p(x_t \mid X_{-t}, Y) = N(\hat{x}_t, D^{-1}) \tag{1c}$$

in which

$$D = \frac{1}{q\sigma_t^2} + \frac{1}{p} \tag{1d}$$

and

$$\hat{x}_t = \left[\frac{f(y_t)}{q\sigma_t^2} + \frac{g(X_{-t})}{p} \right] D^{-1} \tag{1e}$$

The Gibbs Sampler

A moment's reflection leads to the realization that $\hat{x}_t$ in equation (1e) can only be evaluated if $\hat{X}_{-t}$ is known. Thus, it is clear that an iterative procedure is called for that can begin with a set of starting values and converge to the optimum solution. This is provided by the Gibbs sampler (Carlin et al., 1992).

The desired solution is provided by $p(X \mid Y)$, which is the posterior distribution of the complete climate data given the tree data. The Gibbs sampler begins with a set of arbitrary starting values $X^0 = (x_1^0, \ldots, x_T^0)$, and then makes successive drawings from the conditional posterior distributions described by (1c) as follows:

$$x_1^1 \text{ from } p(x_1 \mid X_{-1}^0, Y)$$
$$x_2^1 \text{ from } p(x_2 \mid x_1^1, x_3^0, \ldots, x_T^0, Y)$$
$$x_3^1 \text{ from } p(x_3 \mid x_1^1, x_2^1, x_4^0, \ldots, x_T^0, Y)$$
$$\vdots$$
$$x_T^1 \text{ from } p(x_T \mid X_{-T}^1, Y)$$

This completes a cycle that takes X^0 to X^1. It has been proven (Geman and Geman, 1984) that X^K, for suitably large K, can be treated as a random sample from $p(X \mid Y)$, which is the posterior distribution we seek. In the example application to follow, we let $K = 1,000$ and simultaneously generate 1,000 independent outcomes at each stage of the Gibbs sampler. This makes it easy to compute the mean and variance of the reconstructed series. Simply take the 1000 independent climate reconstructions that were generated after the Kth pass of the Gibbs sampler and average them to compute a mean climate reconstruction, $\hat{X}$. Similarly, the variance for each time, t, is computed from the 1,000 time t reconstructions. We chose 1,000 for both K and the number of independent reconstructions principally because 1,000 seemed more than adequate for both values. However, this did require several hours of computation at a Unix workstation.

Hurricane Signal Detection

We have chosen several methods to illustrate that hurricanes have left their imprint on tree-ring records of baldcypress growing near the Gulf of Mexico which are 1) standardized (high frequency) ring-width series and coincidence of known hurricanes, 2) large deviations or influential data points in a growth-climate model that coincide with known hurricanes, and 3) standardized ring-width variation of old-growth vs second-growth (young) trees.

Over forty hurricanes have made landfall along the coastal area between the Atchafalaya River, Louisiana east to the Mobile River in Alabama since 1831 (Ludlum, 1963). We first noted evidence that baldcypress may contain a hurricane signal during the initial measurement of our tree-ring material from old-growth baldcypress located in the Tangipahoa River basin, Louisiana. We began to notice that many, but not all, old trees had small ring-widths the year of a hurricane and also the year after a hurricane strike (Figure 38.1). A small ring-width the year after a hurricane is more frequent than the year of the hurricane. Accurate hurricane path information for the study area is available from 1871 to 1986 (Neumann et al., 1987). (Historical accounts of hurricanes prior to 1871 back to the late 1700s are available from Ludlum (1963); the accuracy of these reports is excellent for near the coast but diminishes as the storms move inland.) For this study, we were interested in hurricanes that passed near the city of New Orleans, Louisiana. A review of hurricane paths since 1871 from Neumann et al. (1987) shows that

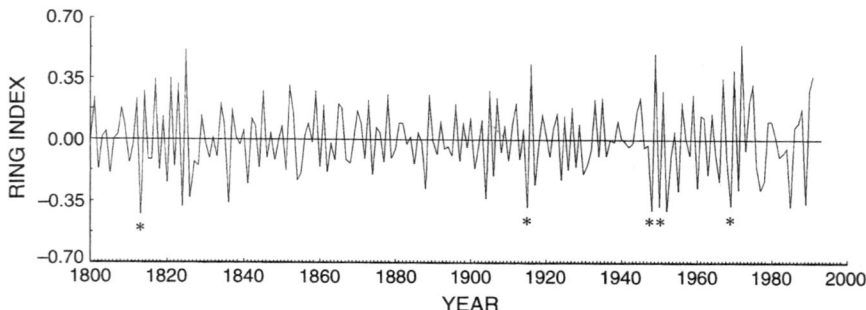

Figure 38.1. Standardized ring-width series from the Tangipahoa River basin. Large negative ring-width indices coincide with known hurricanes in 1812, 1915, 1947, 1948, and 1969.

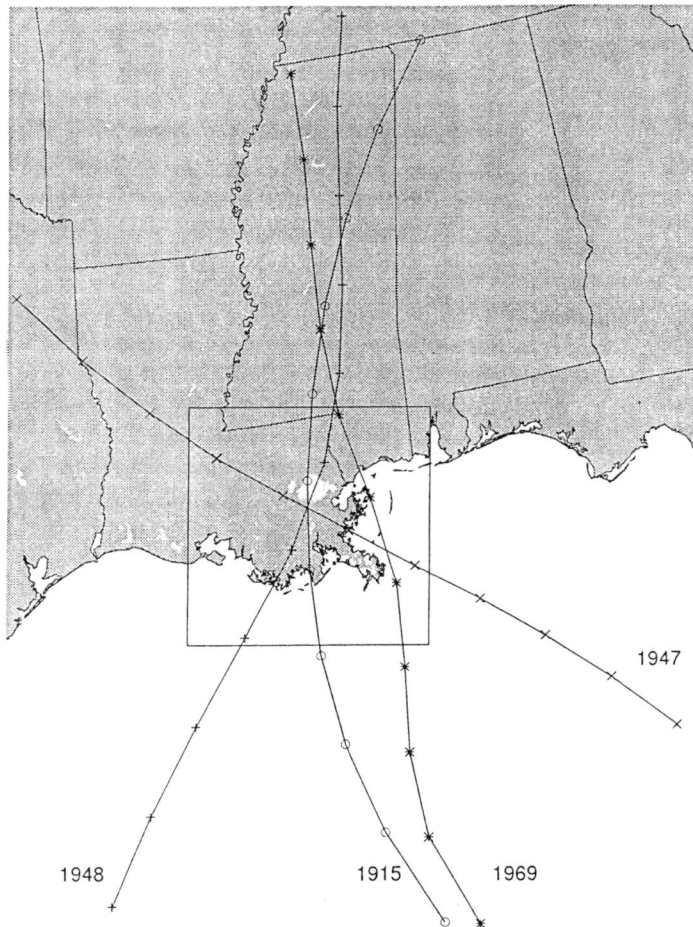

Figure 38.2. Paths of hurricanes in the vicinity of New Orleans and the Tangipahoa River and Pearl River study sites since 1915.

four hurricanes passed over New Orleans and the baldcypress forests in the Tangipahoa River study area (Figure 38.2).

In an attempt to evaluate the existence of a hurricane signal in baldcypress ring-width series we chose the following logic. We first developed a climate-tree-growth model using ordinary least squares (OLS), then looked at several regression diagnostics that provided information about influential data points and leverage points. If the regression diagnostics indicated that certain years had large residuals or leverage points, then we investigated whether these years were associated with known hurricane strikes. Prior studies have shown that standardized ring-widths of baldcypress are positively correlated with the PDSI for the month of June and spring precipitation on a wide regional basis (Stahle and Cleaveland, 1992), and we focused our efforts on investigating whether similar relationships existed for near-coastal baldcypress.

Ring widths from nineteen old-growth (> 400 years) baldcypress were measured, crossdated and then standardized using the inverse-hypersine transformation:

$$h(i) = \ln(r(i) + \sqrt{[r(i)^2 + 1]})$$

in which h is standardized ring-width, r is the actual ring-width and i is an index for years. First differences of the inverse-hypersines are nearly identical with first differences of natural logarithms, however, the transformation avoids the problem of a zero value for a ring-width returning a value of minus infinity (Van Deusen, 1990). We then developed OLS models using monthly average temperature, precipitation, and PDSI from 1895 to 1991, fit to the standardized ring-width data.

Regression Diagnostics

A valuable regression diagnostic statistic is the COVRATIO statistic, based on the ratio of the covariance matrix derived from all the data, $\sigma^2(X^TX)^{-1}$, with the covariance matrix that results when row i has been deleted, $\sigma^2[X^T(i)X(i)]^{-1}$ (Belsley et al., 1980). Because the two matrices differ only by the inclusion of the i'th row in the sum of squares and cross-products, values of this ratio near unity can be taken to indicate that the two covariance matrices are close, and the estimated coefficients are insensitive to deletion of the observation. Values farthest removed from unity indicate that the data for that specific year influence the estimated coefficients and therefore warrant further investigation.

Results and Discussion

Example Application of the Baysian Climate Reconstruction

We used cross-dated baldcypress tree-ring data from the Tangipahoa River basin just north of Lake Ponchartrain near New Orleans, Louisiana to reconstruct the PDSI for the month of June. June PDSI yields the highest R^2 value of any simple linear regression that regresses ring-widths on monthly temperature, precipitation, or PDSI with these data.

The following equation was used for calibrating the climate data to the tree-ring data:

$$f(y_t) = a + by_t \tag{2a}$$

in which the a and b parameters were estimated from that part of the data when both X and Y are known. Weighted regression was used with the weights being $1/\sigma_t^2$. The y_t chronology values were derived from taking first differences of natural logs for each core in the data set and then averaging. The q-parameter in equation (1a) was estimated as follows:

$$q = \frac{\hat{e}'W\hat{e}}{n-2} \quad (2b)$$

in which $\hat{e}$ is a vector of regression residuals, W is a diagonal matrix containing the weights and n is the number of known climate values.

The prior model for climate was an AR(2) model that was fit to the known PDSI values for the month of June.

$$g(X_{-t}) = \lambda_0 + \lambda_1 x_{t-1} + \lambda_2 x_{t-2} \quad (2c)$$

The λ-parameters in (2c) are estimated from the known PDSI values using OLS and the p-parameter in equation (1b) is the resulting mean squared error.

We initialized the Gibbs sampler with PDSI predictions from the calibration equation (2a). The usual practice for climate reconstruction would be to use these as the final values. Our approach calls for generating random variates distributed as $N(0,1)$ and converting these to $N(\hat{x}_t, D^{-1})$ by sequentially stepping through the sequence called for by the Gibbs sampler. As mentioned earlier, convergence was assumed after 1,000 iterations, and 1,000 independent samples were generated at each step of the Gibbs sampler. The programming required here is trivial and allows us to easily compute valid confidence intervals.

The reconstruction was also performed over the period of known climate to indicate the Bayesian procedure's performance vs the traditional reconstruction from the calibration equation. In fact, there was only a slight improvement with the Bayes procedure (Figure 38.3); the sum of squared deviations for the Bayes procedure vs the squared deviations for the calibration equation differed by approximately 2% over the period of known climate. Model predictions from both the traditional and Bayesian procedure underpredict the (year-to-year) extreme observed values. This is an expected and well-known outcome when using OLS procedures for parameter estimation. The 95% prediction interval (CI) computed from the Gibbs sampler, although valid, is not very encouraging (Figure 38.4). The CI indicates that any value between about -2 and $+2$ is included at nearly all years. It is important to note that this CI is somewhat smaller that the CI that would be obtained with the usual calibration equation reconstruction. The prior information has kept the CI from being even larger.

Hurricane Signal

Similar to Stahle et al. (1988) and others, we found that the PDSI for the month of June was consistently the single-most influential variable. The regression model for coastal Louisiana that relates standardized tree growth as a function of climate is:

$$\hat{Y}_t = 0.003965 + 0.06067 \text{ (June PDSI}_t) \quad (3)$$

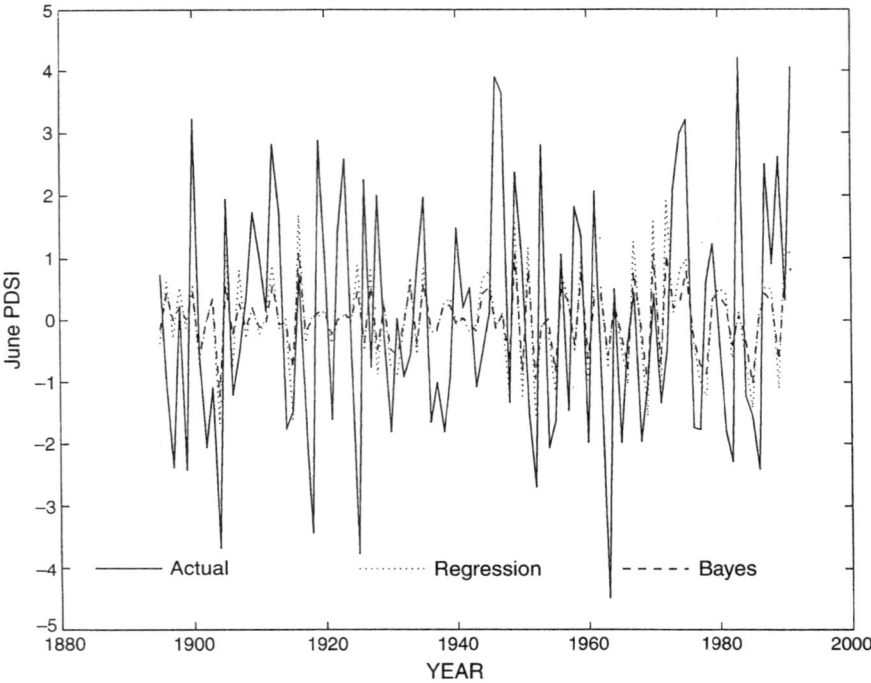

Figure 38.3. Reconstruction over the period of known PDSI for the month of June. The regression procedure results are represented by a dotted line, the Bayes results by a dashed line, and actual PDSI values by a solid line.

in which the subscript t is an index for year. Although the regression is significant at the $p < 0.003$ level, the model R^2 of .10 is unimpressive. We suspect there are several reasons for the relatively weak performance of June PDSI found here as compared with other studies involving baldcypress. First, extended droughts along the coast of Louisiana are rarer than at sites farther inland, as summer afternoon thunderstorms are the norm even during regional inland droughts. The Louisiana (Division six) climate data illustrates the point that long runs of moisture deficits are not apparent in the data available from 1895 to 1991 (Figure 38.5). Second, small ring-widths occur not only for drought years but also for the year of or after a hurricane. This probably causes a confounding of the drought signal and the hurricane signal. Such an interpretation suggests that the climate-growth relationship can change over time. Recent work in dendroecology has shown that climate-growth relationships can be dynamic and change for a number of reasons (Van Deusen, 1987; Peterson and Peterson, 1994).

Large swings between negative and positive studentized residuals occurred for years surrounding 1916, 1970, and during the late 1940s. All of these years followed known hurricane strikes (Figure 38.6). The COVRATIO residuals indicate that 1916, 1970, and the late 1940s are influential data points (Figure 38.7),

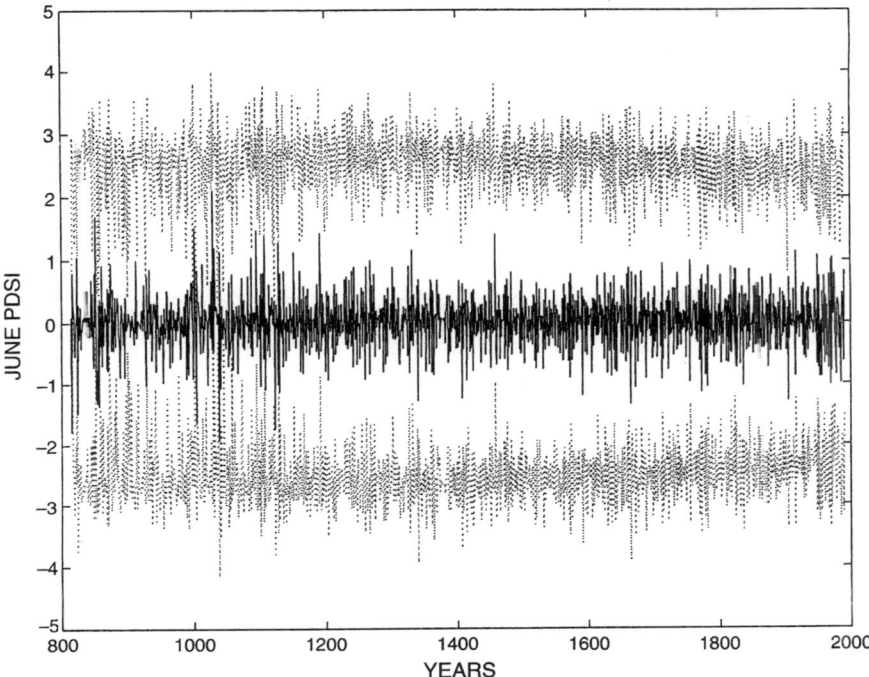

Figure 38.4. Reconstruction of June PDSI over the years 1814–1991 with 95% interval showing the range within which a reasonable climate prediction might fall for each year.

and all coincide with known hurricanes (Figure 38.2). Other such influential data points as 1905 and the mid-1950s appear to be associated with drought (negative PDSI values for June). From the two regression diagnostics, it appears that the most influential data points are associated with summer drought (noted in prior

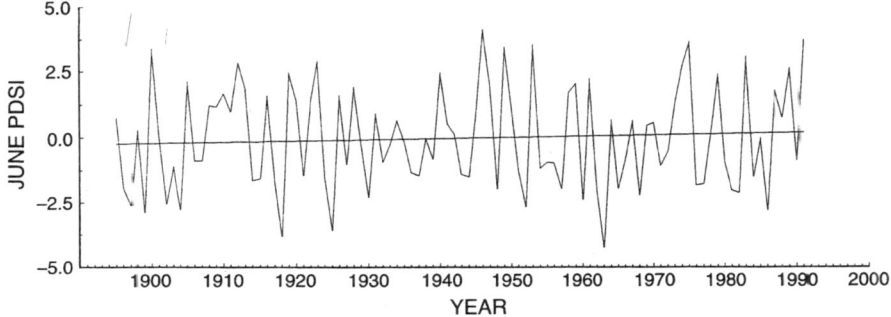

Figure 38.5. June PDSI from 1895–1991 for southeastern Louisiana. The trend line was estimated by fitting a first-degree orthogonal polynomial to the data.

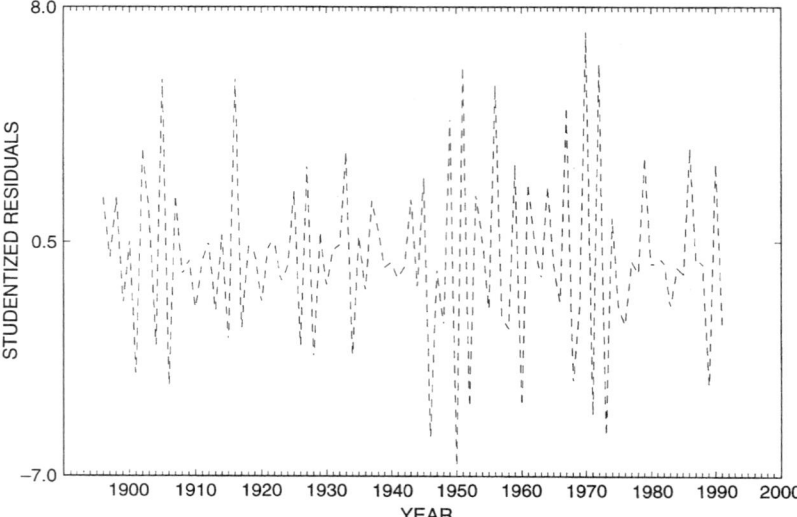

Figure 38.6. Studentized residuals from equation 1. Extreme residuals, both positive and negative, occur in the years that bound known hurricanes (1915, 1969, and the late 1940s)

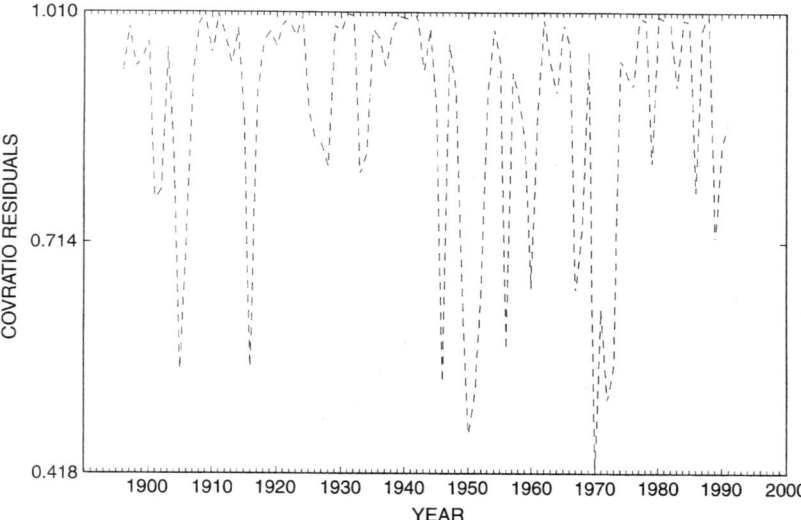

Figure 38.7. COVRATIO residuals from equation 1. Values farthest from unity are data points that heavily influence estimated coefficients. Residuals for the year 1916, 1970, and the late 1940s all coincide with known hurricanes.

baldcypress research) and hurricanes. Both phenomena result in small annual rings. Stahle and Cleaveland (1992) have presented evidence explaining the relationship between reduced ring-widths in baldcypress and drought. We suggest that the mechanism producing small ring-widths following a hurricane is foliage reduction through loss of small and large branches. Francis and Gillespie (1993) related maximum wind gust speeds during Hurricane Hugo to tree damage. They found that large trees were at greater risk than smaller trees, and that crown and bole damage began at speeds of about 60 km/hour (hr) and increased rapidly with gust speeds to about 130 km/hr. Our personal observations of baldcypress conditions in southern Louisiana following Hurricane Andrew in July and August of 1992 were similar to findings of Francis and Gillespie (1993) in that larger, older baldcypress lost individual large branches, while smaller, younger baldcypress appear to be relatively undamaged. Loope et al. (1994) also report that almost all old-growth baldcypress (> 300 years old) from the Corkscrew Swamp in southwest Florida lost upper portions of stems of major branches as a result of recent hurricanes (e.g., Andrew in 1992) and those in the past (Duever, et al., 1984). These observations appear to coincide with the lack of a clear hurricane signal in younger baldcypress when compared to old-growth cypress.

Late Spring and Summer Precipitation

Two approaches have been used by climatologists to derive scenario patterns of climatic changes that might result from an increase in atmospheric CO_2. These are 1) numerical modelling using general circulation models (GCMs), and 2) the use of past warm periods as analogues of the future (Wigley et al., 1980). Both methods have limitations, however, it is noteworthy that there is agreement between the approaches on many of the general results. These results indicate that the greatest temperature changes will occur at high northern latitudes and in the winter.

For southern U.S. forests, regional changes in precipitation patterns are among the most important consequences of climatic warming, nevertheless there is substantial disagreement as to the magnitude and direction of changes. For example, in the lower southeastern United States, studies of the instrument-based climatic record have shown a correlation between increased temperature and increased annual precipitation (Diaz and Quayle, 1980). Conversely, Wigley et al., (1980) suggest that this region would be drier on an annual basis during a warm episode. A more recent analysis by Coleman (1988) suggests a decrease in summer convection activity that would result in reduced summer precipitation from 10 to 20%. Coleman (1988) concludes that deficits greater than these may occur under full-scale climatic warming.

The baldcypress chronologies are positively correlated with the PDSI for June and, therefore, decreases in productivity of baldcypress are probable under the decreased summer precipitation scenario (Wigley et al., 1980; Coleman, 1988). The bulk of rainfall in the lower southeastern United States during June, July, August, and September results from two distinct phenomena with climatically different origins, which are 1) convection thunderstorms, and 2) tropical cyclones. If the suggested decreases in summer precipitation occur from global warming we can expect decreased productivity for baldcypress, and possible reductions in baldcypress habitat because of drier soil conditions.

How this decreased summer precipitation will effect seed production and seedling development is not known. However, it is known that floodwaters spread the scales or cones along streams and this is the most important means of seed dissemination (USDA Forest Service, 1965). Seeds usually fail to germinate on better-drained soils as a result of reduced surface water. Thus, saturated soils are needed for a period of one to three months after seedfall. After germination, seedlings cannot endure submergence by flooding, therefore, provided there is success with germination, there are some possible benefits that could result from decreased summer precipitation.

Summary

The Bayesian reconstruction procedure presented here allows for some prior knowledge about climate to be incorporated into the process. As specified earlier, the weight placed on the prior will tend to increase as the reconstruction moves further back in time. This is because the number of tree-ring observations diminishes and therefore σ_t^2 increases. It is not difficult to extend this approach to reconstruct a vector of climate values with a prior that accounts for intercorrelations among them. However, this extension is beyond the scope of this chapter.

Although the climate reconstruction in the example is somewhat disappointing, this reflects the inexact relationship between the tree-ring data and the climate data, and is not an indictment of the Bayes procedure. The Bayes procedure and the Gibbs sampler produce valid confidence intervals that are easy to produce.

For reconstructing variables other than PDSI, the climate data might need to be standardized to better meet the normal distribution assumption. The Palmer drought severity index by its nature is already standardized. There is clearly a need for further research on Bayesian climate reconstruction and the Gibbs sampler. We believe that the method will show particular promise for vector climate reconstruction because values within the same year have strong correlations that can be modeled with the prior distribution.

The use of the COVRATIO statistic for signal detection was successful in identifying both hurricane and drought events. The COVRATIO results indicate that the most influential data points affecting dendroclimatic parameter estimation are those years associated with hurricane strikes. From 1915 to 1991 there were four hurricanes that passed over the Tangipahoa River study site. Tree rings from old-growth baldcypress contain a hurricane signature that mimicked that of drought (reduced growth). Fortunately, the differences were apparent between drought and hurricane signals. We observed a notable increase in high frequency ring-width variation associated with hurricanes that does not appear around periods of drought. However, climate reconstructions based on tree-ring material from coastal areas with a relatively high frequency of hurricanes should take into account the possibility that depressed ring-widths from hurricanes might mimic those caused by drought.

Crown damage to old-growth trees seems the most probable mechanism by which passing hurricanes are recorded. Old-growth baldcypress from our study suffered from heavy crown damage and subsequent heartrot caused by a fungus that gained entry in the crown and worked downward over the centuries. To date, we have not detected a hurricane signal from second-growth baldcypress. Our observations following Hurricane Andrew in 1992 suggest that in addition to loss of needles, young baldcypress trees experience less crown damage than old-growth trees.

References

Baillie MCL, Munro MAR (1988) Irish tree rings, Santorini and volcanic dust veils. Nature 332:344–346.
Belsley DA, Kuh E, Welsch RE (1980) *Regression diagnostics: Identifying influential data and sources of collinearity.* John Wiley and Sons, New York.
Box GEP, Tiao GC (1973) *Bayesian inference in statistical analysis.* Addison-Wesley, Philippines.
Carlin BP, Gelfand AE, Smith AFM (1992) Hierarchical Bayesian analysis of changepoint problems. Appl Stat 41(2):389–405.
Coleman JM (1988) Climatic warming and increased summer aridity in Florida, U.S.A. Clim Change 12:165–178.
Conner WE, Toliver JR (1990) Long-term trends in the bald-cypress (*Taxodium distichum*) resource in Louisiana (U.S.A.). For Ecol and Manage 33/34:543–557.
Diaz MF, Quayle RG (1980) The climate of the United States since 1895: Spatial and temporal changes. Mon Weat Rev 108:246–266.
Douglass AE (1941) Crossdating and dendrochronology. J For 39:825–831.
Duever MJ, Carlson JE, Riopelle LA (1984) Corkscrew Swamp: A virgin cypress stand. In Ewel KC, Odum T (Eds) *Cypress swamps.* University Presses of Florida, Gainesville, 334–348.
Emanuel KA (1987) The dependence of hurricane intensity on climate. Nature 326:483–485.
Francis JK, Gillespie AJR (1993) Relating gust speed to tree damage in Hurricane Hugo, 1989. J Arbor 19(6):368–373.
Fritts HC (1976) *Tree rings and climate.* Academic Press, London.
Geman S, Geman D (1984) Stochastic relaxation, Gibbs distributions, and the Bayesian restoration of images. IEEE Transactions on Pattern Analysis and Machine Intelligence, PAMI-6:721–741.
Graumlich LJ (1993) A 1000-year record of temperature and precipitation in the Sierra Nevada. Quat Res 39:249–255.
Graybill DA, Idso SB (1993) Detecting the aerial fertilization effect of atmospheric CO_2 enrichment in tree-ring chronologies. Glob Biogeochem Cyc 7:81–95.
Jarrell JD, Elsberry RL (1994) The effect of global climate change on tropical cyclones. Paper presented at the 5th Global Warming Conference, San Francisco, California. 4–7 April.
Loope L, Duever M, Herndon A, Snyder J, Jansen D (1994) Hurricane impacts on uplands and freshwater swamp forests. BioSci 44(4):238–246.
Ludlum DM (1963) *Early American hurricanes 1492–1870.* Lancaster Press, Inc, Boston.

Neumann CJ, Jarvinen BR, Pike AC, Elms JD (1987) *Tropical Cyclones of the North Atlantic ocean, 1871–1986.* Historical Climatology Series 6–2. National Climatic Data Center, Asheville, NC.

Peterson, DW, Peterson DL (1994) Effects of climate on radial growth of subalpine conifers in the North Cascade Mountains. Can J For Res 24:1921–1932.

Putz FE, Sharitz RR (1991) Hurricane damage to old-growth forest in Congaree Swamp National Monument, South Carolina, U.S.A. Can J For Res 21:1765–1770.

Reams GA, Van Deusen PC (1993) Synchronic large-scale disturbances and red spruce growth decline. Can J For Res 23:1361–1374.

Rosson JF, McWilliams WH, Frey PD (1988) Forest resources of Louisiana. USDA For Serv Resour Bull, SO-130.

Schmidt H, von Storch H (1993) German bight storms analyzed. Nature 365:791.

Sheffield RM, Thompson MT (1992) *Hurricane Hugo: Effects on South Carolina's Forest Resource.* Research Paper SE-284. USDA, For Serv, Southeast For Exper Stat, Asheville, NC.

Sheppard PR, Jacoby GC (1987) Dating earthquakes along the San Andreas fault system in California. In Jacoby GC, Hornbeck JW, (Eds) *Proceedings of the international symposium on ecological aspects of tree-ring analysis.* US Department of Commerce, Springfield, VA.

Smith AFM, Gelfand AE (1992) Bayesian statistics without tears: A sampling-resampling perspective. Amer Stat 46:84–88.

Stahle DW, Cook ER, White JWC (1985) Tree-ring dating of baldcypress and potential for millennia-long chronologies in the Southeast. Amer Anti 50:796–802.

Stahle DW, Cleaveland MK, Hehr JG (1988) North Carolina climate changes reconstructed from tree rings: A.D. 372 to 1985. Science 240:1517–1519.

Stahle DW, Cleaveland MK (1992) Reconstruction and analysis of spring rainfall over the southeastern U.S. for the past 1000 years. *Bull Amer Met Soc* 73(12):1947–1961.

Stahle DW, Cleaveland MK (1993) Southern oscillation extremes reconstructed from tree rings of the Sierra Madre Occidental and Southern Great Plains. J Clim 6(1):129–140.

Swetnam TW (1993) Fire history and climate change in giant sequoia groves. Science 262:885–889.

U.S. Department of Agriculture, Forest Service (1965) *Silvics of forest trees of the United States.* USDA, Agriculture Handbook 271. Washington, DC.

Van Deusen PC (1987) Testing for stand dynamics effects on red spruce growth trends. Can J For Res 17:1487–1495.

Van Deusen PC (1990) A dynamic program for cross-dating tree rings. Can J For Res 20:200–205.

Wigley TML, Jones PD, Kelly PM (1980) Scenario for a warm, high-CO_2 world. Nature 283:17–21.

Wilhite LP, Toliver JR (1990) Taxodium distichum. In Burns RM, Honkala BH, (Eds) *Silvics of North America.* U.S. Department of Agriculture, Forest Service Agriculture Handbook 654. Washington, DC.

Williston HL, Shropshire FW, Balmer WE (1980) Cypress management: A forgotten opportunity. USDA For Rep SA-FR-8.

39. Modeling the Differential Sensitivity of Loblolly Pine to Climatic Change Using Tree Rings

Edward R. Cook, Warren L. Nance, Paul J. Krusic, and James Grissom

The Southwide Pine Seed Source Study (SPSSS) was undertaken in 1951 to determine to what extent inherent geographic variation in four southern pine species (loblolly pine, *Pinus taeda* L.; slash pine, *P. elliottii* Engelm. var. *elliottii*; longleaf pine, *P. palustris* Mill.; and shortleaf pine, *P. echinata* Mill.) is related to observable geographic variation in climate and physiography. The study's design was based on the classic common garden test design wherein all geographic sources were planted together at multiple sites across the natural range; and the fundamental objective was to test the widely accepted hypothesis that local seed sources were uniformly better adapted and faster growing than nonlocal seed sources from the same species (complete study details appear in Wells and Wakeley, 1966).

Recently, there has been increasing interest in the SPSSS and other similarly designed studies because these studies offer long-term data that could be useful in assessing genetic sensitivity of tree species to climatic effects. The common garden design allows a comparison of the responses of different individual genotypes and seed sources to the same climatic regime at one common site. Moreover, the existence of many planting sites—all with the same seed sources colocated—provides an opportunity to assess the effect of changing climatic regimes on the same seed sources. Thus, the tree-ring analysis of the SPSSS could provide unique information about the sensitivity of the four southern pine species to future climatic changes resulting from greenhouse warming in the southeastern United States (e.g., Rind et al., 1990).

In this chapter, we examine this potential for one of the SPSSS species: loblolly pine. Specifically, we hope to determine the degree to which different seed sources located at the same plantation differ in their responses to the same local climate regime. Any identified differences could be the result of local adaptations of the seed sources being investigated. In turn, these indicated local adaptations might be used to determine these seed sources that are apt to perform best under various scenarios of future climatic change.

The Southwide Pine Seed Ssource Study Plantations

Originally, the loblolly pine portion of the SPSSS consisted of fifteen provenance plantations containing between eight and fifteen seed sources collected from locations across the natural range of the species. Of the fifteen original plantations, only eight have survived to the present time. Figure 39.1 shows the locations of these eight surviving plantations, along with the locations of the seed sources used. From this map, it is apparent that this subset of original plantations still covers most of the geographic range of loblolly pine. A key to these plantations is also provided in Table 39.1, by both original SPSSS plantation code and geographic name. Similarly, a key to the seed sources is provided in Table 39.2, again by original SPSSS code and geographic name. These SPSSS plantation and seed source codes will be used throughout this paper.

The seed sources and plantings used in the SPSSS were actually divided into two series, Series–1 and Series–2, mainly to avoid the overwhelming task of planting large plots of all fifteen seed sources at all locations (Wells, 1983). The Series–1 plantings are found in all but one of these plantations, the exception being located in northern Mississippi (see Figure 39.1), which only has Series–2 trees. In contrast, Series–2 trees are present at only five of the eight surviving plantations. Thus, four of the eight plantations have both series present.

At each plantation, each seed source was planted in four randomized complete blocks containing 121 trees in an 11 × 11 grid (Wakeley, 1961; Wells and Wakeley, 1966). The inner forty-nine trees laid out in a 7 × 7 grid were used for remeasurement, with the remaining trees were used as border or buffer trees between plots. Over the years, a large amount of natural mortality and some prescribed thinning occurred. Therefore, the number of trees ultimately sampled for tree-ring analysis was a small fraction of the original total planted. This fact should not be viewed as a drawback, however, because the sampled trees were the survivors of a (largely) natural winnowing-out process that occurs in natural, unmanaged forests as well.

Summaries of mortality, growth, and yield have been published several times over the past forty years (Wakeley, 1953, 1959, 1961; Wells and Wakeley, 1966; Nance and Wells, 1981; Wells, 1969, 1983), as well as reports on insect and disease data (Henry, 1959; Henry and Coyne, 1955; Henry and Hepting, 1957; Wells and Switzer, 1975). These data generally show clear genetic differentiation between geographic sources in response to major climatic and physiographic

39. The Differential Sensitivity of Loblolly Pine to Climatic Change

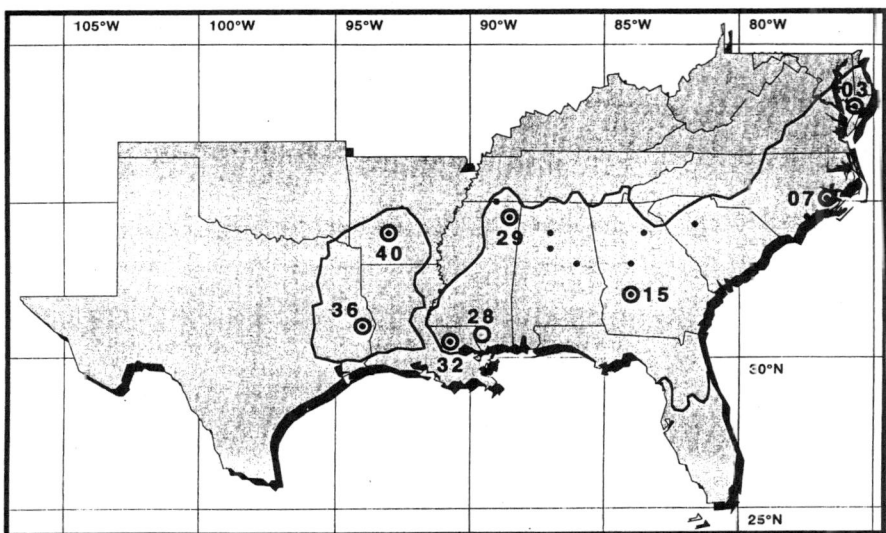

Figure 39.1. A map of the SPSSS plantation and seed source locations. The large open circles are the eight existing plantations sampled for this study. (See Table 39.1 for the plantation codes.) The small, filled circles are the locations of the fifteen seed sources used in the SPSSS. (See Table 39.2 for a listing of those sources.) When a small filled circle falls inside a large open circle, that plantation has a local seed source. The irregular lines on the map delineate the general range boundaries of loblolly pine.

effects, with much smaller amounts of genetic variation within the major climatic and physiographic regions.

The original data collections on the SPSSS were spaced at five-year intervals, which is generally not frequent enough for detection of climatic effects in the southern and southeastern United States. However, it was possible to obtain increment cores from the living trees in the study and obtain direct measurements of annual radial increment. These measurements provided the basis for the data presented and analyzed in this chapter.

Table 39.1. The SPSSS Loblolly Plantations

Plantation Code	Geographic Region
03	Maryland, eastern
07	North Carolina, eastern
15	Georgia, southwestern
28	Mississippi, southern
29	Mississippi, northeastern
32	Louisiana, southeastern
36	Texas, eastern
40	Arkansas, southwestern

Table 39.2. The SPSSS Loblolly Pine Seed Sources[1]

Seed Source	Series	Geographic Region
301	1	Maryland, eastern
303	1,2	North Carolina, southeastern
305	1	North Carolina, eastern
307	2	South Carolina, western
309	1	Georgia, southwestern
311	2	Georgia, northeastern
315	1	Alabama, northern
317	2	Alabama, northeastern
319	1	Alabama, northern
321	2	Mississippi, northeastern
323	1,2	Louisiana, southeastern
325	1	Texas, eastern
327	1,2	Arkansas, southwestern
329	2	Tennessee, western
331	2	Georgia, northwestern

[1] Note the three common seed sources in the two Series.

Climatology of the Southwide Pine Seed Ssource Study Plantations

To place this study in its proper climatological context, comparisons of plantation monthly precipitation and temperature climatologies were made. Figure 39.2 shows the mean monthly maximum and minimum temperature and total monthly precipitation profiles for the eight SPSSS plantations. These monthly profiles are based on 1° × 1° grid-point data from the Richman-Lamb climatological database (Lamb, 1987), covering the period from 1949 to 1988. The grid-points closest to the plantation locations were used. In the case of plantations 28 and 32, the closest grid-point fell roughly equidistant between the two. Consequently, the same climate data were used for each of these plantations.

The maximum temperature profiles (Figure 39.2A) indicate a temperature range of 6 to 16 °C in January, and 30 to 34 °C in July, across all plantations. This indicates generally higher variability in winter maximum temperatures across the plantations, a result consistent with continentality of climate. The profiles also reveal a surprising degree of warm-season concordance. That is, for six of the eight plantations, there is little difference in maximum temperatures during the warm-season months of May to September, the season when the most radial growth of loblolly pine should occur. For the six warmest plantations, warm-season maximum temperatures average at approximately 32 °C. In contrast, the two anomalous plantations, 03 and 07, are the most northerly plantations of the group and have warm-season temperatures that are 3 to 4 °C cooler on average. Only for the cool-season months of November to March is there a clear separation of the plantations into essentially three groups, which are 1) 15, 28, 32, 36; 2) 07, 29, 40; and 3) 03, ranked from warmest to coldest, respectively. This stratification is roughly by latitude, with the warmest cool-seasons occurring at the most south-

39. The Differential Sensitivity of Loblolly Pine to Climatic Change   721

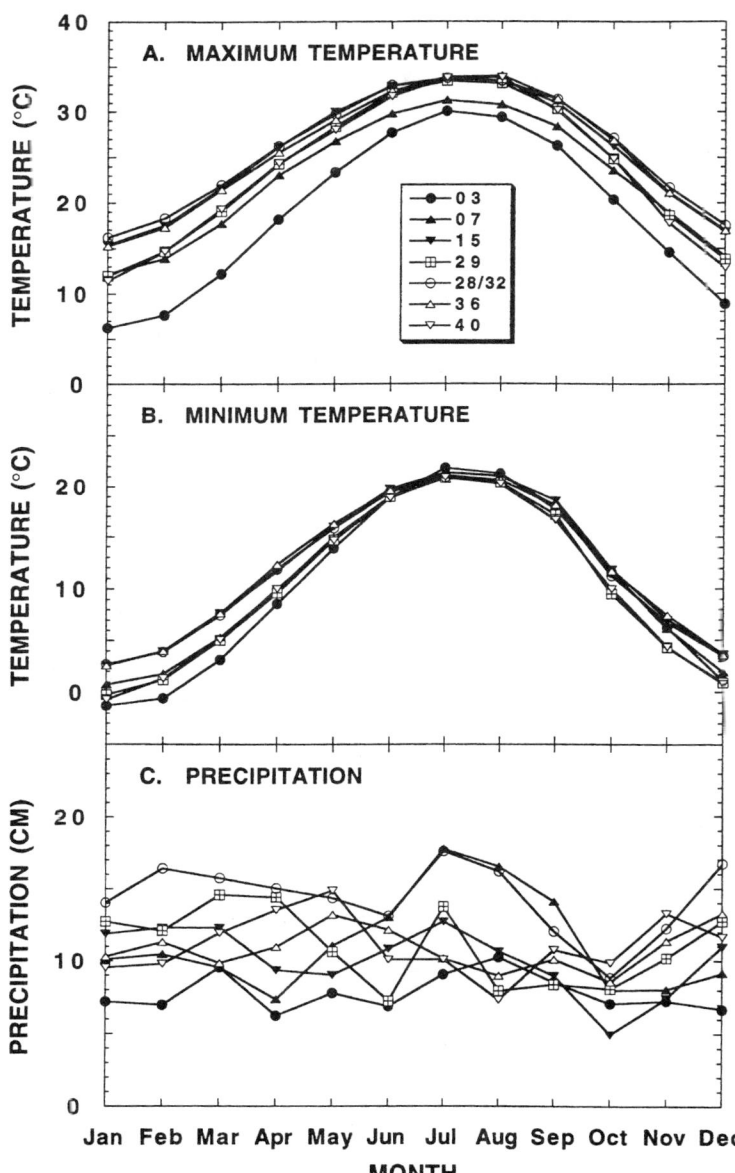

Figure 39.2. The mean monthly maximum and minimum temperature and total monthly precipitation profiles of the eight SPSSS plantations.

erly plantations. As before, plantation 03 is most anomalous, with cool-season temperatures that average 5 to 7 °C below the other plantations. The significance of this phenomenon will become apparent in the climate modeling of the tree-ring series, described later in this chapter.

The minimum temperature profiles (Figure 39.2B) reveal far less variability between plantations. None of the plantations separate out during the growing season months. Only for the cool-season months of January to March is there a clear separation by latitude into the three described groups, but this separation is much smaller for minimum temperatures. Additionally, only in January do any of the plantation minimum temperatures fall marginally below the freezing mark.

The precipitation profiles (Figure 39.2C) indicate that rainfall is evenly distributed throughout the year across all plantations. Only plantation 07 has a regime that is weakly warm-season dominant. Plantation 03 is the driest with an average of about 8 cm/month; plantations 28 and 32 are the wettest with 14 cm/month. All other plantations receive at least 10 cm/month of rainfall.

From this analysis, it is clear that the SPSSS loblolly pine plantations are located in generally warm, moist environments. The most anomalous is plantation 03 located at the northern limit of loblolly pine distribution, which is comparatively cool, dry. Given this exception, the lack of any strong latitude-based differences in climate during the warm-season/growing-season months suggests that loblolly pine "chooses" to grow in a reasonably homogenous regional climate regime (i.e., warm and moist). This means that it could be difficult to find strong differences in the strength of the climatic response in the tree rings, either within or between plantations, because the climate variables influencing growth may be equally limiting to the various seed sources across the range. However, this does not rule out significant differences in which climate variables are most influential on growth because of changing site characteristics (e.g., site hydrology, soil type, fertility) and as the plantation climatologies change geographically (cf. plantations 03 and 07 with the others).

The Southwide Pine Seed Source Study Tree-Ring Database

Between 1952 and 1953, a total of 18,718 loblolly pine trees were planted on the fifteen original SPSSS plantations. Through attrition, by natural and anthropogenic causes, both plantations and trees suffered significant losses. Thus, in the eight plantations surviving today, only 1,634 trees remain. These remaining plantation trees were completely sampled for increment cores between 1990 and 1991.

Two increment cores, diametrically opposed to each other, and passing as near as possible through the pith, were collected from every surviving tree. This was facilitated by the use of a gasoline-powered increment borer that was able to extract a full-diameter, 5-mm core from a tree in less then thirty seconds. Because the objectives of this study emphasized tree growth over the entire period of the plantations since establishment, cores were collected from as low on the stem as possible.

In the lab, the increment cores were processed using standard dendrochronological techniques (e.g., Stokes and Smiley, 1968; Fritts, 1976; Cook and Kairiukstis, 1990). The cores were firmly glued into grooved sticks with the long axis of the cells oriented vertically, sanded to a high polish, and the ring widths carefully cross-dated (Krusic et al., 1987). After measurement to a precision of ± 001 mm, the ring-width series were checked for cross-dating quality using program COFECHA (Holmes, 1982). Each seed source collection and plantation was processed independently of all others to ensure that the dating and measuring procedures were unbiased. Although some tree-ring data were available prior to 1960, a combination of planting shock and juvenile growth effects made the pre-1960 ring-widths highly erratic between trees. Therefore, all analyses presented here only used tree-ring data since 1960.

Trees of each seed source, from all eight plantations, were separated into two distinct stand-canopy classes. Dominant and codominant trees of a seed source were grouped as one class, and those remaining were grouped into a subdominant or suppressed class. The selection criterion for the purposes of this partition was tree diameter, with the five largest diameter trees from each plot considered the dominant-codominant trees. (Tree heights were not available at the time of sampling.) This number was justified by noting that it approximated a stocking level of 40 trees/hectare, which is typical for stands of mature loblolly pine. In the plots that had more than five surviving trees, those smaller than the five largest were considered subdominant or suppressed. Only the dominant-codominant trees were used in the subsequent tree-ring analyses on the basis that these are the ultimate survivors that truly matter. Another rationale for deleting the subdominant-suppressed trees was the way in which radial growth became extremely compressed in some of those trees for some years, which was not the case in the dominant-codominant trees growing on the same plot. Consequently, there was concern that the climate signal in the tree rings of the subdominant-suppressed trees might be confounded by competition-related effects.

Table 39.3 has the tally of cores and trees per plantation that fell into the dominant-codominant category used here. The total number of trees is 1,537, which is 94% of the total trees available. Therefore, little information was lost, in any event, by deleting the trees considered subdominant or suppressed.

Table 39.3. Southwide Pine Seed Source Study Loblolly Pine Plantation Series–1 and Series–2 Dominant-Codominant Core Collection

Plantation code	#Cores	#Trees
03	280	218
07	456	289
15	278	170
28	273	161
29	223	126
32	419	249
36	137	72
40	454	252
Total	2,520	1,537

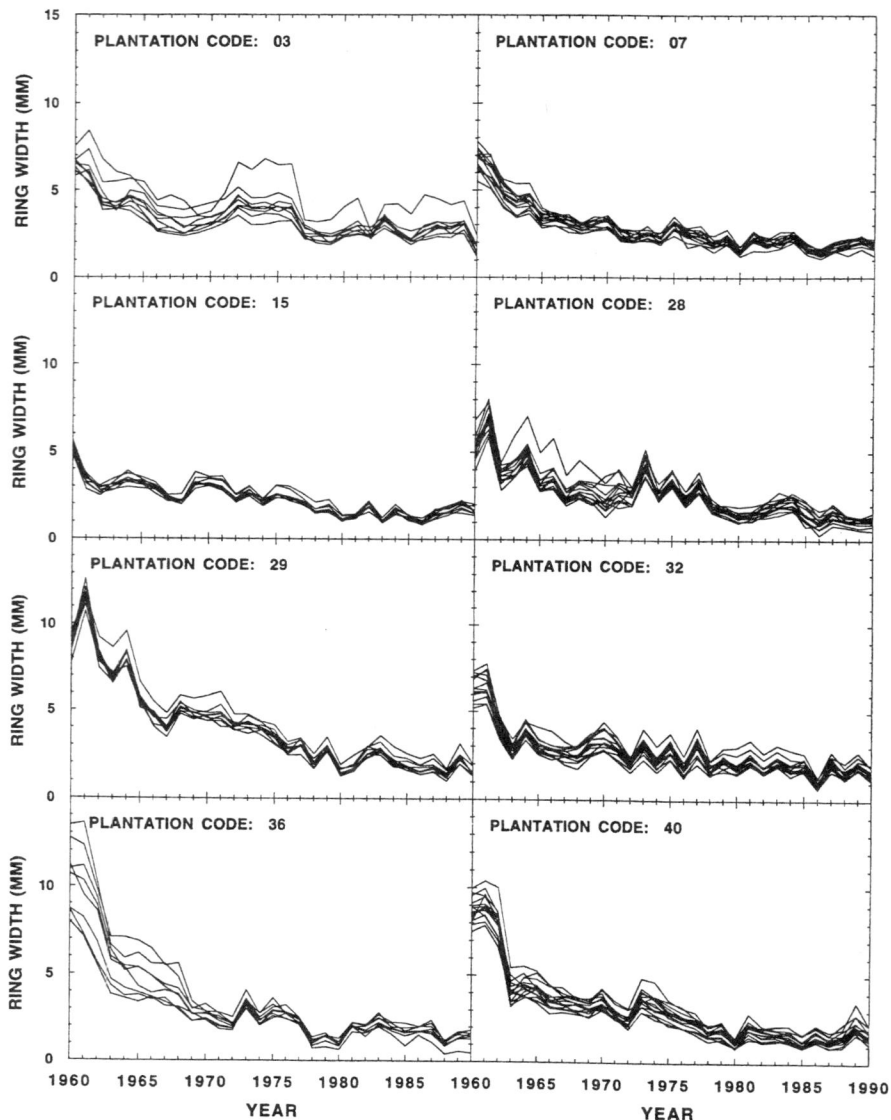

Figure 39.3. The seed source mean ring-width chronologies for each of the eight SPSSS plantations. Note the generally high level of conformity between seed sources at each plantation.

The Southwide Pine Seed Source Study Loblolly Pine Tree-Ring Chronologies

Figure 39.3 shows plots of the mean dominant-codominant ring-width chronologies for all seed sources present at the eight plantations. The overlays of the seed source chronologies are intended to illustrate the degree of homogeneity in the overall trajectory of radial growth within each plantation. With the exceptions of plantation 36, which shows considerable variation between seed sources up to 1970 but excellent convergence thereafter, and the odd seed sources in plantations 03, 28, and 29, the within-plantation seed source mean ring-width chronologies are remarkably similar. The odd behavior of individual seed sources in plantations 03, 28, and 29 may be the result of a combination of genetic and silvicultural factors that affected changes in stocking level and, consequently, growth rate over time. For example, the anomalous seed source in plantation 03 maintained a higher radial growth rate presumably because it had the lowest stocking level of any of the seed sources. This could have occurred from a combination of a higher rate of natural self-thinning and prescribed thinning. Regardless, such factors could wholly obscure any differences in growth resulting from seed source-related differential responses to climate. Consequently, it is necessary to remove absolute growth-rate effects from the tree-ring data.

The removal of absolute growth-rate effects was accomplished by modeling the trajectory of each individual ring-width series with a modified negative exponential curve of the form:

$$G_t = ae^{-bt} + k \tag{1}$$

in which G_t is the growth-curve estimate, a is the intercept, b is the slope, k is the asymptotic growth rate for over-mature trees, and t is time in years (Fritts et al., 1969). An examination of the mean ring-width plots in Figure 39.3 indicates that this model is reasonable for estimating the curvilinear growth trends apparent in the data. So, a modified negative exponential curve was fit to each individual ring-width series and the growth trend removed as:

$$I_t = R_t/G_t \tag{2}$$

in which R_t is the actual ring width, G_t is the growth curve value, and I_t is the resultant tree-ring index, for all years $t = 1,n$. This process of detrending and transforming the tree ring into dimensionless indices is known as "standardization" (Fritts, 1976) because it tends to equalize the growth variations of trees over time regardless of age, size, or absolute growth rates. Tree-ring indices have a defined mean of 1.0 and typically fall in the range of 0 to 2.

Figure 39.4 shows the mean tree-ring index chronologies for the eight plantations, as in Figure 39.3. The growth trends apparent in the raw data are clearly gone, along with much of the scatter in some of the plantations. For example, the scatter in the mean ring-widths of plantation 36 prior to 1970 is now gone. The

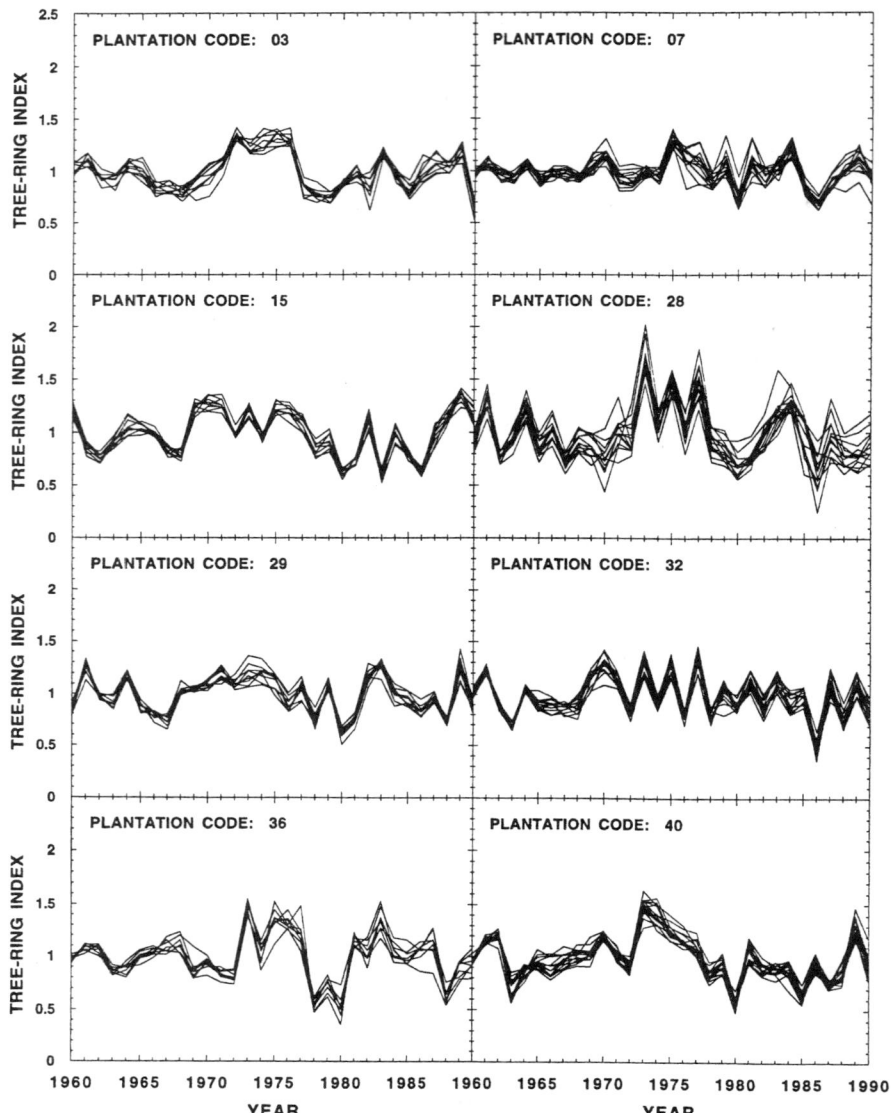

Figure 39.4. The seed source standardized tree-ring chronologies for each of the eight SPSSS plantations. These series were obtained after removing the long-term trends in radial growth from the individual ring-width series. Note that most of the differences between seed sources indicated in Figure 39.3 are now greatly reduced.

seed source anomalies in the ring-width series from plantations 03 and 29 are now also gone. The only clear inflation of differences from ring-widths to indices is evident in plantation 28, principally after 1984. This is related, in part, to the rapid and highly variable growth in the mid-1970s, which caused the end-fitting of the negative exponential curve to be more variable.

The tree-ring indices in Figure 39.4 will be used to ascertain the degree to which differential climate responses exist both within and between plantations. However, before proceeding with the climate modeling, a comparison of certain descriptive statistics will be done. In dendrochronology, four descriptive statistics are frequently computed for interpretive purposes. They are 1) mean sensitivity, 2) standard deviation, 3) serial correlation, and 4) mean between-series correlation.

Mean sensitivity (ms) is a measure of high frequency or year-to-year variability in tree-ring series. It is computed as:

$$ms = \frac{1}{n-1} \sum_{t=1}^{n-1} \left| \frac{2(x_{t+1} - x_t)}{x_{t+1} + x_t} \right| \tag{3}$$

in which, x_t is the tree-ring value for year t. Mean sensitivity has the interesting property that it assumes effectively the same value whether computed from raw ring-widths or from the same series after standardization to tree-ring indices. This is because it emphasizes the high frequency component of the time series only. The numerator is a first-difference operator, which is insensitive to all but the year-to-year changes in growth. Traditionally, ms has been used as a qualitative tool for estimating the relative sensitivity of a tree-ring series to climatic, environmental influences. High ms values are indicative of trees that are highly "sensitive" to yearly changes in growth-limiting influences. For our purposes, it is used to compare within- and between-plantation tree-ring variability in an effort to see if any unusual differences in "sensitivity" can be found.

Standard deviation (sd) is a classical statistical measure of variability. It is computed as:

$$sd = \sqrt{\frac{1}{n-1} \sum_{t=1}^{n} (x_t - \bar{x})^2} \tag{4}$$

in which $\bar{x}$ is the arithmetic mean of series x. Different from ms, sd measures variability in a tree-ring series at all time-scales and therefore, it is sensitive to low frequency, multiyear changes in growth as well, which is not the case for ms. In general, $sd > ms$ when positive autocorrelation is present in the series, as is usually the case with tree rings.

Serial correlation (r_1) is a measure of the year-to-year persistence in growth. As such, it is an expression of the physiological preconditioning (Fritts, 1976) that a tree goes through when climatic and environmental influences during one year affect the potential for growth in subsequent years. It is computed as:

$$r_1 = \frac{\sum_{t=2}^{n}(x_t - \bar{x})(x_{t-1} - \bar{x})}{\sum_{t=1}^{n}(x_t - \bar{x})^2} \qquad (5)$$

In tree-ring series, r_1 is usually positive and in the range $0 < r_1 < 1$, meaning that above-average growth in one year tends to promote above-average growth the following year, and vice versa. In reality, r_1 is only a rough, first-order estimate of chronology persistence. It is well-known that tree-ring chronologies often have more complex persistence structures that are well-modeled as higher-order, autoregressive-moving average processes (Box and Jenkins, 1976). However, as a simple descriptive statistic of persistence in tree-ring chronologies, r_1 is sufficient for our purposes.

Mean sensitivity (*ms*), *sd*, and r_1 are roughly related in the following ways. When r_1 goes up, *ms* goes down, with the converse also true, across the domain $-1 < r_1 < 1$. Also, *sd* is a complex interaction between *ms* and r_1 as each contributes to different aspects of the overall variability expressed in *sd*, especially when $r_1 > 0$.

The mean between-series correlation ($\bar{r}_{bt}$) is a measure of the strength of the common signal between trees. It is computed as:

$$\bar{r}_{bt} = \frac{\sum_{i=1}^{m-1}\sum_{j=i+1}^{m} r_{ij}}{m(m-1)/2} \qquad (6)$$

in which r_{ij} is the correlation between tree i and j and m is the number of trees. When more than one tree-ring series is available per tree, the number of between-tree correlations is increased accordingly in computing $\bar{r}_{bt}$. Because tree-ring series are cross-dated before being used in mean chronologies, $\bar{r}_{bt}$ is always in the range $0 < \bar{r}_{bt} < 1$. The mean between-series correlation is an unbiased estimator of the percent variance in common between tree-ring series (Wigley et al., 1986), and, in this sense, is a measure of the strength of the common climatic-environmental signal contained in the record. In the context of this chapter, it serves two purposes. First, it indicates the homogeneity of the within-seed source common signals in each plantation. Second, it indicates the similarity of the common signal strength between plantations. At times, $\bar{r}_{bt}$ has been advocated as an indirect measure of the strength of the climatic signal in tree-ring series. This is based on the argument that as climate becomes more limiting to growth, $\bar{r}_{bt}$ should increase because the trees will be forced to grow more similarly. Although heuristically appealing, the use of $\bar{r}_{bt}$ for this purpose is often disappointing when compared to the "goodness-of-fit" of climate models based on meteorological data.

39. The Differential Sensitivity of Loblolly Pine to Climatic Change   729

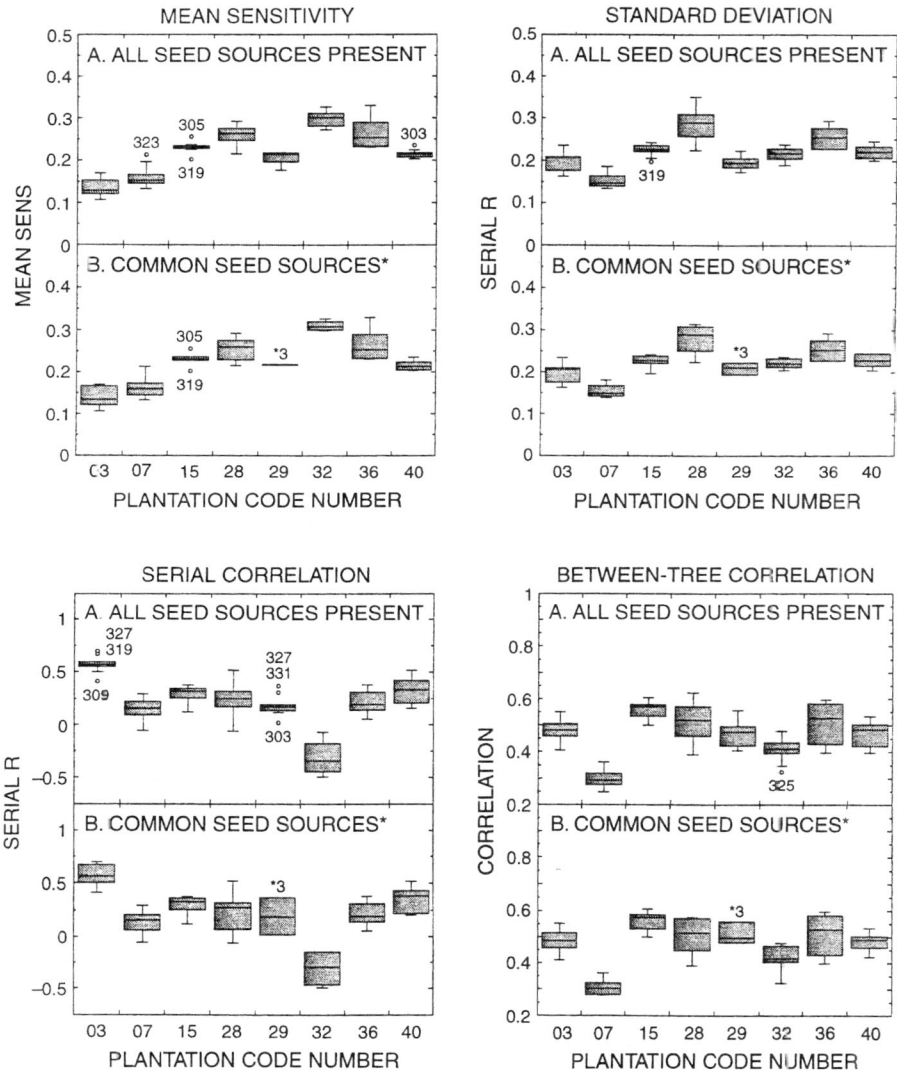

Figure 39.5. Boxplots of tree-ring chronology statistics described in the text. The boxplots were constructed for all seed sources present and for only those in common among all plantations. Note that there is not much difference between the boxplots.

These statistics are compactly displayed as a series of boxplots for all plantations in Figure 39.5. Each boxplot locates the median data value as the horizontal line through the box. The top of the box is the upper quartile (UQ; the data value halfway between the median and maximum value), and the bottom of the box is the lower quartile (LQ; the data value halfway between the median and minimum value). The length of the box is the interquartile distance (IQD) or UQ-LQ, which contains 50% of the sample values. The lines extending above UQ and below LQ are the limits of the data that do not exceed UQ + 1.5 × IQD and LQ − 1.5 × IQD, espectively. The open dots are regarded as outliers that exceed the upper or lower 1.5 × IQD limits. For each statistic, boxplots were made for all seed sources present at each plantation (a), and for only the eight seed sources common to all plantations (b). The exception to the latter is plantation 29, which only has three of the eight common seed sources because it is made up of only Series–2 trees. The common seed sources are 301, 303, 305, 309, 319, 323, 325, and 327. The common seed source boxplots may provide clearer comparisons between plantations by keeping the seed sources constant.

The boxplots indicate considerable variability in the four statistics both within and between plantations. The variation in seed source statistics within each plantation appears to be consistent with the level of visual scatter seen in some of the seed source tree-ring chronologies shown in Figure 39.4. For example, all of the boxplots for plantation 15 are uniformly narrow, which is consistent with the excellent visual agreement between the seed source chronologies (Figure 39.4). Conversely, all the boxplots for plantation 36 are consistently wider and probably reflect the lesser agreement between seed source chronologies for that plantation. Other plantations give more ambiguous results, witness the narrow *ms* and wide r_1 boxplots for plantation 32, even with apparently excellent agreement between chronologies. Given the random variability associated with the estimation of such statistics based on only thirty-one observations, it is not possible to determine if such effects are related to differences in seed source genetics.

In contrast, the boxplot comparisons between the plantations indicate the possibility for some plantation-level differentiation. For example, the *ms* statistics reveal that plantation 03 and, to a lesser degree, plantation 07 have less year-to-year ring-width variability than the other plantations. This may be related to the somewhat cooler maximum temperatures at these sites described earlier, which may reduce the development and severity of internal moisture stress in the trees. In contrast, plantation 32 has the highest *ms,* although it does not stand out greatly. The *sd* results are less clear cut, with only plantation 07 maintaining somewhat lower overall variability compared to the rest. Plantation 28 is marginally the most variable as a result in part of the mid-1970s growth pattern described earlier. However, the *sd* results are partly confounded by variations in r_1 across the plantations. The clearest example is plantation 03, where *sd* increased relative to plantation 07 even though it has the lowest *ms*. This result occurred because plantation 03 has the highest r_1 among all plantations and is, again, an anomalous unit. As will be shown later, this is almost certainly caused by a distinctly different response to climate compared to the other plantations. In terms of r_1, the other odd plantation is 32. In this case, $r_1 < 0$, a highly unusual result in dendrochronology. The reason for this phenomenon is unknown. The $\bar{r}_{bt}$ boxplots are reasonably uniform across plantations. Only plantation 07 has an anomalously low $\bar{r}_{bt}$, meaning that there is unusually high variability between trees, perhaps caused by high variability between the four plots per seed source. This result thus suggests that the North Carolina plantation plots are not homogeneous with regards to local site conditions. Otherwise, the SPSSS plantations appear to have comparable levels of plot homogeneity.

These classical dendrochronological statistics have revealed some evidence for differentiation between plantations that is at least consistent with some differences in the plantation climatologies (i.e., plantations 03 and plantation 07). Plantation 32 is also anomalous, but for reasons that are not presently explicable.

A More Detailed Look for Seed Source Differences

Although the boxplot results do not, in general, suggest strong differences between seed sources, it is still worth looking more carefully for these effects. The boxplots are rather blunt statistical tools that may be obscuring some true, albeit small, differences. First, we will examine the degree of similarity between the seed source chronologies using principal components analysis (PCA; Cooley and Lohnes, 1971). This will be followed by a very detailed linear modeling exercise using a mixed-effects analysis of variance (ANOVA) model that explicitly utilizes all components of the original randomized complete box design of the SPSSS experiment.

Principal Components Analysis

Principal components analysis was carried out on the seed source chronologies of each plantations. Based on the visual similarities of the chronologies in Figure 39.4, it was anticipated that the majority of the variance would be in common. However, PCA has the capability of decomposing the total variance into orthogonal modes of unique covariance, which could be seed source related. Thus, even though the first dominant mode may explain the majority of the variance in the seed source chronologies, it is possible that significant higher-order seed source modes might also be present.

The results of the PCAs confirmed the visual similarities between the seed source chronologies. In every case, the first PC, which accounts for the most common mode of variation among all series, explained 81.1 to 94.4% of the total variance. In contrast, the second PC, which accounts for the next most common mode of variation among all series, explained only 1.9 to 5.6% of the total variance, a result not statistically significant ($p < .10$) using a Monte Carlo testing procedure (Preisendorfer et al., 1981). All remaining higher-order PCs were similarly not significant. The plantation with the highest common seed source signal was plantation 15 (94.4%), followed by 29 (92.6%), 40 (92%), 32 (91.5%), 03 (90.9%), 36 (90.6%), 28 (85.2%), and 07 (81.1%). All of these figures are markedly higher than the $\bar{r}_{bt}$ results in Figure 39.5., in which the average over all plantations is 45.4%, with a range of 24.9 to 62.3%. Thus, there is considerably more variability between trees within provenances than between mean seed source chronologies within plantations. This fact would seem to work against finding seed source level differences in the SPSSS loblolly pine tree-ring data.

Analysis of Variance

The SPSSS employed a randomized complete block design. Specifically, each block consists of eight or nine seed sources composing the particular Series represented at the plantation. Each provenance within a block consists of the surviving individual trees of the forty-nine (7 × 7) planted spaces. Finally, each individual tree with a provenance consists of the one or two radial tree-ring cores sampled from that tree. This rigorous experimental design facilitates a detailed ANOVA components within and between seed sources using ANOVA techniques. In so doing, the inherent error structure of the randomized complete block design can be properly exploited.

The SPSSS was actually composed of two separate plantings: Series–1 and Series–2. Because the Series–1 planting was the most successful in terms of survival rate and is also present in seven of the eight existing plantations, the decision was made to only use those tree-ring series in the ANOVA. As before, to avoid the possible bias of suppressed trees in the results, the data from only the five largest dominant-codominant trees per plot were used.

The ANOVA was formulated to test for differences between seed sources resulting from climate. The SAS general linear model procedure (SAS, 1985), which allows for unbalanced experimental designs, was used for this purpose. All treatments and their interactions were assumed to be random except for provenances, which were assumed to be fixed. This test was conducted on the tree-ring series after they were first transformed to stabilize the variance, detrended to remove long-term growth trends, and prewhitened to remove autocorrelation. The ANOVA proceeded in a sequential fashion. First, the variance resulting from the endogenous treatments implicit in the randomized complete block design were isolated as sources of variation in the model (Table 39.4). Hence, the original error structure of the experimental design was explicitly evaluated before any climate effects on radial growth were tested. The incorporation of climate effects in the model was designed to maximize the correlation with the tree-ring index, and therefore to maximize information from the available climate data.

The climate index was formulated as a multiple linear regression model predicting tree-ring index, given that the variance resulting from all design components had been factored out. Symbollically, therefore:

$$CI = TRI - (Block + Provenance + Plot + Tree + Radius) \qquad (7)$$

in which CI = climate index, TRI = tree-ring index, Plot = Block × Provenance, Tree = tree within Plot, and Radius = radial growth series within Tree. The CI was generated using a stepwise regression technique, with exogenous variables of monthly temperature, precipitation, and Palmer Drought Severity Index (PDSI) for both present and prior growing seasons. The method selected exactly six variables that were maximally correlated with the TRI. The single CI for each plantation represented a common climatic signal among all tree-ring series and provenances. This CI was entered into the linear model as a covariate to remove the common climatic signal before testing for interactions between CI and seed source.

39. The Differential Sensitivity of Loblolly Pine to Climatic Change

Table 39.4. Example of Analysis of Variance Results for Plantation 28

Factor	% Model SS	Prob > F
Block	< 1.0%	0.95
Provenance	3.5%	0.16
Plot (block × prov)	1.3%	0.32
Tree within plot	15.9%	0.0001**
Radius within tree	4.4%	0.0001**
Climate index (CI)	69.3%	0.0001**
CI × block	< 1.0%	0.65
CI × provenance	1.4%	0.0001**
CI × plot	< 1.0%	0.23
CI × tree within plot	2.7%	0.30
CI × radius within tree	< 1.0%	0.99

** significant at the 1% level

The ANOVA just described was applied to all seven plantations containing Series–1 plantings. This represents all but plantation 29 in northern Mississippi. In five of the seven plantations, a significant CI × Provenance interaction remained in the residual tree-ring chronologies after the design variables and common climatic signals were removed. The two plantations not showing a significant CI × Provenance interaction were 40 and 32. The negative result for Plantation 40 was unexpected given its extreme western location. Regardless, these results suggest a differential response of these loblolly pine provenances to the same set of climatic conditions. Table 39.4 provides a detailed breakdown of the model results for plantation 28 as an example. The variance accounted for by the CI × Provenance interactions, although statistically significant in most cases, always accounted for less then 2% and usually < 1% of the total variance of the overall ANOVA models of the plantations. The statistical significance of such small percentages is caused by the very large degrees of freedom available for each test (e.g., 3,712 for plantation 28). Hence, the practical significance of these results is probably not meaningful.

The results thus far suggest little evidence for strong seed source differences within the plantations. Small, yet statistically significant, differences between seed sources can be found in some of the plantations. However, these differences typically account for only < 2% of the total variance among all seed sources, which gives them little operational significance. Consequently, the climate modeling described next will be based on pooling the common variance among seed sources using PCA.

Climate Response Models for the Southwide Plant Seed Source Study Plantations

Using the climate data described earlier, simple correlation analyses were carried out on the time series scores of the first tree-ring PC from each plantation. The

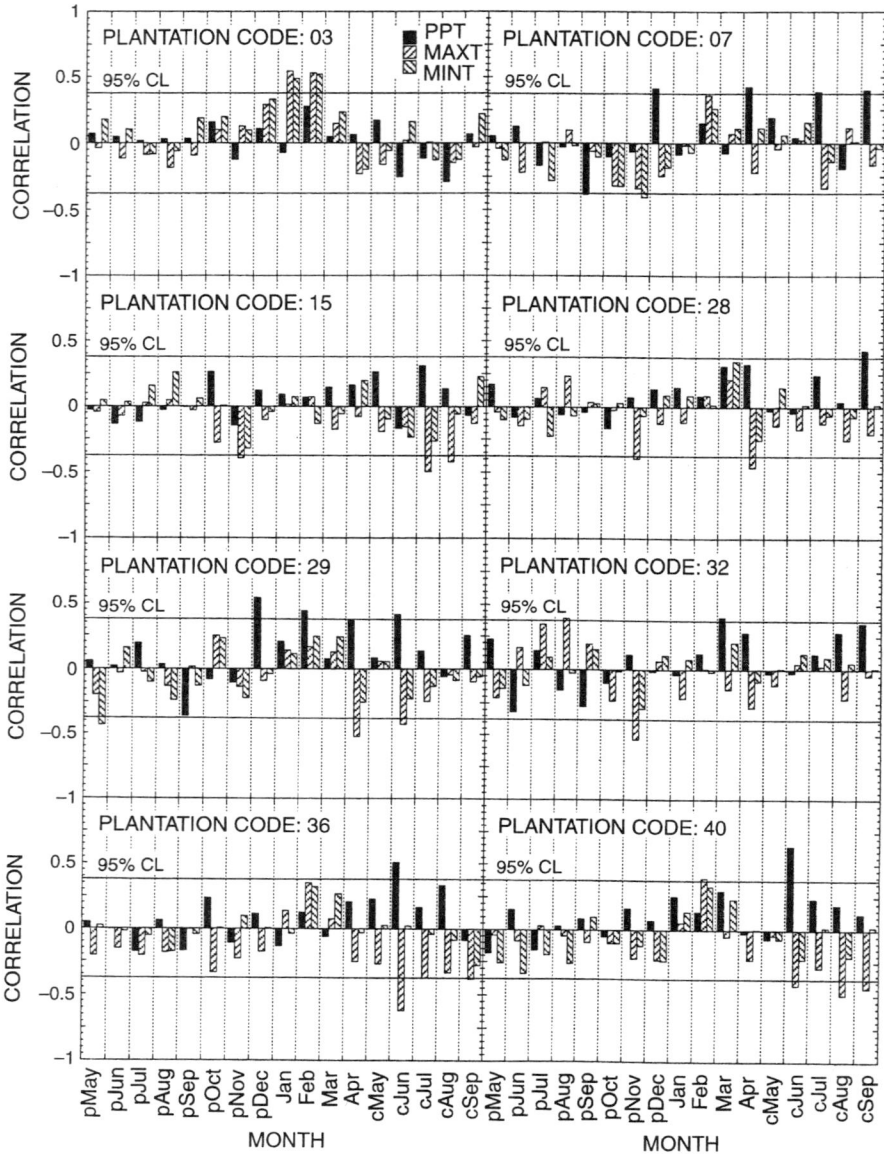

Figure 39.6. Correlations between loblolly pine tree rings and monthly climate over the period from 1960 to 1988. In each case, the tree-ring series used for correlation with climate was the first principal component of the plantation seed source chronologies. This represents the most common mode of covariance between seed sources and is the only orthogonal mode that is statistically significant.

correlations were estimated over the common period of 1960 to 1988 using a dendroclimatic year (Fritts, 1976) extending from the prior May to the current September of growth. The extension of the correlation analyses back into the prior growing season allowed for the possibility of climatic preconditioning on growth the following year. This is a very common phenomenon in tree-ring response functions (Fritts, 1976).

Figure 39.6 shows the correlation analysis results for each plantation. Most months do not show any correlation between climate and loblolly pine radial growth, especially during the prior growing season months. However, certain features are relevant to our purposes. For example, the two most western plantations (36 and 40), indicate a very high sensitivity to rainfall and maximum temperatures in June of the current growing season. The positive correlation with rainfall and negative correlation with maximum temperature during that month is a classic "drought sensitivity" response. That is, overall radial growth is less when June is dry and hot, particularly during the daylight hours when the trees are photosynthetically active. In the subsequent months of July to September, the sensitivity to both precipitation and temperature diminishes, although the signs of the correlations remain consistent with drought sensitivity and are sometimes statistically significant. Plantation 29, which is also a westerly, continental-interior site, also shows a drought response during current-June, but it is somewhat weaker compared to plantations 36 and 40. Precipitation during the earlier months of December, February, and April also appears to be influential on radial growth, but it is difficult to interpret this collective relationship causally other than to say that it may be related to soil-moisture recharge. Together, these results suggest that drought, particularly when it peaks in June, is an important growth-limiting factor to loblolly pine as it approaches its western range limit and probably contributes strongly to the lack of establishment and survival of this species beyond that limit. Although not terribly surprising, this conclusion is obviously relevant to concerns about possible increasing drought frequency in the southeastern United States resulting from greenhouse warming (Rind et al., 1990) and its consequent impact on forests. Loblolly pine would appear to be highly vulnerable in this regard.

The drought sensitivity of loblolly pine diminishes quickly for the plantations at more coastal and easterly locations. Plantations 28 and 32, which are still westerly but more coastal, show no sensitivity to June climatic conditions. At best, there is a weak dependence on spring climate and prior-November maximum temperatures at these plantations, but none are strong enough to warrant much attention. Plantations 07 and 15 indicate a later current growing season (mainly July) drought response, but this response is weaker than that found in plantations 36 and 40.

Plantation 03 has an unusual climate response that stands out from the others, that being significant correlations with January and February temperatures. This unusual result may be related to the comparatively cold January and February temperatures that this plantation experiences (see Figure 39.2). Given that this is the only plantation with mean maximum January and February temperatures

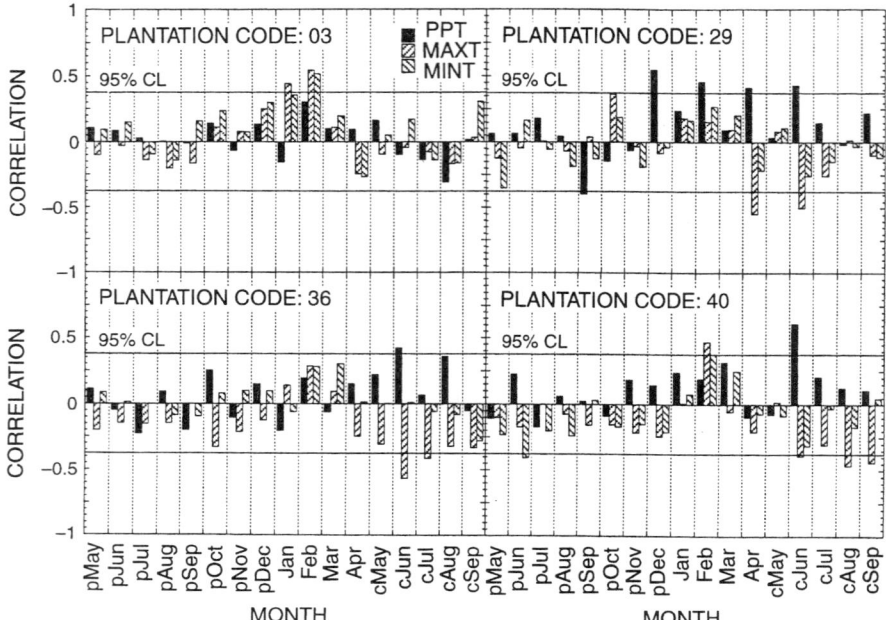

Figure 39.7. Correlations between loblolly pine trees of the local seed source and monthly climate for four extreme locations in the range of loblolly pine where adaptations of local seed sources to climate might be most evident. A comparison of Figure 39.7 with the relevant plantations in Figure 39.6 shows that there is no discernible local seed source adaptations.

below 10 °C, it is possible that this represents a threshold effect whereby loblolly pine is increasingly sensitive to winter injury, either through direct freezing or desiccation. Consequently, this finding may help explain how climate influences the northern range limit of this tree species.

We also examined the local seed source response to climate at each plantation to see if there was any evidence for local adaptations to the climatic environment. Figure 39.7 shows the results for four of the most extreme plantations: 03, 29, 36, and 40. By comparing Figure 39.7 with Figure 39.6, it is apparent that the local seed sources are not differentially adapted to climate in any obvious way, a result that is wholly consistent with similarity of the tree-ring chronologies themselves. Again, there seems to be little evidence for any meaningful differences between the seed sources.

Summary

This study has used the tree rings from a long-term common garden experiment to determine if there is any differential sensitivity of loblolly pine to climatic effects

39. The Differential Sensitivity of Loblolly Pine to Climatic Change

at the seed source level. Based on the ANOVA results, there does appear to be a very weak differential sensitivity to climate at most of the SPSSS plantations. However, this effect is very small in terms of explained variance and is, for all practical purposes, meaningless. This essentially negative result was surprising given the clear differences in seed source performance over geographic space when viewed in terms of mortality, growth, and yield (Wakeley, 1953, 1959, 1961; Wells and Wakeley, 1966; Nance and Wells, 1981; Wells, 1969, 1983). However, it must be pointed out that the level of year-to-year variance in growth provided by the tree rings is typically a small fraction of that resulting from changes in absolute growth, especially during the juvenile and early maturation phases when growth rates are changing rapidly because of intense competitive pressures. Most of the SPSSS five-year remeasurements were made during that very active phase of plantation establishment and maturation.

The lack of any clear differential sensitivity to climate at the seed source level may be the result of the high level of noise or random variability between trees within seed sources, as pointed out earlier. It is difficult to know whether this noise is caused by truly random within- and between-plot effects or to the inherent genetic variability of the seed sources used. It is probably a combination of both effects. Regardless, the net effect was that the within-plantation seed source chronologies were practically identical after the within- and between-tree effects were averaged out over plots. As a consequence, the climate modeling could only be relied upon to provide a plantation-level expression of the response of loblolly pine to climate.

Perhaps the most practically useful results of this study have come from the plantation-level climatic response functions. There is a clear indication of increasing drought sensitivity of loblolly pine as it approaches the western limits of its range. This is indicated especially well for plantations 36 and 40 in east Texas and southwest Arkansas, respectively. Interestingly, the critical month in both cases is June when loblolly pines are especially sensitive to moisture availability and evapotranspiration demand. Any increase in drought frequency and severity resulting from greenhouse warming, especially during late spring and early summer, would have a devastating impact on these plantations and, by extension, loblolly pines growing elsewhere in this part of the range. For the more interior-range plantations (15, 28, and 32), overall climate sensitivity appears to be much weaker. However, given the lack of any meaningful seed source differences in climate response, it is clear that these plantations would also be vulnerable to any increase in drought frequency and severity as well. This conclusion is also supported by independent climatic response function analyses of loblolly pine tree-ring chronologies from Alabama (Jordan and Lockaby, 1990) and Georgia (Grissino-Mayer et al., 1989). In both cases, strong statistical evidence for growing-season drought sensitivity was found.

The climate response of plantation 03 in eastern Maryland is equally interesting for a different reason. In this case, the cardinal climate variables influencing radial growth are January and February maximum and minimum temperatures, with growing-season climate variables having little or no influence on growth. This odd response may be related to a threshold effect in which maximum temperatures below 10 °C have a strong impact on radial growth potential. This result suggests that loblolly pine at the northern limit of its range may actually benefit from greenhouse warming during the winter, at least up to some level. However, from the analyses of the other plantations, it is clear that this benefit would occur only if the warming does not exceed ~ 4to 5 °C during the winter months. Above that, the benefit would probably be lost and the Maryland trees would begin to respond more similarly to other plantations. It also suggests that loblolly pine will have the potential to move northward from its present northern range limit, either naturally or by artificial means, if future warming occurs. This movement would probably not go much beyond the ~ 5 to 6 °C January and February maximum temperature isotherm as it too moves northward, however.

References

Box GEP, Jenkins GM (1976) *Time series analysis: Forecasting and control*. Holden-Day, San Francisco.

Cook ER, Kairiukstis LA (1990) *Methods of dendrochronology: Applications in the environmental sciences*. Kluwer Academic Publishers, Dordrecht, The Netherlands.

Cooley WW, Lohnes PR (1971) *Multivariate data analysis*. John Wiley and Sons, New York.

Fritts HC (1976) *Tree rings and climate*. Academic Press, London.

Fritts HC, Mosimann JE, Bottorff CP (1969) A revised computer program for standardizing tree-ring series. Tree-Ring Bull 29:15–20.

Grissino-Mayer HD, Rosenberger MS, Butler DR (1989) Climatic response in tree rings of loblolly pine from north Georgia. Phys Geog 10(1):32–43.

Henry BW (1959) Diseases and insects in the Southwide Pine Seed Source Study. In Proceedings of fifth southern conference on forest tree improvement. North Carolina State University, Raleigh, North Carolina.

Henry BW, Coyne JF (1955) Occurrence of pests in Southwide Pine Seed Source Study. In Proceedings of third southern conference on forest tree improvement. USDA Forest Service, Southern Forest Experiment Station, New Orleans, Louisiana.

Henry BW, Hepting GH (1957) Pest occurrences in 35 of the Southwide Pine Seed Source Study plantations during the first three years. US For Serv South Southeast For Exper Sta, Asheville, North Carolina.

Holmes RL (1982) Computer-assisted quality control in tree-ring dating and measurement. Tree-Ring Bull 44:69–75.

Jordan DN, Lockaby BG (1990) Time series modelling of relationships between climate and long-term radial growth of loblolly pine. Can J For Res 20:738–742.

Krusic PJ, Kenney M, Hornbeck J (1987) Preparing increment cores for ring width measurement. Nor J Appl For 4(2):104–105.

Lamb PJ (1987) On the development of regional climatic scenarios for policy-oriented climatic-impact assessment. Bull Am Met Soc 68(9):1116–1123.

Nance WL, Wells OO (1981) Estimating volume potential in genetic tests using growth and yield models. In Proceedings of 16th southern conference on forest tree improvement. Virginia Polytechnic Institute and State University, Blacksburg, Virginia.

Preisendorfer RW, Zwiers FW, Barnett TP (1981) *Foundations of principal components selection rules*. SIO Reference Series 81–4, Scripps Institution of Oceanography, La Jolla, CA.

Rind DR, Goldberg R, Hansen J, Rosensweig C, Ruedy R (1990) Potential evapotranspiration and the likelihood of future drought. *J Geophys Res* 95(D7):9983–10,004.

SAS Institute Inc. (1985) *SAS user's guide: Statistics,* Version 5 Ed. Cary, NC.
Stokes MA, Smiley TL (1968) *An Introduction to tree-ring dating.* University of Chicago Press, Chicago, IL.
Wakeley PC (1953) Progress in study of pine races. South Lumber 187(2345):137–140.
Wakeley PC (1959) Five-year results of the Southwide Pine Seed Source Study. In Proceedings of fifth southern conference on forest tree improvement. North Carolina State College, Raleigh, North Carolina.
Wakeley PC (1961) Results of the Southwide Pine Seed Source Study through 1960–61. In Proceedings of sixth southern conference on forest tree improvement. School of Forestry, University of Florida, Gainesville, Florida.
Wells OO (1969) Results of the Southwide Pine Seed Source Study through 1968–1969. In Proceedings of 10th southern conference on forest tree improvement. Texas Forest Service, Texas A&M University, College Station, Texas.
Wells OO (1983) Southwide Pine Seed Source Study—Loblolly pine at twenty-five years. South J Appl For 7(2):63–71.
Wells OO, Wakeley PC (1966) Geographic variation in survival, growth, and fusiform-rust infection of planted loblolly pine. For Sci Mono 11.
Wells OO, Switzer GL (1975) Selecting populations of loblolly pine for rust resistance and fast growth. In Proceedings of 13th southern conference on forest tree improvement. USDA Forest Service, Macon, Georgia.
Wigley TML, Briffa KR, Jones PD (1986) On the average value of correlated time series, with applications in dendroclimatology and hydrometeorology. J Clim Appl Met 23:201–213.

40. Global Change and Disturbance in Southern Forest Ecosystems

Matthew P. Ayres and Gregory A. Reams

Perturbation and Disturbance in Forest Ecosystems

Global change is apt to introduce a variety of perturbations in forests of the southern United States (Figure 40.1). The consequences will vary depending upon characteristics of the perturbations and the ecosystem. Perturbations of high intensity but low frequency (e.g., fires, hurricanes, and regional epidemics of southern pine beetles) can "result in the sudden mortality of biomass in a community" and be described as disturbances (Huston 1994). At the other extreme, perturbations of low intensity but high frequency (e.g., changes in average temperature, elevated carbon dioxide (CO_2), and atmospheric nitrogen deposition) tend to exert sustained but low intensity pressures on ecosystems and are sometimes referred to as "stress" (Underwood, 1989; Winner, 1994; Milchunas and Lauenroth, 1995), but the effects are not necessarily negative (Teskey, see Chapter 8).

Global change is also apt to produce perturbations of intermediate intensity and frequency, such as changes in climatic extremes (e.g., occasional droughts, hard freezes, and hot spells), bouts of ozone exposure, and mild epidemics of pests and pathogens. Populations, communities, and ecosystems may differ in their resistance to the effects of perturbations, and their resilience in recovering from perturbations (Cottingham and Carpenter, 1994; Larsen, 1995). For example, populations with an evolutionary history of perturbation may tend to be more resistant and resilient in the face of perturbations (Bazzaz, 1983; Wilson and Keddy, 1986; Miao and Bazzaz, 1990; Clark, 1991; Parker et al., 1993).

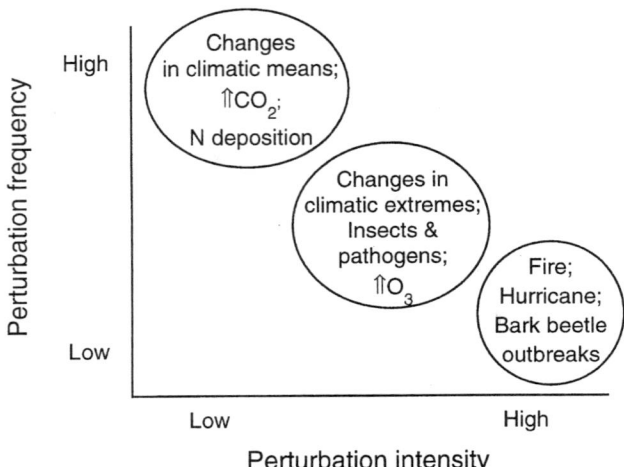

Figure 40.1. Global change could produce or influence a variety of perturbations in forests of the southern United States. The ecological and economic effects will probably vary with the frequency and intensity of perturbations.

Ecosystems with high diversity may be impacted to a lesser degree by climatic extremes than ecosystems with low diversity (Tilman, 1996). Ecosystems with high productivity also tend to recover more quickly from perturbations than those with low productivity (Moore et al., 1993; Huston, 1994). Perturbations are a natural feature of most ecosystems and are not intrinsically deleterious from an ecological perspective (Lorimer, 1980; Glitzenstein et al., 1986; Frelich and Lorimer, 1991; Attiwill, 1994). Nevertheless, natural and anthropogenic patterns of perturbations are a dominant consideration in forest management becuase they exert strong effects on spatial and temporal variability in resources and landscape structure (Turner, 1987; Dayton et al., 1992; Mladenoff et al., 1993 Robertson et al., 1993). One of the prevailing theories for explaining patterns of biodiversity postulates that maximum diversity is attained at intermediate levels of disturbance (Connell 1978; Huston, 1979, Luken et al., 1992; Huston, 1994).

Future Climate for the Southern United States

Climatologists have used two approaches to predict the climate changes that may result from an increase in greenhouse gases, which are 1) numerical modelling using general circulation models (GCMs), and 2) the analysis of past warm periods as analogues of the future (Wigley et al., 1980). Both methods have limitations, but there is agreement between the approaches on many important results. For example, both approaches predict that the greatest temperature increases will occur at high northern latitudes and in the winter. For forests in the southern United States, regional changes in precipitation patterns are among the most important possible consequences of climate change. However, consensus has been slow to emerge regarding the expected magnitude and direction of changes in precipitation. Some studies of the instrument-based climatic record have shown a positive correlation between temperature and annual precipitation in the southeastern United States (Diaz and Quayle, 1980) indicating that rising temperatures may be associated with increased precipitation. However, Wigley et al., (1980) suggest that this region tends to be drier during warm episodes. Recent analyses of instrument-based data suggest that a decrease in summertime convection activity will result in reductions of 10 to 20% in summer precipitation (Coleman, 1988). Thus, present-day projections suggest regional decreases in precipitation and increases in drought frequency (Rind et al., 1990), but there remains considerable uncertainty in these projections.

Potential Impacts Resulting from Changes in Temperature and Precipitation

Climatic variability is one of the dominant exogenous factors determining the productivity and community structure of forests. Interest in the potential impact of climatic change on the forests of the southern United States has led to studies of tree-ring data from mature loblolly pine (*Pinus taeda* L.) and old-growth baldcypress (*Taxodium distidium* L. Rich) trees located throughout the region. For both loblolly pine and baldcypress, relationships between annual ring-width and monthly climate variables suggest that reduced summer precipitation will result in reduced annual growth (Cook et al., see Chapters 39; Reams and Van Deusen, see Chapter 38). In particular, growth is less when June is hot and dry. In the subsequent summer months, the sensitivity to both precipitation and temperature diminishes, although prolonged summer drought can further limit growth (Jordan and Lockaby, 1990). These patterns are consistent with our understanding of seasonal patterns in the growth and differentiation of loblolly pine (Lorio, 1986, 1993; Wilken et al., see Chapter 32). The highest growth rates in loblolly pine are typically attained during spring and early summer, during the time of earlywood production. Potential cambial growth rates in the bole are reduced after the midsummer ontogenetic transition from the production of earlywood to latewood, and this ontogenetic transition tends to occur earlier in years when soil-water deficits occur earlier (caused by warm temperatures and low precipitation in the early summer). Hydrological models (e.g., McNulty et al., see Chapter 22) that predict soil water availability during June through August should provide a valuable tool for evaluating the consequences of specific climate scenarios for the regional productivity of southern pine forests.

The bulk of summer rainfall in the southeastern United States results from convection thunderstorms and tropical cyclones. Hence, atmospheric changes that alter patterns of convection thunderstorms and cyclones are apt to influence forest productivity. Increases in drought frequency and severity resulting from greenhouse warming would probably lead to increased tree mortality from drought in loblolly pine forests in east Texas and southwest Arkansas (Cook et al., see Chapter 39) where water limitations are already severe under the present climate. Loblolly pines in eastern sections of their present-day range would also be affected but to a lesser degree because of the higher baseline precipitation. Similarly, most scenarios of decreased summer precipitation and decreased soil moisture would lead to decreases in the productivity and distribution of baldcypress (Reams and Van Deusen, see Chapter 38).

Although forest productivity may decline in some regions as a result of climate change, forest productivity may increase in other areas. Knowledge of past trends in climate and associated geographic shifts of vegetation provide one basis for predicting changes in forest distribution and composition that may result from climate change. Results from matching the paleoclimate and paleopollen records suggest that increases in the distribution and abundance of southern pines relative to oaks may be associated with increases in January temperature (Overpeck and Bartlein, 1989; Devall and Parresol, see Chapter 36). Dendroclimatic studies provide further evidence that the geographic distribution of loblolly pine may shift northward as a result of greenhouse warming. Cook et al., (see Chapter 39) found that radial growth of loblolly pine was negatively correlated with cold January and February temperatures. Both the recent records of tree-ring and paleobotany record indicate that loblolly pine has the potential to move northward from its present distribution limits. Northern distribution limits for loblolly pine may be approximated by the 5 °C January and February maximum temperature isotherm (Cook et al., see Chapter 39). Relatively little is known about the probable responses to global change of southern tree species other than loblolly pine and baldcypress. However, it has been suggested that distribution shifts in native tree species could be constrained by the increased abundance of introduced plant species that thrive at relatively high temperatures (Hogenbirk and Wein, 1992).

Potential Impacts Resulting from Changes in the Frequency and Pattern of Perturbations

Greenhouse warming may lead to more frequent and intense storms (Schmidt and von Storch, 1993). Emanuel (1987) showed that a doubling of atmospheric CO_2 could lead to an increase of 40 to 50% in the destructive potential of hurricanes. Jarrell and Elsberry (1994) suggested that global warming might increase the frequency of tropical cyclones because it would expand the area of oceans with temperatures above the 26 °C threshold required for the formation of tropical cyclones. Hurricanes and other large-scale disturbances strongly influence the structure, composition, and successional processes of many forests (Putz and Sharitz, 1991; Reams and Van Deusen, 1993). A single hurricane can affect 400,000 hectares of forest (Spurr, 1956; Sheffield and Thompson, 1992), and alter species composition and community dynamics for many decades (Putz and

Sharitz, 1991). An increase in hurricane frequency could favor such species as baldcypress, tupelo (*Nyssa* sp.), and other hardwoods that are more resistant to wind damage than loblolly pine and have a greater ability to resprout after suffering trunk damage (Putz and Sharitz, 1991).

There is increasing support for the hypothesis that climatic warming will lead to general increases in climatic variation and the frequency of biologically important climate disturbance such as hurricanes, ice-storms, and droughts (Gates, 1990; Overpeck et al., 1990; Knox, 1993; Meehl and Washington, 1993; Overpeck, 1996). Changes in environmental variance may be at least as important as changes in the mean environmental conditions (Wigley, 1985). For example, changes in seasonal patterns in soil water and year-to-year variance in soil water, could have large effects on tree growth, herbivore outbreaks, losses to pathogens, and fire frequency in southern forests (even in the absence of changes in average soil water; see Wilkens et al., Chapter 32). In general, variance is of particular relevance to any process that exhibits nonlinear responses to environmental variables. Ecologists are just beginning to incorporate year-to-year climatic variation into their modeling scenarios (Cohen and Pastor, 1991; Goulden et al., 1996; McNulty et al., see Chapter 33). Physiologists have made great progress in characterizing tree responses to average conditions (see Chapters 4 to 15), but generally lack a theoretical framework or empirical basis for evaluating the effects of environmental variance. We hypothesize that patterns of environmental variance will be particularly important for biological processes that involve acclimatization (e.g., adjustments of photosynthetic biochemistry, needle morphology, and root to shoot ratios) and that the most important time-scales of environmental variance will be a function of the time-scale at which trees acclimatize (days to years depending upon the physiological system).

Fire

General circulation models suggest that a doubling of CO_2 levels will produce more episodes of high-pressure systems located over the southeastern United States. Such weather systems can block the transport of Gulf moisture northward over the eastern states, and create the potential for more fire-weather episodes in the southeastern United States (Heilman et al., see Chapter 37). In the longleaf and slash pine belt, conventional silvicultural management involves the use of fire to expose the soil for seed fall, control brush and hardwood competition, control brown spot disease, and release seedlings from the grass stage. Burning is also used as a site preparation technique for establishing loblolly pine. In general, southern pines, especially longleaf pine (*Pinus plustris* Mill.), seem to be more resistant to fire than hardwoods, suggesting that pines may be favored by increased fire. The longleaf pine community, which dominated much of the southeast from the mid-Holocene to the late 1800s, appears to have been sustained by fires and includes many species that are now rare but would probably increase in abundance and distribution with increases in fire frequency (Stout and Marion, 1993; Ware et al., 1993).

Forest Pests and Pathogens

There are numerous pathways by which global change could impact the effects of forest pests and pathogens. Changes in CO_2, temperature, water availability, nutrient availability, and cloud cover can all impact the resistance and resilience of trees to herbivores (Ayres, 1993). Physiological models of tree allocation and defense provide useful tools for predicting the local and regional effects of global change on plant-pest interactions (Reeve et al., 1995; Wilkens et al., see Chapter 32; Britton et al., see Chapter 34). Elevated levels of CO_2 tend to reduce leaf nitrogen and herbivores often respond with reduced growth or increased consumption (Johnson and Lincoln, 1991; Lindroth et al., 1993). Increased cloud cover tends to reduce plant secondary metabolism and increase susceptibility to herbivory (Larsson et al., 1986; Bryant et al., 1987; Mole et al., 1988). Changes in temperature can also impact secondary plant metabolism and plant-nitrogen content (Laine and Henttonen, 1987). Some of the most dramatic effects of global change on plant-pest interactions in forests of the southeastern United States are apt to result from changes in water availability, nutrient availability, and their interactions. Soil-water availability will probably change in many forests as a result of changes in precipitation. Soil-nutrient availability will also probably change as a result of changes in mineralization rates, litter quality, and atmospheric nitrogen deposition (Pastor and Post, 1988; Bazazz, 1990; Olliger et al., 1993). Although it remains difficult to predict the net effect of these changes, they are apt to be of great importance for plant-pest interactions.

A review of published data suggests changes in nutrient availability may have a greater effect on plant-pest interactions than any other parameters of global change (Ayres, 1993). Studies of loblolly pine and dogwood indicate that increased nutrient availability leads to reduced resistance to southern pine beetles and dogwood anthracnose, respectively (Wilken et al., see Chapters 32; Britton et al., see Chapter 34). These results are consistent with emerging physiological models that predict that plants experiencing moderate deficits of water or nutrients will respond to increases in water or nutrients with increased growth, reduced secondary metabolism, and increased susceptibility to pests (see Figure 32.1 in Wilkens et al., Chapter 32, Reeve et al., 1995). However, these same models predict that plant responses are nonlinear and that increases in water or nutrient availability can lead to increased secondary metabolism in plants under extreme deficits of water or nutrients. Therefore, a reduction of 20% in summer precipitation could increase bark beetle outbreaks in east Texas where pines are already under extreme water limitations, could also reduce bark beetle outbreaks in Georgia where water limitations are less severe.

There are other mechanisms by which global change could alter the impacts of forest pests in southern forests. Changes in temperature can have strong direct effects on the physiology, distribution, and population dynamics of herbivores and pathogens with consequences for their host populations and the forest community (Beal, 1933; Wagner et al., 1984; Moser and Thompson, 1986; Allen et al., 1993; Ayres, 1993; Ayres and Scriber, 1994). For the most part, however, the direct effects of temperature on herbivores and pathogens has not been explored in global change research programs. Additional effects could also be produced by changes in thunderstorm activity because trees that are struck by lightning frequently act as foci for the initiation of bark beetle infestations. Increased frequencies or intensity of any kind of disturbance could lead to elevated numbers of pests by promoting the abundance of fast-growing plant species, which tend to be poorly defended against pests (Coley et al., 1985). Natural or anthropogenic effects on landscape structure can also influence the epidemiology of forest pests (Menges and Loucks, 1984; Showalter and Turchin, 1993).

Summary

Global change seems certain to alter spatial and temporal patterns of perturbation and disturbance in southern forests. It is probable that perturbations will become less frequent or less severe in some areas, but more frequent or more severe in other areas. Some climatologists expect that global change will lead to an overall increase in the frequency of extreme climatic events (perturbations). However, even if the average frequency and severity of perturbations remains the same across the southeastern United States as a whole, changes in the spatial and temporal patterns of perturbation will probably lead to increased ecological impacts because we expect that 1) a given level of perturbation will produce the greatest disturbance in forest communities with a limited history of perturbation, and 2) forests with a history of frequent perturbations will acquire different characteristics if perturbations are relaxed. By analogous reasoning, changes in the spatial and temporal patterns of perturbation will probably introduce unavoidable social impacts and economic costs. For example, a large, expensive infrastructure exists within the U.S. Forest Service (i.e., Forest Health) and the private sector (i.e., timber salvage operators) to control infestations of southern pine beetles. The personnel, equipment, and expertise of this infrastructure tend to be concentrated in areas where the frequency of bark beetle infestations is greatest. If bark beetle infestations become less frequent in these areas and more frequent in others, the average efficacy of control is likely to decrease and the average losses associated with infestations will probably increase because the control expertise will be located elsewhere and because such risk mitigation measures as frequent thinning of pine forests, tend to be less common in areas without a sustained history of bark beetle outbreaks. Similarly, if hurricane tracks change (even with no change in hurricane frequency), there will be increased impacts and costs because landowners in formerly high risk areas will tend to be overinsured and landowners in new high risk areas will tend to be underinsured.

Other ecological and socioeconomic consequences may be even more difficult to remedy. Parks, preserves, and conservation easements have value based upon the contemporary distribution of organisms and cannot easily be expanded or relocated if the selected sites become unsuitable for valued organisms because of climatic changes. Because of agriculture, losses of forest due to climate change in some areas will not be easily matched by gains in other areas. Municipalities and school districts make long term investments that are dependent on a stable or growing tax base; global change can threaten these investments because in many areas of the southeastern U.S. timberland is a large portion of the tax base and forest product revenues are a large portion of the local economy.

In some regions, perturbations associated with global change are likely to alter optimal land-use strategies (e.g., forest vs. range vs. agriculture, pine vs hardwood, loblolly pine vs longleaf pine, long rotations vs. short rotations, etc.). Assuming that present land-use strategies are approximately optimal, environmental changes that alter optimal land-use introduce unavoidable costs. These costs will be minimized by (1) early recognition of the changes, (2) accurate predictions of the consequences of changes in land-use strategies, and (3) timely transitions, when appropriate, from the status quo to appropriate new strategies. Research such as that reported in this volume represents progress toward steps 1 and 2 of this process. Step 2 in particular, will require further significant advances in scientific knowledge. Step 3 hinges upon effective decisionmaking by public and private landowners.

Knowledge Gaps

Research published here and elsewhere indicates substantial growth in our understanding of relationships between forest disturbance and global change. Furthermore, we are now in a better position to identify priorities for future research. We conclude with a list of what seems to be the key knowledge gaps that limit our ability to predict future patterns of disturbance, and their consequences, in southern forests:

1. We lack adequate knowledge of the direction and magnitude of regional changes in precipitation, and seasonal patterns in precipitation.
2. We lack adequate knowledge of the direction and magnitude of regional changes in mineralization rates, litter quality, and atmospheric elemental deposition. What will be the net effect on nutrient availability for plants? Will there be changes in the relative importance of nitrogen and phosphorus as limiting nutrients in southern forests?
3. We lack adequate knowledge of regional patterns (present and future) in the effects of herbivores and pathogens on forests of the southeastern United States Rapid progress could be made by the refinement and parameterization of physiological models of environmental effects on plant secondary metabolism and resistance to insects and disease. This could be facilitated by expanding existing tree growth models (see Chapters 16–26) to include pools and fluxes for secondary metabolism.
4. We lack adequate knowledge of the way that stand-level responses can modify tree-level environments. For example, models based on physiological responses of individual trees predict that increased temperature leads to in-

creased transpiration and, therefore, decreased soil water and the more rapid onset of water deficits. In contrast, stand-level models for loblolly pine (McNulty et al., see Chapter 22) predict that projected temperature increases would lead to reductions in the leaf area index of the pine stand and, as a consequence, reduced water deficits in individual trees. Scaling up from physiological responses to ecosystem properties is a major challenge for basic and applied science in forests of the southeastern U.S. and elsewhere.
5. We lack adequate knowledge of the direct effects of climatic change, especially temperature, on keystone species of animals and fungi (e.g., southern pine beetle, gypsy moth, fusiform rust)
6. We lack adequate knowledge of the effects of changes in patterns of perturbations on biodiversity, and effects of biodiversity on responses to perturbations.

References

Allen JC, Foltz JL, Dixon WN, Liebhold AM, Colbert JJ, Regniere J, Gray DR, Wilder JW, Christie I (1993) Will the gypsy moth become a pest in Florida? Flor Entom 76:102–113.

Attiwill PM (1994) The disturbance of forest ecosystems: The ecological basis for conservative management. For Ecol Manage 63:247–300.

Ayres MP (1993) Global change, plant defense, and herbivory. In Kareiva PM, Kingsolver JG, Huey RB (Eds) *Biotic interactions and global change.* Sinauer Associates. Sunderland, MA.

Ayres MP, Scriber JM (1994) Local adaptation to regional climates in Papilio canadensis (Lepidoptera: Papilionidae). Ecol Mono 64:465–482.

Bazzaz FA (1983) Characteristics of populations in relation to disturbance in natural and man-modified ecosystems. In Mooney HA, Godron M (Eds) *Disturbance and ecosystems: Components of response.* Heidelberg, New York.

Bazzaz FA (1990) The response of natural ecosystems to the rising global CO_2 levels. Ann Rev Ecol System 21:167–196.

Beal JA (1933) Temperature extremes as a factor in the ecology of the southern pine beetle. J For 31:329–336.

Bryant JP, Chapin FS, III, Reichardt PB, Clausen TP (1987) Response of winter chemical defense in Alaska paper birch and green alder to manipulation of plant carbon/nutrient balance. Oecologia 72:510–514.

Clark JS (1991) Disturbance and tree life history on the shifting mosaic landscape. Ecol 72:1102–1118.

Cohen Y, Pastor J (1991) The responses of a forest model to serial correlations of global warming. Ecol 72:1161–1165.

Coleman JM (1988) Climatic warming and increased summer aridity in Florida, U.S.A. Clim Change 12:165–178.

Coley PD, Bryant JP, Chapin FS. III (1985) Resource availability and plant antiherbivore defense. Science 230:895–899.

Connell JH (1978) Diversity in tropical rainforests and coral reefs. Science 199 1302–1309.

Cottingham KL, Carpenter SR (1994) Predictive indices of ecosystem resilience in models of north temperate lakes. Ecol 75:2127–2138.

Dayton PK, Tegner MJ, Parnell PE, Edwards PB (1992) Temporal and spatial patterns of disturbance and recovery in a kelp forest community. Ecol Mono 62:421–445.

Diaz MF, Quayle RG (1980) The climate of the United States since 1895: Spatial and temporal changes. Mon Wea Rev 108:246–266.

Emanuel KA (1987) The dependence of hurricane intensity on climate. Nature 326:483–485.
Frelich LE, Lorimer CG (1991) Natural disturbance regimes in hemlock-hardwood forests of the upper Great Lakes region. Ecol Mono 61:145–164.
Gates DM (1990) Climate change and forests. Tree Physiol 7:1–5.
Glitzenstein JS, Harcombe PA, Streng DR (1986) Disturbance, succession, and maintenance of species diversity in an east Texas forest. Ecol Mono 56:243–258.
Goulden ML, Munger JW, Fan S, Daube BC, Wofsky SC (1996) Exchange of carbon dioxide by a deciduous forest: Response to interannual climate variability. Science 271:1576–1578.
Hogenbirk JC, Wein RW (1992) Temperature effects on seedling emergence from boreal wetland soils: Implications for climate change. Aqua Bot 45:361–373.
Huston MA (1979) A general hypothesis of species diversity. Am Natur 113:81–101.
Huston MA (1994) Biological diversity: The coexistence of species on changing landscapes. Cambridge University Press, Cambridge, England.
Jarrell JD, Elsberry RL (1994) The effect of global climate on tropical cyclones. Paper presented at the 5th Global Warming Conference, San Francisco, CA.
Johnson RH, Lincoln DE (1991) Sagebrush carbon allocation patterns and grasshopper nutrition: The influence of CO_2 enrichment and soil mineral nutrition. Oecologia 87:127–134.
Jordan DN, Lockaby BG (1990) Time series modelling of relationships between climate and long-term radial growth of loblolly pine. Can J For Res 20:738–742.
Knox JC (1993) Large increases in flood magnitude in response to modest changes in climate. Nature 361:430–432.
Laine K, Henttonen H (1987) Phenolic/nitrogen ratios in the blueberry Vaccinium myrtillus in relation to temperature and microtine density in Finnish Lapland. Oikos 50:389–395.
Larsen JB (1995) Ecological stability of forests and sustainable silviculture. For Ecol Manage 73:85–96.
Larsson S, Wiren A, Lundgren L, Ericsson T (1986) Effects of light and nutrient stress on leaf phenolic chemistry in *Salix dasyclados* and susceptibility to *Galerucella lineola* (Coleoptera). Oikos 47:205–210.
Lindroth RL, Kinney KK, Platz CL (1993) Responses of deciduous trees to elevated atmospheric carbon dioxide: Productivity, phytochemistry, and insect performance. Ecol 74:763–777.
Lorimer CG (1980) Age structure and disturbance history of a southern Appalachian virgin forest. Ecol 61:1169–1184.
Lorio PL, Jr. (1986) Growth-differentiation balance: A basis for understanding southern pine beetle–tree interactions. For Ecol Manage 14:259–273.
Lorio PL, Jr. (1993) Environmental stress and whole-tree physiology. In Schowalter TD, Filip GM (Eds) *Beetle-pathogen interactions in conifer forests*. Academic Press, London.
Luken JO, Hinton AC, Baker DG (1992) Response of woody plant communities in power-line corridors to frequent anthropogenic disturbance. Ecol Appl 2:356–362.
Meehl GA, Washington WM (1993) South Asian summer monsoon variability in a model with doubled atmospheric carbon dioxide concentration. Science 260:1101–1104.
Menges ES, Loucks OL (1984) Modeling a disease-caused patch disturbance: Oak wilt in the midwestern United States. Ecol 65:487–498.
Miao SL, Bazzaz FA (1990) Responses to nutrient pulses of two colonizers requiring different disturbance frequencies. Ecol 71:2166–2178.
Milchunas DG, Lauenroth WK (1995) Inertia in plant community structure: State changes after cessation of nutrient-enrichment stress. Ecol Appl 5:452–458.
Mladenoff DJ, White MA, Pastor J, Crow TR (1993) Comparing spatial pattern in unaltered old-growth and disturbed forest landscapes. Ecol Appl 3:294–306.

Mole S, Ross JAM, Waterman PG (1988) Light-induced variation in phenolic levels in foliage of rain-forest plants. I. Chemical changes. J Chem Ecol 14:1–21.
Moore JC, De Ruiter PC, Hunt HW (1993) Influence of productivity on the stability of real and model ecosystems. Science 261:906–908.
Moser JC, Thompson WA (1986) Temperature thresholds related to flight of *Dendroctonus frontalis* Zimm. (Col.: Scolytidae). Agron 6:905–910.
Ollinger SV, Aber JD, Lovett GM, Millham SE, Lathrop RG, Ellis JM (1993) A spatial model of atmospheric deposition for the northeastern U.S. Ecol Appl 3:459–472.
Overpeck JT (1996) Warm climate surprises. Science 271:1820–1821.
Overpeck JT, Bartlein PJ (1989) Assessing the response of vegetation to future climate change: Ecological response surfaces and paleoecological model validation. In Smith JB, Tirpak DA (Eds) *The potential effects of global climate change on the United States—Report to Congress.* Volume D—Forests. USEPA, Washington, DC.
Overpeck JT, Rind D, Goldberg R (1990) Climate-induced changes in forest disturbance and vegetation. Nature 343:51–53.
Parker IM, Mertens SK, Schemske DW (1993) Distribution of seven native and two exotic plants in a tallgrass prairie in southeastern WisconsIn The importance of human disturbance. Am Midlandnat 130:43–55.
Pastor J, Post WM (1988) Response of northern forests to CO_2-induced climate change. Nature 334:55–58.
Putz FE, Sharitz RR (1991) Hurricane damage to old-growth forest in Congaree Swamp National Monument, South Carolina, U.S.A. Can J For Res 21:1765–1770.
Reams GA, Van Deusen PC (1993) Synchronic large-scale disturbances and red spruce growth decline. Can J For Res 23:1361–1374.
Reeve JR, Ayres MP, Lorio PL, Jr. (1995) Host suitability, predation, and bark beetle population dynamics. In Cappuccino N, Price PW (Eds) *Population dynamics: New approaches and synthesis.* Academic Press, San Diego, CA.
Rind DR, Goldberg R, Hansen J, Rosensweig C, Ruedy R (1990) Potential evapotranspiration and the likelihood of future drought. J Geophys Res 95(D7):9983–10004.
Robertson GP, Crum JR, Ellis BG (1993) The spatial variability of soil resources following long-term disturbance. Oecologia 96:451–456.
Schmidt H, von Storch H (1993) German Bight storms analyzed. Nature 365:791.
Sheffield RM and Thompson MT (1992) Hurricane Hugo: Effects on South Carolina's forest resource. USDA, For Serv, Southeast For Exper Stat Res Pap SE-284.
Showalter TD, Turchin P (1993) Southern pine beetle infestation development: Interaction between pine and hardwood basal areas. For Science 39:201–210.
Spurr SH (1956) Natural restocking of forests following the 1938 hurricane in central New England. Ecol 37(3):443–451.
Stout IJ, Marion WR (1993) Pine flatwoods and xeric pine forests of the southern (lower) coastal plain. In Martin WH, Boyce SG, Echternacht AC (Eds) *Biodiversity of the southeastern United States: Lowland terrestrial communities.* John Wiley and Sons, Inc., New York.
Tilman D (1996) Biodiversity: Population versus ecosystem stability. Ecol 77:350–363.
Turner MG (1987) *Landscape heterogeneity and disturbance.* Springer-Verlag, New York.
Underwood AJ (1989) The analysis of stress in natural populations. Biol J Linnean Soc 37:51–78.
Wagner TL, Gagne JA, Sharpe PJH, Coulson RN (1984) A biophysical model of southern pine beetle, *Dendroctonus frontalis* Zimmermann (Coleoptera: Scolytidae), development. Ecol Model 21:125–147.
Ware S, Frost C, Doerr PD (1993) Southern mixed hardwood forest: The former longleaf pine forest. In Martin WH, Boyce SG, Echternacht AC (Eds) *Biodiversity of the southeastern United States: Lowland terrestrial communities.* John Wiley and Sons, Inc., New York.

Wigley TML (1985) Impact of extreme events. Nature 316:106–107.
Wigley TML, Jones PD, Kelly PM (1980) Scenario for a warm, high-CO_2 world. Nature 283:17–21.
Wilson SD, Keddy PA (1986) Species competitive ability and position along a natural stress/disturbance gradient. Ecol 67:1236–1242.
Winner WE (1994) Mechanistic analysis of plant responses to air pollution. Ecol Appl 4:651–661.

Section 6. Socioeconomic Impacts of Global Change

41. Evaluation of Effects of Forestry and Agricultural Policies on Forest Carbon and Markets

Ralph J. Alig, Darius M. Adams, and Bruce A. McCarl

Possible global warming has prompted examination of alternative policy measures for reducing excessive carbon dioxide (CO_2) in the atmosphere caused by emissions. These measures include forestry-based strategies for sequestering additional increments of carbon (e.g., Sampson and Hair, 1992; Haynes et al., 1994; Alig et al, 1997). Forests are a dominant part of the landscape in most of the southern United States, storing most of the carbon residing in terrestrial ecosystems. Forestry and agriculture cause the largest changes in the region's vegetation; possible future changes in the area, cover types, and ages of forests are important considerations for those examining policies for sequestering more carbon in forests.

We used a linked model of the forestry and agricultural sectors to evaluate the impacts of forestry and agricultural policies on southern forests and forest carbon. Global change mitigation policies through forest management of carbon involve increasing the standing inventory of forest biomass, thereby sequestering carbon. We examined a forestry policy aimed at increasing forest inventory and carbon, and two other policies that indirectly affect forest carbon storage. The latter two policies more closely resemble past natural resources policies. In this chapter, we first review prior research and summarize historical land exchanges between forestry and agricultural uses in the southern United States[1] We then describe a

---

[1] Agricultural uses here refer to cropland, pastureland, and rangeland.

model that projects changes in land uses, forest types, timber management, and other variables that impact or influence future forest resource conditions and markets. Third, we describe major assumptions and projections for a base case and three alternative scenarios. Finally, we discuss policy implications of our research findings and future research needs.

Prior Research

Our research for the Southern Global Change Program (SGCP) was designed to investigate regional and national economic impacts of selected global change policies (Alig et al., 1994). Our application addressed gaps in prior regional and national forest resource modeling (e.g., Haynes et al., 1994), which are the endogenous modeling of 1) land-use changes between the forest and agricultural sectors, and 2) investments in timber-management intensification on existing forestland. Earlier models also used a period-by-period approach (Adams and Haynes, 1996), in contrast to determining intertemporally optimum outcomes (i.e., present and future market conditions are linked) (Alig and Adams, 1996).

Models used in past applications in global change assessments have not accounted for full land-base interaction between the forestry and agriculture sectors. In general, prior models assume that the area of land that would shift from competing uses will rise (along a "land supply function") as land rents in the use under study increase. Land prices in alternative uses are exogenous. Forest sector and multisector carbon sequestration studies have recognized a portion of this interactive nature of land markets using a land-supply function (Adams et al., 1993; Alig, 1986; Moulton and Richards, 1990; Parks and Hardie, 1995), but limitations remain. First, the land-supply relations have generally been static or have been taken either as some form of "equilibrium" or long-term relation, or simply assumed to be fixed, with no explicit recognition of the intertermporal dynamics of land-use decisions and rent determination. Second, land prices have been variable only in the competing sector, with only a one-way flow allowed (e.g., from agriculture to forestry).

In an earlier paper (Alig et al., 1995), we compared projections from our intertemporal optimization model—the Forest and Agricultural Sector Optimization Model (FASOM)—with those from models used in the 1993 Resource Planning Act (RPA) Assessment Update (Haynes et al., 1995). Although both types of models can reflect interregional differences in the forest resource situation, timber processing, and other factors, the FASOM projections differ in several key ways. The FASOM model projects that 1) significantly more near-term adjustments will occur in response to market signals (e.g., millions more hectares (ha) of tree planting in the South by nonindustrial private forest (NIPF) owners and responses by forest industry), 2) higher levels of near-term timber harvest in anticipation of increased long-term supply resulting from investments in forest management, and 3) more stable longer-term levels of forestry prices (Alig et al., 1995).

Modeling Approach

We applied the FASOM model of the forestry and agriculture sectors (Adams et al., 1996 a,b) to investigate forest carbon differences and economic impacts under selected scenarios. The FASOM model is constructed as a multiperiod, price-endogenous, spatial equilibrium market model. Because of the dynamic nature of the market decisions being considered, the objective function maximizes the discounted sum of producers' and consumers' surplus in the agriculture and forest sectors over the projection horizon. A FASOM solution reflects price and quantity equilibria established in each period with perfect knowledge of market conditions in all periods (e.g., land-price equilibrium between sectors). Producers are assumed to hold rational expectations about future prices and maximize the net present value of timber and agricultural investments. Equivalently, FASOM's accounting of the interactive nature of land markets means that land migrates into the sector that promises the highest net present value of future returns considering the costs of use conversion and land-movement limits.

The forestry and agriculture sectors are linked through the land-transfer activities and constraints. We employ a nine-decade projection period, though our discussion of results and policies focuses on the fifty years from 1990 to 2039. Exogenous model elements are held constant after the fifth decade in the forest sector.

The agricultural sector depicts crop and livestock production and secondary processing using key water, labor, and forage inputs, as well as primary product trade. The agricultural component of FASOM is adapted from the Agricultural Sector Model (Chang et al., 1992; McCarl et al., 1993), aggregated to regions matching those used in forestry. More than 200 production possibilities represent agricultural production options in each decade.

The forest sector portrays the planting and harvesting of timber (logs) on private lands in U.S. regions and foreign trade in logs. Logs are differentiated by six product classes (sawtimber, pulpwood, and fuelwood from both softwoods and hardwoods), and interproduct substitution is permitted from sawlogs to pulpwood to fuelwood (Adams et al., 1996a). Forest land areas are differentiated by "existing" and "new" activities, depending on whether their associated timber stands were present in the initial timber inventory at the start of the projection or were created during the course of the projection.

The basic form of the forest sector model is a "model II" even-aged harvest scheduling structure (Johnson and Schuerman, 1977). A mathematical description is given in Adams et al., (1997). The FASOM model considers only the log market portion of the forest sector. Log demand is derived from the markets for such processed products as lumber, plywood, and paper. Logs are differentiated by six product classes, hardwood and softwood sawlogs, pulpwood, and fuelwood. Empirical demand functions for softwood and hardwood sawtimber and pulpwood are derived from solutions of the Timber Assessment Market Model (TAMM) solidwood and National Acidic Precipitation Assessment Program (NAPAP) pulpwood models by summing regionally derived demand relations (Adams and

Haynes, 1996; Ince, 1994). Substitution is permitted between sawlogs and pulpwood, pulpwood and fuelwood, also between residues generated in sawlog processing and pulpwood. Log trade with regions outside the United States is recognized by including price-sensitive, product-specific demand (export) or supply (import) functions for each region as appropriate, based on historical or anticipated offshore trading patterns.

To project changes in forest carbon stocks, we need to model the basic afforestation, regeneration, intermediate stand treatments, harvest, and other treatments applied to timberland in the southern United States. The FASOM model describes private timberland in terms of location by region, private ownership (forest industry and nonindustrial private), forest type[2] (species composition), site quality (potential for wood volume growth), management intensity (the type of timber management regime applied to the area, if any), and ten-year age-class. Each stratum for private timberlands in FASOM is characterized in terms of the number of timberland ha and the growing stock volume per unit area (in cubic meters per ha) it contains, drawn from data used in the 1993 RPA Timber Assessment Update (Haynes et al., 1995). The FASOM model simulates the growth of existing and regenerated stands by means of timber yield tables that give the net wood volume per ha in unharvested stands for strata by age cohort.[3] Harvest of a ha of timberland involves the simultaneous production of some mix of softwood and hardwood timber volume (e.g., Birdsey, 1992), which is then translated into hardwood and softwood products (e.g., sawlogs, pulpwood, and fuelwood). The FASOM model's determination of optimum timing of final harvest is also important in terms of species shifts, because harvesting facilitates conversion to other species or management types (e.g., planted pine areas in the region). Changes in forest carbon stocks are accounted for in FASOM for five carbon pools which are 1) tree, 2) woody debris, 3) soil, 4) forest floor, and 5) understory (Adams et al., 1996b).

Among the FASOM regions, the U.S. South is a key timber supply region, with more than 85 million ha of forestland[4] (Powell et al., 1993), 29% of the U.S. total. The southern United States contains a large proportion of the nation's opportunities to increase forest growth on timberland and forest carbon sequestration

---

[2]Forest types in the FASOM model represent existing vegetation, in contrast to potential natural vegetation depicted in some global change models. Similarly, simulation of forest growth is empirically based, using recent rates based on measurements from more than 70,000 forest survey plots.

[3]Potential changes in climate are outside the scope of this study and could affect future areas of forest cover types and forest productivity (Burton et al., 1995). Our projections of timber inventory follow those from the 1993 RPA Assessment Update, which assumed a future in which the climate follows historical trends.

[4]The South spans from Virginia to Texas (USDA Forest Service 1988), and in FASOM consists of the Southeast and Southcentral regions. Forest land consists of areas at least 10% stocked by forest trees of any size, including land that formerly had such tree cover and that will be naturally or artificially regenerated. Ninety-four percent of southern forestland is classified as timberland, reflecting the general suitability and availability of forestland for timber production.

(Alig and Wear, 1992). Eight to 12 million ha of those opportunities involve marginal and environmentally sensitive agricultural land that could be converted to forests (USDA Forest Service, 1988; Moulton and Richards, 1990). Private owners control approximately 90% of the region's timberland, a much higher percentage than in competing timber supply regions in the West. Forest industry owns 20% and NIPF owners have 70%, with 10% in public ownership. Half of the United States private timberland is in the southeast region.

Description of Scenarios

The FASOM model is well-suited to investigate outcomes under different forestry and land-use policies. To better illuminate the policy scenarios we examine in the next section, we will first review some important historical background on southern land-use changes involving forestry. Although some land in the southern United States has repeatedly shifted between agricultural and forestry uses, areas in both uses have been converted to developed uses (Alig et al., 1990).[5] Since 1950, forest area has decreased by 11 million ha (Figure 41.1), agricultural area has increased by 4 million ha, and area of urban and other special uses has increased by 6 million ha or 22% (Daugherty, 1995). The potential for further reallocation of land between uses is substantial, as land capability class (LCC) data indicate that substantial areas of land are suitable for use in either forest or

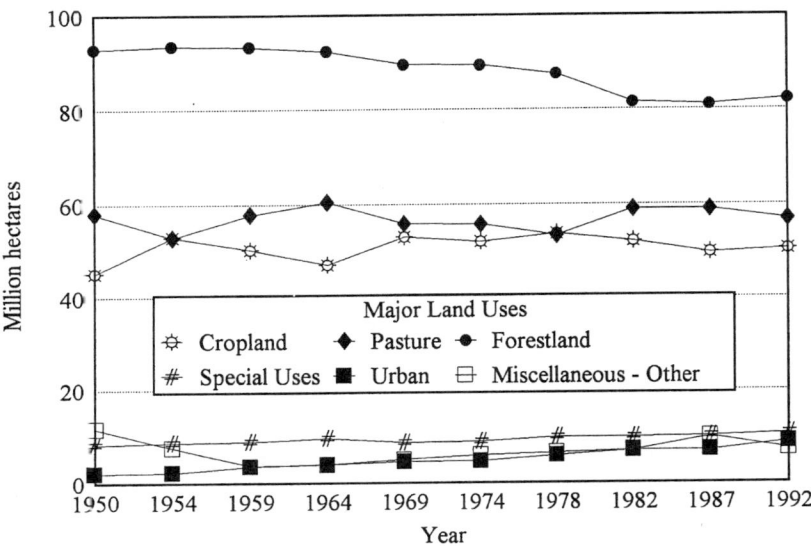

Figure 41.1. Trends in land use in the U.S. South, 1950–1992 (source: Daugherty, 1995).

---

[5]Over the long historical time period, forest area is about two-thirds of what was forested nationally in 1630; most of the land converted from forest has been shifted to agriculture.

agricultural uses (USDA Natural Resource Conservation Service, 1996). At the extremes of the LCC spectrum, there are large areas of land that could be shifted to another use. Approximately 10 million ha of southern forest land are in LCC I and LCCII, which are prime farmland; a further 13 million ha are in LCC III, which are suited to cultivated agricultural crops. Conversely, approximately 8 million ha of cropland and pastureland are in LCC's V–VIII, land with marginal crop productivity in many cases. The land class LCC IV—lands designated as not suitable for continuous cropping—contains more than 3 million ha of cropland in the region, 12 million ha of forestland, and 5 million ha of pasture.

In the FASOM model, we identify NIPF timberland area in the region that could be converted to cropland and pastureland, and also agricultural land that could be shifted to timberland. Land is converted from forestry to agriculture when the present value of expected land rents in agricultural uses exceeds that of timber growing, and from agriculture to forestry when the reverse is true (Alig, 1986). Agricultural programs have influenced forestry by encouraging the cropping of southern forest-land, given that so much land is suitable for both land uses. Such agricultural programs have affected not only the forest land base but the forest type as well because land removed may be of any forest type, but land returned to forestry via programs is generally planted to pine in the region (Lee and Alig, 1991). Tree species composition and level of timber management intensity are free to vary, in line with optimum adjustments, each decade within that fixed industrial land base.

The base case (BASE) assumptions for the agriculture sector are derived from Chang et al., (1992) and McCarl et al., (1997). The forest sector assumptions are derived from the USDA Forest Service 1993 RPA Assessment Update (Haynes et al., 1995). The continued conversion of timberland to developed uses, especially around such metropolitan areas as Atlanta, is projected to result in at least another 3 million ha being converted in the region by 2030 (Alig and Wear, 1992).[6] Land-use competition between forestry and agriculture will determine whether there will be further reductions in the timberland base or if forestry can gain area to offset some of the losses to urban and developed uses.

To investigate the outcomes of land use competition between the forestry and agriculture sectors under different policies, we examined three alternative scenarios. These are outlined in the following sections.

Forest Carbon Program

Projections in the BASE and those of several other studies (Birdsey, 1992; Turner et al., 1993) show a decreasing rate of carbon (C) uptake by forests over time. Policy-makers are interested, however, in the costs and implications of achieving constant or increasing rates of forest C sequestration in the face of greenhouse gas emissions that are apt to grow as a result of rising population and energy use. We

---

[6]Forest industry is assumed to gain some timberland from NIPF owners in the future, but the amount is offset by losses to urban and developed and other nonforest uses, leading to no net change.

developed a "direct policy" scenario in which a C goal is expressed as a series of decadal C-flux targets. The national inventory or stock is approximately 24 gigatonnes, and the forest carbon program (FCARB) scenario specifies a national C-flux (or net increase) target of at least 1.1 gigatonnes between 1990 and 2000, and a growing increment that expands by .1 gigatonnes each decade thereafter (1.1 + 2, 1.1 + .3, and so forth). In contrast, the baseline carbon flux is positive but becomes progressively smaller between 2010 to 2039. An increasing flux amount is consistent with a hypothetical scenario of increasing C emissions as both population and economic activity increase in the future. The model will determine the least-cost manner of attaining the target; no restrictions were placed on how the target could be met. The resulting solution can be considered a least-social-cost allocation of land and investments to meet the C target. Least social cost is defined as the minimum loss in the net present value (NPV) of the welfare of producers and consumers in the agriculture and forest sectors.

Permanent Conservation Reserve Program

Fundamental drivers of cycles in southern softwood forest area in the past have been timber management investment and shifts in land between forestry and agriculture. The Conservation Reserve Program (CRP) of the 1985 and the 1990 farm bills resulted in the direct planting of approximately 1 million ha to trees, largely on marginal cropland in the southern United States. At the same time, the diversion of other cropland to grass cover under the CRP acted to raise rents for cropland and lower them for pastureland, indirectly influencing the area in forests. In this "indirect policy" scenario we assume that the CRP program continues throughout the projection period but at a reduced level (hence, PERMCRP). In contrast to the CRP grass ha being released for possible conversion back to crop use by 2000 in the BASE, our scenario maintains 8.5 million ha nationwide in CRP grass cover. Present-day CRP area is about 14 million ha, so this scenario represents a 40% reduction in CRP area. This would act to increase cropland rents and cause some conversion of forest land to crop use, including land in the region. Such shifts in land use between forestry and agriculture on NIPF ownerships would impact forest carbon storage and timber markets.

No Land Transfers Between Forestry and Agriculture

Under the no land tranfers between forestry and agriculture (NOAG) scenario, no NIPF timberland is assumed to be transferred to or from the agricultural sector over the projection period. In this exploratory research we do not specify policy vehicles but investigate the implications of the assumption of cessation of land transfers between the forestry and agricultural sectors, and simulate outcomes to examine the sensitivity of forest resource and forest carbon projections to land-use changes involving the agricultural sector. The simulation of indirect effects on forest carbon storage from a policy centered on other objectives is of interest because multiple-purpose policies are increasingly used in recognition that forest stocks have many values. Under all scenarios, some NIPF timberland is assumed to be lost to urban and developed uses (Alig et al., 1990), therefore the area of NIPF timberland is also reduced in each future decade in this scenario.

Projections

The results for the BASE are presented here, focusing on the forest sector results, and for the three scenarios. The FASOM model is designed to produce national-level economic projections, including economic welfare measures of the distribution of effects across consumers and producers in forestry and agriculture. Forest resource projections include regional area changes for forest types, shifts in timber management intensity, and levels of softwood and hardwood inventory volumes.

Land Use

Forestry gains 2 million ha of land from agriculture in the southern United States between 1990 and 2039 in the BASE (Table 41.1). Forestry gains land from both crop and pasture uses, with the majority of the net gain from pastureland. Most of the net gain is realized in the 1990s decade, with reverse transfers to agriculture dominating after 2010. The total amount of land transfers in both directions over the projection period is approximately 18 million ha, an area equal in size to the combined timberland area of Alabama and Georgia.

The pattern of projected land transfers are similar under the forest carbon scenario, except that net exchanges in each decade are larger. Forestry has a net gain of about twice the amount of land in the first decade compared to the BASE, or approximately 4 million ha. The model allocates more land into forestry in the first decade to ensure that forest carbon targets will be met in later decades, given the multidecade production process of forestry.

The PERMCRP scenario results in the smallest net gain of land from agriculture, 1.1 million ha for the years 1990 to 2039. Decreased availability of land for crop agriculture in this scenario increases competition for forestland that could be converted to crop use, relative to the BASE in which 8.5 million ha of CRP land are released after the year 2000 for possible use in crop agriculture. Patterns vary over time too, as agriculture has a subsequent net gain of land from forestry for three decades starting in the year 2000.

Forest Cover

Changes in forest cover areas can be viewed in many cases as shifts in timber management intensity, and forest cover type changes in FASOM are driven by expected net returns from such conversions. The FASOM model endogenously selects optimum timber management investment for softwood and hardwood hectares on private timberland. The BASE and policy scenarios all indicate numerous opportunities for shifting more area to the commercially preferred softwood types and installing plantations (Table 41.2). There is some variation in the use of plantations and conversion of forest types across scenarios, as production technology allows variation in the combination of an array of other inputs with the

Table 41.1. Projections of NIPF Timberland Area (Million Hectares) in the South by Decade, and Net Change with Agriculture for the Base Case and Three Policy Scenarios, 1990–2039

End of Decade	Base		NOAG		FCARB		PERMCRP	
	Area[1]	Net Change with Ag.	Area	Net Change with Ag.	Area	Net Change with Ag.	Area	Net Change with Ag.
1990	57.63	2.38[2]	55.25	0.00	59.14	3.89	57.76	2.70
2000	57.90	0.35	55.17	0.00	59.78	0.71	56.98	−0.68
2010	56.73	−1.10	55.10	0.00	58.60	−1.11	55.74	−0.92
2020	56.77	0.11	55.03	0.00	58.80	0.28	55.64	−0.03
2030	56.94	0.25	54.95	0.00	60.15	1.43	56.08	0.25

[1] Although the initial area estimates in the model (not shown above) are for the year 1990, the projections above are by decade; for example, the first row is for the projection at the end of the 1990 decade, or 1999.
[2] Positive number indicates net shift to forest use.

Table 41.2. FASOM Projections for Distribution of Timberland Area (Million Hectares) in the South by Planted and Non-Planted Management Intensity Groups, for the BASE and Three Policy Scenarios, 1990 to 2039

Owner	Base		NOAG		FCARB		PERMCRP	
Decade	Planted	Natural Regen	Planted	Natural Regen	Planted	Natural Regen	Planted	Natural Regen
Forest industry								
1990	6.73	8.96	6.91	8.78	5.91	9.78	6.73	8.96
2000	9.36	6.33	11.29	4.41	8.59	7.10	11.28	4.42
2010	9.82	5.87	11.82	3.87	9.09	6.60	11.70	4.00
2020	9.67	6.02	10.85	4.85	9.09	6.60	11.75	3.94
2030	9.67	6.02	10.87	4.82	9.09	6.60	11.75	3.94
NIPF								
1990	15.72	41.90	14.48	40.77	14.61	44.52	15.61	42.15
2000	24.21	33.70	23.70	31.44	23.47	36.31	23.21	33.77
2010	24.86	31.87	25.90	29.20	24.18	34.41	23.89	31.85
2020	25.60	31.17	27.40	27.62	24.64	34.16	24.51	31.13
2030	28.47	28.47	29.76	25.20	26.03	34.12	28.34	27.73

land base. Use of plantations rather than natural methods to regenerate stands represents a significant alteration in the mix of land and nonland inputs in production.

In terms of forest cover changes, softwood area increases under all scenarios (Figure 41.2). The softwood area in 2039 is largest in the forest carbon scenario, in which faster-growing pine plantations are a key part of the requirements for meeting forest carbon targets. The largest area of pine plantations receiving the highest intensity of intermediate timber management is also projected for the FCARB scenario.

The smallest amount of softwood area is projected for the PERMCRP scenario, in which a smaller amount of land is gained from agriculture for conversion to plantations relative to the BASE. In response to NIPF owners planting a reduced amount of softwoods, forest industry owners increase their planting of softwoods under the PERMCRP scenario.

The area in southern pine plantations on forest industry lands either peaks or plateaus in approximately 2010 across all three scenarios. In contrast, the areas of NIPF pine plantations increase throughout the projection period across scenarios. The forest industry relies to an important degree upon hardwood conversions to increase pine plantation area. The availability of such convertible hardwoods stands on industry lands drops sharply after 2010. The NIPF ownership is able to augment hardwood conversions as a source of pine plantations with land transfers from agriculture, which necessarily enter the forest sector as pine plantations.

Pine plantations are projected to dominate forest industry land by 2039 across all scenarios, although the projected plantation proportions are smaller on NIPF lands (Table 41.2). In FASOM, although plantations compose the two highest timber management intensity classes—medium and high—naturally regenerated stands are in the lower two classes (Adams et al., 1996b). The BASE 2039 areas in plantations and naturally regenerated stands on NIPF southern lands are identical. The NIPF plantation areas slightly exceed the naturally regenerated areas in 2039 in the NOAG and PERMCRP scenarios, but the naturally regenerated areas comprise 57% of the total FCARB area. Thus, the FCARB strategy relies both upon retention of existing, naturally regenerated forests, as well as adding afforestation plantations.

The southern United States presently has more area in hardwood types than softwood ones, and the NIPF hardwood inventory is the largest among owner–species combinations in the region. A key projected land-base adjustment involving this NIPF aggregate is the conversion of more than 6 million ha of hardwoods to planted pine during the first decade. As we will see later, conversion to faster-growing pine plantations allows large increases in timber inventory and forest carbon within several decades.

In addition to conversion to pine plantations, changes in NIPF hardwood areas are also influenced by conversions to nonforest uses. Conversion of NIPF hardwood area to pine plantations is concentrated in the first two projection decades, although conversion to other uses (e.g., urban and developed uses) is a more constant activity over the projection period. When comparing the BASE vs

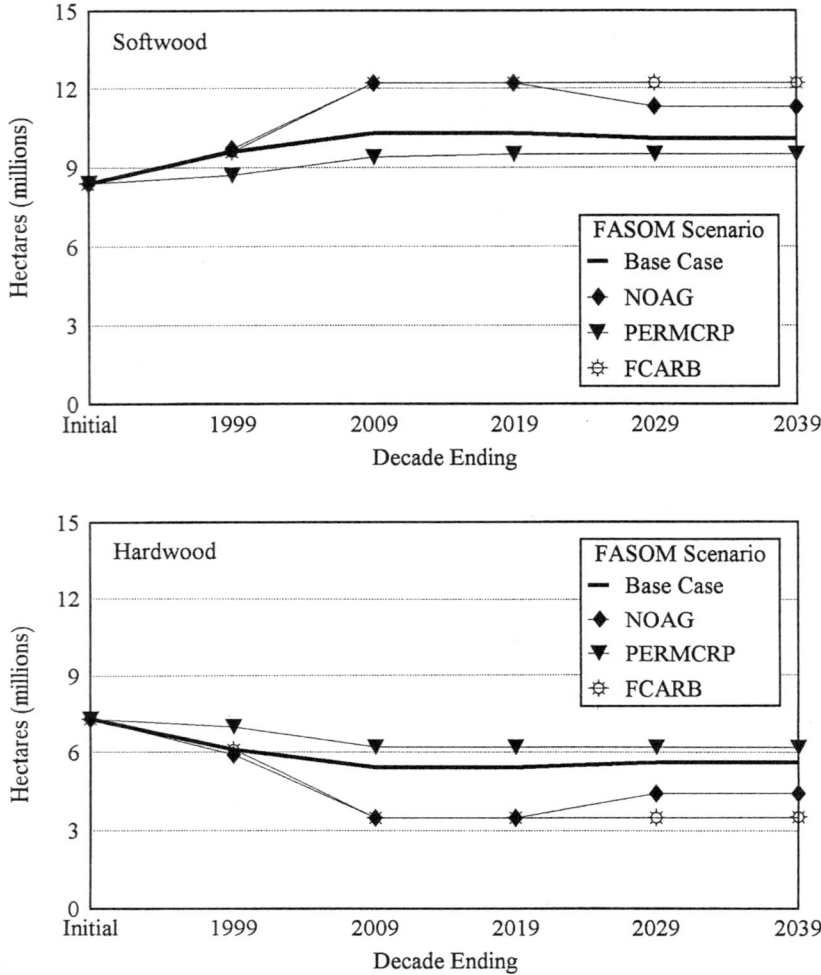

Figure 41.2a. Projected total area of timberland for forest industry ownerships in the U.S. South for the BASE and three alternative scenarios, 1990 to 2039.

NOAG projections, another influence of land-use changes on forest type areas is indicated by the smallest projected NIPF hardwood area for the NOAG scenario (Figure 41.2). Shutting off the net conversion of agricultural land to forestry in the NOAG scenario leads to higher log prices relative to the BASE, which heightens the financial stimulus for harvesting hardwood stands and converting them to planted softwood types.

Changes in hardwood areas on forest industry lands, in which the total land area is fixed over time, largely represent symmetrical changes with respect to those for

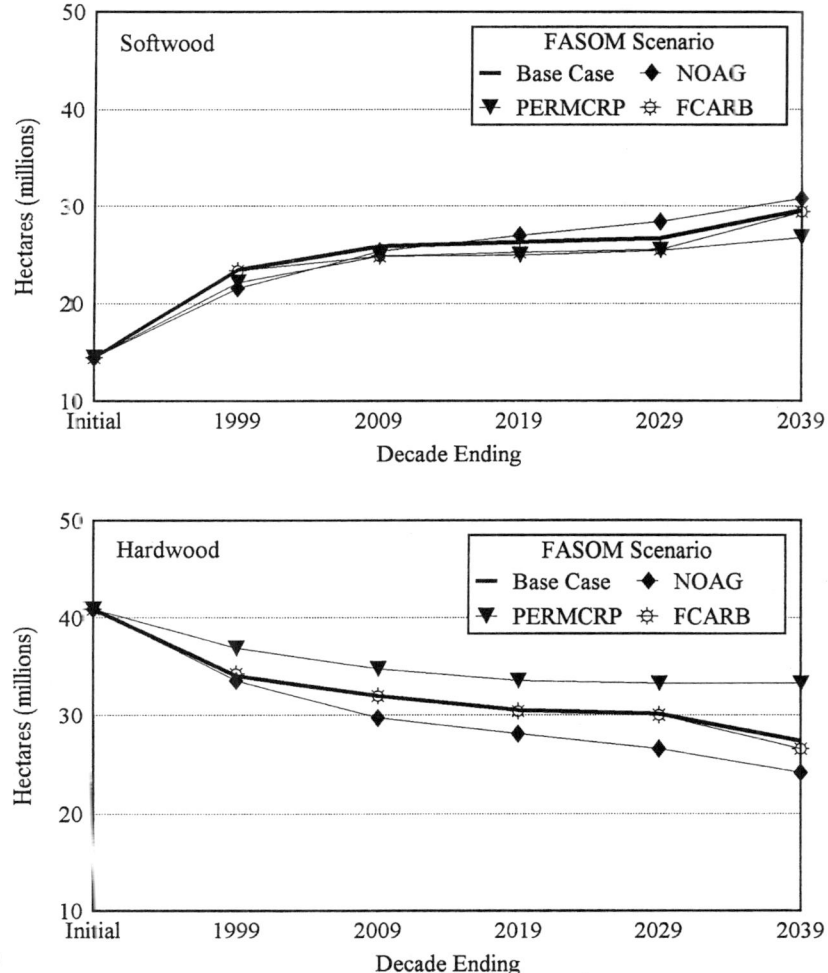

Figure 41.2b. Projected total area of timberland for nonindustrial private forest ownerships in the U.S. South for the BASE and three alternative scenarios, 1990 to 2039.

softwoods (Figure 41.2a). Forest industry converts many hardwood stands to softwood plantations, especially in the first part of the projection period.

Forest Carbon

Figure 41.3 shows the projected national forest C stocks for the BASE and the three scenarios. The BASE national level of forest C stock is projected to increase by about one-fifth by 2039. The forest carbon scenario with a growing C-flux each

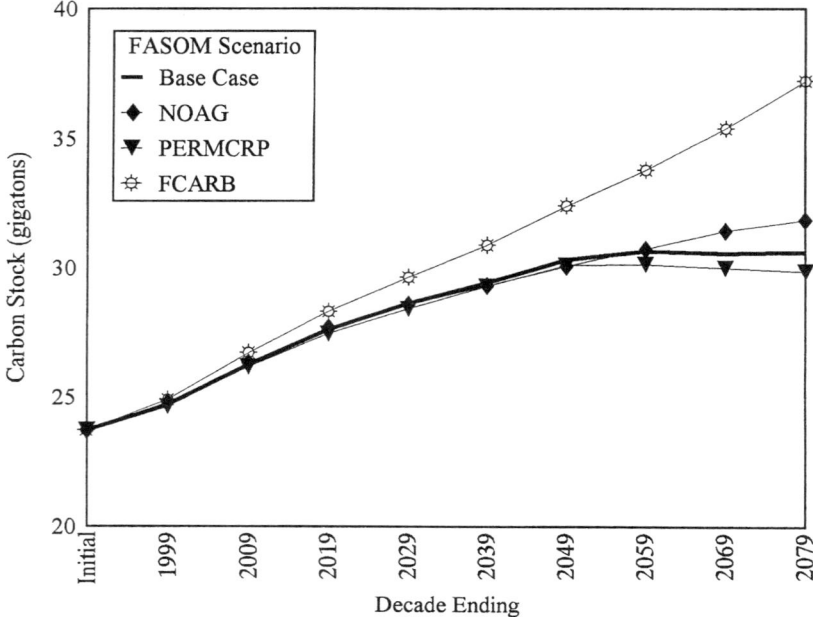

Figure 41.3. Projected stocks of U.S. forest carbon for the base case and three alternative scenarios, 1990 to 2039.

decade leads to 5% more forest carbon in 2039 than the BASE. The south-central region contributes the most to the attainment of the forest carbon goal, through additional pine plantations and land transfers from agriculture. By 2039, the FCARB level of forest carbon in the south-central region is 7% higher than in the BASE.

The southern forest C stock is projected to increase at a higher rate than the national pool because most of the nation's fast-growing plantations are projected to be in the southern region. The age-class structure of the region's forests has important implications for sustaining relatively rapid growth in forest C stocks. The region has many young softwood stands, where stand growth is relatively rapid, especially in plantations. Overall, although southern softwood timber volumes are projected in the BASE to increase by more than 70% by 2039, hardwood volumes increase by less than 10% (Figure 41.4). The FCARB scenario has the highest softwood inventory levels, with the 2039 level more than double the initial inventory level. Projected growth in forest carbon is higher than for the timber inventory because carbon stored in forest products allows for a faster build up.

Scenarios that lead to a smaller forest C stock than the BASE are cessation of land transfers between forestry and agriculture (NOAG) or conversion of more forest land to cropland (PERMCRP). However, in both cases the differences by 2039 are only about 1% because timber management intensification is able to offset some of the timberland area differences.

The present age-class distribution of forests does impose certain limitations for supply responses and opportunities for increasing forest carbon in the short-term. Timber harvest is projected to come from private timber inventories that are sharply compressed into the youngest classes, and an increasing fraction of harvest comes from ages at the merchantability margin, with a projected rise in log prices over the next 10 to 20 years. The age-class "gap" on NIPF timberland is represented by a large drop in area of trees 15 to 25 years of age, as harvest, land-area change, and planting activity have clearly impacted the age-class structure (Adams and Haynes, 1991). Shifting age-class structures affect southern softwood growth, as questions have arisen about the ability of the forest resources in the South to sustain growth and harvest over the next several decades (Cubbage et al., 1995, Sheffield et al., 1985). Long-term, private timber supply depends critically on the extent of investment or management intensity, with substantial financial opportunities for intensified timber management, including expanding the area of fast-growing pine plantations.

Economic Welfare

Table 41.3 shows the present value of selected welfare components for the BASE and the three scenarios (at the national level and producers surplus for the South and other regions). All scenarios have a net social cost compared to the BASE, as indicated by the combined welfare estimates in the last row of Table 41.3. However, the percentage of differences are relatively small, reflecting the arrays of possible adjustments in response to policies.[7] The carbon target scenario has a net social welfare cost of $12 billion, reflecting constraints to increase forest C storage.

Forestry producers would gain under all the scenarios relative to the BASE, although consumers of forest products would have welfare losses. The largest producer gain is when land transfers with agriculture are eliminated, leading to reduced forest area, and higher product prices. Softwood pulpwood and sawtimber prices under the NOAG scenario are 10% and 2% higher than BASE prices by 2039, respectively. These are the largest price differences relative to BASE prices for any of the alternative scenarios, except for a 3% increase in 2039 sawtimber prices under the FCARB scenario. The FCARB scenario results in more timberland area than the BASE, but timber harvest is lower than in the BASE because of the forest carbon constraints. This leads to higher product prices than in the BASE and a 2% increase in forestry producer surplus. None of the policy changes cause more than a 5% change in forestry production levels, as a result of the price-inelastic nature of the product demands within FASOM (Adams et al., 1996a). However, effects on producers are largest for the southern United States, which has the majority of opportunities for changes in private timber investment and land uses.

---

[7]We do not attempt an assessment of overall social welfare impacts, because, for example, we do not account for the increased costs of maintaining the CRP program.

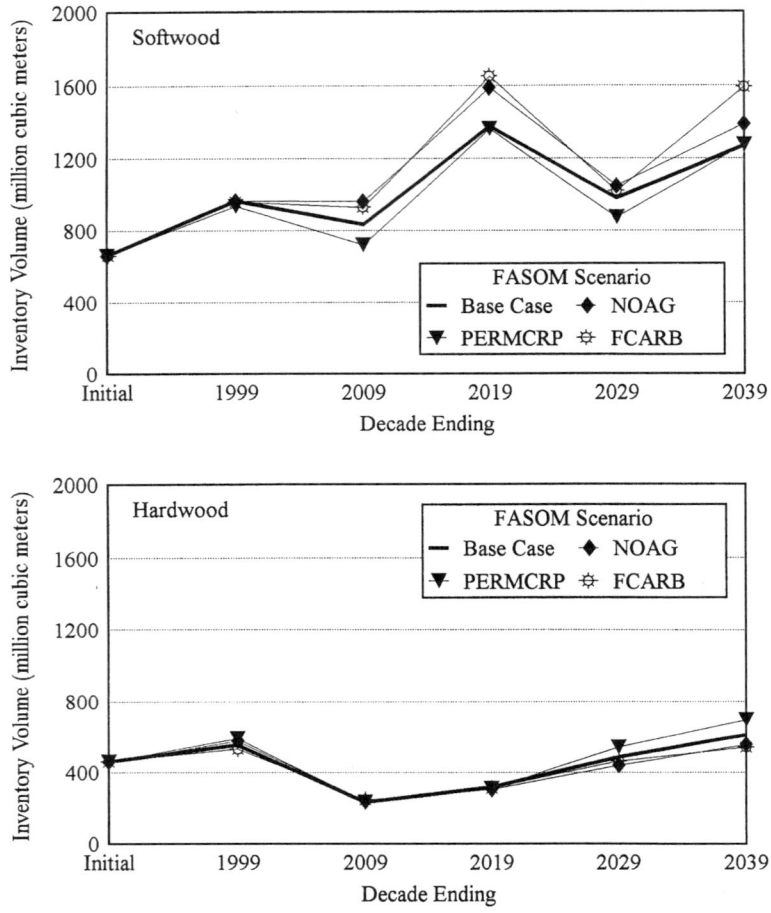

Figure 41.4a. Projected total stock of growing stock inventory for forest industry. Ownerships in the U.S. South for the BASE and three alternative scenarios, 1990 to 2039.

Agricultural welfare is impacted the most by the FCARB scenario, which has the largest net shift of land from agriculture to forestry. Regionally, the impacts are larger in the region, which has most of the land transfers.

The opportunities for expanding timber production and forest carbon sequestration in the South and the attendant welfare implications contrast with other studies that project recent trends in timber removals. For example, Cubbage et al., (1995) state that there is considerable doubt regarding the ability of southern forests to continue producing high levels of timber, as projected in the 1993 RPA Assessment Update (Haynes et al., 1995). The assertions by Cubbage et al. (1995) are based on an examination of southern forest survey data and studies of timberland availability. However, they qualify their prediction by saying that substantial enhancements in timber management intensity, prompted largely by even more substantial price increases, may be the only way that high levels of southern

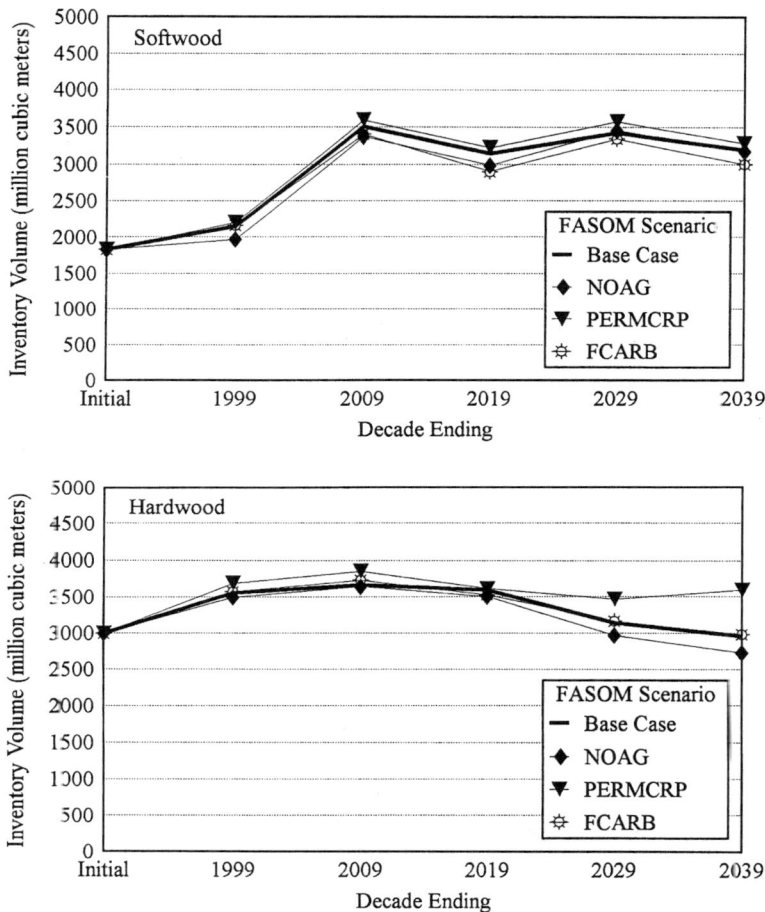

Figure 41.4b. Projected total stock of growing stock inventory for nonindustrial private forest ownerships in the U.S. South for the BASE and three alternative scenarios, 1990 to 2039.

timber could be produced. Our projections suggest it is possible to significantly increase levels of timber production in the region without substantial increases in forestry product prices. Such increases in timber inventory and forest carbon depend in part on expanded investment levels in timber management, especially by NIPF owners (Alig and Wear 1992).

Summary

Our results suggest that sequestration of forest carbon can potentially be significantly increased at the same time that southern timber production is increased,

Table 41.3. Present Value of Selected Welfare Components in FASOM for the Base Case (10^{12} dollars) and Percent Changes from BASE for Three Policy Scenarios

Welfare Component[1]	BASE	Percent Changes From Base		
		NOAG	FCARB	PERMCRP
Forest Sector Domestic Consumer Surplus	2.388	−1.3	−0.2	−0.5
Forest Sector Domestic Producer Surplus	0.229	11.0	2.2	4.1
Southern Producers Surplus	.089	19.9	2.6	6.1
Other Regions Producers Surplus	.140	5.4	1.8	2.8
Forest Sector Net Surplus	2.753	−0.3	0.0	−0.1
Agriculture Sector Net Surplus	37.414	−0.0	−0.1	0.0
Combined Net Surplus for Both Sectors	40.132	−0.0	−0.0	−0.0
Net Social Welfare Cost of Policy (10^{12} dollars)		0.009	0.012	0.012

[1] All benefits and costs are discounted at 4% for the period from 1990 to 2080.

without large increases in timber prices. Land transfers between the sectors, especially in the southern United States, are important in meeting a hypothetical global change policy of an increasing forest carbon flux in each future decade. Projections with the FASOM model indicate that generally economic adjustments act to buffer market-related impacts from policies affecting southern forest ecosystems and use of forest resources. In particular, owners respond rapidly by adjusting land allocation, harvest, and forest investment levels. Investment in forestry may be utilized to maintain or increase productivity, for example, replanting harvested areas back to the same species, or to enhance productivity through conversion of hardwoods to softwoods or through use of genetically improved stock in reforestation. The projected adjustments by private owners demonstrate that an array of resource and market conditions could be impacted by global change policies, and that other policies or developments (e.g., continued conversion of forest land to other uses) can also significantly impact costs and conditions for joint production of forest carbon.

The BASE results are generally robust to the policy changes caused in part by the flexibility imparted by the large and diverse private forest resource and the associated opportunities for additional investment in softwood timber management. Most of the timber investment opportunities are on NIPF lands, where owners have historically opted for relatively few investments in increasing the growth of forests. In contrast, industrial owners control approximately only 20% of the total timberland in the southern United States, but have invested in over 60% of the pine plantations in the region.

In the longer term, the largest area-based changes are projected for the NIPF ownership, and NIPF timber management intensification could also act to mitigate timber supply-based impacts. The FASOM model projects a small net gain in private timberland area in the region between 1990 and 2039 in the BASE. A net gain from agriculture is almost offset by projected conversion of forestland to urban and developed uses. This is consistent with historical trends. There has been only a small net change in timberland area over the last decade despite the afforestation of about 1 million ha of erodible cropland to tree cover through CRP. Land-use changes on NIPF lands also have important implications for forest industry. This is most clearly seen under the permanent CRP scenario, in which increased competition for NIPF forestland leads to higher prices, and, in response, the forest industry intensifies timber management on its fixed land base and increases the area in pine plantations. A major source of the additional plantation area is hardwood ha that are converted to pine, potentially leading to significant environmental effects (e.g., wildlife habitat changes) in addition to changes in forest carbon stocks. However, attaining the carbon target would also require large areas of afforestation, much larger than has been observed historically during one decade, even during such major tree planting programs as the Soil Bank or CRP programs.

In the FASOM model, any change in future condition is anticipated in the optimum (from a net social welfare viewpoint), and investment and land use are free to vary over time. The structure of FASOM provides a useful platform for future research to examine some of these questions of "stickiness" in product, land, and capital markets (e.g., Alig et al., 1996), including imperfections in land markets, limits on investment borrowing or capital budgets, and uncertainty regarding future market conditions.

References

Adams RC, Adams DM, Callaway JM, Chang C, McCarl BA (1993) Sequestering carbon on agricultural land: Social costs and impacts on timber markets. Contemp Pol Iss XI:76–87.

Adams DM, Alig RJ, Callaway JM, McCarl BA, Winnett S (1996b) The forest and agricultural sector model (FASOM): Model structure and policy applications. USDA For Serv Pac Northwest Res Stat Res Pap PNW-495. Portland, OR.

Adams DM, Alig RJ, McCarl BA, Callaway JM, Winnett S (1996a) An analysis of the impacts of public timber harvest policies on private forest management in the U.S. Forest Science 42(3):343–358.

Adams DM, Haynes RW (1991) Softwood timber supply and the future of the southern forest economy. So J Appl For 15:31–37.

Adams DM, Haynes RW (1996) The 1993 timber assessment market model: Structure, projections, and policy simulations. USDA For Serv Pac Northwest Res Stat Gen Tech Rep PNW GTR-368. Portland, OR.

Alig RJ (1986) Econometric analysis of the factors influencing forest area trends in the Southeast. For Sci 32(1):119–134.

Alig RJ, Adams DM (1996) Timber supply analyses in the U.S.: TAMM, FASCM, and related models. In Proceedings of conference, large-scale forestry scenario models: experiences and requirements. European Forest Institute, Joensuu, Finland.

Alig RJ, Adams DM, Haynes RW (1994) Regional changes in land uses and cover types: Modeling links between forestry and agriculture. In Proceedings of the 24th annual southern forest economics workshop, Univ GA, Warnell School of Forest Resources, May, 1994. Savannah, GA.

Alig RJ, Adams DM, Haynes RW (1995) Regional changes in land uses and cover types: Modeling links between forestry and agriculture. In USDA For Serv Southeast For Exper Stat Gen Tech Rep SE-92. Asheville, NC.

Alig RJ, Adams DM, Chmelik, J, Bettinger, P (1996) Private forest investment and long-run sustainable harvest volumes. In Proceedings of planted forests: Contributions to sustainable societies, June 1995, Portland, Oregon.

Alig RJ, Adams DM, McCarl BA, Callaway JM, Winnett S (1997) Assessing effects of mitigation strategies for global climate change with an intertemporal model of the U.S. forest and agriculture sectors. Environmental and Resource Econmics 9:259–274.

Alig RJ, Hohenstein WG, Murray BC, and Haight RG (1990) Changes in area of timberland area in the United States, 1952–2040, by ownership, forest type, region, and state. USDA For Serv Southeast For Exper Stat Gen Tech Rep SE-64. Asheville, NC.

Alig RJ, Wear DN (1992) U.S. private timberlands, 1952–2040. J For 90(5):31–37.

Birdsey RA (1992) Carbon storage in trees and forests. In Sampson N, Hair D (Ed) *Forests and global change: Vol I*. American Forestry Association, Washington DC.

Burton DA, McCarl BA, Adams DM, Alig RJ, Callaway JM, Winnett SM (1995) An exploratory study of the economic impacts of climate change on southern forests: Preliminary results. In USDA For Serv Southeast For Exper Stat Gen Tech Report SE-92. Asheville, NC.

Chang C, McCarl BA, Mjelde J, Richardson J (1992) Sectoral implications of farm program modifications. Am J Agr Econ 74: 38–49.

Cubbage F, Harris T Jr, Wear DN, Abt R, Pacheco G (1995) Timber supply in the South: Where is all the wood? J For 93 (7): 16–20.

Daugherty AB (1995) Major uses of land in the United States, 1992. USDA Econ Res Serv Agric Econ Rep No. 723. Washington, DC.

Haynes RW, Adams DM, Mills J (1995) The 1993 RPA Timber Assessment Update. USDA For Serv Rocky Mtn For Range Exper Stat Gen Tech Rep RM-259. Ft. Collins, CO.

Haynes RW, Alig RJ, Moore E (1994) Alternative simulations of forestry scenarios involving carbon sequestration options: Investigation of impacts on regional and national timber markets. USDA For Serv Pac Northwest Res Stat Gen Tech Rep PNW-335. Portland, OR.

Ince P (1994) Recycling and long-range timber outlook. USDA For Serv Rocky Mtn For Range Exper Stat Gen Tech Rep RM-242, Ft. Collins, CO.

Johnson KN, Schuerman L (1977) Techniques for prescribing optimal timber harvests and investments under different objectives. For Science Mono 18.

Lee KJ, Alig RJ (1991) Public policies and the southern forest landscape. In Chang J (Ed) Proceedings of the 1991 southern forest economics workers (SOFEW) workshop, Feb. 20–22, 1991. Washington, DC.

McCarl BA, Chang C, Atwood J, Nayda W (1997) The U.S. agricultural sector model. Texas A&M University, College Station, TX.

Moulton RJ, Richards KR (1990) Costs of sequestering carbon through treeplanting and forest management in the United States. USDA For Serv Gen Tech Rep WO-58. Washington DC.

Parks P, Hardie I (1995) Least-cost forest carbon reserves: Cost-effective subsidies to convert marginal agricultural land to forests. Land Econ 71:122–36.

Powell DJ, Faulkner P, Zhu Z, MacCleery D (1993) Forest resources of the United States, 1992. USDA Forest Service General Tech. Report RM-234. Rocky Mt. Forest and Range Exp. Station, Ft. Collins. CO.

Sampson N, Hair D (1992) *Forests and global change: Vol I—Opportunities for increasing forest cover.* American Forestry Association, Washington, DC.

Sheffield R, Cost N, Bechtold W, McClure J (1985) Pine growth reductions in the Southeast. USDA Forest Serv Southeast For Exper Stat Res Bull SE-83. Asheville, NC.

Turner D, Lee J, Koerper G, Barker J (1993) The forestland carbon budget of the U.S.: Current status and evaluation. Report to EPA. ManTech, Corvallis, OR.
USDA Forest Service (1988) The South's fourth forest: Alternatives for the future. USDA For Serv For Res Rep No 24. Washington, DC.
USDA National Resource Conservation Service (1996) The 1992 Natural Resources Inventory in the United States. NRCS Report, Washington, DC.

42. Economic Dimensions of Climate Change Impacts on Southern Forests

Diana M. Burton,[1] Bruce A. McCarl,[1] Claudio N.M. de Sousa,[1] Darius M. Adams, Ralph J. Alig, and Steven M. Winnett

Global climate change is apt to impact forests and forestry in the United States. Regions in the United States may be differentially affected. Forecasts of climatic change and of the biological response to climate change in trees and forests, however, are still largely hypothetical. Though much research is being done on aspects of climate change and biological response, conclusions are still preliminary.

This chapter reports a portion of the results from an examination of potential economic impacts of climate change on the forest products sector nationally and in the southern United States. A full presentation of the experiment is in Burton et al. (1997). This chapter focuses on the outer dimensions of the economic impacts. Hypothetical cases of extreme biological response to climate change are considered, representing probable boundaries on the range of forest response to stresses induced by global change. These scenarios should establish outer limits for economic impacts. For this exploratory study, scenarios are not based directly on climatic descriptors, but on alternative changes in forest-growth rates. Sectoral economic impacts are assessed in terms of softwood production and forest inventory, softwood and hardwood price levels and the economic welfare of timber consumers and producers.

---

[1] Senior authorship not assigned.

Literature

Much has been written in recent years about the potential effects of climate change. Several studies have explored the possible impacts of climate change on the agricultural sector (e.g., Adams et al., 1990; Kane et al., 1992). In general, these studies used hypothesized climatic scenarios, often based on general circulation models (GCMs). The climate scenarios were combined with information and assumptions in a biological model of annual crop response to temperature, precipitation, and atmospheric composition changes to determine yields. Resulting economic impacts were determined through the effects of these changes in supply on market equilibria and economic welfare. Articles have considered impacts at the farm level (e.g., Kaiser et al., 1993), the national level (e.g., Adams et al., 1990), and the impacts on international markets (e.g., Reilly and Hohmann, 1993; Tobey et al., 1992). Additionally, Mendelsohn et al., (1994) examined the impacts of climate change on agricultural land prices.

In forestry, the extent of knowledge concerning the biological response to climate change is more limited. The annual nature of most agricultural crops vs the multidecade growing cycle for trees naturally leads to relatively less information about climate change effects on forests. Short-term experimental results have not yet been validated for the life of the tree. In particular, little is yet known about how the response of an individual tree to climatic change generalizes to the stand and forest levels. An understanding of the response of the whole forest over relevant time frames is critical for the analysis of changes in forest sector supply and the resulting economic consequences of climate change.

A second category of prior studies has examined carbon sequestration in forests as a way of mitigating global climate change. Among the many various studies, the economic impacts of large tree-planting programs in the United States have been investigated by Adams et al., (1993); Haynes et al., (1994); Alig et al., (1994); and in Canada by Van Kooten et al., (1992).

The remaining literature consists of a few studies that utilize varying assumptions to examine the potential impacts of climate change on the forestry sector. Van Kooten and Arthur (1989) studied hypothetical increases in the biomass of certain forests in Canada under three alternative growth assumptions for the rest of Canada and the United States. They concluded that, under free trade assumptions, the overall economic welfare implications of climate change may be negative for Canada. In a similar study, Van Kooten (1990) examined the impact of 5% and 7.5% increases in harvests in Canada together with increases or decreases of similar magnitudes for U.S. harvests. He concluded that consumers in both countries benefit but that producers lose, and that the overall impact is positive for Canada only when harvests in the United States decline.

Burton et al., (1994) explored the scenario in which regions of the United States are differentially affected by climate change, depending on the realized pattern of warming and precipitation changes. They considered forest-yield changes of ± 5% and ± 10% and found that if net forest growth increased nationally as a result of climate changes, the northern United States would benefit relatively more than the southern United States. If net forest yields decreased nationally, however, the southern United States would benefit relatively more economically. If yields in the southern United States declined while rising elsewhere, the southern United States will lose economic welfare.

The economic impacts of mitigating climatic changes through tree planting was evaluated by de Steiguer (1994), who found that the increased inventory resulted in net benefits to both producers and consumers. Alig et al., (1994) considered market-level interactions and the linkage between the forestry and agricultural sectors when examining the impacts of afforestation programs under global change policies. They found smaller net effects than in static studies (e.g., Parks and Hardie, 1995) because of countervailing land transfers back to agriculture.

Global Climate Change

The change in global climate considered in this chapter is that which is induced primarily by increasing concentrations of atmospheric carbon dioxide (CO_2). Climatic descriptors used in the literature are generally obtained from GCMs. These models are large and complex and disagree on predicted climate for a doubling of CO_2 levels. Adams et al., (1990) and Kaiser et al., (1993) have commented on the disparity of predictions and the sensitivity of results to the climatic descriptor chosen. Robock et al., (1993) formally compared the predictions of five GCMs for use in impact analysis and found results to be highly sensitive to the GCM used.

However, there is general agreement that the warming effects will probably be greatest in the northern and temperate latitudes. In the United States, the southeastern area may experience a rise in temperature with either no change or a slight decrease in precipitation. The northern latitudes and other areas in the United States may experience a rise in temperature with either no change or a slight increase in precipitation (Adams et al., 1990; Melillo et al., 1993; Rind et al., 1992). In addition to these possibilities, a potential fertilization effect may result from increased concentrations of CO_2 (Sedjo and Solomon, 1989).

The FASOM Model

The model used for this study is the Forest and Agricultural Sector Optimization Model (FASOM) (Adams et al., 1996; Adams et al., 1994). The forestry and agricultural parts of FASOM can work independently; only the forestry model is used for this study. The FASOM model is a large, multisector, regional, multiperiod, nonlinear mathematical programming optimization model of stumpage and log markets written in GAMS (Brooke et al., 1988). The FASOM model maximizes the present value of the sum of the economic surpluses over the ten forecasted decades. Only the first six decades are reported in this chapter; therefore any distortions induced by terminal conditions are largely eliminated.

For these projections, predefined FASOM regions of the contiguous United States are aggregated. The southern region includes the southeastern and south-central United States. The northern region includes the Pacific northwestern states, the Rocky Mountains, the Lake states and the northeastern United States. The remaining states in the Corn Belt, Great Plains, Southwest, and California are considered as an aggregate southwest region. The northern and southern regions are specified along approximate latitudinal and regional lines.

Important exogenous variables in FASOM are public cut, forest product international trade flows and forest product demands. These variables are generated in the Timber Assessment Market Model (TAMM) (Adams and Haynes, 1996) and are inputs into FASOM. Variables endogenous to FASOM include the areas harvested, and production volume for industrial and nonindustrial private forests, market prices, and timber harvesting decisions. The softwood and hardwood products considered are sawtimber and pulpwood. Per-acre yield estimates and inventory data are derived from Aggregate Timberland Assessment System (ATLAS) data sets (Mills and Kincaid, 1992). This timber yield information is empirically based and reflects recent inventories and inventory events, such as fires.

Scenarios

This study compares three scenarios of extreme growth-rate changes induced by global climate change to the FASOM base scenario. One scenario considers a 50% increase in decadal growth rates in both the northern and southern regions of the United States. Another scenario examines a 50% decrease in decadal growth rates in both the northern and southern regions. To explore the possibility that temperature increases and precipitation changes may facilitate forest growth in the northern region and inhibit it in the southern latitudes, a third scenario investigates a 50% growth-rate increase in the northern region and a 50% growth-rate decrease in the southern region during each decade. For simplicity, because timber production is relatively low in the Southwest (when compared to the southern and northern regions), no change is assumed for the southwest region.

Growth-rate changes are compounded over the six decades. Sousa (1994) described the detailed calculations of growth-rate changes in the context of the FASOM model. Yield tables in FASOM are adjusted to reflect growth-rate changes. In essence, the shape of the yield curve is changed by applying the percentage change of the scenario to the derivative of the yield curve.

The Base Scenario

The FASOM base scenario provides a benchmark for comparison. Figures 42.1, 42.2, and 42.3 summarize production, inventory, price, and economic welfare information for the base scenario over six decades. The base FASOM run is detailed in Adams et al., (1996). Projections are shown for both the nation and for

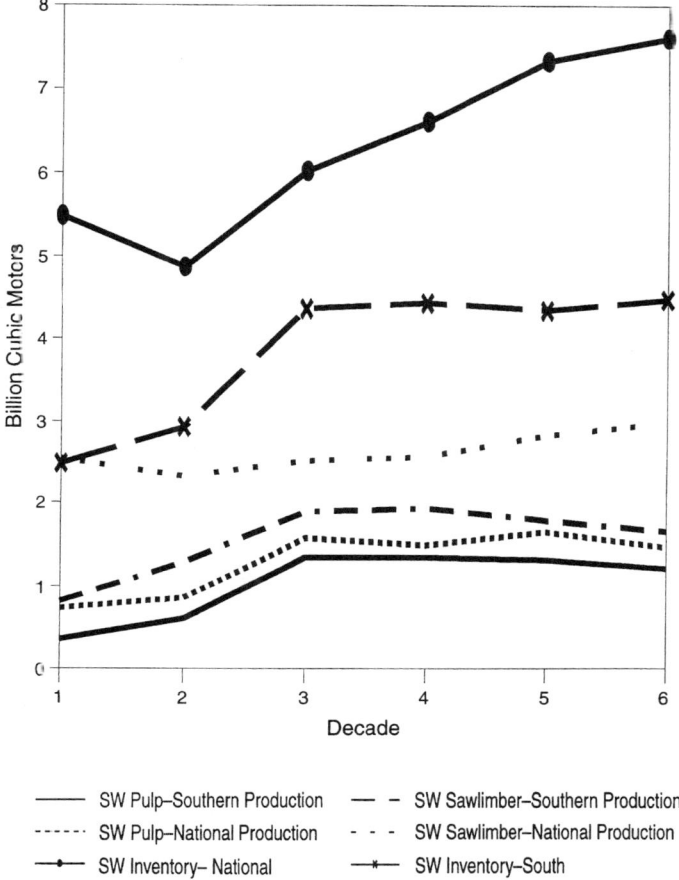

Figure 42.1. Base scenario—softwood production and inventory.

the southern region. Figure 42.1 shows volume of softwood pulp and sawtimber production for the nation and southern region. These are decade totals, representing ten years of harvests. Both national and southern pulp production rise for the first three decades and then level off. Southern sawtimber production shows a similar pattern. However, national sawtimber production decreases at first and then increases gradually over the six decades. Softwood inventory is shown in this figure as well. National softwood inventory falls initially and then rises; southern softwood inventory rises and then levels off.

Because the FASOM model clears a national market, only a national real price is reported for each market category in 1990 dollars per cubic meter. Production trends are reflected in the prices for softwood pulp and sawtimber in Figure 42.2. Softwood pulp prices increase initially as production rises, then decrease as production continues to rise, finally stabilizing as production levels off. Softwood

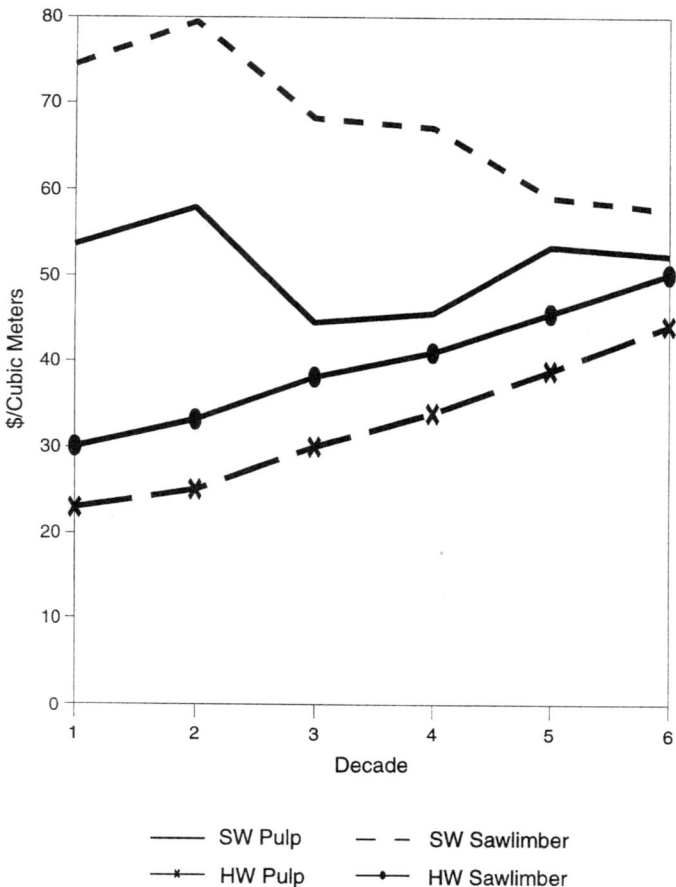

Figure 42.2. Base scenario—Pulp and sawtimber prices.

sawtimber prices rise as production is initially fairly flat, and then prices fall as production increases gradually. Hardwood pulp and sawtimber prices both increase steadily over the six decades.

Economic welfare is set forth in 1990 dollars as national consumer surplus, national producer surplus, and southern producer surplus for each decade totaled across all markets that were modeled. Each surplus comes from markets that clear a decade of production and is consequently large. Price and production patterns are reflected in the economic welfare measures. Real consumer surplus rises after an initial decline. Real national producer surplus, or profits, declines until the third decade, and then rises sharply for the remaining three decades. Southern producer surplus rises more sharply in the initial decade, falls, and then rises gradually again for the remaining decades.

42. Economic Dimensions of Climate Change Impacts

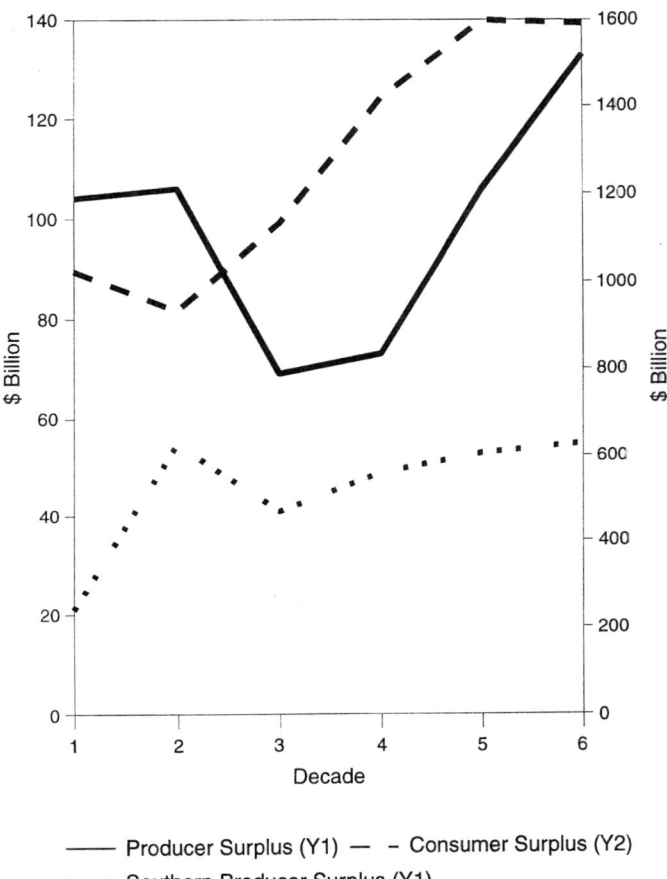

—— Producer Surplus (Y1)   — — Consumer Surplus (Y2)
- - - Southern Producer Surplus (Y1)

Figure 42.3. Base scenario—Economic welfare.

Fifty Percent National Growth-Rate Increase Scenario

Figures 42.4, 42.5, and 42.6 present results from the 50% national growth-rate increase scenario, an extreme upward adjustment in tree-growth rates resulting from climate change. This growth-rate change takes place in both the northern and southern regions of the United States.

National and southern pulp production rise initially, level off, and then decline at the end of the six decades. A similar upward pattern is evident for southern sawtimber. However, national sawtimber production decreases initially and then increases. Shown in the same figure, national softwood inventory falls initially in response to the large change in growth rates and then rises. Southern softwood inventories decrease strongly toward the end of the period as landowners respond

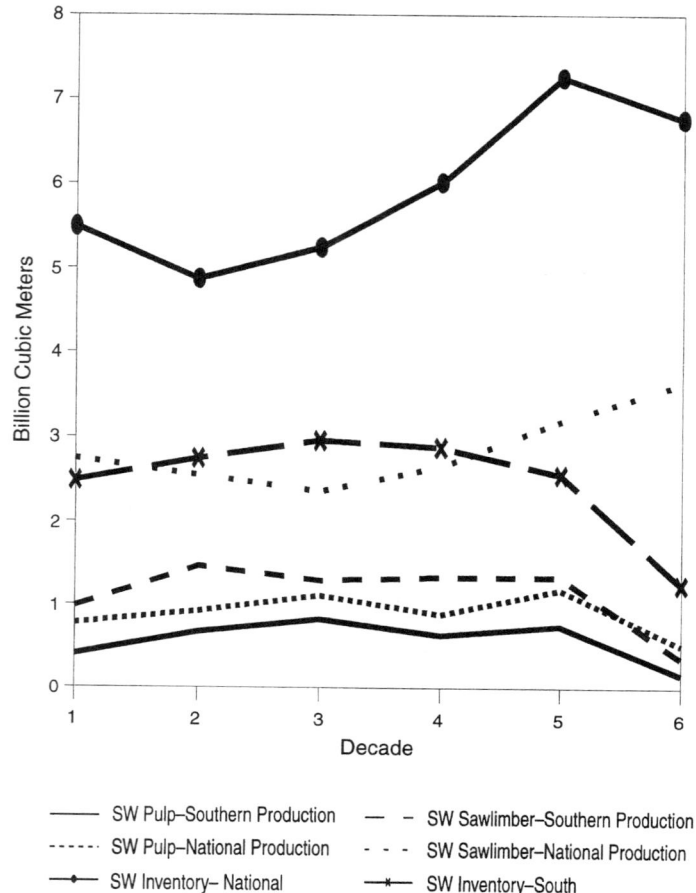

Figure 42.4. Fifty percent national growth-rate increase scenario— Softwood production and inventory.

to the increased growth rates and falling softwood prices with reduced management intensity.

Reflecting the great increase in resource availability from the rise in growth rates, prices fall for softwood sawtimber (Figure 42.5). Softwood pulp prices gradually increase and then fall off. With such high growth rates, sawtimber becomes cheaper to grow and can be sold for less. Hardwood pulp and sawtimber prices rise gradually over the six decades.

In Figure 42.6, consumer surplus increases across the last five decades as consumers benefit from more plentiful wood. Producer surplus falls nationally for the first four decades and then rises slightly. Southern producers initially gain surplus as they produce more but their surplus falls over the remaining five

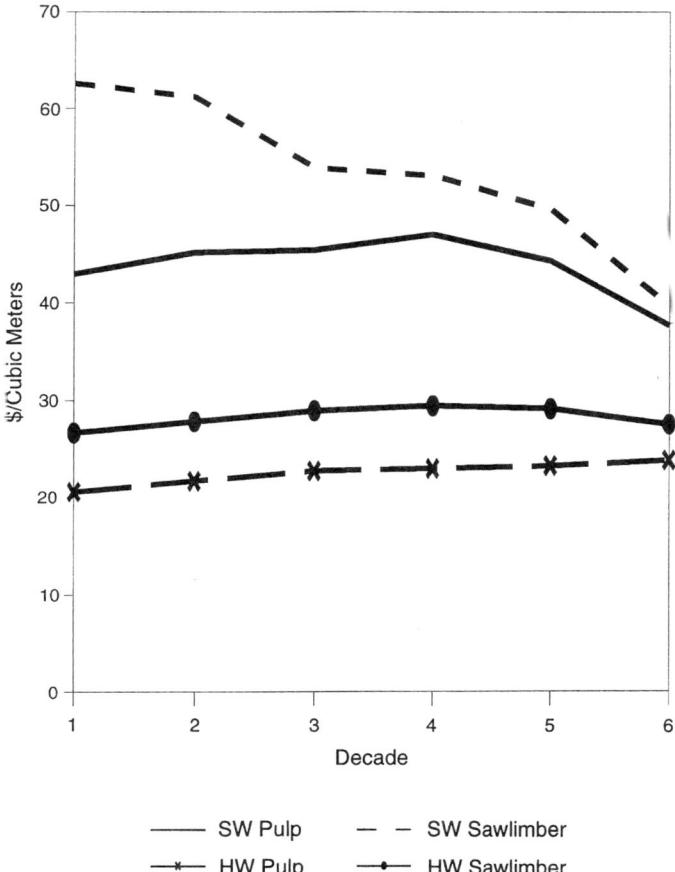

Figure 42.5. Fifty percent national growth-rate increase scenario—Pulp and sawtimber prices.

decades. Southern producer surplus is actually negative in the sixth decade of the simulation, indicating that costs are higher than revenues.

Compared to the base scenario, the 50% national growth-rate increase scenario shows some important changes. Softwood pulp production is much decreased while sawtimber production rises; the increased growth rates mean reduced costs to grow sawtimber nationally. In the southern region, the increased growth rates result in decreased softwood production as the south loses comparative advantage in fast-growing softwood. At the end of six decades, southern softwood production under this scenario is roughly 20% of the levels in the base scenario. Inventories are similarly affected as management of timberland in the southern region is largely abandoned and wood prices fall. Market share shifts northward, and

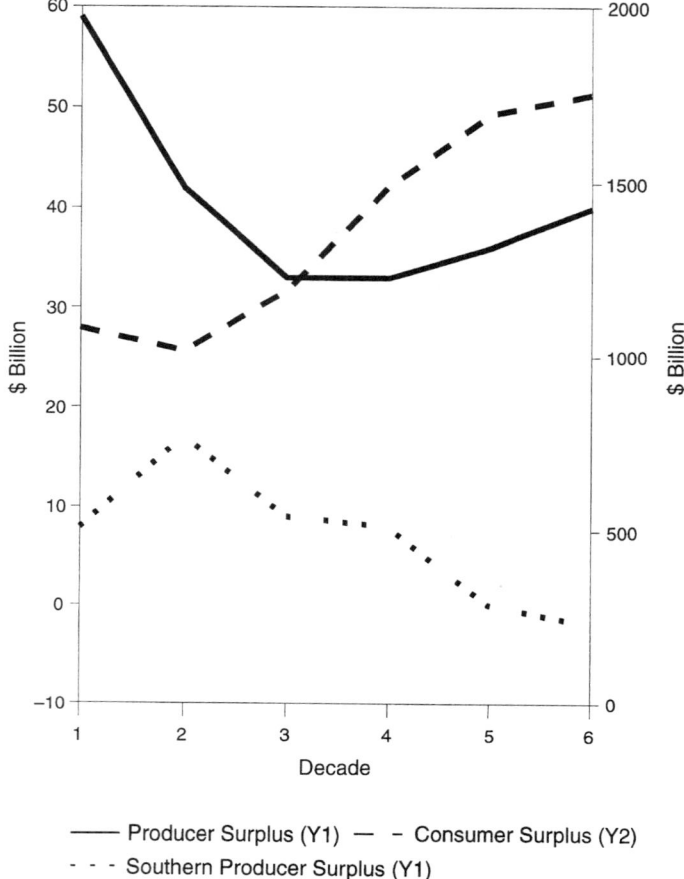

Figure 42.6. Fifty percent national growth-rate increase scenario—Economic welfare.

southern producers lose substantially over time. Consumers benefit more under this scenario than under the base scenario. Although producers lose profits nationally because of the increased abundance of the timber resource, southern producers lose relatively more than producers in other regions.

Fifty Percent National Growth-Rate Decline Scenario

Figures 42.7, 42.8, and 42.9 present results from the 50% national growth-rate decline scenario. Southern production of softwood pulp and sawtimber and national pulp production all rise over the six decades in Figure 42.7. In contrast, national sawtimber production generally falls. National softwood inventory generally increases, and southern softwood inventory rises more strongly over the six decades of the simulation.

Figure 42.8 shows that prices are generally up in this scenario, with only the

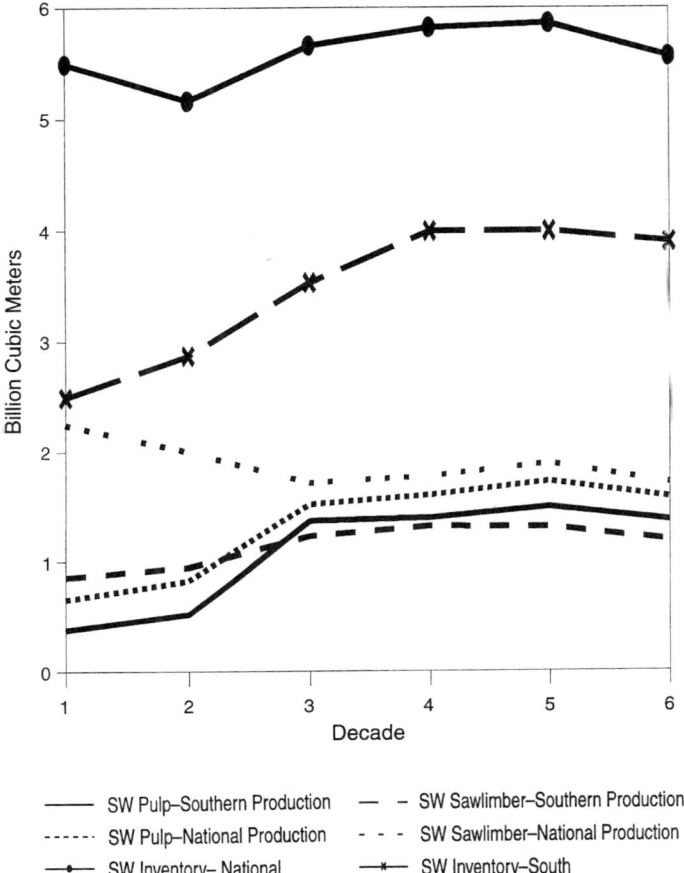

Figure 42.7. Fifty percent national growth-rate decline scenario—Softwood production and inventory.

softwood pulp price taking a downward direction in the third decade. These prices reflect the increasing scarcity of wood under this extreme growth-rate decline. With the increased prices come increased producer surplus both nationally and in the southern region, as shown in Figure 42.9, although southern producers gain less relatively. Consumers have to pay more for the increasingly scarce wood resource, and consumer surplus is lower by the sixth decade.

Compared to the base scenario, the steep decline in growth rates compounded over six decades results in some drastic changes. Because it is relatively more expensive to grow sawtimber, production of sawtimber is reduced compared to the base scenario, although not as much in the southern region. Softwood pulp markets increase production. Softwood inventories are down, with the declining growth rate only partially offset by increased management intensity. Reflecting the increased wood scarcity, prices are much higher than in the base

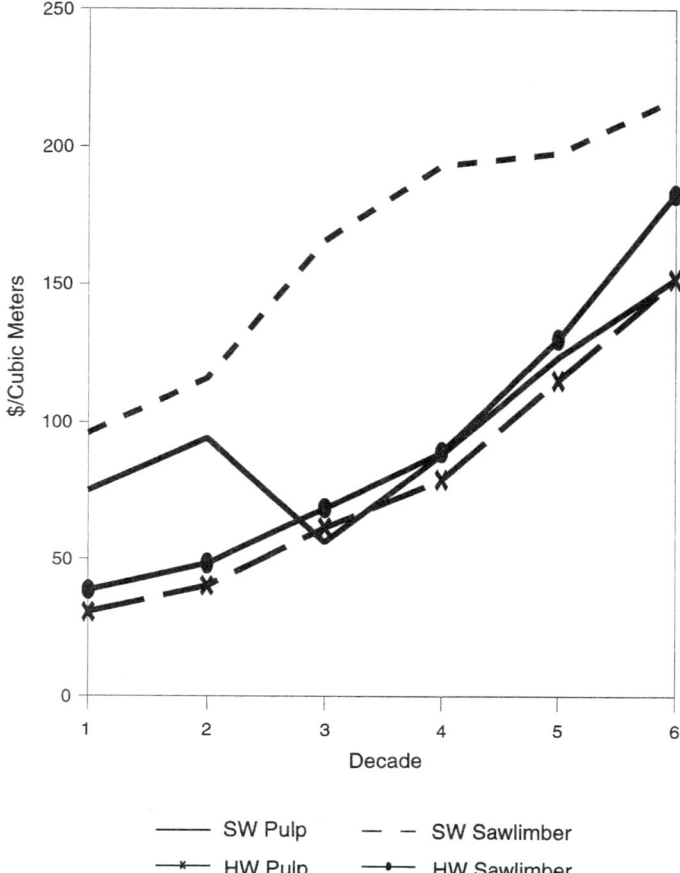

Figure 42.8. Fifty percent national growth-rate decline scenario—Pulp and sawtimber prices.

scenario, roughly three times the base levels by the end of the simulation. Consequently, consumers of wood lose surplus relative to the base scenario, and by the sixth decade consumer surplus is roughly half of that in the base scenario. In contrast, producers, gain greatly, seeing profits at five and six times the base level nationally and even higher in the southern region.

Southern Decline Scenarios

If latitudes in the United States are differentially affected, the northern region may gain forest growth at the expense of the southern region. In the extreme case, there might be a 50% increase in northern growth rates with a 50% decline in southern

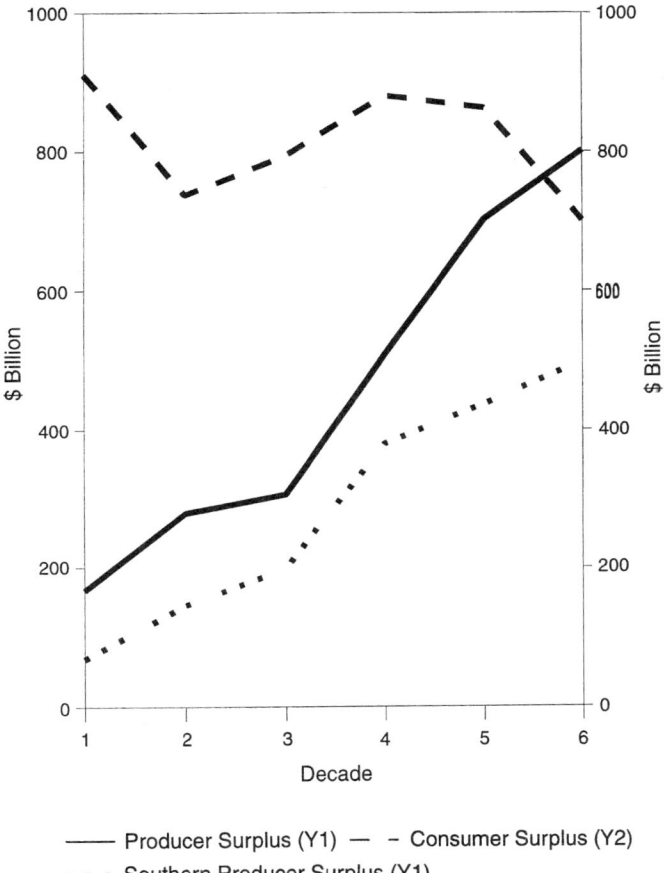

Figure 42.9. Fifty percent national growth-rate decline scenario—Economic welfare.

growth rates, as the timber growing area of the United States shifts to the north. Figures 42.10, 42.11, and 42.12 present results for this scenario.

Production changes are more variable in this scenario (Figure 42.10). Nationally, production of softwood pulp rises then falls, and production of softwood sawtimber falls and then rises. Southern softwood pulp production increases and then decreases to very low levels at the end of the simulation period. Southern sawtimber production presents a similar pattern. National softwood inventory decreases initially, then increases to end the six decades at a substantially higher level. Southern softwood inventory levels rise before falling off steeply toward the end of the simulation period. The southern region, under this extreme scenario, basically withdraws from the softwood business by the end of the six decades.

Price patterns are mixed (Figure 42.11). Softwood prices fall over the six decades, although hardwood sawtimber prices generally rise, with hardwood

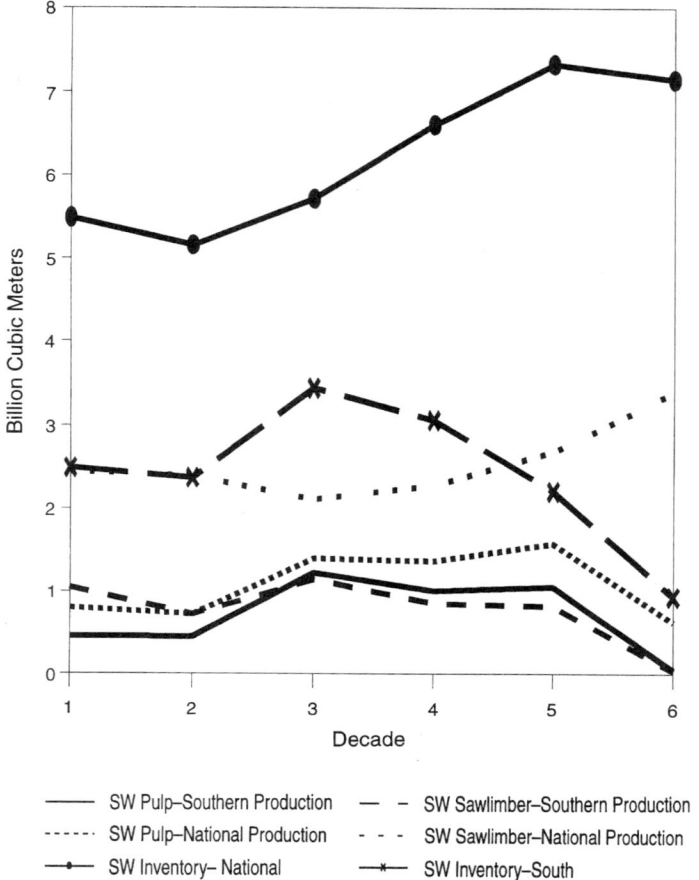

Figure 42.10. Fifty percent southern growth-rate decline scenario—Softwood production and inventory.

sawtimber prices rising and then falling. Reflecting these patterns, consumer surplus decreases and then increases for the remaining decades (Figure 42.12). National producer surplus rises, falls and then gradually rises. Southern producer surplus increases and then decreases to near zero by the end of the simulation.

Compared to the base scenario, this scenario shows that the southern region basically withdraws from the softwood business. With increased growth rates in the northern United States and decreased growth rates in the southern United States, the southern region can no longer compete in these markets. Softwood pulp production is at 4% of the base scenario level, and softwood sawtimber production is at 2% of the base level by the sixth decade. In this extreme scenario, only the hardwood markets remain viable in the southern region. Because prices

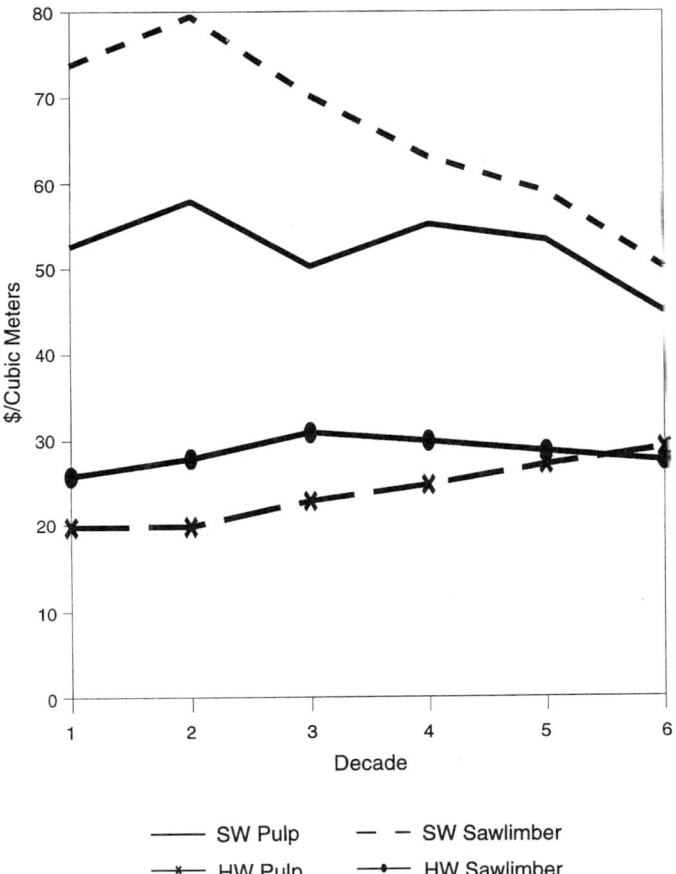

Figure 42.11. Fifty percent southern growth-rate decline scenario—Pulp and sawtimber prices.

are generally lower than under the base scenario consumers gain under this scenario and producers lose. However, as expected, southern producers lose much more than producers do nationally.

Summary

The nature of climate-induced biological change will greatly affect economic impacts. To fully dimension the economic impacts, a variety of scenarios must be examined. This chapter presents the results from the extreme scenarios, which represent the probable boundaries of economic impacts.

In summary, these three extreme case scenarios show that, at the outer limits, the impacts of climate change are apt to be felt more strongly by producers than

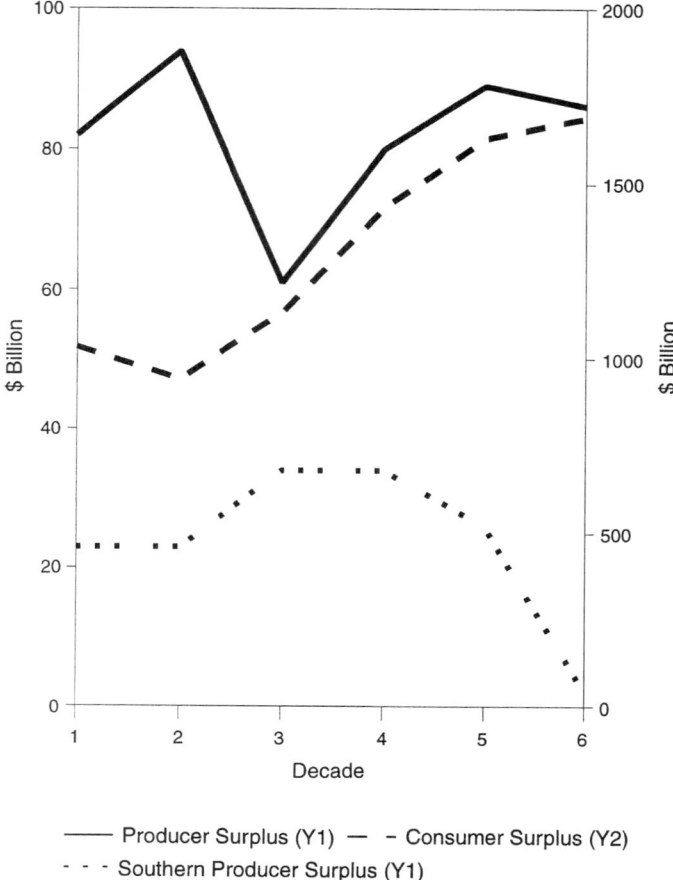

Figure 42.12. Fifty percent southern growth-rate decline scenario—Economic welfare.

consumers and by southern producers more so than producers in other regions. In particular, under two of these scenarios, southern producers see their profits eliminated and their presence in softwood markets drastically reduced. Although these scenarios are indeed extreme, scenarios with more moderate changes in growth rates show a similar pattern of impact in which southern producers could be substantially negatively impacted (Burton et al., 1996) compared to producers elsewhere.

These scenario results are generally consistent with those from earlier studies to the extent that they are comparable. Van Kooten (1990) also found that the impacts on producers are apt to be negative and that consumers will probably benefit.

The sensitivity of these results to certain aspects of the model should be considered. The scenarios evaluated in this study are constructed so that climate change effects are fully implemented in the first decade and the growth-rate changes compound thereafter. In reality climate changes will probably occur gradually with resultant phased-in effects. Effects that occur over time give greater opportunity for economic agents to adjust behavior to mitigate potential negative effects. This is an obvious area for further research.

In this study, the quantity of land in forestry is fixed. In reality, hectares would be expected to come into forestry when profits are high and to move out of forestry when profits sag. Alig et al., (1994) also investigated land shifts between forestry and agriculture.

This study incorporates perfect foresight on the part of economic agents and the mathematical programming model permits the resulting changes (sometimes large changes) to be realized immediately as economic agents adjust their behaviors to scenario conditions. Hence, some major shifts occur in the first decade or two in this model, although in reality such changes probably would take more time. Additionally, this is a full information model, in which all economic agents know everything about the model for the full projection timespan. In reality, this is not the case. To the extent that true behavior may be only partially adaptive in any time period and may be based on inaccurate or incomplete information, the results from these simulations can be seen as representing outer dimensions of potential economic impacts.

Lastly, the model is constructed using the best available data from the USDA Forest Service and other UDSA sources. Any uncertainty in this data will be reflected in the simulation results, as are the model assumptions.

In conclusion, although these scenarios are designed to explore climate change effects at their extremes, the impact on timber producers appears to be potentially very large when compared with the possible effects on consumers. Even considering the limitations of the study, these extreme cases show that the impacts of climate change have large potential economic dimensions. It is clearly in the interests of producers, and southern producers in particular, to try to narrow the amount of uncertainty around climate change impact forecasts. Better climate models and enhanced understanding of the biological response of whole forests to climate change are necessary as prerequisites to more focused economic impact estimates. This information could be quite valuable to forestry producers.

References

Adams RM, Adams DM, Callaway JM, Chang C-C, McCarl BA (1993) Sequestering carbon on agricultural land: Social cost and impacts on timber markets. Contemp Pol Iss XI:76–87.

Adams DM, Alig R, Callaway JM, McCarl BA, Winnett SM (1994) Forest and agricultural sector optimization model: Model description. Final report to USEPA, Climate Change Division, Washington, DC.

Adams DM, Alig R, Callaway JM, McCarl BA, Winnett SM (1996) An analysis of the impacts of public timber harvest policies on private forest management in the U.S. For Science 42:343–358.

Adams DM, Haynes RW (1996) The 1993 timber assessment market model: Structure, projections, and policy simulations. USDA For Serv PNW-368. Portland, OR.

Adams RM, Rosenzweig C, Peart RM, Ritchie JT, McCarl BA, Glyer JD, Curry RB, Jones JW, Boote KJ, Allen LH Jr (1990) Global climate change and US agriculture. Nature 345:219–224.

Alig R, Adams D, Haynes R (1994) Regional changes in land uses and cover types: Modeling links between forestry and agriculture. In Proceedings of the Southern Forest Economics Workshop, March, 1994, Savannah, GA.

Brooke A, Kendrick D, Meeraus A (1988) GAMS: A user's guide. The Scientific Press. South San Francisco, CA.

Burton D, McCarl B, Adams D, Alig R, Callaway J, Winnett S (1994) An exploratory study of the economic impacts of climate change on southern forests: Preliminary results. In:Proceedings of the Southern Forest Economics Workshop, March 1994, Savannah, GA.

Burton, D, McCarl B, Sousa CNM de, Adams DM, Alig R, Winnett S (1997) Economic impacts of climate change on southern forests. Fac Pap Series 97-18, Dep Agri Econ, Texas A&M University, College Station, Texas.

de Steiguer J (1994) Timber market impacts of the climate change action plan. In Proceedings of the Southern Forest Economics Workshop, March, 1994, Savannah, GA.

Haynes R, Alig R, Moore E, (1994) Alternative simulations of forestry scenarios involving carbon sequestration. USDA, For Serv, PNW-335. Portland, OR.

Kaiser HM, Riha SJ, Wilks DS, Rossiter DG, Sampath R (1993) A farm-level analysis of economic and agronomic impacts of gradual climate warming. Am J Agri Econ 75:387–398.

Kane S, Reilly J, Tobey J (1992) An empirical study of the economic effects of climate change on world agriculture. Clim Change 21:17–35.

Melillo JM, McGuire AD, Kicklighter DW, Moore B III, Vorosmarty CJ, Schloss AL (1993) Global climate change and terrestrial net primary production. Nature 363:234–240.

Mendelsohn R, Nordhaus W, Shaw D (1994) The impact of global warming on agriculture: A Ricardian analysis. Am Econ Rev 84(4):753–771.

Mills J, Kincaid J (1992) The aggregate timberland assessment system—ATLAS: A comprehensive timber projection model. Portland, OR: U.S. Department of Agriculture, Forest Service, Gen. Tech. Rep. PNW-281. Pacific Northwest Station.

Parks PJ, Hardie IW (1995) Least-cost forest carbon reserves: Cost-effective subsidies to convert marginal agricultural land to forests. Land Econ 71:122–36.

Reilly J, Hohmann N (1993) Climate change and agriculture: The role of international trade. Am Econ Rev 83(2):306–312.

Rind D, Rosenzweig C, Goldberg R (1992) Modeling the hydrological cycle in assessments of climate change. Nature 358:119–122.

Robock A, Turco RP, Harwell MA, Ackerman TP, Andressen R, Chang H-S, Sivakumar MVK (1993) Use of general circulation model output in the creation of climate change scenarios for impact analysis. Clim Change 23:293–335.

Sedjo RA, Solomon AM (1989) Climate and forests. In Rosenberg NJ et al., (Eds) Greenhouse warming: Abatement and adaptation. Resources for the Future, Washington, DC.

Sousa C (1994) Modeling economic impacts of climate change on U.S. forests. MS thesis, Dep Agri Econ, Texas A&M University, College Station, Texas.

Tobey J, Reilly J, Kane S (1992) Economic implications of global climate change for world agriculture. J Agri Res Econ 17(1):195–204.

Van Kooten GC (1990) Climate change impacts on forestry: economic issues. Can J Agri Econ 38:701–710.

Van Kooten GC, Arthur LM (1989) Assessing economic benefits of climate change on Canada's boreal forest. Can J For Res 19:464–470.

Van Kooten GC, Arthur LM, Wilson WR (1992) Potential to sequester carbon in Canadian forests: Some economic considerations. Can Pub Pol XVIII:2:127–138.

43. Assessing Present Biological Information for Valuating the Economic Impacts of Climate Change on Softwood Stumpage Supply in the South

James T. Gunter, Donald G. Hodges, and James L. Regens

The uncertainty inherent in projecting forest responses to climatic change has limited attempts to evaluate the potential economic impact of global climate change on forestry in the United States (Hodges, et al., 1992; Bazazz, 1990; Smith and Tirpak, 1989; DeLaune et al., 1987). This study compares the possible changes to southern timber supply resulting from the northward shift of softwood forests to the potential growth increases due to elevated atmospheric carbon dioxide (CO_2) concentrations. The study describes a potential approach to obtain the following information:

- Estimates of the potential effects of rapid, anthropogenically induced climate change on the softwood stumpage supply of the southern United States resulting from shifts in the distribution of southern pines;
- Estimates of the potential effects of an enriched CO_2 atmosphere on softwood stumpage supply; and
- Comparison of the effects of changes in softwood forest distribution with the potential growth responses, thus identifying biological information that will likely have the greatest impact on an economic valuation.

Our results provide useful information to policy-makers attempting to allocate research resources. The supply changes resulting from projected southern softwood forest migrations and the potential growth responses of these forests to an enriched CO_2 environment reveal the sensitivity of economic assessment to the underlying biological response scenarios. By comparing the magnitude of potential supply changes, research priorities can be determined that will decrease the uncertainty associated with the forest response scenarios used in future economic impact assessments.

Uncertainty is introduced because of the long-time horizon inherent in climate change scenarios, coupled with limited information about temperature and precipitation regimes, as well as atmospheric CO_2 levels. Determining changes in the distribution of forest tree species as a consequence of increased levels of CO_2 and other trace gases typically involves extrapolating the effects of climate change on species ranges from climatological projections based on general circulation models (GCMs). These models do not agree on the moisture available for forest growth in the region, which dictates the composition and distribution of southern forests. For example, because of drought, the Geophysical Fluid Dynamics Laboratory's (GFDL) climatic projections force significant changes in the composition and distribution of forests in the South (Botkin et al., 1989; Urban and Shugart, 1989; Woodman and Furiness, 1989).

As a result of drought in the southern United States, many species are projected to either shift northward or to become locally extinct. Grasslands and savanna are projected to cover much of the southern United States (Solomon et al., 1984). The most commonly studied species in terms of climatic impact is loblolly pine, primarily because of its commercial importance. At the aggregate level, the range of loblolly pine is projected to migrate northward and increase in area (Woodman and Furiness, 1989; Miller et al., 1987). The northward shift, however, is projected to lessen the overall loblolly pine site index, as the range shifts onto poorer sites. As a result, timber yields could be reduced over much of the range. Most climate scenarios project that southern pines will be eliminated or greatly reduced in Georgia, Mississippi, and South Carolina as a result of migration. Urban and Shugart (1989) evaluated the change in an oak–pine forest type in eastern Tennessee and concluded that the warmer, drier conditions also could result in several hardwood and pine species being eliminated. Conversely, several researchers have noted the possibility of increased tree productivity as a result of enhanced CO_2 levels (Pastor and Post, 1988; Rogers et al., 1983; Funsch et al., 1970). Thus, it is theoretically possible that increasing levels of CO_2 in the atmosphere could enhance tree growth significantly. Several recent studies, however, suggest that tree response could vary significantly by species and age. The results indicate that the concentrations do indeed vary by level, suggesting that growth responses could vary among species as well as among seedlings, saplings, and mature trees (Graumlich, 1991; Bazazz and Williams, 1991; Bazazz et al., 1990).

Methods

Two scenarios were considered for our study. The first utilized the results of a forest gap-dynamic model to determine the extent of the northward shift of the present southern softwood forests and the species that would probably replace

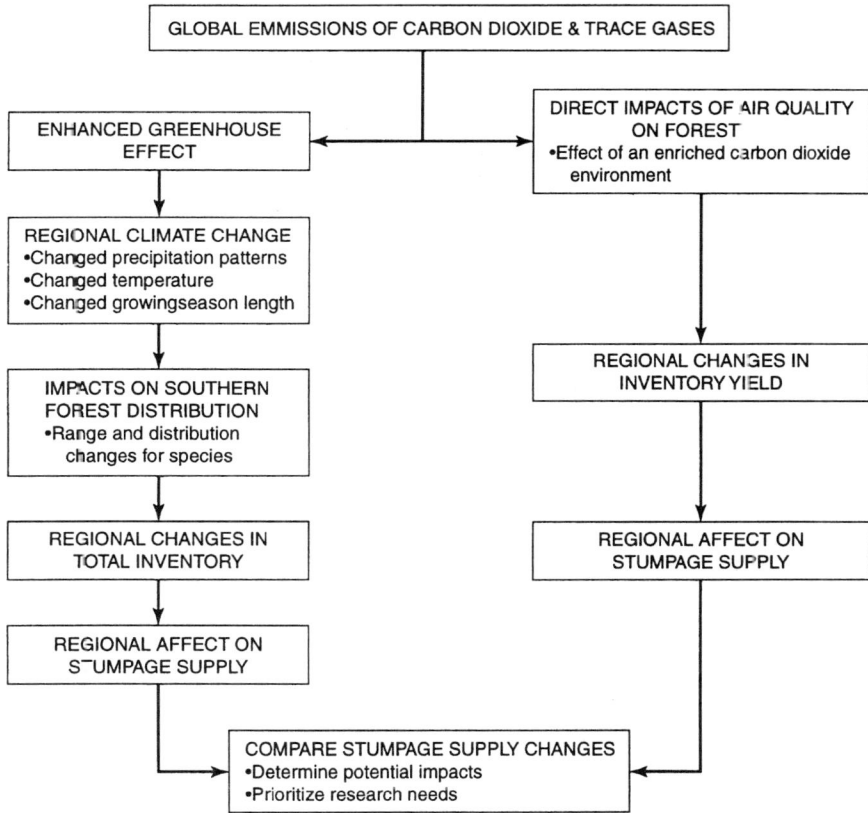

Figure 43.1. A framework for assessing and prioritizing the research needs required to reduce the uncertainty associated with an economic impact assessment of climatic change on the southern forest industry.

present-day forests when they become extinct. Species composition and distribution were assumed to change, although yields in areas predicted to be suitable for southern softwood forest remained constant. The second considered a range of potential changes in forest yield that were determined from the literature. Yields increased although species composition and distribution remained constant. Because the gap-dynamic model did not consider the effects of CO_2 fertilization on species composition and distribution, no scenario was formulated to consider the interactive effects of forest migration in an enriched CO_2 environment.

Figure 43.1 presents a framework similar to Hodges et al. (1992) that outlines the analytic procedure used in our study. First, a climate scenario and the resulting distribution of southern forests were needed. The projections developed for the year 2080 by Solomon et al. (1984) were selected for our analysis because of the comprehensive regional approach used. The biological model used by Solomon et al. (1984) did not include all southern or nonindigenous species that may survive climatic change in the South. As a result, the projections were viewed as a worst case scenario. These projections were employed to estimate the regional changes in total inventory arising from the changes in forest distribution. Using an econometric model modified from the regional softwood stumpage supply and demand model developed by Newman (1986), the effect of the regional inventory changes on softwood stumpage supply was quantified. All steps in this process, as well as the procedure employed to estimate the supply effects of growth response to climate change, are described in the following sections.

Climate Scenario for the Year 2080

Solomon et al. (1984) performed an initial analysis of the effect of climatic change on unmanaged forests in eastern North America. The climate scenario used was generated by multiple GCMs for five points spread throughout the southern United States. It was determined that:

- July temperatures would increase an average of 2 °C.
- January temperatures would increase an average of 1.5 °C.
- July precipitation would decrease an average of 0.5 mm per day.
- January precipitation would decrease an average of 1.5 millimeters per day.

This scenario is similar to those used by other researchers (see Urban and Shugart, 1989), and was used to explore the possible impacts of climatic change on softwood timber supply in the South.

Potential Southern Forests Distribution in the Year 2080

Using a gap-dynamic model, Solomon et al. (1984) projected forest distribution under the climatic conditions described. Their results were among the most comprehensive found for the southern United States and included loblolly pine (*Pinus taeda* L.), a species with great commercial importance. The model did not include longleaf or slash pines (*Pinus elliotti* Engelm.), or bald cypress (*Taxodium distichum* (L.) Rich.), all of which are also commercially important in the South. The simulation indicated that forests in the extreme South and at the western margins would be replaced by nonforest vegetation (see Urban and Shugart, 1989). Urban and Shugart (1989) projected shortleaf pine forests would be replaced by loblolly pine at higher elevations and to the North, and the southern extremes would be replaced by prairie and savanna. Miller et al. (1987) also determined that climatic conditions on the southern-most sites would not be favorable for loblolly pine growth with the range extending at higher elevations and to the North. The literature indicates a loss of loblolly pine and other forest types in the southern United States, although the lands capable of supporting southern forests would increase to the North.

Regional Inventory Changes

The USDA Forest Service Forest Inventory and Analysis (FIA) units periodically inventory the forest resources of every state, with each southern state being subdivided into four to six survey units. These survey units served as the primary unit of analysis for projecting regional inventory changes. Three suitability classes were defined for these units based on the results of Solomon et al. (1984). These were 1) land becoming unsuitable, 2) land remaining suitable, and 3) land becoming suitable for southern forests.

Impacted survey units (Figure 43.2) were identified by overlaying a map of the potential forest vegetation in the year 2080 with a map of the USDA Fcrest Serice Forest Inventory Analysis (FIA) survey units of the southern United States. The survey units were assigned to the suitability class in which the majority of the area occurred. The total land area of a suitability class was the sum of the total land areas (forest and nonforest land) for each survey unit within that suitability class. Total softwood inventory (volume of wood in cubic feet) was the sum of the 1987 cubic fcot volumes of softwood inventory present in each survey unit within a suitability class. Dividing the total softwood inventory for a suitability class by the total land area in the suitability class produced the average softwood volume per acre for the class.

The average volume of softwood inventory per acre weighted the importance of the land in a suitability class for timber production. Because forest industries develop close to the resource base, loss of land that historically produced high

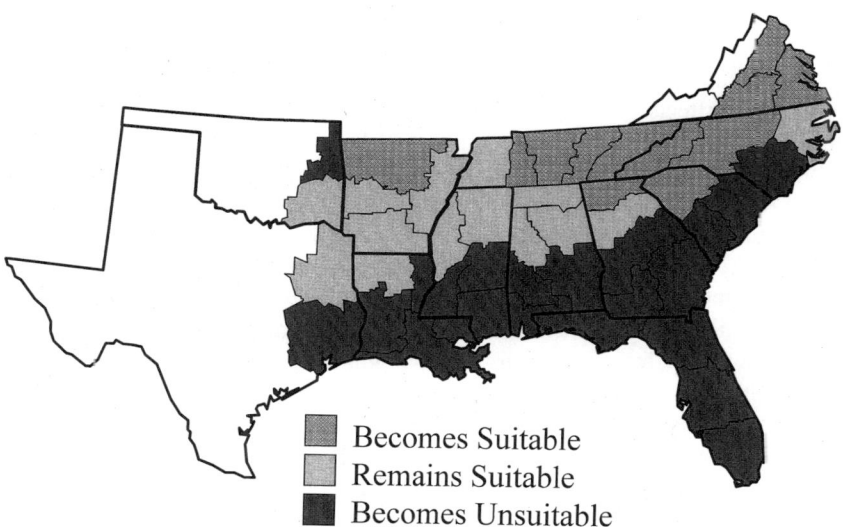

Figure 43.2. A map of the potential suitability of the South for southern softwood forest in the year 2080 by USDA Forest Service FIA survey units (adapted from Solomon et al., 1984)

volumes of softwood inventory would affect southern forest industries more than the loss of historically less productive lands.

Changes in Growth Response and Yield

The gap-dynamic model that simulated possible forest responses to climatic change did not incorporate the possible physiological effects of a CO_2-enriched atmosphere on forests. Photosynthesis, respiration, water use efficiency, nutrient use efficiency, disease and insect susceptibility, and wildfire could affect net forest yield. An enriched CO_2 atmosphere is apt to increase net forest primary production, although the magnitudes of increases vary from 20% to 182% (Melillo et al., 1993b; Unninayar and Bergman, 1993). The effect net forest yield is largely unknown because of the influence of wildfires, insects and diseases. The physiological effects of an enriched CO_2 atmosphere could significantly alter the net volume of softwood timber in the South, change forest management costs, and greatly influence the timber industry.

Bazzaz and Fajer (1992) describe some of the possible affects of a CO_2-enriched atmosphere on plant and forest growth. Because of increases in photosynthetic efficiencies, large increases in photosynthetic rates may occur in nutrient-rich environments that receive adequate light and moisture. If nutrients, sunlight, or moisture are inadequate, only a small rate increases may occur. Even if photosynthetic rates increase, proportional increases in forest growth may not result. Plants allocate sugars, water, and other resources to the roots, stems, leaves, flowers, and seeds. The allocation of resources within a plant affects its growth. Competition appears to limit the effects of CO_2-fertilization for reasons not fully understood. Many of the affects of a CO_2-enriched environment will be plant-specific.

Melillo et al. (1993) developed the Terrestrial Ecosystem Model (TEM) to consider the affects of increasing atmospheric CO_2 on the net amount of carbon sequestered by plants. The net amount of carbon sequestered by plants (net primary production or NPP) indicates the amount of energy available in an ecosystem for plant and animal use. Increases in forest growth are related to increases in the amount of carbon sequestered. The TEM model is a process-based and considers the effects of climate, soils, elevation, vegetation, and water availability to generate monthly estimates of nitrogen and carbon fluxes and pool sizes in an ecosystem. The model considers only mature, undisturbed vegetation and does not consider the effects of land use or changing vegetation distributions.

Assuming that atmospheric CO_2 concentrations would rise to 625 ppm, future climate scenarios from four GCMs were used with TEM to predict changes in NPP. When compared to present-day conditions (historical climatic conditions with atmospheric CO_2 concentrations of 312.5 ppm), global NPP is predicted to rise 20 to 26% with an average of 24%. Globally, temperate coniferous forest NPP will increase 18 to 27% with an average of 21%. The predicted changes for the eastern United States forest range from a 5% decrease to a 50% increase in NPP with two of four GCMs estimating a 5 to 25% increase (see Melillo et al., 1993b for more details).

Climatic changes could increase forest susceptibility to insect and disease attacks (Parry and Duinker, 1990). Air pollution and climatic changes contribute to decline cycles, and predispose trees to insect and disease attacks (Manion, 1981; Parry and Duinker, 1990). Insect feeding may increase because of decreases in plant tissue nutritional quality in an enriched CO_2 environment (Bazzaz, 1990; Melillo et al., 1993a), although a decrease in plant tissue nutritional quality could adversely affect insect health (Bazzaz, 1990).

Forest decline would result in high fuel loads, and would increase the risk and hazard associated with wildfire (Parry and Duinker, 1990). Decreasing summer precipitation would also increase fire risk and hazard. Parry and Duinker (1990) suggest that even if no decrease in summer precipitation occurs, the increase in evapotranspiration that would accompany higher temperatures would result in drier forests, and increase the susceptibility to wildfire. Although increases in forest productivity are possible, yields may potentially decline from insect, disease, and wildfire, and forest management costs may increase in an effort to control such problems.

Uncertainties about the interactions between changing plant physiology and wildfires, insects, and diseases limit our ability to confidently project the effect of an enriched CO_2 environment on forests. As a result, it remains uncertain whether net forest yields will increase or decrease. Because of this uncertainty, scenarios of 10 and 20% decreases, and 10 and 20% increases in timber yield were considered.

Changes to Softwood Stumpage Supply

The changes in future softwood stumpage supply were estimated based on the projected changes in regional inventories and growth. Approximately 47% of the southern United States' total land area could potentially become unsuitable for forests based on Solomon et al. (1984). For the purposes of our analysis, supply impacts were evaluated by decreasing the total land area capable of supporting southern forests by 5, 10, 15, 25, 35, and 45%. These incremental changes reflect the degree of uncertainty surrounding changes in regional inventory, and also allow for comparisons with the supply impacts of growth changes. It is important to note that inventory was only subtracted from the "becomes unsuitable" class. Inventory remained constant in the other two suitability classes. The mean inventory per acre in each suitability class was multiplied by the acreage capable of supporting forest in each suitability class.

The effect of changes in softwood stumpage supply resulting from an CO_2-enriched atmosphere was quantified by decreasing or increasing softwood inventory by 10 and 20% from the 1987 volume. Changes of 10 and 20% were used because the values fall within the range of yield changes described by the literature cited. Using a supply equation, the potential impacts of an enriched CO_2 atmosphere on softwood stumpage supply were quantified.

To facilitate the analysis, several assumptions were made, which were:

1. Doubling atmospheric CO_2 levels would result in the climate described.
2. The possible impacts of climatic change on southern forests presented would be the worst case scenario.
3. All forests within a suitability class were equally affected. Forests in the northern part of a class were affected the same as forests in the southern part.
4. If climatic conditions dictated land no longer supported southern forests, softwood inventory was not present. Managed forests were affected the same as unmanaged forests.
5. As southern forests replaced higher-elevation forests and forests further to the North, species changes occurred but the volume of softwood inventory remained constant. For example, white and pitch pines would be replaced by loblolly and shortleaf pines, but the volume of softwood fiber the area is capable of supporting remained constant. Much more information is needed to assume otherwise.
6. The effects of an enriched CO_2 atmosphere would be reflected in all forests. Softwood timber inventory may decrease or increase by 10 and 20% on the land area that remains suitable for producing softwood timber.

Assumption 6 allowed for possible losses of inventory caused by climatic change to offset or intensify softwood inventory changes resulting from an enriched CO_2 atmosphere.

Econometric Model

Newman (1986) provides a partial equilibrium model of the softwood pulpwood and solid wood stumpage market for the southern United States. The model is estimated using three stage least squares. The linear functional form of the system equations provided the best estimates of stumpage supply and demand. The supply and demand equations for the pulpwood stumpage market are:

$$S^s_{ppt} = a_{p0} + a_{p1} P_{pt} + a_{p2} I_t + a_{p3} P_{swt} + e_1 \quad (1)$$

$$S^d_{ppt} = b_{p0} + b_{p1} P_{pt} + b_{p2} F_{ppt} + b_{p3} W_{ppt} + b_{p4} r_{ppt} + b_{p5} S_{PARt-1} + e_2 \quad (2)$$

in which S^s_{ppt} = quantity of softwood pulpwood stumpage consumed in time t (thousands of cubic feet); p_{pt} = weight average regional price of southern pine pulpwood stumpage in time t (1967 dollars); I_t = standing softwood inventory in time t (thousands of cubic feet); p_{swt} = weight average regional price of southern pine sawtimber stumpage in time t (1967 dollars); F_{ppt} = the real producer price index of paper in time t (1967 = 100); w_{ppt} = real national hourly wages of production workers in the paper and allied products industries in time t (1967 dollars); r_{ppt} = the real user cost of capital for the pulp and paper industry in time t (percent of 1967 dollars); and S_{PARt-1} = quantity of pulpwood stumpage and residuals consumed in the previous time period (thousands of cubic feet).

For the solidwood stumpage market, the supply and demand equations were:

$$S^s_{swt} = a_{s0} + a_{s1} p_{swt} + a_{s2} I_t + a_{s3} p_{ppt} + e_3 \qquad (3)$$

$$S^d_{swt} = b_{s0} + b_{s1} p_{swt} + b_{s2} F_{swt} + b_{s3} w_{swt} + b_{s4} r_{swt} + b_{s5} S_{swt-1} + e_4 \qquad (4)$$

in which S^s_{swt} = quantity of softwood sawtimber stumpage consumed in time t (thousands of cubic feet); p_{swt} = weight average regional price of southern pine sawtimber stumpage in time t (1967 dollars); I_t = standing softwood inventory in time t (thousands of cubic feet); p_{ppt} = weight average regional price of southern pine pulpwood stumpage in time t (1967 dollars); F_{swt} = weighted real producer price index for softwood lumber and veneer in time t (1967 = 100); w_{sw} = national mean hourly wages of production workers in the softwood lumber and veneer industries in time t (1967 dollars); r_{swt} = the user cost of capital for the softwood lumber industry in time t (percent of 1967 dollars); and S_{swt-1} = quantity of stumpage consumed in the prior time period (thousands of cubic feet).

We modified the model developed by Newman (1986) to investigate the potential responses of softwood stumpage supply to regional climatic change. The softwood stumpage supply and demand equations for the solidwood and pulpwood industries were combined into a single system of supply and demand equations for the entire southern forest products industry. The costs of managing timberlands are included in the stumpage supply equation. This model is then reestimated for the years 1964 through 1985.

The system of equations for softwood timber supply and demand are:

$$STUMPAGE_t = a1 + a2\ STUMP\$_t + a3\ INV_t + a4\ MNGNT\$ + e_t \qquad (5)$$

$$STUMPAGE_t = b1 + b2\ STUMP\$_t + b3\ GOODS\$_t + b4\ WAGE\$_t + \\ b5\ PK_t + b6\ LAGRAW_t + m_t \qquad (6)$$

Table 43 1 lists the variables, definitions, and expected signs for equations 5 and 6. It is important to note the stumpage supply equation (5) includes a proxy for the costs of managing timberlands.

Assumptions

Except for assumption 5, the same assumptions regarding the southern forest industry made by Newman (1986) are relevant and apply to our study. They are:

- The exchange of timber between the South and other regions in the country did not occur in quantities that would affect the stumpage market.
- The stumpage market in the South was purely competitive.
- Expected stumpage prices did not affect inventories.
- The quantity of stumpage demanded equaled the quantity supplied.
- Money distributed through the FIP and the ACP reflect the costs of timber management in the South.
- *Ceteris paribus* was applied.

No significant cross-regional trade of unprocessed stumpage occurs because hauling costs increase the price of harvested timber. As the distance from a mill

Table 43.1. Variables, Definitions, Units, Expected Signs, and Applicable Equations, for the Econometric Model Used to Investigate the Potential Impacts of Climate Change and an Enriched Carbon Dioxide Environment on Southern Softwood Stumpage Supply in the Year 2080

Variable	Definition	Units	Expected sign & equation
STUMPAGE$_t$	Total softwood stumpage consumed by the lumber, veneer, pulp, and paper industries in time t	Thousands of cubic feet	dependent variable
STUMP$\$_t$	Weighted price of stumpage consumed by the southern forest products industry	1982 dollars/cubic foot	+ supply − demand
INV$_t$	Total live tree volume of softwood stumpage in the South in time t	Thousands of cubic feet	+ supply
MNGNT$\$_t$	Annual sum of forestry incentives program and agricultural conservation program expenditures on forest regeneration and timber stand improvement—proxy for the costs of timber management in the South in time t	1982 dollars/acre	− supply
GOODS$_t$	Weighted annual mean producer price index for the solid wood and pulp and paper industries—proxy for the price of final goods in time t	Percent of 1982 dollars	+ demand
WAGE$\$_t$	Weighted national hourly wage of production workers in the solid wood, pulp, and paper industries in time t	1982 dollars per hour	? supply
PK$_t$	Costs of capital of the pulp and paper industry in the South	Percent of 1982 dollars	? supply
LAGRAW$_t$	Lagged quantity of softwood stumpage plus residuals consumed by the southern forest products industry in time t	Thousands of cubic feet	+ supply

increases, more must be paid for harvested wood as an incentive for loggers to haul wood. Also, as the distance from the mill increases, stumpage buyers pay reduced amounts to landowners because hauling costs significantly increase harvesting costs. As a result, land owners are less apt to sell timber (Newman, 1986).

The stumpage market is competitive within the southern United States. Hauling costs typically prevent nonregional firms from purchasing southern stumpage. Nevertheless, the forest industry is highly concentrated in the South, and as a result, stumpage prices are competitive regionally. The bias that may result from the violation of this assumption is small (Newman, 1986).

Expected prices do not affect forest inventory because inventory changes occur over long periods. Thus, the influence of stumpage prices cannot be precisely measured. Violations of this assumption should result in little bias (Newman, 1986).

The quantity of stumpage supplied equals the quantity demanded is the market-clearing assumption. This assumption allows the use of simultaneous equation systems.

Money distributed through the Forestry Incentive Program (FIP) and the Agricultural Conservation Program (ACP) reflect the costs of timber management in the South. The FIP and ACP subsidize forest regeneration and timber-stand improvement on nonindustrial private forest (NIPF) lands. Although other factors may influence increases in the per-acre program funding levels, forest landowner and industry would support increases in funding levels as management costs increase. As a result, increased program expenditures reflect increased timber management costs.

Sources of Bias

Real national weighted mean hourly wages for lumber and solid wood products and paper products were used in the analysis. Regional data were not available for the time series modeled. This may bias the results, if southern wages significantly differ from the national averages. Bias in terms of wages is an unknown factor and caution should be used when interpreting the coefficients for wages (Newman, 1986).

The user costs of capital are representative of the user cost of capital for the southern pulpwood industry (Abt 1994). Although data exist for the pulpwood and solidwood industries, accurately weighting the data to reflect the entire southern industry is difficult. The user cost of capital for the solidwood industry was not considered because the pulpwood industry was assumed to be more capital-intensive than the solidwood industry. The user cost of capital will probably be low because the solidwood industry was omitted. As a result, the coefficient for the user cost of capital is probably underestimated.

The quantity of stumpage consumed by the lumber and veneer industries is calculated using the quantities of products produced. A conversion factor computed for a single year would not reflect technological advances. Because the conversion factor for computing the volume of raw materials needed to produce the quantity of final goods is for the years 1975 to 1982 (Newman, 1986), values may be underestimated before 1975 and overestimated after 1982. The bias that may result was expected to be small.

Results

The potential change in the distribution of loblolly pine could substantially affect future softwood supply. More than 150 million acres of land in the South could become unsuitable for forests. Table 43.2 lists the mean estimated volume per acre

Table 43.2. Projected Forest Suitability Class by the Year 2080, Mean 1987 Softwood Inventory Per Acre, Area, and Percent Area of Total Area in the South

Projected forest suitability class by 2080	Estimated softwood inventory volume[1] (cubic feet/acre)	Approximate total land area (thousands of acres)	Percent of total area in the South
Unsuitable	402.7	151,432	47.4
Constant	298.7	93,393	29.3
Suitable	260.5	74,365	23.3
Total	961.9	319,190	100.0

[1] Calculated by softwood live tree vollume divided by total land area

of land able to support southern softwood forests, the total land area, and the percentage of land area for each suitability class. It was estimated that 47.4% of the total regional land base may become unsuitable for southern softwood forests and this area contains more volume per acre (mean of 402.7 cubic feet/acre[1]) than the areas projected to remain or to become suitable for southern forests. Northward migration of southern forests clearly would have a negative impact on southern forest industries.

Table 43.3 indicates that a 45% loss of the area suitable for southern forests would result in a 84.1% loss of softwood stumpage in the region. With a 25% reduction in land area suitable for southern softwood forests, more than 45% of the supply could be lost. This scenario implies that every 5% area reduction resulted in more than a 9% decline in future softwood supply. If 10% or more of the land area becomes unable to support southern softwood forests, it is improbable that inventory losses will be offset by yield increases that result from an enriched CO_2 environment. Uncertainty in economic calculations could be most efficiently reduced by improving projections of the distribution of the southern forests of the future.

Table 43.4 presents the estimated softwood stumpage supply with an inventory yield decrease or increase resulting from an enriched CO_2 atmosphere. A 20%

Table 43.3. Estimated Softwood Stumpage Supply Changes in the Year 2080 Resulting From Northwind Migration of Southern Softwood Forests Because of Climatic Change

Estimated loss of total land area able to support southern softwood forests	Estimated stumpage supply	Percent change from baseline
Baseline	3,241,406	—
5%	2,938,482	−9.3
10%	2,635,558	−18.7
15%	2,332,635	−28.0
25%	1,726,787	−46.7
35%	1,120,940	−65.4
45%	515,092	−84.1

Table 43.4. Estimated Softwood Stumpage Supply Changes in Year 2080 from Growth Response of Southern Softwood Forests to an Enriched Carbon Dioxide Environment

Estimated change in southern softwood industry	Estimated stumpage supply	Percent change from baseline
−20	2,220,957	−31.5
−10	2,731,182	−15.7
Baseline	3,241,406	—
+10	3,751,631	15.7
+20	4,261,855	31.5

change in inventory volume from the 1987 volume could result in a 1.02 billion cubic feet increase or decrease in the South's softwood stumpage supply. This figure represents 31.5% of the present supply. Similarly, a 10% change in inventory volume would affect the future supply by 15.7%. The direction of net inventory change is unclear from the literature because of the effects of fires, insects, and diseases, although an increase in yield from an enriched CO_2 environment is possible.

Summary

As anthropogenically induced climatic changes force southern softwood forests northward, the southern United States most productive forest lands may become unproductive causing a loss of up to 84% of the region's softwood stumpage supply. The estimated effects of effects of a CO_2-enriched environment suggest that net yield increases, if they occur, would not offset more than a 10% northward migration of southern forests. If very large shifts in forest distribution are expected, potential changes in inventory yield will probably be of minor economic consequence. Again, this suggests that refining the estimates of potential forest migration will reduce uncertainty more than obtaining better estimates of the effects of a CO_2-enriched environment. Producing reliable economic impact assessments require refined estimates of potential forest migration. After future forest distribution and stocking are confidently estimated, improved estimates of inventory yield responses to CO_2 fertilization would further enhance impact-assessment accuracy.

References

Apt RC, Burnet J, Murray BC, Roberts DG (1994) Productivity, growth, and price trends in North American sawmilling industries: an inter-regional comparison. Can J For Res 24:139–148.

Bazzaz FA (1990) The response of natural ecosystems to rising global CO_2 levels. Ann Rev Ecol Syst 1:167–196.

Bazzaz FA, Coleman JS, Morse SR (1990) Growth responses of seven major co-occurring tree species of the northeastern United States to elevated CO_2. Can J For Res 20:1479–1484.

Bazzaz FA, Fajer ED (1992) Plant life in a CO_2-rich world. Sci Am: January, 68–74.
Bazzaz FA, Williams WE (1991) Atmospheric CO_2 concentrations within a mixed forest: Implications for seedling growth. Ecol 72:12–16.
Botkin DB, Nisbet RA, Reynales TE. (1989) Effects of climate change on forests of the Great Lake states. In Smith JB, Tirpak DA (Eds) The potential effects of global climate change on the United States. EPA-230-05-89-050. pp. 2:1–31, USEPA, Washington, DC.
DeLaune RD, Patrick WH, Pezesliki SR (1987) Foreseeable flooding and death of coastal wetland forests. Envir Cons 14:129–133.
Funsch RW, Mattson RH, Mowry GR (1970) CO_2-supplemented atmosphere increases growth of *Pinus strobus*. For Sci 16:459–460.
Graumlich LJ (1991) Subalpine tree growth, climate, and increasing CO_2: An assessment of recent growth trends. Ecol 72:1-1 1.
Hodges DG, Cubbage FW, Regens JL (1992) Regional forest migrations and potential economic effects. Envir Toxic Chem II:1129–1136.
Manion PD (1981) *Tree disease concepts*. Prentice-Hall Inc., Edgewood Cliffs, NJ.
Melillo JM, Callaghan TV, Woodwar FI, Salati E, Sinha SK (1993). Effects on ecosystem. In Houghton JT, Jenkins GJ, Ephraus JJ (Eds) *The Scientific Assessment*. University of Cambridge Press, Cambridge, London.
Melillo JM, McGuire AD, Kicklighter DW, Moore B III, Vorosmarty CJ, and Schloss AL (1993). Global climate change and terrestrial net primary production. Nature 363:234–239.
Miller WF, Dougherty PM, Switzer GL (1987) Effect of rising carbon dioxide and potential climate change on loblolly pine distribution, growth, survival and productivity. In Shands WE, Hoffman JS (Eds) *The greenhouse effect, climate change, and U.S. forest*. The Conservation Foundation, Washington DC.
Newman DH (1986) An econometric analysis of aggregate gains from technical change in southern softwood forestry. Ph.D dissertation. Dep For, Duke University, Durham, NC.
Parry ML, Duinker PN (1990) Agriculture and forestry, In McG Tegat WJ, Sheldon GW, Griffiths DC (Eds) *Climate change: The IPPC impacts assessment*. Australian Government Printing Service, Canberra, Australia.
Pastor J, Post WM (1988) Response of northern forests to CO_2-induced climate change. Nature (London) 334:55–58.
Rogers HH, Bingham GE, Cure JD, Smith JM, Surano KA. (1983) Responses of selected plant species to elevated, carbon dioxide in the field. J Environ Qual 12:569–574.
Smith JB, Tirpak DA (Eds) (1989) The potential effects of global climate change on the United States. EPA-230-05-89-050. USEPA, Washington, DC.
Solomon AM, Tharp ML, West DC, Taylor GE, Web JW, Trimble JL (1984) Response of unmanaged forest to CO_2-induced climate change: available information, initial test, and data requirements. DOE/NBB/0053 TROO9, US D E, Off Ener Res, Washington, DC.
Unninayar S, Bergman KH (1993) Modeling the earth system in the mission to planet earth era. NASA. Washington, DC.
Urban DL, Shugart HH. (1989) Forest responses to climatic change: a simulation study for the southeastern forest. In Smith JB, Tirpak DK (Eds) *The potential effects of global climate change on the United States: Appendix D—Forest*. EPA-23005-89-050, USEPA, Washington, DC.
Woodman JN, Furiness CS (1989) Potential effects of climate change on U.S. forests: case studies of California and the Southeast. In Smith JB, Tirpak DA (Eds) *The potential effects of global climate change on the United States*. EPA-230-05-89-050, USEPA, Washington, DC.

44. An Integrated Assessment of Climate Change on Timber Markets of the Southern United States

Joseph E. de Steiguer and Steven G. McNulty

There is growing public concern that continued emissions of greenhouse gases could cause the global climate to change (Gore, 1992). Altered global climate could, in turn, have impacts on the earth's natural systems and, ultimately, on human welfare (Office of Technology Assessment, 1991). Economic assessments of these potential welfare impacts are useful to government officials who ultimately may need to evaluate the costs and benefits of global change legislation.

The purpose of this chapter was to examine the potential economic impacts of climate change on pine timber markets of the southern United States. Southern pine forests are commercially important as they account for approximately one-half of the softwood timber volume harvested in the United States (Haynes, 1990). The three specific objectives of the study were 1) to develop scenarios of climate change using historic climate data and general circulation models (GCMs), 2) to use the climate scenarios to predict changes in the growth and merchantable inventory of southern pine forests from eastern Texas to Virginia, and 3) to estimate the economic impact of this inventory change on timber producers and consumers in the southern pine sawtimber and pulpwood markets.

Review of Prior Economic Studies

The literature on the economic impacts of global change in timber markets was examined to identify methods and results that might be applicable to the present study. Botkin and Nisbet (1990) have estimated that global warming could have major impacts on commercial forestry, timber supply, recreation, and wildlife that depend upon forest habitats, as well as on water supply and erosion rates. The same authors state that losses from increased fire incidence and insect damage could also occur. de Steiguer (1992, 1993) has discussed global climate change damage to forests as economic externalities, which are the unintentional economic side effects of resource consumption. Sedjo and Solomon (1989) have projected a forest area decrease of 6% as a result of global change. Cline (1992) estimated that economic losses in the lumber industry in the United States could reach $4 billion per year. Hodges et al. (1992) have estimated that losses to the forestry sector in the southern United States could total $300 million with an additional $100 million spent for management costs. Adams et al. (1994) have developed FASOM, which is a forest and agriculture sector model that can be used to examine the impacts of climate change on economic welfare as well as carbon accumulation. The model offers some advantages over earlier models because it examines the shift in productivity between the forest and agriculture sectors. Van Kooten and Arthur (1989) explored the effects of global change on the timber markets of Canada and the United States and found that gains in welfare were experienced principally by the United States. de Steiguer (1994) used the Southern Pine Aggregate Market Model (SPAMM) to examine the economic impacts of tree planting to sequester carbon. Sohngen et al. (1996) developed dynamic global change scenarios for U.S. forests that predicted increases in economic surplus.

Study Methods

This study analyzed five climate change scenarios in an integrated assessment framework that included the following three components: 1) a southern pine tree physiology model, 2) a regional forest projection system, and 3) a pine timber market model for the southern United States. The study compared changes in southern pine inventories under each of the five doubled-carbon dioxide (CO_2) climate scenarios to the inventory under historic "normal" climate conditions. The study, therefore, compared steady-state conditions and did not attempt to examine the dynamic nature of the climate change process. The study methods are presented in the following sequence: 1) climate change scenarios, 2) forest productivity modeling, 3) regional forest projections, and 4) timber market modeling.

Climate Change Scenarios

Precipitation and air temperature were the only variables considered in the climate change scenarios. Carbon dioxide-fertilization effects on forest growth were not examined. Two types of climate change scenarios were developed to assess altered temperature and precipitation patterns on southern pine productivity. The first, called the minimum climate change (MCC) scenario, increased the historic (1951 to 1984) monthly average minimum and maximum temperature by 2 °C and increased total monthly precipitation by 20%.

A second group of climate change scenarios were obtained using GCM projections and historic weather data. The GCMs used in the study were the Oregon State University (OSU), Goddard Institute for Space Studies (GISS), General Fluid Dynamics Laboratory (GFDL), and United Kingdom Meteorological Office (UKMO) models. The spatial scale for which the models were developed varied from the OSU GCM at 4.0° × 5.0° to the GISS GCM at 7.8° × 10.0°. Each GCM predicts for each of its grid cells the change in monthly temperature and precipitation that occurs with a simulated doubling of atmospheric levels of CO_2. Predicted temperature changes from each of the four GCMs were added to historic (1951 to 1984) average monthly minimum and maximum air temperatures. Predicted proportional changes in precipitation were multiplied by historic monthly precipitation. These calculations yielded thirty-five years of temperature and precipitation change projections for simulations with the tree physiology model.

Forest Productivity Modeling

The PnET-IIS model, a physiologically based, monthly time step model, predicts changes in forest hydrology and forest growth for forest tree species across the eastern United States. (Aber et al., 1995) Model predictions of forest growth with this model have been well-correlated prior to this study with average annual site basal area growth measured in twelve pine stands located from eastern Texas to eastern Virginia ($r^2 = 0.66$, $P < 0.005$) (McNulty et al., 1996).

The PnET-IIS model uses site-specific, soil-water-holding capacity (SWHC), vegetation process parameters for the separate tree species, and four monthly climate parameters (i.e., minimum and maximum air temperatures, total precipitation, and solar radiation) to predict net primary productivity (NPP). Net primary production is defined as annual gross photosynthesis minus growth and maintenance respiration for leaf, wood, and root compartments. Annual gross photosynthesis is a function of gross photosynthesis per unit leaf area and total leaf area. Changes in water availability and plant-water demand place limitations on the amount of leaf area produced. As vapor pressure diminishes and air temperatures increase, leaf area, and gross photosynthesis decrease.

Southern pine respiration is related to the length of time that the trees have to acclimate to changes in air temperature and the total change in air temperature. As the length of acclimation time increases, gross foliar respiration rates decrease, especially at air temperature greater than 30 °C (Strain et al., 1976). The PnET-IIS model calculates temperature change as the difference between the present and prior months' minimum and maximum air temperatures. The optimum temperature for net photosynthesis varied from 23 to 27 °C, and the maximum air temperature for gross photosynthesis varied from 30 to 43 °C. As temperatures increase beyond the optimum photosynthetic temperature, the respiration rate increased, and gross photosynthesis increased slightly or decreased, so proportionally less net carbon per unit leaf area was fixed.

The PnET-IIS model uses constant generalized species-dependent process

Table 44.1. The PnET–IIS Model Default Parameters and Parameters Used in Sensitivity Analysis

Parameter name	Parameter Abbreviation	Model default value
Light extinction coefficient	k	0.5
Foliar retention time (years)		2.0
Leaf specific weight (g)		9.0
NetPsnMaxA (slope)		2.4
NetPsnMaxB (intercept)		0.0
Light half saturation (J m^2 sec^{-1})	HS	70.0
Vapor deficit efficiency constant	VPDK	0.03
Base leaf respiration fraction		0.10
Water use efficiency constant	WUE C	10.9
Canopy evaporation fraction		0.15
Soil-water release constant	F	0.04
Maximum air temperature for photosynthesis (°C)	TMAX	variable
Optimum air temperature for photosynthesis (°C)	TOPT	variable
Change in historic air temperature (°C)	DTEMP	0.0
Change in historic precipitation (% difference)	DPPT	0.0

coefficients (Table 44.1), site-specific soils, and climate data. Soils series data were derived from a geographic information system (GIS)-based soils atlas compiled by the Soil Conservation Service (Marx, 1988). The soil series were hand-digitized from maps at a scale between 1:500,000 to 1:1,500,000, depending on the state. Soil information associated with each series included SWHC to a depth of 102 cm. All other soil parameter values were held constant across all sites and years (Table 44.1).

The Forest Health Atlas (Marx, 1988) provided cooperator and first-order station data, which was originally acquired from the National Climatic Data Center (NCDC). Cooperator station data included average minimum and maximum monthly air temperature and total monthly precipitation; first-order station records included relative humidity. After checking for accuracy, the database was interpolated on a 0.5° × 0.5° across the southern United States (Marx, 1988). The gridded databases of minimum and maximum air temperature, relative humidity, and precipitation were compiled into a single database and run through a program to calculate monthly solar radiation (Nikolov and Zeller, 1992) at a 0.5° × 0.5° grid. Solar radiation values were then combined with average monthly maximum and minimum air temperatures and total monthly precipitation and input into PnET–IIS to obtain predictions of changes in forest growth.

Converting Net Primary Productivity to Regional Changes in Forest Growth

The PnET–IIS model predictions of biological productivity under the climate change scenarios were converted into regional estimates of merchantable inventory change for use in the SPAMM economic market analysis. These changes in merchantable inventory can come from two sources. First, the geographic extent of pine forests may change. This was calculated by changing the total present-day 61.8 million acres of southern pine forests (USDA, 1988) proportional to the ratio of the number of GCM grid cells that were shown to be without pine production following the PnET–IIS climate simulations vs those that originally had pine production. Second, the merchantable growth of the residual stand may change. In these calculations, it was assumed that the present merchantable forest inventory and growth data obtained from the USDA Forest Service Forest Inventory and Analysis (FIA) database was representative of a historically normal climate. Furthermore, the changes from existing merchantable forest growth and inventory were assumed to be proportional to the ratio of PnET–IIS predicted changes in total biological productivity under the various climate scenarios to the PnET–IIS predicted total biological productivity under historic normal climates. Although present FIA estimates for total pine merchantable inventory are 102 billion cubic feet, while annual growth of these forests has been 5.4 billion cubic feet (USDA, 1988).

Timber Market Model

The SPAMM model calculated changes in timber producer and consumer surpluses, and also changes in timber prices and annual harvest levels in southern pine solidwood and pulpwood markets. Measurement of changes in these four economic indicators constituted the timber market economic assessment for the study. A graphical representation of the SPAMM model is represented in (Figure 44.1). If the market is free of global change effects, timber supply (schedule S) and timber demand (schedule D) prevail. Market equilibrium occurs when timber demand D is equal to supply S and quantity (q^*) clears the market at price (p^*). Producers surplus accrues to timber growers in the amount of a + b + c. Mill owners receive a consumer surplus in the amount equal to area d + e + f + g.

The timber market supply schedule in Figure 44.1 represents an aggregation of all individual agent's supply functions. Market supply is a negative function of timber price. Timber supply is a function of timber production costs, which are, in part, related to the amount of merchantable timber inventory. Timber inventory is thus used as a proxy for the cost of supplying timber (Jackson, 1983). Changes in the standing inventory will change production costs. These cost changes are represented by parallel shifts in the entire supply function. Increases in inventory, and therefore supply costs, causes a downward shift of the supply function relative to its original position on the price (i.e., y) axis. This result can be confirmed intuitively by observing that the price of a given quantity of timber decreases with a downward shift (i.e., an increase) in supply. Conversely, decreases in inventory and supply costs causes an upward shift of the supply function relative to its beginning position on the price axis. Again, this can be confirmed intuitively by observing that the price of a given quantity of timber increases with an upward shift (i.e., a decrease) in supply.

A decrease in timber inventory caused by global change will result in an

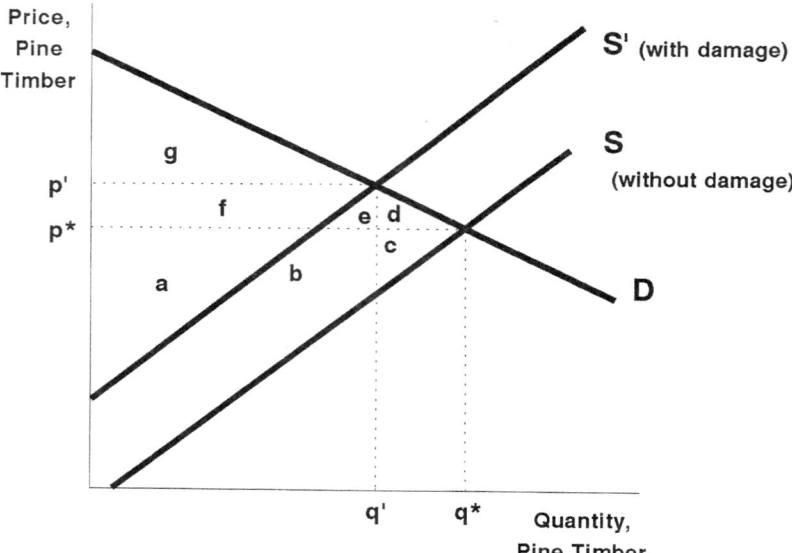

Figure 44.1. A decrease in timber inventory caused by global change will result in an upward shift in the supply function from S to S'. Mill owners would have a surplus equal to area g, and the value equal to area f has been transferred to growers who also retain area a. Area b + c + d + e represents the southern pine timber market welfare loss resulting from global warming.

upward shift in the supply function, as represented by S' (Figure 44.1). The intersection of the new timber supply schedule S' and the demand curve D sets a lower equilibrium harvest quantity (q') at the higher price (p'). Mill owners would have a surplus equal to area g, and the value equal to area f has been transferred to growers who also retain area a. Areas b + c + d + e represent the contribution by southern pine timber markets to total social welfare losses resulting from global warming. Any increase in forest productivity will cause a downward shift of the original supply curve S. Any increase in the original areas of the welfare triangles resulting from a downward supply shift would measure the economic benefits from southern pine forestry caused by climate change.

A working version of the SPAMM model was programmed on a personal computer using the following inverse supply and demand equations from Newman (1987):

$$\text{Sawtimber demand: } P_d = 939.7 - .0003162 Q_d \tag{1}$$

$$\text{Sawtimber supply: } P_s = -239.82 + .0003255 Q_s \tag{2}$$

$$\text{Pulpwood demand: } P_d = 253.7 - .00011 Q_d \tag{3}$$

$$\text{Sawtimber supply: } P_s = -289.8 + .0002032 Q_s \tag{4}$$

in which: Ps = supply price, Pd = demand price, Qs = quantity of timber supplied, Qd = quantity of timber demanded.

Shifts in the timber supply curve under each of the five climate change scenarios was accomplished in the following manner. The historic regional annual productivity was subtracted from each of the five PnET–IIs simulated changes in regional productivity. These results, in billions of ft^3, were divided by the total pine inventory (102 billion ft^3) and then multiplied by 100 to obtain the annual percentage change in regional pine inventory resulting from climate change. The percentage change in productivity was apportioned between solidwood and pulpwood market using recent relative wood consumption shares of 66% and 34% for the two markets respectively (Haynes, 1990). The percentage changes in inventory were multiplied by inventory elasticities (i.e., the ratio of the percentage change in the quantity of timber harvested to the percentage change in forest inventory) to obtain percentage changes in harvest quantity. For this study, the sawtimber inventory elasticity = .387, and pulpwood inventory elasticity = 1.198 (Newman, 1987). The new harvest quantities were substituted into the sawtimber and pulpwood supply equations and solved for the new y-intercept. This procedure provided the newly shifted supply functions, one for each climate change scenario in the pulpwood and sawtimber markets.

With the new supply functions, changes in producer and consumer surplus were calculated in 1991 dollars by computing the area of the welfare triangles using procedures from Holmes (1992). The amounts were not discounted even though global change impacts probably will occur in the future. D'Arge et al. (1982) suggested that discounting may not be appropriate for global change-related losses because it implies that the welfare of future generations is of reduced importance than that of present generations.

Study Results

Climate Change Scenarios

Each of the climate change scenarios predicted increases in average monthly precipitation across the southern United States, with the exception of the UKMO model (Table 44.2). The latter showed a very slight decrease in average monthly precipitation. The MCC scenario yielded the largest percentage increase in average monthly precipitation (1.2%). Each of the climate scenarios indicated increases in average monthly air temperatures across the southern United States (Table 44.2). The UKMO model was the largest at 6.6 °C; while the MCC scenario was the smallest at 2.0 °C.

Projections of Southern Pine Timber Volume

The annual productivity under historic ambient conditions was 5.4 billion ft^3 (Table 44.3). The MCC scenario yielded a volume of 5.1 billion ft^3. The four GCM scenarios yield the following volumes in billion ft^3: 1) OSU 4.0, 2) GISS

Table 44.2. Average Total Monthly Precipitation (cm) Across the Southern United States (1951 to 1980)

	Jan.	Feb.	March	April	May	June	July	Aug.	Sept.	Oct.	Nov.	Dec.	Average (s.e.)
Average	10.9	10.7	12.9	10.9	11.9	11.2	12.8	11.2	10.6	8.2	9.2	11.5	11.0 (0.4)
Model						Percentage Change From Historic Values							
MCC	1.20	1.20	1.20	1.20	1.20	1.20	1.20	1.20	1.20	1.20	1.20	1.20	1.20 (0.00)
OSU	0.88	0.98	0.75	0.83	1.00	1.00	1.13	1.35	1.27	1.20	0.95	1.05	1.03 (0.05)
GISS	0.85	1.31	0.91	1.01	1.22	1.27	1.31	1.05	1.00	0.84	0.76	0.82	1.03 (0.05)
GFDL	1.29	1.06	0.99	1.31	1.22	0.70	1.31	0.78	1.04	1.03	1.03	1.16	1.08 (0.05)
UKMO	0.81	1.05	1.09	1.18	1.08	0.99	0.91	0.95	1.03	0.80	0.93	1.06	0.99 (0.03)

Average Monthly Air Temperature Across the Southern United States (°C) (1951 to 1980)

	Jan.	Feb.	March	April	May	June	July	Aug.	Sept.	Oct.	Nov.	Dec.	Average (s.e.)
Average	6.4	8.2	12.2	17.1	21.1	24.7	26.4	26.0	23.2	17.7	12.1	8.2	16.9 (2.2)
Model						Additive Change (°C)							
MCC	+2.0	+2.0	+2.0	+2.0	+2.0	+2.0	+2.0	+2.0	+2.0	+2.0	+2.0	+2.0	+2.0 (0.0)
OSU	+5.0	+3.2	+4.0	+3.8	+3.2	+3.7	+3.5	+3.1	+3.6	+3.7	+2.3	+3.0	+3.5 (0.2)
GISS	+3.8	+3.8	+5.8	+4.2	+3.8	+3.8	+3.5	+3.2	+5.3	+4.6	+5.4	+4.1	+4.3 (0.2)
GFDL	+5.6	+4.7	+4.3	+3.3	+3.6	+3.8	+3.7	+3.9	+4.8	+5.2	+2.6	+1.9	+4.0 (0.3)
UKMO	+6.7	+6.6	+7.1	+6.5	+5.6	+6.1	+6.7	+6.9	+6.7	+6.7	+6.6	+7.2	+6.6 (0.1)

Prediction use the United Kingdon Meterological (UKMO) general circulation model (GCM), the General Fluid Dynamics Laboratory (GFDL) GCM, the Oregon State University (OSU) GCM, and the Goddard Institutiet of space Studies (GISS) GCM. As a comparison, the minimum climate change scenario (MCC) uses a constant increase in air temperature and percentage increase in precipitation. All models were run in conjunction with historic (1951 to 1984) climate data through PnET–IIS.

Table 44.3. Estimated Changes in the Annual Growth, Total Acreage, and Total Annual Productivity of Pine in the Southern United States, for Historic Climate and Five Climate Change Scenarios

Climate scenario	Annual growth (ft^3/ac)	Change in annual growth (%)	Total acreage (10^6 ac)	Annual change in acreage (%)	Total annual growth (10^9 ft^3)	Annual Change in Growth (%)
Ambient	87	—	61.8	—	5.4	—
MCC	82	−6 %	61.8	0 %	5.1	−6 %
OSU	65	−25	60.8	−2	4.0	−26
GISS	62	−29	60.6	−2	3.8	−30
GFDL	57	−34	58.1	−6	3.3	−39
UKMO	47	−46	35.6	−42	1.7	−69

Note: For the southern pine forests: total acreage = 61.8 million, total volume (or inventory) = 102 billion ft^3, annual harvest = 5.4 billion ft^3, annual growth = 5.4 billion ft^3.

3.8, 3) GFDL 3.3, and 4) UKMO 1.7. Thus, for the MCC climate scenario and all four of the GCM climate scenarios, the total annual forest productivity of southern pine was shown to decrease when compared forest productivity under historic ambient climate conditions.

The annual change in total forest productivity as a percentage of regional timber inventory ranged from −0.3% for the MCC scenario to −3.6% for the UKMO scenario (Table 44.4). These values were used with the inventory elasticities to shift the timber market-supply curves. The negative signs indicate a decrease in timber supply that will result in a loss in economic welfare in southern pine timber markets.

Timber Market Assessment

In the solidwood market, the decreases in total economic surplus ranged from approximately $2.7 million per year under the MCC scenario to $32.2 million per year for the UKMO scenario (Table 44.5). The decreases in consumer surplus ranged from $1.3 million per year with MCC scenario to $15.9 million with UKMO. In the solidwood market, the OSU and GISS scenarios yielded consumer surplus decreases of approximately of $6 million per year, and the GFDL scenario predicted a $9 million consumer surplus decrease per year. The changes in solidwood market producer surplus for each climate scenario were about the same magnitude as the consumer surplus changes (Table 44.5).

In the pulpwood market, the decreases in total economic surplus ranged from about $1.2 million per year under the MCC scenario to $17.7 million per year for the UKMO scenario (Table 44.6). The decreases in consumer surplus ranged from $453 thousand per year with MCC to $6.2 million with UKMO. The OSU, GISS and GFDL scenarios yielded consumer surplus decreases of approximately $2 to $3 million per year. Producer surplus in the pulpwood market decreased $836 thousand per year with MCC scenario. Producer surplus decreases with UKMO

Table 44.4. Changes in Total Annual Growth of Pine Forests as a Percentage of Total Pine Inventory for the Southern United States, for Ambient Climate and Five Climate Change Scenarios

Climate scenario	Annual percentage change in pine inventory
Ambient	—
MCC	−0.3%
OSU	−1.4
GISS	−1.6
GFDL	−2.1
UKMO	−3.6

Table 44.5. Annual Changes in Economic Surplus, Stumpage Price and Annual Harvest for the Southern Pine Solidwood Market Under Five Climate Change Scenarios in 1991 Dollars

Climate scenario	Change in producers surplus (thousands of dollars)	Change in consumers surplus (thousands of dollars)	Change in total surplus (thousands of dollars)	Change in stumpage price (%)	Change in annual harvest (%)
MCC	−1,363	−1,324	−2,687	.12	−.08
OSU	−6,129	−5,954	−12,083	.56	−.35
GISS	−6,809	−6,615	−13,424	.63	−.39
GFDL	−9,530	−9,257	−18,787	.88	−.54
UKMO	−16,321	−15,855	−32,176	1.51	−.93

Table 44.6. Annual Changes in Economic Surplus, Stumpage Price and Annual Harvest for the Southern Pine Pulpwood Market Under Five Climate Change Scenarios in 1991 Dollars

Climate scenario	Change in producers surplus (thousands of dollars)	Change in consumers surplus (thousands of dollars)	Change in total surplus (thousands of dollars)	Change in stumpage price (%)	Change in annual harvest (%)
MCC	−836	−453	−1,289	.36	−.12
OSU	−4,176	−2,260	−6,436	1.82	−.60
GISS	−4,988	−2,701	−7,689	2.16	−.72
GFDL	−5,790	−3,134	−8,294	2.49	−.84
UKMO	−11,494	−6,222	−17,716	4.94	−1.68

Table 44.7. Combined Annual Changes in Economic Surplus for Southern Pine Solidwood and Pulpwood Markets under Five Climate Change Scenarios in 1991 Dollars

Climate scenario	Change in producer surplus (thousands of dollars)	Change in consumer surplus (thousands of dollars)	Change in total surplus (thousands of dollars)
MCC	−2,199	−1,777	−3,976
OSU	−10,305	−8,214	−18,519
GISS	−11,797	−9,316	−21,113
GFDL	−15,320	−12,391	−27,711
UKMO	−27,815	−22,077	−49,892

was about $11 million per year. The OSU, GISS and GFDL scenarios yielded producer surplus decreases in the pulpwood market that ranged from about $4 to $6 million per year.

Combined annual surplus decreases for both the solidwood and pulpwood markets (Table 44.7) ranged from about $4 million for MCC scenario to $50 million for UKMO. Although the OSU and GISS scenarios yielded total surplus changes of about $20 million a year, the GFDL scenario was about $27 million per year.

Summary

The annual economic impacts of global climate change on southern timber markets were negative in each case. The changes in precipitation and temperature contributed to a loss in forest productivity that, in turn, caused economic losses. The economic impacts were negative to both timber producers and timber consumers.

However, although global change did cause annual economic losses, these do not appear to be particularly large in relative terms. For example, the annual losses in economic surplus predicted by the UKMO model were $50 million per year. This represents only about 1% of the $5 billion total annual southern pine timber market surplus. (In Figure 44.1, total surplus is represented by the triangular area bounded by demand function D, supply function S, and the price axis.) The MCC scenario predicted annual economic losses of less than one-tenth of 1% of total annual timber market surplus. The annual losses predicted by the GISS model were about four-tenths of 1% of total timber market surplus.

Future versions of the PnET–IIs and SPAMM models could provide improved estimates of global change impacts on southern forests. Changes to future versions of the model would include the addition of a CO_2 component to the PnET–IIS model. The present version does not have the capability of analyzing CO_2 effects on forest growth. The probable effect of a CO_2 model component would be to increase tree growth and thereby reduce damages or, perhaps, even show a gain in tree growth. Another improvement would be to make the physiology and economic models capable of dynamic analysis. That is, permitting the models to analyze the transitional impacts of global change over a long-time period. In this manner, the cumulative, rather than static, effects of global change could be analyzed. The results of such a dynamic analysis might be a more dramatic effect of global change on southern forests.

References

Adams DM, Alig R, Callaway JM, McCarl B 1994. Forest and Agricultural optimization model: Model description. Final Report to USEPA. RCG Hagler Bailly, Boulder, CO.

Aber JD, Ollinger SV, Federer CA, Reich PM, Goulden ML, Kicklighter DW, Melillo JM, Lathrop RG Jr, Ellis JM (1995) Predicting the effects of climate change on water yield and forest production in the northeastern U.S. Climate Res 5:207–222.

Botkin D, Nisbet R (1990) The response of forests to global warming and CO_2 fertilization. Report to USEPA, January.

Cline WR (1992) *The economics of global warming.* Institute for International Economics, Washington, DC.

D'Arge RC, Schultze WD, Brookshire DS (1982) Carbon dioxide and intergenerational choice. Am Econ Rev 72(2):25(1)–256.

de Steiguer JE (1992) Greenhouse gases, forests and environmental externalities. In Proceedings of IUFRO Centennial meeting, Berlin.

de Steiguer JE (1993) Socio-economic impact of global change and air pollution-related forestry damage. In Schlaepfer R (Ed) *Long-term implications of climate change and air pollution on forest ecosystems.* IUFRO World Series Vol. 4, Vienna.

de Steiguer JE (1994) "Timber market impacts of the climate change action plan, In Proceedings of the 24th Annual Southern Forest Economics Workshop, March 27–29, 1994, Savannah, GA.

Gore A (1992) *Earth in the balance.* Houghton Mifflin Company, Boston.

Haynes RW (1990) An analysis of the timber situation in the United States: (1989–2040. USDA For Serv Gen Tech Rep RM-199. Fort Collins, CO.

Hodges DG, Cubbage FW, Regen JL (1992) Regional forest migrations and potential economic effects. Envir Toxic Chem (11:1)29–(1136)

Holmes TP (1992) Economic effects of air pollution damage to U.S. forests. In de Steiguer JE (Ed) The economic impact of air pollution on timber markets: studies from North America and Europe. USDA For Serv, Southeast For Exper Stat, 19–26.

Jackson DH (1983) Sub-regional timber demand analysis: remarks and an approach for protection. For Ecol Manag 5:109–118.

Marx DH (1988) Southern forest atlas project. In The 8(1st)nnual meeting of The Association Dedicated to Air Pollution Control and Hazardous Waste Management (APCA), Dallas, TX, 1988.

McNulty SG, Vose JM, Swank WT (1996) Loblolly pine hydrology and productivity across the southern United States. For Ecol Manag 86:241–251.

Newman DH (1987) An econometric analysis of the southern softwood stumpage market: 1950UN–1980. For Science 33(4):932–945.

Nikolov NT, Zeller KF (1992) A solar radiation algorithm for ecosystem dynamic models. Ecol Model 6(1:1)9–(168.)EP

Office of Technology Assessment (1991) *Changing by degrees: Steps to reduce greenhouse gases.* Congress of the US, Washington, DC.

Sedjo R, Solomon A (1989) Climate and forests. In Rodenberg NJ (Ed) *Greenhouse warming: Abatement and adaptation.* Resources for the Future, Washington, DC.

Sohngen B, Sedjo R, Mendelsohn R, Lyon K (1996) Analyzing the economic impact of climate change on global timber markets. Discussion Paper 96-08.Resources for the Future. Washington, DC.

Strain BR, Higginbotham KO, Mulroy JC (1976) Temperature preconditioning and photosynthetic capacity of *Pinus taeda* L. Photosyn, 10:47–53.

U.S. Department of Agriculture, Forest Service (1988) The South's fourth forest: Alternatives to the future, Report 24. US Gov Print Off, Washington, DC.

Van Kooten CG, Louise A (1989) Assessing economic benefits of climate change in Canada's boreal forests. Can J For Res 19:463–47

45. Integrating Local and Global Objectives in Forest Management: A Value-Based, Multiscalar Approach

R. Gordon Dailey, Jr. and Bryan G. Norton

Forest management practices have often been criticized for inadequate consideration of the public's environmental concerns in the pursuit of economic resource use. With issues ranging from the preservation of such endangered species as the red-cockaded woodpecker to the siltation and pollution of rivers and lakes, and now including concerns about the long-term impact of global change on forests, resource managers face many new and difficult challenges in the management of public forest land. New tools and techniques will be needed to meet the multiple and apparently conflicting demands for the forest resource in facing these challenges. In particular, because the accumulated effects of localized management practices are manifested on increasingly larger scales, there is a need for a new approach to forest management that can account for both the multiple values and concerns of the public as well as the potentially large-scale impacts from forest management practices.

In *Thinking Like a Mountain,* Aldo Leopold (1949) acknowledged the short-sightedness of wolf eradication as a resource management practice in wilderness areas. From the irruption of deer and the consequent devastation of the mountainside that followed the extirpation of this predator from inaccessible areas, he saw how the wolf was part of a complex interrelationship that unfolded on the mountain's time frame, not that of humans. Such large-scale, longer-term impacts from resource management continue to be an issue for land managers. Increasing calls for "whole ecosystem management" are clear expressions of the need to consider these large-scale dynamics in resource management if we are to maintain the functions of ecological systems that support many different features of the natural environment (Norton, 1991).

Coordination of forest management activities with other resource management efforts is becoming increasingly important. Such forestry practices as clearcutting can significantly impact aquatic systems. A system of analysis that can identify and account for these indirect effects and larger-scale considerations in local-level forest management will be needed to ensure the success of other resource management programs.

Another problem is that wildlife preservation strategies have typically been static and limited in scope (Noss and Harris, 1986). Although these strategies have generally focused on individual species, the factors that affect those species and their habitat are part of larger-scale ecological dynamics. Noss and Harris (1986) suggest that the emphasis on endangered species and the assumption of an ecological "equilibrium" system is insufficient for the maintenance of species diversity over the long term. The problem with such narrowly focused and static approaches is that these approaches fail to recognize that individual plant or animal communities, and even whole ecosystems, are part of larger dynamic processes that represent the context within which they function (Leopold, 1949; Noss and Harris, 1986; Norton, 1992; see generally Costanza et al., 1992). These dynamic patterns, however, exist on many scales, each providing a slower-changing context for the faster-changing processes on smaller scales. To effectively manage for multiple and indirectly related public objectives, such as water quality and wildlife preservation, forest and other resource managers must recognize these larger-level processes and incorporate them into their planning and management efforts.

Furthermore, many of the impacts on publicly valued resources may result from activities that take place on both public and private land. For example, erosion within a watershed and its impact on public water resources is not limited by property lines. In the southern United States, where a significant portion of the forest land is privately held, consideration of private land management activities is particularly important. Public land management must either be coordinated with the management of private land or public forest managers must be able to mitigate the effects of private activities on public resources. Ecological models that characterize these relationships and a management system that can incorporate these considerations are needed to both preserve and use public resources in the future.

Forest managers will also require a management and planning process that can address the effects on forests from such exogenous factors as acid rain, air pollutants, and global climate change. For example, managers require ecological models that can indicate how ground-level ozone affects the growth rates and composition of the forest. Conversely, such large-scale concerns as global climate change may mean that forest managers will be called upon to manage the forest proactively as a carbon-sink in addition to managing for the demands of the timber industry and recreational use.

Perhaps most important, from our perspective, is the need to recognize that public values may themselves exist on different scales. The problem is that when compared in terms of the present, protecting timber harvesting jobs and preservation of wildlife seem to represent competing value positions. However, they may not necessarily imply "either-or" scenarios or trade-offs. The point that is often overlooked is that both of these forest "uses" require the continued functioning of larger, landscape-level ecological systems if they are to be sustained into the future. Neither economic productivity nor aesthetic and instrumentally valued natural features can be protected without protecting the complex, organized system that provides the ecological context on which all these values depend (Norton, 1991). A system that can manage such public resources as forests at these various scales is necessary to minimize trade-offs and maintain the ecological integrity of natural systems that support publicly valued uses and features of those resources.

The Concept: Multiscalar Management

To meet these challenges and help avoid politically determined restrictions on economic resource use, we propose a multiscalar system of resource management. The idea is to develop a scientifically informed, public value-based constraint system on economic resource management designed to achieve specified public goals. A two-stage model for such a management system has been suggested by Norton (1992); see Figure 45.1. In this risk-decision space, potential impacts from

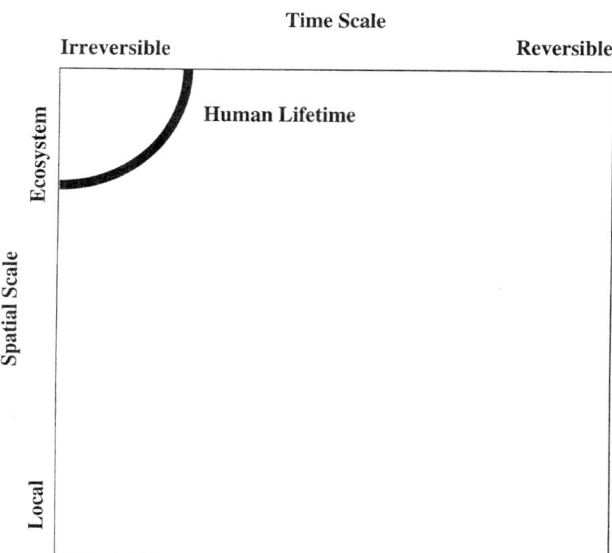

Figure 45.1. An ecologically sensitive risk decision square, which plots irreversibility of expected impacts against the spatial scale of those impacts. The time required to reverse an impact provides a temporal dimension, while classifications of policies according to the probable scale of their environmental impacts provides a spatial dimension.

various management practices are located by the degree of reversibility of the impact (horizontal axis) and the scale of the impact (vertical axis). Activities whose impacts fall in the large lower-right area are sufficiently addressed by traditional economic approaches. The multiscalar approach, however, supplements traditional economic methods of resource management by placing constraints on practices that have large-scale and potentially irreversible impacts on publicly valued resources or features of the landscape. These would be management activities with impacts falling in the upper-left corner of Figure 45.1. However, restrictions are not placed on all activities with a large-scale or irreversible impact. In our approach, public values are used to define the type and scale of impacts that fall into the upper corner: Public values define where the demarcation line falls.

Hierarchy Theory

Following arguments that suggest that evolutionary processes favor nested, hierarchical organization (Simon, 1962), hierarchy theory has been applied to the functioning relationships of ecological systems (Allen and Starr, 1982; O'Neill et al., 1986). Hierarchy theory represents one way of characterizing complex, ecological phenomena that are interrelated but may occur at different scales. In a hierarchical structure, any given system operates as both a part of a larger system, and at the same time as an independent whole unto itself. As a whole, a system exerts some constraint on its component subsystems that are themselves both wholes and parts. As a part, that same system operates as an integrated component of a larger whole (Allen and Starr, 1982). Through this type of structure, the stability of ecological systems is maintained. This stability results from the inherent constraint system. It is, however, a dynamic stability. The self-assertiveness of a constraining environmental system within a larger environment essentially protects the smaller-scale subsystem, that it controls (Allen and Starr, 1982). Disturbances can be incorporated by a system when it exerts control over some factor that is not controlled at lower levels of organization (O'Neill et al., 1986). For example, the forest as a whole minimizes local temperature variations through evapotransportation, although individual trees cannot do this (O'Neill et al., 1986). Under certain conditions, however, this constraint system can be broken with effects that propagate up the hierarchy (Johnson, 1993). Leopold's (1939) examples of the effects of predator eradication and the problems of early German monoculture forestry illustrate these types of larger-scale effects that result from changes in local-level activities.

For forest management in the future, hierarchy theory provides a conceptual basis for incorporating economic resource management at the local level within the need to maintain regenerative systems at larger scales. In a scalar system of management local activities can be constrained by a hierarchy of longer-term, higher-level considerations. That is, in a hierarchical structure, management at the local level may vary considerably according to local desires as long as they do not interfere with objectives established for management at larger scales and over longer time frames.

In our approach, the goals for resource management at each scale are defined by ecological dynamics at those scales that are associated with the maintenance of publicly valued natural features and resource uses. Basically, management at any particular scale focuses on sustaining valued natural resources or features through time by ensuring, via scalar constraints, that management activities and exogenous impacts remain within the bounds of higher-level ecological dynamics operating in longer time frames. By recognizing ecological constraints, multiscalar forest management can help to avoid the larger-scale impacts that can result from exceeding a critical threshold at a lower level.

Hierarchy theory also provides a useful conceptual framework for modeling ecological systems in order to address scale-related resource management issues (Johnson, 1993). This is important because the descriptive scientific models traditionally used to characterize various ecological systems may not be sufficient for resource management in the future. Because there is no single or "fundamental" hierarchical description of nature (Allen and Starr, 1982; O'Neill, 1986; Johnson, 1993), there is no single model that could be used to evaluate the many ecological impacts of different forest management strategies. Because a particular system may be described in several ways, it has been suggested that an appropriate hierarchical description should be based on the phenomena that we are interested in observing (O'Neill et al., 1986). Because certain characteristic properties "emerge" depending on the scale of observation (Allen and Starr, 1982), and ecological systems can be described in several different ways depending on what we are interested in observing, the development of models for ecological resource management is necessarily a prescriptive science (Haskell et al., 1992). This is particularly important for issues of scale in resource management because the correct scale on which to address a management problem is determined by what society wants to accomplish with that system (Norton, 1992). That is, an ecological system can and should be modeled differently depending on that which we are interested in managing.

Thus, in our approach, public values are used not only to define resource management at different scales, but they are also used to define the ecological models of the systems to be managed. As a framework for organizing ecological systems, hierarchy theory helps scientists, forest managers, and the public identify relevant ecological dynamics and variables operating at each scale. Through an iterative and ongoing process of interaction with the public and forest managers, scientific study can then be guided toward developing hierarchical models that inform public understanding of ecosystem functioning in a way that helps to define resource management to achieve public goals (Norton and Ulanowicz, 1992).

The Advantages of a Multiscalar Approach

A multiscalar approach to forest management has many advantages. This approach can incorporate multiple public values into a multiscaled management strategy, and it also provides a framework though which ecological models can be developed to identify appropriate constraints for resource management at various scales. An important feature of our approach is that it is designed in such a way that forests can continue to be managed for economic use at a local (lower) level within constraints imposed by ecological considerations at higher levels, and therefore trade-offs between seemingly conflicting values can be minimized.

With its focus on management at different scales, this approach also provides a system by which to account for many of the factors that complicate contemporary forest management. Because it recognizes the importance of regenerative landscape-level processes in maintaining all publicly valued natural "goods," scalar management can help to avoid large-scale or irreversible impacts and minimize trade-offs. Through its emphasis on multiscaled public values, this approach provides guidance to forest managers in response to the impacts of exogenous factors. For example, air pollution has been documented to have several different impacts on forests leading to decline (MacKenzie and El-Ashry, 1989). Forest managers could adjust their management practices in order to mitigate these impacts on timber production or mitigate further impacts on aquatic systems by minimizing their contribution. Whatever impact is of concern, ecological models that characterize the relevant variables and dynamics can help forest managers determine how their activities influence publicly valued resources and identify appropriate strategies.

A multiscalar approach can also help to coordinate forest management with other efforts to maintain publicly valued environmental resources. Although maintaining the viability of fish populations has not generally been the responsibility of forest managers, the management practices on public forest lands nonetheless affect river and lake systems through soil erosion and nutrient runoff (Morris et al., 1992). Given a publicly expressed concern for the health of rivers and lakes, our approach requires the formulation of a hierarchical model of ecological relationships appropriate for considering these impacts in forest management. In a similar fashion, forest management could also serve as a means to proactively address global warming by increasing net primary productivity in order to enhance the role of the forest as a CO_2 sink (Morris et al., 1992). Models that can characterize the impact on the atmosphere and the ecological relationships associated with carbon fixation would facilitate such activities.

Applying Multiscalar Public Values to Forest Management

After we have accepted the role of public values in defining not only the management but also the ecological science upon which it relies, the problem of developing a value-based management strategy becomes one of translating these values into management objectives at different scales. To do this, we must associate different values with ecological systems at appropriate scales as noted. Before this can be done, however, several questions must be answered, which include: Whose values are to be used? How are they to be assessed? How can they be translated into management objectives? How are the values of different communities to be integrated?

"Bottom-Up" Valuation

Answering the question, "whose values are to be used?" may seem very simple: National values should be applied to national forests, state values to state forests, and local values to local or regional parks. We do not believe this is so. Aggregating values, such as aggregating impacts over a large area, can lead to suboptimum solutions to resource management problems.

Wilkinson and Anderson (1987) note that in the past, public forest management and planning was a decentralized activity in which local managers used their knowledge of the specific characteristics of the local forests that they managed in order to both protect and use the resource. During the 1920s, timber management plans determined the amount of timber that could be harvested from "working circles," which were areas large enough to support local forest-based industries. Such efforts also included explicit concern for protection of the local watershed and recreation was increasingly considered in the plans. However, as private timber lands were cleared in the post-war building boom of the 1950s, this relatively uncontroversial planning framework began to break down as increasing demands were placed on the national forests for all their resources. The result of the controversy over multiple-use forest management was several pieces of legislation that established national planning along with a range of legal standards for local forest planning and management. The need for federal action arose from lumber industry demands, as well as increasing demands for the preservation of existing wilderness areas.

The change that occurred was the imposition of national-level values on local-level forest management. Wilkinson and Anderson (1987) note that the unit planning framework that replaced multiple-use plans in the early 1970s was intended to ensure greater consistency between national and local land-use priorities. In this new framework, "area guides" advised forest planners of an area's relative ability to achieve national objectives for various resources (Wilkinson and Anderson, 1987). In essence, local resources were no longer being managed to meet locally defined needs as expressed by local values.

To see how this may affect ecological systems it is important to examine how cultural institutions define the relationship between humans and their environment. In his ecological history of New England, William Cronon (1983) illustrates how different social institutions, such as various schemes of ownership and shared use, result in different landscape patterns. As cultural patterns changed from the shared-use rights of access recognized by Native Americans, to the individual property ownership of Europeans, the demand driving decisions about land use were effectively internationalized, and demand for certain products, given world markets, was effectively limitless. Social decisions including the types of infrastructure built, were determined not only by the local communities that became established in colonial New England but by all distant places to which those communities sold their goods. The landscape of New England therefore increasingly met not only the needs of its inhabitants for food and shelter but also the demands of markets abroad for cattle, corn, fur, timber, and other goods whose "values" became expressions of the colonists' socially determined needs (Cronon, 1983). A similar phenomenon occurred in the southern forests as the region sought to finance industrial development, and thus increase their wealth, through the exploitation of its forests (Williams, 1989).

Although both the Native Americans and the colonists shaped the land to meet their own purposes, they did so in fundamentally different ways. The colonists, with their experiences of timber as a scarce resource in England, saw the forests as an abundant resource to be exploited. With no prior experience within the landscape of the New World, they did not see the forest as an integral part of their lives. In contrast to the colonists however, the Native Americans had a much closer understanding of the land, through a longer historical association with it, such that they were able to maintain its functioning and coexist within it.

The point is that local forest land must be managed based on the values expressed by local citizens who are familiar with their surrounding landscape and its unique features. Centralized planning, as mandated by federal laws, runs the risk of failing to adequately reflect the relationship of local residents to their landscape even as it tries to balance between preservation and harvesting interests in the aggregate. This is again a problem of scale, but one of the scale of management. Thus, we believe that local values must form the basis for managing local forests; from these local values the larger scales of management must be derived.

Assessing Scalar Public Values

With local values as a starting point, the task of assessing these values requires attention. Sagoff (1988) hypothesizes that citizens may make different choices depending on the context in which a question regarding resource use is asked. In particular, he argued that individuals are more apt to advocate preservation of a natural area if the question is posed to them as a citizen rather than in a market context in which they will think as consumers. We can apply the insight by explicitly asking citizens to conceptualize three levels of values: 1) individual, 2) intergenerational, and 3) evolutionary. These three scales help citizens to visualize management problems in a multiscalar context—a decision context that one might say unfolds on three "time horizons."

To assess "individual" values for local forest management in the short-term, members of a local community can be asked to state their preferences in terms of the present. In this context, it is probable that they will assume the present rules of economics and express values based on present worth, discounting benefits that accrue in the future. In this case, citizens might express a preference for timber-harvesting jobs or recreation-related jobs over increased protection of natural resources. The values expressed in this context should be consistent with those assessed using traditional economic methods.

Next, in the context of a "sustainability convention" (analogous to a constitutional convention) community members would participate in an iterative process of public value articulation. One important aspect of this process, which should involve scientists, stakeholder groups, and interested citizens, is to focus on what, exactly, citizens of this generation view as essential elements in a "bequest package" for future generations (Norton, 1995). For example, we heard a citizen activist say, "We like our hardwood forests. We are comfortable with them. If the chip mills are allowed in they'll level the hardwood forests and replant with monoculture plantations." Similarly, citizens may express a desire to preserve a certain vista or ensure that future generations will be able to fish in a particular river, lake, or stream. These are the values that may not be captured by traditional methods because they represent values that operate on a longer time-scale than short-term economic concerns.

On a third level, community members may be asked what values they associate with the very long term. Values at this scale might include a desire to ensure the survival of the human species or, alternatively, a more inclusive value that reflects concern for the evolutionary potential of other species and natural systems. Minimizing the impact or reducing the possibility of global climate change might be a value expressed in this context. These values, as well as these others described, can be integrated into management goals at various scales, through a multiscalar management system. Figure 45.2 which shows a simple correction between the scales of cultural and social values and the scale of ecological processes.

Temporal Horizon of Human Concern	Time Scales	Temporal Dynamics in Nature
Individual/Economic	0-5 years	Human economics
Community, intergenerational bequests	up to 200 years	Ecological dynamics/ Interaction of species in communities
Species survival and our genetic successors	indefinite time	Global physical systems

Figure 45.2. Correlation of human concerns and natural system dynamics at different temporal scales.

However, the process of assessing these values may not be as simple as gathering citizens together and asking them questions about their economic and environmental values. The problem is that noneconomic values associated with particular places and environmental features may not be explicitly recognized by individuals. Edward Relph (1985) suggests that the landscapes within which we live are "inconspicuous backgrounds" to our everyday experiences. He notes that landscapes are usually unobtrusive backgrounds to other more important concerns, but occasionally they are brought forward into our awareness "in certain affective states in which we may be predisposed to notice the world around us." This implicit nature of landscape values is what necessitates iterative, ongoing processes of public involvement and community value articulation. Ecosystem management projects, for example, can embody such involvement as an essential part of the goal-setting process.

Mugerauer (1985) suggests that "environmental hermeneutics" can be employed to discover the subtle relationship between a community and its local environment. This involves an understanding of the local language and the meaning of the environment within that language, including its historical dimensions. This does not mean that anthropologists or linguists must be dispatched to assess the value of the environment and its specific features for a local community. It merely implies that a more interactive approach should be employed in order to draw out these values. An ongoing discussion between local residents, forest managers and scientists can educate the citizens about ecological relationships, and draw out these values, bringing these "background" features to light.

An interactive approach to assessing local values has other advantages as well. In a highly mobile society such as ours, citizens have little time to develop experiences with a particular place as they move from one location to another. An interactive valuation process can educate residents about their local place and surrounding landscape and bring about a recognition of the landscape and values associated with maintaining larger-level processes. This process allows them to make more educated decisions about the values and goals they choose for the management of public forest lands.

This interactive value assessment process provides two key functions for multiscalar management, which are (1) the identification of the values to emphasize in the management of the local area, and (2) assistance in articulating values that emerge on longer time-scales. These values can then be used to determine both economic and environmental resource management goals that are supported by the local residents.

Translating Values into Scalar Management Objectives

An interactive approach is not only necessary from a value assessment perspective, but it also provides the means by which to determine appropriate hierarchical models and management goals for forest managers. As was noted earlier, because specific management goals cannot be understood prior to scientific understanding, developing goals and hierarchical models that characterize the dynamics that affect valued natural features should be an experimental and interactive process. Biologists and ecologists must be part of this process in order to help the public and forest managers to define the appropriate spatiotemporal scale on which to define the system to be managed (Norton, 1992). Through an interactive and experimental process, the boundaries of the system and the ecological dynamics relevant to those objectives can be identified and appropriate goals for management can then be defined (Norton and Ulanowicz, 1992).

For example, to maximize job opportunities for a timber community, silvicultural techniques for economically sustainable timber management have been well developed. The forest can be maintained in several different ways, including the use of even-aged stands with either clear cutting or old-growth, uneven-aged stands with selective cuts, as well as other types of practices. The boundaries of the system to be managed would probably be defined at the stand level, which is about 100 hectares or less (Morris et al., 1992). At this scale, site factors play an important role and represent the relatively stable context within which the trees of a stand grow. These include soil conditions (especially nutrient flows), topography, aspect on a slope, and climate (Spurr and Barnes, 1980).

Conversely, the value of maximizing recreational attractions could translate into the management goal of establishing and maintaining a "wilderness" area. Because wilderness areas are not confined to a single forest stand but represent a collection of various forest habitats, the scale of management for these systems must necessarily be larger. The boundaries of the area to be managed could be set at the range of the largest ranging species (Noss and Harris, 1986). In this case, diversity within the management unit is an important management goal. This might involve minimum technical interference in natural forest development in some areas so that plant and animal associations develop and natural forest succession proceeds uninterrupted by humans (Morris et al., 1992). Management may also involve some protection from such impacts as recreational use and, in more interventionist practices, fire management.

In our approach, intergenerational values can be thought of as social constraints on shorter-term preferences. These values are associated with longer-time horizons and will generally correspond to management objectives at larger spatiotemporal scales. These higher-level objectives then provide the basis for constraints on management practices at lower levels because in order to maintain a natural "good" expressed by these values over this longer time frame, the ecological processes of the relevant system at this scale must be maintained. This means that after an intergenerational value is translated into a relevant hierarchical system and management goal, the lower-level components that contribute to that system should be constrained by appropriate limits.

Thus, a larger-scale management program can be established, as defined by intergenerational value choices, in order to monitor and coordinate the activities of lower level forest management. For example, the major outputs from catchment basins include water and its associated load of dissolved nutrients and particulate matter (Nelson, 1970). The transport of soluble nutrients is of the most concern for the health of aquatic systems because of the impact of increased concentrations of phosphorous and nitrogen in terms of stream and lake eutrophication (Morris et al., 1992). However, sedimentation from soil erosion and rates of water yield can also have a significant impact (Morris et al., 1992). These factors represent the principal issues of concern in terms of managing the forested watershed to minimize aquatic impacts.

When combined with assessments of the present health of the aquatic systems of concern, these variables can be converted into appropriate management goals that would then have to be incorporated into local management plans. This does not mean that uniform restrictions on, say fertilizer use or clear cutting, would be placed on all lands within the watershed. Instead a watershed-level hydrologic model could be used to identify areas in which higher levels of fertilizer will proabably not impact the aquatic system, and areas where they are more apt to do so. This means that economic forest management, in accordance with the local values in specific areas, could continue as long as the practices did not violate the watershed-scale requirements for maintenance of the aquatic systems.

More important, studying natural dynamics and experimenting with conservation practices provides both scientific progress and increasing public knowledge of the role of ecosystem dynamics in maintaining publicly valued resources and natural features. Through such an approach, hierarchical models and management objectives can be sorted out according to their usefulness in clarifying, explaining, and achieving public goals (Norton, 1992).

Integrating Values at Higher Levels

Before a multiscalar management strategy can be fully defined, the longer-term intergenerational values of local citizens must be integrated with the values of citizens in other local areas. On the local level, the basic principles of economics may be used for determining management activities because they are compatible with the values at this scale. However, local practices defined by these values also operate within higher-level constraints according to our approach. Because intergenerational values are associated with the longer term, the corresponding ecological systems and dynamic processes that maintain a resource or natural feature in that time horizon, generally occur on larger scales, which may require management of practices outside of the boundaries of the local community.

In such a case, citizens from all communities with intergenerational values that require landscape-level management must be brought together so that they can achieve a consensus on what the management goals for higher levels should be. In this manner, the higher-level constraints that are imposed upon local management activities through landscape-level management will not be forced from the "top-down" but will be based on local values with knowledge of the particulars of local ecosystems. Using the process described, it might be determined, for example, that maintaining viable fish populations in rivers and lakes is important to the residents of several communities. By agreement, residents within an area might conclude that they wish to maintain one particular fork of a river or, conversely, they might wish to maintain all the rivers in their area. If they choose to maintain all the rivers in the area, then the management plan will necessarily involve a larger scale. In each case, however, landscape-level management would apply certain constraints on forest management practices within the corresponding watershed. A similar integration could theoretically be performed for global-level concerns, although the difficulties of doing so are quite large.

Summary

In conclusion, what we have proposed is an alternative to traditional approaches to forest management. It is intended to articulate constraints on economic forest management when those practices may interfere with the maintenance of resources and natural features that are valued by the public for the longer term. It is a multiscalar management system based on locally assessed values. The system proceeds by iteration as managers create an interactive and ongoing process involving forest managers and ecological scientists, as well as the public. Through this system of ecologically informed scalar management, we believe that our approach provides a framework for addressing multiple public values—a system that may be able to meet many of the challenges facing forest managers today. Perhaps most important in our approach is that we seek to facilitate a system of forest management that can maximize public values by developing ecological models that characterize the variables and dynamics relevant to publicly valued resources and natural features at various scales.

This approach may also provide a better means by which to address issues of such international concern as global climate change. Instead of relying on "one-size-fits-all" solutions or restrictions on certain activities, the development of an explicitly value-based and scientifically informed process may help to define more appropriate responses for particular countries to take in order to achieve overall goals that are agreed upon by the international community.

Because our proposed methodology is multiscalar, it provides more robust models for ecosystem management. By developing ecologically informed and sensitive models, the proposed approach can show how activities of local communities, regional economies, and national goals can to integrated as nested subsystems of larger, even global models. In this way, the model takes into account local variation in ecological conditions, and also suggests how these submodels can be integrated at larger scales. The interactive process continues both at local level, with the goal of articulating local adaptions and protecting the distinctive character of local communities, and at the larger levels, at which discussions among regions and nations encourage integration across boundaries.

References

Allen TFH, Starr TB (1982) *Hierarchy: Perspective for ecological complexity.* University of Chicago Press, Chicago, IL.
Costanza R, Norton BG, Haskell B (Eds) (1992) *Ecosystem health: New goals for environmental management.* Island Press, Washington, DC.

Cronon, W (1983) *Changes in the land indians, colonists, and the ecology of New England.* Hill and Wang, New York.

Haskell BD, Norton BG, Costanza R (1992) What is ecosystem health and why should we worry about it? In Costanza R, Norton BG, Haskell BD (Eds) *Ecosystem health: New goals for environmental management.* Island Press, Washington, DC.

Johnson, AR (1993) Spatiotemporal hierarchies in ecological theory and modeling. In 2nd international conference on integrating geographic information systems and environmental modeling, Sept. 26–30, 1993, Breckenridge, CO.

Leopold A (1939) A biotic view of land. J For 37:727–730.

Leopold A (1949) *A Sand County almanac and sketches here and there.* Oxford University Press, London.

MacKenzie JJ, El-Ashry M (eds) (1989) *Air pollution's toll on forests and crops.* Yale University Press, New Haven, CT.

Morris LA, Bush PB, Clark JS (1992) Ecological impacts and risks associated with forest management. In Cairns J, Niederlehner BR, Orvos DR (Eds). *Predicting ecosystem risk.* Princeton Scientific Publishing Co., Princeton, NJ.

Mugerauer R (1985) Language and the emergence of the environment. In Seamons D, Mugerauer R. (Eds) *Dwelling, place and environment: Towards a phenomenology of person and world.* Columbia University Press, New York.

Nelson DJ (1970) Measurement and sampling of outputs from watersheds In Reichle, DE (Ed) *Analysis of temporate forest ecosystems.* Springer-Verlag, New York.

Norton BG (1991) *Toward unity among environmentalists.* Oxford University Press, New York.

Norton BG (1995) Evaluating ecosystem states: two competing paradigms. Ecological Economics 14(2):113–128.

Norton BG (1992) A new paradigm for environmental management. In: *Ecosystem health: New goals for environmental management* (eds). Costanza, R, Norton, B, Haskell, B. p.23–41. Island Press: Washington, DC.

Norton BG, Ulanowicz RE (1992) Scale and biodiversity policy: A hierarchical approach. Ambio 21(3):244–249.

Noss RF, Harris LD (1986) Nodes, networks, and MUMs: Preserving diversity at all scales. Envir Manage 10:299–309.

O'Neill RV, DeAngelis DL, Waide JB, Allen TFH (1986) *A hierarchical concept of ecosystems.* Princeton University Press, Princeton, NJ.

Relph E (1985) Geographical experiences and being-in-the-world: The phenomenological origins of geography. In Seamon D, Mugerauer R (Eds) *Dwelling, place and environment: Toward a phenomenology of person and world.* Columbia University Press, New York.

Sagoff M (1988) Environmental protection and property rights. Forum for Applied Research and Public Policy 3(3):75–84.

Simon HA (1962) The architecture of complexity. Proceedings of the American Philosophical Society 106:467–482.

Spurr SH, Barnes BV (1980) *Forest ecology* (third edition). John Wiley and Sons, New York.

Wilkinson CF, Anderson HM (1987) *Land resource planning in the national forests.* Island Press, Washington, DC.

Williams M (1989) *Americans and their forests: A historical geography.* Cambridge University Press, Cambridge, England.

46. Sensitivity of Protection Value to Forest Condition in the Southern Appalachian Spruce–Fir Forest

Dylan H. Jenkins, Jay Sullivan, Niki S. Nicholas, Gregory Amacher, and Dixie Watts Reaves

The southern Appalachian Mountains have a history of natural and anthropogenic disturbance. Wide-scale clear cutting of the region's high-elevation red spruce (*Picea rubens* Sargent) and Fraser fir (*Abies fraseri* (Pursh) Poiret) forests began in the late 1800s and continued until the early decades of this century. By the 1920s, most of the accessible spruce–fir southern Appalachian had been mined for timber, reducing the spruce–fir forests from one-half to one-tenth their pre-European settlement extent (Korstian, 1937; Saunders, 1979). Occupying nearly 20,000 hectares, the rugged high-elevation forests of the Great Smoky Mountains National Park comprise the largest expanse of uncut spruce and fir. Nearly all of the remaining southern spruce–fir forests have been cut at least once and occur in and around the Black and Balsam Mountains in North Carolina, and Mount Rogers in Virginia (Dull et al., 1988; Pyle and Schafale, 1988).

Although logging has long since ceased in the region's higher elevations, the southern spruce–fir forests now suffer from a number of relatively recent disturbances. The present Fraser fir population is under attack from the balsam woolly adelgid (*Adelges piceae* Ratz.), an insect introduced to this country in 1908 from Europe (Kotinsky, 1916). First detected in 1957 on Mount Mitchell in the Black Mountains of North Carolina (Speers, 1958), the adelgid has spread throughout the entire southern Appalachian Fraser fir population (Nicholas et al., 1992a). To an extent, such climate changes as increased ambient air temperature and humidity are favorable for adelgid reproduction. Elevated ambient carbon dioxide (CO_2) levels and increased droughts could both increase the rate of insect population growth and its impact to the region's fir population (Hollingsworth and Hain, 1994; Eager, 1984), as well as impacting fir growth rates even without adelgid infestation (Samuelson and Seiler, 1992). Even without long-term increases in atmospheric temperature, mature Fraser fir are highly susceptible to adelgid attack, and the continued integrity of the spruce–fir ecosystem is uncertain (Nicholas et al., 1992a). Further, Adams et al. (1985), McLaughlin et al. (1987), and LeBlanc et al. (1992) report finding significant decreases in red spruce diameter growth over the past twenty years. Severe deterioration of red spruce crowns has also been recorded during the mid- to late-1980's (Zedaker et al., 1989; Nicholas and Zedaker, 1990). The deterioration of high elevation spruce–fir forests is thought to be exacerbated by air pollution from regional urban and industrial centers (Eager and Adams, 1992; Johnson and Taylor, 1989; White 1984).

Although timber production no longer plays a significant role in the high-elevation region, southern spruce–fir forests are now a key producer of scientific, biologic, and recreational resources. Southern spruce–fir forests are a premier feature of the Great Smoky Mountains National Park, the North Carolina State Park System, the Jefferson and Pisgah National Forests, and the Blue Ridge Parkway. Much of the southern leg of the Appalachian Trail runs through the spruce–fir stands off the southern Appalachians. Additionally, the spruce–fir forest is home to several threatened and endangered plant species, as well as seventeen species or subspecies of plants and animals endemic to the southern Appalachian Mountains (White, 1984). Isolated from related northern vegetation, southern Appalachian spruce–fir forests occur as a series of individual stands that occupy the region's highest mountain tops and ridges. Located at the extreme southern edge of the forest's range, southern spruce–fir forests and their associated amenities are especially sensitive to natural and anthropogenic perturbations. Hence, the southern spruce–fir ecosystem may be a valuable laboratory for understanding how natural and human-caused disturbances influence natural resource values.

Forest Condition and Economic Value for Forest Protection

This chapter seeks to develop a pilot model that links forest conditions that might be associated with global change to the economic value that households place on protecting the southern spruce–fir forest. Contingent valuation (CV) is used to reveal the value that southeastern U.S. households place on protecting southern Appalachian spruce–fir forest quality. Numerous studies have used CV to estimate values for changes in forest condition. For example, in a study conducted by Crocker (1985), recreationists were surveyed to determine the economic values of alternative air pollution-induced health states of southern California national forests. Walsh et al. (1990) surveyed Colorado residents to determine the value that residents place on preservation of forest quality in the national forests of Colorado. Although these models are useful for predicting values for nonmarket good conditions as such, they are limited in their ability to monitor fluctuations in economic value given changes in the individual factors, for example, basal area, degree of insect infestation, downed woody debris, and so forth, which constitute a forest's condition.

Holmes and Kramer (1995) conducted a study of the economic value of protecting different areas of the remaining spruce–fir forests in the southern Appalachian. Respondents were asked their willingness to pay (WTP) to protect remaining undamaged forests along roads and trails versus their WTP to protect all remaining undamaged spruce–fir forests. The principle difference between the Holmes and Kramer (1995) study and this study is the resource being valued. Holmes and Kramer varied the area of protection, whereas this study varied initial forest quality in forest protection scenarios. Alternative damage scenarios were used to examine how household WTP for forest protection is influenced by the initial condition of the forest being protected. That is, we attempt to assess whether households would be willing to pay a different amount to protect a forest that is altered somewhat by an external factor, such as global change, than they are willing to pay to protect a forest in pristine condition.

Willingness to Pay for Forest Quality Changes

Contingent valuation (CV) is a nonmarket valuation method that is an accepted technique for revealing the economic values of natural resources. Many environmental goods, including southern Appalachian spruce–fir forest quality, may be categorized as nonmarket, that is, public, goods. Nonmarket goods are not traded as such, and consequently, their economic value (i.e., prices) are not explicitly revealed in conventional markets. The following discussion is adapted from Fisher (1994) and briefly outlines the theoretical framework supporting the use of CV in eliciting economic values for nonmarket goods.

Contingent valuation is based on consumer theory and the concept of utility. A basic assumption of consumer theory is that an individual maximizes the satisfaction or utility received from goods consumed, subject to his or her budget or income constraint. An individual's utility function may be represented as follows:

$$u = f(m,z,y) \qquad (1)$$

in which the utility (u) an individual receives is derived by consuming both market goods (m) and non-market goods (z) using income (y), and can be maximized subject to the individual's budget constraint ($mp = y$), when mp is the product of market goods consumed and associated prices. From the first-order conditions of utility maximization, demand functions for every market good an individual consumes may be estimated as:

$$m_i = g_i(p,z,y) \; i = 1, \ldots, n, \qquad (2)$$

in which I indexes the ith market good, and n is the number of different market goods consumed. An individual's indirect utility function may now be defined as:

$$h(p,z,y) = u^*[g(p,z,y),z], \qquad (3)$$

so that utility is represented as a function of market prices, income, and environmental goods consumed.

All other factors remaining constant, suppose that the level of one nonmarket commodity, denoted z_f for forest quality, is decreased (the decrease in spruce–fir quality is represented by subscripts denoting forest health condition so that $z_f^0 > z_f^1$, in which z_f^0 is the original undegraded state of the forest and z_f^1 is the forest in a damaged state). The individual's satisfaction as a function of forest condition can now be compared as:

$$[u^0 = h(p, z_f^0, y)] \neq [u^1 = h(p, z_f^1, y)]. \tag{4}$$

Thus our main empirical hypothesis is that, given a perceived change in forest condition, the individual experiences a change in satisfaction when forest health declines. Using the individual's indirect utility function, 3, and the assumption in 4, and recalling that an individual's goal is to maintain a maximized level of satisfaction through goods consumed, an individual's WTP to prevent a decline in forest quality (z_f^0 to z_f^1) is now represented as:

$$h(p, z_f^0, y - WTP) = h(p, z_f^1, y), \tag{5}$$

in which an individual's WTP is the amount of income that a person who values forest health would forego to prevent the original forest condition z_f^0 from degrading to z_f^1. In summary, if forest health is one of the goods an individual values, then the individual may be willing to pay some amount of their income to prevent the forest from degrading to any level below the forest's original condition.

Study Design

The sampling frame for this study includes all households in telephone directories within the seven-state southern Appalachian region (i.e., North Carolina, South Carolina, West Virginia, Virginia, Tennessee, Kentucky, and Georgia). Households were sampled proportionally to population per zip code. To assess the values that both visitors and households who have never visited the region hold for southern spruce–fir forests, households were sampled over entire states rather than from mountain region counties.

Each household was mailed a CV questionnaire built around a referendum type WTP question. As opposed to other types of WTP questions in which respondents are asked to choose their WTP for a nonmarket good from a given set of bid values (payment card) or are asked to write in their WTP value (open ended), referendum-type questions present each household with one value (i.e., price) for the resource they are being asked to value. Respondents are then asked to either accept or reject the given resource price. Each household is randomly assigned one value from a prespecified set of resource prices. In the empirical analysis, the mean value (i.e., mean WTP) for the resource is a function of the probability of each household accepting their given price.

46. Sensitivity of Protection Value to Forest Condition   841

Relative to other value elicitation formats (e.g., open-ended and payment card), referendum-type WTP questions most closely resemble the market and political processes to which individuals are accustomed (Diamond and Hausman, 1993). That is, individuals do not normally bargain for prices, rather they are presented with a price and then choose either to buy or not to buy. In a similar sense, political referencums generally present voters with a specific tax change or bond price for which individuals may vote either for or against. Given the similarity of referendum questions to political and market processes, CV values elicited using yes or no WTP formats may be influenced by yea-saying, which can be defined as the phenomena of individuals accepting their given bid without fully considering their true value for the good (Mitchell and Carson, 1989). More simply, referendum questions may be "too easy" to answer. In a study comparing forest protection, values estimated using open-ended vs dichotomous choice (i.e., referendum-type, WTP questions), Holmes and Kramer (1995) identified yea-saying as a cause of significant value variance between the two elicitation formats that is, the referencum format yielded significantly greater WTP estimates.

Making a complete link between changes in many individual forest-stand attributes and economic value for the spruce–fir forest was beyond the available resources of this study. Therefore, this pilot study was designed to examine the influence of a change in initial stand condition on a household's value for spruce–fir forest protection. To test the null hypothesis that WTP for forest protection programs does not vary depending on the forest's initial condition, 1,000 households were randomly assigned one of two different forest protection scenarios. Households receiving the first scenario were asked to value a protection program for a fir forest in an "unimpacted" state (i.e., 5% dead basal area). Households receiving the second scenario were asked to value a protection program for an "impacted" fir forest that was described as already beginning to show signs of damage from balsam woolly adelgid attack and air pollution (i.e., 30% dead basal area). Each household received two photos with a written description of their protection program scenario; the first photo corresponded to the initial forest condition (either 5% or 30% dead basal area), and the second photo depicted a future forest condition without implementation of a forest protection program (75% dead basal area). Aside from the difference in protection scenarios, the two survey versions were identical. Using survey responses, a test was conducted to determine if a household's WTP for forest quality is influenced by the initial condition of the forest.

Survey Design

The questionnaire for this study borrowed heavily from the Holmes and Kramer (1995) forest quality survey, and consequently, we did not use focus groups to guide the design of survey format or question wording. However, the spruce–fir photos taken as part of this study were used in a related analysis to develop scenic beauty estimates (SBE) (e.g., Daniel and Boster, 1976), which aided us in selecting photos for the WTP section of the CV questionnaire. That analysis used approximately eighty subjects, who registered their preferences (on a ten-point scale) for eighty-five photos of spruce–fir study plots. Photos represented the possible range of forest quality conditions now occurring in the southern Appalachian spruce–fir forests. Based on a scenic preference index derived from subjects' photo-based responses, three photos (two initial conditions and one end condition) were chosen to represent two perceptibly different forest protection scenarios. The three photos represented the midpoint (30% dead basal area), and two extremes (5% and 75% dead basal area) of the scenic beauty index derived for our photos. Given the construction of SBE indices, it is not possible to identify the significance of differences between SBE values. However, using photo-based responses, Daniel et al. (1989) found a high degree of correlation between campers' WTP and scenic beauty values for the same variations in forest condition.

To identify potential difficulties that final sample respondents may have encountered when responding to the survey instrument, a draft questionnaire was mailed to 150 randomly sampled households within the study region. Pretest WTP questions used an open-ended format to establish the range of referendum bids used in the final sample. Based on pretest responses, minor changes were made to survey format and question wording, and the range of referendum bids for the final sample was established. Final survey design and implementation closely followed the Dillman (1978) method. The initial mailing was sent to all final sample households in March 1995, followed by a postcard reminder. A second and third survey packet were mailed to nonrespondents.

Survey Questions

The survey booklet gathered information on household environmental attitudes, familiarity with spruce–fir forest quality and protection, recreation activities, and demographic characteristics. To establish household attitudes regarding forest protection, respondents were asked to rate the importance of several reasons for protecting spruce–fir forests. Information gathered on household recreation activities included a household's past and planned future visits to the southern Appalachian Mountains, and whether respondents had ever noticed large forested areas with dead or dying evergreen trees. Demographic questions included age and sex of respondent, number of people in the household, years of education, and income. Two additional questions were asked to establish a household's time and financial contributions to conservation and environmental organizations.

To aid respondents in assessing their WTP for forest protection, households were given information on southern Appalachian spruce–fir forest features, history, health, and decline, and were provided a map showing the study region and the location of three popular spruce–fir forest sites, that is, Mount Mitchell, North Carolina, Great Smoky Mountains National Park, and Mount Rogers National Recreation Area. Respondents were also given a description of a spruce–fir forest protection program that included two color photographs (one photo of an initial forest condition and another depicting a damaged forest without protection). The photo-based forest protection scenarios were designed to cue respondents on the physical attributes of the forest protection program they were being asked to value.

The WTP question immediately followed the photo and protection program description, and was worded as follows:

> *Based on the two color photographs provided, please answer the following questions:* Would your household pay $X *each year* in additional taxes to provide protection programs for the southern Appalachian spruce–fir forests? That is, your household would be paying to prevent the spruce–fir forests in PHOTO A from becoming like the forests in PHOTO B.

Each household received one of nine possible prices to which they either responded yes, they would pay the given amount for the forest protection program, or no, they would not pay.

Households that responded that they were not willing to pay for forest protection were asked a follow-up question to establish the reason for their negative response. The purpose of the follow-up question on "no" response was to check for protest bids, that is, households that indicated they would not be willing to pay either because "people should not have to pay to protect forest quality," or because "[they] objected to the question." The concern with protest bids is that it is unknown to the researcher whether the respondent truly does not hold as much value for the nonmarket good as the given price, or if the respondent is protesting the method of payment (in this case, additional taxes), or some other aspect of the survey (Diamond et al., 1993).

Households that accepted their resource price were also asked a follow-up question about the percentage of the program price they would assign to each of four reasons for protecting spruce–fir forests, which were 1) use of forests for themselves, 2) use of forests for others (including future generations), 3) protection of the forests even if no one uses them, and 4) other. Past CV studies have shown that nonuse values (i.e., the value placed on a resource so that others, including future generations, may use the resource (bequest value), or for the knowledge that the resource is protected (existence value)), may represent a large part of some respondent's total economic value for a nonmarket resource (Bishop and Heberlein, 1979; Walsh et al., 1990, Holmes and Kramer, 1996).

Biologic Data

Biologic and visual data for this study were collected from a series of plots developed during the mid 1980s as a component of the Spruce-Fir Research Cooperative in the National Acidic Precipitation Assessment Program's Forest Response Program (Nicholas et al., 1992b). From June 21 to July 21, 1994, a total of thirty-three plots near Mount Mitchell, North Carolina and Mount Rogers, Virginia were visited to photosample and to gather data on stem mortality (i.e., balsam woolly adelgid (BWA) infestation).

Photosampling

Photos were taken toward the plot center from each plot corner (a total of eight photos per plot). Human focal length was mimicked using a 35 mm Nikon 6006 camera with a 50 mm lens. All plots were photographed using 200 ASA Kodak Ektachrome(R) slide film. Photographs were taken at a height of approximately 1.6 meters parallel with plot slope from plot corner to plot center. To minimize the influence of subjective compositional judgments, all photographs were taken on automatic focus and aperture settings. Most photographs were taken between the hours of 10:00 am and 3:00 pm, however, hours of photography were extended beyond 3:00 pm (but not later than 5:30 pm) contingent on the presence of favorable lighting conditions, for example, a slope with southwestern aspect. From the approximately 260 photos taken, eighty-five were used to estimate a scenic beauty index for the study plots, and three were chosen to create the two forest protection scenarios.

Basal Area and Stem Mortality

Live vs dead basal area varies widely between plots, and has been shown in scenic beauty studies to be a dominant attribute on which individuals cue to form their aesthetic judgements for forest scenes (Buhyoff et al., 1982). Changes in this forest attribute are one of the primary manifestations of natural and anthropogenic disturbances in the southern Appalachians, and further decline of the region's Fraser fir is expected (Nicholas et al., 1992b). Consequently, forest protection scenarios and photo descriptions were worded to reflect the degree of forest damage in terms of stem mortality.

To estimate these basal area parameters for our plots, each 20-m^2 plot was divided into four quadrants, and one sampling point per quadrant was located 7m toward the plot center from each corner along diagonal transects connecting opposing plot corners. Live and dead basal area parameters were recorded for the four following tree categories: 1) spruce, 2) fir, 3) other softwoods, and 4) hardwoods. Based on the scenic beauty index estimated from plot photos, fir-dominant plots were separated into either an unimpacted or an impacted category. Unimpacted fir plots were defined as having less than 30% of the plot's total basal area composed of standing dead fir.

Empirical Results

Descriptive Statistics

Households were randomly assigned one of two forest protection scenarios: either a protection program for a forest that had very little visual evidence of BWA/air pollution-induced tree mortality (5% dead basal area) or heavy tree mortality (30% dead basal area). Overall response rate for sample households was 40.3% of delivered surveys, although response rates for the 5% damage and 30% damage survey versions were 42.4% and 38.2%, respectively. Based on the recommendations of Freeman (1986), forty-four protest bids were identified (twenty-one for the 5% damage scenario and twenty-three for the 30% damage scenario) and these were excluded from the observations used in the final model estimations.

Difference between samples were tested using two sample t-tests (Table 46.1). Past visitation to the southern Appalachians differs between the 5% and 30% damage samples at the 0.05 significance level. Although the difference in visitation habits is significant, both groups appear to have more than a moderate familiarity with the resource they were being asked to value; more than half of the households in both groups had either read or heard about an increase in spruce–fir forest mortality. Additionally, more than 80% of households in either sample had visited the southern Appalachians sometime in the past.

In their study measuring the contribution of existence and bequest values to total economic value, Walsh et al. (1990) found that nonuse or public benefit values constitute approximately 72.6% of a Colorado resident's WTP value for forest protection. Similarly, bequest and existence values represented nearly 86% of the total value households place on protecting southern Appalachian spruce–fir forests (Haefele et al., 1991). Nonuse values are the set of benefits derived from a good other than value derived from its direct use or consumption. Nonuse values may include bequest value (valuing a good for future generations) and existence value (the satisfaction gained from simply knowing a good exists) (Mitchell and Carson, 1989). Consistent with results from prior studies, nonuse values in this study constitute approximately 79% of the total value households place on protecting spruce–fir forest quality. Recognizing that individuals may not be able to accurately decompose their total WTP for forest protection among different value components (Holmes and Kramer, 1995), the use and nonuse values reported here are for comparative purposes only.

Model Specification

Variables used to explain valuation behavior for forest protection programs in the final models include bid amount, household income, and trips to the southern Appalachian Mountains (Table 46.2). Bid amount recorded the program protection price each household received. Based on the range of WTP responses received from the open-ended pretest survey, households in the final sample were given one of nine bid levels for forest protection, that is, 1, 2, 5, 10, 25, 50, 100, 200, or 400 dollars. Economic theory suggests that a household's budget constraint (i.e., income) is an important explanatory variable in predicting the amount a household will pay for a commodity. From ten income categories, respondents were asked to check the level that best described the total income, before taxes, received by the respondent and other adult household family members in 1994. Household familiarity with the southern Appalachian spruce–fir forest was measured by recording the number of recreational visits to the region in the past three years.

Following the independent subsample design (i.e., households were asked to value only one forest protection scenario) two models explaining household valu-

Table 46.1. Descriptive Statistics of Selected Characteristics of Respondents[a]

Characteristic	Descriptive statistics[b]		t-Value[d]
	Initial condition[c]		
	5% Damage	30% Damage	
Read/heard about increase in damage to spruce-fir forest (yes = 1, no = 0)	0.58 n = 131	0.56 n = 11	0.4215 (0.3262)
Participation in recreation activities 10+ miles from home (no. of days/year)	27.64 (32.27) n = 131	26.87 (37.18) n = 101	0.1686 (0.1337)
Plan to visit southern Appalachians in future (yes = 1, no = 0)	0.86 n = 131	0.88 n = 101	−0.3836 (0.2984)
Visited southern Appalachians in past (yes = 1, no = 0)	0.84 n = 131	0.93 n = 101	−2.0613** (0.9596)
No. of recreational trips to the southern Appalachians in past three years	4.99 (9.70) n = 131	4.47 (6.14) n = 101	0.4708 (0.3581)
If bid amount accepted, percentage of bid amount assigned to following reason for protecting forest:			
Use for self (use value)	16.40 (13.51) n = 78	20.15 (17.76) n = 64	−1.4276 (0.8444)
Use for future generations (bequest value)	33.63 (22.79) n = 78	33.83 (20.48) n = 64	−0.0491 (0.0391)
Protection of forests even if no one uses them (existence value)	46.48 (27.03) n = 78	44.06 (25.20) n = 64	0.5476 (0.4152)
Age (years)	47.73 (16.02) n = 131	50.32 (13.67) n = 101	−1.2932 (0.8028)
Sex (proportion female)	0.28 n = 131	0.25 n = 101	−0.5471 (0.4152)
Household (no. people)	2.69 (1.49) n = 131	2.80 (1.21) n = 101	−0.5869 (0.4422)
Education (years)	14.15 (3.20) n = 131	14.88 (3.12) n = 101	−1.7206* (0.9133)
Income ($)	43,989 (29,335) n = 131	50,817 (32,959) n = 101	−1.6654* (0.9028)

[a] Protest responses were not included in the calculation of descriptive statisics.
[b] Descriptive statistics for each characteristic are mean, standard deviation, and sample size, respectively. Standard deviations are not reported for characteristics that have (0/1) responses.
[c] Initial condition refers to initial forest condition presented in each forest protection scenarios, i.e., 5% Damage refers to 5% dead basal area initial forest condition (unimpacted sample); 30% Damage refers to 30% dead basal area initial forest condition (impacted sample).
[d] Probability $|t| \geq x$ in parentheses.
*Significant at $\alpha = .10$.
**Significant at $\alpha = .05$.

Table 46.2. Estimated Logit Model Coefficients[a,b]

	Coefficient[c]		
	Initial Condition[d]		
Variable	5% Damage (n = 131)	30% Damage (n = 101)	Combined (n = 232)
Constant	0.332 (0.802)	0.329 (0.684)	0.440 (1.445)
Bid Amount	-0.110×10^{-1}*** (-4.115)	-0.951×10^{-2}*** (-3.794)	-0.100×10^{-1}*** (-5.594)
Income	0.2183×10^{-4}*** (2.604)	0.1031×10^{-4} (1.329)	0.1667×10^{-4}*** (2.952)
No. of recreational trips to southern Appalachians	-0.3818×10^{-2} (-0.195)	0.1222* (1.837)	0.9740×10^{-2} (0.532)
Model χ^2 statistic	36.626***	28.334***	59.652***
Likelihood ratio test χ^2 statistic[e]		5.42	
Mean WTP	$137.23	$169.38	$150.63

[a] Protest bids not included in observations used for final estimations.
[b] Model coefficients were estimated using the LIMDEP econometric program (Greene, 1992).
[c] t-statistics in parentheses.
[d] Initial condition refers to initial forest condition presented in each forest protection scenario, i.e., 5% DBA is initial condition for unimpacted sample; 30% DBA is initial condition for impacted sample. Combined refers to model estimated using observations from both survey versions.
[e] Critical value is 9.48 at $\alpha = 0.05$.
*Significant at $\alpha = .10$.
**Significant at $\alpha = .05$.
***Significant at $\alpha = .01$.

ation behavior were analyzed. Model parameters were estimated for the following logistic equation:

$$P_i = \frac{1}{1 + e^{-x_i\beta}} \quad (6)$$

in which P_i is the probability of a household accepting their given referendum WTP price for a forest protection program, X_i is the matrix of independent variable observations (i.e., bid price, income, and number of recreation trips), and β is the coefficient matrix.

Coefficient signs in logit models indicate the direction that the probability of a household accepting their WTP referendum value changes for right-hand side variable changes. As seen from the t-ratios, the referendum bid amount that a household was asked to pay for their forest protection program is a significant predictor of the probability of bid acceptance. The negative coefficients for bid amount indicate that, as the given referendum price increases, households are less apt to accept that bid. The significant positive coefficients on income in the 5% damage and combined models give evidence that higher-income households were more apt to accept their referendum bid. These results are consistent with the downward sloping demand curve of economic theory. The variable representing familiarity with the spruce–fir resource (i.e., the number of recreational trips taken to the southern Appalachians in the past three years) was insignificant in explaining household valuation behavior for forest protection in all models at the 0.05 significance level.

Model coefficients were examined jointly for significance of their difference from zero. The X^2 statistics for all models reveal that the null hypothesis that all coefficients are equal to zero can be rejected at the 0.01 significance level, indicating that WTP values vary significantly from zero. A likelihood ratio test was used to examine the difference between the coefficients of the 5% and 30% damage models, testing the null hypothesis that coefficients were identical across models. Based on the calculated X^2 statistic for this test, no significant difference between forest protection value models can be claimed at the 0.05 level. That is, over the range of initial forest damage considered (5% to 30%), a household's probability of accepting their given price, and hence their WTP, for a spruce–fir forest protection program does not appear to be sensitive to the initial condition of the forest.

To calculate mean WTP values for forest protection, we integrated the estimated function (6) numerically, using the referendum bid level that yields a 1% probability of acceptance as the upper limit of integration. The mean WTP for forest protection estimated by this study, $150.63 (both survey versions combined) is somewhat larger than the mean referendum values of $59.22 to $99.57 (depending on the area of forest protected) estimated by Holmes et al. (1992). Discrepancies in WTP values between the two studies may be explained by geographic differences in sampling, integration limits for calculating mean WTP, and other issues. Holmes et al. (1992) sampled households within a 500-mile radius of Asheville, North Carolina, an area that extends into Michigan, New York, Missouri, Louisiana, and Florida. Households in this study were sampled only from states that contain a portion of the southern Appalachian Mountains (i.e., Virginia, West Virginia, North Carolina, South Carolina, Georgia, Kentucky, and Tennessee). Consequently, our sample is probably more familiar with the resource and may have a greater interest in protecting the spruce–fir forests.

Discussion

The phenomenon of individuals yielding statistically similar economic values for dissimilar public goods and for different provision levels of the same public good has been observed in prior CV studies of public good values (Kahneman and Knetsch, 1992; Kahneman et al., 1993; Boyle et al., 1994). As possible explanations for invariant WTP values, these studies point to deficiencies in survey design, respondent misperceptions of the resources being valued, and the propensity of CV to elicit support responses rather than purchase values for environmental goods. Adopting some of the same arguments, possible causes of the condition-invariant protection values reported here include the influence of bias caused by questionnaire wording, respondents' inability to identify with the resource being valued, and an insufficient range of forest conditions tested.

Deficiencies in survey wording and respondent misperception of the good to be valued may have interacted to influence forest protection values. Inadequate or unrealistic portrayal of the spruce–fir resource may have caused respondents to visualize a commodity different than the resource envisioned by the study. To familiarize respondents with the resource they were being asked to value, surveys contained a detailed information section on southern Appalachian spruce–fir forest history, health, and decline. This section was placed immediately before the forest protection scenario description and WTP question. Wording in the forest information section was designed to avoid biased and alarmist language. Graphic aids were included with mail-out surveys to minimize the incidence of respondents perceiving a good different than the one envisioned by the study. Households were provided with an explicit description of their forest protection program, including a map of southern Appalachian spruce–fir region and two color photographs depicting the forest conditions described. Furthermore, the WTP section instructed respondents to focus specifically on the physical forest condition change represented by the two photos. Nevertheless, respondents simply may not have been able to understand or relate to the resource changes that were described. Also, information regarding a forest in decline is, by nature, cause-oriented. As a consequence, when asked to value a forest protection program, the placement and nature of the information section may have unintentionally cued households' value judgements on forest protection as a cause rather than on the specific level of forest protection provided.

This same effect may have been caused by respondents seeking a sense of moral satisfaction, or "warm glow", from the opportunity to value an environmental good (Shavell, 1993). Such a suggestion is plausible to the degree that individuals viewed the survey as an opportunity to register their support for an environmental cause. If survey wording and design or a "warm glow" effect prompted households to perceive forest protection as a cause to be supported rather than a commodity to be purchased, then WTP values for forest protection could be expected to be invariant regardless of the physical condition of the forest. Unfortunately, we have no information on the psychological processes respondents used when answering the WTP question, and have no way of determining the influence of survey wording vs respondent perceptions of the commodity on final WTP estimates.

A third possible explanation for invariant WTP values for different forest conditions is that the range of damage tested may have been too small to elicit a change in economic value for forest protection. Photographs used to depict our forest protection scenarios represented the midpoint and endpoints of the scenic beauty index derived using plot photos, providing evidence that the photos selected to represent forest protection scenarios depicted a perceptible difference in forest conditions. Because the correlation between WTP values and SBE for the photos used in this study is unknown, it could be argued that the range of conditions tested by this study was insufficient to elicit a significant difference in WTP responses between the two protection scenarios. That is, from 5% to 30% damage, WTP for forest protection may be flat relative to SBE values over the same range.

Recent studies comparing WTP and SBE of the same good have found high correlation between the two value metrics. In a comparison of campers' photo-based aesthetic preferences for forest areas vs other campers' photo-based economic judgements for the same sites, Daniel et al. (1989) found a nearly perfect relationship ($\rho = 0.96$) between the two indexes of value. Similarly, in a study comparing the ranking of 16 social issues using five different metrics of preference (including WTP), Kahneman et al. (1993) report between-metric correlations of $\rho = 0.52$ to 0.97. Given the evidence of high correlation between WTP and other measures of preference (including SBE), it is difficult to support the hypothesis that the range of damage that represented the endpoint and midpoint of scenic beauty judgements was insufficient to elicit a change in economic value between forest protection scenarios.

Finally, we cannot rule out the possibility that our results were influenced by nonresponse bias. Nonresponse bias occurs when people who return the questionnaire are different in some important way from the people who do not return the questionnaire. Given the similarities between our respondent average income level ($46,962) and the regional household average ($44,670) (U.S. Department of Commerce, 1994), we have no evidence, at least on the basis of income, that nonresponse bias influenced our sample.

Implications

This pilot study was conducted to estimate the influence of forest condition on the economic value households place on southern Appalachian spruce–fir forest protection. Although a photo-based scenic beauty study revealed differences in aesthetic preference for the scenes used in two forest protection scenarios, results from our CV study suggest that household economic values for spruce–fir forest protection may be insensitive to nonmarginal changes in the physical condition of the forest. Although the results of this study do not allow the conclusion that, via WTP responses, households "purchased moral satisfaction" (Kahneman and Knetsch, 1992), they do indicate that respondents failed to cue on the physical attributes of forest protection as intended by the study. The inability of this CV experiment to capture values for forest protection that are sensitive to the forest's physical condition is problematic in predicting the influence of potential global changes on southern Appalachian spruce–fir forest value. If economic values for forest quality are governed by a household's support for forest protection as a cause, or if households do not focus on the physical condition of the forest when forming their value responses, then WTP for forest protection may not be sensitive to the manifestations of global change on the forest.

For the present study, there is no information on the psychological processes respondents used to formulate their valuation responses. Also, because WTP for forest protection was examined between 5% and 30% dead basal area, conclusions regarding the sensitivity of household forest protection values cannot be made beyond the range tested. Further research is needed to determine the range of damage over which WTP is invariant, the role of the survey instrument in contributing to the apparent invariance of protection values, and the psychological processes individuals use when valuing environmental goods. Future work should also investigate the correlations between aesthetic and economic value preferences for forest scenes and the dynamic WTP values when responses from the general public are disaggregated into different social groups.

References

Adams HS, Stephenson SL, Blasing TJ, Duvick DN (1985) Growth-trend declines of spruce and fir in mid-Appalachian subalpine forest. Envir and Exper Bot 25:315–325.

Bishop RC, Heberlein TA (1979) Measuring values of extra-market goods: Are indirect measures biased? Am J Agri Econ 61(Dec.):926–930.

Boyle KJ, Desvousges WH, Johnson FR, Dunford RW, Hudson SP (1994) An investigation of part-whole biases in contingent-valuation studies. J Envir Econ Manage 27:64–83.

Buhyoff GJ, Wellman JD, Daniel TC (1982) Predicting scenic quality for mountain pine beetle and western spruce budworm damaged forest vistas. For Science 28:827–838.

Crocker TD (1985) On the value of the condition of a forest stock. Land Econ 61(3):244–254.

Daniel TC, Boster RS (1976) Measuring landscape esthetics: The scenic beauty estimation method. USDA For Serv Res Pap 167.

Daniel TC, Brown TC, King DA, Richards MT, Stewart WP (1989) Perceived scenic beauty and contingent valuation of forest campgrounds. For Science 35(1):76–90.

Diamond PA, Hausman JA (1993) On contingent valuation measurement of nonuse values. In Hausman JA (Ed) *Contingent valuation: A critical assessment.* Elsevier Science Publishers, New York.

Diamond PA, Hausman JA, Leonard GK Denning MA (1993) Does contingent measure preferences? Experimental evidence. In Hausman JA (Ed) *Contingent valuation: A critical assessment.* Elsevier Science Publishers, New York.

Dillman DA (1978) Mail and telephone surveys: The total design method. John Wiley and Sons, Inc. New York.

Dull CW, Ward JD, Brown HD, Ryan GW, Clerke WH, Uhler RJ (1988) Evaluation of spruce and fir mortality in the southern Appalachian Mountains. USDA For South Reg R8-PR 13.

Eagar C (1984) Review of the biology and ecology of the balsam wooly aphid in southern Appalachian spruce–fir forests. In White PS (Ed) *The southern Appalachian spruce-fir ecosystem: its biology and threats.* USDI Nat Park Serv, REs Res Manage Rep SER-71.

Eagar C, Adams MB (1992) *Ecology and decline of red spruce in the eastern United States.* Springer-Verlag, New York.

Fisher AS (1994) The conceptual underpinnings of the contingent valuation method. CA Agri Exper Stat, DOE/EPA workshop on using contingent valuation to measure non-market values, May 19–20, 1994, Washington, DC.

Freeman AM (1986) On assessing the state of the arts of the contingent valuation method of valuing environmental changes. In Cummings RG, Brookshire DS, Shultze WD (Eds)

Valuing environmental goods: An assessment of the contingent valuation method. Rowman and Allanheld, Tottwa, NJ.

Haefele MA, Kramer RA, Holmes TP (1991) Estimating the total value of forest quality in the high-elevation spruce–fir forests. In Proceedings of the National conference of the economic value of wilderness. Gen Tech Rep SE-78. USDA For Serv, Southeast For Exper Stat, Asheville, NC.

Hollingsworth RG, Hain FP (1994) Effect of drought stress and infestation by the Balsam Woolly Adelgid (*Homoptera: Adelgidae*) on abnormal wood production in Fraser fir. Can J For Res 24:2295–2297.

Holmes TP, Kramer RA, Haefele MA (1992) Economic valuation of spruce–fir decline in the southern Appalachian Mountains: A comparison of value elicitation methods. In Proceedings of the forestry and the environment: Economic perspectives conference. March 9–11, 1992. Jasper, Alberta, Canada.

Holmes TP, Kramer RA (1995) An independent sample test of yea-saying and starting point bias in dichotomous-choice contingent valuation. J Envir Econ Manage 29:121–132.

Holmes TP, Kramer RA (1996) Contingent valuation of ecosystem health. Ecosys Health 2(1):56–60.

Johnson DW, Taylor GE (1989) Role of air pollution in forest decline in Eastern North America. Water Air Soil Pollut 48:21–43.

Kahneman D, Knetsch JL (1992) Valuing public goods: The purchase of moral satisfaction. J Envir Econ Manage 22:57–70.

Kahneman D, Ritov I, Jacowitz KE, Grant P (1993) Stated willingness-to-pay for public goods: A psychological perspective. Psychol Sci 4(5):310–315.

Kotinsky J (1916) The European fir trunk louse (*Chermes (Dreyfusia) piceae* Ratz.): Apparently long established in the United States. Proceedings of the Entomological Society of Washington 18:14–16.

LeBlanc DC, Nicholas NS, Zedaker SM (1992) Prevalence of individual-tree growth decline in red spruce populations of the southern Appalachian Mountains. Can J For Res 22:905–914.

McLaughlin SB, Downing DJ, Blasing TJ, Cook ER, Adams HS (1987) An analysis of climate and competition as contributors to decline of red spruce in high elevation Appalachian forests of the Eastern United States. Oecologia 72:487–501.

Mitchell RC, Carson RT (1989) Using surveys to value public goods: The contingent valuation method. Resources for the Future. Washington, DC.

Nicholas NS, Zedaker SM (1990) Forest decline and regeneration success of the Great Smoky Mountains spruce–fir. In Smith ER (Ed) Proceedings of the first annual southern Appalachian man and the biosphere conference. (Abstract) Nov. 5–6, 1990, Gatlingburg, TN. TVA/LR/NRM-90/8, Tennessee Valley Authority, Norris, TN.

Nicholas NS, Zedaker SM, Eagar C (1992a) A comparison of overstory community structure in three southern Appalachian Spruce-Fir Forests. Bull Torrey Bot Club 119(3):316–332.

Nicholas NS, Zedaker SM, Eagar C, Bonner FT (1992b) Seedling recruitment and stand regeneration in spruce–fir forests of the Great Smoky Mountains. Bull Torrey Bot Club 119(3):289–299.

Pyle C, Schafale MP (1988) Land use history of three spruce–fir forest sites in Southern Appalachia. Journal of Forest History 32:4–21.

Samuelson LJ, Seiler JR (1992) Fraser fir seedling gas exchange and growth responses to elevated CO_2. Environ Exper Bot 32:351–356.

Shavell S (1993) Contingent valuation of the nonuse value of natural resources: Implications for public policy and the liability system? In Hausman JA (Ed) *Contingent valuation: A critical assessment.* Elsevier science Publishers, New York.

Speers DM (1958) The balsam woolly aphid in the southeast. J For 56:515–516.

U.S. Department of Commerce (1994) Statistical abstract of the United States, 114th Edition, Washington, DC.

Walsh RG, Bjonback RD, Aiken RA, Rosenthal DH (1990) Estimating the public benefits of protecting forest quality. J Environ Manage 30:175–189.
White, PS (1984) The southern Appalachian spruce–fir ecosystem, an introduction. In White PS (Ed) The southern Appalachian spruce–fir ecosystem: Its biology and threats. USDI Nat Park Serv, Res Res Manage Rep SER-71. SE Regional Office, Atlanta, GA.
Zedaker SM, Nicholas NS, Eagar C (1989) Assessment of forest decline in the Southern Appalachian Spruce-Fir Forest, USA. In Bucher JB, Bucher-Wallin I (Eds) Air pollution and forest decline. Proceedings of 14th International Meeting for specialists in air pollution effects on forest ecosystems, IUFRO P2.05, Interlaken, Switzerland, Oct. 2–8, 1988, Birmensdorf.

47. Economics and Global Climate Change

David N. Wear

Global climate change portends important changes in both the structure and the extent of forestlands in the U.S. South. Climate and biological response models suggest that even modest changes in temperature and precipitation could strongly alter net primary productivity, species ranges, and even the area over which trees can persist (VEMAP Members, 1995). Although the collection of these models produces a broad variety of—and in some cases countervailing—climate change forecasts, they clearly raise concern over the future structure and function of ecosystems and ultimately the welfare of people in the region. The objective of this chapter is to explore our present ability to estimate the potential economic effects of climate change and the effects of efforts to mitigate climate change.

The studies reported in this section look generally at how global climate change might influence the benefits that people derive from forests in the southern United States and how human action might be designed to offset or adapt to changes in ecological systems. Under the best of circumstances, it is difficult to describe the complex interactions that define how human welfare is tied to ecological condition. In the case of global climate change this complexity is compounded further by the high degree of uncertainty surrounding forecasts of climate change and ecological response. Because estimates of human impacts must necessarily be derived from these climate and biological forecasts, economic forecasts can be only tenuous.

In spite of this uncertainty, there is considerable value in undertaking the economic analysis of global climate change that extends beyond the relative accuracy of market forecasts. These values derive not from knowledge of "what will happen," but from improved understanding of how things could happen. In this case, we are especially interested in the mechanisms through which values are derived from forests, how the economy might adapt to forest changes, and what the plausible and efficient strategies for mitigating adverse impacts might be. The studies in this section provide first approximations on these issues and yield some useful insights.

A Role for Economics

Economics is classically defined as the study of how people allocate scarce resources among competing uses to achieve human wants. At one level, economics examines the mechanisms of producing and consuming things of value. The scale of this type of inquiry can range from the individual to markets to economies as a whole. At another level, economics addresses how resource allocation might be improved by intervening in the natural course of events in an economy. This is the study of practical policy instruments that society can apply to correct problems in resource allocation and change the provision and distribution of human wants.

The role of economics in studying global climate change should ring clear. We are concerned first with how resulting changes in ecological structure and productivity will influence the options available to people managing and using forests, and then how they will respond to changes in their options. Because human beings are adaptive, shifts in biophysical productivity generally do not "add up" in a simple way to economic impacts. Estimating economic impacts requires knowing something about how people allocate their forest resources when faced with different options and how they derive values from ecological systems in general (Mendelsohn et al., 1994).

Although the mechanisms of global climate change remain uncertain, it would be injudicious to not begin to contemplate policy aimed at mitigating climate change and its potential impacts. Policies give rise to costs that can result in changes in resource allocation both in targeted sectors and in other sectors. Additionally, policies also have opportunity costs (e.g., government could use funding to address other issues, such as crime, poverty, and so forth). Good policy choices require both effectiveness and relative efficiency of various options and insights into the fundamentals of resource allocation.

Resource Allocation Through Timber Markets

Timber products are the most visible goods produced from southern forests. Although timber production, consumption, and values are easily tracked when compared to the many forest goods and services that are not exchanged in readily defined markets, accurately modeling timber markets is far from a trivial undertaking. This is because 1) timber is not a homogenous product, but is several products trading in separate but interrelated timber markets, and 2) the nature of timber growth means that every action taken today (e.g., harvest or stand improvements) will have effects on timber management options in subsequent years.

The supply of timber depends essentially on the ecological relationships that define forest growth and yield coupled with management inputs (Wear and Parks, 1994). To estimate the effects of climate change on timber supply requires some estimates of how these fundamental growth relationships will be affected. The three ways that aggregate forest growth might be affected are 1) primary productivity could be altered, 2) the species mix of forests could change, and 3) the extent of southern forests could shift both in size and in location. All three could have important impacts on eventual timber production and, depending on the scenario, might increase or decrease the economic benefits derived from forests.

In addition to estimates of biological production relationships, we require some insights into how people manage forests. Differences in biological production imply changes in the options faced by forest managers, and therefore, changes in their contemporary management approach and eventually their levels of timber production.

The key technical construct in studying the output decisions of any sector is a production function that defines how various inputs are transformed into outputs. In the case of forestry, we are concerned with a biological production function that maps inputs of land, management, and time to the outputs of various timber products. The input of time makes timber production functions decidedly more complex than, for example, production functions for manufacturing sectors. This is because changes in management plans can take decades to play out completely, and, conversely, because management possibilities are always constrained by the history of management and the condition of the standing forest.

The timber production function is the fundamental conceptual model behind the analysis of timber markets. However, the timber production function as such, is only rarely estimated in empirical economic analysis. Rather, derivative timber supply models that define timber output as a function of price and other variables are usually estimated. In effect, supply functions combine behavioral elements of harvest and management choices with biological models of growth and yield. If correctly specified, timber supply models can describe how changes in biological production lead to changes in timber output.

The economic analysis of changes in timber markets resulting from climate change is built up from this model of interaction. That is, climate change influences fundamental biological productivity, and therefore shifts management options. As a result, derivative timber supply relationships are altered and timber markets adjust accordingly. If, for example, all other variables are equal and productivity increases, then more timber would be supplied at a given price, and the market clearing level of timber production would increase by an amount determined by the interaction of timber supply and demand.

Of course, the magnitude of a supply effect depends on the magnitude of changes in productivity, but it depends on other variables as well. First, it depends on the shape of the underlying timber yield curve. In the southern United States, where growth is rapid and markets are strong for both large and small diameter products, changes in productivity might lead to discontinuous changes in rotation ages and shifts in the resulting product mix. In addition to these productivity changes, because shifts in productivity apply to the land and not to timber alone, climate change might imply a change in optimum land use between agriculture and forestry. Finally, the estimate of a supply impact can depend critically on the length of the period that is analyzed. In the short term, effects on supply may be substantially constrained by the age-class distribution of existing inventories of timber; given more time, however, supply may adjust to a much greater extent.

There are two fundamental approaches to modeling timber supply. One is empirical analysis that fits the supply functions with historical data. Empirical supply functions are based on a tacit, unspecified production function. Because the production function is unspecified, empirical supply functions can provide only limited insights into effects that change the productivity of forests. However, they do provide a means of examining shifts in the scale or area of forests. The other approach to modeling timber supply is a simulation approach that mechanistically builds supply response up from models of growth and yield and individual choice. These models prove more versatile for simulating productivity shifts but generally lack grounding in historical observations.

The four studies in this section that forecast the economic impacts of global climate change on timber markets use two different models. One is a regional econometric model for the southern United States developed by Newman (1987). The papers by de Steiguer and McNulty (Chapter 44), and Gunter et al. (Chapter 43) apply versions of Newman's model to estimates of aggregate changes in inventory resulting from global climate change. They adopt empirical supply functions derived from thirty years of historical data (through 1980) to gauge impacts on the forest sector as a whole.

In contrast, the chapters by Alig et al. (Chapter 41) and Burton et al. (Chapter 42) use a model that derives supply from a mechanistic model of forest growth and individual choices (the Forest and Agricultural Sector Optimization Model—FASOM). So, rather than use a model of supply that is based on a historical and implicit production function, this model directly models the production function and examines how landowners may respond to changes in productivity. Supply effects are derived from the simulation of management choices. Additionally, FASOM directly models shifts in agriculture and forestry land uses, thereby directly addressing interactions with the other major land-using sector in the region. It, therefore, incorporates the mechanisms for long-term adjustments.

The econometric market supply approach used by de Steiguer and McNulty (Chapter 44) and by Gunter et al. (Chapter 43) has the virtue of being grounded in empirical history. That is, estimated supply responses are consistent with historical supply behavior. As constructed, these models project the effects of proportional shifts in the area of forests. Even when shifts in underlying productivity are estimated, they are applied as effective reductions in forest area. This is because such econometric models of supply as Newman's (1987) are strictly functions of total timber inventory and timber prices within a region, and, implicitly, are derived from an historical production function. There is no mechanism for estimating changes in stock to land ratios that are implied by changes in productivity.

In spite of this limitation, estimates from an essentially short-term adjustment model (Newman, 1987) can provide useful first approximations of the potential economic impacts of global climate change. Until we know more about how productivity relationships are effected by climate change, these types of models can inform the policy debate with order-of-magnitude estimates of impacts.

The FASOM model, in contrast, directly addresses the mechanisms of supply response to productivity changes. However, as described in Burton et al. (Chapter 42), there is little specific information on how productivity at a stand level will actually be altered by climate changes. Their logical response to this uncertainty is to analyze a broad range of productivity change scenarios. Again, the result is insight into order-of-magnitude estimates of impacts rather than specific forecasts.

The challenge facing these types of mechanistic simulation models is calibration. In effect, such models as FASOM are sensitive to specification errors, because they are based on very specific models of how individuals behave. In addition to more specific information on productivity changes, these models need continued refinement of their definitions of individual behavior regarding land use and timber management. For example, we know that the timber management decisions of industrial and nonindustrial private landowners are qualitatively different (Newman and Wear, 1993), and, therefore, their responses to productivity changes should be different. Furthermore, industry ownership of land has to be viewed as a strategic choice that depends, at least in part, on the supply behavior on other private landowners in the South, as well as conditions in other regions. These types of interactions are not well understood.

Short-term econometric models and mechanistic models therefore provide different strategies for addressing the economic impacts of biological changes that are known with little certainty. The short-term econometric models exploit the information contained in historical behavior to estimate first approximations of effects. These models can, however, provide only limited insights into long-term phenomena. Mechanistic models can address long-term adjustments, but they require much more information on the nature of change and on the nature of human behavior. These models provide the first steps toward structuring how shifts in biological productivity could be expressed in forest products markets and, perhaps even more important, they could provide frameworks for organizing information and identifying information deficits.

The results from the studies by de Steiguer and McNulty (Chapter 44), and Burton et al. (Chapter 42) highlight the differences between modeling approaches. de Steiguer and McNulty find that declines in southern growth rates result in costs for both producers and consumers. Analysis of similar scenarios by Burton et al. (Chapter 42) also show costs for consumers. However, their approach, by allowing for adjustments in management by landowners, as well as for adjustments between regions, shows enhanced benefits accruing to wood products firms.

Taken together, these modeling efforts highlight how understanding the impacts of climate change will require new insights into social phenomena (e.g., timber supply) in addition to better information on the climate and biological changes resulting from social phenomena.

Nontimber Benefits

People derive much more than timber from southern forests. The value of recreation in forms that range from camping and hiking, to hunting, fishing, and wildlife viewing depend on forest and ecological conditions. The role of forests in watershed protection and the production of clean water is increasingly important in the face of rapid population growth in the southern United States. Forests also provide aesthetic benefits to the people who live among them. Overarching all of these specific uses, however, is the ultimate concern that human habitation and carrying capacity is directly linked to the materials and energy flows that comprise the biosphere.

We know relatively little about the relationships between forest condition and nontimber benefits. Because these goods and services do not trade in markets, we lack simple measures of their values. In addition to this lack of knowledge, even the production of these benefits—with perhaps the exception of some forms of recreation—is not monitored, and information on the fundamental supply relationships, for example, species persistence or water-production models, are not well understood. With very limited knowledge of prices or even the quantities of these goods and services, it is impossible to accurately account for nonmarket benefits. With only limited knowledge of production relationships it is difficult to estimate how climate change might effect these benefits in the future.

These are critical deficits in our understanding of how human well-being is tied to the function of ecological systems. In response, a body of research has emerged to improve knowledge of how values are formed for the nontimber benefits of forests (e.g., Holmes and Kramer, 1996; Sohngen et al., 1993). In particular, work in nonmarket valuation is presently focused on methods used to elicit values using the two principal approaches of 1) contingent valuation, and 2) travel-cost analysis. The former uses questionnaires and other approaches to simulate a market setting for evaluating choices regarding nonmarket goods. The latter method derives the value of nonmarket goods from actual choices among alternatives with different nonmarket attributes.

Although considerable progress has been made in nonmarket valuation, especially over the last five years, the field has not yet produced the data or methodology necessary for compiling a complete accounting of nontimber benefits. However, promising developments in the field of natural resource accounting may provide at least a rough approximation of the nonmarket benefits obtained, not from direct valuation of nontimber goods, but through analysis of forgone opportunities in the production of timber goods (Lee, 1996). That is, the value of nontimber benefits can be inferred from the difference between the market values actually obtained from timber management and the maximum values that could have been obtained from other activities. These accounting methods can provide important insights into the historical contribution of nonmarket goods and services to human welfare but because they do not directly address the formation of value they have limitations for addressing how values may be shifted in response to climate change.

Both nonmarket valuation and natural resource accounting represent broad fields of inquiry in economics in general and forest economics in particular. They hold considerable promise for shedding light on how human values are inextricably tied to the function of ecosystems and natural places.

In lieu of methods that can provide system-level estimates of nontimber benefits, it makes sense to investigate these relationships through case studies. The chapter by Jenkins et al. (Chapter 46) examines how southeastern households value the conditions of spruce–fir forests in the southern Appalachians. In turn, they examine how these conditions and derived values might be impacted by global change. The results provide not only values of these benefits but also the relative values placed on forest protection programs. This study, and similar studies to it, can begin to provide essential insights into the formation of value, effects of damage, and the structure of policy at a local level.

Programs and Policy

Usually the objective of economic analysis is not simple description or valuation. Rather, it is to examine potential program and policy solutions to social problems. Generally, these problems can be linked to market failures or situations in which the market allocation of resources fails to provide for all desired social benefits. Market failure is often defined by externalities, generally in which production or consumption produces uncompensated "side effects" (see Baumol and Oates, 1988, for an extended discussion). For example, an externality arises when productive activity gives rise to water pollution that, in turn imposes costs on downstream users. In the case of climate change, many production activities have altered the atmosphere, imposing costs at a global scale. The essence of the externality problem is that if producers do not realize the benefits or incur the costs resulting from their actions then they will not allocate resources in a way that maximizes human wants, that is, the externality will not be given its appropriate value when weighing alternative choices.

Government can intervene to correct for misallocation in three basic ways. First way is to simply to impose regulations or standards that reduce or eliminate a harmful substance. For example, use or production of such toxic substances as DDT can be banned or restricted to minimum-use levels. A second approach is to use taxes or subsidies to incrementally shift individual resource allocation decisions. For example, a timber severance tax will, on average, prolong timber harvest rotations.

The third general approach to regulation is to apply so-called market-based approaches that create markets for the externality. The best known example of this is the allocation of rights to pollute that can then be traded among polluting firms. The firms can thereby benefit from installing cleaner technologies and selling their pollution permits. Although permit systems may apply to only a limited set of situations, they can often accomplish pollution reductions much more efficiently than direct regulation.

Forests and forestry play an unusual role in the development of policies addressing global climate change. Although forestry in general, at least in temperate forests, is not generally considered a principal cause of climate changes, it is considered an important mechanism for mitigating climate change. This is because trees transform atmospheric carbon dioxide into woody biomass and serve as one of the largest terrestrial sinks of carbon. Therefore, tree-planting programs that expand the net area of forestland are seen as a useful tool for reducing greenhouse gases in the earth's atmosphere. Additionally, changes in the management of existing forests can also influence carbon storage.

The principal policy instrument proposed for increasing the area of forests and, therefore, the storage of carbon in the South is subsidies for tree planting. The study by Alig et al. (Chapter 41) in this section however, indicates that tree-planting programs may be a fairly inefficient means of reducing atmospheric carbon. Although they find that these programs have relatively small economic effects on the agricultural sector, they do not expand carbon pools by significant amounts. Instead, it appears that programs that combine both tree planting and management intensification hold the most promise for increasing carbon stored in terrestrial vegetation.

These policies appear to be fairly effective tools for increasing carbon storage even though they raise interesting distributional issues. Tree-planting programs funded by the government to correct for actions of firms in other sectors (e.g., energy) are in a sense a subsidy to the polluting industry. To the extent that efficiencies are gained by the policy, it may be more equitable to assess the principal sources of greenhouse gases to pay for these types of programs.

In addition to such standard policy instruments as taxation and subsidy, forest policies in the United States are often aimed at extensive public forestlands. Although there are limited opportunities to use public lands as instruments to shape the landscape in the southern United States (90% of land in the South is privately owned), their management has critical bearing on conditions in some local areas such as the southern Appalachians. Dailey and Norton's chapter (Chapter 45) explores how public lands might be used to address ecological functions at various scales. They recognize the crux of the "ecosystem management" problem as defining relevant ecological information at appropriate scales—in effect, viewing public lands in their appropriate contexts, both ecological and social and designing management that complements or compensates for management on neighboring private land.

Summary

The economic studies of climate change reported in this volume provide some important first steps in understanding how human well-being is influenced by the function of ecological systems. They highlight that present knowledge of the climate and biological impacts of global change is a limiting factor for economic analysis. However, they also indicate that our economic models need further development. Concurrent pursuit of biological and social research is warranted and is clearly necessary for understanding global change and for designing programs to mitigate its effects.

Future progress in the economic analysis of climate change depends critically on linking accurate forest-growth models to theoretically plausible models of economic behavior. Scaling from biological process models to stand development and growth and yield is a key step. Better growth models will need to be coupled with better knowledge of how landowners invest in and choose to harvest their forests.

Finally, it bears mentioning, that the analysis of the effects of climate change on forestry in the southern United States cannot be conducted in isolation. Global changes imply effects that cross both regional and sectoral boundaries and, although it is never possible to measure all the derivative impacts, it is important to identify areas in which critical connections and spillovers may occur.

References

Baumol WJ, Oates WE (1988) *The theory of environmental policy.* Second edition. Cambridge University Press, New York.

Holmes TP, Kramer RA (1996) Contingent valuation of ecosystem health. Ecosys Health 2(1):56–60.

Lee KJ (1996) Natural resource accounting: Counting the forest and the trees. In *SOFEW '95, A world of forestry.* Proceedings of the 1995 South For Econ Workshop, New Orleans, LA. April 17–19.

Mendelsohn R, Nordhaus WO, Shaw D (1994) The impact of global warming on agriculture: A Ricardian analysis. Am Econ Rev 84(4):753–771.

Newman DH (1987) An econometric analysis of the Southern softwood stumpage market: 1950–1980. For Sci 33(4):932–945.

Newman DH, Wear DN (1993) Production economics of private forestry: A comparison of industrial and nonindustrial forest owners. Am J Agri Econ 75:674–684.

Sohngen B, Holmes T, Mendelsohn R (1993) Hedonic travel cost analysis of forest attribute values in the southern United States. In *Policy and forestry: Design, evaluation, and spillovers.* Proceedings of the 1993 Southern Forest Economics Workshop. April 21–23. Duke University, Durham, NC.

VEMAP Members (1995) Vegetation/ecosystem modeling and analysis project: Comparing biogeography and biogeochemistry models in a continental-scale study of terrestrial ecosystem responses to climate change and CO_2 doubling. Global Biogeochem Cyc 9(4):407–437.

Wear DN, Parks PJ (1994) The economics of timber supply: An analytical synthesis of modeling approaches. Nat Res Model 8(3):199–223.

Index

A

Abelmoschus esculentus (Okra), chilling survival of, and carbon dioxide level, 9
Abies balsameri (balsam fir), susceptibility to the adelgids, 273–274
Abies fraseri (Fraser fir), 255
Accelerated growing seasons, 119
Acclimation, 285
    to carbon dioxide level changes, 5–6
    exclusion from NPP simulation, 385
    importance of climatic variance to, 745
    in pot-bound specimens and in free stands, 122
    to temperature changes, PnET-IIS model, 811–812
Acer rubrum (red maple)
    competition experiment, 119
    root length density response to carbon dioxide levels, 125
Acid deposition
    effect of
        on calcium supply, 273–274
        on nutrient supply, 267
        on spruce-fir forests, southern Appalachian Mountains, 255–277
    and nutrient flux, 271
    rain
        effects on southeastern forests, 149
        and public policy, United States versus Canada, 58–59
Acidity
    effect on cold tolerance, 258
    effect on reproductive biology, 104
Acid-neutralizing capacity (ANC), leaching of, simulated, 521–522
Acid precipitation, and dogwood anthracnose, mediation of effects by soil, 635–646
Acid rain. *See* Acid deposition, rain; National Acid Precipitation Assessment Program
Adelgid, wooly, 255
    in the southern spruce-fir forests, 837–838
    susceptibility of stressed forests to, 273–274

Age, and net primary productivity, longleaf pine, 242
Age-class distribution, and carbon stock, 769–771
Aggregate Timberland Assessment System (ATLAS), linkage with FASOM, 780
Agricultural Conservation Program (ACP), 805
Agricultural programs, effect on forest base and forest type, 760, 805
Agricultural Sector Model, 757
Agricultural welfare, predictions of, 769–771
Agroforestry, 552
Air pollution, effect on reproductive biology, 104
Allocation, among competing resources, 855–863. *See also* Carbon allocation
Aluminum, toxicity in soil solutions, 258 and mobilization of, 559, 581
Ammonia, uptake from soil solutions, 261
Analysis
 of litter samples, procedures, 528–529
 statistical, litter decomposition study, 529–530
Analysis of variance, seed source study, 732–733
Animals, temperature and precipitation changes affecting, 667–669
Aquatic systems, goals for managing, 834
ARSTAN program, 263–264
Arthropods, 539
 role in litter fragmentation
  effect of temperature on, 537–538
  and nutrient release, 526, 529
Assessment
 policy-relevant, lessons from, 67
 of research, 66–67
 of the timber market, 817–819
 See also Evaluation
Atmospheric deposition
 effects of
  on forest soils, 558–587
  on soil changes, 579–581
 of nitrogen compounds, 593–594
Atmospheric Model Intercomparison Project (AMIP), 21

B

Baldcypress, predicting climatic variation from old-growth, 701–716
Basal-area-growth rate, for SOFOCLES simulation, 440–441
Basal-area increment model
 for climatic variables in a natural stand study, 247–248
 sensitivity of predictions to climate changes, 253
Basal area per hectare (BApha), and net primary productivity, longleaf pine, 242
Basal area projection model, 246–247
Bayesian model, for climate, 704–705
Benchmark, for FASOM, 781–783
Beta distribution, for describing foliage within a crown, MAESTRO, 328
Bias, in predicting southern softwood stumpage, 805
Biodiversity
 dialogue about, in the United States, 67
 effects of global climate change on, 663–681
Biogeography model, 454
 comparison with biogeochemistry model, 472–473
Biological model, generating data from, 797–798
Biologic data
 for a protection value study, 843–844
 quality objectives for variables, 108
Biomass
 aboveground, longleaf pine, 238–241
 change in, from harvesting and species change, simulated, 506
 of dogwood seedlings treated with simulated acid rain, 641
 effect on
  of carbon dioxide levels, 99–100
  of irrigation and fertilization, 156–159
  of ozone levels, 85–88
  of water deficit, 76–77
 foliage, ratio to woody tissue, simulation results, 318–319
 partitioning of production
  effects of nutrition on, 158, 166, 203–204

effects of water on, 158
response to carbon dioxide levels, variation among species, 123–126
root, maintenance respiration as a function of, 324
seedling, effect of ozone on, 260
simulated accumulation of, 516
simulated branch and bole trends, 317–318
BIOMASS model, 150
carbon dioxide level
effects simulated, 489
estimation of changes with, 159
and growth, 488
focus on carbon allocation, 484
modification for the loblolly pine, 368–369
net primary production (NPP) estimates from, 375–376
productivity predictions of, 491
temperature and growth change in, 488
temperature effects simulated by, interaction with carbon dioxide, 491
temperature variation in, 376–378
validation of, 370
Biosolids, municipal, enhancing forest productivity with, 552
Birds, effects of climate changes on, 668–669
Bole surface area, simulated values for thinned and unthinned stands, 317–318
Boxplots, of tree-ring chronology statistics, 729–731
Branch carbon exchange index (BCEI), 187
effect on, of fertilization and thinning, 193–194
Branch chambers, 133, 171
Budbreak date, effect of environmental variables on, 151
Bud ratios, pine and dogwood, 110
Buffer power (b), 301
Burns, effect on productivity of longleaf pines, 242–243. *See also* Wildfires

C

CADIL model, of soil-chemical adsorption, 410

Calcium
availability of, simulated distributions, 518
exchangeable
changes relating to magnesium exchange, simulation, 519–520
simulation, 507, 509, 511–513, 514–516
trends in, 580
leaching of
from leaves of dogwood in simulated acid rain, 636–637
simulated, 519
recycling of, dogwood role in, 644
role in formation of lignin, 272
roles in plants, 261
sequestration in the biomass, 503
uptake of
calcium to aluminum ratio effect on, 258
through newly elongated roots, 201
Calibration
of the MAPSS model, 456, 457
of the NuCM model, 505–507
of tree-ring data, 702
Calvin cycle, 131
Cambial growth
measuring, 595
response to fertilizer, soil moisture, and thinning, 601
timing of, loblolly pine, 743
Canadian Climate-Vegetation Model (CCVM), 368
Canopy
description in the slash pine carbon dynamics mode, 354
interaction with the atmosphere, MAPSS model, 472
Carbohydrates
effect of reserves on growth potential, 267
partitioning of, measurement procedures, 595
in soil organic matter, effect on water infiltration, 544–545
Carbon
in belowground pools, effect of temperature on, 573–574

forest
  managing, 755–756, 767–769, 771–773
  modeling economic impacts of changes in, 757–759
  program for increasing, 760–761
  mineralization of, and dissolved organic matter, 547–548
  partitioning of
    between primary and secondary metabolites, 610–611
    among branches, 203
    labelled carbon studies, 603, 609
    modified BIOMASS model, 369
    and seasonal responses, 201
  sequestration of
    and global climate change, 778
    in soil organic matter, 549–551
Carbon allocation
  predicting from MAESTRO and SPM, 482–483
  in red spruce, 269–271
  and resin-based defenses, 609–612
Carbon assimilation (A_n)
  and carbon dioxide level, measurements, 135–137
  effects of carbon dioxide and temperature on, 137–138, 138–145
  response to temperature change, in oak seedlings, 655
  slash pine carbon dynamics model, 357
  stresses affecting, 131–132
  Unified Transport Model simulation, 410
Carbon balance, and root turnover, 495–496
Carbon budgets
  annual, 430
  and resistance to bark beetles, 591–616
  seasonal, for loblolly pines, 162–163
  for trees, effects of carbon dioxide levels on, 551–552
Carbon dioxide
  in branch chambers, trend, 175
  effect of
    on competition among grasses, 118
    on forest soils, 571–587
    on light-saturated net photosynthesis, 154
    on loblolly pine, 93–101, 117–118
    on loblolly pine, respiration, 337
    on photosynthesis in miniature stands, 120–121
    on sweet gum, 117–118
  evolutionary importance of, C_3 plants, 3
  incorporating effects of into the SOFOCLES model, 440–441
  interaction of
    with air temperature, effect on loblolly pines, 131–148, 341–352
    with air temperature, slash pine carbon dynamics model, 362–363
    with climate, effect on loblolly pines, 367–389
    with the environment, effect on plants, 4–10
  interaction with the environment, effect on loblolly pines, 149
  net crown uptake of, modified MAESTRO simulation, 335
  net needle uptake of, effect of fertilization on, 193
  and net primary production, 800–801
  predicted effects on growth
    overview of model projections, 494
    variation among models, 489
  and secondary plant metabolites, 668
Carbon-flux model, carbon assimilation and dark respiration from, 344
Carbon gain
  net annual, as a simulated function of temperature and carbon dioxide, 337
  response to carbon dioxide levels, 285
  simulated estimation of
    for a stand, 307
    SOFOCLES model, 493–494
  simulated trends for a canopy, linked model, 319–321
  simulated trends for a tree, linked model, 321
Carbon storage
  policies for increasing, 862
  pools in the FASOM model, 758
  in the SPM, BIOMASS, and UTM models, 496

Index

Carbon to nitrogen ratio
  effect of temperature change on, 574–575
  in spruce-fir stands, 558
Cation exchange capacity, of forest soils, 5–6, 580
Cations
  exchangeable
    changing pools of, 503
    in litter, 534–535
    forest budgets of, 580
Cell walls, role of calcium in formation of, 272
CERES-Maize physiological model, for regional crop assessments, 17–18
CERES model, inclusion in a Unified Transport Model, 409–410
Chemical climate, effect on mature forest trees, 207–230
Chemical data, soil, for NuCM calibration, 506
Chlorofluorocarbons (CFCs), political and scientific issues in banning the use of, 59
Chlorophyll, total, effect of carbon dioxide on, 176–179, 180. *See also* Photosynthesis
Chronologies, ring-width, loblolly pine plantations, 724–731
Circulation models. *See* General circulation models
Circulation patterns, wildfires associated with, 684–693
Circumference growth, seasonal dynamics of, 215. *See also* Diameter at breast height
Climate
  chemical and physical, effect on mature forest trees, 207–230
  data for measured sites, 397
  effect on stands of trees, amplification by ozone, 227
  effect on the productivity of loblolly pines, 367–389
  reconstruction, Bayesian application, 708–715
  responses of trees to extremes of, 359–360
  scenario for the year 2080, 798–807
  scenarios for construction of regional patterns for simulation, 372–374, 435, 438
  simulating the effects of, 438
  of the southern United States, 15–54
    effect on wildfire occurrence, 683–699
Climate change
  economic dimensions of, 777–794
  effect on growth, overview of model projections, 495
  forest responses to, simulated, 479–500
  projects effects on forests and water resources, 453–477
  scenarios for simulation, carbon dioxide and temperature changes, 347
  scenarios for the Southern United States, 23–31
  and southern timber markets, 809–821
Climate growth model, predictions of, about red spruce growth, 261–263
Climate index, as a multiple linear regression, 732
Climate response models, for the Southwide plant seed source study plantations, 733–736
Climate-Vegetation Model (CCVM), 368
Climatic variation, predicting from baldcypress old-growth formation, 701–716
Climatology, Southwide Pine Seed Source Study plantations, 720–722
Climax vegetation, as an equilibrium state, 454
Communication
  dialogue in the Joint Climate Project, 67
  lessons in, public policy development issues, 65–67
Competition
  among forest tree species
    amplification by ozone, 227
    effect of thinning and fertilization on, 199–200
    environmental factors affecting, 117–129
    hardwood versus pine biomass changes, 126–127
  FORET model of, 409

in natural stands of longleaf pine, 231–254
among plants, environmental factors affecting, 670
simulation of, 307
Competition index, use in PTAEDA2, 484
Composite index, for vegetation types, 457–459
Congress of the United States, climate change research agenda of, 60–62
Consensus, social, and effect of scientific certainty, 58–59
Conservation, meeting global threat to biodiversity through, 676–677
Conservation Reserve Program (CRP), 761
Constraint, fixed land use as, MAPSS model, 472
Consumer theory, 839–840
Contingent valuation (CV), 838, 839–840, 860–861
Convection thunderstorms, and forest productivity, 743–744
Cores, for tree-ring studies, 703–704
Cornus florida (flowering dogwood)
dogwood anthracnose in, effect of acid precipitation on, 635
and environmental stress, effect on reproductive biology, 103–116
Cornus nuttallii, dogwood anthracnose in, 635
Correlations
longleaf pine variables, 239–240
nonparametric, SOFOCLES model, 444
predicted and actual southern pine beetle populations, 626–627
tree rings and climate, loblolly pines, 734
utilizing in models, 480
Cost-benefit analysis
nontimber benefits of forests, 860–861
policy-relevant-output-per-dollar ratio, 69
Costs, social, and scientific certainty, 58–59
COVRATIO statistic, 708
for detecting hurricane and drought signals, 714–715

residuals from, magnitude in years of hurricanes, 712
Critique
economic dimensions of climate change, 792–793
of knowledge about the results of global change, 748–749
modified MAESTRO model, 331–338
Cross-dating
COFECHA for, 723–724
dendrochronological methods, 704
Crown, physiology of, responses to fertilization, 187
Crown growth
carbon assimilation trends, simulated, 319–321
carbon dioxide uptake, modified MAESTRO simulation, 332
conditions affecting, 201
net carbon gain relationship with yield, linked model, 322
simulated
changes in shape over time, 311
maintenance respiration, modified MAESTRO simulation, 336
transpiration, modified MAESTRO simulation, 334, 336
trends in volume, 311
volume and foliated length, 313–314
ratio, as a surrogate for leaf area, in PTAEDA2, 484
Crown structure
damage to, and hurricane signals in baldcypress, 715
and resistance to bark beetles, 596
simulated properties, Sitka spruce, 324

D

Dark respiration
and carbon metabolism, red spruce, 259, 267
effect of carbon dioxide levels on, 281–282
effect of temperature on, 269
Data selection
major models, 486
PnET-IIS southern pine beetle population prediction, 625
SOFOCLES model, 435

Decision makers, climate information
  needs of, 62–67
Decomposition, litter
  effect of temperature on, 567
  tannins in, 535–537, 538
  temperature effect on, experimental
    data, 531
Dendrochronological methods
  correlation-function analysis, red
    spruce, 263
  cross-dating, 704
  descriptive statistics used for interpretation, 727–731
Dendroclimatic model, 222
  predicting the range of loblolly pines
    from, 744
Dendroctonus frontalis (pine beetles),
    591–616
  predicting population size, from a forest
    ecosystem model, 617–634
Dendroecological studies, of red spruce,
    256–258
Dendrometer, for stem circumference
    measurement, 209
Deposition, atmospheric, simulated effects
  of, 503–524
Dew point depression, association with
    wildfires, 693
Dew point temperature, modeling humidity effects using, 18
Diameter, growth of, in thinned and unthinned stands, 601
Diameter at breast height (DBH), 186
  of longleaf pines, 232
  in a simulated stand, 309
DIFMAS model, 409–410
Digital elevation model (DEM), source
  of data for the MAPSS model,
    459
Discula destructiva, dogwood anthracnose
    caused by, 635
Dissolved organic matter (DOM), in soil
    organic matter, 547–549
Distribution, of vegetation, MAPPS simulations, 463–468
Disturbance regimes versus equilibrium
    scenarios, 473–474
Diurnal changes, in temperature, in a
    GCM scenario, 23

Diversity, and impact of climatic extremes, 742
Dogwood anthracnose
  interaction with acid precipitation, 635–646
  and soil, 637–638
Drainage, predicted versus measured,
  validation of the PnET-II model,
    401
Drought patterns, of the southern United
  States, 15–16, 796
Drought stress
  and carbon dioxide levels, 5–7
  effects on biomass production, 123
  effect of, on oak, 647–660
  and growth patterns, 222
  and response to ozone levels
    shortleaf pine, 73–92
  various species, 223
  and root elongation, 202
  sensitivity of the loblolly pine to, 737
  species tolerances to, predictions of the
    MAPSS model, 493
  tolerance to, and leaf area in acid rain
    conditions, 639
DRYADS model, 409–410
DYNACLIM, cross-dating tree cores
  with, 704
Dynamic response, to climate changes,
    207–230
Dynamics, stand, FORET model, 484

E

Earth and Environmental Sciences, Committee on, 62
Echinocloa crustgalii, survival at low
  temperature, and carbon dioxide
  level, 9
Ecological dynamics, and resource management, 826
Ecological Risk Assessment (ERA)
  framework, 448–450
Econometric model
  for assessing biological information
    about softwood stumpage, 802–805
  market supply approach, 858–859
Economics
  dimensions of climate change

on softwood stumpage in the south, 795–808
in southern forests, 777–794
and global climate change, 855–863
value of protecting spruce-fir forests, 838–853
Economic welfare, present value of selected components, 769–772, 782
Ecophysiological responses, stands of loblolly pines, 185–206
Ecophysiology, Fraser fir, 273–274
Ecosystem
 dogwood as a nutrient source, 644
 forest, perturbation and disturbance in, 741–742
 model of, predicting southern pine beetle infestations from, 617–634
 mycorrhizae in, 673
 problem of managing, 862
 southern spruce-fir, 272–274
 type of, and forest response to soil temperature change, 577–579
Education, of decision makers, 66
Electric Power Research Institute, Integrated Forest Study (IFS) of, 505–507
El niño, and wildfire incidence, 695–696
Elusine indica, chilling survival of, and carbon dioxide level, 9
Embolisms, in oak xylem vessels, 655–658
Emission density, changes in, and tree-ring chemistry patterns, 259
Energy demand, as a constraint on vegetation distribution, 455
Environment
 chemical and physical, 3–14
 soil, and root system growth, 188–189, 194–199
Environmental hermeneutics, 832
Environment and Development, United Nations Conference on, 62, 63
Equilibrium changes, in climate, estimates from general circulation models, 20
Evaluation
 of the correlation between tree-ring widths for individual trees, 263–264

of general circulation models, 21
quality control, litter decomposition measurements, 530, 537
quality of biological response data, 120
oak study, 652
of tree and stand-growth prediction models, 244–247
See also Assessment; Validation
Evapotranspiration
 and ozone exposure, 419
 PnET-IIS model calculation of, 620
Exotic plant species, threat to biodiversity in Florida, 672
Expert opinion, forming simulation variables from, 439–441
Externalities
 in forest consumption, 810
 market failure defined by, 861
Extremes
 effects of, on biodiversity, 675
 in predictions of temperature and precipitation changes, GCMs, 385

F

Fagus grandifolia (beech), response to temperature change, simulated, 670–671
FASOM model, input to Timber Assessment Market Model, 780
Feedback inhibition, whole tree versus experimental isolated branch conditions, 163–164
Fertility, site, effect on respiration, 488
Fertilization
 calcium, effect of, 260
 effect of
  on light-saturated net photosynthesis, 153–155
  on resin-based defenses, 609–610
  on tree height, DBH and volume, 189–190
 interaction with carbon dioxide, effect on photosynthesis, 282
 phosphorus, effect of, 364
 and resistance to bark beetles, 596
Foliage
 density of, in a loblolly pine crown, 308

as an indicator of ozone damage, 77–79, 87
pattern by canopy level, 202
Foliage density, Weibull distribution for describing, 328–329
Forest, density mapping of, 435
FOREST-Biochemical Cycles Model, 368
Forest carbon program (FCARB), 761
Forest gap-dynamic model, 796–797
Forest inventory and analysis data, 799–800
  hardwood estimates, 648–649
  for PnET-II simulation, 393–394, 813
  for SOFOCLES simulation, 435
Forest plantation management, PTAEDA2 model of, 409
Forest productivity models, 811–812
Forest Response Program (FRP), 10
Forestry
  effects on forest carbon and on markets, 755–775, 778
  gains of land from agriculture, predicted, 762
  integrating local and global objectives in management, 823–836
Forestry and agricultural sector optimization model (FASOM), 756, 779–780, 810, 858
Forestry Incentive Program (FIP), 805
Forests
  change in cover, as a shift in timber management, 762
  growth response, and soil temperature, 578
  responses to climate change simulated, 479–500
  soil warming study, 575–577
  southern, distribution in the year 2080, 798
  temperature at the floor of, versus ambient temperature, 530
  types of, MAPSS estimates, 461
FORET model, 429
  adaptation to loblolly pine plantations, 414
  of competition in a forest, 409, 410–411, 484
  effect of ozone on carbon gain in, 490

linkage of, in the Unified Transport Model, 414–415
modified, for a biodiversity study, 667
tree height and ozone exposure, simulated, 420
Fourier transform infrared spectroscopy (FTIR), for analyzing soil organic matter, 545
Free-Air Carbon Dioxide Enrichment (FACE) exposure system
  Amax response to carbon dioxide level change from, 163–164
  measure of stomatal closure as a function of carbon dioxide levels, 153
Fulvic acids, 545

G
Gap-dynamic model, for projecting forest distributions, 798
  growth and yield in, 800–801
Gapon equation, for cation exchange simulation, 505–506, 515
Gas analyzer, infrared, for photosynthesis and stomatal conductance measurement, 172
Gas exchange
  diurnal measurement of, temperature and carbon dioxide interactions, 134–135
  effect of ozone and water deficit on, in seedlings, 79
G'DAY model, resource balance base of, 429
General circulation models (GMCs), 15–54, 460, 627, 778
  attributes of, for an index of climate change study, 373
  climate change scenarios using, 399–401
  combination with historic weather data, 811
  data for MAPSS from, 454, 455
  dependence on, for distribution of forest tree species, 796
  hurricane energy predictions from, 702
  net primary production estimates from, in a changing climate, 371–376
  for predicting climate change, 742–743
  predictions of, 353, 367–368

temperature and water cycle predictions of, 525
use in the SOFOCLES model, 449
variable predictions of, productivity and climate change, 381
Genetic problems, in large animals, effects of climate change on, 669
Geographic information systems (GIS)
creation of, 430–431
SOFOCLES model, 436
data from, for PnET-II simulation, 393, 626
Geophysical Fluid Dynamics Laboratory (GFDL)
estimate of composition and distribution of forests, 796
general circulation model output, 21–23
Geopotential height
association with circulation, and incidence of wildfires, 686
association with El Niño, 695–696
Germination
and environmental stress, flowering dogwood, 112–114
of loblolly pine and dogwood, after exposure to carbon dioxide, 106–107
Gibbs sampler algorithm, 702, 705–706, 709
Global Air Pollution and Climate Change, Dutch National Research Program on, 67
Global change
and disturbance in southern forest ecosystems, 741–752
and economics, 855–863
influence on tree physiology and growth, 279–290
models of, 779
Global Change Research Program (GCRP), United States, 61
Global climate model data, models using, 487
Glycine max (soybean), ozone injury to, effect of carbon dioxide levels, 10
Goddard Institute for Space Studies, general circulation model output, 21–23
Greenhouse effect, 551

and carbon dioxide levels, 283
Gremmeniella abietina, effect on Norway spruce, 636
Gross primary production (GPP), 150
effect of carbon dioxide levels on, model estimation, 159–161
Growing season, climate change during, model comparisons, 25–26
Growth
accelerated growing seasons, experimental systems, 119
analysis of patterns of, 209–213
mature red spruce, 256–258
belowground, and biomass accumulation, 123–125
changes in, and yield, 800–801
effect on
of carbon dioxide levels, loblolly pine, 93–101
of fertilization and thinning, 189–190, 601–603, 604–605
of global change, 279–290
of moisture supply, 212
of ozone levels, 85–87
of water deficit, 76–77
forest
data collection, 396
PnET-IIS simulation, 396–398
predicting from models, 482–494
relating to net primary productivity, 812–813
individual tree, model for climatic variables, 248–249
loblolly pine, effects of environmental variables, 149–167
model for regional development, loblolly pine, 445
model for stand development, loblolly pine, 305–325, 439, 442–445
of oak seedlings, effects of temperature and soil moisture on, 655
patterns of, and resin-based defenses, 609–612
predicted effects of warming and carbon dioxide on, Pipestem model, 350–351
radial, of loblolly pines, 391
root system, effect of soil environment on, 194–199

Growth-differentiation balance, modeling, 621–622
GSROOT software, lateral root persistence, 188

H
Habitat, baldcypress, 702–703
Hardwoods
 conversion of land between softwoods and, 764
 rate of growth, versus softwood, 768
 response of, to hurricane frequency, 744–745
 role in stand development, 126–127
Harvesting
 effects of
  on exchangeable cations, 514–516
  models incorporating, 757
  of loblolly pine forests, and ozone exposure, simulated, 426
  predicted increase in, for hardwood, 548
Heartrot, in baldcypress, 704
Herpetofauna, diversity of, and climate change, 669
Hierarchical information transfer, 407
Hierarchy theory, of evolutionary processes, 826–827
Historic record
 predicting climate change from, 742–743
 of temperature changes, response of vegetation to, 663–665
Holdridge life zones, 357–358
Human activity, reduction in habitat from woody material loss due to, 673
Human responses, research on, and public policy, 65
Humic acids, 545
Humin, 545
Humus, and physical properties of soil, 543
Hurricanes
 destructive potential of, 744
 signal detection in tree-ring data, 706, 709–711
Hydrogen peroxide, effect of, on red spruce physiology, 260
Hydrology
 climate change effects on, prediction from the PnET-II model, 402, 743
 prediction of, in response to climate change, 391–405

I
Image processing workbench (IPW), 459
Individual tree-growth model, for climatic variables, effect on natural stands, 248–249
Initialization, Pipestem model, 343
Insects, population of, and drought stress in plants, 668. *See also* Pests
In situ study, ozone and water deficit effects study, 83–89
Integrated Forest Study (IFS), of the Electric Power Research Institute, 505–507
Integrated Lake-Watershed Acidification (ILWAS) model, integration with NuCM, 505–507
Integration
 of economic resource management and maintenance of regenerative systems, 825
 of levels of values, 834–835
Intergenerational values, social constraint of, on short-term preferences, 833–834
Intergovernmental Panel on Climate Change (IPCC), 18–19, 61
 temperature and precipitation effects of climate variability, 20
Intertemporal optimization model, forestry and agricultural sector optimization model (FASOM), 756
Irrigation, effect on light-saturated net photosynthesis, 154

J
Job opportunities, maximizing, and economically sustainable timber management, 833
Joint Climate Project, dialogue in, 67

K
Kinetics, of nutrient uptake, 295–297

Knowledge, about the results of global change, critique, 748–749
Küchler potential vegetation map, 456
　calibration of the MAPSS model to, 457

L

Land capability class data (LCC data), 759–772
Land markets, incorporating in a model, 756
Landscapes, of daily life, 832
Land-transfer, linking forestry and agriculture through activities of, 757–759
Land use
　deterministic models for, 472
　national objectives in, impact at the local level, 829–830
　policies interacting with forestry, 759–772
　strategies in global change, 748
Langmuir isotherms, use in simulating sulfate adsorption, 505
La Niña, association with drought in the Great Plains, 695
Larrea divaricata, temperature and carbon dioxide level responses of, 9
Lateral roots
　elongation of, 195
　growth of, response to fertilization, 203
　seasonal rate of elongation, effect of thinning and fertilization on, 197, 198
Latewood, resin ducts in, 618–619
Latin hypercube sampling, 411–412
Leaching, 503
　effect of sulfate ion on, for magnesium, 580
　of magnesium, 504
　simulated, 519
Leaf area
　of dogwood seedlings treated with simulated acid rain, 641
　effect of fertilization on, 155–157, 166, 364
　effect of water availability on, 157
　simulated trends in, thinned and unthinned loblolly pine, 311
　specific, of longleaf pines, natural stand study, 243
Leaf area density
　linked model simulation predicting, 313–317
　MAESTRO simulation of, 307
　PnET-II simulation of, 392–393
Leaf area index
　dependence on nitrogen availability, 374–375
　estimation by forest and woodland type, 462
　longleaf pine, 236–237
　MAPSS model, 457–459
　　assumptions about constraints on, 485
　　iterative calculation, 455
　　prediction of, 468
　　simulation, 461
　prediction from a linked simulation, loblolly pine stand, 315, 386, 415–416
　use in the BIOMASS modified model, 369, 370
Leaf senescence, ozone effects on, 412
Least-social-cost allocation of land, for meeting a carbon sequestration target, 761
Legislation
　Global Change Research Act, 61
　National Climate Program Act, 60
Light
　effect on reproductive biology of *Tectonia* and *Pinus*, 104
　interaction with carbon dioxide level, effect on photosynthesis, 7–8
Lightning, and wildfire incidence, 696
Light-saturated net photosynthesis, response to carbon dioxide level changes, 161–163
Light-saturated rate of photosynthesis
　effect of carbon dioxide on, 176, 177
　as a measure for photosynthetic capacity, 172
Lignin, role of calcium in formation of, 272
Linkage process
　between forest conditions and economic value, 838

between forestry and agriculture simulations, 757–759
for the MAPSS model, 460
models of differing scales
    carbon-flux and Pipestem models, 343–344
    into the Unified Transport Model, 409–410
    PTAEDA2-MAESTRO, 308–309
    for the Unified Transport Model, 414–415
    UTM/FORET/PTAEDA2, 496
    See also PTAEDA2/MAESTRO linkage model
Liquidambar styraciflua, effect of carbon dioxide on, 117–118
Liriodendron tulipifera (yellow poplar)
    effect of carbon dioxide level on photosynthetic capacity of, 350
    response to temperature change, simulated, 670–671
Litter
    concentration of lignin, and decomposition rates, 526, 538
    phosphorus in, experimental sites, 530, 533
    tannins in decomposing, 538
        experimental measurements, 535–537
Litter decomposition
    collecting and analyzing samples, spruce-fir study, 561
    critique of study, 539
    and soil solution chemistry, 573–574
    and solid temperature, 565
Litterfall return, of soil nutrients, 579–580
Loblolly pine
    carbon assimilation of, 131–148
        change with carbon dioxide level, 145
    effect of climate change on, probabilistic regional modeling, 429–451
    effect of physical climate on annual stem growth of, 222
    effects of environmental variables on, 169–184
    effects of environmental variables on physiology and growth, 149–167

environmental stress and reproduction biology of, 103–116
exposure to carbon dioxide
    over a life-span, 3–4
    and growth of, 93–101
microclimate effect on short-term litter decomposition, 525–542
nutrient uptake model, 293–304
physiological response of, and ozone exposure, 407–428
productivity of, effects of climate and carbon dioxide levels on, 367–389
PTAEDA2-MAESTRO system simulation, 312
    outcomes for trees, 310
    range of, limitation by moisture and temperature, 671
    reproductive behavior and response to carbon dioxide levels, 109
    responses to temperature and carbon dioxide levels, 9
    simulation, 327–339
    root length density response to carbon dioxide levels, 125
    sensitivity to climatic changes, 717–739
    simulating stand development and growth, 305–325
Loblolly Pine Growth and Yield Research Cooperative, data from, for calibrating Pipestem, 343–344
Logistic equation, 847
Logit model coefficients, 847
Longleaf pine, productivity of natural stands, 231–254
Lycopersicon esculentum, ozone injury to, effect of carbon dioxide levels, 10

M
MAESTRO biological process model, 306–309, 429
    emphases of, 482
    linkage with PTAEDA2, 484
    respiration simulated with, 309
    steady-state carbon-flux model from, 343
    temperature and carbon dioxide response of the loblolly pine, 327–339

Magnesium
  availability of, simulated distributions, 518
  distribution of, simulation, 510, 514–516
  exchangeable, trends in, 580
  leaching of, 504
    from leaves of dogwood in simulated acid rain, 636–637
    simulated, 519
Maintenance respiration
  cost of, 324, 369
  effect of temperature change on, 164
  rates per unit surface area of woody tissue, 155
  simulated trends for a canopy, linked model, 319–321
  simulated trends for a tree, linked model, 321
  slash pine carbon dynamics model, 357–358
Management techniques
  for forest soils, 552
  responses to climate changes, 497–498
Mapped Atmosphere-Plant-Soil System model. *See* MAPSS model
Maps, of historical and double carbon dioxide scenarios, GCM data, 33–53
MAPSS model, 454–474
  emphases in, 485
  PRISM code linkage with, 459
  respiration in, 488
  runoff predicted by, 497
  temperature use in, 491
  validation of, 456
  vegetation type prediction from, 489
Markets, effect of forestry and agricultural policies on, 755–775. *See also* Economics
Mass, dry weights of oak seedlings, effects of temperature and soil moisture, 656
Mass balance
  losses in litter, effects of temperature and of moisture on, 538
  utilizing in models, 480
Maximum influx rate (I_{max}), nutrient kinetics, 296–297

Mean annual increment (MAI), trends from a simulation, linked model, 321–322
Mean between-series correlation, 728
Mean sensitivity (ms), of a tree-ring series, 727
Metabolism, secondary
  and global climate change, 746–747
  response to fertilization, 593–594
  and soil moisture, 593
Michaelis-Menten kinetics, for nutrient uptake at a root surface, 294–295
Microclimate, effect on short-term litter decomposition, 525–542
Microcosms, plexiglass, for litter decomposition studies, 527–528
Mineralization, rates of, ecosystem characteristics affecting, 593
Mineral weathering, simulation of, 505–506
Minigreenhouse structures
  for monitoring soil temperature and litter decomposition, 560
  temperature differential created by, 562–565
Minimum climate change (MCC), for assessing southern pine productivity, 810–811
Mitigation and Adaptation Research Strategies (MARS), 62
Models
  Barber Cushman, 294–295
  CADIL, of soil-chemical adsorption, 410
  carbon-flux, carbon assimilation and dark respiration from, 344
  for climate
    applied to red spruce, 261–263
  for climate and tree growth, with ordinary least squares, 707
  DIFMAS, 409–410
  DRYADS, 409–410
  forest and agricultural sector optimization, 810, 858
  FOREST-Biochemical Cycles Model, 368
  G'DAY, resource balance base of, 429
  for growth response in mature forest trees, 217–221

linked
- for evaluating forestry and agricultural policies, 755
- for simulating stand development and growth processes, 305–325
- Mapped Atmosphere-Plant-Soil system, 454–474
- for nutrient uptake, loblolly pine, 293–304
- for photosynthesis, Farquhar, 132, 283–284, 375
- PnET-II, for forest hydrology and productivity, 392–393
- slash pine plantation carbon dynamics (SPM), 353–366
- stand-level, distribution describing, 441–442
- steady-state, for nutrient uptake, 294–295, 301–302
- Terrestrial Ecosystem Model (TEM), 368
- for the timber market, 813–815
- for tree and stand growth prediction, 244–250
- tree growth and yield, PTAEDA2, 306–309
- Unified Transport Model, for tree physiological processes, 409
- See also BIOMASS model; FORET; General circulation models; MAESTRO biological process model; Pipestem model; PnET-IIS model; PTAEDA2 model

Moisture
- and decomposition of litter, pine species, 526
- effect on reproductive biology, 104
- and growth, of the loblolly pine, 208, 592–593
- in soil, response to temperature changes, 572

Moisture stress index (MSI), 209

Moisture supply index (MSI)
- and circumference growth, loblolly pine, 221
- correlation with ozone exposure, 225
- growth at variable levels of, interaction with ozone levels, 213
- and ozone exposure, 226

Monte Carlo simulation, of stand growth, 432

Multiscalar management, 825–828

Multistakeholder dialogues, for establishing public policy, 59–60

Multivariate analysis
- of interannual tree growth, 225
- of productivity of natural stands, parameter estimates, 252

Mycorrhizae, responses to climate change, 672–673

N

National Acid Precipitation Assessment Program (NAPAP), 10, 68, 448, 757–758

National Center for Atmospheric Research (NCAR), general circulation model output summaries, 21

National Climate Program (NCP), origin of, 60

National Crop Loss Assessment Program (NCLAN), 10

National Oceanic and Atmospheric Administration (NOAA), carbon dioxide data, 3–4

Natural system dynamics, correlation with human concerns, 831

Needles
- measuring the fall of, 235–238
- morphology and chemistry, effects of thinning and fertilizer on, 596
- ozone effects on retention of, 412
- production of, longleaf pine, 235–238

Netherlands
- climate-related research supporting policy development, 68
- dialogue about climate change in, 67

Net primary production (NPP), 150
- aboveground, longleaf pine, 241–243
- and change in forest respiration rate, 269
- effect of carbon dioxide levels on, loblolly pine, 159–161
- effect on southern pine beetle populations, 631
- estimates from the modified BIOMASS model, 375–376

880 Index

models used for assessing, 368, 811–812
PnET-II model prediction, 401–402, 403
PnET-II validation, 396
PnET-IIS model prediction, 619–620
regional changes in, predictions of two GCMs, 629
response to carbon dioxide levels, 800–801
and stemwood yield index, 370–371
Terrestrial Ecosystem Model prediction of, with carbon dioxide level change, 383, 800–801
and yield, effect of climate changes on, 383

Nitrate ion
deposition of, in the Smoky Mountains, 258–259
leaching of, effect of soil temperature on, 579

Nitrification, in minigreenhouses, spruce-fir study, 566

Nitrogen
availability of
effect on the southeastern forests, 149, 576
interaction with carbon dioxide, 121, 385–386
and leaf area index, 374–375, 386
and model predictions, 497
simulated distributions, 508, 518
cycling of, effects of temperature in a red spruce stand, 557–569
deficiency of, in forest ecosystems, 581
distribution between soil pools and plant pools, 574–575
foliage concentration of, and light-saturated net photosynthesis, 153
interaction with carbon dioxide, 581–583
leaching of, simulated, 521–522
leaf
interaction with ozone, 282
leaf content, of dogwood seedlings treated with simulated acid rain, 641
in litter
experimental measurements, 533–534
mineralization of, 539
mineralization of

forest floor measurements, 561–562, 565
higher altitude southern Appalachians, 558–559
and respiration, 488
in spruce-fir stands, 558–559

Nitrogen deposition
effect on soil-solution chemistry, 260
and leaching, 504

Nitrogen oxides, utilization by red spruce, 260

No land transfers between forestry and agriculture (NOAG), scenario for simulation, 761–762

Nontimber benefits of forests, 860–861
Nonuse values, defined, 845
Nuclear magnetic resonance (NMR), for soil organic matter analysis, 546
NuCM (nutrient cycling model), 504–523
Nutrient cycling model (NuCM), 504–523
Nutrient flux, effect on red spruce and Fraser fir, 271–272

Nutrients
availability of
and growth, 165
and plant-pest interactions, 746–747
cycling of, effects of atmospheric deposition on, simulated, 503–524
interaction with carbon dioxide levels, 5–6, 385
effect on photosynthesis, 283
limiting, simulation, 514
mineral
in air-dried litter, 528
availability of, 201
in decomposing litter, 538
optimum, for fertilizer, 165–166
and productivity, model predictions, 492
supply of, and uptake, 301–302
uptake of
and forest soil characteristics, 579–581
loblolly pine, 293–304

O

Oak, effects of temperature and drought stress on, 647–660

Oleoresin
    as a deterrent to bark beetles, 593
    production of, model of soil water stress, 621–623
Ordinary least squares (OLS), for a climate-tree-growth model, 707
Oregon State University, general circulation model output, 21–23
Organic matter, soil
    effects of temperature on decomposition of, 557–569
    and soil productivity, 543–556
    See also Soil decomposition
Oxygenase, temperature specificity of, 8
Ozone
    and circumference growth, loblolly pine, 221
    and drought stress
        effect on the shortleaf pine, 73–92
        studies of various species, 223
    effects of
        on growth, 286
        on growth, in a mature stand, 216–217
        on growth, models considering, 490, 492–493
        on growth, overview of model projections, 494–495
        on productivity, FORET model, 496
        on productivity, PTAEDA2 model, 496
        on productivity, UTM model, 496
        on root growth, 302
        on the southeastern forests, 149
        on stem growth, 221–222
    exposure to, and stem growth, 213
    exposure data
        for SOFOCLES simulation, 435
        interpolating for SOFOCLES simulation, 436–438
    interactions with temperature and moisture, empirical model, 217–221
    physiological responses of loblolly pine to, 407–428
    sensitivity of seedlings to, 208
    threshold for adverse effects of exposure to, 225

    treating pine seedlings with, 74–75

P
Palmer Drought Severity Index, 212, 222, 234, 344
    for climate reconstruction, 702
Parasites, obligate versus facultative, in preferences in host condition, 635–646
Partitioning, of soil organic matter, 545
Pearson correlation coefficients, basal-area projection model and climatic variables, 250
Pectins, role of calcium in formation of, 272
Perspective, international
    driving policy, 63
    and research, 65
Perturbations, in climate from global change, 747
Pests
    adelgid, wooly, 255, 273–274
    climate and bark beetle outbreaks, 591–616, 627–630
    interactions with plants, in a changing climate, 746–747, 801
    pine-tip moth, 96
    southern pine beetle, predicting infestations with a forest ecosystem model, 617–634
    susceptibility of stressed trees to, 647
    See also Insects
pH, soil-solution, and change in exchangeable calcium ion, 516
Phloem-specific mass, in thinned and unthinned stands, 603, 606–608
Phosphorus
    atmospheric deposition of, 519
    effects of fertilization with, 364
    in fertilizer, effect on root growth, 200
    as a limiting nutrient, 514, 517–518, 581, 582
    critique of the NuCM simulation of, 521–523
    in litter
        experimental sites, 530, 533
        mineralization of, 539

Photorespiration, effect of carbon dioxide levels on, 131
Photosynthesis
  and the canopy environment, 202
  carbon dioxide concentration as a limitation on, 169, 800–801
  carbon dioxide levels and, 279–280
  in dogwood exposed to synthetic acid rain, 643
  effects of environmental variables on the rate of, loblolly pine, 151–155
  effects of temperature on, 224
  model versus measurement, 145
  effects of temperature and soil moisture on, oak seedlings, 653–655
  interactions among variables affecting, 120–123, 385–386
  measuring, 595
  midday net, P_{net}, 638
  models emphasizing, 482
  and oleoresin production, 623
  optimum temperature for
    C_3 and C_4 plants, 8
  PnET-II model, 392–393
  and ozone levels, 224
  simulation, 416–417
  ratio to dark respiration, 259–260
  relative, ozone effects on, 413
  in thinned stands
    and fertilization, 596
    and resin production, 611
  water deficit effect on, 74
  water use efficiency of, and carbon dioxide levels, 5
Photosynthetically active radiation (PAR), 17–18
  branch chamber, above and below a sample branch, 175
  slash pine carbon dynamics model, 354
Photosynthetic carbon reduction cycle, 582
Photosynthetic photon flux density (PPFD), 187
  canopy position and, 202
  typical measurements, 191
Photosynthetic tissue, ratio to woody tissue, simulated calculation, 318–319

Physical climate, effect on mature forest trees, 207–230
Physiological year, defined for simulation, 344
Physiology
  red spruce, and response to climate changes, 267–272
  tree, influence of global change on, 279–290
Phytophthora cinnamomi, predicted increase in range of, 648
Phytotoxicity, quantifying, for ozone exposure, 225
Picea rubens (red spruce)
  range of, 255
  responses to physical climate, 261–264
  seedling growth study, 94
  soil temperature effects, 557–569
Picea sitchensis (Sitka spruce), crown-structure properties, MAESTRO simulation, 324, 350
Pigment content, of oak leaves, effects of temperature and soil moisture on, 655–658
Pine inventory, changes in, comparison of models, 818
Pine-tip moth (*Rhyacionia spp.*), effect on observed growth response to carbon dioxide, 96
Pinus banksiana, effects of temperature and moisture on decomposition of, 525
Pinus contorta, effects of temperature and moisture on decomposition of, 525
Pinus eldarica, effect of carbon dioxide level on photosynthesis in, 281
Pinus elliottii (slash pine)
  fine root data for, 369
  simulation base, 309
Pinus palustris (longleaf pine), harvesting, 150
Pinus ponderosa
  effect of carbon dioxide levels on photosynthesis by, 281
  foliage production patterns in, 202
  seedlings of, effect of temperature and carbon dioxide on growth of, 9
Pinus radiata (Monterey pine)

carbon dioxide evolution from soil and forest floor, 526
effect of carbon dioxide levels on photosynthesis by, 281
effect of phosphorus levels on growth of, 582
effect of phosphorus levels on photosynthesis by, 364
Pinus resinosa (red pine)
    foliage production pattern in, 202
    scleroderris canker severity in, 636
Pinus sylvestris (Scots pine)
    effect of carbon dioxide on photosynthesis in, 281
    effect of temperature and carbon dioxide levels on growth of, 10
    effects of temperature and moisture on litter decomposition, 525
    scleroderris canker severity in, 636
Pinus taeda. See Loblolly pine
Pioneer species, adaptability of, 675
Pipestem model
    carbon dioxide level effects simulated by, 489
    carbon dioxide scenarios for, 343–344
    emphases in, 483–484
    overview, 342–343
    productivity predictions of, 491–492
    temperature effects simulated by, interaction with carbon dioxide, 491
Pitcher plant, extinction in the southeast, predicted, 671
Plantation attributes, loblolly pine, with and without ozone exposure, 423
Plants
    biodiversity of, responses to climate change, 669–671
    flowering dates of, response to climate change, 675
PnET-II model
    hydrologic data, validation of, 398–399
    optimum temperature for photosynthesis, 392–393
    validation of, 394–396
PnET-IIS model
    carbon dioxide level effects simulated by, 489
    predicting southern pine beetle population from, 618, 623–624

productivity prediction of, with temperature change, 811–812
refinement of, for population of southern pine beetles, 631
runoff predicted by, 497
simplification involved in, 484–485
structure of, and predictive outputs, 619–621
temperature responses simulated by, 488
temperature use in, 491
validation of, 620–621
Pollution, air
    interaction with carbon dioxide levels in affecting growth, 10
    phytotoxic, 73–74
Population models, southern pine beetle, 619–626
    based on prior year's population, 623–624
    effects of soil water stress and NPP on, 631
    PnET-IIS prediction, 621
Populus tremuloides (aspen), effects of temperature and moisture on decomposition of, 525
Potassium
    in decomposing litter, 539
    as a limiting nutrient, 581
Precipitation
    of Antlers, Oklahoma, 173–174
    changing patterns of, effect on southern pine beetle outbreaks, 627
    correlation with red spruce growth, 267
    interaction with carbon dioxide levels, GMC calculations, 20
    predicting changes in, 743
    region-wide, from general circulation models, 30
    relationship to carbon dioxide levels, assumed, 372
    scenarios from the general circulation model studies, 24–27, 627
    in the southern United States, 15–16
    average effective, 696
    Southwide Pine Seen Source Study plantations, 721, 722
    spring and summer, baldcypress chronologies, 713–714

and temperature in the southern United States, 16–17
and tree-ring growth, red spruce, 267
trends for the southern United States, 816
use of
  in MAPSS and PnET-IIS, 491
  in PnET-IIS, 493
  variation in, modified BIOMASS model, 376–378
  within-region, from general circulation models, 30
Predawn plant water potential, 79, 81
  effect of ozone on, 224
Predawn xylem pressure potential (PXPP), 638
  of dogwood seedlings in simulated acid rain, 641
  of oak seedlings, temperature and soil moisture effects on, 653
Predictions
  of biodiversity effects of climate change, 674–676
  of climate change, 742–743
  key processes influencing, 496–497
  about temperature and precipitation change effects, 665–667
Principal components analysis (PCA), of seed source chronologies, 731
PRISM code
  linkage with simulation models, 411
  linkage with the MAPSS model, 459
Probabilistic modeling, of climate change
  regional, 429–451
  stand-level, 439–442
Production function, biological, 857
Productivity
  and climate change
    impact of migration of growing sites, 473
    modified BIOMASS simulation, 379–381, 391–405
    of natural stands of longleaf pine, 231–254
  PTAEDA2 model for predicting, 429
  response to temperature, carbon dioxide, and ozone, 487–494
  of soil, and soil organic matter, 543–556

Projected leaf area index (PLAI), natural stands, longleaf pine, 243–244
Projected needle surface area (PNSA), 187
PROSPER model, of the diurnal cycle of soil-water-plant relationships, 409–410
Pseudotsuga menziesii (Douglas fir), biomass partitioning by, 203–204
PTAEDA2/MAESTRO linkage model
  effects of ozone on carbon gain in, 490
  effects of ozone on productivity, simulation predictions, 492–493
  predictions of effects of carbon dioxide level changes, 489
  productivity predictions of, 491
PTAEDA2 model
  effect of ozone on carbon gain in, 490
  focus of, on density of forest plantation growth, 484
  of forest plantation management, 409, 411
  linkage of, in the Unified Transport Model, 414–415
  link to FORET, ozone exposure effects, 421–422
  of tree growth and yield, 306–309
Public policy
  developing
    for global climate change research, 55–70
    for social problems, 861
  implications of, research and certainty in, 66
  research in, 60
Public values, scales of, 824–825

Q

Quercus rubra (red oak), seedlings of, sensitivity to ozone exposure, 425

R

Rainfall
  annual, effect on growth of mature trees, 222
  experimental sites, litter decomposition study, 532
  modelling, 459

Raphanus sativus, ozone injury to, effect of carbon dioxide levels, 10
Recreational opportunities, scale of management, 833
Regenerative landscape-level processes, recognition of, in multiscalar management, 828
Regional analysis system, constructing, 434–439
Regional longleaf pine growth study (RLGS), 232–234
Regional variation, in the impact of global climate change, 778–779
Relative growth rates, in the modified BIOMASS model, 369
Relative humidity, anomalies in, and wildfire incidence, 689, 691, 692
Reproductive biology, and environment stress, in the loblolly pine and flowering dogwood, 103–116
Research
  and certainty, exploring public policy implications, 66
  social science, and public policy, 60
Resin-based defenses, and growth patterns, 609–612
Resin flow, in thinned and unthinned stands, 601, 603, 606–607
Resource allocation, through timber markets, 856–860
Respiration
  calculation in various models, 488
  construction, modified BIOMASS model, 369
  effect on growth, overview of model projections, 495
  leaf, ozone effects on, 412
  maintenance
    simulated trends for a canopy, linked model, 319–321
    simulated trends for trees, linked model, 321
  MAPSS model, 488
  root, effect of soil temperature on, 572
  simulated
    MAESTRO, 309
    PnET-II, 392–393
    response to ozone, 416–417

Response, dynamic, to climate changes, 207–230
Reversibility, of economic resource management practices, and constraints, 826
Rhizosphere, carbon sequestration in, 550–551
Rhizotrons, for measuring long lateral new root length, 188, 195
Ribulose-1,5-biphosphate carboxylase/oxygenase. *See* Rubisco
Richman-Lamb historical database, of temperature and precipitation in the Southern United States, 17
Risk, relative, in global climate change assessments, 66
Risk decision square, ecologically sensitive, 825
Root function, assessing, 76
Root-length density (RLD)
  response to carbon dioxide, interactions with water and nitrogen level, 125
  response to nutrients, 298
Root/shoot ratio
  responses to carbon dioxide, 125
  responses to carbon dioxide, 123
  response to temperature and soil moisture changes, 655
Roots/root system
  fine, production of, 165–166, 369
  effect of ozone on growth of, 302
  growth of
    and acid deposition, 260
    and nutrient uptake, 297–299
    and response to carbon dioxide levels, 582
    and soil environment, 194–199
  length of, and nutrient uptake, 297–299
  nutrient uptake at the surface of, 294–295, 297–299
  soil organic matter surrounding, 549–550
  turnover of, effect on carbon balance, 495–496
  water flux in, effect of ozone and soil moisture, 82
Rubisco
  efficiency of, and salt tolerance, 7

involvement in carbon dioxide response
of pine seedlings, 350
kinetic constraints of, and photosynthetic potential, 4
kinetic restraints of, and photosynthetic potential, 131
response to carbon dioxide levels, 582
Runoff, prediction from models, 469, 497
Ryegrass, effects of light and carbon dioxide concentration on growth of, 7

S

Salinity
increasing, and climate change, 676
tolerance to, and carbon dioxide level, 7
Saplings, effect of aluminum on, 259
SAS general linear model procedure, for testing interactions among tree-ring series, 732
Scale
from leaf level to stand level, light distribution issue, 328
linking models with differences in, 305
of management, 830
and rate of change, 824
Scaling, of experimental study observations to stand level, 164–165
Scaling up, from physiological to landscape scale, responses to ozone, 407–428
Scenarios
for atmospheric deposition, 507–512
for change in soil organic matter, 551–552
for changes in forestry with global climate change, 780–791
climate
for the year 2080, 798–807
climate change, and the southern timber market, 810–811, 815
defined, 23
effects of climate change
on net primary productivity, 628
on soil water stress, 628
fifty percent national growth rate
decline, 786–788
increase, 783–786
for solar radiation general circulation models, 28

southern loss of forest, 788–791
Science
emphases for governmental research in, 56
and public policy, 56–58
questions answered by research in, policy relevance of, 59–60
Scleroderris canker, effects of acid rain on, 636
Sea level, rise in, effect on biodiversity, 674–675
Seasonal distribution, of wildfires, 685
Second generation climate projections, four general circulation models, 22
Sedge, effects of light and carbon dioxide concentration on growth of, 7
Seedling growth study, using *picea rubens*, 94
Seedlings
container-grown
extrapolation from to forest-level systems, 551
study of water stress on, 74–83
container-grown oak, study of temperature and soil moisture effects, 649–659
effects on competition among species, 119
study of water stress and simulated acid rain on, 640
Sensitivity analysis
of environmental variables, 409
modified BIOMASS model, 376, 381–383
of nutrient uptake, 295–297
Serial correlation, 727–728
Shoot/root ratio, of dogwood seedlings treated with simulated acid rain, 641
Shortleaf pine
drought stress and response to ozone levels, 73–92, 224
effect of physical climate on annual stem growth of, 222
Simulated acid rain (SAR), 636–637, 638–639
Simulation
predictions from, of carbon dioxide response, 164–165

of a stand from the time of planting, 306–307
Site index (SI)
and bark beetle infestation, 593
generation of, FORET simulation, 411
and net primary productivity, longleaf pine, 242
simulated effects of, 311
as a summary variable for the Unified Transport Model, 422
updating in a linked model, 306
Snow and sea ice, interaction with carbon dioxide levels, GMC calculations, 20
Social welfare cost, of increasing forest carbon storage, 769–772
SOFOCLES (SOuthern FOrest CLimate Effects Synthesis), 431–450
effects of ozone on productivity, simulation predictions, 492–493
emphases of, 482
on uncertainty in, 485
productivity predictions of, effect of temperature, 493–494
Softwood
conversion of land between hardwood and, 765–767
rate of growth of, versus hardwood, 768
supply and demand equations for, 802–803, 814–815
Softwood stumpage supply, 801–802
Soil
effects on dogwood anthracnose, versus foliar effects, 637–638
fertility of, effect on resin-based defenses, 609–610
productivity of, and organic matter, 543–556
Soil moisture, 195
data in stands of mature trees, 212
effect of thinning and fertilization on, 196
interaction with carbon dioxide levels, GMC calculations, 20
and net primary production, modified BIOMASS model, 376–379
and resistance to bark beetles, 592–593

and wildfire incidence, 694–695
Soil moisture stress/soil water stress
influence on oleoresin production, 622
and ozone exposure, 226
from the PnET-IIS model WATSTRS module, 620
regional changes in, two GCMs, 630
Soil organic matter (SOM)
decomposition of, and soil temperature, 572
defined, 543
Soil solutions
chemistry of, southern spruce-fir forest, 258, 559
collection and analysis of, southern spruce-fir forest, 561
Soil supply, and nutrient uptake, 299–300
Soil-water-holding capacity (SWHC)
site-specific, 392–393
use by MAPSS and PnET-IIS, 491
Solar and Meterological Surface Observation Network (SAMSON), global radiation model data from, 17–18
Solar radiation
data for models, 17–18
general circulation model comparisons, 28
global, from general circulation models, 30
within-region, from general circulation models, 31
Southern Global Change Program (SGCP), 279, 431, 479–480
charter of, 341–342
intertemporally optimum outcomes incorporated in, 756
research on productivity and atmospheric change, 583–584
Southern Pine Aggregate Market Model (SPAMM), 810, 813–815
Southern pine beetle, predicting infestations with a forest ecosystem model, 617–634
Southwide Pine Seed Source Study (SPSSS), 717–739
climate response models for, 733–736
loblolly pine sites, 718–720
Soybean, effects of light and carbon dioxide concentration on growth of, 7

Species changes, effects of, on soil
　　characteristics, 579–581
Species ranges
　　change in, and nutrient cycling, simulated, 503–524
　　predictions of the MAPSS model simulations, 493
Specification errors, model sensitivity to, 859
Specific leaf area, of longleaf pines, natural stand study, 243
SPM model, 482
　　carbon dioxide level effects simulated by, 489
　　productivity predictions of, 492
　　temperature effects simulated by, 488
　　　　interaction with carbon dioxide, 491
Spruce-fir forests
　　acid deposition and global change in southern Appalachian Mountains, 255–277
　　mining of, 837
　　sensitivity of protection value to forest condition in, 837–853
Standard deviation (sd), for tree-ring series data, 727
Standardization, of tree-ring data, 725–727
Stand dynamics, FORET model, 484
Stands
　　carbon gain in, effect of carbon dioxide, 159–161
　　environment of
　　　　and ecophysiological responses of loblolly pines, 185–206
　　　　factors affecting dynamics of growth, 210
　　growth projections for loblolly pine, 341–352
　　miniature
　　　　biomass studies under water stress, 123
　　　　to represent forest canopies, 120
　　mixed versus monoculture, photosynthetic responses of species, 122–123
　　model for development of, loblolly pine, 305–325, 439

resource availability in, thinned and unthinned, 592
Statistics, descriptive, respondents in a values study, 846
Steady state simulation, climate change effect on timber markets, 810–815
Stem growth
　　multiplier of, as a summary variable for UMT input, 422
　　sensitivity of
　　　　to carbon dioxide levels, 353–366
　　　　to optimum temperature for carbon assimilation, 360–361
　　　　to temperature, 353–366
　　　　to water stress, 410–411
　　slash pine carbon dynamics model, 355, 359
　　variation in
　　　　and environmental change, 207
　　　　and ozone exposure, 213, 407, 418
　　　　seasonal over two years, 214
Stemwood production
　　correlation with peak leaf area, effect of water and nutrients on, 158–159
　　effect of fertilization on, 155–156, 157
　　yield index, relating net primary production to, 370–371, 376
Stomatal conductance
　　and carbon dioxide levels, 493
　　in dogwood seedlings under simulated acid rain and drought stress, 639
　　effect of carbon dioxide levels on, 98–100, 121, 146, 152, 176, 177, 178, 180, 284–285, 385, 455
　　in oak seedlings, effects of temperature and soil moisture on, 654
　　and ozone levels, 413–414
　　sensitivity to vapor pressure and air temperature, 358–359
　　temperature response function, modified MAESTRO simulation, 333
Stomatal regulation of water loss, effect of ozone on, 223
Streamflow data, U.S. Geological survey, 399
Stress, response to
　　and nutrient availability, 302
　　and southern pine been damage, 618
Strobili counts, loblolly pine, 110–111

Sulfate ion
  adsorption of, simulation in NuCM, 505
  deposition of, in the Smoky Mountains, 258–259
  leaching of, simulated, 521–522
Sulfur, deposition of
  and leaching, 504
  simulation, 508
Summary functions, for the SOFOCLES model, 439–441
Supercritical fluid extraction (SFE), for analyzing soil organic matter, 546, 548, 550
Supply and demand equations, for softwood, 802–803, 814–815
Surface area
  foliage, simulated maintenance respiration as a function of, 324
  foliage ratio to woody tissue, simulated values, 318–319
  woody tissue, simulated maintenance respiration as a function of, 324
Surface-pressure patterns, anomalies in, and wildfire incidence, 692
Survey, of willingness to pay for environmental values, 841–843
Sweetgum, root length density response to carbon dioxide levels, 125

T

Tannins, in decomposing litter, 538
  experimental measurements of, 535–537
Tax base, and social programs, effect of global change on, 748
Temperature
  anomalies in, and wildfire incidence, 588, 690, 692
  branch, and canopy position, 192
  branch chamber versus ambient, 174
  and carbon dioxide levels
    diurnal effect on carbon assimilation, 136–137
    effect on loblolly pines, 145
    loblolly pine stands, 341–352
  and circumference growth, loblolly pine, 221
  and drought stress, effect on oak, 647–660
  effect of
    on dark respiration, 269–271
    on growth, overview of model projections, 494
    on herbivores and pathogens, 746–747
    on litter decomposition, experimental data, 531
    on photosynthesis, in the loblolly pine, 224
    on productivity, 487–489
    on reproductive biology, 104
    on stand growth, Pipestem projections, 348–349
  forest floor versus ambient, 565
  from general circulation models
    region-wide, 29
    within-region, 30
  general circulation model scenarios, 23–24
  historic data, for the MAPSS model, 459
  interaction with carbon dioxide levels
    effect on photosynthesis, 8–10, 283
    GMC calculations, 20
    SOFOCLES simulation, 444
  and litter decomposition, 565
  lower canopy, effect of thinning on, 191
  and nutrient release from soil, 267
  optimum
    for carbon assimilation, sensitivity of stem growth to, 360–362
    for gross photosynthesis, 392–393
  and precipitation, in the southern United States, 16–17
  and reproductive structure formation, 114
  sensitivity to, in northern forests, 258
  of soil
    effects of thinning on, 196
    effects on organic matter and nitrogen cycling, 557–569
    effects on soil quality, 571–587
    measuring, 188
    in thinned plots, 195
  Southwide Pine Seen Source Study plantations, 721, 722
  and tree-ring growth, red spruce, 265

trends for the southern United States, 816
use of, in MAPSS and PnET-IIS, 491
variation in, modified BIOMASS model, 376–378
Terrestrial Ecosystem Model (TEM), 368, 578
prediction of net primary production with carbon dioxide level changes, 383, 800–801
Thinking Like a Mountain (Leopold), 823
Thinning, effect of
on DBH and volume, 189
on lateral root elongation, 195
on a natural study, 246
on productivity and on root growth, 199–204
on resin production, 611–612
on and resistance to bark beetles, 596, 597–598
Thirty-year average rainfall (TYAR), maintaining in an experimental program, 84
Timber Assessment Market Model (TAMM), 757–758. *See also* Southern Pine Aggregate Market Model (SPAMM)
Timberland area, predicting
planted and non-planted management intensity, 764
three policy scenarios, 763
Timber production function, 857
Timber volume, projections of, 815–817
Time horizons
for decisions about values, 830–832
and modeling approaches, 859
Time replication, isolating variables affected by, 246
Tobacco, ozone injury to, effect of carbon dioxide levels, 10
Total leaf area (TLA)
as a composite index for MAPSS, 457
prediction from MAPSS, 468–469
Toxodium distichum (baldcypress), 702
Transpiration
annual, PnET-II model, 393, 620
and water use effect, model predictions, 469
and wildfire incidence, 694–695

Transpiration rate
and carbon dioxide level, diurnal variation, 136–137
effect of carbon dioxide levels on, 98–99
interaction with temperature, 137–138, 143
loblolly pine, 146
Travel-cost analysis, 860–861
Tree density, and net primary productivity, in Apalachicola National Forest, 242
Tree-growth model, for climatic variables, effect on natural stands, 248–249
Tree-ring data, climate reconstruction from, 701–702, 722–724
loblolly pine study, 717–739
Tree-ring growth
comparison between predicted and actual, red spruce, 262–263
and temperature, red spruce, 265
Tree-ring patterns
changes in, due to air pollution, 264–265
and changes in the chemical environment, red spruce, 259, 263–264
standardization of, 725–727
Trees, population changes from climate shifts, 670
Tridestronia oblongifolia, temperature and carbon dioxide level responses of, 9
Trifolium repens (white clover), ozone injury to, effect of carbon dioxide levels, 10
Triticum aestivum, ozone injury to, effect of carbon dioxide levels, 10

U

Ultraviolet radiation, effect on reproductive biology, 104
Uncertainty
estimating, regional model, 436–438
expressing in a model, 433–434, 442
about interactions in the ecosystem, 801
in long-time projections, 796
in MAPSS vegetation pattern predictions, 469–473

in scaling up tree-level data, 430, 439–440
scientific, and policy decisions, 58–59, 63–64, 445–450
Uncertainty analysis, 407
Unified Transport Model, 416–422
Unified Transport Model (UTM), 409–410
  adaptation for simulation of ozone effects, 412–414
  deterministic output from, 415–416
  effect of ozone on carbon gain in, 490
  effects of ozone on productivity, simulation predictions, 492–493
  for tree physiological processes, 409
  uncertainty analysis, 416–422
United Kingdom Meteorological Office (UKMO), general circulation model output, 21–23
Uptake, cation, interaction with leaching, 503–504
UPTAKE model
  focus on nutrients, 483
  productivity predictions of, 492
Utility function, individual, 839–840

V

Validation
  errors in predictions from simulations, 495–496
  of hydrologic data, PnET-II model, 398–399
  of the MAPSS model, 456
  of models, 481
  of the modified BIOMASS model, 370
  of the PnET-II model, 394–396, 620–621
Valuation, bottom-up, for forest management, 829–830
Values
  multiscalar public, applying to forest management, 828–835
  nonuse, defined, 845
  public
    and ecological model choice, 827
    scalar, assessing, 830–832
    scales of, 824–825
    and public policy decisions, 56–58

and scalar management objectives, 832–834
  study of, descriptive statistics about respondents, 846
Vapor pressure data, for the MAPSS model, 459–460
Vapor pressure deficit (VPD)
  general circulation models, 29
  region-wide, from general circulation models, 30
  stomatal response to, model, 18
  within-region, from general circulation models, 31
Variability, of growth rates
  mature tree measurements, 213
  model predictions of, 489
Variables, southern softwood stumpage supply study, 804
Variance
  analysis of, seed source study, 732–733
  environmental, incorporating in models, 745
  unbiased estimator of, between-series correlation, 728
Vegetation
  classification and analysis of, MAPSS model, 457–459
  climax, as an equilibrium state, 454
  influence on wildfire incidence, 694–695
  types of, survey and MAPSS estimated compared, 463
Vegetation/Ecosystem Model Analysis Project (VEMAP), model intercomparison and model linkage project, 460
VELMEX system, for measuring cores from red spruce, 263–264

W

Water
  availability of, and growth, 165
  as a constraint on vegetation distribution, 455
  effect of deficit on growth allocation, 74
  interaction with carbon dioxide levels, and photosynthesis, 6–7, 121, 169–170, 284–285

soil-water content, 195
Water balance, whole tree, effect of ozone on, 221–222
Water potentials
 effect of, on growth, 209
 effect of ozone and soil moisture on, 81
Water quality, and climate change, 675
Water stress, 593
Water use efficiency (WUE)
 and carbon dioxide levels, 6, 99–100, 121, 144, 455
  MAPSS model, 489
  PnET-IIS model, 489
 in the MAPSS model, 456, 489, 493
  effect on total leaf area, 468–469
  effect on vegetation distribution, 463–468
  simulation as stomatal conductance, 485
Weather data, for simulation, 345
Weibull distribution, for foliage density description, MAESTRO, 328–329
Wetlands
 biodiversity in, responses to climate change, 671
 coastal, effect of climate change on, 676
Wildfires, 745
 and regional climate change, 683–699
 risks of, in forest decline, 801
Wildlife preservation strategies, 824
Willingness to pay (WTP), 839–840
 invariance of, protection value study, 847–849
Wind jet, low-level, accompanying wildfires, 693
Wind speed data, for the MAPSS model, 460
Wood, structural changes in, red spruce forest, 272–273
Woodland types, and MAPSS vegetation classes, 458
Woody tissue weight, linked simulation model predicting, 317
World Meteorological Organization (WMO), Atmospheric Model Intercomparison Project, 21

Y

Yield, and net primary production, 383

Ecological Studies

Volumes published since 1992

Volume 89
Plantago: A Multidisciplinary Study (1992)
P.J.C. Kuiper and M. Bos (Eds.)

Volume 90
Biogeochemistry of a Subalpine Ecosystem: Loch Vale Watershed (1992)
J. Baron (Ed.)

Volume 91
Atmospheric Deposition and Forest Nutrient Cycling (1992)
D.W. Johnson and S.E. Lindberg (Eds.)

Volume 92
Landscape Boundaries: Consequences for Biotic Diversity and Ecological Flows (1992)
A.J. Hansen and F. di Castri (Eds.)

Volume 93
Fire in South African Mountain Fynbos: Ecosystem, Community, and Species Response at Swartboskloof (1992)
B.W. van Wilgen et al. (Eds.)

Volume 94
The Ecology of Aquatic Hyphomycetes (1992)
F. Bärlocher (Ed.)

Volume 95
Palms in Forest-Ecosystems of Amazonia (1992)
F. Kahn and J.-J. DeGranville

Volume 96
Ecology and Decline of Red Spruce in the Eastern United States (1992)
C. Eagar and M.B. Adams (Eds.)

Volume 97
The Response of Western Forests to Air Pollution (1992)
R.K. Olson, D. Binkley, and M. Böhm (Eds.)

Volume 98
Plankton Regulation Dynamics (1993)
N. Walz (Ed.)

Volume 99
Biodiversity and Ecosystem Function (1993)
E.-D. Schulze and H.A. Mooney (Eds.)

Volume 100
Ecophysiology of Photosynthesis (1994)
E.-D. Schulze and M.M. Caldwell (Eds.)

Volume 101
Effects of Land-Use Change on Atmospheric CO_2 Concentrations: South and South East Asia as a Case Study (1993)
V.H. Dale (Ed.)

Volume 102
Coral Reef Ecology (1993)
Y.I. Sorokin (Ed.)

Volume 103
Rocky Shores: Exploitation in Chile and South Africa (1993)
W.R. Siegfried (Ed.)

Volume 104
Long-Term Experiments With Acid Rain in Norwegian Forest Ecosystems (1993)
G. Abrahamsen et al. (Eds.)

Volume 105
Microbial Ecology of Lake Plußsee (1993)
J. Overbeck and R.J. Chrost (Eds.)

Volume 106
Minimum Animal Populations (1994)
H. Remmert (Ed.)

Volume 107
The Role of Fire in Mediterranean-Type Ecosystems (1994)
J.M. Moreno and W.C. Oechel (Eds.)

Volume 108
Ecology and Biogeography of Mediterranean Ecosystems in Chile, California, and Australia (1994)
M.T.K. Arroyo, P.H. Zedler, and M.D. Fox (Eds.)

Volume 109
Mediterranean Type Ecosystems: The Function of Biodiversity (1994)
G.W. Davis and D.M. Richardson (Eds.)

Volume 110
Tropical Montane Cloud Forests (1994)
L.S. Hamilton, J.O. Juvik, and
F.N. Scatena (Eds.)

Volume 111
Peatland Forestry: Ecology and Principles
(1995)
E. Paavilainen and J. Päivänen

Volume 112
Tropical Forests: Management and Ecology
(1995)
A.E. Lugo and C. Lowe (Eds.)

Volume 113
**Arctic and Alpine Biodiversity:
Patterns, Causes and Ecosystem
Consequences** (1995)
F.S. Chapin III and C. Körner (Eds.)

Volume 114
**Crassulacean Acid Metabolism:
Biochemistry, Ecophysiology and Evolution**
(1995)
K. Winter and J.A.C. Smith (Eds.)

Volume 115
**Islands: Biological Diversity and Ecosystem
Function** (1995)
P.M. Vitousek, H. Andersen, and
L. Loope (Eds.)

Volume 116
**High-Latitude Rainforests and
Associate Ecosystems of the West Coast of
the Americas: Climate, Hydrology, Ecology
and Conservation** (1995)
R.G. Lawford, P.B. Alaback, and
E.R. Fuentes (Eds.)

Volume 117
**Anticipated Effects of a Changing Global
Environment on Mediterranean-Type
Ecosystems** (1995)
J.M. Moreno and W.C. Oechel (Eds.)

Volume 118
**Impact of Air Pollutants on Southern Pine
Forests** (1995)
S. Fox and R.A. Mickler (Eds.)

Volume 119
Freshwaters of Alaska: Ecological Synthesis
(1997)
A.M. Milner and M.W. Oswood (Eds.)

Volume 120
**Landscape Function and Disturbance
in Arctic Tundra** (1996)
J.F. Reynolds and J.D. Tenhunen (Eds.)

Volume 121
**Biodiversity and Savanna Ecosystem
Processes: A Global Perspective** (1996)
O.T. Solbrig, E. Medina, and J.F. Silva
(Eds.)

Volume 122
**Biodiversity and Ecosystem Processes in
Tropical Forests** (1996)
G.H. Orians, R. Dirzo, and J.H. Cushman
(Eds.)

Volume 123
**Marine Benthic Vegetation: Recent
Changes and the Effects of Eutrophication**
(1996)
W. Schramm and P.H. Nienhuis (Eds.)

Volume 124
**Global Change and Arctic Terrestrial
Ecosystems** (1996)
W.C. Oechel (Ed.)

Volume 125
**Ecology and Conservation of Great Plains
Vertebrates** (1997)
F.L. Knopf and F.B. Samson (Eds.)

Volume 126
**The Central Amazon Floodplain: Ecology
of a Pulsing System** (1997)
W.J. Junk (Ed.)

Volume 127
**Forest Design and Ozone: A Comparison
of Controlled Chamber and Field
Experiments** (1997)
H. Sanderman, A.R. Wellburn, and
R.L. Heath (Eds.)

Volume 128
**The Productivity and Sustainability of
Southern Forest Ecosystems in a Changing
Environment** (1998)
R.A. Mickler and S. Fox (Eds.)

Volume 129
Pelagic Nutrient Cycles (1997)
T. Andersen

Volume 130
Vertical Food Web Interactions (1997)
K. Dettner, G. Bauer, and W. Völkl (Eds.)

Volume 131
The Structuring Role of Submerged Macrophytes in Lakes (1998)
E. Jeppesen, M. Søndergaard,
M. Søndergaard, and K. Christoffersen (Eds.)

Volume 132
Vegetation of the Tropical Pacific Islands (1998)
D. Mueller-Dombois and F.R. Fosberg